ENGAGE today's students

For the fourth edition of *College Physics: A Strategic Approach*, we expand our focus from HOW students learn physics to WHY students study physics. We now make connections to biology and other sciences throughout the text to keep students engaged, presenting content that is relevant to today's students. This new edition is one of the best college physics book on the market for non-physics majors.

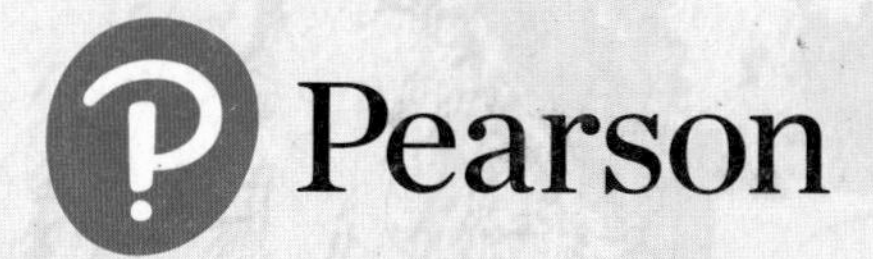

More connections to life science

Build students' problem-solving skills in a context they care about while using real-life data and examples to keep their interest piqued.

13.7 The Circulatory System BIO

The Arteries and Capillaries

In the human body, blood pumped from the heart to the body starts its journey in a single large artery, the aorta. The flow then branches into smaller blood vessels, the large arteries that feed the head, the trunk, and the limbs. These branch into still smaller arteries, which then branch into a network of much smaller arterioles, which branch further into the capillaries. FIGURE 13.37 shows a schematic outline of the circulation, with average values for the diameters of the individual vessels, the total cross-section area of all of each type of vessel considered together, and the pressure in these vessels, assuming that the person is lying down so that there is no pressure change due to differences in elevation.

This preserved section of blood vessels shows the tremendous increase in number and in total area as blood vessels branch from large arteries to arterioles. One large artery gives rise to thousands of smaller vessels.

NEW! Topics of interest to life science students, such as the nature of the drag force at different scales and qualitative and quantitative descriptions of diffusion, provide current coverage of relevant topics based on the evolving consensus in the introductory physics for the life sciences community.

NEW! Material stressing the application of physics to life sciences includes structural color in animals and plants, the electric sense of different animals, the circulatory system (13.7) and on forces and torques in the body (8.5).

8.5 Forces and Torques in the Body BIO

Let's take your foot as the object of interest. When you stand on tiptoe, your foot pivots about your ankle. As shown in FIGURE 8.27, the forces on one foot are an upward force on your toes from the floor, a downward force on your ankle from the lower leg bone, and an upward force on the heel of your foot from your Achilles tendon. Suppose a 61 kg woman stands on one foot, on tiptoe, with the sole of her foot making a 25° angle with the floor; the distances are as shown in Figure 8.27. What is the magnitude of the tension force in the tendon? By what fraction does this force exceed the woman's weight? What is the magnitude of the force in the ankle joint?

FIGURE 8.27 Forces on the foot when standing on tiptoe.

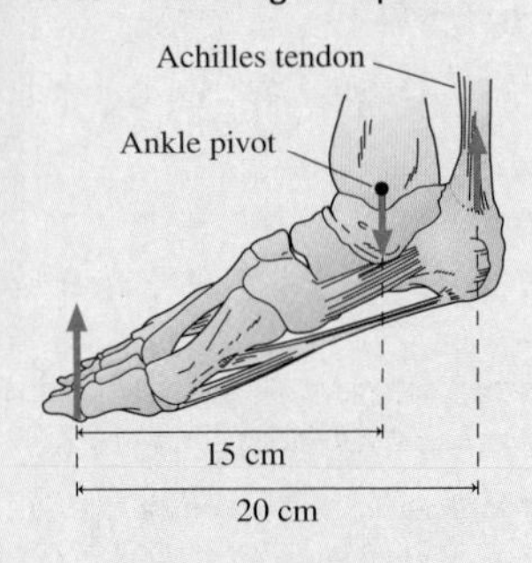

create relevance to students' lives

EXAMPLE 2.16 **Finding the height of a leap** 

A springbok is an antelope found in southern Africa that gets its name from its remarkable jumping ability. When a springbok is startled, it will leap straight up into the air—a maneuver called a "pronk." A particular springbok goes into a crouch to perform a pronk. It then extends its legs forcefully, accelerating at 35 m/s^2 for 0.70 m as its legs straighten. Legs fully extended, it leaves the ground and rises into the air.

a. At what speed does the springbok leave the ground?
b. How high does it go?

STRATEGIZE This is a two-part problem. In the first phase of its motion, the springbok accelerates upward, reaching some maximum speed just as it leaves the ground. As soon as it does so, the springbok is subject to only the force of gravity, so it is in free fall. For both phases, we will use the constant-acceleration equations from Synthesis 2.1.

NEW! STRATEGIZE step in Examples shows students the "big picture" view before delving into the details. Classroom testing of this addition has shown it to be popular with students and effective in teaching problem-solving skills.

NEW! End-of-chapter problem sets now include real-life data and examples, helping students build transferable skills for their future courses and careers.

8. | A hippo's body is 4.0 m long with front and rear feet BIO located as in Figure P8.8. The hippo carries 60% of its weight on its front feet. How far from its tail is the hippo's center of gravity?

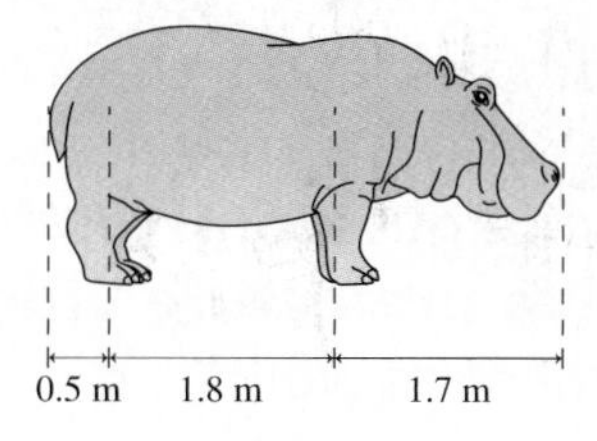

FIGURE P8.8

NEW! Learning Objectives, keyed to relevant end-of-chapter problems, help students check their understanding and guide them in choosing appropriate problems to optimize their study time.

Learning Objectives After studying this chapter, you should be able to:

- Use motion diagrams to interpret motion. *Conceptual Question 2.3; Problems 2.1, 2.2, 2.59*
- Use and interpret motion graphs. *Conceptual Questions 2.5, 2.13; Problems 2.4, 2.18, 2.19, 2.22, 2.62*
- Calculate the velocity of an object. *Conceptual Question 2.9; Problems 2.8, 2.15, 2.57*
- Solve problems about an object in uniform motion. *Problems 2.9, 2.10, 2.11, 2.13, 2.58*
- Calculate the acceleration of an object. *Problems 2.25, 2.27, 2.32, 2.33, 2.72*
- Determine and interpret the sign of acceleration. *Conceptual Questions 2.2, 2.8; Problem 2.50*
- Use the problem-solving approach to solve problems of motion with constant acceleration and free fall. *Problems 2.36, 2.40, 2.41, 2.47, 2.52, 2.75*

Prepare students for engagement

NEW! What the Physics? Videos bring relatable content to engage students with what they are learning and promote curiosity for natural phenomena. These short videos present visually stimulating physical phenomena, pause throughout to address misconceptions, and ask conceptual questions about the physics at hand. Quantitative questions follow some of the videos and will be assignable in Mastering™ Physics and embedded in the eText.

Prelecture Videos, presented by co-author Brian Jones, expand on the Chapter Previews, giving context, examples, and a chance for students to practice the concepts they are studying via short multiple-choice questions. **NEW! Qualitative and Quantitative prelecture videos** now available with assessment as well!

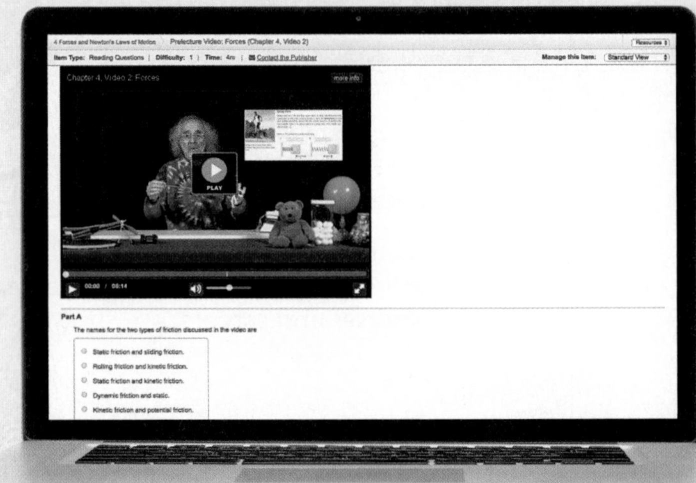

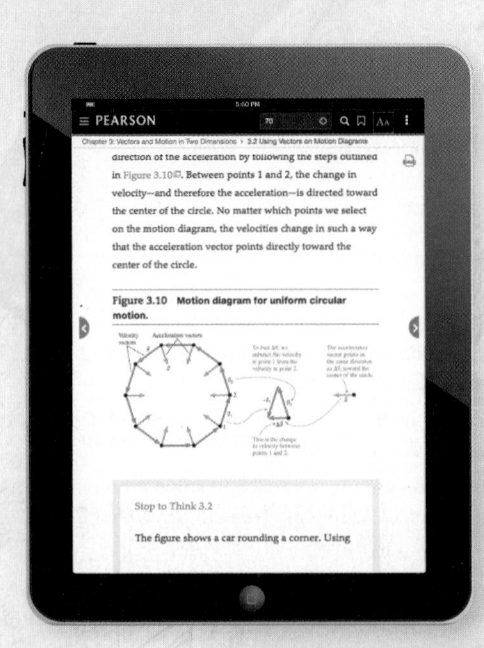

NEW! eText, optimized for mobile, seamlessly integrates videos and other rich media with the text and gives students access to their textbook anytime, anywhere. eText is available with Mastering Physics when packaged with new books, or as an upgrade students can purchase online.

in lecture with interactive media

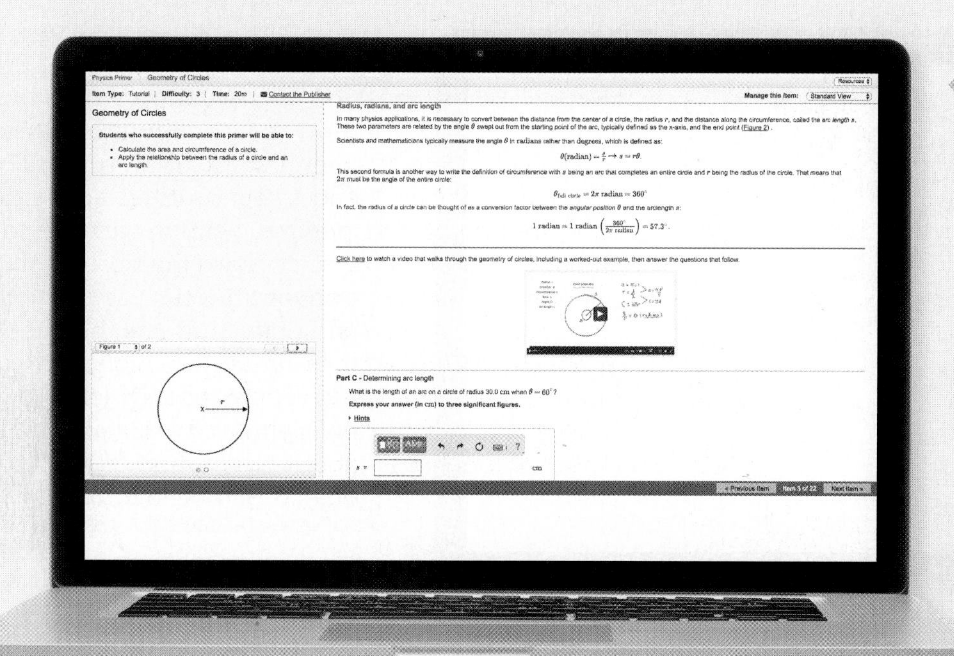

NEW! The Physics Primer relies on videos, hints, and feedback to refresh students' math skills in the context of physics and prepare them for success in the course. These tutorials can be assigned before the course begins as well as throughout the course. They ensure students practice and maintain their math skills, while tying together mathematical operations and physics analysis.

Dynamic Study Modules (DSMs) help students study effectively on their own by continuously assessing their activity and performance in real time and adapting to their level of understanding. The content focuses on definitions, units, and the key relationships for topics across all of mechanics and electricity and magnetism.

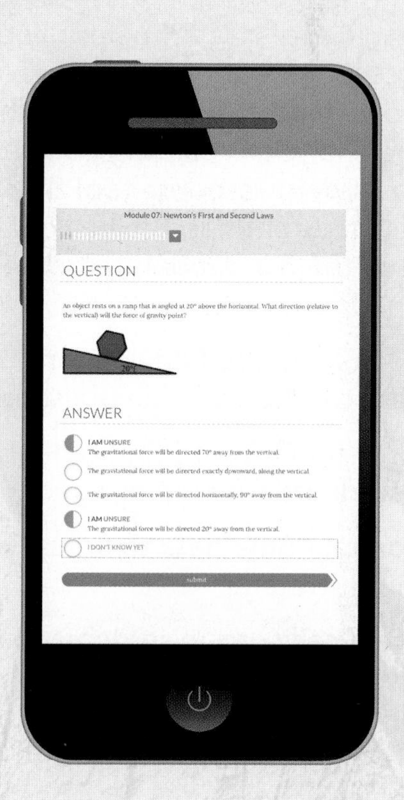

Enhance students' understanding when

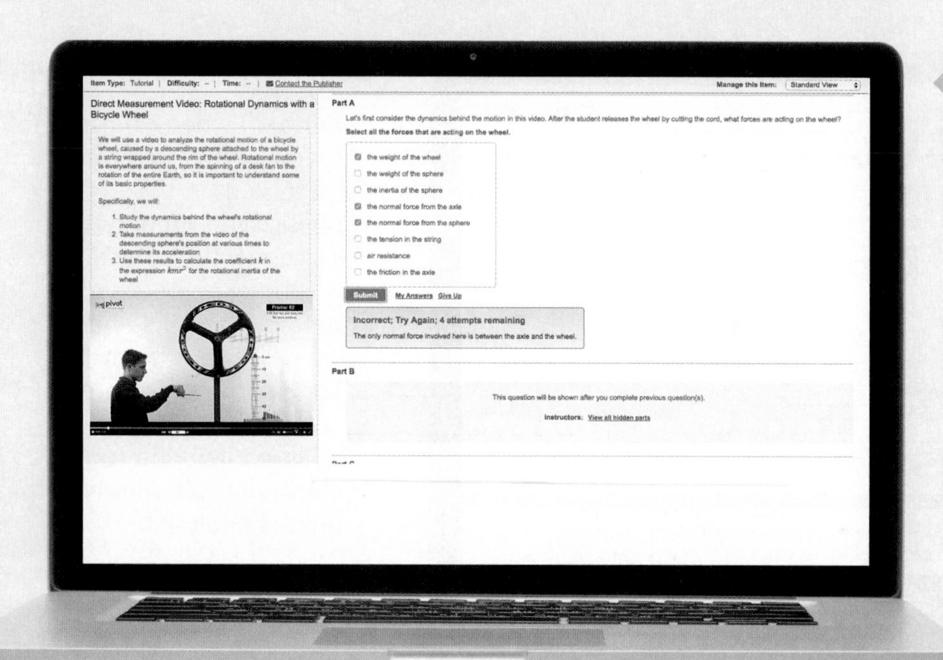

NEW! Direct Measurement Videos are short videos that show real situations of physical phenomena. Grids, rulers, and frame counters appear as overlays, helping students to make precise measurements of quantities such as position and time. Students then apply these quantities along with physics concepts to solve problems and answer questions about the motion of the objects in the video.

Learning Catalytics™ helps generate class discussion, customize lectures, and promote peer-to-peer learning with real-time analytics. Learning Catalytics acts as a student response tool that uses students' smartphones, tablets, or laptops to engage them in more interactive tasks and thinking:

- **NEW!** Upload a full PowerPoint® deck for easy creation of slide questions.
- Monitor responses to find out where your students are struggling.
- Rely on real-time data to adjust your teaching strategy.
- Automatically group students for discussion, teamwork, and peer-to-peer learning.

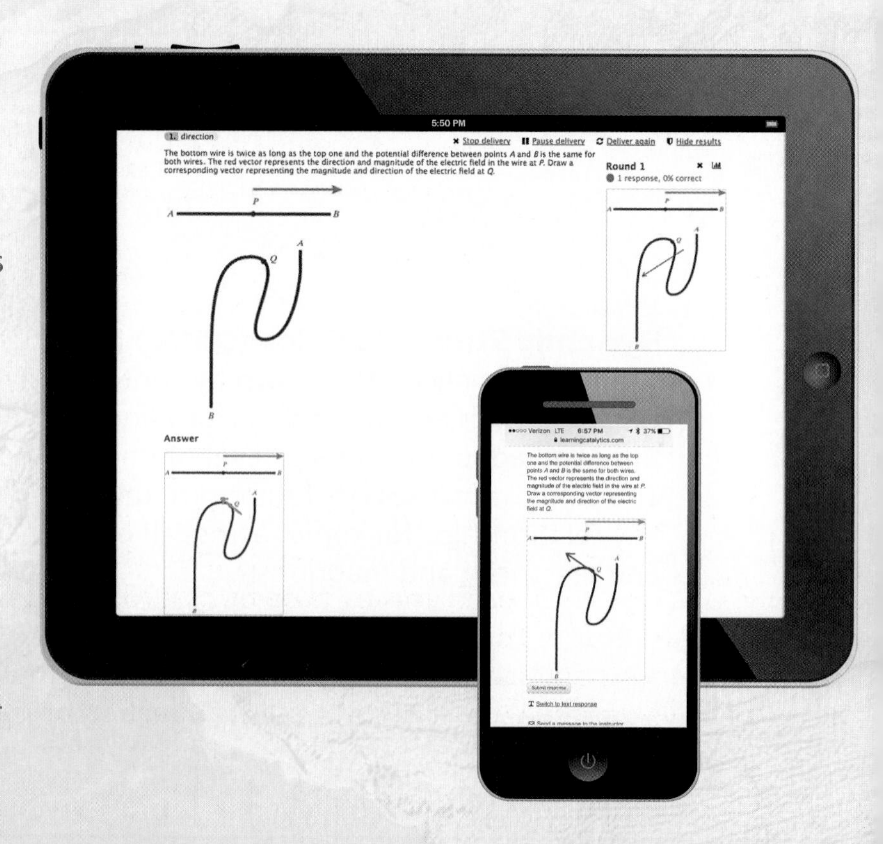

they apply what they've learned

10.6 Potential Energy

17. Below we see a 1 kg object that is initially 1 m above the ground and rises to a height of 2 m. Anjay and Brittany each measure its position but use a different coordinate system to do so. Fill in the table to show the initial and final gravitational potential energies and ΔU as measured by Anjay and Brittany.

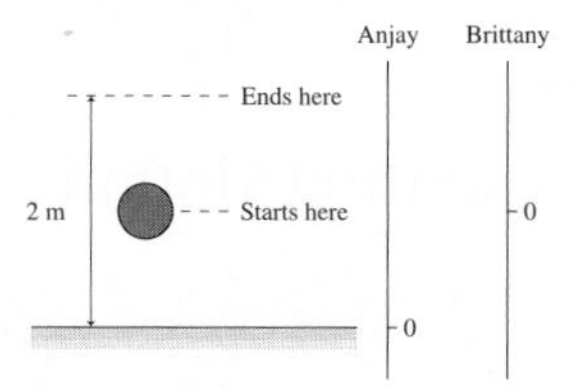

	U_i	U_f	ΔU
Anjay			
Brittany			

18. Three balls of equal mass are fired simultaneously with *equal* speeds from the same height above the ground. Ball 1 is fired straight up, ball 2 is fired straight down, and ball 3 is fired horizontally. Rank in order, from largest to smallest, their speeds v_1, v_2, and v_3 as they hit the ground.

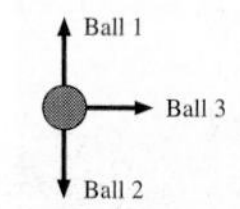

Order:

Explanation:

19. Below are shown three frictionless tracks. Ablock is released from rest at the position shown on the left. To which point does the block make it on the right before reversing direction and sliding back? Point B is the same height as the starting position.

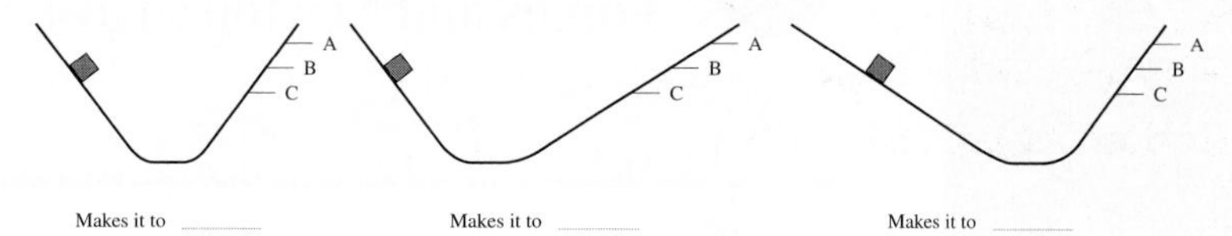

Makes it to ______ Makes it to ______ Makes it to ______

A key component of *College Physics: A Strategic Approach* is the accompanying ***Student Workbook.*** The workbook bridges the gap between textbook and homework problems by providing students the opportunity to learn and practice skills prior to using those skills in quantitative end-of-chapter problems, much as a musician practices technique separately from performance pieces. The workbook exercises, which are keyed to each section of the textbook, focus on developing specific skills, ranging from identifying forces and drawing free-body diagrams to interpreting field diagrams.

Instructor tools help shape your course more efficiently

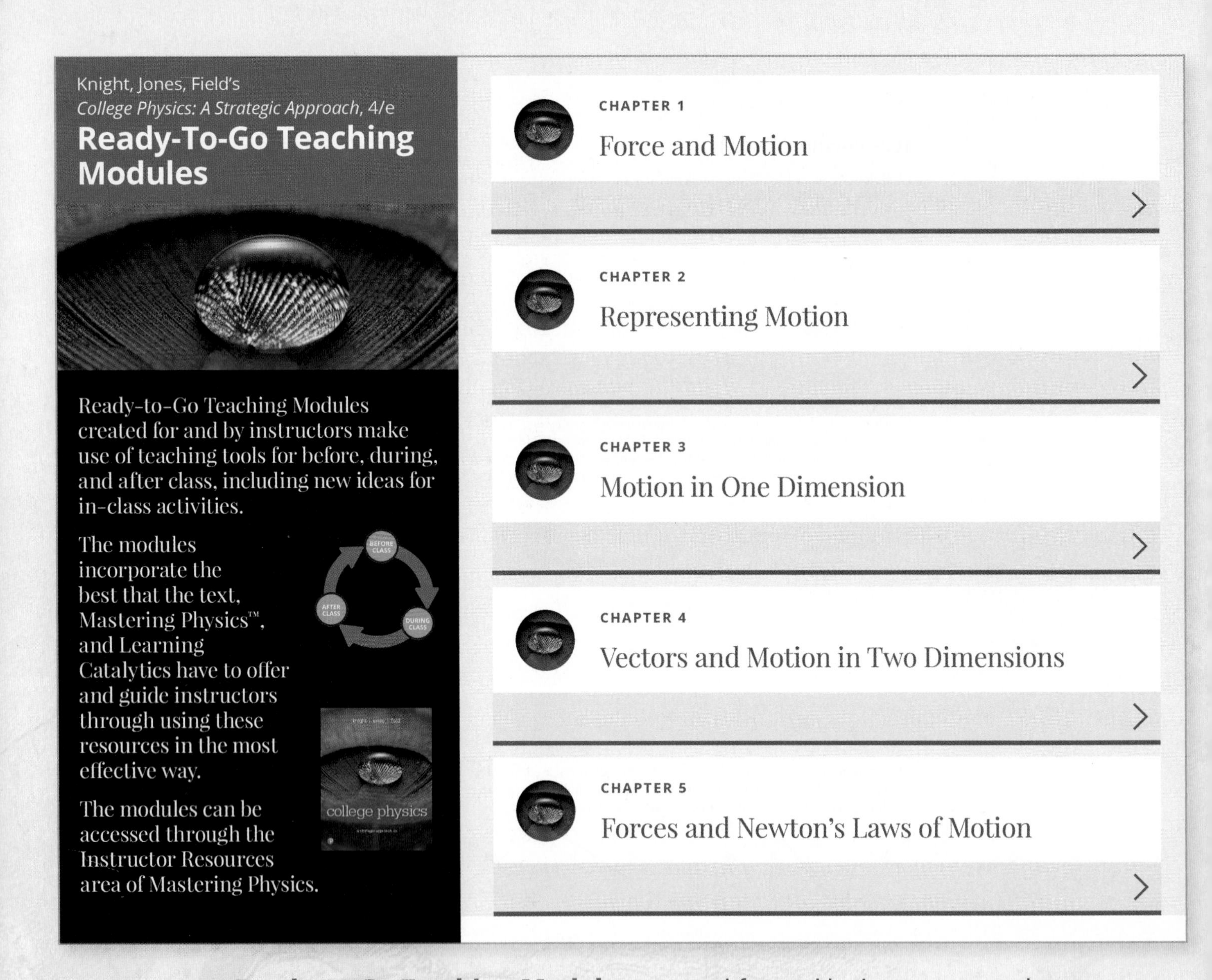

Knight, Jones, Field's
College Physics: A Strategic Approach, 4/e
Ready-To-Go Teaching Modules
Ready-to-Go Teaching Modules created for and by instructors make use of teaching tools for before, during, and after class, including new ideas for in-class activities.
The modules incorporate the best that the text, Mastering Physics™, and Learning Catalytics have to offer and guide instructors through using these resources in the most effective way.
The modules can be accessed through the Instructor Resources area of Mastering Physics.
BEFORE CLASS
DURING CLASS
AFTER CLASS
knight | jones | field
college physics
CHAPTER 1
Force and Motion
CHAPTER 2
Representing Motion
CHAPTER 3
Motion in One Dimension
CHAPTER 4
Vectors and Motion in Two Dimensions
CHAPTER 5
Forces and Newton's Laws of Motion

NEW! Ready-to-Go Teaching Modules, created for and by instructors, make use of teaching tools for before, during, and after class, including new ideas for in-class activities. The modules incorporate the best that the text, Mastering Physics, and Learning Catalytics have to offer and guide instructors through using these resources in the most effective way. The modules can be accessed through the Instructor Resources Area of Mastering Physics and as pre-built, customizable assignments.

Table of Problem-Solving Approaches

Table of Synthesis Boxes

Table of Math Relationships

Note for users of the two-volume edition:
Volume 1 (pp. 1–583)
includes Chapters 1–16.
Volume 2 (pp. 584–1093)
includes Chapters 17–30.

Brief Contents

PART I Force and Motion

PART II Conservation Laws

PART III Properties of Matter

PART IV Oscillations and Waves

PART V Optics

PART VI Electricity and Magnetism

PART VII Modern Physics

college physics

a strategic approach 4e

volume 1

randall d. knight
California Polytechnic State University, San Luis Obispo

brian jones
Colorado State University

stuart field
Colorado State University

330 Hudson Street, NY NY 10013

Editor in Chief, Director Physical Science Courseware Portfolio: Jeanne Zalesky
Editor, Physics Courseware Portfolio Analyst: Darien Estes
Senior Content Producer: Martha Steele
Senior Content Developer: David Hoogewerff
Managing Producer: Kristen Flathman
Courseware Director, Content Development: Jennifer Hart
Senior Analyst, Content Development, Science: Suzanne Olivier
Courseware Editorial Assistant: Kristen Stephens
Rich Media Content Producer: Dustin Hennessey
Full-Service Vendor: Nesbitt Graphics/Cenveo® Publisher Services
Copyeditor: Carol Reitz
Compositor: Nesbitt Graphics/Cenveo® Publisher Services
Art and Design Director: Mark Ong/Side By Side Studios
Interior/Cover Designer: tani hasegawa
Rights & Permissions Project Manager: Katrina Mohn
Rights & Permissions Management: Ben Ferrini
Photo Researcher: Dena Digilio Betz
Manufacturing Buyer: Stacey J. Weinberger/LSC Communications
Product Marketing Manager, Physical Sciences: Elizabeth Bell
Cover Photo Credit: MirageC/Getty

Library of Congress Cataloging in Publication Data on File

ISBN 10: 0-134-61045-8; ISBN 13: 978-0-134-61045-0 (Volume 1)
1 17

About the Authors

Randy Knight taught introductory physics for 32 years at Ohio State University and California Polytechnic State University, where he is Professor Emeritus of Physics. Professor Knight received a Ph.D. in physics from the University of California, Berkeley and was a post-doctoral fellow at the Harvard-Smithsonian Center for Astrophysics before joining the faculty at Ohio State University. It was at Ohio State that he began to learn about the research in physics education that, many years later, led to *Five Easy Lessons: Strategies for Successful Physics Teaching* and this book, as well as *Physics for Scientists and Engineers: A Strategic Approach.* Professor Knight's research interests are in the fields of laser spectroscopy and environmental science. When he's not in front of a computer, you can find Randy hiking, sea kayaking, playing the piano, or spending time with his wife Sally and their five cats.

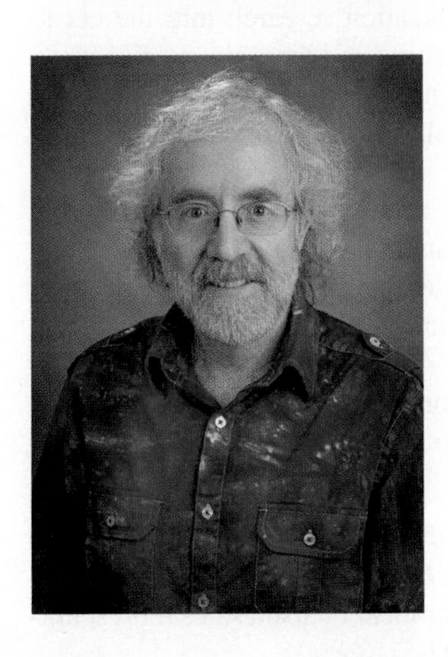

Brian Jones has won several teaching awards at Colorado State University during his 30 years teaching in the Department of Physics. His teaching focus in recent years has been the College Physics class, including writing problems for the MCAT exam and helping students review for this test. In 2011, Brian was awarded the Robert A. Millikan Medal of the American Association of Physics Teachers for his work as director of the Little Shop of Physics, a hands-on science outreach program. He is actively exploring the effectiveness of methods of informal science education and how to extend these lessons to the college classroom. Brian has been invited to give workshops on techniques of science instruction throughout the United States and in Belize, Chile, Ethiopia, Azerbaijan, Mexico, Slovenia, Norway, and Namibia. Brian and his wife Carol have dozens of fruit trees and bushes in their yard, including an apple tree that was propagated from a tree in Isaac Newton's garden.

Stuart Field has been interested in science and technology his whole life. While in school he built telescopes, electronic circuits, and computers. After attending Stanford University, he earned a Ph.D. at the University of Chicago, where he studied the properties of materials at ultralow temperatures. After completing a postdoctoral position at the Massachusetts Institute of Technology, he held a faculty position at the University of Michigan. Currently at Colorado State University, Stuart teaches a variety of physics courses, including algebra-based introductory physics, and was an early and enthusiastic adopter of Knight's *Physics for Scientists and Engineers.* Stuart maintains an active research program in the area of superconductivity. Stuart enjoys Colorado's great outdoors, where he is an avid mountain biker; he also plays in local ice hockey leagues.

Preface to the Instructor

In 2006, we published *College Physics: A Strategic Approach,* a new algebra-based physics textbook for students majoring in the biological and life sciences, architecture, natural resources, and other disciplines. As the first such book built from the ground up on research into how students can more effectively learn physics, it quickly gained widespread critical acclaim from professors and students alike. For this fourth edition, we have continued to build on the research-proven instructional techniques introduced in the first edition while working to make the book more useful for instructors, more relevant to the students who use it, and more connected to the other subjects they study.

Objectives

Our primary goals in writing *College Physics: A Strategic Approach* are:

- To provide students with a textbook that's a more manageable size, less encyclopedic in its coverage, and better designed for learning.
- To integrate proven techniques from physics education research into the classroom in a way that accommodates a range of teaching and learning styles.
- To help students develop both quantitative reasoning skills and solid conceptual understanding, with special focus on concepts well documented to cause learning difficulties.
- To help students develop problem-solving skills and confidence in a systematic manner using explicit and consistent tactics and strategies.
- To motivate students by integrating real-world examples that are relevant to their majors—especially from biology, sports, medicine, the animal world—and that build upon their everyday experiences.
- To utilize proven techniques of visual instruction and design from educational research and cognitive psychology that improve student learning and retention and address a range of learner styles.

A more complete explanation of these goals and the rationale behind them can be found in Randy Knight's paperback book, *Five Easy Lessons: Strategies for Successful Physics Teaching*. Please request a copy from your local Pearson sales representative if it is of interest to you (ISBN 978-0-805-38702-5).

What's New to This Edition

In previous editions of the text, we focused on *how* students learn physics. Each chapter was built from the ground up to present concepts and problem-solving strategies in an engaging and effective manner. In this edition, we are focusing on *why* students learn physics. This is a question our students often ask. Why should a biology major take physics? A student planning a career in medicine? This book is for a physics course, but it's a course that will generally be taken by students in other fields.

The central goal of this edition is to make the text more relatable to the students who will use it, to add examples, explanations, and problems that show physics at work in contexts the students will find engaging. We've considered extensive feedback from scores of instructors and thousands of students as we worked to enhance and improve the text, figures, and end-of-chapter problems. Instructors need not be specialists in the life sciences or other fields to appreciate the new material. We've done the work to connect physics to other disciplines so that instructors can use this material to engage their students while keeping their focus on the basic physics.

Making the text more relatable meant making significant changes throughout the book. These edits aren't cosmetic add-ons; they reflect a thorough reworking of each chapter. Changes include:

- Guided by an evolving consensus in the Introductory Physics for the Life Sciences community, we have included **new sections** on the nature of the drag force at different scales, qualitative and quantitative descriptions of diffusion, and other topics of interest to life science students.
- We have added a great deal of **new material** that stresses the application of physics to life science topics. For example, we have expanded our treatment of vision and vision correction, included new material on structural color in animals and plants and the electric sense of different animals, and added new sections on the circulatory system and on forces and torques in the body.
- We have made **new connections** between physics topics and other courses that students are likely to take. For example, a new section connects the concept of the conservation of energy to topics from chemistry, including ionization energy and the role of catalysts in reactions. We have continued this approach when we introduced the concept of electric potential energy.
- Hundreds of **new end-of-chapter questions and problems** show physics at work in realistic, interesting situations. We have replaced problems that are artificial and abstract with problems that use real data from research in life science fields, problems that show the physics behind modern technologies, and problems that use physics to explore everyday phenomena. We have used the wealth of data from Mastering™ Physics to make sure that we have problems of a wide range of difficulties for each topic and problem-solving approach. A rigorous blind-solving and accuracy cross-checking process has been used to check all new problems to be sure that they are clearly worded and correct in all details, that they are accompanied by carefully worked out solutions.
- **New examples** throughout the book use the concepts of the chapters to explore realistic situations of interest to the students—from how bees use electric fields to locate promising flowers to how a study of force and torque in the jaw explains why dogs have long snouts and cats don't.
- We have changed the **photos and captions** at the starts of the chapters and parts of the text to better interest and engage students. The questions that are raised at the starts of the chapters aren't rhetorical; they are questions that will be answered in the flow of the chapter.

We have also made a number of changes to make the text an even more effective tool for students:

- A new **STRATEGIZE** step in examples shows students the "big picture" view before we delve into the details. Classroom testing of this addition has shown it to be quite popular with students, and quite effective in teaching problem-solving skills.
- **Key Concept figures** encourage students to actively engage with key or complex figures by asking them to reason with a related STOP TO THINK question.
- Additional **STOP TO THINK questions** provide students with more crucial practice and concept checks as they go through the chapters. The solutions to these questions have been moved to a more prominent location.
- We now provide **Learning Objectives** keyed to relevant end-of-chapter problems to help students check their understanding and guide them in choosing appropriate problems to optimize their study time.
- **Streamlined text and figures** tighten and focus the presentation to more closely match student needs. We've scrutinized every figure, caption, discussion, and photo in order to enhance their clarity and focus their role.
- Increased emphasis on **critical thinking, modeling, and reasoning,** both in worked examples and in end-of-chapter problems, promotes these key skills. These skills are especially important for students who are taking the MCAT exam.

- Expanded use of **realistic and real-world data** ensures students can make sense of answers that are grounded in the real world. Our examples and problems use real numbers and real data; they test different types of reasoning using equations, ratios, and graphs.

We have made many small changes to the flow of the text throughout, streamlining derivations and discussions, providing more explanation for complex concepts and situations, and reordering and reorganizing material so that each section and each chapter have a clearer focus. We have updated our treatment of entropy and the second law to better match current thinking. We have reordered the presentation of material on motion in two dimensions to be more logical. Every chapter has significant and meaningful changes, making this course especially relevant for today's students.

We know that students increasingly rely on sources of information beyond the text, and instructors are looking for quality resources that prepare students for engagement in lecture. The text will always be the central focus, but we have added additional media elements closely tied to the text that will enhance student understanding. In the Technology Update to the Second Edition, we added Class Videos, Video Tutor Solutions, and Video Tutor Demonstrations. In the Third Edition, we added an exciting new supplement, **Prelecture Videos,** short videos with author Brian Jones that introduce the topics of each chapter with accompanying assessment questions. In the front of this book, you'll find an illustrated walkthrough of the new media available in this technology update for the third edition:

- **NEW! What the Physics? Videos** bring new, relatable content to engage students with what they are learning and promote curiosity for natural phenomena. These short videos present visually stimulating physical phenomena and pause throughout to address misconceptions and ask conceptual questions about the physics at hand. The videos are embedded in the eText as well as assignable in Mastering Physics. Quantitative questions are also available for assignment.
- **NEW! Direct Measurement Videos** are short videos that show real situations of physical phenomena. Grids, rulers, and frame counters appear as overlays, helping students to make precise measurements of quantities such as position and time. Students then apply these quantities along with physics concepts to solve problems and answer questions about the motion of the objects in the video. The problems are assignable in Mastering Physics and can be used to replace or supplement traditional word problems, or as open-ended questions to help develop problem-solving skills.
- **NEW! The Physics Primer** relies on videos, hints, and feedback to refresh students' math skills in the context of physics and prepares them for success in the course. These tutorials can be assigned before the course begins or throughout the course as just-in-time remediation. They ensure students practice and maintain their math skills, while tying together mathematical operations and physics analysis.
- **NEW! Quantitative Prelecture Videos** are assignable, interactive videos that complement the Conceptual Prelecture Videos, giving students exposure to concepts before class and helping them learn how problems for those concepts are worked.
- **NEW! Ready-to-Go Teaching Modules** provide instructors with easy-to-use tools for teaching the toughest topics in physics. These modules demonstrate how your colleagues effectively use all the resources Pearson has to offer to accompany *College Physics: A Strategic Approach,* including, but not limited to, Mastering Physics items. Ready-to-Go Teaching Modules were created for and by instructors to provide easy-to-use assignments for before, during, and after class. Assets also include in-class activities and questions in Learning Catalytics™.
- **Dynamic Study Modules (DSMs)** help students study on their own by continuously assessing their activity and performance in real time. Students complete a set of questions with a unique answer format that repeats each question until students can answer them all correctly and confidently.

- **Dynamic Figure Videos in each chapter** are one-minute videos based on figures from the textbook that depict important, but often challenging, physics principles.
- **Video Tutor Solutions** created by co-author Brian Jones are an engaging and helpful walkthrough of worked examples and select end-of-chapter (EOC) problems designed to help students solve problems for each main topic. Each chapter has seven Video Tutor Solutions.
- **Prep questions aligned with the MCAT exam** are based on the Foundational Concepts and Content Categories outlined by the Association of American Medical Colleges. These 140 new problems are assignable in Mastering Physics and available for self-study in the Study Area.
- **Video Tutor Demonstrations** feature "pause-and-predict" demonstrations of key physics concepts and incorporate assessment with answer-specific feedback.

Textbook Organization

College Physics: A Strategic Approach is a 30-chapter text intended for use in a two-semester course. The textbook is divided into seven parts: Part I: *Force and Motion,* Part II: *Conservation Laws,* Part III: *Properties of Matter,* Part IV: *Oscillations and Waves,* Part V: *Optics,* Part VI: *Electricity and Magnetism,* and Part VII: *Modern Physics.*

Part I covers Newton's laws and their applications. The coverage of two fundamental conserved quantities, momentum and energy, is in Part II, for two reasons. First, the way that problems are solved using conservation laws—comparing an *after* situation to a *before* situation—differs fundamentally from the problem-solving strategies used in Newtonian dynamics. Second, the concept of energy has a significance far beyond mechanical (kinetic and potential) energies. In particular, the key idea in thermodynamics is energy, and moving from the study of energy in Part II into thermal physics in Part III allows the uninterrupted development of this important idea.

Optics (Part V) is covered directly after oscillations and waves (Part IV), but *before* electricity and magnetism (Part VI). Further, we treat wave optics before ray optics. Our motivations for this organization are twofold. First, wave optics is largely just an extension of the general ideas of waves; in a more traditional organization, students will have forgotten much of what they learned about waves by the time they get to wave optics. Second, optics as it is presented in introductory physics makes no use of the properties of electromagnetic fields. The documented difficulties that students have with optics are difficulties with waves, not difficulties with electricity and magnetism. There's little reason other than historical tradition to delay optics. However, the optics chapters are easily deferred until after Part VI for instructors who prefer that ordering of topics.

- **Complete edition,** with Mastering™ Physics and Student Workbook (ISBN 978-0-134-64149-2): Chapters 1–30.
- **Books a la Carte edition,** with Mastering™ Physics and Student Workbook (ISBN 978-0-134-64414-1)
- **Complete edition,** without Mastering™ Physics (ISBN 978-0-134-60903-4): Chapters 1–30.
- **Volume 1** without Mastering™ Physics (ISBN 978-0-134-61045-0): Chapters 1–16.
- **Volume 2** without Mastering™ Physics (ISBN 978-0-134-61046-7): Chapters 17–30.

Split your text the way you split your course! Log on to www.pearsonhighered.com/collections/educator-features.html and create your own splits of *College Physics: A Strategic Approach,* 4e, including *your* choice of chapters.

Instructional Package

College Physics: A Strategic Approach, fourth edition, provides an integrated teaching and learning package of support material for students and instructors.

NOTE For convenience, most instructor supplements can be downloaded from the "Instructor Resources" area of Mastering Physics and the Instructor Resource Center (www.pearson.com/us/higher-education/customers/educators.html).

Supplement	Print	Online	Instructor or Student Supplement	Description
Mastering Physics with Pearson eText (ISBN 0134671023)		✓	Instructor and Student Supplement	This product features all of the resources of Mastering Physics in addition to the new Pearson eText 2.0. Now available on smartphones and tablets, Pearson eText 2.0 comprises the full text, including videos and other rich media.
Instructor's Solutions Manual (ISBN 0134796829)		✓	Instructor Supplement	This comprehensive solutions manual contains complete solutions to all end-of-chapter questions and problems.
TestGen Test Bank (ISBN 0134702409)		✓	Instructor Supplement	The Test Bank contains more than 2,000 high-quality problems, with a range of multiple-choice, true/false, short answer, and regular homework-type questions. Test files are provided in both TestGen® and Word format.
Instructor's Resource Materials	✓	✓	Instructor Supplement	All art, photos, and tables from the book are available in JPEG format and as modifiable PowerPoints™. In addition, instructors can access lecture outlines as well as "clicker" questions in PowerPoint format, editable content for key features, all the instructor's resources listed above, and solutions to the Student Workbook. Materials are accessible to download from the Instructor Resource area of Mastering Physics.
Student's Workbook Standard (CH1–30) (ISBN 0134609891X) Volume 1 (CH1–16) (ISBN 0134724828) Volume 2 (CH17–30) (ISBN 0134724801)	✓		Student Supplement	For a more detailed description of the *Student's Workbook,* see page ix.
Student's Solutions Manual Volume 1 (CH1–16) (ISBN 0134704193) Volume 2 (CH17–30) (ISBN 0134724798)	✓		Student Supplement	These solutions manuals contain detailed solutions to all of the odd-numbered end-of-chapter problems from the textbook.
Ready-to-Go Teaching Modules		✓	Instructor Supplement	Ready-to-Go Teaching Modules provide instructors with easy-to-use tools for teaching the toughest topics in physics. Created by the authors and designed to be used before, during, and after class, these modules demonstrate how to effectively use all the book, media, and assessment resources that accompany *College Physics: A Strategic Approach 4e.*

The Student Workbook

A key component of *College Physics: A Strategic Approach* is the accompanying *Student Workbook*. The workbook bridges the gap between textbook and homework problems by providing students the opportunity to learn and practice skills prior to using those skills in quantitative end-of-chapter problems, much as a musician practices technique separately from performance pieces. The workbook exercises, which are keyed to each section of the textbook, focus on developing specific skills, ranging from identifying forces and drawing free-body diagrams to interpreting field diagrams.

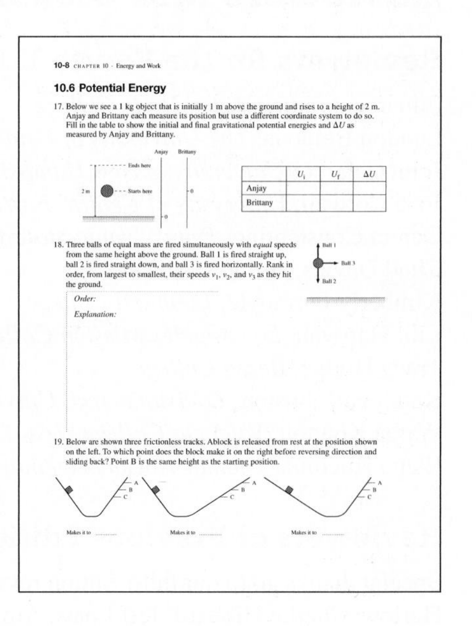
10-8 CHAPTER 10 · Energy and Work

10.6 Potential Energy

17. Below we see a 1 kg object that is initially 1 m above the ground and rises to a height of 2 m. Anjay and Brittany each measure its position but use a different coordinate system to do so. Fill in the table to show the initial and final gravitational potential energies and ΔU as measured by Anjay and Brittany.

	U_i	U_f	ΔU
Anjay			
Brittany			

18. Three balls of equal mass are fired simultaneously with *equal* speeds from the same height above the ground. Ball 1 is fired straight up, ball 2 is fired straight down, and ball 3 is fired horizontally. Rank in order, from largest to smallest, their speeds v_1, v_2, and v_3 as they hit the ground.

Order:

Explanation:

19. Below are shown three frictionless tracks. Ablock is released from rest at the position shown on the left. To which point does the block make it on the right before reversing direction and sliding back? Point B is the same height as the starting position.

The workbook exercises, which are generally qualitative and/or graphical, draw heavily upon the physics education research literature. The exercises deal with issues known to cause student difficulties and employ techniques that have proven to be effective at overcoming those difficulties. Also included are *jeopardy problems* that ask students to work backward from equations to physical situations, enhancing their understanding and critical thinking skills. The workbook exercises can be used in-class as part of an active-learning teaching strategy, in recitation sections, or as assigned homework. More information about effective use of the *Student Workbook* can be found in the *Instructor's Guide* in the Ready-to-Go modules.

Available versions: Standard Edition (ISBN 978-0-134-60989-8): Chapters 1-30, Volume 1 (ISBN 978-0-134-72482-9): Chapters 1–16, and Volume 2 (978-0-134-72480-5): Chapters 17–30.

Acknowledgments

We have relied upon conversations with and, especially, the written publications of many members of the physics education community. Those who may recognize their influence include the late Arnold Arons, Uri Ganiel, Fred Goldberg, Ibrahim Halloun, Paula Heron, David Hestenes, the late Leonard Jossem, Jill Larkin, Priscilla Laws, John Mallinckrodt, Richard Mayer, Lillian McDermott and members of the Physics Education Research Group at the University of Washington, Edward "Joe" Redish, Fred Reif, Rachel Scherr, Bruce Sherwood, David Sokoloff, Ronald Thornton, Sheila Tobias, Alan Van Heuvelen, Carl Wieman, and Michael Wittman.

We are very grateful to Larry Smith for the difficult task of writing the *Instructor Solutions Manual;* to Scott Nutter for writing out the Student Workbook answers; to Wayne Anderson, Jim Andrews, Nancy Beverly, David Cole, Karim Diff, Jim Dove, Marty Gelfand, Kathy Harper, Charlie Hibbard, Robert Lutz, Matt Moelter, Kandiah Manivannan, Ken Robinson, Cindy Schwarz-Rachmilowitz, and Rachel Jones for their contributions to the end-of-chapter questions and problems; to Wayne again for helping with the Test Bank questions; and to Steven Vogel for his careful review of the biological content of many chapters and for helpful suggestions.

We especially want to thank Editor-in-Chief, Director Physical Science Courseware Portfolio, Jeanne Zalesky; Editor, Physics Courseware Portfolio Analyst, Darien Estes; Courseware Director, Content Development, Jennifer Hart; Senior Analyst, Courseware Development, Suzanne Olivier; Senior Content Producer, Martha Steele; and all the other staff at Pearson for their enthusiasm and hard work on this project. Having a diverse author team is one of the strengths of this book, but it has meant that we rely a great deal on Darien to help us keep to a single focus. Special thanks is due Martha for her keen attention to all details, careful scheduling and shepherding of all the elements of this complex project, and gentle prodding to keep everything moving forward.

Rose Kernan and the team at Nesbitt Graphics/Cenveo, copy editor Carol Reitz, and photo researcher Dena Digilio Betz get much credit for making this complex project all come together. In addition to the reviewers and classroom testers listed below, who gave invaluable feedback, we are particularly grateful to Ansel Foxley for his close scrutiny of every word, symbol, number, and figure.

Randy Knight: I would like to thank my Cal Poly colleagues for many valuable conversations and suggestions. I am endlessly grateful to my wife Sally for her love, encouragement, and patience, and to our many cats for nothing in particular other than being cats.

Brian Jones: I would like to thank my fellow AAPT and PIRA members for their insight and ideas, the students and colleagues who are my partners in the Little Shop of Physics, the students in my College Physics classes who teach me so much about the world, and the team at Pearson who help me develop as an educator and writer. Most of all, I thank my wife Carol, my best friend and gentlest editor, whose love makes the journey worthwhile.

Stuart Field: I would like to thank my wife Julie and my children, Sam and Ellen, for their love, support, and encouragement.

Reviewers and Classroom Testers

Reviewers for the Fourth Edition

Eduardo Araujo, *Miami Dade College*
Landon Bellavia, *The University of Findlay*
John Calarco, *University of New Hampshire*
Josh Colwell, *University of Central Florida*
Daniel Constantino, *Pennsylvania State University*
Chad Davies, *Gordon College*
Kimberly Fermoyle, *Elmhurst College*
Lilit Haroyan, *Los Angeles Mission College*
Tracy Hodge, *Berea College*
Kathleen Johnston, *Louisiana Tech University*
Wafaa Khattou, *Valencia College West*
Elena Kuchina, *Thomas Nelson Community College*
Ravi Kumar, *Wayne Community College, Goldsboro*
Jorge Lopez, *University of Texas*
Luiz Manzoni, *Concordia College*
Dario Martinez, *Lone Star College*
John McClain, *Temple College*
Irina Perevalova, *Middle Tennessee State University*
Amy Pope, *Clemson University*
Firouzeh Sabri, *University of Memphis*
Chandralekha Singh, *University of Pittsburgh*
Brenda Skoczelas, *Lake Sumter State College*
Valeriia Starovoitova, *Idaho State University*
Susa Stonedahl, *St. Ambrose University*
Ulrich Zurcher, *Cleveland State University*

Reviewers of Previous Editions

Special thanks go to our third edition review panel: Taner Edis, Marty Gelfand, Jason Harlow, Charlie Hibbard, Jeff Loats, Amy Pope, and Bruce Schumm.

David Aaron, *South Dakota State University*
Susmita Acharya, *Cardinal Stritch University*
Ugur Akgun, *University of Iowa*
Ralph Alexander, *University of Missouri—Rolla*
Kyle Altmann, *Elon University*
Donald Anderson, *Ivy Tech*
Michael Anderson, *University of California—San Diego*
Steve Anderson, *Montana Tech*
James Andrews, *Youngstown State University*
Charles Ardary, *Edmond Community College*
Charles Bacon, *Ferris State University*
John Barry, *Houston Community College*
David H. Berman, *University of Northern Iowa*
Phillippe Binder, *University of Hawaii—Hilo*
Jeff Bodart, *Chipola College*
James Boger, *Flathead Valley Community College*
Richard Bone, *Florida International University*
James Borgardt, *Juniata College*
Daniela Bortoletto, *Purdue University*
Don Bowen, *Stephen F. Austin State University*
Asa Bradley, *Spokane Falls Community College*
Elena Brewer, *SUNY at Buffalo*
Dieter Brill, *University of Maryland*
Hauke Busch, *Augusta State University*
Kapila Castoldi, *Oakland University*
Raymond Chastain, *Louisiana State University*
Michael Cherney, *Creighton University*
Lee Chow, *University of Central Florida*
Song Chung, *William Paterson University*
Alice Churukian, *Concordia College*
Christopher M. Coffin, *Oregon State University*
John S. Colton, *Brigham Young University*
Kristi Concannon, *Kings College*
Teman Cooke, *Georgia Perimeter College at Lawrenceville*
Daniel J. Costantino, *The Pennsylvania State University*
Jesse Cude, *Hartnell College*
Melissa H. Dancy, *University of North Carolina at Charlotte*
Loretta Dauwe, *University of Michigan—Flint*
Mark Davenport, *San Antonio College*
Chad Davies, *Gordon College*
Lawrence Day, *Utica College*
Carlos Delgado, *Community College of Southern Nevada*
David Donovan, *Northern Michigan University*
James Dove, *Metropolitan State University of Denver*
Archana Dubey, *University of Central Florida*
Andrew Duffy, *Boston University*
Taner Edis, *Truman State University*
Ralph Edwards, *Lurleen B. Wallace Community College*
Steve Ellis, *University of Kentucky*
Paula Engelhardt, *Tennessee Technical University*
Davene Eryes, *North Seattle Community College*
Gerard Fasel, *Pepperdine University*
Luciano Fleischfresser, *OSSM Autry Tech*
Cynthia Galovich, *University of Northern Colorado*
Bertram Gamory, *Monroe Community College*
Sambandamurthy Ganapathy, *SUNY at Buffalo*
Delena Gatch, *Georgia Southern University*
Richard Gelderman, *Western Kentucky University*
Martin Gelfand, *Colorado State University*
Terry Golding, *University of North Texas*
Robert Gramer, *Lake City Community College*
William Gregg, *Louisiana State University*
Paul Gresser, *University of Maryland*
Robert Hagood, *Washtenaw Community College*
Jason Harlow, *University of Toronto*
Heath Hatch, *University of Massachusetts*

Carl Hayn, *Santa Clara University*
James Heath, *Austin Community College*
Zvonko Hlousek, *California State University Long Beach*
Greg Hood, *Tidewater Community College*
Sebastian Hui, *Florence-Darlington Technical College*
Eric Hudson, *The Pennsylvania State University*
Joey Huston, *Michigan State University*
David Iadevaia, *Pima Community College—East Campus*
Fred Jarka, *Stark State College*
Ana Jofre, *University of North Carolina—Charlotte*
Daniel Jones, *Georgia Tech*
Erik Jensen, *Chemeketa Community College*
Todd Kalisik, *Northern Illinois University*
Ju H. Kim, *University of North Dakota*
Armen Kocharian, *California State University Northridge*
J. M. Kowalski, *University of North Texas*
Laird Kramer, *Florida International University*
Christopher Kulp, *Eastern Kentucky University*
Richard Kurtz, *Louisiana State University*
Kenneth Lande, *University of Pennsylvania*
Tiffany Landry, *Folsom Lake College*
Todd Leif, *Cloud County Community College*
John Levin, *University of Tennessee—Knoxville*
John Lindberg, *Seattle Pacific University*
Jeff Loats, *Metropolitan State University of Denver*
Rafael López-Mobilia, *The University of Texas at San Antonio*
Robert W. Lutz, *Drake University*
Lloyd Makorowitz, *SUNY Farmingdale*
Colleen Marlow, *Rhode Island College*
Eric Martell, *Millikin University*
Mark Masters, *Indiana University—Purdue*
John McClain, *Temple College*
Denise Meeks, *Pima Community College*
Henry Merrill, *Fox Valley Technical College*
Mike Meyer, *Michigan Technological University*
Karie Meyers, *Pima Community College*
Tobias Moleski, *Nashville State Tech*
April Moore, *North Harris College*
Gary Morris, *Rice University*
Krishna Mukherjee, *Slippery Rock University*
Charley Myles, *Texas Tech University*
Meredith Newby, *Clemson University*
David Nice, *Bryn Mawr*
Fred Olness, *Southern Methodist University*
Charles Oliver Overstreet, *San Antonio College*
Paige Ouzts, *Lander University*
Russell Palma, *Minnesota State University—Mankato*
Richard Panek, *Florida Gulf Coast University*
Joshua Phiri, *Florence-Darling Technical College*
Iulia Podariu, *University of Nebraska at Omaha*
David Potter, *Austin Community College*
Promod Pratap, *University of North Carolina—Greensboro*
Michael Pravica, *University of Nevada, Las Vegas*
Earl Prohofsky, *Purdue University*
Marilyn Rands, *Lawrence Technological University*
Andrew Rex, *University of Puget Sound*
Andrew Richter, *Valparaiso University*
William Robinson, *North Carolina State University*
Phyliss Salmons, *Embry—Riddle Aeronautical University*
Michael Schaab, *Maine Maritime Academy*
Bruce Schumm, *University of California, Santa Cruz*
Mizuho Schwalm, *University of Minnesota Crookston*
Cindy Schwarz, *Vassar College*
Natalia Semushkhina, *Shippensburg University*
Khazhgery (Jerry) Shakov, *Tulane University*
Kathy Shan, *University of Toledo*
Anwar Sheikh, *Colorado Mesa University*
Bart Sheinberg, *Houston Community College*
Marllin Simon, *Auburn University*
Kenneth Smith, *Pennsylvania State University*
Michael Smutko, *Northwestern University*
Jon Son, *Boston University*
Noel Stanton, *Kansas State University*
Donna Stokes, *University of Houston*
Chuck Stone, *North Carolina A&T*
Chun Fu Su, *Mississippi State University*
Jeffrey Sudol, *West Chester University*
Scott Thompson, *Georgia Gwinnett College*
William Tireman, *Northern Michigan University*
Negussie Tirfessa, *Manchester Community College*
Rajive Tiwari, *Belmont Abbey College*
Herman Trivilino, *College of the Mainland*
Dmitri Tsybychev, *Stony Brook University*
Douglas Tussey, *Pennsylvania State University*
Stephen Van Hook, *Pennsylvania State University*
Manuel Valera, *Slippery Rocky University*
Christos Valiotis, *Antelope Valley College*
James Vesenka, *University of New England*
Stamatis Vokos, *Seattle Pacific University*
James Wanliss, *Embry—Riddle Aeronautical University*
Henry Weigel, *Arapahoe Community College*
Luc T. Wille, *Florida Atlantic University*
Courtney Willis, *University of Northern Colorado*
Katherine Wu, *University of Tampa*
Ali Yazdi, *Jefferson State Community College*
David Young, *Louisiana State University*
Hsiao-Ling Zhou, *Georgia State University*
Todd Zimmerman, *University of Wisconsin—Stout*
Ulrich Zurcher, *Cleveland State University*

About the Cover

The cover image isn't just a pretty picture. All of the elements—the lensing by the water droplet, the structure of the feather, the mechanism for the feather colors—make an appearance as applications of the physics concepts that students are learning.

Preface to the Student

One may say the eternal mystery of the world is its comprehensibility.
—Albert Einstein

If you are taking a course for which this book is assigned, you probably aren't a physics major or an engineering major. It's likely that you aren't majoring in a physical science. So why are you taking physics?

It's almost certain that you are taking physics because you are majoring in a discipline that requires it. Someone, somewhere, has decided that it's important for you to take this course. And they are right. There is a lot you can learn from physics, even if you don't plan to be a physicist. We regularly hear from doctors, physical therapists, biologists, and others that physics was one of the most interesting and valuable courses they took in college.

So, what can you expect to learn in this course? Let's start by talking about what physics is. Physics is a way of thinking about the physical aspects of nature. Physics is not about "facts." It's far more focused on discovering *relationships* between facts and the *patterns* that exist in nature than on learning facts for their own sake. Our emphasis will be on thinking and reasoning. We are going to look for patterns and relationships in nature, develop the logic that relates different ideas, and search for the reasons *why* things happen as they do.

The concepts and techniques you will learn will have a wide application. In this text we have a special emphasis on applying physics to understanding the living world. You'll use your understanding of charges and electric potential to analyze the electrical signal produced when your heart beats. You'll learn how sharks can detect this signal to locate prey and, further, how and why this electrical sensitivity seems to allow hammerhead sharks to detect magnetic fields, aiding navigation in the open ocean.

Like any subject, physics is best learned by doing. "Doing physics" in this course means solving problems, applying what you have learned to answer questions at the end of the chapter. When you are given a homework assignment, you may find yourself tempted to simply solve the problems by thumbing through the text looking for a formula that seems like it will work. This isn't how to do physics; if it was, whoever required you to take this course wouldn't bother. The folks who designed your major want you to learn to *reason,* not to "plug and chug." Whatever you end up studying or doing for a career, this ability will serve you well.

How do you learn to reason in this way? There's no single strategy for studying physics that will work for all students, but we can make some suggestions that will certainly help:

- **Read each chapter *before* it is discussed in class.** Class attendance is much more effective if you have prepared.
- **Use the other resources that accompany the text.** The text includes many videos and online tools to help you better master new material.
- **Participate actively in class.** Take notes, ask and answer questions, take part in discussion groups. There is ample scientific evidence that *active participation* is far more effective for learning science than is passive listening.
- **After class, go back for a careful rereading of the chapter.** In your second reading, pay close attention to the details and the worked examples. Look for the *logic* behind each example, not just at what formula is being used.
- **Apply what you have learned to the homework problems at the end of each chapter.** By following the techniques of the worked examples, applying the tactics and problem-solving strategies, you'll learn how to apply the knowledge you are gaining.
- **Form a study group with two or three classmates.** There's good evidence that students who study regularly with a group do better than the rugged individualists who try to go it alone.
- **Don't be afraid to ask questions.** The more you engage with your instructor and other students, the more successful you will be.

We have one final suggestion. As you read the book, take part in class, and work through problems, step back every now and then to appreciate the big picture. You are going to study topics that range from motions in the solar system to the electrical signals in the nervous system that let you tell your hand to turn the pages of this book. It's a remarkable breadth of topics and techniques that is based on a very compact set of organizing principles.

Now, let's get down to work.

Studying for and Taking the MCAT Exam

If you are taking the College Physics course, there's a good chance that you are majoring in the biological sciences. There's also a good chance that you are preparing for a career in the health professions, and so might well be required to take the Medical College Admission Test, the MCAT exam.

The *Chemical and Physical Foundations of Biological Systems* section of the MCAT assesses your understanding of the concepts of this course by testing your ability to apply these concepts to living systems. You will be expected to use what you've learned to analyze situations you've never seen before, making simplified but realistic models of the world. Your reasoning skills will be just as important as your understanding of the universal laws of physics.

Structure of the MCAT Exam

Most of the test consists of a series of passages of technical information followed by a series of questions based on each passage, much like the passage problems at the end of each chapter in this book. Some details:

- **The passages and the questions are *always* integrated.** Understanding the passage and answering the questions will require you to use knowledge from several different areas of physics.
- **Passages will generally be about topics for which you do not have detailed knowledge.** But, if you read carefully, you'll see that the treatment of the passage is based on information you should know well.
- **The test assumes a basic level of background knowledge.** You'll need to have facility with central themes and major concepts, but you won't need detailed knowledge of any particular topic. Such detailed information, if needed, will be provided in the passage.
- **You can't use calculators on the test, so any math that you do will be reasonably simple.** Quickly estimating an answer with ratio reasoning or a knowledge of the scale of physical quantities will be a useful skill.
- **The answers to the questions are all designed to be plausible.** You can't generally weed out the "bad" answers with a quick inspection.
- **The test is given online.** Practicing with Mastering Physics will help you get used to this format.

Preparing for the Test

Because you have used this book as a tool for learning physics, you should use it as a tool for reviewing for the MCAT exam. Several of the key features of the book will be useful for this, including some that were explicitly designed with the MCAT exam in mind.

As you review the chapters:

- Start with the *Chapter Previews,* which provide a "big picture" overview of the content. What are the major themes of each chapter?
- Look for the *Synthesis* boxes that bring together key concepts and equations. These show connections and highlight differences that you should understand and be ready to apply.
- Go through each chapter and review the *Stop to Think* exercises. These are a good way to test your understanding of the key concepts and techniques.
- Each chapter closes with a passage problem that is designed to be "MCAT-exam-like." They'll give you good practice with the "read a passage, answer questions" structure of the MCAT exam.

The passage problems are a good tool, but the passages usually don't integrate topics that span several chapters—a key feature of the MCAT exam. For integrated passages and problems, turn to the *Part Summaries:*

- For each Part Summary, read the *One Step Beyond* passage and answer the associated questions.
- After this, read the passages and answer the questions that end each Part Summary section. These passages and associated problems are—by design—very similar to the passages and questions you'll see on the actual MCAT exam.

Taking the Test: Reading the Passage

As you read each passage, you'll need to interpret the information presented and connect it with concepts you are familiar with, translating it into a form that makes sense based on your background.

The next page shows a passage that was written to very closely match the style and substance of an actual MCAT passage. Blue annotations highlight connections you should make as you read. The passage describes a situation (the mechanics and energetics of sled dogs) that you probably haven't seen before. But the basic physics (friction, energy conversion) are principles that you are familiar with, principles that you have seen applied to related situations. When you read the passage, think about the underlying physics concepts and how they apply to this case.

Translating the Passage
As you read the passage, do some translation. Connect the scenario to examples you've seen before, translate given information into forms you are familiar with, think about the basic physical principles that apply.

Passage X

For travel over snow, a sled with runners that slide on snow is the best way to get around. Snow is slippery, but there is still friction between runners and the ground; the forward force required to pull a sled at a constant speed might be 1/6 of the sled's weight.

As you read this part of the passage, think about the forces involved: For a sled moving at a constant speed, there is no net force. The downward weight force is equal to the upward normal force; the forward pulling force must be equal to the friction force, which is acting opposite the sled's motion. There are many problems like this in Chapter 5.

Part of translating is converting given information into a more usual or more useful form. This is really a statement about the coefficient of kinetic friction.

The pulling force might well come from a dog. In a typical sled, the rope that the dog uses to pull attaches at a slight angle, as in Figure 1. The pulling force is the horizontal component of the tension in the rope.

The force applied to the sled is the tension force in the rope, which is shown at an angle. The horizontal component is the pulling force; you're told this. There is a vertical component of the force as well.

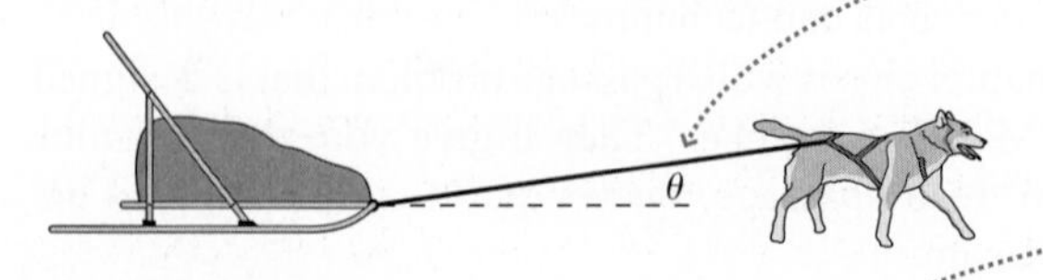

Figure 1

In the data given here, and the description given above, the sled moves at a constant speed—there is no mention of acceleration anywhere in this passage. In such cases, the net force is zero and the kinetic energy of the sled isn't changing.

Sled dogs have great aerobic capacity; a 40 kg dog can provide output power to pull with a 60 N force at 2.2 m/s for hours. The output power is related to force and velocity by $P = F \cdot v$, so they can pull lighter loads at higher speeds.

Notice that the key equation relating power, force, and velocity is given to you. That's to be expected. Any specific information, including equations, constants, and other such details, will generally be given in the passage. The MCAT is a test of reasoning, not recall.

Doing 100 J of work means that a dog must expend 400 J of metabolic energy. The difference must be exhausted as heat; given the excellent insulation provided by a dog's fur, this is mostly via evaporation as it pants. At a typical body temperature, the evaporation of 1.0 l of water carries away 240,000 J, so this is an effective means of cooling.

The concepts of metabolic energy and energy output are treated in Chapter 11. The details here match those in the chapter (as they should!); this corresponds to an efficiency of 25%. 400 J of energy is used by the body; 25% of this, 100 J, is the energy output. This means that 300 J is exhausted as heat.

Chapter 12 discusses means of heat transfer: conduction, convection, radiation, evaporation. This paragraph gives biological details about dogs that you can interpret as follows: A dog's fur limits transfer by conduction, convection, and radiation; evaporation of water by a panting dog must take up the slack.

The specific data for energy required to evaporate water are given. If you need such information to answer questions, it will almost certainly be provided. As we noted above, this is a test of reasoning, not recall.

FIGURE MCAT-EXAM.1 Interpreting a passage.

Taking the Test: Answering the Questions

The passages on the MCAT exam seem complicated at first, but, as we've seen, they are about basic concepts and central themes that you know well. The same is true of the questions; they aren't as difficult as they may seem at first. As with the passage, you should start by translating the questions, identifying the physical concepts that apply in each case. You then proceed by reasoning, determining the solution to the question, using your understanding of these basic concepts. The practical suggestions below are followed by a detailed overview of the solutions to the questions based on the passage on the previous page.

You Can Answer the Questions in Any Order

The questions test a range of skills and have a range of difficulties. Many questions will involve simple reading comprehension; these are usually quite straightforward. Some require sophisticated reasoning and (slightly) complex mathematical manipulations. Start with the easy ones, ones that you can quickly solve. Save the more complex ones for later, and skip them if time is short.

Take Steps to Simplify or Eliminate Calculations

You won't be allowed to use a calculator on the exam, so any math that you do will be reasonably straightforward. To rapidly converge on a correct answer choice, there are some important "shortcuts" that you can take.

- **Use ratio reasoning.** What's the relationship between the variables involved in a question? You can use this to deduce the answer with only a very simple calculation, as we've seen many times in the book. For instance, suppose you are asked the following question:

  A model rocket is powered by chemical fuel. A student launches a rocket with a small engine containing 1.0 g of combustible fuel. The rocket reaches a speed of 10 m/s. The student then launches the rocket again, using an engine with 4.0 g of fuel. If all other parameters of the launch are kept the same, what final speed would you expect for this second trial?

  This is an energy conversion problem: Chemical energy of the fuel is converted to kinetic energy of the rocket. Kinetic energy is related to the speed by $K = \frac{1}{2}mv^2$. The chemical energy—and thus the kinetic energy—in the second trial is increased by a factor of 4. Since $K \sim v^2$, the speed must increase by a factor of 2, to 20 m/s.
- **Simplify calculations by liberally rounding numbers.** You can round off numbers to make calculations more straightforward. Your final result will probably be close enough to choose the correct answer from the list given. For instance, suppose you are asked the following question:

  A ball moving at 2.0 m/s rolls off the edge of a table that's 1.2 m high. How far from the edge of the table does the ball land?

  A. 2 m B. 1.5 m C. 1 m D. 0.5 m

  We know that the vertical motion of the ball is free fall; so the vertical distance fallen by the ball in a time Δt is $\Delta y = -\frac{1}{2}gt^2$. The time to fall 1.2 m is $\Delta t = \sqrt{2(1.2\text{ m})/g}$. Rather than complete this calculation, we estimate the results as follows: $\Delta t = \sqrt{2.4/9.8} \approx \sqrt{1/4} = 1/2 = 0.5\text{s}$. During this free-fall time, the horizontal motion is constant at 2.0 m/s, so we expect the ball to land about 1 m away. Our quick calculation shows us that the correct answer is choice C—no other answer is close.
- **For calculations using values in scientific notation, compute either the first digits or the exponents, not both.** In some cases, a quick calculation can tell you the correct leading digit, and that's all you need to figure out the correct answer. In other cases, you'll find possible answers with the same leading digit but very different exponents or decimal places. In this case, all you need is a simple order-of-magnitude estimate to decide on the right result.
- **Where possible, use your knowledge of the expected scale of physical quantities to quickly determine the correct answer.** For instance, suppose a question asks you to find the photon energy for green light of wavelength 550 nm. Visible light has photon energies of about 2 eV, or about 3×10^{-19} J, and that might be enough information to allow you to pick out the correct answer with no calculation.
- **Beware of "distractors," answers that you'll get if you make common mistakes.** For example, Question 4 on the next page is about energy conversion. The dog is keeping the sled in motion, so it's common for students to say that the dog is converting chemical energy in its body into kinetic energy. However, the kinetic energy isn't changing. The two answer choices that involve kinetic energy are common, but incorrect, choices. Be aware that the questions are constructed to bring out such misconceptions and that these tempting, but wrong, answer choices will be provided.

One Final Tip: Look at the Big Picture

The MCAT exam tests your ability to look at a technical passage about which you have some background knowledge and quickly get a sense of what it is saying, enough to answer questions about it. Keep this big picture in mind:

- **Don't get bogged down in technical details of the particular situation.** Focus on the basic physics.
- **Don't spend too much time on any one question.** If one question is taking too much time, make an educated guess and move on.
- **Don't get confused by details of notation or terminology.** For instance, different people use different symbols for physical variables. In this text, we use the symbol K for kinetic energy; others use E_K.

Finally, don't forget the most important aspect of success on the MCAT exam: The best way to prepare for this or any test is simply to understand the subject. As you prepare for the test, focus your energy on reviewing and refining your knowledge of central topics and techniques, and practice applying your knowledge by solving problems like you'll see on the actual MCAT.

Translating

Look at the questions and think about the physics principles that apply, how they connect to concepts you know and understand.

Tips

- Numerical choices are presented in order; that's the usual practice on the test. Estimate the size of the answer, and think about where it falls.
- For questions with sentences as choices, decide on the solution before you look at the choices; this will save time reading.

Reasoning

Think about the question and the range of possible answers, and converge to a solution with as few steps as possible—time is limited!

This is a question about the size of the friction force. You are told that it takes a force that's about 1/6 of the sled's weight to pull it forward on snow. You can estimate the friction coefficient from this information.

1. What is the approximate coefficient of kinetic friction for a sled on snow?

A. 0.35
B. 0.25
C. 0.15
D. 0.05

For an object on level ground, the normal force equals the weight force. If the sled is moving at a constant speed, the pulling force equals the friction force. This implies that $\mu = f_k/n = f_{pull}/w = 1/6$. Two of the answer choices convert easily to fractions: 0.25 = 1/4; 0.05 = 1/20. 1/6 is between these, so C must be our choice. (Indeed, 1/6 = 0.167, so 0.15 is pretty close.)

If the speed is constant, there is no net force. We are told that the pulling force is the horizontal component of the tension force, not the tension force itself. Because there is no net force, this horizontal component is equal to the friction force, which is directed backward. So this is really a question about the friction force.

2. If a rope pulls at an angle, as in Figure 1, how will this affect the pulling force necessary to keep the sled moving at a constant speed?

A. This will reduce the pulling force.
B. This will not change the pulling force.
C. This will increase the pulling force.
D. It will increase or decrease the pulling force, depending on angle.

A vertical component of the tension force will reduce the normal force, reducing the friction force—and thus the pulling force.

We assume that the output power is the same for the two cases—this is implied in the passage.

3. A dog pulls a 40 kg sled at a maximum speed of 2 m/s. What is the maximum speed for an 80 kg sled?

A. 2 m/s
B. 1.5 m/s
C. 1.0 m/s
D. 0.5 m/s

Doubling the weight doubles the normal force, which doubles the friction force. This will double the necessary pulling force as well. Given the expression for power given in the passage, this means the maximum speed will be halved.

This is a question about energy transformation. For such questions, think about changes. What forms of energy are *changing*? We know that thermal energy is part of the picture because some of the chemical energy is converted to thermal energy in the dog's body.

4. As a dog pulls a sled at constant speed, chemical energy in the dog's body is converted to

A. kinetic energy
B. thermal energy
C. kinetic energy and thermal energy
D. kinetic energy and potential energy

Choice B is correct, but A and C are clever distractors. It's tempting to choose an answer that includes kinetic energy. The sled is in motion, after all! But don't be swayed. The kinetic energy isn't changing, and friction to the sled converts any energy the dog supplies into thermal energy.

Increasing speed increases power, as the passage told us. But the energy to pull the sled is not the *power*, it's the *work*, and we know that the work is $W = F\Delta x$. This is a question about work and energy, not about power.

5. A dog pulls a sled for a distance of 1.0 km at a speed of 1 m/s, requiring an energy output of 60,000 J. If the dog pulls the sled at 2 m/s, the necessary energy is

A. 240,000 J
B. 120,000 J
C. 60,000 J
D. 30,000 J

Doubling the speed doubles the power, but it doesn't change the force; that's fixed by friction. The distance is the same as well, and so is the work done, the energy required. Since the speed doubles, it's tempting to think the energy doubles, though. This "obvious" but incorrect solution is one of the choices—expect such situations on the actual MCAT.

The passage tells us that the dog uses 400 J of metabolic energy to do 100 J of work. 300 J, or 75%, must be exhausted to the environment. We can assume the same efficiency here.

6. A dog uses 100,000 J of metabolic energy pulling a sled. How much energy must the dog exhaust by panting?

A. 100,000 J
B. 75,000 J
C. 50,000 J
D. 25,000 J

If 75% of the energy must be exhausted to the environment, that's 75,000 J.

FIGURE MCAT-EXAM.2 Answering the questions for the passage of Figure MCAT-EXAM.1.

Real-World Applications

Applications of biological or medical interest are marked BIO in the list below, including MCAT-style Passage Problems. Other end-of-chapter problems of biological or medical interest are marked BIO in the chapter.

Chapter 21

Chapter 22

Chapter 23

Chapter 24

Chapter 25

Chapter 26

Part VI Summary

Chapter 27

Chapter 28

Chapter 29

Chapter 30

Part VII Summary

Detailed Contents

PART II Conservation Laws

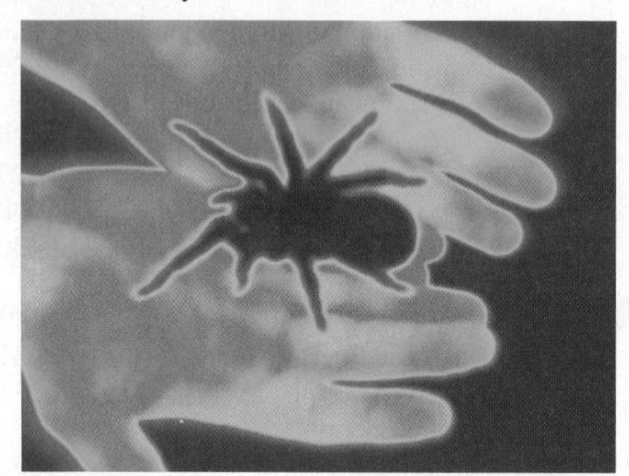

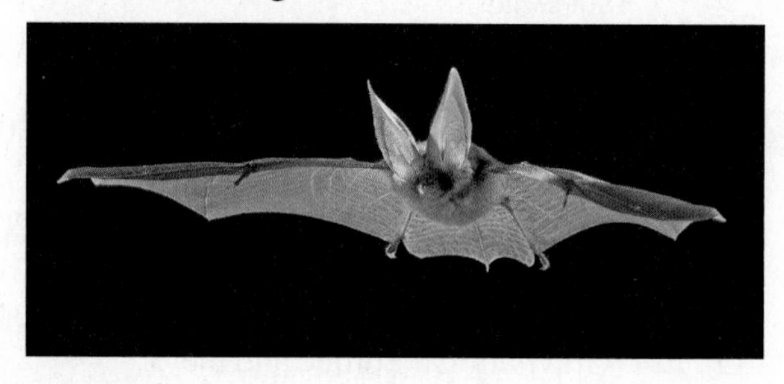

PART

I

Force and Motion

The cheetah is the fastest land animal, able to run at speeds exceeding 60 miles per hour. Nonetheless, the rabbit has an advantage in this chase. It can *change* its motion more quickly and will likely escape. How can you tell, by looking at the picture, that the cheetah is changing its motion?

OVERVIEW

The Science of Physics

Physics is the foundational science that underlies biology, chemistry, earth sciences, and all other fields that attempt to understand our natural world. Physicists couple careful experimentation with deep theoretical insights to build powerful and predictive models of how the world works. A key aspect of physics is that it is a *unifying* discipline: A relatively small number of key concepts can explain a vast array of natural phenomena. In this text, we have organized the chapters into parts according to seven of these unifying principles. Each of the seven parts of this text opens with an overview that gives you a look ahead, a glimpse of where your journey will take you in the next few chapters. It's easy to lose sight of the big picture while you're busy negotiating the terrain of each chapter. In Part I, the big picture is, in a word, *change*.

Why Things Change

Simple observations of the world around you show that most things change. Some changes, such as aging, are biological. Others, such as the burning of gasoline in your car, are chemical. We will look at changes that involve *motion* of one form or another—running and jumping, throwing balls, lifting weights.

There are two big questions we must tackle to study how things change by moving:

- **How do we describe motion?** How should we measure or characterize the motion if we want to analyze it mathematically?
- **How do we explain motion?** Why do objects have the particular motion they do? Why, when you toss a ball upward, does it go up and then come back down rather than keep going up? What are the "laws of nature" that allow us to predict an object's motion?

Two key concepts that will help answer these questions are *force* (the "cause") and *acceleration* (the "effect"). Our basic tools will be three laws of motion worked out by Isaac Newton. Newton's laws relate force to acceleration, and we will use them to explain and explore a wide range of problems. As we learn to solve problems dealing with motion, we will learn basic techniques that we can apply in all the parts of this text.

Simplifying Models

Another key aspect of physics is the importance of models. Suppose we want to analyze a ball moving through the air. Is it necessary to analyze the way the atoms in the ball are connected? Or the details of how the ball is spinning? Or the small drag force it experiences as it moves? These are interesting questions, of course. But if our task is to understand the motion of the ball, we need to simplify!

We can conduct a perfectly fine analysis of the ball's motion if we treat the ball as a single particle moving through the air. This is a *model* of the situation. A model is a simplified description of reality that is used to reduce the complexity of a problem so it can be analyzed and understood. Model building is a major part of the strategy that we will develop for solving problems throughout the text. Learning how to simplify a situation is the essence of successful modeling—and successful problem solving.

1 Representing Motion

As this falcon moves in a graceful arc through the air, the direction of its motion and the distance between each of its positions and the next are constantly changing. What language should we use to describe this motion?

LOOKING AHEAD ▸▸

Chapter Preview

Each chapter starts with a preview outlining the major topics and what you'll be learning for each topic.

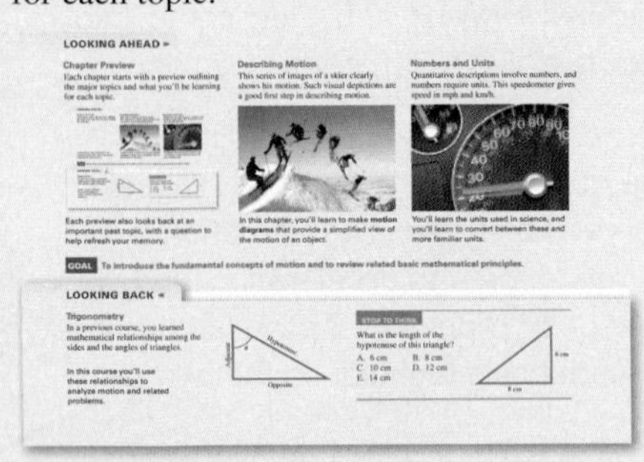

Each preview also looks back at an important past topic, with a question to help refresh your memory.

Describing Motion

This series of images of a skier clearly shows his motion. Such visual depictions are a good first step in describing motion.

In this chapter, you'll learn to make **motion diagrams** that provide a simplified view of the motion of an object.

Numbers and Units

Quantitative descriptions involve numbers, and numbers require units. This speedometer gives speed in mph and km/h.

You'll learn the units used in science, and you'll learn to convert between these and more familiar units.

GOAL **To introduce the fundamental concepts of motion and to review related basic mathematical principles.**

LOOKING BACK ◂◂

Trigonometry

In a previous course, you learned mathematical relationships among the sides and the angles of triangles.

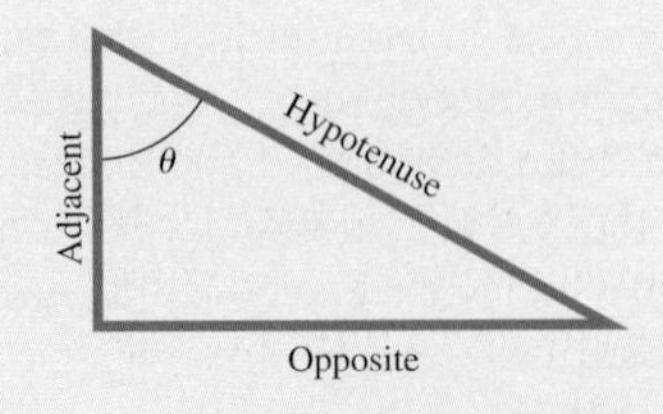

In this course you'll use these relationships to analyze motion and related problems.

STOP TO THINK

What is the length of the hypotenuse of this triangle?

A. 6 cm B. 8 cm
C. 10 cm D. 12 cm
E. 14 cm

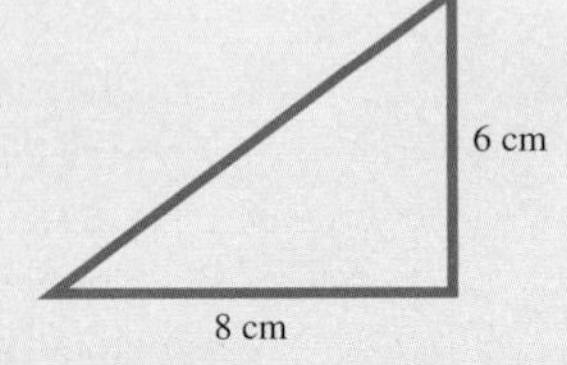

1.1 Motion: A First Look

Motion is a theme that will appear in one form or another throughout this entire text. You have a well-developed intuition about motion based on your experiences, but we'll find that some of the most important aspects of motion can be rather subtle. We need to develop some tools to help us explain and understand motion, so rather than jumping immediately into a lot of mathematics and calculations, this first chapter focuses on *visualizing* motion and becoming familiar with the *concepts* needed to describe a moving object.

One key difference between physics and other sciences is how we set up and solve problems. We'll often use a two-step process to solve motion problems. The first step is to develop a simplified *representation* of the motion so that key elements stand out. For example, the photo of the falcon at the start of the chapter allows us to observe its position at many successive times. We will begin our study of motion by considering this sort of picture. The second step is to analyze the motion with the language of mathematics. The process of putting numbers on nature is often the most challenging aspect of the problems you will solve. In this chapter, we will explore the steps in this process as we introduce the basic concepts of motion.

Types of Motion

As a starting point, let's define **motion** as the change of an object's position or orientation with time. Examples of motion are easy to list. Bicycles, baseballs, cars, airplanes, and rockets are all objects that move. The path along which an object moves, which might be a straight line or might be curved, is called the object's **trajectory.**

FIGURE 1.1 shows four basic types of motion that we will study in this text. In this chapter, we will focus on the first type of motion in the figure, motion along a straight line, or *straight-line motion*. In later chapters, we will learn about *circular motion*, which is the motion of an object along a circular path; *projectile motion*, the motion of an object through the air; and *rotational motion*, the spinning of an object about an axis.

FIGURE 1.1 Four basic types of motion.

Straight-line motion

Circular motion

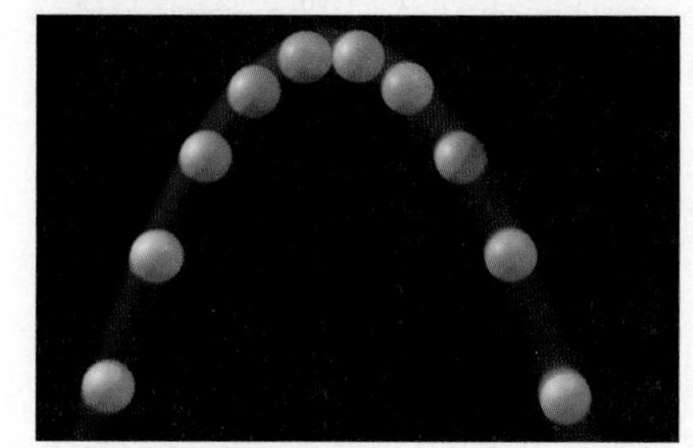

Projectile motion

Rotational motion

Making a Motion Diagram

An easy way to study motion is to record a video of a moving object with a stationary camera. A video camera takes images at a fixed rate, typically 30 images every second. Each separate image is called a *frame*. As an example, FIGURE 1.2 shows several frames from a video of a car going past, with the camera in a fixed position. Not surprisingly, the car is in a different position in each frame.

FIGURE 1.2 Several frames from the video of a car.

FIGURE 1.3 A motion diagram of the car shows all the frames simultaneously.

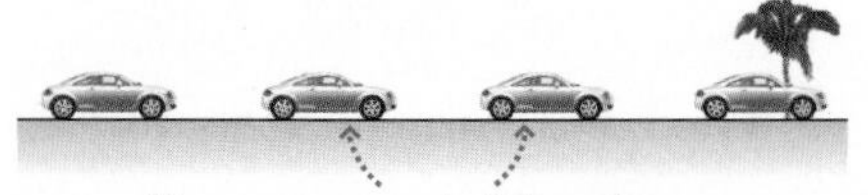
The same amount of time elapses between each image and the next.

Suppose we now edit the video by layering the frames on top of each other. We end up with the picture in FIGURE 1.3. This composite image, showing an object's positions at several *equally spaced instants of time,* is called a **motion diagram.** As simple as motion diagrams seem, they will turn out to be powerful tools for analyzing motion.

Now let's take our camera out into the world and make some motion diagrams. The following table illustrates how a motion diagram shows important features of different kinds of motion.

Examples of motion diagrams

Images that are *equally spaced* indicate an object moving with *constant speed.*

A skateboarder rolling down the sidewalk.

An *increasing distance* between the images shows that the object is *speeding up.*

A sprinter starting the 100 meter dash.

A *decreasing distance* between the images shows that the object is *slowing down.*

A car stopping for a red light.

A more complex motion diagram shows changes in speed and direction.

A basketball free throw.

We have defined several concepts (constant speed, speeding up, and slowing down) in terms of how the moving object appears in a motion diagram. These are called **operational definitions,** meaning that the concepts are defined in terms of a particular procedure or operation. For example, we could answer the question Is the airplane speeding up? by checking whether the images in the plane's motion diagram are getting farther apart. Many of the concepts in physics will be introduced as operational definitions. This reminds us that physics is an experimental science.

STOP TO THINK 1.1 Which car is going faster, A or B? Assume there are equal intervals of time between the frames of both videos.

Car A
Car B

NOTE ▶ Each chapter in this text has several *Stop to Think* questions. These questions are designed to see if you've understood the basic ideas that have just been presented. The answers are given at the end of the chapter, but you should make a serious effort to think about these questions before turning to the answers. ◀

1.2 Models and Modeling

The real world is messy and complicated. Our goal in studying physics is to brush aside many of the real-world details in order to discern patterns that occur over and over. For example, a swinging pendulum, a vibrating guitar string, a sound wave, and jiggling atoms in a crystal are all very different—yet they share a common core characteristic: Each is an example of an *oscillating system*, something that moves back and forth around an equilibrium position. If we focus on understanding a very simple oscillating system, such as a block (generically, a "mass") attached to a spring, we'll automatically understand quite a bit about the many real-world examples of oscillations.

Stripping away the details to focus on essential features is a process called *modeling*. A **model** is a highly simplified picture of reality, but one that still captures the essence of what we want to study. Thus a mass attached to a spring is a simple but realistic model of many oscillating systems.

Models allow us to make sense of complex situations by providing a framework for thinking about them. One could go so far as to say that developing and testing models is at the heart of the scientific process. Albert Einstein once said, "Physics should be as simple as possible—but not simpler." We want to find the simplest model that allows us to understand the phenomenon we're studying, but we can't make the model so simple that key aspects of the phenomenon get lost.

We'll develop and use many models throughout this text; they'll be one of our most important thinking tools. These models will be of two types:

- *Descriptive models:* What are the essential characteristics and properties of a phenomenon? How do we describe it in the simplest possible terms? For example, the mass-on-a-spring model of an oscillating system is a descriptive model.
- *Explanatory models:* Why do things happen as they do? Explanatory models, based on the laws of physics, have predictive power. They allow us to test—against experimental data—whether a model provides an adequate explanation of our observations. For example, the *charge model* that we will introduce in Chapter 20 helps us explain and predict a wide range of experimental outcomes related to electric forces.

When we solve physics problems, one of the most important steps is choosing an appropriate model for the system we are studying. In the worked examples in this text, in the first "Strategize" step, we'll point out the model being used, when appropriate.

The Particle Model

For many objects, the motion of the object *as a whole* is not influenced by the details of the object's size and shape. To describe the object's motion, all we really need to keep track of is the motion of a single point: You could imagine looking at the motion of a dot painted on the side of the object.

In fact, for the purposes of analyzing the motion, we can often consider the object *as if* it were just a single point. We can also treat the object *as if* all of its mass were concentrated into this single point. An object that can be represented as a mass at a single point in space is called a **particle.**

If we treat an object as a particle, we can represent the object in each frame of a motion diagram as a simple dot. FIGURE 1.4 shows how much simpler motion diagrams appear when the object is represented as a particle. Note that the dots have been numbered 0, 1, 2, . . . to tell the sequence in which the frames were exposed. These diagrams still convey a complete understanding of the object's motion.

In representing the car in Figure 1.4 as a particle, we have discarded many of the details of the car, such as the shape of its body and the motion of its wheels, which are unimportant in understanding its overall motion. In other words, we have developed a model for moving objects, the **particle model,** that allows us to see

FIGURE 1.4 Simplifying a motion diagram using the particle model.

(a) Motion diagram of a car stopping

(b) Same motion diagram using the particle model

The same amount of time elapses between each frame and the next.

0 1 2 3

Numbers show the order in which the frames were taken.

A single dot is used to represent the object.

A video to support a section's topic is embedded in the eText.

Video Figure 1.4

FIGURE 1.5 The particle model for two falling objects.

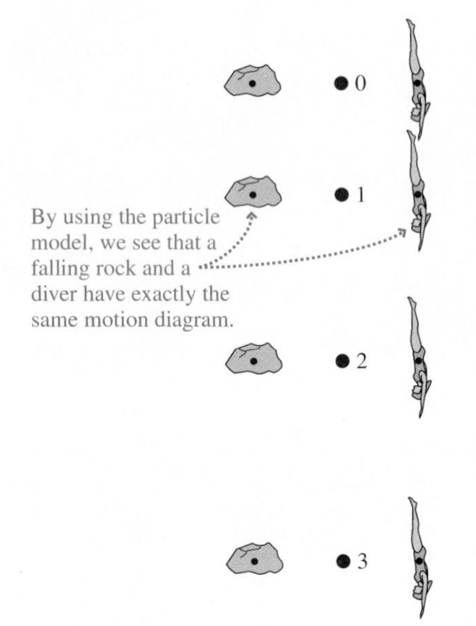

connections that are very important but that are obscured or lost by examining all the parts of an extended, real object. Consider the motion of the rock and the diver shown in FIGURE 1.5. These two very different objects have exactly the same motion diagram. As we will see, all objects falling under the influence of gravity move in exactly the same manner if no other forces act. The simplification of the particle model has revealed something about the physics that underlies both of these situations.

STOP TO THINK 1.2 Three motion diagrams are shown. Which is a dust particle settling to the floor at constant speed, which is a ball dropped from the roof of a building, and which is a descending rocket slowing to make a soft landing on Mars?

A. 0● 1● 2● 3● 4● 5●

B. 0● 1● 2● 3● 4● 5●

C. 0● 1● 2● 3● 4● 5●

1.3 Position and Time: Putting Numbers on Nature

To develop our understanding of motion further, we need to be able to make quantitative measurements: We need to use numbers. As we analyze a motion diagram, it is useful to know where the object is (its *position*) and when the object was at that position (the *time*). We'll start by considering the motion of an object that can move only along a straight line. Examples of this **one-dimensional** or "1-D" motion are a car moving along a long, straight road; an airplane taxiing down a runway; and an elevator moving up and down a shaft.

Position and Coordinate Systems

FIGURE 1.6 Describing your position.

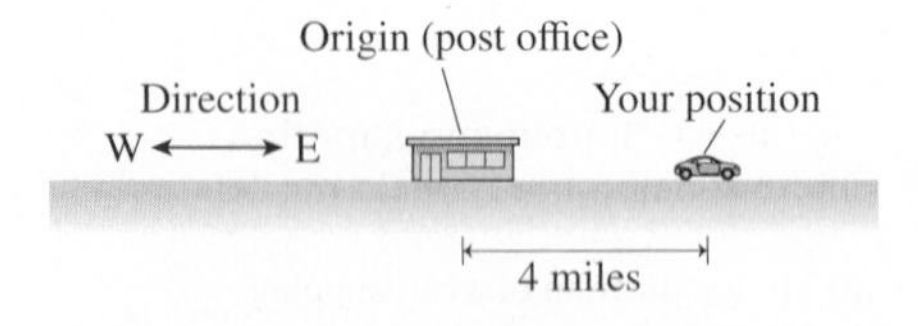

Suppose you are driving along a long, straight country road, as in FIGURE 1.6, and your friend calls and asks where you are. You might reply that you are 4 miles east of the post office, and your friend would then know just where you were. Your location at a particular instant in time (when your friend phoned) is called your **position.** Notice that to know your position along the road, your friend needed three pieces of information. First, you had to give her a reference point (the post office) from which all distances are to be measured. We call this fixed reference point the **origin.** Second, she needed to know how far you were from that reference point or origin—in this case, 4 miles. Finally, she needed to know which side of the origin you were on: You could be 4 miles to the west of it or 4 miles to the east.

We will need these same three pieces of information in order to specify any object's position along a line. We first choose our origin, from which we measure the distance to the object. The position of the origin is arbitrary, and we are free to place it where we like. Usually, however, there are certain points (such as the well-known post office) that are more convenient choices than others.

In order to specify how far our object is from the origin, we lay down an imaginary axis along the line of the object's motion. Like a ruler, this axis is marked off in equally spaced divisions of distance, perhaps in inches, meters, or miles, depending on the problem at hand. We place the zero mark of this ruler at the origin, allowing us to locate the position of our object by reading the ruler mark where the object is.

This gauge's vertical scale measures the depth of snow when it falls. It has a natural origin at the level of the road.

Finally, we need to be able to specify which side of the origin our object is on. To do this, we imagine the axis extending from one side of the origin with increasing positive markings; on the other side, the axis is marked with increasing *negative* numbers. By reporting the position as either a positive or a negative number, we know on what side of the origin the object is.

These elements—an origin and an axis marked in both the positive and negative directions—can be used to unambiguously locate the position of an object. We call this a **coordinate system.** We will use coordinate systems throughout this text, and we will soon develop coordinate systems that can be used to describe the positions of objects moving in more complex ways than just along a line. FIGURE 1.7 shows a coordinate system that we can use to locate various objects along the country road discussed earlier.

FIGURE 1.7 The coordinate system used to describe objects along a country road.

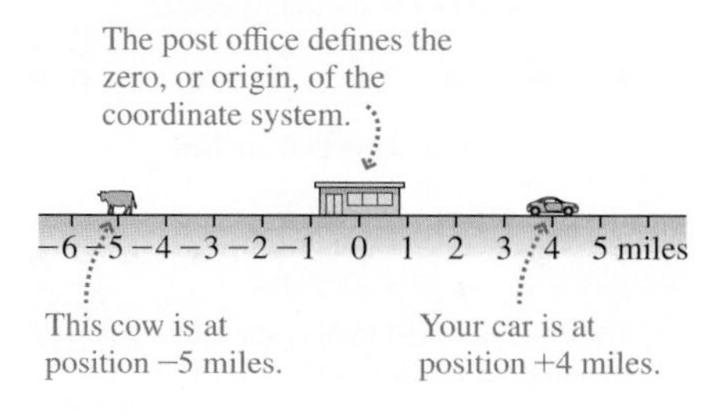

Although our coordinate system works well for describing the positions of objects located along the axis, our notation is somewhat cumbersome because we keep needing to say things like "the car is at position +4 miles." A better notation, and one that will become particularly important when we study motion in two dimensions, is to use a symbol such as x or y to represent the position along the axis. Then we can say "the cow is at $x = -5$ miles." The symbol that represents a position along an axis is called a **coordinate.** The introduction of symbols to represent positions (and, later, velocities and accelerations) also allows us to work with these quantities mathematically.

FIGURE 1.8 shows how we would set up a coordinate system for a sprinter running a 50 meter race (we use the standard abbreviation "m" for meters). For horizontal motion like this we usually use the coordinate x to represent the position.

FIGURE 1.8 A coordinate system for a 50 meter race.

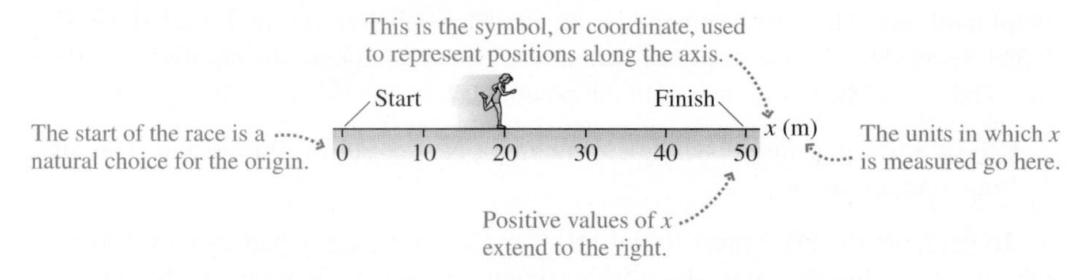

Motion along a straight line need not be horizontal. As shown in FIGURE 1.9, a rock falling vertically downward and a skier skiing down a straight slope are also examples of straight-line or one-dimensional motion.

FIGURE 1.9 Examples of one-dimensional motion.

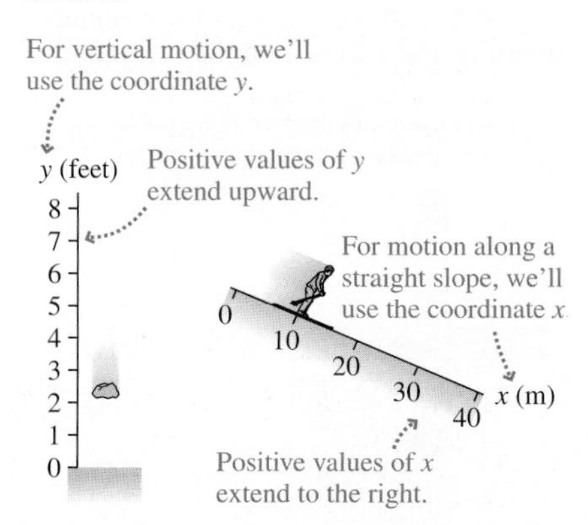

Time

The pictures in Figure 1.9 show the position of an object at just one instant of time. But a full motion diagram represents how an object moves as time progresses. So far, we have labeled the dots in a motion diagram by the numbers 0, 1, 2, . . . to indicate the order in which the frames were taken. But to fully describe the motion, we need to indicate the *time,* as read off a clock or a stopwatch, at which each frame of a video was made. This is important, as we can see from the motion diagram of a stopping car in FIGURE 1.10. If the frames were taken 1 second apart, this motion diagram shows a leisurely stop; if 1/10 of a second apart, it represents a screeching halt.

FIGURE 1.10 Is this a leisurely stop or a screeching halt?

For a complete motion diagram, we thus need to label each frame with its corresponding time (symbol t) as read off a clock. But when should we start the clock? Which frame should be labeled $t = 0$? This choice is much like choosing the origin $x = 0$ of a coordinate system: You can pick any arbitrary point in the motion and label it "$t = 0$ seconds." This is simply the instant you decide to start your clock or stopwatch, so it is the origin of your time coordinate. A video frame labeled "$t = 4$ seconds" means it was taken 4 seconds after you started your clock. We typically choose $t = 0$ to represent the "beginning" of a problem, but the object may have been moving before then.

FIGURE 1.11 The motion diagram of a car that travels at constant speed and then brakes to a halt.

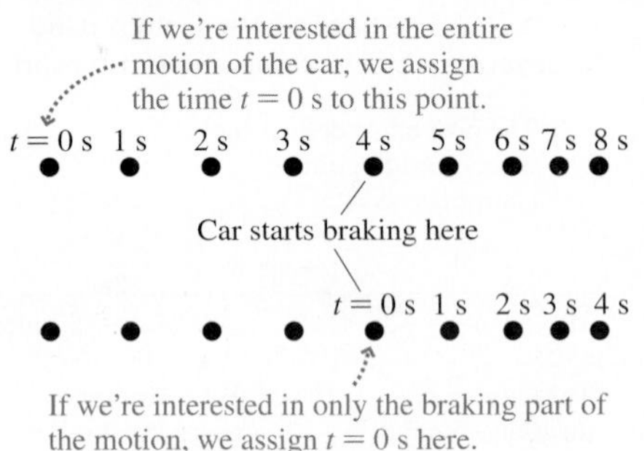

To illustrate, FIGURE 1.11 shows the motion diagram for a car moving at a constant speed and then braking to a halt. Two possible choices for the frame labeled $t = 0$ seconds are shown; our choice depends on what part of the motion we're interested in. Each successive position of the car is then labeled with the clock reading in seconds (abbreviated by "s").

Changes in Position: Displacement

Now that we've seen how to measure position and time, let's return to the problem of motion. To describe motion we'll need to measure the *changes* in position that occur with time. Consider the following:

Sam is standing 50 feet (ft) east of the corner of 12th Street and Vine. He then walks to a second point 150 ft east of Vine. What is Sam's change of position?

FIGURE 1.12 Sam undergoes a displacement Δx from position x_i to position x_f.

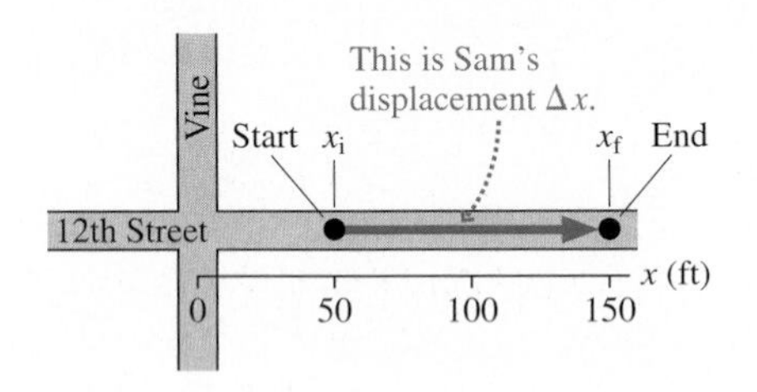

FIGURE 1.12 shows Sam's motion on a map. We've placed a coordinate system on the map, using the coordinate x. We are free to place the origin of our coordinate system wherever we wish, so we have placed it at the intersection. Sam's initial position is then at $x_i = 50$ ft. The positive value for x_i tells us that Sam is east of the origin.

NOTE ▶ We will label special values of x or y with subscripts. The value at the start of a problem is usually labeled with a subscript "i," for *initial,* and the value at the end is labeled with a subscript "f," for *final.* For cases having several special values, we will usually use subscripts "1," "2," and so on. ◀

Sam's final position is $x_f = 150$ ft, indicating that he is 150 ft east of the origin. You can see that Sam has changed position, and a *change* of position is called a **displacement**. His displacement is the distance labeled Δx in Figure 1.12. The Greek letter delta (Δ) is used in math and science to indicate the *change* in a quantity. Thus Δx indicates a change in the position x.

NOTE ▶ Δx is a *single* symbol. You cannot cancel out or remove the Δ in algebraic operations. ◀

To get from the 50 ft mark to the 150 ft mark, Sam clearly had to walk 100 ft, so the change in his position—his displacement—is 100 ft. We can think about displacement in a more general way, however. Displacement is the *difference* between a final position x_f and an initial position x_i. Thus we can write

$$\Delta x = x_f - x_i = 150 \text{ ft} - 50 \text{ ft} = 100 \text{ ft}$$

FIGURE 1.13 A displacement is a signed quantity. Here Δx is a negative number.

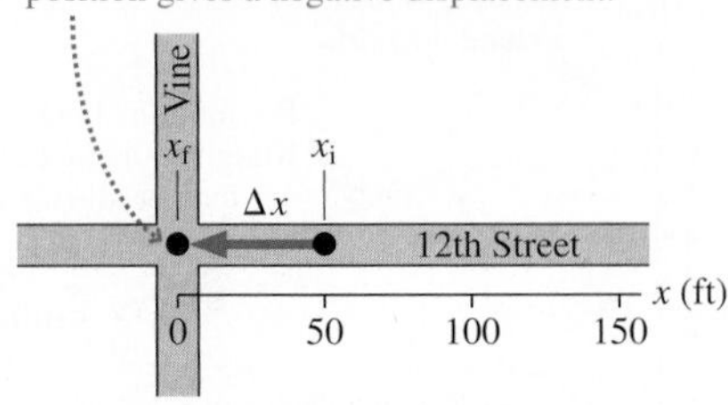

NOTE ▶ A general principle, used throughout this text, is that the change in any quantity is the final value of the quantity minus its initial value. ◀

Displacement is a *signed quantity;* that is, it can be either positive or negative. If, as shown in FIGURE 1.13, Sam's final position x_f had been at the origin instead of the 150 ft mark, his displacement would have been

$$\Delta x = x_f - x_i = 0 \text{ ft} - 50 \text{ ft} = -50 \text{ ft}$$

The negative sign tells us that he moved to the *left* along the x-axis, or 50 ft *west.*

Changes in Time

A displacement is a change in position. In order to quantify motion, we'll need to also consider changes in *time,* which we call **time intervals.** We've seen how we can label each frame of a motion diagram with a specific time, as determined by our stopwatch. FIGURE 1.14 shows the motion diagram of a bicycle moving at a constant speed, with the times of the measured points indicated.

FIGURE 1.14 The motion diagram of a bicycle moving to the right at a constant speed.

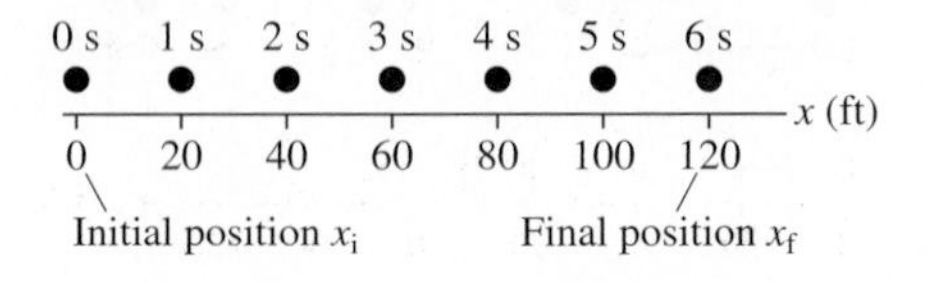

The displacement between the initial position x_i and the final position x_f is

$$\Delta x = x_f - x_i = 120 \text{ ft} - 0 \text{ ft} = 120 \text{ ft}$$

Similarly, we define the time interval between these two points to be

$$\Delta t = t_f - t_i = 6\text{ s} - 0\text{ s} = 6\text{ s}$$

A time interval Δt measures the elapsed time as an object moves from an initial position x_i at time t_i to a final position x_f at time t_f. Note that, unlike Δx, Δt is always positive because t_f is always greater than t_i.

EXAMPLE 1.1 How long a ride?

Emily is enjoying a bicycle ride on a country road that runs east-west past a water tower. At noon, Emily is 3 miles (mi) east of the water tower. A half-hour later, she is 2 mi west of the water tower. What is her displacement during that half-hour?

STRATEGIZE We will use the particle model to represent Emily. She is riding along a straight east-west road, so we will make use of a one-dimensional coordinate system to describe her motion.

PREPARE Although it may seem like overkill for such a simple problem, you should start by making a drawing, like the one in FIGURE 1.15, with the x-axis along the road. We choose our coordinate system so that increasing x means moving to the east. Distances are measured with respect to the water tower, so it is a natural origin for the coordinate system. Once the coordinate system is established, we can show Emily's initial and final positions and her displacement between the two.

FIGURE 1.15 A drawing of Emily's motion.

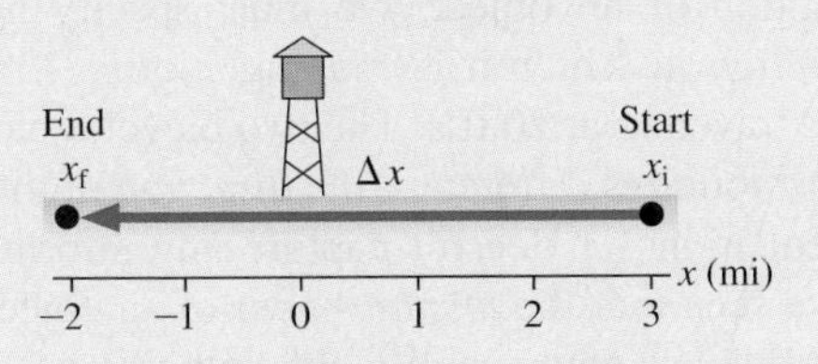

SOLVE We've specified values for Emily's initial and final positions in our drawing. We can thus compute her displacement:

$$\Delta x = x_f - x_i = (-2\text{ mi}) - (3\text{ mi}) = -5\text{ mi}$$

ASSESS Once we've completed the solution to the problem, we need to go back to see if it makes sense. Emily is moving to the west, so we expect her displacement to be negative—and it is. We can see from our drawing in Figure 1.15 that she has moved 5 miles from her starting position, so our answer seems reasonable. As part of the Assess step, we also check our answers to see if they make physical sense. Emily travels 5 miles in a half-hour, quite a reasonable pace for a cyclist.

NOTE ▶ All of the numerical examples in the text are worked out with the same four-step process: Strategize, Prepare, Solve, Assess. It's tempting to cut corners, especially for the simple problems in these early chapters, but you should take the time to do all of these steps now, to practice your problem-solving technique. We'll have more to say about our general problem-solving approach in Chapter 2. ◀

STOP TO THINK 1.3 Sarah starts at a positive position along the x-axis. She then undergoes a negative displacement. Her final position

A. Is positive. B. Is negative. C. Could be either positive or negative.

1.4 Velocity

We all have an intuitive sense of whether something is moving very fast or just cruising slowly along. To make this intuitive idea more precise, let's start by examining the motion diagrams of some objects moving along a straight line at a *constant* speed, objects that are neither speeding up nor slowing down. This motion at a constant speed is called **uniform motion.** As we saw for the skateboarder in Section 1.1, for an object in uniform motion, successive frames of the motion diagram are *equally spaced*, so the object's displacement Δx is the *same* between successive frames.

To see how an object's displacement between successive frames is related to its speed, consider the motion diagrams of a bicycle and a car, traveling along the same street as shown in FIGURE 1.16. Clearly the car is moving faster than the bicycle: In any 1 second time interval, the car undergoes a displacement $\Delta x = 40$ ft, while the bicycle's displacement is only 20 ft.

FIGURE 1.16 Motion diagrams for a car and a bicycle.

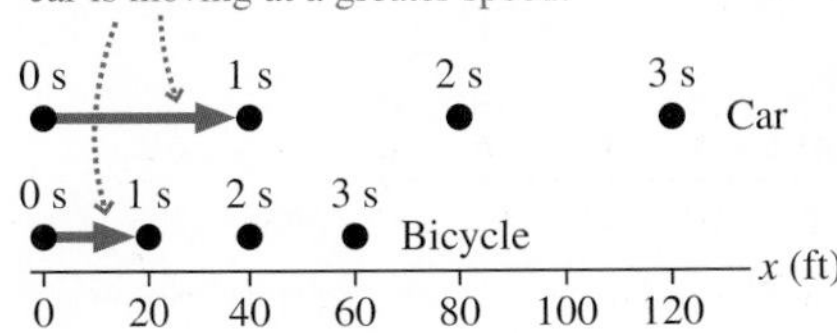

The greater the distance traveled by an object in a given time interval, the greater its speed. This idea leads us to define the speed of an object as

$$\text{speed} = \frac{\text{distance traveled in a given time interval}}{\text{time interval}} \tag{1.1}$$

Speed of an object in uniform motion

For the bicycle, this equation gives

$$\text{speed} = \frac{20 \text{ ft}}{1 \text{ s}} = 20 \frac{\text{ft}}{\text{s}}$$

while for the car we have

$$\text{speed} = \frac{40 \text{ ft}}{1 \text{ s}} = 40 \frac{\text{ft}}{\text{s}}$$

The speed of the car is twice that of the bicycle, which seems reasonable.

NOTE ▸ The division gives units that are a fraction: ft/s. This is read as "feet per second," just like the more familiar "miles per hour." ◂

FIGURE 1.17 Two bicycles traveling at the same speed, but with different velocities.

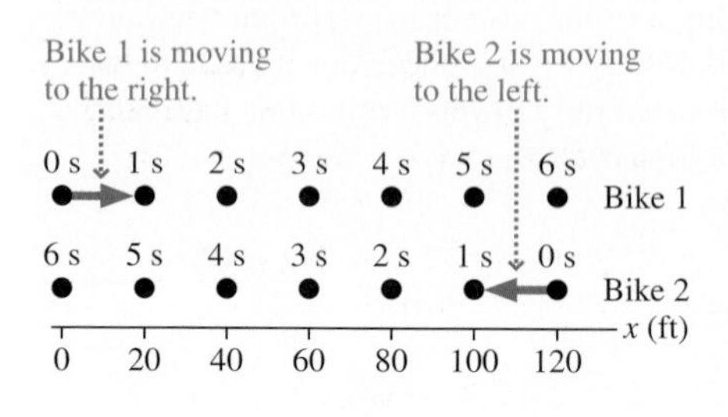

To fully characterize the motion of an object, we must specify not only the object's speed but also the *direction* in which it is moving. FIGURE 1.17 shows the motion diagrams of two bicycles traveling at 20 ft/s. The two bicycles have the same speed, but something about their motion is different—the *direction* of their motion.

The "distance traveled" in Equation 1.1 doesn't capture any information about the direction of travel. But we've seen that the *displacement* of an object does contain this information. We can introduce a new quantity, the **velocity,** as

$$\text{velocity} = \frac{\text{displacement}}{\text{time interval}} = \frac{\Delta x}{\Delta t} \tag{1.2}$$

Velocity of a moving object

The velocity of bicycle 1 in Figure 1.17, computed using the 1 second time interval between the $t = 0$ s and $t = 1$ s positions, is

$$v = \frac{\Delta x}{\Delta t} = \frac{x_1 - x_0}{1 \text{ s} - 0 \text{ s}} = \frac{20 \text{ ft} - 0 \text{ ft}}{1 \text{ s}} = +20 \frac{\text{ft}}{\text{s}}$$

while the velocity of bicycle 2, during the same time interval, is

$$v = \frac{\Delta x}{\Delta t} = \frac{x_1 - x_0}{1 \text{ s} - 0 \text{ s}} = \frac{100 \text{ ft} - 120 \text{ ft}}{1 \text{ s}} = -20 \frac{\text{ft}}{\text{s}}$$

NOTE ▸ We have used x_0 for the position at time $t = 0$ seconds and x_1 for the position at time $t = 1$ second. The subscripts serve the same role as before—identifying particular positions—but in this case the positions are identified by the time at which each position is reached. ◂

The two velocities have opposite signs because the bicycles are traveling in opposite directions. **Speed measures only how fast an object moves, but velocity tells us both an object's speed *and its direction*.** In this text, we'll use a positive velocity to indicate motion to the right or, for vertical motion, upward. We'll use a negative velocity for an object moving to the left, or downward.

NOTE ▸ Learning to distinguish between speed, which is always a positive number, and velocity, which can be either positive or negative, is one of the most important tasks in the analysis of motion. ◂

The velocity as defined by Equation 1.2 is actually what is called the *average* velocity. On average, over each 1 s interval bicycle 1 moves 20 ft, but we don't know if it was moving at exactly the same speed at every moment during this time interval. In Chapter 2, we'll develop the idea of *instantaneous* velocity, the velocity of an object at a particular instant in time. Since our goal in this chapter is to *visualize* motion with motion diagrams, we'll somewhat blur the distinction between average and instantaneous quantities, refining these definitions in Chapter 2.

The "Per" in Miles Per Hour

The units for speed and velocity are a unit of distance, such as feet, meters, or miles, divided by a unit of time, such as seconds or hours. Thus we could measure velocity in units of mi/h (or mph) or m/s, pronounced "miles *per* hour" or "meters *per* second." The word "per" will often arise in physics when we consider the ratio of two quantities. What do we mean, exactly, by "per"?

If a car moves with a speed of 23 m/s, we mean that it travels 23 meters *for each* second of elapsed time. The word "per" thus associates the number of units in the numerator (23 m) with *one* unit of the denominator (1 s). We'll see many other examples of this idea as the text progresses. You may already know a bit about *density;* you can look up the density of gold and you'll find that it is 19.3 g/cm^3 ("grams *per* cubic centimeter"). This means that there are 19.3 grams of gold *for each* cubic centimeter of the metal.

EXAMPLE 1.2 Finding the speed of a seabird BIO

Albatrosses are seabirds that spend most of their lives flying over the ocean looking for food. With a stiff tailwind, an albatross can fly at high speeds. Satellite data on one particularly speedy albatross showed it 60 miles east of its roost at 3:00 PM and then, at 3:15 PM, 80 miles east of its roost. What was its velocity?

STRATEGIZE We will assume that the albatross is flying at a constant speed; that is, it is in uniform motion. Using the particle model, we can represent its motion on a coordinate system.

PREPARE The statement of the problem provides us with a natural coordinate system: We can measure distances with respect to the roost, with distances to the east as positive. With this coordinate system, the motion of the albatross appears as in FIGURE 1.18. The motion takes place between 3:00 and 3:15, a time interval of 15 minutes. If we want our final velocity to be in the familiar units of miles per hour, or mph, we need this time interval in hours (abbreviated "h"). So we make the conversion 15 min = 0.25 h.

FIGURE 1.18 The motion of an albatross at sea.

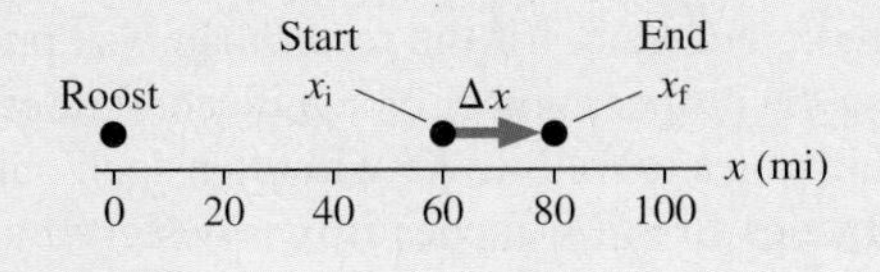

SOLVE We know the initial and final positions, and we know the time interval, so we can calculate the velocity:

$$v = \frac{\Delta x}{\Delta t} = \frac{x_f - x_i}{0.25\text{ h}} = \frac{20\text{ mi}}{0.25\text{ h}} = 80\text{ mph}$$

ASSESS The velocity is positive, which makes sense because Figure 1.18 shows that the motion is to the right. A speed of 80 mph is certainly fast, but the problem said it was a "particularly speedy" albatross, so our answer seems reasonable. (Indeed, albatrosses have been observed to fly at such speeds in the very fast winds of the Southern Ocean. This problem is based on real observations, as will be our general practice in this text.)

STOP TO THINK 1.4 Jane starts from her house to take a stroll in her neighborhood. After walking for 2 hours at a steady pace, she has walked 4 miles and is 2 miles from home. For this time interval, what was her speed?

A. 4 mph B. 3 mph C. 2 mph D. 1 mph

1.5 A Sense of Scale: Significant Figures, Scientific Notation, and Units

Physics attempts to explain the natural world, from the very small to the exceedingly large. And in order to understand our world, we need to be able to *measure* quantities both minuscule and enormous. A properly reported measurement has three elements. First, we can measure our quantity with only a certain precision. To make this precision clear, we need to make sure that we report our measurement with the correct number of *significant figures.*

Second, writing down the really big and small numbers that often come up in physics can be awkward. To avoid writing all those zeros, scientists use *scientific notation* to express numbers both big and small.

Finally, we need to choose an agreed-upon set of *units* for the quantity. For speed, common units include meters per second and miles per hour. For mass, the kilogram is the most commonly used unit. Every physical quantity that we can measure has an associated set of units.

Measurements and Significant Figures

FIGURE 1.19 The precision of a measurement depends on the instrument used to make it.

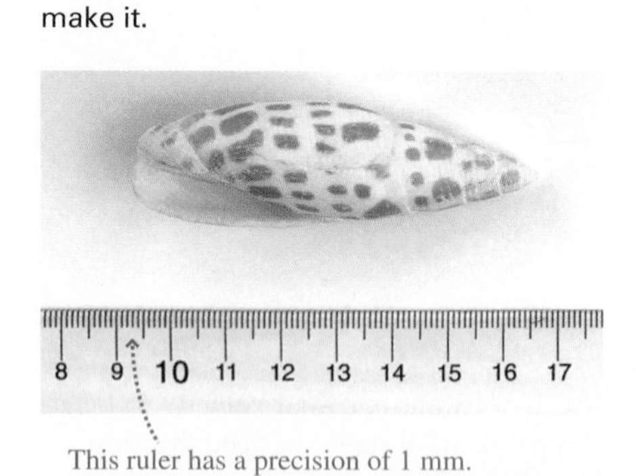
8 9 10 11 12 13 14 15 16 17
This ruler has a precision of 1 mm.

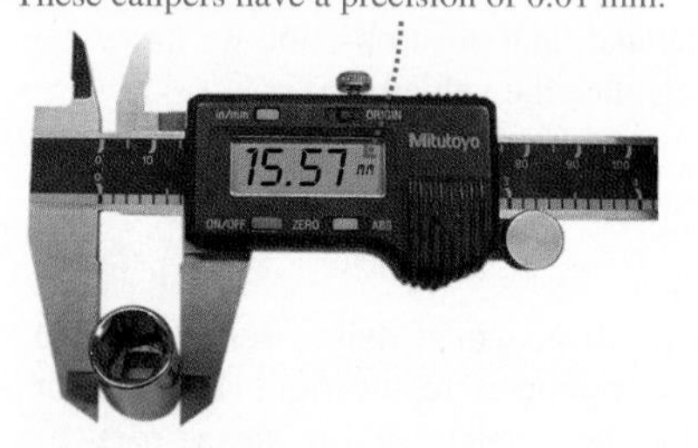
These calipers have a precision of 0.01 mm.
15.57
Mitutoyo

When we measure any quantity, such as the length of a bone or the weight of a specimen, we can do so with only a certain *precision.* If you make a measurement with the ruler shown in FIGURE 1.19, you probably can't be more accurate than about ± 1 mm, so the ruler has a precision of 1 mm. The digital calipers shown can make a measurement to within ± 0.01 mm, so it has a precision of 0.01 mm. The precision of a measurement can also be affected by the skill or judgment of the person performing the measurement. A stopwatch might have a precision of 0.001 s, but, due to your reaction time, your measurement of the time of a sprinter would be much less precise.

It is important that your measurement be reported in a way that reflects its actual precision. Suppose you use a ruler to measure the length of a particular frog. You judge that you can make this measurement with a precision of about 1 mm, or 0.1 cm. In this case, the frog's length should be reported as, say, 6.2 cm. We interpret this to mean that the actual value falls between 6.15 cm and 6.25 cm and thus rounds to 6.2 cm. If you reported the frog's length as simply 6 cm, you would be saying less than you know; you would be withholding information. If you reported the number as 6.213 cm, however, anyone reviewing your work would interpret this to mean that the actual length falls between 6.2125 cm and 6.2135 cm, a precision of 0.001 cm. In this case, you would be claiming to have more information than you really possessed.

The way to state your knowledge precisely is through the proper use of **significant figures.** You can think of a significant figure as a digit that is reliably known. A measurement such as 6.2 cm has *two* significant figures, the 6 and the 2. The next decimal place—the hundredths—is not reliably known and is thus not a significant figure. Similarly, a time measurement of 34.62 s has four significant figures, implying that the 2 in the hundredths place is reliably known.

Walter Davis's best long jump on this day was reported as 8.24 m. This implies that the actual length of the jump was between 8.235 m and 8.245 m, a spread of only 0.01 m, which is 1 cm. Does this claimed accuracy seem reasonable?

When we perform a calculation such as adding or multiplying two or more measured numbers, we can't claim more accuracy for the result than was present in the initial measurements. Determining the proper number of significant figures is straightforward, but there are a few definite rules to follow. We will often spell out such technical details in what we call a "Tactics Box." A Tactics Box is designed to teach you particular skills and techniques. Each Tactics Box will include the icon to designate exercises in the *Student Workbook* that you can use to practice these skills.

TACTICS BOX 1.1 Using significant figures

❶ When you multiply or divide several numbers, or when you take roots, the number of significant figures in the answer should match the number of significant figures of the *least* precisely known number used in the calculation:

Three significant figures

$$3.73 \times 5.7 = 21$$

Two significant figures

Answer should have the *lower* of the two, or two significant figures.

Continued

❷ When you add or subtract several numbers, the number of decimal places in the answer should match the *smallest* number of decimal places of any number used in the calculation:

$$\begin{array}{r} 18.54 \\ +106.6 \\ \hline 125.1 \end{array}$$

18.54 — Two decimal places
+106.6 — One decimal place
125.1 — Answer should have the *lower* of the two, or one decimal place.

❸ **Exact numbers** have no uncertainty and, when used in calculations, do not change the number of significant figures of measured numbers. Examples of exact numbers are π and the number 2 in the relation $d = 2r$ between a circle's diameter and radius.

There is one notable exception to these rules:

- It is acceptable to keep one or two extra digits during *intermediate* steps of a calculation to minimize round-off errors in the calculation. But the *final* answer must be reported with the proper number of significant figures.

Exercise 15 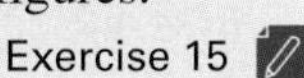

EXAMPLE 1.3 Using significant figures when measuring the velocity of a car

To measure the velocity of a car, clocks A and B are set up at two points along the road, as shown in FIGURE 1.20. Clock A is precise to 0.01 s, while clock B is precise to only 0.1 s. The distance between these two clocks is carefully measured to be 124.5 m. The two clocks are automatically started when the car passes a trigger in the road; each clock stops automatically when the car passes that clock. After the car has passed both clocks, clock A is found to read $t_A = 1.22$ s, and clock B to read $t_B = 4.5$ s. The time from the less-precise clock B is correctly reported with fewer significant figures than that from A. What is the velocity of the car, and how should it be reported with the correct number of significant figures?

FIGURE 1.20 Measuring the velocity of a car.

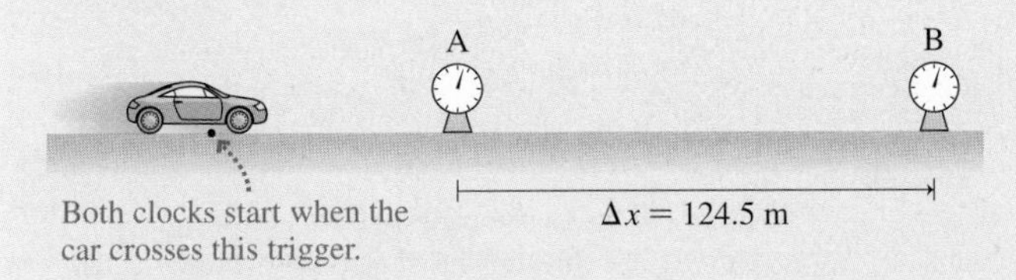
A
B
Both clocks start when the car crosses this trigger.
$\Delta x = 124.5$ m

STRATEGIZE To find the car's velocity with the correct precision, we will need to take into account the significant figures in the measured quantities and then apply the rules of Tactics Box 1.1.

PREPARE To calculate the velocity, we need the displacement Δx and the time interval Δt as the car moves between the two clocks. The displacement is given as $\Delta x = 124.5$ m; we can calculate the time interval as the difference between the two measured times.

SOLVE The time interval is

This number has one decimal place. This number has two decimal places.

$$\Delta t = t_B - t_A = (4.5 \text{ s}) - (1.22 \text{ s}) = 3.3 \text{ s}$$

By rule 2 of Tactics Box 1.1, the result should have *one* decimal place.

We can now calculate the velocity with the displacement and the time interval:

The displacement has four significant figures.

$$v = \frac{\Delta x}{\Delta t} = \frac{124.5 \text{ m}}{3.3 \text{ s}} = 38 \text{ m/s}$$

The time interval has two significant figures. By rule 1 of Tactics Box 1.1, the result should have *two* significant figures.

ASSESS Our final value has two significant figures. Suppose you had been hired to measure the speed of a car this way, and you reported 37.72 m/s. It would be reasonable for someone looking at your result to assume that the measurements you used to arrive at this value were correct to four significant figures and thus that you had measured time to the nearest 0.001 second. Our correct result of 38 m/s has all of the accuracy that you can claim, but no more!

Scientific Notation

It's easy to write down measurements of ordinary-sized objects: Your height might be 1.72 meters, the weight of an apple 0.34 pound. But the radius of a hydrogen atom is 0.000 000 000 053 m, and the distance to the moon is 384,000,000 m. Keeping track of all those zeros is quite cumbersome.

Beyond requiring you to deal with all the zeros, writing quantities this way makes it unclear how many significant figures are involved. For the distance to the moon, how many of those digits are significant? Three? Four? All nine?

Writing numbers using **scientific notation** avoids both these problems. A value in scientific notation is a number with one digit to the left of the decimal point and zero or more to the right of it, multiplied by a power of ten. This solves the problem of writing so many zeros and makes the number of significant figures immediately apparent. In scientific notation, writing the distance to the moon as 3.84×10^8 m indicates that three digits are significant; writing it as 3.8×10^8 m indicates that only two digits are.

Even for smaller values, scientific notation can clarify the number of significant figures. Suppose a distance is reported as 1200 m. How many significant figures does this measurement have? It's ambiguous, but using scientific notation can remove any ambiguity. If this distance is known to within 1 m, we can write it as 1.200×10^3 m, showing that all four digits are significant; if it is accurate to only 100 m or so, we can report it as 1.2×10^3 m, indicating two significant figures.

TACTICS BOX 1.2 Using scientific notation

To convert a number into scientific notation:

❶ For a number greater than 10, move the decimal point to the left until only one digit remains to the left of the decimal point. The remaining number is then multiplied by 10 to a power; this power is given by the number of spaces the decimal point was moved. Here we convert the radius of the earth to scientific notation:

We move the decimal point until there is only one digit to its left, counting the number of steps.

Since we moved the decimal point 6 steps, the power of ten is 6.

$$6\,370\,000 \text{ m} = 6.37 \times 10^6 \text{ m}$$

The number of digits here equals the number of significant figures.

❷ For a number less than 1, move the decimal point to the right until it passes the first digit that isn't a zero. The remaining number is then multiplied by 10 to a negative power; the power is given by the number of spaces the decimal point was moved. For the diameter of a red blood cell we have:

We move the decimal point until it passes the first digit that is not a zero, counting the number of steps.

Since we moved the decimal point 6 steps, the power of ten is −6.

$$0.000\,007\,5 \text{ m} = 7.5 \times 10^{-6} \text{ m}$$

The number of digits here equals the number of significant figures.

Exercise 16 

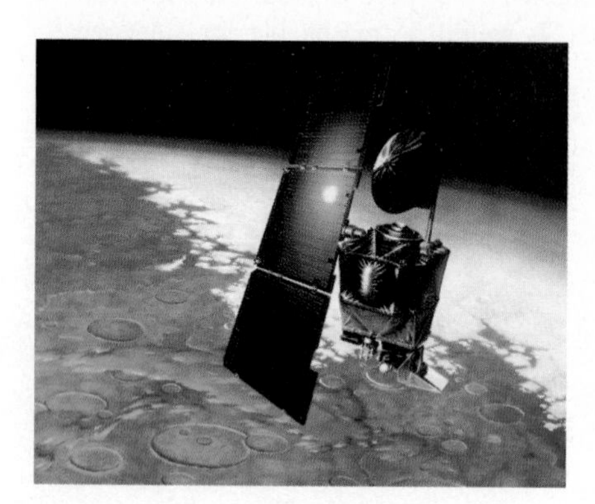

◀ **The importance of units** In 1999, the \$125 million Mars Climate Orbiter burned up in the Martian atmosphere instead of entering a safe orbit from which it could perform observations. The problem was faulty units! An engineering team had provided critical data on spacecraft performance in English units, but the navigation team assumed these data were in metric units. As a consequence, the navigation team had the spacecraft fly too close to the planet, and it burned up in the atmosphere.

Units

As we have seen, in order to measure a quantity we need to give it a numerical value. But a measurement is more than just a number—it requires a *unit* to be given. You can't go to the deli and ask for "three quarters of cheese." You need to use a unit—here, one of weight, such as pounds—in addition to the number.

In your daily life, you probably use the English system of units, in which distances are measured in inches, feet, and miles. These units are well adapted for daily life, but they are rarely used in scientific work. Given that science is an international discipline, it is also important to have a system of units that is recognized around the world. For these reasons, scientists use a system of units called *le Système Internationale d'Unités,* commonly referred to as **SI units.** We often refer to these as *metric units* because the meter is the basic standard of length.

The three basic SI quantities, shown in TABLE 1.1, are time, length (or distance), and mass. The SI units for these quantities are meters, seconds, and kilograms, respectively. Other quantities needed to understand motion can be expressed as combinations of these basic units. For example, speed and velocity are expressed in meters per second or m/s. This combination is a ratio of the length unit (the meter) to the time unit (the second).

TABLE 1.1 Common SI units

Quantity	Unit	Abbreviation
Time	second	s
Length	meter	m
Mass	kilogram	kg

Using Prefixes

We will have many occasions to use lengths, times, and masses that are either much less or much greater than the standards of 1 meter, 1 second, and 1 kilogram. We will do so by using *prefixes* to denote various powers of ten. For instance, the prefix "kilo" (abbreviation k) denotes 10^3, or a factor of 1000. Thus 1 km equals 1000 m, 1 MW (megawatt) equals 10^6 watts, and 1 μV (microvolt) equals 10^{-6} volts. TABLE 1.2 lists the common prefixes that will be used frequently throughout this text. A more extensive list of prefixes is shown at the front of the text.

Although prefixes make it easier to talk about quantities, those given with prefixed units are usually converted to base SI units of seconds and meters before any calculations are done. Thus 23.0 cm should be converted to 0.230 m before starting calculations. (The exception is the kilogram, which is already the base SI unit.)

TABLE 1.2 Common prefixes for powers of 10

Prefix	Abbreviation	Power of 10
mega-	M	10^6
kilo-	k	10^3
centi-	c	10^{-2}
milli-	m	10^{-3}
micro-	μ	10^{-6}
nano-	n	10^{-9}

Unit Conversions Between Measurement Systems

Although SI units are our standard, we cannot entirely forget that the United States still uses English units. Even after repeated exposure to metric units in classes, most of us "think" in English units. Thus it remains important to be able to convert back and forth between SI units and English units. TABLE 1.3 shows some frequently used conversions that will come in handy.

TABLE 1.3 Useful unit conversions

1 inch (in) = 2.54 cm
1 foot (ft) = 0.305 m
1 mile (mi) = 1.609 km
1 mile per hour (mph) = 0.447 m/s
1 m = 39.37 in
1 km = 0.621 mi
1 m/s = 2.24 mph

One effective method of performing unit conversions begins by noticing that since, for example, 1 mi = 1.609 km, the ratio of these two distances—*including their units*—is equal to 1, so that

$$\frac{1 \text{ mi}}{1.609 \text{ km}} = \frac{1.609 \text{ km}}{1 \text{ mi}} = 1$$

A ratio of values equal to 1 is called a **conversion factor.** The following Tactics Box shows how to make a unit conversion.

TACTICS BOX 1.3 Making a unit conversion

❶ Start with the quantity you wish to convert.

❷ Multiply by the appropriate conversion factor. Because this conversion factor is equal to 1, multiplying by it does not change the value of the quantity—only its units.

❺ Remember to convert your final answer to the correct number of significant figures!

$$60 \text{ mi} = 60 \text{ mi} \times \frac{1.609 \text{ km}}{1 \text{ mi}} = 96.54 \text{ km} = 97 \text{ km}$$

❸ You can cancel the original unit (here, miles) because it appears in both the numerator and the denominator.

❹ Calculate the answer; it is in the desired units. Remember, 60 mi and 96.54 km are the same distance; they are simply in different units.

Exercise 17

More complicated conversions can be done with several successive multiplications of conversion factors, as we see in the example on the next page.

EXAMPLE 1.4 **Can a bicycle go that fast?**

In Section 1.4, we calculated the speed of a bicycle to be 20 ft/s. Is this a reasonable speed for a bicycle?

STRATEGIZE In order to determine whether or not this speed is reasonable, we will convert it to more familiar units. For speed, the unit you are most familiar with is likely miles per hour.

PREPARE We need the following unit conversions:

$$1 \text{ mi} = 5280 \text{ ft} \qquad 1 \text{ hour } (1 \text{ h}) = 60 \text{ min} \qquad 1 \text{ min} = 60 \text{ s}$$

SOLVE We then multiply our original value by successive factors of 1 in order to convert the units:

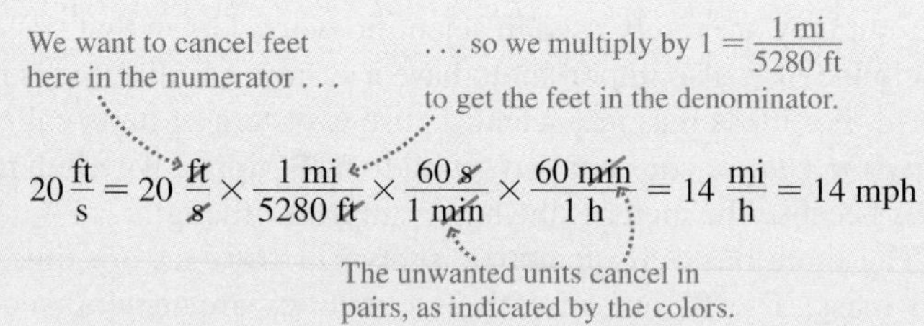

We want to cancel feet here in the numerator . . .
. . . so we multiply by 1 = 1 mi / 5280 ft to get the feet in the denominator.
The unwanted units cancel in pairs, as indicated by the colors.

$$20\,\frac{\text{ft}}{\text{s}} = 20\,\frac{\text{ft}}{\text{s}} \times \frac{1 \text{ mi}}{5280 \text{ ft}} \times \frac{60 \text{ s}}{1 \text{ min}} \times \frac{60 \text{ min}}{1 \text{ h}} = 14\,\frac{\text{mi}}{\text{h}} = 14 \text{ mph}$$

ASSESS Our final result of 14 miles per hour (14 mph) is a very reasonable speed for a bicycle, which gives us confidence in our answer. If we had calculated a speed of 140 miles per hour, we would have suspected that we had made an error because this is quite a bit faster than the average bicyclist can travel!

◄ The man has a mass of 70 kg. What is the mass of the elephant standing next to him? By thinking about the relative dimensions of the two, you can make a reasonable one-significant-figure *estimate*.

Estimation

Precise calculations are appropriate when we have precise data, but there are many times when just a rough estimate is sufficient. Suppose you saw a rock fall off a cliff and wanted to know how fast it was going when it hit the ground. By doing a mental comparison with the speeds of familiar objects, such as cars and bicycles, you might judge that the rock was traveling at about 20 mph.

This is a one-significant-figure estimate: You can probably distinguish 20 mph from either 10 mph or 30 mph, but you certainly cannot distinguish 20 mph from 21 mph just from an observation. A one-significant-figure estimate or calculation, such as this estimate of speed, is called an **order-of-magnitude estimate.** An order-of-magnitude estimate is indicated by the symbol $\sim$, which indicates even less precision than the "approximately equal" symbol $\approx$. You would report your estimate of the speed of the falling rock as $v \sim 20$ mph.

It's a useful skill to make reliable order-of-magnitude estimates on the basis of known information (or information found on the Internet), simple reasoning, and common sense. It may help to convert from SI units to more familiar units to make such estimates. You can also do this to assess problem solutions given in SI units. TABLE 1.4 lists some approximate conversion factors to apply in such cases.

TABLE 1.4 Some approximate conversion factors

Quantity	SI unit	Approximate conversion
Mass	kg	1 kg ≈ 2 lb
Length	m	1 m ≈ 3 ft
	cm	3 cm ≈ 1 in
	km	5 km ≈ 3 mi
Speed	m/s	1 m/s ≈ 2 mph
	km/h	10 km/h ≈ 6 mph

EXAMPLE 1.5 **How fast do you walk?**

Estimate how fast you walk, in meters per second.

STRATEGIZE In this example we're asked for an *estimate* of your walking speed, so we'll need to use only rough values obtained from our everyday experience of walking.

PREPARE In order to compute speed, we need a distance and a time. If you walked a mile to campus, how long would this take? You'd probably say 30 minutes or so—half an hour. Let's use this rough number in our estimate.

SOLVE Given this estimate, we compute your speed as

$$\text{speed} = \frac{\text{distance}}{\text{time}} \sim \frac{1 \text{ mile}}{1/2 \text{ hour}} = 2\,\frac{\text{mi}}{\text{h}}$$

But we want the speed in meters per second. Since our calculation is only an estimate, we use an approximate conversion factor from Table 1.4:

$$1\,\frac{\text{mi}}{\text{h}} \sim 0.5\,\frac{\text{m}}{\text{s}}$$

This gives an approximate walking speed of 1 m/s.

ASSESS Is this a reasonable value? Let's do another estimate. Your stride is probably about 1 yard long—about 1 meter. And you take about one step per second; next time you are walking, you can count and see. So a walking speed of 1 meter per second sounds pretty reasonable.

STOP TO THINK 1.5 Rank in order, from the most to the fewest, the number of significant figures in the following numbers. For example, if B has more than C, C has the same number as A, and A has more than D, give your answer as B > C = A > D.

A. 0.43 B. 0.0052 C. 0.430 D. 4.321×10^{-10}

1.6 Vectors and Motion: A First Look

Many physical quantities, such as time, temperature, and mass, can be described completely by a number with a unit. For example, the mass of an object might be 6 kg and its temperature 30°C. When a physical quantity is described by a single number (with a unit), we call it a **scalar quantity.** A scalar can be positive, negative, or zero.

Many other quantities, however, have a directional quality and cannot be described by a single number. To describe the motion of a car, for example, you must specify not only how fast it is moving, but also the *direction* in which it is moving. A **vector quantity** is a quantity that has both a *size* (How far? or How fast?) and a *direction* (Which way?). The size or length of a vector is called its **magnitude.** The magnitude of a vector can be positive or zero, but it cannot be negative.

We graphically represent a vector as an *arrow*, as illustrated for the velocity and force vectors. The arrow is drawn to point in the direction of the vector quantity, and the *length* of the arrow is proportional to the magnitude of the vector quantity.

When we want to represent a vector quantity with a *symbol,* we need somehow to indicate that the symbol is for a vector rather than for a scalar. We do this by drawing an arrow over the letter that represents the quantity. Thus $\vec{r}$ and $\vec{A}$ are symbols for vectors, whereas r and A, without the arrows, are symbols for scalars. In handwritten work you *must* draw arrows over all symbols that represent vectors. This may seem strange until you get used to it, but it is very important because we will often use both r and $\vec{r}$ or both A and $\vec{A}$ in the same problem, and they mean different things!

NOTE ▶ The arrow over the symbol always points to the right, regardless of which direction the actual vector points. Thus we write $\vec{r}$ or $\vec{A}$, never $\overleftarrow{r}$ or $\overleftarrow{A}$. ◀

Scalars and vectors

Scalars

Time, temperature, and mass are all *scalar* quantities. To specify the current time, the temperature outside, or your mass, we need only a single number.

Vectors

The velocity of the race car is a *vector.* To fully specify a velocity, we need to give its magnitude (e.g., 120 mph) *and* its direction (e.g., west).

The force with which the boy pushes on his friend is another example of a vector. To completely specify this force, we must know not only how hard he pushes (the magnitude) but also in which direction.

Displacement Vectors

For motion along a line, we found in Section 1.3 that the displacement is a quantity that specifies not only how *far* an object moves but also the *direction*—to the left or to the right—that the object moves. Since displacement is a quantity that has both a magnitude (How far?) and a direction, it can be represented by a vector, the **displacement vector.** FIGURE 1.21 shows the displacement vector for Sam's trip that we discussed earlier. We've simply drawn an arrow—the vector—from his initial to his final position and assigned it the symbol $\vec{d}_S$. Because $\vec{d}_S$ has both a magnitude and a direction, it is convenient to write Sam's displacement as $\vec{d}_S = (100 \text{ ft, east})$. The first value in the parentheses is the magnitude of the vector (i.e., the size of the displacement), and the second value specifies its direction.

Also shown in Figure 1.21 is the displacement vector $\vec{d}_J$ for Jane, who started on 12th Street and ended up on Vine. As with Sam, we draw her displacement vector as an arrow from her initial to her final position. In this case, $\vec{d}_J = (100 \text{ ft, } 60° \text{ north of east})$.

Jane's trip illustrates an important point about displacement vectors. Jane started her trip on 12th Street and ended up on Vine, leading to the displacement vector

FIGURE 1.21 Two displacement vectors.

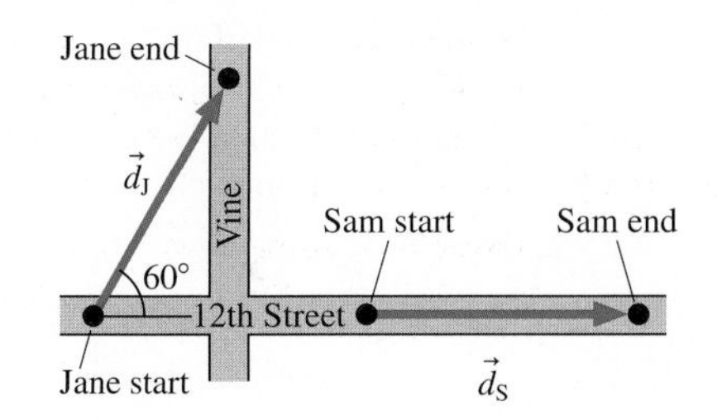

The boat's displacement is the straight-line connection from its initial to its final position.

shown. But to get from her initial to her final position, she didn't have to walk along the straight-line path denoted by $\vec{d}_J$. If she walked east along 12th Street to the intersection and then headed north on Vine, her displacement would still be the vector shown. **An object's displacement vector is drawn from the object's initial position to its final position, regardless of the actual path followed between these two points.**

Vector Addition

Let's consider one more trip for the peripatetic Sam. In FIGURE 1.22, he starts at the intersection and walks east 50 ft; then he walks 100 ft to the northeast through a vacant lot. His displacement vectors for the two legs of his trip are labeled $\vec{d}_1$ and $\vec{d}_2$ in the figure.

FIGURE 1.22 Sam undergoes two displacements.

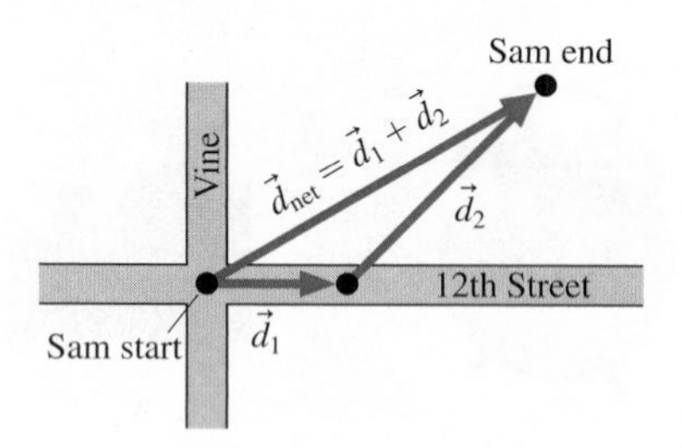

Sam's trip consists of two legs that can be represented by the two vectors $\vec{d}_1$ and $\vec{d}_2$, but we can represent his trip as a whole, from his initial starting position to his overall final position, with the *net* displacement vector labeled $\vec{d}_{net}$. Sam's net displacement is in a sense the *sum* of the two displacements that made it up, so we can write

$$\vec{d}_{net} = \vec{d}_1 + \vec{d}_2$$

Sam's net displacement thus requires the *addition* of two vectors, but vector addition obeys different rules from the addition of two scalar quantities. The directions of the two vectors, as well as their magnitudes, must be taken into account. Sam's trip suggests that we can add vectors together by putting the "tail" of one vector at the tip of the other. This idea, which is reasonable for displacement vectors, in fact is how *any* two vectors are added. Tactics Box 1.4 shows how to add two vectors $\vec{A}$ and $\vec{B}$ to get their **vector sum** $\vec{A} + \vec{B}$.

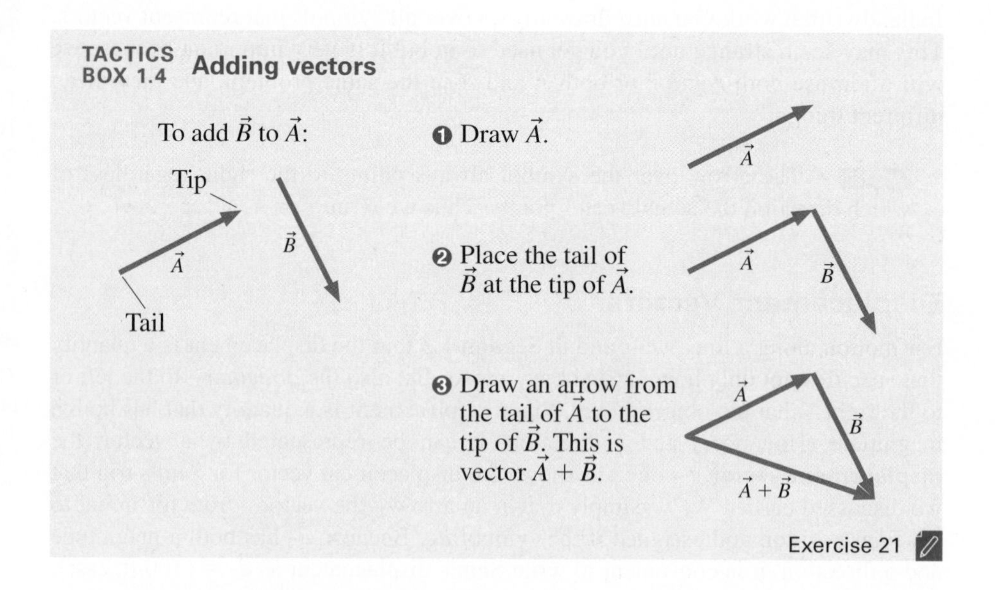

TACTICS BOX 1.4 Adding vectors
To add $\vec{B}$ to $\vec{A}$:
Tip
$\vec{A}$
$\vec{B}$
Tail
❶ Draw $\vec{A}$.
$\vec{A}$
❷ Place the tail of $\vec{B}$ at the tip of $\vec{A}$.
$\vec{A}$
$\vec{B}$
❸ Draw an arrow from the tail of $\vec{A}$ to the tip of $\vec{B}$. This is vector $\vec{A} + \vec{B}$.
$\vec{A}$
$\vec{B}$
$\vec{A} + \vec{B}$
Exercise 21

Vectors and Trigonometry

When we need to add displacements or other vectors in more than one dimension, we'll end up computing lengths and angles of triangles. This is the job of trigonometry. FIGURE 1.23 reviews the basic ideas of trigonometry.

KEY CONCEPT FIGURE 1.23 Relating sides and angles of right triangles using trigonometry.

We specify the sides of a right triangle in relation to one of the angles.

The longest side, opposite to the right angle, is the **hypotenuse**.

This is the side **opposite** to angle θ.

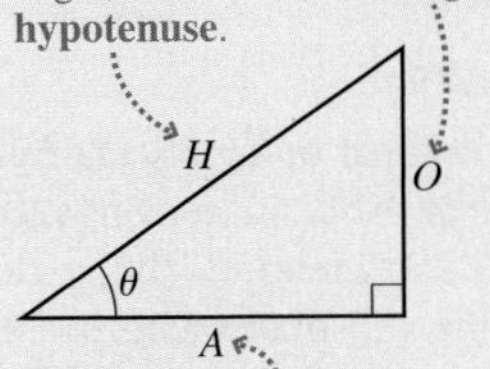

This is the side **adjacent** to angle θ.

The three sides are related by the *Pythagorean theorem*:

$$H = \sqrt{A^2 + O^2}$$

The sine, cosine, and tangent of angle θ are defined as ratios of the side lengths.

$$\sin\theta = \frac{O}{H}$$

$$\cos\theta = \frac{A}{H}$$

$$\tan\theta = \frac{O}{A}$$

We can rearrange these equations in useful ways:

$$O = H\sin\theta$$

$$A = H\cos\theta$$

Given the length of the hypotenuse and one angle, we can find the side lengths.

y is opposite to the angle; use the sine formula.

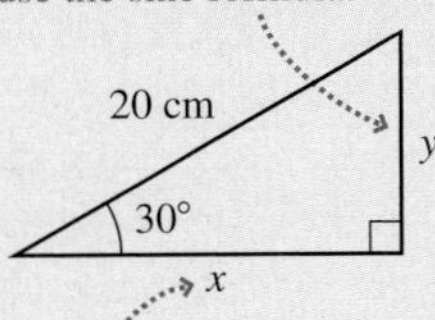

x is adjacent to the angle; use the cosine formula.

$$x = (20\text{ cm})\cos(30°) = 17\text{ cm}$$

$$y = (20\text{ cm})\sin(30°) = 10\text{ cm}$$

Inverse trig functions let us find angles given lengths.

$$\theta = \sin^{-1}\left(\frac{O}{H}\right)$$

$$\theta = \cos^{-1}\left(\frac{A}{H}\right)$$

$$\theta = \tan^{-1}\left(\frac{O}{A}\right)$$

If we are given the lengths of the triangle's sides, we can find angles.

θ is adjacent to the 10 cm side; use the $\cos^{-1}$ formula.

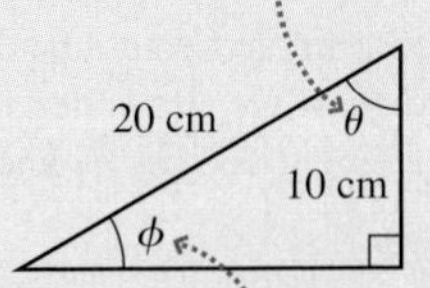

ϕ is opposite to the 10 cm side; use the $\sin^{-1}$ formula.

$$\theta = \cos^{-1}\left(\frac{10\text{ cm}}{20\text{ cm}}\right) = 60°$$

$$\phi = \sin^{-1}\left(\frac{10\text{ cm}}{20\text{ cm}}\right) = 30°$$

STOP TO THINK 1.6 Using the information in Figure 1.23, what is the distance x, to the nearest cm, in the triangle at the right?

A. 26 cm B. 20 cm C. 17 cm D. 15 cm

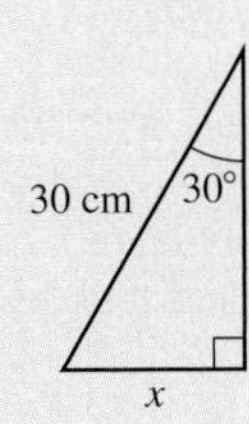

EXAMPLE 1.6 How far north and east?

Suppose Alex is navigating using a compass. She starts walking at an angle 60° north of east and walks a total of 100 m. How far north is she from her starting point? How far east?

STRATEGIZE We'll need to use trigonometry to solve this problem. To do so, we'll need to sketch the situation so that we can identify a right triangle along with its hypotenuse and adjacent and opposite sides.

PREPARE A sketch of Alex's motion is shown in FIGURE 1.24a. We've shown north and east as they are on a map, and we've noted Alex's displacement as a vector, giving its magnitude and direction. FIGURE 1.24b shows a triangle with this displacement as the hypotenuse. Alex's distance north of her starting point is this triangle's opposite side, and her distance east of her starting point is its adjacent side.

SOLVE Because we want to find O and A for a triangle with $\theta = 60°$ and $H = 100$ m, we use the equations $O = H\sin\theta$ and $A = H\cos\theta$, giving

$$\text{distance north of start} = (100\text{ m})\sin(60°) = 87\text{ m}$$

$$\text{distance east of start} = (100\text{ m})\cos(60°) = 50\text{ m}$$

ASSESS Both of the distances we calculated are less than 100 m, as they must be, and the distance east is less than the distance north, as our diagram in Figure 1.24b shows it should be. Our answers seem reasonable. In finding the solution to this problem, we "broke down" the displacement into two different distances, one north and one east. This hints at the idea of the *components* of a vector, something we'll explore in the next chapter.

FIGURE 1.24 An analysis of Alex's motion.

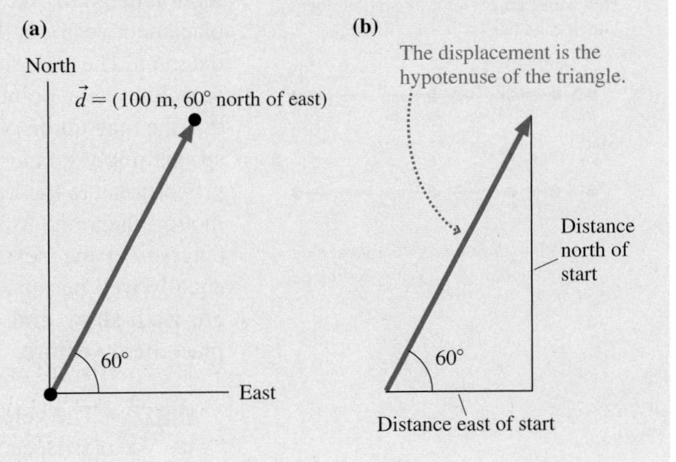

EXAMPLE 1.7 How far away is Anna?

Anna walks 90 m due east and then 50 m due north. What is her displacement from her starting point?

STRATEGIZE Again, we will need to use trigonometry to solve this problem, so we'll draw a right triangle and identify its sides.

PREPARE Let's start with the sketch in FIGURE 1.25a. We set up a coordinate system with Anna's original position as the origin, and then we drew her two subsequent motions as the two displacement vectors $\vec{d}_1$ and $\vec{d}_2$.

FIGURE 1.25 Analyzing Anna's motion.

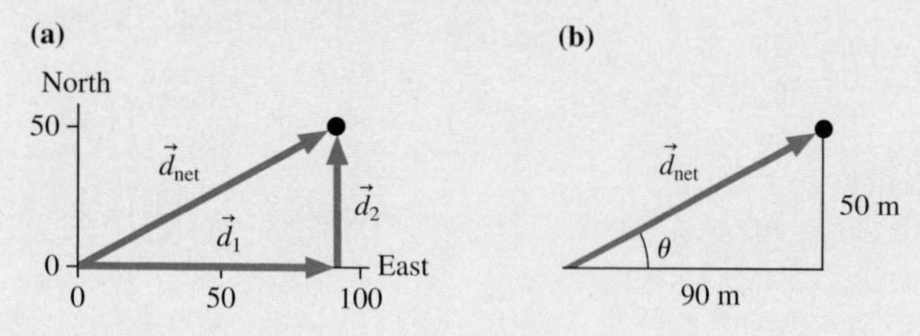

SOLVE We drew the two vector displacements with the tail of one vector starting at the tip of the previous one—exactly what is needed to form a vector sum. The vector $\vec{d}_{net}$ in Figure 1.25a is the vector sum of the successive displacements and thus represents Anna's net displacement from the origin.

Anna's distance from the origin is the length of this vector $\vec{d}_{net}$. FIGURE 1.25b shows that this vector is the hypotenuse of a right triangle with sides 50 m (because Anna walked 50 m north) and 90 m (because she walked 90 m east). We can compute the magnitude of this vector, her net displacement, using the Pythagorean theorem:

$$d_{net} = \sqrt{(50\text{ m})^2 + (90\text{ m})^2} = 103\text{ m} \approx 100\text{ m}$$

We have rounded off to the appropriate number of significant figures, giving us 100 m for the magnitude of the displacement vector. How about the direction? Figure 1.25b identifies the angle that gives the angle north of east of Anna's displacement. In the right triangle, 50 m is the opposite side and 90 m is the adjacent side, so the angle is

$$\theta = \tan^{-1}\left(\frac{50\text{ m}}{90\text{ m}}\right) = \tan^{-1}\left(\frac{5}{9}\right) = 29°$$

Putting it all together, we get a net displacement of

$$\vec{d}_{net} = (100\text{ m, }29°\text{ north of east})$$

ASSESS We can use our drawing to assess our result. If the two sides of the triangle are 50 m and 90 m, a length of 100 m for the hypotenuse seems about right. The angle is certainly smaller than 45°, but not too much smaller, so 29° seems reasonable.

Velocity Vectors

We've seen that a basic quantity describing the motion of an object is its velocity. Velocity is a vector quantity because its specification involves how fast an object is moving (its speed) and also the direction in which the object is moving. We thus represent the velocity of an object by a **velocity vector** $\vec{v}$ that points in the direction of the object's motion, and whose magnitude is the object's speed.

FIGURE 1.26 The motion diagram for a car starting from rest.

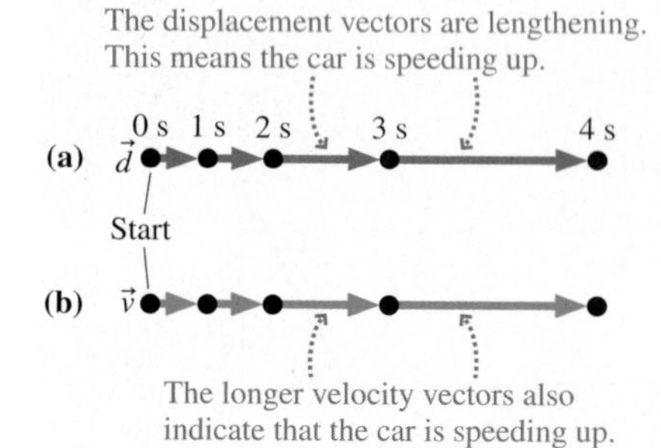

FIGURE 1.26a shows the motion diagram of a car accelerating from rest. We've drawn vectors showing the car's displacement between successive positions in the motion diagram. To draw the velocity vectors, first note that the direction of the displacement vector is the direction of motion between successive points in the motion diagram. The velocity of an object also points in the direction of motion, so the velocity vector points in the same direction as its displacement vector. Next, note that the magnitude of the velocity vector—How fast?—is the object's speed. Higher speeds imply greater displacements, so the length of the velocity vector should be proportional to the length of the displacement vector between successive points on a motion diagram. All this means that the vectors connecting each dot of a motion diagram to the next dot, which we have labeled as displacement vectors, could equally well be identified as velocity vectors, as shown in FIGURE 1.26b. **From now on, we'll show and label velocity vectors on motion diagrams rather than displacement vectors.**

NOTE ▶ The velocity vectors shown in Figure 1.26b are actually *average* velocity vectors. Because the velocity is steadily increasing, it's a bit less than this average at the start of each time interval, and a bit greater at the end. In Chapter 2 we'll refine these ideas as we develop the idea of instantaneous velocity. ◀

EXAMPLE 1.8 **Drawing a ball's motion diagram**

Jake hits a ball at a 60° angle from the horizontal. It is caught by Jim. Draw a motion diagram of the ball that shows velocity vectors rather than displacement vectors.

STRATEGIZE This example is typical of how many problems in physics are worded. The problem does not give a clear statement of where the motion begins or ends. Are we interested in the motion of the ball only during the time it is in the air between Jake and Jim? What about the motion *as* Jake hits it (ball rapidly speeding up) or *as* Jim catches it (ball rapidly slowing down)? Should we include Jim dropping the ball after he catches it? The point is that *you* will often be called on to make a *reasonable interpretation* of a problem statement. In this problem, the details of hitting and catching the ball are complex. The motion of the ball through the air is easier to describe, and it's a motion you might expect to learn about in a physics class. So our *interpretation* is that the motion diagram should start as the ball leaves Jake's bat (ball already moving) and should end the instant it touches Jim's hand (ball still moving).

PREPARE We model the ball as a particle, and sketch a motion diagram that represents the motion of a thrown ball along an arc.

SOLVE FIGURE 1.27 shows the motion diagram of the ball. Notice how, in contrast to the car of Figure 1.26, the ball is already moving as the motion diagram begins. As before, the velocity vectors are shown by connecting the dots with arrows. You can see that the velocity vectors get shorter (ball slowing down), get longer (ball speeding up), and change direction. Each $\vec{v}$ is different, so this is *not* constant-velocity motion.

FIGURE 1.27 The motion diagram of a ball traveling from Jake to Jim.

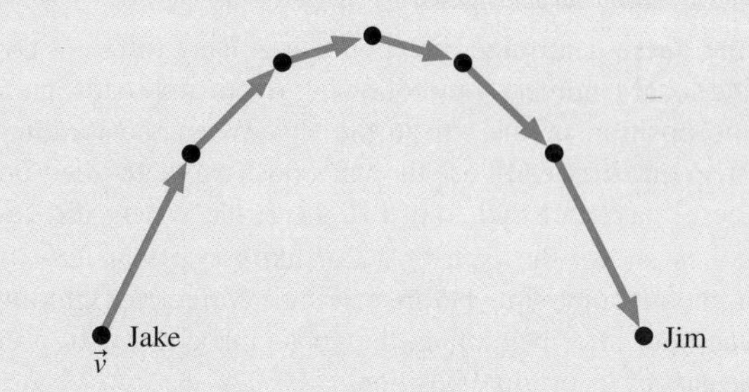

ASSESS We haven't learned enough to make a detailed analysis of the motion of the ball, but it's still worthwhile to do a quick assessment. Does our diagram make sense? Think about the velocity of the ball—we show it moving upward at the start and downward at the end. This does match what happens when you toss a ball back and forth, so our answer seems reasonable.

STOP TO THINK 1.7 $\vec{P}$ and $\vec{Q}$ are two vectors of equal length but different direction. Which vector shows the sum $\vec{P} + \vec{Q}$?

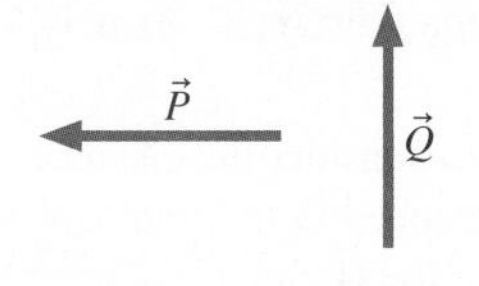

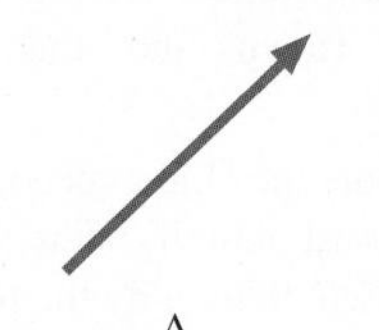

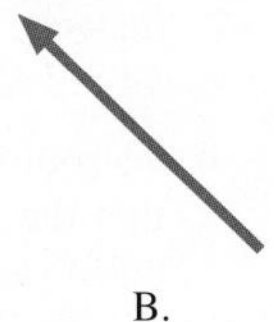

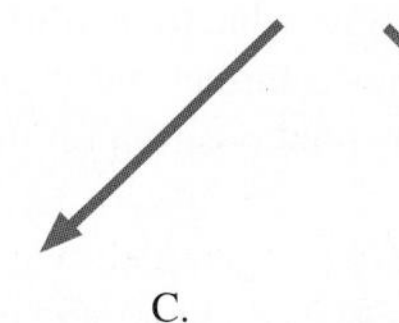

1.7 Where Do We Go from Here?

This first chapter has been an introduction to some of the fundamental ideas about motion and some of the basic techniques that you will use in the rest of the course. You have seen some examples of how to make *models* of a physical situation, thereby focusing on the essential elements of the situation. You have learned some practical ideas, such as how to convert quantities from one set of units to another. The rest of this text—and the rest of your course—will extend these themes.

In each chapter of this text, you'll learn both new principles and more tools and techniques. As you proceed, you'll find that each new chapter depends on those that preceded it. The principles and the problem-solving strategies you learned in this chapter will still be needed in Chapter 30.

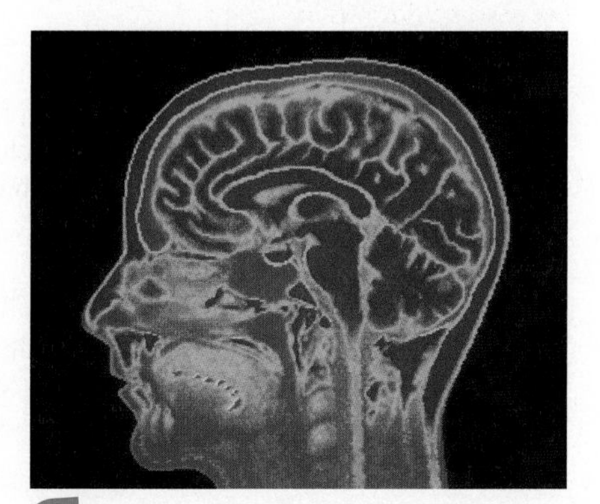

BIO Chapter 28 ends with an integrated example that explores the basic physics of magnetic resonance imaging (MRI), explaining how the interaction of magnetic fields with the nuclei of atoms in the body can be used to create an image of the body's interior.

We'll give you some assistance integrating new ideas with the material of previous chapters. When you start a chapter, the **chapter preview** will let you know which topics are especially important to review. And the last element in each chapter will be an **integrated example** that brings together the principles and techniques you have just learned with those you learned previously. The integrated nature of these examples will also be a helpful reminder that the problems of the real world are similarly complex, and solving such problems requires you to do just this kind of integration.

Our first integrated example is reasonably straightforward because there's not much to integrate yet. The examples in future chapters will be much richer.

INTEGRATED EXAMPLE 1.9 A goose gets its bearings BIO

Migrating geese determine direction using many different tools: by noting local landmarks, by following rivers and roads, and by using the position of the sun in the sky. When the weather is overcast so that they can't use the sun's position to get their bearings, geese may start their day's flight in the wrong direction. FIGURE 1.28 shows the path of a Canada goose that flew in a straight line for some time before making a corrective right-angle turn. One hour after beginning, the goose made a rest stop on a lake due east of its original position.

FIGURE 1.28 Trajectory of a misdirected goose.

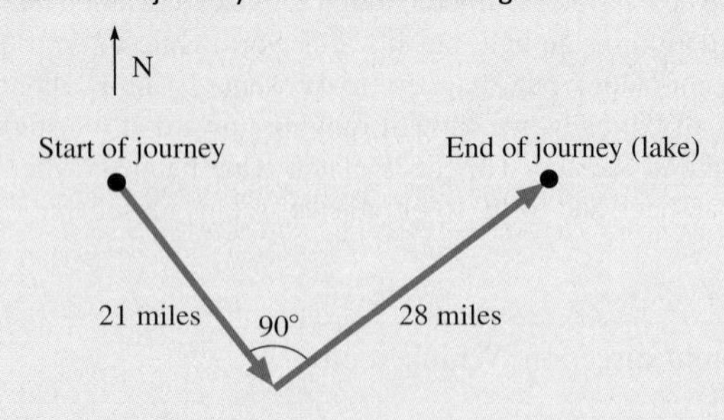

a. How much extra distance did the goose travel due to its initial error in flight direction? That is, how much farther did it fly than if it had simply flown directly to its final position on the lake?
b. What was the flight speed of the goose?
c. A typical flight speed for a migrating goose is 80 km/h. Given this, does your result seem reasonable?

STRATEGIZE In this integrated example, we'll need to pull together all our knowledge about right triangles, speed, and significant figures.

PREPARE Figure 1.28 shows the trajectory of the goose, but it's worthwhile to redraw Figure 1.28 and note the displacement from the start to the end of the journey, the shortest distance the goose could have flown. (The examples in the chapter to this point have used professionally rendered drawings, but these are much more careful and detailed than you are likely to make. FIGURE 1.29 shows a drawing that is more typical of what you might actually do when working problems yourself.) Drawing and labeling the displacement between the starting and ending points in Figure 1.29 show that it is the hypotenuse of a right triangle, so we can use our rules for triangles as we look for a solution.

FIGURE 1.29 A typical student sketch shows the motion and the displacement of the goose.

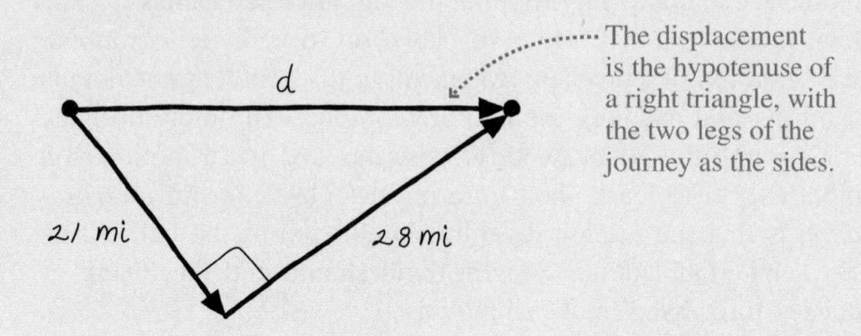

SOLVE

a. The minimum distance the goose *could* have flown, if it flew straight to the lake, is the hypotenuse of a triangle with sides 21 mi and 28 mi. This straight-line distance is

$$d = \sqrt{(21 \text{ mi})^2 + (28 \text{ mi})^2} = 35 \text{ mi}$$

The actual distance the goose flew is the sum of the distances traveled for the two legs of the journey:

$$\text{distance traveled} = 21 \text{ mi} + 28 \text{ mi} = 49 \text{ mi}$$

The extra distance flown is the difference between the actual distance flown and the straight-line distance—namely, 14 miles.

b. To compute the flight speed, we need to consider the distance that the bird actually flew. The flight speed is the total distance flown divided by the total time of the flight:

$$v = \frac{49 \text{ mi}}{1.0 \text{ h}} = 49 \text{ mi/h}$$

c. To compare our calculated speed with a typical flight speed, we must convert our solution to km/h, rounding off to the correct number of significant figures:

$$49 \frac{\text{mi}}{\text{h}} \times \frac{1.61 \text{ km}}{1.00 \text{ mi}} = 79 \frac{\text{km}}{\text{h}}$$

A calculator will return many more digits, but the original data had only two significant figures, so we report the final result to this accuracy.

ASSESS In this case, an assessment was built into the solution of the problem. The calculated flight speed matches the expected value for a goose, which gives us confidence that our answer is correct. As a further check, our calculated net displacement of 35 mi seems about right for the hypotenuse of the triangle in Figure 1.29.

SUMMARY

GOAL To introduce the fundamental concepts of motion and to review related basic mathematical principles.

IMPORTANT CONCEPTS

Motion Diagrams

The **particle model** represents a moving object as if all its mass were concentrated at a single point. Using this model, we can represent motion with a **motion diagram,** where dots indicate the object's positions at successive times. In a motion diagram, the time interval between successive dots is always the same.

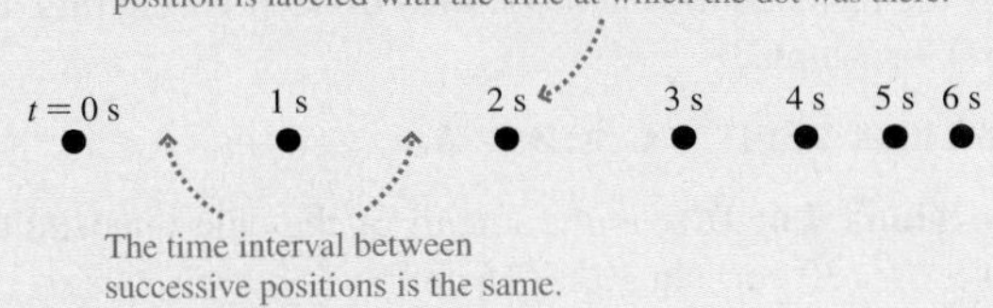

Each dot represents the position of the object. Each position is labeled with the time at which the dot was there.
t = 0 s
1 s
2 s
3 s
4 s
5 s
6 s
The time interval between successive positions is the same.

Scalars and Vectors

Scalar quantities have only a magnitude and can be represented by a single number. Temperature, time, and mass are scalars.

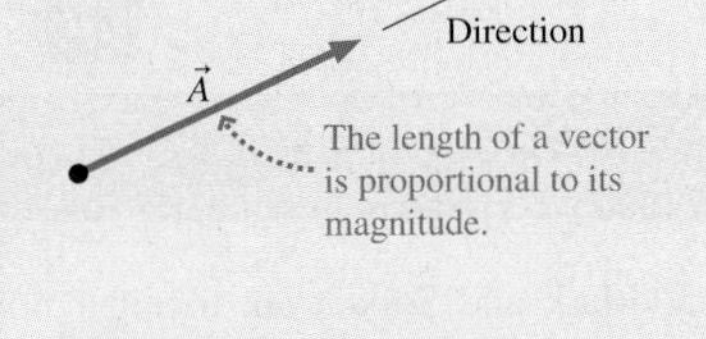

The length of a vector is proportional to its magnitude.

A **vector** is a quantity described by both a magnitude and a direction. Velocity and displacement are vectors.

Velocity vectors can be drawn on a motion diagram by connecting successive points with a vector.

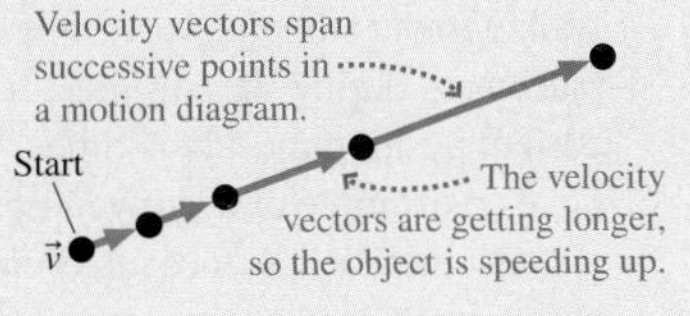

Describing Motion

Position locates an object with respect to a chosen coordinate system. It is described by a **coordinate**.

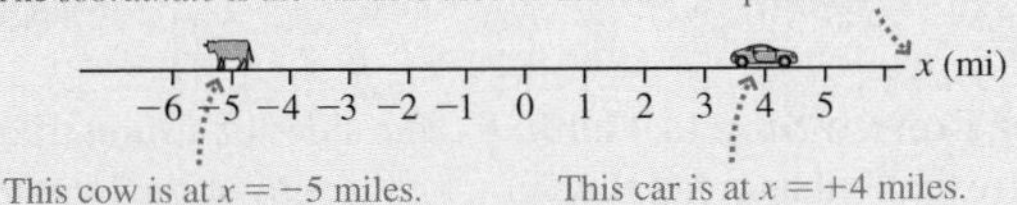

−6 −5 −4 −3 −2 −1 0 1 2 3 4 5
This cow is at x = −5 miles.
This car is at x = +4 miles.

A change in position is called a **displacement**. For motion along a line, a displacement is a signed quantity. The displacement from x_i to x_f is $\Delta x = x_f - x_i$.

Time is measured from a particular instant to which we assign $t = 0$. A **time interval** is the elapsed time between two specific instants t_i and t_f. It is given by $\Delta t = t_f - t_i$.

Velocity is the ratio of the displacement of an object to the time interval during which this displacement occurs:

$$v = \frac{\Delta x}{\Delta t}$$

Units

Every measurement of a quantity must include a **unit.**
The standard system of units used in science is the **SI system.** Common SI units include:

- Length: meters (m)
- Time: seconds (s)
- Mass: kilograms (kg)

APPLICATIONS

Working with Numbers

In **scientific notation,** a number is expressed as a decimal number between 1 and 10 multiplied by a power of ten. In scientific notation, the diameter of the earth is 1.27×10^7 m.

A **prefix** can be used before a unit to indicate a multiple of 10 or 1/10. Thus we can write the diameter of the earth as 12,700 km, where the k in km denotes 1000.

We can perform a **unit conversion** to convert the diameter of the earth to a different unit, such as miles. We do so by multiplying by a conversion factor equal to 1, such as 1 = 1 mi/1.61 km.

Significant figures are reliably known digits. The number of significant figures for:

- **Multiplication, division, and powers** is set by the value with the fewest significant figures.
- **Addition and subtraction** is set by the value with the smallest number of decimal places.

An **order-of-magnitude estimate** is an estimate that has an accuracy of about one significant figure. Such estimates are usually made using rough numbers from everyday experience.

Learning Objectives After studying this chapter, you should be able to:

- Draw and interpret motion diagrams to represent motion. *Conceptual Questions 1.2, 1.13; Problems 1.1, 1.2, 1.3*
- Describe motion in terms of position, velocity, and time. *Conceptual Question 1.8; Problems 1.5, 1.6, 1.7, 1.8, 1.9*
- Calculate the speed and velocity of an object. *Conceptual Question 1.5; Problems 1.11, 1.12, 1.13, 1.15*
- Use scientific notation. *Problems 1.23, 1.24, 1.25*
- Express quantities with the appropriate units and the proper number of significant figures. *Problems 1.19, 1.20, 1.21, 1.22*
- Perform unit conversions. *Conceptual Question 1.17; Problems 1.16, 1.17, 1.18*
- Describe motion using vectors and trigonometry. *Conceptual Question 1.16; Problems 1.27, 1.28, 1.32, 1.35, 1.38*

STOP TO THINK ANSWERS

Chapter Preview Stop to Think: C. The sides of a right triangle are related by the Pythagorean theorem. The length of the hypotenuse is thus $\sqrt{(6\text{ cm})^2 + (8\text{ cm})^2} = 10\text{ cm}$. Note that this triangle is a version of a 3-4-5 right triangle; the lengths of the sides are in this ratio.

Stop to Think 1.1: B. The images of B are farther apart, so B travels a greater distance than does A during the same intervals of time.

Stop to Think 1.2: A. Dropped ball. B. Dust particle. C. Descending rocket.

Stop to Think 1.3: C. Depending on her initial positive position and how far she moves in the negative direction, she could end up on either side of the origin.

Stop to Think 1.4: C. Her speed is given by Equation 1.1. Her speed is the distance traveled (4 miles) divided by the time interval (2 hours), or 2 mph.

Stop to Think 1.5: D > C > B = A.

Stop to Think 1.6: D. x is the length of the side opposite the 30° angle, so $x = (30\text{ cm})\sin 30° = 15\text{ cm}$.

Stop to Think 1.7: B. The vector sum is found by placing the tail of one vector at the tip of the other vector.

Video Tutor Solution Chapter 1

QUESTIONS

Conceptual Questions

1. A softball player slides into second base. Use the particle model to draw a motion diagram of the player from the time he begins to slide until he reaches the base. Number the dots in order, starting with zero.
2. A car travels to the left at a steady speed for a few seconds, then brakes for a stop sign. Use the particle model to draw a motion diagram of the car for the entire motion described here. Number the dots in order, starting with zero.
3. BIO The bush baby, a small African mammal, is a remarkable jumper. Although only about 8 inches long, it can jump, from a standing start, straight up to a height of over 7 feet! Use the particle model to draw a motion diagram for a bush baby's jump, from its start until it reaches its highest point.
4. A ball is dropped from the roof of a tall building and students in a physics class are asked to sketch a motion diagram for this situation. A student submits the diagram shown in Figure Q1.4. Is the diagram correct? Explain.

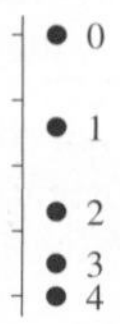

FIGURE Q1.4

5. Mark and Sofia walk together down a long, straight road. They walk without stopping for 4 miles. At this point Sofia says their displacement during the trip must have been 4 miles; Mark says their current position must be 4 miles. Who, if either, is correct? Explain.
6. Give an example of a trip you might take in your car for which the distance traveled as measured on your car's odometer is not equal to the displacement between your initial and final positions.
7. Write a sentence or two describing the difference between speed and velocity. Give one example of each.
8. The motion of a skateboard along a horizontal axis is observed for 5 s. The initial position of the skateboard is negative with respect to a chosen origin, and its velocity throughout the 5 s is also negative. At the end of the observation time, is the skateboard closer to or farther from the origin than initially? Explain.
9. You are standing on a straight stretch of road and watching the motion of a bicycle; you choose your position as the origin. At one instant, the position of the bicycle is negative and its velocity is positive. Is the bicycle getting closer to you or farther away? Explain.
10. Two friends watch a jogger complete a 400 m lap around the track in 100 s. One of the friends states, "The jogger's velocity was 4 m/s during this lap." The second friend objects, saying, "No, the jogger's speed was 4 m/s." Who is correct? Justify your answer.

Problem difficulty is labeled as | (straightforward) to ||||| (challenging). INT Problems labeled integrate significant material from earlier chapters; BIO are of biological or medical interest.

The eText icon indicates when there is a video tutor solution available for the chapter or for a specific problem. To launch these videos, log into your eText through Mastering™ Physics or log into the Study Area.

11. A softball player hits the ball and starts running toward first base. Draw a motion diagram, using the particle model, showing her velocity vectors during the first few seconds of her run.
12. A child is sledding on a smooth, level patch of snow. She encounters a rocky patch and slows to a stop. Draw a motion diagram, using the particle model, showing her velocity vectors.
13. A skydiver jumps out of an airplane. Her speed steadily increases until she deploys her parachute, at which point her speed quickly decreases. She subsequently falls to earth at a constant rate, stopping when she lands on the ground. Draw a motion diagram, using the particle model, that shows her position at successive times and includes velocity vectors.
14. Your roommate drops a tennis ball from a third-story balcony. It hits the sidewalk and bounces as high as the second story. Draw a motion diagram, using the particle model, showing the ball's velocity vectors from the time it is released until it reaches the maximum height on its bounce.
15. A car is driving north at a steady speed. It makes a gradual 90° left turn without losing speed, then continues driving to the west. Draw a motion diagram, using the particle model, showing the car's velocity vectors as seen from a helicopter hovering over the highway.
16. Three displacement vectors have lengths 1 m, 2 m, and 4 m. Could they possibly add together to get a vector of length zero?
17. Your friend Travis claims to have set the new world speed record for riding a unicycle. His top speed, he says, was 55 m/s. Do you believe him? Explain.

Multiple-Choice Questions

18. | A student walks 1.0 mi west and then 1.0 mi north. Afterward, how far is she from her starting point?
A. 1.0 mi B. 1.4 mi C. 1.6 mi D. 2.0 mi
19. | You throw a rock upward. The rock is moving upward, but it is slowing down. If we define the ground as the origin, the position of the rock is ______ and the velocity of the rock is ______.
A. positive, positive B. positive, negative
C. negative, positive D. negative, negative
20. | Which of the following motions could be described by the motion diagram of Figure Q1.20?
A. A hockey puck sliding across smooth ice.
B. A cyclist braking to a stop.
C. A sprinter starting a race.
D. A ball bouncing off a wall.

FIGURE Q1.20 5 4 3 2 1 0

21. | Which of the following motions is described by the motion diagram of Figure Q1.21?
A. An ice skater gliding across the ice.
B. An airplane braking to a stop after landing.
C. A car pulling away from a stop sign.
D. A pool ball bouncing off a cushion and reversing direction.

FIGURE Q1.21 0 1 2 3 4 5

22. | A bird flies 3.0 km due west and then 2.0 km due north. What is the magnitude of the bird's displacement?
A. 2.0 km B. 3.0 km C. 3.6 km D. 5.0 km
23. || Weddell seals make holes in sea ice so that they can swim down to forage on the ocean floor below. Measurements for one seal showed that it dived straight down from such an opening, reaching a depth of 0.30 km in a time of 5.0 min. What was the speed of the diving seal? BIO
A. 0.60 m/s B. 1.0 m/s C. 1.6 m/s D. 6.0 m/s
E. 10 m/s
24. || A bird flies 3.0 km due west and then 2.0 km due north. Another bird flies 2.0 km due west and 3.0 km due north. What is the angle between the net displacement vectors for the two birds?
A. 23° B. 34° C. 56° D. 90°
25. || Hicham El Guerrouj of Morocco holds the world record in the 1500 m running race. He ran the final 400 m in a time of 51.9 s. What was his average speed in mph over the last 400 m?
A. 14.2 mph B. 15.5 mph
C. 17.2 mph D. 23.9 mph
26. | Compute 3.24 m + 0.532 m to the correct number of significant figures.
A. 3.7 m B. 3.77 m
C. 3.772 m D. 3.7720 m
27. || An American football field is 109.7 m long and 48.8 m wide. To the correct number of significant figures, what is its area?
A. 5351 m^2 B. 5.35×10^3 m^2
C. 5351.17 m^2 D. 5400 m^2
28. | The earth formed 4.57×10^9 years ago. What is this time in seconds?
A. 1.67×10^{12} s B. 4.01×10^{13} s
C. 2.40×10^{15} s D. 1.44×10^{17} s
29. ||| An object's average density ρ is defined as the ratio of its mass to its volume: $\rho = M/V$. The earth's mass is 5.94×10^{24} kg, and its volume is 1.08×10^{12} km^3. What is the earth's average density?
A. 5.50×10^3 kg/m^3 B. 5.50×10^6 kg/m^3
C. 5.50×10^9 kg/m^3 D. 5.50×10^{12} kg/m^3

PROBLEMS

Section 1.1 Motion: A First Look

1. | A car skids to a halt to avoid hitting an object in the road. Draw a motion diagram of the car from the time the skid begins until the instant the car stops.
2. | A man rides a bike along a straight road for 5 min, then has a flat tire. He stops for 5 min to repair the flat, but can't fix it. He walks the rest of the way, which takes him another 10 min. Use the particle model to draw a motion diagram of the man for the entire motion described here. Number the dots in order, starting with zero.
3. | Amanda has just entered an elevator. The elevator rises and stops at the third floor. Use the particle model to draw a motion diagram of Amanda during her entire ride on the elevator. Number the dots in order, starting from zero. (Be sure to consider how the elevator speeds up and slows down.)

Watch Video Solution Problem 1.25

Section 1.2 Models and Modeling

Section 1.3 Position and Time: Putting Numbers on Nature

4. | Figure P1.4 shows Sue along the straight-line path between her home and the cinema. What is Sue's position x if
   a. Her home is the origin?
   b. The cinema is the origin?

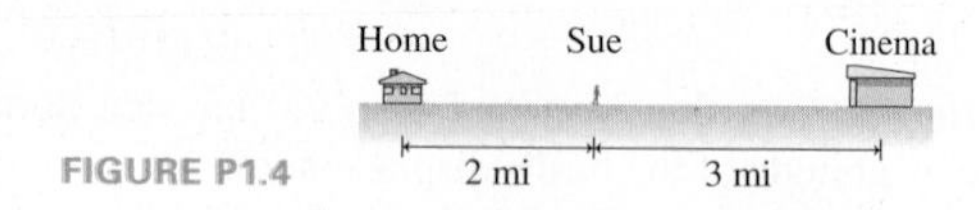

FIGURE P1.4

5. | Figure P1.4 shows Sue along the straight-line path between her home and the cinema. Now Sue walks home. What is Sue's displacement if
   a. Her home is the origin?
   b. The cinema is the origin?
6. | Logan observes a paramecium under a microscope. The eyepiece of the microscope has a horizontal scale marked in mm. The paramecium starts at the 65 mm mark and ends up at the 42 mm mark. What is the paramecium's displacement? BIO
7. | Keira starts at position $x = 23$ m along a coordinate axis. She then undergoes a displacement of -45 m. What is her final position?
8. | A car travels along a straight east-west road. A coordinate system is established on the road, with x increasing to the east. The car ends up 14 mi west of the origin, which is defined as the intersection with Mulberry Road. If the car's displacement was -23 mi, what side of Mulberry Road did the car start on? How far from the intersection was the car at the start?
9. | Foraging bees often move in straight lines away from and toward their hives. Suppose a bee starts at its hive and flies 500 m due east, then flies 400 m west, then 700 m east. How far is the bee from the hive? BIO

Section 1.4 Velocity

10. | A security guard walks at a steady pace, traveling 110 m in one trip around the perimeter of a building. It takes him 240 s to make this trip. What is his speed?
11. || List the following items in order of decreasing speed, from greatest to least: (i) A wind-up toy car that moves 0.15 m in 2.5 s. (ii) A soccer ball that rolls 2.3 m in 0.55 s. (iii) A bicycle that travels 0.60 m in 0.075 s. (iv) A cat that runs 8.0 m in 2.0 s.
12. || Figure P1.12 shows the motion diagram for a horse galloping in one direction along a straight path. Not every dot is labeled, but the dots are at equally spaced instants of time. What is the horse's velocity
    a. During the first 10 seconds of its gallop?
    b. During the interval from 30 s to 40 s?
    c. During the interval from 50 s to 70 s?

70 s 50 s 30 s 10 s
x (m)
50 150 250 350 450 550 650

FIGURE P1.12

13. || It takes Harry 35 s to walk from $x = -12$ m to $x = -47$ m. What is his velocity?
14. | A dog trots from $x = -12$ m to $x = 3$ m in 10 s. What is its velocity?
15. || In Michael Johnson's world-record 400 m sprint, he ran the first 100 m in 11.20 s; then he reached the 200 m mark after a total time of 21.32 s had elapsed, reached the 300 m mark after 31.76 s, and finished in 43.18 s. BIO
    a. During what 100 m segment was his speed the highest?
    b. During this segment, what was his speed in m/s?

Section 1.5 A Sense of Scale: Significant Figures, Scientific Notation, and Units

16. || Convert the following to SI base units:
    a. 9.12 μs b. 3.42 km
    c. 44 cm/ms d. 80 km/h
17. | Convert the following to SI units:
    a. 8.0 in b. 66 ft/s c. 60 mph
18. | Convert the following to SI units:
    a. 1.0 hour b. 1.0 day c. 1.0 year
19. || How many significant figures does each of the following numbers have?
    a. 6.21 b. 62.1
    c. 0.620 d. 0.062
20. | How many significant figures does each of the following numbers have?
    a. 0.621 b. 0.006200
    c. 1.0621 d. 6.21×10^3
21. | Compute the following numbers to three significant figures.
    a. 33.3×25.4 b. $33.3 - 25.4$
    c. $\sqrt{33.3}$ d. $333.3 \div 25.4$
22. || If you make multiple measurements of your height, you are likely to find that the results vary by nearly half an inch in either direction due to measurement error and actual variations in height. You are slightly shorter in the evening, after gravity has compressed and reshaped your spine over the course of a day. One measurement of a man's height is 6 feet and 1 inch. Express his height in meters, using the appropriate number of significant figures. BIO

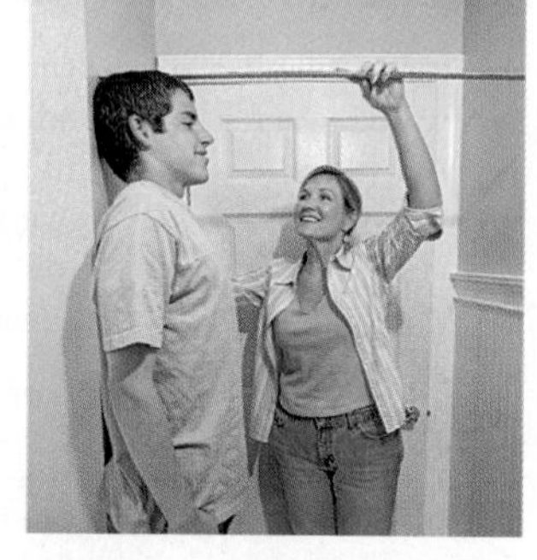

23. | Mount Everest has a height of 29,029 ft above sea level. Express this height in meters, giving your result in scientific notation with the correct number of significant figures.
24. ||| Blades of grass grow from the bottom, so, as growth occurs, the top of the blade moves upward. During the summer, when your lawn is growing quickly, estimate this speed, in m/s. Make this estimate from your experience noting, for instance, how often you mow the lawn and what length you trim. Express your result in scientific notation. BIO
25. || Estimate the average speed, in m/s, with which the hair on your head grows. Make this estimate from your own experience noting, for instance, how often you cut your hair and how much you trim. Express your result in scientific notation. BIO

Section 1.6 Vectors and Motion: A First Look

26. | Loveland, Colorado, is 18 km due south of Fort Collins and 31 km due west of Greeley. What is the distance between Fort Collins and Greeley?

27. || A city has streets laid out in a square grid, with each block 135 m long. If you drive north for three blocks, then west for two blocks, how far are you from your starting point?
28. | Joe and Max shake hands and say goodbye. Joe walks east 0.55 km to a coffee shop, and Max flags a cab and rides north 3.25 km to a bookstore. How far apart are their destinations?
29. || In downtown Chicago, the east-west blocks are 400 ft long while the north-south blocks are 280 ft long. Because of the many one-way streets, it can be challenging to get around. Veronica starts at the corner of Dearborn and Ohio Streets. She drives four blocks north to Superior, two blocks east to Wabash, then a block south to get to her destination at Wabash and Huron. What is the straight-line distance from her starting point?
30. || A butterfly flies from the top of a tree in the center of a garden to rest on top of a red flower at the garden's edge. The tree is 8.0 m taller than the flower, and the garden is 12 m wide. Determine the magnitude of the butterfly's displacement.
31. ||| A garden has a circular path of radius 50 m. John starts at the easternmost point on this path, then walks counterclockwise around the path until he is at its southernmost point. What is John's displacement? Use the (magnitude, direction) notation for your answer.
32. ||| Luis is visiting a public garden that has a large, circular path. When he has walked one-quarter of the distance around the path, the magnitude of his displacement is 180 m. What is the diameter of the path?
33. | Migrating geese tend to travel at approximately constant speed, flying in segments that are straight lines. A goose flies 32 km south, then turns to fly 20 km west. Afterward, how far is the goose from its original position? BIO
34. || A circular test track for cars in England has a circumference of 3.2 km. A car travels around the track from the southernmost point to the northernmost point.
    a. What distance does the car travel?
    b. What is the car's displacement from its original position?
35. || Black vultures excel at gliding flight; they can move long distances through the air without flapping their wings while undergoing only a modest drop in height. A vulture in a typical glide in still air moves along a path tipped 3.5° below the horizontal. If the vulture moves a horizontal distance of 100 m, how much height does it lose? BIO
36. ||| Figure P1.36 shows a map of Olivia's trip to a coffee shop. She gets on her bike at Loomis and then rides south 0.8 mi to Broadway. She turns east onto Broadway, rides 0.8 mi to where Broadway turns, and then continues another 1.2 mi to the shop. What is the total displacement of her trip, in (magnitude, direction) form?

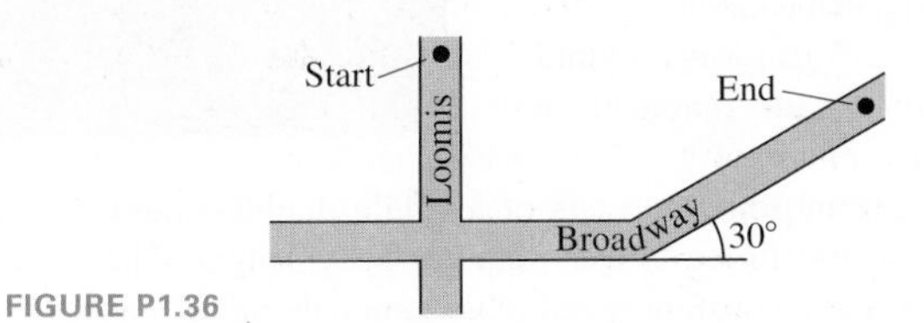

FIGURE P1.36

37. || The Great Pyramid of Giza is 139 m tall, with a slope of 51.8°. If you were to climb the pyramid from base to top (which is forbidden!), what distance along the face of the pyramid would you travel?
38. | A hiker is climbing a steep 10° slope. Her pedometer shows that she has walked 1500 m along the slope. How much elevation has she gained?
39. ||||| A ball on a porch rolls 60 cm to the porch's edge, drops 40 cm, continues rolling on the grass, and eventually stops 80 cm from the porch's edge. What is the magnitude of the ball's net displacement, in centimeters?
40. || A kicker punts a football from the very center of the field to the sideline 43 yards downfield. What is the net displacement of the ball? (A football field is 53 yards wide.)

Problems 41 and 42 relate to the gliding flight of flying squirrels. These squirrels glide from tree to tree at a constant speed, moving in a straight line tipped below the vertical and steadily losing altitude as they move forward. Short and long glides have different profiles.

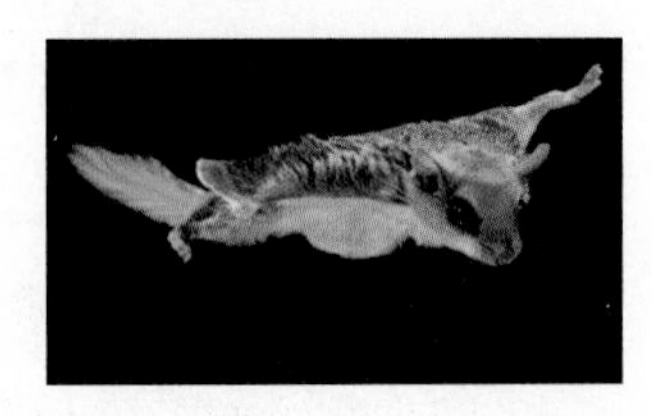

41. || A squirrel completing a short glide travels in a straight line tipped 40° below the horizontal. The squirrel starts 9.0 m above the ground on one tree and glides to a second tree that is a horizontal distance of 3.5 m away. BIO
    a. What is the length of the squirrel's glide path?
    b. What is the squirrel's height above the ground when it lands?
42. || A squirrel in a typical long glide covers a horizontal distance of 16 m while losing 8.0 m of elevation. During this glide, BIO
    a. What is the angle of the squirrel's path below the horizontal?
    b. What is the total distance covered by the squirrel?

General Problems

Problems 43 through 49 are motion problems similar to those you will learn to solve in Chapter 2. For now, simply *interpret* the problem by drawing a motion diagram showing the object's position and its velocity vectors. **Do *not* solve these problems** or do any mathematics.

43. || In a typical greyhound race, a dog accelerates to a speed of 20 m/s over a distance of 30 m. It then maintains this speed. What would be a greyhound's time in the 100 m dash? BIO
44. || Billy drops a watermelon from the top of a three-story building, 10 m above the sidewalk. How fast is the watermelon going when it hits?
45. || A skateboarder starts from rest at the top of a ramp. He rolls down the ramp and then continues rolling on the smooth, horizontal floor.
46. || A speed skater moving across frictionless ice at 8.0 m/s hits a 5.0-m-wide patch of rough ice. She slows steadily, then continues on at 6.0 m/s. What is her acceleration on the rough ice?
47. || The giant eland, an African antelope, is an exceptional jumper, able to leap 1.5 m off the ground. To jump this high, with what speed must the eland leave the ground? BIO
48. || A ball rolls along a smooth horizontal floor at 10 m/s, then starts up a 20° ramp. How high does it go before rolling back down?
49. || A motorist is traveling at 20 m/s. He is 60 m from a stop light when he sees it turn yellow. His reaction time, before stepping on the brake, is 0.50 s. What steady deceleration while braking will bring him to a stop right at the light?

Watch Video Solution Problem 1.61

Problems 50 through 54 show a motion diagram. For each of these problems, write a one or two sentence "story" about a *real object* that has this motion diagram. Your stories should talk about people or objects by name and say what they are doing. Problems 43 through 49 are examples of motion short stories.

50. |

51. |

52. |

53. |

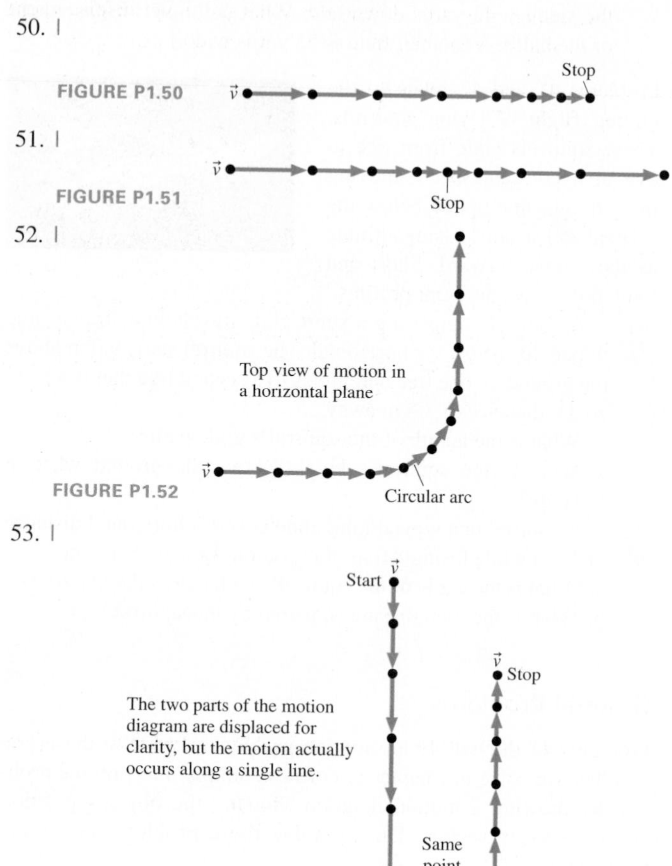
Stop
$\vec{v}$
$\vec{v}$
Stop
Top view of motion in a horizontal plane
$\vec{v}$
Circular arc
Start
$\vec{v}$
$\vec{v}$
Stop
The two parts of the motion diagram are displaced for clarity, but the motion actually occurs along a single line.
Same point

FIGURE P1.50

FIGURE P1.51

FIGURE P1.52

FIGURE P1.53

54. |

$\vec{v}$

FIGURE P1.54

55. || Estimate the length of a human lifetime, in seconds.

56. ||| On a highway trip, Joseph drives the first 25 miles at 55 mph, and the next 15 miles at 70 mph. What is his average speed for this trip?

57. ||| Evan is just leaving his house to visit his grandmother. Normally, the trip takes him 25 minutes on the freeway, going 55 mph. But tonight he's running 5 minutes late. How fast will he need to drive on the freeway to make up the 5 minutes?

58. ||| Gretchen runs the first 4.0 km of a race at 5.0 m/s. Then a stiff wind comes up, so she runs the last 1.0 km at only 4.0 m/s. If she later ran the same course again, what constant speed would let her finish in the same time as in the first race?

59. |||| If you swim with the current in a river, your speed is increased by the speed of the water; if you swim against the current, your speed is decreased by the water's speed. The current in a river flows at 0.52 m/s. In still water you can swim at 1.78 m/s. If you swim downstream a certain distance, then back again upstream, how much longer, in percent, does it take compared to the same trip in still water?

60. || The end of Hubbard Glacier in Alaska advances by an average of 105 feet per year. What is the speed of advance of the glacier in m/s?

61. | The earth completes a circular orbit around the sun in one year. The orbit has a radius of 93,000,000 miles. What is the speed of the earth around the sun in m/s? Report your result using scientific notation.

62. || BIO The Greenland shark is thought to be the longest-living vertebrate on earth. Early estimates of its maximum age were based on the fact that such sharks grow in length only about a centimeter per year, and yet an adult shark can reach a length of 15 feet. Estimate how long a 15 foot shark might have lived. (A newborn shark is about 1 foot long.)

63. ||| The winner of the 2016 Keystone (Colorado) Uphill/Downhill mountain bike race finished in a total time of 47 minutes and 25 seconds. The uphill leg was 4.6 miles long, and on this leg his average speed was 8.75 mph. The downhill leg was 6.9 miles. What was his average speed on this leg?

64. ||| Shannon decides to check the accuracy of her speedometer. She adjusts her speed to read exactly 70 mph on her speedometer and holds this steady, measuring the time between successive mile markers separated by exactly 1.00 mile. If she measures a time of 54 s, is her speedometer accurate? If not, is the speed it shows too high or too low?

65. || The Nardo ring is a circular test track for cars. It has a circumference of 12.5 km. Cars travel around the track at a constant speed of 100 km/h. A car starts at the easternmost point of the ring and drives for 15 minutes at this speed.
   a. What distance, in km, does the car travel?
   b. What is the magnitude of the car's displacement, in km, from its initial position?
   c. What is the speed of the car in m/s?

66. || BIO Motor neurons in mammals transmit signals from the brain to skeletal muscles at approximately 25 m/s. Estimate how much time in ms (10^{-3} s) it will take for a signal to get from your brain to your hand.

67. ||| BIO Satellite data taken several times per hour on a particular albatross showed travel of 1200 km over a time of 1.4 days.
   a. Given these data, what was the bird's average speed in mph?
   b. Data on the bird's position were recorded only intermittently. Explain how this means that the bird's actual average speed was higher than what you calculated in part a.

68. || BIO The bacterium *Escherichia coli* (or *E. coli*) is a single-celled organism that lives in the gut of healthy humans and animals. Its body shape can be modeled as a 2-μm-long cylinder with a 1 μm diameter, and it has a mass of 1×10^{-12} g. Its chromosome consists of a single double-stranded chain of DNA 700 times longer than its body length. The bacterium moves at a constant speed of 20 μm/s, though not always in the same direction. Answer the following questions about *E. coli* using SI base units (unless specifically requested otherwise) and correct significant figures.

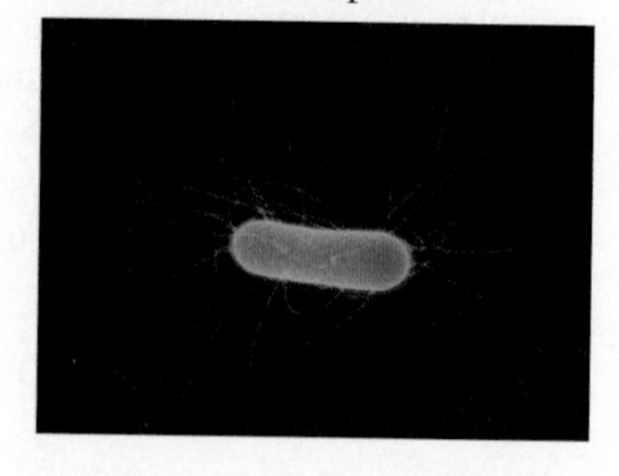

   a. What is its length?
   b. Diameter?
   c. Mass?
   d. What is the length of its DNA, in millimeters?
   e. If the organism were to move along a straight path, how many meters would it travel in one day?

69. || The bacterium *Escherichia coli* (or *E. coli*) is a single-celled organism that lives in the gut of healthy humans and animals. When grown in a uniform medium rich in salts and amino acids, it swims along zig-zag paths at a constant speed changing direction at varying time intervals. Figure P1.69 shows the positions of an *E. coli* as it moves from point A to point J. Each segment of the motion can be identified by two letters, such as segment BC. During which segments, if any, does the bacterium have the same
BIO
   a. Displacement? b. Speed? c. Velocity?

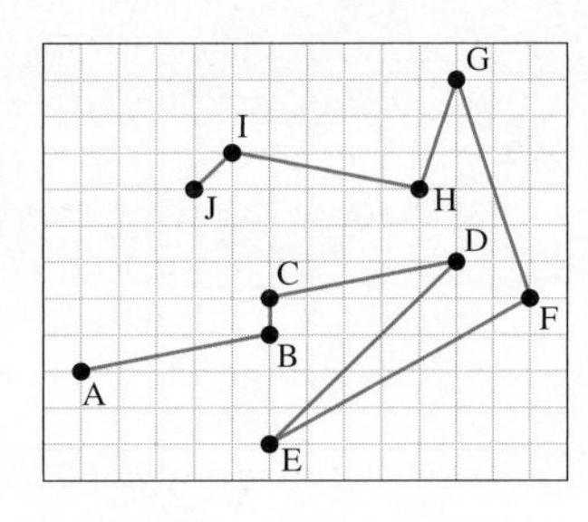

FIGURE P1.69

70. ||| The sun is 30° above the horizon. It makes a 52-m-long shadow of a tall tree. How high is the tree?
71. ||| Weddell seals foraging in open water dive toward the ocean bottom by swimming forward in a straight-line path tipped below the horizontal. The tracking data for one seal showed it taking 4.0 min to descend 360 m below the surface while moving 920 m horizontally.
BIO
   a. What was the angle of the seal's path below the horizontal?
   b. What distance did the seal cover in making this dive?
   c. What was the seal's speed, in m/s?
72. ||| Erica is participating in a road race. The first part of the race is on a 5.2-mile-long straight road oriented at an angle of 25° north of east. The road then turns due north for another 4.0 mi to the finish line. In miles, what is the straight-line distance from the starting point to the end of the race?
73. ||| Whale sharks swim forward while ascending or descending. They swim along a straight-line path at a shallow angle as they move from the surface to deep water or from the depths to the surface. In one recorded dive, a shark started 50 m below the surface and swam at 0.85 m/s along a path tipped at a 13° angle above the horizontal until reaching the surface.
BIO
   a. What was the horizontal distance between the shark's starting and ending positions?
   b. What was the total distance that the shark swam?
   c. How much time did this motion take?
74. ||| Starting from its nest, an eagle flies at constant speed for 3.0 min due east, then 4.0 min due north. From there the eagle flies directly to its nest at the same speed. How long is the eagle in the air?
75. ||| John walks 1.00 km north, then turns right and walks 1.00 km east. His speed is 1.50 m/s during the entire stroll.
   a. What is the magnitude of his displacement, from beginning to end?
   b. If Jane starts at the same time and place as John, but walks in a straight line to the endpoint of John's stroll, at what speed should she walk to arrive at the endpoint just when John does?

MCAT-Style Passage Problems

Growth Speed

The images of trees in Figure P1.76 come from a catalog advertising fast-growing trees. If we mark the position of the top of the tree in the successive years, as shown in the graph in the figure, we obtain a motion diagram much like ones we have seen for other kinds of motion. The motion isn't steady, of course. In some months the tree grows rapidly; in other months, quite slowly. We can see, though, that the average speed of growth is fairly constant for the first few years.

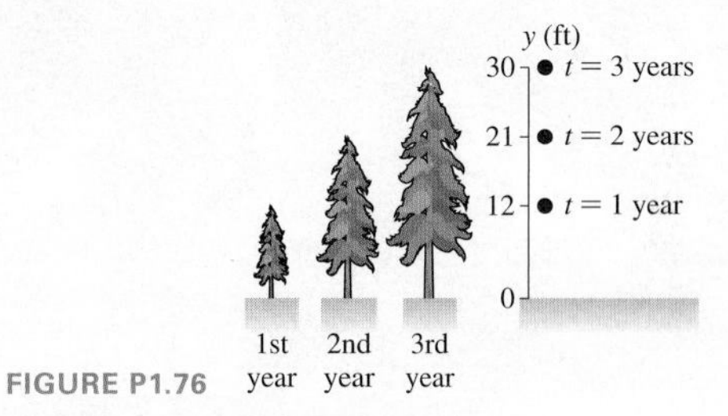

FIGURE P1.76

76. | What is the tree's speed of growth, in feet per year, from $t = 1$ yr to $t = 3$ yr?
   A. 12 ft/yr B. 9 ft/yr
   C. 6 ft/yr D. 3 ft/yr
77. | What is this speed in m/s?
   A. 9×10^{-8} m/s B. 3×10^{-9} m/s
   C. 5×10^{-6} m/s D. 2×10^{-6} m/s
78. | At the end of year 3, a rope is tied to the very top of the tree to steady it. This rope is staked into the ground 15 feet away from the tree. What angle does the rope make with the ground?
   A. 63° B. 60°
   C. 30° D. 27°

2 Motion in One Dimension

A horse can run at 35 mph, much faster than a human. And yet, surprisingly, a man can win a race against a horse if the length of the course is right. When, and how, can a man outrun a horse?

LOOKING AHEAD ▸

Uniform Motion

Successive images of the Segway rider are the same distance apart, so his velocity is constant. This is **uniform motion**.

You'll learn to describe motion in terms of quantities such as distance and velocity, an important first step in analyzing motion.

Acceleration

A cheetah is capable of running at very high speeds but, more important, it is capable of a rapid *change* in speed—a large **acceleration.**

You'll use the concept of acceleration to solve problems of changing velocity, such as races or predators chasing prey.

Free Fall

When the diver jumps, his motion—both going up and coming down—is determined by gravity alone. We call this **free fall.**

How long will it take this diver to reach the water? This is the type of free-fall problem you'll learn to solve.

GOAL To describe and analyze motion along a line.

LOOKING BACK ◂

Motion Diagrams

As you saw in Section 1.6, a good first step in analyzing motion is to draw a motion diagram, marking the position of an object at successive times.

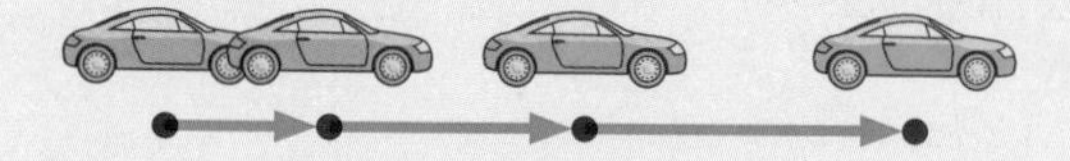

In this chapter, you'll learn to create motion diagrams for different types of motion along a line. Drawing pictures like this is a good starting point for solving problems.

STOP TO THINK

A bicycle is moving to the left with increasing speed. Which of the following motion diagrams illustrates this motion?

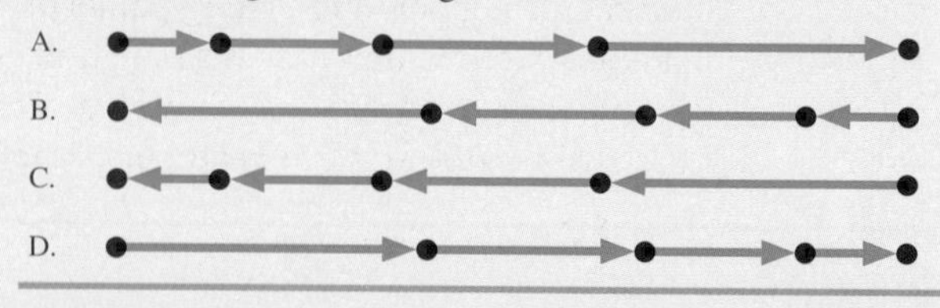

2.1 Describing Motion

In this chapter, we'll focus on how to *describe* motion using several different representations, including motion diagrams, graphs, and mathematical equations. We will defer a treatment of *why* objects move as they do until Chapter 4. The branch of physics that deals with the description of motion is **kinematics,** from the Greek word *kinema,* meaning "movement." You know this word through its English variation *cinema*—motion pictures!

Representing Position

As we saw in Chapter 1, *kinematic variables* such as position and velocity are measured with respect to a coordinate system, an axis that *you* impose on a system. We will use an x-axis to analyze both horizontal motion and motion on a ramp; a y-axis will be used for vertical motion. We will adopt the convention that the positive end of an x-axis is to the right and the positive end of a y-axis is up. This convention is illustrated in FIGURE 2.1.

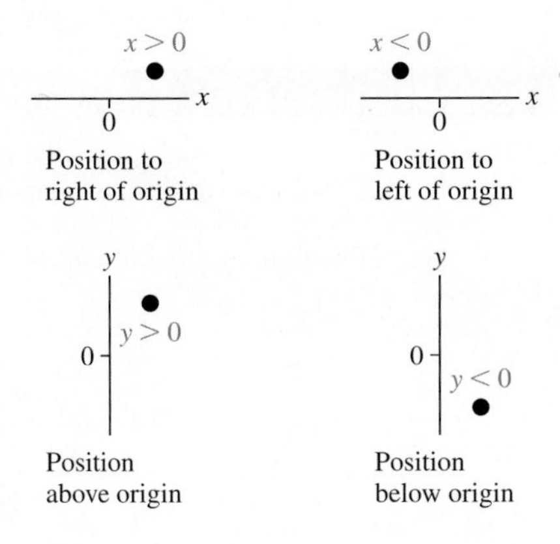

FIGURE 2.1 Sign conventions for position.

NOTE ▶ The conventions illustrated in Figure 2.1 aren't absolute. In most cases, we are free to define the coordinate system, and doing so in this standardized way makes sense. In some cases, though, we'll want to make a different choice. ◀

Now, let's look at a practical problem. FIGURE 2.2 is a motion diagram of a straightforward situation, a student walking to school. She is moving horizontally, so we use the variable x to describe her motion. We have set the origin of the coordinate system, $x = 0$, at her starting position, and we measure her position in meters. We have included velocity vectors connecting successive positions on the motion diagram, as we saw we could do in Chapter 1.

FIGURE 2.2 The motion diagram of a student walking to school and a coordinate axis for making measurements.

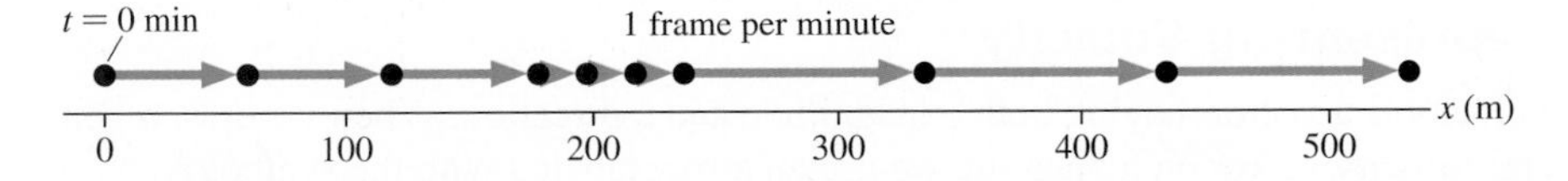

TABLE 2.1 Measured positions of a student walking to school

Time t (min)	Position x (m)	Time t (min)	Position x (m)
0	0	5	220
1	60	6	240
2	120	7	340
3	180	8	440
4	200	9	540

The motion diagram shows that she leaves home at a time we choose to call $t = 0$ min, and then makes steady progress for a while. Beginning at $t = 3$ min there is a pericd in which the distance traveled during each time interval becomes shorter—perhaps she slowed down to speak with a friend. Then, at $t = 6$ min, the distances traveled within each interval are longer—perhaps, realizing she is running late, she begins walking more quickly.

Every dot in the motion diagram of Figure 2.2 represents the student's position at a particular time. For example, the student is at position $x = 120$ m at $t = 2$ min. TABLE 2.1 lists her position for every point in the motion diagram.

The motion diagram of Figure 2.2 is one way to represent the student's motion. Presenting the data as in Table 2.1 is a second way to represent this motion. A third way to represent the motion is to use the data to make a graph. FIGURE 2.3 is a graph of the positions of the student at different times; we say it is a graph of x versus t for the student.

FIGURE 2.3 A graph of the student's motion.

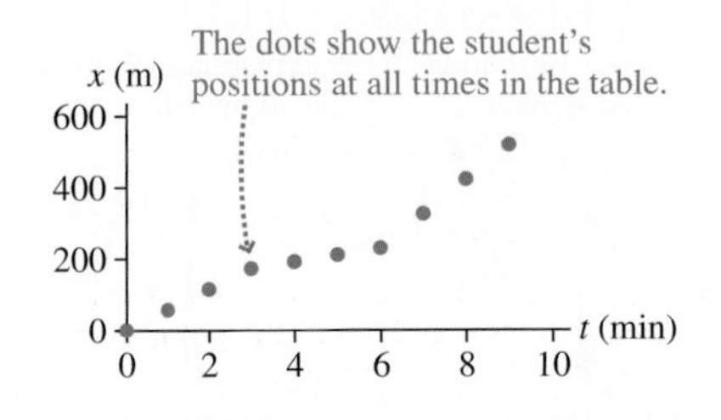

NOTE ▶ A graph of "a versus b" means that a is graphed on the vertical axis and b on the horizontal axis. We say that such a graph represents a "as a function of" b. ◀

We can flesh out the graph of Figure 2.3, though. We can assume that the student moved *continuously* through all intervening points of space, so we can represent her motion as a continuous curve that passes through the measured points, as shown in FIGURE 2.4. Such a continuous curve that shows an object's position as a function of time is called a **position-versus-time graph** or, sometimes, just a *position graph.*

FIGURE 2.4 Extending the graph of Figure 2.3 to a position-versus-time graph.

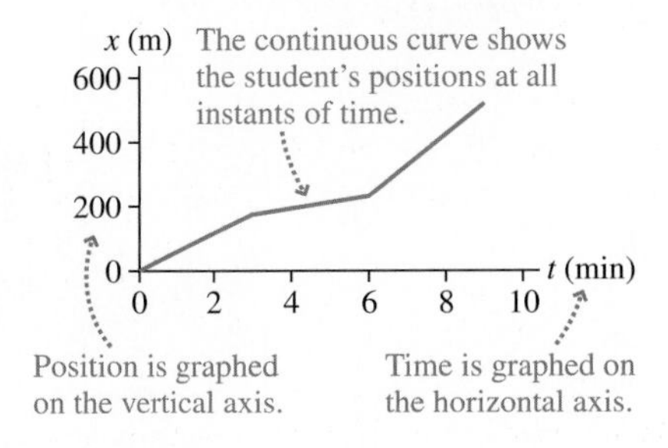

NOTE ▶ A graph is *not* a "picture" of the motion. The student is walking along a straight line, but the graph itself is not a straight line. Further, we've graphed her position on the vertical axis even though her motion is horizontal. A graph is an *abstract representation* of motion. ◀

CONCEPTUAL EXAMPLE 2.1 **Interpreting a car's position-versus-time graph**

The graph in FIGURE 2.5 represents the motion of a car along a straight road. Describe (in words) the motion of the car.

FIGURE 2.5 Position-versus-time graph for the car.

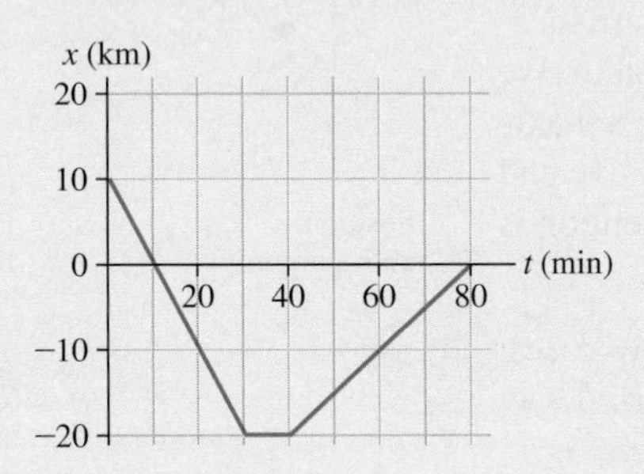

FIGURE 2.6 Looking at the position-versus-time graph in detail.

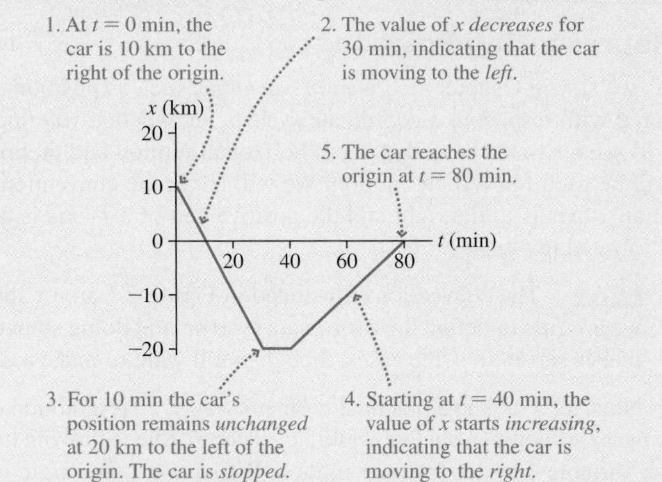

1. At t = 0 min, the car is 10 km to the right of the origin.
2. The value of x decreases for 30 min, indicating that the car is moving to the left.
5. The car reaches the origin at t = 80 min.
x (km)
20
10
0
−10
−20
20
40
60
80
t (min)
3. For 10 min the car's position remains unchanged at 20 km to the left of the origin. The car is stopped.
4. Starting at t = 40 min, the value of x starts increasing, indicating that the car is moving to the right.

REASON The vertical axis in Figure 2.5 is labeled "x (km)," so the car's position is measured in kilometers. Our convention for motion along the x-axis given in Figure 2.1 tells us that x increases as the car moves to the right and x decreases as the car moves to the left. As FIGURE 2.6 explains in detail, the graph thus shows that the car travels to the left for 30 minutes, stops for 10 minutes, then travels to the right for 40 minutes. It ends up 10 km to the left of where it began.

ASSESS The car travels to the left for 30 minutes and to the right for 40 minutes. Nonetheless, it ends up to the left of where it started. This means that the car was moving faster when it was moving to the left than when it was moving to the right. We can deduce this fact from the graph as well, as we will see in the next section.

Representing Velocity

Velocity is a vector, having both a magnitude and a direction. When we draw a general velocity vector on a diagram, we use an arrow labeled with the symbol $\vec{v}$.

FIGURE 2.7 Sign conventions for velocity.

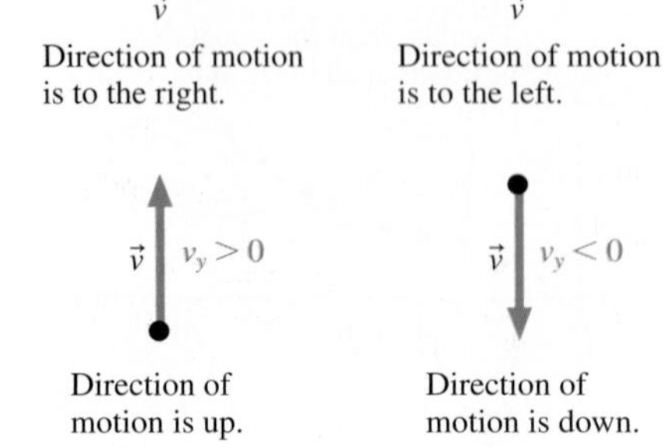

For motion in one dimension, however, velocity vectors are restricted to point only forward or backward for horizontal motion, or up or down for vertical motion. This restriction lets us simplify our notation for velocity vectors in one dimension. When we solve problems for motion along an x-axis, we represent the velocity with the simple variable v_x. As FIGURE 2.7 shows, we adopt the convention that for an object moving to the right, v_x is positive, whereas for motion to the left, v_x is negative.

For vertical motion along the y-axis, we use the symbol v_y to represent the velocity. The sign conventions for this vertical motion are also shown in Figure 2.7.

We use the symbol v, with no subscript, to represent the *speed* of an object. **Speed is the *magnitude* of the velocity vector** and is thus always positive.

In Chapter 1 we defined an object's velocity as $\Delta x/\Delta t$, where $\Delta x = x_f - x_i$ is the *displacement*, or change in position, as the object moves from an initial position x_i to a final position x_f, and Δt is the interval of time during which the motion occurs. For motion along a horizontal line, we can write

$$v_x = \frac{\Delta x}{\Delta t} \tag{2.1}$$

This agrees with the sign conventions in Figure 2.7. If Δx is positive, x is increasing, the object is moving to the right, and Equation 2.1 gives a positive value for velocity. If Δx is negative, x is decreasing, the object is moving to the left, and Equation 2.1 gives a negative value for velocity.

Equation 2.1 is the first of many *kinematic equations* we'll see in this chapter. We'll often specify equations in terms of the coordinate x, but if the motion is vertical, in which case we use the coordinate y, the equations can be easily adapted. For example, Equation 2.1 for motion along a vertical axis becomes

$$v_y = \frac{\Delta y}{\Delta t} \qquad (2.2)$$

From Position to Velocity

How is an object's velocity related to its position-versus-time graph? To find out, let's take another look at the motion diagram of the student walking to school. As we see in FIGURE 2.8, where we have repeated the motion diagram of Figure 2.2, her motion has three clearly defined phases. In each phase her speed is constant (because the velocity vectors have the same length) but the speed varies from phase to phase.

FIGURE 2.8 Revisiting the motion diagram of the student walking to school.

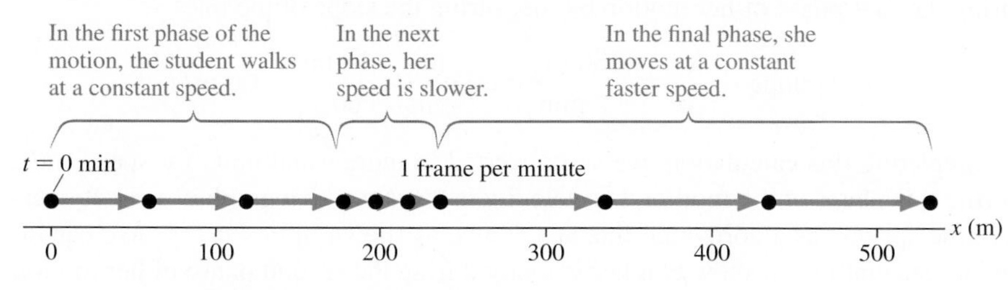

In the first phase of the motion, the student walks at a constant speed.
In the next phase, her speed is slower.
In the final phase, she moves at a constant faster speed.
t = 0 min
1 frame per minute
x (m)
0
100
200
300
400
500
Throughout the motion, she moves toward the right, in the direction of increasing x. Her velocity is always positive.

Just as her motion has three different phases, her position-versus-time graph (redrawn in FIGURE 2.9a) has three clearly defined segments with three different *slopes*. We can see that there's a relationship between her speed and the slope of the graph: **A faster speed corresponds to a steeper slope.**

The correspondence is actually deeper than this. Let's look at the slope of the third segment of the position-versus-time graph, as shown in FIGURE 2.9b. The slope of a graph is defined as the ratio of the "rise," the vertical change, to the "run," the horizontal change. For the segment of the graph shown, the slope is

$$\text{slope of graph} = \frac{\text{rise}}{\text{run}} = \frac{\Delta x}{\Delta t}$$

This ratio has a physical meaning—it's the velocity, exactly as we defined it in Equation 2.1. We've shown this correspondence for one particular graph, but it is a general principle: **The slope of an object's position-versus-time graph is the object's velocity at that point in the motion.** This principle also holds for negative slopes, which correspond to negative velocities. We can associate the slope of a position-versus-time graph, a *geometrical* quantity, with velocity, a *physical* quantity.

FIGURE 2.9 Interpreting the slope of the position graph for the student walking to school.

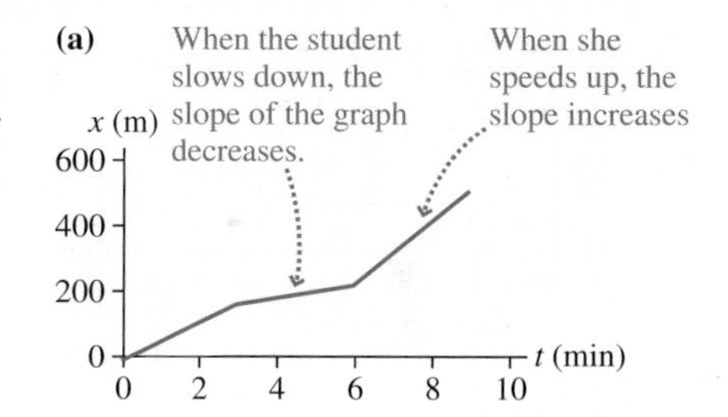

When she speeds up, the slope increases

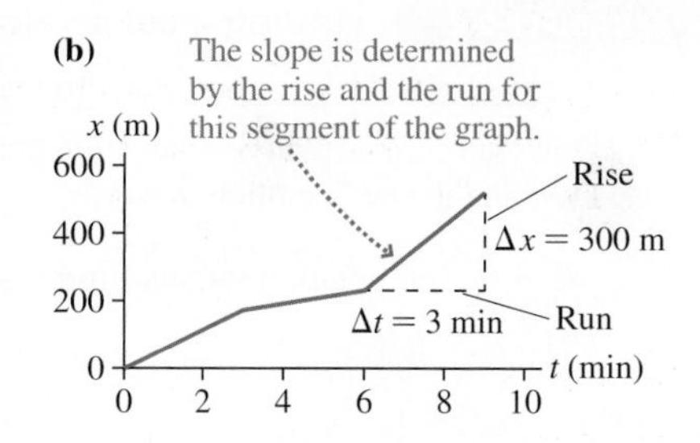

TACTICS BOX 2.1 Interpreting position-versus-time graphs

Information about motion can be obtained from position-versus-time graphs as follows:

1. Determine an object's *position* at time t by reading the graph at that instant of time.
2. Determine the object's *velocity* at time t by finding the slope of the position graph at that point. Steeper slopes correspond to faster speeds.
3. Determine the *direction of motion* by noting the sign of the slope. Positive slopes correspond to positive velocities and, hence, to motion to the right (or up). Negative slopes correspond to negative velocities and, hence, to motion to the left (or down).

Exercises 3,4

NOTE ▶ The slope is a ratio of intervals, $\Delta x/\Delta t$, not a ratio of coordinates; that is, the slope is *not* simply x/t. ◀

NOTE ▶ We are distinguishing between the actual slope—the slope on the graph—and the *physically meaningful* slope. If you were to use a ruler to measure the rise and the run of the graph, you could compute the actual slope of the line as drawn on the page. That is not the slope we are referring to when we equate the velocity with the slope of the line. Instead, we find the *physically meaningful* slope by measuring the rise and run using the scales along the axes. The "rise" Δx is some number of meters; the "run" Δt is some number of seconds. The physically meaningful rise and run include units, and the ratio of these units gives the units of the slope. ◀

FIGURE 2.10 Deducing the velocity-versus-time graph from the position-versus-time graph.

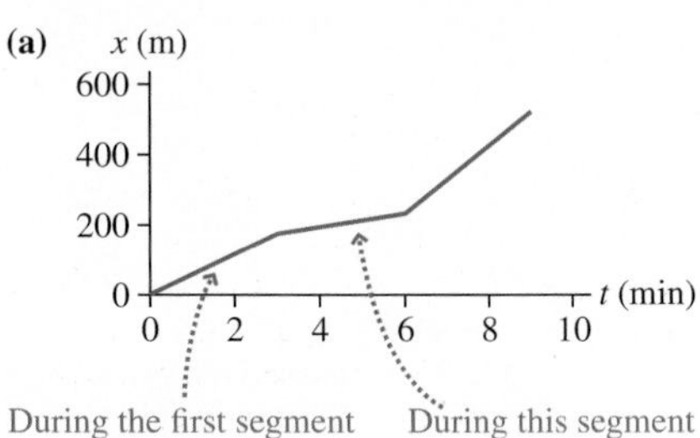

During the first segment of the motion, the slope is positive, a constant 60 m/min = 1.0 m/s . . .

During this segment of the motion, the slope decreases but is still positive . . .

(b) . . . so the velocity is positive, a constant 1.0 m/s.

. . . so the velocity is positive, but with a smaller magnitude.

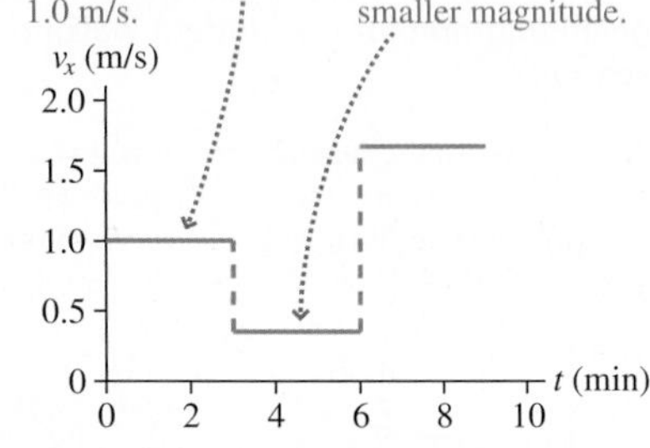

We can now use the approach of Tactics Box 2.1 to analyze the student's position-versus-time graph, redrawn in FIGURE 2.10a. We can determine her velocity during the first phase of her motion by measuring the slope of the line:

$$v_x = \text{slope} = \frac{\Delta x}{\Delta t} = \frac{180 \text{ m}}{3.0 \text{ min}} = 60 \frac{\text{m}}{\text{min}} \times \frac{1 \text{ min}}{60 \text{ s}} = 1.0 \text{ m/s}$$

In completing this calculation, we've converted to more usual units for speed, m/s. During this phase of the motion, her velocity is constant, so a graph of velocity versus time appears as a horizontal line at 1.0 m/s, as shown in FIGURE 2.10b. We can do similar calculations to show that her velocity during the second phase of her motion is +0.33 m/s, and then increases to +1.7 m/s during the final phase. We combine this information to create the **velocity-versus-time graph** shown in Figure 2.10b.

An inspection of the velocity-versus-time graph shows that it matches our understanding of the student's motion: There are three phases of the motion, each with constant speed. In each phase, the velocity is positive because she is always moving to the right. The second phase is slow (low velocity) and the third phase is fast (high velocity). All of this can be clearly seen on the velocity-versus-time graph, which is yet another way to represent her motion.

NOTE ▶ The velocity-versus-time graph in Figure 2.10b includes vertical segments in which the velocity changes instantaneously. Such rapid changes are an idealization; it actually takes at least a small amount of time to change velocity. ◀

EXAMPLE 2.2 Finding a car's velocity graph from its position graph

FIGURE 2.11 gives the position-versus-time graph of a car.

a. Draw the car's velocity-versus-time graph.
b. Describe the car's motion in words.

FIGURE 2.11 The position-versus-time graph of a car.

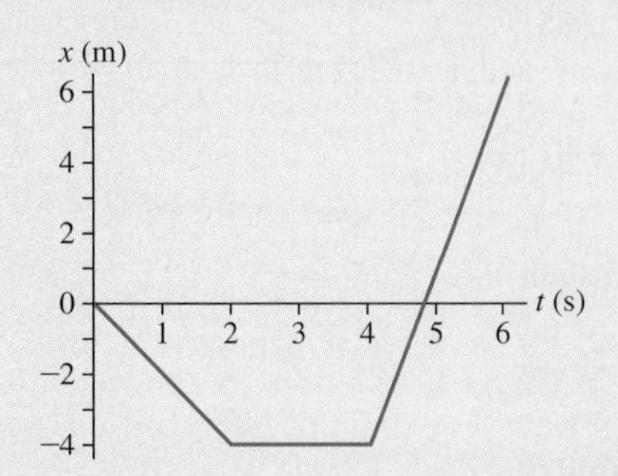

STRATEGIZE We will use the steps from Tactics Box 2.1 to understand the car's motion and to draw its velocity-versus-time graph based on its position graph.

PREPARE Figure 2.11 is a graphical representation of the motion. The car's position-versus-time graph is a sequence of three straight lines. Each of these straight lines represents uniform motion at a constant velocity. We can determine the car's velocity during each interval of time by measuring the slope of the line.

SOLVE

a. From $t = 0$ s to $t = 2$ s ($\Delta t = 2$ s) the car's displacement is $\Delta x = -4 \text{ m} - 0 \text{ m} = -4$ m. The velocity during this interval is

$$v_x = \frac{\Delta x}{\Delta t} = \frac{-4 \text{ m}}{2 \text{ s}} = -2 \text{ m/s}$$

The car's position does not change from $t = 2$ s to $t = 4$ s ($\Delta x = 0$ m), so $v_x = 0$ m/s. Finally, the displacement between $t = 4$ s and $t = 6$ s ($\Delta t = 2$ s) is $\Delta x = 10$ m. Thus the velocity during this interval is

$$v_x = \frac{10 \text{ m}}{2 \text{ s}} = 5 \text{ m/s}$$

These velocities are represented graphically in FIGURE 2.12.

FIGURE 2.12 The velocity-versus-time graph for the car.

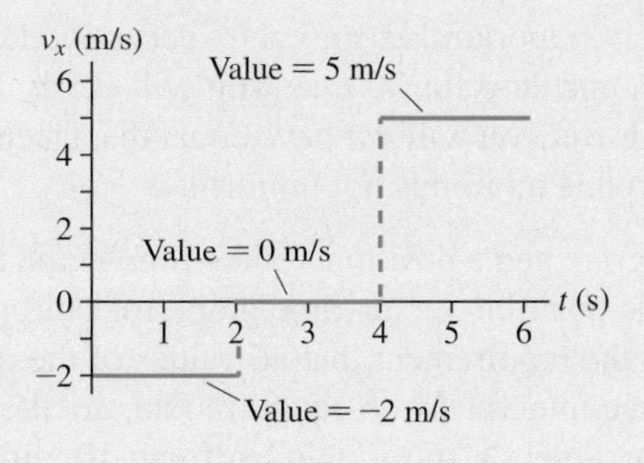

v_x (m/s)
Value = 5 m/s
Value = 0 m/s
Value = −2 m/s
t (s)

b. The velocity-versus-time graph of Figure 2.12 shows the motion in a way that we can describe in a straightforward manner: The car backs up for 2 s at 2 m/s, sits at rest for 2 s, then drives forward at 5 m/s for 2 s.

ASSESS Notice that the velocity graph and the position graph look completely different. They should! The value of the velocity graph at any instant of time equals the *slope* of the position graph. Since the position graph is made up of segments of constant slope, the velocity graph should be made up of segments of constant *value*, as it is. This gives us confidence that the graph we have drawn is correct.

From Velocity to Position

We've now seen how to move between different representations of uniform motion. There's one last issue to address: If you have a graph of velocity versus time, how can you determine the position graph?

Suppose you leave a lecture hall and begin walking toward your next class, which is down the hall to the east. You then realize that you left your textbook at your seat. You turn around and run back to the lecture hall to retrieve it. A velocity-versus-time graph for this motion appears in FIGURE 2.13a. There are two clear phases to the motion: walking away from the lecture hall (velocity +1.0 m/s) and running back (velocity −3.0 m/s). How can we deduce your position-versus-time graph?

As before, we can analyze the graph segment by segment, as shown in Figure 2.13. For the first segment, the velocity graph in Figure 2.13a indicates motion with a constant velocity of +1.0 m/s. This tells us that the corresponding position graph must be a straight line with a positive slope of +1.0 m/s, as shown in the position graph of FIGURE 2.13b. For the second segment, where the velocity is −3.0 m/s, as in Figure 2.13a, the position graph must be a line with a negative slope of −3.0 m/s, also shown in the position graph of Figure 2.13b.

The position graph makes sense: It shows 15 seconds of slowly increasing position (walking away from the lecture hall) and then 5 seconds of rapidly decreasing position (running back). And you end up back where you started.

There's one important detail that we didn't talk about in the preceding paragraph: How did we know that the position graph started at $x = 0$ m? The velocity graph tells us the *slope* of the position graph, but it doesn't tell us where the position graph should start. Although you're free to select any point you choose as the origin of the coordinate system, here it seems reasonable to set $x = 0$ m at your starting point in the lecture hall; as you walk away, your position increases.

FIGURE 2.13 Deducing a position graph from a velocity-versus-time graph.

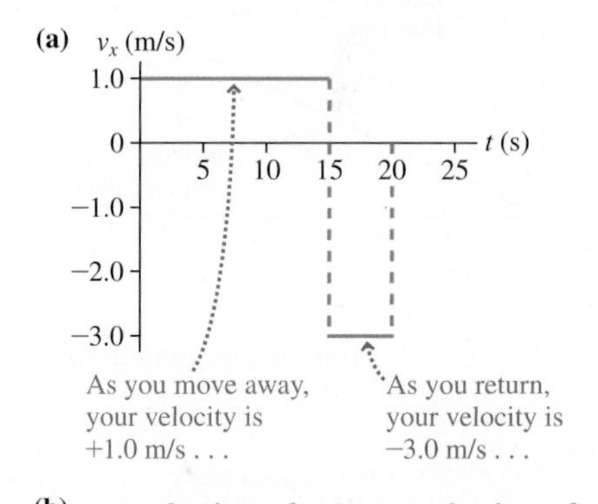

(a)
v_x (m/s)
t (s)
As you move away, your velocity is +1.0 m/s . . .
As you return, your velocity is −3.0 m/s . . .

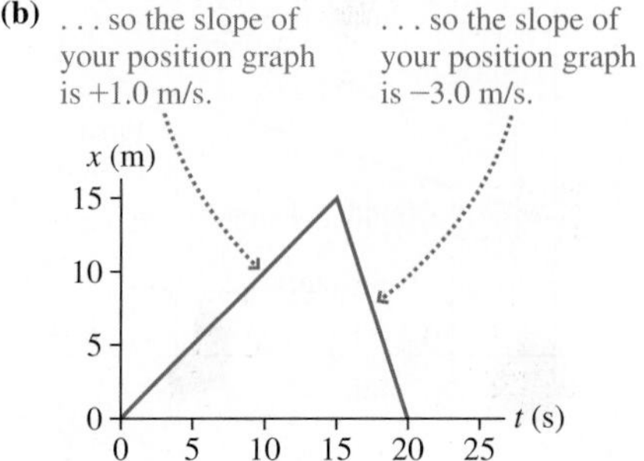

(b)
. . . so the slope of your position graph is +1.0 m/s.
. . . so the slope of your position graph is −3.0 m/s.
x (m)
t (s)

STOP TO THINK 2.1 Which position-versus-time graph best describes the motion diagram at left?

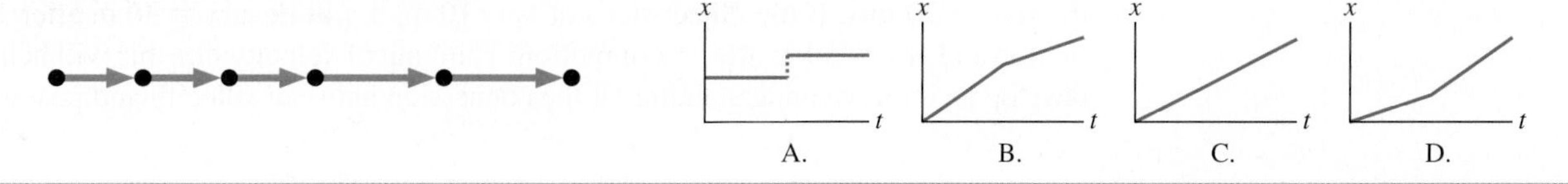

x
t
A.
B.
C.
D.

2.2 Uniform Motion

If you drive your car on a straight road at a perfectly steady 60 miles per hour (mph), you will cover 60 mi during the first hour, another 60 mi during the second hour, yet another 60 mi during the third hour, and so on. This is an example of what we call *uniform motion.* **Straight-line motion in which equal displacements occur during any successive equal-time intervals is called uniform motion or constant-velocity motion.**

NOTE ▶ The qualifier "any" is important. If for each successive hour of a trip you drive 120 mph for 30 min and stop for 30 min, you will cover 60 mi during each successive 1 hour interval. But you will *not* have equal displacements during successive 30 min intervals, so this motion is not uniform. ◀

FIGURE 2.14 Motion diagram and position-versus-time graph for uniform motion.

Uniform motion

The displacements between successive frames are the same. Dots are equally spaced. v_x is constant.

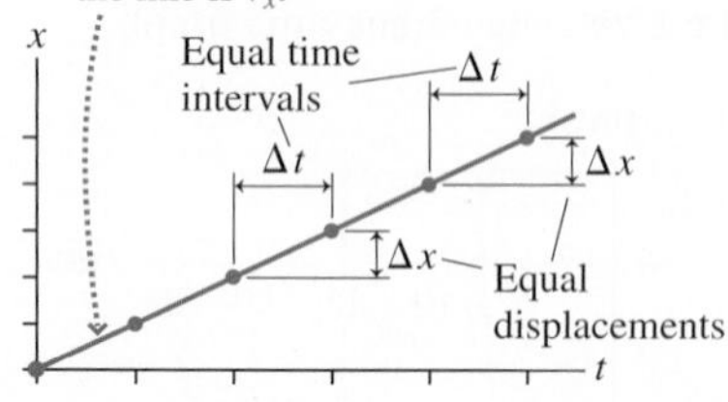

FIGURE 2.14 shows a motion diagram and a position-versus-time graph for an object in uniform motion. Notice that the position-versus-time graph for uniform motion is a straight line. This follows from the requirement that all values of the displacement Δx corresponding to the same time interval Δt be equal. In fact, an alternative definition of uniform motion is: **An object's motion is uniform if and only if its position-versus-time graph is a straight line.**

Equations of Uniform Motion

We've just learned that an object in uniform motion along the *x*-axis will have a linear (straight-line) position-versus-time graph like the one shown in FIGURE 2.15. Recall from Chapter 1 that we denote the object's initial position as x_i at time t_i. The term "initial" refers to the starting point of our analysis or the starting point in a problem. The object may or may not have been in motion prior to t_i. We use the term "final" for the ending point of our analysis or the ending point of a problem, and denote the object's final position x_f at the time t_f. As we've seen, the object's velocity v_x along the *x*-axis can be determined by finding the slope of the graph:

$$v_x = \frac{\text{rise}}{\text{run}} = \frac{\Delta x}{\Delta t} = \frac{x_f - x_i}{t_f - t_i} \tag{2.3}$$

FIGURE 2.15 Position-versus-time graph for an object in uniform motion.

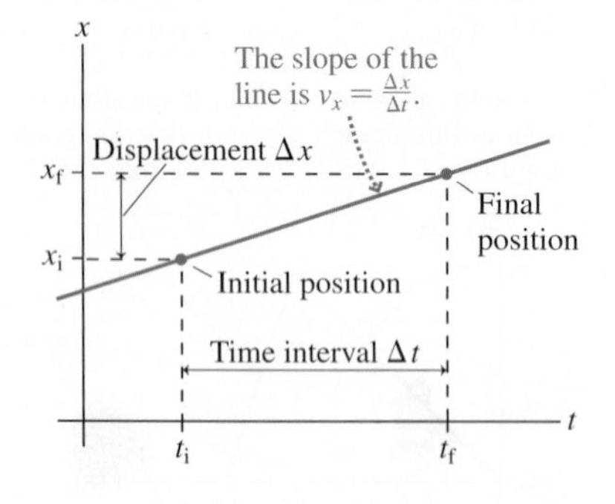

Here, the displacement $\Delta x = x_f - x_i$ is the change in position that occurs during the time interval $\Delta t = t_f - t_i$. Equation 2.3 can be rearranged to give

$$x_f = x_i + v_x\,\Delta t \tag{2.4}$$

Position equation for an object in uniform motion (v_x is constant)

where Δt is the interval of time in which the object moves from position x_i to position x_f. Equation 2.4 applies to any time interval Δt during which the velocity is constant. We can also write this in terms of the object's displacement Δx:

$$\Delta x = v_x\,\Delta t \tag{2.5}$$

The velocity of an object in uniform motion tells us the amount by which its position changes during each second. An object with a velocity of 20 m/s *changes* its position by 20 m during every second of motion: by 20 m during the first second of its motion, by another 20 m during the next second, and so on. We say that position is changing at the *rate* of 20 m/s. If the object starts at $x_i = 10$ m, it will be at $x = 30$ m after 1 s of motion and at $x = 50$ m after 2 s of motion. Thinking of velocity like this will help you develop an intuitive understanding of the connection between velocity and position.

Mathematical Relationships

Physics may seem densely populated with equations, but most equations follow a few basic forms. FIGURE 2.16 shows three graphs: a mathematical equation, the kinetic energy of a moving object versus its speed, and the potential energy of a spring versus how far the spring is compressed.

All of these graphs have the same overall appearance. The three expressions differ in their variables, but all three equations have the same **mathematical**

FIGURE 2.16 Three graphs with the same mathematical relationship.

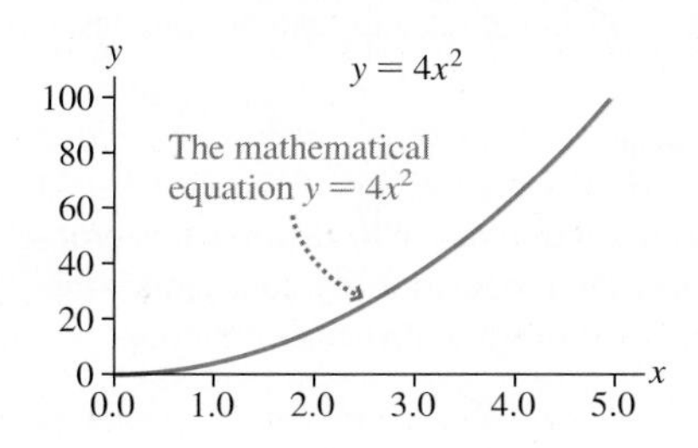

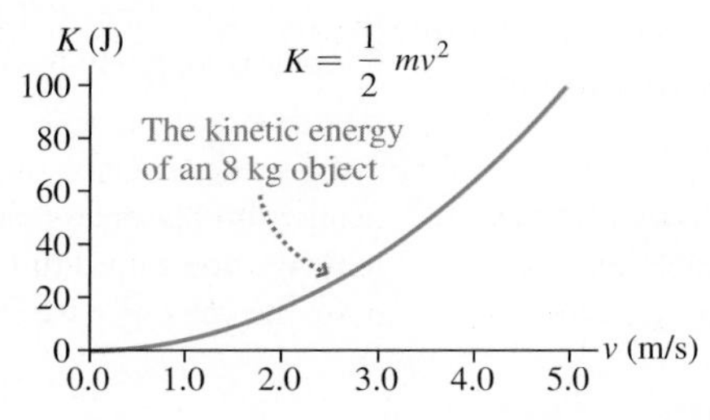

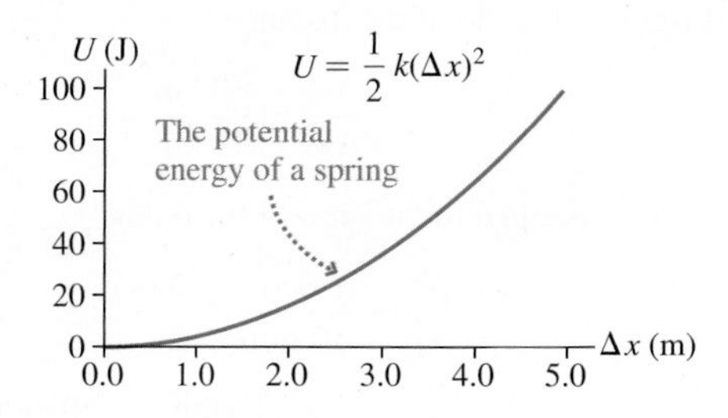

relationship. We'll use only a handful of different mathematical relationships in this text. As we meet each relationship for the first time, we will give an overview of its most important properties. When you see the relationship again in a new equation, we'll insert an icon, such as , that refers back to the overview so that you can remind yourself of the key details.

For instance, the mathematical form of Equation 2.5 is a type that we will see often: The displacement Δx is *proportional* to the time interval Δt. The following proportional relationships overview gives the details.

Proportional relationships

We say that y is **proportional** to x if they are related by an equation of the form

$$y = Cx$$

y is proportional to x

We call C the **proportionality constant.** A graph of y versus x is a straight line that passes through the origin.

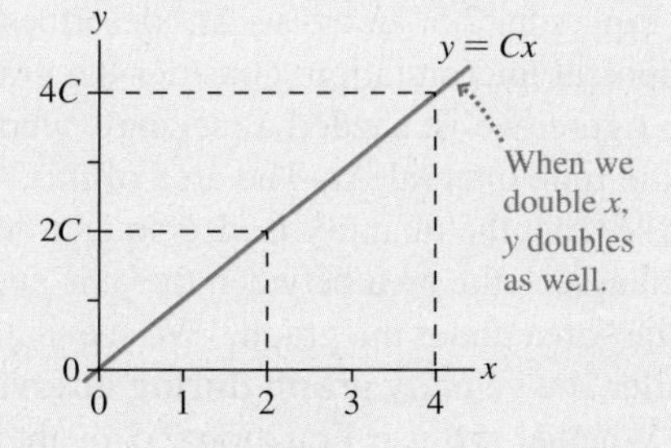

SCALING If x has the initial value x_1, then y has the initial value $y_1 = Cx_1$. Changing x from x_1 to x_2 changes y from y_1 to y_2. The ratio of y_2 to y_1 is

$$\frac{y_2}{y_1} = \frac{Cx_2}{Cx_1} = \frac{x_2}{x_1}$$

The ratio of y_2 to y_1 is exactly the same as the ratio of x_2 to x_1. If y is proportional to x, which is often written $y \propto x$, then x and y change by the same factor:

- If you double x, you double y.
- If you decrease x by a factor of 3, you decrease y by a factor of 3.

If two variables have a proportional relationship, we can draw important conclusions from ratios without knowing the value of the proportionality constant C. We can often solve problems in a very straightforward manner by looking at such ratios. This is an important skill called *ratio reasoning.*

Exercise 11 

EXAMPLE 2.3 If a train leaves Cleveland at 2:00 ...

A train is moving due west at a constant speed. A passenger notes that it takes 10 minutes to travel 12 km. How long will it take the train to travel 60 km?

STRATEGIZE For an object in uniform motion, Equation 2.5 shows that the distance traveled Δx is proportional to the time interval Δt, so this is a good problem to solve using ratio reasoning.

PREPARE We are comparing two cases: the time $\Delta t_1 = 10$ min it takes to travel the distance $\Delta x_1 = 12$ km, and the (unknown) time Δt_2 it will take to travel $\Delta x_2 = 60$ km. Ratio reasoning tells us that $\Delta x_2/\Delta x_1 = \Delta t_2/\Delta t_1$.

Continued

SOLVE The ratio of the distances is

$$\frac{\Delta x_2}{\Delta x_1} = \frac{60\text{ km}}{12\text{ km}} = 5$$

This is equal to the ratio of the times:

$$\frac{\Delta t_2}{\Delta t_1} = \frac{\Delta t_2}{10\text{ min}} = \frac{\Delta x_2}{\Delta x_1} = 5$$

$$\Delta t_2 = 5 \times (10\text{ min}) = 50\text{ min}$$

It takes 10 minutes to travel 12 km, so it will take 50 minutes—5 times as long—to travel 60 km.

ASSESS For an object in steady motion, it makes sense that 5 times the distance requires 5 times the time. We can see that using ratio reasoning is a straightforward way to solve this problem. We don't need to know the proportionality constant (in this case, the velocity); we just used ratios of distances and times.

A Second Way to Find Position from Velocity

Earlier, we saw that we could deduce an object's position graph from its velocity graph by drawing a position graph in which the slopes everywhere matched the velocity graph. But there's another way to understand the relationship between velocity and position graphs—by looking at what we call the *area under the graph.* Let's look at an example.

Suppose a car is in uniform motion at 12 m/s. How far does it travel—that is, what is its displacement—during the time interval between $t = 1.0$ s and $t = 3.0$ s?

Equation 2.5, $\Delta x = v_x \Delta t$, describes the displacement mathematically; for a graphical interpretation, consider the graph of velocity versus time in FIGURE 2.17. In the figure, we've shaded a rectangle whose height is the velocity v_x and whose base is the time interval Δt. The area of this rectangle is $v_x \Delta t$. Looking at Equation 2.5, we see that the quantity is also equal to the displacement of the car. The area of this rectangle is the area between the axis and the line representing the velocity; we call it the "area under the graph." We see that **the displacement Δx is equal to the area under the velocity graph during interval Δt.**

FIGURE 2.17 Displacement is the area under a velocity-versus-time graph.

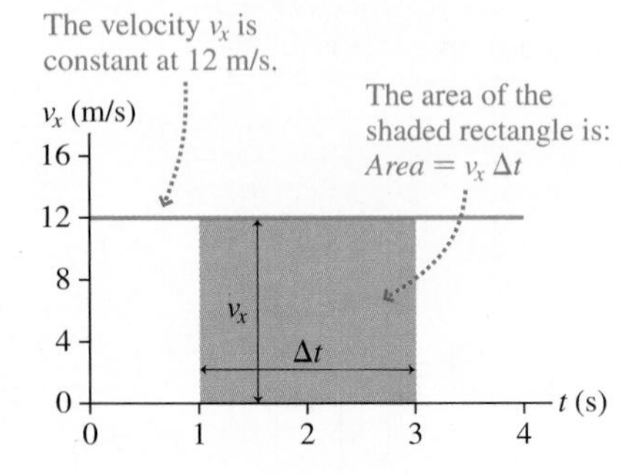

Whether we use Equation 2.5 or the area under the graph to compute the displacement, we get the same result:

$$\Delta x = v_x \Delta t = (12\text{ m/s})(2.0\text{ s}) = 24\text{ m}$$

Although we've shown that the displacement is the area under the graph only for uniform motion, where the velocity is constant, we'll soon see that this result applies to any one-dimensional motion.

NOTE ▸ Wait a minute! The displacement $\Delta x = x_f - x_i$ is a length. How can a length equal an area? Recall that earlier, when we found that the velocity is the slope of the position graph, we made a distinction between the *actual* slope and the *physically meaningful* slope? The same distinction applies here. The velocity graph does indeed bound a certain area on the page. That is the actual area—say, in square inches of paper—but it is *not* the area to which we are referring. Once again, we need to measure the quantities we are using, v_x and Δt, by referring to the scales on the axes. Δt is some number of seconds, while v_x is some number of meters per second. When these are multiplied, the *physically meaningful* area has units of meters, appropriate for a displacement. ◂

STOP TO THINK 2.2 Four objects move with the velocity-versus-time graphs shown. Which object has the largest displacement between $t = 0$ s and $t = 2$ s?

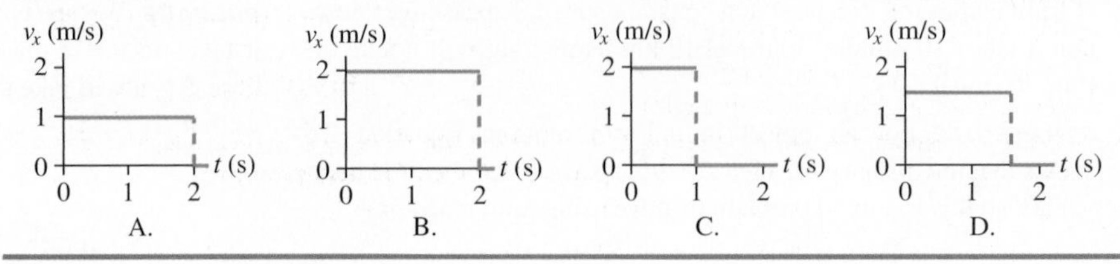

2.3 Instantaneous Velocity

The objects we've studied so far have moved with a constant, unchanging velocity or, like the car in Example 2.1, have a velocity that changes abruptly from one constant value to another. This is not very realistic. Real moving objects speed up and slow down, with their velocity changing smoothly. Suppose you're sitting at a red light at the start of a freeway ramp. When the light turns green, you increase your speed steadily from 0 mph to 60 mph until you merge onto the freeway.

A car's speed increases smoothly as it enters the freeway.

Perhaps as you speed down the ramp you glance at your speedometer and notice that, at that specific instant, it reads 40 mph. The speedometer indicates how fast you're moving *at a particular instant* in time. An object's velocity—its speed and direction—at a specific instant of time t is called the object's **instantaneous velocity.**

But what does it mean to have a velocity "at an instant"? An instantaneous velocity with a speed of 40 mph means that the rate at which your car's position is changing—at that exact instant—is such that it would travel a distance of 40 miles in 1 hour if it continued at that rate without change. If you speed up to pass a car that is traveling at a steady 40 mph, then at the very moment that your instantaneous velocity is 40 mph, your speed will match that of the other car—but an instant later, you'll be moving faster than 40 mph.

The velocity we introduced in Section 1.4 is really the *average velocity*; it is the velocity *averaged* over a *finite* time interval, such as 1 s or 1 min. **From now on, though, the word "velocity" will always mean instantaneous velocity**—the velocity at a single instant of time.

For uniform motion, an object's position-versus-time graph is a straight line and the object's velocity is the slope of that line. In contrast, FIGURE 2.18 shows that the position-versus-time graph for a car entering a freeway is a *curved* line. The displacement Δx during equal intervals of time gets greater as the car speeds up. Even so, we can use the slope of the position graph to measure the car's velocity. We can say that

$$\text{instantaneous velocity } v_x \text{ at time } t = \text{slope of position graph at time } t \qquad (2.6)$$

But how do we determine the slope of a curved line at a particular point? The following table shows how.

FIGURE 2.18 Position-versus-time graph for a car entering a freeway.

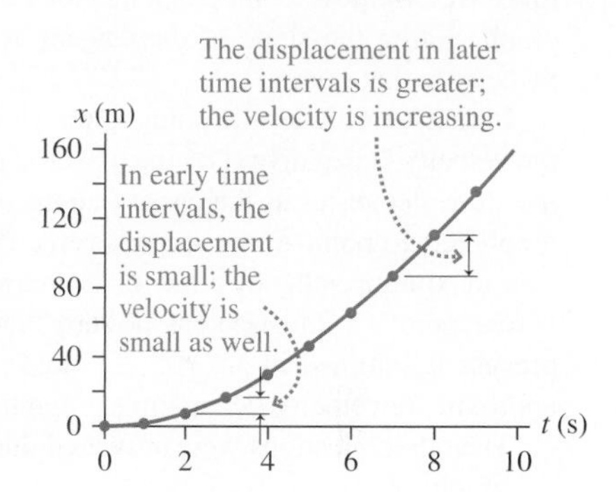

Finding the instantaneous velocity

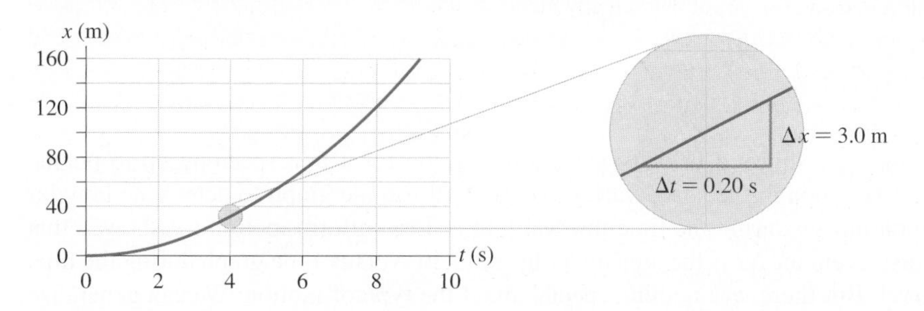

If the velocity changes, the position graph is a curved line. But we can compute a slope at a point by considering a small segment of the graph. Let's look at the motion in a very small time interval right around $t = 4.0$ s. This is highlighted with a circle, and we show a closeup in the next graph at the right.

In this magnified segment of the position graph, the curve isn't apparent. It appears to be a line segment. We can find the slope by calculating the rise over the run, just as before:

$$v_x = (3.0\text{ m})/(0.20\text{ s}) = 15\text{ m/s}$$

This is the slope at $t = 4.0$ s and thus the velocity at this instant of time.

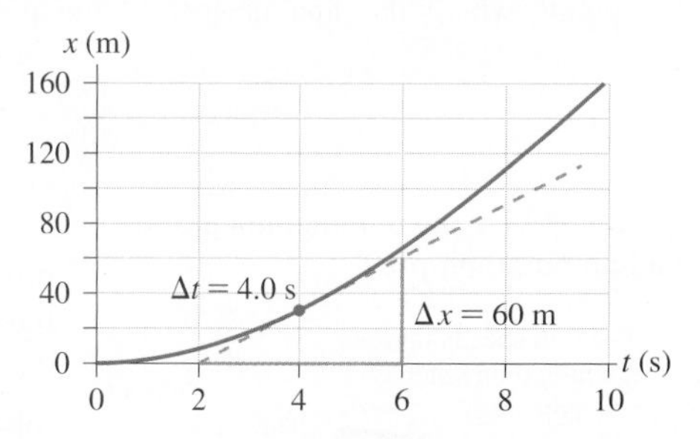

Graphically, the slope of the curve at a point is the same as the slope of a straight line drawn *tangent* to the curve at that point. Calculating rise over run for the tangent line, we get

$$v_x = (60\text{ m})/(4.0\text{ s}) = 15\text{ m/s}$$

This is the same value we obtained from the close-up view. **The slope of the tangent line is the instantaneous velocity at that instant of time.**

CONCEPTUAL EXAMPLE 2.4 Analyzing a hockey player's position graph

A hockey player moves in a straight line along the length of the ice in a game. We measure position from the center of the rink. FIGURE 2.19 shows a position-versus-time graph for his motion.

a. Sketch an approximate velocity-versus-time graph.
b. At which point or points is the player moving the fastest?
c. Is the player ever at rest? If so, at which point or points?

FIGURE 2.19 The position-versus-time graph for a hockey player.

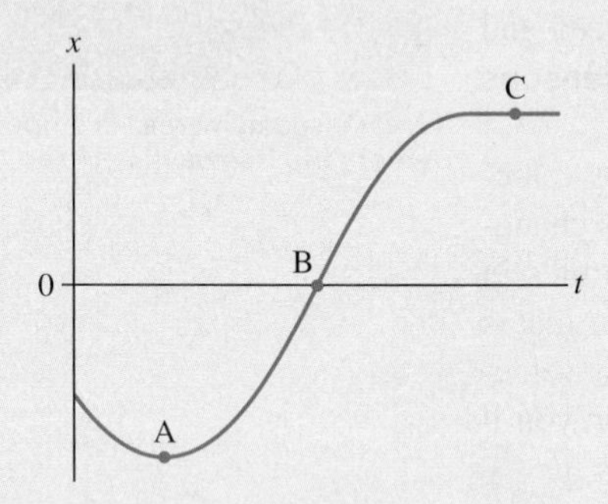

FIGURE 2.20 Finding a velocity graph from a position graph.

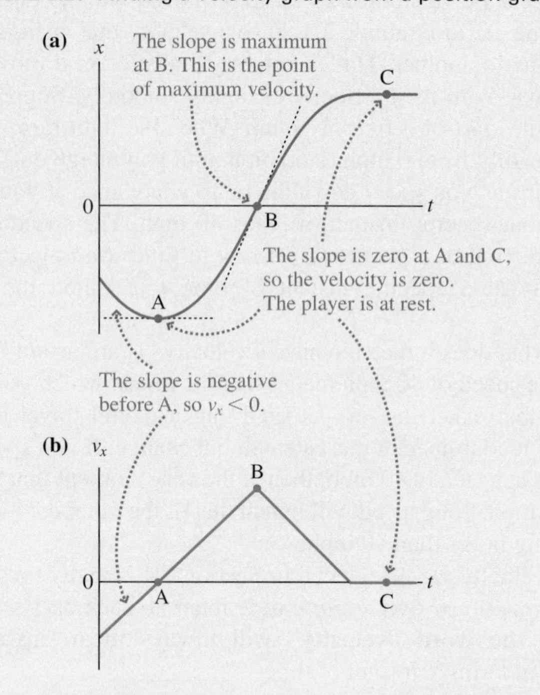

REASON a. The velocity at a particular instant of time is the slope of the tangent line to the position-versus-time graph at that time. We can move point-by-point along the position-versus-time graph, noting the slope of the tangent at each point to find the velocity at that point.

Initially, to the left of point A, the slope is negative and thus the velocity is negative (i.e., the player is moving to the left). But the slope decreases as the curve flattens out, and by the time the graph gets to point A, the slope is zero. The slope then increases to a maximum value at point B, decreases back to zero a little before point C, and remains at zero thereafter. This reasoning process is outlined in FIGURE 2.20a, and FIGURE 2.20b shows the approximate velocity-versus-time graph that results.

The other questions were answered during the construction of the graph:

b. The player moves the fastest at point B where the slope of the position graph is the steepest.
c. If the player is at rest, $v_x = 0$. Graphically, this occurs at points where the line tangent to the position-versus-time graph is horizontal and thus has zero slope. Figure 2.20 shows that the slope is zero at point A and for a small range of times near point C. At point A, the velocity is only instantaneously zero—the player is reversing direction, changing from moving to the left to moving to the right. Near point C, he has stopped moving and stays at rest.

ASSESS The best way to check our work is to look at different segments of the motion and see if the velocity and position graphs match. Until point A, x is decreasing. The player is moving to the left, so the velocity should be negative, which our graph shows. Between points A and C, x is increasing, so the velocity should be positive, which is also a feature of our graph. The steepest slope is at point B, so this should be the high point of our velocity graph, as it is.

FIGURE 2.21 Velocity-versus-time graph for a lion pursuing prey.

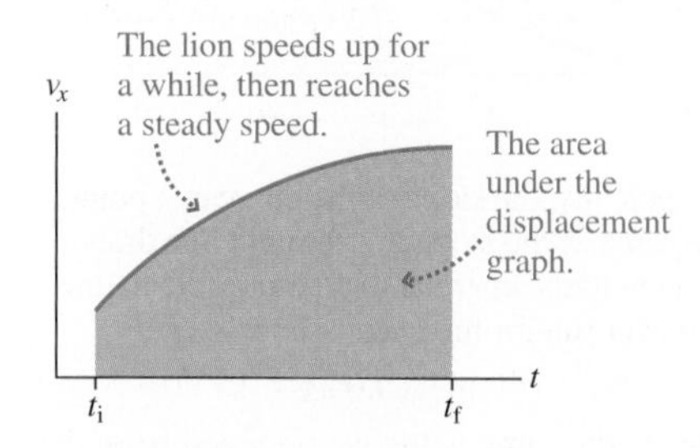

FIGURE 2.21 shows a velocity-versus-time graph for a lion speeding up to pursue prey. Even though the speed varies, we can still use the graph to determine how far the lion moves during the time interval t_i to t_f. For uniform motion we showed that the displacement Δx is the area under the velocity-versus-time graph during the time interval. But there was nothing special about the type of motion: We can generalize this idea to the case of an object whose velocity varies. If we draw a velocity graph for the motion, the object's displacement is given by

$$x_f - x_i = \text{area under the velocity graph between } t_i \text{ and } t_f \qquad (2.7)$$

The area under the graph in Figure 2.21 tells us how far the lion ran during this segment of the chase.

In many cases, as in the next example, the area under the graph is a simple shape whose area we can easily compute. If the shape is complex, however, we can approximate the area using a number of simpler shapes that closely match it.

EXAMPLE 2.5 **Calculating the displacement of a car during a rapid start**

FIGURE 2.22 shows the velocity-versus-time graph of a car pulling away from a stop. How far does the car move during the first 3.0 s?

STRATEGIZE The question How far? indicates that we need to find a displacement Δx rather than a position x. Graphically, the displacement is given by the area under the velocity-versus-time graph.

PREPARE In Figure 2.22 we have shaded the area we need to find. It is the area between the straight line of the velocity graph and the t-axis, between $t_i = 0$ s and $t_f = 3.0$ s.

FIGURE 2.22 Velocity-versus-time graph for the car of Example 2.5.

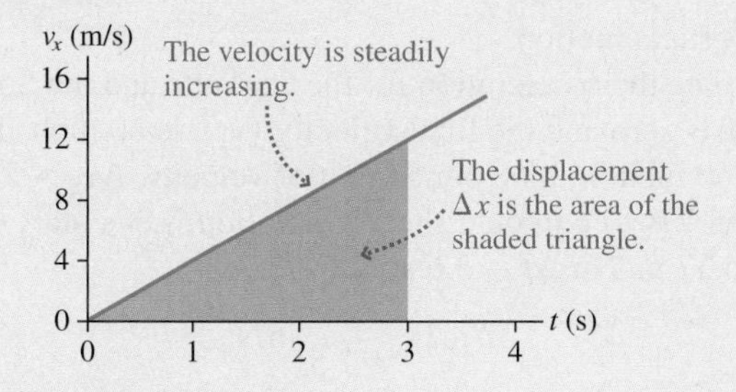

SOLVE The graph in this case is an angled line, so the area is that of a triangle:

$$\Delta x = \text{area of triangle between } t = 0 \text{ s and } t = 3.0 \text{ s}$$
$$= \tfrac{1}{2} \times \text{base} \times \text{height} = \tfrac{1}{2} \times 3.0 \text{ s} \times 12 \text{ m/s} = 18 \text{ m}$$

The car moves 18 m during the first 3 seconds as its velocity changes from 0 to 12 m/s.

ASSESS The physically meaningful area is a product of a time in s and a velocity in m/s, so Δx has the proper units of m. Let's check the numbers to see if they make physical sense. The final velocity, 12 m/s, is about 25 mph. Pulling away from a stop, you might expect to reach this speed in about 3 s—at least if you have a reasonably sporty vehicle! Another check is to realize that if the car had moved at a constant 12 m/s (the final velocity) during these 3 s, the distance would be 36 m. The actual distance traveled during the 3 s is 18 m—half of 36 m. This makes sense, as the velocity was 0 m/s at the start of the problem and increased steadily to 12 m/s.

STOP TO THINK 2.3 Which velocity-versus-time graph goes with the position-versus-time graph on the left?

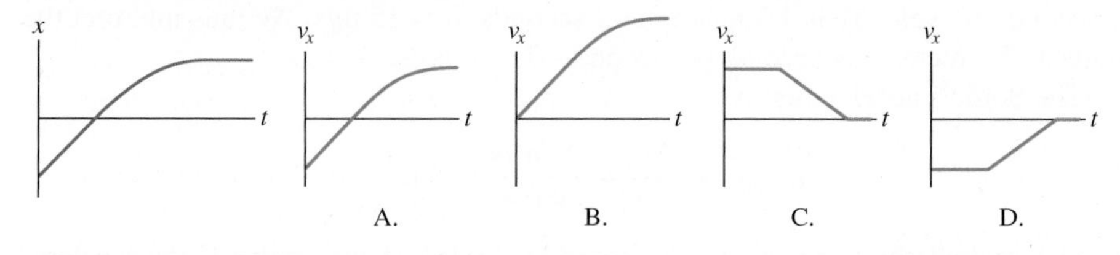

2.4 Acceleration

The goal of this chapter is to describe motion. We've seen that velocity describes the rate at which an object changes position. We need one more motion concept to complete the description, one that will describe an object whose velocity is changing.

As an example, let's look at a frequently quoted measurement of car performance, the time it takes the car to go from 0 to 60 mph. TABLE 2.2 shows this time for two different cars, a sporty Corvette and a compact Sonic with a much more modest engine.

TABLE 2.2 Performance data for vehicles

Vehicle	Time to go from 0 to 60 mph
2016 Chevy Corvette	3.6 s
2016 Chevy Sonic	9.0 s

Let's look at motion diagrams for the Corvette and the Sonic in FIGURE 2.23. We can see two important facts about the motion. First, the lengths of the velocity vectors are increasing, showing that the speeds are increasing. Second, the velocity vectors for the Corvette are increasing in length more rapidly than those of the Sonic. The quantity we seek is one that measures how rapidly an object's velocity vectors change in length.

When we wanted to measure changes in position, the ratio $\Delta x/\Delta t$ was useful. This ratio, which we defined as the velocity, is the *rate of change of position.* Similarly, we

FIGURE 2.23 Motion diagrams for the Corvette and Sonic.

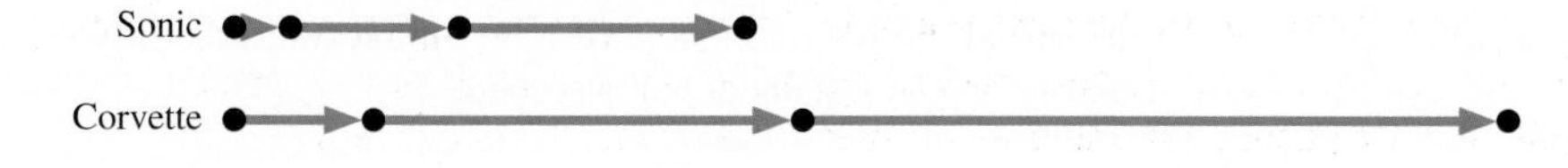

can measure how rapidly an object's velocity changes with the ratio $\Delta v_x / \Delta t$. Given our experience with velocity, we can say a couple of things about this new ratio:

- The ratio $\Delta v_x / \Delta t$ is the *rate of change of velocity.*
- The ratio $\Delta v_x / \Delta t$ is the *slope of a velocity-versus-time graph.*

We will define this ratio as the **acceleration,** for which we use the symbol a_x:

$$a_x = \frac{\Delta v_x}{\Delta t} \tag{2.8}$$

Definition of acceleration as the rate of change of velocity

Cushion kinematics When a car hits an obstacle head-on, the damage to the car and its occupants can be reduced by making the acceleration as small as possible. As we can see from Equation 2.8, acceleration can be reduced by making the *time* for a change in velocity as long as possible. This is the purpose of the yellow crash cushion barrels you may have seen in work zones on highways—to lengthen the time of a collision with a barrier.

Similarly, $a_y = \Delta v_y / \Delta t$ for vertical motion.

As an example, let's calculate the acceleration for the Corvette and the Sonic. For both, the initial velocity $(v_x)_i$ is zero and the final velocity $(v_x)_f$ is 60 mph. Thus the *change* in velocity is $\Delta v_x = 60$ mph. In m/s, our SI unit of velocity, $\Delta v_x = 27$ m/s.

Now we can use Equation 2.8 to compute the acceleration. Let's start with the Corvette, which speeds up to 27 m/s in $\Delta t = 3.6$ s:

$$a_{\text{Corvette } x} = \frac{\Delta v_x}{\Delta t} = \frac{27 \text{ m/s}}{3.6 \text{ s}} = 7.5 \frac{\text{m/s}}{\text{s}}$$

Here's the meaning of this final figure: Every second, the Corvette's velocity changes by 7.5 m/s. In the first second of motion, the Corvette's velocity increases by 7.5 m/s; in the next second, it increases by another 7.5 m/s, and so on. Thus after 1 second, the velocity is 7.5 m/s; after 2 seconds, it is 15 m/s. We thus interpret the units as 7.5 meters per second, per second—7.5 (m/s)/s.

The Sonic's acceleration is

$$a_{\text{Sonic } x} = \frac{\Delta v_x}{\Delta t} = \frac{27 \text{ m/s}}{9.0 \text{ s}} = 3.0 \frac{\text{m/s}}{\text{s}}$$

In each second, the Sonic changes its speed by 3.0 m/s. This is only 2/5 the acceleration of the Corvette! The reason the Corvette is capable of greater acceleration has to do with what causes the motion. We will explore the reasons for acceleration in Chapter 4. For now, we will simply note that the Corvette is capable of much greater acceleration, something you would have suspected.

NOTE ▶ It is customary to abbreviate the acceleration units (m/s)/s as m/s^2, which we say as "meters per second squared." For example, the Sonic has an acceleration of 3.0 m/s^2. When you use this notation, keep in mind its *meaning* as "(meters per second) per second." ◀

EXAMPLE 2.6 Animal acceleration BIO

Lions, like most predators, are capable of very rapid starts. From rest, a lion can sustain an acceleration of 9.5 m/s^2 for up to one second. How much time does it take a lion to go from rest to a typical recreational runner's top speed of 10 mph?

STRATEGIZE The lion's speed increases by 9.5 m/s^2 each second. Once we know the runner's speed in m/s, we will calculate the time it would take for the lion to reach that speed.

PREPARE We start by converting to SI units. The speed the lion must reach is

$$v_f = 10 \text{ mph} \times \frac{0.45 \text{ m/s}}{1.0 \text{ mph}} = 4.5 \text{ m/s}$$

The lion can accelerate at 9.5 m/s^2, changing its speed by 9.5 m/s per second, for only 1.0 s—long enough to reach 9.5 m/s. It will take the lion less than 1.0 s to reach 4.5 m/s, so we can use $a_x = 9.5 \text{ m/s}^2$ in our solution.

SOLVE We know the acceleration and the desired change in velocity, so we can rearrange Equation 2.8 to find the time:

$$\Delta t = \frac{\Delta v_x}{a_x} = \frac{4.5 \text{ m/s}}{9.5 \text{ m/s}^2} = 0.47 \text{ s}$$

ASSESS The lion changes its speed by 9.5 meters per second in one second. So it's reasonable (if a bit intimidating) that it will reach 4.5 m/s in just under half a second.

From Velocity to Acceleration

Let's use the values we have computed for acceleration to make a table of velocities for the Corvette and the Sonic we considered earlier. TABLE 2.3 uses the idea that the Sonic's velocity increases by 3.0 m/s every second while the Corvette's velocity increases by 7.5 m/s every second. The data in Table 2.3 are the basis for the velocity-versus-time graphs in FIGURE 2.24. As you can see, **an object undergoing constant acceleration has a straight-line velocity graph.**

TABLE 2.3 Velocity data for the Sonic and the Corvette

Time (s)	Velocity of Sonic (m/s)	Velocity of Corvette (m/s)
0	0	0
1	3.0	7.5
2	6.0	15.0
3	9.0	22.5
4	12.0	30.0

FIGURE 2.24 Velocity-versus-time graphs for the two cars.

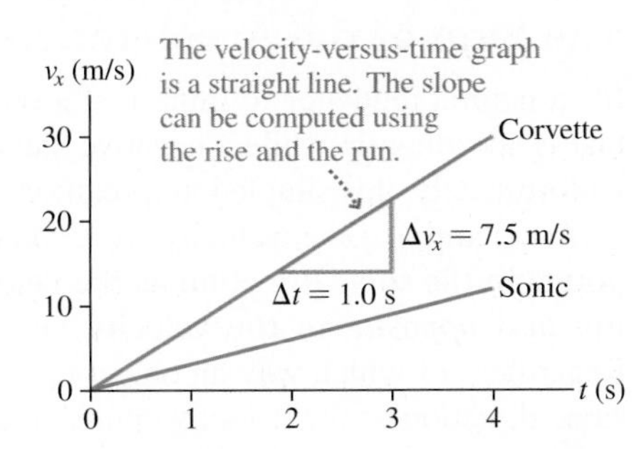

The slope of either of these lines—the rise over the run—is $\Delta v_x / \Delta t$. Comparing this with Equation 2.8, we see that the equation for the slope is the same as that for the acceleration. That is, **an object's acceleration is the slope of its velocity-versus-time graph:**

$$\text{acceleration } a_x \text{ at time } t = \text{slope of velocity graph at time } t \qquad (2.9)$$

The Sonic has a smaller acceleration, so its velocity graph has a smaller slope.

CONCEPTUAL EXAMPLE 2.7 Analyzing a car's velocity-versus-time graph

FIGURE 2.25a is a graph of velocity versus time for a car. Sketch a graph of the car's acceleration versus time.

REASON The graph can be divided into three sections:

- An initial segment, in which the velocity increases at a steady rate
- A middle segment, in which the velocity is constant
- A final segment, in which the velocity decreases at a steady rate

In each section, the acceleration is the slope of the velocity-versus-time graph. Thus the initial segment has constant, positive acceleration, the middle segment has zero acceleration, and the final segment has constant, *negative* acceleration. The acceleration graph appears in FIGURE 2.25b.

ASSESS This process is analogous to finding a velocity graph from the slope of a position graph. It is important to understand that the zero acceleration in the middle segment does *not* mean that the velocity there is zero. In this segment the velocity is *constant*, which means that it is *not changing* and thus the car is not accelerating.

In the first and last segments, the velocity is changing, and so the car does have a nonzero acceleration. In the first segment, the acceleration is positive; in the last segment, it is negative. What does the *sign* of the acceleration tell us? We will address this issue in the next section.

FIGURE 2.25 Finding an acceleration graph from a velocity graph.

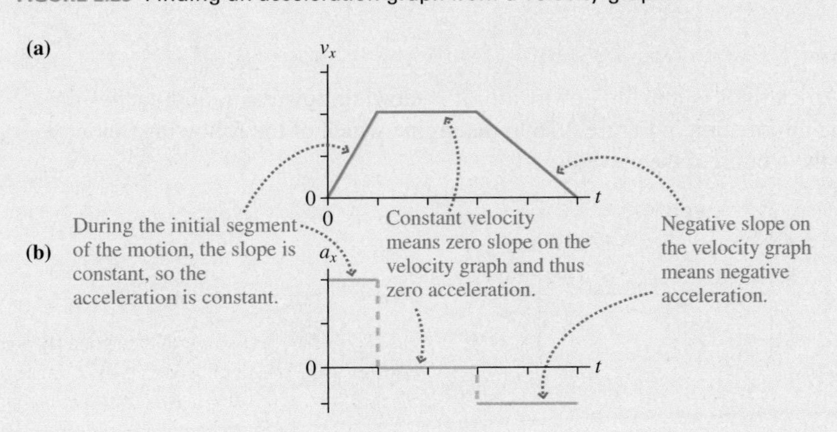

STOP TO THINK 2.4 A particle moves with the velocity-versus-time graph shown here. At which labeled point is the magnitude of the acceleration the greatest?

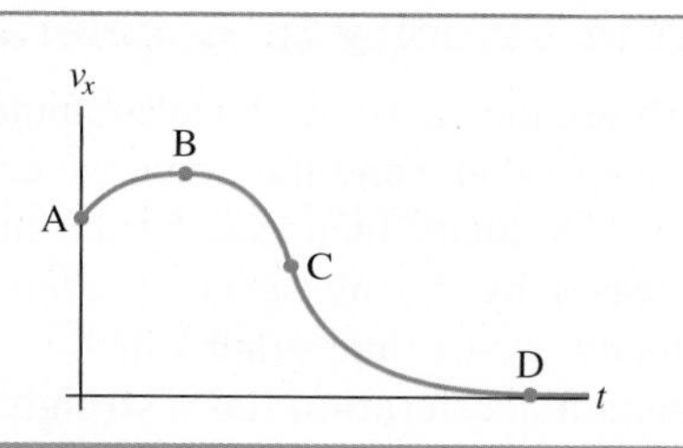

The Sign of the Acceleration

A video to support a section's topic is embedded in the eText.

Video Motion Along a Straight Line

It's a natural tendency to think that a positive value of a_x or a_y describes an object that is speeding up while a negative value describes an object that is slowing down. Unfortunately, this simple interpretation *is not correct.*

Acceleration, like velocity, is a vector. Specifically, **the acceleration vector points in the *same* direction as the velocity vector for an object that is speeding up, and *opposite* to the velocity vector for an object that is slowing down.** Regardless of which way an object moves, an acceleration vector that points in the same direction as the velocity "pulls" the velocity vectors to make them longer and longer—speeding up—while an acceleration vector that points opposite to the velocity "pushes against" the velocity vectors to make them shorter and shorter—slowing down. You can see this in the four situations shown in FIGURE 2.26.

Just as we did with velocity, we can simplify our analysis of one-dimensional motion by using an ordinary variable a_x, which can be positive or negative, to represent the one-dimensional acceleration. The sign convention for a_x (and a_y) is exactly

KEY CONCEPT **FIGURE 2.26** Determining the sign of the acceleration.

The object is moving to the right, so $v_x > 0$. Because it is speeding up, its acceleration vector points in the same direction as its velocity (i.e., to the right), so $a_x > 0$.

$\vec{v}$

$\vec{a}$

Positive a_x, positive slope

v_x

t

The object is moving to the left, so $v_x < 0$. Because it is slowing down, its acceleration vector points opposite to its velocity (i.e., to the right), so $a_x > 0$.

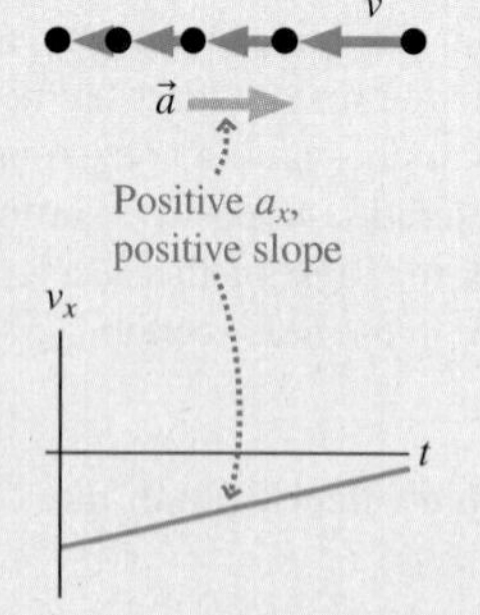

The object is moving to the right, so $v_x > 0$. Because it is slowing down, its acceleration vector points opposite to its velocity (i.e., to the left), so $a_x < 0$.

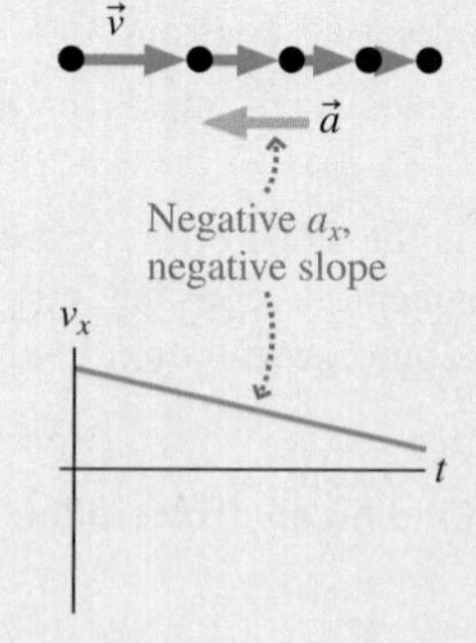

The object is moving to the left, so $v_x < 0$. Because it is speeding up, its acceleration vector points in the same direction as its velocity (i.e., to the left), so $a_x < 0$.

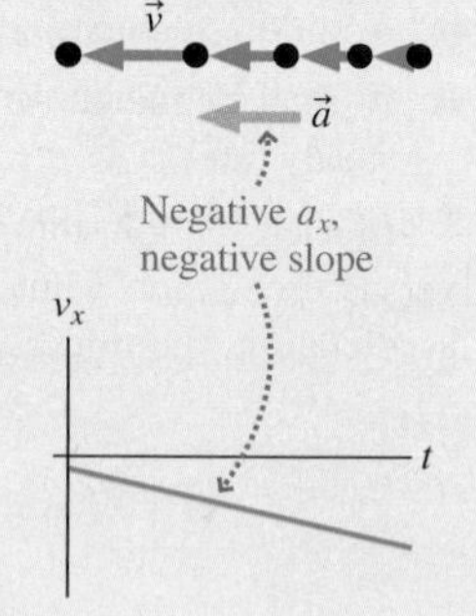

STOP TO THINK 2.5 An elevator is moving downward. It is slowing down as it approaches the ground floor. Adapt the information in Figure 2.26 to determine which of the following velocity graphs best represents the motion of the elevator.

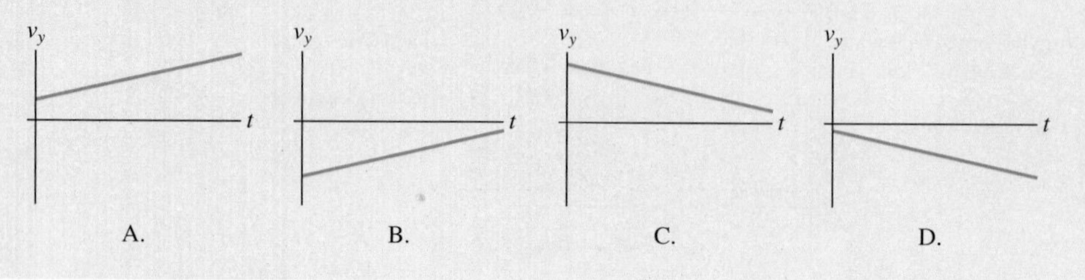

the same as the sign convention for velocity that was shown in Figure 2.7: a_x (or a_y) is positive when the acceleration vector points to the right (or up), negative when the acceleration vector points to the left (or down).

Notice that the first two situations in Figure 2.26, where the acceleration vectors point to the right, both have positive values of the acceleration a_x even though one shows an object speeding up and the other an object slowing down. **The sign of a_x is based on the direction the acceleration vector points,** not on whether the object is speeding up or slowing down.

Figure 2.26 illustrates two more ideas: First, our convention for the sign of the acceleration is consistent with what you just saw about an object's acceleration being the slope of its velocity graph. Second, **an object is speeding up if v_x and a_x have the same sign, and slowing down if they have opposite signs.**

2.5 Motion with Constant Acceleration

For uniform motion—motion with constant velocity—we found in Equation 2.3 a simple relationship between position and time. It's no surprise that there are also simple relationships that connect the various kinematic variables in constant-acceleration motion. We will start with a concrete example, the launch of a Saturn V rocket like the one that carried the Apollo astronauts to the moon in the 1960s and 1970s. FIGURE 2.27 shows one frame from a video of a rocket lifting off the launch pad. The red dots show the positions of the top of the rocket at equally spaced intervals of time in earlier frames of the video. This is a motion diagram for the rocket, and we can see that the velocity is increasing. The graph of velocity versus time in FIGURE 2.28 shows that the velocity is increasing at a fairly constant rate. We can approximate the rocket's motion as having constant acceleration.

We can use the slope of the graph in Figure 2.28 to determine the acceleration of the rocket:

$$a_y = \frac{\Delta v_y}{\Delta t} = \frac{27 \text{ m/s}}{1.5 \text{ s}} = 18 \text{ m/s}^2$$

This acceleration is more than double the acceleration of the Corvette we discussed earlier, and it goes on for a long time—the first phase of the launch lasts over 2 minutes! How fast is the rocket moving at the end of this acceleration, and how far has it traveled? To answer questions like these, we first need to work out some basic kinematic equations for motion with constant acceleration.

FIGURE 2.27 The red dots show the positions of the top of the Saturn V rocket at equally spaced intervals of time during liftoff.

FIGURE 2.28 A graph of the rocket's velocity versus time.

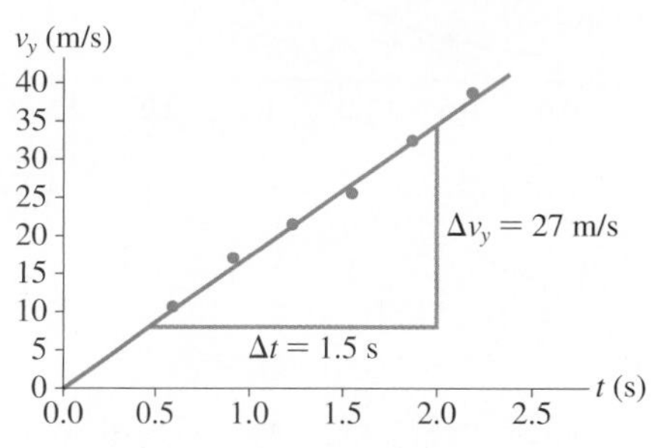

Constant-Acceleration Equations

Consider an object whose acceleration a_x remains constant during the time interval $\Delta t = t_f - t_i$. At the beginning of this interval, the object has initial velocity $(v_x)_i$ and initial position x_i. Note that t_i is often zero, but it need not be. FIGURE 2.29a shows the acceleration-versus-time graph. It is a horizontal line between t_i and t_f, indicating a *constant* acceleration.

FIGURE 2.29 Acceleration and velocity graphs for motion with constant acceleration.

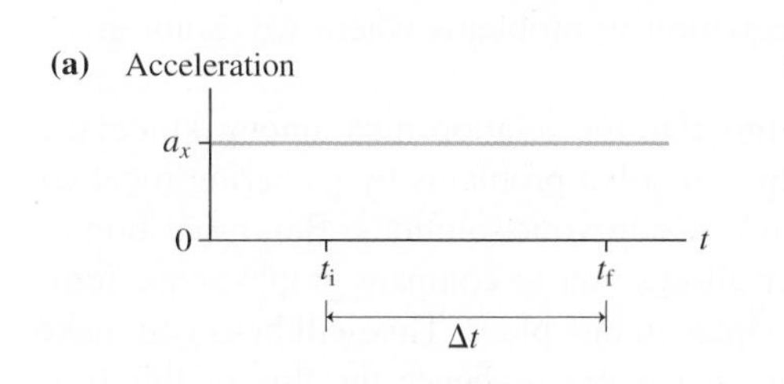

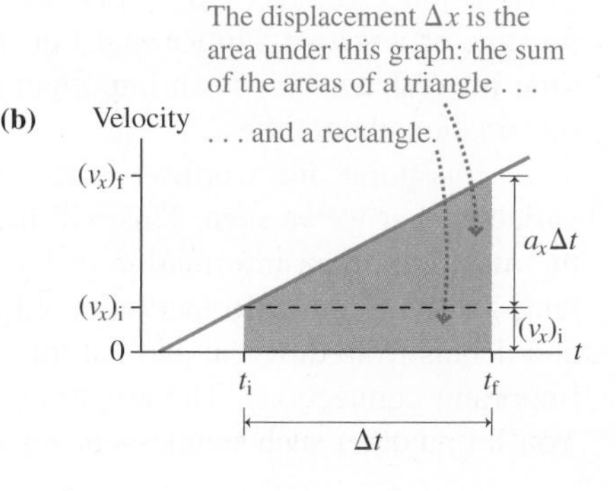

The object's velocity is changing because the object is accelerating. We can use the acceleration to find $(v_x)_f$ at a later time t_f. We defined acceleration as

$$a_x = \frac{\Delta v_x}{\Delta t} = \frac{(v_x)_f - (v_x)_i}{\Delta t} \tag{2.10}$$

which is rearranged to give

$$(v_x)_f = (v_x)_i + a_x \Delta t \tag{2.11}$$

Velocity equation for an object with constant acceleration

NOTE ▶ We have expressed this equation for motion along the x-axis, but it is a general result that will apply to any axis. ◀

The velocity-versus-time graph for this constant-acceleration motion, shown in FIGURE 2.29b, is a straight line with value $(v_x)_i$ at time t_i and with slope a_x.

We would also like to know the object's position x_f at time t_f. As you learned earlier, the displacement Δx during a time interval Δt is the area under the velocity-versus-time graph. This area is shown shaded in Figure 2.29b. The shaded area can be divided into a rectangle of area $(v_x)_i \Delta t$ and a triangle of area $\frac{1}{2}(a_x \Delta t)(\Delta t) = \frac{1}{2}a_x(\Delta t)^2$. Adding these gives

$$x_f = x_i + (v_x)_i \Delta t + \tfrac{1}{2}a_x(\Delta t)^2 \tag{2.12}$$

Position equation for an object with constant acceleration

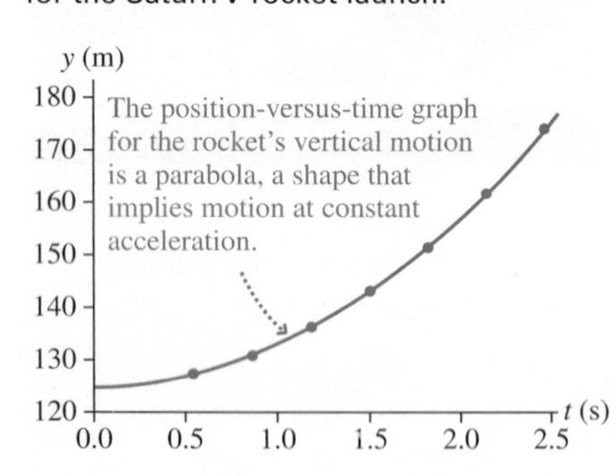

FIGURE 2.30 Position-versus-time graph for the Saturn V rocket launch.

where $\Delta t = t_f - t_i$ is the elapsed time. The fact that the time interval Δt appears in the equation as $(\Delta t)^2$ causes the position-versus-time graph for constant-acceleration motion to have a parabolic shape. For the rocket launch of Figure 2.27, a graph of the position of the top of the rocket versus time appears as in FIGURE 2.30.

Equations 2.11 and 2.12 are two of the basic kinematic equations for motion with constant acceleration. They allow us to predict an object's position and velocity at a future instant of time. We need one more equation to complete our set, a direct relationship between displacement and velocity. To derive this relationship, we first use Equation 2.11 to write $\Delta t = ((v_x)_f - (v_x)_i)/a_x$. We can substitute this into Equation 2.12 to obtain

$$(v_x)_f^{\,2} = (v_x)_i^{\,2} + 2a_x \Delta x \tag{2.13}$$

Relating velocity and displacement for constant-acceleration motion

In Equation 2.13, $\Delta x = x_f - x_i$ is the *displacement* (not the distance!), so it can be positive or negative. Notice that Equation 2.13 does not require that we know the time interval Δt. This is an important equation in problems where we're not given information about times.

At this point, it's worthwhile to summarize the relationships among kinematic variables that we've seen. This will help you solve problems by gathering together the most important information that you'll use in your solutions. But, more important, gathering this information together allows you to compare graphs, equations, and details from different parts of the chapter in one place. This will help you make important connections. The emphasis is on *synthesis*—hence the title of this box. You'll find other such synthesis boxes in most chapters.

SYNTHESIS 2.1 Describing motion in one dimension

We describe motion in terms of position, velocity, and acceleration.

For **all motion:**

Velocity is the rate of change of position, in m/s. $v_x = \dfrac{\Delta x}{\Delta t}$

Acceleration is the rate of change of velocity, in m/s². $a_x = \dfrac{\Delta v_x}{\Delta t}$

For **uniform motion:**
- acceleration is zero
- velocity is constant
- position changes steadily

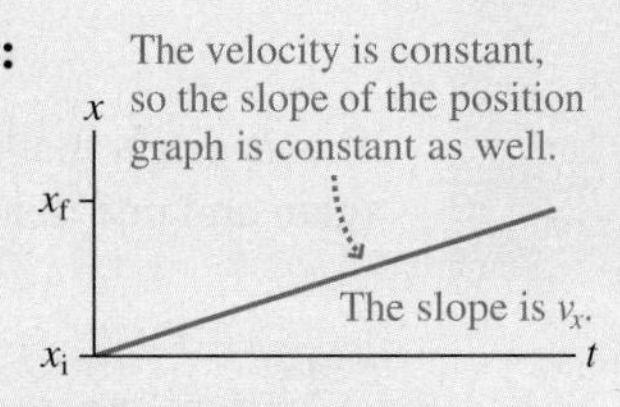

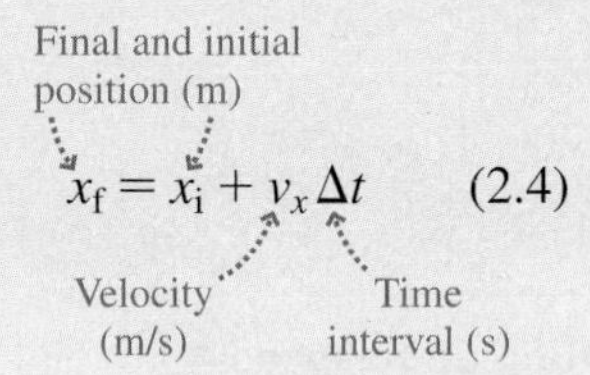

$$x_f = x_i + v_x \Delta t \quad (2.4)$$

For **motion with constant acceleration:**
- acceleration is steady; it does not change

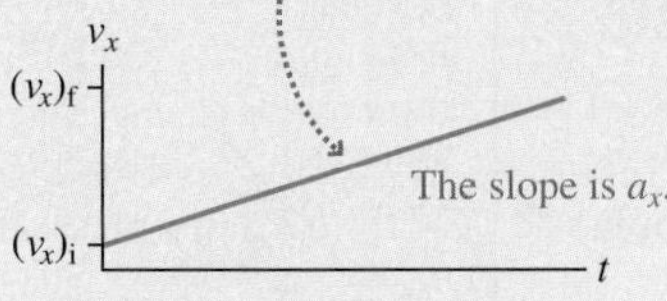

The slope is a_x.

- velocity changes steadily

Final and initial velocity (m/s)

$$(v_x)_f = (v_x)_i + a_x \Delta t \quad (2.11)$$

Acceleration (m/s²) Time interval (s)

- the position changes as the square of the time interval

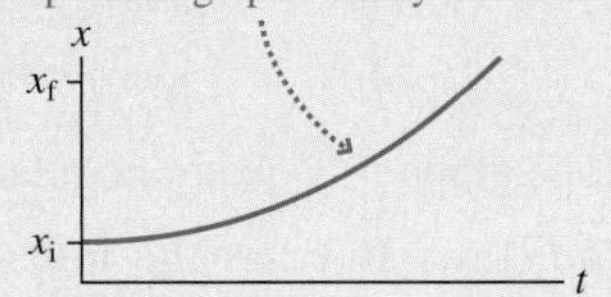

The velocity steadily increases, so the slope of the position graph steadily increases.

The slope is v_x.

Final and initial position (m) Time interval (s)

$$x_f = x_i + (v_x)_i \,\Delta t + \tfrac{1}{2}a_x(\Delta t)^2 \quad (2.12)$$

Initial velocity (m/s) Acceleration (m/s²)

- we can also express the change in velocity in terms of **displacement, not time**

This gives us a third equation, which is useful for many kinematics problems.

Final and initial velocity (m/s)

$$(v_x)_f^2 = (v_x)_i^2 + 2a_x \Delta x \quad (2.13)$$

Acceleration (m/s²) Change in position (m)

EXAMPLE 2.8 Coming to a stop in a car

As you drive in your car at 15 m/s (just a bit under 35 mph), you see a child's ball roll into the street ahead of you. You hit the brakes and stop as quickly as you can. In this case, you come to rest in 1.5 s. How far does your car travel as you brake to a stop?

STRATEGIZE The problem states that your car begins to slow down when you hit the brakes; we'll model this as constant-acceleration motion. We know the initial and final speeds, and we want to find the distance traveled. These observations suggest that we use Equation 2.12 of Synthesis 2.1.

PREPARE The problem gives us a description of motion in words. To help us visualize the situation, FIGURE 2.31 illustrates the key features of the motion with a motion diagram and a velocity graph. The graph is based on the car slowing from 15 m/s to 0 m/s in 1.5 s.

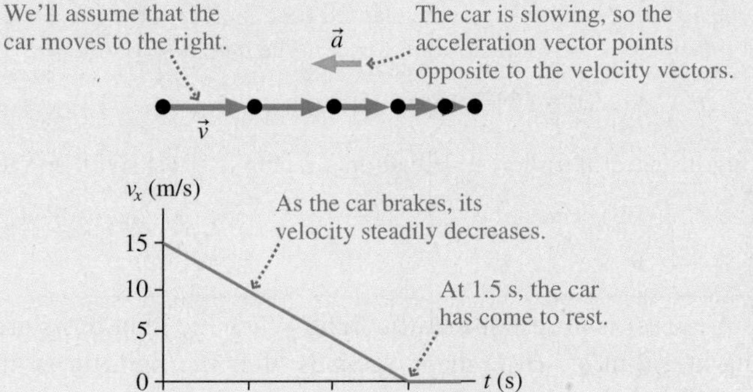

FIGURE 2.31 Motion diagram and velocity graph for a car coming to a stop.

SOLVE We've assumed that your car is moving to the right, so its initial velocity is $(v_x)_i = +15$ m/s. After you come to rest, your final velocity is $(v_x)_f = 0$ m/s. We use the definition of acceleration from Synthesis 2.1:

$$a_x = \frac{\Delta v_x}{\Delta t} = \frac{(v_x)_f - (v_x)_i}{\Delta t} = \frac{0\text{ m/s} - 15\text{ m/s}}{1.5\text{ s}} = -10\text{ m/s}^2$$

Now that we know the acceleration, we can compute the distance that the car moves as it comes to rest using Equation 2.12:

$$x_f - x_i = (v_x)_i\Delta t + \tfrac{1}{2}a_x(\Delta t)^2$$
$$= (15\text{ m/s})(1.5\text{ s}) + \tfrac{1}{2}(-10\text{ m/s}^2)(1.5\text{ s})^2 = 11\text{ m}$$

ASSESS 11 m is a little over 35 feet. That's a reasonable distance for a quick stop while traveling at about 35 mph.

We found that the acceleration a_x is negative. This makes sense from two perspectives. First, as we learned in Figure 2.26, an object moving to the right and slowing down has a negative acceleration. Second, the slope of the velocity graph in Figure 2.31 is negative, again indicating a negative acceleration.

Getting up to speed BIO A bird must have a minimum speed to fly. Generally, the larger the bird, the faster the takeoff speed. Small birds can get moving fast enough to fly with a vigorous jump, but larger birds may need a running start. This swan must accelerate for a long distance in order to achieve the high speed it needs to fly, so it makes a frenzied dash across the frozen surface of a pond. Swans require a long, clear stretch of water or land to become airborne.

As we've noted, for motion at constant acceleration, the position changes as the square of the time interval. If the initial velocity $(v_x)_i$ is zero, Equation 2.12 from Synthesis 2.1 can be written as

$$x_f - x_i = \Delta x = \frac{1}{2}a_x(\Delta t)^2$$

This is a new mathematical relationship—a *quadratic relationship*—that we will see again and one that we can use as the basis of reasoning to solve problems.

Quadratic relationships

Two quantities are said to have a **quadratic relationship** if y is proportional to the square of x. We write the mathematical relationship as

$$y = Ax^2$$

y is proportional to x^2

y
$y = Ax^2$
4A
A
0
0
1
2
x
Doubling x causes y to change by a factor of 4.

The graph of a quadratic relationship is a parabola.

SCALING If x has the initial value x_1, then y has the initial value $y_1 = A(x_1)^2$. Changing x from x_1 to x_2 changes y from y_1 to y_2. The ratio of y_2 to y_1 is

$$\frac{y_2}{y_1} = \frac{A(x_2)^2}{A(x_1)^2} = \left(\frac{x_2}{x_1}\right)^2$$

The ratio of y_2 to y_1 is the square of the ratio of x_2 to x_1. If y is a quadratic function of x, a change in x by some factor changes y by the square of that factor:

- If you increase x by a factor of 2, you increase y by a factor of $2^2 = 4$.
- If you decrease x by a factor of 3, you decrease y by a factor of $3^2 = 9$.

Generally, we can say that:

Changing x by a factor of c changes y by a factor of c^2.

Exercise 19

EXAMPLE 2.9 Finding the displacement of a drag racer

A drag racer, starting from rest, travels 6.0 m in 1.0 s. Suppose the car continues this acceleration for an additional 4.0 s. How far from the starting line will the car be?

STRATEGIZE We assume that the acceleration is constant. Because the initial position and velocity are zero, the displacement will then scale as the square of the time; we can then use ratio reasoning to solve the problem.

PREPARE After 1.0 s, the car has traveled 6.0 m; after another 4.0 s, a total of 5.0 s will have elapsed.

SOLVE The initial elapsed time was 1.0 s, so the elapsed time increases by a factor of 5. The displacement thus increases by a factor of 5^2, or 25. The total displacement is

$$\Delta x = 25(6.0 \text{ m}) = 150 \text{ m}$$

ASSESS This is a big distance in a short time, but drag racing is a fast sport, so our answer makes sense.

STOP TO THINK 2.6 A cyclist is at rest at a traffic light. When the light turns green, he begins accelerating at 1.2 m/s^2. How many seconds after the light turns green does he reach his cruising speed of 6.0 m/s?

A. 1.0 s B. 2.0 s C. 3.0 s D. 4.0 s E. 5.0 s

2.6 Solving One-Dimensional Motion Problems

The big challenge when solving a physics problem is to translate the words into symbols that can be manipulated, calculated, and graphed. This translation from words to symbols is the heart of problem solving in physics. Ambiguous words and phrases must be clarified, the imprecise must be made precise, and you must arrive at an understanding of exactly what the question is asking.

PROBLEM-SOLVING APPROACH

The first step in solving a seemingly complicated problem is to break it down into a series of smaller steps. In worked examples in the text, we use a problem-solving approach that consists of four steps: *strategize, prepare, solve,* and *assess*. Each of these steps has important elements that you should follow when you solve problems on your own.

STRATEGIZE The Strategize step of the solution is where you address the *big-picture* questions about the problem. Here, you take a step back from the details of the problem to ask:

- **What kind of problem is this?** From reading the problem statement, try to categorize the problem in terms of what you've learned in the chapter. If, for instance, the problem refers to a bicyclist riding at a constant 7.0 m/s, this suggests the problem is about uniform motion.
- **What's the correct general approach?** What principles, strategies, and tactics that you've learned are relevant in solving this problem? For example, if you're given a position-versus-time graph and are asked to find the velocity, the principle that the velocity is related to the slope of the position graph is likely to be important.
- **What should the answer look like?** Is a numerical answer asked for? Do you need a graph or a sketch?

PREPARE The Prepare step of a solution is where you identify important elements of the problem and collect information. It's tempting to jump right to the Solve step, but a skilled problem solver spends the most time on preparation, which includes:

- **Drawing a picture.** This is often the most important part of a problem. The picture lets you model the problem and identify the important elements. As you add information to your picture, the outline of the solution will take shape. For the problems in this chapter, a picture could be a motion diagram or a graph—or perhaps both.
- **Collecting necessary information.** The problem's statement may give you some values of variables. Other information may be implied, or looked up in a table, or estimated or measured.
- **Doing preliminary calculations.** Some calculations, such as unit conversions, are best done in advance.

SOLVE The Solve step of a solution is where you actually do the mathematics or reasoning necessary to arrive at the answer. This is the part of the problem-solving approach that you likely think of as "solving problems." The Strategize and Prepare steps help you be certain you understand the problem before you start putting numbers in equations.

ASSESS The Assess step of your solution is very important. Once you have an answer, you should check to see whether it makes sense. Ask yourself:

- **Does my solution answer the question that was asked?** Make sure you have addressed all parts of the question and clearly written down your solutions.
- **Does my answer have the correct units and number of significant figures?**
- **Does the value I computed make physical sense?** In this book all calculations use physically reasonable numbers. If your answer seems unreasonable, go back and check your work.
- **Can I estimate what the answer should be to check my solution?**
- **Does my final solution make sense in the context of the material I am learning?**

The Pictorial Representation

Many physics problems, including one-dimensional motion problems, have several variables and other pieces of information to keep track of. The best way to tackle such problems is to draw a picture, as we noted when we introduced a general problem-solving approach. But what kind of picture should you draw?

In this section, we will begin to draw **pictorial representations** as an aid to solving problems. A pictorial representation shows all of the important details that we need to keep track of and will be very important in solving motion problems.

Dinner at a distance BIO A chameleon's tongue is a powerful tool for catching prey. Certain species can extend the tongue to a distance of over 1 ft in less than 0.1 s! A study of the kinematics of the motion of the chameleon tongue reveals that the tongue has a period of rapid acceleration followed by a period of constant velocity. This knowledge is a very valuable clue in the analysis of the evolutionary relationships between chameleons and other animals.

TACTICS BOX 2.2 Drawing a pictorial representation

1. **Sketch the situation.** Not just any sketch: Show the object at the *beginning* of the motion, at the *end,* and at any point where the character of the motion changes. Very simple drawings are adequate.
2. **Establish a coordinate system.** Select your axes and origin to match the motion.
3. **Define symbols.** Use the sketch to define symbols representing quantities such as position, velocity, acceleration, and time. *Every* variable used later in the mathematical solution should be defined on the sketch.

We will generally combine the pictorial representation with a **list of values.** In this list, you should:

4. *List the known information.* Make a table of the quantities whose values you can determine from the problem statement or that you can find quickly with simple geometry or unit conversions.
5. *Identify the desired unknowns.* What quantity or quantities will allow you to answer the question?

Exercise 21 

EXAMPLE 2.10 Drawing a pictorial representation

Complete a pictorial representation and a list of values for the following problem: A rocket sled accelerates at 50 m/s^2 for 5 s. What are the total distance traveled and the final velocity?

STRATEGIZE We'll prepare the pictorial representation and list of values according to the steps of Tactics Box 2.2.

PREPARE FIGURE 2.32a shows a pictorial representation as drawn by an artist in the style of the figures in this book. This is certainly neater and more artistic than the sketches you will make when solving problems yourself! FIGURE 2.32b shows a sketch like one you might actually draw. It's less formal, but it contains all of the important information you need to solve the problem. The circled numbers in the sketch correspond to the steps in Tactics Box 2.2.

NOTE ▶ Throughout this book we will illustrate select examples with actual hand-drawn figures so that you have them to refer to as you work on your own pictures for homework and practice. ◀

Let's look at how these pictures were constructed. The motion has a clear beginning and end; these are the points in the motion that we've sketched. A coordinate system has been chosen with the origin at the starting point. The quantities x, v_x, and t are needed at both points, so these have been defined on the sketch and distinguished by subscripts. The acceleration is associated with an interval between these points. Values for two of these quantities are given in the problem statement. Others, such as $x_i = 0$ m and $t_i = 0$ s, are inferred from our choice of coordinate system. The value $(v_x)_i = 0$ m/s is part of our *interpretation* of the problem. Finally, we identify x_f and $(v_x)_f$ as the quantities that will answer the question. We now understand quite a bit about the problem and would be ready to start a quantitative analysis.

ASSESS We didn't *solve* the problem; that was not our purpose. Constructing a pictorial representation and a list of values is part of a systematic approach to interpreting a problem and getting ready for a mathematical solution.

FIGURE 2.32 Constructing a pictorial representation and a list of values.

(a) Artist's version
Pictorial representation

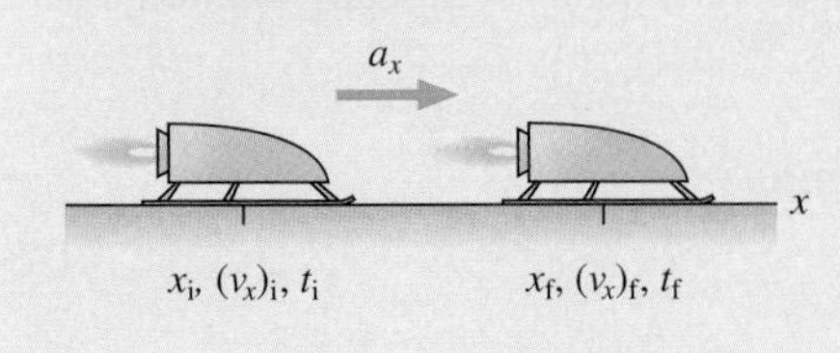

List of values

Known
$x_i = 0$ m
$(v_x)_i = 0$ m/s
$t_i = 0$ s
$a_x = 50$ m/s^2
$t_f = 5$ s

Find
x_f, $(v_x)_f$

(b) Student sketch

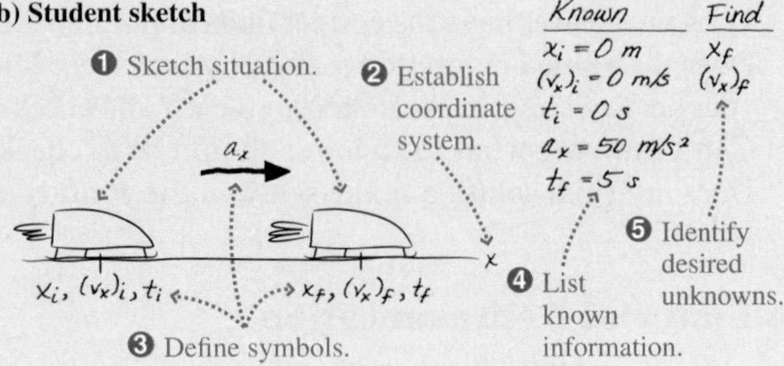

Video Motion with Constant Acceleration

The Visual Overview

The pictorial representation and the list of values are a very good complement to the motion diagram and other ways of looking at a problem that we have seen. As we translate a problem into a form we can solve, we will combine these elements into what we will term a **visual overview.** The visual overview will consist of some or all of the following elements:

- A *motion diagram.* A good approach for solving a motion problem is to start by drawing a motion diagram.
- A *pictorial representation,* as defined in Tactics Box 2.2.
- A *list of values,* also described in Tactics Box 2.2. This list should sum up all of the important values in the problem.
- A *graphical representation.* For motion problems, it is often quite useful to include a graph of position and/or velocity.

Future chapters will add other elements to this visual overview of the physics.

EXAMPLE 2.11 Kinematics of a rocket launch

A Saturn V rocket is launched straight up with a constant acceleration of 18 m/s^2. After 150 s, how fast is the rocket moving and how far has it traveled?

STRATEGIZE We are given the acceleration and the time interval, suggesting that this is a constant-acceleration problem. We will find the velocity from Equation 2.11 and the position from Equation 2.12.

PREPARE FIGURE 2.33 shows a visual overview of the rocket launch that includes a motion diagram, a pictorial representation, and a list of values. The visual overview shows the whole problem in a nutshell. The motion diagram illustrates the motion of the rocket. The pictorial representation (produced according to Tactics Box 2.2) shows the axis, identifies the important points of the motion, and defines the variables. Finally, we include a list of values that gives the known and unknown quantities (again according to Tactics Box 2.2). In the visual overview we have taken the statement of the problem in words and made it much more precise. The overview contains everything we need to know about the problem.

SOLVE Our first task is to find the final velocity. Our list of values includes the initial velocity, the acceleration, and the time interval, so we can use Equation 2.11 of Synthesis 2.1 to find the final velocity:

$$(v_y)_f = (v_y)_i + a_y \Delta t = 0 \text{ m/s} + (18 \text{ m/s}^2)(150 \text{ s})$$
$$= 2700 \text{ m/s}$$

The distance traveled is found using Equation 2.12 of Synthesis 2.1:

$$y_f = y_i + (v_y)_i \Delta t + \tfrac{1}{2} a_y (\Delta t)^2$$
$$= 0 \text{ m} + (0 \text{ m/s})(150 \text{ s}) + \tfrac{1}{2}(18 \text{ m/s}^2)(150 \text{ s})^2$$
$$= 2.0 \times 10^5 \text{ m} = 200 \text{ km}$$

ASSESS The acceleration is very large, and it goes on for a long time, so the large final velocity and large distance traveled seem reasonable.

FIGURE 2.33 Visual overview of the rocket launch.

Motion diagram
Pictorial representation
List of values
$\vec{v}$
$\vec{a}$
The motion diagram for the rocket shows the full range of the motion.
y
$y_f, (v_y)_f, t_f$
$y_i, (v_y)_i, t_i$
0
The pictorial representation identifies the two important points of the motion, the start and the end, and shows that the rocket accelerates between them.
Known
$y_i = 0$ m
$(v_y)_i = 0$ m/s
$t_i = 0$ s
$a_y = 18$ m/s^2
$t_f = 150$ s
Find
$(v_y)_f$ and y_f
The list of values makes everything concrete. We define the start of the problem to be at time 0 s, when the rocket has a position of 0 m and a velocity of 0 m/s. The end of the problem is at time 150 s. We are to find the position and velocity at this time.

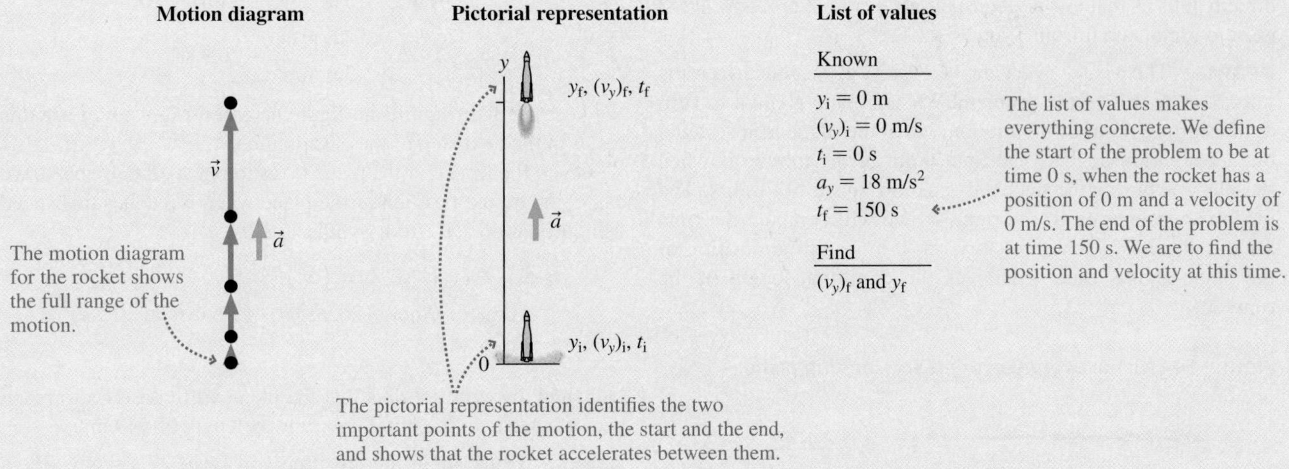

Problem-Solving Approach for Motion with Constant Acceleration

Earlier in this section, we introduced a general problem-solving approach. In this and future chapters we will adapt this general approach to specific types of problems.

PROBLEM-SOLVING APPROACH 2.1 Motion with constant acceleration

Problems involving constant acceleration—speeding up, slowing down, vertical motion, horizontal motion—can all be treated with the same problem-solving approach.

STRATEGIZE Identify the problem as one involving constant-acceleration motion: Look for statements that give the acceleration or indicate that the speed or velocity of an object is changing. Free-fall problems, discussed in the next section, are always constant-acceleration problems. Solve constant-acceleration problems using the ideas in Synthesis 2.1.

PREPARE Draw a visual overview of the problem. This should include a motion diagram, a pictorial representation, and a list of values; a graphical representation may be useful for certain problems.

SOLVE The mathematical solution is based on the three constant-acceleration equations in Synthesis 2.1.

- Though the equations are phrased in terms of the variable x, it's customary to use y for motion in the vertical direction.
- Use the equation that best matches what you know and what you need to find. For example, if you know acceleration and time and are looking for a change in velocity, the first equation is the best one to use.
- Uniform motion with constant velocity has $a = 0$.

ASSESS Is your result believable? Does it have proper units? Does it make sense?

Exercise 25

EXAMPLE 2.12 Calculating the minimum length of a runway

A fully loaded Boeing 747 with all engines at full thrust accelerates at 2.6 m/s^2. Its minimum takeoff speed is 70 m/s. How much time will the plane take to reach its takeoff speed? What minimum length of runway does the plane require for takeoff?

STRATEGIZE The acceleration of the plane is given, which directly tells us that this is a constant-acceleration problem. We'll need to use material from Synthesis 2.1.

PREPARE The visual overview of FIGURE 2.34 summarizes the important details of the problem. We set x_i and t_i equal to zero at the starting point of the motion, when the plane is at rest and the acceleration begins. The final point of the motion is when the plane achieves the necessary takeoff speed of 70 m/s. The plane is accelerating to the right, so we will compute the time for the plane to reach a velocity of 70 m/s and the position of the plane at this time, giving us the minimum length of the runway.

FIGURE 2.34 Visual overview for an accelerating plane.

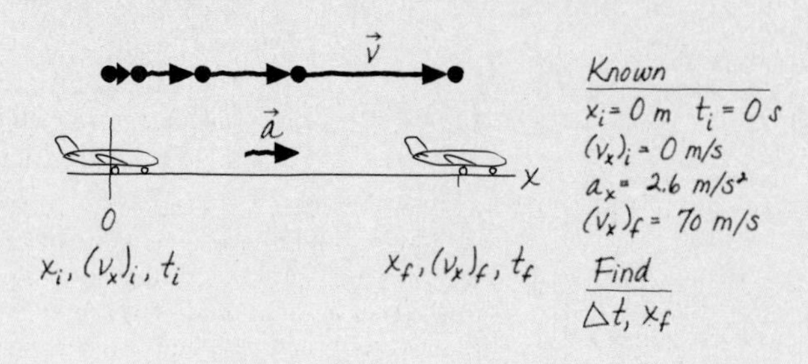

SOLVE First we solve for the time required for the plane to reach takeoff speed. We can use Equation 2.11 from Synthesis 2.1 to compute this time:

$$(v_x)_f = (v_x)_i + a_x \Delta t$$

$$70 \text{ m/s} = 0 \text{ m/s} + (2.6 \text{ m/s}^2)\, \Delta t$$

$$\Delta t = \frac{70 \text{ m/s}}{2.6 \text{ m/s}^2} = 26.9 \text{ s}$$

We keep an extra significant figure here because we will use this result in the next step of the calculation.

Given the time that the plane takes to reach takeoff speed, we can compute the position of the plane when it reaches this speed using Equation 2.12 from Synthesis 2.1:

$$\begin{aligned} x_f &= x_i + (v_x)_i \,\Delta t + \tfrac{1}{2} a_x (\Delta t)^2 \\ &= 0 \text{ m} + (0 \text{ m/s})(26.9 \text{ s}) + \tfrac{1}{2}(2.6 \text{ m/s}^2)(26.9 \text{ s})^2 \\ &= 940 \text{ m} \end{aligned}$$

Our final answers are thus that the plane will take 27 s to reach takeoff speed, with a minimum runway length of 940 m.

ASSESS Think about the last time you flew; 27 s seems like a reasonable time for a plane to accelerate on takeoff. Actual runway lengths at major airports are 3000 m or more, a few times greater than the minimum length, because they have to allow for emergency stops during an aborted takeoff. (If we had calculated a distance far greater than 3000 m, we would know we had done something wrong!)

EXAMPLE 2.13 Finding the braking distance of a car on the highway

A car is traveling at a speed of 30 m/s, a typical highway speed, on wet pavement. The driver sees an obstacle ahead and decides to stop. From this instant, it takes him 0.75 s to begin applying the brakes. Once the brakes are applied, the car experiences an acceleration of $-6.0\ \text{m/s}^2$. How far does the car travel from the instant the driver notices the obstacle until it stops?

STRATEGIZE The wording of this problem makes it clear that, while it is braking, the car is experiencing constant acceleration. We'll once more use the information from Synthesis 2.1.

PREPARE This problem is more involved than previous problems we have solved, so we will take more care with the visual overview in FIGURE 2.35. In addition to a motion diagram and a pictorial representation, we include a graphical representation. Notice that there are two different phases of the motion: a constant-velocity phase before braking begins, and a steady slowing down once the brakes are applied. We will need to do two different calculations, one for each phase. Consequently, we use numerical subscripts rather than the simple i and f.

SOLVE From t_1 to t_2 the velocity stays constant at 30 m/s. This is uniform motion, so we compute the position at time t_2 using Equation 2.4 from Synthesis 2.1:

$$x_2 = x_1 + (v_x)_1(t_2 - t_1) = 0\ \text{m} + (30\ \text{m/s})(0.75\ \text{s}) = 22.5\ \text{m}$$

Starting at t_2, the velocity begins to decrease at a steady $-6.0\ \text{m/s}^2$ until the car comes to rest at t_3. We can compute this time interval using Equation 2.11 from Synthesis 2.1, $(v_x)_3 = (v_x)_2 + a_x\,\Delta t$:

$$\Delta t = t_3 - t_2 = \frac{(v_x)_3 - (v_x)_2}{a_x} = \frac{0\ \text{m/s} - 30\ \text{m/s}}{-6.0\ \text{m/s}^2} = 5.0\ \text{s}$$

We can compute the position at time t_3 using Equation 2.12 from Synthesis 2.1. We take point 2 as the initial point and point 3 as the final point for this phase of the motion and use $\Delta t = t_3 - t_2$:

$$x_3 = x_2 + (v_x)_2\,\Delta t + \tfrac{1}{2}a_x(\Delta t)^2$$
$$= 22.5\ \text{m} + (30\ \text{m/s})(5.0\ \text{s}) + \tfrac{1}{2}(-6.0\ \text{m/s}^2)(5.0\ \text{s})^2$$
$$= 98\ \text{m}$$

x_3 is the position of the car at the end of the problem—and so the car travels 98 m before coming to rest.

ASSESS The numbers for the reaction time and the acceleration on wet pavement are reasonable ones for an alert driver in a car with good tires. The final distance is quite large—more than the length of a football field.

FIGURE 2.35 Visual overview for a car braking to a stop.

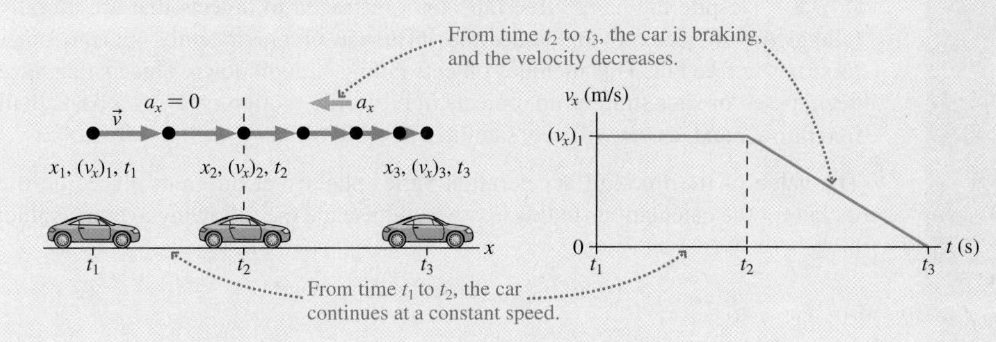

2.7 Free Fall

If you drop a hammer and a feather, you know what will happen. The hammer quickly strikes the ground, and the feather drifts slowly downward and lands some time later. But if you do this experiment on the moon, the result is strikingly different: Both the hammer and the feather experience the exact same acceleration, undergo the exact same motion, and strike the ground at the same time.

The moon lacks an atmosphere, and so objects falling to its surface experience no air resistance. They are acted upon by only one force—gravity. If an object moves under the influence of gravity only, and no other forces, we call the resulting motion **free fall.** Many experiments have shown that **all objects in free fall, regardless of their mass, have the same acceleration.** Thus, if you drop two objects and they are both in free fall, they hit the ground at the same time.

On the earth, air resistance is a factor. But when you drop a heavy object like a hammer, air resistance can be ignored, so we make only a slight error in treating the hammer *as if* it were in free fall. Motion with air resistance is a problem we will

Free-falling feather Apollo 15 lunar astronaut David Scott performed a classic experiment on the moon, simultaneously dropping a hammer and a feather from the same height. Both hit the ground at the exact same time—something that would not happen in the atmosphere of the earth!

study in Chapter 5. Until then, we will restrict our attention to situations in which air resistance can be ignored, and we will assume that falling objects are in free fall.

FIGURE 2.36a shows the motion diagram for an object that was released from rest and falls freely. Since the acceleration is the same for all objects, the diagram and graph would be the same for a falling baseball or a falling boulder! FIGURE 2.36b shows the object's velocity graph. The velocity changes at a steady rate. The slope of the velocity-versus-time graph is the free-fall acceleration $a_{\text{free fall}}$.

Video Figure 2.36

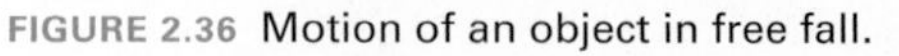

FIGURE 2.36 Motion of an object in free fall.

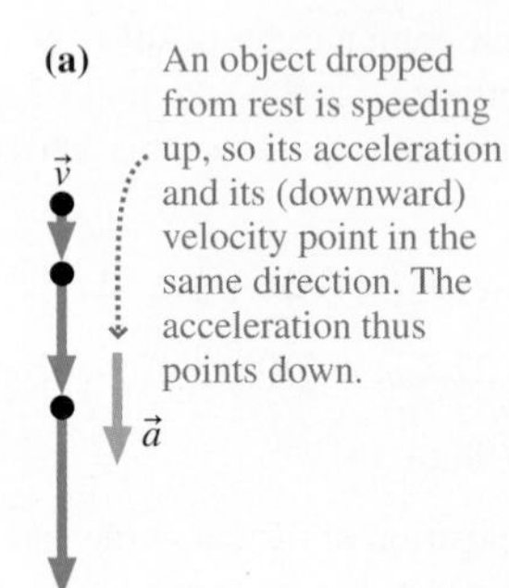

(a)
$\vec{v}$
An object dropped from rest is speeding up, so its acceleration and its (downward) velocity point in the same direction. The acceleration thus points down.
$\vec{a}$

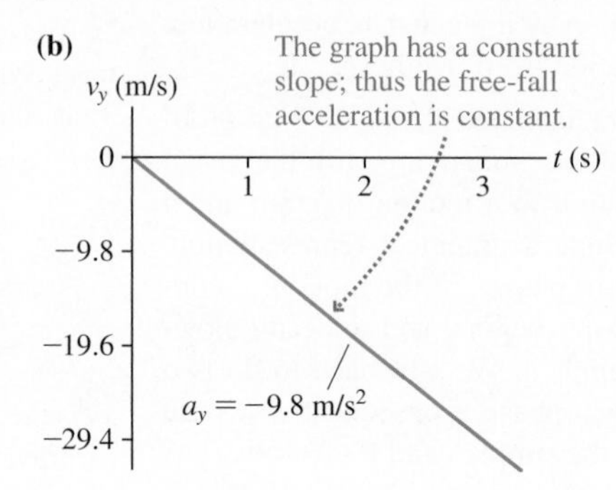

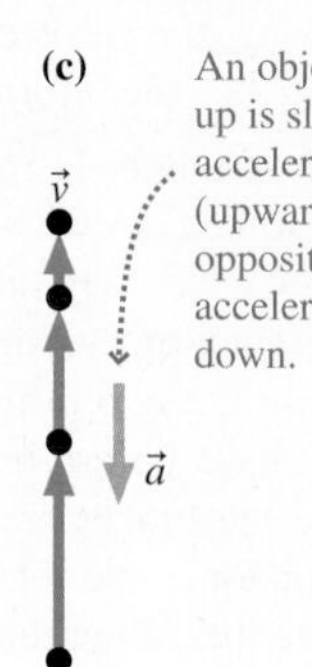

Instead of dropping the object, suppose we throw it upward. What happens then? You know that the object will move up and that its speed will decrease as it rises. This is illustrated in the motion diagram of FIGURE 2.36c, which shows a surprising result: Even though the object is moving up, its acceleration still points down. In fact, **the free-fall acceleration always points down,** no matter what direction an object is moving.

Video Free Fall

NOTE ► Despite the name, free fall is not restricted to objects that are literally falling. Any object moving under the influence of gravity only, and no other forces, is in free fall. This includes objects falling straight down, objects that have been tossed or shot straight up, objects in projectile motion (such as a basketball free throw), and, as we will see, satellites in orbit. ◄

Some of the children on this trampoline are moving up and some are moving down, but all are in free fall—and so are accelerating downward at $9.8\ \text{m/s}^2$.

The value of the free-fall acceleration varies slightly at different places on the earth, but for the calculations in this text we will use the the following average value:

$$\vec{a}_{\text{free fall}} = (9.80\ \text{m/s}^2,\ \text{vertically downward}) \tag{2.14}$$

Standard value for the acceleration of an object in free fall

The magnitude of the **free-fall acceleration** has the special symbol g:

$$g = 9.80\ \text{m/s}^2$$

We will generally work with two significant figures and so will use $g = 9.8\ \text{m/s}^2$.

Several points about free fall are worthy of note:

- g, by definition, is *always* positive. **There will never be a problem that uses a negative value for g.**
- The velocity graph in Figure 2.36b has a negative slope. Even though a falling object speeds up, it has *negative* acceleration. Thus g is *not* the object's acceleration, simply the *magnitude* of the acceleration. The one-dimensional acceleration is
$$a_y = -g$$
- Because free fall is motion with constant acceleration, we can use the kinematic equations for constant acceleration with $a_y = -g$.
- Once an object is acted upon by only the force of gravity, it is in free fall with $a_y = -9.8\ \text{m/s}^2$. It doesn't matter how the object entered free fall: Once in the air, a football that was punted straight up has the *same* acceleration of $-g$ as a stone dropped off a bridge.

- g is not called "gravity." Gravity is a force, not an acceleration. g is the *free-fall acceleration.*
- $g = 9.80\ \text{m/s}^2$ only on earth. Other planets have different values of g. You will learn in Chapter 6 how to determine g for other planets.
- We will sometimes compute acceleration of objects not in free fall in units of g. An acceleration of $9.8\ \text{m/s}^2$ is an acceleration of $1g$; an acceleration of $19.6\ \text{m/s}^2$ is $2g$. Generally, we can compute

$$\text{acceleration (in units of } g\text{, or } g\text{'s)} = \frac{\text{acceleration (in units of m/s}^2\text{)}}{9.8\ \text{m/s}^2} \quad (2.15)$$

This allows us to express accelerations in units that have a definite physical reference.

EXAMPLE 2.14 **Analyzing a rock's fall**

A heavy rock is dropped from rest at the top of a cliff and falls 100 m before hitting the ground. How long does the rock take to fall to the ground, and what is its velocity when it hits?

STRATEGIZE This is a free-fall problem, so it is a constant-acceleration problem with $a_y = -g$. We will use the constant-acceleration equations from Synthesis 2.1.

PREPARE FIGURE 2.37 shows a visual overview with all necessary data. We have placed the origin at the ground, so that $y_i = 100$ m.

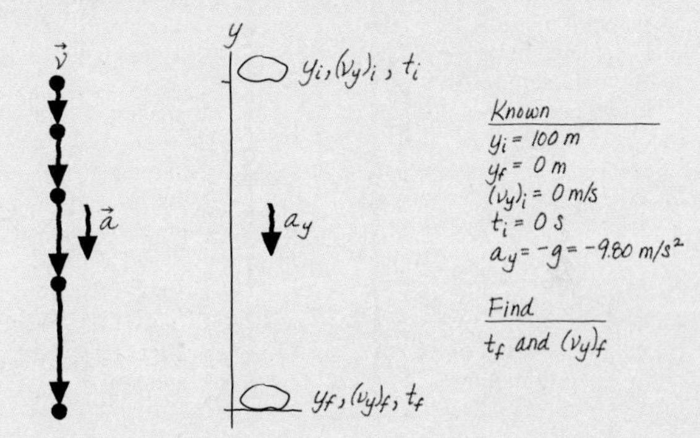

FIGURE 2.37 Visual overview of a falling rock.

SOLVE The first question in the problem statement involves a relationship between time and distance, a relationship expressed by Equation 2.12 in Synthesis 2.1. Using $(v_y)_i = 0$ m/s and $t_i = 0$ s, we find

$$y_f = y_i + (v_y)_i\,\Delta t + \tfrac{1}{2}a_y(\Delta t)^2 = y_i - \tfrac{1}{2}g(\Delta t)^2 = y_i - \tfrac{1}{2}gt_f^{\,2}$$

We can now solve for t_f:

$$t_f = \sqrt{\frac{2(y_i - y_f)}{g}} = \sqrt{\frac{2(100\ \text{m} - 0\ \text{m})}{9.80\ \text{m/s}^2}} = 4.52\ \text{s}$$

Now that we know the fall time, we can use Equation 2.11 to find $(v_y)_f$:

$$(v_y)_f = (v_y)_i - g\,\Delta t = -gt_f = -(9.80\ \text{m/s}^2)(4.52\ \text{s})$$
$$= -44.3\ \text{m/s}$$

ASSESS Are the answers reasonable? Well, 100 m is about 300 feet, which is about the height of a 30-floor building. How long does it take an object to fall 30 floors? Four or five seconds seems pretty reasonable. How fast would the object be going at the bottom? Using an approximate version of our conversion factor 1 m/s ≈ 2 mph, we find that 44.3 m/s ≈ 90 mph. That also seems like a pretty reasonable speed for something that has fallen 30 floors. Suppose we had made a mistake. If we had misplaced a decimal point, we could have calculated a speed of 443 m/s, or about 900 mph! This is clearly *not* reasonable. If we had misplaced the decimal point in the other direction, we would have calculated a speed of 4.3 m/s ≈ 9 mph. This is another unreasonable result, because this is slower than a typical bicycling speed.

CONCEPTUAL EXAMPLE 2.15 **Analyzing the motion of a ball tossed upward**

Draw a motion diagram and a velocity-versus-time graph for a ball tossed straight up in the air from the point that it leaves the hand until just before it is caught.

REASON You know what the motion of the ball looks like: The ball goes up, and then it comes back down again. This complicates the drawing of a motion diagram a bit because the ball retraces its route as it falls. A literal motion diagram would show the upward motion and downward motion on top of each other, leading to confusion. We can avoid this difficulty by horizontally separating the upward motion and downward motion diagrams. This will not affect our conclusions because it does not change any of the vectors. The motion diagram and velocity-versus-time graph appear as in FIGURE 2.38.

ASSESS The highest point in the ball's motion, where it reverses direction, is called a *turning point.* What are the velocity and the acceleration at this point? We can see from the motion diagram that the velocity vectors are pointing upward but getting shorter as the ball approaches the top. As it starts to fall, the velocity vectors are pointing downward and getting longer. There must be a moment—just an instant as $\vec{v}$ switches from pointing up to pointing down—when the velocity is zero. Indeed, the ball's velocity *is* zero for an instant at the precise top of the motion! We can also see on the velocity graph

Continued

that there is one instant of time when $v_y = 0$. This is the turning point.

But what about the acceleration at the top? You might expect the acceleration to be zero at the highest point. But recall that the velocity at the top point is changing—from up to down. If the velocity is changing, there *must* be an acceleration. The slope of the velocity graph at the instant when $v_y = 0$—that is, at the highest point—is no different than at any other point in the motion. The ball is still in free fall with acceleration $a_y = -g$!

Another way to think about this is to note that zero acceleration would mean no change of velocity. When the ball reached zero velocity at the top, it would hang there and not fall if the acceleration were also zero!

FIGURE 2.38 Motion diagram and velocity graph of a ball tossed straight up in the air.

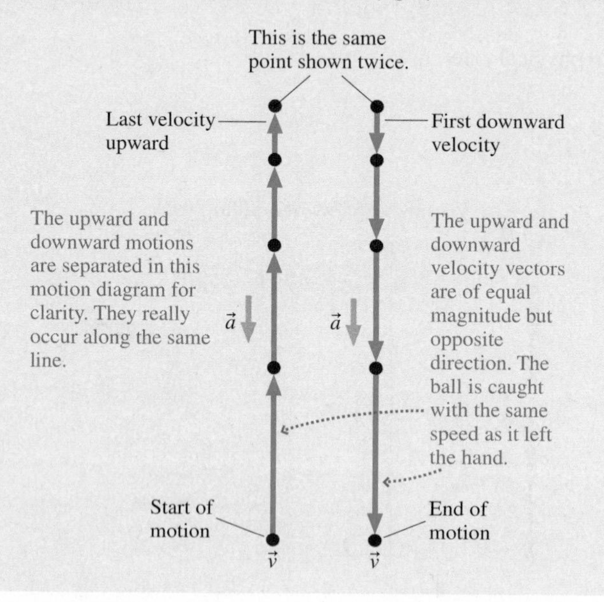
This is the same point shown twice.
Last velocity upward
First downward velocity
The upward and downward motions are separated in this motion diagram for clarity. They really occur along the same line.
The upward and downward velocity vectors are of equal magnitude but opposite direction. The ball is caught with the same speed as it left the hand.
$\vec{a}$
$\vec{a}$
Start of motion
End of motion
$\vec{v}$
$\vec{v}$

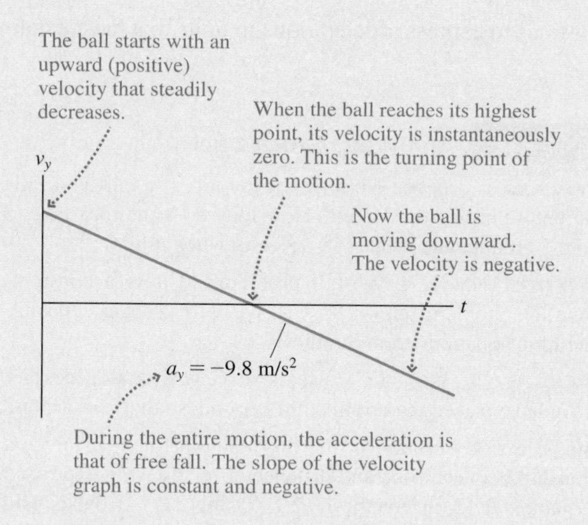
The ball starts with an upward (positive) velocity that steadily decreases.
When the ball reaches its highest point, its velocity is instantaneously zero. This is the turning point of the motion.
v_y
Now the ball is moving downward. The velocity is negative.
t
$a_y = -9.8\text{ m/s}^2$
During the entire motion, the acceleration is that of free fall. The slope of the velocity graph is constant and negative.

EXAMPLE 2.16 Finding the height of a leap 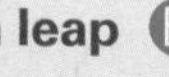

A springbok is an antelope found in southern Africa that gets its name from its remarkable jumping ability. When a springbok is startled, it will leap straight up into the air—a maneuver called a "pronk." A particular springbok goes into a crouch to perform a pronk. It then extends its legs forcefully, accelerating at 35 m/s^2 for 0.70 m as its legs straighten. Legs fully extended, it leaves the ground and rises into the air.

a. At what speed does the springbok leave the ground?
b. How high does it go?

STRATEGIZE This is a two-part problem. In the first phase of its motion, the springbok accelerates upward, reaching some maximum speed just as it leaves the ground. As soon as it does so, the springbok is subject to only the force of gravity, so it is in free fall. For both phases, we will use the constant-acceleration equations from Synthesis 2.1.

PREPARE We begin with the visual overview shown in FIGURE 2.39, where we've identified the two different phases of the motion: the springbok pushing off the ground and the springbok rising into the air. We'll treat these as two separate problems that we solve in turn. We will "re-use" the variables y_i, y_f, $(v_y)_i$, and $(v_y)_f$ for the two phases of the motion.

For the first part of our solution, in Figure 2.39a we choose the origin of the y-axis at the position of the springbok deep in its crouch. The final position is the top extent of the push, at the instant the springbok leaves the ground. We want to find the velocity at this position because that's how fast the springbok is moving as it leaves the ground. Figure 2.39b essentially starts over—we have defined a new vertical axis with its origin at the ground, so the highest point of the springbok's motion is its distance above the ground. The table of values shows the key piece of information for this second part of the problem: The initial velocity for part b is the final velocity from part a.

After the springbok leaves the ground, this is a free-fall problem because the springbok is moving under the influence of gravity only. We want to know the height of the leap, so we are looking for the height at the top point of the motion. This is a turning point of the motion, with the instantaneous velocity equal to zero. Thus y_f, the height of the leap, is the springbok's position at the instant $(v_y)_f = 0$.

SOLVE a. For the first phase, pushing off the ground, we have information about displacement, initial velocity, and acceleration, but we don't know anything about the time interval. Equation 2.13 from Synthesis 2.1 is perfect for this type of situation. We can use it to solve for the velocity with which the springbok lifts off the ground:

$$(v_y)_f^2 = (v_y)_i^2 + 2a_y\Delta y$$
$$= (0\text{ m/s})^2 + 2(35\text{ m/s}^2)(0.70\text{ m}) = 49\text{ m}^2/\text{s}^2$$
$$(v_y)_f = \sqrt{49\text{ m}^2/\text{s}^2} = 7.0\text{ m/s}$$

The springbok leaves the ground with a speed of 7.0 m/s.

b. Now we are ready for the second phase of the motion, the vertical motion after leaving the ground. Equation 2.13 is again appropriate because again we don't know the time. Because $y_i = 0$, the springbok's displacement is $\Delta y = y_f - y_i = y_f$, the height of the vertical leap. From part a, the initial velocity is $(v_y)_i = 7.0$ m/s, and the final velocity is $(v_y)_f = 0$. This is free-fall motion, with $a_y = -g$; thus

$$(v_y)_f^2 = 0 = (v_y)_i^2 - 2g\Delta y = (v_y)_i^2 - 2gy_f$$

which gives

$$(v_y)_i^2 = 2gy_f$$

Solving for y_f, we get a jump height of

$$y_f = \frac{(7.0\text{ m/s})^2}{2(9.8\text{ m/s}^2)} = 2.5\text{ m}$$

ASSESS 2.5 m is a remarkable leap—a bit over 8 ft—but these animals are known for their jumping ability, so this seems reasonable.

FIGURE 2.39 A visual overview of the springbok's leap.

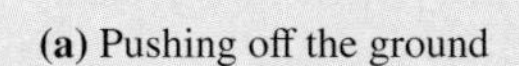

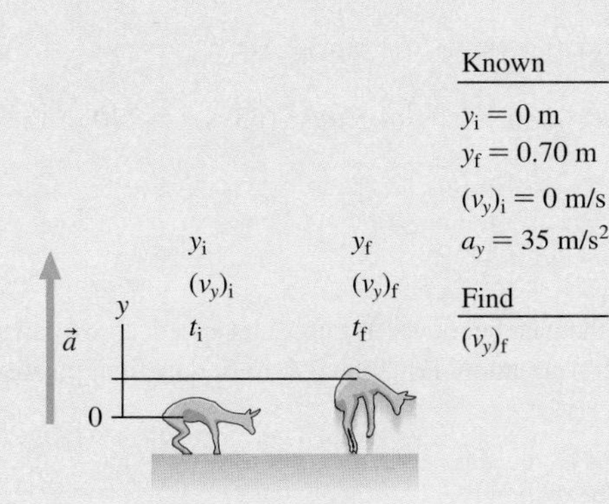

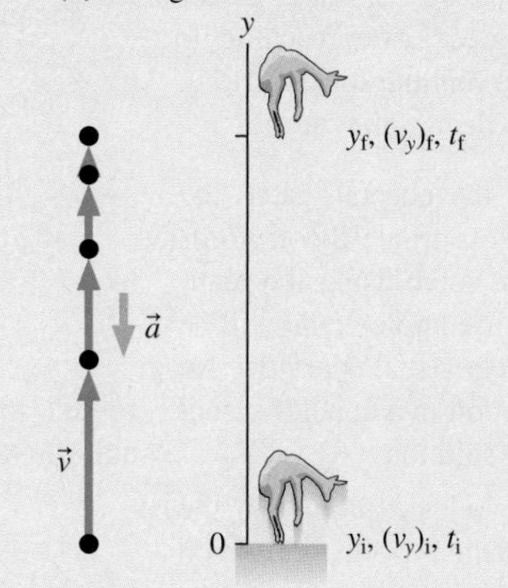

The caption accompanying the photo at the start of the chapter suggested a question about animals and their athletic abilities: Who is the winner in a race between a horse and a man? The surprising answer is "It depends." Specifically, the winner depends on the length of the race.

Some animals are capable of high speed; others are capable of great acceleration. Horses can run much faster than humans, but, when starting from rest, humans are capable of much greater initial acceleration. FIGURE 2.40 shows velocity and position graphs for an elite male sprinter and a thoroughbred racehorse. The horse's maximum velocity is about twice that of the man, but the man's initial acceleration—the slope of the velocity graph at early times—is greater than that of the horse. As the second graph shows, a man could win a short race. For a longer race, the horse's higher maximum velocity will put it in the lead. The men's world-record time for the mile is a bit under 4 min, but a horse can easily run a mile in less than 2 min.

For a race of many miles, another factor comes into play: energy. A very long race is less about velocity and acceleration than about endurance—the ability to continue expending energy for a long time. In such endurance trials, humans often win. We will explore such energy issues in Chapter 11.

FIGURE 2.40 BIO Velocity-versus-time and position-versus-time graphs for a sprint between a man and a horse.

(a) Velocity-versus-time graph

The velocity-versus-time graph for the man starts with a steeper slope. Initially, the man is running at a higher velocity, but . . .

. . . the horse has a higher *maximum* velocity.

v_x (m/s), Horse, Man, t (s)

(b) Position-versus-time graph

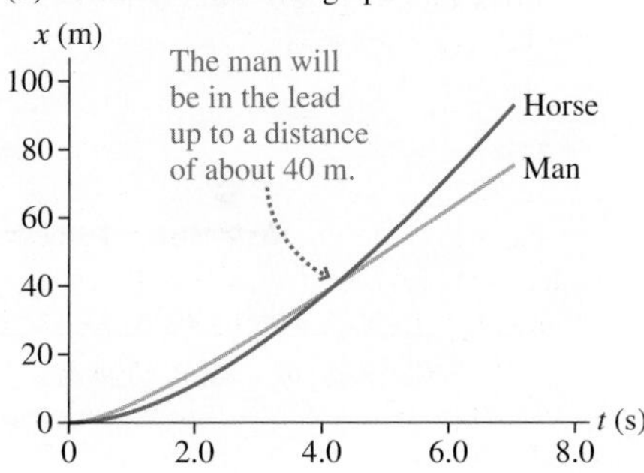

STOP TO THINK 2.7 A volcano ejects a chunk of rock straight up at a velocity of $v_y = 30$ m/s. Ignoring air resistance, what will be the velocity v_y of the rock when it falls back into the volcano's crater?

A. > 30 m/s B. 30 m/s C. 0 m/s D. −30 m/s E. < −30 m/s

INTEGRATED EXAMPLE 2.17 Speed versus endurance 

Cheetahs have the highest top speed of any land animal, but they usually fail in their attempts to catch their prey because their endurance is limited. They can maintain their maximum speed of 30 m/s for only about 15 s before they need to stop.

Thomson's gazelles, their preferred prey, have a lower top speed than cheetahs, but they can maintain this speed for a few minutes. When a cheetah goes after a gazelle, success or failure is a simple matter of kinematics: Is the cheetah's high speed enough to allow it to reach its prey before the cheetah runs out of steam? The following problem uses realistic data for such a chase.

A cheetah has spotted a gazelle. The cheetah leaps into action, reaching its top speed of 30 m/s in a few seconds. At this instant, the gazelle, 160 m from the running cheetah, notices the danger and heads directly away. The gazelle accelerates at 4.5 m/s^2 for 6.0 s, then continues running at a constant speed. After reaching its maximum speed, the cheetah can continue running for only 15 s. Does the cheetah catch the gazelle, or does the gazelle escape?

STRATEGIZE The example asks, "Does the cheetah catch the gazelle?" Our most challenging task is to translate these words into a mathematical problem that we can solve using the techniques of this chapter. For a problem of this complexity, it will be particularly important to prepare a complete visual overview. An overview lays out all the relevant information in a concise visual form, helping to guide your mathematical solution.

PREPARE This example consists of two related problems, the motion of the cheetah and the motion of the gazelle, for which we'll use the subscripts "C" and "G." Let's take our starting time, $t_1 = 0$ s, as the instant that the gazelle notices the cheetah and begins to run. We'll take the position of the cheetah at this instant as the origin of our coordinate system, so $x_{1C} = 0$ m and $x_{1G} = 160$ m—the gazelle is 160 m away when it notices the cheetah. We've used this information to draw the visual overview in FIGURE 2.41, which includes motion diagrams and velocity graphs for the cheetah and the gazelle. The visual overview sums up everything we know about the problem.

With a clear picture of the situation, we can now rephrase the problem this way: Compute the position of the cheetah and the position of the gazelle at $t_3 = 15$ s, the time when the cheetah needs to break off the chase. If $x_{3G} > x_{3C}$, then the gazelle stays out in front and escapes. If $x_{3G} \le x_{3C}$, the cheetah wins the race—and gets its dinner.

SOLVE The cheetah is in uniform motion for the entire duration of the problem, so we can use Equation 2.4 of Synthesis 2.1 to solve for its position at $t_3 = 15$ s:

$$x_{3C} = x_{1C} + (v_x)_{1C}\Delta t = 0\text{ m} + (30\text{ m/s})(15\text{ s}) = 450\text{ m}$$

The gazelle's motion has two phases: one of constant acceleration and then one of constant velocity. We can solve for the position and the velocity at t_2, the end of the first phase, using Equations 2.11 and 2.12 of Synthesis 2.1. Let's find the velocity first:

$$(v_x)_{2G} = (v_x)_{1G} + (a_x)_G\Delta t = 0\text{ m/s} + (4.5\text{ m/s}^2)(6.0\text{ s}) = 27\text{ m/s}$$

The gazelle's position at t_2 is

$$x_{2G} = x_{1G} + (v_x)_{1G}\Delta t + \tfrac{1}{2}(a_x)_G(\Delta t)^2$$

$$= 160\text{ m} + 0 + \tfrac{1}{2}(4.5\text{ m/s}^2)(6.0\text{ s})^2 = 240\text{ m}$$

The gazelle has a head start; it begins at $x_{1G} = 160$ m.

Δt is the time for this phase of the motion, $t_2 - t_1 = 6.0$ s.

From t_2 to t_3 the gazelle moves at a constant speed, so we can use the uniform motion equation, Equation 2.4, to find its final position:

The gazelle begins this phase of the motion at $x_{2G} = 240$ m.

Δt for this phase of the motion is $t_3 - t_2 = 9.0$ s.

$$x_{3G} = x_{2G} + (v_x)_{2G}\ \Delta t = 240\text{ m} + (27\text{ m/s})(9.0\text{ s}) = 480\text{ m}$$

x_{3C} is 450 m; x_{3G} is 480 m. The gazelle is 30 m ahead of the cheetah when the cheetah has to break off the chase, so the gazelle escapes.

ASSESS Does our solution make sense? Let's look at the final result. The numbers in the problem statement are realistic, so we expect our results to mirror real life. The speed for the gazelle is close to that of the cheetah, which seems reasonable for two animals known for their speed. And the result is the most common occurrence—the chase is close, but the gazelle gets away.

FIGURE 2.41 Visual overview for the cheetah and for the gazelle.

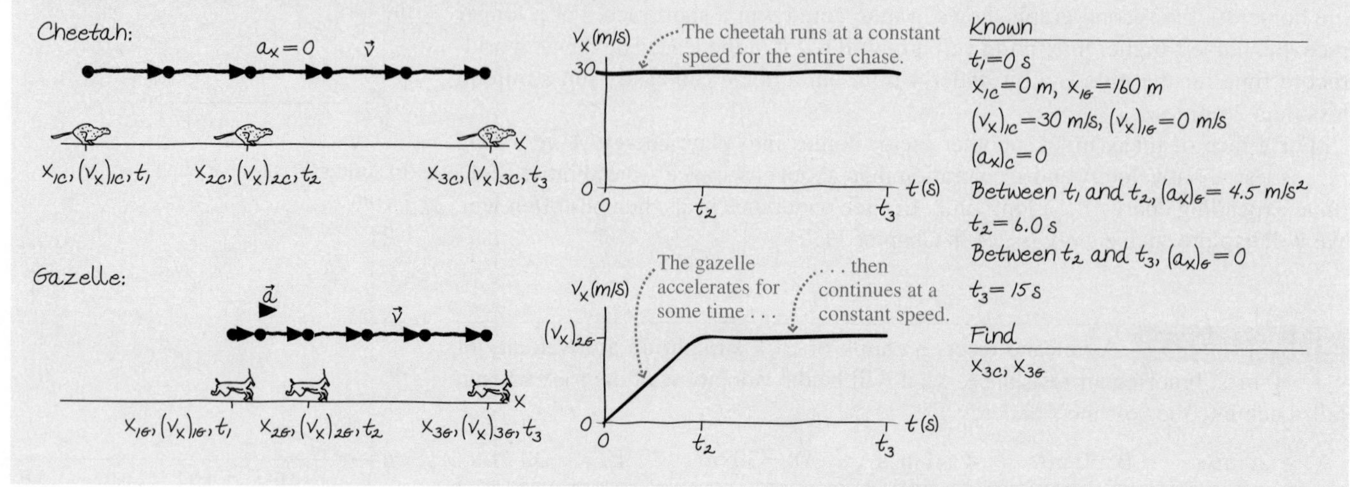

SUMMARY

GOAL To describe and analyze motion along a line.

GENERAL STRATEGIES

Problem-Solving Approach

Our general problem-solving approach has four parts:

STRATEGIZE Think about the big picture: What kind of problem is this? What general approach should be used? What should the answer look like?

PREPARE Set up the problem:

- Draw a picture.
- Collect necessary information.
- Do preliminary calculations.

SOLVE Do the necessary mathematics or reasoning.

ASSESS Check your answer to see if it is complete in all details and makes physical sense.

Visual Overview

A visual overview consists of several parts that completely specify a problem. This may include any or all of the elements below:

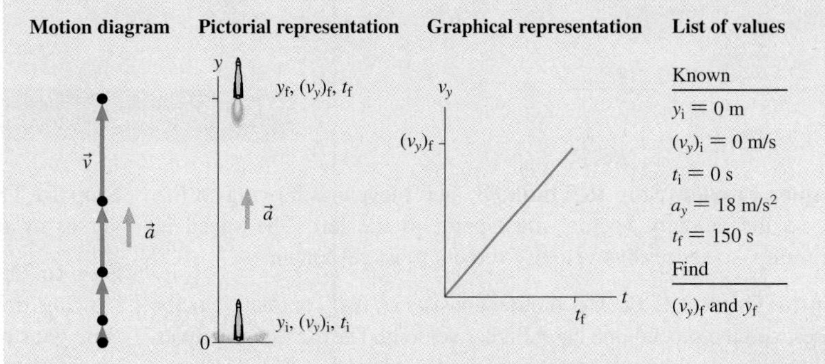

IMPORTANT CONCEPTS

Velocity is the rate of change of position:

$$v_x = \frac{\Delta x}{\Delta t}$$

Acceleration is the rate of change of velocity:

$$a_x = \frac{\Delta v_x}{\Delta t}$$

The units of acceleration are m/s^2.

An object is speeding up if v_x and a_x have the same sign, slowing down if they have opposite signs.

A **position-versus-time graph** plots position on the vertical axis against time on the horizontal axis.

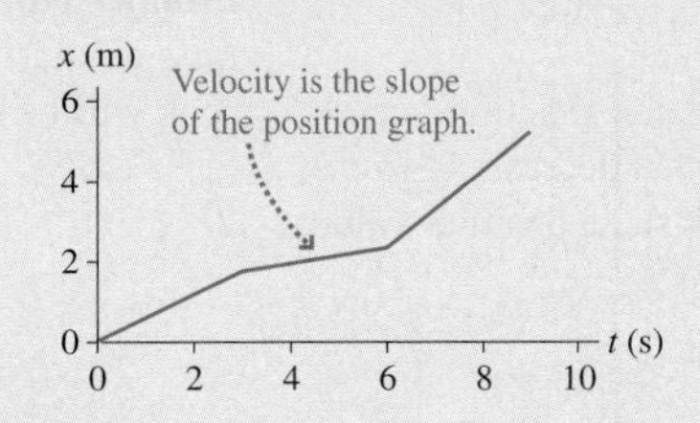

A **velocity-versus-time graph** plots velocity on the vertical axis against time on the horizontal axis.

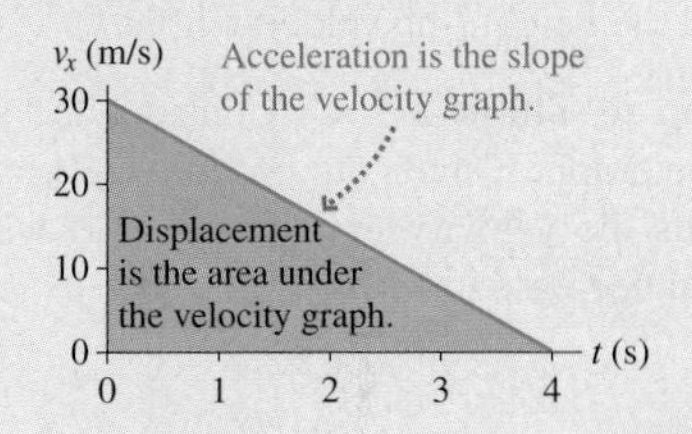

APPLICATIONS

Uniform Motion

An object in uniform motion has a constant velocity. Its velocity graph is a horizontal line; its position graph is linear.

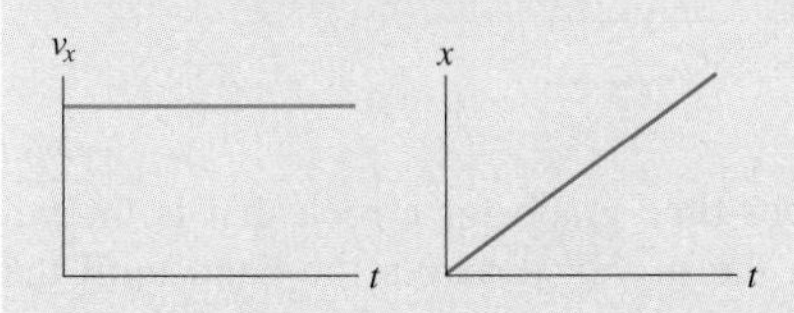

Kinematic equation for uniform motion:

$$x_f = x_i + v_x\,\Delta t$$

Uniform motion is a special case of constant-acceleration motion, with $a_x = 0$.

Motion with Constant Acceleration

An object with constant acceleration has a constantly changing velocity. Its velocity graph is linear; its position graph is a parabola.

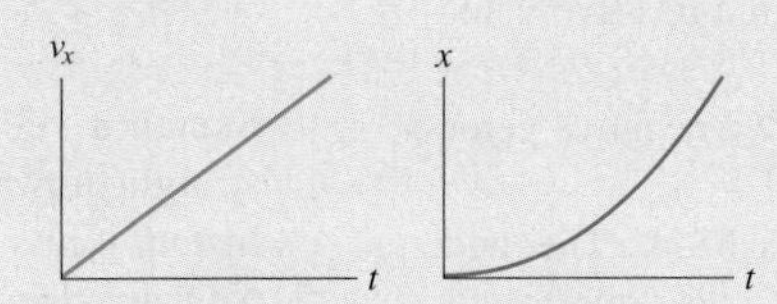

Kinematic equations for motion with constant acceleration:

$$(v_x)_f = (v_x)_i + a_x\,\Delta t$$

$$x_f = x_i + (v_x)_i\,\Delta t + \tfrac{1}{2}a_x(\Delta t)^2$$

$$(v_x)_f{}^2 = (v_x)_i{}^2 + 2a_x\,\Delta x$$

Free Fall

Free fall is a special case of constant-acceleration motion. The acceleration has magnitude $g = 9.80$ m/s^2 and is always directed vertically downward whether an object is moving up or down.

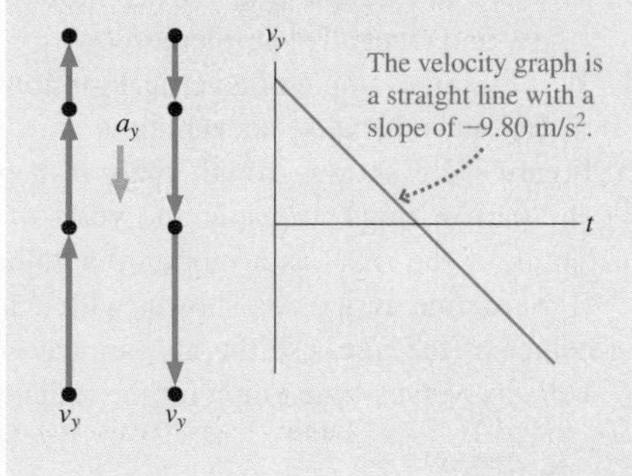

Learning Objectives After studying this chapter, you should be able to:

- Use motion diagrams to interpret motion. *Conceptual Question 2.3; Problems 2.1, 2.2, 2.59*
- Use and interpret motion graphs. *Conceptual Questions 2.5, 2.13; Problems 2.4, 2.18, 2.19, 2.22, 2.62*
- Calculate the velocity of an object. *Conceptual Question 2.9; Problems 2.8, 2.15, 2.57*
- Solve problems about an object in uniform motion. *Problems 2.9, 2.10, 2.11, 2.13, 2.58*
- Calculate the acceleration of an object. *Problems 2.25, 2.27, 2.32, 2.33, 2.72*
- Determine and interpret the sign of acceleration. *Conceptual Questions 2.2, 2.8; Problem 2.50*
- Use the problem-solving approach to solve problems of motion with constant acceleration and free fall. *Problems 2.36, 2.40, 2.41, 2.47, 2.52, 2.75*

STOP TO THINK ANSWERS

Chapter Preview Stop to Think: B. The bicycle is moving to the left, so the velocity vectors must point to the left. The speed is increasing, so successive velocity vectors must get longer.

Stop to Think 2.1: D. The motion consists of two constant-velocity phases, and the second one has a higher velocity. The correct graph has two straight-line segments, with the second one having a steeper slope.

Stop to Think 2.2: B. The displacement is the area under a velocity-versus-time graph. In all four cases, the graph is a straight line, so the area under the graph is a rectangle. The area is the product of the length and the height, so the largest displacement belongs to the graph with the largest product of the length (the time interval, in s) and the height (the velocity, in m/s).

Stop to Think 2.3: C. Consider the slope of the position-versus-time graph. It starts out positive and constant, then decreases to zero. Thus the velocity graph must start with a constant positive value, then decrease to zero.

Stop to Think 2.4: C. Acceleration is the slope of the velocity-versus-time graph. The largest magnitude of the slope is at point C.

Stop to Think 2.5: B. The elevator is moving down, so $v_y < 0$. It is slowing down, so the magnitude of the velocity is decreasing. As time goes on, the velocity graph should get closer to the origin. This means that the acceleration is positive, and the slope of the graph is positive.

Stop to Think 2.6: E. An acceleration of 1.2 m/s^2 corresponds to an increase of 1.2 m/s every second. At this rate, the cruising speed of 6.0 m/s will be reached after 5.0 s.

Stop to Think 2.7: D. The final velocity will have the same *magnitude* as the initial velocity, but the velocity is negative because the rock will be moving downward.

Video Tutor Solution Chapter 2

QUESTIONS

Conceptual Questions

1. A person gets in an elevator on the ground floor and rides it to the top floor of a building. Sketch a velocity-versus-time graph for this motion.
2. a. Give an example of a vertical motion with a positive velocity and a negative acceleration.
   b. Give an example of a vertical motion with a negative velocity and a negative acceleration.
3. Figure Q2.3 shows growth rings in a tree's trunk. The wide and narrow rings correspond to years of fast and slow growth. Think of the rings as a motion diagram for the tree's growth. If we define an axis as shown, with x measured out from the center of the tree, use the appearance of the rings to sketch a velocity-versus-time graph for the radial growth of the tree. BIO

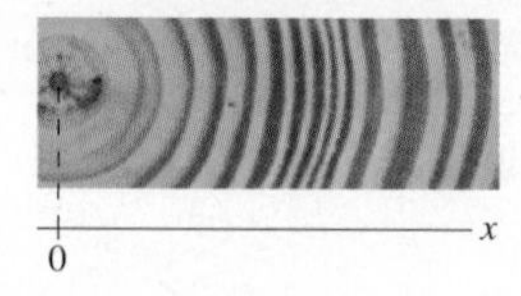

FIGURE Q2.3

4. Sketch a velocity-versus-time graph for a rock that is thrown straight upward, from the instant it leaves the hand until the instant it hits the ground.
5. You are driving down the road at a constant speed. Another car going a bit faster catches up with you and passes you. Draw a position graph for both vehicles on the same set of axes, and note the point on the graph where the other vehicle passes you.

Problem difficulty is labeled as | (straightforward) to ||||| (challenging). Problems labeled BIO are of biological or medical interest.

The eText icon indicates when there is a video tutor solution available for the chapter or for a specific problem. To launch these videos, log into your eText through Mastering™ Physics or log into the Study Area.

6. Figure Q2.6 shows the velocity-versus-time graphs for two objects A and B. Students Zach and Victoria are asked to tell stories that correspond to the motion of the objects. Zach says, "The graph could represent two cars traveling in opposite directions that pass each other." Victoria says, "No, I think they could be two rocks thrown vertically from a bridge; rock A is thrown upward and rock B is thrown downward." Which student, if either, is correct? Explain.

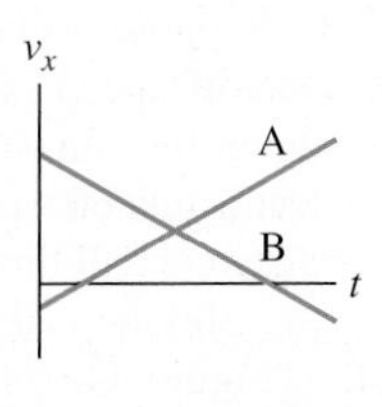

FIGURE Q2.6

7. BIO Certain animals are capable of running at great speeds; other animals are capable of tremendous accelerations. Speculate on which would be more beneficial to a predator—large maximum speed or large acceleration.
8. A ball is thrown straight up into the air. At each of the following instants, is the ball's acceleration a_y equal to g, $-g$, 0, $< g$, or $> g$?
   a. Just after leaving your hand?
   b. At the very top (maximum height)?
   c. Just before hitting the ground?
9. Janelle stands on a balcony, two stories above Michael. She throws one ball straight up and one ball straight down, but both with the same initial speed. Eventually each ball passes Michael. Which ball, if either, is moving faster when it passes Michael? Explain.
10. Figure Q2.10 shows an object's position-versus-time graph. The letters A to E correspond to various segments of the motion in which the graph has constant slope.
   a. Write a realistic motion short story for an object that would have this position graph.
   b. In which segment(s) is the object at rest?
   c. In which segment(s) is the object moving to the right?
   d. Is the speed of the object during segment C greater than, equal to, or less than its speed during segment E? Explain.

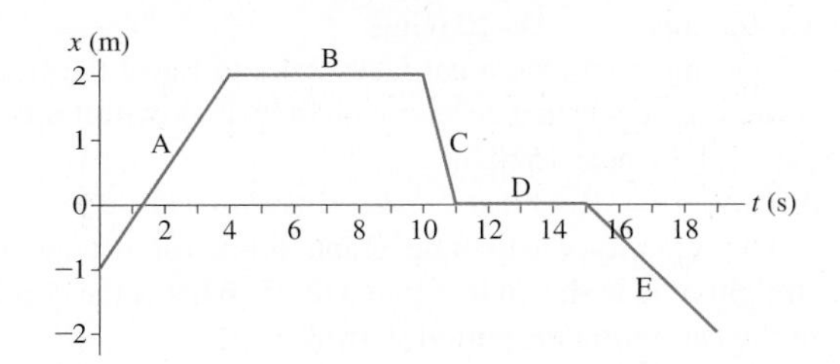

FIGURE Q2.10

11. Figure Q2.11 shows the position graph for an object moving along the horizontal axis.
   a. Write a realistic motion short story for an object that would have this position graph.
   b. Draw the corresponding velocity graph.

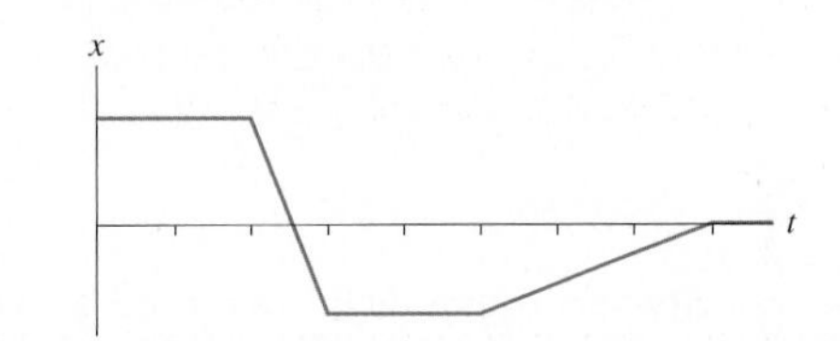

FIGURE Q2.11

12. Figure Q2.12 shows the position-versus-time graphs for two objects, A and B, that are moving along the same axis.
   a. At the instant $t = 1$ s, is the speed of A greater than, less than, or equal to the speed of B? Explain.
   b. Do objects A and B ever have the *same* speed? If so, at what time or times? Explain.

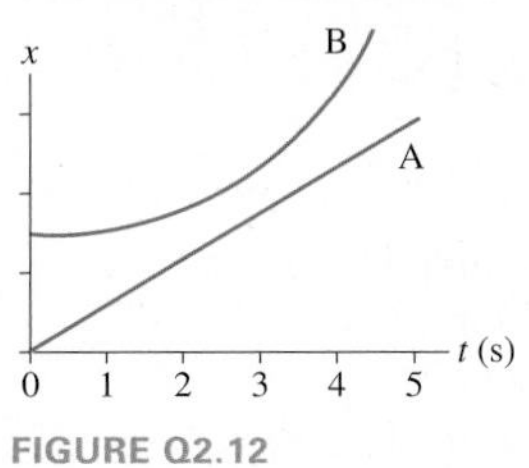

FIGURE Q2.12

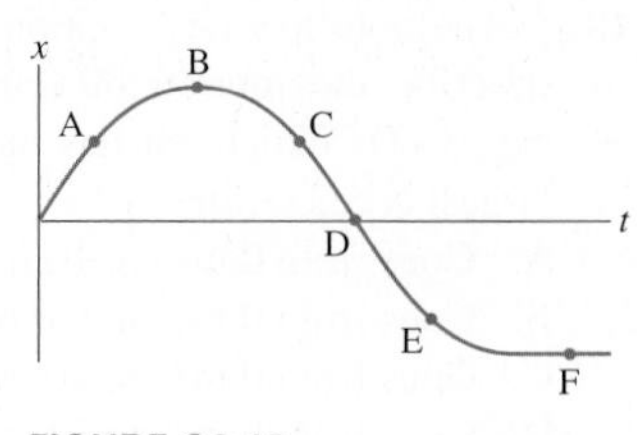

FIGURE Q2.13

13. Figure Q2.13 shows a position-versus-time graph. At which lettered point or points is the object
   a. Moving the fastest?
   b. Moving to the left?
   c. Speeding up?
   d. Slowing down?
   e. Turning around?
14. Figure Q2.14 is the velocity-versus-time graph for an object moving along the x-axis.
   a. During which segment(s) is the velocity constant?
   b. During which segment(s) is the object speeding up?
   c. During which segment(s) is the object slowing down?
   d. During which segment(s) is the object standing still?
   e. During which segment(s) is the object moving to the right?

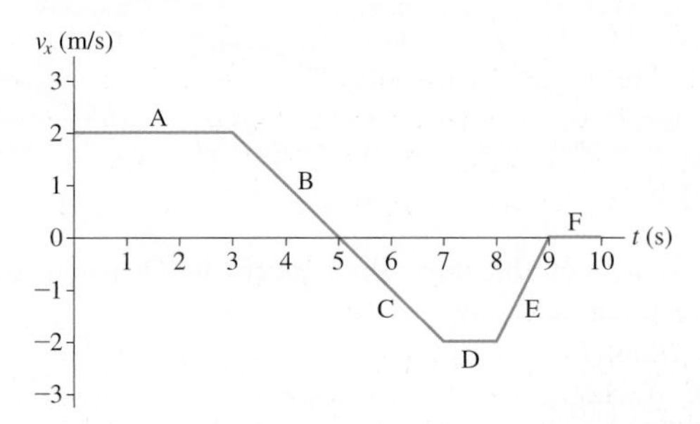

FIGURE Q2.14

Multiple-Choice Questions

15. | Figure Q2.15 shows the position graph of a car traveling on a straight road. At which labeled instant is the speed of the car greatest?

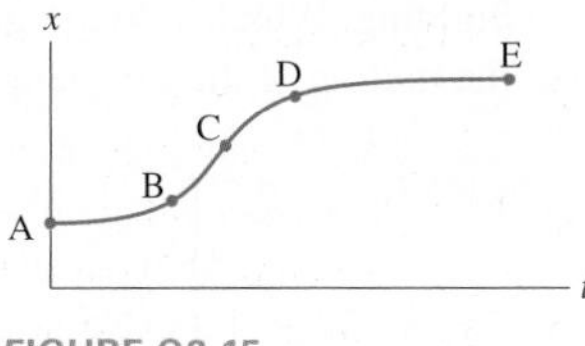

FIGURE Q2.15

16. | Figure Q2.16 shows the position graph of a car traveling on a straight road. The velocity at instant 1 is _________ and the velocity at instant 2 is _________.
   A. positive, negative
   B. positive, positive
   C. negative, negative
   D. negative, zero
   E. positive, zero

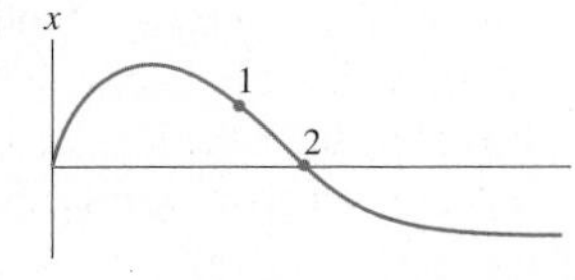

FIGURE Q2.16

17. | Figure Q2.17 shows an object's position-versus-time graph. What is the velocity of the object at $t = 6$ s?
A. 0.67 m/s B. 0.83 m/s
C. 3.3 m/s D. 4.2 m/s
E. 25 m/s

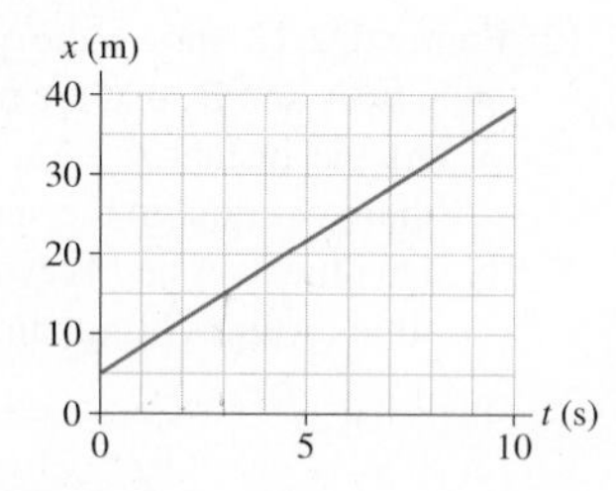

FIGURE Q2.17

18. | The following options describe the motion of four cars A–D. Which car has the largest acceleration?
A. Goes from 0 m/s to 10 m/s in 5.0 s
B. Goes from 0 m/s to 5.0 m/s in 2.0 s
C. Goes from 0 m/s to 20 m/s in 7.0 s
D. Goes from 0 m/s to 3.0 m/s in 1.0 s

19. | A car is traveling at $v_x = 20$ m/s. The driver applies the brakes, and the car slows with $a_x = -4.0$ m/s^2. What is the stopping distance?
A. 5.0 m
B. 25 m
C. 40 m
D. 50 m

20. || Velocity-versus-time graphs for three drag racers are shown in Figure Q2.20. At $t = 5.0$ s, which car has traveled the farthest?
A. Andy
B. Rachel
C. Carl
D. All have traveled the same distance

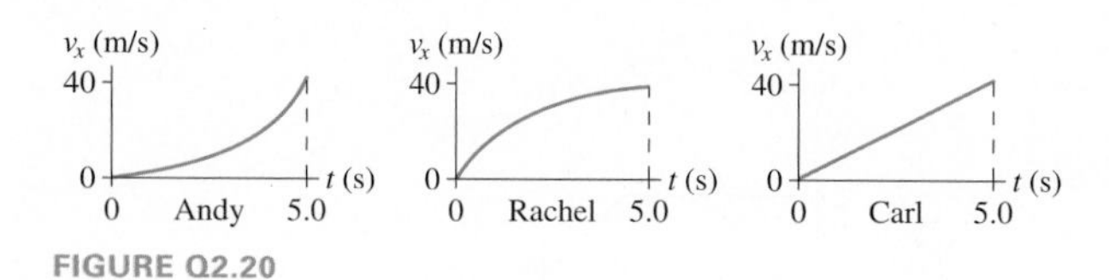

FIGURE Q2.20

21. | Which of the three drag racers in Question 20 had the greatest acceleration at $t = 0$ s?
A. Andy
B. Rachel
C. Carl
D. All had the same acceleration

22. || Chris is holding two softballs while standing on a balcony. She throws ball 1 straight up in the air and, at the same instant, releases her grip on ball 2, letting it drop over the side of the building. Which velocity graph in Figure Q2.22 best represents the motion of the two balls?

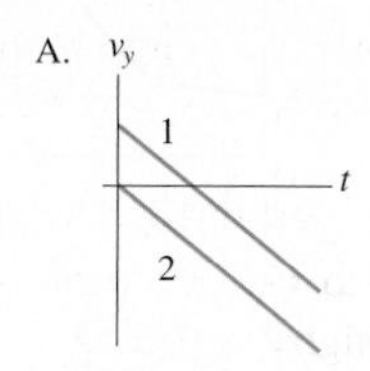

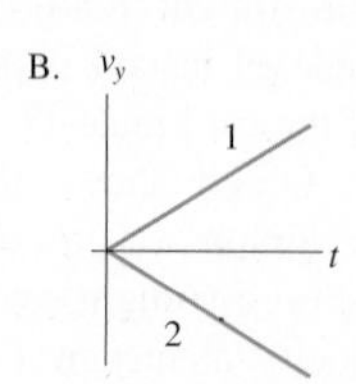

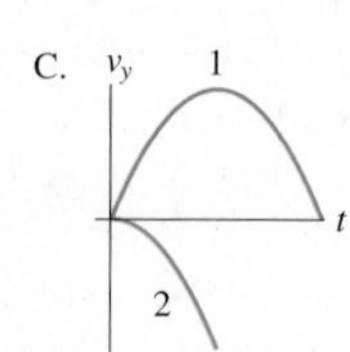

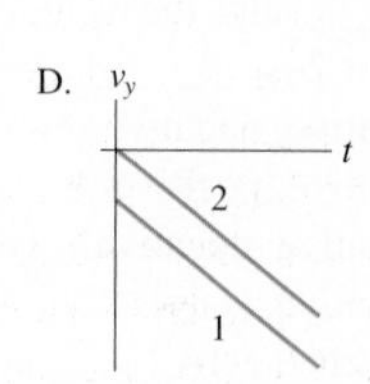

FIGURE Q2.22

23. || Suppose a plane accelerates from rest for 30 s, achieving a takeoff speed of 80 m/s after traveling a distance of 1200 m down the runway. A smaller plane with the same acceleration has a takeoff speed of 40 m/s. Starting from rest, after what distance will this smaller plane reach its takeoff speed?
A. 300 m B. 600 m C. 900 m D. 1200 m

24. || Figure Q2.24 shows a motion diagram with the clock reading (in seconds) shown at each position. From $t = 9$ s to $t = 15$ s the object is at the same position. After that, it returns along the same track. The positions of the dots for $t \geq 16$ s are offset for clarity. Which graph best represents the object's *velocity?*

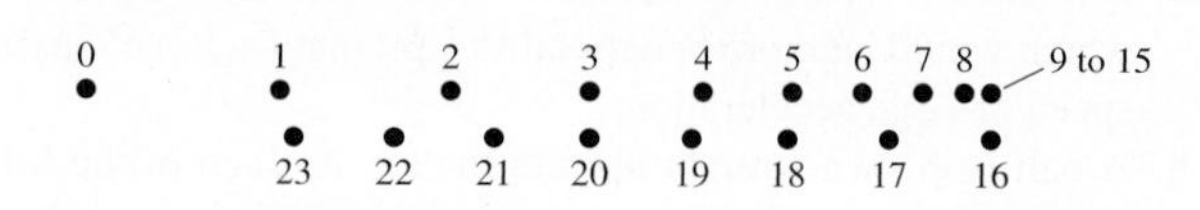

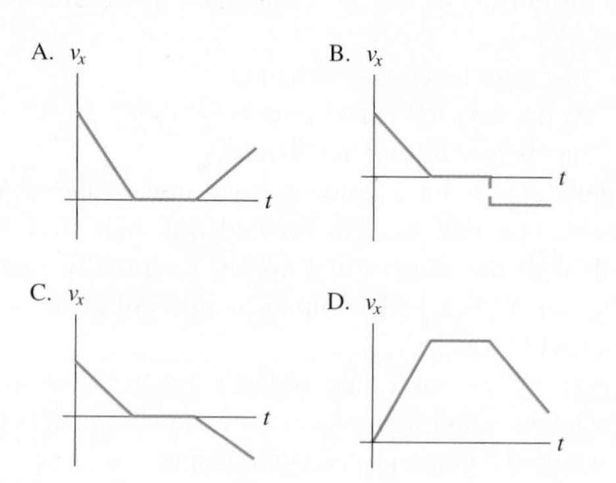

FIGURE Q2.24

25. || Nate throws a ball straight up to Kayla, who is standing on a balcony 3.8 m above Nate. When she catches it, the ball is still moving upward at a speed of 2.8 m/s. With what initial speed did Nate throw the ball?
A. 7.0 m/s B. 12.3 m/s
C. 9.1 m/s D. 10.6 m/s

26. | Starting from rest, a car takes 2.4 s to travel the first 15 m. Assuming a constant acceleration, how long will it take the car to travel the next 15 m?
A. 0.67 s B. 1.0 s C. 1.8 s D. 3.6 s

27. | The velocity-versus-time graph for a car driving down a straight road is shown in Figure Q2.27. What is the acceleration of the car during the period shown?
A. 1.0 m/s^2 B. 2.5 m/s^2
C. 3.8 m/s^2 D. 5.0 m/s^2

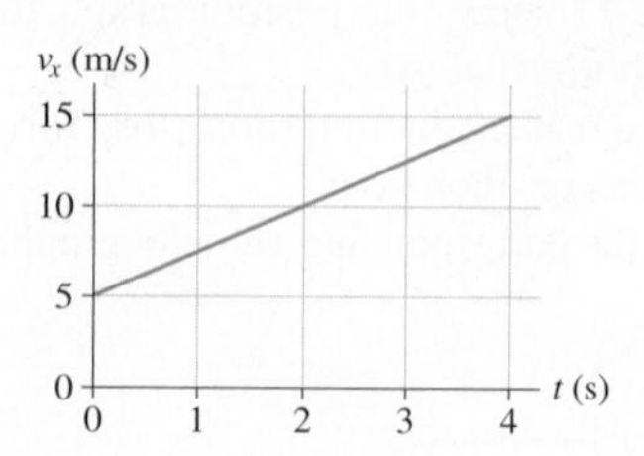

FIGURE Q2.27

28. || The velocity-versus-time graph for a car driving down a straight road is shown in Figure Q2.27. How far does the car travel during the time interval from $t = 0$ s to $t = 4.0$ s?
A. 10 m B. 20 m C. 40 m D. 60 m

PROBLEMS

Section 2.1 Describing Motion

1. ||| Figure P2.1 shows a motion diagram of a car traveling down a street. The camera took one frame every second. A distance scale is provided.
   a. Use the scale to determine the x-value of the car at each dot. Place your data in a table, similar to Table 2.1, showing each position and the instant of time at which it occurred.
   b. Make a graph of x versus t, using the data in your table. Because you have data only at certain instants of time, your graph should consist of dots that are not connected together.

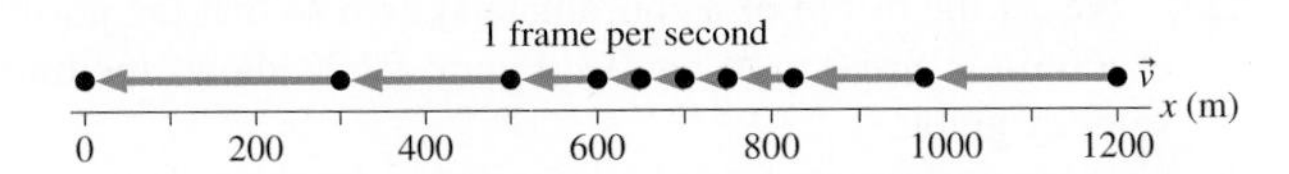

FIGURE P2.1

2. | For each motion diagram in Figure P2.2, determine the sign (positive or negative) of the position and the velocity.

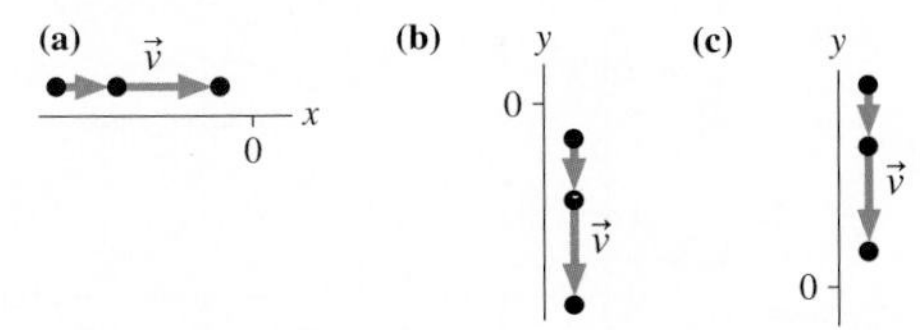

FIGURE P2.2

3. || The position graph of Figure P2.3 shows a dog slowly sneaking up on a squirrel, then putting on a burst of speed.
   a. For how many seconds does the dog move at the slower speed?
   b. Draw the dog's velocity-versus-time graph. Include a numerical scale on both axes.

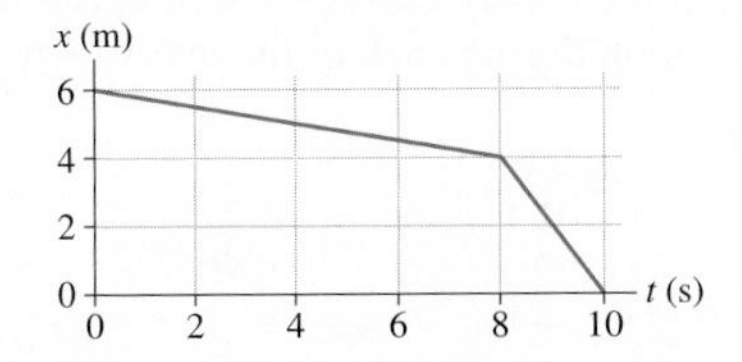

FIGURE P2.3

4. || A rural mail carrier is driving slowly, putting mail in mailboxes near the road. He overshoots one mailbox, stops, shifts into reverse, and then backs up until he is at the right spot. The velocity graph of Figure P2.4 represents his motion.
   a. Draw the mail carrier's position-versus-time graph. Assume that $x = 0$ m at $t = 0$ s.
   b. What is the position of the mailbox?

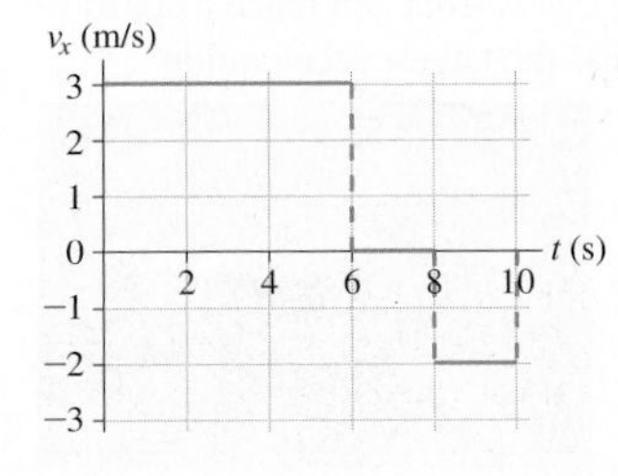

FIGURE P2.4

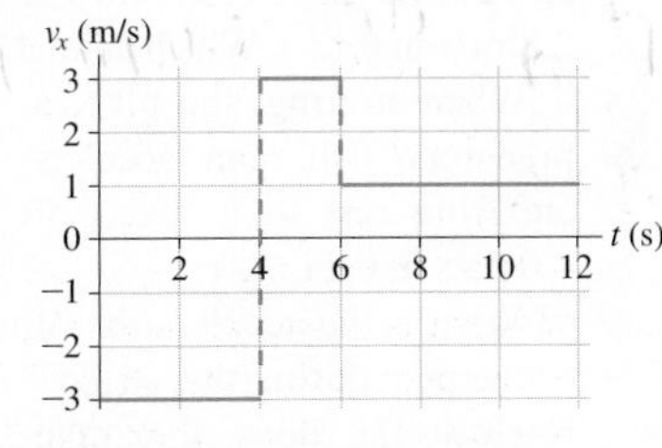

FIGURE P2.5

5. || For the velocity-versus-time graph of Figure P2.5:
   a. Draw the corresponding position-versus-time graph. Assume that $x = 0$ m at $t = 0$ s.
   b. What is the object's position at $t = 12$ s?
   c. Describe a moving object that could have these graphs.
6. || Starting at 48th Street, Dylan rides his bike due east on Meridian Road with the wind at his back. He rides for 20 min at 15 mph. He then stops for 5 min, turns around, and rides back to 48th Street; because of the headwind, his speed is only 10 mph.
   a. How long does his trip take?
   b. Assuming that the origin of his trip is at 48th Street, draw a position-versus-time graph for his trip.
7. ||| An elevator in a high-rise building goes up and down at the same speed. Starting at the ground floor, Rafael and Monica ride up five floors, a vertical rise of 20 m. The elevator stops for 10 s as Monica gets off. Rafael then goes back down two floors. Rafael's entire trip takes 24 s. Taking the origin to be at the ground floor, draw position-versus-time and velocity-versus-time graphs for Rafael's trip.
8. || A bicyclist has the position-versus-time graph shown in Figure P2.8. What is the bicyclist's velocity at $t = 10$ s, at $t = 25$ s, and at $t = 35$ s?

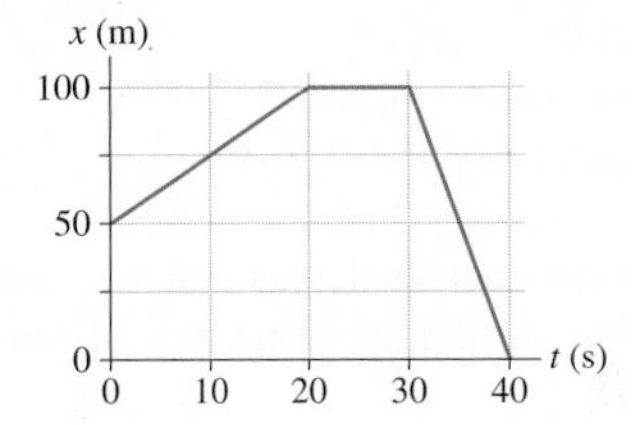

FIGURE P2.8

Section 2.2 Uniform Motion

9. | In major league baseball, the pitcher's mound is 60 feet from the batter. If a pitcher throws a 95 mph fastball, how much time elapses from when the ball leaves the pitcher's hand until the ball reaches the batter?
10. | In college softball, the distance from the pitcher's mound to the batter is 43 feet. If the ball leaves the bat at 100 mph, how much time elapses between the hit and the ball reaching the pitcher?
11. || Alan leaves Los Angeles at 8:00 AM to drive to San Francisco, 400 mi away. He travels at a steady 50 mph. Beth leaves Los Angeles at 9:00 AM and drives a steady 60 mph.
    a. Who gets to San Francisco first?
    b. How long does the first to arrive have to wait for the second?
12. || Richard is driving home to visit his parents. 125 mi of the trip are on the interstate highway where the speed limit is 65 mph. Normally Richard drives at the speed limit, but today he is running late and decides to take his chances by driving at 70 mph. How many minutes does he save?
13. ||| In a 5.00 km race, one runner runs at a steady 12.0 km/h and another runs at 14.5 km/h. How long does the faster runner have to wait at the finish line to see the slower runner cross?
14. |||| In an 8.00 km race, one runner runs at a steady 11.0 km/h and another runs at 14.0 km/h. How far from the finish line is the slower runner when the faster runner finishes the race?

15. ‖ Figure P2.15 shows actual data from Usain Bolt's 2009 word-record run in the 100 m sprint. From this graph, estimate his top speed in m/s and in mph.

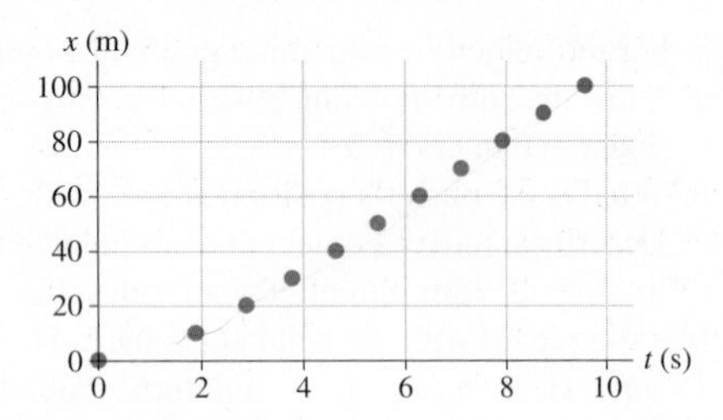

FIGURE P2.15

16. ‖ While running a marathon, a long-distance runner uses a stopwatch to time herself over a distance of 100 m. She finds that she runs this distance in 18 s. Answer the following by considering ratios, without computing her velocity.
   a. If she maintains her speed, how much time will it take her to run the next 400 m?
   b. How long will it take her to run a mile at this speed?

Section 2.3 Instantaneous Velocity

17. | Figure P2.17 shows the position graph of a particle.
   a. Draw the particle's velocity graph for the interval $0\text{ s} \le t \le 4\text{ s}$.
   b. Does this particle have a turning point or points? If so, at what time or times?

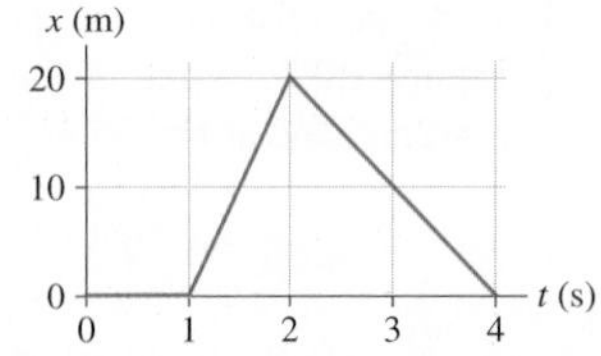

FIGURE P2.17

18. ‖ A somewhat idealized graph of the speed of the blood in
BIO the ascending aorta during one beat of the heart appears as in Figure P2.18.
   a. Approximately how far, in cm, does the blood move during one beat?
   b. Assume similar data for the motion of the blood in your aorta, and make a rough estimate of the distance from your heart to your brain. Estimate how many beats of the heart it takes for blood to travel from your heart to your brain.

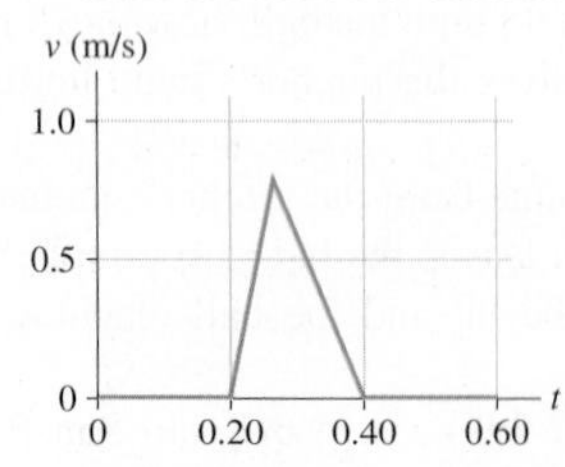

FIGURE P2.18

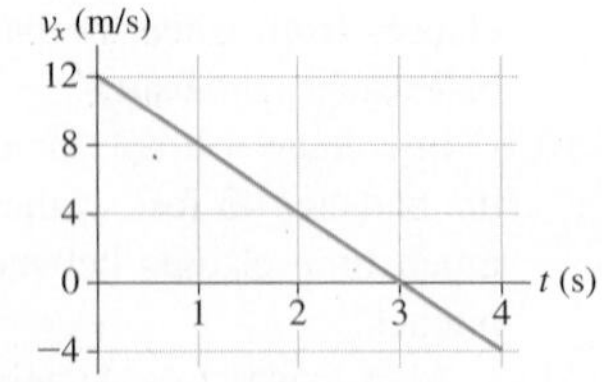

FIGURE P2.19

19. ||| A car starts from $x_i = 10$ m at $t_i = 0$ s and moves with the velocity graph shown in Figure P2.19.
   a. What is the car's position at $t = 2$ s, 3 s, and 4 s?
   b. Does this car ever change direction? If so, at what time?

20. ‖ Figure P2.20 shows a graph of actual position-versus-time data for a particular type of drag racer known as a "funny car."
   a. Estimate the car's velocity at 2.0 s.
   b. Estimate the car's velocity at 4.0 s.

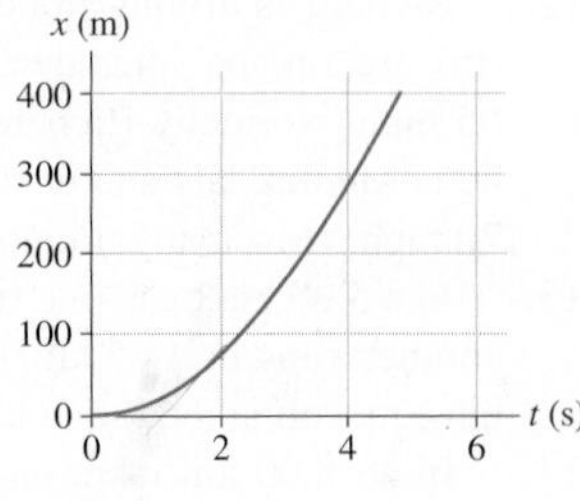

FIGURE P2.20

Section 2.4 Acceleration

21. ‖ Figure P2.21 shows the velocity graph of a bicycle. Draw the bicycle's acceleration graph for the interval $0\text{ s} \le t \le 4\text{ s}$. Give both axes an appropriate numerical scale.

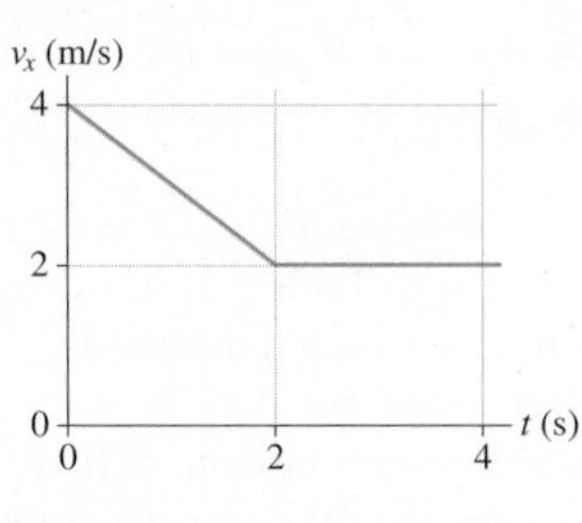

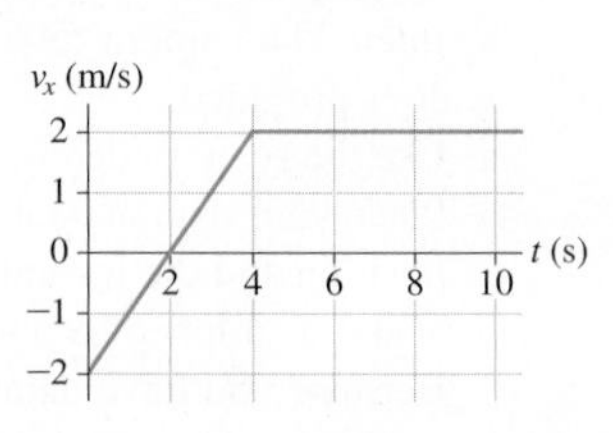

FIGURE P2.21

FIGURE P2.22

22. ||| We set the origin of a coordinate system so that the position of a train is $x = 0$ m at $t = 0$ s. Figure P2.22 shows the train's velocity graph.
   a. Draw position and acceleration graphs for the train.
   b. Find the acceleration of the train at $t = 3.0$ s.

23. ||| An object has the acceleration graph shown in Figure P2.23. Its velocity at $t = 0$ s is $v_x = 2.0$ m/s. Draw the object's velocity graph.

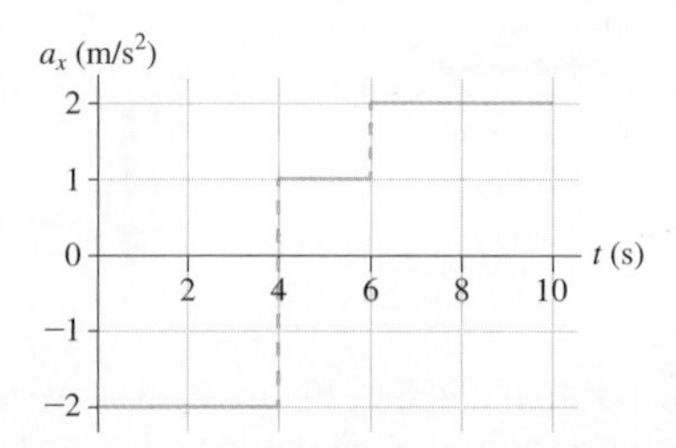

FIGURE P2.23

24. ‖ Figure P2.18 showed data for the speed of blood in the aorta.
BIO Determine the magnitude of the acceleration for both phases, speeding up and slowing down.

25. ‖ Figure P2.25 is a somewhat simplified velocity graph for Olympic sprinter Carl Lewis starting a 100 m dash. Estimate his acceleration during each of the intervals A, B, and C.

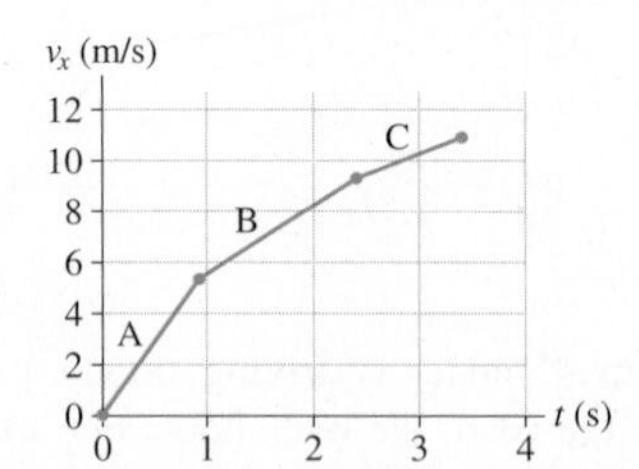

FIGURE P2.25

26. | Small frogs that are good jumpers are capable of remarkable
BIO accelerations. One species reaches a takeoff speed of 3.7 m/s in 60 ms. What is the frog's acceleration during the jump?

27. | A Thomson's gazelle can reach a speed of 13 m/s in 3.0 s. A lion
BIO can reach a speed of 9.5 m/s in 1.0 s. A trout can reach a speed of 2.8 m/s in 0.12 s. Which animal has the largest acceleration?

28. ||| When striking, the pike, a
BIO predatory fish, can accelerate from rest to a speed of 4.0 m/s in 0.11 s.
   a. What is the acceleration of the pike during this strike?
   b. How far does the pike move during this strike?

Section 2.5 Motion with Constant Acceleration

29. || a. What constant acceleration, in SI units, must a car have to go from zero to 60 mph in 10 s?
    b. What fraction of g is this?
    c. How far has the car traveled when it reaches 60 mph? Give your answer both in SI units and in feet.
30. || When jumping, a flea rapidly extends its legs, reaching a takeoff speed of 1.0 m/s over a distance of 0.50 mm.
BIO
    a. What is the flea's acceleration as it extends its legs?
    b. How long does it take the flea to leave the ground after it begins pushing off?
31. ||| In a car crash, large accelerations of the head can lead to severe injuries or even death. A driver can probably survive an acceleration of $50g$ that lasts for less than 30 ms, but in a crash with a $50g$ acceleration lasting longer than 30 ms, a driver is unlikely to survive. Imagine a collision in which a driver's head experienced a $50g$ acceleration.
BIO
    a. What is the highest speed that the car could have had such that the driver survived?
    b. What is the shortest survivable distance over which the driver's head could have come to rest?
32. || Light-rail passenger trains that provide transportation within and between cities speed up and slow down with a nearly constant (and quite modest) acceleration. A train travels through a congested part of town at 5.0 m/s. Once free of this area, it speeds up to 12 m/s in 8.0 s. At the edge of town, the driver again accelerates, with the same acceleration, for another 16 s to reach a higher cruising speed. What is the final speed?
33. ||| A cross-country skier is skiing along at a zippy 8.0 m/s. She stops pushing and simply glides along, slowing to a reduced speed of 6.0 m/s after gliding for 5.0 m. What is the magnitude of her acceleration as she slows?
34. || A small propeller airplane can comfortably achieve a high enough speed to take off on a runway that is 1/4 mile long. A large, fully loaded passenger jet has about the same acceleration from rest, but it needs to achieve twice the speed to take off. What is the minimum runway length that will serve? **Hint:** You can solve this problem using ratios without having any additional information.
35. || Formula One racers speed up much more quickly than normal passenger vehicles, and they also can stop in a much shorter distance. A Formula One racer traveling at 90 m/s can stop in a distance of 110 m. What is the magnitude of the car's acceleration as it slows during braking?
36. || Figure P2.36 shows a velocity-versus-time graph for a particle moving along the x-axis. At $t = 0$ s, assume that $x = 0$ m.
    a. What are the particle's position, velocity, and acceleration at $t = 1.0$ s?
    b. What are the particle's position, velocity, and acceleration at $t = 3.0$ s?

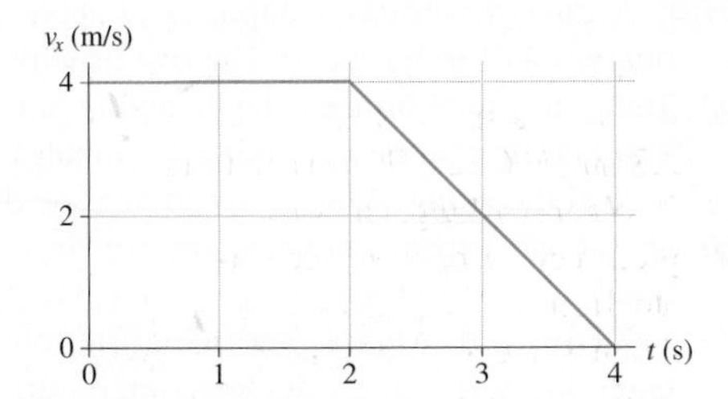

FIGURE P2.36

Section 2.6 Solving One-Dimensional Motion Problems

37. || A driver has a reaction time of 0.50 s, and the maximum deceleration of her car is 6.0 m/s^2. She is driving at 20 m/s when suddenly she sees an obstacle in the road 50 m in front of her. Can she stop the car in time to avoid a collision?
38. || Chameleons catch insects with their tongues, which they can rapidly extend to great lengths. In a typical strike, the chameleon's tongue accelerates at a remarkable 250 m/s^2 for 20 ms, then travels at constant speed for another 30 ms. During this total time of 50 ms, 1/20 of a second, how far does the tongue reach?
BIO
39. ||| You're driving down the highway late one night at 20 m/s when a deer steps onto the road 35 m in front of you. Your reaction time before stepping on the brakes is 0.50 s, and the maximum deceleration of your car is 10 m/s^2.
    a. How much distance is between you and the deer when you come to a stop?
    b. What is the maximum speed you could have and still not hit the deer?
40. || Upon impact, bicycle helmets compress, thus lowering the potentially dangerous acceleration experienced by the head. A new kind of helmet uses an airbag that deploys from a pouch worn around the rider's neck. In tests, a headform wearing the inflated airbag is dropped onto a rigid platform; the speed just before impact is 6.0 m/s. Upon impact, the bag compresses its full 12.0 cm thickness, slowing the headform to rest. What is the acceleration, in g's, experienced by the headform? (An acceleration greater than $60g$ is considered especially dangerous.)
BIO

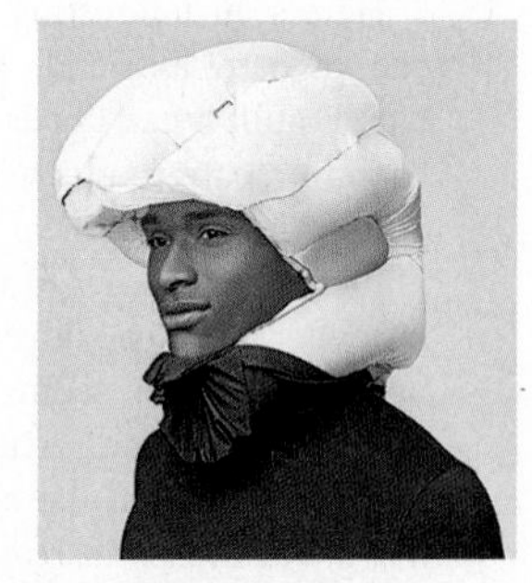

41. || A car is traveling at a steady 80 km/h in a 50 km/h zone. A police motorcycle takes off at the instant the car passes it, accelerating at a steady 8.0 m/s^2.
    a. How much time elapses before the motorcycle is moving as fast as the car?
    b. How far is the motorcycle from the car when it reaches this speed?
42. ||| The velocity-versus-time graph for the vertical jump of a green leafhopper, a small insect, is shown in Figure P2.42. This insect is unusual because it jumps with nearly constant acceleration.
BIO
    a. Estimate the leafhopper's acceleration.
    b. About how far does it move during this phase of its jump?

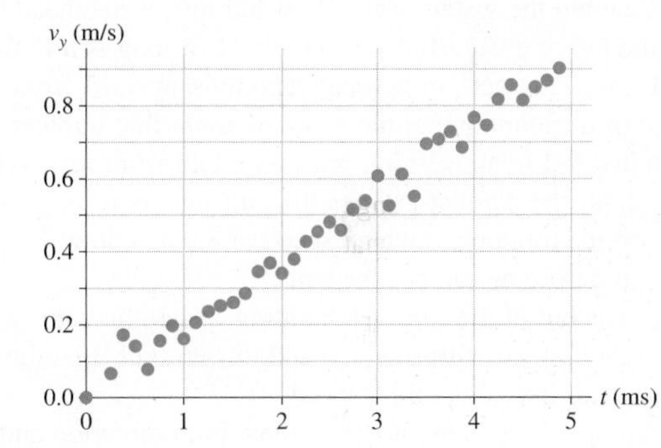

FIGURE P2.42

43. ||| A simple model for a person running the 100 m dash is to assume the sprinter runs with constant acceleration until reaching top speed, then maintains that speed through the finish line. If a sprinter reaches his top speed of 11.2 m/s in 2.14 s, what will be his total time?

Section 2.7 Free Fall

44. ||| Scientists have investigated how quickly hoverflies start beating their wings when dropped both in complete darkness and in a lighted environment. Starting from rest, the insects were dropped from the top of a 40-cm-tall box. In the light, those flies that began flying 200 ms after being dropped avoided hitting the bottom of the box 80% of the time, while those in the dark avoided hitting only 22% of the time. BIO
    a. How far would a fly have fallen in the 200 ms before it began to beat its wings?
    b. How long would it take for a fly to hit the bottom if it never began to fly?
45. || Here's an interesting challenge you can give to a friend. Hold a \$1 (or larger!) bill by an upper corner. Have a friend prepare to pinch a lower corner, putting her fingers near but not touching the bill. Tell her to try to catch the bill when you drop it by simply closing her fingers. This seems like it should be easy, but it's not. After she sees that you have released the bill, it will take her about 0.25 s to react and close her fingers—which is not fast enough to catch the bill. How much time does it take for the bill to fall beyond her grasp? The length of a bill is 16 cm. BIO

46. ||| In the preceding problem we saw that a person's reaction time is generally not quick enough to allow the person to catch a \$1 bill dropped between the fingers. The 16 cm length of the bill passes through a student's fingers before she can grab it if she has a typical 0.25 s reaction time. How long would a bill need to be for her to have a good chance of catching it? BIO
47. | A gannet is a seabird that fishes by diving from a great height. If a gannet hits the water at 32 m/s (which they do), what height did it dive from? Assume that the gannet was motionless before starting its dive.
48. |||| Steelhead trout migrate upriver to spawn. Occasionally they need to leap up small waterfalls to continue their journey. Fortunately, steelhead are remarkable jumpers, capable of leaving the water at a speed of 8.0 m/s. BIO
    a. What is the maximum height that a steelhead can jump?
    b. Leaving the water vertically at 8.0 m/s, a steelhead lands on the top of a waterfall 1.8 m high. How long is it in the air?
49. ||| In a circus act, an acrobat rebounds upward from the surface of a trampoline at the exact moment that another acrobat, perched 9.0 m above him, releases a ball from rest. While still in flight, the acrobat catches the ball just as it reaches him. If he left the trampoline with a speed of 8.0 m/s, how long is he in the air before he catches the ball?
50. | A student at the top of a building of height h throws ball A straight upward with speed v_0 and throws ball B straight downward with the same initial speed.
    a. Compare the balls' accelerations, both direction and magnitude, immediately after they leave her hand. Is one acceleration larger than the other? Or are the magnitudes equal?
    b. Compare the final speeds of the balls as they reach the ground. Is one larger than the other? Or are they equal?
51. || Excellent human jumpers can leap straight up to a height of 110 cm off the ground. To reach this height, with what speed would a person need to leave the ground? BIO
52. || A football is kicked straight up into the air; it hits the ground 5.2 s later.
    a. What was the greatest height reached by the ball? Assume it is kicked from ground level.
    b. With what speed did it leave the kicker's foot?
53. ||||| In an action movie, the villain is rescued from the ocean by grabbing onto the ladder hanging from a helicopter. He is so intent on gripping the ladder that he lets go of his briefcase of counterfeit money when he is 130 m above the water. If the briefcase hits the water 6.0 s later, what was the speed at which the helicopter was ascending?
54. ||| Spud Webb was, at 5 ft 8 in, one of the shortest basketball players to play in the NBA. But he had an amazing vertical leap; he could jump to a height of 1.1 m off the ground, so he could easily dunk a basketball. For such a leap, what was his "hang time"—the time spent in the air after leaving the ground and before touching down again? BIO
55. |||| A rock climber stands on top of a 50-m-high cliff overhanging a pool of water. He throws two stones vertically downward 1.0 s apart and observes that they cause a single splash. The initial speed of the first stone was 2.0 m/s.
    a. How long after the release of the first stone does the second stone hit the water?
    b. What was the initial speed of the second stone?
    c. What is the speed of each stone as it hits the water?

General Problems

56. ||| Actual velocity data for a lion pursuing prey are shown in Figure P2.56. Estimate: BIO
    a. The initial acceleration of the lion.
    b. The acceleration of the lion at 2 s and at 4 s.
    c. The distance traveled by the lion between 0 s and 8 s.

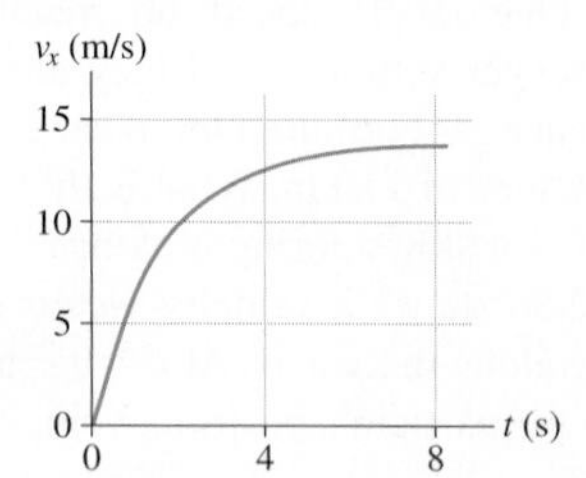

FIGURE P2.56

57. || A truck driver has a shipment of apples to deliver to a destination 440 miles away. The trip usually takes him 8 hours. Today he finds himself daydreaming and realizes 120 miles into his trip that he is running 15 minutes later than his usual pace at this point. At what speed must he drive for the remainder of the trip to complete the trip in the usual amount of time?
58. || Jenny and Alyssa are members of the cross-country team. On a training run, Jenny starts off and runs at a constant 3.8 m/s. Alyssa starts 15 s later and runs at a constant 4.0 m/s. At what time after Jenny's start does Alyssa catch up with Jenny?

59. ‖ Figure P2.59 shows the motion diagram, made at two frames of film per second, of a ball rolling along a track. The track has a 3.0-m-long sticky section.
    a. Use the scale to determine the positions of the center of the ball. Place your data in a table, similar to Table 2.1, showing each position and the instant of time at which it occurred.
    b. Make a graph of x versus t for the ball. Because you have data only at certain instants of time, your graph should consist of dots that are not connected together.
    c. What is the *change* in the ball's position from $t = 0$ s to $t = 1.0$ s?
    d. What is the *change* in the ball's position from $t = 2.0$ s to $t = 4.0$ s?
    e. What is the ball's velocity before reaching the sticky section?
    f. What is the ball's velocity after passing the sticky section?
    g. Determine the ball's acceleration on the sticky section of the track.

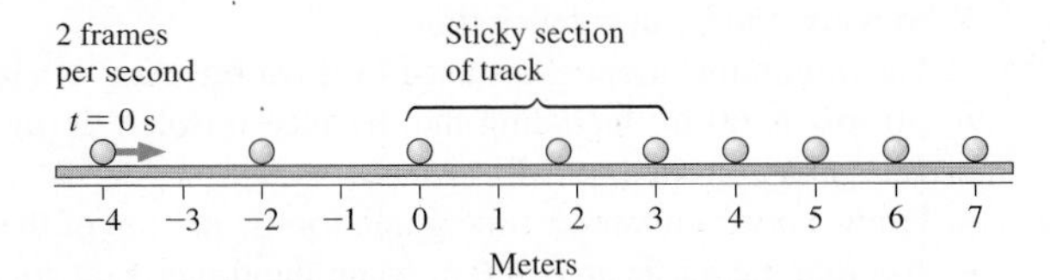

FIGURE P2.59

60. ||| In a 5000 m race, the athletes run $12\frac{1}{2}$ laps; each lap is 400 m. Kara runs the race at a constant pace and finishes in 17.5 min. Hannah runs the race in a blistering 15.3 min, so fast that she actually passes Kara during the race. How many laps has Hannah run when she passes Kara?

61. ‖ The takeoff speed for an Airbus A320 jetliner is 80 m/s. Velocity data measured during takeoff are as shown in the table.
    a. What is the jetliner's acceleration during takeoff, in m/s^2 and in g's?
    b. At what time do the wheels leave the ground?
    c. For safety reasons, in case of an aborted takeoff, the length of the runway must be three times the takeoff distance. What is the minimum length runway this aircraft can use?

t (s)	v_x (m/s)
0	0
10	23
20	46
30	69

62. |||| Does a real automobile have constant acceleration? Measured data for a Porsche 944 Turbo at maximum acceleration are as shown in the table.
    a. Convert the velocities to m/s, then make a graph of velocity versus time. Based on your graph, is the acceleration constant? Explain.
    b. Estimate how far the car traveled in the first 10 s.
    c. Draw a smooth curve through the points on your graph, then use your graph to *estimate* the car's acceleration at 2.0 s and 8.0 s. Give your answer in SI units. **Hint:** Remember that acceleration is the slope of the velocity graph.

t (s)	v_x (mph)
0	0
2	41
4	66
6	83
8	97
10	110

63. ‖ Scientists have studied two species of sand lizards, the Mojave fringe-toed lizard and the western zebra-tailed lizard, to understand the extent to which the different structure of the two species' toes is related to their preferred habitats—fine sand for the Mojave lizard and coarse sand for the zebra-tailed lizard. Figure P2.63 shows a somewhat simplified velocity-versus-time graph for the Mojave fringe-toed lizard.

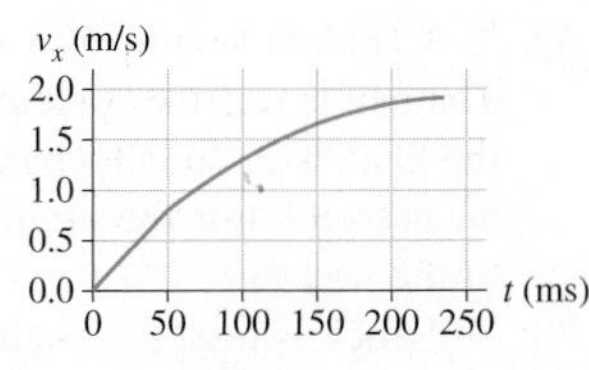

FIGURE P2.63

    a. Estimate the maximum acceleration of the lizard in both m/s^2 and g's.
    b. Estimate its acceleration at $t = 150$ ms.
    c. Estimate how far it travels in the first 50 ms.

64. ||| You are driving to the grocery store at 20 m/s. You are 110 m from an intersection when the traffic light turns red. Assume that your reaction time is 0.70 s and that your car brakes with constant acceleration.
    a. How far are you from the intersection when you begin to apply the brakes?
    b. What acceleration will bring you to rest right at the intersection?
    c. How long does it take you to stop?

65. | When you blink your eye, the upper lid goes from rest with your eye open to completely covering your eye in a time of 0.024 s.
BIO
    a. Estimate the distance that the top lid of your eye moves during a blink.
    b. What is the acceleration of your eyelid? Assume it to be constant.
    c. What is your upper eyelid's final speed as it hits the bottom eyelid?

66. ||| A bush baby, an African primate, is capable of a remarkable vertical leap. The bush baby goes into a crouch and extends its legs, pushing upward for a distance of 0.16 m. After this upward acceleration, the bush baby leaves the ground and travels upward for 2.3 m. What is the acceleration during the pushing-off phase? Give your answer in m/s^2 and in g's.
BIO

67. |||| When jumping, a flea reaches a takeoff speed of 1.0 m/s over a distance of 0.50 mm.
BIO
    a. What is the flea's acceleration during the jump phase?
    b. How long does the acceleration phase last?
    c. If the flea jumps straight up, how high will it go? (Ignore air resistance for this problem; in reality, air resistance plays a large role, and the flea will not reach this height.)

68. ||| Certain insects can achieve seemingly impossible accelerations while jumping. The click beetle accelerates at an astonishing 400g over a distance of 0.60 cm as it rapidly bends its thorax, making the "click" that gives it its name.
BIO

    a. Assuming the beetle jumps straight up, at what speed does it leave the ground?
    b. How much time is required for the beetle to reach this speed?
    c. Ignoring air resistance, how high would it go?

Watch Video Solution Problem 2.75

69. IIII A student standing on the ground throws a ball straight up. The ball leaves the student's hand with a speed of 15 m/s when the hand is 2.0 m above the ground. How long is the ball in the air before it hits the ground? (The student moves her hand out of the way.)
70. IIII A rock is tossed straight up with a speed of 20 m/s. When it returns, it falls into a hole 10 m deep.
   a. What is the rock's velocity as it hits the bottom of the hole?
   b. How long is the rock in the air, from the instant it is released until it hits the bottom of the hole?
71. IIII In springboard diving, the diver strides out to the end of the board, takes a jump onto its end, and uses the resultant spring-like nature of the board to help propel him into the air. Assume that the diver's motion is essentially vertical. He leaves the board, which is 3.0 m above the water, with a speed of 6.3 m/s.
   a. How long is the diver in the air, from the moment he leaves the board until he reaches the water?
   b. What is the speed of the diver when he reaches the water?
72. III Haley is driving down a straight highway at 75 mph. A construction sign warns that the speed limit will drop to 55 mph in 0.50 mi. What constant acceleration (in m/s) will bring Haley to this lower speed in the distance available?
73. IIII A car starts from rest at a stop sign. It accelerates at 2.0 m/s^2 for 6.0 seconds, coasts for 2.0 s, and then slows down at a rate of 1.5 m/s^2 for the next stop sign. How far apart are the stop signs?
74. III Chameleons can rapidly project their very long tongues to catch nearby insects. The tongue of the tiny Rosette-nosed chameleon has the highest acceleration of a body part of any amniote (reptile, bird, or mammal) ever measured. In a somewhat simplified model of its tongue motion, the tongue, starting from rest, first undergoes a constant-acceleration phase with an astounding magnitude of 2500 m/s^2. This acceleration brings the tongue up to a final speed of 5.0 m/s. It continues at this speed for 22 ms until it hits its target. BIO
   a. How long does the acceleration phase last?
   b. What is the total distance traveled by the chameleon's tongue?
75. III Heather and Jerry are standing on a bridge 50 m above a river. Heather throws a rock straight down with a speed of 20 m/s. Jerry, at exactly the same instant of time, throws a rock straight up with the same speed. Ignore air resistance.
   a. How much time elapses between the first splash and the second splash?
   b. Which rock has the faster speed as it hits the water?
76. II A Thomson's gazelle can run at very high speeds, but its acceleration is relatively modest. A reasonable model for the sprint of a gazelle assumes an acceleration of 4.2 m/s^2 for 6.5 s, after which the gazelle continues at a steady speed. BIO
   a. What is the gazelle's top speed?
   b. A human would win a very short race with a gazelle. The best time for a 30 m sprint for a human runner is 3.6 s. How much time would the gazelle take for a 30 m race?
   c. A gazelle would win a longer race. The best time for a 200 m sprint for a human runner is 19.3 s. How much time would the gazelle take for a 200 m race?
77. III We've seen that a man's higher initial acceleration means that he can outrun a horse in very short race. A simple—but plausible—model for a sprint by a man and a horse uses these assumptions: The man accelerates at 6.0 m/s^2 for 1.8 s and then runs at a constant speed. A horse accelerates at 5.0 m/s^2 but continues accelerating for 4.8 s and then continues at a constant speed. A man and a horse are competing in a 200 m race. The man is given a 100 m head start, so he begins 100 m from the finish line. How much time does the man take to complete the race? How much time does the horse take? Who wins the race? BIO
78. II A pole-vaulter is nearly motionless as he clears the bar, set 4.2 m above the ground. He then falls onto a thick pad. The top of the pad is 80 cm above the ground, and it compresses by 50 cm as he comes to rest. What is his acceleration as he comes to rest on the pad?
79. II A Porsche challenges a Honda to a 400 m race. Because the Porsche's acceleration of 3.5 m/s^2 is larger than the Honda's 3.0 m/s^2, the Honda gets a 100-m head start—it is only 300 m from the finish line. Assume, somewhat unrealistically, that both cars can maintain these accelerations the entire distance. Who wins, and by how much time?
80. IIIII The minimum stopping distance for a car traveling at a speed of 30 m/s is 60 m, including the distance traveled during the driver's reaction time of 0.50 s.
   a. Draw a position-versus-time graph for the motion of the car. Assume the car is at $x_i = 0$ m when the driver first sees the emergency situation ahead that calls for a rapid halt.
   b. What is the minimum stopping distance for the same car traveling at a speed of 40 m/s?
81. IIIII A rocket is launched straight up with constant acceleration. Four seconds after liftoff, a bolt falls off the side of the rocket. The bolt hits the ground 6.0 s later. What was the rocket's acceleration?

MCAT-Style Passage Problems

Free Fall on Different Worlds

Objects in free fall on the earth have acceleration $a_y = -9.8\ m/s^2$. On the moon, free-fall acceleration is approximately 1/6 of the acceleration on earth. This changes the scale of problems involving free fall. For instance, suppose you jump straight upward, leaving the ground with velocity v_i and then steadily slowing until reaching zero velocity at your highest point. Because your initial velocity is determined mostly by the strength of your leg muscles, we can assume your initial velocity would be the same on the moon. But considering the final equation in Synthesis 2.1 we can see that, with a smaller free-fall acceleration, your maximum height would be greater. The following questions ask you to think about how certain athletic feats might be performed in this reduced-gravity environment.

82. I If an astronaut can jump straight up to a height of 0.50 m on earth, how high could he jump on the moon?
   A. 1.2 m B. 3.0 m C. 3.6 m D. 18 m
83. I On the earth, an astronaut can safely jump to the ground from a height of 1.0 m; her velocity when reaching the ground is slow enough to not cause injury. From what height could the astronaut safely jump to the ground on the moon?
   A. 2.4 m B. 6.0 m C. 7.2 m D. 36 m
84. I On the earth, an astronaut throws a ball straight upward; it stays in the air for a total time of 3.0 s before reaching the ground again. If a ball were to be thrown upward with the same initial speed on the moon, how much time would pass before it hit the ground?
   A. 7.3 s B. 18 s C. 44 s D. 108 s

3 Vectors and Motion in Two Dimensions

Once the leopard jumps, its trajectory is fixed by the initial speed and angle of the jump. How can we work out where the leopard will land?

LOOKING AHEAD ➤

Vectors and Components

The dark green vector represents the basketball's velocity. The light green **component vectors** show its horizontal and vertical velocities.

You'll learn how to find components of vectors and how to use these components to solve problems.

Projectile Motion

A leaping fish's parabolic arc is an example of **projectile motion.** The details are the same for a fish or a basketball.

You'll see how to solve projectile motion problems, determining how long an object is in the air and how far it travels.

Circular Motion

These riders move in a circle at a constant speed, but they have an acceleration because the direction of their motion is constantly changing.

You'll learn how to determine the magnitude and the direction of the acceleration for an object in circular motion.

GOAL To learn more about vectors and to use vectors as a tool to analyze motion in two dimensions.

LOOKING BACK ◄

Free Fall

You learned in Section 2.7 that an object tossed straight up is in free fall. The acceleration is the same whether the object is going up or coming back down.

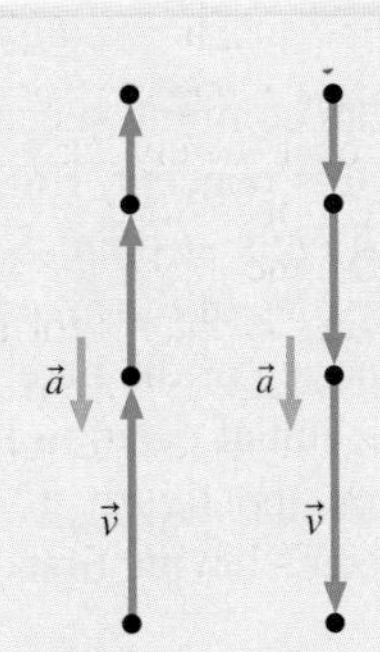

For an object in projectile motion, the vertical component of the motion is also free fall. You'll use your knowledge of free fall to solve projectile motion problems.

STOP TO THINK

A player kicks a soccer ball straight up into the air. The ball takes 2.0 s to reach its highest point. Approximately how fast was the ball moving when it left the player's foot?

A. 5 m/s B. 10 m/s
C. 15 m/s D. 20 m/s

3.1 Using Vectors

In Chapter 2, we solved problems in which an object moved in a straight-line path. In this chapter, objects move in curving paths, so that their motion is in two dimensions. A ball arcing through the air is an example of this two-dimensional motion, because it is moving horizontally and vertically at the same time. Since the direction of motion will be so important, we need to develop an appropriate mathematical language to describe it—the language of vectors.

FIGURE 3.1 The velocity vector $\vec{v}$ has both a magnitude and a direction.

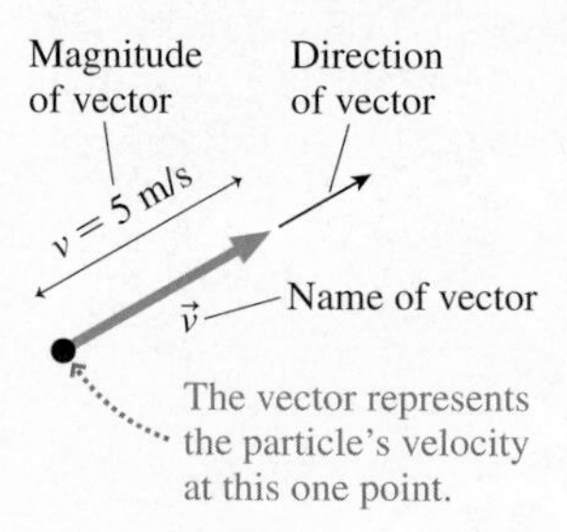

We introduced the concept of a vector in ◀◀ SECTION 1.6. In the next few sections we will develop techniques for working with vectors as a tool for studying motion in two dimensions. As we learned, a vector is a quantity with both a size (magnitude) and a direction. FIGURE 3.1 shows how to represent a particle's velocity as a vector $\vec{v}$. The particle's speed at this point is 5 m/s *and* it is moving in the direction indicated by the arrow. Recall that the magnitude of a vector is represented by the letter symbol of the vector, but without an arrow. In this case, the particle's speed—the magnitude of the velocity vector $\vec{v}$—is $v = 5$ m/s. The magnitude of a vector, a *scalar* quantity, cannot be a negative number.

NOTE ▶ Although the vector arrow is drawn across the page, from its tail to its tip, this arrow does *not* indicate that the vector "stretches" across this distance. Instead, the arrow tells us the value of the vector quantity only at the one point where the tail of the vector is placed. ◀

FIGURE 3.2 Displacement vectors.

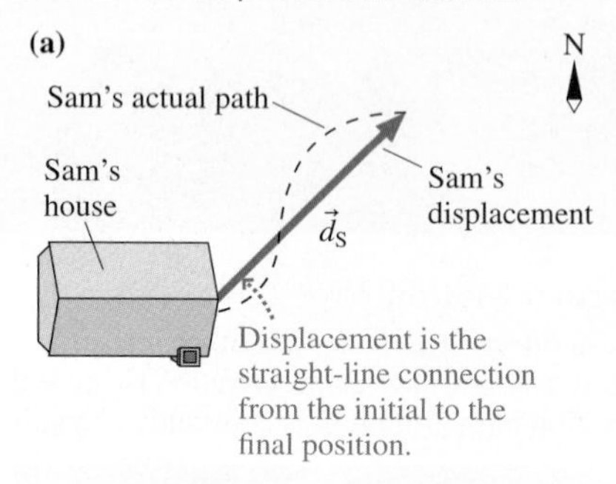

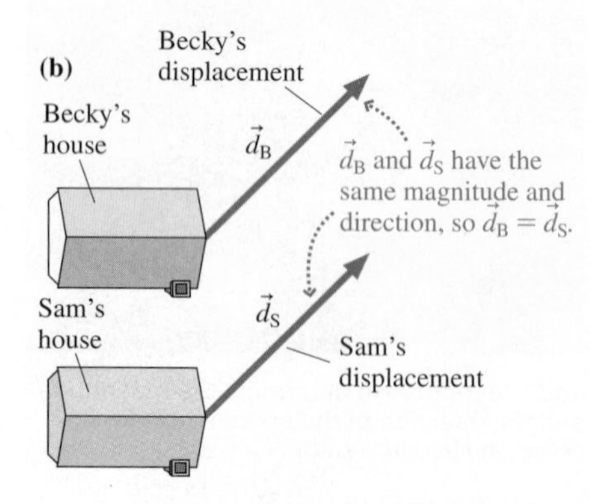

We saw in Chapter 1 that the displacement of an object is a vector drawn from its initial position to its position at some later time. Because displacement is an easy concept to think about, we will use it to introduce some of the properties of vectors. However, all the properties we will discuss in this chapter **(addition, subtraction, multiplication, components) apply to all types of vectors, not just to displacement.**

Suppose that Sam, our old friend from Chapter 1, starts from his front door, walks across the street, and ends up 200 ft to the northeast of where he started. Sam's displacement, which we will label $\vec{d}_S$, is shown in FIGURE 3.2a. The displacement vector is a straight-line connection from his initial to his final position, not necessarily his actual path. The dashed line indicates a possible route Sam might have taken, but his displacement is the vector $\vec{d}_S$ from his initial to his final position.

To describe a vector we must specify both its magnitude and its direction. We can write Sam's displacement as

$$\vec{d}_S = (200 \text{ ft, northeast})$$

where the first number specifies the magnitude and the second item gives the direction. The magnitude of Sam's displacement is $d_S = 200$ ft, the distance between his initial and final points.

Sam's next-door neighbor Becky also walks 200 ft to the northeast, starting from her own front door. Becky's displacement $\vec{d}_B = (200 \text{ ft, northeast})$ has the same magnitude and direction as Sam's displacement $\vec{d}_S$. Because vectors are defined by their magnitude and direction, **two vectors are equal if they have the same magnitude and direction.** This is true regardless of the individual starting points of the vectors. Thus the two displacements in FIGURE 3.2b are equal to each other, and we can write $\vec{d}_B = \vec{d}_S$.

FIGURE 3.3 The net displacement $\vec{C}$ resulting from two displacements $\vec{A}$ and $\vec{B}$.

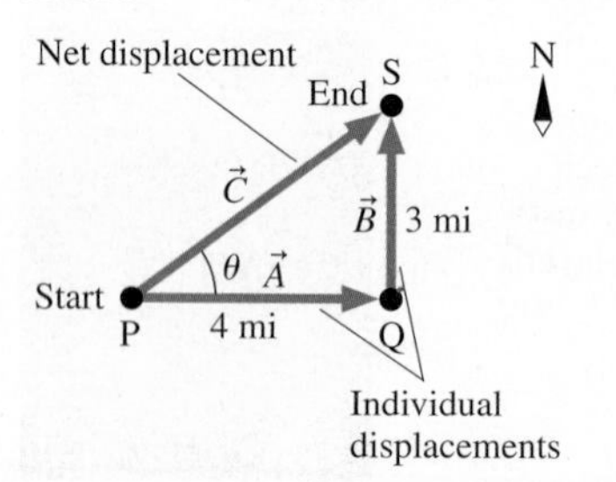

Vector Addition

As we saw in Chapter 1, we can combine successive displacements by vector addition. Let's review and extend this concept. FIGURE 3.3 shows the displacement of a hiker who starts at point P and ends at point S. She first hikes 4 miles to the east, then 3 miles to the north. The first leg of the hike is described by the displacement vector $\vec{A} = (4 \text{ mi, east})$. The second leg of the hike has displacement $\vec{B} = (3 \text{ mi, north})$. By definition, a vector from her initial position P to her final position S is also a displacement. This is vector $\vec{C}$ on the figure. $\vec{C}$ is the *net displacement* because it describes the net result of the hiker's having first displacement $\vec{A}$, then displacement $\vec{B}$.

The word "net" implies addition. The net displacement $\vec{C}$ is an initial displacement $\vec{A}$ *plus* a second displacement $\vec{B}$, or

$$\vec{C} = \vec{A} + \vec{B} \tag{3.1}$$

The sum of two vectors is called the **resultant vector.** Vector addition is commutative: $\vec{A} + \vec{B} = \vec{B} + \vec{A}$. You can add vectors in any order you wish.

Look back at ◀◀ TACTICS BOX 1.4 to review the three-step procedure for adding two vectors. This *tip-to-tail* method for adding vectors, which is used to find $\vec{C} = \vec{A} + \vec{B}$ in Figure 3.3, is called *graphical addition.*

There is a second method for adding vectors that is sometimes more convenient than the tip-to-tail method. FIGURE 3.4a shows the addition of two vectors $\vec{D}$ and $\vec{E}$ using the tip-to-tail rule. But recall that a vector such as $\vec{E}$ is not fixed in place on the diagram; it can be "slid" to any other location as long as its magnitude and direction remain unchanged. In particular, it can be moved so that its tail coincides with the tail of $\vec{D}$, as shown in FIGURE 3.4b. In this case, you can see that the vector sum $\vec{D} + \vec{E}$ can be found as the diagonal of the parallelogram defined by $\vec{D}$ and $\vec{E}$. This method is called the *parallelogram rule* of vector addition. It is especially useful when two vectors are already drawn with their tails touching.

FIGURE 3.4 Two vectors can be added using the tip-to-tail rule or the parallelogram rule.

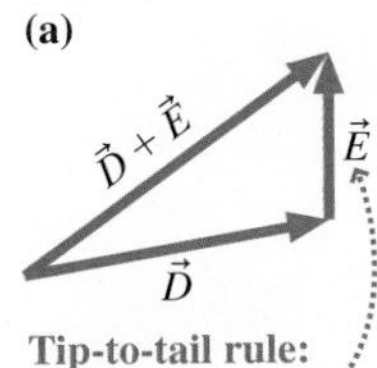
(a)
$\vec{D}+\vec{E}$
$\vec{E}$
$\vec{D}$
Tip-to-tail rule:
Place the tail of $\vec{E}$ at the tip of $\vec{D}$.

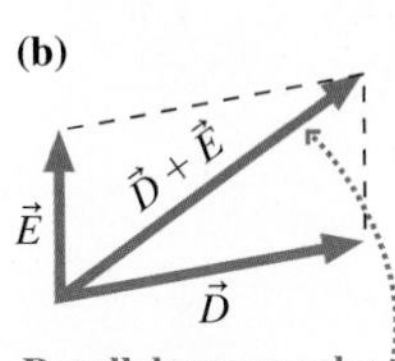
(b)
$\vec{D}+\vec{E}$
$\vec{E}$
$\vec{D}$
Parallelogram rule:
Find the diagonal of the parallelogram formed by $\vec{D}$ and $\vec{E}$.

Multiplication by a Scalar

The hiker in Figure 3.3 started with displacement $\vec{A}_1 = (4 \text{ mi, east})$. Suppose a second hiker walks twice as far to the east. The second hiker's displacement will then certainly be $\vec{A}_2 = (8 \text{ mi, east})$. The words "twice as" indicate a multiplication, so we can say

$$\vec{A}_2 = 2\vec{A}_1$$

Multiplying a vector by a positive scalar gives another vector of *different magnitude* but pointing in the *same direction.*

Suppose a vector $\vec{A}$ has magnitude A and direction θ. Now we let $\vec{B} = c\vec{A}$, where c is a positive scalar constant. Then vector $\vec{B}$ is stretched or compressed by the factor c (i.e., it has magnitude $B = cA$), but $\vec{B}$ points in the *same* direction as $\vec{A}$. This is illustrated in FIGURE 3.5.

FIGURE 3.5 Multiplication of a vector by a positive scalar.

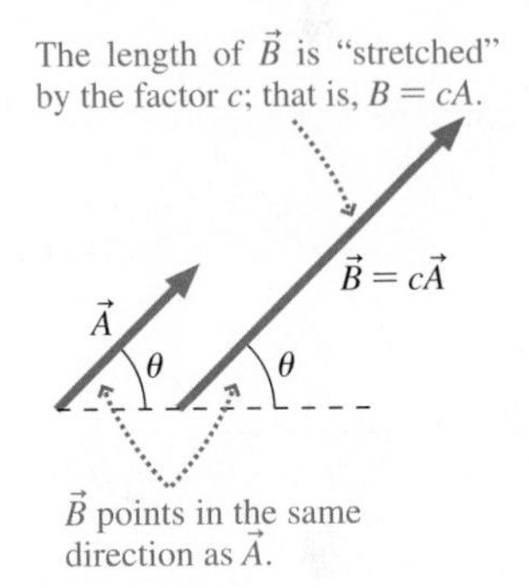
The length of $\vec{B}$ is "stretched" by the factor c; that is, $B = cA$.
$\vec{B} = c\vec{A}$
$\vec{A}$
θ
θ
$\vec{B}$ points in the same direction as $\vec{A}$.

What happens if we multiply $\vec{A}$ by zero? We get a vector that has zero length or magnitude. This vector is known as the **zero vector,** denoted $\vec{0}$. The direction of the zero vector is not defined; you cannot describe the direction of an arrow of zero length!

Finally, what happens if we multiply a vector by a negative number? Consider the vector $-\vec{A}$, which is equivalent to multiplying $\vec{A}$ by -1. Because

$$\vec{A} + (-\vec{A}) = \vec{0} \tag{3.2}$$

the vector $-\vec{A}$ must be such that, when it is added to $\vec{A}$, the resultant is the zero vector $\vec{0}$. In other words, the *tip* of $-\vec{A}$ must return to the *tail* of $\vec{A}$, as shown in FIGURE 3.6. This will be true only if $-\vec{A}$ is equal in magnitude to $\vec{A}$ but opposite in direction. Thus we can conclude that

$$-\vec{A} = (A, \text{direction opposite } \vec{A}) \tag{3.3}$$

FIGURE 3.6 Vector $-\vec{A}$.

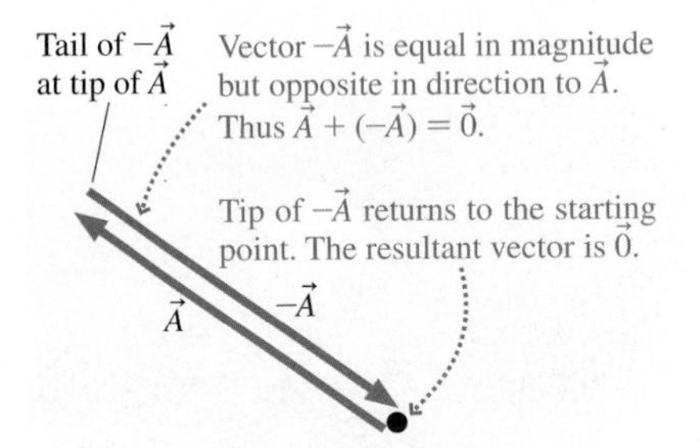
Tail of $-\vec{A}$ at tip of $\vec{A}$
Vector $-\vec{A}$ is equal in magnitude but opposite in direction to $\vec{A}$. Thus $\vec{A} + (-\vec{A}) = \vec{0}$.
Tip of $-\vec{A}$ returns to the starting point. The resultant vector is $\vec{0}$.
$\vec{A}$
$-\vec{A}$

Multiplying a vector by -1 reverses its direction without changing its length.

FIGURE 3.7 shows vectors $\vec{A}$, $2\vec{A}$, and $-3\vec{A}$. Multiplication by 2 doubles the length of the vector but does not change its direction. Multiplication by -3 stretches the length by a factor of 3 *and* reverses the direction.

FIGURE 3.7 Vectors $\vec{A}$, $2\vec{A}$, and $-3\vec{A}$.

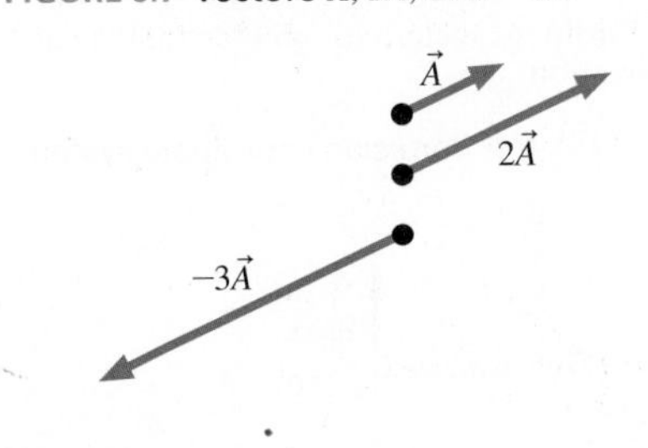
$\vec{A}$
$2\vec{A}$
$-3\vec{A}$

Vector Subtraction

How might we *subtract* vector $\vec{B}$ from vector $\vec{A}$ to form the vector $\vec{A} - \vec{B}$? With numbers, subtraction is the same as the addition of a negative number. That is, $5 - 3$ is the same as $5 + (-3)$. Similarly, $\vec{A} - \vec{B} = \vec{A} + (-\vec{B})$. We can use the rules for vector addition and the fact that $-\vec{B}$ is a vector opposite in direction to $\vec{B}$ to form rules for vector subtraction.

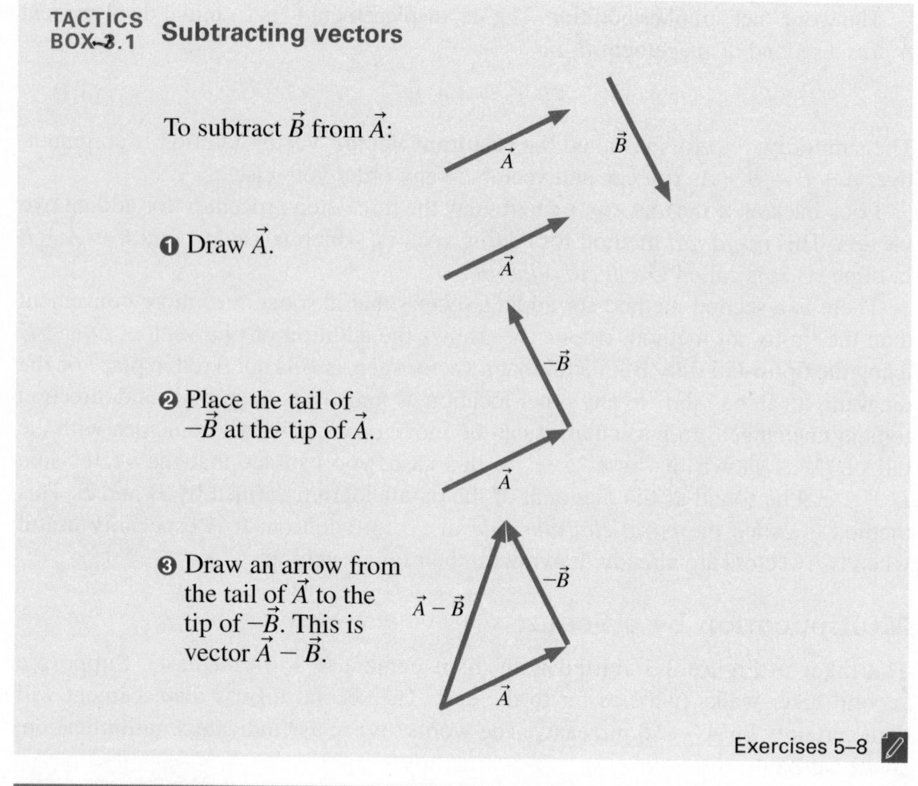

TACTICS BOX 3.1 Subtracting vectors
To subtract $\vec{B}$ from $\vec{A}$:
$\vec{A}$
$\vec{B}$
❶ Draw $\vec{A}$.
$\vec{A}$
❷ Place the tail of $-\vec{B}$ at the tip of $\vec{A}$.
$-\vec{B}$
$\vec{A}$
❸ Draw an arrow from the tail of $\vec{A}$ to the tip of $-\vec{B}$. This is vector $\vec{A} - \vec{B}$.
$\vec{A} - \vec{B}$
$-\vec{B}$
$\vec{A}$
Exercises 5–8

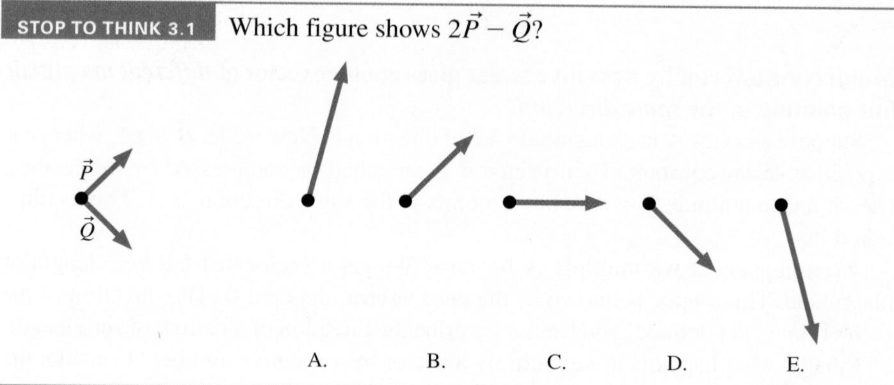

STOP TO THINK 3.1 Which figure shows $2\vec{P} - \vec{Q}$?

3.2 Coordinate Systems and Vector Components

A coordinate system for an archaeological excavation.

We've just learned how to add and subtract vectors graphically. But working with vectors graphically is not an especially good way to find quantitative results. In this section we will introduce a *coordinate description* of vectors that will be the basis for doing vector calculations.

Coordinate Systems

As we saw in Chapter 1, the world does not come with a coordinate system attached to it. A coordinate system is an artificially imposed grid that you place on a problem in order to make quantitative measurements. The right choice of coordinate system will make a problem easier to solve. We will generally use **Cartesian coordinates,** the familiar rectangular grid with perpendicular axes, as illustrated in FIGURE 3.8.

FIGURE 3.8 A Cartesian coordinate system.

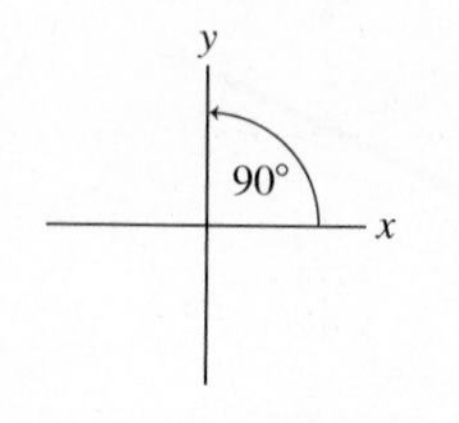

Coordinate axes have a positive end and a negative end, separated by zero at the origin where the two axes cross. When you draw a coordinate system, it is important to label the axes. This is done by placing x and y labels at the *positive* ends of the axes, as in Figure 3.8.

Component Vectors

FIGURE 3.9 shows a vector $\vec{A}$ and an *xy*-coordinate system that we've chosen. Once the directions of the axes are known, we can define two new vectors *parallel to the axes* that we call the **component vectors** of $\vec{A}$. Vector $\vec{A}_x$, called the *x-component vector,* is the projection of $\vec{A}$ along the *x*-axis. Vector $\vec{A}_y$, the *y-component vector,* is the projection of $\vec{A}$ along the *y*-axis.

You can see, using the parallelogram rule, that $\vec{A}$ is the vector sum of the two component vectors:

$$\vec{A} = \vec{A}_x + \vec{A}_y \tag{3.4}$$

In essence, we have "broken" vector $\vec{A}$ into two perpendicular vectors that are parallel to the coordinate axes. We say that we have **decomposed,** or **resolved,** vector $\vec{A}$ into its component vectors.

NOTE ▶ It is not necessary for the tail of $\vec{A}$ to be at the origin. All we need to know is the *orientation* of the coordinate system so that we can draw $\vec{A}_x$ and $\vec{A}_y$ parallel to the axes. ◀

FIGURE 3.9 Component vectors $\vec{A}_x$ and $\vec{A}_y$ are drawn parallel to the coordinate axes such that $\vec{A} = \vec{A}_x + \vec{A}_y$.

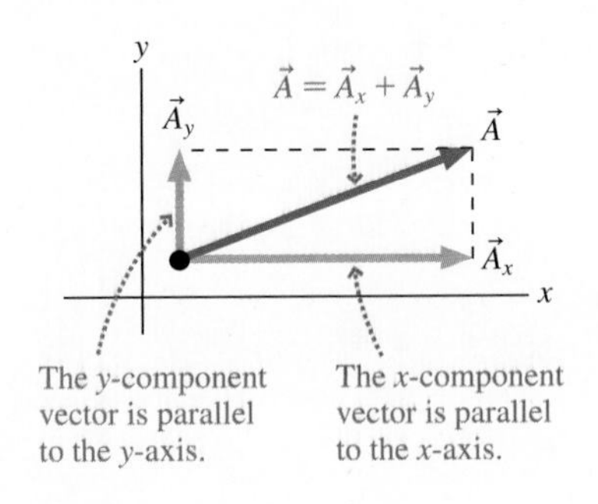
y
$\vec{A} = \vec{A}_x + \vec{A}_y$
$\vec{A}_y$
$\vec{A}$
$\vec{A}_x$
x
The y-component vector is parallel to the y-axis.
The x-component vector is parallel to the x-axis.

Components

As we learned from studying velocity in one-dimensional motion, we give v_x a positive sign if the velocity vector $\vec{v}$ points toward the positive end of the *x*-axis and a negative sign if $\vec{v}$ points in the negative *x*-direction. Thus the *sign* of v_x indicates the *direction* of the velocity. We can extend this idea to vectors in general.

Suppose we have a vector $\vec{A}$ that has been decomposed into component vectors $\vec{A}_x$ and $\vec{A}_y$ parallel to the coordinate axes. We can describe each component vector with a single number (a scalar) called the **component.** The *x-component* and *y-component* of vector $\vec{A}$, denoted A_x and A_y, are determined as follows:

TACTICS BOX 3.2 Determining the components of a vector

1. The absolute value $|A_x|$ of the *x*-component A_x is the magnitude of the component vector $\vec{A}_x$.
2. The *sign* of A_x is positive if $\vec{A}_x$ points in the positive *x*-direction, negative if $\vec{A}_x$ points in the negative *x*-direction.
3. The *y*-component A_y is determined similarly.

Exercises 12–14

In other words, the component A_x tells us two things: how big $\vec{A}_x$ is and toward which end of the axis $\vec{A}_x$ points. FIGURE 3.10 shows three examples of determining the components of a vector.

NOTE ▶ $\vec{A}_x$ and $\vec{A}_y$ are component *vectors*—they have a magnitude and a direction. A_x and A_y are simply *components.* The components A_x and A_y are scalars—just numbers (with units) that can be positive or negative. ◀

Much of physics is expressed in the language of vectors. We will frequently need to decompose a vector into its components or to "reassemble" a vector from its components, moving back and forth between the graphical and the component representations of a vector.

Let's start with the problem of decomposing a vector into its *x*- and *y*-components. FIGURE 3.11a shows a vector $\vec{A}$ at an angle θ above horizontal. It is *essential* to use a picture or diagram such as this to define the angle you are using to describe a vector's direction. $\vec{A}$ points to the right and up, so Tactics Box 3.2 tells us that the components A_x and A_y are both positive.

FIGURE 3.10 Determining the components of a vector.

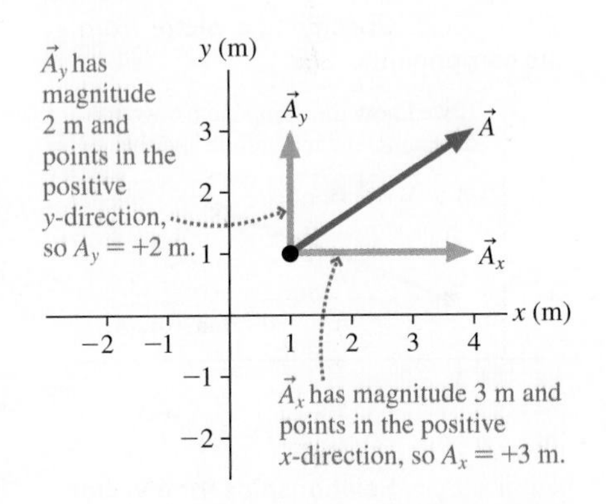
y (m)
$\vec{A}_y$ has magnitude 2 m and points in the positive y-direction, so $A_y = +2$ m.
$\vec{A}_y$
$\vec{A}$
$\vec{A}_x$
x (m)
$\vec{A}_x$ has magnitude 3 m and points in the positive x-direction, so $A_x = +3$ m.

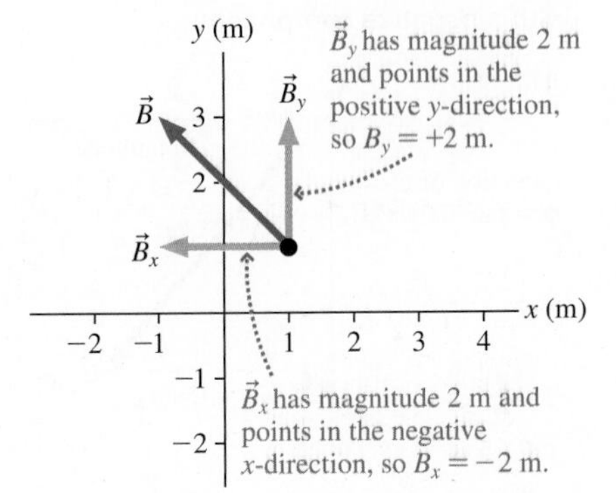
y (m)
$\vec{B}_y$ has magnitude 2 m and points in the positive y-direction, so $B_y = +2$ m.
$\vec{B}$
$\vec{B}_y$
$\vec{B}_x$
x (m)
$\vec{B}_x$ has magnitude 2 m and points in the negative x-direction, so $B_x = -2$ m.

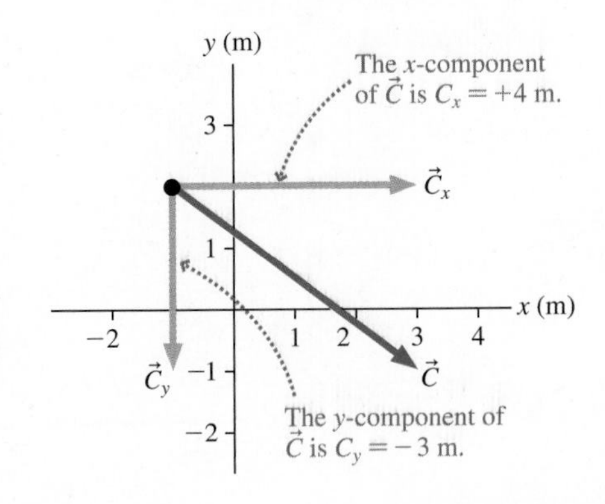
y (m)
The x-component of $\vec{C}$ is $C_x = +4$ m.
$\vec{C}_x$
$\vec{C}_y$
$\vec{C}$
x (m)
The y-component of $\vec{C}$ is $C_y = -3$ m.

FIGURE 3.11 Finding the components of a vector.

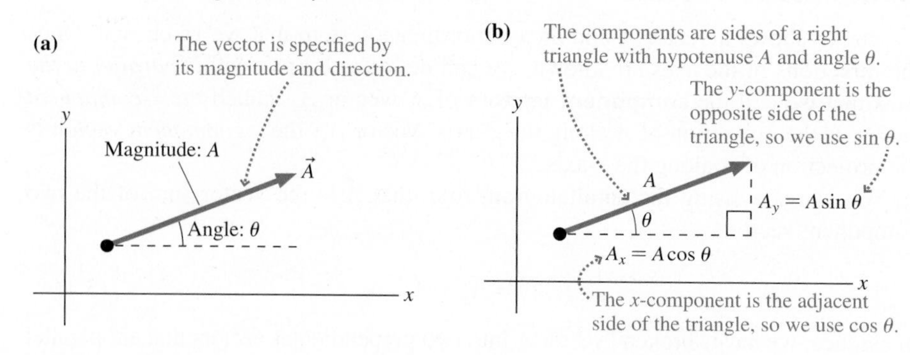

We can find the components using trigonometry, as illustrated in FIGURE 3.11b. For this case, we find that

$$A_x = A\cos\theta$$
$$A_y = A\sin\theta \qquad (3.5)$$

where A is the magnitude, or length, of $\vec{A}$. These equations convert the length and angle description of vector $\vec{A}$ into the vector's components, but they are correct *only* if θ is measured from horizontal.

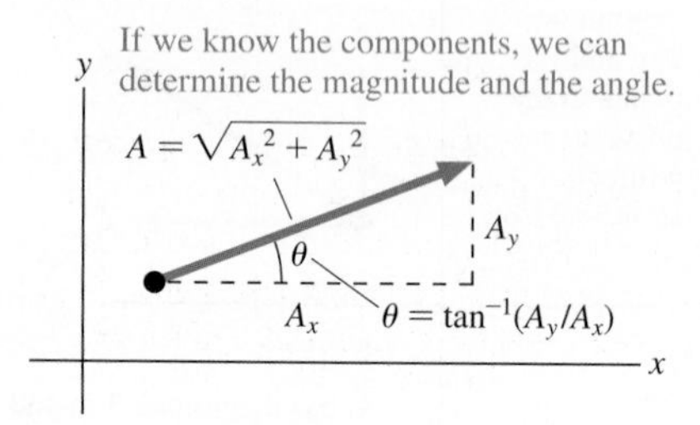

FIGURE 3.12 Specifying a vector from its components.

Alternatively, if we are given the x- and y-components of a vector, we can determine the length and angle of the vector, as shown in FIGURE 3.12. Because A in Figure 3.12 is the hypotenuse of a right triangle, its length is given by the Pythagorean theorem:

$$A = \sqrt{A_x^2 + A_y^2} \qquad (3.6)$$

Similarly, the tangent of angle θ is the ratio of the opposite side to the adjacent side, so

$$\theta = \tan^{-1}\left(\frac{A_y}{A_x}\right) \qquad (3.7)$$

Equations 3.6 and 3.7 can be thought of as the "inverse" of Equations 3.5.

Vectors may not always point to the right and up as in Figure 3.11, and the angle defining their direction may not always be measured from the x-axis. For example, FIGURE 3.13 shows vector $\vec{C}$ pointing down and to the right. In this case, the component vector $\vec{C}_y$ is pointing *down,* in the negative y-direction, so the y-component C_y is a *negative* number. In addition, the angle ϕ is drawn measured from the y-axis, so the components of $\vec{C}$ are

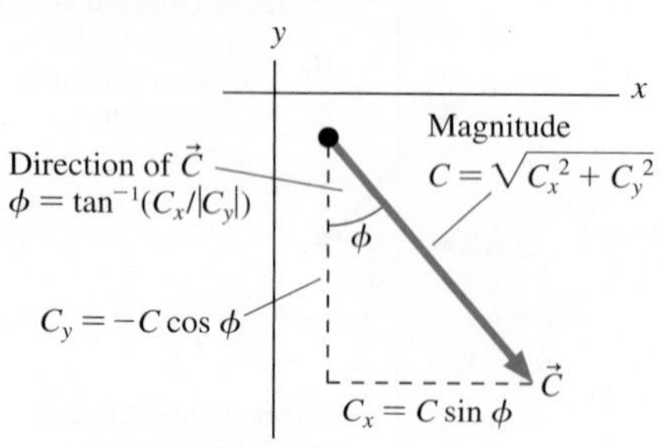

FIGURE 3.13 Relationships for a vector with a negative component.

$$C_x = C\sin\phi$$
$$C_y = -C\cos\phi \qquad (3.8)$$

The roles of sine and cosine are reversed from those in Equations 3.5 because the angle ϕ is measured with respect to vertical, not horizontal.

NOTE ▶ Whether the x- and y-components use the sine or cosine depends on how you define the vector's angle. As noted previously, you *must* draw a diagram to define the angle that you use, and you must be sure to refer to the diagram when computing components. Don't use Equations 3.5 or 3.8 as general rules—they aren't! They appear as they do because of how we defined the angles. ◀

Next, let's look at the "inverse" problem for this case: determining the length and direction of the vector given the components. The signs of the components don't matter for determining the length; the Pythagorean theorem always works to find the length or magnitude of a vector because the squares eliminate any concerns over the signs. The length of the vector in Figure 3.13 is simply

$$C = \sqrt{C_x^2 + C_y^2} \qquad (3.9)$$

When we determine the direction of the vector from its components, we must consider the signs of the components. Finding the angle of vector $\vec{C}$ in Figure 3.13 requires the length of C_y *without* the minus sign, so vector $\vec{C}$ has direction

$$\phi = \tan^{-1}\left(\frac{C_x}{|C_y|}\right) \tag{3.10}$$

Notice that the roles of x and y differ from those in Equation 3.7.

EXAMPLE 3.1 Finding the components of an acceleration vector

Find the x- and y-components of the acceleration vector $\vec{a}$ shown in FIGURE 3.14.

FIGURE 3.14 An acceleration vector $\vec{a}$.

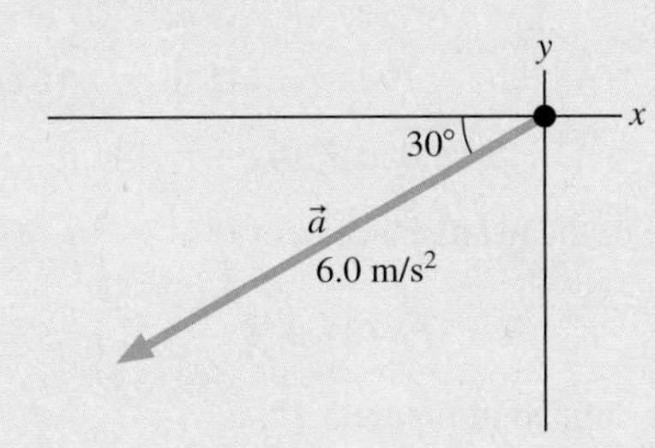

FIGURE 3.15 The components of the acceleration vector.

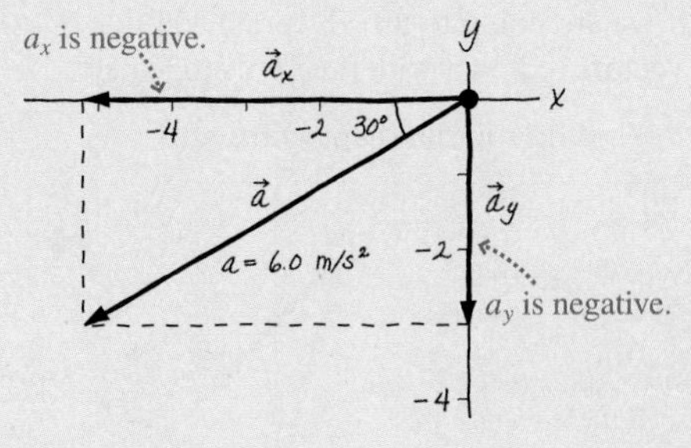

STRATEGIZE We can find the components from the vector's magnitude and by using trigonometry, paying special attention to how the angle is defined in the problem.

PREPARE Making a sketch is crucial to setting up this problem. FIGURE 3.15 shows the original vector $\vec{a}$ decomposed into component vectors parallel to the axes.

SOLVE The acceleration vector $\vec{a} = (6.0\text{ m/s}^2$, 30° below the negative x-axis) points to the left (negative x-direction) and down (negative y-direction), so the components a_x and a_y are both negative:

$$a_x = -a\cos 30° = -(6.0\text{ m/s}^2)\cos 30° = -5.2\text{ m/s}^2$$

$$a_y = -a\sin 30° = -(6.0\text{ m/s}^2)\sin 30° = -3.0\text{ m/s}^2$$

ASSESS The magnitude of the y-component is smaller than that of the x-component, as seems to be the case in Figure 3.15, a good check on our work. The units of a_x and a_y are the same as the units of vector $\vec{a}$. Notice that we had to insert the components' minus signs manually after observing that the vector points down and to the left.

STOP TO THINK 3.2 What are the x- and y-components C_x and C_y of vector $\vec{C}$?

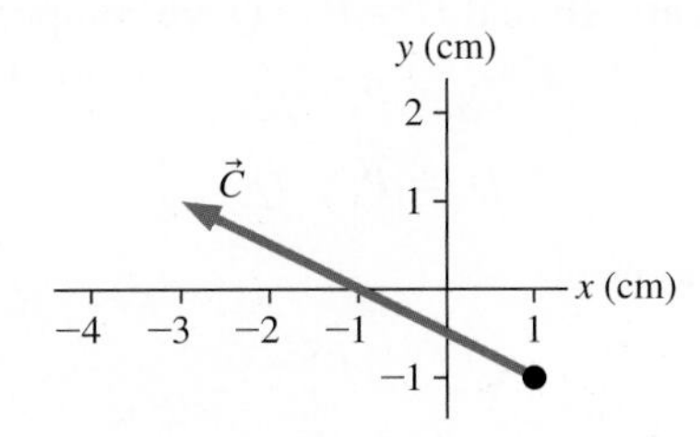

Working with Components

We've seen how to add vectors graphically, but there's an easier way: using components. To illustrate, let's look at the vector sum $\vec{C} = \vec{A} + \vec{B}$ for the vectors shown in FIGURE 3.16. You can see that the component vectors of $\vec{C}$ are the sums of the component vectors of $\vec{A}$ and $\vec{B}$. The same is true of the components: $C_x = A_x + B_x$ and $C_y = A_y + B_y$.

In general, if $\vec{D} = \vec{A} + \vec{B} + \vec{C} + \cdots$, then the x- and y-components of the resultant vector $\vec{D}$ are

$$\begin{aligned} D_x &= A_x + B_x + C_x + \cdots \\ D_y &= A_y + B_y + C_y + \cdots \end{aligned} \tag{3.11}$$

This method of vector addition is called *algebraic addition*.

FIGURE 3.16 Using components to add vectors.

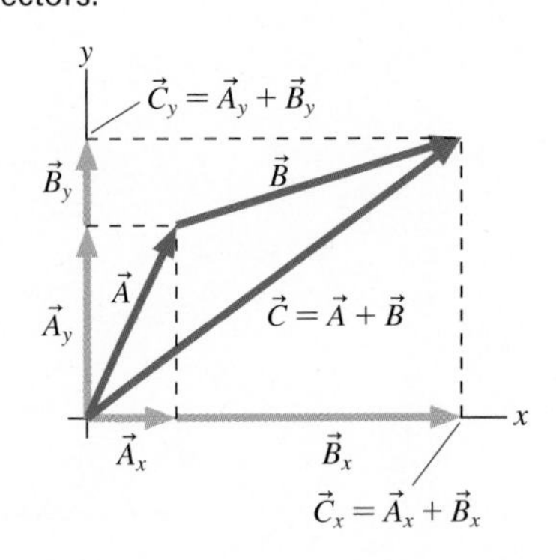

EXAMPLE 3.2 Using algebraic addition to find a bird's displacement

A bird flies 100 m due east from a tree, then 200 m northwest (that is, 45° north of west). What is the bird's net displacement?

STRATEGIZE To find the bird's net displacement using algebraic addition, we will first need to find the components of each of its two individual displacements. Then we will find the components of the net displacement by adding the individual components.

PREPARE FIGURE 3.17a shows the displacement vectors $\vec{A} = (100 \text{ m, east})$ and $\vec{B} = (200 \text{ m, northwest})$ and also the net displacement $\vec{C}$. We draw vectors tip-to-tail if we are going to add them graphically, but it's usually easier to draw them all from the origin if we are going to use algebraic addition. FIGURE 3.17b shows the vectors redrawn with their tails together.

FIGURE 3.17 Finding the net displacement.

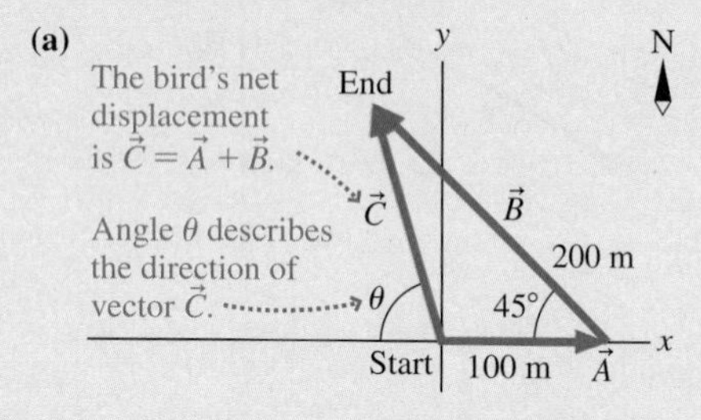

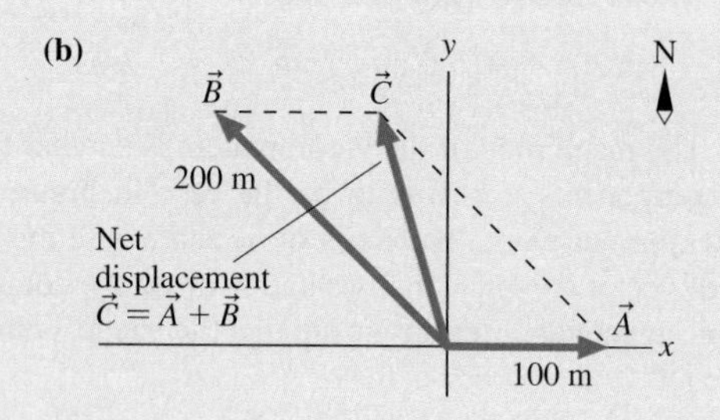

SOLVE To add the vectors algebraically we must know their components. From the figure these are seen to be

$$A_x = 100 \text{ m}$$
$$A_y = 0 \text{ m}$$
$$B_x = -(200 \text{ m}) \cos 45° = -141 \text{ m}$$
$$B_y = (200 \text{ m}) \sin 45° = 141 \text{ m}$$

We learned *from the figure* that $\vec{B}$ has a negative x-component. Adding $\vec{A}$ and $\vec{B}$ by components gives

$$C_x = A_x + B_x = 100 \text{ m} - 141 \text{ m} = -41 \text{ m}$$
$$C_y = A_y + B_y = 0 \text{ m} + 141 \text{ m} = 141 \text{ m}$$

The magnitude of the net displacement $\vec{C}$ is

$$C = \sqrt{C_x^{\,2} + C_y^{\,2}} = \sqrt{(-41 \text{ m})^2 + (141 \text{ m})^2} = 147 \text{ m}$$

The angle θ, as defined in Figure 3.17, is

$$\theta = \tan^{-1}\left(\frac{C_y}{|C_x|}\right) = \tan^{-1}\left(\frac{141 \text{ m}}{41 \text{ m}}\right) = 74°$$

Thus the bird's net displacement is $\vec{C} = (147 \text{ m}, 74° \text{ north of west})$.

ASSESS The final values of C_x and C_y match what we would expect from the sketch in Figure 3.17. The geometric addition was a valuable check on the answer we found by algebraic addition.

Vector subtraction and the multiplication of a vector by a scalar are also easily performed using components. To find $\vec{D} = \vec{P} - \vec{Q}$ we compute

$$D_x = P_x - Q_x$$
$$D_y = P_y - Q_y \tag{3.12}$$

Similarly, $\vec{T} = c\vec{S}$ has components

$$T_x = cS_x$$
$$T_y = cS_y \tag{3.13}$$

The next few chapters will make frequent use of *vector equations*. For example, you will learn that the equation to calculate the net force on a car skidding to a stop is

$$\vec{F} = \vec{n} + \vec{w} + \vec{f} \tag{3.14}$$

Equation 3.14 is really just a shorthand way of writing the two simultaneous equations:

$$F_x = n_x + w_x + f_x$$
$$F_y = n_y + w_y + f_y \tag{3.15}$$

In other words, a vector equation is interpreted as meaning: Equate the x-components on both sides of the equals sign, then equate the y-components. Vector notation allows us to write these two equations in a more compact form.

Tilted Axes

Although we are used to having the x-axis horizontal, there is no requirement that it has to be that way. In Chapter 1, we saw that for motion on a slope, it is often most convenient to put the x-axis along the slope. When we add the y-axis, this gives us a tilted coordinate system such as that shown in FIGURE 3.18.

Finding components with tilted axes is no harder than what we have done so far. Vector $\vec{C}$ in Figure 3.18 can be decomposed into component vectors $\vec{C}_x$ and $\vec{C}_y$, with $C_x = C\cos\theta$ and $C_y = C\sin\theta$.

FIGURE 3.18 A coordinate system with tilted axes.

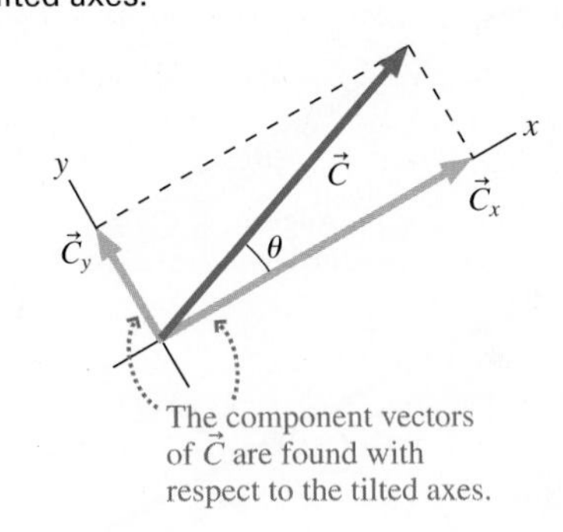

STOP TO THINK 3.3 Angle ϕ that specifies the direction of $\vec{C}$ is computed as

A. $\tan^{-1}(C_x/C_y)$.
B. $\tan^{-1}(C_x/|C_y|)$.
C. $\tan^{-1}(|C_x|/|C_y|)$.
D. $\tan^{-1}(C_y/C_x)$.
E. $\tan^{-1}(C_y/|C_x|)$.
F. $\tan^{-1}(|C_y|/|C_x|)$.

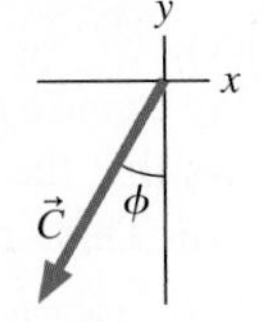

3.3 Motion on a Ramp

In this section, we will examine the problem of motion on a ramp, or *incline*. There are three reasons to look at this problem. First, it will provide good practice using vectors to analyze motion. Second, it is a simple problem for which we can find an exact solution. Third, this seemingly abstract problem has real and important applications.

A video to support a section's topic is embedded in the eText.

Video Motion on a Ramp

Constant-Velocity Motion on a Ramp

We begin with a constant-velocity example to give us some practice with vectors and components before moving on to the more general case of accelerated motion.

EXAMPLE 3.3 Finding the height gained by a car moving up a slope

A car drives up a steep 10° slope at a constant speed of 15 m/s. After 10 s, how much height has the car gained?

STRATEGIZE The vertical motion of the car is determined by the vertical component of the velocity.

PREPARE FIGURE 3.19 is a visual overview, with x- and y-axes defined. The velocity vector $\vec{v}$ points up the slope. We decompose $\vec{v}$ into component vectors $\vec{v}_x$ and $\vec{v}_y$ as shown.

FIGURE 3.19 Visual overview of a car moving up a slope.

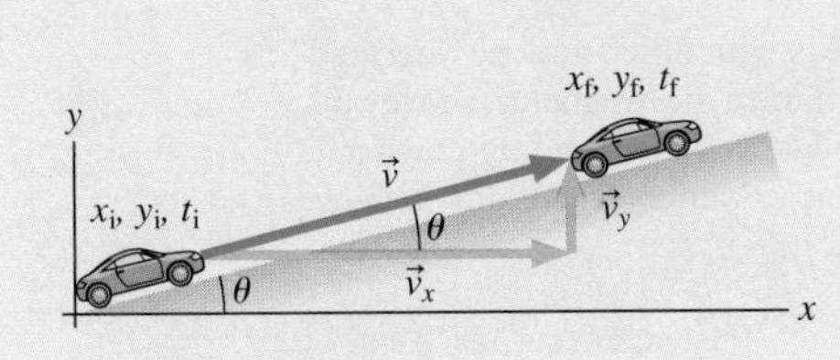

SOLVE The velocity component we need is v_y, which describes the vertical motion of the car. Using the rules for finding components outlined previously, we find

$$v_y = v\sin\theta = (15\text{ m/s})\sin 10° = 2.6\text{ m/s}$$

Because the velocity is constant, the car's vertical displacement (i.e., the height gained) during 10 s is

$$\Delta y = v_y\,\Delta t = (2.6\text{ m/s})(10\text{ s}) = 26\text{ m}$$

ASSESS The car is traveling at a pretty good clip—15 m/s is a bit faster than 30 mph—up a steep slope, so it should climb a respectable height in 10 s. 26 m, or about 80 ft, seems reasonable.

Accelerated Motion on a Ramp

FIGURE 3.20a shows a crate sliding down a frictionless (i.e., smooth) ramp tilted at angle θ. The crate accelerates due to the action of gravity, but it is *constrained* to accelerate parallel to the surface. What is the acceleration?

A motion diagram for the crate is drawn in FIGURE 3.20b. The velocity vectors point in the direction of motion, downhill and parallel to the ramp's surface. Because the crate is speeding up, its acceleration points in the same direction as its velocity. We can take advantage of the properties of vectors to find the crate's acceleration.

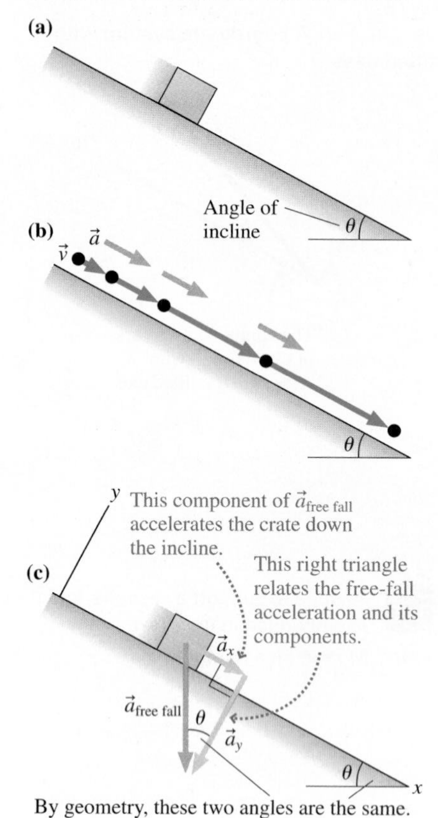
(a)
Angle of incline
θ
(b)
$\vec{a}$
$\vec{v}$
θ
(c)
y
This component of $\vec{a}_{\text{free fall}}$ accelerates the crate down the incline.
This right triangle relates the free-fall acceleration and its components.
$\vec{a}_x$
$\vec{a}_{\text{free fall}}$
θ
$\vec{a}_y$
θ
x
By geometry, these two angles are the same.

FIGURE 3.20 Acceleration on an incline.

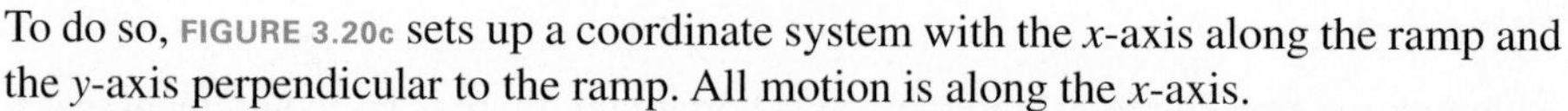

To do so, FIGURE 3.20c sets up a coordinate system with the x-axis along the ramp and the y-axis perpendicular to the ramp. All motion is along the x-axis.

If the incline suddenly vanished, the object would have a free-fall acceleration $\vec{a}_{\text{free fall}}$ straight down. As Figure 3.20c shows, this acceleration vector can be decomposed into two component vectors: a vector $\vec{a}_x$ that is *parallel* to the incline and a vector $\vec{a}_y$ that is *perpendicular* to the incline. The vector addition rules studied earlier in this chapter tell us that $\vec{a}_{\text{free fall}} = \vec{a}_x + \vec{a}_y$.

The motion diagram shows that the object's actual acceleration $\vec{a}_x$ is parallel to the incline. The surface of the incline somehow "blocks" the other component of the acceleration $\vec{a}_y$, through a process we will examine in Chapter 5, but $\vec{a}_x$ is unhindered. It is this component of $\vec{a}_{\text{free fall}}$, parallel to the incline, that accelerates the object.

We can use trigonometry to work out the magnitude of this acceleration. Figure 3.20c shows that the three vectors $\vec{a}_{\text{free fall}}$, $\vec{a}_y$, and $\vec{a}_x$ form a right triangle with angle θ as shown. This angle is the same as the angle of the incline. By definition, the magnitude of $\vec{a}_{\text{free fall}}$ is g. This vector is the hypotenuse of the right triangle. The vector we are interested in, $\vec{a}_x$, is opposite angle θ. Thus the value of the acceleration along a frictionless slope is

$$a_x = \pm g \sin\theta \tag{3.16}$$

NOTE ▸ The correct sign depends on the direction in which the ramp is tilted. The acceleration in Figure 3.20 is $+g \sin\theta$, but upcoming examples will show situations in which the acceleration is $-g \sin\theta$. ◂

Let's look at Equation 3.16 to verify that it makes sense. A good way to do this is to consider **limiting cases** in which the angle is at one end of its range. In these cases, the physics is clear and we can check our result. Let's look at two such possibilities:

1. Suppose the angle of the ramp is lowered until the ramp is perfectly horizontal, with $\theta = 0°$. An object sliding on a (frictionless) horizontal surface would move with a constant velocity and so have zero acceleration. Equation 3.16 gives $a_x = 0$ when $\theta = 0°$, in agreement with our expectations.
2. Now suppose the ramp is tilted until it becomes vertical, with $\theta = 90°$. You know what happens—the object will be in free fall, parallel to the vertical surface. Equation 3.16 gives $a_x = \pm g$ when $\theta = 90°$, again in agreement with our expectations.

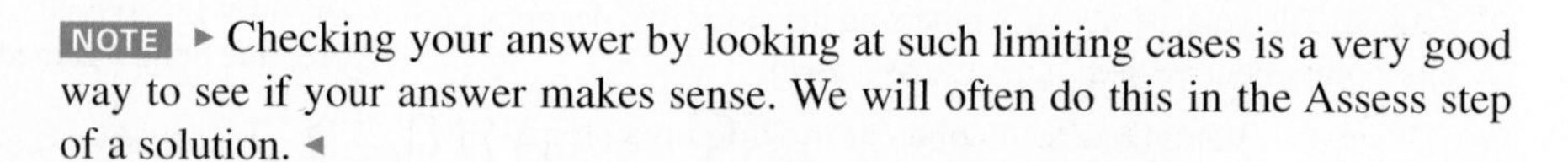

NOTE ▸ Checking your answer by looking at such limiting cases is a very good way to see if your answer makes sense. We will often do this in the Assess step of a solution. ◂

◂ **Extreme physics** A speed skier, on wide skis with little friction, wearing an aerodynamic helmet and crouched low to minimize air resistance, moves in a straight line down a steep slope—pretty much like an object sliding down a frictionless ramp. There is a maximum speed that a skier could possibly achieve at the end of the slope.

EXAMPLE 3.4 Maximum possible speed for a skier

The Willamette Pass ski area in Oregon was the site of the 1993 U.S. National Speed Skiing Competition. The skiers started from rest and then accelerated down a stretch of the mountain with a reasonably constant slope, aiming for the highest possible speed at the end of this run. During this acceleration phase, the skiers traveled 360 m while dropping a vertical distance of 170 m. What is the fastest speed a skier could achieve at the end of this run?

STRATEGIZE The fastest possible run is one with no friction or air resistance, meaning the acceleration down the slope is given by Equation 3.16. The problem is then one-dimensional motion with constant acceleration.

PREPARE We make the visual overview in FIGURE 3.21. The motion diagram shows the acceleration of the skier and the pictorial representation gives an overview of the problem including the dimensions of the slope. As before, we put the x-axis along the slope.

FIGURE 3.21 Visual overview of a skier accelerating down a slope.

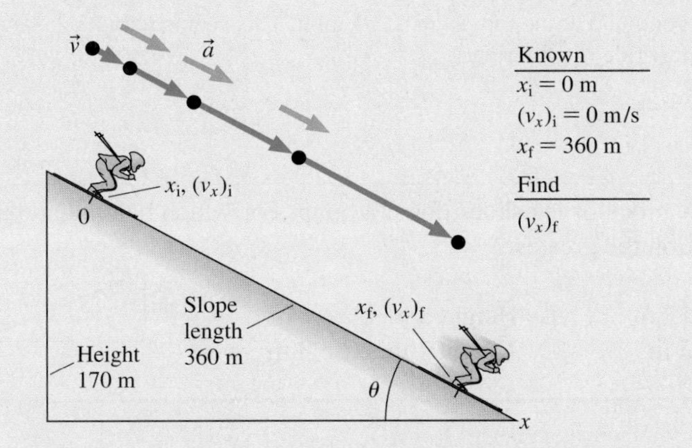

SOLVE The acceleration is in the positive x-direction, so we use the positive sign in Equation 3.16. What is the angle in Equation 3.16? Figure 3.21 shows that the 360-m-long slope is the hypotenuse of a triangle of height 170 m, so we use trigonometry to find

$$\sin \theta = \frac{170 \text{ m}}{360 \text{ m}}$$

which gives $\theta = \sin^{-1}(170/360) = 28°$. Equation 3.16 then gives

$$a_x = +g \sin \theta = (9.8 \text{ m/s}^2)(\sin 28°) = 4.6 \text{ m/s}^2$$

For linear motion with constant acceleration, we can use Equation 2.13 in Synthesis 2.1: $(v_x)_f{}^2 = (v_x)_i{}^2 + 2a_x \Delta x$. The initial velocity $(v_x)_i$ is zero; thus

This is the distance along the slope, the length of the run.

$$(v_x)_f = \sqrt{2a_x \Delta x} = \sqrt{2(4.6 \text{ m/s}^2)(360 \text{ m})} = 58 \text{ m/s}$$

This is the fastest that any skier could hope to be moving at the end of the run. Any friction or air resistance would decrease this speed.

ASSESS The final speed we calculated is 58 m/s, which is about 130 mph, reasonable because we expect a high speed for this sport. In the competition noted, the actual winning speed was 111 mph, not much slower than the result we calculated. Obviously, efforts to minimize friction and air resistance are working!

Skis on snow have very little friction, but there are other ways to reduce the friction between surfaces. For instance, a roller coaster car rolls along a track on low-friction wheels. No drive force is applied to the cars after they are released at the top of the first hill; the speed changes due to gravity alone. The cars speed up as they go down hills and slow down as they climb.

EXAMPLE 3.5 **Speed of a roller coaster**

A classic wooden coaster has cars that go down a big first hill, gaining speed. The cars then ascend a second hill with a slope of 30°. If the cars are going 25 m/s at the bottom and it takes them 2.0 s to climb this hill, how fast are they going at the top?

STRATEGIZE The coaster's velocity is directed up the ramp. But because the cars are slowing, the acceleration points opposite to the velocity, or down the hill. The acceleration is constant and given by Equation 3.16; we will use the kinematic equations from Synthesis 2.1 to solve for the speed at the top.

PREPARE The visual overview in FIGURE 3.22 includes a motion diagram, a pictorial representation, and a list of values. We've done this with a sketch such as you might draw for your homework. Notice how the motion diagram of Figure 3.22 differs from that of Figure 3.21 in Example 3.4: The motion is along the x-axis, as before, but the acceleration vector points in the negative x-direction, so the component a_x is negative.

FIGURE 3.22 The coaster's speed decreases as it goes up the hill.

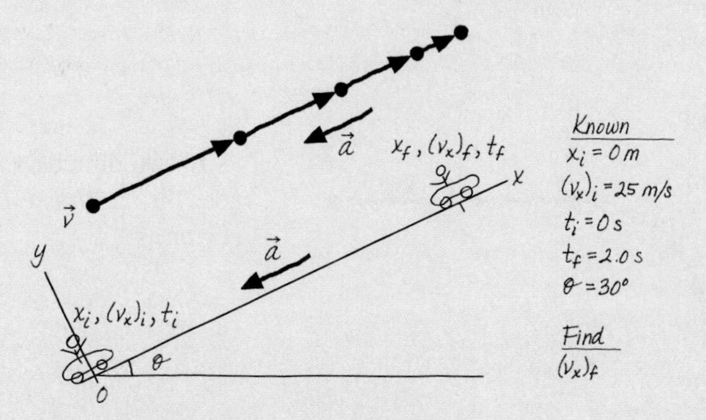

SOLVE To determine the final speed, we need to know the acceleration. We will assume that there is no friction or air resistance, so the magnitude of the roller coaster's acceleration is given by Equation 3.16 using the minus sign, as noted:

$$a_x = -g \sin \theta = -(9.8 \text{ m/s}^2) \sin 30° = -4.9 \text{ m/s}^2$$

Continued

The speed at the top of the hill can then be computed using Equation 2.11 of Synthesis 2.1:

$$(v_x)_f = (v_x)_i + a_x \Delta t = 25 \text{ m/s} + (-4.9 \text{ m/s}^2)(2.0 \text{ s}) = 15 \text{ m/s}$$

ASSESS The speed is less at the top of the hill than at the bottom, as it should be, but the coaster is still moving at a pretty good clip at the top—almost 35 mph. This seems reasonable—a fast ride is a fun ride.

STOP TO THINK 3.4 A block of ice slides down a ramp. For which height and base lengths is the acceleration the greatest?

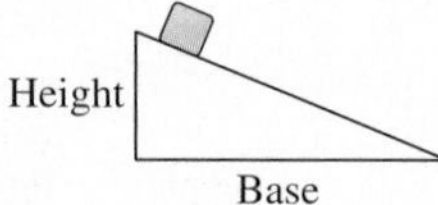

A. Height 4 m, base 12 m
B. Height 3 m, base 6 m
C. Height 2 m, base 5 m
D. Height 1 m, base 3 m

3.4 Motion in Two Dimensions

So far, we've studied in detail only one-dimensional motion, such as a sprinter running a 100 m race or a skier accelerating down a slope. For such motion we can represent the object's displacement, velocity, and acceleration by vectors that are directed along the single axis of motion.

Now we'll examine the more complex case of *two-dimensional motion.* In this kind of motion, the object moves not along a single line, but in an entire *plane.* For example, a skater gliding around a rink moves in the plane defined by the horizontal ice surface. A rider on a Ferris wheel moves in a vertical plane. As an object undergoes two-dimensional motion, its displacement, velocity, and acceleration vectors all change, and point in many different directions.

FIGURE 3.23 Motion diagram for a car going around a curve.

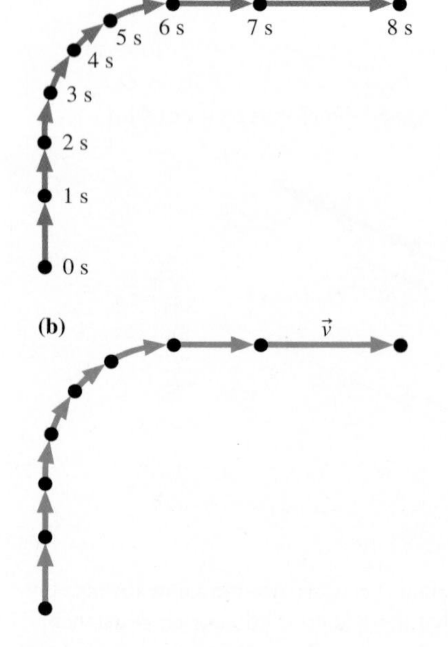

FIGURE 3.23a shows a motion diagram for a car, as seen from above, rounding a curve in the road. This is two-dimensional motion in the plane of the road. The motion diagram shows that the car slows as it approaches the curve, rounds the curve slowly, and then speeds up as it exits the curve. We have drawn displacement vectors $\vec{d}$ connecting the dots that represent the car's positions at successive times.

We can also represent the car's velocity vectors on a motion diagram. In Chapter 2, we defined velocity for one-dimensional motion as an object's displacement—its change in position—divided by the time interval in which the change occurs:

$$v_x = \frac{\Delta x}{\Delta t} = \frac{x_f - x_i}{\Delta t}$$

In two dimensions, an object's displacement is a vector. Suppose an object undergoes displacement $\vec{d}$ during the time interval Δt. Then we define the object's velocity *vector* as

$$\vec{v} = \frac{\vec{d}}{\Delta t} = \left(\frac{d}{\Delta t}, \text{ same direction as } \vec{d}\right) \tag{3.17}$$

Definition of velocity in two or more dimensions

The velocity vector is simply the displacement vector multiplied by the scalar $1/\Delta t$. Consequently, **the velocity vector points in the same direction as the displacement.** As a result, we represent the velocity on a motion diagram by vectors that extend from one dot to the next, just as the displacement vectors do. Using this idea, we draw the velocity vectors for the car's motion as in FIGURE 3.23b.

EXAMPLE 3.6 A kayaker goes over a waterfall

A kayaker paddles off the edge of a waterfall, arcing down to the water below. The partial motion diagram in FIGURE 3.24 shows his position at 0.1 s intervals. Estimate the kayaker's velocity during the time interval from $t = 0.5$ s to $t = 0.6$ s.

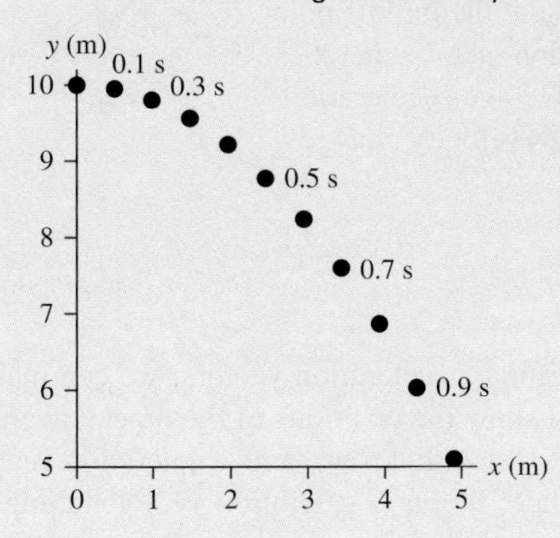

FIGURE 3.24 Partial motion diagram for a kayaker.

STRATEGIZE The problem asks for the kayaker's velocity, so we'll need to specify both his speed and his direction. We'll first find his displacement from the motion diagram and then use Equation 3.17 to find his velocity.

PREPARE During this time interval, we estimate the distance he traveled to be 0.8 m, and the angle of his motion looks to be about 45° below the horizontal.

SOLVE From our estimates, the kayaker's displacement during the stated time interval is

$$\vec{d} = (0.8 \text{ m}, 45° \text{ below horizontal})$$

Then from Equation 3.17 his velocity is

$$\vec{v} = \left(\frac{0.8 \text{ m}}{0.1 \text{ s}}, 45° \text{ below horizontal}\right)$$
$$= (8 \text{ m/s}, 45° \text{ below horizontal})$$

ASSESS The kayaker's *speed* of 8 m/s $\approx$ 18 mph seems reasonable for a daredevil kayaker.

Acceleration in Two Dimensions

In ◀◀ SECTION 2.4 we defined an object's acceleration in one dimension as the change in its velocity divided by the time interval in which the change occurs, so that $a_x = \Delta v_x / \Delta t$. In two dimensions, we need to use a vector to describe acceleration. The vector definition of acceleration is a straightforward extension of the one-dimensional version:

$$\vec{a} = \frac{\vec{v}_f - \vec{v}_i}{t_f - t_i} = \frac{\Delta \vec{v}}{\Delta t} \qquad (3.18)$$

Definition of acceleration in two or more dimensions

From its definition, we see that **$\vec{a}$ points in the same direction as $\Delta\vec{v}$,** the change in velocity. Because velocity in two dimensions is a vector, as an object moves, its velocity vector can change in two possible ways:

1. The magnitude of $\vec{v}$ can change, which indicates that the object's *speed* has changed.
2. The direction of $\vec{v}$ can change, which indicates that the object's *direction of motion* has changed.

In Chapter 2 we saw how to compute an acceleration vector for the first case, in which an object speeds up or slows down while moving in a straight line. Now we will examine the second case, in which an object's direction of motion changes.

Suppose an object has initial velocity $\vec{v}_i$ and, a short time Δt later, final velocity $\vec{v}_f$. During this time interval, the velocity changes by an amount $\Delta\vec{v} = \vec{v}_f - \vec{v}_i$. Because the velocity *changes,* the object must undergo an *acceleration* during the time interval Δt, as Equation 3.18 shows. Equation 3.18 further shows that the acceleration vector points in the *same direction* as the change in velocity $\Delta\vec{v} = \vec{v}_f - \vec{v}_i$. This means that in order to find the acceleration, we have to subtract two velocity vectors.

Tactics Box 3.1 showed how to perform vector subtraction. Tactics Box 3.3 shows how to use vector subtraction to find the acceleration vector.

TACTICS BOX 3.3 Finding the acceleration vector

To find the acceleration as the velocity changes from $\vec{v}_i$ to $\vec{v}_f$:

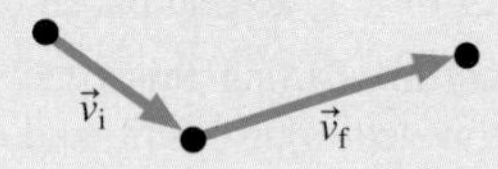

❶ Draw the velocity vector $\vec{v}_f$.

❷ Draw $-\vec{v}_i$ at the tip of $\vec{v}_f$.

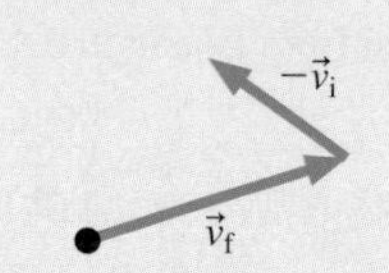

❸ Draw $\Delta\vec{v} = \vec{v}_f - \vec{v}_i$
$= \vec{v}_f + (-\vec{v}_i)$
This is the direction of $\vec{a}$.

❹ In the original motion diagram, draw a vector at the middle point in the direction of $\Delta\vec{v}$; label it $\vec{a}$. This is the average acceleration as the object's velocity changes from $\vec{v}_i$ to $\vec{v}_f$.

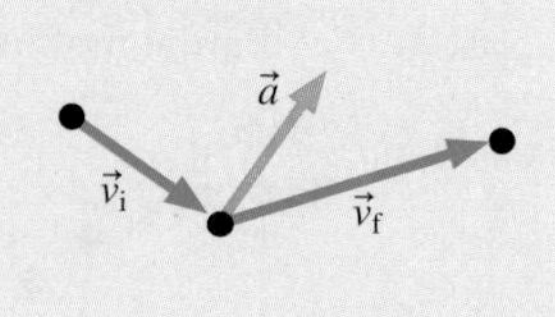

Exercises 27, 28

Now that we know how to determine acceleration vectors, we can make a complete motion diagram with dots that show the positions of the object, average velocity vectors found by connecting the dots with arrows, and acceleration vectors found using Tactics Box 3.3. Note that there is *one* acceleration vector linking each *two* velocity vectors, and $\vec{a}$ is drawn at the dot between the two velocity vectors it links.

EXAMPLE 3.7 Skiing through a valley

A skier skis down a long, straight hill, through a short valley, and then back up a hill on the other side. Draw a complete motion diagram of her motion.

STRATEGIZE This is two-dimensional motion. As the skier skis down through the valley, both her speed and the direction of her velocity change. Both these changes contribute to her acceleration, which we will determine using the steps of Tactics Box 3.3.

PREPARE In FIGURE 3.25 we have drawn dots to represent her positions at equally spaced times.

SOLVE The velocity vectors of the skier's motion are found by connecting successive dots in the motion diagram with arrows. Her motion down the hill is just one-dimensional motion on a ramp; because she is speeding up, her acceleration vectors point in the same direction as her velocity vectors—that is, downhill. Similarly, for her straight-line motion up the hill on the other side she is slowing, so her acceleration vectors point in the direction opposite to her velocity.

As the skier moves through the valley, where both her speed and the direction of her velocity are changing, we must use the steps of Tactics Box 3.3 to find her acceleration. We need to use vector subtraction to find the direction of $\Delta\vec{v}$ and thus of $\vec{a}$. The procedure is shown at two points in the motion diagram.

ASSESS This example illustrates a new and counterintuitive result: For two-dimensional motion, the acceleration need not point in the same (or opposite) direction as the velocity! In particular, at the very bottom of the valley, the skier is neither speeding up nor slowing down, so her speed is, at this instant in time, constant. But she still has an acceleration, directed at right angles to her velocity. We'll see more examples of this interesting behavior of acceleration as we go through this chapter.

FIGURE 3.25 Motion diagram for a skier skiing through a valley.

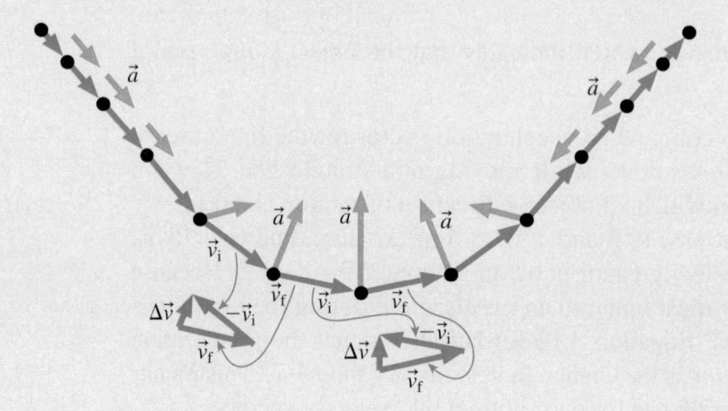

As Example 3.7 shows, the relationship between velocity and acceleration can sometimes be subtle. But the example illustrates some important ideas. First, as shown in Tactics Box 3.3, we can always find the acceleration graphically by taking the difference between the two velocity vectors on either side of one dot in a motion diagram. Second, even if the object has a constant speed (as at the bottom of the valley), it will have an acceleration if its path is curving; the acceleration in this case points at right angles to the velocity, directed toward the "inside" of the curve. Finally, if the object is moving along a straight-line segment of its path, just as in Chapter 2 the acceleration points in the direction of the velocity (object speeding up) or in the direction opposite to the velocity (object slowing down).

STOP TO THINK 3.5 The positions of four cars on four different roads are indicated by the dots. Which cars could possibly have the acceleration shown?

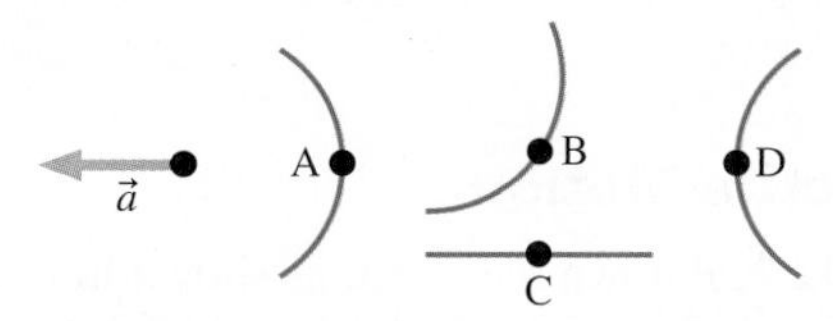

3.5 Projectile Motion

Balls flying through the air, long jumpers, and cars doing stunt jumps are all examples of the two-dimensional motion that we call **projectile motion.** Projectile motion is an extension to two dimensions of the free-fall motion we studied in ◀◀ SECTION 2.7. **A projectile is an object that moves in two dimensions under the influence of gravity and nothing else.** Although real objects are also influenced by air resistance, the effect of air resistance is small for reasonably dense objects moving at modest speeds, so we can ignore it for the cases we consider in this chapter. As long as we can neglect air resistance, any projectile will follow the same type of path. Because the form of the motion will always be the same, the strategies we develop to solve one projectile problem can be applied to others as well.

Video Video Dropped and Thrown Balls

Video Figure 3.26

FIGURE 3.26 The motions of two balls launched at the same time.

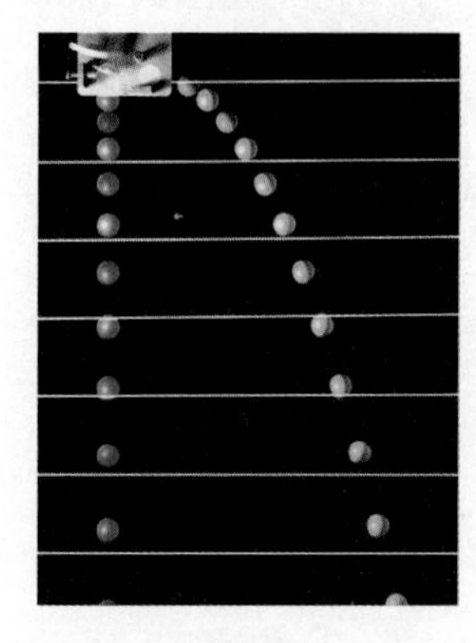

FIGURE 3.26 is a strobe photograph of two falling balls. The red ball was released from rest and undergoes ordinary free fall. At the same instant the red ball was released, the yellow ball was shot horizontally. Note that the *vertical* motions of the two balls are identical—at the same time after release, they are at the same height—and they hit the floor simultaneously. Evidently the vertical motion of the yellow ball is not affected by the fact that it is moving horizontally. Thus the vertical motion of each ball is free fall, which is the same for all objects, as we learned in Chapter 2. Furthermore, a careful look at the *horizontal* motion of the yellow ball shows that it is uniform motion; the horizontal motion continues as if the ball were not falling.

So, for an object in projectile motion, the initial horizontal velocity has *no* influence over the vertical motion, and vice versa. This is a general rule for *any* projectile motion: **The horizontal and vertical components of an object undergoing projectile motion are independent of each other.**

CONCEPTUAL EXAMPLE 3.8 Time and distance for balls rolled off the table

Two balls are rolling toward the edge of a table, with ball 1 rolling twice as fast as ball 2. Both balls leave the table edge at the same time. Which ball hits the ground first? Which ball goes farther?

REASON The vertical motion of both balls is the same—free fall—and they both fall from the same height. Both balls are in the air for the same time interval and hit the ground at the exact same time. During this time interval, the two balls continue moving horizontally at the speed with which they left the table. Ball 2 has twice the horizontal speed of ball 1; it will therefore go twice as far.

ASSESS This result makes sense. The vertical and horizontal motions are independent of each other, so we can analyze them separately. If you drop two objects from the same height, they hit the ground at the same time, so both balls should land at the same time. And as they drop, the one that was moving faster horizontally will continue to do so—thus it will go farther.

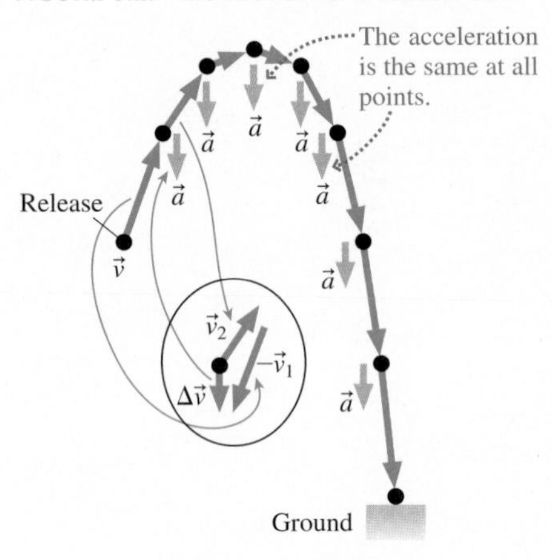

FIGURE 3.27 The motion of a tossed ball.

In Chapter 2, you learned that for an object in free fall, the acceleration is always directed vertically downward, with magnitude g. FIGURE 3.27 shows that this statement is also true for a ball tossed into the air at an angle—that is, for projectile motion. We can use the procedure of Tactics Box 3.3 to find the ball's acceleration. The inset in the figure shows this procedure at one point in the ball's motion; we can see that the ball's acceleration vector points straight down. In fact, a careful analysis would find that *at every point in the motion* the acceleration vector points straight down and has magnitude $g = 9.80\text{ m/s}^2$. This should not be surprising. If we ignore the tiny effect of air resistance, a projectile is an object moving under the influence of gravity only, so it is in free fall with acceleration $\vec{a}_{\text{free fall}} = (9.80\text{ m/s}^2,\text{ vertically downward})$.

As the projectile moves, the free-fall acceleration will change the *vertical* component of the velocity, but there will be no change to the *horizontal* component of the velocity. **The vertical component of acceleration a_y for all projectile motion is the familiar $-g$ of free fall, while the horizontal component a_x is zero.**

Analyzing Projectile Motion

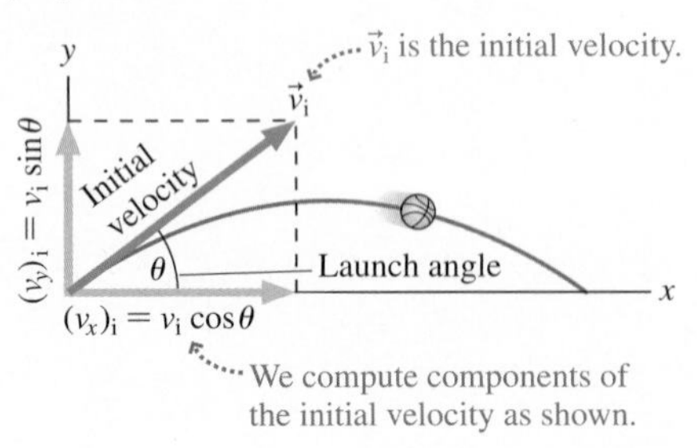

FIGURE 3.28 The launch and motion of a projectile.

Suppose you toss a basketball down the court, as shown in FIGURE 3.28. To study the resulting projectile motion, we've established a coordinate system with the x-axis horizontal and the y-axis vertical. In projectile motion, the angle of the initial velocity above the horizontal (i.e., above the x-axis) is called the **launch angle.** As you learned in Section 3.2, the initial velocity vector can be expressed in terms of its x- and y-components, which are shown in the figure.

NOTE ▶ The components $(v_x)_i$ and $(v_y)_i$ are not always positive. A projectile launched at an angle *below* the horizontal (such as a ball thrown downward from the roof of a building) has *negative* values for θ and $(v_y)_i$. However, the *speed* v_i is always positive. ◀

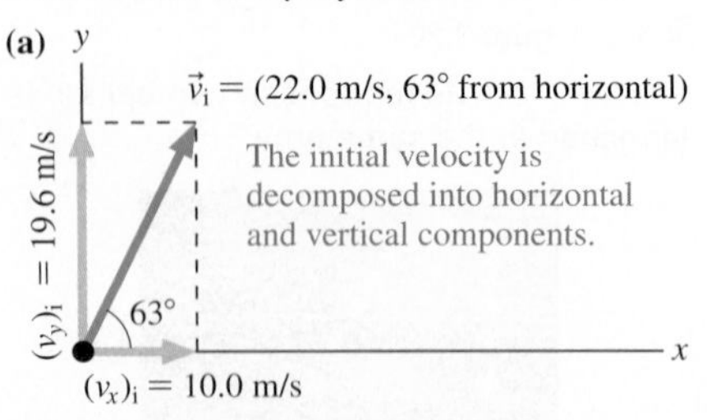

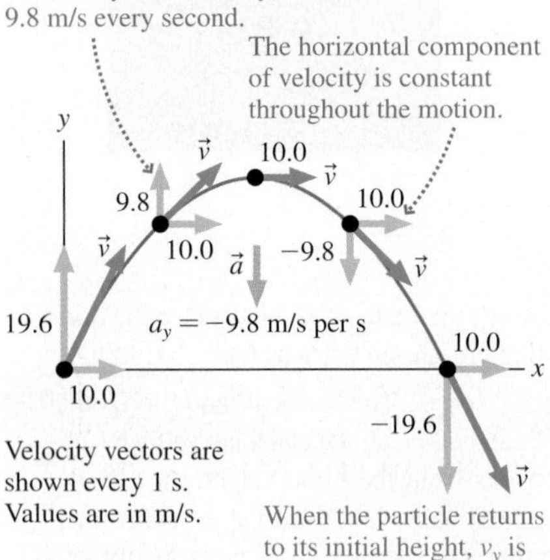

FIGURE 3.29 The velocity and acceleration vectors of a projectile.

Once the basketball leaves your hand, its subsequent motion is determined by the initial components of the velocity and the acceleration. To see how this plays out in practice, let's look at a specific case with definite numbers. FIGURE 3.29 shows a projectile launched at a speed of 22.0 m/s at an angle of 63° from the horizontal. In Figure 3.29a, the initial velocity vector v_i is broken into its horizontal and vertical components. In Figure 3.29b, the velocity vector and its component vectors are shown every 1.0 s. Because there is no horizontal acceleration ($a_x = 0$) the value of v_x never changes. In contrast, v_y decreases by 9.8 m/s every second. (This is what it *means* to accelerate at $a_y = -9.8\text{ m/s}^2 = (-9.8\text{ m/s})$ per second.) There is nothing *pushing* the projectile along the curve. As the ball moves, the downward acceleration changes the velocity vector as shown, causing it to tip downward as the motion proceeds. At the end of the arc, when the ball is at the same height as it started, v_y is -19.6 m/s, the negative of its initial value. **The ball finishes its motion moving downward at the same speed as it started moving upward,** just as we saw in the case of one-dimensional free fall in Chapter 2.

You can see from Figure 3.29 that **projectile motion is made up of two independent motions: uniform motion at constant velocity in the horizontal direction and free-fall motion in the vertical direction.** The two motions are independent, but because they occur together they must be analyzed together, as we'll see.

In Chapter 2, we saw kinematic equations for constant-velocity and constant-acceleration motion. We can adapt these equations to this current case. The horizontal motion is constant-velocity motion at $(v_x)_i$, and the vertical motion is constant-acceleration motion with initial velocity $(v_y)_i$ and an acceleration of $a_y = -g$.

Let's summarize and synthesize everything we've learned to this point about projectile motion, including the equations from Chapter 2 for the horizontal and vertical motion. We'll use this information to solve some problems in the next section.

SYNTHESIS 3.1 **Projectile motion**

The horizontal and vertical components of projectile motion are independent, but must be analyzed together.

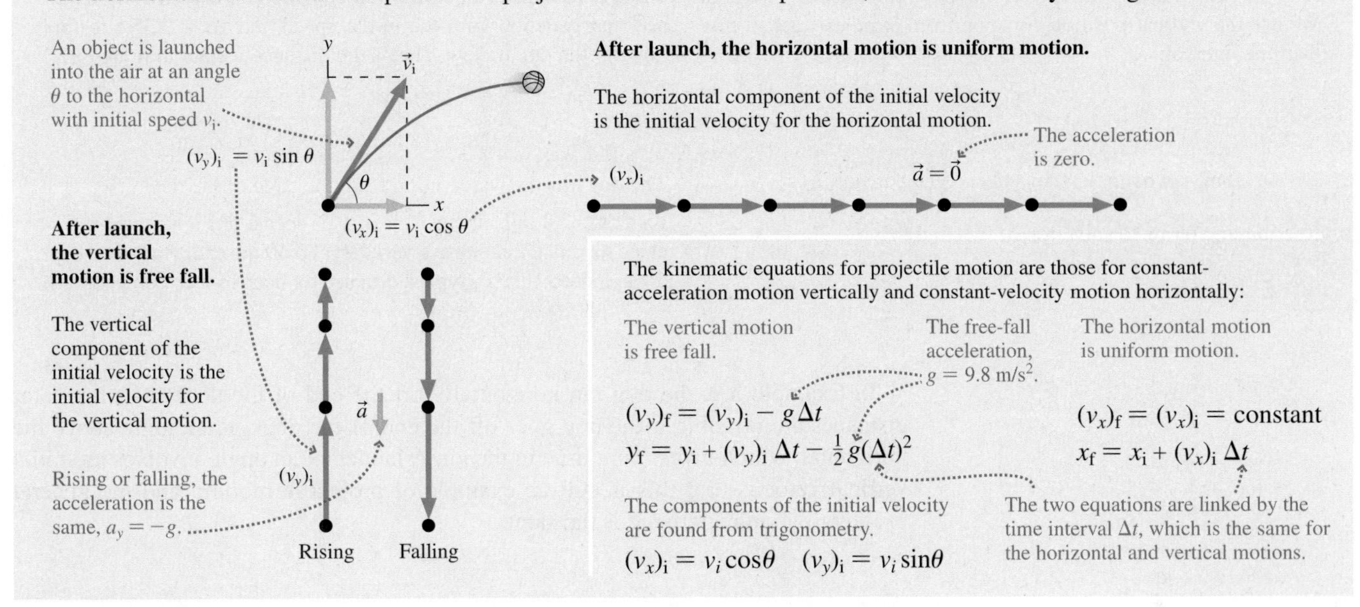

An object is launched into the air at an angle θ to the horizontal with initial speed v_i.
$(v_y)_i = v_i \sin\theta$
$(v_x)_i = v_i \cos\theta$
After launch, the vertical motion is free fall.
The vertical component of the initial velocity is the initial velocity for the vertical motion.
Rising or falling, the acceleration is the same, $a_y = -g$.
Rising
Falling
After launch, the horizontal motion is uniform motion.
The horizontal component of the initial velocity is the initial velocity for the horizontal motion.
The acceleration is zero.
$\vec{a} = \vec{0}$

The kinematic equations for projectile motion are those for constant-acceleration motion vertically and constant-velocity motion horizontally:

The vertical motion is free fall.

The free-fall acceleration, $g = 9.8\ \text{m/s}^2$.

The horizontal motion is uniform motion.

$$(v_y)_f = (v_y)_i - g\,\Delta t$$
$$y_f = y_i + (v_y)_i\,\Delta t - \tfrac{1}{2}g(\Delta t)^2$$

$$(v_x)_f = (v_x)_i = \text{constant}$$
$$x_f = x_i + (v_x)_i\,\Delta t$$

The components of the initial velocity are found from trigonometry.

$$(v_x)_i = v_i\cos\theta \quad (v_y)_i = v_i\sin\theta$$

The two equations are linked by the time interval Δt, which is the same for the horizontal and vertical motions.

STOP TO THINK 3.6 A 100 g ball rolls off a table and lands 2 m from the base of the table. A 200 g ball rolls off the same table with the same speed. How far does it land from the base of the table?

A. <1 m
B. 1 m
C. Between 1 m and 2 m
D. 2 m
E. Between 2 m and 4 m
F. 4 m

Video What the Physics? Projectile Paths

Video Ball Fired Upward from Moving Cart

3.6 Projectile Motion: Solving Problems

Now that we have a good idea of how projectile motion works, we can use that knowledge to solve some true two-dimensional motion problems.

EXAMPLE 3.9 **Dock jumping**

In the sport of dock jumping, dogs run at full speed off the end of a dock that sits a few feet above a pool of water. The winning dog is the one that lands farthest from the end of the dock. If a dog runs at 8.5 m/s (a pretty typical speed for this event) straight off the end of a dock that is 0.61 m (2 ft, a standard height) above the water, how far will the dog go before splashing down?

STRATEGIZE We can treat this as a projectile motion problem, so the dog's vertical motion is free fall and its horizontal motion is uniform motion. The relevant equations are given in Synthesis 3.1.

PREPARE We make a visual overview of the situation as in FIGURE 3.30. We have chosen to put the origin of the coordinate system at the base of the dock. The dog runs horizontally off the end of the dock, so the initial components of the velocity are $(v_x)_i = 8.5$ m/s and $(v_y)_i = 0$ m/s.

We know that the horizontal and vertical motions are independent. The fact that the dog is falling toward the water doesn't affect its horizontal motion. When the dog leaves the end of the dock, it will continue to move horizontally at 8.5 m/s. The vertical motion is free fall. The jump ends when the dog hits the water—that is, when it has dropped by 0.61 m. We are ultimately interested in how far the dog goes, but to make this determination we'll need to find the time interval Δt that the dog is in the air.

FIGURE 3.30 Visual overview for Example 3.9.

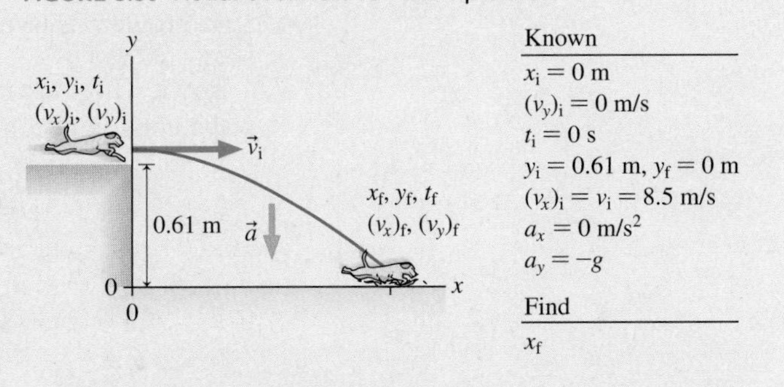

Continued

SOLVE We'll start by solving for the time interval Δt, the time the dog is in the air. This time is determined by the vertical motion, which is free fall with an initial velocity $(v_y)_i = 0$ m/s. We use the vertical-position equation from Synthesis 3.1 to find the time interval:

$$y_f = y_i + (v_y)_i \Delta t - \frac{1}{2} g (\Delta t)^2$$

$$0 \text{ m} = 0.61 \text{ m} + (0 \text{ m/s})\Delta t - \frac{1}{2}(9.8 \text{ m/s}^2)(\Delta t)^2$$

Rearranging terms to solve for Δt, we find that

$$\Delta t = 0.35 \text{ s}$$

This is how long it takes the dog's vertical motion to reach the water. During this time interval, the dog's horizontal motion is uniform motion at the initial velocity. We can use the horizontal-position equation with the initial speed and $\Delta t = 0.35$ s to find how far the dog travels. This is the distance we are looking for:

$$x_f = x_i + (v_x)_i \Delta t$$
$$= 0 \text{ m} + (8.5 \text{ m/s})(0.35 \text{ s}) = 3.0 \text{ m}$$

The dog hits the water 3.0 m from the edge of the dock.

ASSESS 3.0 m is about 10 feet. This seems like a reasonable distance for a dog running at a very fast clip off the end of a 2-foot-high dock. Indeed, this is a typical distance for dogs in such competitions.

In Example 3.9, the dog ran horizontally off the end of the dock. Much greater distances are possible if the dog goes off the end of the dock at an angle above the horizontal, which gives more time in the air. A launch at an angle involves an initial vertical velocity, but this is still an example of projectile motion, and the general problem-solving approach is the same.

PROBLEM-SOLVING APPROACH 3.1

Projectile motion problems

STRATEGIZE We will solve projectile motion problems by considering the horizontal and vertical motions as separate but related problems.

PREPARE There are a number of steps that you should go through in setting up the solution to a projectile motion problem:

- Make simplifying assumptions. Whether the projectile is a car or a basketball, the motion will be the same.
- Draw a visual overview including a pictorial representation showing the beginning and ending points of the motion.
- Establish a coordinate system with the x-axis horizontal and the y-axis vertical. In this case, you know that the horizontal acceleration will be zero and the vertical acceleration will be free fall: $a_x = 0$ and $a_y = -g$.
- Draw a vector representing the initial velocity, and find its x- and y-components in terms of the initial speed and the launch angle.
- Define symbols and write down a list of known values. Identify what the problem is trying to find.

SOLVE There are two sets of kinematic equations for projectile motion, one for the horizontal component and one for the vertical:

Horizontal	Vertical
$x_f = x_i + (v_x)_i \Delta t$	$y_f = y_i + (v_y)_i \Delta t - \frac{1}{2} g (\Delta t)^2$
$(v_x)_f = (v_x)_i = \text{constant}$	$(v_y)_f = (v_y)_i - g \Delta t$

Δt is the same for the horizontal and vertical components of the motion. Find Δt by solving for the vertical or the horizontal component of the motion; then use that value to complete the solution for the other component.

ASSESS Check that your result has the correct units, is reasonable, and answers the question.

EXAMPLE 3.10 A free kick

In a soccer free kick, a player kicks a stationary ball toward the goal that is 18 m away. He kicks the ball at an angle of 22° from the horizontal at a speed of 23 m/s. How long does the ball take to reach the goal? And how far off the ground is the ball when it reaches the goal?

STRATEGIZE This is a projectile motion problem with the initial velocity at an angle above the horizontal. This means that we will have both vertical and horizontal components of the initial velocity when we apply the equations of Problem-Solving Approach 3.1.

PREPARE Following Problem-Solving Approach 3.1, we prepare the visual overview shown in FIGURE 3.31. In choosing our axes, we've placed the origin at the point where the ball is kicked. The initial velocity vector is tilted at 22° above the horizontal, so the components of the initial velocity are

$$(v_x)_i = v_i \cos\theta = (23 \text{ m/s})(\cos 22°) = 21.3 \text{ m/s}$$

$$(v_y)_i = v_i \sin\theta = (23 \text{ m/s})(\sin 22°) = 8.6 \text{ m/s}$$

FIGURE 3.31 Visual overview of a kicked soccer ball.

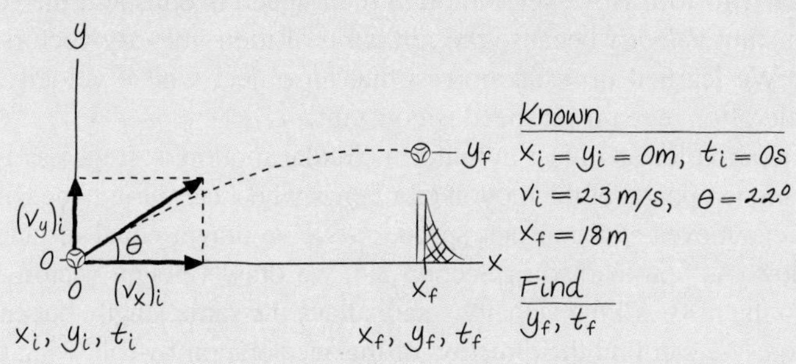

SOLVE The problem-solving approach suggests using one component of the motion to solve for Δt. We can do so using the kinematic equation for the horizontal position,

$$x_f = x_i + (v_x)_i \Delta t$$

We know that $x_i = 0$ m, so we can solve this equation for Δt to get

$$\Delta t = \frac{x_f}{(v_x)_i} = \frac{18 \text{ m}}{21.3 \text{ m/s}} = 0.845 \text{ s}$$

where we have kept one extra significant figure for our calculations.

We can now use the kinematic equation for the vertical position to find the height of the ball when it reaches the goal:

$$y_f = y_i + (v_y)_i \Delta t - \tfrac{1}{2}g(\Delta t)^2$$
$$= 0 \text{ m} + (8.6 \text{ m/s})(0.845 \text{ s}) - \tfrac{1}{2}(9.8 \text{ m/s}^2)(0.845 \text{ s})^2 = 3.8 \text{ m}$$

The ball is 3.8 m off the ground when it reaches the goal.

ASSESS A height of 3.8 m—about 12 ft—seems reasonable for the height of a kicked ball. Unfortunately, the height of the goal itself is only 2.44 m, so this kick sails easily over the crossbar.

The Range of a Projectile

When the quarterback throws a football down the field, how far will it go? What will be the **range** for this particular projectile motion—the horizontal distance traveled?

For a projectile, its initial speed and its launch angle are the two variables that determine the range. A higher speed means a greater range, of course. But how does the angle figure in?

FIGURE 3.32 shows the trajectory that a projectile launched at 100 m/s will follow for different launch angles. At very small or very large angles, the range is quite small. If you throw a ball at a 75° angle, it will do a great deal of up-and-down motion, but it won't achieve much horizontal travel. If you throw a ball at a 15° angle, the ball won't be in the air long enough to go very far. These cases both have the same range, as Figure 3.32 shows.

If the angle is too small or too large, the range is shorter than it could be. The "just right" case that gives the maximum range when the landing is at the same elevation as the launch is a launch angle of 45°, as Figure 3.32 shows.

For real-life projectiles, such as golf balls and baseballs, the optimal angle may be less than 45° because of air resistance. Up to this point we've ignored air resistance, but for small objects traveling at high speeds, air resistance is critical. Aerodynamic forces come into play, causing the projectile's trajectory to deviate from a parabola. The maximum range for a golf ball comes at an angle much less than 45°, as you no doubt know if you have ever played golf.

FIGURE 3.32 Trajectories of a projectile for different launch angles, assuming air resistance can be ignored.

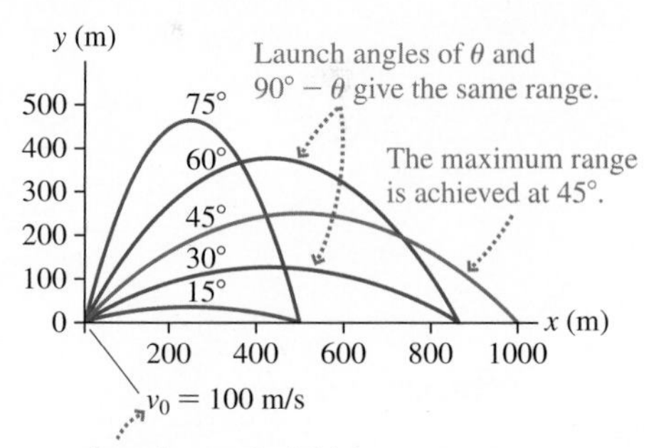

► **A long long jump** A 45° angle gives the greatest range for a projectile, so why do long jumpers take off at a much shallower angle? Two of the assumptions that lead to the 45° optimal angle don't apply here. The athlete changes the position of his legs in the air—he doesn't land at the same height as that from which he took off. Also, athletes can't keep the same launch speed for different angles—they can jump faster at smaller angles. The gain from the faster speed outweighs the effect of the smaller angle.

STOP TO THINK 3.7 A baseball player is taking batting practice. He hits two successive pitches at different angles but at exactly the same speed. Ball 1 and ball 2 follow the paths shown. Which ball is in the air for a longer time? Assume that you can ignore air resistance for this problem.

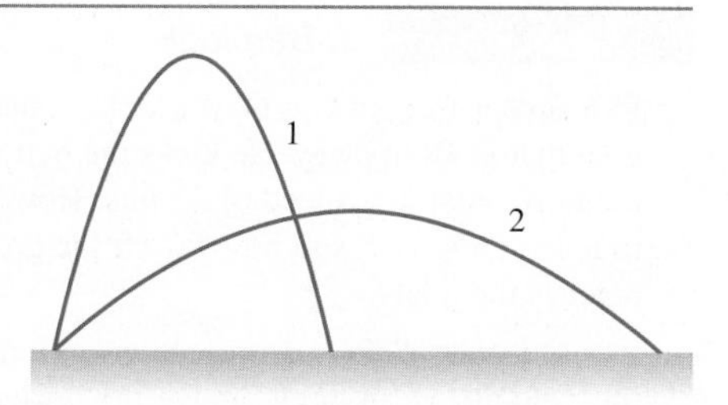
1
2

A. Ball 1
B. Ball 2
C. Both balls are in the air for the same amount of time.

3.7 Circular Motion

The London Eye Ferris wheel.

The 32 cars on the London Eye Ferris wheel move at about 0.5 m/s in a vertical circle with a 65 m radius. The cars have a *constant speed,* and so we say that the cars are in **uniform circular motion.** However, although their speed is constant, the cars do not move with constant *velocity* because the *direction* of their velocity vectors is constantly changing. We learned in ◀◀ SECTION 3.4 that an object whose velocity is changing has an acceleration, even if its speed is constant.

Let's find the acceleration for an object in uniform circular motion. FIGURE 3.33 is a motion diagram showing ten points of the motion of a Ferris wheel car during one complete revolution. The car moves at a constant speed, so we've drawn equal distances between successive dots. As you learned in Section 3.4, we draw velocity vectors by connecting each dot to the next. All the velocity vectors have the same length, but each has a different direction. We can find the direction of the acceleration by following the steps listed in Tactics Box 3.3, where we found that the acceleration points in the same direction as the *change* in velocity $\Delta\vec{v}$. Figure 3.33 shows how to do this for two successive velocities $\vec{v}_1$ and $\vec{v}_2$. We see that $\Delta\vec{v}$—and therefore the acceleration—is directed

KEY CONCEPT FIGURE 3.33 Motion diagram for uniform circular motion.

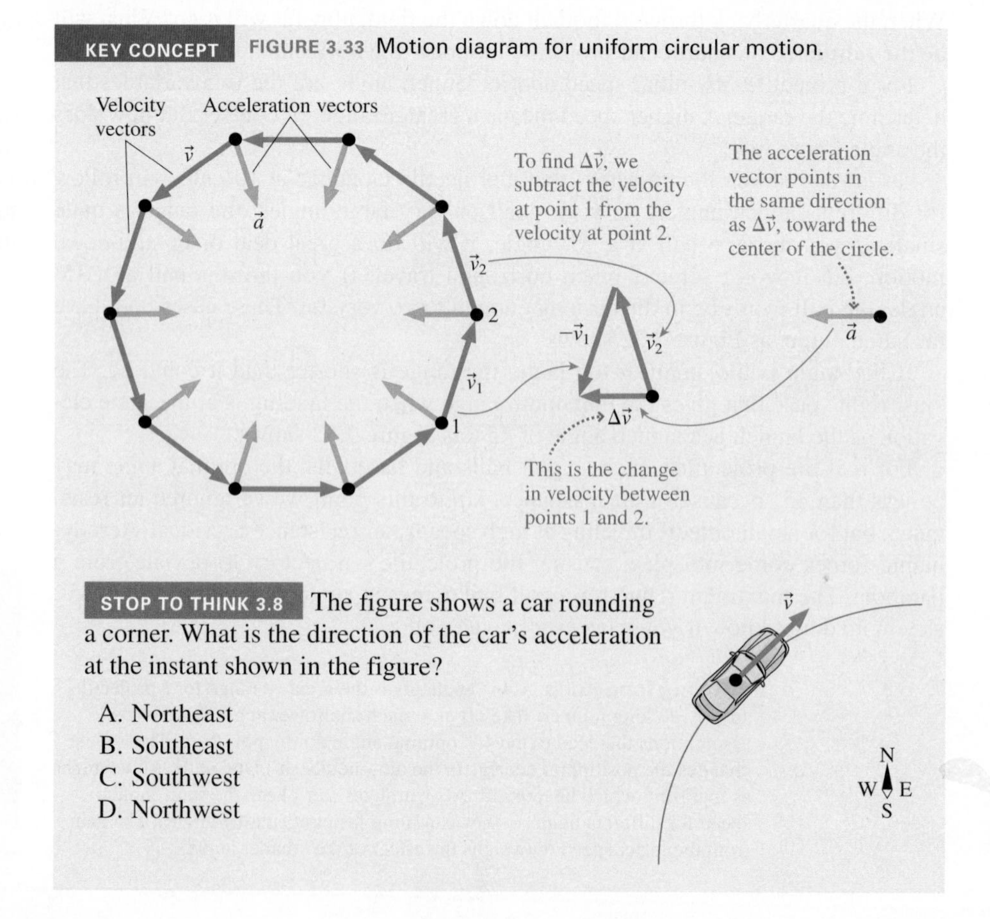
Velocity vectors
Acceleration vectors
$\vec{v}$
$\vec{a}$
$\vec{v}_2$
2
$\vec{v}_1$
1
To find $\Delta\vec{v}$, we subtract the velocity at point 1 from the velocity at point 2.
$-\vec{v}_1$
$\vec{v}_2$
$\Delta\vec{v}$
This is the change in velocity between points 1 and 2.
The acceleration vector points in the same direction as $\Delta\vec{v}$, toward the center of the circle.
$\vec{a}$
$\vec{v}$
N
W
E
S

STOP TO THINK 3.8 The figure shows a car rounding a corner. What is the direction of the car's acceleration at the instant shown in the figure?

A. Northeast
B. Southeast
C. Southwest
D. Northwest

toward the center of the circle. No matter which points we select on the motion diagram, the velocities change in such a way that **the acceleration vector $\vec{a}$ points toward the center of the circle.** This is a key result for uniform circular motion.

The acceleration that always points directly toward the center of a circle is called **centripetal acceleration.** The word "centripetal" comes from a Greek root meaning "center seeking." FIGURE 3.34 summarizes what we've learned about centripetal acceleration.

FIGURE 3.34 The velocity and acceleration vectors for circular motion.

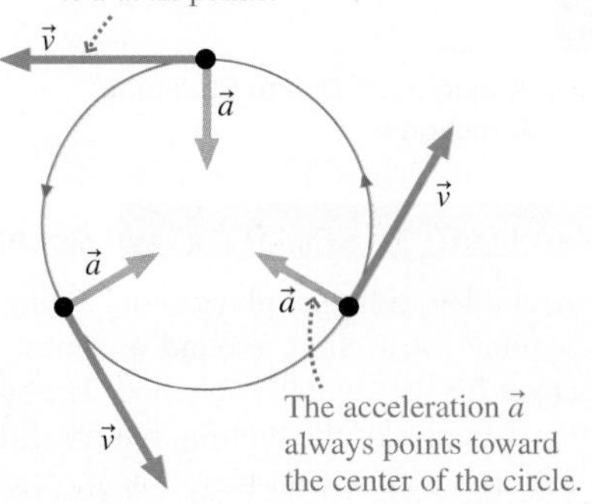
The velocity $\vec{v}$ is always tangent to the circle and perpendicular to $\vec{a}$ at all points.
$\vec{v}$
$\vec{a}$
$\vec{v}$
$\vec{a}$
$\vec{a}$
$\vec{v}$
The acceleration $\vec{a}$ always points toward the center of the circle.

NOTE ▸ Centripetal acceleration is not a new type of acceleration. It is simply the name given to the acceleration of an object moving in a circular path. ◂

To complete our description of circular motion, we need to find a quantitative relationship between the magnitude of the acceleration a and the speed v. Let's return to the case of the Ferris wheel. As shown in FIGURE 3.35a, during a time Δt in which a car on the Ferris wheel moves along the circle from point 1 to point 2, the car moves through an angle θ and undergoes a displacement $\vec{d}$. We've chosen a relatively large angle θ for our drawing so that angular relationships can be clearly seen, but for a small angle the displacement is essentially identical to the actual distance traveled along the circular arc, and we'll make this approximation.

FIGURE 3.35 Changing position and velocity for an object in circular motion.

(a) As the car moves from point 1 to point 2, the displacement is $\vec{d}$.

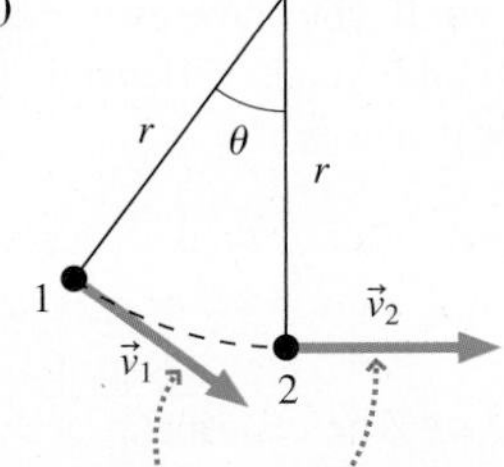
(b)
r
θ
r
1
$\vec{v}_2$
$\vec{v}_1$
2
The magnitude of the velocity is constant, but the direction changes.

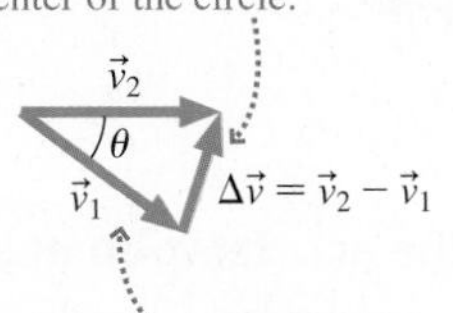
(c) The change in velocity is a vector pointing toward the center of the circle.
$\vec{v}_2$
θ
$\vec{v}_1$
$\Delta\vec{v} = \vec{v}_2 - \vec{v}_1$

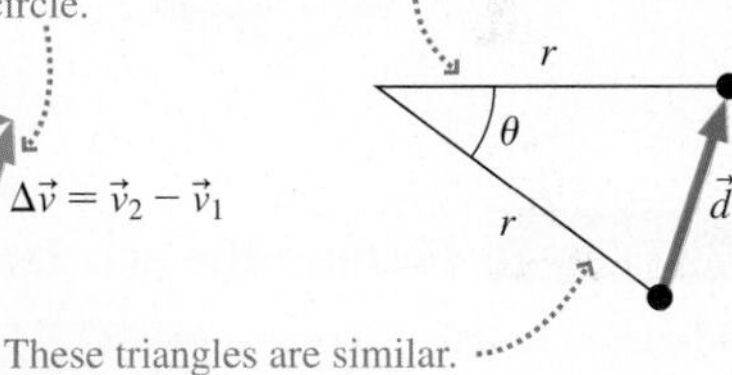
This triangle is the same as in part a, but rotated.
r
θ
r
$\vec{d}$
These triangles are similar.

FIGURE 3.35b shows how the velocity changes as the car moves, and FIGURE 3.35c shows the vector calculation of the change in velocity. The triangle we use to make this calculation is geometrically *similar* to the one that shows the displacement, as Figure 3.35c shows. This is a key piece of information: You'll remember from geometry that similar triangles have equal ratios of their sides, so we can write

$$\frac{\Delta v}{v} = \frac{d}{r} \tag{3.19}$$

where Δv is the magnitude of the velocity-change vector $\Delta\vec{v}$. We've used the unsubscripted speed v for the length of a side of the first triangle because it is the same for velocities $\vec{v}_1$ and $\vec{v}_2$.

Now we're ready to compute the acceleration. The displacement is just the speed v times the time interval Δt, so we can write

$$d = v\Delta t$$

We can substitute this for d in Equation 3.19 to obtain

$$\frac{\Delta v}{v} = \frac{v\Delta t}{r}$$

which we can rearrange as

$$\frac{\Delta v}{\Delta t} = \frac{v^2}{r}$$

We recognize the left-hand side of the equation as the acceleration, so this becomes

$$a = \frac{v^2}{r}$$

Combining this magnitude with the direction we noted previously, we can write the centripetal acceleration as

Video Acceleration Due to Changing Direction

$$\vec{a} = \left(\frac{v^2}{r}, \text{ toward center of circle}\right) \qquad (3.20)$$

Centripetal acceleration of an object moving in a circle of radius r at speed v

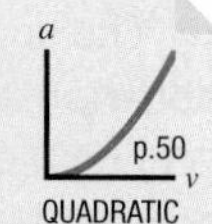

CONCEPTUAL EXAMPLE 3.11 Acceleration on a swing

A child is riding a playground swing. The swing rotates in a segment of a circle around a central point where the rope or chain for the swing is attached. The speed isn't changing at the lowest point of the motion, but the direction is—this is circular motion, with an acceleration directed upward, as shown in FIGURE 3.36. More acceleration will mean a more exciting ride.

FIGURE 3.36 A child at the lowest point of motion on a swing.

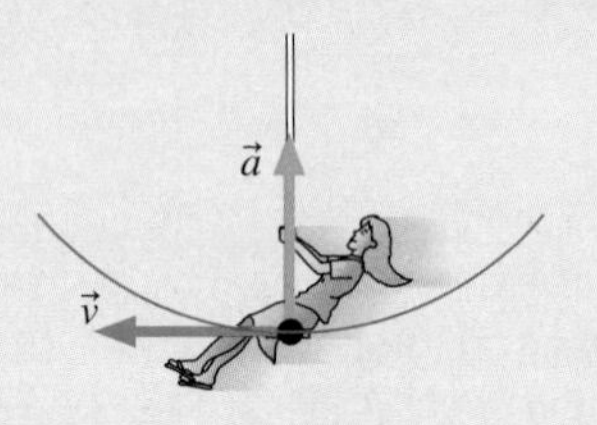

What change could the child make to increase the acceleration she experiences?

REASON The acceleration the child experiences is the "changing direction" acceleration of circular motion, given by Equation 3.20. The acceleration depends on the speed and the radius of the circle. The radius of the circle is determined by the length of the chain or rope, so the only easy way to change the acceleration is to change the speed, which she could do by swinging higher. Because the acceleration is proportional to the square of the speed, doubling the speed means a fourfold increase in the acceleration.

ASSESS If you have ever ridden a swing, you know that the acceleration you experience is greater the faster you go—so our answer makes sense.

EXAMPLE 3.12 Finding the acceleration of speed skaters

World-class female short-track speed skaters can cover the 500 m of a race in 45 s. The most challenging elements of the race are the turns, which are very tight, with a radius of approximately 11 m. Estimate the magnitude of the skater's centripetal acceleration in a turn.

STRATEGIZE The skaters are moving in a circular path, so they have a centripetal acceleration directed inward, toward the center of the curve. We'll need to know their speed, which we can find from the distance around the track and the time to complete the race.

PREPARE The centripetal acceleration depends on two quantities: the radius of the turn (given as approximately 11 m) and the speed. The speed varies during the race, but we can make a good estimate of the speed by using the total distance and time:

$$v \simeq \frac{500 \text{ m}}{45 \text{ s}} = 11 \text{ m/s}$$

SOLVE We can use these values to estimate the magnitude of the acceleration:

$$a = \frac{v^2}{r} \simeq \frac{(11 \text{ m/s})^2}{11 \text{ m}} = 11 \text{ m/s}^2$$

ASSESS This is a large acceleration—a bit more than g—but the photo shows the skaters leaning quite hard into the turn, so such a large acceleration seems quite reasonable.

STOP TO THINK 3.9 Which of the following particles has the greatest centripetal acceleration?

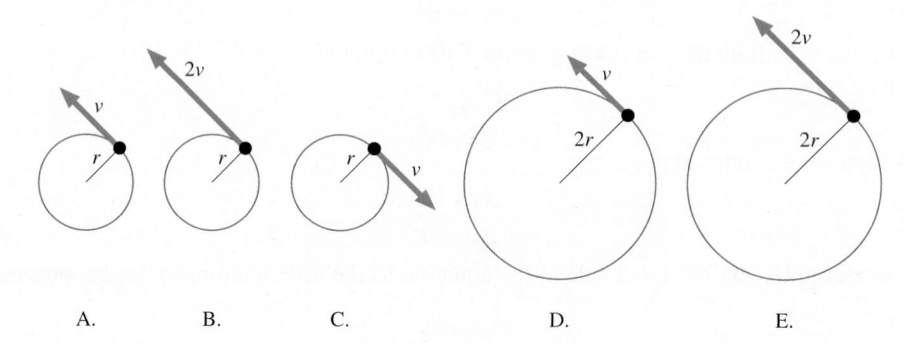

3.8 Relative Motion

You've now dealt many times with problems that say something like "A car travels at 30 m/s" or "A plane travels at 300 m/s." But, as we will see, we may need to be a bit more specific.

In FIGURE 3.37, Amy, Bill, and Carlos are watching a runner. According to Amy, the runner's velocity is $v_x = 5$ m/s. But to Bill, who's riding a bicycle alongside, the runner is lifting his legs up and down but going neither forward nor backward relative to Bill. As far as Bill is concerned, the runner's velocity is $v_x = 0$ m/s. Driving his car, Carlos sees the runner receding in his rearview mirror, in the *negative* x-direction, getting 10 m farther away from him every second. According to Carlos, the runner's velocity is $v_x = -10$ m/s. Which is the runner's *true* velocity?

FIGURE 3.37 Amy, Bill, and Carlos each measure the velocity of the runner. The velocities are shown relative to Amy.

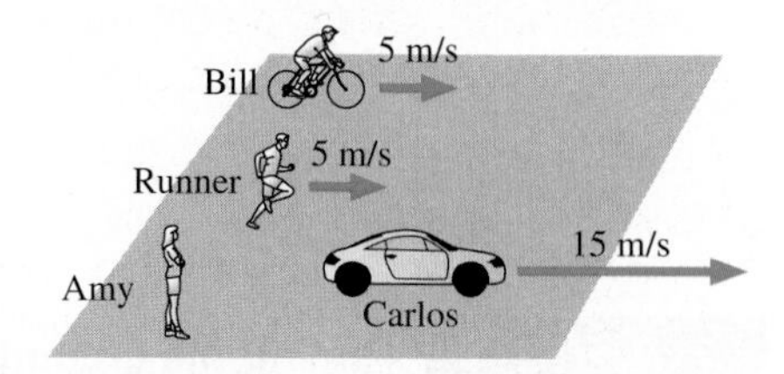

As we'll see, there is no single true velocity for an object; we can define an object's velocity only *relative* to the observer who is measuring it. The runner's velocity *relative to Amy* is 5 m/s; that is, his velocity is 5 m/s in a coordinate system attached to Amy and in which Amy is at rest. The runner's velocity relative to Bill is 0 m/s, and the velocity relative to Carlos is -10 m/s. These are all valid descriptions of the runner's motion.

Relative Velocity

Suppose we know that the runner's velocity relative to Amy is 5 m/s; we will call this velocity $(v_x)_{RA}$. The second subscript "RA" means "**R**unner relative to **A**my." We also know that the velocity of **C**arlos relative to **A**my is 15 m/s; we write this as $(v_x)_{CA} = 15$ m/s. It is equally valid to compute Amy's velocity relative to Carlos. From Carlos's point of view, Amy is moving to the left at 15 m/s. We write **A**my's velocity relative to **C**arlos as $(v_x)_{AC} = -15$ m/s. Note that $(v_x)_{AC} = -(v_x)_{CA}$.

Given the runner's velocity relative to Amy and Amy's velocity relative to Carlos, we can compute the runner's velocity relative to Carlos by combining the two velocities we know. The subscripts as we have defined them are our guide for this combination:

$$(v_x)_{RC} = (v_x)_{RA} + (v_x)_{AC} \tag{3.21}$$

The "A" appears on the right of the first expression and on the left of the second; when we combine these velocities, we "cancel" the A to get $(v_x)_{RC}$.

Throwing for the gold An athlete throwing the javelin does so while running. It's harder to throw the javelin on the run, but there's a very good reason to do so. The distance of the throw will be determined by the velocity of the javelin with respect to the ground—which is the sum of the velocity of the throw plus the velocity of the athlete. A faster run means a longer throw.

Generally, we can add two relative velocities in this manner, by "canceling" subscripts as in Equation 3.21. In Chapter 27, when we learn about relativity, we will have a more rigorous scheme for computing relative velocities, but this technique will serve our purposes at present.

EXAMPLE 3.13 Speed of a seabird

Researchers doing satellite tracking of albatrosses in the Southern Ocean observed a bird maintaining sustained flight speeds of 35 m/s—nearly 80 mph! This seems surprisingly fast until you realize that this particular bird was flying with the wind, which was moving at 23 m/s. What was the bird's airspeed—its speed relative to the air? This is a truer measure of its flight speed.

STRATEGIZE We are asked how fast the bird appears to be moving relative to someone drifting along at the same speed as the wind. We will use Equation 3.21 to find this relative velocity.

PREPARE FIGURE 3.38 shows the wind and the albatross moving to the right, so all velocities will be positive. We've shown the velocity $(v_x)_{bw}$ of the **b**ird with respect to the **w**ater, which is the measured flight speed, and the velocity $(v_x)_{aw}$ of the **a**ir with respect to the **w**ater, which is the known wind speed. We want to find the bird's airspeed—the speed of the **b**ird with respect to the **a**ir.

FIGURE 3.38 Relative velocities for the albatross and the wind for Example 3.13.

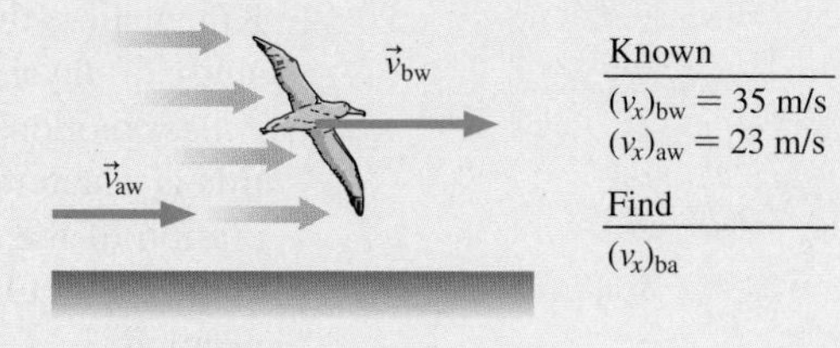

Continued

SOLVE We've noted three different velocities that are important in the problem: $(v_x)_{bw}$, $(v_x)_{aw}$, and $(v_x)_{ba}$. We can combine these in the usual way:

$$(v_x)_{bw} = (v_x)_{ba} + (v_x)_{aw}$$

Then, to solve for $(v_x)_{ba}$, we can rearrange the terms:

$$(v_x)_{ba} = (v_x)_{bw} - (v_x)_{aw} = 35 \text{ m/s} - 23 \text{ m/s} = 12 \text{ m/s}$$

ASSESS 12 m/s—about 25 mph—is a reasonable airspeed for a bird. And it's slower than the observed flight speed, which makes sense because the bird is flying with the wind.

This technique for finding relative velocities also works for two-dimensional situations, another good exercise in working with vectors.

EXAMPLE 3.14 Finding the ground speed of an airplane

Cleveland is approximately 300 miles east of Chicago. A plane leaves Chicago flying due east at 500 mph. The pilot forgot to check the weather and doesn't know that the wind is blowing to the south at 100 mph. What is the plane's velocity relative to the ground?

STRATEGIZE We will generalize the idea of relative velocity to the case where the velocities are not directed along a single line by using *vectors* to represent the velocities in Equation 3.21.

PREPARE FIGURE 3.39 is a visual overview of the situation. We are given the velocity of the **p**lane relative to the **a**ir ($\vec{v}_{pa}$) and the velocity of the **a**ir relative to the **g**round ($\vec{v}_{ag}$); the velocity of the **p**lane relative to the **g**round will be the vector sum of these velocities:

$$\vec{v}_{pg} = \vec{v}_{pa} + \vec{v}_{ag}$$

This vector sum is shown in Figure 3.39.

FIGURE 3.39 The wind causes a plane flying due east in the air to move to the southeast relative to the ground.

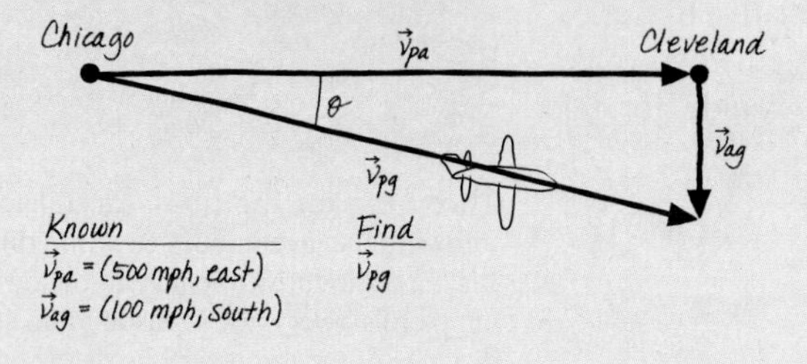

SOLVE The plane's speed relative to the ground is the hypotenuse of the right triangle in Figure 3.39; thus:

$$v_{pg} = \sqrt{v_{pa}^2 + v_{ag}^2} = \sqrt{(500 \text{ mph})^2 + (100 \text{ mph})^2} = 510 \text{ mph}$$

The plane's direction can be specified by the angle θ measured from due east:

$$\theta = \tan^{-1}\left(\frac{100 \text{ mph}}{500 \text{ mph}}\right) = \tan^{-1}(0.20) = 11°$$

The velocity of the plane relative to the ground is thus

$$\vec{v}_{pg} = (510 \text{ mph}, 11° \text{ south of east})$$

ASSESS The good news is that the wind is making the plane move a bit faster relative to the ground. The bad news is that the wind is making the plane move in the wrong direction!

STOP TO THINK 3.10 The water in a river flows downstream at 3.0 m/s. A boat is motoring upstream against the flow at 5.0 m/s relative to the water. What is the boat's speed relative to the riverbank?

A. 8.0 m/s B. 5.0 m/s C. 3.0 m/s D. 2.0 m/s

What Comes Next: Forces

Kinematics, the mathematical description of motion, is a good place to start our study of physics because motion is very visible and very familiar. But what actually *causes* motion? The skaters in Example 3.12 were "leaning quite hard into the turn"; this is a statement about the forces acting on them, the forces that cause the acceleration of the turn. In the next chapter, we'll explore the nature of forces and the connection to motion, giving us the tools to treat a much broader range of problems.

INTEGRATED EXAMPLE 3.15 World-record jumpers BIO

Frogs, with their long, strong legs, are excellent jumpers. And thanks to the good folks of Calaveras County, California, who have a jumping frog contest every year in honor of a Mark Twain story, we have very good data as to just how far a determined frog can jump. The current record holder is Rosie the Ribeter, a bullfrog that made a leap of 2.2 m from a standing start. This compares favorably with the world record for a human, which is a mere 3.7 m.

Typical data for a serious leap by a bullfrog look like this: The frog goes into a crouch, then rapidly extends its legs by 15 cm as it pushes off, leaving the ground at an angle of 30° to the horizontal. It's in the air for 0.49 s before landing at the same height from which it took off. Given this leap, what is the acceleration while the frog is pushing off? How far does the frog jump?

STRATEGIZE The problem really has two parts: the leap through the air and the acceleration required to produce this leap. We'll need to analyze the leap—the projectile motion—first, which will give us the frog's launch speed and the distance of the jump. Once we know the velocity with which the frog leaves the ground, we can calculate its acceleration while pushing off the ground.

PREPARE Let's start with a visual overview of the two parts, as shown in FIGURE 3.40. Notice that the second part of the problem uses a different x-axis, tilted as we did earlier for motion on a ramp.

SOLVE The "flying through the air" part shown in Figure 3.40a is projectile motion. The frog lifts off at a 30° angle with a speed v_i. The x- and y-components of the initial velocity are

$$(v_x)_i = v_i \cos 30^\circ$$
$$(v_y)_i = v_i \sin 30^\circ$$

The vertical motion can be analyzed as we did in Example 3.10. The kinematic equation is

$$y_f = y_i + (v_y)_i \Delta t - \tfrac{1}{2}g(\Delta t)^2$$

We know that $y_f = y_i = 0$, so this reduces to

$$(v_y)_i = \tfrac{1}{2}g\Delta t = \tfrac{1}{2}(9.8 \text{ m/s}^2)(0.49 \text{ s}) = 2.4 \text{ m/s}$$

We know the y-component of the velocity and the angle, so we can find the magnitude of the velocity and the x-component:

$$v_i = \frac{(v_y)_i}{\sin 30^\circ} = \frac{2.4 \text{ m/s}}{\sin 30^\circ} = 4.8 \text{ m/s}$$

$$(v_x)_i = v_i \cos 30^\circ = (4.8 \text{ m/s}) \cos 30^\circ = 4.2 \text{ m/s}$$

The horizontal motion is uniform motion, so the frog's horizontal position when it returns to the ground is

$$x_f = x_i + (v_x)_i \Delta t = 0 + (4.2 \text{ m/s})(0.49 \text{ s}) = 2.1 \text{ m}$$

This is the length of the jump.

Now that we know how fast the frog is going when it leaves the ground, we can calculate the acceleration necessary to produce this jump—the "pushing off the ground" part shown in Figure 3.40b. We've drawn the x-axis along the direction of motion, as we did for problems of motion on a ramp. We know the displacement Δx of the launch but not the time, so we can use Equation 2.13 from Synthesis 2.1:

$$(v_x)_f^2 = (v_x)_i^2 + 2a_x \Delta x$$

The initial velocity is zero, the final velocity is $(v_x)_f = 4.8$ m/s, and the displacement is the 15 cm (or 0.15 m) stretch of the legs during the jump. Thus the frog's acceleration while pushing off is

$$a_x = \frac{(v_x)_f^2}{2\,\Delta x} = \frac{(4.8 \text{ m/s})^2}{2(0.15 \text{ m})} = 77 \text{ m/s}^2$$

ASSESS A 2.1 m jump is an impressive jump, but it's less than the record for a frog, so the final result for the distance seems reasonable. Such a long jump must require a large acceleration during the pushing-off phase, which is what we found.

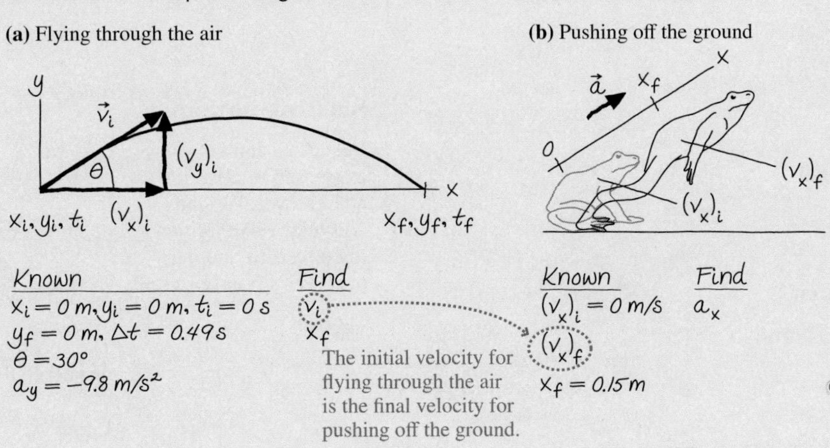

FIGURE 3.40 A visual overview for the leap of a frog.

SUMMARY

GOAL **To learn more about vectors and to use vectors as a tool to analyze motion in two dimensions.**

GENERAL PRINCIPLES

Projectile Motion

A projectile is an object that moves through the air under the influence of gravity and nothing else.

The path of the motion is a parabola.

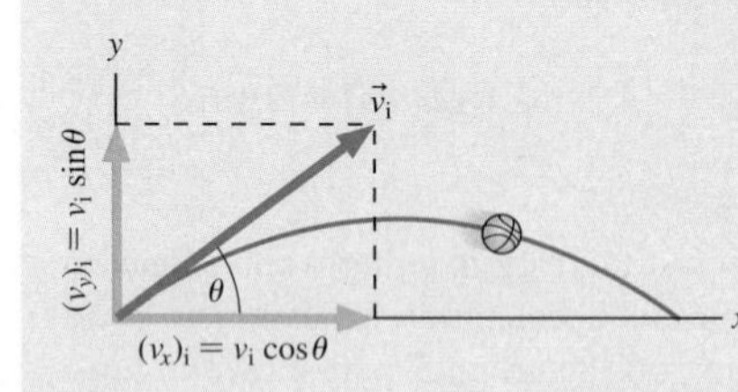

The motion consists of two pieces:

1. Vertical motion with free-fall acceleration, $a_y = -g$
2. Horizontal motion with constant velocity

Kinematic equations:

$$x_f = x_i + (v_x)_i\,\Delta t$$

$$(v_x)_f = (v_x)_i = \text{constant}$$

$$y_f = y_i + (v_y)_i\,\Delta t - \tfrac{1}{2}g(\Delta t)^2$$

$$(v_y)_f = (v_y)_i - g\,\Delta t$$

Circular Motion

An object moving in a circle at a constant speed has a velocity that is constantly changing direction, and so experiences an acceleration:

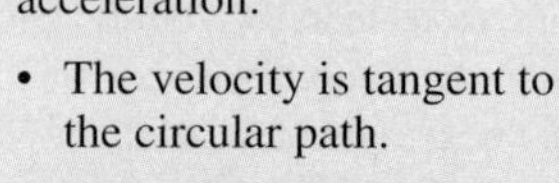

- The velocity is tangent to the circular path.
- The acceleration points toward the center of the circle and has magnitude

$$a = \frac{v^2}{r}$$

IMPORTANT CONCEPTS

Vectors and Components

A vector can be decomposed into x- and y-**components.**

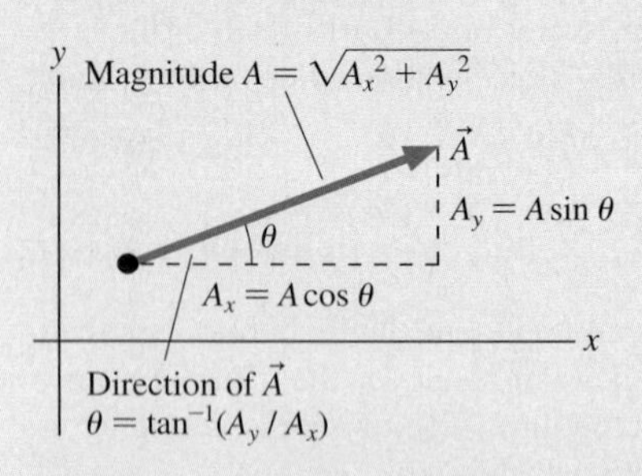

The magnitude and direction of a vector can be expressed in terms of its components.

The sign of the components depends on the direction of the vector:

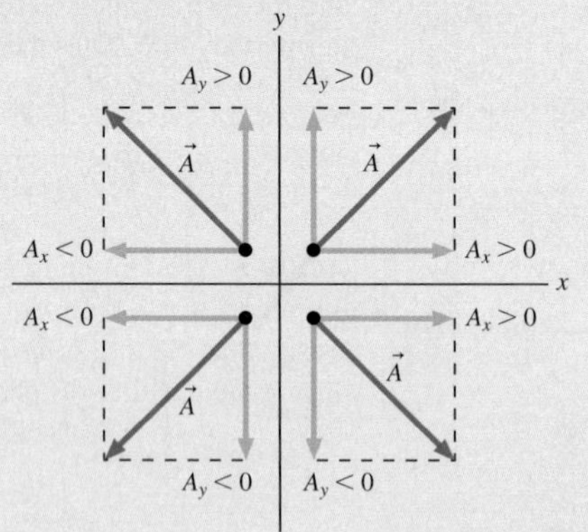

The Acceleration Vector

We define the acceleration vector as

$$\vec{a} = \frac{\vec{v}_f - \vec{v}_i}{t_f - t_i} = \frac{\Delta\vec{v}}{\Delta t}$$

We find the acceleration vector on a motion diagram as follows:

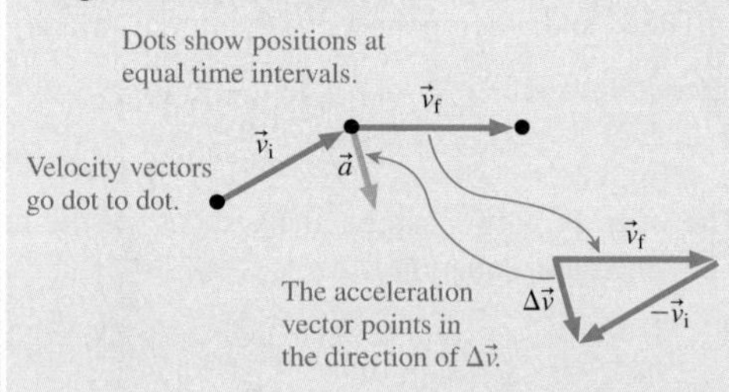

APPLICATIONS

Motion on a ramp

An object sliding down a ramp will accelerate parallel to the ramp:

$$a_x = \pm g\sin\theta$$

The correct sign depends on the direction in which the ramp is tilted.

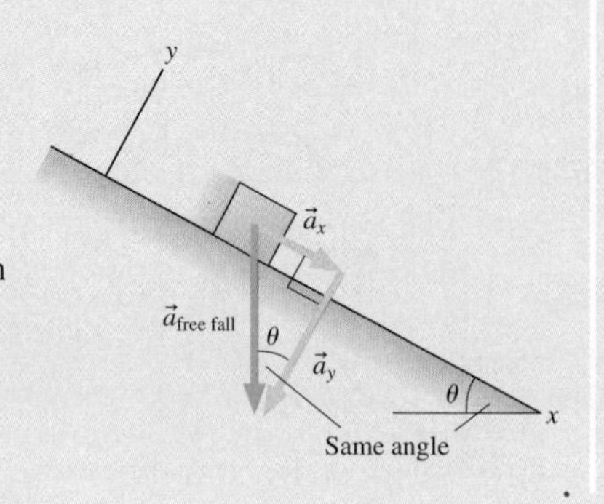

Relative motion

Velocities can be expressed relative to an observer. We can add relative velocities to convert to another observer's point of view.

c = car, r = runner, g = ground

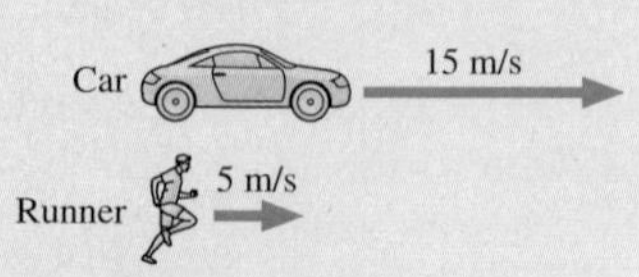

The speed of the car with respect to the runner is

$$(v_x)_{cr} = (v_x)_{cg} + (v_x)_{gr}$$

Learning Objectives
After studying this chapter, you should be able to:

- Add and subtract vectors graphically. *Conceptual Questions 3.2, 3.3; Problems 3.1, 3.2, 3.3, 3.4*
- Use vectors and motion diagrams to find acceleration. *Conceptual Question 3.12; Problems 3.21, 3.22, 3.23, 3.24, 3.25*
- Perform vector calculations using vector components. *Problems 3.5, 3.6, 3.9, 3.13, 3.52*
- Solve problems about an object moving on a ramp. *Conceptual Question 3.11; Problems 3.16, 3.17, 3.18, 3.19, 3.20, 3.59*
- Understand and use relative velocity. *Conceptual Question 3.19; Problems 3.45, 3.46, 3.47, 3.49, 3.76*
- Solve problems for projectiles that follow parabolic trajectories. *Conceptual Questions 3.6, 3.9; Problems 3.27, 3.28, 3.31, 3.34, 3.66*
- Calculate and use centripetal acceleration. *Conceptual Questions 3.15, 3.17; Problems 3.37, 3.38, 3.40, 3.41, 3.43*

STOP TO THINK ANSWERS

Chapter Preview Stop to Think: D. The ball took 2.0 s to reach the highest point, at which time it is (momentarily) at rest. As it rises, its vertical velocity is changing at $a_y = -g = -9.8\text{ m/s}^2 \approx -10\text{ m/s}^2$. The change in speed is $\Delta v_y \approx (-10\text{ m/s}^2)(2.0\text{ s}) = -20\text{ m/s}$, so the initial speed is approximately 20 m/s.

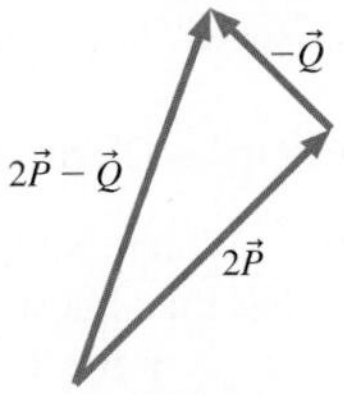

Stop to Think 3.1: A. The graphical construction of $2\vec{P} - \vec{Q}$ is shown at right.

$-\vec{Q}$, $2\vec{P} - \vec{Q}$, $2\vec{P}$

Stop to Think 3.2: B. From the axes on the graph, we can see that the x- and y-components are -4 cm and $+2$ cm, respectively.

Stop to Think 3.3: C. Vector $\vec{C}$ points to the left and down, so both C_x and C_y are negative. C_x is in the numerator because it is the side opposite ϕ.

Stop to Think 3.4: B. The angle of the slope is greatest in this case, leading to the greatest acceleration.

Stop to Think 3.5: A, B, and C. If car A has a constant speed, then at the point shown its acceleration will be to the left; this is like the valley of Example 3.7 turned counterclockwise. Car B can also have the acceleration shown; imaging tilting the valley slightly counterclockwise. Car C is undergoing linear motion, so its acceleration will point to the left if it is either moving to the left and speeding up, or moving to the right and slowing down. But car D's acceleration can never point as shown: In the valley example, the acceleration vectors always point to the inside of the curve, never to the outside, no matter whether the skier is speeding up or slowing down.

Stop to Think 3.6: D. Mass does not appear in the kinematic equations, so the mass has no effect. The balls follow the same path.

Stop to Think 3.7: A. The time in the air is determined by the vertical component of the velocity. Ball 1 has a higher vertical velocity, so it will be in the air for a longer time.

Stop to Think 3.8: B. The acceleration is directed toward the center of the circle. At the instant shown, the acceleration is down and to the right—to the southeast.

Stop to Think 3.9: B. The magnitude of the acceleration is v^2/r. Acceleration is greatest for the combination of highest speed and smallest radius.

Stop to Think 3.10: D. Let's define upstream as positive and downstream as negative. Then the speed of the boat with respect to the water is $v_{bw} = +5$ m/s, and the speed of the water with respect to the riverbank is $v_{wr} = -3$ m/s. Thus the speed of the boat with respect to the riverbank is

$$v_{br} = v_{bw} + v_{wr}$$
$$= 5\text{ m/s} - 3\text{ m/s} = 2\text{ m/s}$$

Video Tutor Solution Chapter 3

QUESTIONS

Conceptual Questions

1. a. Can a vector have nonzero magnitude if a component is zero? If no, why not? If yes, give an example.
   b. Can a vector have zero magnitude and a nonzero component? If no, why not? If yes, give an example.
2. If $\vec{C} = \vec{A} + \vec{B}$ can $C = 0$? Can $C < 0$? For each question, show how or explain why not.
3. Suppose two vectors have unequal magnitudes. Can their sum be $\vec{0}$? Explain.
4. At the instant shown, is the particle in Figure Q3.4 speeding up, slowing down, or traveling at a constant speed?

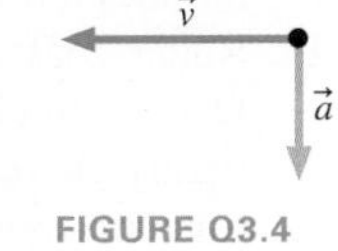

$\vec{v}$, $\vec{a}$

FIGURE Q3.4

5. For a projectile, which of the following quantities are constant during the flight: x, y, v_x, v_y, v, a_x, a_y? Which of the quantities are zero throughout the flight?

Problem difficulty is labeled as | (straightforward) to ||||| (challenging). INT Problems labeled integrate significant material from earlier chapters; BIO are of biological or medical interest.

The eText icon indicates when there is a video tutor solution available for the chapter or for a specific problem. To launch these videos, log into your eText through Mastering™ Physics or log into the Study Area.

6. A baseball player throws a ball at a 40° angle to the ground. The ball lands on the ground some distance away.
   a. Is there any point on the trajectory where $\vec{v}$ and $\vec{a}$ are parallel to each other? If so, where?
   b. Is there any point where $\vec{v}$ and $\vec{a}$ are perpendicular to each other? If so, where?
7. An athlete performing the long jump tries to achieve the maximum distance from the point of takeoff to the first point of touching the ground. After the jump, rather than land upright, she extends her legs forward as in the photo. How does this affect the time in the air? How does this give the jumper a longer range?

8. If you kick a football, at what angle to the ground should you kick the ball for the maximum range—that is, the greatest distance down the field? At what angle to the ground should you kick the ball for the maximum "hang time"—that is, the maximum time in the air?
9. Carla holds a ball 1.5 m above the ground. Daniel, leaning out of a car window, also holds a ball 1.5 m above the ground. Daniel drives past Carla at 40 mph and, just as he passes her, both release their balls at the same instant. Whose ball hits the ground first? Explain.
10. A cart that is rolling at a constant velocity on a level table fires a ball straight up out of a vertical tube.
    a. When the ball comes back down, will it land in front of the launching tube, behind the launching tube, or directly into the tube? Explain.
    b. Does your answer change if the cart is accelerating in the forward direction? If so, how?
11. If you go to a ski area, you'll likely find that the beginner's slope has the smallest angle. Use the concept of acceleration on a ramp to explain why this is so.
12. In an amusement-park ride, cars rolling along at high speed suddenly head up a long, straight ramp. They roll up the ramp, reverse direction at the highest point, then roll backward back down the ramp. In each of the following segments of the motion, which way does the acceleration vector point?
    a. As the cars roll up the ramp.
    b. At the highest point on the ramp.
    c. As the cars roll back down the ramp.
13. There are competitions in which pilots fly small planes low over the ground and drop weights, trying to hit a target. A pilot flying low and slow drops a weight; it takes 2.0 s to hit the ground, during which it travels a horizontal distance of 100 m. Now the pilot does a run at the same height but twice the speed. How much time does it take the weight to hit the ground? How far does it travel before it lands?
14. Roller coaster loops are rarely perfectly circular. Instead, they are tightly curved at the top, where the cars are moving more slowly, and they have a gentler curve at the bottom, where the cars are moving much faster. Explain why the loops are designed this way.

15. You are cycling around a circular track at a constant speed. Does the magnitude of your acceleration change? The direction?
16. An airplane has been directed to fly in a clockwise circle, as seen from above, at constant speed until another plane has landed. When the plane is going north, is it accelerating? If so, in what direction does the acceleration vector point? If not, why not?
17. When you go around a corner in your car, your car follows a path that is a segment of a circle. To turn safely, you should keep your car's acceleration below some safe upper limit. If you want to make a "tighter" turn—that is, turn in a circle with a smaller radius—how should you adjust your speed? Explain.
18. A person trying to throw a ball as far as possible will run forward during the throw. Explain why this increases the distance of the throw.
19. Anna is running to the right, as shown in Figure Q3.19. Balls 1 and 2 are thrown toward her by friends standing on the ground. According to Anna, both balls are approaching her at the same speed. Which ball was thrown with the faster speed? Or were they thrown with the same speed? Explain.

FIGURE Q3.19

Multiple-Choice Questions

20. || Which combination of the vectors shown in Figure Q3.20 has the largest magnitude?
    A. $\vec{A}+\vec{B}+\vec{C}$
    B. $\vec{B}+\vec{A}-\vec{C}$
    C. $\vec{A}-\vec{B}+\vec{C}$
    D. $\vec{B}-\vec{A}-\vec{C}$

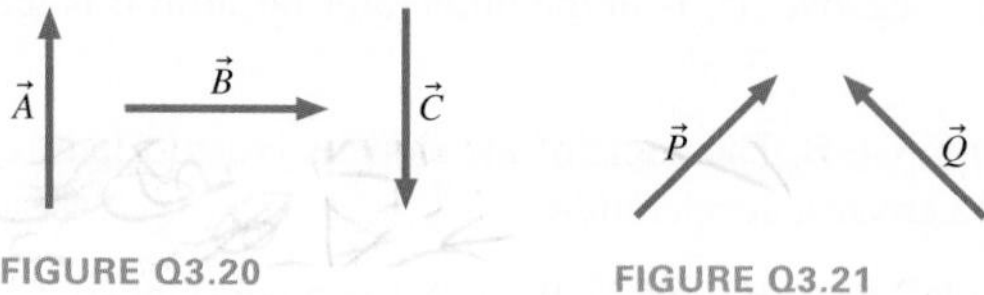

FIGURE Q3.20 FIGURE Q3.21

21. || Two vectors appear as in Figure Q3.21. Which combination points directly to the left?
    A. $\vec{P}+\vec{Q}$ B. $\vec{P}-\vec{Q}$ C. $\vec{Q}-\vec{P}$ D. $-\vec{Q}-\vec{P}$
22. | The gas pedal in a car is sometimes referred to as "the accelerator." Which other controls on the vehicle can be used to produce acceleration?
    A. The brakes.
    B. The steering wheel.
    C. The gear shift.
    D. All of the above.
23. | A car travels at constant speed along the curved path shown from above in Figure Q3.23. Five possible vectors are also shown in the figure; the letter E represents the zero vector. Which vector best represents
    a. The car's velocity at position 1?
    b. The car's acceleration at position 1?
    c. The car's velocity at position 2?
    d. The car's acceleration at position 2?
    e. The car's velocity at position 3?
    f. The car's acceleration at position 3?

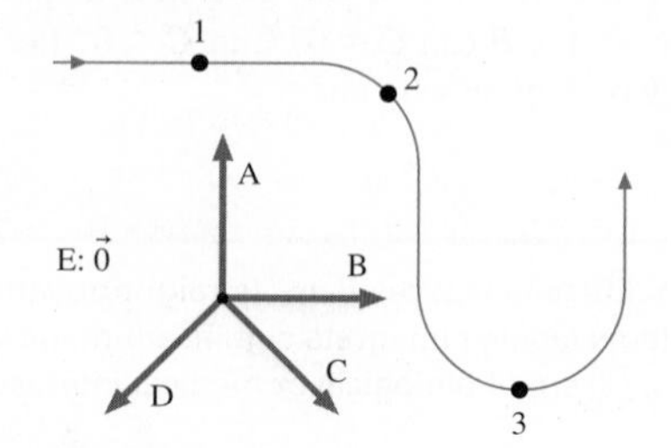

FIGURE Q3.23

24. | A ball is fired from a cannon at point 1 and follows the trajectory shown in Figure Q3.24. Air resistance may be neglected. Five possible vectors are also shown in the figure; the letter E represents the zero vector. Which vector best represents
    a. The ball's velocity at position 2?
    b. The ball's acceleration at position 2?
    c. The ball's velocity at position 3?
    d. The ball's acceleration at position 3?

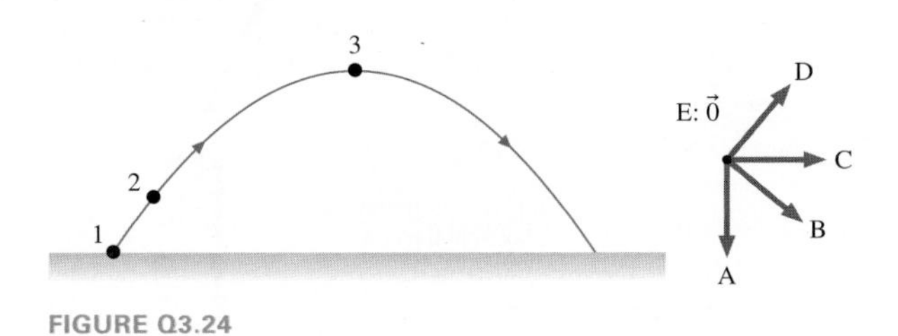

FIGURE Q3.24

25. | A ball thrown at an initial angle of 37.0° and initial velocity of 23.0 m/s reaches a maximum height *h*, as shown in Figure Q3.25. With what initial speed must a ball be thrown *straight up* to reach the same maximum height *h*?
    A. 13.8 m/s B. 17.3 m/s C. 18.4 m/s D. 23.0 m/s

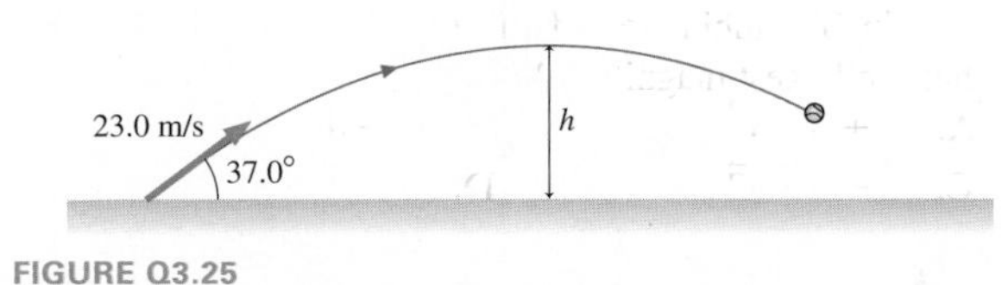

FIGURE Q3.25

26. || At a football game, an air gun fires T-shirts into the crowd. The gun is fired at an angle of 40° from the horizontal with an initial speed of 28 m/s. A fan who is sitting 60 m horizontally from the gun, but high in the stands, catches a T-shirt.
    a. How long does it take for the T-shirt to reach the fan?
       A. 1.3 s B. 2.0 s C. 2.8 s D. 3.4 s E. 3.7 s
    b. At what height *h* is the fan from the ground?
       A. 3.2 m B. 4.3 m C. 6.6 m D. 9.2 m E. 12.0 m
27. || Liam throws a water balloon horizontally at 8.2 m/s out of a window 10 m from the ground. How far from the base of the building does the balloon land?
    A. 4.2 m B. 8.0 m C. 12 m D. 15 m E. 18 m
28. | A football is kicked at an angle of 30° with a speed of 20 m/s. To the nearest second, how long will the ball stay in the air?
    A. 1 s B. 2 s C. 3 s D. 4 s
29. | A football is kicked at an angle of 30° with a speed of 20 m/s. To the nearest 5 m, how far will the ball travel?
    A. 15 m B. 25 m C. 35 m D. 45 m
30. || BIO An ultracentrifuge spins biological samples at very high speeds in order to separate their constituent microscopic components. One such instrument spins at 120,000 rpm, and the samples experience a centripetal acceleration of $6.6 \times 10^6\ m/s^2$ What is the radius of the circle around which the samples travel?
    A. 2.6 cm B. 4.2 cm C. 7.8 cm D. 12 cm E. 17 cm
31. | Formula One race cars are capable of remarkable accelerations when speeding up, slowing down, and turning corners. At one track, cars round a corner that is a segment of a circle of radius 95 m at a speed of 68 m/s. What is the approximate magnitude of the centripetal acceleration, in units of *g*?
    A. 1*g* B. 2*g* C. 3*g* D. 4*g* E. 5*g*

PROBLEMS

Section 3.1 Using Vectors

1. || Trace the vectors in Figure P3.1 onto your paper. Then use graphical methods to draw the vectors (a) $\vec{A} + \vec{B}$ and (b) $\vec{A} - \vec{B}$.
2. ||| Trace the vectors in Figure P3.2 onto your paper. Then use graphical methods to draw the vectors (a) $\vec{A} + \vec{B}$ and (b) $\vec{A} - \vec{B}$.

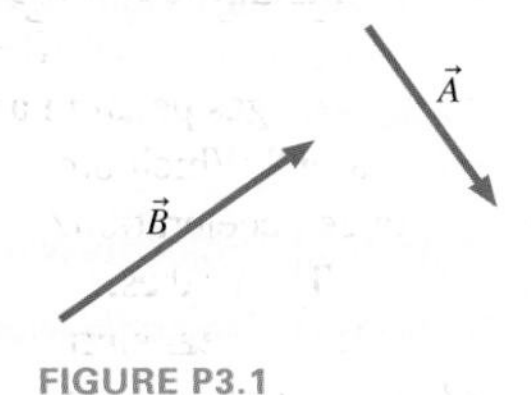

FIGURE P3.1

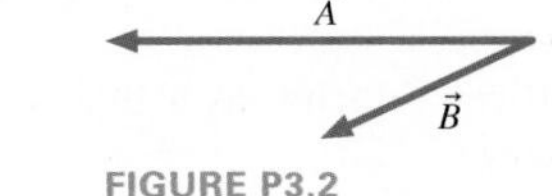

FIGURE P3.2

FIGURE P3.3

3. | Trace the vectors in Figure P3.3 onto your paper. Then draw the vector $\vec{C}$ such that $\vec{A} + \vec{B} + \vec{C} = 0$.
4. || Two vectors $\vec{A}$ and $\vec{B}$ are at right angles to each other. The magnitude of $\vec{A}$ is 1. What should be the length of $\vec{B}$ so that the magnitude of their vector sum is 2?

Section 3.2 Coordinate Systems and Vector Components

5. || A position vector with magnitude 10 m points to the right and up. Its *x*-component is 6.0 m. What is the value of its *y*-component?
6. ||| A velocity vector 40° above the positive x-axis has a *y*-component of 10 m/s. What is the value of its *x*-component?
7. || A cannon tilted upward at 30° fires a cannonball with a speed of 100 m/s. At that instant, what is the component of the cannonball's velocity parallel to the ground?
8. || a. What are the *x*- and *y*-components of vector $\vec{E}$ of Figure P3.8 in terms of the angle θ and the magnitude *E*?
    b. For the same vector, what are the *x*- and *y*-components in terms of the angle ϕ and the magnitude *E*?

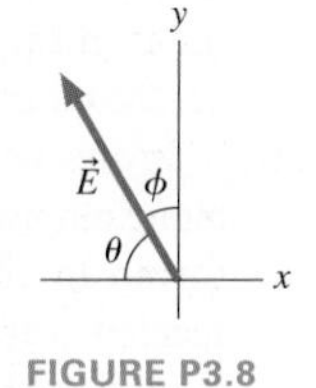

FIGURE P3.8

9. | Draw each of the following vectors, then find its *x*- and *y*-components.
    a. $\vec{d} = (100\ m, 45°\ below\ +x\text{-axis})$
    b. $\vec{v} = (300\ m/s, 20°\ above\ +x\text{-axis})$
    c. $\vec{a} = (5.0\ m/s^2, -y\text{-direction})$
10. || Draw each of the following vectors, then find its *x*- and *y*-components.
    a. $\vec{d} = (2.0\ km, 30°\ left\ of\ +y\text{-axis})$
    b. $\vec{v} = (5.0\ cm/s, -x\text{-direction})$
    c. $\vec{a} = (10\ m/s^2, 40°\ left\ of\ -y\text{-axis})$
11. | Each of the following vectors is given in terms of its *x*- and *y*-components. Draw the vector, label an angle that specifies the vector's direction, then find the vector's magnitude and direction.
    a. $v_x = 20\ m/s$, $v_y = 40\ m/s$
    b. $a_x = 2.0\ m/s^2$, $a_y = -6.0\ m/s^2$

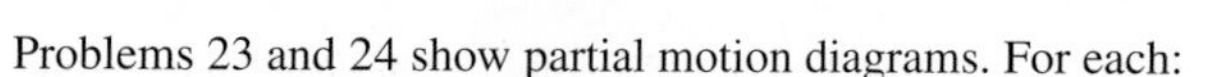

12. | Each of the following vectors is given in terms of its x- and y-components. Draw the vector, label an angle that specifies the vector's direction, then find the vector's magnitude and direction.
    a. $v_x = 10$ m/s, $v_y = 30$ m/s
    b. $a_x = 20$ m/s^2, $a_y = 10$ m/s^2
13. || A wildlife researcher is tracking a flock of geese. The geese BIO fly 4.0 km due west, then turn toward the north by 40° and fly another 4.0 km. How far west are they of their initial position? What is the magnitude of their displacement?
14. || Jack and Jill ran up the hill at 3.0 m/s. The horizontal component of Jill's velocity vector was 2.5 m/s.
    a. What was the angle of the hill?
    b. What was the vertical component of Jill's velocity?
15. | Josh is climbing up a steep 34° slope, moving at a steady 0.75 m/s along the ground. How many meters of elevation does he gain in one minute of this climb?

Section 3.3 Motion on a Ramp

16. ||| You begin sliding down a 15° ski slope. Ignoring friction and air resistance, how fast will you be moving after 10 s?
17. |||| A car traveling at 30 m/s runs out of gas while traveling up a 5.0° slope. How far will it coast before starting to roll back down?
18. || In the Soapbox Derby, young participants build non-motorized cars with very low-friction wheels. Cars race by rolling down a hill. The track at Akron's Derby Downs, where the national championship is held, begins with a 55-ft-long section tilted 13° below horizontal.

    a. What is the maximum possible acceleration of a car moving down this stretch of track?
    b. If a car starts from rest and undergoes this acceleration for the full 55 ft, what is its final speed in m/s?
19. ||| A piano has been pushed to the top of the ramp at the back of a moving van. The workers think it is safe, but as they walk away, it begins to roll down the ramp. If the back of the truck is 1.0 m above the ground and the ramp is inclined at 20°, how much time do the workers have to get to the piano before it reaches the bottom of the ramp?
20. || In the winter sport of bobsledding, athletes push their sled along a horizontal ice surface and then hop on the sled as it starts to careen down the steeply sloped track. In one event, the sled reaches a top speed of 9.2 m/s before starting down the initial part of the track, which is sloped downward at an angle of 6.0°. What is the sled's speed after it has traveled the first 100 m?

Section 3.4 Motion in Two Dimensions

21. | A car goes around a corner in a circular arc at constant speed. Draw a motion diagram including positions, velocity vectors, and acceleration vectors.
22. |||| Figure 3.33 showed the motion diagram for a rider on a Ferris wheel that was turning at a constant speed. The inset to the figure showed how to find the acceleration vector at the rightmost point. Use a similar analysis to find the rider's acceleration vector at the leftmost position of the motion diagram, then at one of the highest positions and at one of the lowest positions. Use a ruler so that your analysis is accurate.

Problems 23 and 24 show partial motion diagrams. For each:
   a. Complete the motion diagram by adding acceleration vectors.
   b. Write a physics *problem* for which this is the correct motion. Be imaginative! Don't forget to include enough information to make the problem complete and to state clearly what is to be found.

23. ||
24. ||

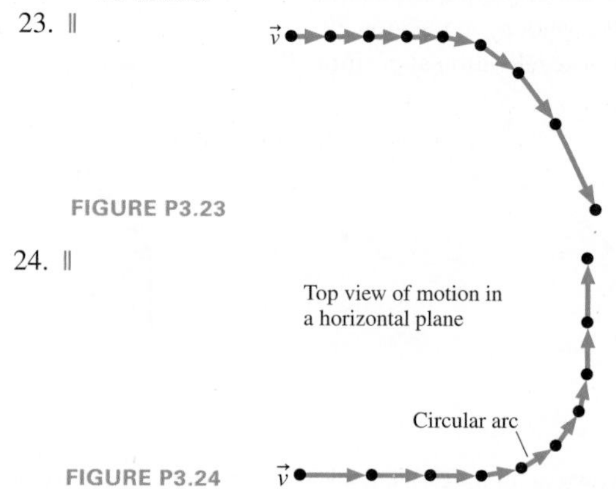

FIGURE P3.23

FIGURE P3.24

Answer Problems 25 and 26 by choosing one of the eight labeled acceleration vectors or selecting option I: $\vec{a} = \vec{0}$.

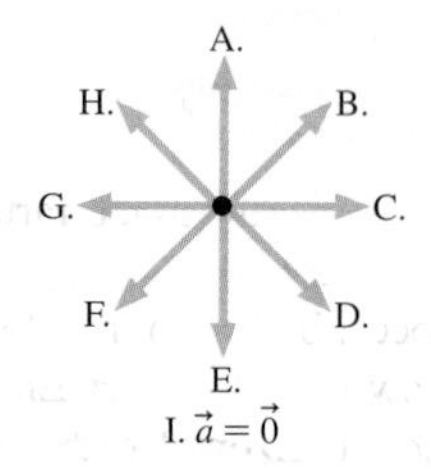

25. || At this instant, the particle has steady speed and is curving to the right. What is the direction of its acceleration?
26. || At this instant, the particle is speeding up and curving upward. What is the direction of its acceleration?

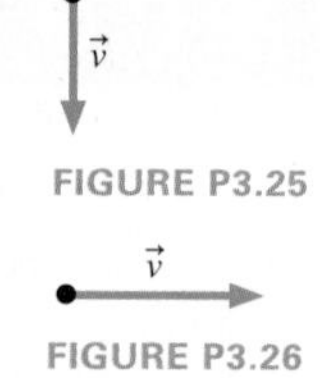

FIGURE P3.25

FIGURE P3.26

Section 3.5 Projectile Motion

Section 3.6 Projectile Motion: Solving Problems

27. ||| A ball is thrown horizontally from a 20-m-high building with a speed of 5.0 m/s.
    a. Make a sketch of the ball's trajectory.
    b. Draw a graph of v_x, the horizontal velocity, as a function of time. Include units on both axes.
    c. Draw a graph of v_y, the vertical velocity, as a function of time. Include units on both axes.
    d. How far from the base of the building does the ball hit the ground?
28. || A ball with a horizontal speed of 1.25 m/s rolls off a bench 1.00 m above the floor.
    a. How long will it take the ball to hit the floor?
    b. How far from a point on the floor directly below the edge of the bench will the ball land?
29. || A pipe discharges storm water into a creek. Water flows horizontally out of the pipe at 1.5 m/s, and the end of the pipe is 2.5 m above the creek. How far out from the end of the pipe is the point where the stream of water meets the creek?

30. ‖ On a day when the water is flowing relatively gently, water in the Niagara River is moving horizontally at 4.5 m/s before shooting over Niagara Falls. After moving over the edge, the water drops 53 m to the water below. If we ignore air resistance, how much time does it take for the water to go from the top of the falls to the bottom? How far does the water move horizontally during this time?
31. ||| A running mountain lion can make a leap 10.0 m long, reaching a maximum height of 3.0 m.
BIO
    a. What is the speed of the mountain lion just as it leaves the ground?
    b. At what angle does it leave the ground?
32. ||| A rifle is aimed horizontally at a target 50 m away. The bullet hits the target 2.0 cm below the aim point.
    a. What was the bullet's flight time?
    b. What was the bullet's speed as it left the barrel?
33. ||| A gray kangaroo can bound across a flat stretch of ground with each jump carrying it 10 m from the takeoff point. If the kangaroo leaves the ground at a 20° angle, what are its (a) takeoff speed and (b) horizontal speed?
BIO
34. ||| On the Apollo 14 mission to the moon, astronaut Alan Shepard hit a golf ball with a golf club improvised from a tool. The free-fall acceleration on the moon is 1/6 of its value on earth. Suppose he hit the ball with a speed of 25 m/s at an angle 30° above the horizontal.
    a. How long was the ball in flight?
    b. How far did it travel?
    c. Ignoring air resistance, how much farther would it travel on the moon than on earth?
35. ‖ Emily throws a soccer ball out of her dorm window to Allison, who is waiting below to catch it. If Emily throws the ball at an angle of 30° below horizontal with a speed of 12 m/s, how far from the base of the dorm should Allison stand to catch the ball? Assume the vertical distance between where Emily releases the ball and Allison catches it is 6.0 m.
36. ‖ A soccer player takes a free kick from a spot that is 20 m from the goal. The ball leaves his foot at an angle of 32°, and it eventually hits the crossbar of the goal, which is 2.4 m from the ground. At what speed did the ball leave his foot?

Section 3.7 Circular Motion

37. | Racing greyhounds are capable of rounding corners at very high speeds. A typical greyhound track has turns that are 45-m-diameter semicircles. A greyhound can run around these turns at a constant speed of 15 m/s. What is its acceleration in m/s^2 and in units of g?
BIO
38. ‖ To withstand "g-forces" of up to 10 g's, caused by suddenly pulling out of a steep dive, fighter jet pilots train on a "human centrifuge." 10 g's is an acceleration of 98 m/s^2. If the length of the centrifuge arm is 12 m, at what speed is the rider moving when she experiences 10 g's?
BIO
39. ‖ The moon completes one (circular) orbit of the earth in 27.3 days. The distance from the earth to the moon is 3.84×10^8 m. What is the moon's centripetal acceleration?
40. ‖ In a roundabout (or traffic circle), cars go around a 25-m-diameter circle. If a car's tires will skid when the car experiences a centripetal acceleration greater than $0.60g$, what is the maximum speed of the car in this roundabout?
41. ‖ A particle rotates in a circle with centripetal acceleration $a = 8.0\ m/s^2$. What is a if
    a. The radius is doubled without changing the particle's speed?
    b. The speed is doubled without changing the circle's radius?
42. ‖ Entrance and exit ramps for freeways are often circular stretches of road. As you go around one at a constant speed, you will experience a constant acceleration. Suppose you drive through an entrance ramp at a modest speed and your acceleration is 3.0 m/s^2. What will be the acceleration if you double your speed?

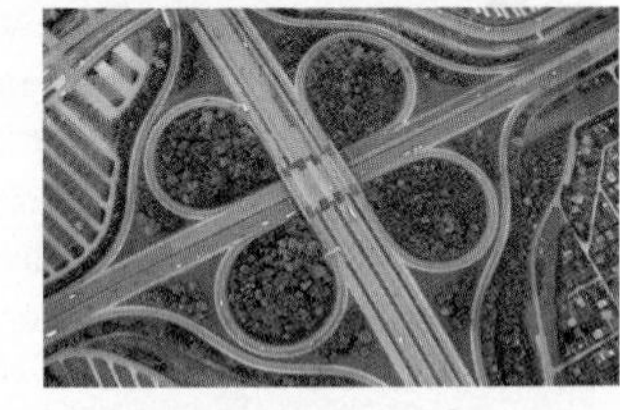

43. ‖ A peregrine falcon in a tight, circular turn can attain a centripetal acceleration 1.5 times the free-fall acceleration. If the falcon is flying at 20 m/s, what is the radius of the turn?
BIO

Section 3.8 Relative Motion

44. ‖ An airplane cruises at 880 km/h relative to the air. It is flying from Denver, Colorado, due west to Reno, Nevada, a distance of 1200 km, and will then return. There is a steady 90 km/h wind blowing to the east. What is the difference in flight time between the two legs of the trip?
45. | Anita is running to the right at 5 m/s, as shown in Figure P3.45. Balls 1 and 2 are thrown toward her at 10 m/s by friends standing on the ground. According to Anita, what is the speed of each ball?

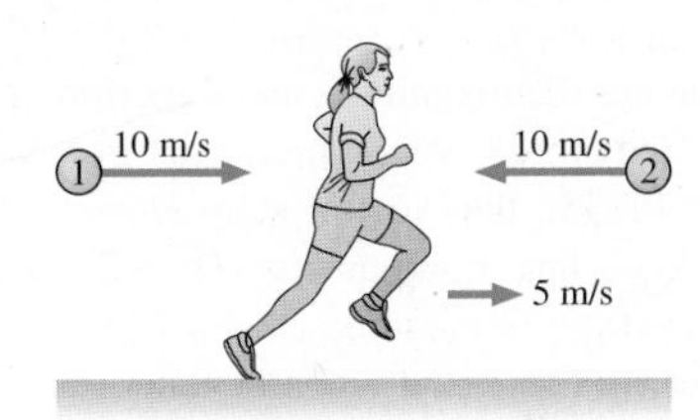

FIGURE P3.45

46. ||| In the 2016 Olympics in Rio, after the 50 m freestyle competition, a problem with the pool was found. In lane 1 there was a gentle 1.2 cm/s current flowing in the direction that the swimmers were going, while in lane 8 there was a current of the same speed but directed opposite to the swimmers' direction. Suppose a swimmer could swim the 50 m in 25.0 s in the absence of any current. What would be her time in lane 1? In lane 8? How does the difference in these times compare to the actual 0.06 s difference in times between the gold medal winner and the fourth-place finisher?
47. | Anita is running to the right at 5 m/s, as shown in Figure P3.47. Balls 1 and 2 are thrown toward her by friends standing on the ground. According to Anita, both balls are approaching her at 10 m/s. According to her friends, with what speeds were the balls thrown?

FIGURE P3.47

48. || Two children who are bored while waiting for their flight at the airport decide to race from one end of the 20-m-long moving sidewalk to the other and back. Phillippe runs on the sidewalk at 2.0 m/s (relative to the sidewalk). Renee runs on the floor at 2.0 m/s. The sidewalk moves at 1.5 m/s relative to the floor. Both make the turn instantly with no loss of speed.
   a. Who wins the race?
   b. By how much time does the winner win?
49. |||| A boat takes 3.0 h to travel 30 km down a river, then 5.0 h to return. How fast is the river flowing?

General Problems

50. | Suppose $\vec{C} = \vec{A} + \vec{B}$ where vector $\vec{A}$ has components $A_x = 5$, $A_y = 2$ and vector $\vec{B}$ has components $B_x = -3$, $B_y = -5$.
   a. What are the x- and y-components of vector $\vec{C}$?
   b. Draw a coordinate system and on it show vectors $\vec{A}$, $\vec{B}$, and $\vec{C}$.
   c. What are the magnitude and direction of vector $\vec{C}$?
51. | Suppose $\vec{D} = \vec{A} - \vec{B}$ where vector $\vec{A}$ has components $A_x = 5$, $A_y = 2$ and vector $\vec{B}$ has components $B_x = -3$, $B_y = -5$.
   a. What are the x- and y-components of vector $\vec{D}$?
   b. Draw a coordinate system and on it show vectors $\vec{A}$, $\vec{B}$, and $\vec{D}$.
   c. What are the magnitude and direction of vector $\vec{D}$?
52. || Suppose $\vec{E} = 2\vec{A} + 3\vec{B}$ where vector $\vec{A}$ has components $A_x = 5$, $A_y = 2$ and vector $\vec{B}$ has components $B_x = -3$, $B_y = -5$.
   a. What are the x- and y-components of vector $\vec{E}$?
   b. Draw a coordinate system and on it show vectors $\vec{A}$, $\vec{B}$, and $\vec{E}$.
   c. What are the magnitude and direction of vector $\vec{E}$?
53. || For the three vectors shown in Figure P3.53, the vector sum $\vec{D} = \vec{A} + \vec{B} + \vec{C}$ has components $D_x = 2$ and $D_y = 0$.
   a. What are the x- and y-components of vector $\vec{B}$?
   b. Write $\vec{B}$ as a magnitude and a direction.

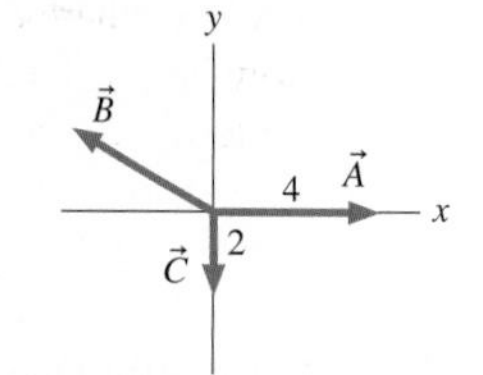

FIGURE P3.53

54. || Let $\vec{A} = (3.0\text{ m}, 20°\text{ south of east})$, $\vec{B} = (2.0\text{ m, north})$, and $\vec{C} = (5.0\text{ m}, 70°\text{ south of west})$.
   a. Draw and label $\vec{A}$, $\vec{B}$, and $\vec{C}$ with their tails at the origin. Use a coordinate system with the x-axis to the east.
   b. Write the x- and y-components of vectors $\vec{A}$, $\vec{B}$, and $\vec{C}$.
   c. Find the magnitude and the direction of $\vec{D} = \vec{A} + \vec{B} + \vec{C}$.
55. || To get to his office from home, Greg walks 5 blocks north and then 3 blocks east. After work he meets some friends at a café; to get there he walks 2 blocks south and 5 blocks west. All blocks are 660 feet long. What is the straight-line distance from the café to his home?
56. ||| A pilot in a small plane encounters shifting winds. He flies 26.0 km northeast, then 45.0 km due north. From this point, he flies an additional distance in an unknown direction, only to find himself at a small airstrip that his map shows to be 70.0 km directly north of his starting point. What were the length and direction of the third leg of his trip?
57. ||| In punting a football, the kicker tries to maximize both the distance of the kick and its "hang time"—the time that the ball is in the air. A kicker gets off a great punt with a hang time of 5.0 s that lands 50 yards from the kicker.
   a. What is the speed of the ball as it leaves the kicker's foot?
   b. What is the angle of the ball's initial velocity?

Watch Video Solution Problem 3.53

58. || The bacterium *Escherichia coli* (or *E. coli*) is a single-celled BIO organism that lives in the gut of healthy humans and animals. When grown in a uniform medium rich in salts and amino acids, these bacteria swim along zig-zag paths at a constant speed of 20 μm/s. Figure P3.58 shows the trajectory of an *E. coli* as it moves from point A to point E. Each segment of the motion can be identified by two letters, such as segment BC.
   a. For each of the four segments in the bacterium's trajectory, calculate the x- and y-components of its displacement and of its velocity.
   b. Calculate both the total distance traveled and the magnitude of the net displacement for the entire motion.
   c. What are the magnitude and the direction of the bacterium's average velocity for the entire trip?

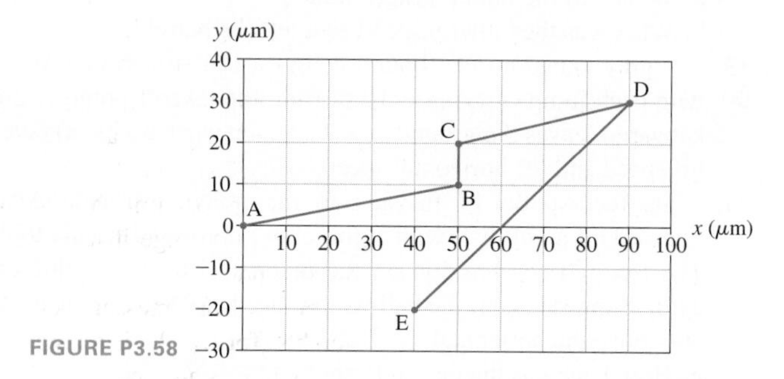

FIGURE P3.58

59. |||| A skier gliding across the snow at 3.0 m/s suddenly starts down a 10° incline, reaching a speed of 15 m/s at the bottom. Friction between the snow and her freshly waxed skis is negligible.
   a. What is the length of the incline?
   b. How long does it take her to reach the bottom?
60. ||| As shown in Figure P3.60, a skier speeds along a flat patch of snow, and then flies horizontally off the edge at 12.0 m/s. He eventually lands on a straight, sloped section that is at an angle of 45° below the horizontal. How long is he in the air?

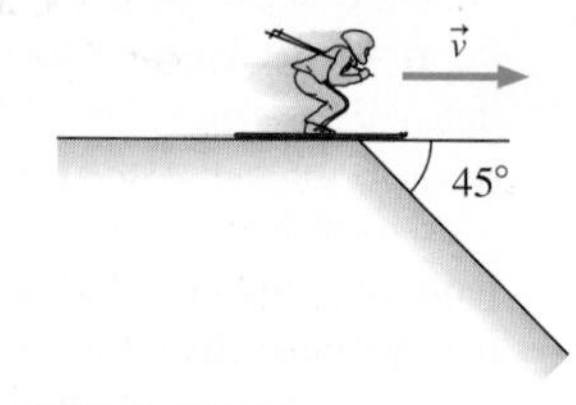

FIGURE P3.60

61. || A physics student on Planet Exidor throws a ball, and it follows the parabolic trajectory shown in Figure P3.61. The ball's position is shown at 1.0 s intervals until $t = 3.0$ s. At $t = 1.0$ s, the ball's velocity has components $v_x = 2.0$ m/s, $v_y = 2.0$ m/s.

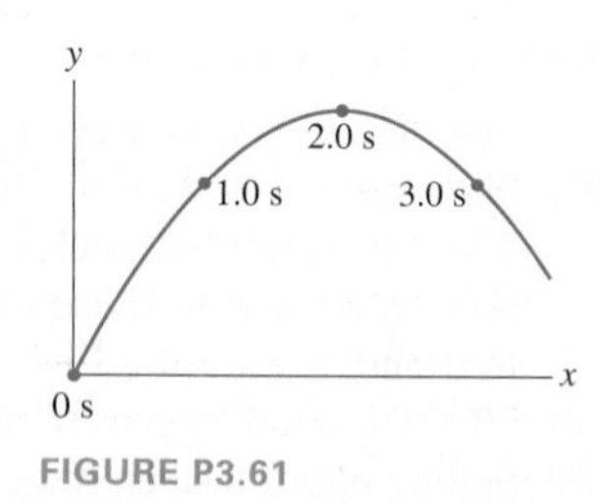

FIGURE P3.61

   a. Determine the x- and y-components of the ball's velocity at $t = 0.0$ s, 2.0 s, and 3.0 s.
   b. What is the value of g on Planet Exidor?
   c. What was the ball's launch angle?
62. ||| The archerfish uses a remarkable method for catching insects BIO sitting on branches or leaves above the waterline. The fish rises to the surface and then shoots out a stream of water precisely aimed to knock the insect off its perch into the water, where the archerfish gobbles it up. Scientists have measured the speed of the water stream exiting the fish's mouth to be 3.7 m/s. An archerfish spots an insect sitting 19 cm above the waterline and a horizontal distance of 30 cm away. The fish aims its stream at an angle of 39° from the waterline. Does the stream hit its mark?

63. ||| In 1780, in what is now referred to as "Brady's Leap," Captain Sam Brady of the U.S. Continental Army escaped certain death from his enemies by running horizontally off the edge of the cliff above Ohio's Cuyahoga River, which is confined at that spot to a gorge. He landed safely on the far side of the river. It was reported that he leapt 22 ft across while falling 20 ft. Tall tale, or possible?

FIGURE P3.63

   a. What is the minimum speed with which he'd need to run off the edge of the cliff to make it safely to the far side of the river?
   b. The world-record time for the 100 m dash is approximately 10 s. Given this, is it reasonable to expect Brady to be able to run fast enough to achieve Brady's leap?

64. ||| The longest recorded pass in an NFL game traveled 83 yards in the air from the quarterback to the receiver. Assuming that the pass was thrown at the optimal 45° angle, what was the speed at which the ball left the quarterback's hand?

65. ||| A spring-loaded gun, fired vertically, shoots a marble 6.0 m straight up in the air. What is the marble's range if it is fired horizontally from 1.5 m above the ground?

66. || Small-plane pilots regularly compete in "message drop" competitions, dropping heavy weights (for which air resistance can be ignored) from their low-flying planes and scoring points for having the weights land close to a target. A plane 60 m above the ground is flying directly toward a target at 45 m/s.
   a. At what distance from the target should the pilot drop the weight?
   b. The pilot looks down at the weight after she drops it. Where is the plane located at the instant the weight hits the ground—not yet over the target, directly over the target, or past the target?

67. ||| Paintball guns were originally developed to mark trees for logging. A forester aims his gun directly at a knothole in a tree that is 4.0 m above the gun. The base of the tree is 20 m away. The speed of the paintball as it leaves the gun is 50 m/s. How far below the knothole does the paintball strike the tree?

68. ||| Trained dolphins are capable of a vertical leap of 7.0 m straight up from the surface of the water—an impressive feat. Suppose you could train a dolphin to launch itself out of the water at this same speed but at an angle. What maximum horizontal range could the dolphin achieve?
BIO INT

69. ||| A tennis player hits a ball 2.0 m above the ground. The ball leaves his racquet with a speed of 20 m/s at an angle 5.0° above the horizontal. The horizontal distance to the net is 7.0 m, and the net is 1.0 m high. Does the ball clear the net? If so, by how much? If not, by how much does it miss?

70. ||| The shot put is a track-and-field event in which athletes throw a heavy ball—the shot—as far as possible. The best athletes can throw the shot as far as 23 m. Athletes who use the "glide" technique push the shot outward in a reasonably straight line, accelerating it over a distance of about 2.0 m. What acceleration do they provide to the shot as they push on it? Assume that the shot is launched at an angle of 37°, a reasonable value for an excellent throw. You can assume that the shot lands at the same height from which it is thrown; this simplifies the calculation considerably, and makes only a small difference in the final result.
INT

71. ||| Water at the top of Horseshoe Falls (part of Niagara Falls) is moving horizontally at 9.0 m/s as it goes off the edge and plunges 53 m to the pool below. If you ignore air resistance, at what angle is the falling water moving as it enters the pool?

72. || A supply plane needs to drop a package of food to scientists working on a glacier in Greenland. The plane flies 100 m above the glacier at a speed of 150 m/s. How far short of the target should it drop the package?

73. |||| A BMX bicycle rider takes off from a ramp at a point 1.8 m above the ground. The ramp is angled at 40° from the horizontal, and the rider's speed is 6.7 m/s when he leaves the ramp. How far from the end of the ramp does he land?

74. ||||| Ships A and B leave port together. For the next two hours, ship A travels at 20 mph in a direction 30° west of north while ship B travels 20° east of north at 25 mph.
   a. What is the distance between the two ships two hours after they depart?
   b. What is the speed of ship A as seen by ship B?

75. ||| A flock of ducks is trying to migrate south for the winter, but they keep being blown off course by a wind blowing from the west at 12 m/s. A wise elder duck finally realizes that the solution is to fly at an angle to the wind. If the ducks can fly at 16 m/s relative to the air, in what direction should they head in order to move directly south?

76. ||| A kayaker needs to paddle north across a 100-m-wide harbor. The tide is going out, creating a tidal current flowing east at 2.0 m/s. The kayaker can paddle with a speed of 3.0 m/s.
   a. In which direction should he paddle in order to travel straight across the harbor?
   b. How long will it take him to cross?

77. |||| A plane has an airspeed of 200 mph. The pilot wishes to reach a destination 600 mi due east, but a wind is blowing at 50 mph in the direction 30° north of east.
   a. In what direction must the pilot head the plane in order to reach her destination?
   b. How long will the trip take?

78. ||| The Gulf Stream off the east coast of the United States can flow at a rapid 3.6 m/s to the north. A ship in this current has a cruising speed of 10 m/s. The captain would like to reach land at a point due west from the current position.
   a. In what direction with respect to the water should the ship sail?
   b. At this heading, what is the ship's speed with respect to land?

79. ||| A ball thrown horizontally at 25 m/s travels a horizontal distance of 50 m before hitting the ground. From what height was the ball thrown?

80. ||| A sports car is advertised as capable of "reaching 60 mph in 5 seconds flat, cornering at 0.85g, and stopping from 70 mph in only 168 feet." In which of those three situations is the magnitude of the car's acceleration the largest? In which is it the smallest?
INT

81. || A Ford Mustang can accelerate from 0 to 60 mph in a time of 5.6 s. A Mini Cooper isn't capable of such a rapid start, but it can turn in a very small circle 34 ft in diameter. How fast would you need to drive the Mini Cooper in this tight circle to match the magnitude of the Mustang's acceleration?
INT

82. || The "Screaming Swing" is a carnival ride that is—not surprisingly—a giant swing. It's actually two swings moving in opposite directions. At the bottom of its arc, a rider in one swing is moving at 30 m/s with respect to the ground in a 50-m-diameter circle. The rider in the other swing is moving in a similar circle at the same speed, but in the exact opposite direction.
    a. What is the acceleration, in m/s^2 and in units of g, that riders experience?
    b. At the bottom of the ride, as they pass each other, how fast do the riders move with respect to each other?
83. || On an otherwise straight stretch of road near Moffat, Colorado, the road suddenly turns. This bend in the road is a segment of a circle with radius 110 m. Drivers are cautioned to slow down to 40 mph as they navigate the curve.
    a. If you heed the sign and slow to 40 mph, what will be your acceleration going around the curve at this constant speed? Give your answer in m/s^2 and in units of g.
    b. At what speed would your acceleration be double that at the recommended speed?

MCAT-Style Passage Problems

Riding the Water Slide

A rider on a water slide goes through three different kinds of motion, as illustrated in Figure P3.84. Use the data and details from the figure to answer the following questions.

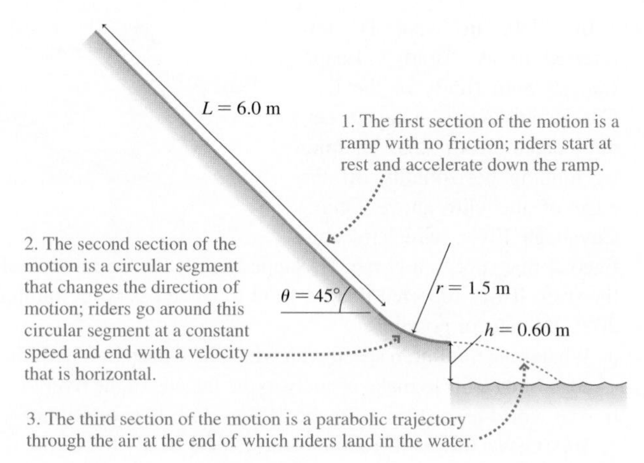

$L = 6.0$ m
1. The first section of the motion is a ramp with no friction; riders start at rest and accelerate down the ramp.
2. The second section of the motion is a circular segment that changes the direction of motion; riders go around this circular segment at a constant speed and end with a velocity that is horizontal.
$\theta = 45°$
$r = 1.5$ m
$h = 0.60$ m
3. The third section of the motion is a parabolic trajectory through the air at the end of which riders land in the water.

FIGURE P3.84

84. | At the end of the first section of the motion, riders are moving at what approximate speed?
    A. 3 m/s B. 6 m/s C. 9 m/s D. 12 m/s
85. | Suppose the acceleration during the second section of the motion is too large to be comfortable for riders. What change could be made to decrease the acceleration during this section?
    A. Reduce the radius of the circular segment.
    B. Increase the radius of the circular segment.
    C. Increase the angle of the ramp.
    D. Increase the length of the ramp.
86. | What is the vertical component of the velocity of a rider as he or she hits the water?
    A. 2.4 m/s B. 3.4 m/s C. 5.2 m/s D. 9.1 m/s
87. | Suppose the designers of the water slide want to adjust the height h above the water so that riders land twice as far away from the bottom of the slide. What would be the necessary height above the water?
    A. 1.2 m B. 1.8 m C. 2.4 m D. 3.0 m
88. | During which section of the motion is the magnitude of the acceleration experienced by a rider the greatest?
    A. The first. B. The second.
    C. The third. D. It is the same in all sections.

4 Forces and Newton's Laws of Motion

We don't normally think of turtles as speedy, but the snake-necked turtle is an ambush predator, catching prey by surprise with a very rapid acceleration of its head. How is it able to achieve this feat?

LOOKING AHEAD ▸▸

Forces

A force is a push or a pull. It is an interaction between two objects, the **agent** (here, the woman) and the **object** (the car).

In this chapter, you'll learn how to identify different forces, and you'll learn their properties.

Forces and Motion

Forces cause objects to *accelerate*. A forward acceleration of the sled requires a forward force.

A larger acceleration requires a larger force. You'll learn this connection between force and motion, part of *Newton's second law*.

Reaction Forces

The hammer exerts a downward force on the nail. Surprisingly, the nail exerts an equal force on the hammer, directed upward.

You'll learn how to identify and reason with **action/reaction pairs** of forces according to *Newton's third law*.

GOAL **To establish the connection between force and motion.**

LOOKING BACK ◂◂

Acceleration

You learned in Chapters 2 and 3 that acceleration is a vector that points in the direction of the change in velocity.

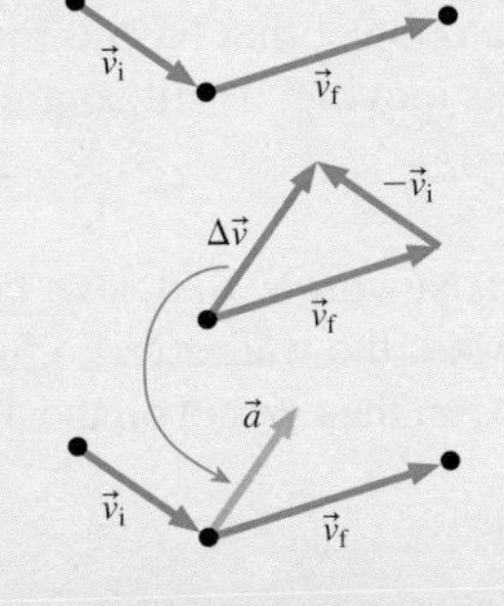

If the velocity is changing, there is an acceleration. And so, as you'll learn in this chapter, there must be a net force.

STOP TO THINK

A swan is landing on an icy lake, sliding across the ice and gradually coming to a stop. As the swan slides, the direction of the acceleration is

A. To the left.
B. To the right.
C. Upward.
D. Downward.

Interstellar coasting A nearly perfect example of Newton's first law is the pair of Voyager space probes launched in 1977. Both spacecraft long ago ran out of fuel and are now coasting through the frictionless vacuum of space. Although not entirely free of influence from the sun's gravity, they are now so far from the sun and other stars that gravitational influences are very nearly zero, and the probes will continue their motion for billions of years.

4.1 Motion and Forces

The snake-necked turtle in the photo at the beginning of the chapter can accelerate its head forward at 40 m/s^2 to capture prey. In Chapters 1 through 3, we've learned how to *describe* this and other types of motion with pictures, graphs, and equations, and you know enough about the scale of things to see that this acceleration—which is about $4g$—is quite impressive. But until now we've said nothing to *explain* the motion, to say how the turtle is able to achieve this feat. In this chapter, we will turn our attention to the *cause* of motion—**forces.** This topic is called **dynamics,** which joins with kinematics to form **mechanics,** the general science of motion. We'll begin our study of dynamics qualitatively in this chapter and then add quantitative detail over the next four chapters.

What Causes Motion?

Let's start with a basic question: Do you need to keep pushing on something—to keep applying a force—to keep it going? Your daily experience might suggest that the answer is Yes. If you slide your textbook across the desk and then stop pushing, the book will quickly come to rest. Other objects will continue to move for a longer time: When a hockey puck is sliding across the ice, it keeps going for a long time, but it, too, comes to rest at some point. Now let's take a closer look to see whether this idea holds up to scrutiny.

FIGURE 4.1 shows a series of motion experiments. Tyler slides down a hill on his sled and then out onto a horizontal patch of smooth snow, as shown in Figure 4.1a. Even if the snow is smooth, the friction between the sled and the snow will soon cause the sled to come to rest. What if Tyler slides down the hill onto some very slick ice, as in Figure 4.1b? The friction is much less, so the sled could slide for quite a distance before stopping. Now, *imagine* the situation in Figure 4.1c, where the sled slides on idealized *frictionless* ice. In this case, the sled, once started in its motion, would continue in motion forever, moving in a straight line with no loss of speed.

FIGURE 4.1 A sled sliding on increasingly smooth surfaces.

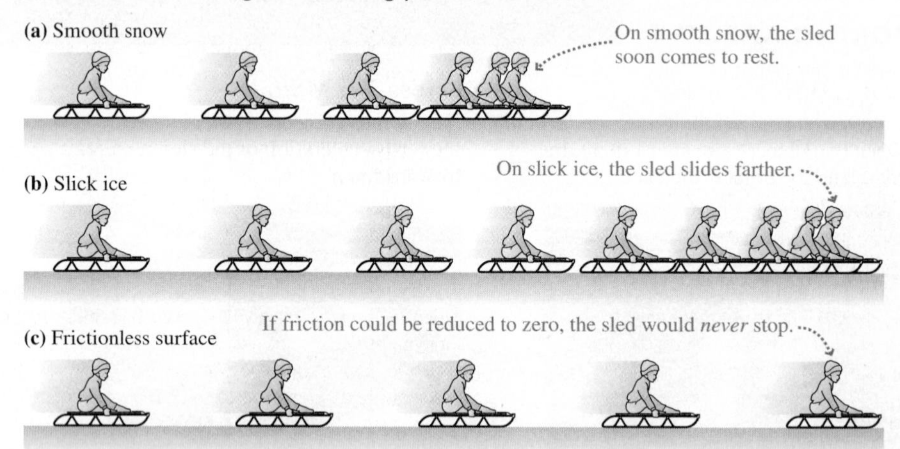

(a) Smooth snow
On smooth snow, the sled soon comes to rest.
(b) Slick ice
On slick ice, the sled slides farther.
(c) Frictionless surface
If friction could be reduced to zero, the sled would *never* stop.

In the absence of friction, **if the sled is moving, it will stay in motion.** It's also true that if the sled were sitting still, it wouldn't start moving on its own; if the sled is at rest, it will stay at rest. Careful experiments done over the past few centuries, notably by Galileo and then by Isaac Newton, verify that this is in fact the way the world works. What we have concluded for the sled is actually a general rule that applies to other similar situations. We call the generalization **Newton's first law** of motion:

> **Newton's first law** Consider an object that has no forces acting on it. If it is at rest, it will remain at rest. If it is moving, it will continue to move in a straight line at a constant speed.

As an important application of Newton's first law, consider the crash test of FIGURE 4.2. As the car contacts the wall, the wall exerts a force on the car and the car begins to slow. But the wall is a force on the *car,* not on the dummy. In accordance with

FIGURE 4.2 Newton's first law tells us: Wear your seatbelts!

At the instant of impact, the car and the dummy are moving at the same speed.

The car slows as it hits, but the dummy continues at the same speed . . .

. . . until it hits the now-stationary dashboard. Ouch!

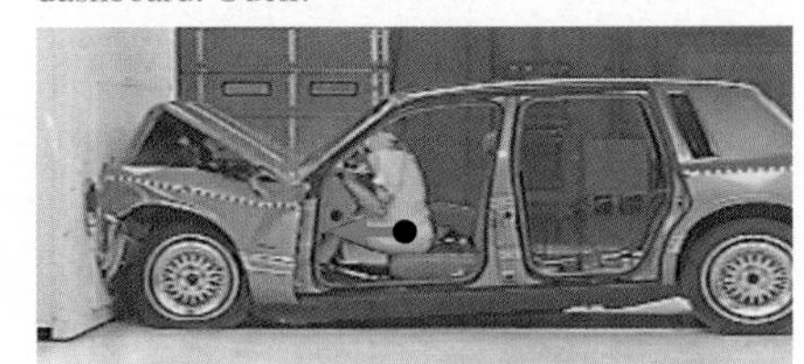

Newton's first law, the unbelted dummy continues to move straight ahead at its original speed. Sooner or later, a force will act to bring the dummy to rest. The only questions are when and how large the force will be. In the case shown, the dummy comes to rest in a short, violent collision with the dashboard of the stopped car. Seatbelts and air bags slow a dummy—or a driver—at a much lower rate and provide a much gentler stop.

Forces

We started with the question What causes motion? Our answer is Newton's first law, which tells us that an object in motion needs *no* cause—or *force*—to keep moving in a straight line forever. But this law does not explain in any detail exactly what a force *is*. The concept of force is best introduced by looking at examples of some common forces and considering the basic properties shared by all forces. Let's begin by examining the properties that all forces have in common, as presented in the table below.

What is a force?

A force is a push or a pull.

Our commonsense idea of a **force** is that it is a *push* or a *pull*. We will refine this idea as we go along, but it is an adequate starting point. Notice our careful choice of words: We refer to "*a* force" rather than simply "force." We want to think of a force as a very specific *action*, so that we can talk about a single force or perhaps about two or three individual forces that we can clearly distinguish—hence the concrete idea of "a force" acting on an object.

A force acts on an object.

Implicit in our concept of force is that **a force acts on an object.** In other words, pushes and pulls are applied *to* something—an object. From the object's perspective, it has a force *exerted* on it. Forces do not exist in isolation from the object that experiences them.

A force requires an agent.

Every force has an **agent,** something that acts or pushes or pulls; that is, a force has a specific, identifiable *cause*. As you throw a ball, it is your hand, while in contact with the ball, that is the agent or the cause of the force exerted on the ball. *If* a force is being exerted on an object, you must be able to identify a specific cause (i.e., the agent) of that force. Conversely, a force is not exerted on an object *unless* you can identify a specific cause or agent. Note that an agent can be an inert, inanimate object such as a tabletop or a wall. Such agents are the cause of many common forces.

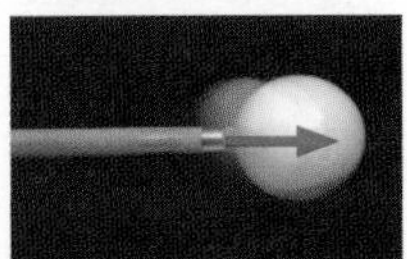

A force is a vector.

If you push an object, you can push either gently or very hard. Similarly, you can push either left or right, up or down. To quantify a push, we need to specify both a magnitude *and* a direction. It should thus come as no surprise that a force is a vector quantity. The general symbol for a force is the vector symbol $\vec{F}$. The size or strength of such a force is its magnitude F.

A force can be either a contact force . . .

There are two basic classes of forces, depending on whether the agent touches the object or not. **Contact forces** are forces that act on an object by touching it at a point of contact. The bat must touch the ball to hit it. A string must be tied to an object to pull it. The majority of forces that we will examine are contact forces.

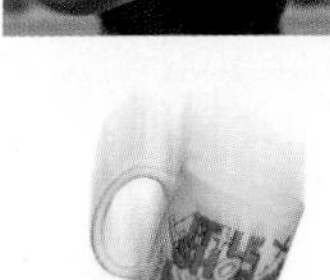

. . . or a long-range force.

Long-range forces are forces that act on an object without physical contact. Magnetism is an example of a long-range force. You have undoubtedly held a magnet over a paper clip and seen the paper clip leap up to the magnet. A coffee cup released from your hand is pulled to the earth by the long-range force of gravity.

There's one more important aspect of forces. If you push against a door (the object) to close it, the door pushes back against your hand (the agent). If a tow rope pulls on a car (the object), the car pulls back on the rope (the agent). In general, if an agent exerts a force on an object, the object exerts a force on the agent. We really need to think of a force as an *interaction* between two objects. Although the interaction perspective is a more exact way to view forces, it adds complications that we would like to avoid for now. Our approach will be to start by focusing on how a single object responds to forces exerted on it. Later, in Section 4.7, we'll return to the larger issue of how two or more objects interact with each other.

Force Vectors

We can use a simple diagram to visualize how forces are exerted on objects. Because we are using the particle model, in which objects are treated as points, the process of drawing a force vector is straightforward:

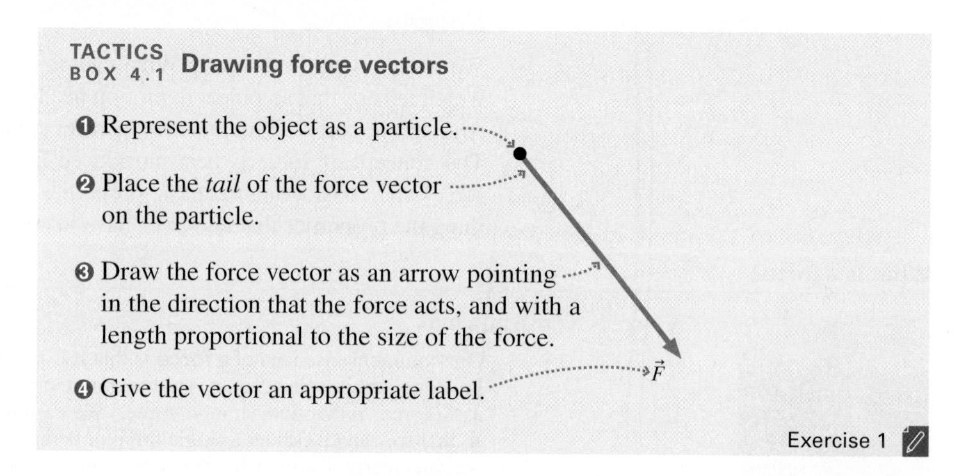

TACTICS BOX 4.1 Drawing force vectors
❶ Represent the object as a particle.
❷ Place the *tail* of the force vector on the particle.
❸ Draw the force vector as an arrow pointing in the direction that the force acts, and with a length proportional to the size of the force.
❹ Give the vector an appropriate label.
$\vec{F}$
Exercise 1

Step 2 may seem contrary to what a "push" should do (it may look as if the force arrow is *pulling* the object rather than *pushing* it), but recall that moving a vector does not change it as long as the length and angle do not change. The vector $\vec{F}$ is the same regardless of whether the tail or the tip is placed on the particle. Our reason for using the tail will become clear when we consider how to combine several forces.

FIGURE 4.3 shows three examples of force vectors. One is a pull, one a push, and one a long-range force, but in all three the *tail* of the force vector is placed on the particle that represents the object. Although the generic symbol for a force is $\vec{F}$, as the figure shows we often use special symbols for certain forces that arise frequently, such as $\vec{T}$ for the tension force in a rope, $\vec{F}_{sp}$ for the force of a spring, and $\vec{w}$ for the force of gravity (an object's *weight*).

FIGURE 4.3 Three force vectors.

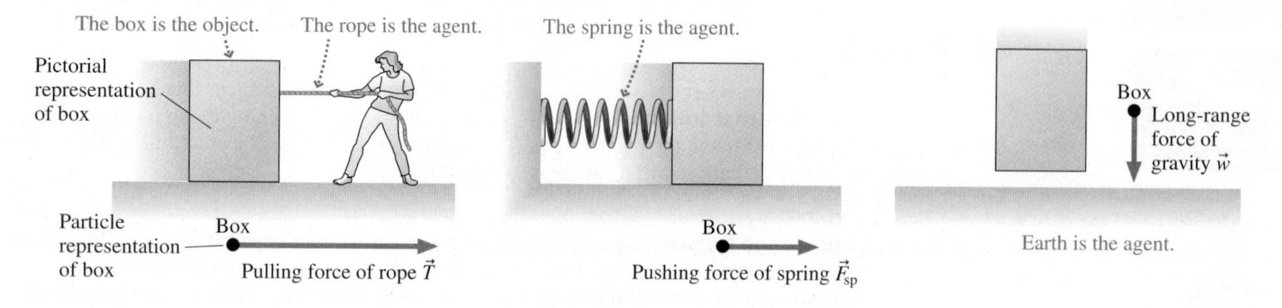

Combining Forces

Force is a vector quantity. We saw in ◀◀ SECTION 3.1 how to find the *vector sum* of two vectors. FIGURE 4.4a shows a top view of a box being pulled by two ropes, each exerting a force on the box. How will the box respond? Experiments show that when several forces $\vec{F}_1, \vec{F}_2, \vec{F}_3, \ldots$ are exerted on an object, they combine to form a **net force** that is the vector sum of all the forces:

$$\vec{F}_{\text{net}} = \vec{F}_1 + \vec{F}_2 + \vec{F}_3 + \cdots \tag{4.1}$$

That is, the single force $\vec{F}_{\text{net}}$ causes the exact same motion of the object as this combination of original forces $\vec{F}_1, \vec{F}_2, \vec{F}_3, \ldots$. FIGURE 4.4b shows the net force on the box.

NOTE ▶ It is important to realize that the net force $\vec{F}_{\text{net}}$ is not a new force acting *in addition* to the original forces $\vec{F}_1, \vec{F}_2, \vec{F}_3, \ldots$. Instead, we should think of the original forces being *replaced* by $\vec{F}_{\text{net}}$. ◀

FIGURE 4.4 Two forces applied to a box.

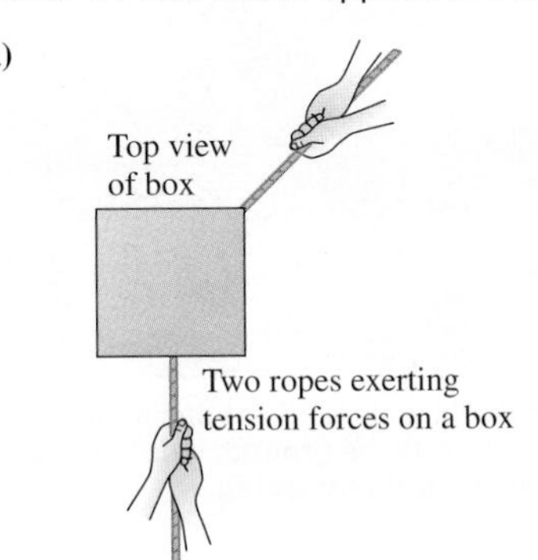

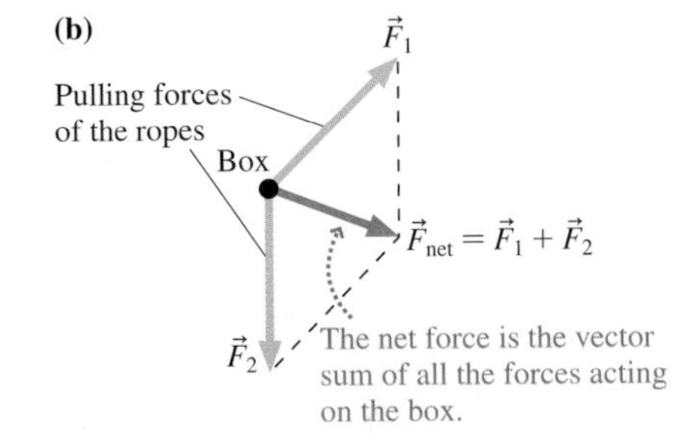

STOP TO THINK 4.1 Two of the three forces exerted on an object are shown. The net force points directly to the left. Which is the missing third force?

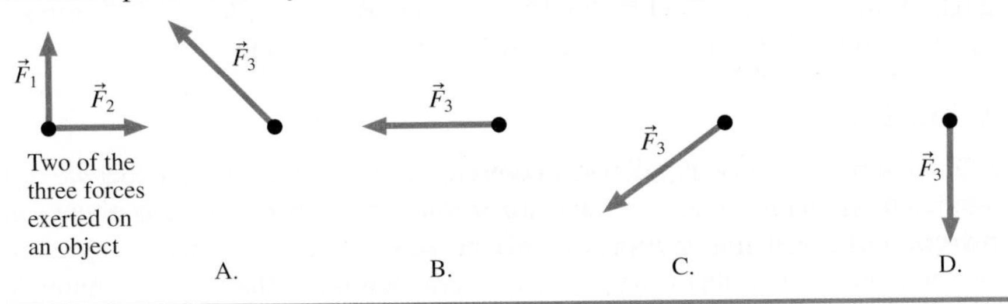

4.2 A Short Catalog of Forces

There are many forces we will deal with over and over. This section will introduce you to some of them and to the symbols we use to represent them.

Weight

A falling rock is pulled toward the earth by the long-range force of gravity. Gravity is what keeps you in your chair, keeps the planets in their orbits around the sun, and shapes the large-scale structure of the universe. We'll have a thorough look at gravity in Chapter 6. For now we'll concentrate on objects on or near the surface of the earth (or other planet).

The gravitational pull of the earth on an object on or near the surface of the earth is called **weight.** The symbol for weight is $\vec{w}$. Weight is the only long-range force we will encounter in the next few chapters. The agent for the weight force is the *entire earth* pulling on an object. The weight force is in some ways the simplest force we'll study. As FIGURE 4.5 shows, **an object's weight vector always points vertically downward,** no matter how the object is moving.

NOTE ▶ We often refer to "the weight" of an object. This is an informal expression for w, the magnitude of the weight force exerted on the object. Note that **weight is not the same thing as mass.** We will briefly examine mass later in the chapter, and we'll explore the connection between weight and mass in Chapter 5. ◀

FIGURE 4.5 Weight always points vertically downward.

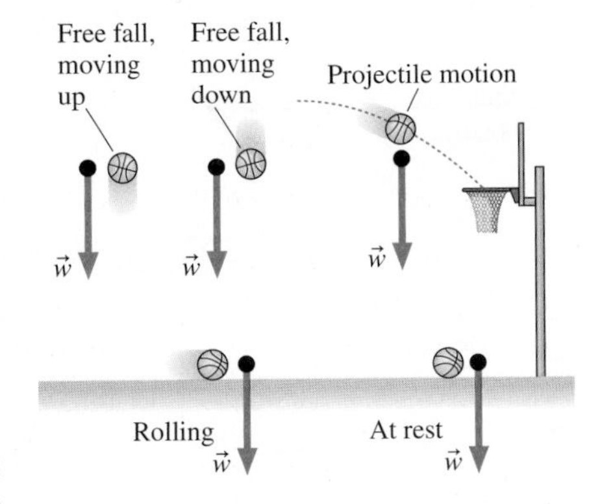

Spring Force

When the flexible blade of this athlete's prosthesis hits the ground, it compresses just like an ordinary spring.

Springs exert one of the most basic contact forces. A spring can either push (when compressed) or pull (when stretched). FIGURE 4.6 shows the **spring force.** In both cases, pushing and pulling, the tail of the force vector is placed on the particle in the force diagram. There is no special symbol for a spring force, so we simply use a subscript label: $\vec{F}_{sp}$.

FIGURE 4.6 The spring force is parallel to the spring.

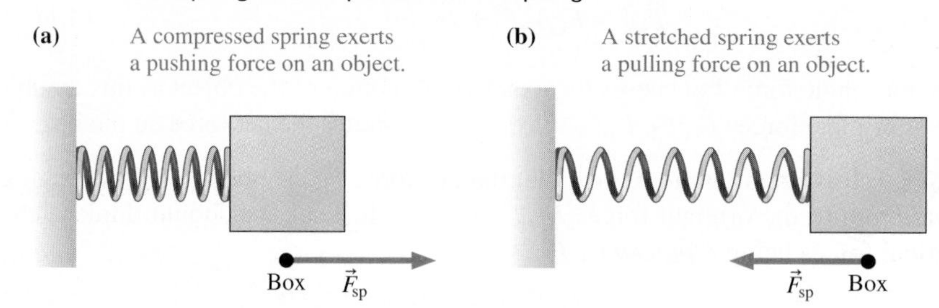
(a)
A compressed spring exerts a pushing force on an object.
Box
$\vec{F}_{sp}$
(b)
A stretched spring exerts a pulling force on an object.
$\vec{F}_{sp}$
Box

Although you may think of a spring as a metal coil that can be stretched or compressed, this is only one type of spring. Hold a ruler, or any other thin piece of wood or metal, by the ends and bend it slightly. It flexes. When you let go, it "springs" back to its original shape. This is just as much a spring as is a metal coil.

Tension Force

FIGURE 4.7 Tension is parallel to the rope.

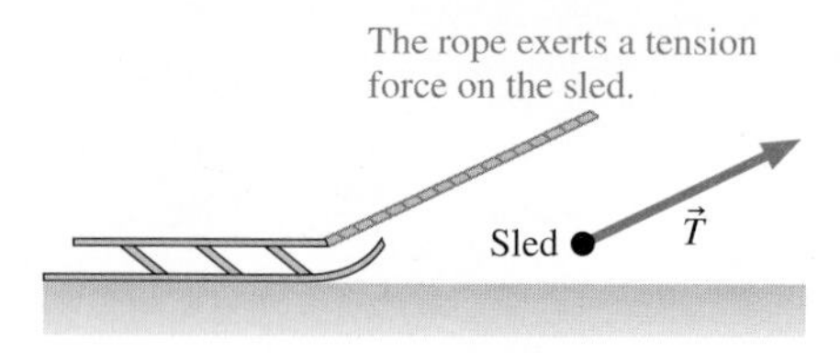
The rope exerts a tension force on the sled.
Sled
$\vec{T}$

When a string or rope or wire pulls on an object, it exerts a contact force that we call the **tension force,** represented by $\vec{T}$. **The direction of the tension force is always in the direction of the string or rope,** as you can see in FIGURE 4.7. When we speak of "the tension" in a string, this is an informal expression for T, the size or magnitude of the tension force. Note that the tension force can only *pull* in the direction of the string; if you try to *push* with a string, it will go slack and be unable to exert a force.

We can think about the tension force using a microscopic picture. If you were to use a very powerful microscope to look inside a rope, you would "see" that it is made of *atoms* joined together by *molecular bonds.* Molecular bonds are not rigid connections between the atoms. They are more accurately thought of as tiny *springs* holding the atoms together, as in FIGURE 4.8. Pulling on the ends of a string or rope stretches the spring-like molecular bonds ever so slightly. The tension within a rope and the tension force experienced by an object at the end of the rope are really the net spring force exerted by billions and billions of microscopic springs.

FIGURE 4.8 An atomic model of tension.

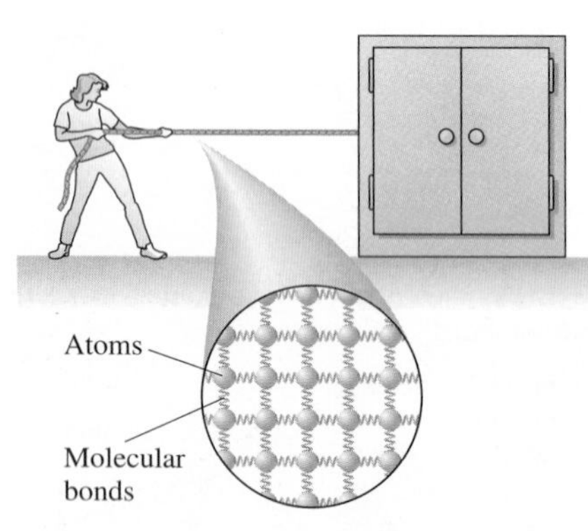
Atoms
Molecular bonds

This atomic-level view of tension introduces a new idea: a microscopic **atomic model** for understanding the behavior and properties of **macroscopic** (i.e., containing many atoms) objects. We will frequently use atomic models to obtain a deeper understanding of our observations.

The atomic model of tension also helps to explain one of the basic properties of ropes and strings. When you pull on a rope tied to a heavy box, the rope in turn exerts a tension force on the box. If you pull harder, the tension force on the box becomes greater. How does the box "know" that you are pulling harder on the other end of the rope? According to our atomic model, when you pull harder on the rope, its microscopic springs stretch a bit more, increasing the spring force they exert on each other—and on the box they're attached to.

Normal Force

If you sit on a bed, the springs in the mattress compress and, as a consequence of the compression, exert an upward force on you. Stiffer springs would show less compression but would still exert an upward force. The compression of extremely stiff springs might be measurable only by sensitive instruments. Nonetheless, the springs would compress ever so slightly and exert an upward spring force on you.

FIGURE 4.9 shows a book resting on top of a sturdy table. The table may not visibly flex or sag, but—just as you do to the bed—the book compresses the molecular "springs" in the table. The compression is very small, but it is not zero. As a consequence, the compressed molecular springs *push upward* on the book. We say that "the table" exerts the upward force, but it is important to understand that the pushing is *really* done by molecular springs. Similarly, an object resting on the ground compresses the molecular springs holding the ground together and, as a consequence, the ground pushes up on the object.

FIGURE 4.9 An atomic model of the force exerted by a table.

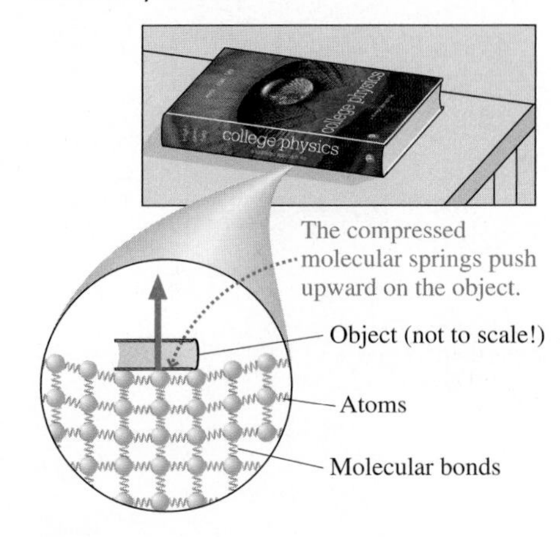
college physics
The compressed molecular springs push upward on the object.
Object (not to scale!)
Atoms
Molecular bonds

A video to support a section's topic is embedded in the eText.

Video Figure 4.9

Video What the Physics? Pushing Down to Move Upward

We can extend this idea. Suppose you place your hand on a wall and lean against it, as shown in FIGURE 4.10. Does the wall exert a force on your hand? As you lean, you compress the molecular springs in the wall and, as a consequence, they push outward *against* your hand. So the answer is Yes, the wall does exert a force on you. It's not hard to see this if you examine your hand as you lean: You can see that your hand is slightly deformed, and becomes more so the harder you lean. This deformation is direct evidence of the force that the wall exerts on your hand. Consider also what would happen if the wall suddenly vanished. Without the wall there to push against you, you would topple forward.

FIGURE 4.10 The wall pushes outward against your hand.

The compressed molecular springs in the wall press outward against her hand.

The force the table surface exerts is vertical, while the force the wall exerts is horizontal. In all cases, the force exerted on an object that is pressing against a surface is in a direction *perpendicular* to the surface. Mathematicians refer to a line that is perpendicular to a surface as being *normal* to the surface. In keeping with this terminology, we define the **normal force** as the force exerted by a surface (the agent) against an object that is pressing against the surface. The symbol for the normal force is $\vec{n}$.

We're not using the word "normal" to imply that the force is an "ordinary" force or to distinguish it from an "abnormal force." A surface exerts a force *perpendicular* (i.e., normal) to itself as the molecular springs press *outward.* FIGURE 4.11 shows an object on an inclined surface, a common situation. Notice how the normal force $\vec{n}$ is perpendicular to the surface.

FIGURE 4.11 The normal force is perpendicular to the surface.

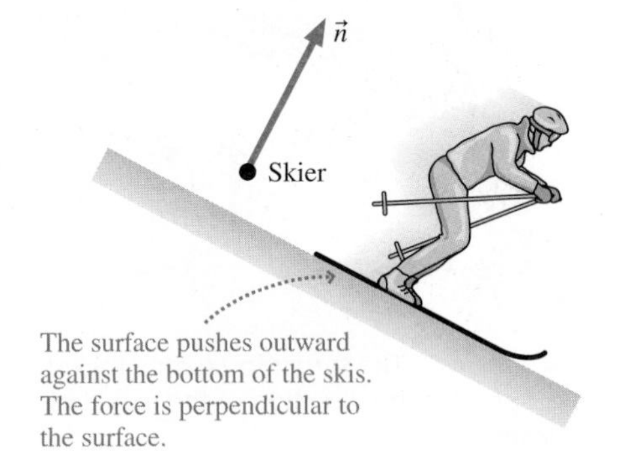
$\vec{n}$
Skier
The surface pushes outward against the bottom of the skis. The force is perpendicular to the surface.

The normal force is a very real force arising from the very real compression of molecular bonds. It is in essence just a spring force, but one exerted by a vast number of microscopic springs acting at once. The normal force is responsible for the "solidness" of solids. It is what prevents you from passing right through the chair you are sitting in and what causes the pain and the lump if you bang your head into a door. Your head can then tell you that the force exerted on it by the door was very real!

Friction

You've certainly observed that a rolling or sliding object, if not pushed or propelled, slows down and eventually stops. You've probably discovered that you can slide better across a sheet of ice than across asphalt. And you also know that most objects stay in place on a table without sliding off even if the table is tilted a bit. The force responsible for these sorts of behavior is **friction**. The symbol for friction is $\vec{f}$.

Friction, like the normal force, is exerted by a surface. Unlike the normal force, however, **the frictional force is always *parallel* to the surface,** not perpendicular to it. (In many cases, a surface will exert *both* a normal and a frictional force.) On a microscopic level, friction arises as atoms from the object and atoms on the surface run into each other. The rougher the surface is, the more these atoms are forced into close proximity and, as a result, the larger the friction force. We will develop a simple model of friction in the next chapter that will be sufficient for our needs. For now, it is useful to distinguish between two kinds of friction:

- *Kinetic friction,* denoted $\vec{f}_k$, acts as an object *slides* across a surface. Kinetic friction is a force that always "opposes the motion," meaning that the friction force $\vec{f}_k$ on a sliding object points in the direction opposite to the direction of the object's motion.

- *Static friction,* denoted $\vec{f}_s$, is the force that keeps an object "stuck" on a surface and prevents its motion relative to the surface. Finding the direction of $\vec{f}_s$ is a little trickier than finding the direction of $\vec{f}_k$. Static friction points opposite the direction in which the object *would* move if there were no friction; that is, it points in the direction necessary to *prevent* motion.

Examples of kinetic and static friction are shown in FIGURE 4.12.

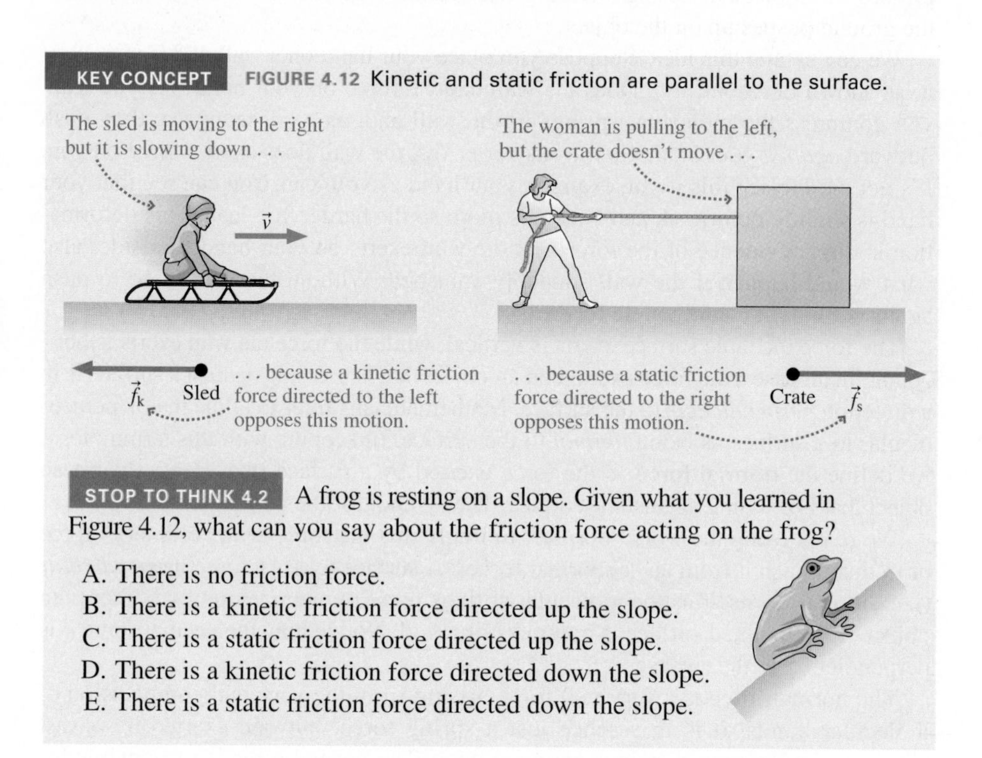

KEY CONCEPT
FIGURE 4.12 Kinetic and static friction are parallel to the surface.
The sled is moving to the right but it is slowing down . . .
$\vec{v}$
The woman is pulling to the left, but the crate doesn't move . . .
$\vec{f}_k$
Sled
. . . because a kinetic friction force directed to the left opposes this motion.
. . . because a static friction force directed to the right opposes this motion.
Crate
$\vec{f}_s$

STOP TO THINK 4.2 A frog is resting on a slope. Given what you learned in Figure 4.12, what can you say about the friction force acting on the frog?

A. There is no friction force.
B. There is a kinetic friction force directed up the slope.
C. There is a static friction force directed up the slope.
D. There is a kinetic friction force directed down the slope.
E. There is a static friction force directed down the slope.

FIGURE 4.13 Air resistance is an example of drag.

Air resistance is a significant force on falling leaves. It points opposite the direction of motion.

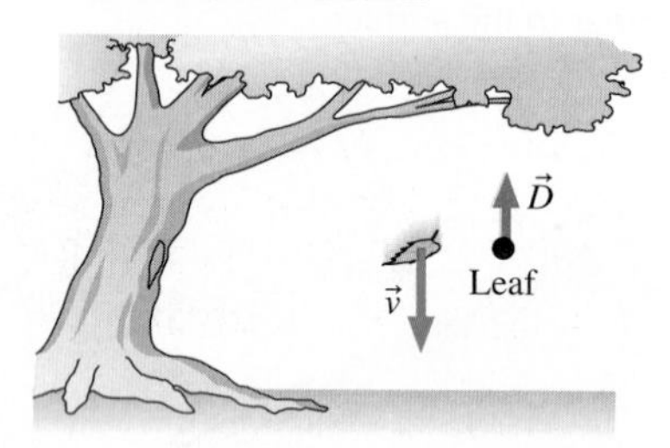

FIGURE 4.14 The thrust force on a rocket is opposite the direction of the expelled gases.

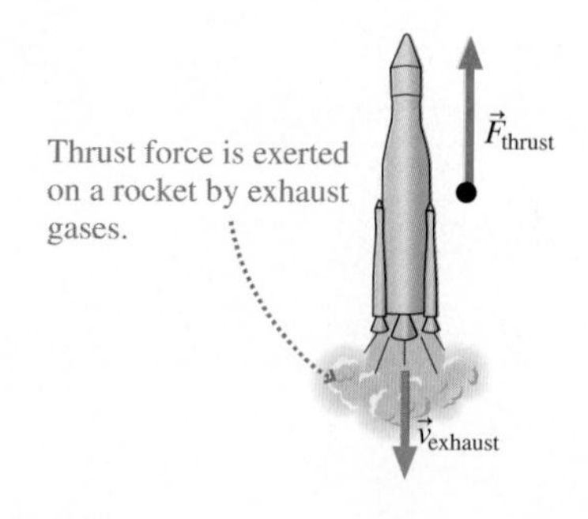

Drag

Friction at a surface is one example of a *resistive force,* a force that opposes or resists motion. Resistive forces are also experienced by objects moving through *fluids*—gases (like air) and liquids (like water). This kind of resistive force—the force of a fluid on a moving object—is called **drag** and is symbolized as $\vec{D}$. Like kinetic friction, **drag points opposite the direction of motion.** FIGURE 4.13 shows an example of drag.

Drag can be a large force for objects moving at high speeds or in dense fluids. Hold your arm out the window as you ride in a car and feel how hard the air pushes against your arm. Note also how the air resistance against your arm increases rapidly as the car's speed increases. For a small particle moving in water, such as a swimming *Paramecium,* drag can be the dominant force.

On the other hand, for objects that are heavy and compact, moving in air, and with a speed that is not too great, the drag force of air resistance is fairly small. To keep things as simple as possible, **you can neglect air resistance in all problems unless a problem explicitly asks you to include it.** The error introduced into calculations by this approximation is generally pretty small.

Thrust

A jet airplane obviously has a force that propels it forward; likewise for the rocket in FIGURE 4.14. This force, called **thrust,** occurs when a jet or rocket engine expels gas molecules at high speed. Thrust is a contact force, with the exhaust gas being the agent that pushes on the engine. The process by which thrust is generated is rather subtle and requires an appreciation of Newton's third law, introduced later in this chapter. For now, we need only consider that **thrust is a force opposite the direction in which the exhaust gas is expelled.** There's no special symbol for thrust, so we call it $\vec{F}_{\text{thrust}}$.

It's not just rocket science BIO Rockets are propelled by thrust, but many animals are as well. Scallops are shellfish with no feet and no fins, but they can escape from predators or move to new territory by using a form of jet propulsion. A scallop forcibly ejects water from the rear of its shell, resulting in a thrust force that moves it forward.

Electric and Magnetic Forces

Electricity and magnetism, like gravity, exert long-range forces. The forces of electricity and magnetism act on charged particles. We will study electric and magnetic forces in detail in Part VI of this text. These forces—and the forces inside the nucleus, which we will also see later in the text—won't be important for the dynamics problems we consider in the next several chapters.

STOP TO THINK 4.3 A boy is using a rope to pull a sled to the right. What are the directions of the tension force and the friction force on the sled, respectively?

A. Right, right B. Right, left
C. Left, right D. Left, left

4.3 Identifying Forces

Video Identifying Forces

A typical physics problem describes an object that is being pushed and pulled in various directions. Some forces are given explicitly, while others are only implied. In order to proceed, it is necessary to determine all the forces that act on the object. It is also necessary to avoid including forces that do not really exist. Now that you have learned the properties of forces and seen a catalog of typical forces, we can develop a step-by-step method for identifying each force in a problem. A list of the most common forces we'll come across in the next few chapters is given in TABLE 4.1.

NOTE ► Occasionally, you'll see labels for forces that aren't included in Table 4.1. For instance, if you push a book across a table, we might simply refer to the force you apply as $\vec{F}_{\text{hand}}$. ◄

Tactics Box 4.2 will help you correctly identify the forces acting on an object. It's followed by two examples that explicitly illustrate the steps of the Tactics Box.

TACTICS BOX 4.2 **Identifying forces**

❶ **Identify the object of interest.** This is the object whose motion you wish to study.

❷ **Draw a picture of the situation.** Show the object of interest and all other objects—such as ropes, springs, and surfaces—that touch it.

❸ **Draw a closed curve around the object.** Only the object of interest is inside the curve; everything else is outside.

❹ **Locate every point on the boundary of this curve where other objects touch the object of interest.** These are the points where *contact forces* are exerted on the object.

❺ **Name and label each contact force acting on the object.** There is at least one force at each point of contact; there may be more than one. When necessary, use subscripts to distinguish forces of the same type.

❻ **Name and label each long-range force acting on the object.** For now, the only long-range force we'll consider is weight.

Exercises 4–8 

TABLE 4.1 Common forces and their notation

Force	Notation
General force	$\vec{F}$
Weight	$\vec{w}$
Spring force	$\vec{F}_{\text{sp}}$
Tension	$\vec{T}$
Normal force	$\vec{n}$
Static friction	$\vec{f}_{\text{s}}$
Kinetic friction	$\vec{f}_{\text{k}}$
Drag	$\vec{D}$
Thrust	$\vec{F}_{\text{thrust}}$

CONCEPTUAL EXAMPLE 4.1 Identifying forces on a bungee jumper

A bungee jumper has leapt off a bridge and is nearing the bottom of her fall. What forces are being exerted on the bungee jumper?

REASON FIGURE 4.15 Forces on a bungee jumper.

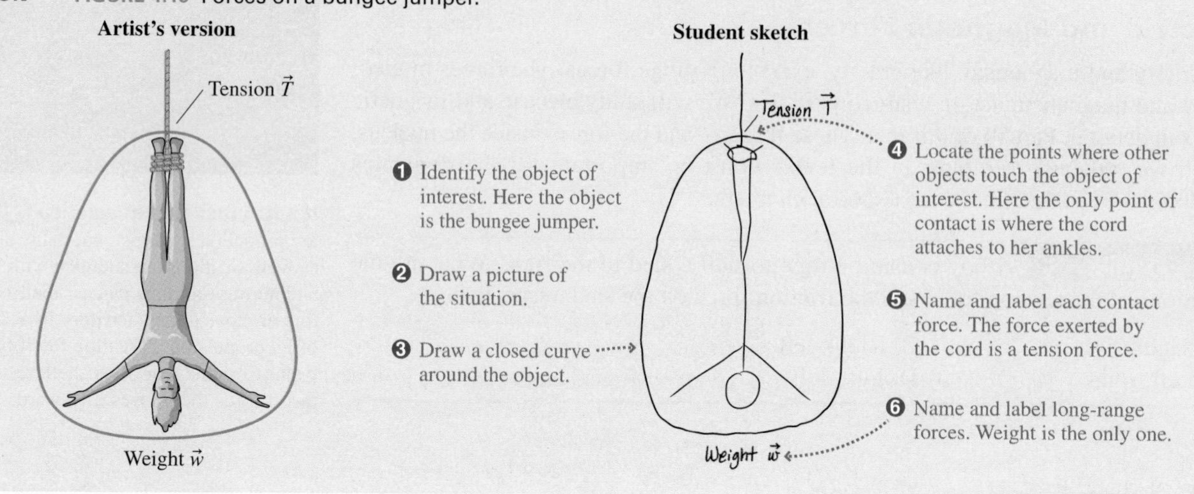

Artist's version
Tension $\vec{T}$
Weight $\vec{w}$
Student sketch
Tension $\vec{T}$
Weight $\vec{w}$
1 Identify the object of interest. Here the object is the bungee jumper.
2 Draw a picture of the situation.
3 Draw a closed curve around the object.
4 Locate the points where other objects touch the object of interest. Here the only point of contact is where the cord attaches to her ankles.
5 Name and label each contact force. The force exerted by the cord is a tension force.
6 Name and label long-range forces. Weight is the only one.

CONCEPTUAL EXAMPLE 4.2 Identifying forces on a skier

A skier is being towed up a snow-covered hill by a tow rope. What forces are being exerted on the skier?

REASON FIGURE 4.16 Forces on a skier.

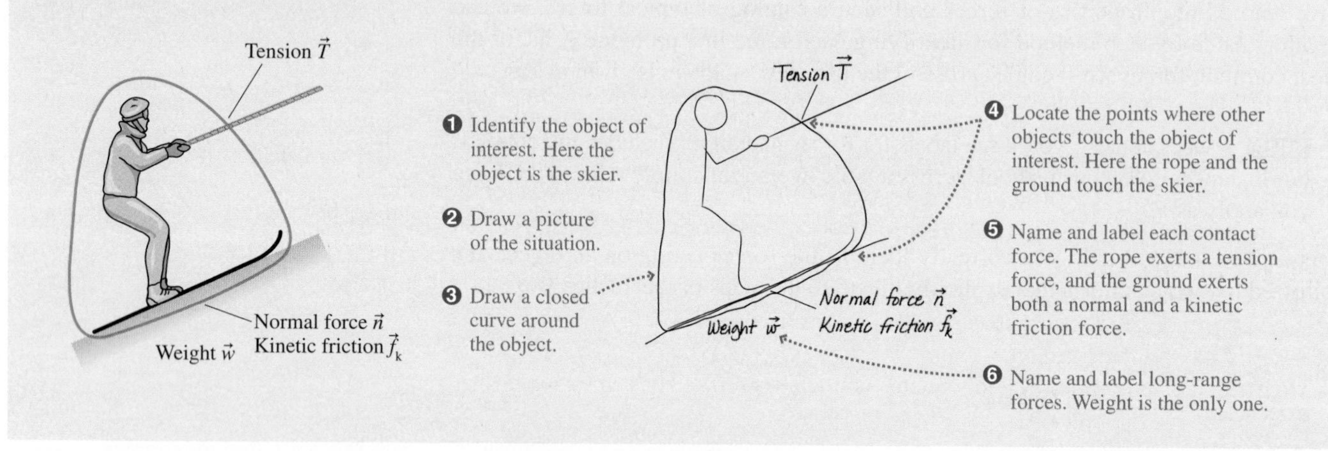

Tension $\vec{T}$
Normal force $\vec{n}$
Weight $\vec{w}$
Kinetic friction $\vec{f}_k$
Tension $\vec{T}$
Normal force $\vec{n}$
Weight $\vec{w}$
Kinetic friction $\vec{f}_k$
1 Identify the object of interest. Here the object is the skier.
2 Draw a picture of the situation.
3 Draw a closed curve around the object.
4 Locate the points where other objects touch the object of interest. Here the rope and the ground touch the skier.
5 Name and label each contact force. The rope exerts a tension force, and the ground exerts both a normal and a kinetic friction force.
6 Name and label long-range forces. Weight is the only one.

NOTE ▶ You might have expected two friction forces and two normal forces in Example 4.2, one on each ski. Keep in mind, however, that we're working within the particle model, which represents the skier by a single point. A particle has only one contact with the ground, so there is a single normal force and a single friction force. The particle model is valid if we want to analyze the motion of the skier as a whole, but we would have to go beyond the particle model to find out what happens to each ski. ◀

CONCEPTUAL EXAMPLE 4.3 Identifying forces on a rocket

A rocket is flying upward through the air, high above the ground. Air resistance is not negligible. What forces are being exerted on the rocket?

REASON

FIGURE 4.17 Forces on a rocket.

STOP TO THINK 4.4 You've just kicked a rock, and it is now sliding across the ground about 2 meters in front of you. Which of these are forces acting on the rock? Include all that apply.

A. Gravity, acting downward
B. The normal force, acting upward
C. The force of the kick, acting in the direction of motion
D. Friction, acting opposite the direction of motion
E. Air resistance, acting opposite the direction of motion

4.4 What Do Forces Do?

The fundamental question is: How does an object move when a force is exerted on it? The only way to answer this question is to do experiments. To do experiments, however, we need a way to reproduce the same force again and again, and we need a standard object so that our experiments are repeatable.

FIGURE 4.18 shows how you can use your fingers to stretch a rubber band to a certain length—say, 10 centimeters—that you can measure with a ruler. We'll call this the *standard length.* You know that a stretched rubber band exerts a force because your fingers *feel* the pull. Furthermore, this is a reproducible force. If you use the same rubber band stretched to the standard length, it will exert the same force. We'll call the magnitude of this force the *standard force* F. Not surprisingly, two identical rubber bands, each stretched to the standard length, exert twice the force of one rubber band; three rubber bands exert three times the force; and so on.

FIGURE 4.18 A reproducible force.

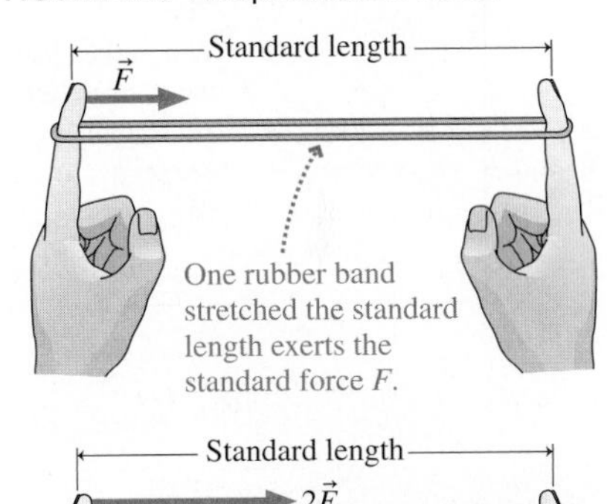
Standard length
$\vec{F}$
One rubber band stretched the standard length exerts the standard force F.

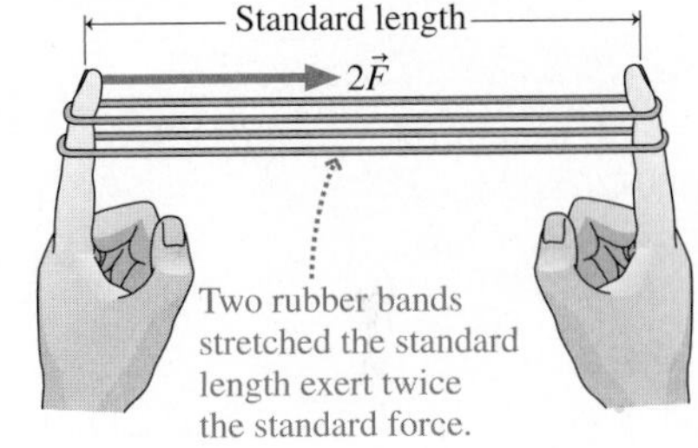
Standard length
$2\vec{F}$
Two rubber bands stretched the standard length exert twice the standard force.

We'll also need several identical standard objects to which the force will be applied. As we learned in Chapter 1, the SI unit of mass is the kilogram (kg). For our standard objects, we will make ourselves several identical blocks, each with a mass of 1 kg.

Now we're ready to start a virtual experiment. First, place one of the 1 kg blocks on a frictionless surface. (In a real experiment, we can nearly eliminate friction by floating the block on a cushion of air.) Second, attach a rubber band to the block and stretch the band to the standard length. Then the block experiences the same force F as your finger did. As the block starts to move, in order to keep the pulling force constant you must *move your hand* in just the right way to keep the length of the rubber band—and thus the force—*constant.* FIGURE 4.19 shows the experiment being carried out. Once the motion is complete, you can use motion diagrams and kinematics to analyze the block's motion.

FIGURE 4.19 Measuring the motion of a 1 kg block that is pulled with a constant force.

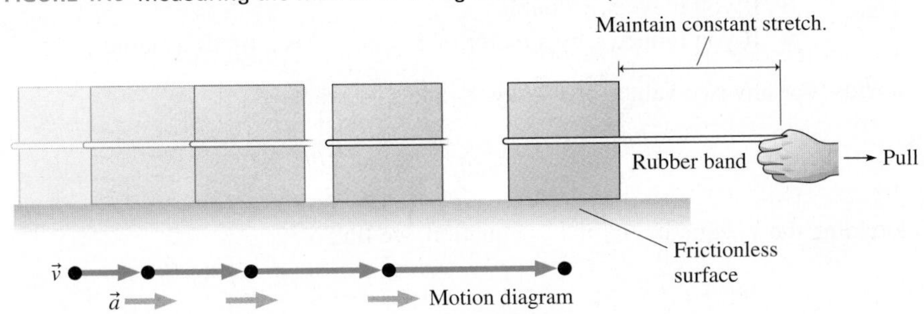
Maintain constant stretch.
Rubber band
Pull
Frictionless surface
$\vec{v}$
$\vec{a}$
Motion diagram

The motion diagram in Figure 4.19 shows that the velocity vectors are getting longer, so the velocity is increasing: The block is *accelerating.* Furthermore, a close inspection of the motion diagram shows that the acceleration vectors are all the same length. This is the first important finding of this experiment: **An object pulled with a constant force moves with a constant acceleration.** This finding could not have been anticipated in advance. It's conceivable that the object would speed up for a

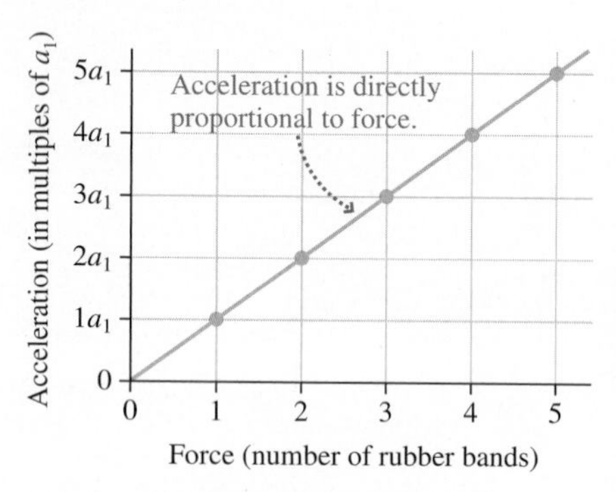

FIGURE 4.20 Graph of acceleration versus force.

FIGURE 4.21 Graph of acceleration versus number of blocks.

The acceleration of one block is a_1.

With two blocks, the acceleration is 1/2 that with only one block.

With three blocks, the acceleration is 1/3 as much.

Acceleration

a_1 $a_1/2$ $a_1/3$ $a_1/4$ 0

0 1 2 3 4

Number of blocks

while and then move with a steady speed. Or that it would continue to speed up, but the *rate* of increase, the acceleration, would steadily decline. But these descriptions do not match what happens. Instead, the object continues *with a constant acceleration* for as long as you pull it with a constant force. We'll call this constant acceleration of *one* block pulled by *one* band a_1.

What happens if you increase the force by using several rubber bands? To find out, use two rubber bands. Stretch both to the standard length to double the force to $2F$, then measure the acceleration. Measure the acceleration due to three rubber bands, then four, and so on. FIGURE 4.20 is a graph of the results. Force is the independent variable, the one you can control, so we've placed force on the horizontal axis to make an acceleration-versus-force graph. The graph reveals our second important finding: **Acceleration is directly proportional to force.**

The final question for our virtual experiment is: How does the acceleration of an object depend on the mass of the object? To find out, glue two of the 1 kg blocks together, so that we have a block with twice as much matter as a 1 kg block—that is, a 2 kg block. Now apply the same force—a single rubber band—as you applied to the single 1 kg block. FIGURE 4.21 shows that the acceleration is *one-half* as great as that of the single block. If we glue three blocks together, making a 3 kg object, we find that the acceleration is only *one-third* of the 1 kg block's acceleration. In general, we find that the acceleration is proportional to the *inverse* of the mass of the object. So our third important result is: **Acceleration is *inversely proportional* to an object's mass.**

Inversely proportional relationships

Two quantities are said to be **inversely proportional** to each other if one quantity is proportional to the *inverse* of the other. Mathematically, this means that

$$y = \frac{A}{x}$$

y is inversely proportional to x

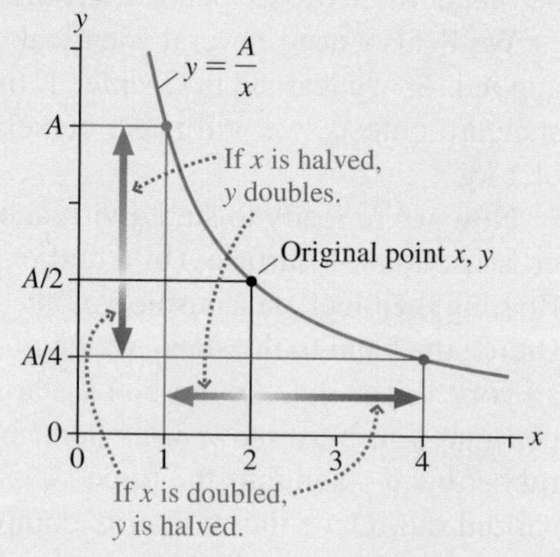

Here, A is a proportionality constant. This relationship is sometimes written as $y \propto 1/x$.

SCALING
- If you double x, you halve y.
- If you triple x, y is reduced by a factor of 3.
- If you halve x, y doubles.
- If you reduce x by a factor of 3, y becomes 3 times as large.

RATIOS For any two values of x—say, x_1 and x_2—we have

$$y_1 = \frac{A}{x_1} \quad \text{and} \quad y_2 = \frac{A}{x_2}$$

Dividing the y_1 equation by the y_2 equation, we find

$$\frac{y_1}{y_2} = \frac{A/x_1}{A/x_2} = \frac{A}{x_1}\frac{x_2}{A} = \frac{x_2}{x_1}$$

That is, the ratio of y-values is the inverse of the ratio of the corresponding values of x.

LIMITS
- As x gets very large, y approaches zero.
- As x approaches zero, y gets very large.

Exercises 10, 11

You're familiar with this idea: It's much harder to get your car rolling by pushing it than to get your bicycle rolling, and it's harder to stop a heavily loaded grocery cart than to stop a skateboard. This tendency to resist a change in velocity (i.e., to resist speeding up or slowing down) is called **inertia.** Thus we can say that more massive objects have more inertia.

Feel the difference Because of its high sugar content, a can of regular soda has a mass about 4% greater than that of a can of diet soda. If you try to judge which can is more massive by simply holding one in each hand, this small difference is almost impossible to detect. If you *move* the cans up and down, however, the difference becomes subtly but noticeably apparent: People evidently are more sensitive to how the mass of each can resists acceleration than they are to the cans' weights alone.

EXAMPLE 4.4 Finding the mass of an unknown block

When a rubber band is stretched to pull on a 1.0 kg block with a constant force, the acceleration of the block is measured to be 3.0 m/s^2. When a block with an unknown mass is pulled with the same rubber band, using the same force, its acceleration is 5.0 m/s^2. What is the mass of the unknown block?

STRATEGIZE Because acceleration is inversely proportional to mass, we will use ratio reasoning to solve this problem.

PREPARE We denote the mass of the unknown block by m.

SOLVE We can use the result of the Inversely proportional relationships box to write

$$\frac{3.0\ \text{m/s}^2}{5.0\ \text{m/s}^2} = \frac{m}{1.0\ \text{kg}}$$

or

$$m = \frac{3.0\ \text{m/s}^2}{5.0\ \text{m/s}^2} \times (1.0\ \text{kg}) = 0.60\ \text{kg}$$

ASSESS With the same force applied, the unknown block had a *larger* acceleration than the 1.0 kg block. It makes sense, then, that its mass—its resistance to acceleration—is *less* than 1.0 kg.

Video Newton's Second Law

STOP TO THINK 4.5 Two rubber bands stretched to the standard length cause an object to accelerate at 2 m/s^2. Suppose another object with twice the mass is pulled by four rubber bands stretched to the standard length. What is the acceleration of this second object?

A. 1 m/s^2 B. 2 m/s^2 C. 4 m/s^2 D. 8 m/s^2 E. 16 m/s^2

4.5 Newton's Second Law

Video Newton's Second Law Application

We can now summarize the results of our experiments. We've seen that **a force causes an object to accelerate. The acceleration a is directly proportional to the force F and inversely proportional to the mass m.** We can express both these relationships in equation form as

$$a = \frac{F}{m} \tag{4.2}$$

Note that if we double the size of the force F, the acceleration a will double, as we found experimentally. And if we triple the mass m, the acceleration will be only one-third as great, again agreeing with our experiments.

Equation 4.2 tells us the magnitude of an object's acceleration in terms of its mass and the force applied. But our experiments also had another important finding: The *direction* of the acceleration was the same as the direction of the force. We can express this fact by writing Equation 4.2 in *vector* form as

$$\vec{a} = \frac{\vec{F}}{m} \tag{4.3}$$

Finally, our experiment was limited to looking at an object's response to a *single* applied force acting in a single direction. Realistically, an object is likely to be subjected to several distinct forces $\vec{F}_1, \vec{F}_2, \vec{F}_3, \ldots$ that may point in different directions. What happens then? Experiments show that the acceleration of the object is determined by the *net force* acting on it. Recall from Figure 4.4 and Equation 4.1 that the net force is the *vector sum* of all forces acting on the object. So if several forces are acting, we use the net force in Equation 4.4.

Newton was the first to recognize these connections between force and motion. This relationship is known today as **Newton's second law.**

Newton's second law An object of mass m subjected to forces $\vec{F}_1, \vec{F}_2, \vec{F}_3, \ldots$ will undergo an acceleration $\vec{a}$ given by

$$\vec{a} = \frac{\vec{F}_{net}}{m} \tag{4.4}$$

INVERSE p. 116

where the net force $\vec{F}_{net} = \vec{F}_1 + \vec{F}_2 + \vec{F}_3 + \cdots$ is the vector sum of all forces acting on the object. **The acceleration vector $\vec{a}$ points in the same direction as the net force vector $\vec{F}_{net}$.**

HONDA
HERSHEY'S
7-ELEVEN

Size matters? Race car driver Danica Patrick was the subject of controversial comments by drivers who thought her relatively small mass of 45 kg gave her an unfair advantage. Because every driver's car must have the same mass, Patrick's overall racing mass was lower than any other driver's, so her car could be expected to have a slightly greater acceleration.

While some relationships are found to apply only in special circumstances, others seem to have universal applicability. Those equations that appear to apply at all times and under all conditions have come to be called "laws of nature." Newton's first and second laws are laws of nature; you will meet others as we go through this text.

We can rewrite Newton's second law in the form

$$\vec{F}_{net} = m\vec{a} \tag{4.5}$$

which is how you'll see it presented in many textbooks and how, in practice, we'll often use the second law. Equations 4.4 and 4.5 are mathematically equivalent, but Equation 4.4 better describes the central idea of Newtonian mechanics: A force applied to an object causes the object to accelerate and the acceleration is in the direction of the net force.

NOTE ► When several forces act on an object, be careful not to think that the strongest force "overcomes" the others to determine the motion on its own. It is $\vec{F}_{net}$, the sum of *all* the forces, that determines the acceleration $\vec{a}$. ◄

CONCEPTUAL EXAMPLE 4.5 Acceleration of a wind-blown basketball

You drop a basketball while a stiff breeze is blowing to the right. In what direction does the ball accelerate?

REASON Wind is just air in motion. If the air is moving to the *right* with respect to the ball, then the ball is moving to the *left* with respect to the air. There will be a drag force opposite the velocity of the ball relative to the air, to the right. So, as FIGURE 4.22a shows, two forces are acting on the ball: its weight $\vec{w}$ directed downward and the drag force $\vec{D}$ directed to the right. Newton's second law tells us that the direction of the acceleration is the same as the direction of the net force $\vec{F}_{net}$. In FIGURE 4.22b we find $\vec{F}_{net}$ by graphical vector addition of $\vec{w}$ and $\vec{D}$. We see that $\vec{F}_{net}$ and therefore $\vec{a}$ point downward and to the right.

FIGURE 4.22 A basketball falling in a strong breeze.

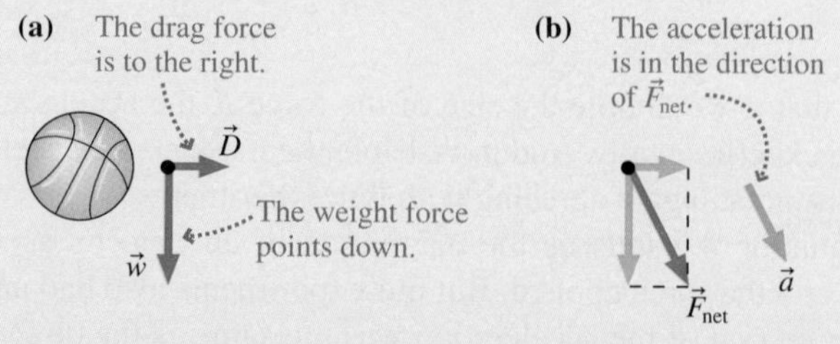
(a)
The drag force is to the right.
$\vec{D}$
The weight force points down.
$\vec{w}$
(b)
The acceleration is in the direction of $\vec{F}_{net}$.
$\vec{a}$
$\vec{F}_{net}$

ASSESS This makes sense on the basis of your experience. Weight pulls the ball down, and the wind pushes the ball to the right. The net result is an acceleration down and to the right.

The Unit of Force

Because $\vec{F}_{net} = m\vec{a}$, the units of force must be the unit of mass (kg) multiplied by the unit of acceleration (m/s^2); thus the units of force are kg · m/s^2. This unit of force is called the **newton**:

$$1 \text{ newton} = 1 \text{ N} = 1 \frac{\text{kg} \cdot \text{m}}{\text{s}^2}$$

The abbreviation for the newton is N. TABLE 4.2 lists some typical forces in newtons.

The newton is a *secondary unit,* meaning that it is defined in terms of the *primary units* of kilograms, meters, and seconds.

The unit of force in the English system is the *pound* (abbreviated lb). Although the definition of the pound has varied, it is now defined in terms of the newton:

$$1 \text{ pound} = 1 \text{ lb} = 4.45 \text{ N}$$

You very likely associate pounds with kilograms rather than with newtons. Everyday language often confuses the ideas of mass and weight, but we're going to need to make a clear distinction between them. We'll have more to say about this in the next chapter.

TABLE 4.2 Approximate magnitude of some typical forces

Force	Approximate magnitude (newtons)
Weight of a U.S. nickel	0.05
Weight of ¼ cup of sugar	0.5
Weight of a 1 pound object	5
Weight of a typical house cat	50
Weight of a 110 pound person	500
Propulsion force of a car	5000
Thrust force of a small jet engine	50,000
Pulling force of a locomotive	500,000

EXAMPLE 4.6 Racing down the runway

A Boeing 737—a small, short-range jet with a mass of 51,000 kg—sits at rest at the start of a runway. The pilot turns the pair of jet engines to full throttle, and the thrust accelerates the plane down the runway. After traveling 940 m, the plane reaches its takeoff speed of 70 m/s and leaves the ground. What is the thrust of each engine?

STRATEGIZE If we assume that the plane undergoes a constant acceleration (a reasonable assumption), we can use kinematics to find the magnitude of that acceleration. Then we will use Newton's second law to find the force—the thrust—that produced this acceleration.

PREPARE FIGURE 4.23 is a visual overview of the airplane's motion.

FIGURE 4.23 Visual overview of the accelerating airplane.

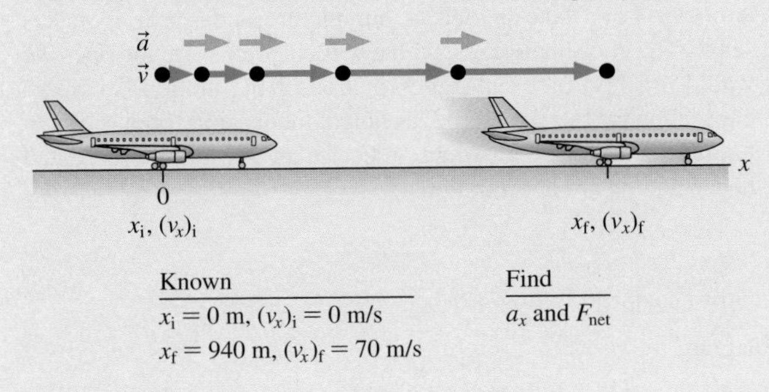

SOLVE We don't know how much time it took the plane to reach its takeoff speed, but we do know that it traveled a distance of 940 m. We can solve for the acceleration by using Equation 2.13 of Synthesis 2.1 :

$$(v_x)_f^2 = (v_x)_i^2 + 2a_x\Delta x$$

The displacement is $\Delta x = x_f - x_i = 940$ m, and the initial velocity is 0. We can rearrange the equation to solve for the acceleration:

$$a_x = \frac{(v_x)_f^2}{2\,\Delta x} = \frac{(70 \text{ m/s})^2}{2(940 \text{ m})} = 2.61 \text{ m/s}^2$$

We've kept an extra significant figure because this isn't our final result—we are asked to solve for the thrust. We complete the solution by using Newton's second law:

$$F = ma_x = (51{,}000 \text{ kg})(2.61 \text{ m/s}^2) = 133{,}000 \text{ N}$$

The thrust of each engine is half of this total force:

$$\text{Thrust of one engine} = 67{,}000 \text{ N} = 67 \text{ kN}$$

ASSESS An acceleration of about ¼g seems reasonable for an airplane: It's zippy, but it's not a thrill ride. And the final value we find for the thrust of each engine is close to the value given in Table 4.2. This gives us confidence that our final result makes good physical sense.

STOP TO THINK 4.6 Three forces act on an object. In which direction does the object accelerate?

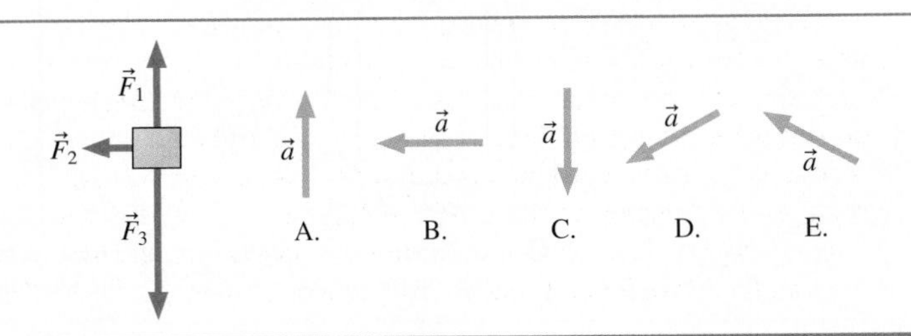

4.6 Free-Body Diagrams

When we solve a dynamics problem, it is useful to assemble all of the information about the forces that act on an object (or "body") into a single diagram called a **free-body diagram.** A free-body diagram represents the object as a particle and shows *all* of the forces that act on the object. Learning how to draw a correct free-body diagram is a very important skill, one that in the next chapter will become a critical part of our approach to solving motion problems.

TACTICS BOX 4.3 Drawing a free-body diagram

1. **Identify all forces acting on the object.** This step was described in Tactics Box 4.2.
2. **Draw a coordinate system.** Use the axes defined in your pictorial representation (see Tactics Box 2.2). If those axes are tilted, for motion along an incline, then the axes of the free-body diagram should be similarly tilted.
3. **Represent the object as a dot at the origin of the coordinate axes.** This is the particle model.
4. **Draw vectors representing each of the identified forces.** This was described in Tactics Box 4.1. Be sure to label each force vector.
5. **Draw and label the *net force* vector $\vec{F}_{net}$.** Draw this vector beside the diagram, not on the particle. Then check that $\vec{F}_{net}$ points in the same direction as the acceleration vector $\vec{a}$ on your motion diagram. Or, if appropriate, write $\vec{F}_{net} = \vec{0}$.

Exercises 17–22

EXAMPLE 4.7 Forces on an elevator

An elevator, suspended by a cable, speeds up as it moves upward from the ground floor. Draw a free-body diagram of the elevator.

STRATEGIZE We'll follow the steps of Tactics Box 4.3. We note that the elevator is moving upward and its speed is increasing. That means the acceleration is directed upward, so that $\vec{F}_{net}$ must be directed upward as well.

PREPARE In FIGURE 4.24 we illustrate the steps listed in Tactics Box 4.3. Because the net force is directed upward, the magnitude of the (upward) tension force $\vec{T}$ must be greater than that of the (downward) weight force $\vec{w}$. We have shown this in the free-body diagram by drawing the tension vector longer than the weight vector.

ASSESS Let's take a look at our picture and see if it makes sense. The coordinate axes, with a vertical *y*-axis, are the ones we would use in a pictorial representation of the motion, so we've chosen the correct axes. And, as noted, the tension force is drawn longer than the weight, which indicates an upward net force and hence an upward acceleration.

FIGURE 4.24 Free-body diagram of an elevator accelerating upward.

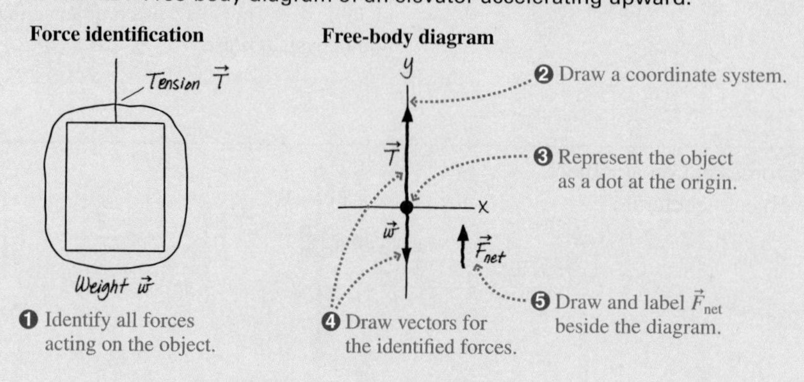
Force identification
Tension $\vec{T}$
Weight $\vec{w}$
1 Identify all forces acting on the object.
Free-body diagram
y
x
$\vec{T}$
$\vec{w}$
$\vec{F}_{net}$
2 Draw a coordinate system.
3 Represent the object as a dot at the origin.
4 Draw vectors for the identified forces.
5 Draw and label $\vec{F}_{net}$ beside the diagram.

EXAMPLE 4.8 Forces on the world's fastest car

The Thrust SSC is the world's fastest car.

The speed record for the world's fastest car is held by the jet-powered Thrust SSC, whose top speed exceeds the speed of sound. At high speeds, air drag is significant, but when the car is just starting to accelerate from rest, friction and drag are negligible compared to the jet thrust. Draw a visual overview—motion diagram, force identification picture, and free-body diagram—for the car as it starts to move forward from rest.

STRATEGIZE To draw the visual overview, we'll use the steps listed in Tactics Box 4.3.

PREPARE We treat the car as a particle. The visual overview is shown in FIGURE 4.25.

ASSESS The motion diagram tells us that the acceleration is in the positive x-direction. According to the rules of vector addition, this can be true only if the upward-pointing $\vec{n}$ and the downward-pointing $\vec{w}$ are equal in magnitude and thus cancel each other. The vectors have been drawn accordingly, and this leaves the net force vector pointing toward the right, in agreement with $\vec{a}$ from the motion diagram.

FIGURE 4.25 Visual overview for a carbon dioxide racer.

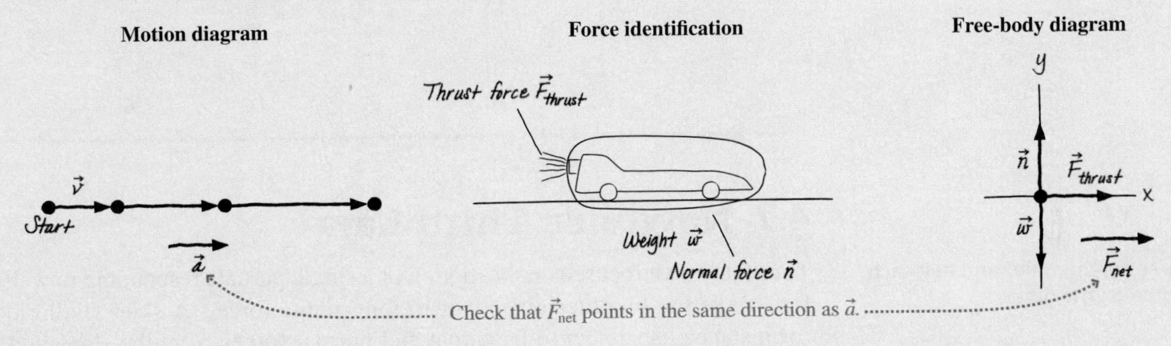

EXAMPLE 4.9 Forces on a towed skier

A tow rope pulls a skier up a snow-covered hill at a constant speed. Draw a full visual overview of the skier.

STRATEGIZE This is example 4.2 again, with the additional information that the skier is moving at a constant speed. Because his acceleration is zero, the net force acting on him must be zero as well.

PREPARE If we were doing a kinematics problem, the pictorial representation would use a tilted coordinate system with the x-axis parallel to the slope, so we use these same tilted coordinate axes for the free-body diagram. The full visual overview is shown in FIGURE 4.26.

ASSESS We have shown $\vec{T}$ pulling parallel to the slope and $\vec{f}_k$, which opposes the direction of motion, pointing down the slope. The normal force $\vec{n}$ is perpendicular to the surface and thus along the y-axis. Finally, and this is important, the weight $\vec{w}$ is *vertically* downward, *not* along the negative y-axis.

The skier moves in a straight line with constant speed, so $\vec{a} = \vec{0}$. Newton's second law then tells us that $\vec{F}_{net} = m\vec{a} = \vec{0}$. Thus we have drawn the vectors such that the forces add to zero. We'll learn more about how to do this in Chapter 5.

FIGURE 4.26 Visual overview for a skier being towed at a constant speed.

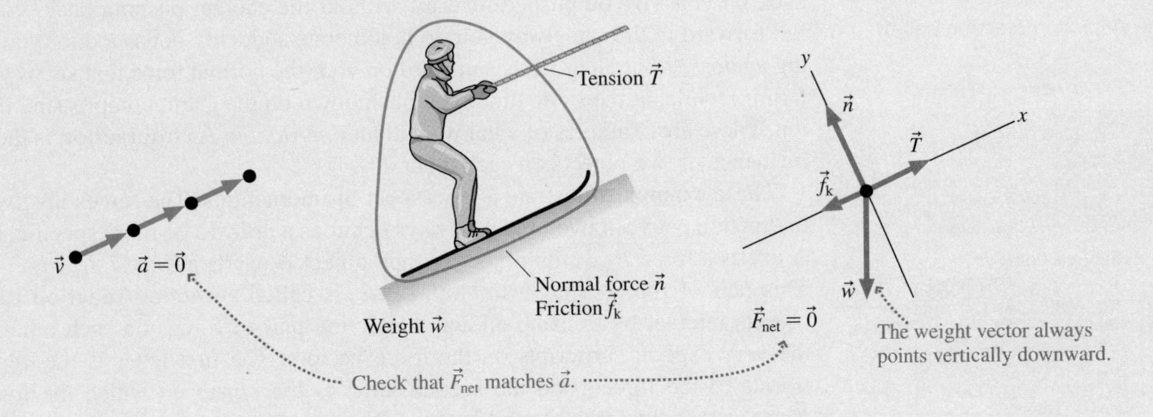

Free-body diagrams will be our major tool for the next several chapters. Careful practice with the workbook exercises and homework in this chapter will pay immediate benefits in the next chapter. Indeed, it is fair to say that a problem is more than half solved when you correctly complete the free-body diagram.

STOP TO THINK 4.7 An elevator suspended by a cable is moving upward and slowing to a stop. Which free-body diagram is correct?

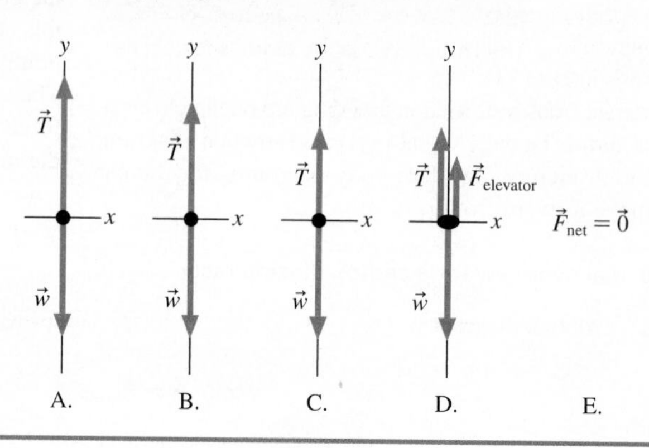

4.7 Newton's Third Law

Thus far, we've focused on the motion of a single particle responding to well-defined forces exerted by other objects or to long-range forces. A skier sliding downhill, for instance, is subject to frictional and normal forces from the slope and the pull of gravity on his body. Once we have identified these forces, we can use Newton's second law to calculate the acceleration, and hence the overall motion, of the skier.

But motion in the real world often involves two or more objects *interacting* with each other. Consider the hammer and nail in FIGURE 4.27. As the hammer hits the nail, the nail pushes back on the hammer. A bat and a ball, your foot and a soccer ball, and the earth–moon system are other examples of interacting objects. We need to consider how the forces on these interacting objects are related to each other.

FIGURE 4.27 The hammer and nail each exert a force on the other.

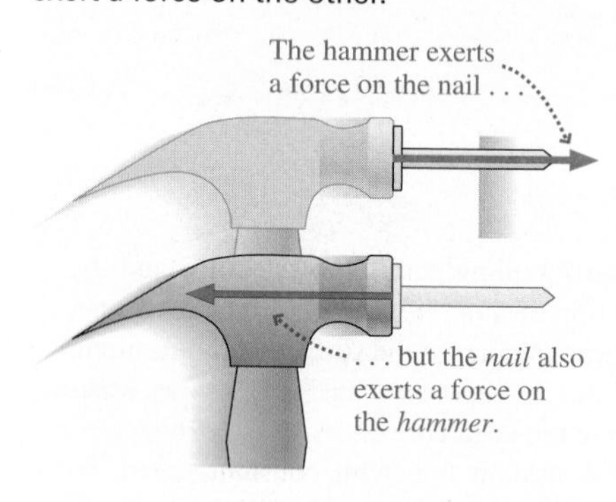

Interacting Objects

Think about the hammer and nail in Figure 4.27 again. The hammer certainly exerts a force on the nail as it drives the nail forward. At the same time, the nail exerts a force on the hammer. If you are not sure that it does, imagine hitting the nail with a glass hammer. It's the force of the nail on the hammer that would cause the glass to shatter.

Indeed, any time that object A pushes or pulls on object B, object B pushes or pulls back on object A. As you push on a filing cabinet to move it, the cabinet pushes back on you. (If you pushed forward without the cabinet pushing back, you would fall forward in the same way you do if someone suddenly opens a door you're leaning against.) Your chair pushes upward on you (the normal force that keeps you from falling) while, at the same time, you push down on the chair, compressing the cushion. These are examples of what we call an *interaction.* An **interaction** is the mutual influence of two objects on each other.

These examples illustrate a key aspect of interactions: The forces involved in an interaction between two objects always occur as a *pair.* To be more specific, if object A exerts a force $\vec{F}_{\text{A on B}}$ on object B, then object B exerts a force $\vec{F}_{\text{B on A}}$ on object A. This pair of forces, shown in FIGURE 4.28, is called an **action/reaction pair.** Two objects interact by exerting an action/reaction pair of forces on each other. Notice the very explicit subscripts on the force vectors. The first letter is the *agent*—the source of the force—and the second letter is the *object* on which the force acts. $\vec{F}_{\text{A on B}}$ is thus the force exerted *by* A *on* B.

FIGURE 4.28 An action/reaction pair of forces.

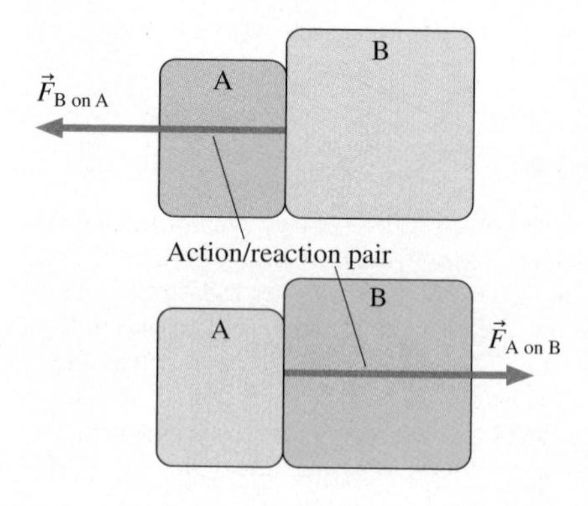

NOTE ▸ The name "action/reaction pair" is somewhat misleading. The forces occur simultaneously, and we cannot say which is the "action" and which the "reaction." Neither is there any implication about cause and effect: The action does not cause the reaction. **An action/reaction pair of forces exists as a pair, or not at all.** For action/reaction pairs, the labels are the key: Force $\vec{F}_{\text{A on B}}$ is paired with force $\vec{F}_{\text{B on A}}$. ◂

Reasoning with Newton's Third Law

Two objects always interact via an action/reaction pair of forces. Newton was the first to recognize how the two members of an action/reaction pair of forces are related to each other. Today we know this as **Newton's third law:**

Newton's third law Every force occurs as one member of an action/reaction pair of forces.

- The two members of an action/reaction pair act on two *different* objects.
- The two members of an action/reaction pair point in *opposite* directions and are *equal in magnitude.*

Newton's third law is often stated: "For every action there is an equal but opposite reaction." While this is a catchy phrase, it lacks the preciseness of our preferred version. In particular, it fails to capture an essential feature of the two members of an action/reaction pair—that each acts on a *different* object. This is shown in FIGURE 4.29, where a hammer hitting a nail exerts a force $\vec{F}_{\text{hammer on nail}}$ on the nail, and by the third law, the nail must exert a force $\vec{F}_{\text{nail on hammer}}$ to complete the action/reaction pair.

Figure 4.29 also illustrates that these two forces point in *opposite directions.* This feature of the third law is also in accord with our experience. If the hammer hits the nail with a force directed to the right, then the force of the nail on the hammer is directed to the left. If the force of me on my chair pushes down, the force of the chair on me pushes up.

Finally, Figure 4.29 shows that, according to Newton's third law, the two members of an action/reaction pair have *equal* magnitudes, so that $F_{\text{hammer on nail}} = F_{\text{nail on hammer}}$. This is something new, and it is by no means obvious.

FIGURE 4.29 Newton's third law.

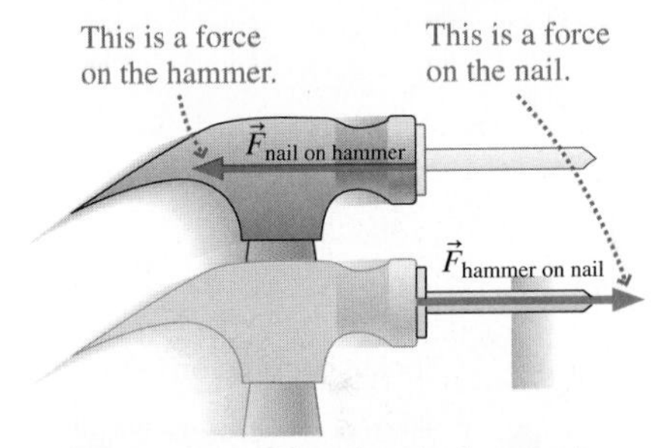
Each force in an action/reaction pair acts on a *different* object.
This is a force on the hammer.
This is a force on the nail.
$\vec{F}_{\text{nail on hammer}}$
$\vec{F}_{\text{hammer on nail}}$
The members of the pair point in *opposite directions*, but are of *equal magnitude*.

CONCEPTUAL EXAMPLE 4.10 Forces in a collision

A 10,000 kg truck has a head-on collision with a 1000 kg compact car. During the collision, is the force of the truck on the car greater than, less than, or equal to the force of the car on the truck?

REASON Newton's third law tells us that the magnitude of the force of the car on the truck must be equal to that of the truck on the car! How can this be, when the car is so small compared to the truck? The source of puzzlement in problems like this is that Newton's third law equates the sizes of the *forces* acting on the two objects, not their *accelerations*. The acceleration of each object depends not only on the force applied to it but also, according to Newton's second law, on its mass. The car and the truck do in fact feel forces of equal strength from the other, but the car, with its smaller mass, undergoes a much greater acceleration than the more massive truck.

ASSESS This is a type of question for which your intuition may need to be refined. When you think about questions of this sort, be sure to separate the *effects* (the accelerations) from the *causes* (the forces themselves). Because two interacting objects can have very different masses, their accelerations can be very different. Don't let this dissuade you from realizing that the interaction forces are of the same strength.

Video Weighing a Hovering Magnet

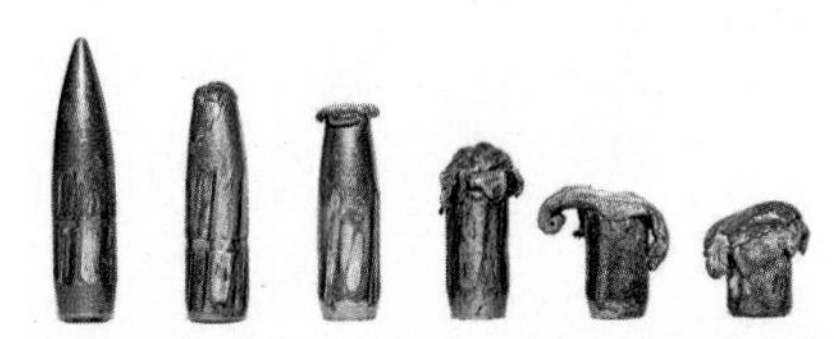

Revenge of the target We normally think of the damage that the force of a bullet inflicts on its target. But according to Newton's third law, the target exerts an equal force on the bullet. The bullet on the left has not been fired. The bullets on the right have been fired into a test target at increasing speeds and were clearly damaged by this interaction.

A runner starts from rest and then begins to move down the track. Because he's accelerating, there must be a force on him in the forward direction. The *energy* to put his body into motion comes from inside his body (we'll consider this type of problem in detail in Chapter 11). But where does the *force* come from?

FIGURE 4.30 Pushing backward to move forward.

(a)

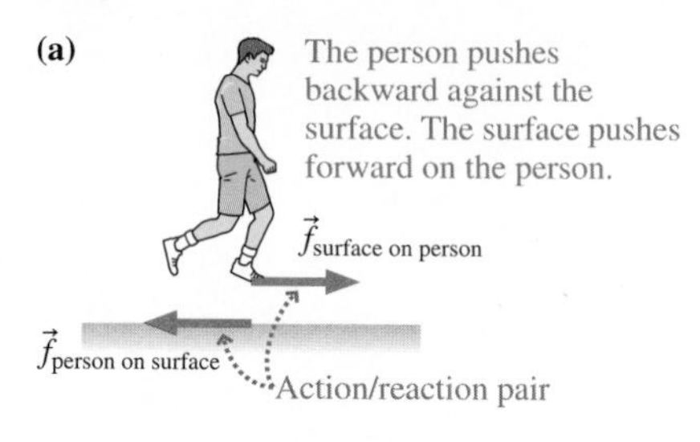
The person pushes backward against the surface. The surface pushes forward on the person.
$\vec{f}_{\text{surface on person}}$
$\vec{f}_{\text{person on surface}}$
Action/reaction pair

(b)

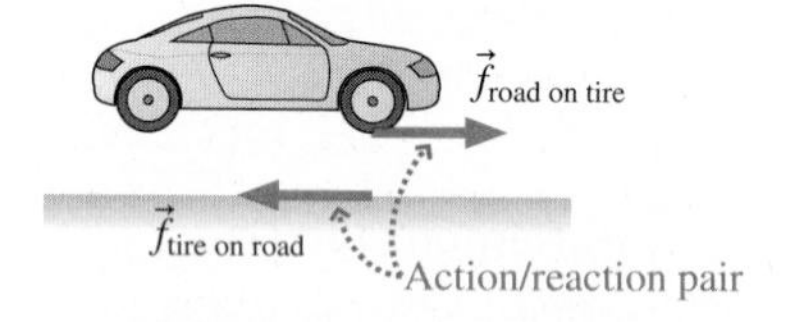
The tire pushes backward against the road. The road pushes forward on the tire.
$\vec{f}_{\text{road on tire}}$
$\vec{f}_{\text{tire on road}}$
Action/reaction pair

If you tried to walk across a frictionless floor, your foot would slip and slide *backward*. In order for you to walk, the floor needs to have friction so that your foot *sticks* to the floor as you straighten your leg, moving your body forward. The friction that prevents slipping is *static* friction. Static friction, you will recall, acts in the direction that prevents slipping, so the static friction force $\vec{f}_{\text{surface on person}}$ has to point in the *forward* direction to prevent your foot from slipping backward. As shown in FIGURE 4.30a, it is this forward-directed static friction force that propels you forward! The force of your foot on the floor, $\vec{f}_{\text{person on surface}}$, is the other half of the action/reaction pair, and it points in the opposite direction as you push backward against the floor. So, when the runner starts down the track—or when you start walking across the floor—it is the static friction force between the ground and the runner that provides the acceleration. This may seem surprising, but imagine that the race was held on an icy pond. The runners would have much more trouble getting started.

NOTE ▶ A counterintuitive notion for many students is that it is *static* friction that pushes you forward. You are moving, so how can this be static friction? It's true that your body is in motion, but your feet are not sliding, so this is not kinetic friction, but static. ◀

Similarly, the car in FIGURE 4.30b uses static friction to propel itself. The car uses its motor to turn the tires, causing the tires to push backward against the road ($\vec{f}_{\text{tire on road}}$). The road surface responds by pushing the car forward ($\vec{f}_{\text{road on tire}}$). Again, the forces involved are *static* friction forces. The tire is rolling, but the bottom of the tire, where it contacts the road, is instantaneously at rest. If it weren't, you would leave one giant skid mark as you drove and would burn off the tread within a few miles.

FIGURE 4.31 Rocket propulsion.

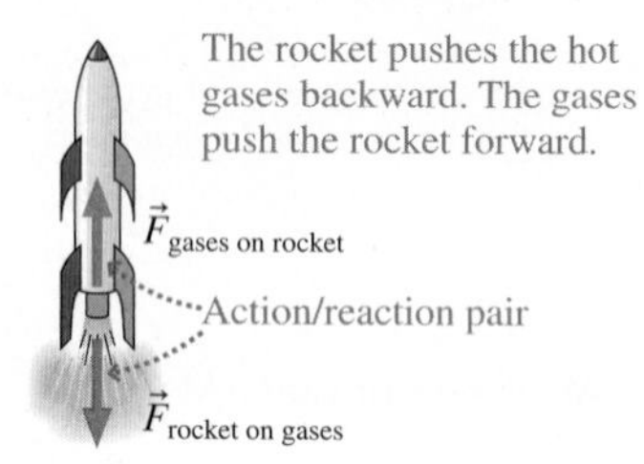
The rocket pushes the hot gases backward. The gases push the rocket forward.
$\vec{F}_{\text{gases on rocket}}$
Action/reaction pair
$\vec{F}_{\text{rocket on gases}}$

Rocket motors provide propulsion as well, but there is a difference from the earlier cases: A rocket doesn't push against the ground or even the atmosphere. Instead, the rocket engine pushes hot, expanding gases out of the back of the rocket, as shown in FIGURE 4.31. In response, the exhaust gases push the rocket forward with the force we've called *thrust*. That's why rocket propulsion works in the vacuum of space.

Now let's consider the opposite case—a pull *in* rather than a push *out*. The chapter opened with a photo of a snake-necked turtle. The muscles in the turtle's neck aren't strong enough to snap its head forward with the observed acceleration. Instead, the turtle uses a different approach. The turtle opens its mouth and forcefully pulls water into its throat. The turtle's head and the water form an action/reaction pair: The water is pulled backward, and this results in a forward force on the turtle's head. This is just the reverse of what happens in a rocket, and it is a surprisingly effective technique, enabling the turtle to strike more rapidly than many predatory fish.

Now we've assembled all the pieces we need in order to start solving problems in dynamics. We have seen what forces are and how to identify them, and we've learned how forces cause objects to accelerate according to Newton's second law. We've also found how Newton's third law governs the interaction forces between two objects. Our goal in the next several chapters is to apply Newton's laws to a variety of problems involving straight-line and circular motion.

STOP TO THINK 4.8 A small car is pushing a larger truck that has a dead battery. The mass of the truck is greater than the mass of the car. The car and the truck are moving to the right and are speeding up. Which of the following statements is true?

A. The car exerts a force on the truck, but the truck doesn't exert a force on the car.
B. The car exerts a larger force on the truck than the truck exerts on the car.
C. The car exerts the same amount of force on the truck as the truck exerts on the car.
D. The truck exerts a larger force on the car than the car exerts on the truck.
E. The truck exerts a force on the car, but the car doesn't exert a force on the truck.

INTEGRATED EXAMPLE 4.11 Pulling an excursion train

An engine slows as it pulls two cars of an excursion train up a mountain. Draw a visual overview (motion diagram, force identification diagram, and free-body diagram) for the car just behind the engine. Ignore friction.

STRATEGIZE We will treat the train car as a particle. We'll prepare the visual overview using the steps of Tactics Box 4.2 to identify the forces, and Tactics Box 4.3 to draw the free-body diagram.

PREPARE Because the train is slowing down, the motion diagram consists of a series of particle positions that become closer together at successive times; the corresponding velocity vectors become shorter and shorter. To identify the forces acting on the car we use the steps of Tactics Box 4.2. Finally, we can draw a free-body diagram using Tactics Box 4.3.

SOLVE Finding the forces acting on car 1 can be tricky. The engine exerts a forward force $\vec{F}_{\text{engine on 1}}$ on car 1 where the engine touches the front of car 1. At its back, car 1 touches car 2, so car 2 must also exert a force on car 1. The direction of this force can be understood from Newton's third law. Car 1 exerts an uphill force on car 2 in order to pull it up the mountain. Thus, by Newton's third law, car 2 must exert an oppositely directed *downhill* force on car 1. This is the force we label $\vec{F}_{\text{2 on 1}}$. The three diagrams that make up the full visual overview are shown in FIGURE 4.32.

ASSESS Correctly preparing the three diagrams illustrated in this example is critical for solving problems using Newton's laws. The motion diagram allows you to determine the direction of the acceleration and hence of $\vec{F}_{\text{net}}$. Using the force identification diagram, you will correctly identify all the forces acting on the object and, just as important, not add any extraneous forces. And by properly drawing these force vectors in a free-body diagram, you'll be ready for the quantitative application of Newton's laws that is the focus of Chapter 5.

FIGURE 4.32 Visual overview for a slowing train car being pulled up a mountain.

Motion diagram

Because the train is slowing down, its acceleration vector points in the direction opposite to its motion.

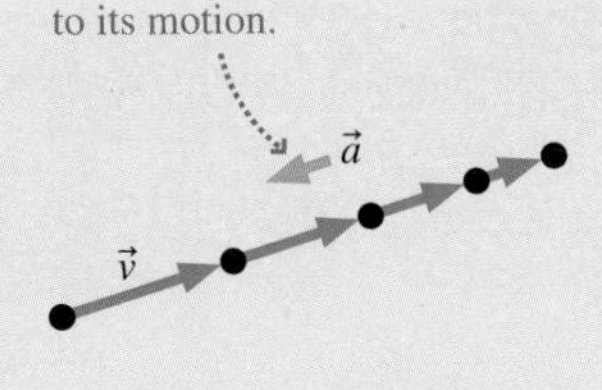

Force identification
(Numbered steps from Tactics Box 4.2)

❶ The object of interest is car 1.
❷ Draw a picture.
❸ Draw a closed curve around the object.
❹ Locate the points where the object touches other objects.
❺ Name and label each contact force.
❻ Weight is the only long-range force.

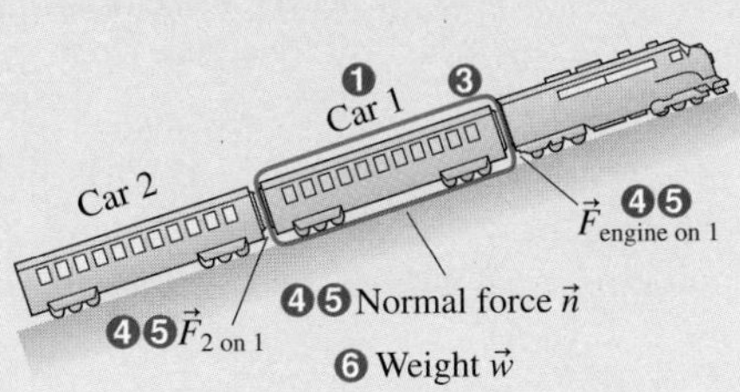

Free-body diagram
(Numbered steps from Tactics Box 4.3)

❶ Identify all forces (already done).
❷ Draw a coordinate system. Because the motion here is along an incline, we tilt our x-axis to match.
❸ Represent the object as a dot at the origin.
❹ Draw vectors representing each identified force.
❺ Draw the net force vector. Check that it points in the same direction as $\vec{a}$.

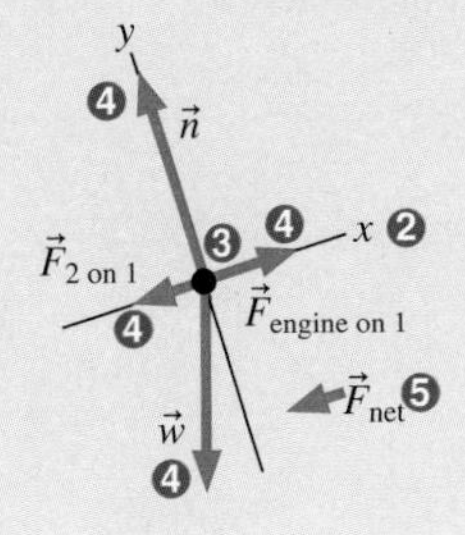

SUMMARY

GOAL To establish the connection between force and motion.

GENERAL PRINCIPLES

Newton's First Law

Consider an object with no force acting on it. If it is at rest, it will remain at rest. If it is in motion, then it will continue to move in a straight line at a constant speed.

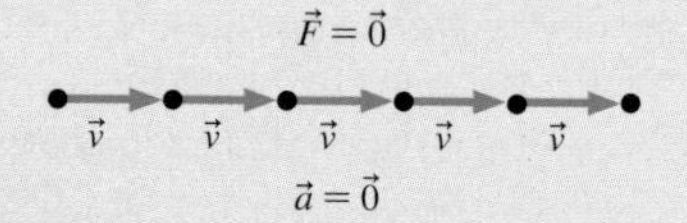

The first law tells us that an object that experiences no force will experience no acceleration.

Newton's Second Law

An object with mass m will undergo acceleration

$$\vec{a} = \frac{\vec{F}_{\text{net}}}{m}$$

where the net force $\vec{F}_{\text{net}} = \vec{F}_1 + \vec{F}_2 + \vec{F}_3 + \cdots$ is the vector sum of all the individual forces acting on the object.

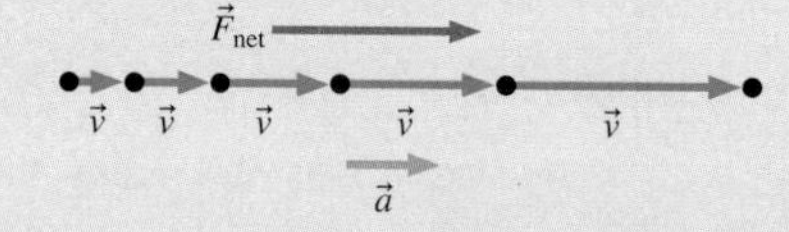

The second law tells us that a net force causes an object to accelerate. This is the relationship between force and motion. The acceleration points in the direction of $\vec{F}_{\text{net}}$.

Newton's Third Law

Every force occurs as one member of an **action/reaction** pair of forces. The two members of an action/reaction pair:

- act on two *different* objects.
- point in opposite directions and are equal in magnitude:

$$\vec{F}_{\text{A on B}} = -\vec{F}_{\text{B on A}}$$

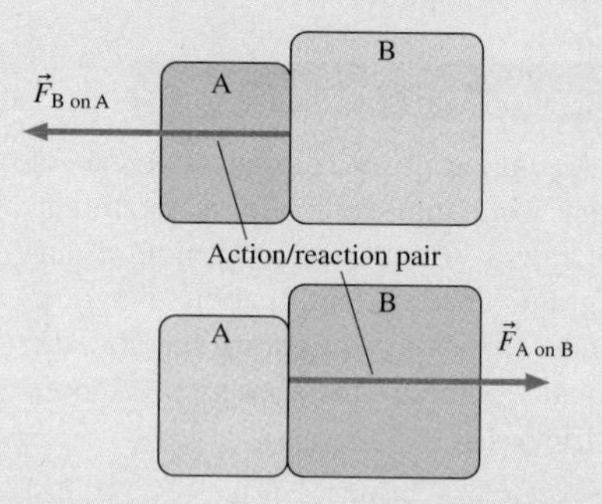

IMPORTANT CONCEPTS

Force is a push or a pull on an object.

- Force is a vector, with a magnitude and a direction.
- A force requires an agent.
- A force is either a contact force or a long-range force.

The SI unit of force is the **newton** (N). A 1 N force will cause a 1 kg mass to accelerate at 1 m/s^2.

Net force is the vector sum of all the forces acting on an object.

$\vec{F}_1$ $\vec{F}_2$ $\vec{F}_3$ = $\vec{F}_{\text{net}} = \vec{F}_1 + \vec{F}_2 + \vec{F}_3$

Mass is the property of an object that determines its resistance to acceleration.

If the same force is applied to objects A and B, then the ratio of their accelerations is related to the ratio of their masses as

$$\frac{a_A}{a_B} = \frac{m_B}{m_A}$$

The mass of objects can be determined in terms of their accelerations.

APPLICATIONS

Identifying Forces

Forces are identified by locating the points where other objects touch the object of interest. These are the points where contact forces are exerted. In addition, objects feel a long-range weight force.

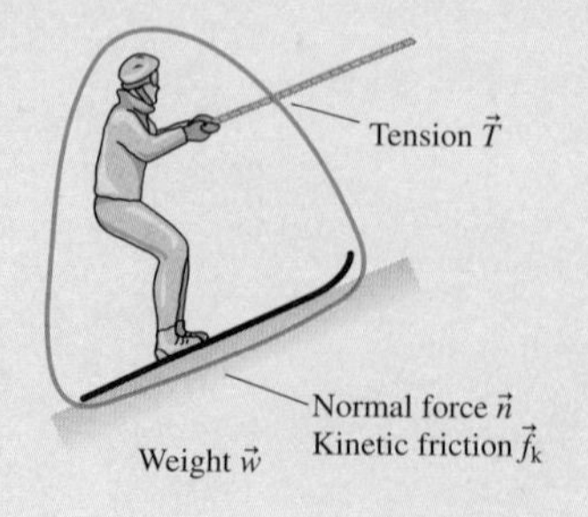

Free-Body Diagrams

A free-body diagram represents the object as a particle at the origin of a coordinate system. Force vectors are drawn with their tails on the particle. The net force vector is drawn beside the diagram.

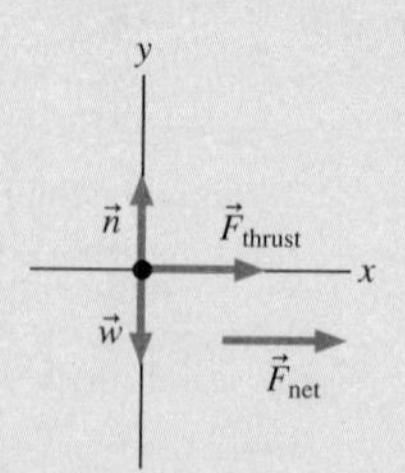

Learning Objectives After studying this chapter, you should be able to:

- Recognize and identify the forces acting on an object. *Conceptual Questions 4.14, 4.20; Problems 4.8, 4.9, 4.10, 4.11, 4.12*
- Combine multiple forces acting on an object. *Problems 4.4, 4.5, 4.6, 4.28*
- Draw a free-body diagram. *Problems 4.34, 4.35, 4.36, 4.37, 4.38*
- Understand the connection between force and motion. *Conceptual Questions 4.1, 4.3; Problems 4.2, 4.3, 4.20, 4.23, 4.24*
- Use Newton's second law. *Problems 4.14, 4.17, 4.21, 4.22, 4.25*
- Identify action/reaction pairs of forces on interacting objects. *Conceptual Questions 4.11, 4.17; Problems 4.43, 4.44, 4.45, 4.46*

STOP TO THINK ANSWERS

Chapter Preview Stop to Think: B. The swan's velocity is to the left. Its speed is decreasing, so the acceleration is opposite the velocity, or to the right.

Stop to Think 4.1: C.

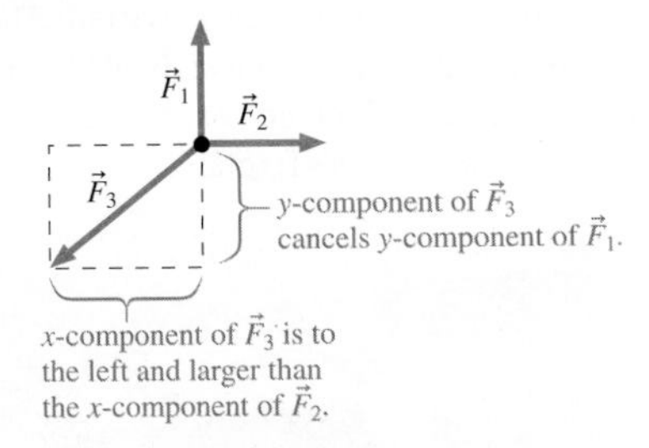

Stop to Think 4.2: C. The frog isn't moving, so a static friction force is keeping it at rest. If there was no friction, the weight force would cause the frog to slide down the slope. The static friction force opposes this, so it must be directed up the slope.

Stop to Think 4.3: B. The tension force is the force pulling the sled to the right. The friction force is opposing the motion, so it is directed to the left.

Stop to Think 4.4: A, B, and D. Friction and the normal force are the only contact forces. Nothing is touching the rock to provide a "force of the kick." We've agreed to ignore air resistance unless a problem specifically calls for it.

Stop to Think 4.5: B. Acceleration is proportional to force, so doubling the number of rubber bands doubles the acceleration of the original object from 2 m/s^2 to 4 m/s^2. But acceleration is also inversely proportional to mass. Doubling the mass cuts the acceleration in half, back to 2 m/s^2.

Stop to Think 4.6: D.

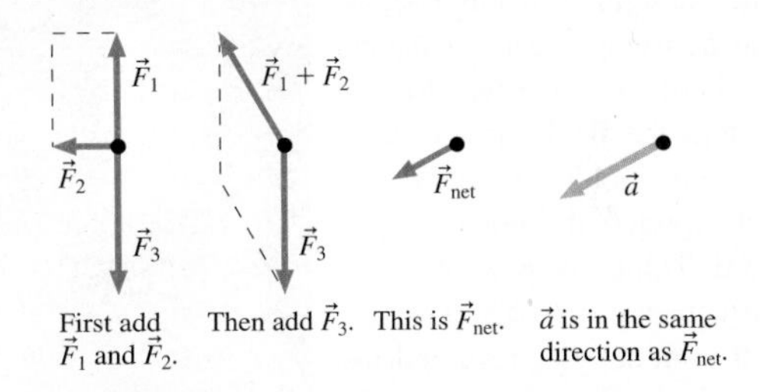

Stop to Think 4.7: C. The acceleration vector points downward as the elevator slows. $\vec{F}_{net}$ points in the same direction as $\vec{a}$, so $\vec{F}_{net}$ also points downward. This will be true if the tension is less than the weight: $T < w$.

Stop to Think 4.8: C. Newton's third law says that the force of A on B is *equal* and opposite to the force of B on A. This is always true. The mass of the objects isn't relevant, nor is the fact that the car and truck are accelerating.

Video Tutor Solution Chapter 4

QUESTIONS

Conceptual Questions

1. If an object is not moving, does that mean that there are no forces acting on it? Explain.
2. An object moves in a straight line at a constant speed. Is it true that there must be no forces of any kind acting on this object? Explain.
3. If you know all of the forces acting on a moving object, can you tell in which direction the object is moving? If the answer is Yes, explain how. If the answer is No, give an example.
4. Three arrows are shot horizontally. They have left the bow and are traveling parallel to the ground as shown in Figure Q4.4. Air resistance is negligible. Rank in order, from largest to smallest, the magnitudes of the *horizontal* forces F_1, F_2, and F_3 acting on the arrows. Some may be equal. State your reasoning.

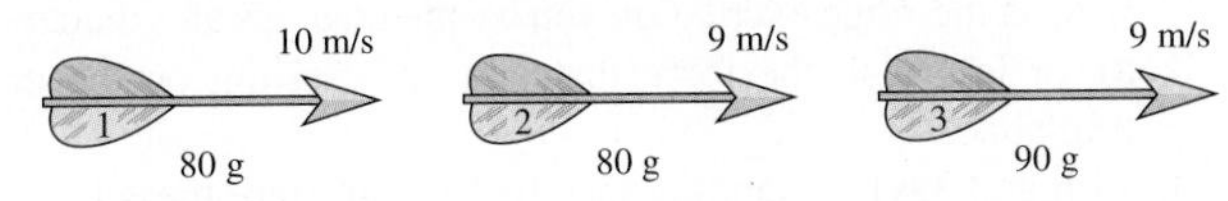

FIGURE Q4.4

Problem difficulty is labeled as | (straightforward) to ||||| (challenging). INT Problems labeled integrate significant material from earlier chapters; BIO are of biological or medical interest.

The eText icon indicates when there is a video tutor solution available for the chapter or for a specific problem. To launch these videos, log into your eText through Mastering™ Physics or log into the Study Area.

5. When your car accelerates away from a stop sign, you feel like you're being pushed back into your seat. Can you identify the force that is pushing you back? If not, why do you feel like you're being pushed back?
6. Internal injuries in vehicular accidents may be due to what is BIO called the "third collision." The first collision is the vehicle hitting the external object. The second collision is the person hitting something on the inside of the car, such as the dashboard or windshield. This may cause external lacerations. The third collision, possibly the most damaging to the body, is when organs, such as the heart or brain, hit the ribcage, skull, or other confines of the body, bruising the tissues on the leading edge and tearing the organ from its supporting structures on the trailing edge.
   a. Why is there a third collision? In other words, why are the organs still moving after the second collision?
   b. If the vehicle was traveling at 60 mph before the first collision, would the organs be traveling faster than, equal to, or slower than 60 mph just before the third collision?
7. Here's a great everyday use of the physics described in this chapter. If you are trying to get ketchup out of the bottle, the best way to do it is to turn the bottle upside down and give the bottle a sharp *upward* smack, forcing the bottle rapidly upward. Think about what subsequently happens to the ketchup, which is initially at rest, and use Newton's first law to explain why this technique is so successful.

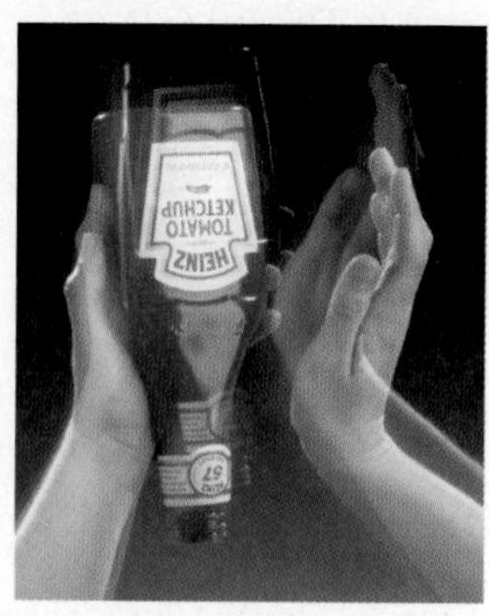

8. a. Give an example of the motion of an object in which the frictional force on the object is directed opposite to the motion.
   b. Give an example of the motion of an object in which the frictional force on the object is in the same direction as the motion.
9. Suppose you are an astronaut in deep space, far from any source of gravity. You have two objects that look identical, but one has a large mass and the other a small mass. How can you tell the difference between the two?
10. Jonathan accelerates away from a stop sign. His eight-year-old daughter sits in the passenger seat. On whom does the back of the seat exert a greater force?
11. Suppose you are an astronaut on a spacewalk, far from any source of gravity. You find yourself floating alongside your spacecraft but 10 m away, with no propulsion system to get back to it. In your tool belt you have a hammer, a wrench, and a roll of duct tape. How can you get back to your spacecraft?
12. You're pushing horizontally on a large crate, but it won't budge. According to Newton's third law, the crate pushes back on you with the same magnitude of force that you exert on it. Suddenly the crate breaks free and you can now push it along the floor. Now is the force exerted on you by the crate greater than, equal to, or less than the force that you are exerting on the crate? Explain.
13. Josh and Taylor, standing face-to-face on frictionless ice, push off each other, causing each to slide backward. Josh is much bigger than Taylor. After the push, which of the two is moving faster?
14. A person sits on a sloped hillside. Is it ever possible to have the static friction force on this person point down the hill? Explain.
15. Walking without slipping requires a static friction force BIO between your feet (or footwear) and the floor. As described in this chapter, the force on your foot as you push off the floor is forward while the force exerted by your foot on the floor is backward. But what about your *other* foot, the one moved during a stride? What is the direction of the force on that foot as it comes into contact with the floor? Explain.
16. Figure 4.30b showed a situation in which the force of the road on the car's tire points forward. In other situations, the force points backward. Give an example of such a situation.
17. Two teams are engaged in a tug-of-war. According to Newton's third law, the red team pulls on the blue team with the same magnitude of force that the blue team pulls on the red team. How, then, can one team win?
18. A baseball pitcher throws a 100 mph fastball to batter Brandon Bopp. Bopp hits a 450 foot home run out of the park. When the ball contacts the bat, which force was greater: the force of the ball on the bat or the force of the bat on the ball?
19. The tire on this drag racer is severely twisted: The force of the road on the tire is quite large (most likely several times the weight of the car) and is directed forward as shown. Is the car speeding up or slowing down? Explain.

20. Suppose that, while in a squatting position, you stand on your hands, and then you pull up on your feet with a great deal of force. You are applying a large force to the bottoms of your feet, but no matter how strong you are, you will never be able to lift yourself off the ground. Use your understanding of force and motion to explain why this is not possible.

Multiple-Choice Questions

21. | A block has acceleration a when pulled by a string. If two identical blocks are glued together and pulled with twice the original force, their acceleration will be
   A. $(1/4)a$ B. $(1/2)a$ C. a D. $2a$ E. $4a$
22. | A 5.0 kg block has an acceleration of 0.20 m/s^2 when a force is exerted on it. A second block has an acceleration of 0.10 m/s^2 when subject to the same force. What is the mass of the second block?
   A. 10 kg B. 5.0 kg C. 2.5 kg D. 7.5 kg
23. | Tennis balls experience a large drag force. A tennis ball is hit so that it goes straight up and then comes back down. The direction of the drag force is
   A. Always up. B. Up and then down.
   C. Always down. D. Down and then up.
24. || A group of students is making model cars that will be propelled by model rocket engines. These engines provide a nearly constant thrust force. The cars are light—most of the weight comes from the rocket engine—and friction and drag are very small. As the engine fires, it uses fuel, so it is much lighter at the end of the run than at the start. A student ignites the engine in a car, and the car accelerates. As the fuel burns and the car continues to speed up, the magnitude of the acceleration will
   A. Increase. B. Stay the same. C. Decrease.

25. | A person gives a box a shove so that it slides up a ramp, then reverses its motion and slides down. The direction of the force of friction is
    A. Always down the ramp.
    B. Up the ramp and then down the ramp.
    C. Always down the ramp.
    D. Down the ramp and then up the ramp.
26. || Craig is trying to push a heavy crate up a wooden ramp, but the crate won't budge. The direction of the static friction force acting on the crate is
    A. Up the ramp.
    B. Down the ramp.
    C. The friction force is zero.
    D. There is not enough information to tell.
27. | As shown in the chapter, scallops use jet propulsion to move from one place to another. Their shells make them denser than water, so they normally rest on the ocean floor. If a scallop wishes to remain stationary, hovering a fixed distance above the ocean floor, it must eject water ______ so that the thrust force on the scallop is ______.
BIO INT
    A. upward, upward B. upward, downward
    C. downward, upward D. downward, downward
28. | A train engine pulls two identical cars behind it. Car 1 is attached directly to the engine, and car 2 is attached to car 1. The engine and cars are speeding up. The force of the engine on car 1 is ______ the force of car 1 on car 2.
    A. greater than B. less than C. equal to
    D. There is no force on either car.
29. | You're pushing a heavy filing cabinet across your office. You are tiring quickly and the cabinet is slowing down. The force that you exert on the cabinet is ______ the force that the cabinet exerts on you.
    A. greater than B. less than C. equal to
30. || Dave pushes his four-year-old son Thomas across the snow on a sled. As Dave pushes, Thomas speeds up. Which statement is true?
    A. The force of Dave on Thomas is larger than the force of Thomas on Dave.
    B. The force of Thomas on Dave is larger than the force of Dave on Thomas.
    C. Both forces have the same magnitude.
    D. It depends on how hard Dave pushes on Thomas.
31. | Figure Q4.31 shows block A sitting on top of block B. A constant force $\vec{F}$ is exerted on block B, causing block B to accelerate to the right. Block A rides on block B without slipping. Which statement is true?
    A. Block B exerts a friction force on block A, directed to the left.
    B. Block B exerts a friction force on block A, directed to the right.
    C. Block B does not exert a friction force on block A.

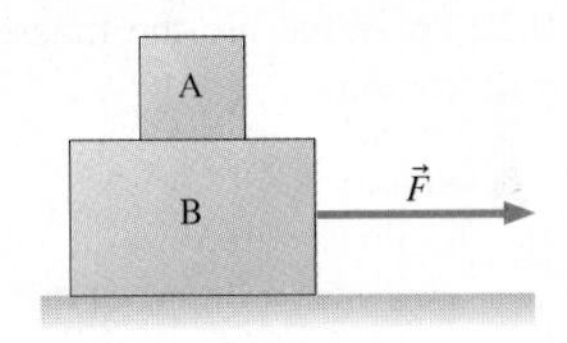

FIGURE Q4.31

PROBLEMS

Section 4.1 Motion and Forces

1. | Whiplash injuries during an automobile accident are caused by the inertia of the head. If someone is wearing a seatbelt, her body will tend to move with the car seat. However, her head is free to move until the neck restrains it, causing damage to the neck. Brain damage can also occur.
BIO

   Figure P4.1 shows two sequences of head and neck motion for a passenger in an auto accident. One corresponds to a head-on collision, the other to a rear-end collision. Which is which? Explain.

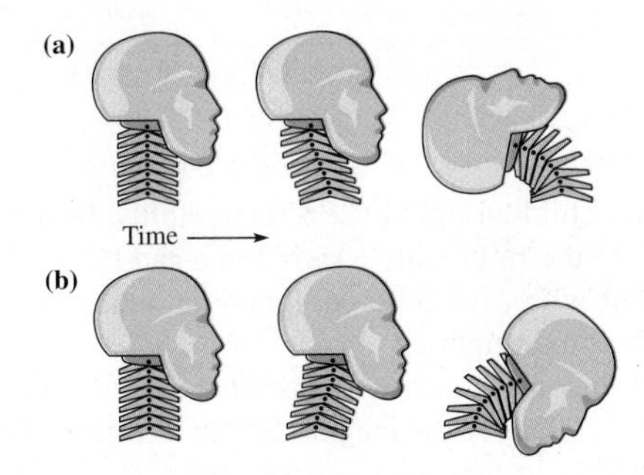

FIGURE P4.1

2. | An automobile has a head-on collision. A passenger in the car experiences a compression injury to the brain. Is this injury most likely to be in the front or rear portion of the brain? Explain.
BIO
3. | In a head-on collision, an infant is much safer in a child safety seat when the seat is installed facing the rear of the car. Explain.

Problems 4 through 6 show two forces acting on an object at rest. Redraw the diagram, then add a third force that results in a net force of zero. Label the new force $\vec{F}_3$.

4. || 5. || 6. ||

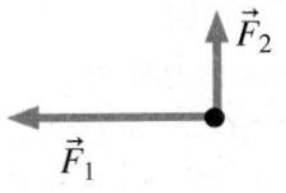

FIGURE P4.4

FIGURE P4.5

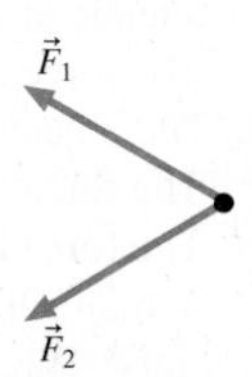

FIGURE P4.6

Watch Video Solution Problems 4.11, 4.15, and 4.23

Section 4.2 A Short Catalog of Forces

Section 4.3 Identifying Forces

7. | A girl is swinging on a rope. Identify the forces acting on the girl at the very top of her swing.
8. | You're exploring a cave. Standing in a low cavern, you steady yourself by pushing straight up on the rock ceiling. Identify the forces acting on you.
9. | A baseball player is sliding into second base. Identify the forces on the baseball player.
10. | A jet plane is speeding down the runway during takeoff. Air resistance is not negligible. Identify the forces on the jet.
11. | A skier is sliding down a 15° slope. Friction is not negligible. Identify the forces on the skier.
12. | A falcon is hovering above the ground, then suddenly pulls BIO in its wings and begins to fall toward the ground. Air resistance is not negligible. Identify the forces on the falcon.

Section 4.4 What Do Forces Do?

13. |||| Figure P4.13 shows an acceleration-versus-force graph for three objects pulled by identical rubber bands. The mass of object 2 is 0.20 kg. What are the masses of objects 1 and 3? Explain your reasoning.

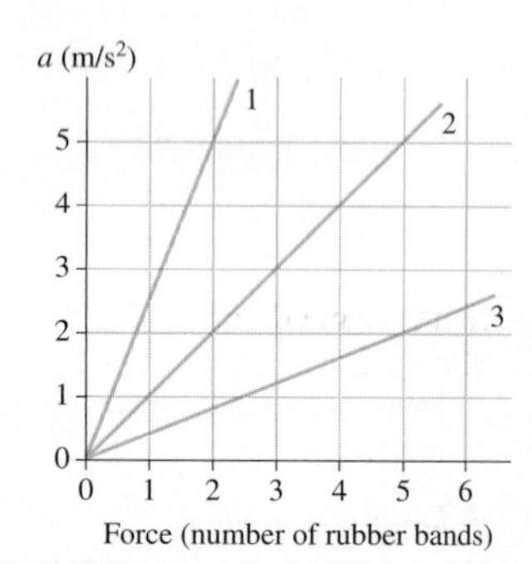

FIGURE P4.13

14. | A constant force applied to object A causes it to accelerate at 5 m/s². The same force applied to object B causes an acceleration of 3 m/s². Applied to object C, it causes an acceleration of 8 m/s².
    a. Which object has the largest mass?
    b. Which object has the smallest mass?
    c. What is the ratio of mass A to mass B (m_A/m_B)?
15. || A compact car has a maximum acceleration of 4.0 m/s² when it carries only the driver and has a total mass of 1200 kg. What is its maximum acceleration after picking up four passengers and their luggage, adding an additional 400 kg of mass?
16. || A student builds a rocket-propelled cart for a science project. Its acceleration is not quite high enough to win a prize, so he uses a larger rocket engine that provides 33% more thrust, although doing so increases the mass of the cart by 12%. By what percentage does the cart's acceleration increase?
17. | A constant force is applied to an object, causing the object to accelerate at 8.0 m/s². What will the acceleration be if
    a. The force is doubled?
    b. The object's mass is doubled?
    c. The force and the object's mass are both doubled?
    d. The force is doubled and the object's mass is halved?
18. ||| A man pulling an empty wagon causes it to accelerate at 1.4 m/s². What will the acceleration be if he pulls with the same force when the wagon contains a child whose mass is three times that of the wagon?
19. The Lamborghini Huracán has an initial acceleration of $0.75g$. Its mass, with driver, is 1510 kg. If an 80 kg passenger rode along, what would the car's acceleration be?

Section 4.5 Newton's Second Law

20. || Scallops eject water from their shells to provide a thrust BIO INT force. The graph shows a smoothed graph of actual data for the initial motion of a 25 g scallop speeding up to escape a predator. What is the magnitude of the net force needed to achieve this motion? How does this force compare to the 0.25 N weight of the scallop?

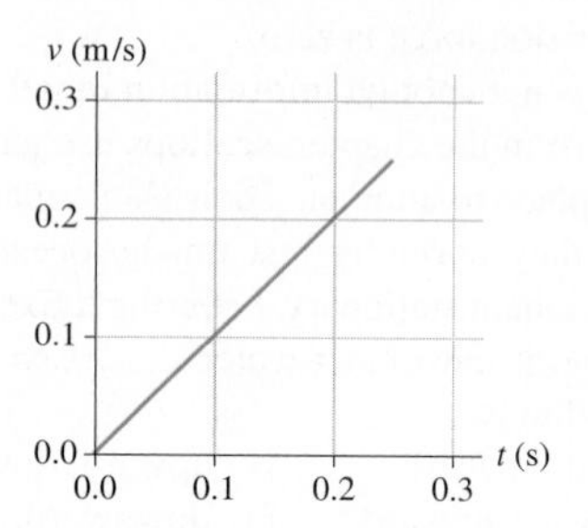

FIGURE P4.20

21. | Figure P4.21 shows an object's acceleration-versus-force graph. What is the object's mass?

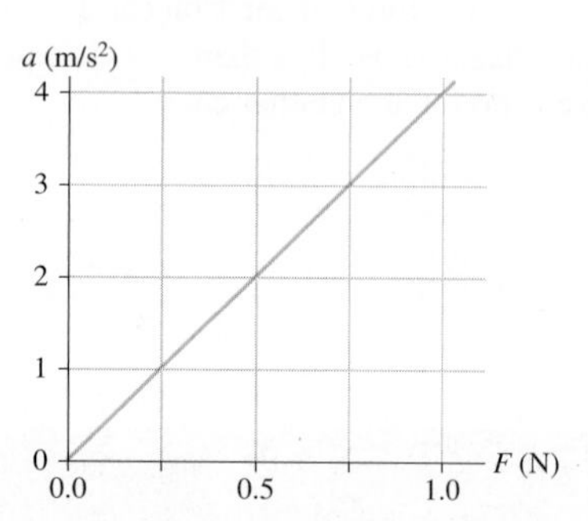

FIGURE P4.21

22. || In a crash test, a car is driven into a solid wall at a speed of 35 mph. Figure P4.22 shows the car's acceleration as it crashes into the wall. What is the maximum force experienced by the 1650 kg car?

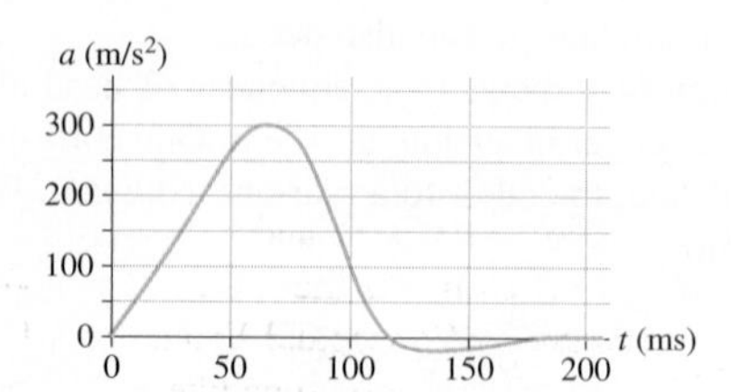

FIGURE P4.22

23. || Two children fight over a 200 g stuffed bear. The 25 kg boy pulls to the right with a 15 N force and the 20 kg girl pulls to the left with a 17 N force. Ignore all other forces on the bear (such as its weight).
    a. At this instant, can you say what the velocity of the bear is? If so, what are the magnitude and direction of the velocity?
    b. At this instant, can you say what the acceleration of the bear is? If so, what are the magnitude and direction of the acceleration?

24. || In t-ball, young players use a bat to hit a stationary ball off a stand. The 140 g ball has about the same mass as a baseball, but it is larger and softer. In one hit, the ball leaves the bat at 12 m/s after being in contact with the bat for 2.0 ms. Assume constant acceleration during the hit.
INT
   a. What is the acceleration of the ball?
   b. What is the net force on the ball during the hit?
25. || A 55 kg ice skater is gliding along at 3.5 m/s. Five seconds later her speed has dropped to 2.9 m/s. What is the magnitude of the kinetic friction acting on her skates?
INT
26. || The IKAROS spacecraft, launched in 2010, was designed to test the feasibility of *solar sails* for spacecraft propulsion. These large, ultralight sails are pushed on by the force of light from the sun, so the spacecraft doesn't need to carry any fuel. The force on IKAROS's sails was measured to be 1.12 mN. If this were the only force acting on the 290 kg spacecraft, by how much would its speed increase after 6 months of flight?
27. || The *head injury criterion* (HIC) is used to assess the likelihood of head injuries arising from various types of collisions; an HIC greater than about 1000 s is likely to result in severe injuries or even death. The criterion can be written as HIC $= (a_{avg}/g)^{2.5}\Delta t$, where a_{avg} is the average acceleration during the time Δt that the head is being accelerated, and g is the free-fall acceleration. Figure P4.27 shows a simplified graph of the net force on a crash dummy's 4.5 kg head as it hits the airbag during a automobile collision. What is the HIC in this collision?
BIO

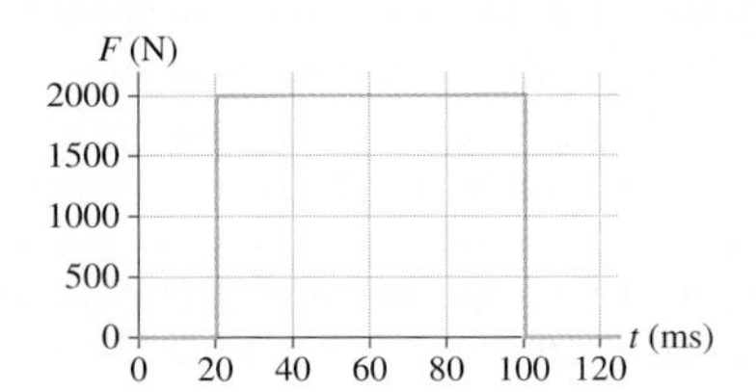

FIGURE P4.27

28. || The particle in Figure P4.28 is acted on by force $\vec{F}_1$ and a second force that is not shown. The particle's acceleration is also shown. Which of the forces $\vec{F}_2 - \vec{F}_4$ could be the second force? (There may be more than one correct answer.)

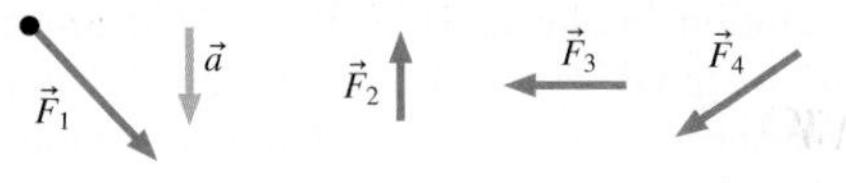

FIGURE P4.28

Section 4.6 Free-Body Diagrams

Problems 29 through 31 show a free-body diagram. For each problem, (a) redraw the free-body diagram and (b) write a short description of a real object for which this is the correct free-body diagram. Use the situations described in Conceptual Examples 4.1, 4.2, and 4.3 as models of what a description should be like.

29. | 30. |

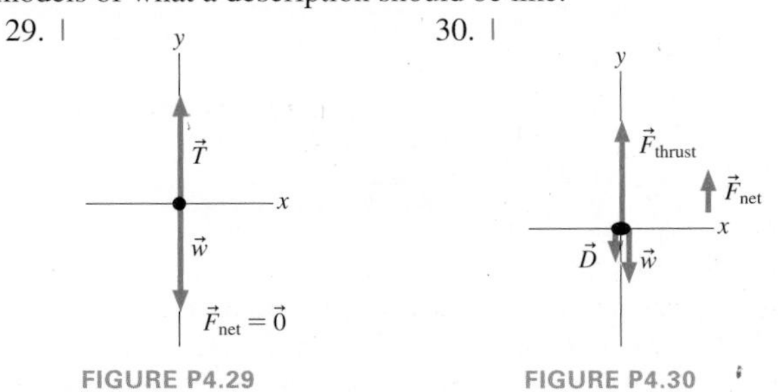

FIGURE P4.29 FIGURE P4.30

31. |

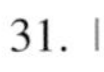

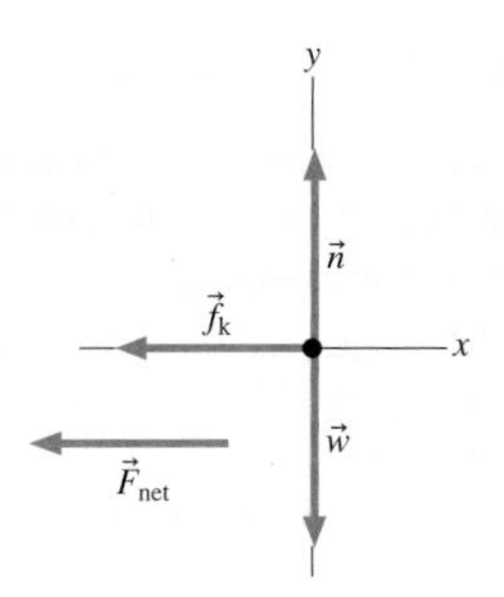

FIGURE P4.31

Problems 32 through 42 describe a situation. For each problem, identify all the forces acting on the object and draw a free-body diagram of the object.

32. You are leaning against a wall.
33. || Your car is accelerating from a stop.
34. || Your car is skidding to a stop from a high speed.
35. || Your physics textbook is sliding across the table.
36. || An ascending elevator, hanging from a cable, is coming to a stop.
37. | You are driving on the highway, and you come to a steep downhill section. As you roll down the hill, you take your foot off the gas pedal. You can ignore friction, but you can't ignore air resistance.
38. || You hold a picture motionless against a wall by pressing on it, as shown in Figure P4.38.

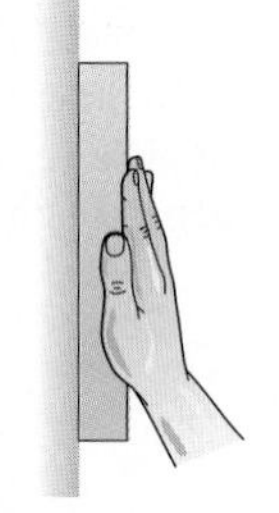

FIGURE P4.38

39. | A box is being dragged across the floor at a constant speed by a rope pulling horizontally on it. Friction is not negligible.
40. | A skydiver has his parachute open and is floating downward through the air at a constant speed.
41. || A cannonball has just been shot out of a cannon aimed 45° above the horizontal. Drag cannot be neglected.
42. || The light turns green and you accelerate rapidly forward so that you and your passenger feel "thrown back" into your seat. Consider your passenger to be the object.

Section 4.7 Newton's Third Law

43. || Three ice skaters, numbered 1, 2, and 3, stand in a line, each with her hands on the shoulders of the skater in front. Skater 3, at the rear, pushes on skater 2. Identify all the action/reaction pairs of forces between the three skaters. Draw a free-body diagram for skater 2, in the middle. Assume the ice is frictionless.
44. | A girl stands on a sofa. Identify all the action/reaction pairs of forces between the girl and the sofa.
45. || A car is skidding to a stop on a level stretch of road. Identify all the action/reaction pairs of forces between the car and the road surface. Then draw a free-body diagram for the car.
46. ||| Squid use jet propulsion for rapid escapes. A squid pulls water into its body and then rapidly ejects the water backward to propel itself forward. A 1.5 kg squid (not including water mass) can accelerate at 20 m/s^2 by ejecting 0.15 kg of water.
BIO
   a. What is the magnitude of the thrust force on the squid?
   b. What is the magnitude of the force on the water being ejected?
   c. What acceleration is experienced by the water?

General Problems

47. | Redraw the motion diagram shown in Figure P4.47, then draw a vector beside it to show the direction of the net force acting on the object. Explain your reasoning.
INT

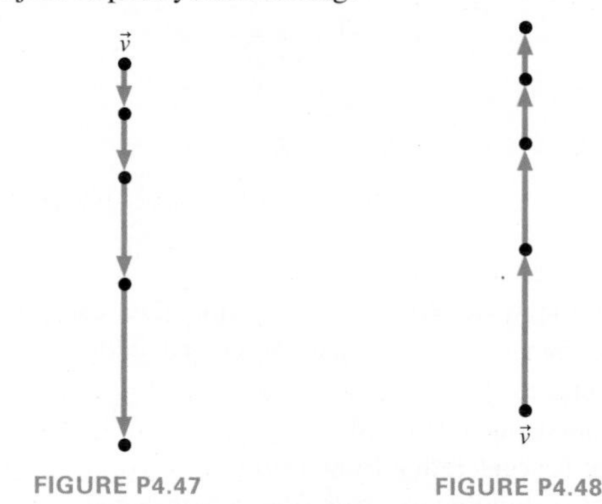

FIGURE P4.47 FIGURE P4.48

48. | Redraw the motion diagram shown in Figure P4.48, then draw a vector beside it to show the direction of the net force acting on the object. Explain your reasoning.
INT

49. | Redraw the motion diagram shown in Figure P4.49, then draw a vector beside it to show the direction of the net force acting on the object. Explain your reasoning.
INT

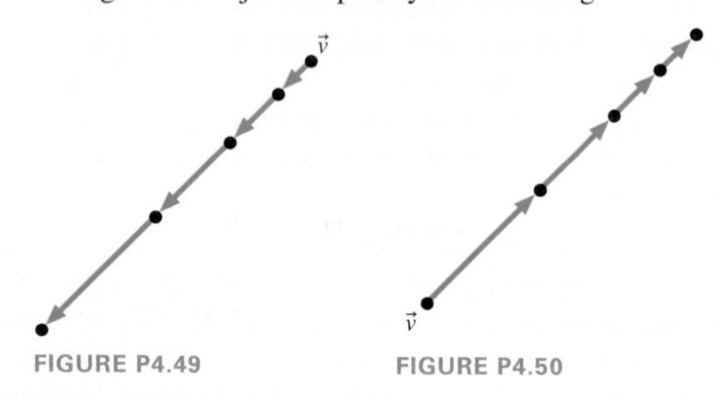

FIGURE P4.49 FIGURE P4.50

50. | Redraw the motion diagram shown in Figure P4.50, then draw a vector beside it to show the direction of the net force acting on the object. Explain your reasoning.
INT

51. ||| A student draws the flawed free-body diagram shown in Figure P4.51 to represent the forces acting on a car traveling at constant speed on a level road. Identify the errors in the diagram, then draw a correct free-body diagram for this situation.

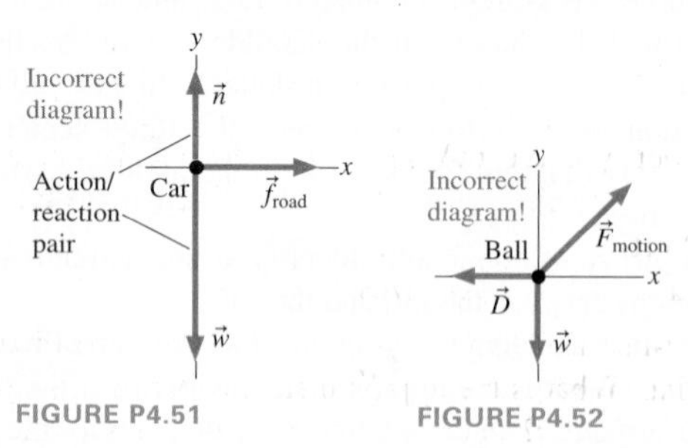

FIGURE P4.51 FIGURE P4.52

52. ||| A student draws the flawed free-body diagram shown in Figure P4.52 to represent the forces acting on a golf ball that is traveling upward and to the right a very short time after being hit off the tee. Air resistance is assumed to be relevant. Identify the errors in the diagram, then draw a correct free-body diagram for this situation.

Problems 53 through 65 describe a situation. For each problem, draw a motion diagram, a force identification diagram, and a free-body diagram.

53. || An elevator, suspended by a single cable, has just left the tenth floor and is speeding up as it descends toward the ground floor.
54. ||| A rocket is being launched straight up. Air resistance is not negligible.
55. ||| A jet plane is speeding down the runway during takeoff. Air resistance is not negligible.
56. || You've slammed on the brakes and your car is skidding to a stop while going down a 20° hill.
57. || A cave explorer is being lowered into a vertical shaft by a rope. He is slowing down as he nears the bottom.
58. || A basketball player is getting ready to jump, pushing off the ground and accelerating upward.
59. || A bale of hay sits on the bed of a trailer. The trailer is starting to accelerate forward, and the bale is slipping toward the back of the trailer.
60. ||| In the process of nailing up a heavy framed poster, a student pushes the poster straight in toward the wall; the poster is sliding downward at a constant speed.
61. ||| A spring-loaded gun shoots a plastic ball. The trigger has just been pulled and the ball is starting to move down the barrel. The barrel is horizontal.
62. || A person on a bridge throws a rock straight down toward the water. The rock has just been released.
63. ||| A gymnast has just landed on a trampoline. She's still moving downward as the trampoline stretches.
64. ||| A heavy box is in the back of a truck. The truck is accelerating to the right. Apply your analysis to the box.
65. || A bag of groceries is on the back seat of your car as you stop for a stop light. The bag does not slide. Apply your analysis to the bag.
66. || A car has a mass of 1500 kg. If the driver applies the brakes while on a gravel road, the maximum friction force that the tires can provide without skidding is about 7000 N. If the car is moving at 20 m/s, what is the shortest distance in which the car can stop safely?
67. ||| Researchers have measured the acceleration of racing greyhounds as a function of their speed; a simplified version of their results is shown in Figure P4.67. The acceleration at low speeds is constant and is limited by the fact that any greater acceleration would result in the dog pitching forward because of the force acting on its hind legs during its power stroke. At higher speeds, the dog's acceleration is limited by the maximum power its muscles can provide.
BIO INT
    a. What is the agent of the force that causes the dog to accelerate?
    b. If the dog's mass is 32 kg, what is the average force acting on it during its initial acceleration phase?
    c. How far does the dog run in the first 4.0 s?

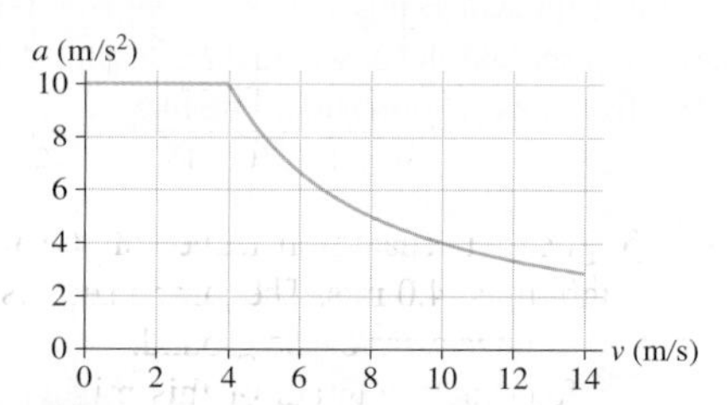

FIGURE P4.67

68. IIII In researching the forces that act on flying birds, scientists studied the motion of cockatiels' bodies as they fly in horizontal flight. As the birds beat their wings, their bodies move up and down. Figure P4.68 shows the vertical acceleration of a cockatiel's body during one wing beat, consisting of an upstroke followed by a downstroke.
BIO
a. If the bird's mass is 87 g, what is the maximum net vertical force exerted on the bird?
b. The bird is held in level flight due to the force exerted on it by the air as the bird beats its wings. What is the maximum value of this force due to the air?

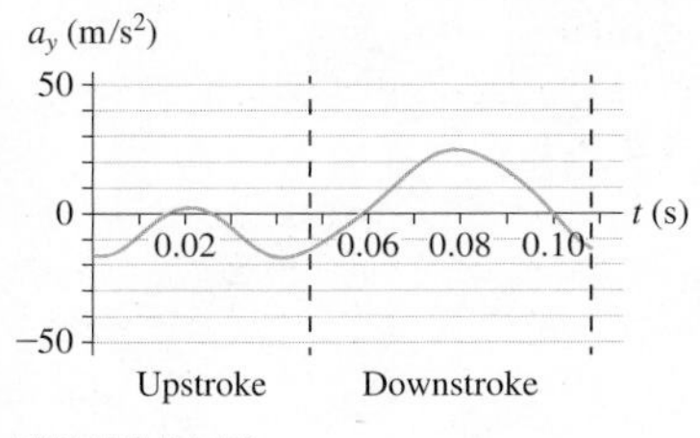

FIGURE P4.68

69. IIII The fastest pitched baseball was clocked at 47 m/s. If the pitcher exerted his force (assumed to be horizontal and constant) over a distance of 1.0 m, and a baseball has a mass of 145 g,
BIO
a. Draw a free-body diagram of the ball during the pitch.
b. What force did the pitcher exert on the ball during this record-setting pitch?
c. Estimate the force in part b as a fraction of the pitcher's weight.

70. III The trap-jaw ant, found throughout tropical South America, catches its prey by *very* rapidly closing its mandibles around its victim. Figure P4.70 shows the speed of one of its mandible jaws versus time in *micro*seconds.
BIO INT
a. What is the maximum acceleration of the ant's mandible?
b. The mass of a trap-jaw ant's mandible has been estimated to be about 1.3×10^{-7} kg. Estimate the maximum force exerted by one mandible.

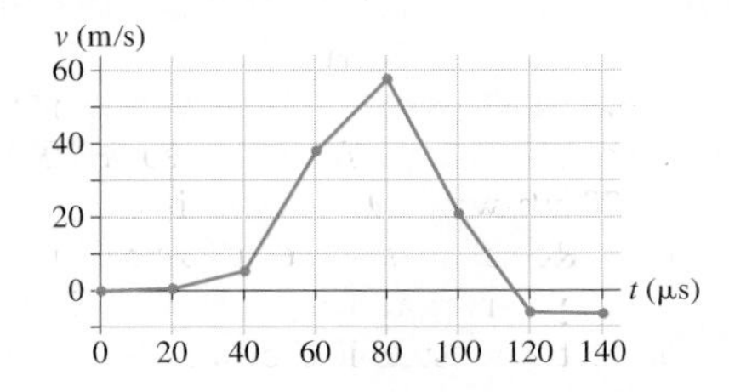

FIGURE P4.70

71. II The jumping ability of the African desert locust was measured by placing the insect on a *force plate,* a platform that can accurately measure the force that acts on it. When the locust jumped straight up, its acceleration was measured to follow the curve in Figure P4.71. What was the maximum force that this 0.50 g locust exerted on the force plate?
BIO

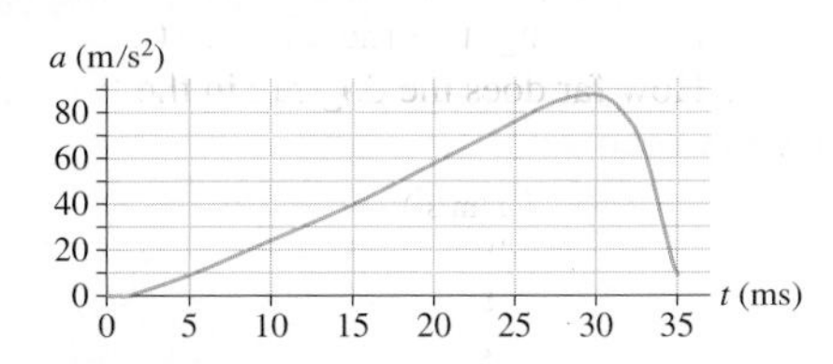

FIGURE P4.71

72. I The froghopper, champion leaper of the insect world, can jump straight up at 4.0 m/s. The jump itself lasts a mere 1.0 ms before the insect is clear of the ground.
BIO
a. Draw a free-body diagram of this mighty leaper while the jump is taking place.
b. While the jump is taking place, is the force that the ground exerts on the froghopper greater than, less than, or equal to the insect's weight? Explain.

73. II A beach ball is thrown straight up, and some time later it lands on the sand. Is the magnitude of the net force on the ball greater when it is going up or when it is on the way down? Or is it the same in both cases? Explain. Air resistance should not be neglected for a large, light object.

MCAT-Style Passage Problems

A Simple Solution for a Stuck Car

If your car is stuck in the mud and you don't have a winch to pull it out, you can use a piece of rope and a tree to do the trick. First, you tie one end of the rope to your car and the other to a tree, then pull as hard as you can on the middle of the rope, as shown in Figure P4.74a. This technique applies a force to the car much larger than the force that you can apply directly. To see why the car experiences such a large force, look at the forces acting on the center point of the rope, as shown in Figure P4.74b. The sum of the forces is zero, thus the tension is much greater than the force you apply. It is this tension force that acts on the car and, with luck, pulls it free.

74. I The sum of the three forces acting on the center point of the rope is assumed to be zero because
A. This point has a very small mass.
B. Tension forces in a rope always cancel.
C. This point is not accelerating.
D. The angle of deflection is very small.

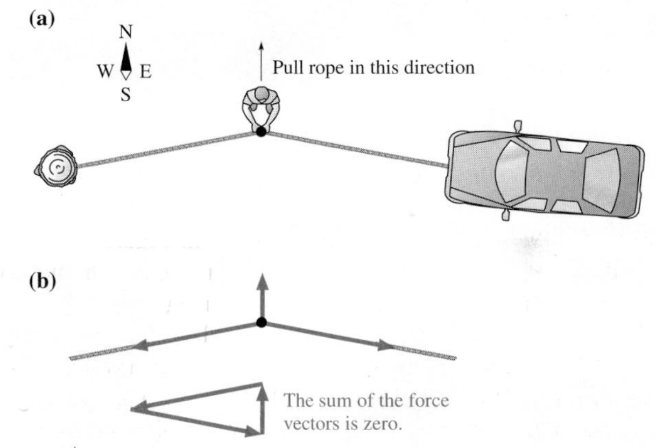

FIGURE P4.74

75. I When you are pulling on the rope as shown, what is the approximate direction of the tension force on the tree?
A. North B. South
C. East D. West

76. I Assume that you are pulling on the rope but the car is not moving. What is the approximate direction of the force of the mud on the car?
A. North B. South
C. East D. West

77. I Suppose your efforts work, and the car begins to move forward out of the mud. As it does so, the force of the car on the rope is
A. Zero.
B. Less than the force of the rope on the car.
C. Equal to the force of the rope on the car.
D. Greater than the force of the rope on the car.

5 Applying Newton's Laws

The sifaka is a type of lemur, found only in Madagascar, that is capable of spectacular leaps from tree to tree. Why does the sifaka spread out its arms and legs this way while making a leap?

LOOKING AHEAD ▸

Working with Forces

In this chapter you'll learn expressions for the different forces we've seen, and you'll learn how to use them to solve problems.

You'll learn how a balance between weight and drag forces leads to a maximum speed for a skydiver.

Equilibrium Problems

The boy is pushing as hard as he can, but the sofa isn't going anywhere. It's in **equilibrium**—the sum of the forces on it is zero.

You'll learn to solve equilibrium problems by using the fact that there is no net force.

Dynamics Problems

Newton's laws allow us to relate the forces acting on an object to its motion, and so to solve a wide range of **dynamics** problems.

This skier is picking up speed. You'll see how her acceleration is determined by the forces acting on her.

GOAL **To use Newton's laws to solve equilibrium and dynamics problems.**

LOOKING BACK ◂

Free-Body Diagrams

In Section 4.6 you learned to draw a free-body diagram showing the magnitudes and directions of the forces acting on an object.

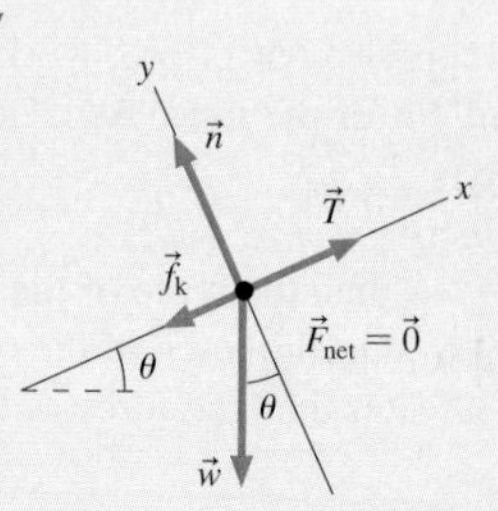

In this chapter, you'll use free-body diagrams as an essential problem-solving tool for single objects and interacting objects.

STOP TO THINK

An elevator is suspended from a cable. It is moving upward at a steady speed. Which is the correct free-body diagram for this situation?

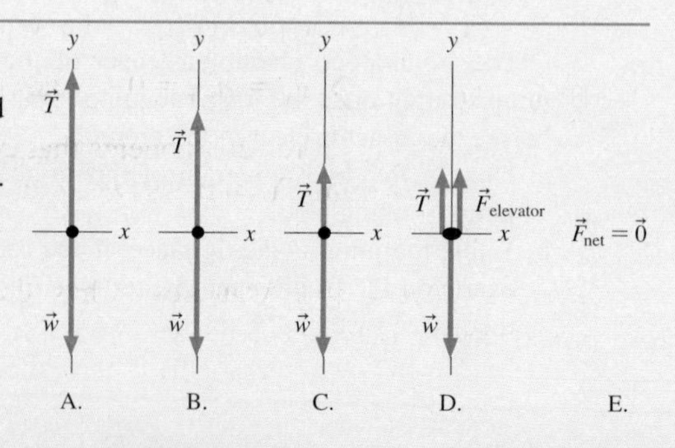

5.1 Equilibrium

In this chapter, we will use Newton's laws to solve force and motion problems. Before turning to a general treatment of such motion, let's examine the two simplest cases, in which the object is either at rest or moving in a straight line at a constant speed. Note that in both cases the object is not accelerating, so $\vec{a} = \vec{0}$.

We say that an object at rest is in **static equilibrium**; an object that is moving in a straight line at a constant speed is said to be in **dynamic equilibrium.** Because $\vec{a} = \vec{0}$ for both cases, Newton's second law $\vec{F}_{\text{net}} = m\vec{a}$ tells us that for an object to be in equilibrium, the net force $\vec{F}_{\text{net}}$ acting on it must be zero: $\vec{F}_{\text{net}} = \vec{0}$. Recall that $\vec{F}_{\text{net}}$ is the vector sum

$$\vec{F}_{\text{net}} = \vec{F}_1 + \vec{F}_2 + \vec{F}_3 + \cdots$$

where $\vec{F}_1$, $\vec{F}_2$, and so on are the individual forces, such as tension or friction, acting on the object. We found in ◀◀ SECTION 3.2 that vector sums can be evaluated in terms of the x- and y-components of the vectors; that is, the x-component of the net force is $(F_{\text{net}})_x = F_{1x} + F_{2x} + F_{3x} + \cdots$. If we restrict ourselves to problems where all the forces are in the xy-plane, then the equilibrium requirement $\vec{F}_{\text{net}} = \vec{0}$ is a shorthand way of writing two simultaneous equations:

$$(F_{\text{net}})_x = F_{1x} + F_{2x} + F_{3x} + \cdots = 0$$

$$(F_{\text{net}})_y = F_{1y} + F_{2y} + F_{3y} + \cdots = 0$$

Recall from your math classes that the Greek letter Σ (sigma) stands for "the sum of." It will be convenient to abbreviate the sum of the x-components of all forces as

$$F_{1x} + F_{2x} + F_{3x} + \cdots = \sum F_x$$

With this notation, Newton's second law for an object in equilibrium, with $\vec{a} = \vec{0}$, can be written as the two equations

$$\sum F_x = ma_x = 0 \quad \text{and} \quad \sum F_y = ma_y = 0 \tag{5.1}$$

In equilibrium, the sums of the x- and y-components of the force are zero.

These equations are the basis for a general approach for solving equilibrium problems.

This human tower is in equilibrium because the net force on each man is zero.

PROBLEM-SOLVING APPROACH 5.1 Equilibrium problems

STRATEGIZE If an object is in equilibrium, the net force acting on it must be zero. We will use this fact to find the forces that keep it in equilibrium.

PREPARE First check that the object is in equilibrium: Does $\vec{a} = \vec{0}$?

- An object at rest is in static equilibrium.
- An object moving at a constant velocity is in dynamic equilibrium.

Then identify all forces acting on the object and show them on a free-body diagram. Determine which forces you know and which you need to solve for.

SOLVE An object in equilibrium must satisfy Newton's second law for the case where $\vec{a} = \vec{0}$. In component form, the requirement is

$$\sum F_x = ma_x = 0 \quad \text{and} \quad \sum F_y = ma_y = 0$$

You can find the force components that go into these sums directly from your free-body diagram. From these two equations, solve for the unknown forces in the problem.

ASSESS Check that your result has the correct units, is reasonable, and answers the question.

The following examples show how to apply this Problem-Solving Approach to both static and dynamic equilibrium problems.

Static Equilibrium

EXAMPLE 5.1 Finding the forces on an orangutan

An orangutan weighing 500 N hangs from a vertical rope. What is the tension in the rope?

STRATEGIZE The orangutan is at rest, so it is in static equilibrium. The net force on it must then be zero. We will use this fact to find the tension.

PREPARE FIGURE 5.1 identifies the forces acting on the orangutan: the upward force of the tension in the rope and the downward, long-range force of gravity (the orangutan's weight $\vec{w}$).

FIGURE 5.1 The forces on an orangutan.

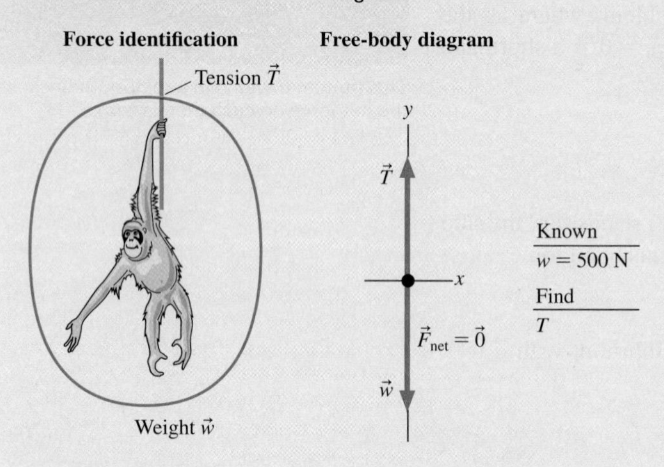

These forces are then shown on a free-body diagram, where it's noted that equilibrium requires $\vec{F}_{net} = \vec{0}$.

SOLVE Neither force has an x-component, so we need to examine only the y-components of the forces. In this case, the y-component of Newton's second law is

$$\sum F_y = T_y + w_y = ma_y = 0$$

You might have been tempted to write $T_y - w_y$ because the weight force points down. But remember that T_y and w_y are *components* of vectors and can thus be positive (for a vector such as $\vec{T}$ that points up) or negative (for a vector such as $\vec{w}$ that points down). The fact that $\vec{w}$ points down is taken into account when we *evaluate* the components—that is, when we write them in terms of the *magnitudes* T and w of the vectors $\vec{T}$ and $\vec{w}$.

Because the tension vector $\vec{T}$ points straight up, in the positive y-direction, its y-component is $T_y = T$. Because the weight vector $\vec{w}$ points straight down, in the negative y-direction, its y-component is $w_y = -w$. This is where the signs enter. With these components, Newton's second law becomes

$$T - w = 0$$

This equation is easily solved for the tension in the rope:

$$T = w = 500 \text{ N}$$

ASSESS It's not surprising that the tension in the rope equals the weight of the orangutan. That gives us confidence in our solution.

EXAMPLE 5.2 Readying a wrecking ball

A wrecking ball weighing 2500 N hangs from a cable. Prior to swinging, it is pulled back to a 20° angle by a second, horizontal cable. What is the tension in the horizontal cable?

STRATEGIZE Because the ball is not yet moving, this is a static equilibrium problem; the net force acting on the ball must be zero.

PREPARE In FIGURE 5.2, we start by identifying all the forces acting on the ball: a tension force from each cable and the ball's weight. We've used different symbols $\vec{T}_1$ and $\vec{T}_2$ for the two different tension forces. We then construct a free-body diagram for these three forces, noting that $\vec{F}_{net} = m\vec{a} = \vec{0}$. We're looking for the magnitude T_1 of the tension force $\vec{T}_1$ in the horizontal cable.

FIGURE 5.2 Visual overview of a wrecking ball just before release.

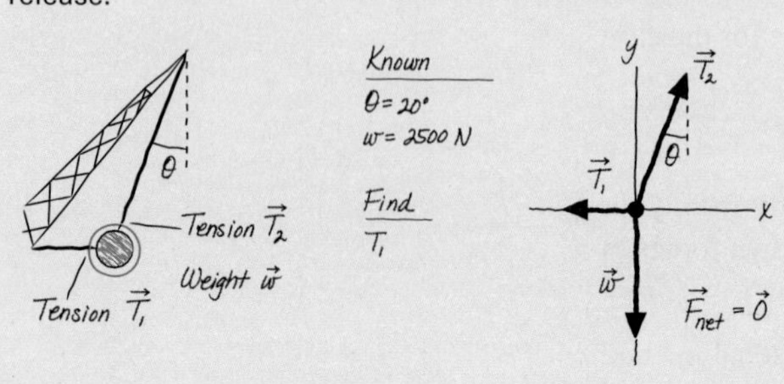

SOLVE The requirement of equilibrium is $\vec{F}_{net} = m\vec{a} = \vec{0}$. In component form, we have the two equations:

$$\sum F_x = T_{1x} + T_{2x} + w_x = ma_x = 0$$
$$\sum F_y = T_{1y} + T_{2y} + w_y = ma_y = 0$$

As always, we *add* the force components together. Now we're ready to write the components of each force vector in terms of the magnitudes and directions of those vectors. We learned how to do this in Section 3.3. With practice you'll learn to read the components directly off the free-body diagram, but to begin it's worthwhile to organize the components into a table.

Force	Name of x-component	Value of x-component	Name of y-component	Value of y-component
$\vec{T}_1$	T_{1x}	$-T_1$	T_{1y}	0
$\vec{T}_2$	T_{2x}	$T_2 \sin\theta$	T_{2y}	$T_2 \cos\theta$
$\vec{w}$	w_x	0	w_y	$-w$

We see from the free-body diagram that $\vec{T}_1$ points along the negative x-axis, so $T_{1x} = -T_1$ and $T_{1y} = 0$. We need to be careful with our trigonometry as we find the components of $\vec{T}_2$. Remembering that the side adjacent to the angle is related to the cosine, we see that the vertical (y) component of $\vec{T}_2$ is $T_2 \cos\theta$. Similarly, the horizontal (x) component is $T_2 \sin\theta$. The weight vector points straight down, so its y-component is $-w$. Notice that negative signs enter as we evaluate the components of the vectors, *not* when we write Newton's second law. This is a critical aspect of solving force and motion problems. With these components, Newton's second law now becomes

$$-T_1 + T_2 \sin\theta + 0 = 0 \quad \text{and} \quad 0 + T_2 \cos\theta - w = 0$$

We can rewrite these equations as

$$T_1 = T_2 \sin\theta \quad \text{and} \quad T_2 \cos\theta = w$$

These are two simultaneous equations with two unknowns: T_1 and T_2. To eliminate T_2 from the two equations, we solve the second equation for T_2, giving $T_2 = w/\cos\theta$. Then we insert this expression for T_2 into the first equation to get

$$T_1 = \frac{w}{\cos\theta}\sin\theta = \frac{\sin\theta}{\cos\theta}w = w\tan\theta = (2500\text{ N})\tan 20° = 910\text{ N}$$

where we made use of the fact that $\tan\theta = \sin\theta/\cos\theta$.

ASSESS It seems reasonable that to pull the ball back to this modest angle, a force substantially less than the ball's weight is required.

CONCEPTUAL EXAMPLE 5.3 Forces in static equilibrium

A rod is free to slide on a frictionless sheet of ice. One end of the rod is lifted by a string. If the rod is at rest, which diagram in FIGURE 5.3 shows the correct angle of the string?

FIGURE 5.3 Which is the correct angle of the string?

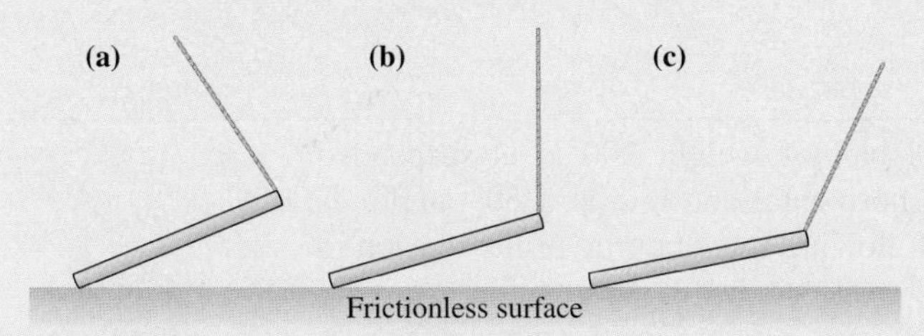

FIGURE 5.4 Free-body diagrams for three angles of the string.

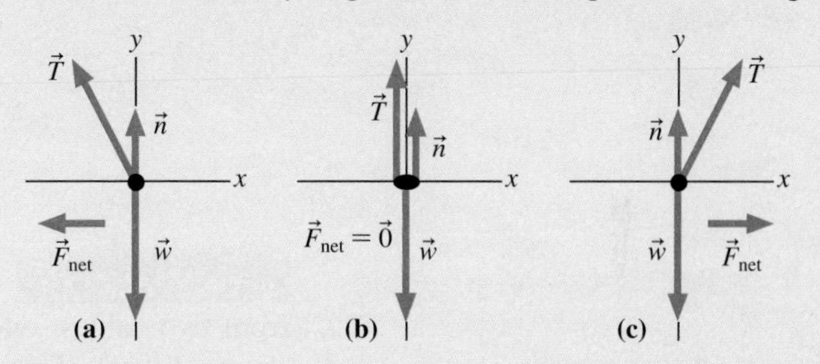

REASON Let's start by identifying the forces that act on the rod. In addition to the weight force, the string exerts a tension force and the ice exerts an upward normal force. What can we say about these forces? If the rod is to hang motionless, it must be in static equilibrium with $\sum F_x = ma_x = 0$ and $\sum F_y = ma_y = 0$. FIGURE 5.4 shows free-body diagrams for the three string orientations. Remember that tension always acts along the direction of the string and that the weight force always points straight down. The ice pushes up with a normal force perpendicular to the surface, but frictionless ice cannot exert any horizontal force. If the string is angled, we see that its horizontal component exerts a net force on the rod. Only in case b, where the tension and the string are vertical, can the net force be zero.

ASSESS If friction were present, the rod could in fact hang as in cases a and c. But without friction, the rods in these cases would slide until they came to rest as in case b.

Dynamic Equilibrium

EXAMPLE 5.4 Finding the tension in a rope while towing a car

A car with a mass of 1500 kg is being towed at a steady speed by a rope held at a 20° angle from the horizontal. A friction force of 320 N opposes the car's motion. What is the tension in the rope?

STRATEGIZE The car is moving in a straight line at a constant speed ($\vec{a} = \vec{0}$) so it is in dynamic equilibrium and must have $\vec{F}_{net} = m\vec{a} = \vec{0}$.

PREPARE FIGURE 5.5 shows the three contact forces acting on the car—the tension force $\vec{T}$, friction $\vec{f}$, and the normal force $\vec{n}$— and the long-range force of gravity $\vec{w}$. These four forces are shown on the free-body diagram.

FIGURE 5.5 Visual overview of a car being towed.

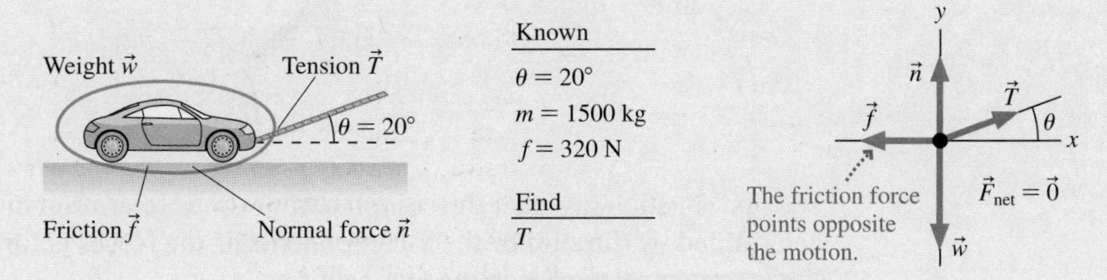

Continued

SOLVE This is still an equilibrium problem, even though the car is moving, so our problem-solving procedure is unchanged. With four forces, the requirement of equilibrium is

$$\sum F_x = n_x + T_x + f_x + w_x = ma_x = 0$$
$$\sum F_y = n_y + T_y + f_y + w_y = ma_y = 0$$

We can again determine the horizontal and vertical components of the forces by "reading" the free-body diagram. The results are shown in the table.

Force	Name of x-component	Value of x-component	Name of y-component	Value of y-component
$\vec{n}$	n_x	0	n_y	n
$\vec{T}$	T_x	$T\cos\theta$	T_y	$T\sin\theta$
$\vec{f}$	f_x	$-f$	f_y	0
$\vec{w}$	w_x	0	w_y	$-w$

With these components, Newton's second law becomes

$$T\cos\theta - f = 0$$
$$n + T\sin\theta - w = 0$$

The first equation can be used to solve for the tension in the rope:

$$T = \frac{f}{\cos\theta} = \frac{320\text{ N}}{\cos 20°} = 340\text{ N}$$

to two significant figures. It turned out that we did not need the y-component equation in this problem. We would need it if we wanted to find the normal force $\vec{n}$.

ASSESS Had we pulled the car with a horizontal rope, the tension would need to exactly balance the friction force of 320 N. Because we are pulling at an angle, however, part of the tension in the rope pulls *up* on the car instead of in the forward direction. Thus we need a little more tension in the rope when it's at an angle, so our result seems reasonable.

STOP TO THINK 5.1 A ball of weight 200 N is suspended from two cables, one horizontal and one at a 60° angle, as shown. Which of the following must be true of the tension in the angled cable?

A. $T > 200$ N B. $T = 200$ N C. $T < 200$ N

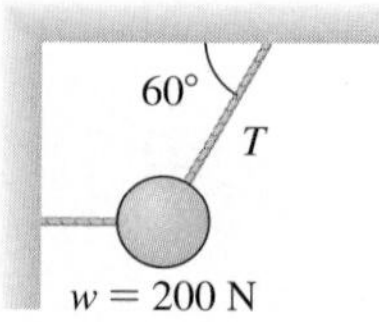

5.2 Dynamics and Newton's Second Law

Newton's second law is the essential link between force and motion. The essence of Newtonian mechanics can be expressed in two steps:

- The forces acting on an object determine its acceleration $\vec{a} = \vec{F}_{\text{net}}/m$.
- The object's motion can be found by using $\vec{a}$ in the equations of kinematics.

A video to support a section's topic is embedded in the eText.
Video Forces in Jumping

We want to develop an approach to solve a variety of problems in mechanics, but first we need to write the second law in terms of its components. To do so, let's first rewrite Newton's second law in the form

$$\vec{F}_{\text{net}} = \vec{F}_1 + \vec{F}_2 + \vec{F}_3 + \cdots = m\vec{a}$$

where $\vec{F}_1, \vec{F}_2, \vec{F}_3$, and so on are the forces acting on an object. To write the second law in component form merely requires that we use the x- and y-components of the acceleration. Thus Newton's second law, $\vec{F}_{\text{net}} = m\vec{a}$, is

$$\sum F_x = ma_x \quad \text{and} \quad \sum F_y = ma_y \tag{5.2}$$

Newton's second law in component form

The first equation says that **the component of the acceleration in the x-direction is determined by the sum of the x-components of the forces acting on the object.** A similar statement applies to the y-direction.

PROBLEM-SOLVING APPROACH 5.2

Dynamics problems

Dynamics problems use Newton's second law as the connection between forces and kinematics.

STRATEGIZE There are two basic kinds of dynamics problems. If the forces on the object are known, you can use Newton's second law to find the acceleration and then, from kinematics, the object's position and velocity. In other cases, you can calculate the object's acceleration from kinematics; using the acceleration, you can apply Newton's second law to find the forces acting on the object. In both kinds of problems the approach to the solution is the same.

PREPARE Sketch a visual overview consisting of:

- A list of values that identifies known quantities and what the problem is trying to find.
- A force identification diagram to help you identify all the forces acting on the object.
- A free-body diagram that shows all the forces acting on the object.

If you'll need to use kinematics to find velocities or positions, you'll also need to sketch:

- A motion diagram to determine the direction of the acceleration.
- A pictorial representation that establishes a coordinate system, shows important points in the motion, and defines symbols.

It's OK to go back and forth between these steps as you visualize the situation.

SOLVE Write Newton's second law in component form as

$$\sum F_x = ma_x \qquad \text{and} \qquad \sum F_y = ma_y$$

You can find the components of the forces directly from your free-body diagram. Depending on the problem, either:

- Solve for the acceleration, then use kinematics to find velocities and positions.
- Use kinematics to determine the acceleration, then solve for unknown forces.

ASSESS Check that your result has the correct units, is reasonable, and answers the question.

Exercise 22

EXAMPLE 5.5 **Putting a golf ball**

A golfer putts a 46 g ball with a speed of 3.0 m/s. Friction exerts a 0.020 N retarding force on the ball, slowing it down. Will her putt reach the hole, 10 m away?

STRATEGIZE We are given the friction force on the ball, which is an indication that we should first find the ball's acceleration using Newton's second law, then use kinematics to solve for the distance the ball rolls.

PREPARE FIGURE 5.6 is a visual overview of the problem. We've collected the known information, drawn a sketch, and identified what we want to find. The motion diagram shows that the ball is

FIGURE 5.6 Visual overview of a golf putt.

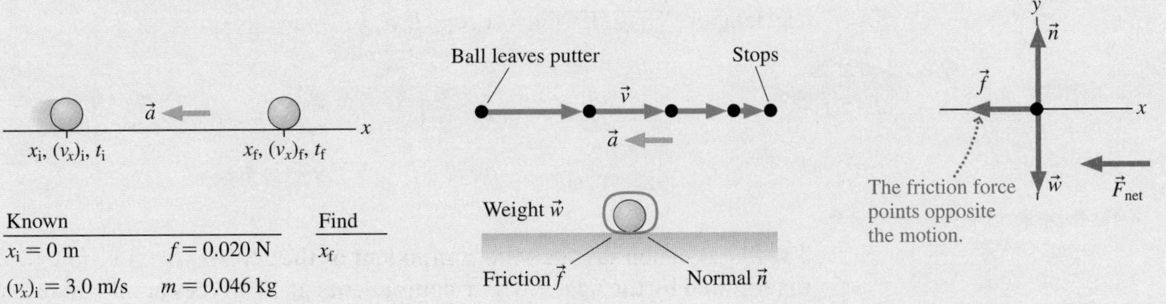

Continued

slowing down as it rolls to the right, so the acceleration vector points to the left. Next, we identify the forces acting on the ball and show them on a free-body diagram. Note that the net force points to the left, as it must because the acceleration points to the left.

SOLVE Newton's second law in component form is

$$\sum F_x = n_x + f_x + w_x = 0 - f + 0 = ma_x$$
$$\sum F_y = n_y + f_y + w_y = n + 0 - w = ma_y = 0$$

We've written the equations as sums, as we did with equilibrium problems, then "read" the values of the force components from the free-body diagram. The components are simple enough in this problem that we don't really need to show them in a table. It is particularly important to notice that we set $a_y = 0$ in the second equation. This is because the ball does not move in the y-direction, so it can't have any acceleration in the y-direction. This will be an important step in many problems.

The first equation is $-f = ma_x$, from which we find

$$a_x = -\frac{f}{m} = \frac{-(0.020\text{ N})}{0.046\text{ kg}} = -0.435\text{ m/s}^2$$

To avoid rounding errors we keep an extra significant figure in this intermediate step in the calculation. The negative sign shows that the acceleration is directed to the left, as expected.

Now that we know the acceleration, we can use kinematics to find how far the ball will roll before stopping. We don't have any information about the time it takes for the ball to stop, so we'll use the kinematic equation $(v_x)_f^2 = (v_x)_i^2 + 2a_x(x_f - x_i)$. This gives

$$x_f = x_i + \frac{(v_x)_f^2 - (v_x)_i^2}{2a_x} = 0\text{ m} + \frac{(0\text{ m/s})^2 - (3.0\text{ m/s})^2}{2(-0.435\text{ m/s}^2)}$$
$$= 10.3\text{ m}$$

If her aim is true, the ball will just make it into the hole.

ASSESS It seems reasonable that a ball putted on grass with an initial speed of 3 m/s—about jogging speed—would travel roughly 10 m.

EXAMPLE 5.6 Towing a car with acceleration

A car with a mass of 1500 kg is being towed by a rope held at a 20° angle to the horizontal. A friction force of 320 N opposes the car's motion. What is the tension in the rope if the car goes from rest to 12 m/s in 10 s?

STRATEGIZE As you may have recognized, this problem is almost identical to Example 5.4. The difference is that the car is now accelerating, so it is no longer in equilibrium. We will first use kinematics to find the car's acceleration. Then we will use this acceleration in Newton's second law to find the tension in the rope.

PREPARE FIGURE 5.7 is a visual overview of the problem showing that the car's acceleration is directed to the right. The force identification diagram is the same as in Example 5.4, but the free-body diagram now indicates that there is a net force directed to the right, in the same direction as the acceleration.

SOLVE Newton's second law in component form is

$$\sum F_x = n_x + T_x + f_x + w_x = ma_x$$
$$\sum F_y = n_y + T_y + f_y + w_y = ma_y = 0$$

We've again used the fact that $a_y = 0$ for motion that is purely along the x-axis. The components of the forces were worked out in Example 5.4. With that information, Newton's second law in component form is

$$T\cos\theta - f = ma_x$$
$$n + T\sin\theta - w = 0$$

Because the car speeds up from rest to 12 m/s in 10 s, we can use kinematics to find the acceleration:

$$a_x = \frac{\Delta v_x}{\Delta t} = \frac{(v_x)_f - (v_x)_i}{t_f - t_i} = \frac{(12\text{ m/s}) - (0\text{ m/s})}{(10\text{ s}) - (0\text{ s})} = 1.2\text{ m/s}^2$$

We can now use the first Newton's-law equation above to solve for the tension. We have

$$T = \frac{ma_x + f}{\cos\theta} = \frac{(1500\text{ kg})(1.2\text{ m/s}^2) + 320\text{ N}}{\cos 20°} = 2300\text{ N}$$

ASSESS The tension is substantially greater than the 340 N found in Example 5.4. It takes much more force to accelerate the car than to keep it rolling at a constant speed.

FIGURE 5.7 Visual overview of a car being towed.

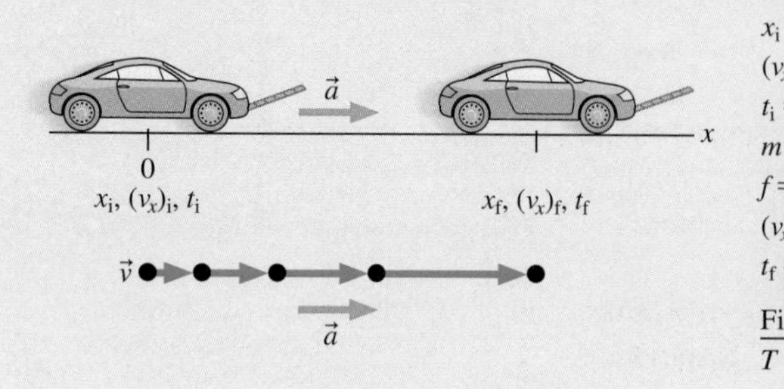

Known
$x_i = 0$ m
$(v_x)_i = 0$ m/s
$t_i = 0$ s, $\theta = 20°$
$m = 1500$ kg
$f = 320$ N
$(v_x)_f = 12$ m/s
$t_f = 10$ s

Find
T

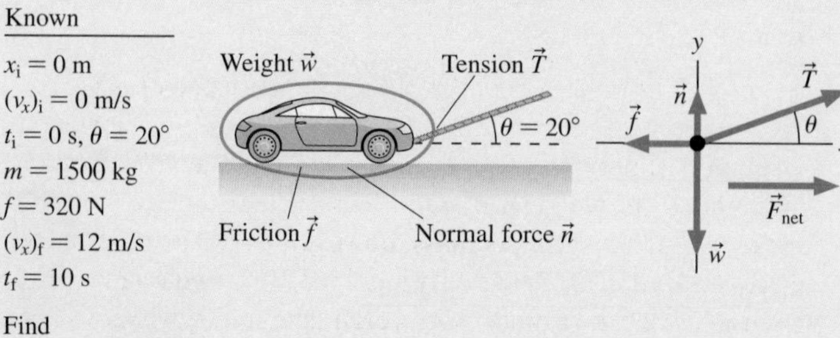

These first examples have shown all the details of our problem-solving approach. Our purpose has been to demonstrate how the approach is put into practice. Future examples will be briefer, but the basic *procedure* will remain the same.

STOP TO THINK 5.2 A Martian lander is approaching the surface. It is slowing its descent by firing its rocket motor. Which is the correct free-body diagram for the lander?

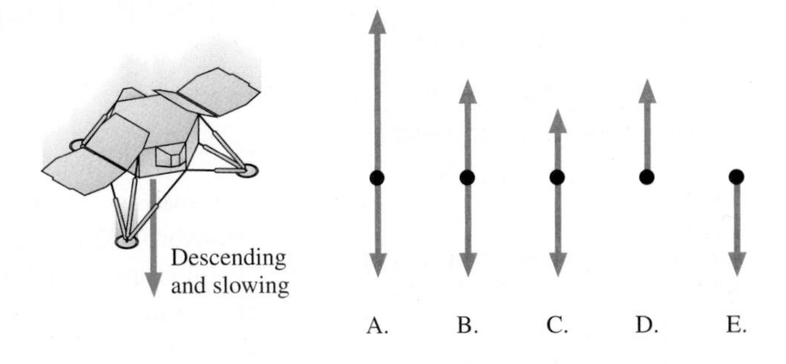

5.3 Mass and Weight

When the doctor asks what you weigh, what does she really mean? We do not make much distinction in our ordinary use of language between the terms "weight" and "mass," but in physics their distinction is of critical importance.

Mass, you'll recall from Chapter 4, is a quantity that describes an object's inertia, its tendency to resist being accelerated. Loosely speaking, it also describes the amount of matter in an object. Mass, measured in kilograms, is an intrinsic property of an object; it has the same value wherever the object may be and whatever forces might be acting on it.

Weight, however, is a *force*. Specifically, it is the gravitational force exerted on an object by a planet. Weight is a vector, not a scalar, and the vector's direction is always straight down. Weight is measured in newtons.

Mass and weight are not the same thing, but they are related. FIGURE 5.8 shows the free-body diagram of an object in free fall. The *only* force acting on this object is its weight $\vec{w}$, the downward pull of gravity. The object is in free fall, so, as we saw in ◀◀ SECTION 2.7, the acceleration is vertical, with $a_y = -g$, where g is the free-fall acceleration, 9.80 m/s^2. Newton's second law for this object is thus

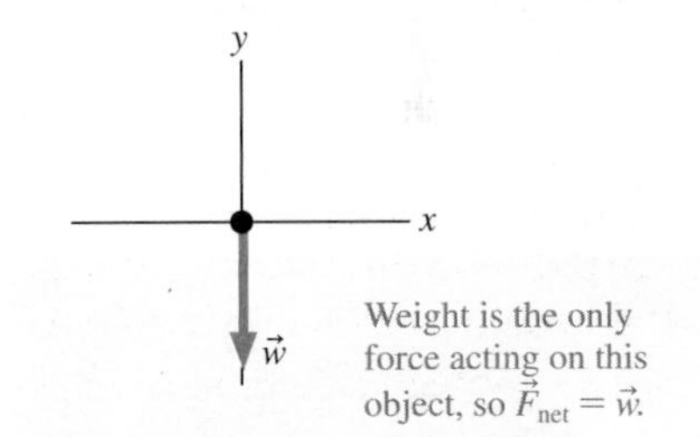

FIGURE 5.8 The free-body diagram of an object in free fall.

$$\sum F_y = -w = -mg$$

which tells us that

$$w = mg \tag{5.3}$$

The magnitude of the weight force, which we call simply "the weight," is directly proportional to the mass, with g as the constant of proportionality.

NOTE ▶ Although we derived the relationship between mass and weight for an object in free fall, the weight of an object is *independent* of its state of motion. Equation 5.3 holds for an object at rest on a table, sliding horizontally, or moving in any other way. ◀

Because an object's weight depends on g, and the value of g varies from planet to planet, weight is not a fixed, constant property of an object. The value of g at the surface of the moon is about one-sixth its earthly value, so an object on the moon would have only one-sixth its weight on earth. The object's weight on Jupiter would be greater than its weight on earth. Its mass, however, would be the same. The amount of matter has not changed, only the gravitational force exerted on that matter.

So, when the doctor asks what you weigh, she really wants to know your *mass*. That's the amount of matter in your body. You can't really "lose weight" by going to the moon, even though you would weigh less there!

On the moon, astronaut John Young jumped 2 feet straight up, despite his spacesuit that weighs 370 pounds on earth. On the moon, where $g = 1.6$ m/s^2, he and his suit together weighed only 90 pounds.

We need to make a clarification here. When you give your weight, you most likely give it in pounds, which is the unit of force in the English system. (We noted in Chapter 4 that the pound is defined as 1 lb = 4.45 N.) You might then "convert" this to kilograms. But a kilogram is a unit of mass, not a unit of force. An object that weighs 1 pound, meaning $w = mg = 4.45$ N, has a mass of

$$m = \frac{w}{g} = \frac{4.45 \text{ N}}{9.80 \text{ m/s}^2} = 0.454 \text{ kg}$$

This calculation is different from converting, for instance, feet to meters. Both feet and meters are units of length, and it's always true that 1 m = 3.28 ft. When you "convert" from pounds to kilograms, you are determining the mass that has a certain weight—two fundamentally different quantities—and this calculation depends on the value of g. But we are usually working on the earth, where we assume that $g = 9.80 \text{ m/s}^2$. In this case, a given mass always corresponds to the same weight, and we can use the relationships listed in TABLE 5.1.

TABLE 5.1 Mass, weight, force

Conversion between force units:
1 pound = 4.45 N
1 N = 0.225 pound
Correspondence between mass and weight, assuming $g = 9.80 \text{ m/s}^2$:
1 kg $\leftrightarrow$ 2.20 lb
1 lb $\leftrightarrow$ 0.454 kg = 454 g

EXAMPLE 5.7 Typical masses and weights

What are the weight, in N, and the mass, in kg, of a 90 pound gymnast, a 150 pound professor, and a 240 pound football player?

STRATEGIZE Weight and mass are related by Equation 5.3.

PREPARE We can use the conversions and correspondences in Table 5.1.

SOLVE We use the correspondence between mass and weight just as we use the conversion factor between different forces:

$$w_{\text{gymnast}} = 90 \text{ lb} \times \frac{4.45 \text{ N}}{1 \text{ lb}} = 400 \text{ N} \qquad m_{\text{gymnast}} = 90 \text{ lb} \times \frac{0.454 \text{ kg}}{1 \text{ lb}} = 41 \text{ kg}$$

$$w_{\text{prof}} = 150 \text{ lb} \times \frac{4.45 \text{ N}}{1 \text{ lb}} = 670 \text{ N} \qquad m_{\text{prof}} = 150 \text{ lb} \times \frac{0.454 \text{ kg}}{1 \text{ lb}} = 68 \text{ kg}$$

$$w_{\text{player}} = 240 \text{ lb} \times \frac{4.45 \text{ N}}{1 \text{ lb}} = 1070 \text{ N} \qquad m_{\text{player}} = 240 \text{ lb} \times \frac{0.454 \text{ kg}}{1 \text{ lb}} = 110 \text{ kg}$$

ASSESS We can use the information in this problem to assess the results of future problems. If you get an answer of 1000 N, you now know that this is approximately the weight of a football player, which can help with your assessment.

Apparent Weight

The weight of an object is the force of gravity on that object. You may never have thought about it, but gravity is not a force that you can feel or sense directly. Your *sensation* of weight—how heavy you *feel*—is due to *contact forces* supporting you. As you read this, your sensation of weight is due to the normal force exerted on you by the chair in which you are sitting. The chair's surface touches you and activates nerve endings in your skin. You sense the magnitude of this force, and this is your sensation of weight. When you stand, you feel the contact force of the floor pushing against your feet. If you are hanging from a rope, you feel the friction force between the rope and your hands.

Physics students can't jump If you ride in an elevator and try to jump up into the air just as the elevator starts to rise, you'll feel like you can hardly get off the ground because your apparent weight is greater than your actual weight. On a fast elevator with a large acceleration, it's like trying to jump with an extra 30 or 40 pounds!

Let's define your **apparent weight** w_{app} in terms of the force you feel:

$$w_{\text{app}} = \text{magnitude of supporting contact forces} \tag{5.4}$$

Definition of apparent weight

If you are in equilibrium, your weight and apparent weight are generally the same. But if you undergo an acceleration, this is not necessarily the case. For instance, you feel "heavy" when an elevator you are riding in suddenly accelerates upward, and you feel lighter than normal as the upward-moving elevator brakes to a halt. Your true weight $w = mg$ has not changed during these events, but your *sensation* of your weight has.

Let's look at the details for this case. Imagine a man standing in an elevator as it accelerates upward. As FIGURE 5.9 shows, the only forces acting on the man are the upward normal force of the floor and the downward weight force. Because the man has an acceleration $\vec{a}$, according to Newton's second law there must be a net force acting on the man in the direction of $\vec{a}$.

Looking at the free-body diagram in Figure 5.9, we see that the y-component of Newton's second law is

$$\sum F_y = n_y + w_y = n - w = ma_y \tag{5.5}$$

where m is the man's mass. Solving Equation 5.5 for n gives

$$n = w + ma_y \tag{5.6}$$

The normal force is the contact force supporting the man, so, given the definition of Equation 5.4, we can rewrite Equation 5.6 as

$$w_{app} = w + ma_y$$

For the case shown in Figure 5.9, where the elevator is accelerating upward, $a_y = +a$ so that $w_{app} = w + ma$. The apparent weight is greater than the man's weight w, so he feels heavier than normal. If, instead, the elevator has a downward acceleration, then $a_y = -a$ so that $w_{app} = w - ma$. The apparent weight is less than the man's weight, and the man feels lighter than normal.

The apparent weight isn't just a sensation, though. You can measure it with a scale. When you stand on a bathroom scale, the scale reading is the upward force of the scale on you. If you aren't accelerating (usually a pretty good assumption!), the upward force of the scale on you equals your weight, so the scale reading is your true weight. But if you are accelerating, this correspondence may not hold. In the example above, if the man were standing on a bathroom scale in the elevator, the contact force supporting him would be the upward force of the scale. This force is equal to the scale reading, so if the elevator is accelerating upward, the scale shows an increased weight: The scale reading is equal to w_{app}.

Apparent weight can be measured by a scale, so it's no surprise that apparent weight has real, physical implications. Astronauts are nearly crushed by their apparent weight during a rocket launch when a is much greater than g. Much of the thrill of amusement park rides, such as roller coasters, comes from rapid changes in your apparent weight.

FIGURE 5.9 A man in an accelerating elevator.

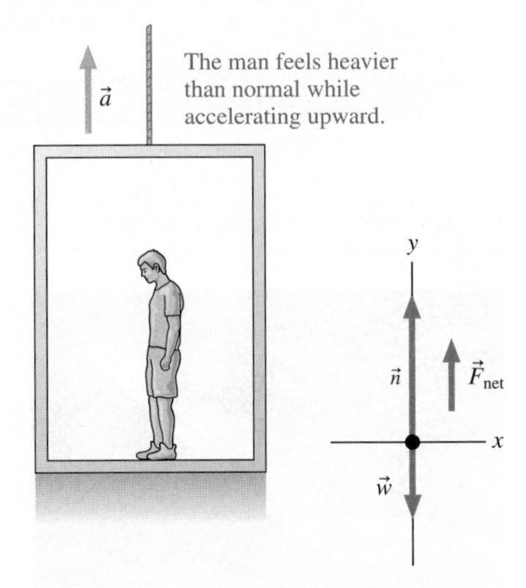
The man feels heavier than normal while accelerating upward.
$\vec{a}$
y
$\vec{n}$
$\vec{F}_{net}$
x
$\vec{w}$

EXAMPLE 5.8 Finding a rider's apparent weight in an elevator

Anjay's mass is 70 kg. He is standing on a scale in an elevator that is moving at 5.0 m/s. As the elevator slows to a stop, the scale reads 750 N. Before it stopped, was the elevator moving up or down? How long did the elevator take to come to rest?

STRATEGIZE The scale reads Anjay's apparent weight, which differs from his actual weight by ma_y. If his apparent weight is greater than his actual weight, then his acceleration is upward; if less, his acceleration is downward.

PREPARE The scale reading as the elevator comes to rest, 750 N, is Anjay's apparent weight. Anjay's actual weight is

$$w = mg = (70 \text{ kg})(9.80 \text{ m/s}^2) = 686 \text{ N}$$

This is an intermediate step in the calculation, so we are keeping an extra significant figure.

SOLVE Because Anjay's apparent weight w_{app} is greater than his actual weight w, his acceleration $a_y = (w_{app} - w)/m$ is *positive*. You learned in Chapter 2 that if an object is slowing, as is the case here, its velocity and acceleration vectors point in opposite directions, so the elevator's velocity must be *negative*—it is moving down.

We can use kinematics to find the stopping time. For this step we need the acceleration, which is

$$a_y = \frac{w_{app} - w}{m} = \frac{750 \text{ N} - 686 \text{ N}}{70 \text{ kg}} = 0.91 \text{ m/s}^2$$

Then we use the kinematic equation

$$(v_y)_f = (v_y)_i + a_y \Delta t$$

Continued

The elevator is initially moving downward, so $(v_y)_i = -5.0$ m/s, and it then comes to a halt, so $(v_y)_f = 0$. We know the acceleration, so the time interval is

$$\Delta t = \frac{(v_y)_f - (v_y)_i}{a_y} = \frac{0 - (-5.0 \text{ m/s})}{0.91 \text{ m/s}^2} = 5.5 \text{ s}$$

ASSESS Think back to your experiences riding elevators. If the elevator is moving downward and then comes to rest, you "feel heavy." This gives us confidence that our analysis of the motion is correct. And 5.0 m/s is a pretty fast elevator: At this speed, the elevator will be passing more than one floor per second. If you've been in a fast elevator in a tall building, you know that 5.5 s is reasonable for the time it takes for the elevator to slow to a stop.

Weightlessness

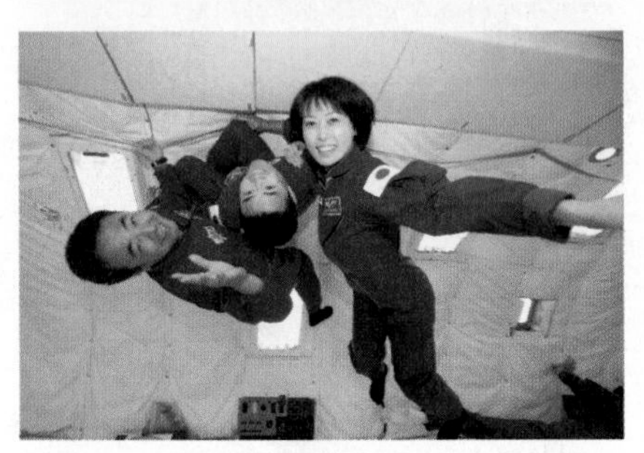

A weightless experience As we learned in Chapter 3, objects undergoing projectile motion are in free fall. This specially adapted plane flies in the same parabolic trajectory as would a projectile with no air resistance. Objects inside, such as these passengers, thus move along a free-fall trajectory. They feel weightless, and then float with respect to the plane's interior, until the plane resumes normal flight, up to 30 seconds later.

Let's return to the elevator example. Suppose Anjay's apparent weight was 0 N—meaning that the scale read *zero*. Could this happen? Recall that Anjay's apparent weight is the contact force supporting him, so we are saying that the upward force from the scale is zero. If we use $W_{app} = 0$ in the equation for the apparent acceleration, we find

$$0 = w + ma_y = mg + ma_y$$

so that $a_y = -g = -9.8 \text{ m/s}^2$. This is a case we've seen before—it is free fall! This means that **a person (or any object) in free fall has zero apparent weight.**

Think about this carefully. Imagine a man inside an elevator that is in free fall (a frightening case, for sure!). This man is also in free fall, so he is falling at the exact same rate as the elevator. From the man's perspective, he is "floating" inside the elevator. A small push off the floor would send him floating toward the top of the elevator. He is what we call *weightless*.

"Weightless" does *not* mean "no weight." An object that is **weightless** has no *apparent* weight. The distinction is significant. The man's weight is still mg because gravity is still pulling down on him, but he has no *sensation* of weight as he free falls. The term "weightless" is in a sense a poor one because it implies that objects have no weight. As we see, that is not the case.

You've seen videos of astronauts and various objects floating inside the International Space Station as it orbits the earth. If an astronaut tries to stand on a scale, it does not exert any force against her feet and reads zero. She is said to be weightless. But if the criterion to be weightless is to be in free fall, and if astronauts orbiting the earth are weightless, does this mean that they are in free fall? This is a very interesting question to which we shall return in Chapter 6.

STOP TO THINK 5.3 You're bouncing up and down on a trampoline. After you have left the trampoline and are moving upward, your apparent weight is

A. More than your true weight.
B. Less than your true weight.
C. Equal to your true weight.
D. Zero.

5.4 Normal Forces

In Chapter 4 we saw that an object at rest on a table is subject to an upward force due to the table. This force is called the *normal force* because it is always directed normal, or perpendicular, to the surface of contact. As we saw, the normal force has its origin in the atomic "springs" that make up the surface. The harder the object bears down on the surface, the more these springs are compressed and the harder they push back. Thus the normal force *adjusts* itself so that the object stays on the surface without penetrating it. This fact is key in solving for the normal force.

EXAMPLE 5.9 Normal force on a pressed book

A 1.2 kg book lies on a table. You press down on the book from above with a force of 15 N. What is the normal force acting on the book from the table below?

STRATEGIZE The book is in static equilibrium, so the net force acting on it is zero. We will use this fact to find the forces acting on the book.

PREPARE We need to identify the forces acting on the book and prepare a free-body diagram showing these forces. These steps are illustrated in FIGURE 5.10.

FIGURE 5.10 Finding the normal force on a book pressed from above.

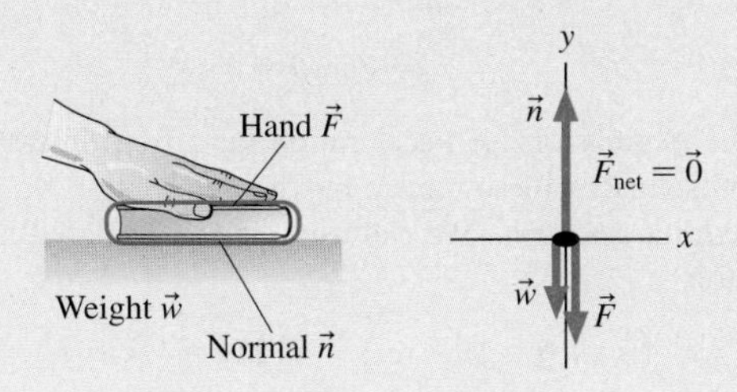

SOLVE Because the book is in static equilibrium, the net force on it must be zero. The only forces acting are in the y-direction, so Newton's second law is

$$\sum F_y = n_y + w_y + F_y = n - w - F = ma_y = 0$$

We learned in the last section that the weight force is $w = mg$. The weight of the book is thus

$$w = mg = (1.2 \text{ kg})(9.8 \text{ m/s}^2) = 12 \text{ N}$$

With this information, we see that the normal force exerted by the table is

$$n = F + w = 15 \text{ N} + 12 \text{ N} = 27 \text{ N}$$

ASSESS The magnitude of the normal force is *larger* than the weight of the book. From the table's perspective, the extra force from the hand pushes the book further into the atomic springs of the table. These springs then push back harder, giving a normal force that is greater than the weight of the book.

A common situation is an object on a ramp or incline. If friction is ignored, there are only two forces acting on the object: gravity and the normal force. However, we need to carefully work out the components of these two forces in order to solve dynamics problems. FIGURE 5.11a shows how. Be sure you avoid the two common errors shown in FIGURE 5.11b.

KEY CONCEPT FIGURE 5.11 The forces on an object on an incline.

(a) Analyzing forces on an incline

The normal force always points perpendicular to the surface.

When we rotate the x-axis to match the surface, the angle between $\vec{w}$ and the negative y-axis is the same as the angle θ of the slope.

The weight force always points straight down.

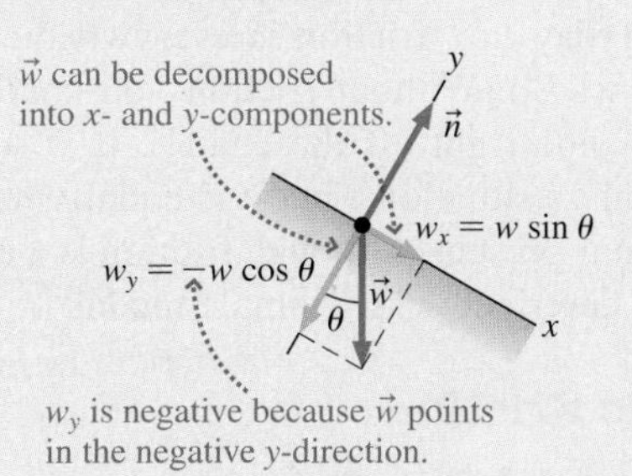

(b) Two common mistakes to avoid

Wrong! The normal force is always perpendicular to the surface of contact!

Correct

Correct

Wrong! The weight always points straight down!

STOP TO THINK 5.4 A mountain biker is climbing a steep 20° slope at a constant speed. The cyclist and bike have a combined weight of 800 N. Referring to Figure 5.11 for guidance, what can you say about the magnitude of the normal force of the ground on the bike?

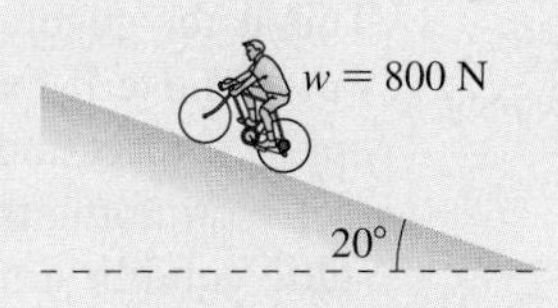

A. $n > 800$ N
B. $n = 800$ N
C. $n < 800$ N

EXAMPLE 5.10 Acceleration of a downhill skier

A skier slides down a steep 27° slope. On a slope this steep, friction is much smaller than the other forces acting on the skier and can be ignored. What is the skier's acceleration?

STRATEGIZE We will use Newton's second law in component form to find the acceleration. We should choose a coordinate system with the x-axis pointing down the slope. This greatly simplifies the analysis because then the skier does not move in the y-direction at all, making $a_y = 0$.

PREPARE FIGURE 5.12 is a visual overview. The free-body diagram is based on the information in Figure 5.11.

FIGURE 5.12 Visual overview of a downhill skier.

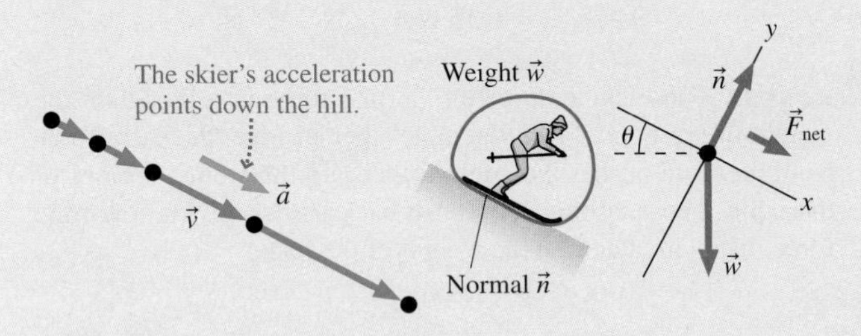

SOLVE We can now use Newton's second law in component form to find the skier's acceleration:

$$\sum F_x = w_x + n_x = ma_x$$
$$\sum F_y = w_y + n_y = ma_y$$

Because $\vec{n}$ points directly in the positive y-direction, $n_y = n$ and $n_x = 0$. Figure 5.11a showed the important fact that the angle between $\vec{w}$ and the negative y-axis is the *same* as the slope angle θ. With this information, the components of $\vec{w}$ are $w_x = w\sin\theta = mg\sin\theta$ and $w_y = -w\cos\theta = -mg\cos\theta$, where we used the fact that $w = mg$. With these components in hand, Newton's second law becomes

$$\sum F_x = w_x + n_x = mg\sin\theta = ma_x$$
$$\sum F_y = w_y + n_y = -mg\cos\theta + n = ma_y = 0$$

In the second equation we used the fact that $a_y = 0$. The m cancels in the first of these equations, leaving us with

$$a_x = g\sin\theta$$

This is the expression for acceleration on a frictionless surface that we presented, without proof, in Chapter 3. Now we've justified our earlier assertion. We can use this to calculate the skier's acceleration:

$$a_x = g\sin\theta = (9.8 \text{ m/s}^2)\sin 27° = 4.4 \text{ m/s}^2$$

ASSESS Our result shows that when $\theta = 0$, so that the slope is horizontal, the skier's acceleration is zero, as it should be. Further, when $\theta = 90°$ (a vertical slope), his acceleration is g, which makes sense because he's in free fall when $\theta = 90°$. Notice that the mass canceled out, so we didn't need to know the skier's mass. We first saw the formula for the acceleration in ◀◀ SECTION 3.4, but now we see the physical reasons behind it.

5.5 Friction

Video Solving Problems Using Newton's Laws

In everyday life, friction is everywhere. Friction is absolutely essential for many things we do. Without friction you could not walk, drive, or even sit down (you would slide right off the chair!). It is sometimes useful to think about idealized frictionless situations, but it is equally necessary to understand a real world where friction is present. Although friction is a complicated force, many aspects of friction can be described with a simple model.

FIGURE 5.13 Static friction keeps an object from slipping.

(a) Force identification

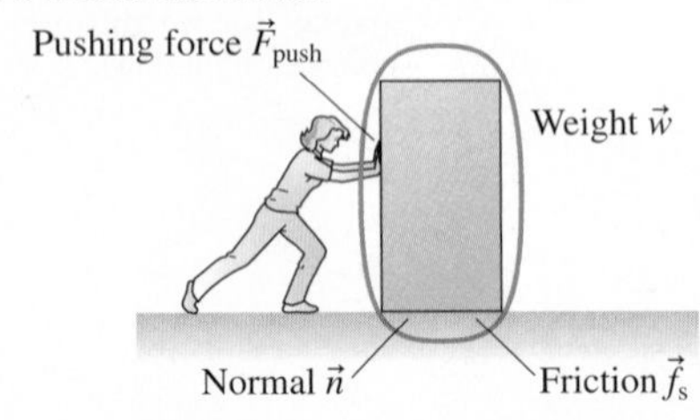

(b) Free-body diagram

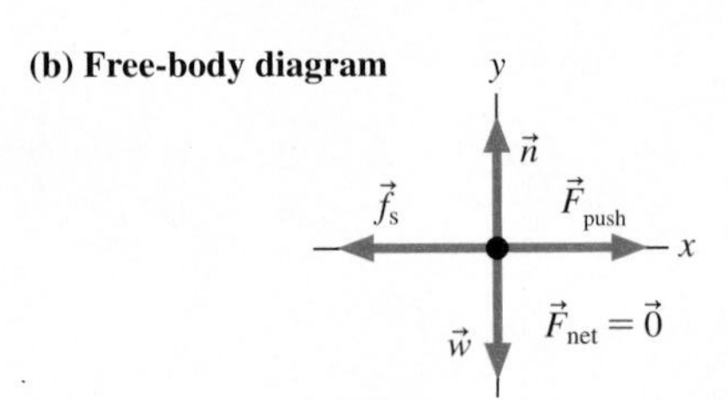

Static Friction

Chapter 4 defined static friction $\vec{f}_s$ as the force that a surface exerts on an object to keep it from slipping across that surface. Consider the woman pushing on the box in FIGURE 5.13a. Because the box is not moving with respect to the floor, the woman's push to the right must be balanced by a static friction force $\vec{f}_s$ pointing to the left. This is the general rule for finding the *direction* of $\vec{f}_s$: Decide which way the object *would* move if there were no friction. The static friction force $\vec{f}_s$ then points in the opposite direction, to prevent motion relative to the surface.

Determining the *magnitude* of $\vec{f}_s$ is a bit trickier. Because the box is at rest, it's in static equilibrium. From the free-body diagram of FIGURE 5.13b, this means that the static friction force must exactly balance the pushing force, so that $f_s = F_{push}$. As shown in FIGURES 5.14a and 5.14b, the harder the woman pushes, the harder the friction force from the floor pushes back. If she reduces her pushing force, the friction force will automatically be reduced to match. Static friction acts in *response* to an applied force.

But there's clearly a limit to how big $\vec{f}_s$ can get. If the woman pushes hard enough, the box will slip and start to move across the floor. In other words, the static friction force has a *maximum* possible magnitude $f_{s\,max}$, as illustrated in FIGURE 5.14c. Experiments with friction show that $f_{s\,max}$ is proportional to the magnitude of the normal force between the surface and the object; that is,

$$f_{s\,max} = \mu_s n \tag{5.7}$$

where μ_s is called the **coefficient of static friction.** The coefficient is a number that depends on the materials from which the object and the surface are made. The higher the coefficient of static friction, the greater the "stickiness" between the object and the surface, and the harder it is to make the object slip. TABLE 5.2 lists some approximate values of coefficients of friction.

NOTE ▸ Equation 5.7 does not say $f_s = \mu_s n$. The value of f_s depends on the force or forces that static friction has to balance to keep the object from moving. It can have any value from zero up to, but not exceeding, $\mu_s n$. ◂

So our rules for static friction are:

- The direction of static friction is such as to oppose motion.
- The magnitude f_s of static friction adjusts itself so that the net force is zero and the object doesn't move.
- The magnitude of static friction cannot exceed the maximum value $f_{s\,max}$ given by Equation 5.7. If the friction force needed to keep the object stationary is greater than $f_{s\,max}$, the object slips and starts to move.

FIGURE 5.14 Static friction acts in response to an applied force.

(a) Pushing gently: friction pushes back gently.

$\vec{f}_s$ $\vec{F}_{push}$

$\vec{f}_s$ balances $\vec{F}_{push}$ and the box does not move.

(b) Pushing harder: friction pushes back harder.

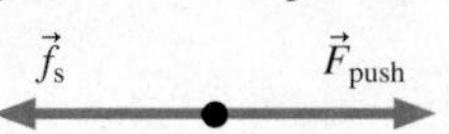

$\vec{f}_s$ grows as $\vec{F}_{push}$ increases, but the two still cancel and the box remains at rest.

(c) Pushing harder still: $\vec{f}_s$ is now pushing back as hard as it can.

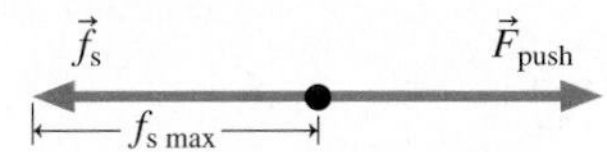

$f_{s\,max}$

Now the magnitude of f_s has reached its maximum value $f_{s\,max}$. If $\vec{F}_{push}$ gets any bigger, the forces will *not* cancel and the box will start to accelerate.

Video Figure 5.14

Kinetic Friction

Once the box starts to slide, as in FIGURE 5.15, the static friction force is replaced by a kinetic (or sliding) friction force $\vec{f}_k$. Kinetic friction is in some ways simpler than static friction: The direction of $\vec{f}_k$ is always opposite to the direction in which an object slides across the surface, and experiments show that kinetic friction, unlike static friction, has a nearly *constant* magnitude, given by

$$f_k = \mu_k n \tag{5.8}$$

where μ_k is called the **coefficient of kinetic friction.** Equation 5.8 also shows that kinetic friction, like static friction, is proportional to the magnitude of the normal force n. Notice that **the magnitude of the kinetic friction force does not depend on how fast the object is sliding.**

TABLE 5.2 Coefficients of friction

Materials	Static μ_s	Kinetic μ_k	Rolling μ_r
Rubber on concrete	1.00	0.80	0.02
Steel on steel (dry)	0.80	0.60	0.002
Steel on steel (lubricated)	0.10	0.05	
Wood on wood	0.50	0.20	
Wood on snow	0.12	0.06	
Ice on ice	0.10	0.03	

FIGURE 5.15 The kinetic friction force is *opposite* to the direction of motion.

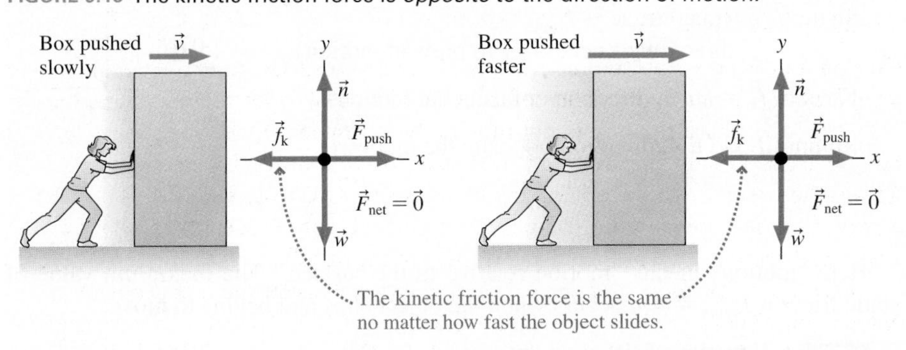
Box pushed slowly
Box pushed faster
The kinetic friction force is the same no matter how fast the object slides.

Table 5.2 includes approximate values of μ_k. You can see that $\mu_k < \mu_s$, which explains why it is easier to keep a box moving than it is to start it moving.

FIGURE 5.16 The bottom of the wheel is stationary.

Rolling Friction

When you are driving, if you slam on the brakes hard enough, your car tires slide against the road surface and leave skid marks. This is kinetic friction because the tire and the road are *sliding* against each other. A wheel *rolling* on a surface also experiences friction, but not kinetic friction: The portion of the wheel that contacts the surface is stationary with respect to the surface, not sliding. The photo in FIGURE 5.16 was taken with a stationary camera. Note that the part of the wheel touching the ground is not blurred, which indicates that this part of the wheel is not moving with respect to the ground.

The interaction between a rolling wheel and the road involves adhesion between and deformation of surfaces and can be quite complicated, but in many cases we can treat it like another type of friction force that opposes the motion, one defined by a **coefficient of rolling friction** μ_r:

$$f_r = \mu_r n \tag{5.9}$$

Rolling friction acts very much like kinetic friction, but values of μ_r (see Table 5.2) are much lower than values of μ_k. It's easier to roll something on wheels than to slide it!

STOP TO THINK 5.5 Rank in order, from largest to smallest, the size of the friction forces $\vec{f}_A$ to $\vec{f}_E$ in the five different situations (one or more friction forces could be zero). The box and the floor are made of the same materials in all situations.

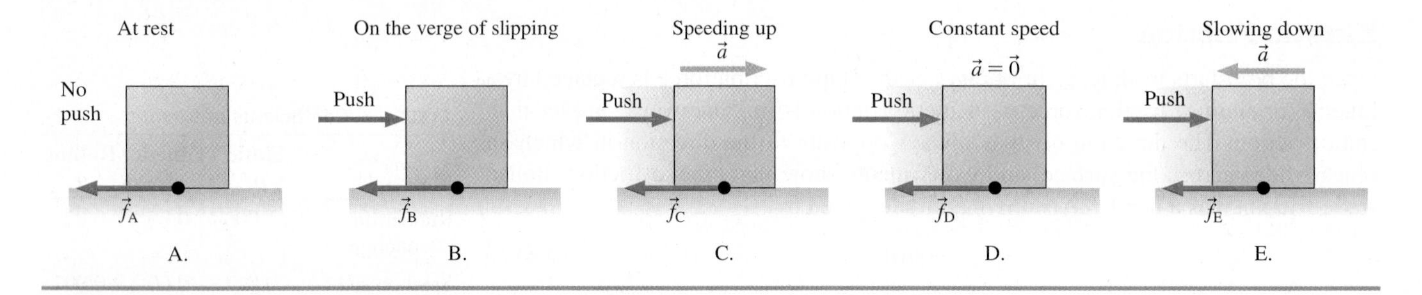

To skid or not to skid If you brake as hard as you can without skidding, the force that stops your car is the static friction force between your tires and the road. This force is bigger than the kinetic friction force, so if you skid, not only can you lose control of your car but you also end up taking a longer distance to stop. Antilock braking systems brake the car as hard as possible without skidding, stopping the car in the shortest possible distance while also retaining control.

Working with Friction Forces

These ideas can be summarized in a *model* of friction:

$$\text{Static: } \vec{f}_s = (\text{magnitude} \le f_{s\,max} = \mu_s n, \text{ direction as necessary to prevent motion})$$
$$\text{Kinetic: } \vec{f}_k = (\mu_k n, \text{ direction opposite the motion}) \tag{5.10}$$
$$\text{Rolling: } \vec{f}_r = (\mu_r n, \text{ direction opposite the motion})$$

p. 39 PROPORTIONAL

Here "motion" means "motion relative to the surface." The maximum value of static friction $f_{s\,max} = \mu_s n$ occurs when the object slips and begins to move.

NOTE ▸ Equations 5.10 are a "model" of friction, not a "law" of friction. These equations provide a reasonably accurate, but not perfect, description of how friction forces act. They are a simplification of reality that works reasonably well, which is what we mean by a "model." They are not a "law of nature" on a level with Newton's laws. ◂

TACTICS BOX 5.1 Working with friction forces

❶ If the object is *not moving* relative to the surface it's in contact with, then the friction force is **static friction.** Draw a free-body diagram of the object. The *direction* of the friction force is such as to oppose sliding of the object relative to the surface. Then use Problem-Solving Approach 5.1 to solve for f_s. If f_s is greater than $f_{s\,max} = \mu_s n$, then static friction cannot hold the object in place. The assumption that the object is at rest is not valid, and you need to redo the problem using kinetic friction.

❷ If the object is *sliding* relative to the surface, then **kinetic friction** is acting. From Newton's second law, find the normal force n. Equation 5.10 then gives the magnitude and direction of the friction force.

❸ If the object is *rolling* along the surface, then **rolling friction** is acting. From Newton's second law, find the normal force n. Equation 5.10 then gives the magnitude and direction of the friction force.

Exercises 20, 21

Video What the Physics? Tablecloth Trick

EXAMPLE 5.11 Finding the force to slide a sofa

Carol wants to move her 32 kg sofa to a different room in the house. She places "sofa sliders," slippery disks with $\mu_k = 0.080$, under the feet of the sofa. She then pushes the sofa at a steady 0.40 m/s across the floor. How much force does she apply to the sofa?

STRATEGIZE The sofa is sliding, so the friction force is kinetic friction. We will use the second step of Tactics Box 5.1.

PREPARE Let's assume the sofa slides to the right. In this case, a kinetic friction force $\vec{f}_k$ opposes the motion by pointing to the left. In FIGURE 5.17 we identify the forces acting on the sofa and construct a free-body diagram.

FIGURE 5.17 Forces on a sofa being pushed across a floor.

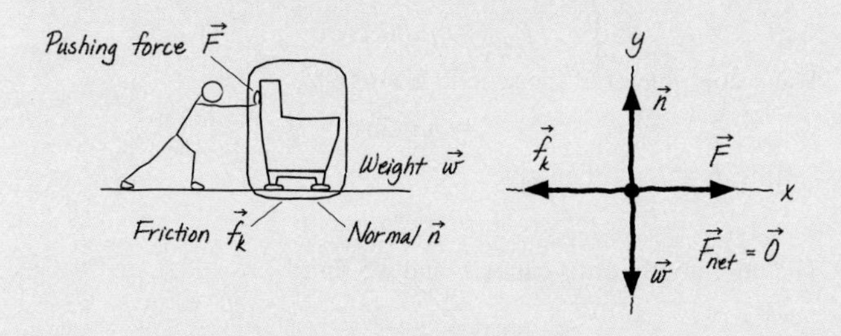

SOLVE The sofa is moving at a constant speed, so it is in dynamic equilibrium with $\vec{F}_{net} = \vec{0}$. This means that the x- and y-components of the net force must be zero:

$$\sum F_x = n_x + w_x + F_x + (f_k)_x = 0 + 0 + F - f_k = 0$$
$$\sum F_y = n_y + w_y + F_y + (f_k)_y = n - w + 0 + 0 = 0$$

In the first equation, the x-component of $\vec{f}_k$ is equal to $-f_k$ because $\vec{f}_k$ is directed to the left. Similarly, in the second equation, $w_y = -w$ because the weight force points down.

From the first equation, we see that Carol's pushing force is $F = f_k$. To evaluate this, we need f_k. Here we can use our model for kinetic friction:

$$f_k = \mu_k n$$

Let's look at the vertical motion first. The second equation ultimately reduces to

$$n - w = 0$$

The weight force $w = mg$, so we can write

$$n = mg$$

This is a common result we'll see again. The force that Carol pushes with is equal to the friction force, and this depends on the normal force and the coefficient of kinetic friction, $\mu_k = 0.080$:

$$F = f_k = \mu_k n = \mu_k mg$$
$$= (0.080)(32\,\text{kg})(9.80\,\text{m/s}^2) = 25\,\text{N}$$

ASSESS The speed with which Carol pushes the sofa does not enter into the answer. This makes sense because the kinetic friction force doesn't depend on speed. The final result of 25 N is a rather small force—only about $5\frac{1}{2}$ pounds—but we expect this because Carol has used slippery disks to move the sofa.

CONCEPTUAL EXAMPLE 5.12 To push or pull a lawn roller?

A lawn roller is a heavy cylinder used to flatten a bumpy lawn, as shown in FIGURE 5.18. Is it easier to push or pull such a roller? Which is more effective for flattening the lawn: pushing or pulling? Assume that the pushing or pulling force is directed along the handle of the roller.

FIGURE 5.18 Pushing and pulling a lawn roller.

Continued

REASON FIGURE 5.19 shows free-body diagrams for the two cases. We assume that the roller is pushed at a constant speed so that it is in dynamic equilibrium with $\vec{F}_{net} = \vec{0}$. Because the roller does not move in the y-direction, the y-component of the net force must be zero. According to our model, the magnitude f_r of rolling friction is proportional to the magnitude n of the normal force. If we *push* on the roller, our pushing force $\vec{F}$ will have a downward y-component. To compensate for this, the normal force must increase and, because $f_r = \mu_r n$, the rolling friction will increase as well. This makes the roller harder to move. If we *pull* on the roller, the now upward y-component of $\vec{F}$ will lead to a *reduced* value of n and hence of f_r. Thus the roller is easier to pull than to push.

However, the purpose of the roller is to flatten the soil. If the normal force $\vec{n}$ of the ground on the roller is greater, then by Newton's third law the force of the roller on the ground will be greater as well. So for smoothing your lawn, it's better to push.

FIGURE 5.19 Free-body diagrams for the lawn roller.

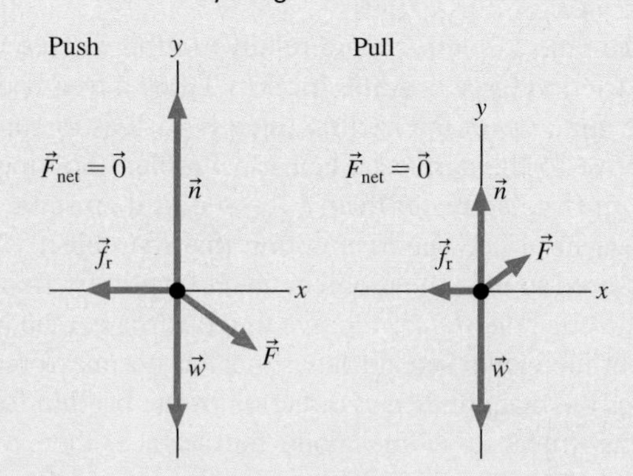

ASSESS You've probably experienced this effect while using an upright vacuum cleaner. The vacuum is harder to push on the forward stroke than when drawing it back.

EXAMPLE 5.13 How to dump a file cabinet

A 50.0 kg steel file cabinet is in the back of a dump truck. The truck's bed, also made of steel, is slowly tilted. What is the magnitude of the static friction force on the cabinet when the bed is tilted 20°? At what angle will the file cabinet begin to slide?

STRATEGIZE Until it slides, the file cabinet is held in place by static friction, so we will use the first step of Tactics Box 5.1. To do so, we need the normal force, which we will find using the fact that the cabinet is in static equilibrium.

PREPARE From Tactics Box 5.1, we know that the file cabinet will slip when the static friction force reaches its maximum possible value $f_{s\,max}$. FIGURE 5.20 shows the visual overview when the truck bed is tilted at angle θ. We can make the analysis easier if we tilt the coordinate system to match the bed of the truck. To prevent the file cabinet from slipping, the static friction force must point *up* the slope.

SOLVE Before it slips, the file cabinet is in static equilibrium. Newton's second law gives

$$\sum F_x = n_x + w_x + (f_s)_x = 0$$
$$\sum F_y = n_y + w_y + (f_s)_y = 0$$

From the free-body diagram we see that f_s has only a negative x-component and that n has only a positive y-component. We also have $w_x = w\sin\theta$ and $w_y = -w\cos\theta$. Thus the second law becomes

$$\sum F_x = w\sin\theta - f_s = mg\sin\theta - f_s = 0$$
$$\sum F_y = n - w\cos\theta = n - mg\cos\theta = 0$$

The x-component equation allows us to determine the magnitude of the static friction force when $\theta = 20°$:

$$f_s = mg\sin\theta = (50.0\text{ kg})(9.80\text{ m/s}^2)\sin 20° = 168\text{ N}$$

This value does not require that we know μ_s. The coefficient of static friction enters only when we want to find the angle at which the file cabinet slips. Slipping occurs when the static friction force reaches its maximum value:

$$f_s = f_{s\,max} = \mu_s n$$

From the y-component of Newton's second law we see that $n = mg\cos\theta$. Consequently,

$$f_{s\,max} = \mu_s mg\cos\theta$$

The x-component of the second law gave

$$f_s = mg\sin\theta$$

Setting $f_s = f_{s\,max}$ then gives

$$mg\sin\theta = \mu_s mg\cos\theta$$

The mg in both terms cancels, and we find

$$\frac{\sin\theta}{\cos\theta} = \tan\theta = \mu_s$$
$$\theta = \tan^{-1}\mu_s = \tan^{-1}(0.80) = 39°$$

ASSESS Steel doesn't slide all that well on unlubricated steel, so a fairly large angle is not surprising. The answer seems reasonable. It is worth noting that $n = mg\cos\theta$ in this example. A common error is to use simply $n = mg$. Be sure to evaluate the normal force within the context of each particular problem.

FIGURE 5.20 Visual overview of a file cabinet in a tilted dump truck.

Known
$\mu_s = 0.80$ $m = 50.0$ kg
$\mu_k = 0.60$

Find
f_s when $\theta = 20°$
θ at which cabinet slips

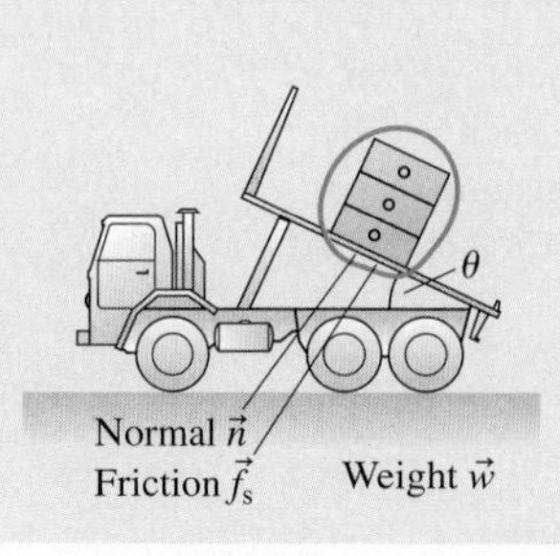

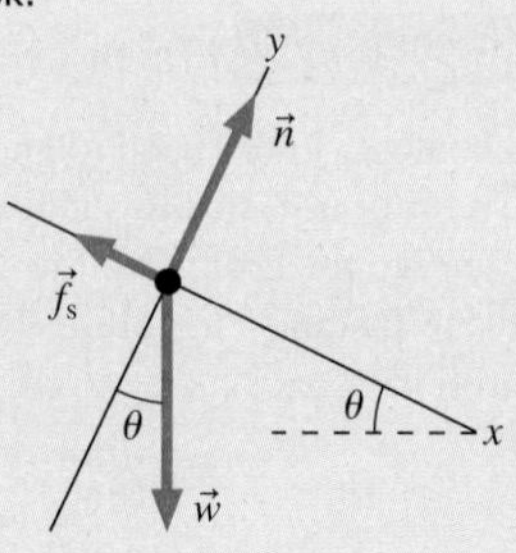

Causes of Friction

It is worth taking a brief pause to look at the *causes* of friction. All surfaces, even those that are quite smooth to the touch, are very rough on a microscopic scale. When two objects are placed in contact, they do not make a smooth fit. Instead, as FIGURE 5.21 shows, the high points on one surface become jammed against the high points on the other surface, while the low points are not in contact at all. Only a very small fraction (typically 10^{-4}) of the surface area is in actual contact. The amount of contact depends on how hard the surfaces are pushed together, which is why friction forces are proportional to n.

For an object to slip, you must push it hard enough to overcome the forces exerted at these contact points. Once the two surfaces are sliding against each other, their high points undergo constant collisions, deformations, and even brief bonding that lead to the resistive force of kinetic friction.

FIGURE 5.21 A microscopic view of friction.

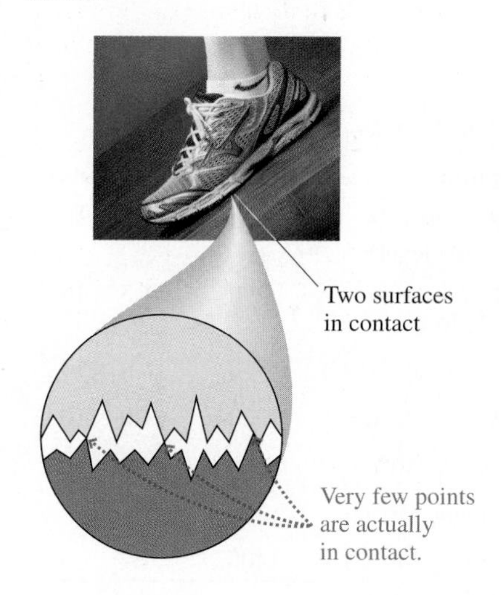
Two surfaces
in contact
Very few points
are actually
in contact.

5.6 Drag

Drag is a force that opposes or retards the motion of an object as it moves through a *fluid*—a substance that flows, such as air or water. You experience drag forces due to the air—sometimes called *air resistance*—whenever you jog, bicycle, or drive your car. In biology, drag forces are extremely important for microorganisms moving in water. The drag force $\vec{D}$:

- Is opposite in direction to the velocity $\vec{v}$, as shown in FIGURE 5.22.
- Increases in magnitude as the object's speed increases.

FIGURE 5.22 The drag force is opposite the direction of motion.

$\vec{D}$

In general, drag is more complex than friction, in part because the drag force has its origin in two rather different physical principles. Fortunately, in many cases of interest only one of these causes of drag is in effect; in such cases we can characterize drag through the use of fairly simple models. To understand which cause of drag applies in a given situation, we need to introduce the *Reynolds number*.

Reynolds Number

We can illustrate the two basic causes of drag with examples. When a thrown baseball moves through the air, most of the drag—the force that resists its motion—occurs because the rapidly moving ball has to push away the air in front of it. This drag is thus due to *inertial forces*, forces that arise because the air's inertia makes it difficult to push through. In contrast, consider the motion of a small sphere slowly falling in a thick fluid such as honey. Here, the drag is mostly the result of *viscous forces* due to the properties of the honey itself: The honey "grabs" the surface of the sphere and tries to hold it back.

Which of these two causes of drag is most important depends on the particular situation. As we'll learn in the following discussion, for an object like the fast-moving baseball in air, where the drag is due to inertial forces, the drag force is proportional to $\rho v^2 L^2$, where ρ (lowercase Greek rho) is the density of the fluid (here, air) through which the object moves, v is its speed, and L is a measure of its size (such as the diameter for a baseball). For an object moving through a thick fluid, where the drag is due to viscous forces, the drag force is proportional to $\eta L v$, where η (lowercase Greek eta) is the **viscosity** of the fluid. The viscosity is larger for thick fluids such as honey than for more free-flowing ones such as water.

To characterize which kind of drag is relevant, we can formulate a number, called the **Reynolds number,** that is the *ratio* of the inertial force to the viscous force. Thus

$$\text{Reynolds number} = Re = \frac{\text{inertial forces}}{\text{viscous forces}} = \frac{\rho v^2 L^2}{\eta L v} = \frac{\rho v L}{\eta} \tag{5.11}$$

Written this way, the Reynolds number is *dimensionless;* that is, it has no units. When the Reynolds number is high, inertial drag is dominant; when the Reynolds number is low, viscous drag is more important.

NOTE ▶ The symbol for the Reynolds number is the two-letter combination Re. It is the only symbol in this text that uses two letters. ◀

In SI units, density is measured in kg/m^3 and viscosity in $\text{Pa}\cdot\text{s}$. Pa is the abbreviation for *pascal*, the SI unit of pressure, which is defined as $1\ \text{Pa} = 1\ \text{N/m}^2$. We'll have a lot more to say about pressure in later chapters. TABLE 5.3 lists the values of density and viscosity for some typical fluids. Viscosity is *very* dependent on temperature—think how quickly honey loses viscosity as you heat it—so the values shown are not appropriate at temperatures other than those listed.

TABLE 5.3 Density and viscosity

Fluid	ρ (kg/m³)	η(Pa · s)
Air (20°C at sea level)	1.20	1.8×10^{-5}
Ethyl alcohol (20°C)	790	1.3×10^{-3}
Olive oil (20°C)	910	8.4×10^{-2}
Water (20°C)	1000	1.0×10^{-3}
Water (40°C)	1000	7.0×10^{-4}
Honey (20°C)	1400	10
Honey (40°C)	1400	1.7

The object's size L is a characteristic or typical size of the object, which could be a height, width, or diameter. This seems a bit odd; for an object that is rectangular, the value of Re depends on whether you select the width or the height for L. However, in practice, how you define L makes very little difference. As you'll see, we want to know whether Re is "high" or "low," values that differ by factors of 1000 or more. A difference of a factor of 2 or 3 will not affect our judgment about the size of Re. Said another way, our use of the Reynolds number needs only approximate values, not precise calculations.

As a simple example, consider a 75-mm-diameter baseball moving through 20°C air at a speed of 20 m/s. We can take the ball's characteristic size L to be its diameter. The Reynolds number for this motion is

$$Re = \frac{\rho v L}{\eta} = \frac{(1.2\ \text{kg/m}^3)(20\ \text{m/s})(0.075\ \text{m})}{1.8 \times 10^{-5}\ \text{Pa}\cdot\text{s}} = 100{,}000$$

As you'll see, this is considered a very high Reynolds number.

TABLE 5.4 Drag Coefficients

Object	C_D
Commercial airliner	0.024
Toyota Prius	0.24
Pitched baseball	0.35
Racing cyclist	0.88
Running person	1.2

Drag at High Reynolds Number

When the Reynolds number is greater than about 1000, it is considered high. A high Reynolds number means, as we've seen, that the drag arises not from viscosity but from inertial drag: The moving object has to push the fluid out of the way. As the example of the baseball suggests, the Reynolds number is high for most ordinary objects—balls, cars, planes—moving through air at ordinary speeds. Re is also high for fish and larger objects moving through water.

For high Reynolds numbers, the drag force for motion through a fluid at speed v is

$$\vec{D} = \left(\tfrac{1}{2}C_D \rho A v^2,\ \text{direction opposite the motion}\right) \quad (5.12)$$

Drag force for high Reynolds number, $Re > 1000$

QUADRATIC p.50

Here ρ is again the density of the fluid, A is the cross-section area of the object (in m²) as defined below, and the dimensionless **drag coefficient** C_D depends on the object's shape. More streamlined, aerodynamic shapes have lower values of C_D. TABLE 5.4 lists the drag coefficients for some common moving objects.

For a high Reynolds number, **the size of the drag force is proportional to the *square* of the object's speed.** If the speed doubles, drag increases by a factor of 4. This model of drag is often called **quadratic drag.**

The cross-section area A is the two-dimensional projection that you see when an object is coming toward you. Many objects can be modeled as spheres or cylinders, so FIGURE 5.23 shows cross-section areas and drag coefficients for a sphere and for two different orientations of a cylinder.

FIGURE 5.23 Cross-section areas and drag coefficients for a sphere and a cylinder.

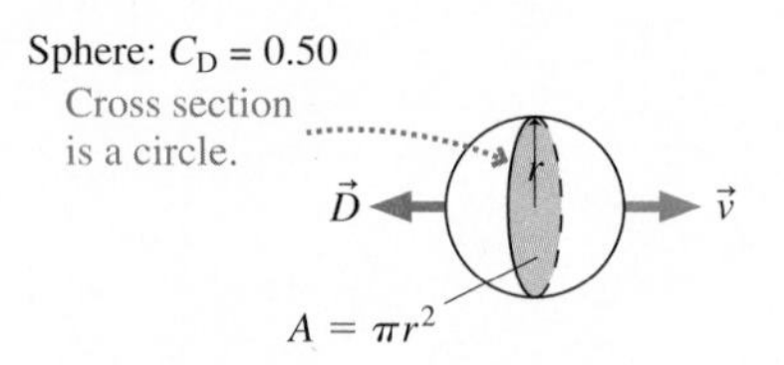

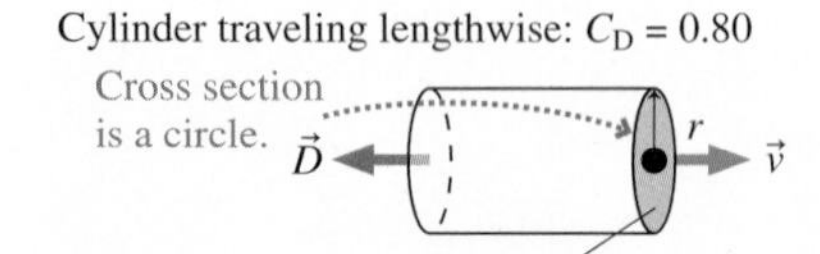

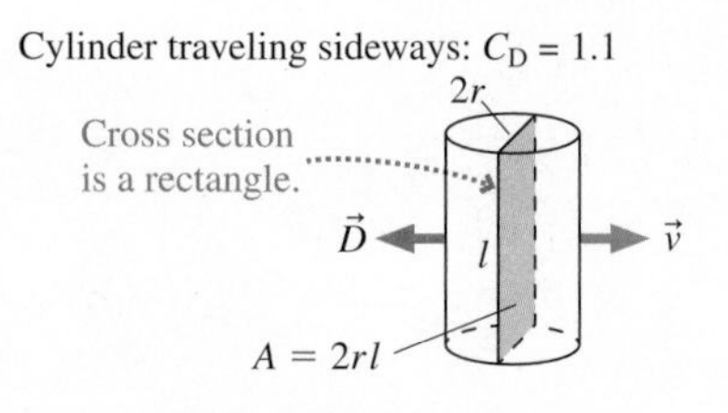

EXAMPLE 5.14 The drag coefficient of a swimming penguin BIO

Biologists have estimated the drag coefficient of a swimming penguin by observing the rate at which a penguin's speed decreases in its glide phase, when it's not actively swimming and is slowing down. In one study, a gliding 4.8 kg Gentoo penguin has an acceleration of -0.52 m/s^2 when its speed is 1.60 m/s. If its frontal area is 0.020 m^2, what is the penguin's drag coefficient?

STRATEGIZE When the penguin is gliding horizontally, the only force acting to slow it down is the drag force. Using Newton's second law, we will find the drag force from the penguin's acceleration and mass. Once the drag force is known, we will use Equation 5.12 to find the drag coefficient.

PREPARE In Equation 5.12, the density ρ is that of the fluid through which the object moves, which is in this case water. From Table 5.3 the density of water is 1000 kg/m^3.

SOLVE Assume that the penguin is moving to the right along the x-axis. Then Newton's second law is

$$F_x = -D = ma_x$$

where the x-component of the drag is negative because it points to the left. We can solve for the magnitude of the drag force as

$$D = -ma_x = -(4.8\text{ kg})(-0.52\text{ m/s}^2) = 2.5\text{ N}$$

Finally, from Equation 5.12 we can solve for the drag coefficient. We have

$$C_D = \frac{2D}{\rho A v^2} = \frac{2(2.5\text{ N})}{(1000\text{ kg/m}^3)(0.020\text{ m}^2)(1.60\text{ m/s})^2} = 0.098$$

ASSESS This drag coefficient is quite a bit better than that of the Toyota Prius in Table 5.4. This is reasonable given that penguins have a highly adapted, streamlined shape.

Terminal Speed

Suppose an object starting from rest is pushed or pulled through a fluid by a constant force $\vec{F}$. Initially the speed is low and the drag force is small, as shown in FIGURE 5.24a, so the net force causes the object to speed up. But the drag force increases as the speed increases, and eventually the object reaches a speed at which the drag force has exactly the same magnitude as $\vec{F}$. Now the net force is zero, as shown in FIGURE 5.24b, so the object can no longer accelerate but will maintain this steady speed for as long as force $\vec{F}$ is applied. The steady, unchanging speed at which drag exactly counterbalances an applied force is called the object's **terminal speed.** In the case of an object fired at an initial speed *greater* than the terminal speed, the very large drag force slows the object until it is moving at the terminal speed.

FIGURE 5.24 An object reaches its terminal speed when the drag force exactly balances the applied force.

(a) At low speeds, D is small and the object accelerates.

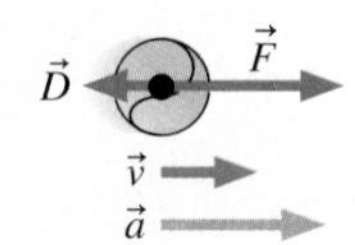

(b) Eventually, v reaches a value such that $D = F$. Then the net force is zero and the object moves at a constant terminal speed v_{term}.

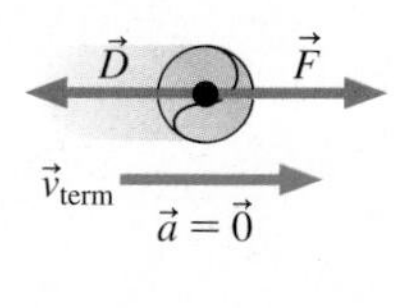

EXAMPLE 5.15 The terminal speeds of a man and a mouse

A 75 kg skydiver and his 0.020 kg pet mouse falling after jumping from a plane are shown in FIGURE 5.25. Find the terminal speed of each.

FIGURE 5.25 A falling skydiver and mouse, and their cross-section areas.

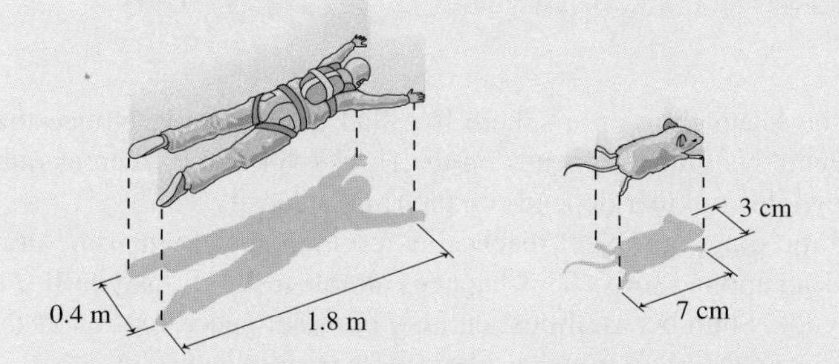

STRATEGIZE The force of gravity is a constant force pulling down on a falling object, while drag due to the air is an upward force. The terminal speed is reached when the force of gravity on each object is equal to the drag force. We will model both the man and the mouse as cylinders falling sideways.

PREPARE From Figure 5.23 we see that for a cylinder $C_D = 1.1$ and that a cylinder's cross-section area as seen from the side is $A = 2rl$. With the dimensions given, we can calculate $A_{man} = 0.72\text{ m}^2$ and $A_{mouse} = 2.1 \times 10^{-3}\text{ m}^2$. We assume that the two skydivers are at sufficiently low altitude that we can use the sea-level value of the density of air.

SOLVE The terminal speed is reached when $D = mg$, or

$$\frac{1}{2}C_D\rho A(v_{term})^2 = mg$$

from which we can find the terminal velocity as

$$v_{term} = \sqrt{\frac{2mg}{C_D\rho A}}$$

Thus our two skydivers have terminal speeds

$$v_{man} = \sqrt{\frac{2(75\text{ kg})(9.8\text{ m/s}^2)}{(1.1)(1.2\text{ kg/m}^3)(0.72\text{ m}^2)}} = 39\text{ m/s}$$

$$v_{mouse} = \sqrt{\frac{2(0.020\text{ kg})(9.8\text{ m/s}^2)}{(1.1)(1.2\text{ kg/m}^3)(2.1 \times 10^{-3}\text{ m}^2)}} = 12\text{ m/s}$$

Continued

ASSESS 39 m/s is about 85 mph. Reported terminal speeds for skydivers falling in the prone position are in the 100–120 mph range, so our simple model of the fall gives a result that is close but a bit too low. We've probably overestimated both ρ, because skydivers are at a high enough altitude that the air density is lower than at sea level, and A, because their legs are actually spread apart and allow air to flow between them. A more realistic, but also more complex, model would give a better prediction. The mouse, though, falls at a much more modest 12 m/s $\approx$ 25 mph. Small animals can usually survive a fall from *any* height. Many tree-dwelling animals, such as the sifaka in the photograph at the beginning of the chapter, can extend flaps of skin to increase their area and thus fall at an even slower speed.

NOTE ▶ Drag forces are fairly small for heavier objects, such as balls or bicycles, moving at not-too-fast speeds through air. To keeps things simple, **you can neglect drag in problems about objects moving through air unless the problem explicitly asks you to include it or asks about the terminal speed.** ◀

Drag at Low Reynolds Number

A 10-μm-diameter dust particle settles to the ground at about 20 mm/s. It's not hard to calculate that the Reynolds number for the falling particle is $Re \approx 0.01$. This is an example of motion at low Reynolds number, which we define to be $Re < 1$. The fluid's viscosity, rather than its density, is the key factor for motion at low Reynolds number.

For low Reynolds number, the drag force for motion through a fluid at speed v is

$$\vec{D} = (b\eta v, \text{ direction opposite the motion}) \tag{5.13}$$

where b is a constant that depends on the size and shape of the object. For low Reynolds number, **the magnitude of the drag force is proportional to the object's speed.** If the speed doubles, drag also increases by a factor of 2. This model of drag is called **linear drag.**

NOTE ▶ Drag is more complex for intermediate values of Reynolds number, $1 < Re < 1000$. We don't consider those situations in this text. ◀

Linear drag is mostly applicable to very small objects, many of which can be reasonably modeled as spheres. That's fortunate because only for spheres does theory give us any guidance for the coefficient b. For a spherical object of radius r, it can be shown that $b = 6\pi r$. Thus

$$\vec{D}_{\text{sphere}} = (6\pi\eta r v, \text{ direction opposite the motion}) \tag{5.14}$$

Stokes' law for low Reynolds number, $Re < 1$

This expression for the linear drag on a sphere is called **Stokes' law.** Notice that linear drag at low Reynolds number depends on the fluid's viscosity, whereas quadratic drag at high Reynolds number depends on the fluid's density.

Just as with quadratic drag, an object reaches its terminal speed when the drag force exactly balances an applied force. This happens almost instantly, as you'll see, so motion at low Reynolds number is almost entirely constant-speed motion at the terminal speed. The force needed to propel a sphere at its terminal speed is

$$F = 6\pi\eta r v_{\text{term}} \tag{5.15}$$

We can find the terminal speed of a small sphere falling in air by using $F = mg$, as we did for high Reynolds number, but this approach doesn't quite work for objects falling in liquids. The reason is that the upward buoyant force of the liquid—a topic we'll study in Chapter 13—cannot be neglected.We'll return to this after we introduce buoyancy.

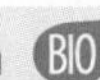

EXAMPLE 5.16 Measuring the mass of a pollen grain BIO

Pollen grains are very light. In one experiment to determine their mass, researchers dropped grains inside a clear glass cylinder and then watched their motion with a microscope. A 40-μm-diameter pollen grain was observed to fall at a rate of 5.3 cm/s. What is the mass in nanograms of this grain?

STRATEGIZE We will model the grain as a sphere. At terminal speed, the force of gravity on the grain is equal to the drag force. Because this is a very small object moving quite slowly, its Reynolds number is very low, so the drag force is linear drag.

PREPARE We'll assume the air in the cylinder was at a room temperature of 20°C and use the value for air viscosity in Table 5.3. In SI units, the terminal speed is $v_{\text{term}} = 0.053$ m/s.

SOLVE We set the force of gravity equal to the linear drag force to get

$$mg = 6\pi\eta r v_{\text{term}}$$

We can solve for the mass as

$$m = \frac{6\pi\eta r v_{\text{term}}}{g} = \frac{6\pi(1.8\times10^{-5}\ \text{Pa}\cdot\text{s})(20\times10^{-6}\ \text{m})(0.053\ \text{m/s})}{9.8\ \text{m/s}^2} = 3.7\times10^{-11}\ \text{kg}$$

We can convert this result to nanograms (ng) as

$$m = 3.7\times10^{-11}\ \text{kg}\times\frac{1000\ \text{g}}{\text{kg}}\times\frac{1\ \text{ng}}{10^{-9}\ \text{g}} = 37\ \text{ng}$$

ASSESS This extremely tiny mass is hard to assess. If we assume that the grain's density is similar to that of water (typical for living objects), then the mass of the 40-um-diameter sphere is about 30 ng, which is close to our result. Our answer is reasonable.

Life at Low Reynolds Number BIO

From protozoa swimming in pond water to bacteria moving in intercellular fluid (essentially water), biology is filled with examples of motion at extremely low Reynolds number. For example, a 1-μm-diameter bacterium moving at a typical speed of 30 μm/s has $Re \approx 3\times10^{-5}$. Life at such low Reynolds number is very much at odds with our everyday experience.

The motion of your body and other everyday objects is at high Reynolds number, where inertia plays a central role. When you apply a force to an object, because of its inertia it takes time for it to gain speed. When you remove the force, drag forces don't cause the object to stop immediately; because of its inertia it takes some time for the object to coast to a stop. Newton's second law, which relates force and motion, is key to understanding motion at high Reynolds number.

Motion at extremely low Reynolds number couldn't be more different. After a force is applied, the object reaches its terminal speed v_{term} *almost instantaneously* and then moves at constant speed. When the force is removed, the object comes to a dead halt *almost instantaneously*. Viscosity is so much more important than inertia that there's no coasting. The only way a tiny organism can move at all is to continuously apply a force—as do the cilia and flagella, the microscopic strands that propel microorganisms.

Suppose a tiny spherical organism of mass m is moving at its terminal speed v_{term} in response to an applied force F. If the applied force is removed, only the drag force remains. How long does it take to stop? We can estimate this time by using kinematics. If the drag force slows the organism with constant acceleration a, then we have

$$\Delta v = v_{\text{f}} - v_{\text{i}} = 0 - v_{\text{term}} = -a\Delta t = -\frac{F_{\text{drag}}}{m}\Delta t = \frac{-6\pi\eta r v_{\text{term}}}{\text{m}}\Delta t$$

from which we can find the time to stop as

$$\Delta t_{\text{stop}} = \frac{m}{6\pi\eta r} \tag{5.16}$$

In addition, during this time interval, the organism coasts a distance $\Delta x_{\text{stop}} = v_{\text{term}}\Delta t_{\text{stop}}$. (The acceleration is not actually constant because the drag force gets smaller as the organism slows down, but these stopping times and distances are still reasonable estimates.)

EXAMPLE 5.17 **A *Paramecium* coasts to a stop** BIO

Paramecia, common unicellular inhabitants of freshwater ponds, are about 100 μm in length and swim at about 200 μm/s. If a *Paramecium* suddenly stops swimming, how long does it take to come to a stop, and how far does it coast while stopping?

STRATEGIZE A *Paramecium*'s swimming speed is its terminal speed v_{term} in response to the pushing force generated by beating its cilia. Although *Paramecia* are somewhat elongated, we'll simplify the situation by modeling the *Paramecium* as a 100-μm-diameter sphere. The density of a *Paramecium* is essentially that of water, so we can calculate its mass from the density of water and the volume of a sphere. We'll assume a moderate water temperature of 20°C.

PREPARE The mass is $m = \frac{4}{3}\pi r^3 \rho$, where ρ is the density. We need to know the density of water and its viscosity at 20°C, both of which are given in Table 5.3.

SOLVE If we use the expression for the *Paramecium*'s mass in Equation 5.16, then the time it takes to come to a stop is

$$\Delta t_{stop} = \frac{\frac{4}{3}\pi r^3 \rho}{6\pi \eta r} = \frac{2r^2\rho}{9\eta} = \frac{2(50 \times 10^{-6}\text{ m})^2(1000\text{ kg/m}^3)}{9(1.0 \times 10^{-3}\text{ Pa}\cdot\text{s})}$$

$$= 5.6 \times 10^{-4}\text{ s} = 0.56\text{ ms}$$

During this time, the *Paramecium* travels distance

$$\Delta x_{stop} = v_{term}\Delta t_{stop} = (200 \times 10^{-6}\text{ m/s})(5.6 \times 10^{-4}\text{ s})$$

$$= 1.1 \times 10^{-7}\text{ m} = 0.11\ \mu\text{m}$$

ASSESS Stopping in less than 1 ms over a distance of 0.11 μm, which is only $\frac{1}{1000}$ the diameter of the organism, is, for all practical purposes, an instantaneous stop. This agrees with our assertion about the motion of microorganisms.

You can swim through water because your hands are able to push the water away from you, but swimming is much more challenging at very low Reynolds number. If you tried to swim through an extremely viscous fluid, like honey, your back-and-forth strokes would cause you to rock back and forth but there would be no net motion.

FIGURE 5.26 A rotating flagellum exerts a continuous, propeller-like push on the fluid.

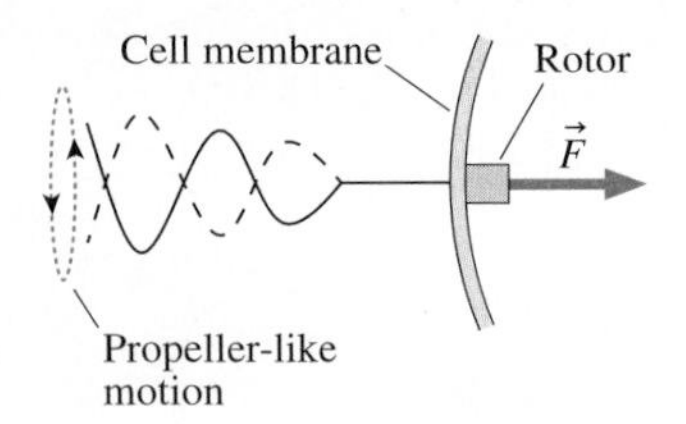

Microorganisms that need to move at very low Reynolds number had to evolve a different form of locomotion, their flagella and cilia. FIGURE 5.26 shows a simple model of a flagellum like the one a bacterium uses; it's basically a rotating corkscrew. The rotary motion—rather than the back-and-forth stroking motion of human swimming—enables the flagellum to apply a continuous push to the fluid. Thus the fluid, by Newton's third law, applies a continuous pushing force to the organism, and, as we've just seen, the organism almost instantaneously reaches its terminal speed and continues swimming at that constant speed.

Protozoa are propelled by cilia. The beating motion of cilia is more complex than the motion of a rotating flagellum, but the end result is the same: a constant push against the fluid. The essential point is that the physics of life at low Reynolds number dictates evolutionary constraints not faced by larger organisms.

STOP TO THINK 5.6 The terminal speed of a Styrofoam ball is 15 m/s. Suppose a Styrofoam ball is shot straight down with an initial speed of 30 m/s. Which velocity graph is correct?

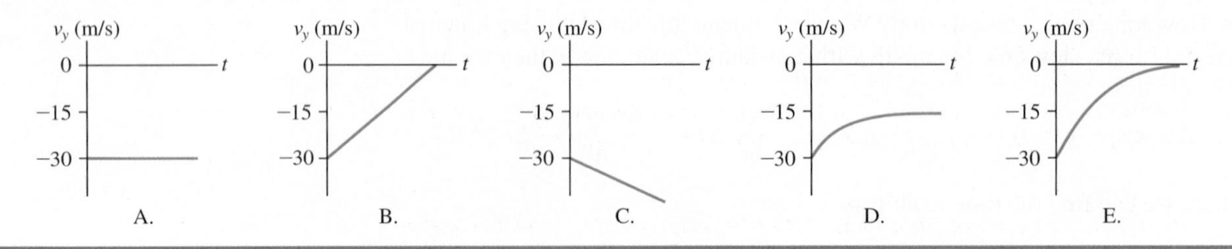

Catalog of Forces Revisited

Before we continue, let's summarize the details of the different forces that we've seen so far, extending the catalog of forces that we presented in ◀◀ SECTION 4.2.

SYNTHESIS 5.1 A catalog of forces

When solving mechanics problems, you'll often use the directions and details of the most common forces, outlined here.

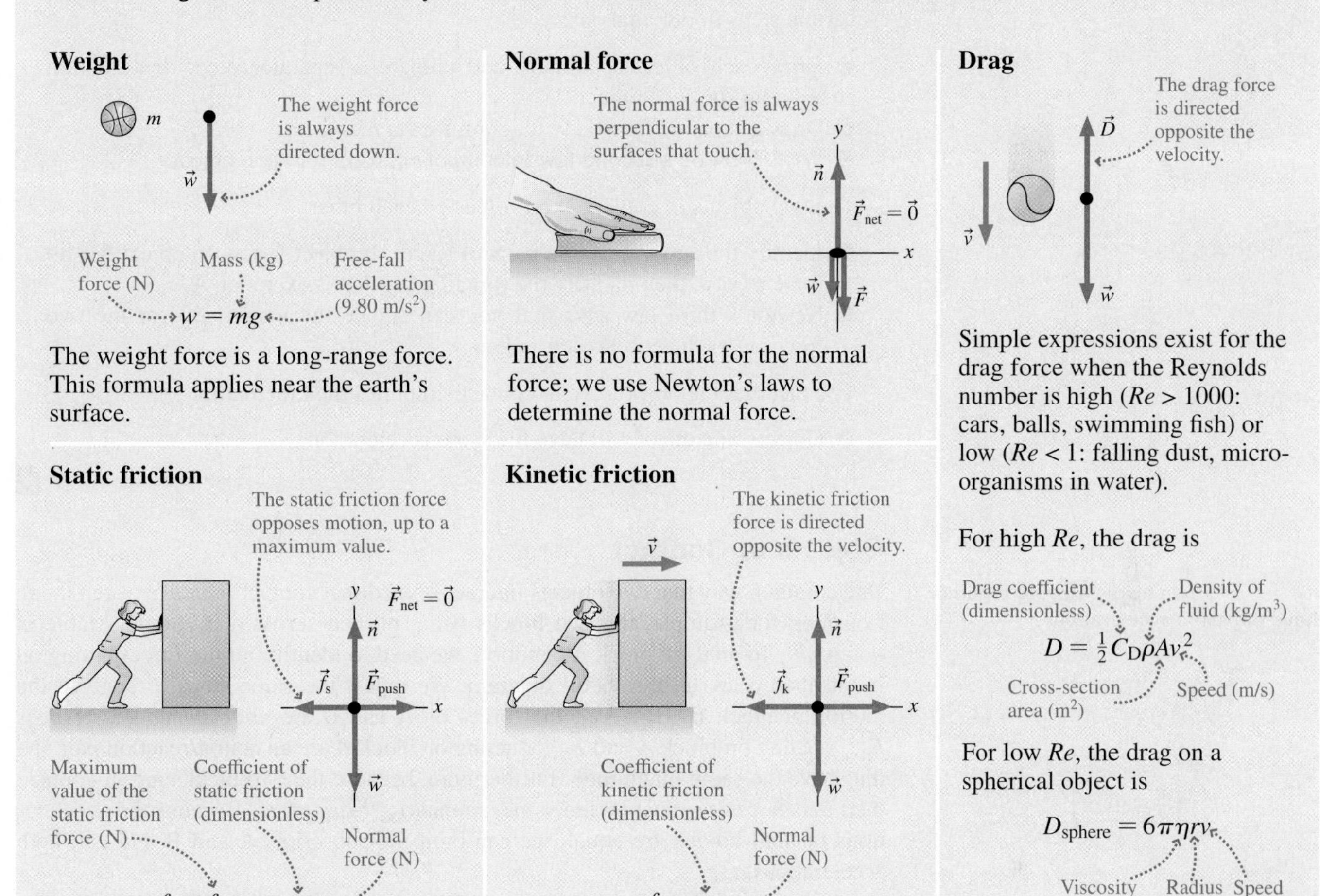

Weight
The weight force is always directed down.
Weight force (N)
Mass (kg)
Free-fall acceleration (9.80 m/s²)
w = mg
The weight force is a long-range force. This formula applies near the earth's surface.
Normal force
The normal force is always perpendicular to the surfaces that touch.
There is no formula for the normal force; we use Newton's laws to determine the normal force.
Drag
The drag force is directed opposite the velocity.
Simple expressions exist for the drag force when the Reynolds number is high (Re > 1000: cars, balls, swimming fish) or low (Re < 1: falling dust, microorganisms in water).
For high Re, the drag is
Drag coefficient (dimensionless)
Density of fluid (kg/m³)
D = ½C_DρAv²
Cross-section area (m²)
Speed (m/s)
For low Re, the drag on a spherical object is
D_sphere = 6πηrv
Viscosity (Pa·s)
Radius (m)
Speed (m/s)
Static friction
The static friction force opposes motion, up to a maximum value.
Maximum value of the static friction force (N)
Coefficient of static friction (dimensionless)
Normal force (N)
f_s ≤ f_s max = μ_s n
Kinetic friction
The kinetic friction force is directed opposite the velocity.
Coefficient of kinetic friction (dimensionless)
Normal force (N)
f_k = μ_k n

5.7 Interacting Objects

Up to this point we have studied the dynamics of a single object subject to forces exerted on it by other objects. In Example 5.11, for instance, the sofa was acted upon by friction, normal, weight, and pushing forces that came from the floor, the earth, and the person pushing. As we've seen, such problems can be solved by an application of Newton's second law after all the forces have been identified.

But in Chapter 4 we found that real-world motion often involves two or more objects interacting with each other. We further found that forces always come in action/reaction *pairs* that are related by Newton's third law. To remind you, Newton's third law states:

- Every force occurs as one member of an action/reaction pair of forces. The two members of the pair always act on *different* objects.
- The two members of an action/reaction pair point in *opposite* directions and are *equal* in magnitude.

Our goal in this section is to learn how to apply the second *and* third laws to interacting objects.

TACTICS BOX 5.2 Working with objects in contact

When two objects are in contact and their motion is linked, we need to duplicate certain steps in our analysis:

1. Draw each object separately and prepare a separate force identification diagram for *each* object.
2. Draw a separate free-body diagram for *each* object.
3. Write Newton's second law in component form for *each* object.

The two objects in contact exert forces on each other:

4. Identify the action/reaction pairs of forces. If object A acts on object B with force $\vec{F}_{\text{A on B}}$, then identify the force $\vec{F}_{\text{B on A}}$ that B exerts on A.
5. Newton's third law says that you can equate the magnitudes of the two forces in each action/reaction pair.

The fact that the objects are in contact simplifies the kinematics:

6. Objects in contact will have the same acceleration.

Exercises 29, 30, 31 

Objects in Contact

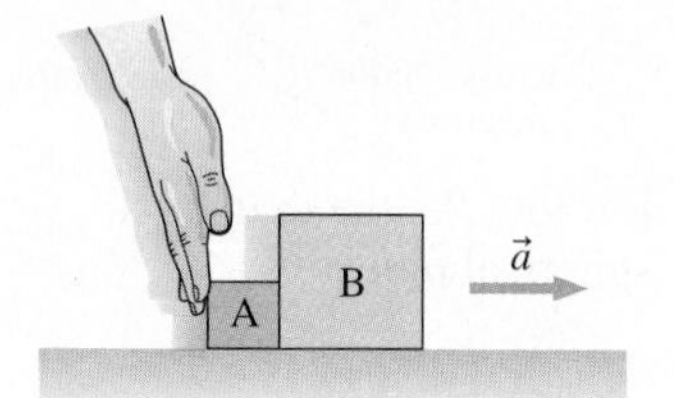

FIGURE 5.27 Two boxes moving together have the same acceleration.

One common way that two objects interact is via direct contact forces between them. Consider, for example, the two blocks being pushed across a frictionless table in FIGURE 5.27. To analyze block A's motion, we need to identify all the forces acting on it and then draw its free-body diagram. We repeat the same steps to analyze the motion of block B. However, the forces on A and B are *not* independent: Forces $\vec{F}_{\text{B on A}}$ acting on block A and $\vec{F}_{\text{A on B}}$ acting on block B are an action/reaction pair and thus have the same magnitude. Furthermore, because the two blocks are in contact, their *accelerations* must be the same, so that $a_{Ax} = a_{Bx} = a_x$. Because the accelerations of both blocks are equal, we can omit the subscripts A and B and call both accelerations a_x.

These observations suggest that we can't solve for the motion of one block without considering the motion of the other block. Solving a motion problem thus means solving two problems in parallel.

NOTE ▶ Two steps are especially important when you draw the free-body diagrams. First, draw a *separate* diagram for each object. They need not have the same coordinate system. Second, show only the forces acting *on* that object. The force $\vec{F}_{\text{A on B}}$ goes on the free-body diagram of object B, but $\vec{F}_{\text{B on A}}$ goes on the diagram of object A. The two members of an action/reaction pair *always* appear on two different free-body diagrams—*never* on the same diagram. ◀

EXAMPLE 5.18 Pushing two blocks

FIGURE 5.28 shows a 5.0 kg block A being pushed with a 3.0 N force. In front of this block is a 10 kg block B; the two blocks move together. What force does block A exert on block B?

STRATEGIZE The two blocks are in contact and are moving together, so we will follow the steps in Tactics Box 5.2.

FIGURE 5.28 Two blocks are pushed by a hand.

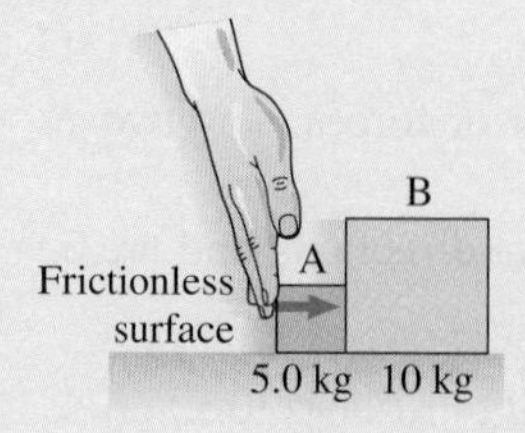

PREPARE The visual overview of FIGURE 5.29 lists the known information and identifies $F_{\text{A on B}}$ as what we're trying to find. Then, following the steps of Tactics Box 5.2, we've drawn *separate* force identification diagrams and *separate* free-body diagrams for the two blocks. Both blocks have a weight force and a normal force, so we've used subscripts A and B to distinguish between them.

The force $\vec{F}_{\text{A on B}}$ is the contact force that block A exerts on B; it forms an action/reaction pair with the force $\vec{F}_{\text{B on A}}$ that block B exerts on A. Notice that force $\vec{F}_{\text{A on B}}$ is drawn acting on block B; it is the force *of* A *on* B. **Force vectors are always drawn on the**

free-body diagram of the object that *experiences* the force, not the object exerting the force. Because action/reaction pairs act in opposite directions, force $\vec{F}_{\text{B on A}}$ pushes backward on block A and appears on A's free-body diagram.

SOLVE We begin by writing Newton's second law in component form for each block. Because the motion is only in the x-direction, we need only the x-component of the second law. For block A,

$$\sum F_x = (F_{\text{H}})_x + (F_{\text{B on A}})_x = m_{\text{A}}a_{\text{A}x}$$

The force components can be "read" from the free-body diagram, where we see $\vec{F}_{\text{H}}$ pointing to the right and $\vec{F}_{\text{B on A}}$ pointing to the left. Thus

$$F_{\text{H}} - F_{\text{B on A}} = m_{\text{A}}a_{\text{A}x}$$

For B, we have

$$\sum F_x = (F_{\text{A on B}})_x = F_{\text{A on B}} = m_{\text{B}}a_{\text{B}x}$$

We have two additional pieces of information: First, Newton's third law tells us that $F_{\text{B on A}} = F_{\text{A on B}}$. Second, the boxes are in contact and must have the same acceleration a_x; that is, $a_{\text{A}x} = a_{\text{B}x} = a_x$. With this information, the two x-component equations become

$$F_{\text{H}} - F_{\text{A on B}} = m_{\text{A}}a_x$$

$$F_{\text{A on B}} = m_{\text{B}}a_x$$

Our goal is to find $F_{\text{A on B}}$, so we need to eliminate the unknown acceleration a_x. From the second equation, $a_x = F_{\text{A on B}}/m_{\text{B}}$. Substituting this into the first equation gives

$$F_{\text{H}} - F_{\text{A on B}} = \frac{m_{\text{A}}}{m_{\text{B}}}F_{\text{A on B}}$$

This can be solved for the force of block A on block B, giving

$$F_{\text{A on B}} = \frac{F_{\text{H}}}{1 + m_{\text{A}}/m_{\text{B}}} = \frac{3.0\text{ N}}{1 + (5.0\text{ kg})/(10\text{ kg})} = \frac{3.0\text{ N}}{1.5} = 2.0\text{ N}$$

ASSESS Force F_{H} accelerates both blocks, a total mass of 15 kg, but force $F_{\text{A on B}}$ accelerates only block B, with a mass of 10 kg. Thus it makes sense that $F_{\text{A on B}} < F_{\text{H}}$.

FIGURE 5.29 A visual overview of the two blocks.

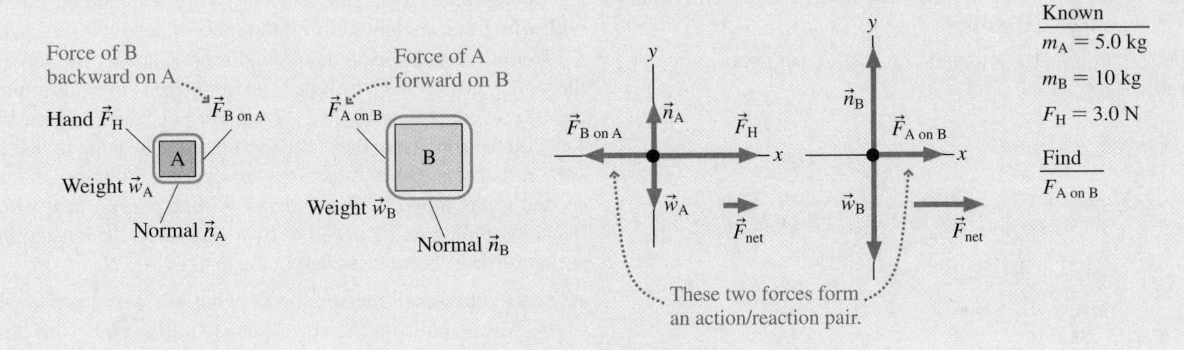

STOP TO THINK 5.7 Boxes P and Q are sliding to the right across a frictionless table. The hand H is slowing them down. The mass of P is larger than the mass of Q. Rank in order, from largest to smallest, the *horizontal* forces on P, Q, and H.

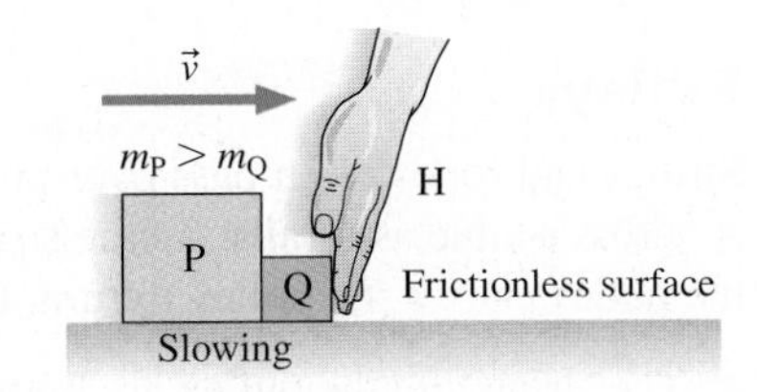

A. $F_{\text{Q on H}} = F_{\text{H on Q}} = F_{\text{P on Q}} = F_{\text{Q on P}}$
B. $F_{\text{Q on H}} = F_{\text{H on Q}} > F_{\text{P on Q}} = F_{\text{Q on P}}$
C. $F_{\text{Q on H}} = F_{\text{H on Q}} < F_{\text{P on Q}} = F_{\text{Q on P}}$
D. $F_{\text{H on Q}} = F_{\text{H on P}} > F_{\text{P on Q}}$

5.8 Ropes and Pulleys

Many objects are connected by strings, ropes, cables, and so on. We can learn several important facts about ropes and tension by considering the box being pulled by a rope in FIGURE 5.30. The rope in turn is being pulled by a hand that exerts a force $\vec{F}$ on the rope.

The box is pulled by the rope, so the box's free-body diagram shows a tension force $\vec{T}$. The *rope* is subject to two horizontal forces: the force $\vec{F}$ of the hand on the rope, and the force $\vec{F}_{\text{box on rope}}$ with which the box pulls back on the rope. In problems

FIGURE 5.30 A box being pulled by a rope.

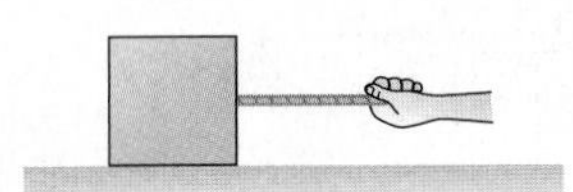

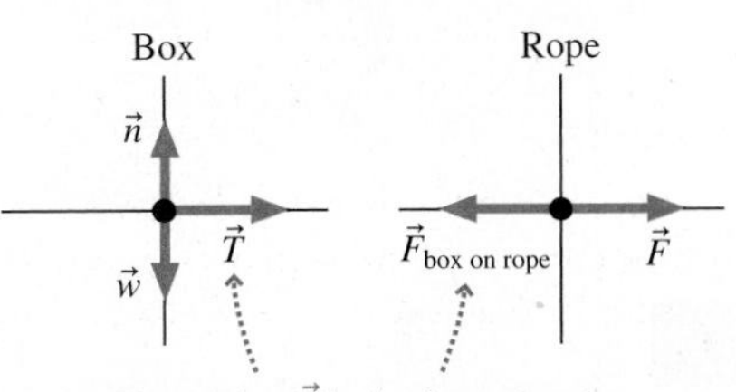

The tension $\vec{T}$ is the force that the rope exerts on the box. Thus $\vec{T}$ and $\vec{F}_{\text{box on rope}}$ are an action/reaction pair and have the same magnitude.

we'll consider, the mass of a string or rope is significantly less than the mass of the objects it pulls on, so we'll make the approximation—called the **massless string approximation**—that $m_{\text{rope}} = 0$. There is no weight force acting on the string, and so no supporting force is necessary. In this case, there are no forces that act along the vertical axis. $\vec{T}$ and $\vec{F}_{\text{box on rope}}$ form an action/reaction pair, so their magnitudes are equal: $F_{\text{box on rope}} = T$. Newton's second law *for the rope* is thus

$$\sum F_x = F - F_{\text{box on rope}} = F - T = m_{\text{rope}} a_x = 0 \tag{5.17}$$

We've used the approximation $m_{\text{rope}} = 0$. In this case, we can say that $T = F$. Generally, **the tension in a massless string or rope equals the magnitude of the force pulling on the end of the string or rope.** As a result:

- A massless string or rope "transmits" a force undiminished from one end to the other: If you pull on one end of a rope with force F, the other end of the rope pulls on what it's attached to with a force of the same magnitude F.
- The tension in a massless string or rope is the same from one end to the other.

CONCEPTUAL EXAMPLE 5.19 **Pulling a rope**

FIGURE 5.31a shows a student pulling horizontally with a 100 N force on a rope that is attached to a wall. In FIGURE 5.31b, two students in a tug-of-war pull on opposite ends of a rope with 100 N each. Is the tension in the second rope larger than, smaller than, or the same as that in the first?

FIGURE 5.31 Pulling on a rope in two ways. Which produces a larger tension?

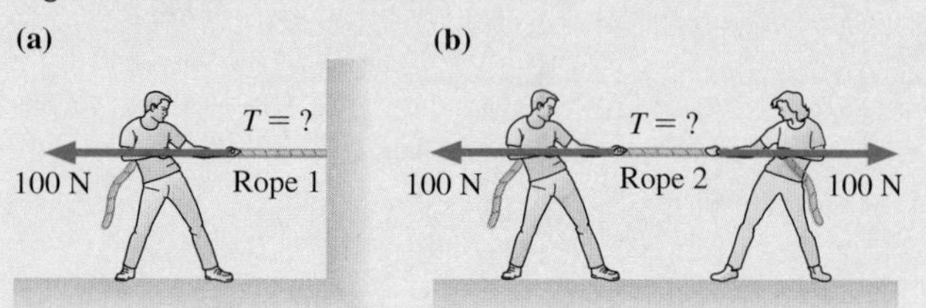

REASON Surely pulling on a rope from both ends causes more tension than pulling on one end. Right? Before jumping to conclusions, let's analyze the situation carefully. We found previously that the force pulling on the end of a rope—here, the 100 N force exerted by the student—and the tension in the rope have the same magnitude. Thus, the tension in rope 1 is 100 N, the force with which the student pulls on the rope.

To find the tension in the second rope, consider the force that the *wall* exerts on the *first* rope. The first rope is in equilibrium, so the 100 N force exerted by the student must be balanced by a 100 N force on the rope from the wall. The first rope is being pulled from *both* ends by a 100 N force—the exact same situation as for the second rope, pulled by the students. A rope doesn't care whether it's being pulled on by a wall or by a person, so the tension in the second rope is the *same* as that in the first, or 100 N.

ASSESS This example reinforces what we just learned about ropes: A rope pulls on the objects at each of its ends with a force equal in magnitude to the tension, and the external force applied to each end of the rope and the rope's tension have equal magnitude.

Pulleys

FIGURE 5.32 An ideal pulley changes the direction in which a tension force acts, but not its magnitude.

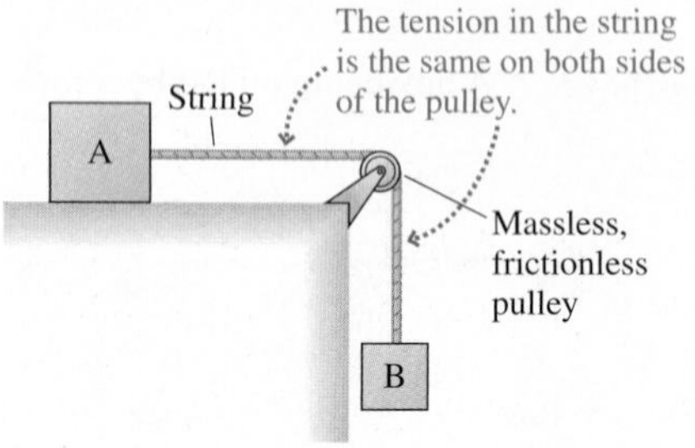

The tension in the string is the same on both sides of the pulley.
String
A
Massless, frictionless pulley
B

Strings and ropes often pass over pulleys. FIGURE 5.32 shows block B dragging block A across a table as it falls. As the string moves, static friction between the string and the pulley causes the pulley to turn. If we assume that

- The string *and* the pulley are both massless, and
- There is no friction where the pulley turns on its axle,

then no net force is needed to accelerate the string or turn the pulley. In this case, **the tension in a massless string is unchanged by passing over a massless, frictionless pulley.** We'll assume such an *ideal pulley* for problems in this chapter.

TACTICS BOX 5.3 **Working with ropes and pulleys**

For massless ropes or strings and massless, frictionless pulleys:

1. If a force pulls on one end of a rope, the tension in the rope equals the magnitude of the pulling force.
2. If two objects are connected by a rope, the tension is the same at both ends.
3. If the rope passes over a pulley, the tension in the rope is unaffected.

Exercises 33, 34 

Video Tension in String between Hanging Weights

EXAMPLE 5.20 Placing a leg in traction BIO

For serious fractures of the leg, the leg may need to have a stretching force applied to it to keep contracting leg muscles from forcing the broken bones together too hard. This is often done using *traction*, an arrangement of a rope, a weight, and pulleys as shown in FIGURE 5.33. The rope must make the same angle θ on both sides of the pulley so that the net force of the rope on the pulley is horizontally to the right, but θ can be adjusted to control the amount of traction. The doctor has specified 50 N of traction for this patient, with a 4.2 kg hanging mass. What is the proper angle θ?

FIGURE 5.33 A leg in traction.

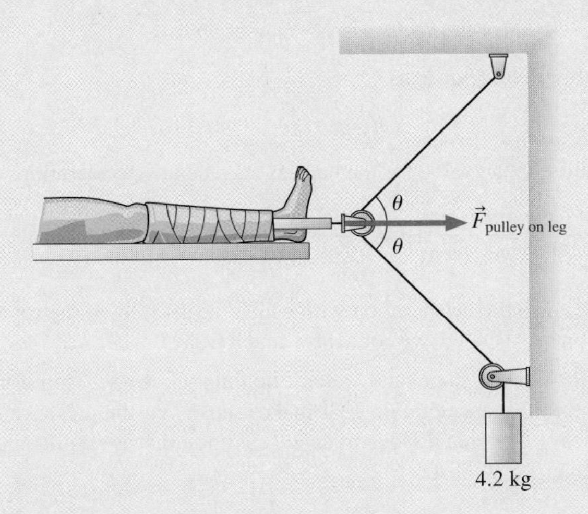
θ
θ
$\vec{F}_{\text{pulley on leg}}$
4.2 kg

STRATEGIZE We will find the tension in the rope from the weight that hangs from it. The rope pulls on the upper pulley with *two tension forces, pointing in the two directions in which the rope leaves the pulley.*

PREPARE The pulley attached to the patient's leg is in static equilibrium, so the net force on it must be zero. FIGURE 5.34 shows a free-body diagram for the pulley, which we assume to be frictionless. Forces $\vec{T}_1$ and $\vec{T}_2$ are the tension forces of the rope as it pulls on the pulley. These forces are equal in magnitude for a frictionless pulley, and their combined pull is to the right. This force is balanced by the force $\vec{F}_{\text{leg on pulley}}$ of the patient's leg pulling to the left. The traction force $\vec{F}_{\text{pulley on leg}}$ forms an action/reaction pair with $\vec{F}_{\text{leg on pulley}}$, so 50 N of traction means that $\vec{F}_{\text{leg on pulley}}$ also has a magnitude of 50 N.

FIGURE 5.34 Free-body diagram for the pulley.

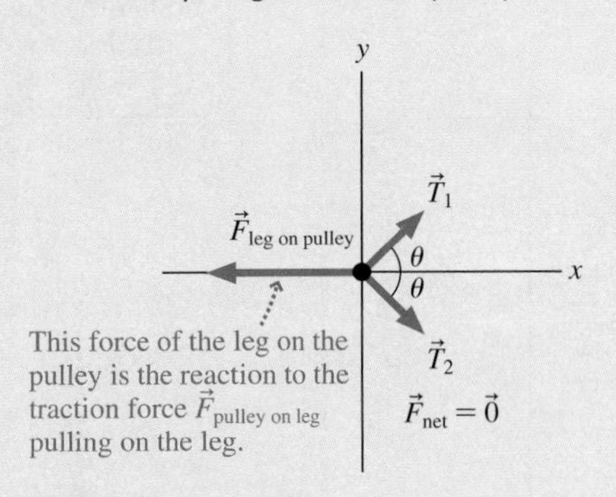
y
x
$\vec{T}_1$
$\vec{T}_2$
$\vec{F}_{\text{leg on pulley}}$
θ
θ
This force of the leg on the pulley is the reaction to the traction force $\vec{F}_{\text{pulley on leg}}$ pulling on the leg.
$\vec{F}_{\text{net}} = \vec{0}$

SOLVE Two important properties of ropes, given in Tactics Box 5.3, are that (1) the tension equals the magnitude of the force pulling on its end and (2) the tension is the same throughout the rope. Thus, if a hanging mass m pulls on the rope with its weight mg, the tension along the entire rope is $T = mg$. For a 4.2 kg hanging mass, the tension is then $T = mg = 41.2$ N.

The pulley, in equilibrium, must satisfy Newton's second law for the case where $\vec{a} = \vec{0}$. Thus

$$\sum F_x = T_{1x} + T_{2x} + (F_{\text{leg on pulley}})_x = ma_x = 0$$

The tension forces both have the same magnitude T, and both are at angle θ from horizontal. The x-component of the leg force is negative because it's directed to the left. Then Newton's law becomes

$$2T\cos\theta - F_{\text{leg on pulley}} = 0$$

so that

$$\cos\theta = \frac{F_{\text{leg on pulley}}}{2T} = \frac{50\text{ N}}{82.4\text{ N}} = 0.607$$

$$\theta = \cos^{-1}(0.607) = 53°$$

ASSESS The traction force would approach $2mg = 82$ N if angle θ approached zero because the two tensions would pull in parallel. Conversely, the traction force would approach 0 N if θ approached 90°. Because the desired traction force is roughly halfway between 0 N and 82 N, an angle near 45° is reasonable.

EXAMPLE 5.21 Lifting a stage set

A 200 kg set used in a play is stored in the loft above the stage. The rope holding the set passes up and over a pulley, then is tied backstage. The director tells a 100 kg stagehand to lower the set. When he unties the rope, the set falls and the unfortunate man is hoisted into the loft. What is the stagehand's acceleration?

STRATEGIZE The acceleration of each object—stagehand and set—is determined by the forces that act on it, so we will start by drawing a separate free-body diagram for each. From Tactics Box 5.3 we know that the tension force acting on each object has the same magnitude. In addition, their accelerations are related because they are connected by the rope.

PREPARE FIGURE 5.35 shows the visual overview. The objects of interest are the stagehand M and the set S, for which we've drawn separate free-body diagrams. Assume a massless rope and a massless, frictionless pulley. Tension forces $\vec{T}_S$ and $\vec{T}_M$ are due to a massless rope going over an ideal pulley, so their magnitudes are the same.

SOLVE From the two free-body diagrams, we can write Newton's second law in component form. For the man we have

$$\sum F_{My} = T_M - w_M = T_M - m_M g = m_M a_{My}$$

Continued

FIGURE 5.35 Visual overview for the stagehand and set.

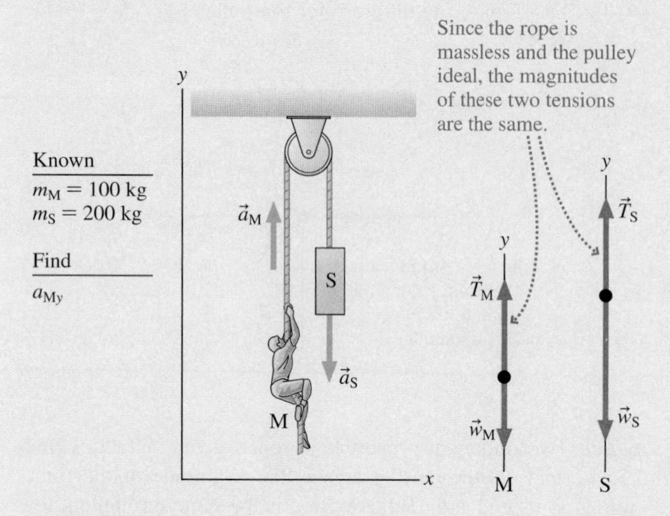

For the set we have

$$\sum F_{Sy} = T_S - w_S = T_S - m_S g = m_S a_{Sy}$$

Only the y-equations are needed. Because the stagehand and the set are connected by a rope, the upward distance traveled by one is the *same* as the downward distance traveled by the other. Thus the *magnitudes* of their accelerations must be the same, but, as Figure 5.35 shows, their *directions* are opposite. We can express this mathematically as $a_{Sy} = -a_{My}$. We also know that the two tension forces have equal magnitudes, which we'll call T. Inserting this information into the above equations gives

$$T - m_M g = m_M a_{My}$$
$$T - m_S g = -m_S a_{My}$$

These are simultaneous equations in the two unknowns T and a_{My}. We can solve for T in the first equation to get

$$T = m_M a_{My} + m_M g$$

Inserting this value of T into the second equation then gives

$$m_M a_{My} + m_M g - m_S g = -m_S a_{My}$$

which we can rewrite as

$$(m_S - m_M)g = (m_S + m_M)a_{My}$$

Finally, we can solve for the hapless stagehand's acceleration:

$$a_{My} = \frac{m_S - m_M}{m_S + m_M} g = \left(\frac{100 \text{ kg}}{300 \text{ kg}}\right) \times 9.80 \text{ m/s}^2 = 3.3 \text{ m/s}^2$$

This is also the acceleration with which the set falls. If the rope's tension was needed, we could now find it from $T = m_M a_{My} + m_M g$.

ASSESS If the stagehand weren't holding on, the set would fall with free-fall acceleration g. It makes sense that the presence of the heavy stagehand leads to an acceleration that is significantly less than g.

EXAMPLE 5.22 A not-so-clever bank robbery

Bank robbers have pushed a 1000 kg safe to a second-story floor-to-ceiling window. They plan to break the window, then lower the safe 3.0 m to their truck. Not being too clever, they stack up 500 kg of furniture, tie a rope between the safe and the furniture, and place the rope over a pulley. Then they push the safe out the window. What is the safe's speed when it hits the truck? The coefficient of kinetic friction between the furniture and the floor is 0.50.

STRATEGIZE We will again need to draw a separate free-body diagram for each object. The tension forces on the safe and the furniture are the same, and their accelerations are related because they are attached by the rope.

PREPARE The visual overview in FIGURE 5.36 establishes a coordinate system and defines the symbols that we need to calculate the safe's motion. The objects of interest are the safe S and the

FIGURE 5.36 Visual overview of the furniture and falling safe.

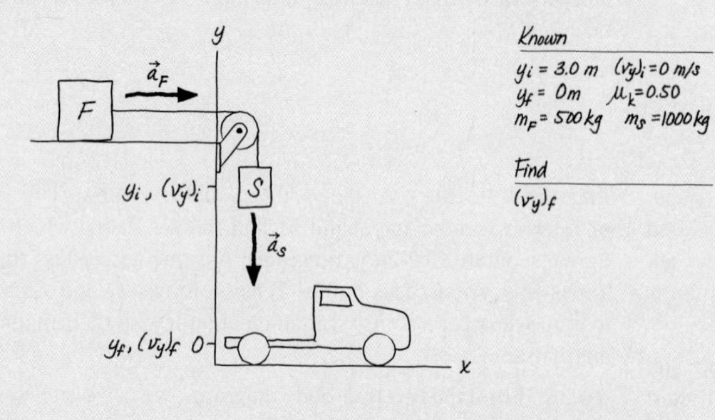

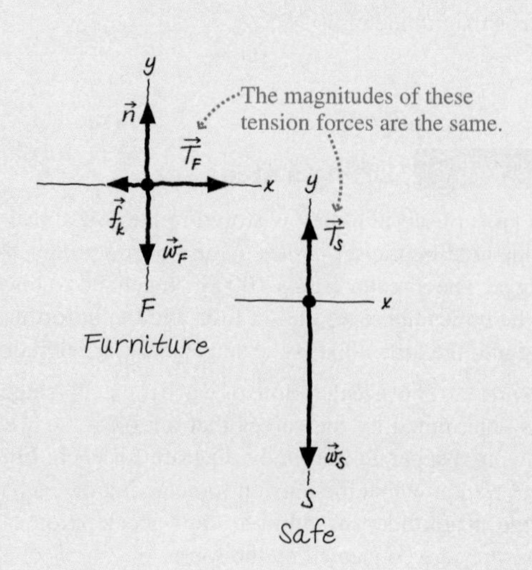

furniture F, which we model as particles. We assume a massless rope and a massless, frictionless pulley. The tension is then the same everywhere in the rope.

SOLVE We can write Newton's second law directly from the free-body diagrams. For the furniture,

$$\sum F_{Fx} = T_F - f_k = T - f_k = m_F a_{Fx}$$

$$\sum F_{Fy} = n - w_F = n - m_F g = 0$$

And for the safe,

$$\sum F_{Sy} = T_S - w_S = T - m_S g = m_S a_{Sy}$$

The safe and the furniture are tied together, so their accelerations have the same magnitude. But as the furniture slides to the right with positive acceleration a_{Fx}, the safe falls in the negative y-direction, so its acceleration a_{Sy} is negative; we can express this mathematically as $a_{Fx} = -a_{Sy}$. We also have made use of the fact that $T_S = T_F = T$. We have one additional piece of information, the model of kinetic friction:

$$f_k = \mu_k n = \mu_k m_F g$$

where we used the y-equation of the furniture to deduce that $n = m_F g$. We substitute this result for f_k into the x-equation of the furniture, then rewrite the furniture's x-equation and the safe's y-equation:

$$T - \mu_k m_F g = -m_F a_{Sy}$$

$$T - m_S g = m_S a_{Sy}$$

We have succeeded in reducing our knowledge to two simultaneous equations in the two unknowns a_{Sy} and T. We subtract the second equation from the first to eliminate T:

$$(m_S - \mu_k m_F)g = -(m_S + m_F)a_{Sy}$$

Finally, we can solve for the safe's acceleration:

$$a_{Sy} = -\left(\frac{m_S - \mu_k m_F}{m_S + m_F}\right)g$$

$$= -\frac{1000\text{ kg} - 0.5(500\text{ kg})}{1000\text{ kg} + 500\text{ kg}} \times 9.80\text{ m/s}^2 = -4.9\text{ m/s}^2$$

Now we need to calculate the kinematics of the falling safe. Because the time of the fall is not known or needed, we can use

$$(v_y)_f{}^2 = (v_y)_i{}^2 + 2a_{Sy}\,\Delta y = 0 + 2a_{Sy}(y_f - y_i) = -2a_{Sy}y_i$$

$$(v_y)_f = \sqrt{-2a_{Sy}y_i} = \sqrt{-2(-4.9\text{ m/s}^2)(3.0\text{ m})} = 5.4\text{ m/s}$$

The value of $(v_y)_f$ is negative, but we only needed to find the speed, so we took the absolute value. It seems unlikely that the truck will survive the impact of the 1000 kg safe!

ASSESS 5.4 m/s, about 12 mph, seems reasonable for a falling safe that is dragging heavy furniture.

Newton's three laws form the cornerstone of the science of mechanics. These laws allowed scientists to understand many diverse phenomena, from the motion of a raindrop to the orbits of the planets. We will continue to develop Newtonian mechanics in the next few chapters because of its tremendous importance to the physics of everyday life. But it's worth keeping in the back of your mind that Newton's laws aren't the ultimate statement about motion. Later in this text we'll reexamine motion and mechanics from the perspective of Einstein's theory of relativity.

STOP TO THINK 5.8 All three 50 kg blocks are at rest. Is the tension in rope 2 greater than, less than, or equal to the tension in rope 1?

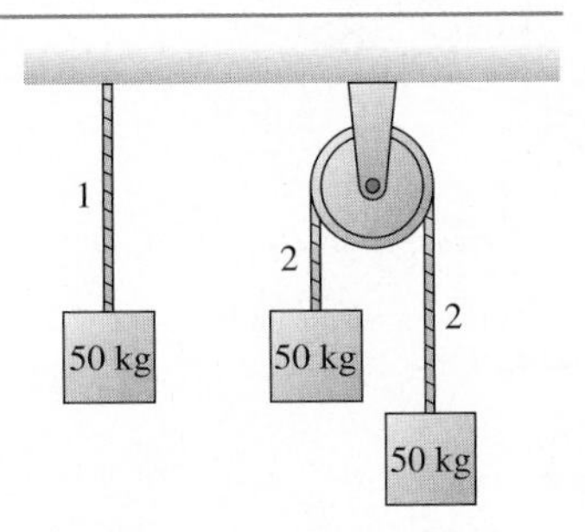

INTEGRATED EXAMPLE 5.23 Stopping distances of a skidding car

A 1500 kg car is traveling at a speed of 30 m/s when the driver slams on the brakes and skids to a halt. Determine the stopping distance if the car is traveling up a 10° slope, down a 10° slope, or on a level road.

STRATEGIZE We will find the acceleration from the known forces and an application of Newton's second law. Then we will use kinematics to find the stopping distance.

PREPARE We represent the car as a particle and we use the model of kinetic friction. We want to solve the problem only once, not three separate times, so we'll leave the slope angle θ unspecified until the end.

FIGURE 5.37 shows the visual overview. We show the car sliding uphill, but these representations work equally well for a level or downhill slide if we let θ be zero or negative, respectively. We use a tilted coordinate system so that the motion is along the x-axis. The car *skids* to a halt, so we use the coefficient of *kinetic* friction for rubber on concrete from Table 5.2.

SOLVE Newton's second law and the model of kinetic friction are

$$\sum F_x = n_x + w_x + (f_k)_x$$
$$= 0 - mg\sin\theta - f_k = ma_x$$
$$\sum F_y = n_y + w_y + (f_k)_y$$
$$= n - mg\cos\theta + 0 = ma_y = 0$$

We've written these equations by "reading" the motion diagram and the free-body diagram. Notice that both components of the weight vector $\vec{w}$ are negative. $a_y = 0$ because the motion is entirely along the x-axis.

The second equation gives $n = mg\cos\theta$. Using this in the friction model, we find $f_k = \mu_k mg\cos\theta$. Inserting this result back into the first equation then gives

$$ma_x = -mg\sin\theta - \mu_k mg\cos\theta$$
$$= -mg(\sin\theta + \mu_k\cos\theta)$$
$$a_x = -g(\sin\theta + \mu_k\cos\theta)$$

This is a constant acceleration. Constant-acceleration kinematics gives

$$(v_x)_f^2 = 0 = (v_x)_i^2 + 2a_x(x_f - x_i) = (v_x)_i^2 + 2a_x x_f$$

which we can solve for the stopping distance x_f:

$$x_f = -\frac{(v_x)_i^2}{2a_x} = \frac{(v_x)_i^2}{2g(\sin\theta + \mu_k\cos\theta)}$$

Notice how the minus sign in the expression for a_x canceled the minus sign in the expression for x_f. Evaluating our result at the three different angles gives the stopping distances:

$$x_f = \begin{cases} 48\text{ m} & \theta = 10° & \text{uphill} \\ 57\text{ m} & \theta = 0° & \text{level} \\ 75\text{ m} & \theta = -10° & \text{downhill} \end{cases}$$

The implications are clear about the danger of driving downhill too fast!

ASSESS 30 m/s ≈ 60 mph and 57 m ≈ 180 feet on a level surface. These are similar to the stopping distances you learned when you got your driver's license, so the results seem reasonable. Additional confirmation comes from noting that the expression for a_x becomes $-g\sin\theta$ if $\mu_k = 0$. This is what you learned in Chapter 3 for the acceleration on a frictionless inclined plane.

FIGURE 5.37 Visual overview for a skidding car.

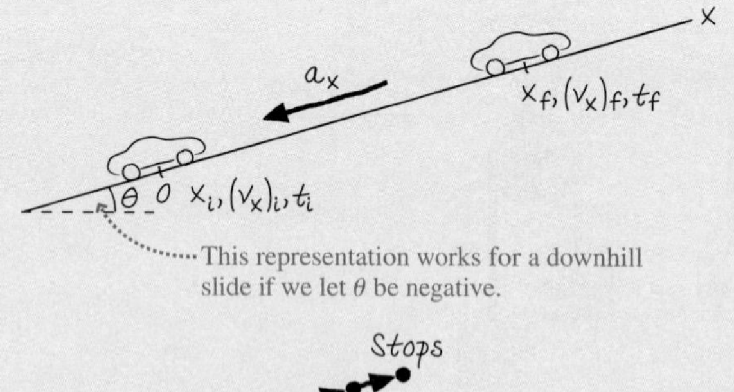

Known
$x_i = 0$ m, $t_i = 0$ s $\quad (v_x)_i = 30$ m/s
$m = 1500$ kg $\quad (v_x)_f = 0$ m/s
$\mu_k = 0.80$
$\theta = -10°, 0°, 10°$

Find
$\Delta x = x_f - x_i = x_f$

Stops

Normal $\vec{n}$
Weight $\vec{w}$ Friction $\vec{f}_k$

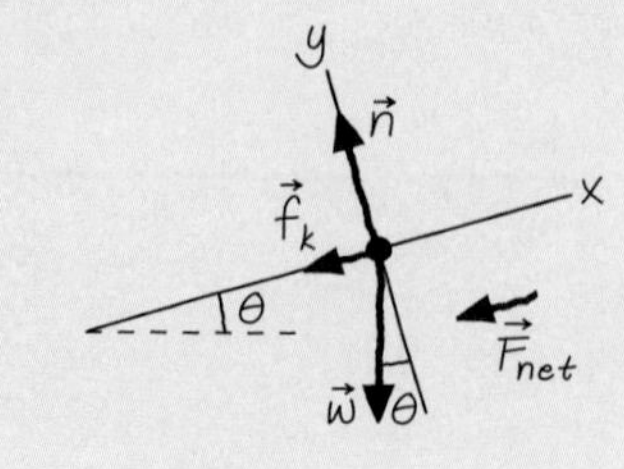

SUMMARY

GOAL To use Newton's laws to solve equilibrium and dynamics problems.

GENERAL STRATEGY

All examples in this chapter follow a four-part approach. You'll become a better problem solver if you adhere to it as you do the homework problems. The *Dynamics Worksheets* in the *Student Workbook* will help you structure your work in this way.

Equilibrium Problems

Object at rest or moving at constant velocity.

STRATEGIZE The net force acting on an object in equilibrium is zero.

PREPARE Make simplifying assumptions.

- Check that the object is either at rest or moving with constant velocity ($\vec{a} = \vec{0}$).
- Identify forces and show them on a free-body diagram.

SOLVE Use Newton's second law in component form:

$$\sum F_x = ma_x = 0$$
$$\sum F_y = ma_y = 0$$

"Read" the components from the free-body diagram.

ASSESS Is your result reasonable?

Dynamics Problems

Object accelerating.

STRATEGIZE Find the acceleration from the forces and Newton's second law, or use kinematics to find the acceleration and, from the second law, the forces.

PREPARE Make simplifying assumptions.

Make a **visual overview:**

- Sketch a pictorial representation.
- Identify known quantities and what the problem is trying to find.
- Identify all forces and show them on a free-body diagram.

SOLVE Use Newton's second law in component form:

$$\sum F_x = ma_x \quad \text{and} \quad \sum F_y = ma_y$$

"Read" the components of the vectors from the free-body diagram. If needed, use kinematics to find positions and velocities.

ASSESS Is your result reasonable?

IMPORTANT CONCEPTS

Specific information about three important forces:

Weight $\vec{w} = (mg, \text{downward})$

Friction $\vec{f}_s = (0 \text{ to } \mu_s n$, direction as necessary to prevent motion)

$\vec{f}_k = (\mu_k n$, direction opposite to the motion)

$\vec{f}_r = (\mu_r n$, direction opposite to the motion)

Drag Drag is directed opposite to the motion.

$D = \frac{1}{2} C_D \rho A v^2$, high *Re*

$D = 6\pi\eta r v$, sphere at low *Re*

Newton's laws are vector expressions. You must write them out by **components:**

$$(F_{net})_x = \sum F_x = ma_x$$
$$(F_{net})_y = \sum F_y = ma_y$$

For equilibrium problems, $a_x = 0$ and $a_y = 0$.

Objects in contact

When two objects interact, you need to draw two separate free-body diagrams.

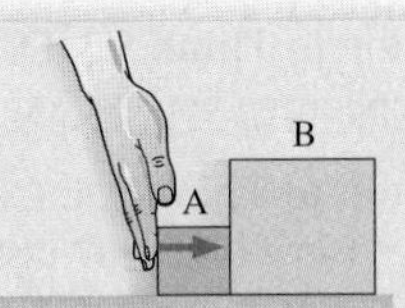

The action/reaction pairs of forces have equal magnitude and opposite directions.

APPLICATIONS

Apparent weight is the magnitude of the contact force supporting an object. It is what a scale reads, and it is your sensation of weight.

Apparent weight equals your true weight $w = mg$ only when the vertical acceleration is zero.

A falling object reaches **terminal speed** when the drag force exactly balances the weight force: $\vec{a} = \vec{0}$.

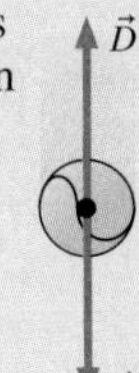

Strings and pulleys

- A string or rope pulls what it's connected to with a force equal to its tension.
- The tension in a rope is equal to the force pulling on the rope.
- The tension in a massless rope is the same at all points in the rope.
- Tension does not change when a rope passes over a massless, frictionless pulley.

$F_{\text{rope on wall}}$ = tension

$F_{\text{hand on rope}}$ = tension

Learning Objectives After studying this chapter, you should be able to:

- Solve problems about objects in equilibrium. *Conceptual Questions 5.1, 5.2; Problems 5.1, 5.2, 5.3, 5.6, 5.7*
- Use free-body diagrams, Newton's second law, and the problem-solving approach to solve dynamics problems. *Conceptual Question 5.6; Problems 5.10, 5.11, 5.12, 5.13, 5.14*
- Work with and distinguish between mass and weight. *Conceptual Questions 5.5, 5.13; Problems 5.16, 5.17, 5.18, 5.19, 5.20*
- Solve problems with sliding and rolling friction; understand how static friction can prevent motion. *Conceptual Questions 5.3, 5.4; Problems 5.24, 5.27, 5.28, 5.32, 5.33*
- Use the linear and quadratic models of drag to solve problems about motion through a fluid and to calculate terminal speeds. *Conceptual Questions 5.8, 5.15; Problems 5.35, 5.36, 5.37, 5.38, 5.40*
- Use Newton's third law to identify forces on and to solve problems about interacting objects. *Problems 5.41, 5.42, 5.43*
- Calculate the tension in ropes and solve problems involving ropes and pulleys. *Conceptual Questions 5.17, 5.20; Problems 5.44, 5.45, 5.46, 5.47, 5.48*

STOP TO THINK ANSWERS

Chapter Preview Stop to Think: B. The elevator is moving at a steady speed, so $a = 0$ and therefore $F_{net} = 0$. There are two forces that act: the downward weight force and the upward tension force. Since $F_{net} = 0$, the magnitudes of these two forces must be equal.

Stop to Think 5.1: A. The ball is stationary, so the net force on it must be zero. Newton's second law for the vertical components of forces is $\sum F_y = T_y - w = 0$. The vertical component of the tension by itself is equal to w, so the tension must be greater than w.

Stop to Think 5.2: A. The lander is descending and slowing. The acceleration vector points upward, and so $\vec{F}_{net}$ points upward. This can be true only if the thrust has a larger magnitude than the weight.

Stop to Think 5.3: D. When you are in the air, there is no contact force supporting you, so your apparent weight is zero: You are weightless.

Stop to Think 5.4: C. The cyclist moves at a constant speed, so the net force is zero. We set the axes parallel to and perpendicular to the slope and compute the components of the weight force as in Figure 5.11. Newton's second law for the vertical components of forces is $\sum F_y = n - w \cos 20° = 0$. For this to be true, it's clear that $n < w$.

Stop to Think 5.5: $f_B > f_C = f_D = f_E > f_A$. Situations C, D, and E are all kinetic friction, which does not depend on either velocity or acceleration. Kinetic friction is less than the maximum static friction that is exerted in B. $f_A = 0$ because no friction is needed to keep the object at rest.

Stop to Think 5.6: D. The ball is shot *down* at 30 m/s, so $v_{0y} = -30$ m/s. This exceeds the terminal speed, so the upward drag force is *greater* than the downward weight force. Thus the ball *slows down* even though it is "falling." It will slow until $v_y = -15$ m/s, the terminal velocity, then maintain that velocity.

Stop to Think 5.7: B. $F_{Q\,on\,H} = F_{H\,on\,Q}$ and $F_{P\,on\,Q} = F_{Q\,on\,P}$ because these are action/reaction pairs. Box Q is slowing down and therefore must have a net force to the left. So from Newton's second law we also know that $F_{H\,on\,Q} > F_{P\,on\,Q}$.

Stop to Think 5.8: Equal to. Each block is hanging in equilibrium, with no net force, so the upward tension force is mg.

Video Tutor Solution Chapter 5

QUESTIONS

Conceptual Questions

1. An object is subject to two forces that do not point in opposite directions. Is it possible to choose their magnitudes so that the object is in equilibrium? Explain.
2. Are the objects described here in static equilibrium, dynamic equilibrium, or not in equilibrium at all?
   a. A girder is lifted at constant speed by a crane.
   b. A girder is lowered by a crane. It is slowing down.
   c. You're straining to hold a 200 lb barbell over your head.
   d. A jet plane has reached its cruising speed and altitude.
   e. A rock is falling into the Grand Canyon.
   f. A box in the back of a truck doesn't slide as the truck stops.
3. Boxes A and B in Figure Q5.3 both remain at rest. Is the friction force on A larger than, smaller than, or equal to the friction force on B? Explain.

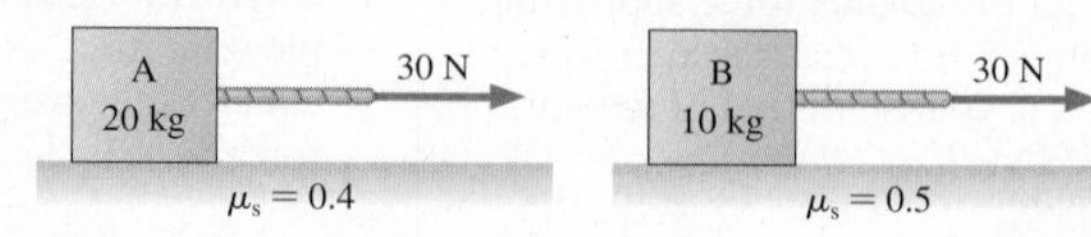

FIGURE Q5.3

Problem difficulty is labeled as | (straightforward) to ||||| (challenging). INT Problems labeled integrate significant material from earlier chapters; BIO are of biological or medical interest.

The eText icon indicates when there is a video tutor solution available for the chapter or for a specific problem. To launch these videos, log into your eText through Mastering™ Physics or log into the Study Area.

4. Boxes A and B in Figure Q5.3 are both sliding. Is the friction force on A larger than, smaller than, or equal to the friction force on B? Explain.
5. An astronaut takes his bathroom scale to the moon and then stands on it. Is the reading of the scale his true weight? Explain.
6. A light block of mass m and a heavy block of mass M are attached to the ends of a rope. A student holds the heavier block and lets the lighter block hang below it, as shown in Figure Q5.6. Then she lets go. Air resistance can be neglected.
   a. What is the tension in the rope while the blocks are falling, before either hits the ground?
   b. Would your answer be different if she had been holding the lighter block initially?

FIGURE Q5.6

7. a. Can the normal force on an object be directed horizontally? If not, why not? If so, provide an example.
   b. Can the normal force on an object be directed downward? If not, why not? If so, provide an example.
8. A ball is thrown straight up. Taking the drag force of air into account, does it take longer for the ball to travel to the top of its motion or for it to fall back down again?
9. You are going sledding with your friends, sliding down a snowy hill. Friction can't be ignored. Riding solo on your sled, you have a certain acceleration. Would the acceleration change if you let a friend ride with you, increasing the mass? Explain.
10. Suppose you are holding a box in front of you and away from your body by squeezing the sides, as shown in Figure Q5.10. Draw a free-body diagram showing all of the forces on the box. What is the force that is holding the box up, the force that is opposite the weight force?

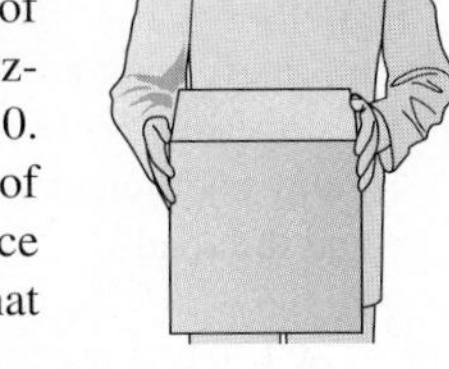
FIGURE Q5.10

11. You are walking up an icy slope. Suddenly your feet slip, and you start to slide backward. Will you slide at a constant speed, or will you accelerate?
12. Three objects move through the air as shown in Figure Q5.12. Rank in order, from largest to smallest, the three drag forces D_1, D_2, and D_3. Some may be equal. Give your answer in the form $A < B = C$ and state your reasoning.

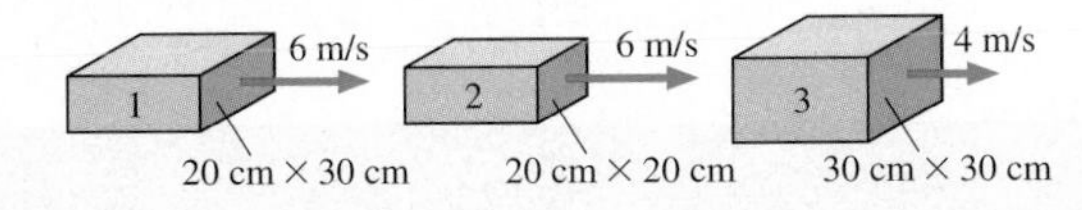

FIGURE Q5.12

13. A trampolinist flies straight up into the air, then back down onto the trampoline, where she comes instantaneously to rest when the trampoline is fully stretched. Rank in order, from largest to smallest, her apparent weight
   a. When she is instantaneously at rest.
   b. Just before she leaves the trampoline.
   c. When she is at the top of her flight.
14. Raindrops can fall at different speeds; some fall quite quickly, others quite slowly. Why might this be true?
15. An airplane moves through the air at a constant speed. The engines' thrust applies a force in the direction of motion, and this force is equal in magnitude and opposite in direction to the drag force. Reducing thrust will cause the plane to fly at a slower—but still constant—speed. Explain why this is so.
16. Is it possible for an object to travel in air faster than its terminal speed? If not, why not? If so, explain how this might happen.

For Questions 17 through 20, determine the tension in the rope at the point indicated with a dot.

- All objects are at rest.
- The strings and pulleys are massless, and the pulleys are frictionless.

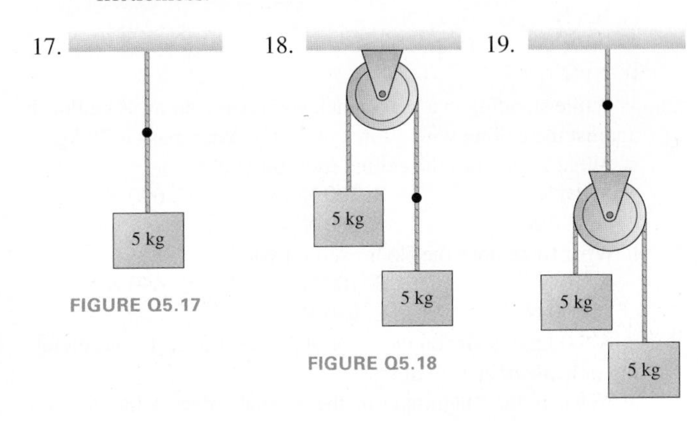

FIGURE Q5.17

FIGURE Q5.18

FIGURE Q5.19

20.

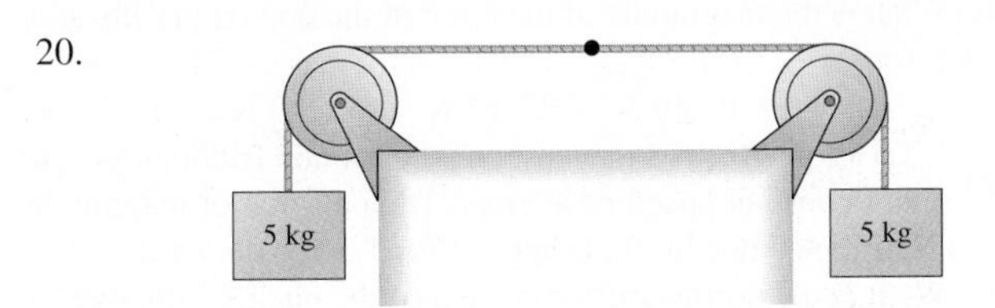

FIGURE Q5.20

21. In Figure Q5.21, block 2 is moving to the right. There is no friction between the floor and block 2, but there is a friction force between blocks 1 and 2. In which direction is the kinetic friction force on block 1? On block 2? Explain.

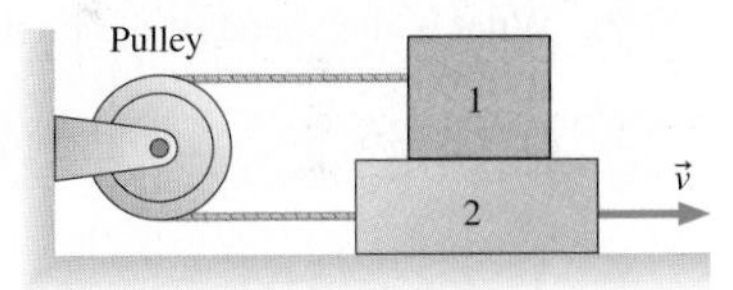

FIGURE Q5.21

Multiple-Choice Questions

22. || The wood block in Figure Q5.22 is at rest on a wood ramp. In which direction is the static friction force on block 1?
   A. Up the slope.
   B. Down the slope.
   C. The friction force is zero.
   D. There's not enough information to tell.

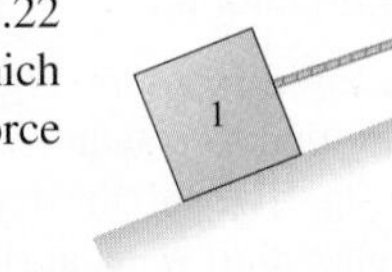

FIGURE Q5.22

23. || Wood block 1 in Figure Q5.22, which has a mass of 1.0 kg, is at rest on a wood ramp. The angle of the ramp is 20° above horizontal. What is the smallest mass of block 2 that will start block 1 sliding uphill?
   a. 0.35 kg b. 0.83 kg c. 12. kg
   d. 1.9 kg e. 2.5 kg

24. || A bicyclist rides down a hill with a 3.0° slope; he attains a terminal speed of 15.0 m/s. What terminal speed would he reach if he went down a 6.0° slope?
a. 17 m/s b. 21 m/s c. 26 m/s
d. 30 m/s e. 33 m/s

25. || A 2.0 kg ball is suspended by two light strings as shown in Figure Q5.25 . What is the tension T in the angled string?
A. 9.5 N B. 15 N
C. 20 N D. 26 N
E. 30 N

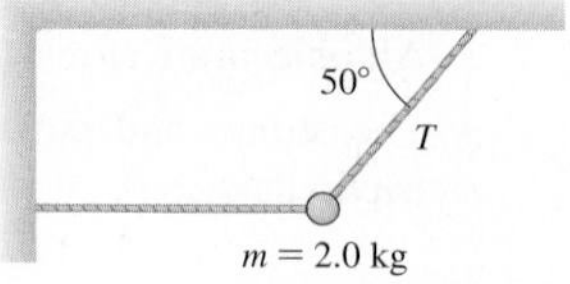

50° T $m = 2.0$ kg

FIGURE Q5.25

26. | While standing in a low tunnel, you raise your arms and push against the ceiling with a force of 100 N. Your mass is 70 kg.
a. What force does the ceiling exert on you?
A. 10 N B. 100 N C. 690 N
D. 790 N E. 980 N
b. What force does the floor exert on you?
A. 10 N B. 100 N C. 690 N
D. 790 N E. 980 N

27. | A 5.0 kg dog sits on the floor of an elevator that is accelerating *downward* at 1.20 m/s^2.
a. What is the magnitude of the normal force of the elevator floor on the dog?
A. 34 N B. 43 N C. 49 N D. 55 N E. 74 N
b. What is the magnitude of the force of the dog on the elevator floor?
A. 4.2 N B. 49 N C. 55 N D. 43 N E. 74 N

28. | A 3.0 kg puck slides due east on a horizontal frictionless surface at a constant speed of 4.5 m/s. Then a force of magnitude 6.0 N, directed due north, is applied for 1.5 s. Afterward,
a. What is the northward component of the puck's velocity?
A. 0.50 m/s B. 2.0 m/s C. 3.0 m/s
D. 4.0 m/s E. 4.5 m/s
b. What is the speed of the puck?
A. 4.9 m/s B. 5.4 m/s C. 6.2 m/s
D. 7.5 m/s E. 11 m/s

29. | Eric has a mass of 60 kg. He is standing on a scale in an elevator that is accelerating downward at 1.7 m/s^2. What is the approximate reading on the scale?
A. 0 N B. 400 N C. 500 N D. 600 N

30. || The two blocks in Figure Q5.30 are at rest on frictionless surfaces. What must be the mass of the right block in order that the two blocks remain stationary?
A. 4.9 kg B. 6.1 kg C. 7.9 kg
D. 9.8 kg E. 12 kg

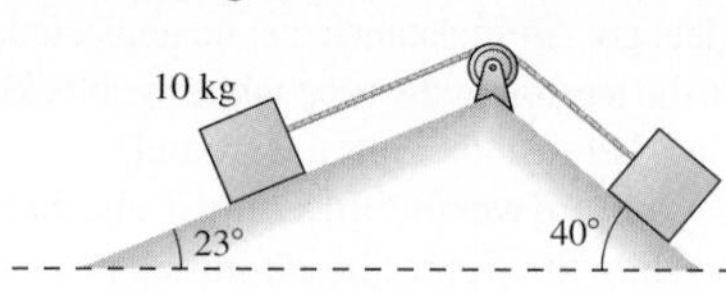

FIGURE Q5.30

31. | A football player at practice pushes a 60 kg blocking sled across the field at a constant speed. The coefficient of kinetic friction between the grass and the sled is 0.30. How much force must he apply to the sled?
A. 18 N B. 60 N C. 180 N D. 600 N

32. | Two football players are pushing a 60 kg blocking sled across the field at a constant speed of 2.0 m/s. The coefficient of kinetic friction between the grass and the sled is 0.30. Once they stop pushing, how far will the sled slide before coming to rest?
A. 0.20 m B. 0.68 m C. 1.0 m D. 6.6 m

33. || Land Rover ads used to claim that their vehicles could climb a slope of 45°. For this to be possible, what must be the minimum coefficient of static friction between the vehicle's tires and the road?
A. 0.5 B. 0.7 C. 0.9 D. 1.0

34. || A truck is traveling at 30 m/s on a slippery road. The driver slams on the brakes and the truck starts to skid. If the coefficient of kinetic friction between the tires and the road is 0.20, how far will the truck skid before stopping?
A. 230 m B. 300 m C. 450 m D. 680 m

PROBLEMS

Section 5.1 Equilibrium

1. | The three ropes in Figure P5.1 are tied to a small, very light ring. Two of the ropes are anchored to walls at right angles, and the third rope pulls as shown. What are T_1 and T_2, the magnitudes of the tension forces in the first two ropes?

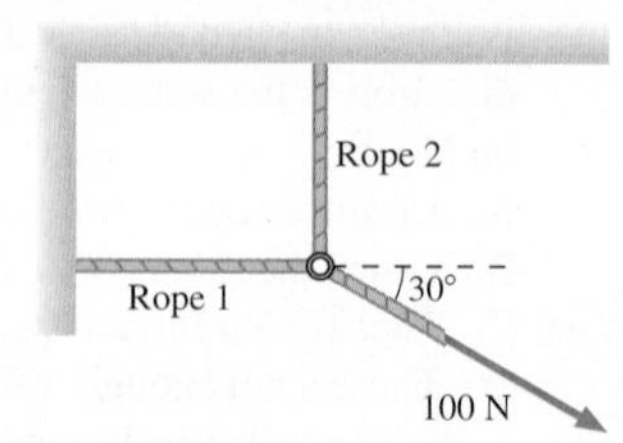

FIGURE P5.1

2. || In the sport of parasailing, a person is attached to a rope being pulled by a boat while hanging from a parachute-like sail. A rider is towed at a constant speed by a rope that is at an angle of 15° from horizontal. The tension in the rope is 2300 N. The force of the sail on the rider is 30° from horizontal. What is the weight of the rider?

3. || In the Skycoaster amusement park ride, riders are suspended from a tower by a long cable. A second cable then lifts them until they reach the starting position indicated in Figure P5.3. The lifting cable is then released, and the riders swing down the circular arc shown. If the four riders have a total mass of 270 kg, what are the tensions in the two cables just before release?

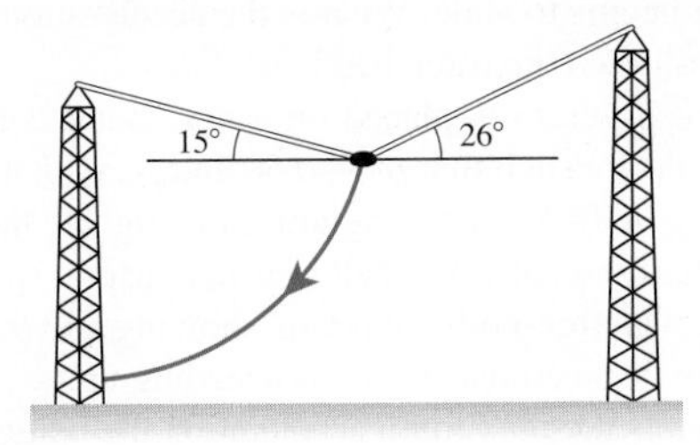

FIGURE P5.3

4. |||| A construction crew would like to support a 1000 kg steel beam with two angled ropes as shown in Figure P5.4. Their rope can support a maximum tension of 5600 N. Is this rope strong enough to do the job?

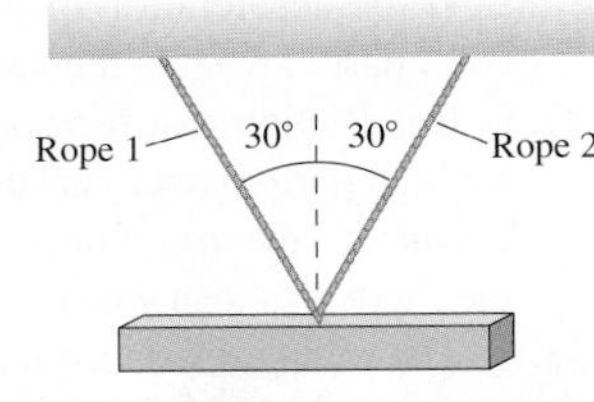

FIGURE P5.4

5. |||| When you bend your knee, the quadriceps muscle is stretched. This increases the tension in the quadriceps tendon attached to your kneecap (patella), which, in turn, increases the tension in the patella tendon that attaches your kneecap to your lower leg bone (tibia). Simultaneously, the end of your upper leg bone (femur) pushes outward on the patella. Figure P5.5 shows how these parts of a knee joint are arranged. What size force does the femur exert on the kneecap if the tendons are oriented as in the figure and the tension in each tendon is 60 N?

BIO

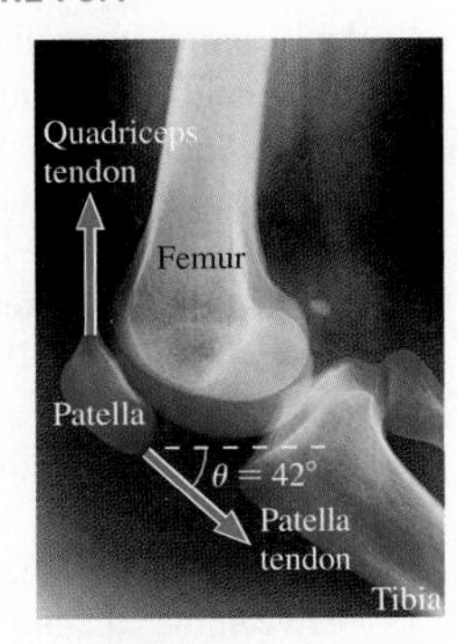

FIGURE P5.5

6. || An early submersible craft for deep-sea exploration was raised and lowered by a cable from a ship. When the craft was stationary, the tension in the cable was 6000 N. When the craft was lowered or raised at a steady rate, the motion through the water added an 1800 N drag force.
   a. What was the tension in the cable when the craft was being lowered to the seafloor?
   b. What was the tension in the cable when the craft was being raised from the seafloor?

7. || Bethany, who weighs 560 N, lies in a hammock suspended by ropes tied to two trees. One rope makes an angle of 45° with the ground; the other makes an angle of 30°. Find the tension in each of the ropes.

8. || A 65 kg student is walking on a slackline, a length of webbing stretched between two trees. The line stretches and so has a noticeable sag, as shown in Figure P5.8. At the point where his foot touches the line, the rope applies a tension force in each direction, as shown. What is the tension in the line?

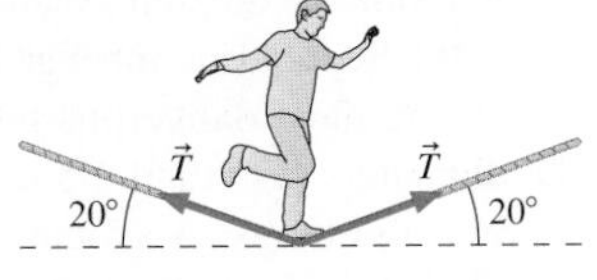

FIGURE P5.8

Section 5.2 Dynamics and Newton's Second Law

9. ||| In the winter sport of curling, two teams alternate sliding 20 kg stones on an icy surface in an attempt to end up with the stone closest to the center of a target painted on the ice. During one turn, a player releases a stone that travels 27.9 m before coming to rest. The friction force acting on the stone is 2.0 N. What was the speed of the stone when the player released it?

INT

10. || The forces in Figure P5.10 are acting on a 2.0 kg object. What is a_x, the x-component of the object's acceleration?

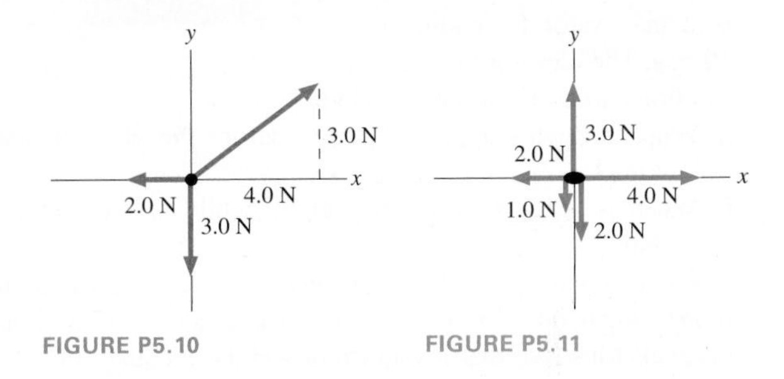

FIGURE P5.10 FIGURE P5.11

11. | The forces in Figure P5.11 are acting on a 2.0 kg object. Find the values of a_x and a_y, the x- and y-components of the object's acceleration.

12. | A horizontal rope is tied to a 50 kg box on frictionless ice. What is the tension in the rope if
   a. The box is at rest?
   b. The box moves at a steady 5.0 m/s?
   c. The box has $v_x = 5.0$ m/s and $a_x = 5.0$ m/s^2?

13. ||| The acceleration of a baseball pitcher's hand as he delivers a pitch is extreme. For a professional player, this acceleration phase lasts only 50 ms, during which the ball's speed increases from 0 to about 90 mph, or 40 m/s. What is the force of the pitcher's hand on the 0.145 kg ball during this acceleration phase?

14. || In a head-on collision, a car stops in 0.10 s from a speed of 14 m/s. The driver has a mass of 70 kg, and is, fortunately, tightly strapped into his seat. What force is applied to the driver by his seat belt during that fraction of a second?

Section 5.3 Mass and Weight

15. | An astronaut's weight on earth is 800 N. What is his weight on Mars, where $g = 3.76$ m/s^2?

16. | A woman has a mass of 55.0 kg.
   a. What is her weight on earth?
   b. What are her mass and her weight on the moon, where $g = 1.62$ m/s^2?

17. || The acceleration of the spacecraft in which the Apollo astronauts took off from the moon was 3.4 m/s^2. On the moon, $g = 1.6$ m/s^2. What was the apparent weight of a 75 kg astronaut during takeoff?

18. || a. How much force does an 80 kg astronaut exert on his chair while sitting at rest on the launch pad?
   b. How much force does the astronaut exert on his chair while accelerating straight up at 10 m/s^2?

Watch Video Solution Problems 5.19, 5.28, and 5.31

19. | It takes the elevator in a skyscraper 4.0 s to reach its cruising speed of 10 m/s. A 60 kg passenger gets aboard on the ground floor. What is the passenger's apparent weight
   a. Before the elevator starts moving?
   b. While the elevator is speeding up?
   c. After the elevator reaches its cruising speed?
20. || Riders on the Power Tower are launched skyward with an acceleration of $4g$, after which they experience a period of free fall. What is a 60 kg rider's apparent weight
   a. During the launch?
   b. During the period of free fall?

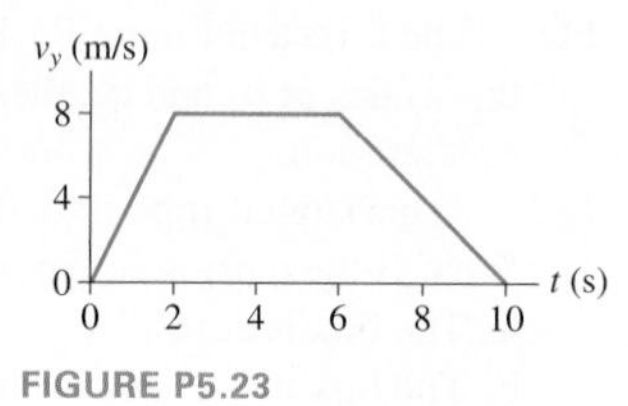

21. || Zach, whose mass is 80 kg, is in an elevator descending at 10 m/s. The elevator takes 3.0 s to brake to a stop at the first floor.
   a. What is Zach's apparent weight before the elevator starts braking?
   b. What is Zach's apparent weight while the elevator is braking?
22. ||| An astronaut lifts off in a rocket from the surface of the moon, where $g = 1.6\ \text{m/s}^2$. What vertical acceleration should his rocket have so that his apparent weight is equal to his true weight on earth?
23. ||| Figure P5.23 shows the velocity graph of a 75 kg passenger in an elevator. What is the passenger's apparent weight at $t = 1.0$ s? At 5.0 s? At 9.0 s?
INT

FIGURE P5.23

Section 5.4 Normal Forces

24. | Mountain goats can easily scale slopes angled at 60° from horizontal. What are the normal force and the static friction force acting on a mountain goat that weighs 900 N and is standing on such a slope?
BIO
25. ||| A 23 kg child goes down a straight slide inclined 38° above horizontal. The child is acted on by his weight, the normal force from the slide, and kinetic friction.
   a. Draw a free-body diagram of the child.
   b. How large is the normal force of the slide on the child?

Section 5.5 Friction

26. ||| A crate pushed along the floor with velocity $\vec{v}_i$ slides a distance d after the pushing force is removed.
   a. If the mass of the crate is doubled but the initial velocity is not changed, what distance does the crate slide before stopping? Explain.
   b. If the initial velocity of the crate is doubled to $2\vec{v}_i$ but the mass is not changed, what distance does the crate slide before stopping? Explain.
27. || Two workers are sliding a 300 kg crate across the floor. One worker pushes forward on the crate with a force of 380 N while the other pulls in the same direction with a force of 350 N using a rope connected to the crate. Both forces are horizontal, and the crate slides with a constant speed. What is the crate's coefficient of kinetic friction on the floor?
28. ||| A 4000 kg truck is parked on a 7.0° slope. How big is the friction force on the truck?
29. ||| A 1000 kg car traveling at a speed of 40 m/s skids to a halt on wet concrete where $\mu_k = 0.60$. How long are the skid marks?
30. ||| A dump truck, whose bed is made of steel, holds an old steel watering trough. The bed of the truck is slowly raised until the trough begins to slide. What is the acceleration of the trough as it slides down the truck bed?
31. ||| A 10 kg crate is placed on a horizontal conveyor belt. The materials are such that $\mu_s = 0.50$ and $\mu_k = 0.30$.
   a. Draw a free-body diagram showing all the forces on the crate if the conveyer belt runs at constant speed.
   b. Draw a free-body diagram showing all the forces on the crate if the conveyer belt is speeding up.
   c. What is the maximum acceleration the belt can have without the crate slipping?
   d. If the acceleration of the belt exceeds the value determined in part c, what is the acceleration of the crate?
32. || It is friction that provides the force for a car to accelerate, so for high-performance cars the factor that limits acceleration isn't the *engine*; it's the *tires*. For typical rubber-on-concrete friction, what is the shortest time in which a car could accelerate from 0 to 60 mph?
33. || The rolling resistance for steel on steel is quite low; the coefficient of rolling friction is typically $\mu_r = 0.002$. Suppose a 180,000 kg locomotive is rolling at 10 m/s (just over 20 mph) on level rails. If the engineer disengages the engine, how much time will it take the locomotive to coast to a stop? How far will the locomotive move during this time?
34. || A vendor at the local art fair ties her tent to the concrete-filled coffee can shown in Figure P5.34. A stiff breeze comes up and the string becomes taut. What is the maximum value that the string tension can have before the can slips? The coefficient of static friction between the can and the ground is 0.60.

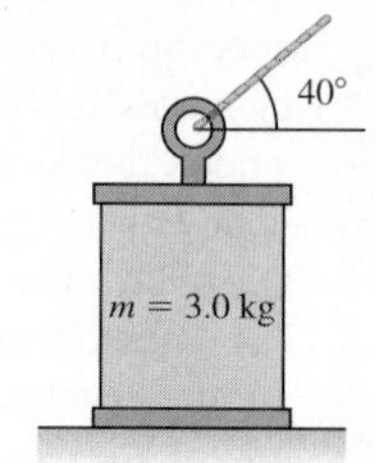

FIGURE P5.34

Section 5.6 Drag

35. ||| Oceanographers use submerged sonar systems, towed by a cable from a ship, to map the ocean floor. In addition to their downward weight, there are buoyant forces and forces from the flowing water that allow them to travel in a horizontal path. One such submersible has a cross-section area of 1.3 m^2, a drag coefficient of 1.2, and, when towed at 5.1 m/s, the tow cable makes an angle of 30° with the horizontal. What is the tension in the cable?
36. || At its widest point, the diameter of a bottlenose dolphin is 0.50 m. Bottlenose dolphins are particularly sleek, having a drag coefficient of only about 0.090.
BIO
   a. What is the drag force acting on such a dolphin swimming at 7.5 m/s?
   b. Using the dolphin's diameter as its characteristic length, what is the Reynolds number as it swims at this speed in 20° C water?
37. ||| The most dangerous particles in polluted air are those with diameters less than 2.5 μm because they can penetrate deeply into the lungs. A 15-cm-tall closed container holds a sample of polluted air containing many spherical particles with a diameter of 2.5 μm and a mass of 1.4×10^{-14} kg. How long does it take for all of the particles to settle to the bottom of the container?
BIO

38. | Running on a treadmill is slightly easier than running outside BIO because there is no drag force to work against. Suppose a 60 kg runner completes a 5.0 km race in 18 minutes. Use the cross-section area estimate of Example 5.15 to determine the drag force on the runner during the race. What is this force as a fraction of the runner's weight?

39. ||| What is the magnitude of the acceleration of a skydiver who is currently falling at one-half his eventual terminal speed?

40. | The air is less dense at higher elevations, so skydivers reach a high terminal speed. The highest recorded speed for a skydiver was achieved in a jump from a height of 39,000 m. At this elevation, the density of the air is only 4.3% of the surface density. Use the data from Example 5.15 to estimate the terminal speed of a skydiver at this elevation.

Section 5.7 Interacting Objects

41. ||| A 1000 kg car pushes a 2000 kg truck that has a dead battery. When the driver steps on the accelerator, the drive wheels of the car push backward against the ground with a force of 4500 N.
    a. What is the magnitude of the force of the car on the truck?
    b. What is the magnitude of the force of the truck on the car?

42. || A 2200 kg truck has put its front bumper against the rear bumper of a 2400 kg SUV to give it a push. With the engine at full power and good tires on good pavement, the maximum forward force on the truck is 18,000 N.
    a. What is the maximum possible acceleration the truck can give the SUV?
    b. At this acceleration, what is the force of the SUV's bumper on the truck's bumper?

43. ||||| Blocks with masses of 1.0 kg, 2.0 kg, and 3.0 kg are lined up in a row on a frictionless table. All three are pushed forward by a 12 N force applied to the 1.0 kg block. How much force does the 2.0 kg block exert on (a) the 3.0 kg block and (b) the 1.0 kg block?

Section 5.8 Ropes and Pulleys

44. ||| What is the tension in the rope of Figure P5.44?

45. |||| A house painter uses the chair-and-pulley arrangement of Figure P5.45 to lift himself up the side of a house. The painter's mass is 70 kg and the chair's mass is 10 kg. If the painter is at rest, what is the tension in the rope?

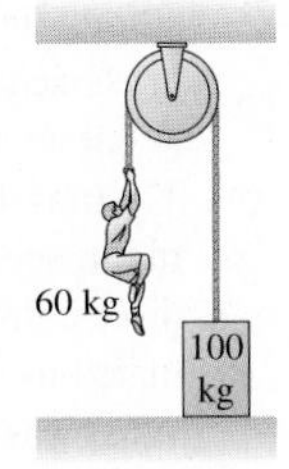

FIGURE P5.44

FIGURE P5.45

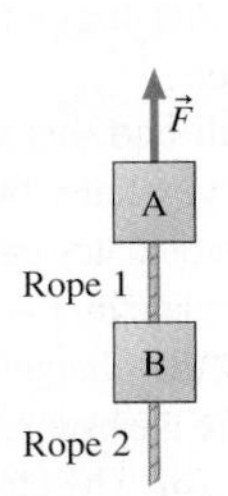

FIGURE P5.46

46. ||| Figure P5.46 shows two 1.00 kg blocks connected by a rope. A second rope hangs beneath the lower block. Both ropes have a mass of 250 g. The entire assembly is accelerated upward at 3.00 m/s^2 by force $\vec{F}$.
    a. What is F?
    b. What is the tension at the top end of rope 1?
    c. What is the tension at the bottom end of rope 1?
    d. What is the tension at the top end of rope 2?

47. || Each of 100 identical blocks sitting on a frictionless surface is connected to the next block by a massless string. The first block is pulled with a force of 100 N.
    a. What is the tension in the string connecting block 100 to block 99?
    b. What is the tension in the string connecting block 50 to block 51?

48. || A Jeep is stuck in the mud. The driver has a winch that can pull on its cable with a force of 40,000 N. The driver loops the cable through a pulley attached to a tree, then attaches the end of the cable to his Jeep, as shown in Figure P5.48. What is the maximum force that can be exerted on the Jeep by this cable arrangement?

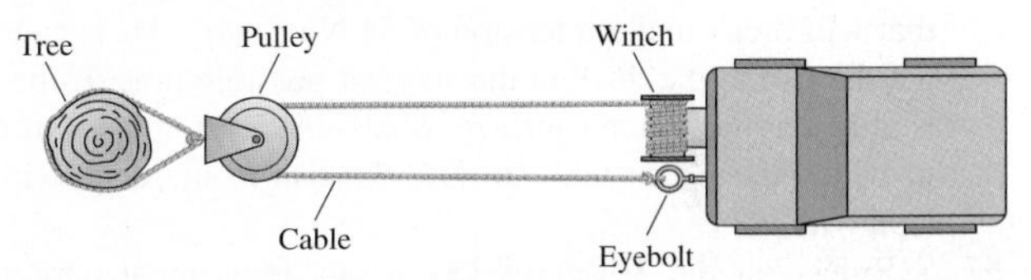

FIGURE P5.48

49. ||| A 500 kg piano is being lowered into position by a crane while two people steady it with ropes pulling to the sides. Bob's rope pulls to the left, 15° below horizontal, with 500 N of tension. Ellen's rope pulls toward the right, 25° below horizontal.
    a. What tension must Ellen maintain in her rope to keep the piano descending vertically at constant speed?
    b. What is the tension in the vertical main cable supporting the piano?

General Problems

50. || Dana has a sports medal suspended by a long ribbon from her rearview mirror. As she accelerates onto the highway, she notices that the medal is hanging at an angle of 10° from the vertical.
    a. Does the medal lean toward or away from the windshield? Explain.
    b. What is her acceleration?

51. || Figure P5.51 shows the INT velocity graph of a 2.0 kg object as it moves along the x-axis. What is the net force acting on this object at $t = 1$ s? At 4 s? At 7 s?

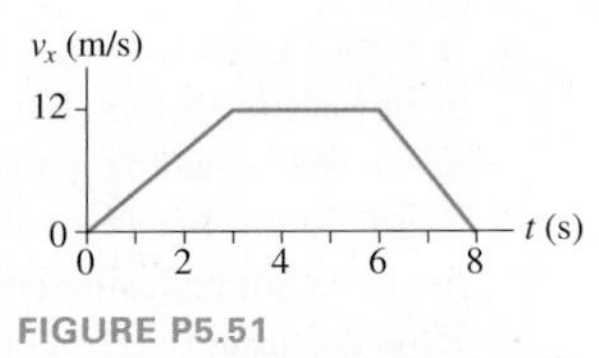

FIGURE P5.51

52. | Your forehead can withstand a force of about 6.0 kN before BIO fracturing, while your cheekbone can only withstand about 1.3 kN.
    a. If a 140 g baseball strikes your head at 30 m/s and stops in 0.0015 s, what is the magnitude of the ball's acceleration?
    b. What is the magnitude of the force that stops the baseball?
    c. What force does the baseball apply to your head? Explain.
    d. Are you in danger of a fracture if the ball hits you in the forehead? In the cheek?

Watch Video Solution Problem 5.61

53. || A 50 kg box hangs from a rope. What is the tension in the rope if
   a. The box is at rest?
   b. The box has $v_y = 5.0$ m/s and is speeding up at 5.0 m/s^2?

54. |||| Scientists have studied how snakes grip and climb ropes. In one study, they found that an important characteristic of a rope is its "compliance"—that is, how easily the rope, while under tension, can be flexed. Figure P5.54 shows how scientists measured a rope's compliance by attaching it to two strings, each supporting an identical mass m. The strings contort the rope so that its middle section lies at angle θ. For $\theta = 30°$ and $m = 100$ g, what are the tensions T_1 and T_2 in the upper and middle parts of the rope?
BIO

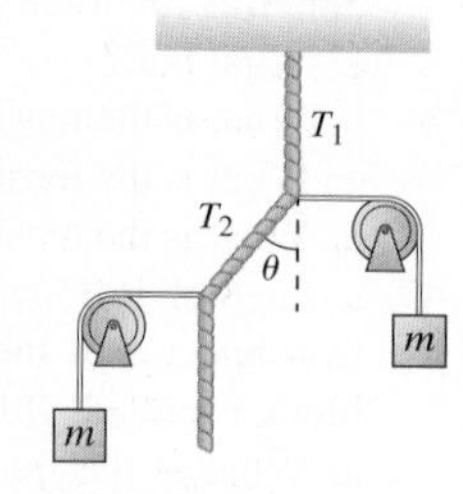

FIGURE P5.54

55. || A 50 kg box hangs from a rope. What is the tension in the rope if
   a. The box moves up at a steady 5.0 m/s?
   b. The box has $v_y = 5.0$ m/s and is slowing down at 5.0 m/s^2?

56. || A fisherman has caught a very large, 5.0 kg fish from a dock that is 2.0 m above the water. He is using lightweight fishing line that will break under a tension of 54 N or more. He is eager to get the fish to the dock in the shortest possible time. If the fish is at rest at the water's surface, what's the least amount of time in which the fisherman can raise the fish to the dock without losing it?
INT

57. ||| Riders on the Tower of Doom, an amusement park ride, experience 2.0 s of free fall, after which they are slowed to a stop in 0.50 s. What is a 65 kg rider's apparent weight as the ride is coming to rest? By what factor does this exceed her actual weight?

58. || Just after launch, the space shuttle takes 8.0 s to reach a speed of 160 km/h. During this phase, what is the apparent weight of a 72 kg astronaut?
INT

59. ||| Seat belts and air bags save lives by reducing the forces exerted on the driver and passengers in an automobile collision. Cars are designed with a "crumple zone" in the front of the car. In the event of an impact, the passenger compartment decelerates over a distance of about 1 m as the front of the car crumples. An occupant restrained by seat belts and air bags decelerates with the car. By contrast, an unrestrained occupant keeps moving forward with no loss of speed (Newton's first law!) until hitting the dashboard or windshield, as we saw in Figure 4.2. These are unyielding surfaces, and the unfortunate occupant then decelerates over a distance of only about 5 mm.
BIO
   a. A 60 kg person is in a head-on collision. The car's speed at impact is 15 m/s. Estimate the net force on the person if he or she is wearing a seat belt and if the air bag deploys.
   b. Estimate the net force that ultimately stops the person if he or she is not restrained by a seat belt or air bag.
   c. How do these two forces compare to the person's weight?

60. ||| Corey, whose mass is 95 kg, stands on a bathroom scale in an elevator. The scale reads 830 N for the first 3.0 s after the elevator starts to move, then 930 N for the next 3.0 s. What is the elevator's velocity 6.0 s after starting?

61. ||| A 20,000 kg rocket has a rocket motor that generates 3.0×10^5 N of thrust.
   a. What is the rocket's initial upward acceleration?
   b. At an altitude of 5.0 km the rocket's acceleration has increased to 6.0 m/s^2. What mass of fuel has it burned?

62. ||| You've always wondered about the acceleration of the elevators in the 101-story-tall Empire State Building. One day, while visiting New York, you take your bathroom scale into the elevator and stand on it. The scale reads 150 lb as the door closes. The reading varies between 120 lb and 170 lb as the elevator travels 101 floors.
   a. What is the magnitude of the acceleration as the elevator starts upward?
   b. What is the magnitude of the acceleration as the elevator brakes to a stop?

63. |||| A 23 kg child goes down a straight slide inclined 38° above horizontal. The child is acted on by his weight, the normal force from the slide, kinetic friction, and a horizontal rope exerting a 30 N force as shown in Figure P5.63. How large is the normal force of the slide on the child?

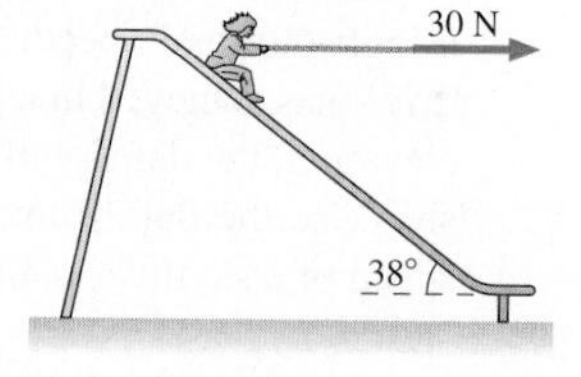

FIGURE P5.63

64. || An impala is an African antelope capable of a remarkable vertical leap. In one recorded leap, a 45 kg impala went into a deep crouch, pushed straight up for 0.21 s, and reached a height of 2.5 m above the ground. To achieve this vertical leap, with what force did the impala push down on the ground? What is the ratio of this force to the antelope's weight?
BIO INT

65. || Josh starts his sled at the top of a 3.0-m-high hill that has a constant slope of 25°. After reaching the bottom, he slides across a horizontal patch of snow. Ignore friction on the hill, but assume that the coefficient of kinetic friction between his sled and the horizontal patch of snow is 0.050. How far from the base of the hill does he end up?

66. |||| The standing vertical jump is a good test of an athlete's strength and fitness. The athlete goes into a deep crouch, then extends his legs rapidly; when his legs are fully extended, he leaves the ground and rises to his highest height. It is the force of the ground on the athlete during the extension phase that accelerates the athlete to the final speed with which he leaves the ground. A good jumper can exert a force on the ground equal to twice his weight. If his crouch is 60 cm deep, how far off the ground does he rise?
BIO INT

67. ||| A 70 kg bicyclist is coasting down a long hill with a 3.5° slope. He's moving quite rapidly, so air drag is important. His cross-section area is 0.32 m^2 and his drag coefficient is 0.88. What speed does he eventually reach, in mph?

68. ||| Many birds can attain very high speeds when diving. Using radar, scientists measured the altitude of a barn swallow in a vertical dive; it dropped 208 m in 3.0 s. The mass of the swallow was estimated to be 0.018 kg, and its cross-section area as 5.6×10^{-4} m^2. What was the drag coefficient for this swallow as it dove?
BIO

69. ||| The drag on a pitched baseball can be surprisingly large. Suppose a 145 g baseball with a diameter of 7.4 cm has an initial speed of 40.2 m/s (90 mph).
INT
   a. What is the magnitude of the ball's acceleration due to the drag force?
   b. If the ball had this same acceleration during its entire 18.4 m trajectory, what would its final speed be?

70. || Small particulates can be removed from the emissions of a coal-fired power plant by *electrostatic precipitation*. The particles are given a small electric charge that draws them toward oppositely charged plates, where they stick. Consider a spherical particulate with a diameter of 1.0 μm. The electric force on this particle is 2.0×10^{-13} N. What is the speed of such a particle? (The electric force is much greater than the particle's weight, which can be ignored.)

71. ||| You probably think of wet surfaces as being slippery. Surprisingly, the opposite is true for human skin, as you can demonstrate by sliding a dry versus a slightly damp fingertip along a smooth surface such as a desktop. Researchers have found that the static coefficient of friction between dry skin and steel is 0.27, while that between damp skin and steel can be as high as 1.4. Suppose a man holds a steel rod vertically in his hand, exerting a 400 N grip force on the rod. What is the heaviest rod he can hold without slipping if
BIO
   a. His hands are dry?
   b. His hands are wet?

72. || Researchers often use *force plates* to measure the forces that people exert against the floor during movement. A force plate works like a bathroom scale, but it keeps a record of how the reading changes with time. Figure P5.72 shows the data from a force plate as a woman jumps straight up and then lands.
BIO INT
   a. What was the vertical component of her acceleration during push-off?
   b. What was the vertical component of her acceleration while in the air?
   c. What was the vertical component of her acceleration during the landing?
   d. What was her speed as her feet left the force plate?
   e. How high did she jump?

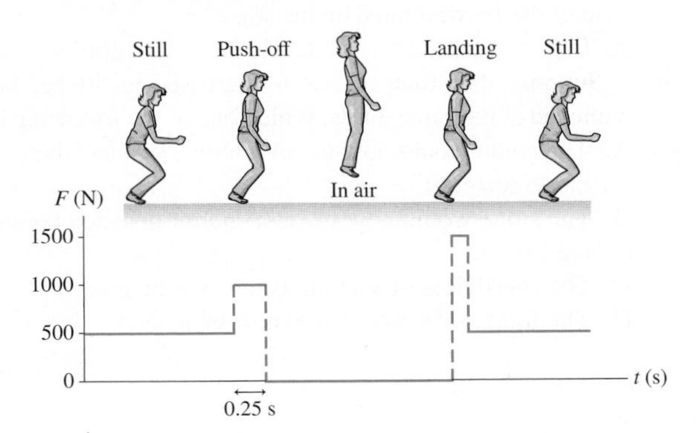

FIGURE P5.72

73. |||| A person with compromised pinch strength in his fingers can only exert a normal force of 6.0 N to either side of a pinch-held object, such as the book shown in Figure P5.73. What is the greatest mass book he can hold onto vertically before it slips out of his fingers? The coefficient of static friction of the surface between the fingers and the book cover is 0.80.

FIGURE P5.73

74. || It's possible for a determined group of people to pull an aircraft. Drag is negligible at low speeds, and the only force impeding motion is the rolling friction of the rubber tires on the concrete runway. In 2000, a team of 60 British police officers set a world record by pulling a Boeing 747, with a mass of 200,000 kg, a distance of 100 m in 53 s. The plane started at rest. Estimate the force with which each officer pulled on the plane, assuming constant pulling force and constant acceleration.
INT

75. ||| A 1.0 kg wood block is pressed against a vertical wood wall by a 12 N force as shown in Figure P5.75. If the block is initially at rest, will it move upward, move downward, or stay at rest?

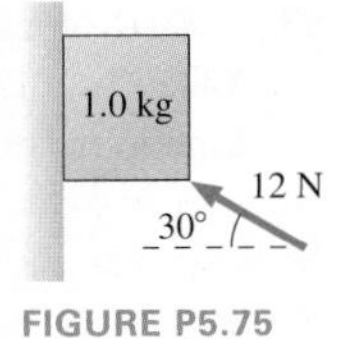

FIGURE P5.75

76. |||| A simple model shows how drawing a bow across a violin string causes the string to vibrate. As the bow moves across the string, static friction between the bow and the string pulls the string along with the bow. At some point, the tension pulling the string back exceeds the maximum static friction force and the string snaps back. This process repeats cyclically, causing the string's vibration. Assume the tension in a 0.33-m-long violin string is 50 N, and the coefficient of static friction between the bow and the string is $\mu_s = 0.80$. If the normal force of the bow on the string is 0.75 N, how far can the string be pulled before it slips if the string is bowed at its center?

77. ||| Two blocks are at rest on a frictionless incline, as shown in Figure P5.77. What are the tensions in the two strings?

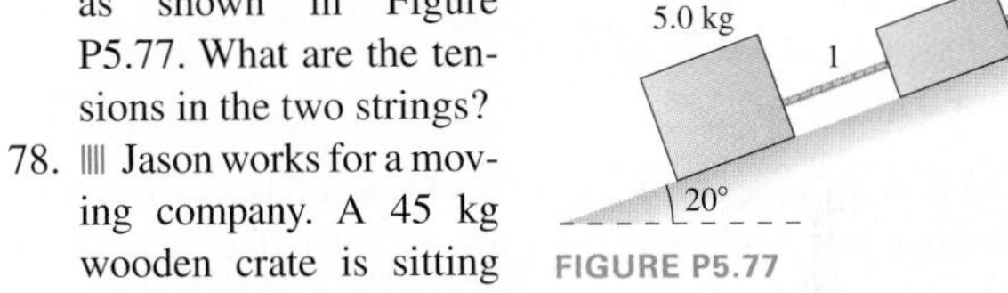

FIGURE P5.77

78. |||| Jason works for a moving company. A 45 kg wooden crate is sitting on the wooden ramp of his truck; the ramp is angled at 11°. What is the magnitude of the force, directed parallel to the ramp, that he needs to exert on the crate to get it to start moving
   a. Up the ramp?
   b. Down the ramp?

79. || Two identical 2.0 kg blocks are stacked as shown in Figure P5.79. The bottom block is free to slide on a frictionless surface. The coefficient of static friction between the blocks is 0.35. What is the maximum horizontal force that can be applied to the lower block without the upper block slipping?

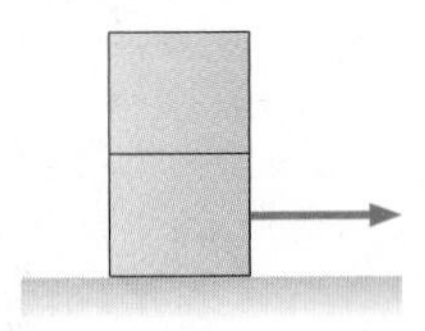
FIGURE P5.79

80. ||| Dana loads luggage into an airplane using a conveyor belt tilted at an angle of 20°. She places a few pieces of luggage on the belt before it starts to move, then she turns the belt on. It takes the belt 0.70 s to reach its top speed of 1.2 m/s. Does the luggage slip? Assume $\mu_s = 0.50$ between the luggage and the belt.

81. |||| In rock climbing, various rope and pulley systems have been devised to help haul up heavy loads, including injured climbers. A rescuer is hauling up an injured climber who weighs 660 N using the rope and pulley system shown in Figure P5.81. (The ropes in this figure are drawn at various angles for clarity, but you can assume they're all vertical.)
   a. What is the tension in rope 1, the rope that the rescuer pulls on?
   b. What is the tension in rope 2?

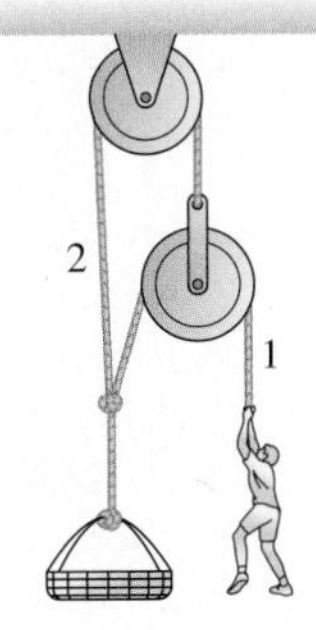

FIGURE P5.81

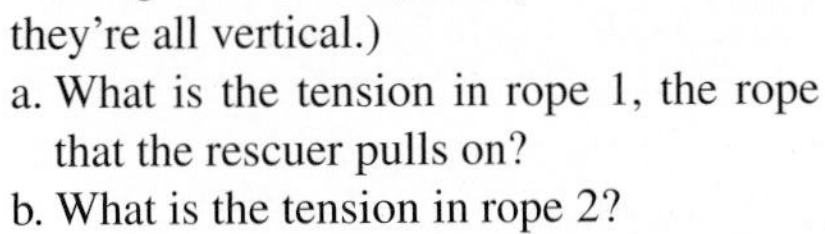

Watch Video Solution Problem 5.83

82. ||| Two blocks are connected by a string as in Figure P5.82. What is the upper block's acceleration if the coefficient of kinetic friction between the block and the table is 0.20?

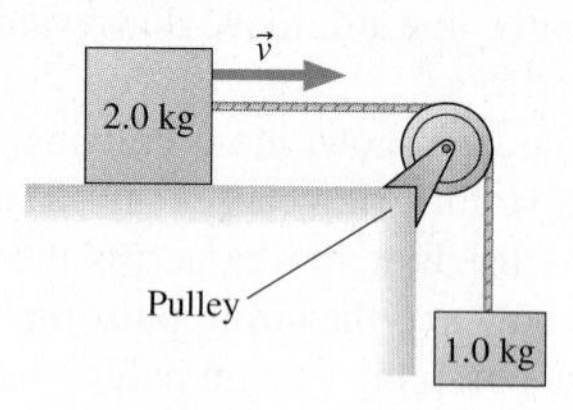

FIGURE P5.82

83. ||| The ramp in Figure P5.83 is frictionless. If the blocks are released from rest, which way does the 10 kg block slide, and what is the magnitude of its acceleration?

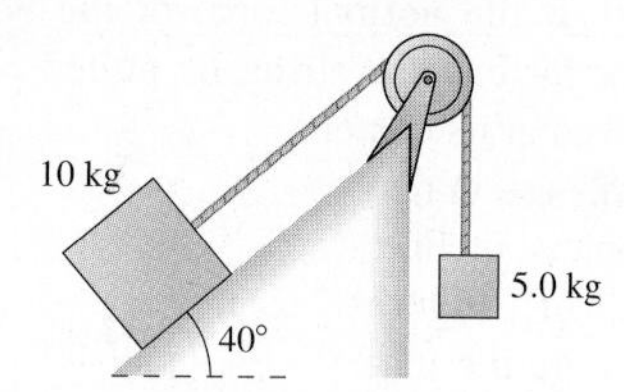

FIGURE P5.83

84. ||| INT The 100 kg block in Figure P5.84 takes 6.0 s to reach the floor after being released from rest. What is the mass of the block on the left?

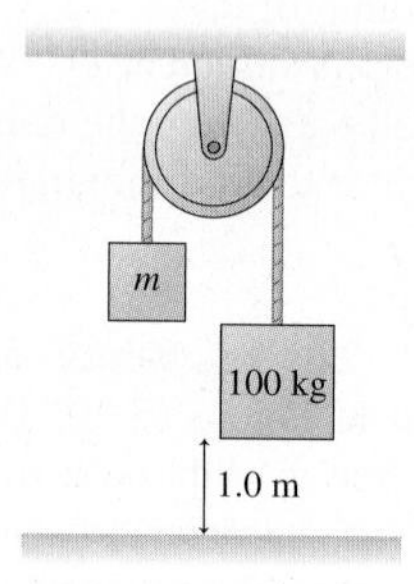

FIGURE P5.84

MCAT-Style Passage Problems

Sliding on the Ice

In the winter sport of curling, players give a 20 kg stone a push across a sheet of ice. The stone moves approximately 40 m before coming to rest. The final position of the stone, in principle, only depends on the initial speed at which it is launched and the force of friction between the ice and the stone, but team members can use brooms to sweep the ice in front of the stone to adjust its speed and trajectory a bit; they must do this without touching the stone. Judicious sweeping can lengthen the travel of the stone by 3 m.

85. | A curler pushes a stone to a speed of 3.0 m/s over a time of 2.0 s. Ignoring the force of friction, how much force must the curler apply to the stone to bring it up to speed?
A. 3.0 N B. 15 N C. 30 N D. 150 N

86. | The sweepers in a curling competition adjust the trajectory of the stone by
A. Decreasing the coefficient of friction between the stone and the ice.
B. Increasing the coefficient of friction between the stone and the ice.
C. Changing friction from kinetic to static.
D. Changing friction from static to kinetic.

87. | Suppose the stone is launched with a speed of 3 m/s and travels 40 m before coming to rest. What is the *approximate* magnitude of the friction force on the stone?
A. 0 N B. 2 N C. 20 N D. 200 N

88. | Suppose the stone's mass is increased to 40 kg, but it is launched at the same 3 m/s. Which one of the following is true?
A. The stone would now travel a longer distance before coming to rest.
B. The stone would now travel a shorter distance before coming to rest.
C. The coefficient of friction would now be greater
D. The force of friction would now be greater.

6 Circular Motion, Orbits, and Gravity

The horses are rounding a corner on a grassy course. Why do they lean into the turn this way?

LOOKING AHEAD ▸

Circular Motion

An object moving in a circle has an acceleration toward the center, so there must be a net force toward the center as well.

How much force does it take to swing the girl in a circle? You'll learn how to solve such problems.

Apparent Forces

These riders feel like they're being pushed against the wall. But this force isn't a real force, it's an **apparent force.**

This apparent force makes the riders "feel heavy." You'll learn to calculate their apparent weight.

Gravity and Orbits

The space station appears to float in space, but gravity is actually pulling down on it quite forcefully.

You'll learn **Newton's law of gravity**, and you'll see how the force of gravity keeps the station in orbit.

GOAL To learn about motion in a circle, including orbital motion under the influence of a gravitational force.

LOOKING BACK ◂

Centripetal Acceleration

In Section 3.7, you learned that an object moving in a circle at a constant speed experiences an acceleration directed toward the center of the circle.

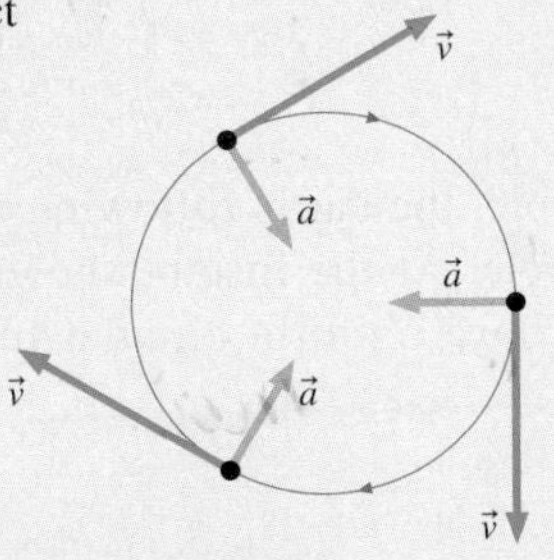

In this chapter, you'll learn how to extend Newton's second law, which relates acceleration to the forces that cause it, to this type of acceleration.

STOP TO THINK

A softball pitcher is throwing a pitch. At the instant shown, the ball is moving in a circular arc at a steady speed. At this instant, the acceleration is

A. Directed up.
B. Directed down.
C. Directed left.
D. Directed right.
E. Zero.

6.1 Uniform Circular Motion

Uniform circular motion at the fair.

The riders on this carnival ride are going in a circle at a constant speed, a type of motion that we've called uniform circular motion. We saw in ◀◀SECTION 3.7 that uniform circular motion results in an acceleration directed toward the center of the circle. According to Newton's second law, this means that there must be a net force directed toward the center of the circle. For these riders, it is the tension in the cables and the force of gravity that combine to provide this net force.

In this chapter, we'll look at objects moving in circles or circular arcs. We'll consider the details of the acceleration and the forces that provide this acceleration, combining our discussion and description of motion from Chapters 1–3 with our treatment of Newton's laws and dynamics from Chapters 4 and 5. For now, we consider objects that move at a *constant* speed; circular motion with *changing* speed will wait until Chapter 7.

Velocity and Acceleration in Uniform Circular Motion

FIGURE 6.1 Velocity and acceleration for uniform circular motion.

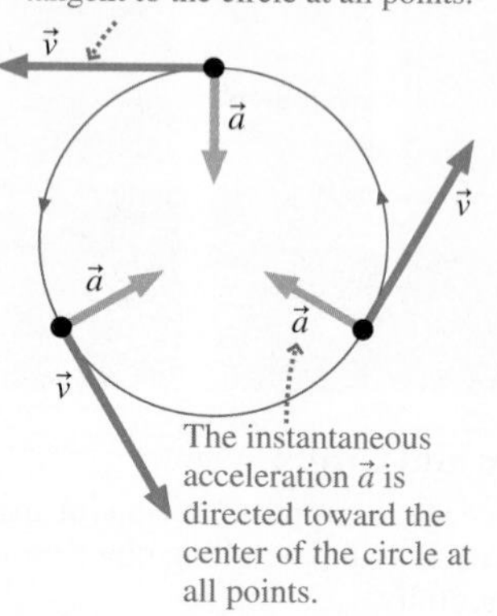
The instantaneous velocity $\vec{v}$ is tangent to the circle at all points.
$\vec{v}$
$\vec{a}$
$\vec{v}$
$\vec{a}$
$\vec{a}$
$\vec{v}$
The instantaneous acceleration $\vec{a}$ is directed toward the center of the circle at all points.

Although the *speed* of a particle in uniform circular motion is constant, its *velocity* is not constant because the *direction* of the motion is always changing. FIGURE 6.1 reminds you of the details: There is an acceleration at every point in the motion, with the acceleration vector $\vec{a}$ pointing toward the center of the circle. We called this the *centripetal acceleration,* and we showed that for uniform circular motion the acceleration is given by

$$a = \frac{v^2}{r} \tag{6.1}$$

Centripetal acceleration for uniform circular motion

a
p.50
v
QUADRATIC

Thus an object's centripetal acceleration depends on both its speed and the distance from the center of its circular path.

CONCEPTUAL EXAMPLE 6.1 **A car rounding a corner**

A car is turning a tight corner at a constant speed. A top view of the motion is shown in FIGURE 6.2. The velocity vector for the car points to the east at the instant shown. What is the direction of the acceleration?

FIGURE 6.2 Top view of a car turning a corner.

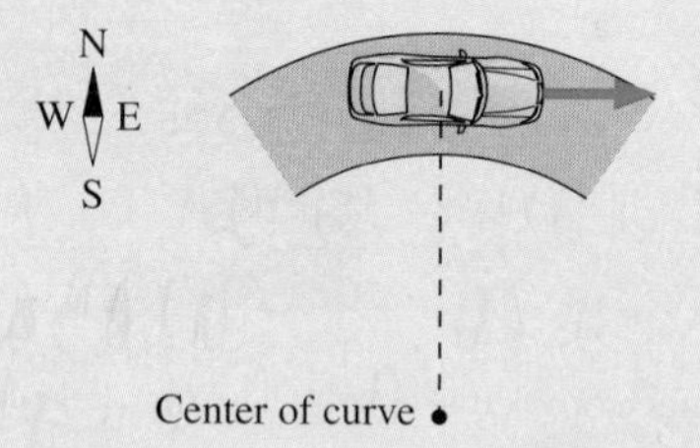
N
W
E
S
Center of curve

REASON The curve that the car is following is a segment of a circle, so this is an example of uniform circular motion. For uniform circular motion, the acceleration is directed toward the center of the circle, which is to the south.

ASSESS This acceleration is due to a change in direction, not a change in speed. And this matches your experience in a car: If you turn the wheel to the right—as the driver of this car is doing—your car then *changes* its motion toward the right, in the direction of the center of the circle.

NOTE ▶ In this example, the car is following a curve that is only a *segment* of a circle, not a *full* circle. At the instant shown, though, the motion follows a circular arc. You can have uniform circular motion without completing a full circle. ◀

Period, Frequency, and Speed

Although an object needn't complete a full circle to be in uniform circular motion, in many of the cases we'll consider, objects will complete multiple full circles of motion, one after another. Since the motion is uniform, each time around the circle is just a repeat of the one before, so the motion is *periodic*.

The time interval it takes an object to go around a circle one time, completing one revolution (abbreviated rev), is called the **period** of the motion. Period is represented by the symbol T.

Rather than specify the time for one revolution, we can specify circular motion by its **frequency,** the number of revolutions per second, for which we use the symbol f. An object with a period of one-half second completes 2 revolutions each second. Similarly, an object can make 10 revolutions in 1 s if its period is one-tenth of a second. This shows that frequency is the inverse of the period:

$$f = \frac{1}{T} \tag{6.2}$$

Although frequency is often expressed as "revolutions per second," *revolutions* are not true units but merely the counting of events. Thus the SI unit of frequency is simply inverse seconds, or s^{-1}. Frequency may also be given in revolutions per minute (rpm) or another time interval, but these usually need to be converted to s^{-1} before calculations are done.

FIGURE 6.3 shows an object moving at a constant speed in a circular path of radius r. We know the time for one revolution—one period T—and we know the distance traveled, so we can write an equation relating the speed to the period and radius:

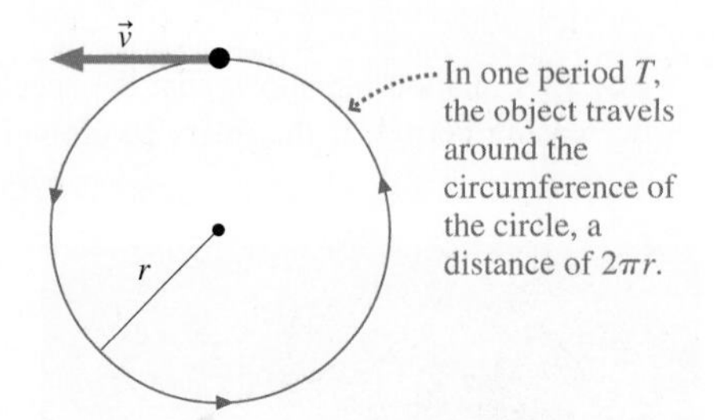

$\vec{v}$
In one period T, the object travels around the circumference of the circle, a distance of $2\pi r$.
r

FIGURE 6.3 Relating frequency and speed.

$$v = \frac{2\pi r}{T} \tag{6.3}$$

Given Equation 6.2 relating frequency and period, we can also write this equation as

$$v = 2\pi f r \tag{6.4}$$

We can combine this equation with Equation 6.1 for acceleration to get an expression for the centripetal acceleration in terms of the frequency or the period for the circular motion:

$$a = \frac{v^2}{r} = (2\pi f)^2 r = \left(\frac{2\pi}{T}\right)^2 r \tag{6.5}$$

EXAMPLE 6.2 A spinning table saw blade

The circular blade of a table saw is 25 cm in diameter and spins at 3600 rpm. How much time is required for one revolution? How fast is one of the teeth at the edge of the blade moving? What is the tooth's acceleration?

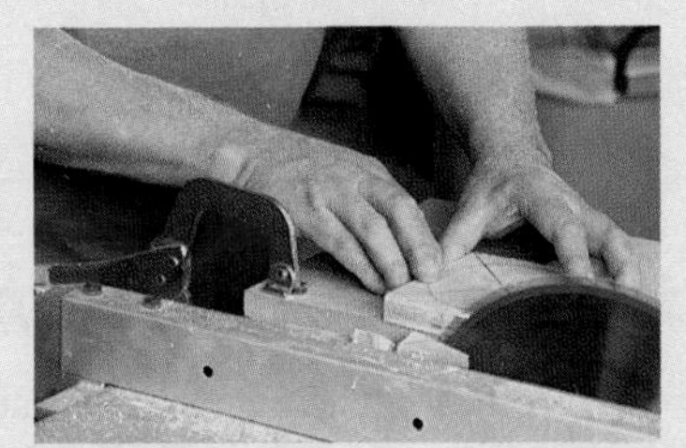

STRATEGIZE Each tooth of the saw blade is undergoing uniform circular motion, so the period, frequency, speed, and acceleration of a tooth are all related by Equations 6.2–6.5.

PREPARE Before we get started, we need to do a couple of unit conversions. The diameter of the blade is 0.25 m, so its radius is 0.125 m. The frequency is given in rpm; we need to convert this to s^{-1}:

$$f = 3600 \frac{\text{rev}}{\text{min}} \times \frac{1 \text{ min}}{60 \text{ s}} = 60 \frac{\text{rev}}{\text{s}} = 60 \text{ s}^{-1}$$

SOLVE The time for one revolution is the period, which is given by Equation 6.2:

$$T = \frac{1}{f} = \frac{1}{60 \text{ s}^{-1}} = 0.017 \text{ s}$$

The speed of the tooth is given by Equation 6.4:

$$v = 2\pi f r = 2\pi(60 \text{ s}^{-1})(0.125 \text{ m}) = 47 \text{ m/s}$$

We can then use Equation 6.5 to find the acceleration:

$$a = (2\pi f)^2 r = (2\pi(60 \text{ s}^{-1}))^2(0.125 \text{ m}) = 1.8 \times 10^4 \text{ m/s}^2$$

ASSESS The speed of the tooth is extremely high: 47 m/s ≈ 100 mph! Still, this seems plausible for a high-speed saw. The acceleration is also remarkable, almost $2000g$.

EXAMPLE 6.3 Finding the period of a carnival ride

In the Quasar carnival ride, passengers travel in a horizontal 5.0-m-radius circle. For safe operation, the maximum sustained acceleration that riders may experience is 20 m/s^2, approximately twice the free-fall acceleration. What is the period of the ride when it is being operated at the maximum acceleration? How fast are the riders moving when the ride is operated at this period?

STRATEGIZE The passengers are in uniform circular motion, so their period, speed, and acceleration are related by Equations 6.4 and 6.5.

PREPARE We will assume that the cars on the ride are in uniform circular motion. The visual overview of FIGURE 6.4 shows a top view of the motion of the ride.

FIGURE 6.4 Visual overview for the Quasar carnival ride.

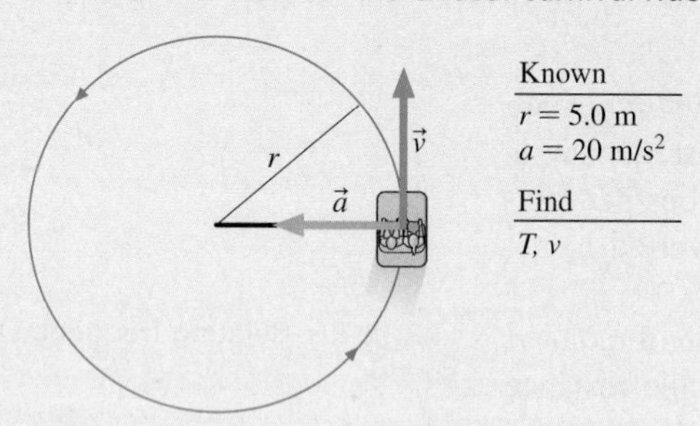

SOLVE Equation 6.5 shows that the acceleration increases with decreasing period: If the riders go around in a shorter time, the acceleration increases. Setting a maximum value for the acceleration means setting a minimum value for the period—as fast as the ride can safely go. We can rearrange Equation 6.5 to find the period in terms of the acceleration. Setting the acceleration equal to the maximum value gives the minimum period:

$$T = 2\pi\sqrt{\frac{r}{a}} = 2\pi\sqrt{\frac{5.0\text{ m}}{20\text{ m/s}^2}} = 3.1\text{ s}$$

We can then use Equation 6.4 to find the speed at which the riders move:

$$v = \frac{2\pi r}{T} = \frac{2\pi(5.0\text{ m})}{3.1\text{ s}} = 10\text{ m/s}$$

ASSESS One rotation in just over 3 seconds seems reasonable for a pretty zippy carnival ride. (The period for this particular ride is actually 3.7 s, so it runs a bit slower than the maximum safe speed.) But in this case we can do a quantitative check on our work. If we use our calculated velocity to find the acceleration using Equation 6.1, we find

$$a = \frac{v^2}{r} = \frac{(10\text{ m/s})^2}{5.0\text{ m}} = 20\text{ m/s}^2$$

This is the acceleration given in the problem statement, so we can have confidence in our work.

Hurling the heavy hammer Scottish games involve feats of strength. Here, a man is throwing a 30 lb hammer for distance. He starts by swinging the hammer rapidly in a circle. You can see from how he is leaning that he is providing a large force directed toward the center to produce the necessary centripetal acceleration. When the hammer is heading in the right direction, the man lets go. With no force directed toward the center, the hammer will stop going in a circle and fly in the chosen direction across the field.

STOP TO THINK 6.1 Rank in order, from largest to smallest, the period of the motion of particles A to D.

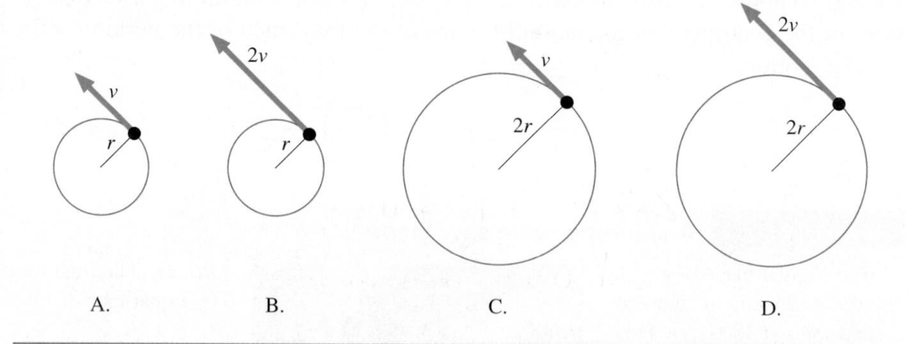

6.2 Dynamics of Uniform Circular Motion

Riders traveling around on a circular carnival ride are accelerating, as we have seen. Consequently, according to Newton's second law, the riders must have a net *force* acting on them.

We've already determined the acceleration of a particle in uniform circular motion—the centripetal acceleration of Equation 6.1. Newton's second law tells us what the net force must be to cause this acceleration:

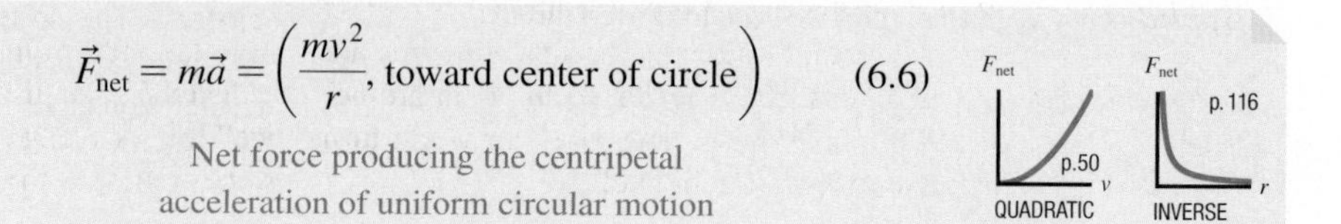

$$\vec{F}_{net} = m\vec{a} = \left(\frac{mv^2}{r}, \text{ toward center of circle}\right) \qquad (6.6)$$

Net force producing the centripetal acceleration of uniform circular motion

A video to support a section's topic is embedded in the eText.

Video Forces in Circular Motion

In other words, **a particle of mass m moving at constant speed v around a circle of radius r must always have a net force of magnitude mv^2/r pointing toward the center of the circle,** as in FIGURE 6.5. It is this net force that causes the centripetal acceleration of circular motion. Without such a net force, the particle would move off in a straight line tangent to the circle.

The force described by Equation 6.6 is not a *new* kind of force. The net force will be due to one or more of our familiar forces, such as tension, friction, or the normal force. Equation 6.6 simply tells us how the net force needs to act—how strongly and in which direction—to cause the particle to move with speed v in a circle of radius r.

In each example of circular motion that we will consider in this chapter, a physical force or a combination of forces directed toward the center produces the necessary acceleration.

FIGURE 6.5 Net force for circular motion.

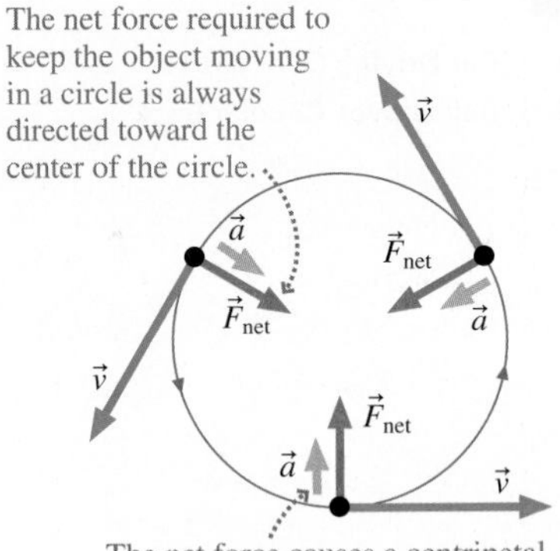
The net force required to keep the object moving in a circle is always directed toward the center of the circle.
$\vec{v}$
$\vec{a}$
$\vec{F}_{net}$
The net force causes a centripetal acceleration.

CONCEPTUAL EXAMPLE 6.4 **Forces on a car, part I**

Engineers design curves on roads to be segments of circles. They also design dips and peaks in roads to be segments of circles with a radius that depends on expected speeds and other factors. A car is moving at a constant speed and goes into a dip in the road. At the very bottom of the dip, is the normal force of the road on the car greater than, less than, or equal to the car's weight?

FIGURE 6.6 Visual overview for the car in a dip.

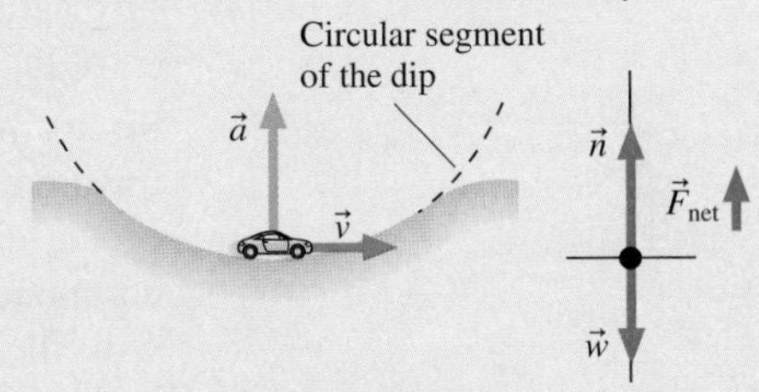
Circular segment of the dip
$\vec{a}$
$\vec{v}$
$\vec{n}$
$\vec{F}_{net}$
$\vec{w}$

REASON FIGURE 6.6 shows a visual overview of the situation. The car is accelerating, even though it is moving at a constant speed, because its direction is changing. When the car is at the bottom of the dip, the center of its circular path is directly above it and so its acceleration vector points straight up. The free-body diagram of Figure 6.6 identifies the only two forces acting on the car as the normal force, pointing upward, and its weight, pointing downward. Which is larger: n or w?

Because $\vec{a}$ points upward, by Newton's second law there must be a net force on the car that also points upward. In order for this to be the case, the free-body diagram shows that the magnitude of the normal force must be *greater* than the weight.

ASSESS You have probably experienced this situation. As you drive through a dip in the road, you feel "heavier" than normal. As discussed in Section 5.3, this is because your apparent weight—the normal force that supports you—is greater than your true weight.

CONCEPTUAL EXAMPLE 6.5 **Forces on a car, part II**

A car is turning a corner at a constant speed, following a segment of a circle. What force provides the necessary centripetal acceleration?

REASON The car moves along a circular arc at a constant speed—uniform circular motion—for the quarter-circle necessary to complete the turn. We know that the acceleration is directed toward the center of the circle. What force or forces can we identify that provide this acceleration?

Imagine you are driving a car on a frictionless road, such as a very icy road. You would not be able to turn a corner. Turning the steering wheel would be of no use. The car would slide straight ahead, in accordance with both Newton's first law and the experience of anyone who has ever driven on ice! So it must be *friction* that causes the car to turn. The top view of the tire in FIGURE 6.7 shows the force on one of the car's tires as it turns a corner. It must be a *static* friction force, not kinetic, because the tires are not skidding: The points where the tires touch the road are not moving relative to the surface. If you skid, your car won't turn the corner—it will continue in a straight line!

FIGURE 6.7 Top views of a car turning a corner.

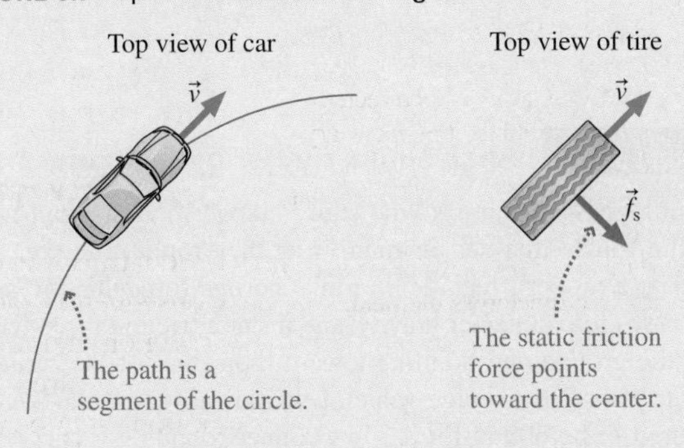
Top view of car
Top view of tire
$\vec{v}$
$\vec{f}_s$
The path is a segment of the circle.
The static friction force points toward the center.

ASSESS This result agrees with your experience. You know that reduced friction makes it harder to turn corners, and you know as well that skidding on a curve is bad news. So it makes sense that static friction is the force at work.

Video Car Driving Over a Rise
Video Ball Leaves Circular Track

PROBLEM-SOLVING APPROACH 6.1 Circular dynamics problems

Circular motion involves an acceleration and thus a net force. We can therefore use techniques very similar to those we've already seen for other Newton's second-law problems.

STRATEGIZE An object undergoing uniform circular motion has an acceleration directed toward the center of the circle. This acceleration, according to Newton's second law, is caused by a net force directed toward the center. The origin of this force is one or more of the forces we're familiar with, such as tension, weight, or friction.

PREPARE Begin your visual overview with a pictorial representation in which you sketch the motion, define symbols, define axes, and identify what the problem is trying to find. There are two common situations:

- If the motion is in a horizontal plane, like a tabletop, draw the free-body diagram with the circle viewed edge-on, the x-axis pointing toward the center of the circle, and the y-axis perpendicular to the plane of the circle.
- If the motion is in a vertical plane, like a Ferris wheel, draw the free-body diagram with the circle viewed face-on, the x-axis pointing toward the center of the circle, and the y-axis tangent to the circle.

SOLVE Newton's second law for uniform circular motion, $\vec{F}_{\text{net}} = (mv^2/r,$ toward center of circle), is a vector equation. Some forces act in the plane of the circle, some act perpendicular to the circle, and some may have components in both directions. In the coordinate system just described, with the x-axis pointing toward the center of the circle, Newton's second law is

$$\sum F_x = \frac{mv^2}{r} \quad \text{and} \quad \sum F_y = 0$$

That is, the net force toward the center of the circle has magnitude mv^2/r while the net force perpendicular to the circle is zero. The components of the forces are found directly from the free-body diagram. Depending on the problem, either:

- Use the net force to determine the speed v, then use circular kinematics to find frequencies or other details of the motion.
- Use circular kinematics to determine the speed v, then solve for unknown forces.

ASSESS Make sure your net force points toward the center of the circle. Check that your result has the correct units, is reasonable, and answers the question.

Exercise 9 

EXAMPLE 6.6 Analyzing a runner on a circular track BIO

Kinesiologists—scientists who study human motion—can measure the forces that act on runners as they round a curve. The forces on a runner's feet are a vertical normal force that, on average, serves to counteract gravity, and a static friction force parallel to the ground and pointing toward the center of the runner's circular path. The average value of this inward-pointing force is measured to be 600 N for a 75 kg runner rounding a curve that has a 2.5 m radius. What is this runner's speed?

STRATEGIZE We saw in Conceptual Example 6.5 that the force that causes the car's centripetal acceleration is the friction force between the road and the tires. The same situation holds for a runner moving in a circle: It is the static friction force between the track and the runner's shoes that causes the runner to move in a circle. Following the steps of Problem-Solving Approach 6.1, we will use Newton's second law to relate this force to the centripetal acceleration, from which we can find the runner's speed.

PREPARE FIGURE 6.8 on the next page shows a visual overview of the problem. The main reason for the pictorial representation on the left is to illustrate the relevant geometry and to define the symbols that will be used. A circular dynamics problem usually does not have starting and ending points like a projectile problem, so subscripts such as x_i or y_f are usually not needed. Here we need to define the runner's speed v and the radius r of the circle.

FIGURE 6.8 A visual overview of the runner moving in a circular path.

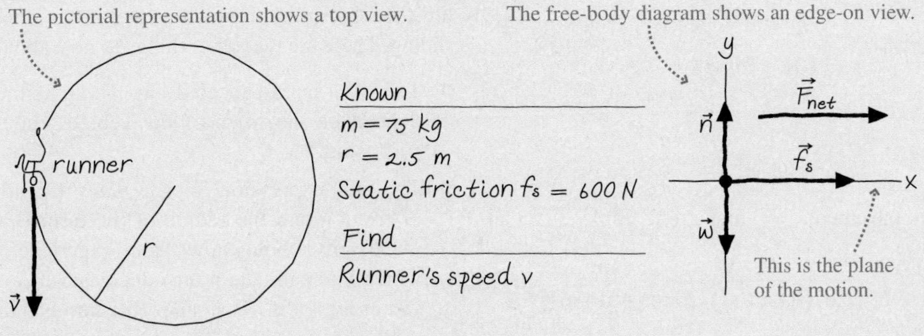

The pictorial representation shows a top view.

The free-body diagram shows an edge-on view.

Known
$m = 75$ kg
$r = 2.5$ m
Static friction $f_s = 600$ N
Find
Runner's speed v

This is the plane of the motion.

The free-body diagram shows the forces in the problem. Because the motion is in a horizontal plane, Problem-Solving Approach 6.1 tells us to draw the free-body diagram looking at the edge of the circle, with the x-axis pointing toward the center of the circle and the y-axis perpendicular to the plane of the circle. Three forces are acting on the runner: the weight force $\vec{w}$, the normal force of the ground $\vec{n}$, and the friction force $\vec{f}_s$.

SOLVE There is no net force in the y-direction, perpendicular to the circle, so $\vec{w}$ and $\vec{n}$ must be equal and opposite. There is a net force in the x-direction, toward the center of the circle, as there must be to cause the centripetal acceleration of circular motion. Only the friction force has an x-component, so Newton's second law is

$$\sum F_x = f_s = \frac{mv^2}{r}$$

We know the mass, the radius of the circle, and the friction force, so we can solve for v:

$$v = \sqrt{\frac{f_s r}{m}} = \sqrt{\frac{(600 \text{ N})(2.5 \text{ m})}{75 \text{ kg}}} = 4.5 \text{ m/s}$$

ASSESS The speed is about 10 mph, a reasonable value for a runner going around a sharp curve.

EXAMPLE 6.7 Finding the maximum speed for a car to turn a corner

What is the maximum speed with which a 1500 kg car can make a turn around a curve of radius 20 m on a level (unbanked) road without sliding? (This radius turn is about what you might expect at a major intersection in a city.)

STRATEGIZE As Conceptual Example 6.5 showed, the force that causes the car's centripetal acceleration is the static friction between the car's tires and the road. If the car's speed is too high, the friction force needed to cause this centripetal force will exceed its maximum possible value and the car will slide.

PREPARE We start with the visual overview in FIGURE 6.9. The car moves along a circular arc at a constant speed—uniform circular motion—during the turn. The static friction force causes the centripetal acceleration, so this force must be directed toward the center of the circular arc. We've drawn the free-body diagram, shown from behind the car, with the friction force pointing in this direction. Because the motion is in a horizontal plane, we've again chosen an x-axis toward the center of the circle and a y-axis perpendicular to the plane of motion.

FIGURE 6.9 Visual overview of a car turning a corner.

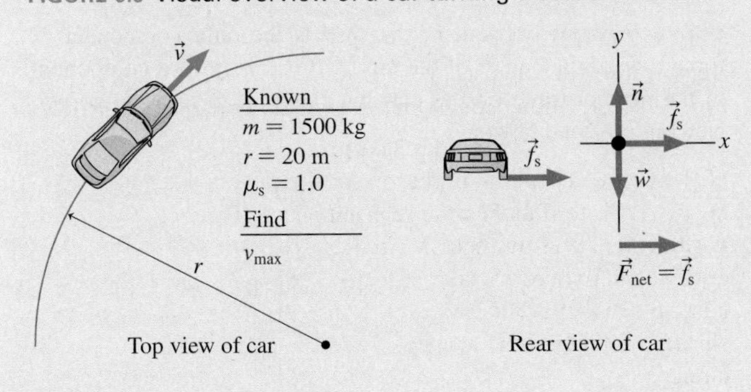

SOLVE The only force in the x-direction, toward the center of the circle, is static friction. Newton's second law along the x-axis is

$$\sum F_x = f_s = \frac{mv^2}{r}$$

This is the same equation as in the previous example, because the force toward the center of the circle in that example is also static friction.

Newton's second law in the y-direction is

$$\sum F_y = n - w = ma_y = 0$$

so that $n = w = mg$.

The car will slide when the static friction force reaches its maximum value. Recall from Equation 5.7 in Chapter 5 that this maximum force is given by

$$f_{s\,max} = \mu_s n = \mu_s mg$$

The maximum speed occurs when the static friction force reaches its maximum value, or when

$$f_{s\,max} = \frac{m{v_{max}}^2}{r}$$

Continued

Using the known value of $f_{s\,max}$, we find

$$\frac{m{v_{max}}^2}{r} = f_{s\,max} = \mu_s mg$$

Rearranging, we get

$${v_{max}}^2 = \mu_s gr$$

For rubber tires on pavement, we find from Table 5.2 that $\mu_s = 1.0$. We then have

$$v_{max} = \sqrt{\mu_s gr} = \sqrt{(1.0)(9.8\ \text{m/s}^2)(20\ \text{m})} = 14\ \text{m/s}$$

ASSESS 14 m/s ≈ 30 mph, which seems like a reasonable upper limit for the speed at which a car can go around a curve without sliding. There are two other things to note about the solution:

- The car's mass canceled out. The maximum speed *does not* depend on the mass of the vehicle, though this may seem surprising.
- The final expression for v_{max} *does* depend on the coefficient of friction and the radius of the turn. Both of these factors make sense. You know, from experience, that the speed at which you can take a turn decreases if μ_s is less (the road is wet or icy) or if r is smaller (the turn is tighter).

FIGURE 6.10 Modifications to increase the maximum speed around a curve.

(a) Wings on a race car.

(b) A banked turn on a racetrack.

Because v_{max} depends on μ_s and because μ_s depends on road conditions, the maximum safe speed through turns can vary dramatically. A car that easily handles a curve in dry weather can suddenly slide out of control when the pavement is wet. Icy conditions are even worse. If you lower the value of the coefficient of friction in Example 6.7 from 1.0 (dry pavement) to 0.1 (icy pavement), the maximum speed for the turn goes down to 4.4 m/s—about 10 mph!

Race cars turn corners at much higher speeds than normal passenger vehicles. One design modification of the *cars* to allow this is the addition of wings, as on the car in FIGURE 6.10a. The wings provide an additional force pushing the car *down* onto the pavement by deflecting air upward. This extra downward force increases the normal force, thus increasing the maximum static friction force and making faster turns possible.

There are also design modifications of the *track* that allow race cars to take corners at high speeds. If the track is banked by raising the outside edge of curved sections, the normal force can provide some of the force necessary to produce the centripetal acceleration, as we will see in the next example. The curves on racetracks may be quite sharply banked (FIGURE 6.10b). Curves on ordinary highways are often banked as well, though at more modest angles suiting the lower speeds.

EXAMPLE 6.8 Finding a car's speed on a banked turn

A curve on a racetrack of radius 70 m is banked at a 15° angle. At what speed can a car take this curve without assistance from friction?

STRATEGIZE With no friction acting, it is the horizontal component of the normal force that causes the centripetal acceleration.

PREPARE After drawing the pictorial representation in FIGURE 6.11, we use the force identification diagram to find that, given that there is no friction acting, the only two forces are the normal force and the car's weight. We can then construct the free-body diagram, making sure that we draw the normal force perpendicular to the road's surface.

FIGURE 6.11 Visual overview for the car on a banked turn.

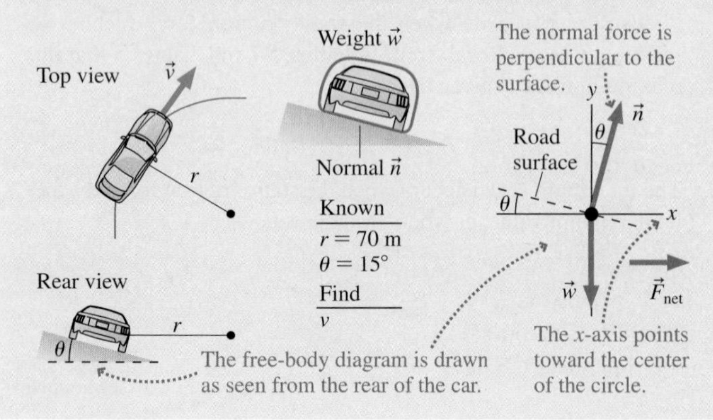

Even though the car is tilted, it is still moving in a *horizontal* circle. Thus, following Problem-Solving Approach 6.1, we choose the x-axis to be horizontal and pointing toward the center of the circle.

SOLVE Without friction, $n_x = n\sin\theta$ is the only component of force toward the center of the circle. It is this inward component of the normal force on the car that causes it to turn the corner. Newton's second law is

$$\sum F_x = n\sin\theta = \frac{mv^2}{r}$$

$$\sum F_y = n\cos\theta - w = 0$$

where θ is the angle at which the road is banked, and we've assumed that the car is traveling at the correct speed v. From the y-equation,

$$n = \frac{w}{\cos\theta} = \frac{mg}{\cos\theta}$$

Substituting this into the x-equation and solving for v give

$$\left(\frac{mg}{\cos\theta}\right)\sin\theta = mg\tan\theta = \frac{mv^2}{r}$$

$$v = \sqrt{rg\tan\theta} = 14 \text{ m/s}$$

ASSESS This is ≈ 30 mph, a reasonable speed. Only at this exact speed can the turn be negotiated without reliance on friction forces.

The friction force provides the necessary centripetal acceleration not only for cars turning corners, but also for bicycles, horses, and, as we saw in Example 6.6, humans. The cyclists in FIGURE 6.12 are going through a tight turn; you can tell by how they lean. The road exerts both a vertical normal force and a horizontal friction force on their tires. A vector sum of these forces points at an angle. The cyclists lean to the side so that the sum of the road forces points along the line of their bikes and their bodies; this keeps them in balance. The horses in the photo that opened the chapter are leaning into the turn for a similar reason.

FIGURE 6.12 Road forces on a cyclist leaning into a turn.

Maximum Walking Speed BIO

Humans and other two-legged animals have two basic gaits: walking and running. At slow speeds, you walk. When you need to go faster, you run. Why don't you just walk faster? There is an upper limit to the speed of walking, and this limit is set by the physics of circular motion.

Think about the motion of your body as you take a walking stride. You put one foot forward, then push off with your rear foot. Your body pivots over your front foot, and you bring your rear foot forward to take the next stride. As you can see in FIGURE 6.13a, the path that your body takes during this stride is the arc of a circle. **In a walking gait, your body is in circular motion as you pivot on your forward foot.**

A force toward the center of the circle is required for this circular motion, as shown in Figure 6.13. FIGURE 6.13b shows the forces acting on the woman's body during the midpoint of the stride: her weight, directed down, and the normal force of the ground, directed up. Newton's second law for the x-axis is

$$\sum F_x = w - n = \frac{mv^2}{r}$$

Because of her circular motion, the net force must point toward the center of the circle, or, in this case, down. In order for the net force to point down, the normal force must be *less* than her weight. Your body tries to "lift off" as it pivots over your foot, decreasing the normal force exerted on you by the ground. The normal force becomes smaller as you walk faster, but n cannot be less than zero. Thus the maximum possible walking speed v_{max} occurs when $n = 0$. Setting $n = 0$ in Newton's second law gives

$$w = mg = \frac{mv_{max}^2}{r}$$

Thus

$$v_{max} = \sqrt{gr} \tag{6.7}$$

The maximum possible walking speed is limited by r, the length of the leg, and g, the free-fall acceleration. This formula is a good approximation of the maximum walking speed for humans and other animals. Giraffes, with their very long legs, can walk at high speeds. Animals such as mice with very short legs have such a low maximum walking speed that they rarely use this gait.

For humans, the length of the leg is approximately 0.7 m, giving $v_{max} \approx 2.6$ m/s ≈ 6 mph. You *can* walk this fast, though it becomes energetically unfavorable to walk at speeds above 4 mph. Most people make a transition to a running gait at about this speed.

FIGURE 6.13 Analysis of a walking stride.

(a) Walking stride

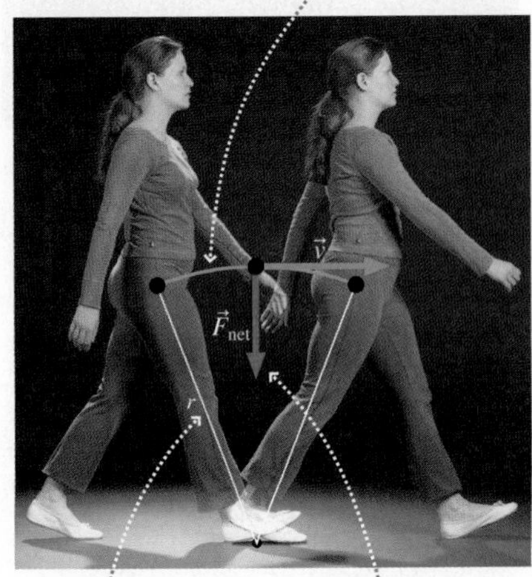
During each stride, her hip undergoes circular motion.
$\vec{F}_{net}$
The radius of the circular motion is the length of the leg from the foot to the hip.
The circular motion requires a force directed toward the center of the circle.

(b) Forces in the stride

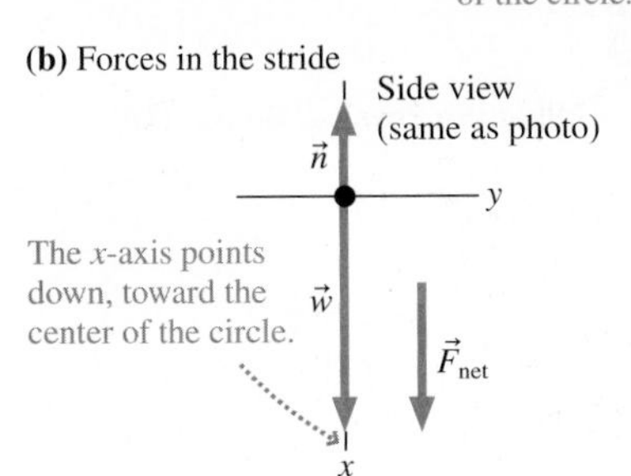
Side view (same as photo)
$\vec{n}$
y
The x-axis points down, toward the center of the circle.
$\vec{w}$
$\vec{F}_{net}$
x

STOP TO THINK 6.2 A block on a string spins in a horizontal circle on a frictionless table. Rank in order, from largest to smallest, the tensions T_A to T_E acting on the blocks A to E.

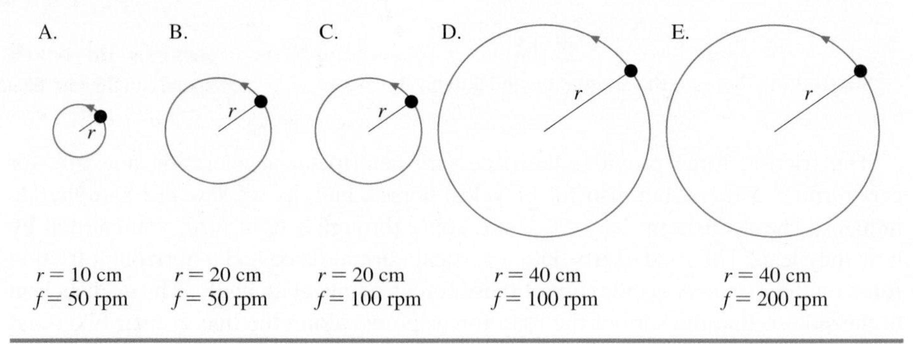

6.3 Apparent Forces in Circular Motion

FIGURE 6.14 Inside the Gravitron, a rotating circular room.

FIGURE 6.14 shows a carnival ride that spins the riders around inside a large cylinder. The people are "stuck" to the inside wall of the cylinder! As you probably know from experience, the riders *feel* that they are being pushed outward, into the wall. But our analysis has found that an object in circular motion must have an *inward* force to create the centripetal acceleration. How can we explain this apparent difference?

Centrifugal Force?

If you are a passenger in a car that turns a corner quickly, you may feel "thrown" by some mysterious force against the door. But is there really such a force? FIGURE 6.15 shows a bird's-eye view of you riding in a car as it makes a left turn. Just before the turn, you were moving in a straight line so, according to Newton's first law, as the car begins to turn, you want to continue moving along that same line. However, the door moves into your path and so runs into you! You feel the force of the door because it is this force, pushing *inward* toward the center of the curve, that is causing you to turn the corner. But you were not "thrown" into the door; the door ran into you.

FIGURE 6.15 Bird's-eye view of a passenger in a car turning a corner.

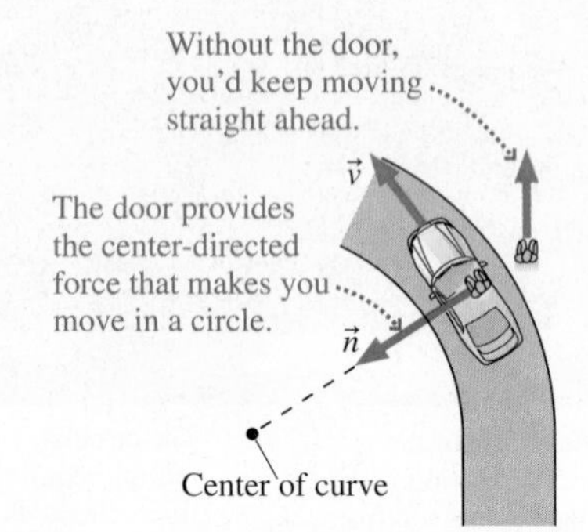

A "force" that *seems* to push an object to the outside of a circle is called a *centrifugal force.* Despite having a name, there really is no such force. What you feel is your body trying to move ahead in a straight line (which would take you away from the center of the circle) as outside forces act to turn you in a circle. The only real forces, those that appear on free-body diagrams, are the ones pushing inward toward the center. **A centrifugal force will never appear on a free-body diagram and never be included in Newton's laws.**

With this in mind, let's revisit the rotating carnival ride. A person watching from above would see the riders in the cylinder moving in a circle with the walls providing the inward force that causes their centripetal acceleration. The riders *feel* as if they're being pushed outward because their natural tendency to move in a straight line is being resisted by the wall of the cylinder, which keeps getting in the way. But feelings aren't forces. The only actual force is the contact force of the cylinder wall pushing *inward.*

Video What the Physics? Shake It Off

Apparent Weight in Circular Motion

Imagine swinging a bucket of water over your head. If you swing the bucket fast enough, the water stays in. But you'll get a shower if you swing too slowly. Why does the water stay in the bucket? Or think about a roller coaster that does a loop-the-loop. How does the car stay on the track when it's upside down? You might have said that there was a centrifugal force holding the water in the bucket and the car on the track, but we have seen that there really isn't a centrifugal force. Analyzing these questions will tell us a lot about forces in general and circular motion in particular.

FIGURE 6.16a shows a roller coaster car going around a vertical loop-the-loop of radius r. If you've ever ridden a roller coaster, you know that your sensation of weight changes as you go over the crests and through the dips. To understand why, let's look at the forces on passengers going through the loop. To simplify our analysis, we will assume that the speed of the car stays constant as it moves through the loop.

FIGURE 6.16b shows a passenger's free-body diagram at the top and the bottom of the loop. Let's start by examining the forces on the passenger at the bottom of the loop. The only forces acting on her are her weight $\vec{w}$ and the normal force $\vec{n}$ of the seat pushing up on her. But recall from ◀◀ SECTION 5.3 that you don't feel the weight force. The force you *feel,* your apparent weight, is the magnitude of the contact force that supports you. Here the seat is supporting the passenger with the normal force $\vec{n}$, so her apparent weight is $w_{app} = n$. Based on our understanding of circular motion, we can say:

- She's moving in a circle, so there must be a net force directed toward the center of the circle—currently directly above her head—to provide the centripetal acceleration.
- The net force points *upward,* so it must be the case that $n > w$.
- Her apparent weight is $w_{app} = n$, so her apparent weight is greater than her true weight $(w_{app} > w)$. Thus she "feels heavy" at the bottom of the circle.

This situation is the same as for the car driving through a dip in Conceptual Example 6.4. To analyze the situation quantitatively, we'll apply the steps of Problem-Solving Approach 6.1. As always, we choose the x-axis to point toward the center of the circle or, in this case, vertically upward. Then Newton's second law is

$$\sum F_x = n_x + w_x = n - w = \frac{mv^2}{r}$$

From this equation, the passenger's apparent weight is

$$w_{app} = n = w + \frac{mv^2}{r} \tag{6.8}$$

Her apparent weight at the bottom is *greater* than her true weight w, which agrees with your experience when you go through a dip or a valley.

Now let's look at the roller coaster car as it crosses the top of the loop. Things are a little trickier here. As Figure 6.16b shows, whereas the normal force of the seat pushes up when the passenger is at the bottom of the circle, it pushes *down* when she is at the top and the seat is above her. It's worth thinking carefully about this diagram to make sure you understand what it is showing.

The passenger is still moving in a circle, so there must be a net force *downward,* toward the center of the circle, to provide her centripetal acceleration. As always, we define the x-axis to be toward the center of the circle, so here the x-axis points vertically downward. Newton's second law gives

$$\sum F_x = n_x + w_x = n + w = \frac{mv^2}{r}$$

Note that w_x is now *positive* because the x-axis is directed downward. We can solve for the passenger's apparent weight:

$$w_{app} = n = \frac{mv^2}{r} - w \tag{6.9}$$

If v is sufficiently large, her apparent weight can exceed the true weight, just as it did at the bottom of the track.

But let's look at what happens if the car goes slower. Notice from Equation 6.9 that, as v decreases, there comes a point when $mv^2/r = w$ and n becomes zero. At that point, the seat is *not* pushing against the passenger at all! Instead, she is able to complete the circle because her weight force alone provides sufficient centripetal acceleration.

FIGURE 6.16 A roller coaster car going around a loop-the-loop.

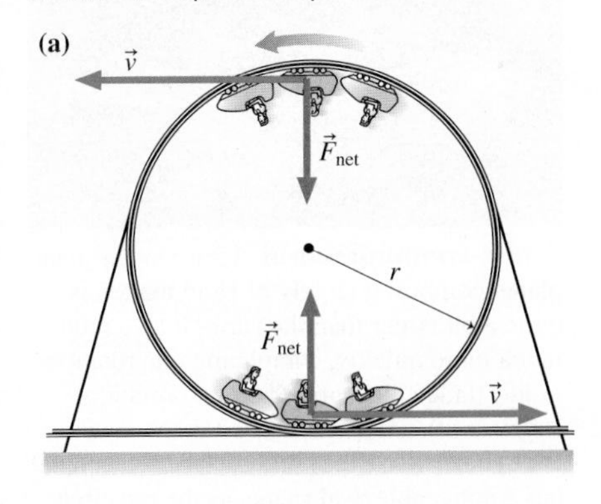

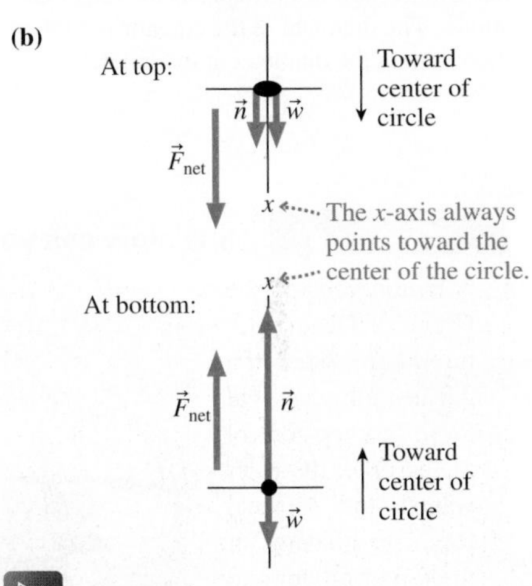

Video Figure 6.16

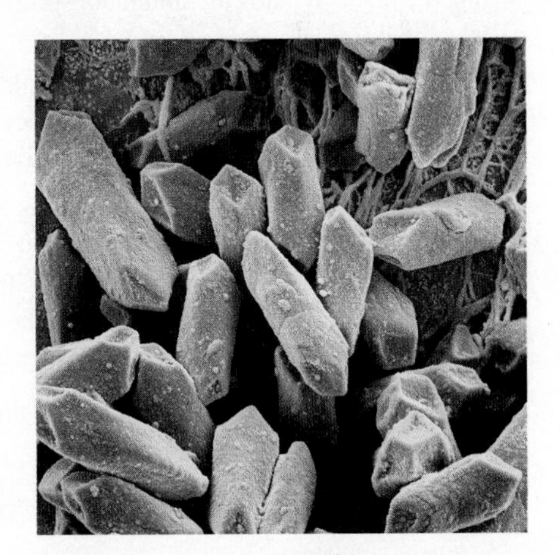

When "down" is up BIO You can tell, even with your eyes closed, what direction is down. Organs in your inner ear contain small crystals of calcium carbonate, called *otoliths.* These crystals are supported by a sensitive membrane. Your brain interprets "down" as the opposite of the direction of the normal force of the membrane on the otoliths. At the top of a loop in a roller coaster, this normal force is directed down, so your inner ear tells you that "down" is up. You are upside down, but it doesn't feel that way.

A fast-spinning world Saturn, a gas giant planet composed largely of fluid matter, is quite a bit larger than the earth. It also rotates much more quickly, completing one rotation in just under 11 hours. The rapid rotation decreases the apparent weight at the equator enough to distort the fluid surface. The planet has a noticeable oval shape, as the red circle shows. The diameter at the equator is 11% greater than the diameter at the poles.

The speed for which $n = 0$ is called the *critical speed* v_c. Because for n to be zero we must have $mv_c^2/r = w$, the critical speed is

$$v_c = \sqrt{\frac{rw}{m}} = \sqrt{\frac{rmg}{m}} = \sqrt{gr} \tag{6.10}$$

What happens if the speed is slower than the critical speed? In this case, Equation 6.9 gives a *negative* value for n if $v < v_c$. But that is physically impossible. The seat can push against the passenger ($n > 0$), but it can't *pull* on her, so the slowest possible speed is the speed for which $n = 0$ at the top. Thus, **the critical speed is the slowest speed at which the car can complete the circle.** If $v < v_c$, the passenger cannot turn the full loop but, instead, will fall from the car as a projectile! (This is why you're always strapped into a roller coaster.)

Water stays in a bucket swung over your head for the same reason. The bottom of the bucket pushes against the water to provide the inward force that causes circular motion. If you swing the bucket too slowly, the force of the bucket on the water drops to zero. At that point, the water leaves the bucket and becomes a projectile following a parabolic trajectory onto your head!

EXAMPLE 6.9 How slow can you go?

A motorcyclist in the Globe of Death, pictured here, rides in a 2.2-m-radius vertical loop. To keep control of the bike, the rider wants the normal force on his tires at the top of the loop to equal or exceed his and the bike's combined weight. What is the minimum speed at which the rider can take the loop?

STRATEGIZE At the top of the loop both the normal force and the weight point downward. Together they provide the net force that determines the centripetal acceleration.

PREPARE The visual overview for this problem is shown in FIGURE 6.17. In accordance with Problem-Solving Approach 6.1, we've chosen the x-axis to point toward the center of the circle.

SOLVE We will consider the forces at the top point of the loop. Because the x-axis points downward, Newton's second law is

$$\sum F_x = w + n = \frac{mv^2}{r}$$

FIGURE 6.17 Riding in a vertical loop around the Globe of Death.

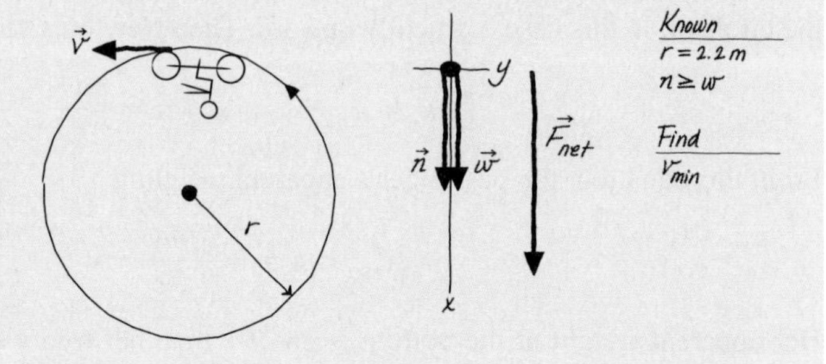

The minimum acceptable speed occurs when $n = w$; thus

$$2w = 2mg = \frac{mv_{min}^2}{r}$$

Solving for the speed, we find

$$v_{min} = \sqrt{2gr} = \sqrt{2(9.8\ \text{m/s}^2)(2.2\ \text{m})} = 6.6\ \text{m/s}$$

ASSESS The minimum speed is ≈ 15 mph, which isn't all that fast; the bikes can easily reach this speed. But normally several bikes are in the globe at one time. The big challenge is to keep all of the riders in the cage moving at this speed in synchrony. The period for the circular motion at this speed is $T = 2\pi r/v \approx 2$ s, leaving little room for error!

Centrifuges BIO

The *centrifuge,* an important biological application of circular motion, is used to separate the components of a liquid that have different densities. Typically these are different types of cells, or the components of cells, suspended in water. You probably know that small particles suspended in water will eventually settle to the bottom. However, the downward motion due to gravity for extremely small objects such as cells is so slow that it could take days or even months for the cells to settle out. It's not practical to wait for biological samples to separate due to gravity alone.

The separation would go faster if the force of gravity could be increased. Although we can't change gravity, we can increase the apparent weight of objects in the sample by spinning them very fast, and that is what the centrifuge shown in FIGURE 6.18 does. The centrifuge produces centripetal accelerations that are thousands of times greater than free-fall acceleration. As the centrifuge effectively increases gravity to thousands of times its normal value, the cells or cell components settle out and separate by density in a matter of minutes or hours.

FIGURE 6.18 The operation of a centrifuge.

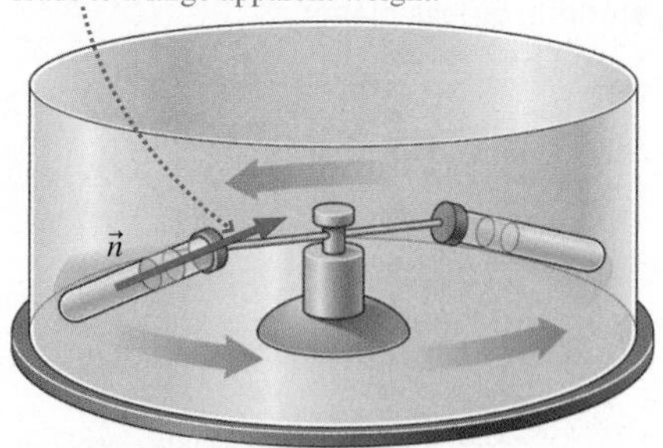
The high centripetal acceleration requires a large normal force, which leads to a large apparent weight.
$\vec{n}$

EXAMPLE 6.10 Analyzing the ultracentrifuge

An 18-cm-diameter ultracentrifuge produces an extraordinarily large centripetal acceleration of 250,000g, where g is the free-fall acceleration due to gravity. What is its frequency in rpm? What is the apparent weight of a sample with a mass of 0.0030 kg?

STRATEGIZE We will use Equation 6.5 for the relationship between the centripetal acceleration and the frequency and radius.

PREPARE The acceleration in SI units is

$$a = 250{,}000(9.80\ \text{m/s}^2) = 2.45 \times 10^6\ \text{m/s}^2$$

The radius is half the diameter, or $r = 9.0\ \text{cm} = 0.090\ \text{m}$.

SOLVE We can rearrange Equation 6.5 to find the frequency given the centripetal acceleration:

$$f = \frac{1}{2\pi}\sqrt{\frac{a}{r}} = \frac{1}{2\pi}\sqrt{\frac{2.45 \times 10^6\ \text{m/s}^2}{0.090\ \text{m}}} = 830\ \text{rev/s}$$

Converting to rpm, we find

$$830\ \frac{\text{rev}}{\text{s}} \times \frac{60\ \text{s}}{1\ \text{min}} = 50{,}000\ \text{rpm}$$

The acceleration is so high that every force is negligible except for the force that provides the centripetal acceleration. The net force is simply equal to the inward force, which is also the sample's apparent weight:

$$w_{\text{app}} = F_{\text{net}} = ma = (3.0 \times 10^{-3}\ \text{kg})(2.45 \times 10^6\ \text{m/s}^2) = 7.4 \times 10^3\ \text{N}$$

The 3 gram sample has an effective weight of about 1700 pounds!

ASSESS Because the acceleration is 250,000g, the apparent weight is 250,000 times the actual weight. This makes sense, as does the fact that we calculated a very high frequency, which is necessary to give the large acceleration.

Human centrifuge BIO If you spin your arm rapidly in a vertical circle, the motion will produce an effect like that in a centrifuge. The motion will assist outbound blood flow in your arteries and retard inbound blood flow in your veins. There will be a buildup of fluid in your hand that you will be able to see (and feel!) quite easily.

STOP TO THINK 6.3 A car is rolling over the top of a hill at constant speed v. At this instant,

A. $n > w$.
B. $n < w$.
C. $n = w$.
D. We can't tell about n without knowing v.

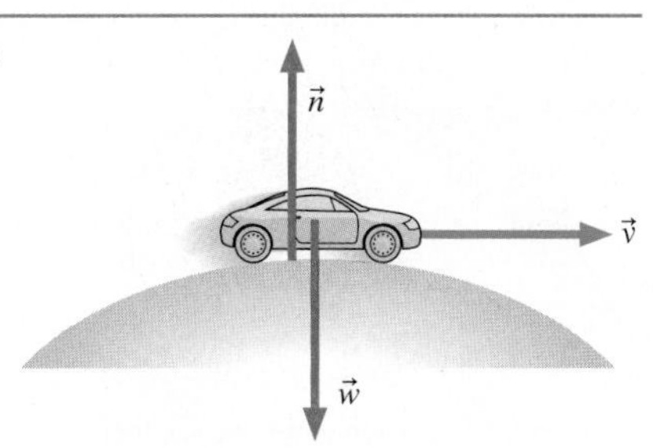
$\vec{n}$
$\vec{v}$
$\vec{w}$

6.4 Circular Orbits and Weightlessness

The International Space Station orbits the earth in a circular path at a speed of over 17,000 miles per hour. What forces act on it? Why does it move in a circle? Before we start considering the physics of orbital motion, let's return, for a moment, to projectile motion. Projectile motion occurs when the only force on an object is gravity. Our analysis of projectiles made an implicit assumption that the earth is flat and that the free-fall acceleration, due to gravity, is everywhere straight down. This is an acceptable approximation for projectiles of limited range, such as baseballs or cannon balls, but there comes a point where we can no longer ignore the curvature of the earth.

Orbital Motion

FIGURE 6.19 Projectiles being launched at increasing speeds from height h on a smooth, airless planet.

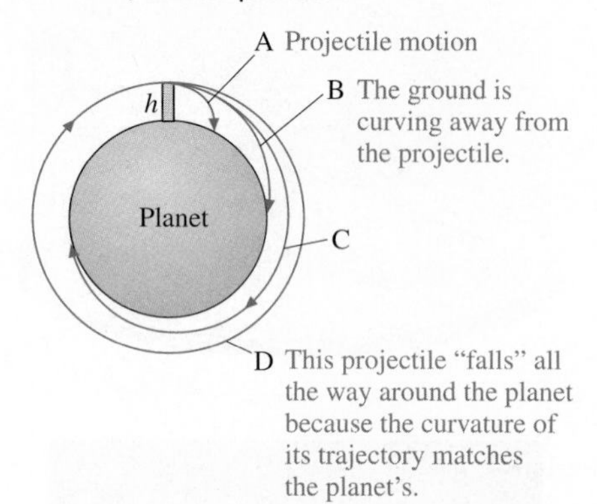
A Projectile motion
h
B The ground is curving away from the projectile.
Planet
C
D This projectile "falls" all the way around the planet because the curvature of its trajectory matches the planet's.

FIGURE 6.19 shows a perfectly smooth, spherical, airless planet with a vertical tower of height h. A projectile is launched from this tower with initial speed v_i parallel to the ground. If v_i is very small, as in trajectory A, the "flat-earth approximation" is valid and the problem is identical to Example 3.9 in which a dog ran off the end of a dock. The projectile simply falls to the ground along a parabolic trajectory.

As the initial speed v_i is increased, it seems from the projectile's perspective that the ground is curving out from beneath it. It is still falling the entire time, always getting closer to the ground, but the distance that the projectile travels before finally reaching the ground—that is, its range—increases because the projectile must "catch up" with the ground that is curving away from it. Trajectories B and C are like this.

If the launch speed v_i is sufficiently large, there comes a point at which the curve of the trajectory and the curve of the earth are parallel. In this case, the projectile "falls" but it never gets any closer to the ground! This is the situation for trajectory D. The projectile returns to the point from which it was launched, at the same speed at which it was launched, making a closed trajectory. Such a closed trajectory around a planet or star is called an **orbit**.

The most important point of this qualitative analysis is that, in the absence of air resistance, **an orbiting projectile is in free fall.** This is, admittedly, a strange idea, but one worth careful thought. An orbiting projectile is really no different from a thrown baseball or a dog jumping off a dock. The only force acting on it is gravity, but its tangential velocity is so great that the curvature of its trajectory matches the curvature of the earth. When this happens, the projectile "falls" under the influence of gravity but never gets any closer to the surface, which curves away beneath it.

When we first studied free fall in Chapter 2, we said that free-fall acceleration is always directed vertically downward. As we see in FIGURE 6.20, "downward" really means "toward the center of the earth." For a projectile in orbit, the direction of the force of gravity changes, always pointing toward the center of the earth.

FIGURE 6.20 The force of gravity is really directed toward the center of the earth.

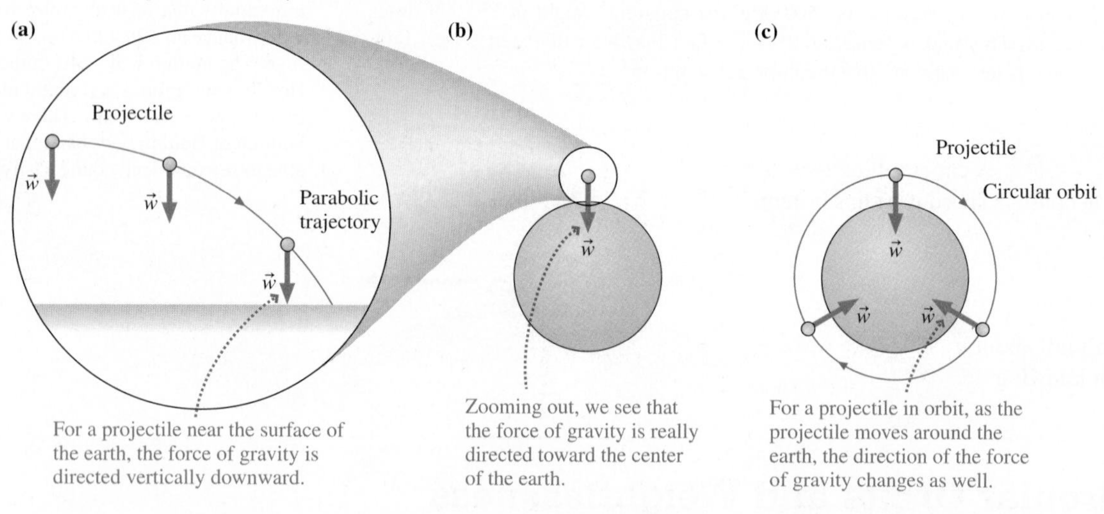
(a)
Projectile
$\vec{w}$
Parabolic trajectory
For a projectile near the surface of the earth, the force of gravity is directed vertically downward.
(b)
Zooming out, we see that the force of gravity is really directed toward the center of the earth.
(c)
Projectile
Circular orbit
For a projectile in orbit, as the projectile moves around the earth, the direction of the force of gravity changes as well.

As you have learned, a force of constant magnitude that always points toward the center of a circle causes the centripetal acceleration of uniform circular motion. Because the only force acting on the orbiting projectile in Figure 6.20 is gravity, and we're assuming the projectile is very near the surface of the earth, we can write

$$a = \frac{F_{net}}{m} = \frac{w}{m} = \frac{mg}{m} = g \qquad (6.11)$$

An object moving in a circle of radius r at speed v_{orbit} will have this centripetal acceleration if

$$a = \frac{(v_{orbit})^2}{r} = g \tag{6.12}$$

That is, if an object moves parallel to the surface with the speed

$$v_{orbit} = \sqrt{gr} \tag{6.13}$$

then the free-fall acceleration provides exactly the centripetal acceleration needed for a circular orbit of radius r. An object with any other speed will not follow a circular orbit.

The earth's radius is $r = R_e = 6.37 \times 10^6$ m. The orbital speed of a projectile just skimming the surface of a smooth, airless earth is

$$v_{orbit} = \sqrt{gR_e} = \sqrt{(9.80\ \text{m/s}^2)(6.37 \times 10^6\ \text{m})} = 7900\ \text{m/s} \approx 18{,}000\ \text{mph}$$

We can use v_{orbit} to calculate the period of the satellite's orbit:

$$T = \frac{2\pi r}{v_{orbit}} = 2\pi\sqrt{\frac{r}{g}} \tag{6.14}$$

For this earth-skimming orbit, $T = 5065\ \text{s} = 84.4$ min.

Of course, actual satellites must orbit at some height above the surface to be above mountains and trees—and most of the atmosphere, so there is little drag. Most people envision the International Space Station orbiting far above the earth's surface, but the average height is just over 250 miles, giving a value of r that is only 6% greater than the earth's radius—not too far from skimming the surface. At this slightly larger value of r, Equation 6.14 gives $T = 87$ min. In fact, the International Space Station orbits with a period of about 93 min, going around the earth more than 15 times each day.

Zero apparent weight in space.

Weightlessness in Orbit

When we discussed *weightlessness* in ◀◀ SECTION 5.3, we saw that it occurs during free fall. We asked whether astronauts and their spacecraft are in free fall. We can now give an affirmative answer: They are, indeed, in free fall. They are falling continuously around the earth, under the influence of only the gravitational force, but never getting any closer to the ground because the earth's surface curves beneath them. Weightlessness in space is no different from the weightlessness in a free-falling elevator. **Weightlessness does *not* occur from an absence of weight or an absence of gravity.** Instead, the astronaut, the spacecraft, and everything in it are "weightless" (i.e., their *apparent* weight is zero) because they are all falling together. We know that the free-fall acceleration doesn't depend on mass, so the astronaut and the station follow exactly the same orbit.

Rotating space stations BIO The weightlessness astronauts experience in orbit has serious physiological consequences. Astronauts who spend time in weightless environments lose bone and muscle mass and suffer other adverse effects. One solution is to introduce "artificial gravity." On a space station, the easiest way to do this would be to make the station rotate, producing an apparent weight. The designers of this space station model for the movie *2001: A Space Odyssey* made it rotate for just that reason.

The Orbit of the Moon

The moon moves in an orbit around the earth that is approximately circular. The force that holds the moon in its orbit, that provides the necessary centripetal acceleration, is the gravitational attraction of the earth. The moon, like all satellites, is simply "falling" around the earth. But if we use the distance to the moon, $r = 3.84 \times 10^8$ m, in Equation 6.14 to predict the period of the moon's orbit, we get a period of approximately 11 hours. This is clearly wrong; you know that the period of the moon's orbit is about one month. What went wrong?

In using Equation 6.14, we assumed that the free-fall acceleration g is the same at the distance of the moon as it is on or near the earth's surface. But if gravity is the force of the earth pulling on an object, it seems plausible that the size of that force, and thus the size of g, should diminish with increasing distance from the earth. And, indeed, the force of gravity does decrease with distance, in a manner we'll explore in the next section.

STOP TO THINK 6.4 A satellite is in a low earth orbit. Which of the following changes would increase the orbital period?

A. Increasing the mass of the satellite
B. Increasing the height of the satellite about the surface
C. Increasing the value of g

6.5 Newton's Law of Gravity

Our current understanding of the force of gravity begins with Isaac Newton. The popular image of Newton coming to a key realization about gravity after an apple fell on his head is at least close to the truth: Newton himself said that the "notion of gravitation" came to him as he "sat in a contemplative mood" and "was occasioned by the fall of an apple."

The important notion that came to Newton is this: *Gravity is a universal force that affects all objects in the universe.* The force that causes the fall of an apple is the same force that keeps the moon in orbit. This is something widely accepted now, but at the time this was a revolutionary idea, and there were some important details for Newton to work out—in particular, the way that the force varies with distance.

Gravity Obeys an Inverse-Square Law

Newton proposed that *every* object in the universe attracts *every other* object with a force that has the following properties:

1. The force is inversely proportional to the square of the distance between the objects.
2. The force is directly proportional to the product of the masses of the two objects.

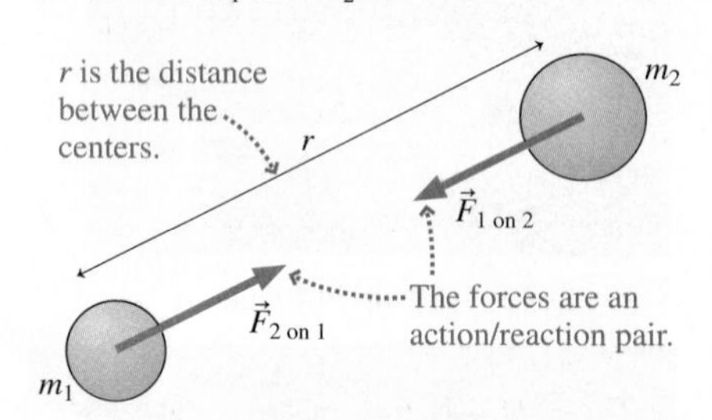
r is the distance between the centers.
r
m_2
$\vec{F}_{1 \text{ on } 2}$
The forces are an action/reaction pair.
$\vec{F}_{2 \text{ on } 1}$
m_1

FIGURE 6.21 The gravitational forces on masses m_1 and m_2.

FIGURE 6.21 shows two spherical objects with masses m_1 and m_2 separated by distance r. Each object exerts an attractive force on the other, a force that we call the **gravitational force.** These two forces form an action/reaction pair, so $\vec{F}_{1 \text{ on } 2}$ is equal in magnitude and opposite in direction to $\vec{F}_{2 \text{ on } 1}$. The magnitude of the forces is given by Newton's law of gravity.

Newton's law of gravity If two objects with masses m_1 and m_2 are a distance r apart, the objects exert attractive forces on each other of magnitude

$$F_{1 \text{ on } 2} = F_{2 \text{ on } 1} = \frac{Gm_1m_2}{r^2} \quad (6.15)$$

INVERSE-SQUARE p.191

The forces are directed along the line joining the two objects.

The constant G is called the **gravitational constant.** In SI units,

$$G = 6.67 \times 10^{-11} \text{ N} \cdot \text{m}^2/\text{kg}^2$$

NOTE ▶ Strictly speaking, Newton's law of gravity applies to *particles* with masses m_1 and m_2. However, it can be shown that the law also applies to the force between two spherical objects if r is the distance between their centers. ◀

As the distance r between two objects increases, the gravitational force between them decreases. Because the distance appears squared in the denominator, Newton's law of gravity is what we call an **inverse-square** law. Doubling the distance between two masses causes the force between them to decrease by a factor of 4. This mathematical form is one we will see again, so it is worth our time to explore it in more detail.

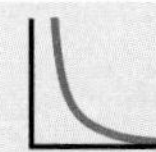

Inverse-square relationships

Two quantities have an **inverse-square relationship** if y is inversely proportional to the *square* of x. We write the mathematical relationship as

$$y = \frac{A}{x^2}$$

y is inversely proportional to x^2.

Here, A is a constant. This relationship is sometimes written as $y \propto 1/x^2$.

SCALING As the graph shows, inverse-square scaling means, for example:

- If you double x, you decrease y by a factor of 4.
- If you halve x, you increase y by a factor of 4.
- If you increase x by a factor of 3, you decrease y by a factor of 9.
- If you decrease x by a factor of 3, you increase y by a factor of 9.

Generally, **if *x increases* by a factor of *C*, *y decreases* by a factor of C^2.** If x *decreases* by a factor of C, y *increases* by a factor of C^2.

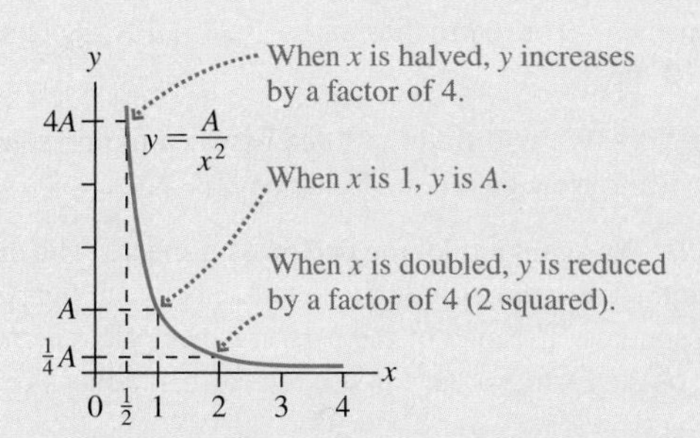

RATIOS For any two values of x—say, x_1 and x_2—we have

$$y_1 = \frac{A}{x_1^2} \quad \text{and} \quad y_2 = \frac{A}{x_2^2}$$

Dividing the y_1-equation by the y_2-equation, we find

$$\frac{y_1}{y_2} = \frac{A/x_1^2}{A/x_2^2} = \frac{A}{x_1^2}\frac{x_2^2}{A} = \frac{x_2^2}{x_1^2}$$

That is, the ratio of y-values is the inverse of the ratio of the squares of the corresponding values of x.

LIMITS As x becomes large, y becomes very small; as x becomes small, y becomes very large.

Exercises 19, 20 

CONCEPTUAL EXAMPLE 6.11 The gravitational force between two spheres

The gravitational force between two giant lead spheres is 0.010 N when the centers of the spheres are 20 m apart. What is the distance between their centers when the gravitational force between them is 0.160 N?

REASON We can solve this problem without knowing the masses of the two spheres. The key is to consider the ratios of forces and distances. Gravity is an inverse-square relationship; the force is related to the inverse square of the distance. The force *increases* by a factor of (0.160 N)/(0.010 N) = 16, so the distance must *decrease* by a factor of $\sqrt{16} = 4$. The distance is thus (20 m)/4 = 5.0 m.

ASSESS This type of ratio reasoning is a very good way to get a quick handle on the solution to a problem.

EXAMPLE 6.12 Finding the gravitational force between two people

You are seated in your physics class next to another student 0.60 m away. Estimate the magnitude of the gravitational force between you. Assume that you each have a mass of 65 kg.

STRATEGIZE We will estimate the force using Newton's law of gravity.

PREPARE We model each of you as a sphere. This is not a particularly good model, but it will do for making an estimate. We then take 0.60 m as the distance between your centers.

SOLVE The gravitational force is given by Equation 6.15:

$$F_{(\text{you})\text{ on }(\text{other student})} = \frac{Gm_{\text{you}}m_{\text{other student}}}{r^2}$$
$$= \frac{(6.67 \times 10^{-11}\ \text{N} \cdot \text{m}^2/\text{kg}^2)(65\ \text{kg})(65\ \text{kg})}{(0.60\ \text{m})^2}$$
$$= 7.8 \times 10^{-7}\ \text{N}$$

ASSESS The force is quite small, roughly the weight of one hair on your head. This seems reasonable; you don't normally sense this attractive force!

There is a gravitational force between all objects in the universe, but the gravitational force between two ordinary-sized objects is extremely small. Only when one (or both) of the masses is exceptionally large does the force of gravity become important. The downward force of the earth on you—your weight—is large because the earth has an enormous mass. And the attraction is mutual: By Newton's third law, you exert an upward force on the earth that is equal to your weight. However, the large mass of the earth makes the *effect* of this force on the earth negligible.

EXAMPLE 6.13 Finding the gravitational force of the earth on a person

What is the magnitude of the gravitational force of the earth on a 60 kg person? The earth has mass 5.98×10^{24} kg and radius 6.37×10^6 m.

STRATEGIZE We will find the force between the person and the earth by using Newton's law of gravity.

PREPARE We again model the person as a sphere. The distance r in Newton's law of gravity is the distance between the *centers* of the two spheres. The size of the person is negligible compared to the size of the earth, so we can use the earth's radius as r.

SOLVE The force of gravity on the person due to the earth can be computed using Equation 6.15:

$$F_{\text{earth on person}} = \frac{GM_e m}{R_e^2}$$
$$= \frac{(6.67 \times 10^{-11}\ \text{N}\cdot\text{m}^2/\text{kg}^2)(5.98 \times 10^{24}\ \text{kg})(60\ \text{kg})}{(6.37 \times 10^6\ \text{m})^2}$$
$$= 590\ \text{N}$$

ASSESS This force is exactly the same as we would calculate using the formula for the weight force, $w = mg$. This isn't surprising, though. Chapter 5 introduced the weight of an object as simply the "force of gravity" acting on it. Newton's law of gravity is a more fundamental law for calculating the force of gravity, but it's still the same force that we earlier called "weight."

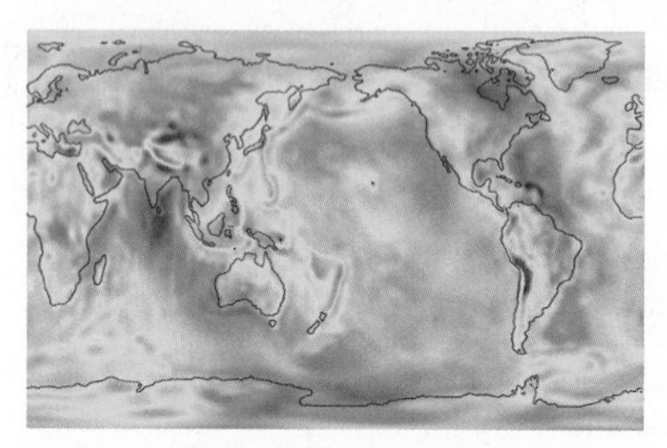

Variable gravity When we calculated the force of the earth's gravity, we assumed that the earth's shape and composition are uniform. In reality, unevenness in density and other factors create small variations in the earth's gravity, as shown in this image. Red means slightly stronger surface gravity; blue means slightly weaker. These variations are important for scientists who study the earth, but they are small enough that we can ignore them for the computations we'll do in this text.

NOTE ► We will use uppercase R and M to represent the large mass and radius of a star or planet, as we did in Example 6.13. ◄

Gravity on Other Worlds

The force of gravitational attraction between the earth and you is responsible for your weight. If you traveled to another planet, your *mass* would be the same but your *weight* would vary, as we discussed in Chapter 5. Indeed, when astronauts ventured to the moon, television images showed them walking—and even jumping and skipping—with ease, even though they were wearing life-support systems with a mass greater than 80 kg, a visible reminder that the weight of objects is less on the moon. Let's consider why this is so.

FIGURE 6.22 shows an astronaut on the moon weighing a rock of mass m. When we compute the weight of an object on the surface of the earth, we use the formula $w = mg$. We can do the same calculation for a mass on the moon, as long as we use the value of g on the moon:

$$w = mg_{\text{moon}} \tag{6.16}$$

This is the "little g" perspective. Falling-body experiments on the moon give the value of g_{moon} as 1.62 m/s^2.

But we can also take a "big G" perspective. The weight of the rock comes from the gravitational attraction of the moon, and we can compute this weight using Equation 6.15. The distance r is the radius of the moon, which we'll call R_{moon}. Thus

$$F_{\text{moon on } m} = \frac{GM_{\text{moon}} m}{R_{\text{moon}}^2} \tag{6.17}$$

Because Equations 6.16 and 6.17 are two names and two expressions for the same force, we can equate the right-hand sides to find that

$$g_{\text{moon}} = \frac{GM_{\text{moon}}}{R_{\text{moon}}^2}$$

We have done this calculation for an object on the moon, but the result is completely general. At the surface of a planet (or a star), the free-fall acceleration g, a consequence of gravity, can be computed as

FIGURE 6.22 An astronaut weighing a mass on the moon.

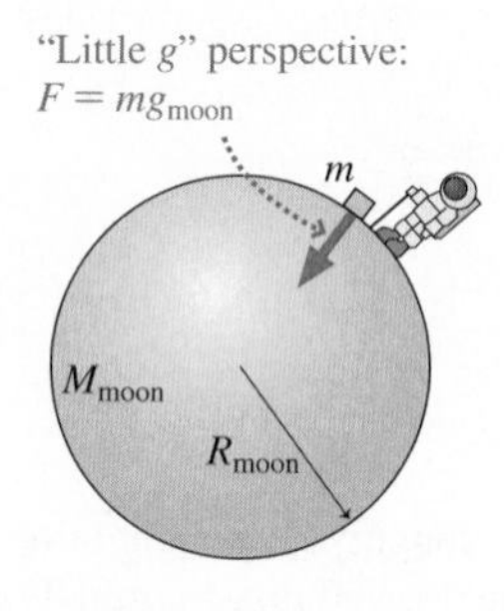

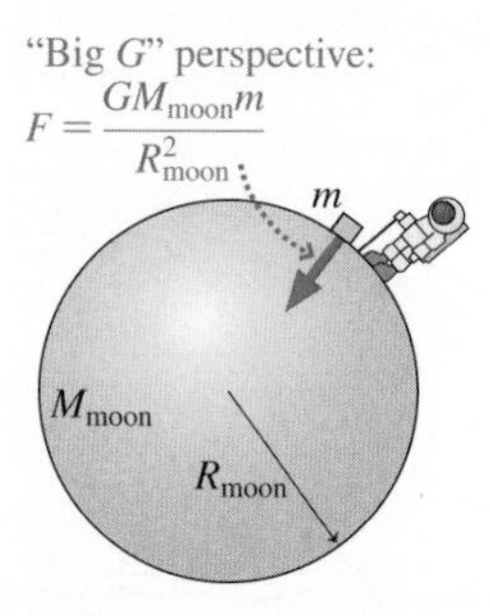

$$g_{\text{planet}} = \frac{GM_{\text{planet}}}{R_{\text{planet}}^2} \tag{6.18}$$

Free-fall acceleration on the surface of a planet

g — M: PROPORTIONAL p. 39; g — R: INVERSE-SQUARE p. 191

If we use values for the mass and the radius of the moon, we compute $g_{moon} = 1.62\ m/s^2$. This means that an object would weigh less on the moon than it would on the earth, where g is $9.80\ m/s^2$. A 70 kg astronaut wearing an 80 kg spacesuit would weigh more than 330 lb on the earth but only 54 lb on the moon.

Equation 6.18 gives g at the surface of a planet. More generally, imagine an object at distance $r > R$ from the center of a planet. Its free-fall acceleration at this distance is

$$g = \frac{GM}{r^2} \tag{6.19}$$

This more general result agrees with Equation 6.18 if $r = R$, but it allows us to determine the "local" free-fall acceleration at distances $r > R$. Equation 6.19 expresses Newton's idea that the size of g should decrease as you get farther from the earth.

As you're flying in a jet airplane at a height of about 10 km, the free-fall acceleration is about 0.3% less than on the ground. At the height of the International Space Station, about 400 km, Equation 6.19 gives $g = 8.7\ m/s^2$, about 13% less than the free-fall acceleration on the earth's surface. If you use this slightly smaller value of g in Equation 6.14 for the period of a satellite's orbit, you'll get the correct period of about 93 minutes. This value of g, only slightly less than the ground-level value, emphasizes the point that an object in orbit is not "weightless" due to the absence of gravity, but rather because it is in free fall.

Walking on the moon BIO We saw (Figure 6.13) that the maximum walking speed depends on leg length and the value of g. The low lunar gravity makes walking very easy but also makes it very slow: The maximum walking speed on the moon is about 1 m/s—a very gentle stroll! Walking at a reasonable pace was difficult for the Apollo astronauts, but the reduced weight made jumping quite easy. Videos from the surface of the moon often showed the astronauts getting from place to place by hopping or skipping—not for fun, but for speed and efficiency.

EXAMPLE 6.14 Finding the speed required to orbit Deimos

Mars has two moons, each much smaller than the earth's moon. The smaller of these two bodies, Deimos, isn't quite spherical, but we can model it as a sphere of radius 6.3 km. Its mass is 1.8×10^{15} kg. At what speed would a projectile move in a very low orbit around Deimos?

SOLVE The free-fall acceleration at the surface of Deimos is small:

$$g_{\text{Deimos}} = \frac{GM_{\text{Deimos}}}{R^2_{\text{Deimos}}} = \frac{(6.67 \times 10^{-11}\ \text{N} \cdot \text{m}^2/\text{kg}^2)(1.8 \times 10^{15}\ \text{kg})}{(6.3 \times 10^3\ \text{m})^2} = 0.0030\ \text{m/s}^2$$

Given this, we can use Equation 6.13 to calculate the orbital speed:

$$v_{\text{orbit}} = \sqrt{gr} = \sqrt{(0.0030\ \text{m/s}^2)(6.3 \times 10^3\ \text{m})} = 4.3\ \text{m/s} \approx 10\ \text{mph}$$

ASSESS This is quite slow. With a good jump, you could easily launch yourself into an orbit around Deimos!

STOP TO THINK 6.5 Rank in order, from largest to smallest, the free-fall accelerations on the surfaces of the following planets.

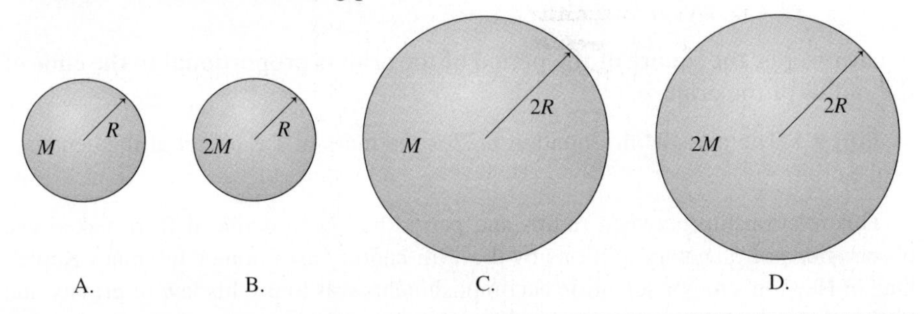

6.6 Gravity and Orbits

The planets of the solar system orbit the sun because the sun's gravitational pull, a force that points toward the center, causes the centripetal acceleration of circular motion. Mercury, the closest planet, experiences the largest acceleration, while Neptune, the most distant, has the smallest.

FIGURE 6.23 The orbital motion of a satellite is due to the force of gravity.

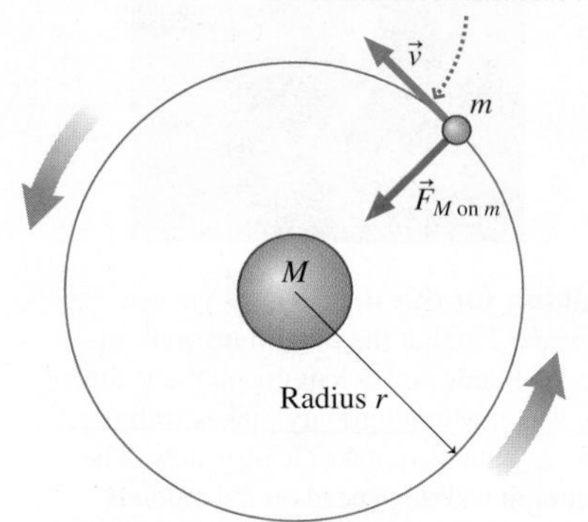
The satellite must have speed $\sqrt{GM/r}$ to maintain a circular orbit of radius r.
$\vec{v}$
m
$\vec{F}_{M \text{ on } m}$
M
Radius r

FIGURE 6.23 shows a large body of mass M such as the sun, with a much smaller body such as the earth of mass m orbiting it. The smaller body is called a **satellite,** even though it may be a planet orbiting the sun. Newton's second law tells us that $F_{M \text{ on } m} = ma$, where $F_{M \text{ on } m}$ is the gravitational force of the large body on the satellite and a is the satellite's acceleration. $F_{M \text{ on } m}$ is given by Equation 6.15, and, because it's moving in a circular orbit, the satellite's acceleration is its centripetal acceleration, mv^2/r. Thus Newton's second law gives

$$F_{M \text{ on } m} = \frac{GMm}{r^2} = ma = \frac{mv^2}{r} \tag{6.20}$$

Solving for v, we find that the speed of a satellite in a circular orbit is

$$v = \sqrt{\frac{GM}{r}} \tag{6.21}$$

Speed of a satellite in a circular orbit of radius r about a star or planet of mass M

A satellite must have this specific speed in order to maintain a circular orbit of radius r about the larger mass M. If the velocity differs from this value, the orbit will become elliptical rather than circular. Notice that the orbital speed does not depend on the satellite's mass m. This is consistent with our previous discoveries that free-fall motion and projectile motion due to gravity are independent of the mass.

For a planet orbiting the sun, the period T is the time to complete one full orbit. The relationship among speed, radius, and period is the same as for any circular motion, $v = 2\pi r/T$. Combining this with the value of v for a circular orbit from Equation 6.21 gives

$$\sqrt{\frac{GM}{r}} = \frac{2\pi r}{T}$$

If we square both sides and rearrange, we find the period of a satellite:

$$T^2 = \left(\frac{4\pi^2}{GM}\right)r^3 \tag{6.22}$$

Relationship between the orbital period T and radius r for a satellite in a circular orbit around an object of mass M

In other words, **the square of the period of the orbit is proportional to the cube of the radius of the orbit.**

NOTE ▸ The mass M in Equation 6.22 is the mass of the object at the center of the orbit. ◂

This relationship between radius and period had been deduced from naked-eye observations of planetary motions by the 17th-century astronomer Johannes Kepler. One of Newton's major scientific accomplishments was to use his law of gravity and his laws of motion to prove what Kepler had deduced from observations. Even today, Newton's law of gravity and equations such as Equation 6.22 are essential tools for the NASA engineers who launch probes to other planets in the solar system.

NOTE ▸ The table inside the back cover of this text contains astronomical information about the sun and the planets that will be useful for many of the end-of-chapter problems. Note that planets farther from the sun have longer periods, in agreement with Equation 6.22. ◂

EXAMPLE 6.15 Locating a geostationary satellite

Communication satellites appear to "hover" over one point on the earth's equator. A satellite that appears to remain stationary as the earth rotates is said to be in a *geostationary orbit.* What is the radius of the orbit of such a satellite?

STRATEGIZE For the satellite to remain stationary with respect to the earth, the satellite's period must be 24 hours.

PREPARE Converted to seconds, the 24-hour period of the satellite is $T = 8.64 \times 10^4$ s.

SOLVE We solve for the radius of the orbit by rearranging Equation 6.22. The mass at the center of the orbit is the earth:

$$r = \left(\frac{GM_e T^2}{4\pi^2}\right)^{\frac{1}{3}}$$

$$= \left(\frac{(6.67 \times 10^{-11}\ \text{N}\cdot\text{m}^2/\text{kg}^2)(5.98 \times 10^{24}\ \text{kg})(8.64 \times 10^4\ \text{s})^2}{4\pi^2}\right)^{\frac{1}{3}}$$

$$= 4.23 \times 10^7\ \text{m}$$

ASSESS This is a high orbit; the radius is about 7 times the radius of the earth. Recall that the radius of the International Space Station's orbit is only about 6% larger than that of the earth.

Gravity on a Grand Scale

A spiral galaxy, similar to our Milky Way galaxy.

Although relatively weak, gravity is a long-range force. No matter how far apart two objects may be, there is a gravitational attraction between them. Consequently, gravity is the most ubiquitous force in the universe. It not only keeps your feet on the ground, but also is at work on a much larger scale. The Milky Way galaxy, the collection of stars of which our sun is a part, is held together by gravity. But why doesn't the attractive force of gravity simply pull all of the stars together?

The reason is that all of the stars in the galaxy are in orbit around the center of the galaxy. The gravitational attraction keeps the stars moving in orbits around the center of the galaxy rather than falling inward, much as the planets orbit the sun rather than falling into the sun. In the nearly 5 billion years that our solar system has existed, it has orbited the center of the galaxy approximately 20 times.

The galaxy as a whole doesn't rotate as a fixed object, though. All of the stars in the galaxy are different distances from the galaxy's center, and so orbit with different periods. Stars closer to the center complete their orbits in less time, as we would expect from Equation 6.22. As the stars orbit, their relative positions shift. Stars that are relatively near neighbors now could be on opposite sides of the galaxy at some later time.

The rotation of a *rigid body* like a wheel is much simpler. As a wheel rotates, all of the points keep the same relationship to each other. The rotational dynamics of such rigid bodies is a topic we will take up in the next chapter.

STOP TO THINK 6.6 Each year, the radius of the moon's orbit increases by about 3.8 cm. How does this change affect the length of a month?

A. A month gets longer.
B. A month gets shorter.
C. The length of a month stays the same.

INTEGRATED EXAMPLE 6.16 A hunter and his sling

A Stone Age hunter stands on a cliff overlooking a flat plain. He places a 1.0 kg rock in a sling, ties the sling to a 1.0-m-long vine, then swings the rock in a horizontal circle around his head. The plane of the motion is 25 m above the plain below. The tension in the vine increases as the rock goes faster and faster. Suddenly, just as the tension reaches 200 N, the vine snaps. If the rock is moving toward the cliff at this instant, how far out on the plain (from the base of the cliff) will it land?

STRATEGIZE This is a two-part problem. First, we will use circular-motion dynamics to calculate the speed of the rock as it moves in a circle. When the vine snaps, the rock will be moving horizontally with this same speed; we can then use projectile-motion kinematics to find how far it goes.

PREPARE We start by modeling the rock as a particle in uniform circular motion. We can use Problem-Solving Approach 6.1 to analyze this part of the motion.

The force identification diagram of FIGURE 6.24a shows that the only contact force acting on the rock is the tension in the vine. Because the rock moves in a horizontal circle, you may be tempted to draw a free-body diagram like FIGURE 6.24b, where $\vec{T}$ is directed along the x-axis. You will quickly run into trouble, however, because in this diagram the net force has a downward y-component that would cause the rock to rapidly accelerate downward. But we know that it moves in a horizontal circle and that the net force must point toward the center of the circle. In this free-body diagram, the weight force $\vec{w}$ points straight down and is certainly correct, so the difficulty must be with $\vec{T}$.

Continued

FIGURE 6.24 Visual overview of a hunter swinging a rock.

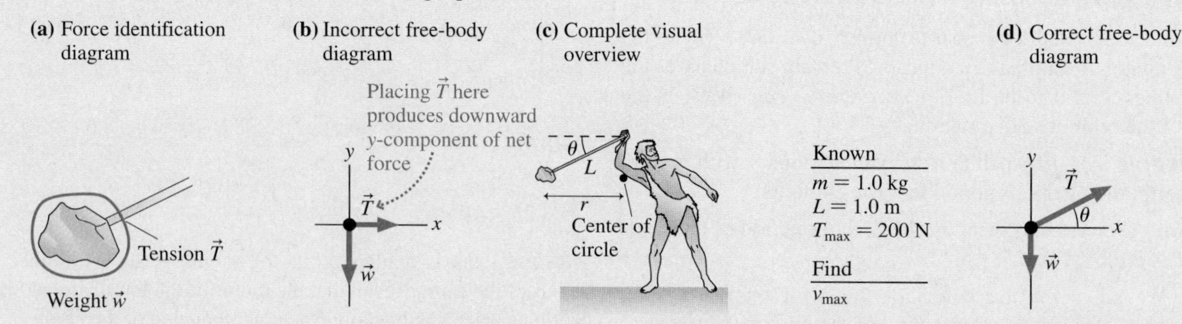

As an experiment, tie a small weight to a string, swing it over your head, and check the angle of the string. You will discover that the string is not horizontal but, instead, is angled downward. The sketch of FIGURE 6.24c labels this angle θ. Notice that the rock moves in a *horizontal* circle, so the center of the circle is not at his hand. The x-axis points horizontally, to the center of the circle, but the tension force is directed along the vine. Thus the correct free-body diagram is the one in FIGURE 6.24d.

Once the vine breaks, the visual overview of the situation is shown in FIGURE 6.25. The important thing to note here is that the initial x-component of velocity is the speed the rock had an instant before the vine broke.

SOLVE From the free-body diagram of Figure 6.24d, Newton's second law for circular motion is

$$\sum F_x = T\cos\theta = \frac{mv^2}{r}$$

$$\sum F_y = T\sin\theta - mg = 0$$

where θ is the angle of the vine below the horizontal. We can use the y-equation to find the angle of the vine:

$$\sin\theta = \frac{mg}{T}$$

$$\theta = \sin^{-1}\left(\frac{mg}{T}\right) = \sin^{-1}\left(\frac{(1.0\text{ kg})(9.8\text{ m/s}^2)}{200\text{ N}}\right) = 2.81°$$

where we've evaluated the angle at the maximum tension of 200 N. The vine's angle of inclination is small but not zero.

Turning now to the x-equation, we find the rock's speed around the circle is

$$v = \sqrt{\frac{rT\cos\theta}{m}}$$

Be careful! The radius r of the circle is not the length L of the vine. You can see in Figure 6.24c that $r = L\cos\theta$. Thus

$$v = \sqrt{\frac{LT\cos^2\theta}{m}} = \sqrt{\frac{(1.0\text{ m})(200\text{ N})(\cos 2.81°)^2}{1.0\text{ kg}}} = 14.1\text{ m/s}$$

Because this is the horizontal speed of the rock just when the vine breaks, the initial velocity $(v_x)_i$ in the visual overview of the projectile motion, Figure 6.25, must be $(v_x)_i = 14.1$ m/s. Recall that a projectile has no horizontal acceleration, so the rock's final position is

$$x_f = x_i + (v_x)_i\,\Delta t = 0\text{ m} + (14.1\text{ m/s})\Delta t$$

where Δt is the time the projectile is in the air. We're not given Δt, but we can find it from the vertical motion. For a projectile, the vertical motion is just free-fall motion, so we have

$$y_f = y_i + (v_y)_i\,\Delta t - \frac{1}{2}g(\Delta t)^2$$

The initial height is $y_i = 25$ m, the final height is $y_f = 0$ m, and the initial vertical velocity is $(v_y)_i = 0$ m/s. With these values, we have

$$0\text{ m} = 25\text{ m} + (0\text{ m/s})\Delta t - \frac{1}{2}(9.8\text{ m/s}^2)(\Delta t)^2$$

Solving this for Δt gives

$$\Delta t = \sqrt{\frac{2(25\text{ m})}{9.8\text{ m/s}^2}} = 2.26\text{ s}$$

Now we can use this time to find

$$x_f = 0\text{ m} + (14.1\text{ m/s})(2.26\text{ s}) = 32\text{ m}$$

The rock lands 32 m from the base of the cliff.

ASSESS The circumference of the rock's circle is $2\pi r$, or about 6 m. At a speed of 14.1 m/s, the rock takes roughly half a second to go around once. This seems reasonable. The 32 m distance is about 100 ft, which seems easily attainable from a cliff over 75 feet high.

FIGURE 6.25 Visual overview of the rock in projectile motion.

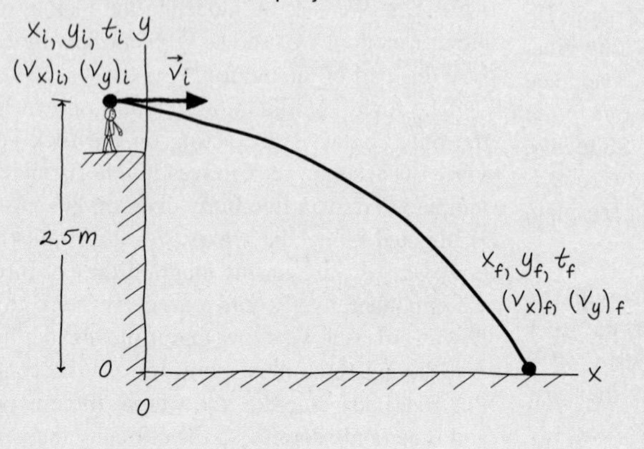

SUMMARY

GOAL To learn about motion in a circle, including orbital motion under the influence of a gravitational force.

GENERAL PRINCIPLES

Uniform Circular Motion

An object moving in a circular path is in uniform circular motion if v is constant.

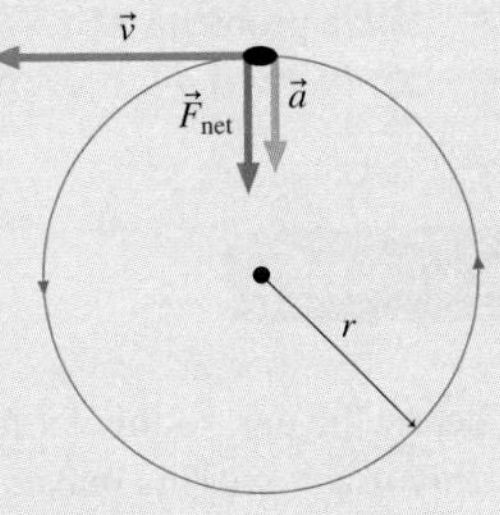

- The speed is constant, but the direction of motion is constantly changing.
- The **centripetal acceleration** is directed toward the center of the circle and has magnitude

$$a = \frac{v^2}{r}$$

- This acceleration requires a net force directed toward the center of the circle. Newton's second law for circular motion is

$$\vec{F}_{net} = m\vec{a} = \left(\frac{mv^2}{r}, \text{ toward center of circle}\right)$$

Universal Gravitation

Two objects with masses m_1 and m_2 that are distance r apart exert attractive gravitational forces on each other of magnitude

$$F_{1 \text{ on } 2} = F_{2 \text{ on } 1} = \frac{Gm_1m_2}{r^2}$$

where the gravitational constant is

$$G = 6.67 \times 10^{-11} \text{ N} \cdot \text{m}^2/\text{kg}^2$$

This is **Newton's law of gravity.** Gravity is an inverse-square law.

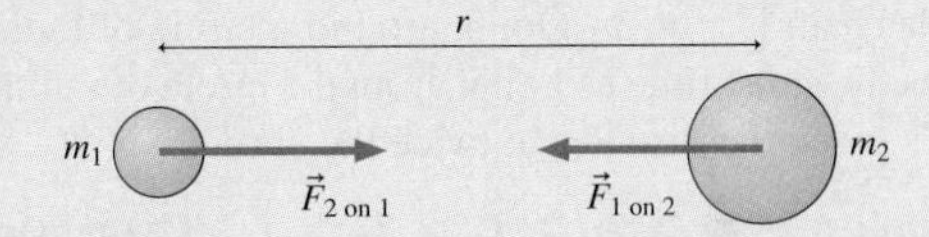

IMPORTANT CONCEPTS

Describing circular motion

For an object moving in a circle of radius r at a constant speed v:

- The **period** T is the time to go once around the circle:

$$T = \text{time for one revolution}$$

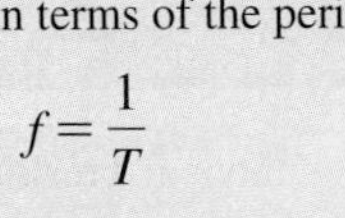

- The **frequency** f is defined as the number of revolutions per second. It is defined in terms of the period:

$$f = \frac{1}{T}$$

- The frequency and period are related to the speed and the radius: $v = 2\pi fr = \dfrac{2\pi r}{T}$

Planetary gravity

The gravitational attraction between a planet and a mass on the surface depends on the two masses and the distance to the center of the planet:

$$F_{\text{planet on } m} = \frac{GM_{planet}m}{R^2_{planet}}$$

We can use this to define a value of the free-fall acceleration at the surface of a planet:

$$g_{planet} = \frac{GM_{planet}}{R^2_{planet}}$$

APPLICATIONS

Apparent weight and weightlessness

Circular motion requires a net force pointing to the center. The apparent weight $w_{app} = n$ is usually not the same as the true weight w. n must be > 0 for the object to be in contact with a surface.

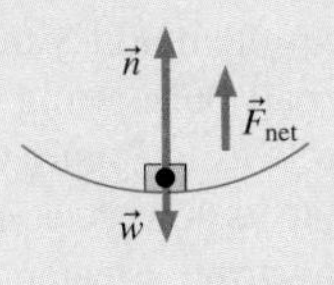

In orbital motion, the net force is provided by gravity alone. An astronaut and his spacecraft are both in free fall, so he feels weightless.

Orbital motion

A **satellite** in a circular orbit of radius r around an object of mass M moves at a speed v given by

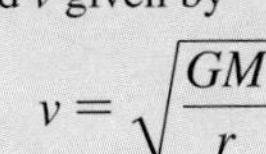

$$v = \sqrt{\frac{GM}{r}}$$

The period and radius are related as follows:

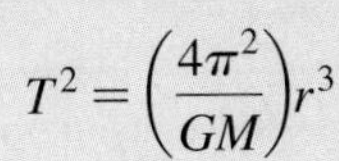

$$T^2 = \left(\frac{4\pi^2}{GM}\right)r^3$$

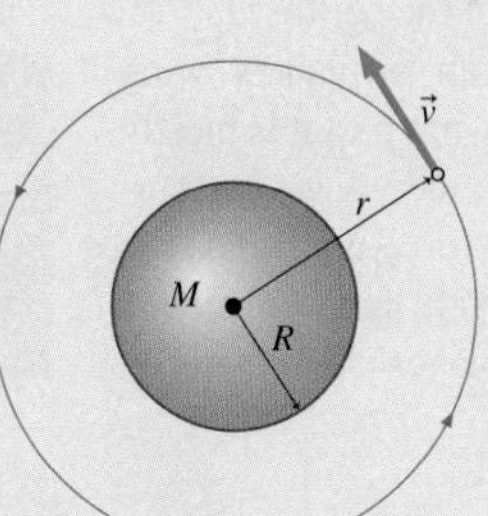

The speed of a satellite in a low orbit is

$$v = \sqrt{gr}$$

The orbital period is

$$T = 2\pi\sqrt{\frac{r}{g}}$$

Learning Objectives After studying this chapter, you should be able to:

- Calculate period, frequency, and speed for objects in circular motion. *Conceptual Questions 6.1, 6.2; Problems 6.1, 6.2, 6.3, 6.9, 6.10*
- Use Newton's laws to solve dynamics problems for objects in uniform circular motion. *Conceptual Questions 6.8, 6.9; Problems 6.16, 6.17, 6.19, 6.20, 6.23*
- Understand the apparent weight of an object in circular motion. *Conceptual Questions 6.4, 6.12; Problems 6.22, 6.26, 6.28, 6.29*
- Analyze the circular orbits of planets and satellites. *Problems 6.7, 6.32, 6.33*
- Use Newton's law of gravity to calculate long-range gravitational forces. *Conceptual Questions 6.18, 6.19; Problems 6.36, 6.38, 6.39, 6.40, 6.41*
- Solve problems about gravity and orbits. *Conceptual Question 6.20; Problems 6.44, 6.47, 6.48, 6.49, 6.50*

STOP TO THINK ANSWERS

Chapter Preview Stop to Think: D. The ball is in uniform circular motion. The acceleration is directed toward the center of the circle, which is to the right at the instant shown.

Stop to Think 6.1: C > D = A > B. Rearranging Equation 6.3 gives $T = \frac{2\pi r}{v}$. For the cases shown, speed is either v or $2v$; the radius of the circular path is r or $2r$. Going around a circle of radius r at a speed v takes the same time as going around a circle of radius $2r$ at a speed $2v$. It's twice the distance at twice the speed.

Stop to Think 6.2: $T_E > T_D > T_C > T_B > T_A$. The tension force provides the centripetal acceleration, and larger acceleration implies larger force. So the question reduces to one about acceleration: Rank the centripetal accelerations for these cases. Equation 6.5 shows that the acceleration is proportional to the radius of the circle and the square of the frequency, and so the acceleration increases steadily as we move from A to E.

Stop to Think 6.3: B. The car is moving in a circle, so there must be a net force toward the center of the circle. The center of the circle is below the car, so the net force must point downward. This can be true only if $w > n$. This makes sense; $n < w$, so the apparent weight is less than the true weight. The riders in the car "feel light"; if you've driven over a rise like this, you know that this is what you feel.

Stop to Think 6.4: B. The period of a satellite doesn't depend on the mass of the satellite. If you increase the height above the surface, you increase the radius of the orbit. This will result in an increased period. Increasing the value of g would cause the period to decrease.

Stop to Think 6.5: B > A > D > C. The free-fall acceleration is proportional to the mass, but inversely proportional to the square of the radius.

Stop to Think 6.6: A. The length of a month is determined by the period of the moon's orbit. Equation 6.22 shows that as the moon gets farther away, the period of the orbit—and thus the length of a month—increases.

Video Tutor Solution Chapter 6

QUESTIONS

Conceptual Questions

1. A cyclist goes around a level, circular track at constant speed. Do you agree or disagree with the following statement? "Since the cyclist's speed is constant, her acceleration is zero." Explain.
2. In uniform circular motion, which of the following quantities are constant: speed, instantaneous velocity, centripetal acceleration, the magnitude of the net force?
3. The force on an airplane's wing from the air flowing past it is mostly perpendicular to the wing's flat surface; this is the force that holds the plane aloft in level flight. Given this fact, why do planes "bank" when making a turn?

4. Tarzan hangs from a vine without it breaking. But when he swings on the same vine starting from a high branch, the vine snaps at the bottom of his swing. Explain why this happens.
5. BIO Large birds like pheasants often walk short distances. Small birds like chickadees never walk. They either hop or fly. Why might this be?
6. When you drive fast on the highway with muddy tires, you can hear the mud flying off the tires into your wheel wells. Why does the mud fly off?
7. A ball on a string moves in a vertical circle as in Figure Q6.7. When the ball is at its lowest point, is the tension in the string greater than, less than, or equal to the ball's weight? Explain. (You may want to include a free-body diagram as part of your explanation.)

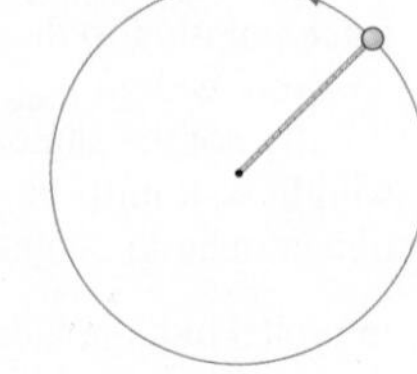

FIGURE Q6.7

Problem difficulty is labeled as | (straightforward) to ||||| (challenging). INT Problems labeled integrate significant material from earlier chapters; BIO are of biological or medical interest.

The eText icon indicates when there is a video tutor solution available for the chapter or for a specific problem. To launch these videos, log into your eText through Mastering™ Physics or log into the Study Area.

8. Give an everyday example of circular motion for which the centripetal acceleration is mostly or completely due to a force of the type specified: (a) Static friction. (b) Tension.
9. Give an everyday example of circular motion for which the centripetal acceleration is mostly or completely due to a force of the type specified: (a) Gravity. (b) Normal force.
10. On a roller coaster loop-the-loop the riders are upside-down at the top of the loop, while on a Ferris wheel the riders are upright at the top. Suppose a Ferris wheel and the loop of a roller coaster have the same diameter. If riders on each have the same magnitude of apparent weight at the top, which rider is moving faster? Explain.
11. A car coasts at a constant speed over a circular hill. Which of the free-body diagrams in Figure Q6.11 is correct? Explain.

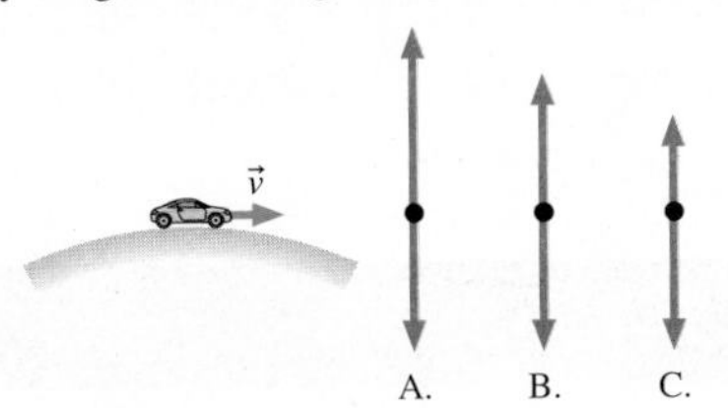

FIGURE Q6.11

12. In Figure Q6.11, at the instant shown, is the apparent weight of the car's driver greater than, less than, or equal to his true weight? Explain.
13. Riding in the back of a pickup truck can be very dangerous. If the truck turns suddenly, the riders can be thrown from the truck bed. Why are the riders ejected from the bed?
14. Playground swings move through an arc of a circle. When you are on a swing, and at the lowest point of your motion, is your apparent weight greater than, less than, or equal to your true weight? Explain.
15. Variation in your apparent weight is desirable when you ride a roller coaster; it makes the ride fun. However, too much variation over a short period of time can be painful. For this reason, the loops of real roller coasters are not simply circles like Figure 6.16a. A typical loop is shown in Figure Q6.15. The radius of the circle that matches the track at the top of the loop is much smaller than that of a matching circle at other places on the track. Explain why this shape gives a more comfortable ride than a circular loop.

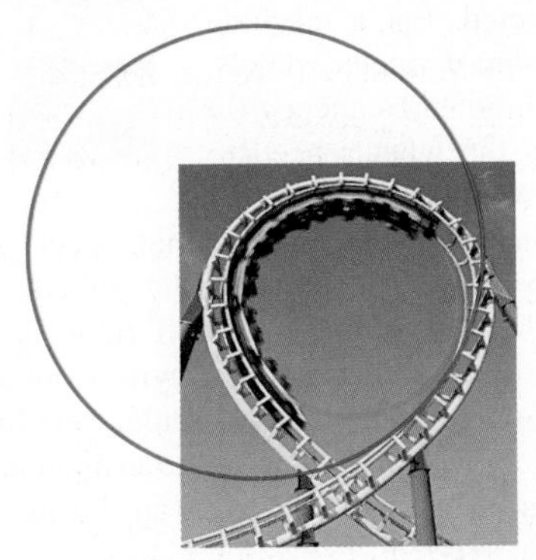

FIGURE Q6.15

16. An object's apparent weight is slightly less when it is at the equator than when it is at the North Pole. Explain why this is so.
17. Why is it impossible for an astronaut inside an orbiting space station to go from one end to the other by walking normally?
18. If every object in the universe feels an attractive gravitational force due to every other object, why don't you feel a pull from someone seated next to you?
19. A mountain climber's weight is slightly less on the top of a tall mountain than at the base, though his mass is the same. Why?
20. The mass of Mars was not known until its two moons were discovered in 1877. Explain how observations of these moons—their distances from Mars and their orbital periods—could be used to find Mars's mass.

Multiple-Choice Questions

21. | A ball on a string moves around a complete circle, once a second, on a frictionless, horizontal table. The tension in the string is measured to be 6.0 N. What would the tension be if the ball went around in only half a second?
    A. 1.5 N B. 3.0 N C. 12 N D. 24 N
22. | As seen from above, a car rounds the curved path shown in Figure Q6.22 at a constant speed. Which vector best represents the net force acting on the car?

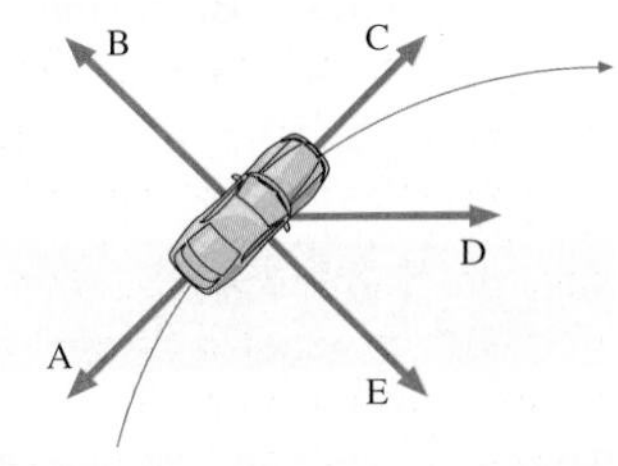

FIGURE Q6.22

23. | As we saw in the chapter, wings on race cars push them into the track. The increased normal force makes large friction forces possible. At one Formula One racetrack, cars turn around a half-circle with diameter 190 m at 68 m/s. For a 610 kg vehicle, the approximate minimum static friction force to complete this turn is
    A. 6000 N B. 15,000 N C. 18,000 N
    D. 24,000 N E. 30,000 N
24. || You swing a weight attached to a string in a vertical circle. At the top of the circle the string just barely goes slack for an instant. At this point the centripetal acceleration of the weight is
    A. Greater than g.
    B. Less than g.
    C. Equal to g.
    D. Can't answer without knowing the weight's speed.
25. || On a snowy day, when the coefficient of friction μ_s between a car's tires and the road is 0.50, the maximum speed that the car can go around a curve is 20 mph. What is the maximum speed at which the car can take the same curve on a sunny day when $\mu_s = 1.0$?
    A. 20 mph B. 24 mph C. 28 mph
    D. 33 mph E. 40 mph
26. | The cylindrical space station in Figure Q6.26, 200 m in diameter, rotates in order to provide artificial gravity of g for the occupants. How much time does the station take to complete one rotation?
    A. 3 s B. 20 s C. 28 s D. 32 s

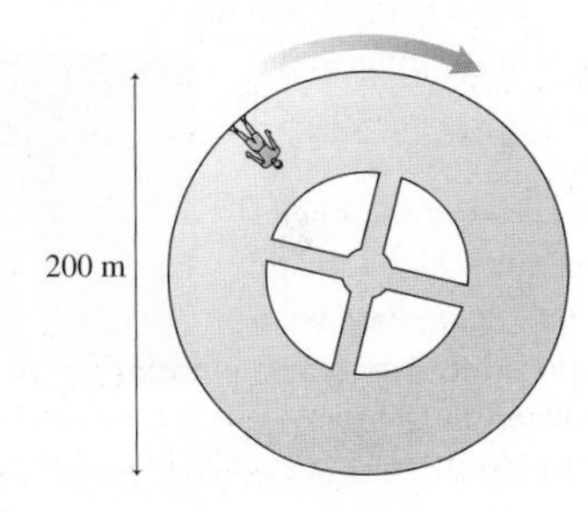

FIGURE Q6.26

27. ‖ The radius of Jupiter is 11 times that of earth, and the free-fall acceleration near its surface is 2.5 times that on earth. If we someday put a spacecraft in low Jupiter orbit, its orbital speed will be
 A. Greater than that for an earth satellite.
 B. The same as that for an earth satellite.
 C. Less than that for an earth satellite.
28. | A newly discovered planet has twice the mass and three times the radius of the earth. What is the free-fall acceleration at its surface, in terms of the free-fall acceleration g at the surface of the earth?
 A. $\frac{2}{9}g$ B. $\frac{2}{3}g$ C. $\frac{3}{4}g$ D. $\frac{4}{3}g$
29. ‖ Suppose one night the radius of the earth doubled but its mass stayed the same. What would be an approximate new value for the free-fall acceleration at the surface of the earth?
 A. 2.5 m/s^2 B. 5.0 m/s^2 C. 10 m/s^2 D. 20 m/s^2
30. | Currently, the moon goes around the earth once every 27.3 days. If the moon could be brought into a new circular orbit with a smaller radius, its orbital period would be
 A. More than 27.3 days.
 B. 27.3 days.
 C. Less than 27.3 days.
31. ‖ Two planets orbit a star. You can ignore the gravitational interactions between the planets. Planet 1 has orbital radius r_1 and planet 2 has $r_2 = 4r_1$. Planet 1 orbits with period T_1. Planet 2 orbits with period
 A. $T_2 = \frac{1}{2}T_1$ B. $T_2 = 2T_1$ C. $T_2 = 4T_1$ D. $T_2 = 8T_1$

PROBLEMS

Section 6.1 Uniform Circular Motion

1. ‖ A 5.0-m-diameter merry-go-round is turning with a 4.0 s period. What is the speed of a child on the rim?
2. | The earth, with a radius of 6.4×10^6 m, rotates on its axis once a day. What is the speed of a person standing on the equator, due to the earth's rotation?
3. | An old-fashioned LP record rotates at $33\frac{1}{3}$ rpm.
 a. What is its frequency, in rev/s?
 b. What is its period, in seconds?
4. | A typical hard disk in a computer spins at 5400 rpm.
 a. What is the frequency, in rev/s?
 b. What is the period, in seconds?
5. | The hammer throw was one of the earliest Olympic events. In this event, a heavy ball attached to a chain is swung several times in a circular path until it is released. The winning athlete is the one who throws the ball the greatest distance. The last complete rotation of 2016 Olympic champion Anita Włodarczyk's final turn took only 0.43 s. The radius of the ball's path, including her extended arms, was 2.1 m.
 a. What was the frequency of this rotation?
 b. What was the speed of the ball?
 c. What was the ball's acceleration, in units of g?
6. ‖ The horse on a carousel is 4.0 m from the central axis.
 a. If the carousel rotates at 0.10 rev/s, how long does it take the horse to go around twice?
 b. How fast is a child on the horse going (in m/s)?
7. ‖| The radius of the earth's very nearly circular orbit around the sun is 1.50×10^{11} m. Find the magnitude of the earth's (a) velocity and (b) centripetal acceleration as it travels around the sun. Assume a year of 365 days.
8. | Modern wind turbines are larger than they appear, and despite their apparently lazy motion, the speed of the blades tips can be quite high—many times higher than the wind speed. A typical modern turbine has blades 56 m long that spin at 13 rpm. At the tip of a blade, what are (a) the speed and (b) the centripetal acceleration?

9. | The California sea lion is capable of making extremely fast, tight turns while swimming underwater. In one study, scientists observed a sea lion making a circular turn with a radius of 0.35 m while swimming at 4.2 m/s.
BIO
 a. What is the sea lion's centripetal acceleration, in units of g?
 b. What percentage is this acceleration of that of an F-15 fighter jet's maximum centripetal acceleration of 9g?
10. | Baseball pitching machines are used to fire baseballs toward a batter for hitting practice. In one kind of machine, the ball is fed between two wheels that are rapidly rotating in opposite directions; as the ball is pulled in between the wheels, it rapidly accelerates up to the speed of the wheels' rims, at which point it is ejected. For a machine with 35-cm-diameter wheels, what rotational frequency (in rpm) do the wheels need to pitch a 90 mph fastball?

11. ‖ Astronauts in the International Space Station must work out every day to counteract the effects of weightlessness. Researchers have investigated if riding a stationary bicycle while experiencing artificial gravity from a rotating platform gives any additional cardiovascular benefit. What frequency of rotation, in rpm, is required to give an acceleration of 1.4g to an astronaut's feet, if her feet are 1.1 m from the platform's rotational axis?
BIO
12. ‖ A typical running track is an oval with 74-m-diameter half circles at each end. A runner going once around the track covers a distance of 400 m. Suppose a runner, moving at a constant speed, goes once around the track in 1 min 40 s. What is her centripetal acceleration during the turn at each end of the track?

Section 6.2 Dynamics of Uniform Circular Motion

13. || Figure P6.13 is a bird's-eye view of particles on a string moving in horizontal circles on a tabletop. All are moving at the same speed. Rank in order, from largest to smallest, the tensions T_1 to T_4.

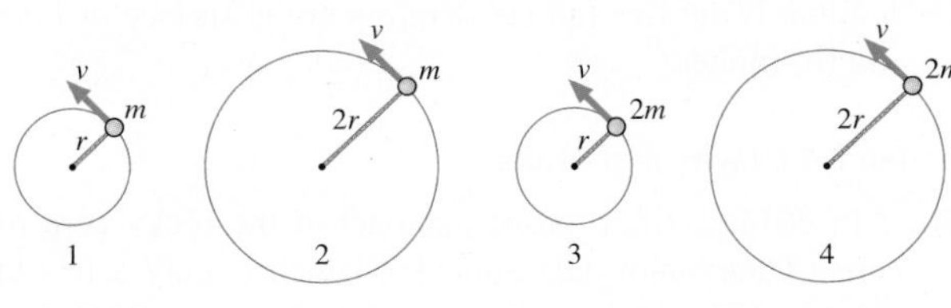

FIGURE P6.13

14. || In short-track speed skating, the track has straight sections and semicircles 16 m in diameter. Assume that a 65 kg skater goes around the turn at a constant 12 m/s.

   a. What is the horizontal force on the skater?
   b. What is the ratio of this force to the skater's weight?

15. ||| In addition to their remarkable top speeds of almost 60 mph, BIO cheetahs have impressive cornering abilities. In one study, the maximum centripetal acceleration of a cheetah was measured to be 18 m/s^2. What minimum value of the coefficient of static friction between the ground and the cheetah's feet is necessary to provide this acceleration?

16. || A cyclist is rounding a 20-m-radius curve at 12 m/s. What is INT the minimum possible coefficient of static friction between the bike tires and the ground?

17. || A 1500 kg car drives around a flat 200-m-diameter circular track at 25 m/s. What are the magnitude and direction of the net force on the car? What causes this force?

18. || A fast pitch softball player does a "windmill" pitch, illustrated BIO in Figure P6.18, moving her hand through a circular arc to pitch a ball at 70 mph. The 0.19 kg ball is 50 cm from the pivot point at her shoulder. At the lowest point of the circle, the ball has reached its maximum speed.
   a. At the bottom of the circle, just before the ball leaves her hand, what is its centripetal acceleration?
   b. What are the magnitude and direction of the force her hand exerts on the ball at this point?

FIGURE P6.18

19. || One kind of baseball pitching machine works by rotating a light and stiff rigid rod about a horizontal axis until the ball is moving toward the target. Suppose a 144 g baseball is held 85 cm from the axis of rotation and released at the major league pitching speed of 85 mph.
   a. What is the ball's centripetal acceleration just before it is released?
   b. What is the magnitude of the net force that is acting on the ball just before it is released?

20. | A wind turbine has 12,000 kg blades that are 38 m long. The blades spin at 22 rpm. If we model a blade as a point mass at the midpoint of the blade, what is the inward force necessary to provide each blade's centripetal acceleration?

21. ||| You're driving your pickup truck around a curve that has a radius of 20 m. How fast can you drive around this curve before a steel toolbox slides on the steel bed of the truck?

22. || The spin cycle of a clothes washer extracts the water in clothing by greatly increasing the water's apparent weight so that it is efficiently squeezed through the clothes and out the holes in the drum. In a top loader's spin cycle, the 45-cm-diameter drum spins at 1200 rpm around a vertical axis. What is the apparent weight of a 1.0 g drop of water?

23. || Gibbons, small Asian apes, move by *brachiation,* swinging BIO below a handhold to move forward to the next handhold. A 9.0 kg gibbon has an arm length (hand to shoulder) of 0.60 m. We can model its motion as that of a point mass swinging at the end of a 0.60-m-long, massless rod. At the lowest point of its swing, the gibbon is moving at 3.5 m/s. What upward force must a branch provide to support the swinging gibbon?

Section 6.3 Apparent Forces in Circular Motion

24. ||| The passengers in a roller coaster car feel 50% heavier than their true weight as the car goes through a dip with a 30 m radius of curvature. What is the car's speed at the bottom of the dip?

25. ||| In the very Dutch sport of Fierljeppen, athletes run up to a long pole and then use it to vault across a canal. At the very top of his arc, a 55 kg vaulter is moving at 2.5 m/s and is 5.1 m from the bottom end of the pole. What vertical force does the pole exert on the vaulter?

26. || A roller coaster car is going over the top of a 15-m-radius circular rise. At the top of the hill, the passengers "feel light," with an apparent weight only 50% of their true weight. How fast is the coaster moving?

27. ||| As a roller coaster car crosses the top of a 40-m-diameter loop-the-loop, its apparent weight is the same as its true weight. What is the car's speed at the top?

28. || Unlike a roller coaster, the seats in a Ferris wheel swivel so that the rider is always seated upright. An 80-ft-diameter Ferris wheel rotates once every 24 s. What is the apparent weight of a 70 kg passenger at (a) the lowest point of the circle and (b) the highest point?

29. ||| You're driving your new sports car at 75 mph over the top of a hill that has a radius of curvature of 525 m. What fraction of your normal weight is your apparent weight as you crest the hill?

30. IIII In a car's suspension system, each wheel is connected to a vertical spring; these springs absorb shocks when the car travels on bumpy roads. In one car, each spring has a spring constant of 5.0×10^4 N/m. If this 1400 kg car is driven at 25 m/s through the bottom of a circular dip in the road that has a radius of 600 m, by how much do these springs compress compared to when the car is driven on a flat road?
INT

31. III A typical laboratory centrifuge rotates at 4000 rpm. Test tubes have to be placed into a centrifuge very carefully because of the very large accelerations.
BIO INT
    a. What is the acceleration at the end of a test tube that is 10 cm from the axis of rotation?
    b. For comparison, what is the magnitude of the acceleration a test tube would experience if stopped in a 1.0-ms-long encounter with a hard floor after falling from a height of 1.0 m?

Section 6.4 Circular Orbits and Weightlessness

32. III A satellite orbiting the moon very near the surface has a period of 110 min. Use this information, together with the radius of the moon from the table on the inside of the back cover, to calculate the free-fall acceleration on the moon's surface.
33. I Many spacecraft have visited Mars over the years. Mars is smaller than Earth and has correspondingly weaker surface gravity. On Mars, the free-fall acceleration is only 3.8 m/s^2. What is the orbital period of a spacecraft in a low orbit near the surface of Mars?

Section 6.5 Newton's Law of Gravity

34. IIII The centers of a 10 kg lead ball and a 100 g lead ball are separated by 10 cm.
    a. What gravitational force does each exert on the other?
    b. What is the ratio of this gravitational force to the weight of the 100 g ball?
35. II The gravitational force of a star on an orbiting planet 1 is F_1. Planet 2, which is twice as massive as planet 1 and orbits at twice the distance from the star, experiences gravitational force F_2. What is the ratio F_2/F_1? You can ignore the gravitational force between the two planets.
36. II The free-fall acceleration at the surface of planet 1 is 20 m/s^2. The radius and the mass of planet 2 are twice those of planet 1. What is the free-fall acceleration on planet 2?
37. IIII What is the ratio of the sun's gravitational force on you to the earth's gravitational force on you?
38. I Just before it landed on the moon, the Apollo 12 lunar lander had a mass of 7200 kg. What rocket thrust was necessary to have the lander touch down with zero acceleration?
INT
39. II In recent years, astronomers have found planets orbiting nearby stars that are quite different from planets in our solar system. Kepler-12b, has a diameter that is 1.7 times that of Jupiter, but a mass that is only 0.43 that of Jupiter. What is the value of g on this large, but low-density, world?
40. III The gravitational constant G was first measured accurately by Henry Cavendish in 1798. He used an exquisitely sensitive balance to measure the force between two lead spheres whose centers were 0.23 m apart. One of the spheres had a mass of 158 kg, while the mass of the other sphere was 0.73 kg. What was the ratio of the gravitational force between these spheres to the weight of the lighter sphere?
41. II a. What is the gravitational force of the sun on the earth?
    b. What is the gravitational force of the moon on the earth?
    c. The moon's force is what percent of the sun's force?
42. II What is the value of g on the surface of Saturn? Explain how such a low value is possible given Saturn's large mass—100 times that of Earth.
43. II What is the free-fall acceleration at the surface of (a) Mars and (b) Jupiter?

Section 6.6 Gravity and Orbits

44. II In 2014, a space probe approached the rocky core of the comet Churyumov–Gerasimenko, which is only a few km in diameter. The probe then entered orbit around the comet at a distance of 30 km. The comet was found to have a mass of 1.0×10^{13} kg. What was the orbital period of the probe around the comet, in earth days?
45. III Mars has two moons, Phobos and Deimos. Phobos orbits Mars at a distance of 9380 km from Mars's center, while Deimos orbits at 23,500 km from the center. What is the ratio of the orbital period of Deimos to that of Phobos?
46. III Mars's moon Phobos orbits the planet at a distance of 9380 km from its center, and it takes 7 hours and 39 minutes to complete one orbit. What is the ratio of Mars's mass to the mass of the earth?
47. II The *dwarf planet* Praamzius is estimated to have a diameter of about 300 km and orbits the sun at a distance of 6.4×10^{12} m. What is its orbital period in years?
48. III Planet X orbits the star Omega with a "year" that is 200 earth days long. Planet Y circles Omega at four times the distance of Planet X. How long is a year on Planet Y?
49. III The International Space Station is in a 250-mile-high orbit. What are the station's orbital period, in minutes, and speed?
50. IIII An earth satellite moves in a circular orbit at a speed of 5500 m/s. What is its orbital period?

In recent years, scientists have discovered hundreds of planets orbiting other stars. Some of these planets are in orbits that are similar to that of earth, which orbits the sun ($M_{\text{sun}} = 1.99 \times 10^{30}$ kg) at a distance of 1.50×10^{11} m, called 1 *astronomical unit* (1 au). Others have extreme orbits that are much different from anything in our solar system. Problems 51–53 relate to some of these planets that follow circular orbits around other stars.

51. III WASP-32b orbits with a period of only 2.7 days a star with a mass that is 1.1 times that of the sun. How many au from the star is this planet?
52. III HD 10180g orbits with a period of 600 days at a distance of 1.4 au from its star. What is the ratio of the star's mass to our sun's mass?
53. III Kepler-42c orbits at a very close 0.0058 au from a small star with a mass that is 0.13 that of the sun. How long is a "year" on this world?

General Problems

54. II How fast must a plane fly along the earth's equator so that the sun stands still relative to the passengers? In which direction must the plane fly, east to west or west to east? Give your answer in both km/h and mph. The radius of the earth is 6400 km.

55. ||| The car in Figure P6.55 travels at a constant speed along the road shown. Draw vectors showing its acceleration at the three points A, B, and C, or write $\vec{a} = \vec{0}$. The lengths of your vectors should correspond to the magnitudes of the accelerations.

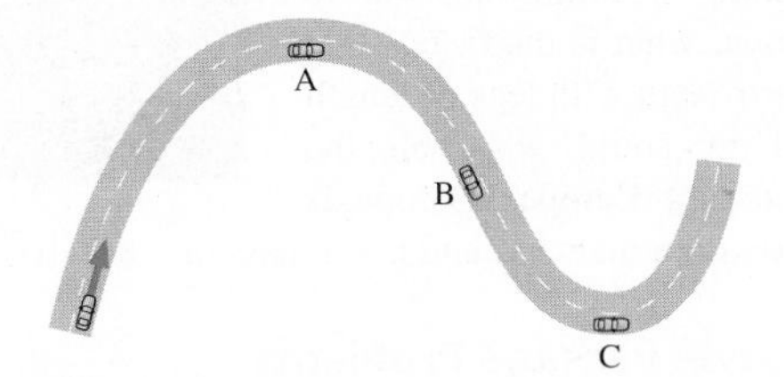

FIGURE P6.55

56. | A 75 kg man weighs himself at the north pole and at the equator. Which scale reading is higher? By how much? Assume the earth is a perfect sphere. Explain why the readings differ.

57. ||| A 1400 kg car drives at 27 m/s over a circular hill that has a radius of 430 m. At the point shown in Figure P6.57, what is the normal force on the car?

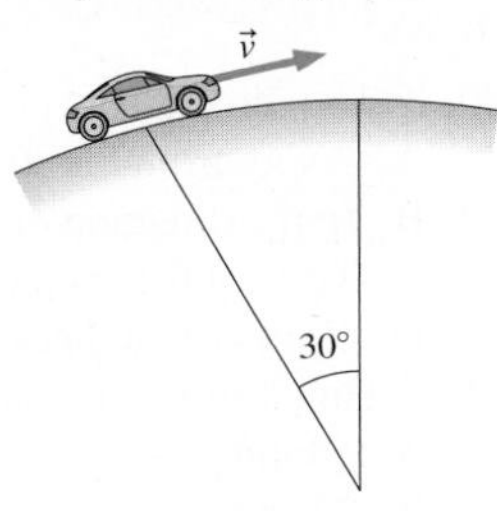

FIGURE P6.57

58. |||| A curve at a racetrack has a radius of 800 m and is banked at an angle of 7.0°. On a rainy day, the coefficient of friction between the cars' tires and the track is 0.50. What is the maximum speed at which a car could go around this curve without slipping?

59. |||| It is well known that runners run more slowly around a
BIO curved track than a straight one. One hypothesis to explain this is that the *total* force from the track on a runner's feet—the magnitude of the *vector* sum of the normal force (that has average value *mg* to counteract gravity) and the inward-directed friction force that causes the runner's centripetal acceleration—is greater when running around a curve than on a straight track. Runners compensate for this greater force by increasing the time their feet are in contact with the ground, which slows them down. For a sprinter running at 10 m/s around a curved track of radius 20 m, how much greater (as a percentage) is the average total force on their feet compared to when they are running in a straight line?

60. ||| Biologists have studied the running ability of the north-
BIO ern quoll, a marsupial indigenous to Australia. In one set of experiments, they studied the maximum speed that quolls could run around a curved path without slipping. One quoll was running at 2.8 m/s around a curve with a radius of 1.2 m when it started to slip. What was the coefficient of static friction between the quoll's feet and the ground in this trial?

61. ||| In a recent study of how mice negotiate turns, the mice ran
BIO around a circular 90° turn on a track with a radius of 0.15 m. The maximum speed measured for a mouse (mass = 18.5 g) running around this turn was 1.29 m/s. What is the minimum coefficient of friction between the track and the mouse's feet that would allow a turn at this speed?

62. ||| You are driving your car through a roundabout that has a radius of 9.0 m. Your physics textbook is lying on the seat next to you. What is the fastest speed at which you can go around the curve without the book sliding? The coefficient of static friction between the book and the seat is 0.30.

63. ||| In the Wall of Death carnival attraction, stunt motorcyclists ride around the inside of a large, 10-m-diameter wooden cylinder that has *vertical* walls. The coefficient of static friction between the riders' tires and the wall is 0.90. What is the minimum speed at which the motorcyclists can ride without slipping down the wall?

64. |||| In the swing carousel amusement park ride, riders sit in chairs that are attached by a chain to a large rotating drum. As the carousel turns, the riders move in a large circle with the chains tilted out from the vertical. In one such carousel, the riders move in a 16.5-m-radius circle and take 8.3 s to complete one revolution. What is the angle of the chains, as measured from the vertical?

65. | A 5.0 g coin is placed 15 cm from the center of a turn-
INT table. The coin has static and kinetic coefficients of friction with the turntable surface of $\mu_s = 0.80$ and $\mu_k = 0.50$. The turntable very slowly speeds up to 60 rpm. Does the coin slide off?

66. ||| In an old-fashioned amusement park ride, passengers stand inside a 3.0-m-tall, 5.0-m-diameter hollow steel cylinder with their backs against the wall. The cylinder begins to rotate about a vertical axis. Then the floor on which the passengers are standing suddenly drops away! If all goes well, the passengers will "stick" to the wall and not slide. Clothing has a static coefficient of friction against steel in the range 0.60 to 1.0 and a kinetic coefficient in the range 0.40 to 0.70. What is the minimum rotational frequency, in rpm, for which the ride is safe?

67. ||| The 0.20 kg puck on the
INT frictionless, horizontal table in Figure P6.67 is connected by a string through a hole in the table to a hanging 1.20 kg block. With what speed must the puck rotate in a circle of radius 0.50 m if the block is to remain hanging at rest?

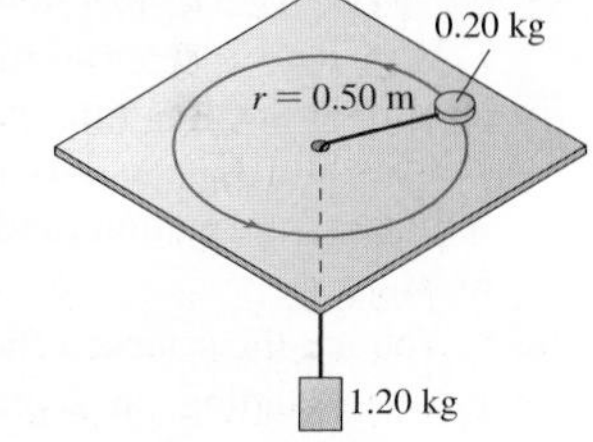

FIGURE P6.67

68. | While at the county fair, you decide to ride the Ferris wheel. Having eaten too many candy apples and elephant ears, you find the motion somewhat unpleasant. To take your mind off your stomach, you wonder about the motion of the ride. You estimate the radius of the big wheel to be 15 m, and you use your watch to find that each loop around takes 25 s.
   a. What are your speed and magnitude of your acceleration?
   b. What is the ratio of your apparent weight to your true weight at the top of the ride?
   c. What is the ratio of your apparent weight to your true weight at the bottom?

69. | A car drives over the top of a hill that has a radius of 50 m. What maximum speed can the car have without flying off the road at the top of the hill?

Watch Video Solution Problems 6.73 and 6.77

70. ||| The ultracentrifuge is an important tool for separating and
BIO analyzing proteins in biological research. Because of the enor
INT mous centripetal accelerations that can be achieved, the apparatus (see Figure 6.18) must be carefully balanced so that each sample is matched by another on the opposite side of the rotor shaft. Any difference in mass of the opposing samples will cause a net force in the horizontal plane on the shaft of the rotor; this force can actually be large enough to destroy the centrifuge. Suppose that a scientist makes a slight error in sample preparation, and one sample has a mass 10 mg greater than the opposing sample. If the samples are 10 cm from the axis of the rotor and the ultracentrifuge spins at 70,000 rpm, what is the magnitude of the net force on the rotor due to the unbalanced samples?
71. |||| A sensitive gravimeter at a mountain observatory finds that the free-fall acceleration is $0.0075\ \text{m/s}^2$ less than that at sea level. What is the observatory's altitude?
72. |||| In 2014, the Rosetta space probe reached the comet Churyumov–
INT Gerasimenko. Although the comet's core is actually far from spherical, in this problem we'll model it as a sphere with a mass of 1.0×10^{13} kg and a radius of 1.6 km. If a rock were dropped from a height of 1.0 m above the comet's surface, how long would it take to hit the surface?
73. ||| Planet Z is 10,000 km in diameter. The free-fall acceleration on Planet Z is $8.0\ \text{m/s}^2$.
    a. What is the mass of Planet Z?
    b. What is the free-fall acceleration 10,000 km above Planet Z's north pole?
74. ||| How long will it take a rock dropped from 2.0 m above the
INT surface of Mars to reach the ground?
75. ||| A 20 kg sphere is at the origin and a 10 kg sphere is at
INT $(x, y) = (20\ \text{cm}, 0\ \text{cm})$. At what point or points could you place a small mass such that the net gravitational force on it due to the spheres is zero?
76. ||| a. At what height above the earth is the free-fall acceleration 10% of its value at the surface?
    b. What is the speed of a satellite orbiting at that height?
77. | Mars has a small moon, Phobos, that orbits with a period of 7 h 39 min. The radius of Phobos' orbit is 9.4×10^6 m. Use only this information (and the value of G) to calculate the mass of Mars.
78. || You are the science officer on a visit to a distant solar system. Prior to landing on a planet you measure its diameter to be 1.80×10^7 m and its rotation period to be 22.3 h. You have previously determined that the planet orbits 2.20×10^{11} m from its star with a period of 402 earth days. Once on the surface you find that the free-fall acceleration is $12.2\ \text{m/s}^2$. What are the masses of (a) the planet and (b) the star?
79. ||||| Europa, a satellite of Jupiter, is
BIO believed to have a liquid ocean of water (with a possibility of life) beneath its icy surface. In planning a future mission to Europa, what is the fastest that an astronaut with legs of length 0.70 m could walk on the surface of Europa? Europa is 3100 km in diameter and has a mass of 4.8×10^{22} kg.

MCAT-Style Passage Problems

Orbiting the Moon

Suppose a spacecraft orbits the moon in a very low, circular orbit, just a few hundred meters above the lunar surface. The moon has a diameter of 3500 km, and the free-fall acceleration at the surface is $1.6\ \text{m/s}^2$.

80. | The direction of the net force on the craft is
    A. Away from the surface of the moon.
    B. In the direction of motion.
    C. Toward the center of the moon.
    D. Nonexistent, because the net force is zero.
81. | How fast is this spacecraft moving?
    A. 53 m/s
    B. 75 m/s
    C. 1700 m/s
    D. 2400 m/s
82. | How much time does it take for the spacecraft to complete one orbit?
    A. 38 min
    B. 76 min
    C. 110 min
    D. 220 min
83. | The material that comprises the side of the moon facing the earth is actually slightly more dense than the material on the far side. When the spacecraft is above a more dense area of the surface, the moon's gravitational force on the craft is a bit stronger. In order to stay in a circular orbit of constant height and speed, the spacecraft could fire its rockets while passing over the denser area. The rockets should be fired so as to generate a force on the craft
    A. Away from the surface of the moon.
    B. In the direction of motion.
    C. Toward the center of the moon.
    D. Opposite the direction of motion.

7 Rotational Motion

Design modifications make this cyclist and her racing bicycle more aerodynamic, which enables her to achieve higher speeds. But there are less visible modifications that permit greater acceleration as well—most important, the design of the tires and wheels. How do you design bicycle wheels to achieve greater acceleration?

LOOKING AHEAD ▸▸

Rotational Kinematics

The spinning roulette wheel isn't going anywhere, but it is moving. This is **rotational motion**.

You'll learn about angular velocity and other quantities we use to describe rotational motion.

Torque

To start something moving, apply a force. To start something rotating, apply a **torque,** as this sailor is doing to the wheel.

You'll see that torque depends on how hard you push and also on *where* you push. A push far from the axle gives a large torque.

Rotational Dynamics

The girl pushes on the outside edge of the merry-go-round, gradually increasing its rotation rate.

You'll learn a version of Newton's second law for rotational motion and use it to solve problems.

GOAL **To understand the physics of rotating objects.**

LOOKING BACK ◂◂

Circular Motion

In Chapter 6, you learned to describe *circular* motion in terms of period, frequency, velocity, and centripetal acceleration.

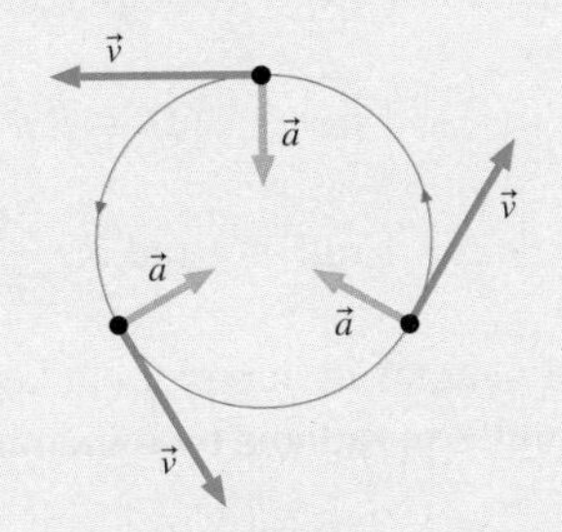

In this chapter, you'll learn to use angular velocity, angular acceleration, and other quantities that describe *rotational* motion.

STOP TO THINK

As an audio CD plays, the frequency at which the disk spins changes. At 210 rpm, the speed of a point on the outside edge of the disk is 1.3 m/s. At 420 rpm, the speed of a point on the outside edge is

A. 1.3 m/s B. 2.6 m/s
C. 3.9 m/s D. 5.2 m/s

Moving blades on a wind turbine.

7.1 Describing Circular and Rotational Motion

The photo shows a spinning wind turbine. The blurring of the photo tells us that the blades are in motion. On a typical wind turbine, each blade takes 6 seconds to go around. But the outer parts of the blades appear more blurred than the parts nearer the hub; therefore these points are moving more rapidly. This means that each point on each blade completes one circle in 6 seconds, but different points move at different speeds. We'll need to extend our description of motion to handle this type of situation.

In this chapter, we'll consider **rotational motion,** the motion of objects that spin about an axis, like the blade assembly of the wind turbine. We'll introduce some new concepts, but we'll start by returning to a topic we first considered in ◀◀ SECTION 6.1—the motion of a particle in uniform circular motion, such as the motion of the tip of a wind turbine blade. We'll extend this treatment to describe the motion of the whole system. Doing this will require us to define some new quantities.

Angular Position

FIGURE 7.1 A particle's angular position is described by angle θ.

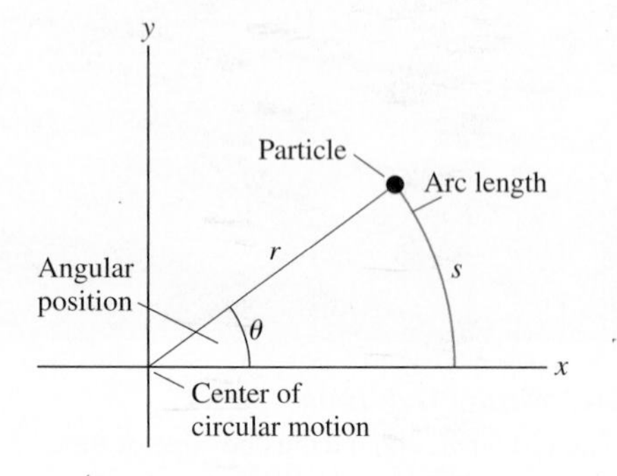

When we describe the motion of a particle as it moves in a circle, it is convenient to use the angle θ from the positive x-axis to describe the particle's location. This is shown in FIGURE 7.1. Because the particle travels in a circle with a fixed radius r, specifying θ completely locates the position of the particle. Thus we call angle θ the **angular position** of the particle.

We define θ to be positive when measured *counterclockwise* from the positive x-axis. An angle measured *clockwise* from the positive x-axis has a negative value. Rather than measure angles in degrees, mathematicians and scientists usually measure angle θ in the angular unit of *radians.* In Figure 7.1, we also show the **arc length** s, the distance that the particle has traveled along its circular path. We define the particle's angle θ in **radians** in terms of this arc length and the radius of the circle:

$$\theta \text{ (radians)} = \frac{s}{r} \tag{7.1}$$

This is a sensible definition of an angle: The farther the particle has traveled around the circle (i.e., the greater s is), the larger the angle θ in radians. The radian, abbreviated rad, is the SI unit of angle. An angle of 1 rad has an arc length s exactly equal to the radius r. An important consequence of Equation 7.1 is that the arc length spanning the angle θ is

$$s = r\theta \tag{7.2}$$

NOTE ▸ Equation 7.2 is valid only if θ is measured in radians, not degrees. This very simple relationship between angle and arc length is one of the primary motivations for using radians. ◂

When a particle travels all the way around the circle—completing one *revolution,* abbreviated rev—the arc length it travels is the circle's circumference $2\pi r$. Thus the angle of a full circle is

$$\theta_{\text{full circle}} = \frac{s}{r} = \frac{2\pi r}{r} = 2\pi \text{ rad}$$

We can use this fact to write conversion factors among revolutions, radians, and degrees:

$$1 \text{ rev} = 360° = 2\pi \text{ rad}$$

$$1 \text{ rad} = 1 \text{ rad} \times \frac{360°}{2\pi \text{ rad}} = 57.3°$$

We will often specify angles in degrees, but keep in mind that the SI unit is the radian. You can visualize angles in radians by remembering that 1 rad is just about 60°.

Angular Displacement and Angular Velocity

For the *linear* motion you studied in Chapters 1 and 2, a particle with a larger velocity undergoes a greater displacement in each second than one with a smaller velocity, as FIGURE 7.2a shows. FIGURE 7.2b shows two particles undergoing uniform *circular* motion. The particle on the left is moving slowly around the circle; it has gone only one-quarter of the way around after 5 seconds. The particle on the right is moving much faster around the circle, covering half of the circle in the same 5 seconds. You can see that the particle to the right undergoes twice the **angular displacement** $\Delta\theta$ during each interval as the particle to the left. Its **angular velocity,** the angular displacement through which the particle moves each second, is twice as large. We use the symbol ω to represent angular velocity. This symbol is a lowercase Greek omega, not an ordinary w. The SI unit of angular velocity is rad/s.

In analogy with linear motion, where $v_x = \Delta x/\Delta t$, we thus define angular velocity as

$$\omega = \frac{\text{angular displacement}}{\text{time interval}} = \frac{\Delta\theta}{\Delta t} \tag{7.3}$$

Angular velocity of a particle in uniform circular motion

FIGURE 7.2 Comparing uniform linear and circular motion.

(a) Uniform linear motion

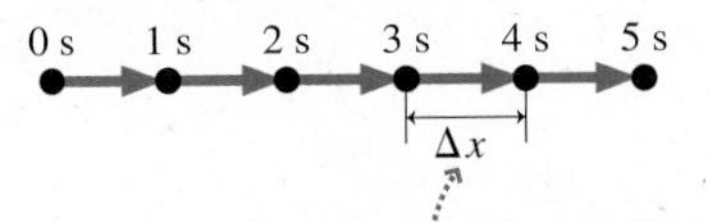

(b) Uniform circular motion

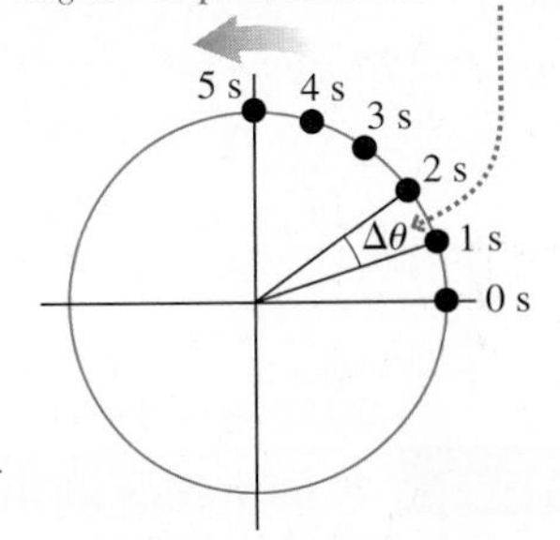

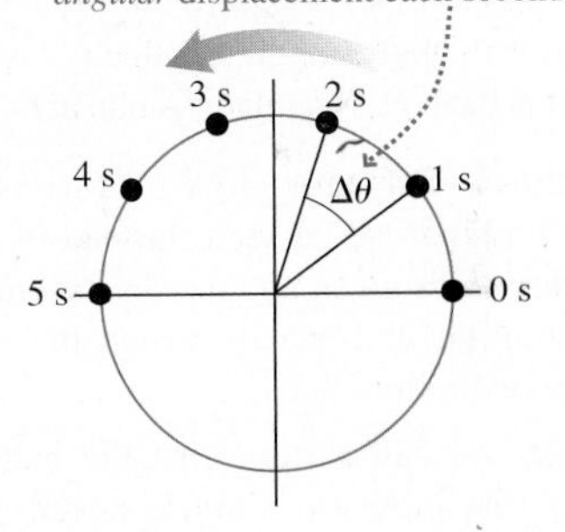

Figure 7.2a shows that the displacement Δx of a particle in uniform linear motion changes by the same amount each second. Similarly, as Figure 7.2b shows, the *angular* displacement $\Delta\theta$ of a particle in uniform *circular* motion changes by the same amount each second. This means that **the angular velocity $\omega = \Delta\theta/\Delta t$ is constant for a particle moving with uniform circular motion.**

EXAMPLE 7.1 Comparing angular velocities

Find the angular velocities of the two particles in Figure 7.2b.

STRATEGIZE This is a straightforward calculation—the angular velocity is simply the change in the angular displacement divided by the time interval. The second particle moves through twice as large an angle as the first, so its angular velocity should be twice that of the first.

PREPARE For uniform circular motion, we can use any angular displacement $\Delta\theta$, as long as we use the corresponding time interval Δt. For each particle, we'll choose the angular displacement corresponding to the motion from $t = 0$ s to $t = 5$ s.

SOLVE The particle on the left travels one-quarter of a full circle during the 5 s time interval. We learned earlier that a full circle corresponds to an angle of 2π rad, so the angular displacement for this particle is $\Delta\theta = (2\pi \text{ rad})/4 = \pi/2$ rad. Thus its angular velocity is

$$\omega = \frac{\Delta\theta}{\Delta t} = \frac{\pi/2 \text{ rad}}{5 \text{ s}} = 0.314 \text{ rad/s}$$

The particle on the right travels halfway around the circle, or π rad, in the 5 s interval. Its angular velocity is

$$\omega = \frac{\Delta\theta}{\Delta t} = \frac{\pi \text{ rad}}{5 \text{ s}} = 0.628 \text{ rad/s}$$

ASSESS The angular velocity of the second particle is double that of the first, as it should be. We should also check the scale of the answers. The angular velocity of the particle on the right is 0.628 rad/s, meaning that the particle travels through an angle of 0.628 rad each second. Because 1 rad ≈ 60°, 0.628 rad is roughly 35°. In Figure 7.2b, the particle on the right appears to move through an angle of about this size during each 1 s time interval, so our answer is reasonable.

Angular velocity, like the velocity v_x of one-dimensional motion, can be positive or negative. The signs for ω noted in FIGURE 7.3 are based on the convention that angles are positive when measured counterclockwise from the positive x-axis.

FIGURE 7.3 Positive and negative angular velocities.

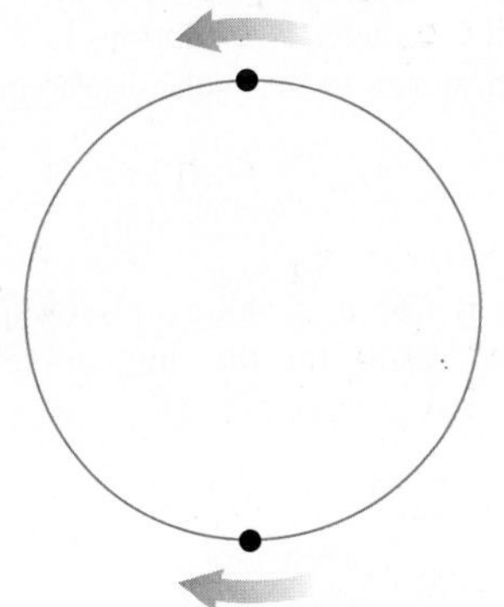

We've already noted how circular motion is analogous to linear motion, with angular variables replacing linear variables. Thus much of what you learned about linear kinematics and dynamics carries over to circular motion. For example, Equation 2.4 gave us a formula for computing a linear displacement during a time interval:

$$x_f - x_i = \Delta x = v_x \, \Delta t$$

You can see from Equation 7.3 that we can write a similar equation for an angular displacement:

$$\theta_f - \theta_i = \Delta\theta = \omega \, \Delta t \quad (7.4)$$

Angular displacement for uniform circular motion

For linear motion, we use the term *speed* v when we are not concerned with the direction of motion, *velocity* v_x when we are. For circular motion, we define the **angular speed** to be the absolute value of the angular velocity, so that it's a positive quantity irrespective of the particle's direction of rotation. Although potentially confusing, it is customary to use the symbol ω for angular speed *and* for angular velocity. If the direction of rotation is not important, we will interpret ω to mean angular speed. In kinematic equations, such as Equation 7.4, ω is always the angular velocity, and you need to use a negative value for clockwise rotation.

EXAMPLE 7.2 Kinematics at the roulette wheel

A small steel ball rolls counterclockwise around the inside of a stationary 30.0-cm-diameter roulette wheel, like the one shown in the chapter preview. The ball completes exactly 2 rev in 1.20 s.

a. What is the ball's angular velocity?
b. What is the ball's angular position at $t = 2.00$ s? Assume $\theta_i = 0$.

STRATEGIZE This problem is analogous to problems of motion along a line that we've seen. Instead of displacement and time, however, we are given angular displacement and time. The steps in the solution will exactly match the steps we would use for motion along a line.

PREPARE We can assume that the ball is in uniform circular motion, so its angular position is steadily changing. We are given the angular displacement for a certain time interval, so we can compute the angular velocity. Then, knowing the angular velocity, we can find the angular position at a future time.

SOLVE

a. The ball's angular velocity is $\omega = \Delta\theta/\Delta t$. We know that the ball completes 2 revolutions in 1.20 s and that each revolution corresponds to an angular displacement $\Delta\theta = 2\pi$ rad. Thus

$$\omega = \frac{2(2\pi \text{ rad})}{1.20 \text{ s}} = 10.47 \text{ rad/s}$$

We'll use this value in subsequent calculations, but our final result for the angular velocity should be reported to three significant figures, $\omega = 10.5$ rad/s. Because the rotation direction is counterclockwise, the angular velocity is positive.

b. The ball moves with constant angular velocity, so its angular position is given by Equation 7.4. Thus the ball's angular position at $t = 2.00$ s is

$$\theta_f = \theta_i + \omega \, \Delta t = 0 \text{ rad} + (10.47 \text{ rad/s})(2.00 \text{ s})$$
$$= 20.94 \text{ rad}$$

If we're interested in where the ball is at $t = 2.00$ s, we can write its angular position as an integer multiple of 2π (representing the number of complete revolutions the ball has made) plus a remainder:

$$\theta_f = 20.94 \text{ rad} = 3.333 \times 2\pi \text{ rad}$$
$$= 3 \times 2\pi \text{ rad} + 0.333 \times 2\pi \text{ rad}$$
$$= 3 \times 2\pi \text{ rad} + 2.09 \text{ rad}$$

In other words, at $t = 2.00$ s, the ball has completed 3 rev and is 2.09 rad = 120° into its fourth revolution. An observer would say that the ball's angular position is $\theta = 120°$.

ASSESS Since the ball completes 2 revolutions in 1.20 s, it seems reasonable that it completes 3.33 revolutions in 2.00 s.

The angular speed ω is closely related to the period T and the frequency f of the motion. If a particle in uniform circular motion moves around a circle once, the angular displacement is $\Delta\theta = 2\pi$ rad; the time interval is, by definition, $\Delta t = T$.

The angular speed, $\omega = \Delta\theta/\Delta t$, can thus be expressed in terms of the period T or the frequency $f = 1/T$:

$$\omega = \frac{2\pi \text{ rad}}{T} = (2\pi \text{ rad})f \qquad (7.5)$$

Angular speed in terms of period and frequency

EXAMPLE 7.3 **Rotations in a car engine**

The crankshaft in your car engine is turning at 3000 rpm. What is the shaft's angular speed in rad/s?

PREPARE We'll need to convert rpm to rev/s and then use Equation 7.5.

SOLVE We convert rpm to rev/s by

$$\left(3000 \frac{\text{rev}}{\text{min}}\right)\left(\frac{1 \text{ min}}{60 \text{ s}}\right) = 50.0 \text{ rev/s}$$

Thus the crankshaft's angular speed is

$$\omega = (2\pi \text{ rad})f = (2\pi \text{ rad})(50.0 \text{ rev/s}) = 314 \text{ rad/s}$$

Angular-Position and Angular-Velocity Graphs

For the one-dimensional motion you studied in Chapter 3, we found that position- and velocity-versus-time graphs were important and useful representations of motion. We can use the same kinds of graphs to represent angular motion. Let's begin by considering the motion of the roulette ball of Example 7.2. We found that it had angular velocity $\omega = 10.5$ rad/s, meaning that its angular *position* changed by $+10.5$ rad every second. This is exactly analogous to the one-dimensional motion problem of a car driving in a straight line with a velocity of 10.5 m/s, so that its position increases by 10.5 m each second. Using this analogy, we can construct the **angular position-versus-time graph** for the roulette ball shown in FIGURE 7.4.

The angular velocity is given by $\omega = \Delta\theta/\Delta t$. Graphically, this is the *slope* of the angular position-versus-time graph, just as the ordinary velocity is the slope of the position-versus-time graph. Thus we can create an **angular velocity-versus-time graph** by finding the slope of the corresponding angular position-versus-time graph.

FIGURE 7.4 Angular position for the ball on the roulette wheel.

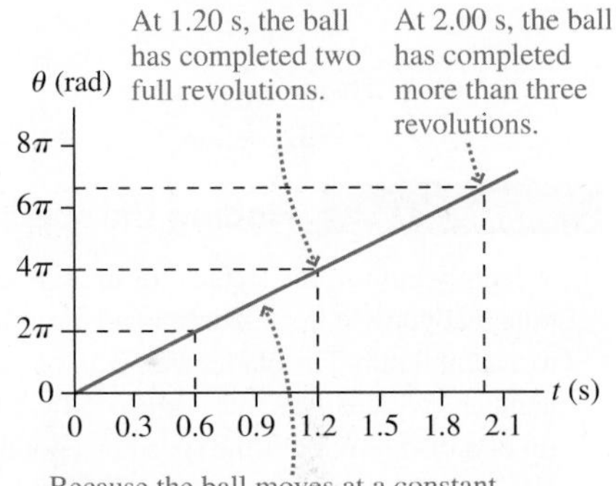

EXAMPLE 7.4 **Graphing a bike ride**

Jake rides his bicycle home from campus. FIGURE 7.5 is the angular position-versus-time graph for a small rock stuck in the tread of his tire. First, draw the rock's angular velocity-versus-time graph, using rpm on the vertical axis. Then interpret the graphs with a story about Jake's ride.

STRATEGIZE There are three portions of the graph in Figure 7.5. For the two horizontal sections, there is no change in the angular position—the tire isn't moving. In between, there is a line with a positive slope. This indicates a steadily increasing angular position, so the tire is rotating at a constant angular velocity.

PREPARE The angular velocity ω is the slope of the angular position-versus-time graph. For the two horizontal segments of the graph, the slope is zero, so the angular speed is zero—as we've noted, the tire isn't moving. For the segment with the positive slope, computing the slope gives us the angular velocity.

FIGURE 7.5 Angular position-versus-time graph for Jake's bike ride.

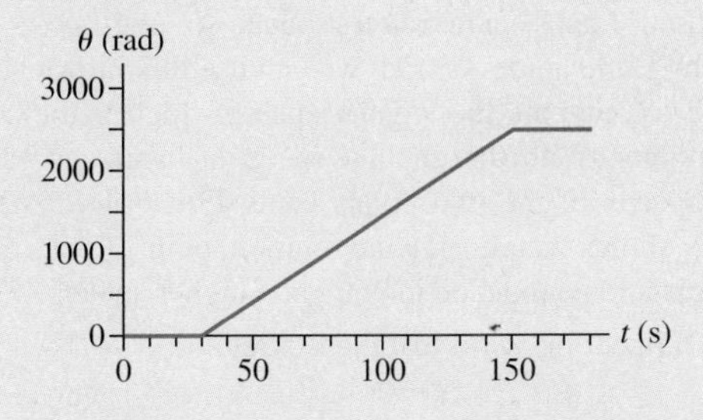

SOLVE Between $t = 30$ s and $t = 150$ s, an interval of 120 s, the rock's angular velocity (the slope of the angular position-versus-time graph) is

$$\omega = \text{slope} = \frac{2500 \text{ rad} - 0 \text{ rad}}{120 \text{ s}} = 20.8 \text{ rad/s}$$

Continued

Then we need to convert this to rpm:

$$\omega = \left(\frac{20.8 \text{ rad}}{1 \text{ s}}\right)\left(\frac{1 \text{ rev}}{2\pi \text{ rad}}\right)\left(\frac{60 \text{ s}}{1 \text{ min}}\right) = 200 \text{ rpm}$$

These values have been used to draw the angular velocity-versus-time graph of FIGURE 7.6. It looks like Jake waited 30 s for the light to change, then pedaled so that the bike wheel turned at a constant angular velocity of 200 rpm. 2.0 min later, he quickly braked to a stop for another 30-s-long red light.

ASSESS 200 rpm is about 3 revolutions per second. This seems reasonable for someone cycling at a pretty good clip.

FIGURE 7.6 Angular velocity-versus-time graph for Jake's bike ride.

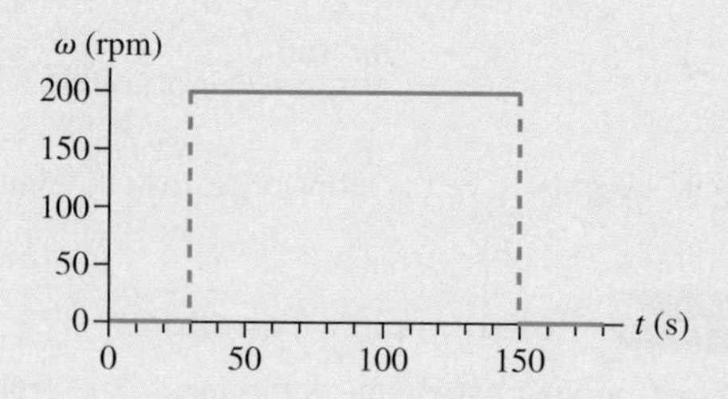

Relating Speed and Angular Speed

Let's think back to the wind turbine example that opened the chapter. Different points on the blades move at different speeds. And we know that points farther from the axis—at larger values of r—move at higher speeds. In Chapter 6 we found that the speed of a particle moving with frequency f around a circular path of radius r is $v = 2\pi fr$. If we combine this result with Equation 7.5 for the angular speed, we find that speed v and angular speed ω are related by

$$v = \omega r \tag{7.6}$$

Relationship between speed and angular speed

NOTE ▶ In Equation 7.6, **ω must be in units of rad/s.** If you are given a frequency in rev/s or rpm, you should convert it to an angular speed in rad/s. ◀

EXAMPLE 7.5 Finding the speed at different points on a wind turbine blade

A large wind turbine (typical of the size and specifications of a turbine that you see in a modern wind farm) has three blades connected to a central hub. The blades are 50 m long and rotate at 10 rpm. What angular speed ω does this correspond to? What is the speed v of the tip of a blade? What is the speed of a point 25 m from the hub?

STRATEGIZE This is a case of uniform rotational motion. The blades and the hub all move at the same angular speed, but points farther from the hub move at a higher speed. A point 50 m from the hub sweeps out a 50-m-radius circle 10 times each minute, so we expect that the speed will be quite high.

PREPARE FIGURE 7.7 shows the blades of the wind turbine, with blue circles illustrating the motion of two points, A and B, during one rotation. Each blade rotates once, so both points rotate through the same angle, 2π rad; we can use this information and the period to compute the angular speed, which is the same for the two points. With the angular speed in hand, we can compute the speeds of the two points using Equation 7.6. The figure clearly shows that, during the rotation, point B goes around a larger circle and so must be moving at a higher speed.

FIGURE 7.7 Rotating wind turbine blades.

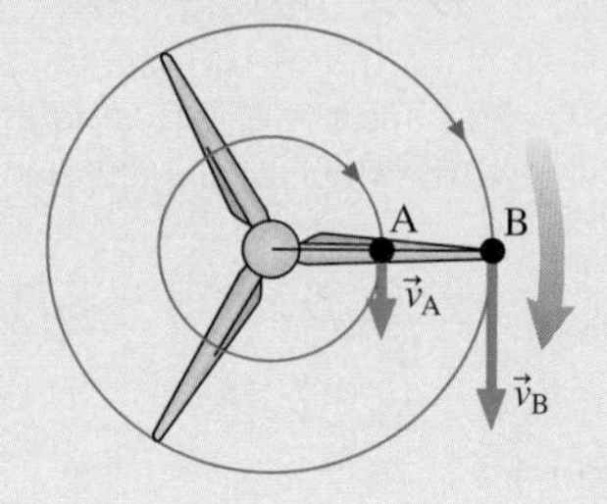

SOLVE The frequency is 10 rpm, so the blades rotate 10 times in 1 minute. The period is thus

$$T = \frac{1}{10} \text{ min} = 6.0 \text{ s}$$

Now we can find the angular speed:

$$\omega = \frac{2\pi \text{ rad}}{T} = \frac{2\pi}{6.0 \text{ s}} = 1.05 \text{ rad/s}$$

We can use the angular speed to find the speeds of points A and B:

$$v_A = \omega r = (1.05 \text{ rad/s})(25 \text{ m}) = 26 \text{ m/s}$$

$$v_B = \omega r = (1.05 \text{ rad/s})(50 \text{ m}) = 52 \text{ m/s}$$

ASSESS The computed angular speed corresponds to a rotation of about 1 rad in 1 second; 1 rad is about 60°, so this will take the blades around a 360° circle in about 6 seconds, which is just right. The tip of the blade travels in a circle of radius 50 m, which has a circumference of just over 300 m. The tip covers this distance in a mere 6.0 s, so we expect that v_B will be just over 50 m/s, exactly what we found.

Now, let's do a reality check: 52 m/s is nearly 120 mph! Do the tips of the blades really move that fast? A quick investigation of manufacturer specifications reveals that they do. This seems surprising because 120 mph is faster than the wind blows except in extreme situations. But the action of the wind on the blades is similar to the action of the air on the wings of a moving plane: It's the lift force that pushes the blades around. An airplane can certainly slice through the air at 120 mph, so the final result makes sense, although it is initially surprising.

▶ **Why do clocks go clockwise?** In the northern hemisphere, the rotation of the earth causes the sun to follow a circular arc through the southern sky, rising in the east and setting in the west. The shadow cast by the sun thus sweeps in an arc from west to east, so the shadow on a sundial—the first practical timekeeping device—sweeps around the top of the dial from left to right. Early clockmakers used the same convention, which is how it came to be clockwise.

STOP TO THINK 7.1 Which particle has angular position $5\pi/2$?

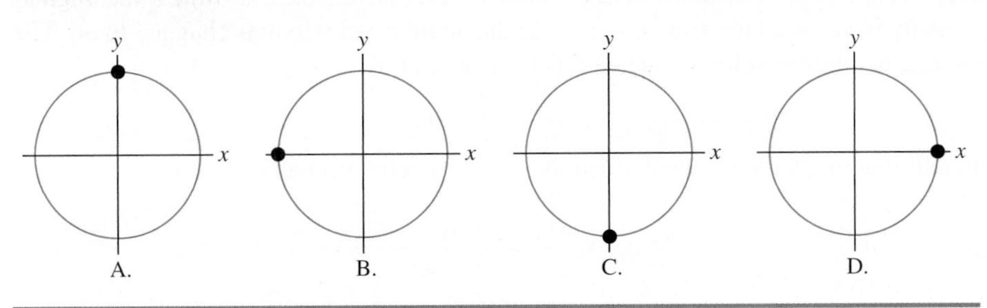

7.2 The Rotation of a Rigid Body

Think about the rotating wind turbine in Example 7.5. Until now, our study of physics has focused almost exclusively on the particle model, in which an entire object is represented as a single point in space. The particle model is adequate for understanding motion in a wide variety of situations, but we need to treat the turbine as an *extended object*—a system of particles for which the size and shape *do* make a difference and cannot be ignored.

A **rigid body** is an extended object whose size and shape do not change as it moves. For example, a bicycle wheel can be thought of as a rigid body. FIGURE 7.8 shows a rigid body as a collection of atoms held together by the rigid "massless rods" of molecular bonds.

Real molecular bonds are, of course, not perfectly rigid. That's why an object seemingly as rigid as a bicycle wheel can flex and bend. Thus Figure 7.8 is really a simplified *model* of an extended object, the **rigid-body model.** The rigid-body model is a very good approximation for many real objects of practical interest, such as wheels and axles.

FIGURE 7.9 illustrates the three basic types of motion of a rigid body: **translational motion, rotational motion,** and **combination motion.** We've already studied translational motion of a rigid body using the particle model. If a rigid body doesn't rotate, this model is often adequate for describing its motion. The rotational motion of a rigid body will be the main focus of this chapter. We'll also discuss an important case of combination motion—that of a *rolling* object—later in this chapter.

FIGURE 7.8 The rigid-body model of an extended object.

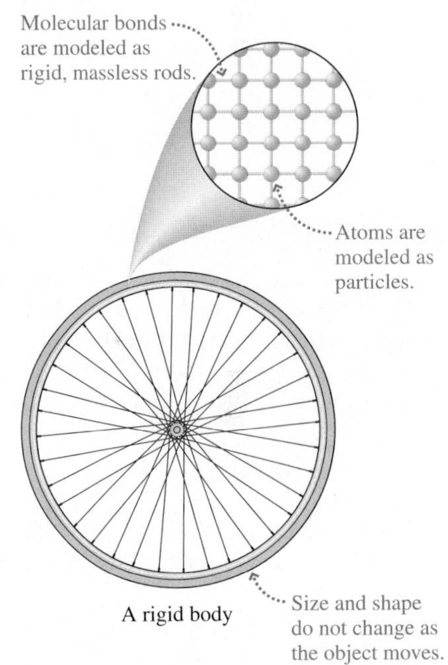

Atoms are modeled as particles.

FIGURE 7.9 Three basic types of motion of a rigid body.

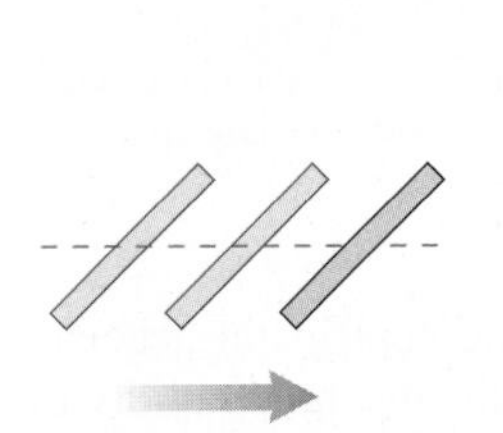

Translational motion: The object as a whole moves along a trajectory but does not rotate.

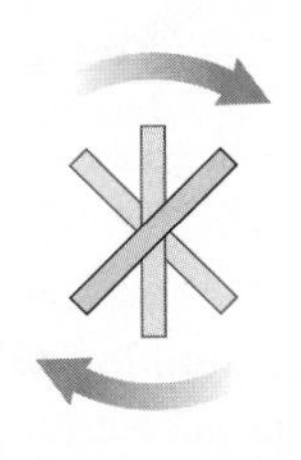

Rotational motion: The object rotates about a fixed point. Every point on the object moves in a circle.

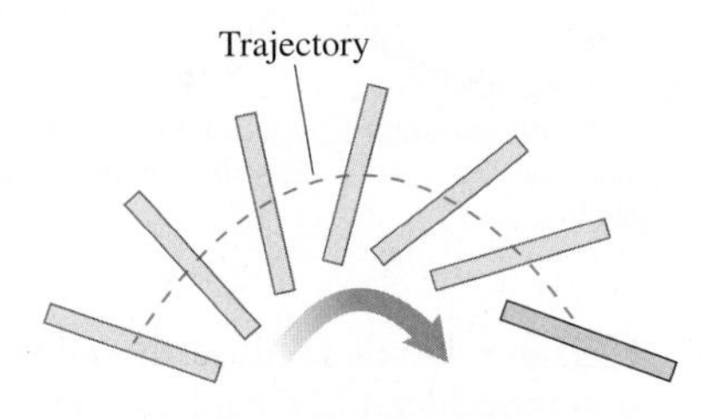

Trajectory
Combination motion: An object rotates as it moves along a trajectory.

FIGURE 7.10 shows a wheel rotating on an axle. The figure shows the motion as the wheel rotates for a time interval Δt. Two points 1 and 2 on the wheel move different distances, so they move at different speeds. But these two points move through the same angle, so they move with the same angular velocity. In general, **every point on a rotating rigid body has the same angular velocity,** so we can refer to the angular velocity ω of the wheel.

FIGURE 7.10 All points on a wheel rotate with the same angular velocity.

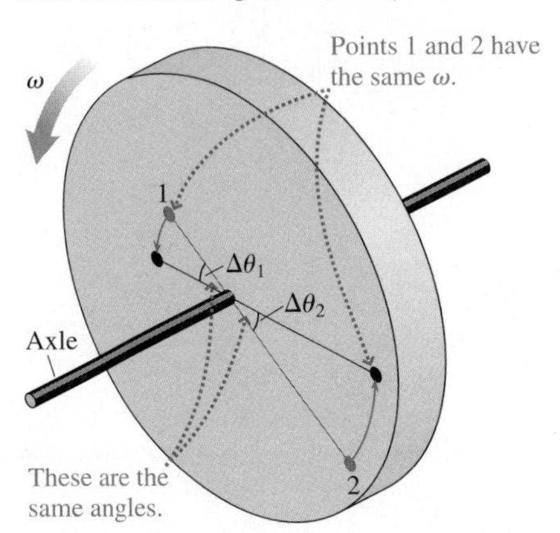

Angular Acceleration

If you push on the edge of a bicycle wheel, it begins to rotate. If you continue to push, it rotates ever faster. Its angular velocity is *changing.* To understand the dynamics of rotating objects, we'll need to be able to describe this case of changing angular velocity—that is, the case of *nonuniform* circular motion.

FIGURE 7.11 A rotating wheel with a changing angular velocity.

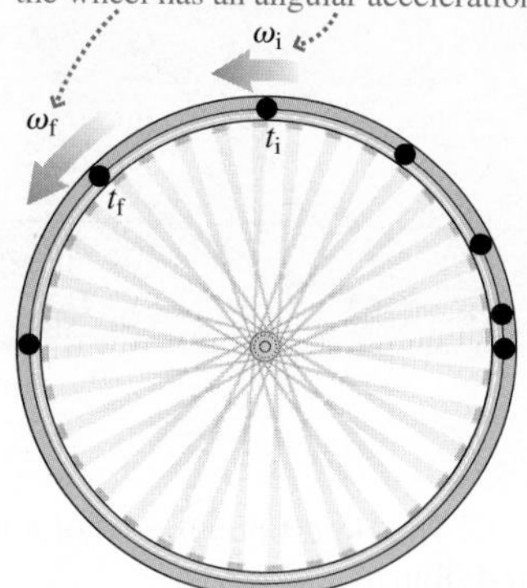
The angular velocity is *changing*, so the wheel has an angular acceleration.
ω_i
ω_f
t_i
t_f

FIGURE 7.11 shows a bicycle wheel whose angular velocity is changing. The dot represents a particular point on the wheel at successive times. At time t_i the angular velocity is ω_i; at a later time $t_f = t_i + \Delta t$ the angular velocity has changed to ω_f. The change in angular velocity during this time interval is

$$\Delta\omega = \omega_f - \omega_i$$

Recall that in Chapter 2 we defined the *linear* acceleration as

$$a_x = \frac{\Delta v_x}{\Delta t} = \frac{(v_x)_f - (v_x)_i}{\Delta t}$$

By analogy, we now define the **angular acceleration** as

$$\alpha = \frac{\text{change in angular velocity}}{\text{time interval}} = \frac{\Delta\omega}{\Delta t} \tag{7.7}$$

Angular acceleration for a particle in nonuniform circular motion

We use the symbol α (Greek alpha) for angular acceleration. Because the units of ω are rad/s, the units of angular acceleration are (rad/s)/s, or rad/s^2. From Equation 7.7, the sign of α is the same as the sign of $\Delta\omega$. FIGURE 7.12 shows how to determine the sign of α. Be careful with the sign of α; just as with linear acceleration, positive and negative values of α can't be interpreted as simply "speeding up" and "slowing down." Like ω, the angular acceleration α is the same for every point on a rotating rigid body.

FIGURE 7.12 Determining the sign of the angular acceleration.

α is *positive* when the rigid body is . . .

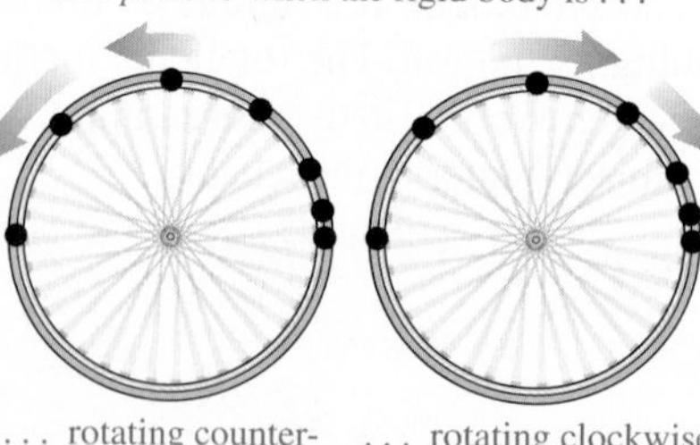

. . . rotating counterclockwise and speeding up.

. . . rotating clockwise and slowing down.

α is *negative* when the rigid body is . . .

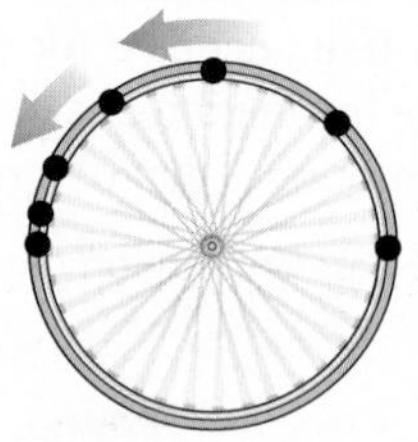

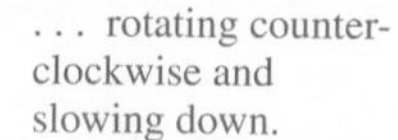

. . . rotating counterclockwise and slowing down.

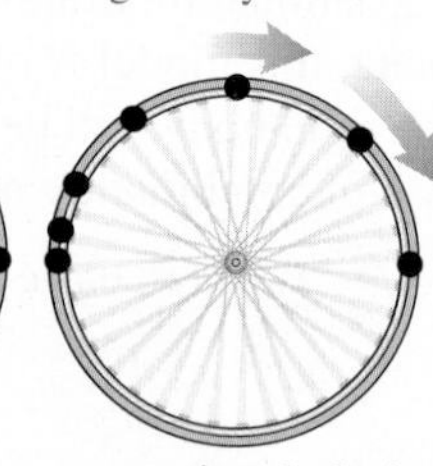

. . . rotating clockwise and speeding up.

NOTE ▸ Don't confuse the angular acceleration with the centripetal acceleration introduced in Chapter 6. The angular acceleration indicates how rapidly the *angular* velocity is changing. The centripetal acceleration is a vector quantity that points toward the center of a particle's circular path; it is nonzero even if the angular velocity is constant. ◂

We noted that the solution to Example 7.2 for uniform circular motion was analogous to a solution to a problem of uniform motion along a line. Instead of position, we have angle; instead of velocity, we have angular velocity. In fact, the equations for circular motion that we use in this chapter are exactly analogous to the equations for motion along a line that you saw in Chapter 2, so we needn't derive any more equations; we can simply adapt ones you've seen before.

SYNTHESIS 7.1 Linear and circular motion

The variables and equations for linear motion have analogs for circular motion.

		Linear motion	Circular motion	
Variables	Position (m)	x	θ	Angle (rad)
	Velocity (m/s)	$v_x = \dfrac{\Delta x}{\Delta t}$	$\omega = \dfrac{\Delta \theta}{\Delta t}$	Angular velocity (rad/s)
	Acceleration (m/s²)	$a_x = \dfrac{\Delta v_x}{\Delta t}$	$\alpha = \dfrac{\Delta \omega}{\Delta t}$	Angular acceleration (rad/s²)
Equations	Constant velocity	$\Delta x = v\Delta t$	$\Delta\theta = \omega\Delta t$	Constant angular velocity
	Constant acceleration	$\Delta v = a\Delta t$	$\Delta\omega = \alpha\Delta t$	Constant angular acceleration
		$\Delta x = v\Delta t + \frac{1}{2}a(\Delta t)^2$	$\Delta\theta = \omega_i\Delta t + \frac{1}{2}\alpha(\Delta t)^2$	

EXAMPLE 7.6 Spinning up a computer disk

The disk in a computer disk drive spins up from rest to a final angular speed of 5400 rpm in 2.00 s. What is the angular acceleration of the disk? At the end of 2.00 s, how many revolutions has the disk made?

STRATEGIZE This problem is clearly analogous to the linear motion problems we saw in Chapter 2. Something starts at rest and then moves with constant angular acceleration for a fixed interval of time. We will solve this problem by referring to Synthesis 7.1 and using equations that correspond to the equations we'd use to solve a similar linear motion problem.

PREPARE The initial angular velocity is $\omega_i = 0$ rad/s. The final angular velocity is $\omega_f = 5400$ rpm. In rad/s this is

$$\omega_f = \frac{5400\text{ rev}}{\text{min}} \times \frac{1\text{ min}}{60\text{ s}} \times \frac{2\pi\text{ rad}}{1\text{ rev}} = 565\text{ rad/s}$$

SOLVE We start by finding the angular acceleration. We can do this using the definition of angular acceleration, the change in angular speed divided by the time interval:

$$\alpha = \frac{\Delta\omega}{\Delta t} = \frac{565\text{ rad/s} - 0\text{ rad/s}}{2.00\text{ s}} = 282.5\text{ rad/s}^2$$

We've kept an extra significant figure for later calculations, but we report our final result as $\alpha = 283$ rad/s^2.

Next, we use the equation for the angular displacement—analogous to the equation for linear displacement—during the period of constant angular acceleration to determine the angle through which the disk moves:

$$\begin{aligned}\Delta\theta &= \omega_i\Delta t + \tfrac{1}{2}\alpha(\Delta t)^2\\ &= (0\text{ rad/s})(2.00\text{ s}) + \tfrac{1}{2}(282.5\text{ rad/s}^2)(2.00\text{ s})^2\\ &= 565\text{ rad}\end{aligned}$$

Each revolution corresponds to an angular displacement of 2π, so we have

$$\begin{aligned}\text{number of revolutions} &= \frac{565\text{ rad}}{2\pi\text{ rad/revolution}}\\ &= 90\text{ revolutions}\end{aligned}$$

The disk completes 90 revolutions during the first 2 seconds.

ASSESS The disk spins up to 5400 rpm, which corresponds to 90 rev/s. If the disk spins at full speed for 2.0 s, it will undergo 180 revolutions. But it spins up from rest to this speed over a time of 2.0 s, so it makes sense that it undergoes half this number of revolutions as it gets up to speed.

Graphs for Rotational Motion with Constant Angular Acceleration

In ◀◀ SECTION 2.5, we studied position, velocity, and acceleration graphs for motion with constant acceleration. We can extend that treatment to the current case of rotational motion. Because of the analogies between linear and angular quantities in Synthesis 7.1, the rules for graphing angular variables are identical to those for linear variables. In particular, **the angular velocity is the slope of the angular position-versus-time graph,** and **the angular acceleration is the slope of the angular velocity-versus-time graph.**

EXAMPLE 7.7 Graphing angular quantities

FIGURE 7.13 shows the angular velocity-versus-time graph for the propeller of a ship.

a. Describe the motion of the propeller.
b. Draw the angular acceleration graph for the propeller.

FIGURE 7.13 The propeller's angular velocity.

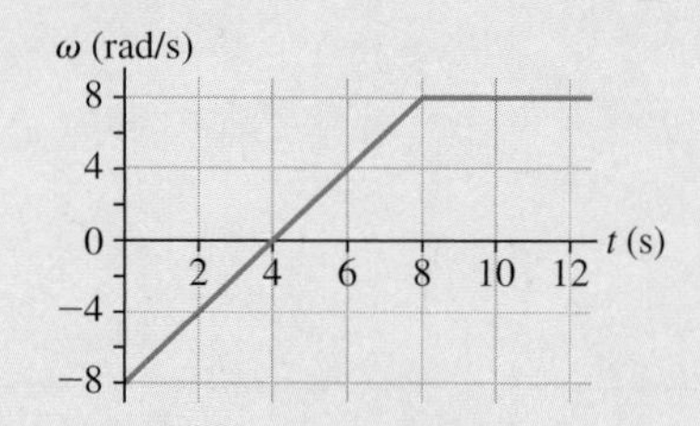

STRATEGIZE We've solved analogous problems for motion in one dimension, and we will use the correspondence between the descriptions of linear and circular motion to guide our solution.

PREPARE The graph shows the angular velocity as a function of time, so we can interpret the graph to describe the motion. We can then use the slope of the graph to determine the angular acceleration.

SOLVE

a. Initially the propeller has a negative angular velocity, so it is turning clockwise. It slows down until, at $t = 4$ s, it is instantaneously stopped. It then speeds up in the opposite direction until it is turning counterclockwise at a constant angular velocity.
b. The angular acceleration graph is the slope of the angular velocity graph. From $t = 0$ s to $t = 8$ s, the slope is

$$\frac{\Delta\omega}{\Delta t} = \frac{\omega_f - \omega_i}{\Delta t} = \frac{(8.0 \text{ rad/s}) - (-8.0 \text{ rad/s})}{8.0 \text{ s}} = 2.0 \text{ rad/s}^2$$

After $t = 8$ s, the slope is zero, so the angular acceleration is zero. This graph is plotted in FIGURE 7.14.

FIGURE 7.14 Angular acceleration graph for a propeller.

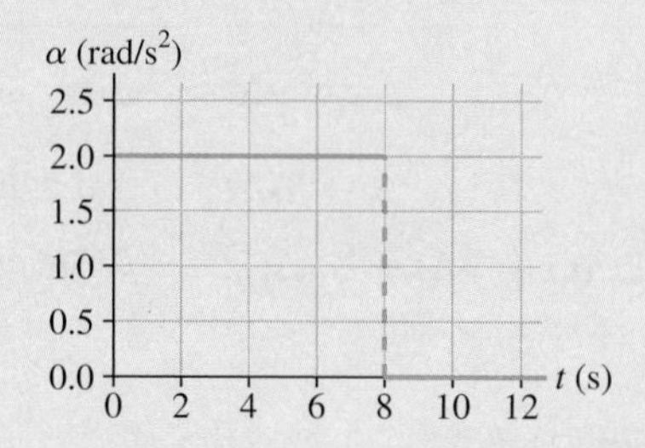

ASSESS A comparison of these graphs with their linear analogs in Figure 2.25 suggests that we're on the right track.

FIGURE 7.15 Uniform and nonuniform circular motion.

(a) Uniform circular motion

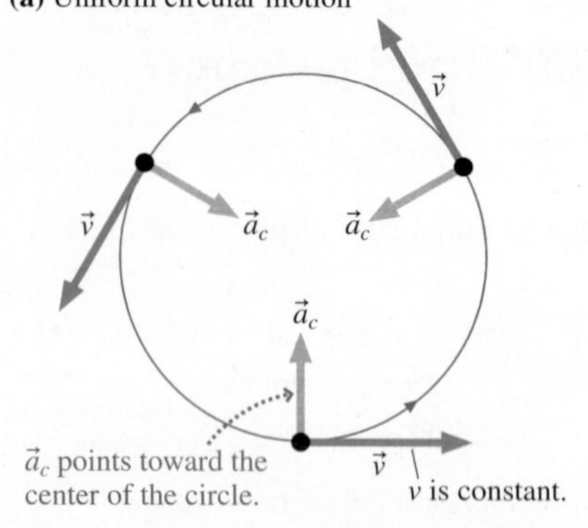

(b) Nonuniform circular motion

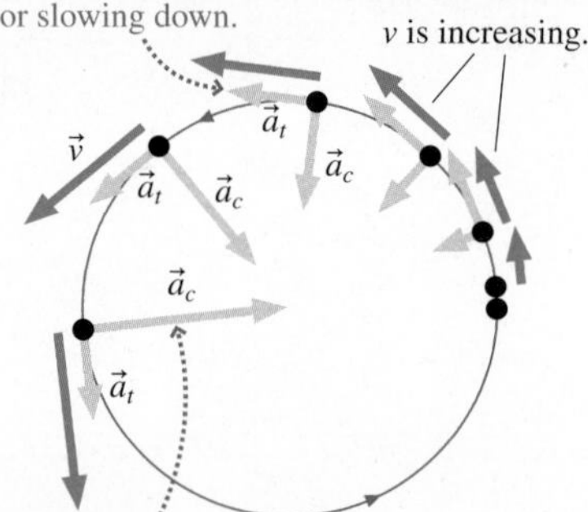

The tangential acceleration $\vec{a}_t$ causes the particle's speed to change. There's a tangential acceleration only when the particle is speeding up or slowing down.
v is increasing.
$\vec{v}$
$\vec{a}_t$
$\vec{a}_c$
$\vec{a}_t$
$\vec{a}_c$
$\vec{a}_c$
$\vec{a}_t$
The centripetal acceleration $\vec{a}_c$ causes the particle's direction to change. As the particle speeds up, a_c gets larger. Circular motion always has a centripetal acceleration.

Tangential Acceleration

As you learned in Chapter 6, and as FIGURE 7.15a reminds you, a particle undergoing uniform circular motion has an acceleration directed inward toward the center of the circle. This centripetal acceleration $\vec{a}_c$ is due to the change in the *direction* of the particle's velocity. Recall that the magnitude of the centripetal acceleration is $a_c = v^2/r = \omega^2 r$.

NOTE ▶ Centripetal acceleration will now be denoted a_c to distinguish it from tangential acceleration a_t, discussed below. ◀

If a particle is undergoing angular acceleration, its angular speed is changing and, therefore, so is its speed. This means that the particle will have another component to its acceleration. FIGURE 7.15b shows a particle whose speed is increasing as it moves around its circular path. Because the *magnitude* of the velocity is increasing, this second component of the acceleration is directed *tangentially* to the circle, in the same direction as the velocity. This component of acceleration is called the **tangential acceleration.**

The tangential acceleration measures the rate at which the particle's speed around the circle increases. Thus its magnitude is

$$a_t = \frac{\Delta v}{\Delta t}$$

We can relate the tangential acceleration to the *angular* acceleration by using the relationship $v = \omega r$ between the speed of a particle moving in a circle of radius r and its angular velocity ω. We have

$$a_t = \frac{\Delta v}{\Delta t} = \frac{\Delta(\omega r)}{\Delta t} = \frac{\Delta\omega}{\Delta t} r$$

or, because $\alpha = \Delta\omega/\Delta t$ from Equation 7.7,

$$a_t = \alpha r \tag{7.8}$$

Relationship between tangential and angular acceleration

We've seen that all points on a rotating rigid body have the same angular acceleration. From Equation 7.8, however, the centripetal and tangential accelerations of a point on a rotating object depend on the point's distance r from the axis, so these accelerations are *not* the same for all points.

STOP TO THINK 7.2 A ball on the end of a string swings in a horizontal circle once every second. State whether the magnitude of each of the following quantities is zero, constant (but not zero), or changing.

a. Velocity
b. Angular velocity
c. Centripetal acceleration
d. Angular acceleration
e. Tangential acceleration

7.3 Torque

We've seen that force is the cause of acceleration. But what about *angular* acceleration? What do Newton's laws tell us about rotational motion? To begin our study of rotational motion, we'll need to find a rotational equivalent of force.

Consider the common experience of pushing open a heavy door. FIGURE 7.16 is a top view of a door that is hinged on the left. Four forces are shown, all of equal strength. Which of these will be most effective at opening the door?

Force $\vec{F}_1$ will open the door, but force $\vec{F}_2$, which pushes straight toward the hinge, will not. Force $\vec{F}_3$ will open the door, but not as effectively as $\vec{F}_1$. What about $\vec{F}_4$? It is perpendicular to the door and it has the same magnitude as $\vec{F}_1$, but you know from experience that pushing close to the hinge is not as effective as pushing at the outer edge of the door.

The ability of a force to cause a rotation thus depends on three factors:

1. The magnitude F of the force
2. The distance r from the pivot—the axis about which the object can rotate—to the point at which the force is applied
3. The angle at which the force is applied

We can incorporate these three factors into a single quantity called the **torque τ** (Greek tau). Loosely speaking, τ measures the effectiveness of a force at causing an object to rotate about a pivot. **Torque is the rotational equivalent of force.** In Figure 7.16, for instance, the torque τ_1 due to $\vec{F}_1$ is greater than τ_4 due to $\vec{F}_4$.

To make these ideas specific, FIGURE 7.17 shows a force $\vec{F}$ applied at one point of a wrench that's loosening a nut. Figure 7.17 defines the distance r from the pivot to the point at which the force is applied; the **radial line,** the line starting at the pivot and extending through this point; and the angle ϕ (Greek phi) measured from the radial line to the direction of the force.

We saw in Figure 7.16 that force $\vec{F}_1$, which was directed perpendicular to the door, was effective in opening it, but force $\vec{F}_2$, directed toward the hinges, had no effect on its rotation. As shown in FIGURE 7.18, this suggests breaking the force $\vec{F}$ applied to the wrench into two component vectors: $\vec{F}_\perp$ directed perpendicular to the radial line, and $\vec{F}_\parallel$ directed parallel to it. Because $\vec{F}_\parallel$ points either directly toward or away from the pivot, it has no effect on the wrench's rotation, and thus contributes nothing to the torque. Only $\vec{F}_\perp$ tends to cause rotation of the wrench, so it is this component of the force that determines the torque.

NOTE ► The perpendicular component $\vec{F}_\perp$ is pronounced "F perpendicular" and the parallel component $\vec{F}_\parallel$ is "F parallel." ◄

We've seen that a force applied at a greater distance r from the pivot has a greater effect on rotation, so we expect a larger value of r to give a greater torque. We also

FIGURE 7.16 The four forces are the same strength, but they have different effects on the swinging door.

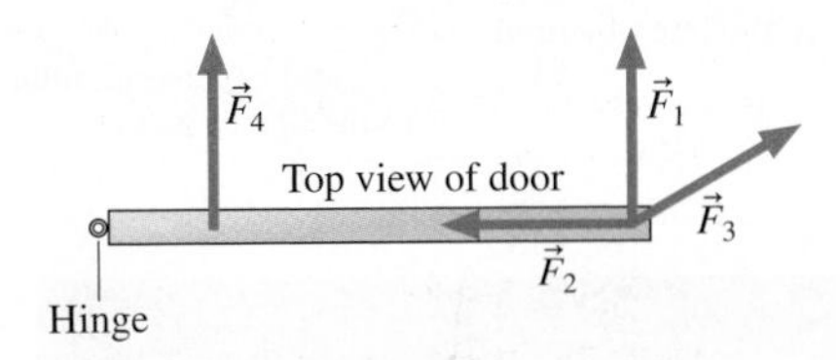
$\vec{F}_4$
$\vec{F}_1$
Top view of door
$\vec{F}_3$
$\vec{F}_2$
Hinge

A video to support a section's topic is embedded in the eText.

Video Torque

FIGURE 7.17 Force $\vec{F}$ exerts a torque about the pivot point.

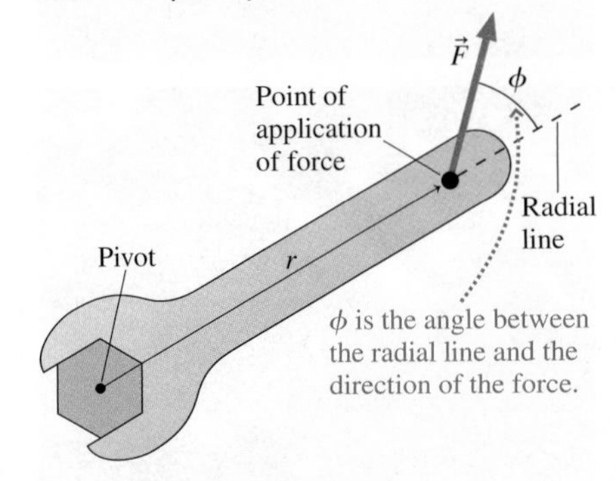
$\vec{F}$
ϕ
Point of application of force
Radial line
Pivot
r
ϕ is the angle between the radial line and the direction of the force.

FIGURE 7.18 Torque is due to the component of the force perpendicular to the radial line.

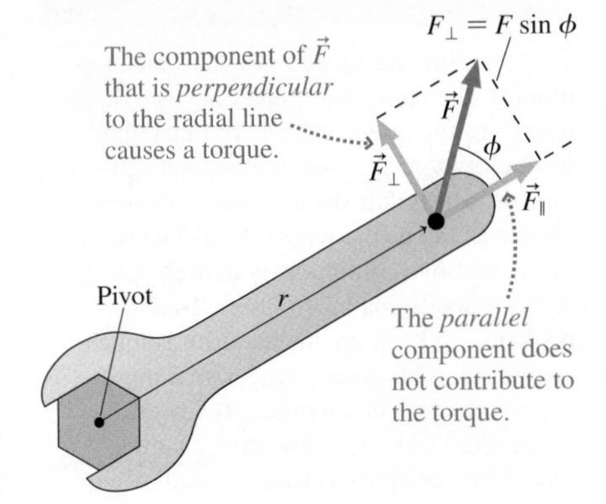
$F_\perp = F \sin\phi$
The component of $\vec{F}$ that is *perpendicular* to the radial line causes a torque.
$\vec{F}$
ϕ
$\vec{F}_\perp$
$\vec{F}_\parallel$
Pivot
r
The *parallel* component does not contribute to the torque.

FIGURE 7.19 You can also calculate torque in terms of the moment arm between the pivot and the line of action.

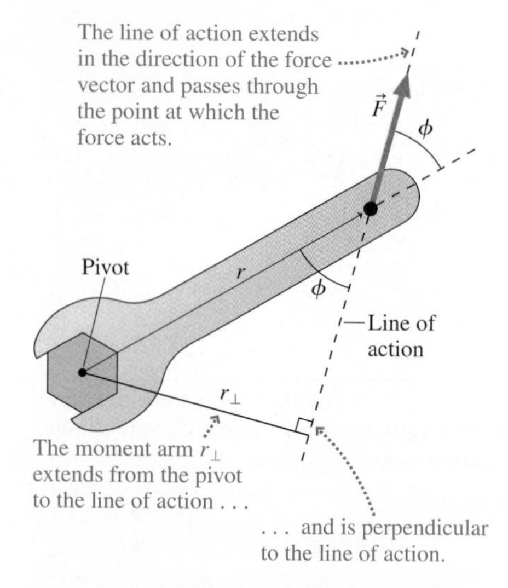
The line of action extends in the direction of the force vector and passes through the point at which the force acts.
$\vec{F}$
ϕ
Pivot
r
ϕ
Line of action
$r_\perp$
The moment arm $r_\perp$ extends from the pivot to the line of action . . .
. . . and is perpendicular to the line of action.

saw that only $\vec{F}_\perp$ contributes to the torque. Both these observations are contained in our first expression for torque:

$$\tau = rF_\perp \tag{7.9}$$

Torque due to a force with perpendicular component $F_\perp$ acting at a distance r from the pivot

From this equation, we see that the SI unit of torque is the newton-meter, abbreviated N · m.

FIGURE 7.19 shows an alternative way to calculate torque. The line that is in the direction of the force, and passes through the point at which the force acts, is called the *line of action.* The perpendicular distance from this line to the pivot is called the **moment arm** (or *lever arm*) $r_\perp$. You can see from the figure that $r_\perp = r\sin\phi$. Further, Figure 7.18 showed that $F_\perp = F\sin\phi$. We can then write Equation 7.9 as $\tau = rF\sin\phi = F(r\sin\phi) = Fr_\perp$. Thus an equivalent expression for the torque is

$$\tau = r_\perp F \tag{7.10}$$

Torque due to a force F with moment arm $r_\perp$

Torque versus speed To start and stop quickly, the basketball player needs to apply a large torque to her wheel. To make the torque as large as possible, the handrim—the outside wheel that she actually grabs—is almost as big as the wheel itself. The racer needs to move continuously at high speed, so his wheel spins much faster. To allow his hands to keep up, his handrim is much smaller than his chair's wheel, making its linear velocity correspondingly lower. The smaller radius means, however, that the torque he can apply is lower as well.

CONCEPTUAL EXAMPLE 7.8 Where do you put the pedals?

It is hard to get going if you try to start your bike with the pedal at the highest point. Why is this?

REASON Aided by the weight of the body, the greatest force can be applied to the pedal straight down. But with the pedal at the top, this force is exerted almost directly toward the pivot, causing only a small torque. We could say either that the perpendicular component of the force is small or that the moment arm is small.

ASSESS If you've ever climbed a steep hill while standing on the pedals, you know that you get the greatest forward motion when one pedal is completely forward with the crank parallel to the ground. This gives the maximum possible torque because the force you apply is entirely perpendicular to the radial line, and the moment arm is as long as it can be.

We've seen that Equation 7.9 can be written as $\tau = rF_\perp = r(F\sin\phi)$, and Equation 7.10 as $\tau = r_\perp F = (r\sin\phi)F$. This shows that both methods of calculating torque lead to the same expression for torque—namely:

$$\tau = rF\sin\phi \tag{7.11}$$

where ϕ is the angle between the radial line and the direction of the force as illustrated in Figure 7.17.

NOTE ▸ Torque differs from force in a very important way. Torque is calculated or measured *about a particular point.* To say that a torque is 20 N · m is meaningless without specifying the point about which the torque is calculated. Torque can be calculated about any point, but its value depends on the point chosen because this choice determines r and ϕ. In practice, we usually calculate torques about a hinge, pivot, or axle. ◂

Equations 7.9–7.11 are three different ways of thinking about—and calculating—the torque due to a force. Equation 7.11 is the most general equation for torque, but

Equations 7.9 and 7.10 are generally more useful for practical problem solving. All three equations calculate the *same* torque, and all will give the same value.

EXAMPLE 7.9 Torque to open a stuck door

Ryan is trying to open a stuck door. He pushes it at a point 0.75 m from the hinges with a 240 N force directed 20° away from being perpendicular to the door. There's a natural pivot point, the hinges. What torque does Ryan exert? How could he exert more torque?

STRATEGIZE This is a straightforward torque problem. But even for a problem as straightforward as this, we will start by drawing the visual overview in FIGURE 7.20. The figure shows the door, identifies the pivot point (the hinge, in this case), and shows the force, where it is applied and at what angle.

FIGURE 7.20 Ryan's force exerts a torque on the door.

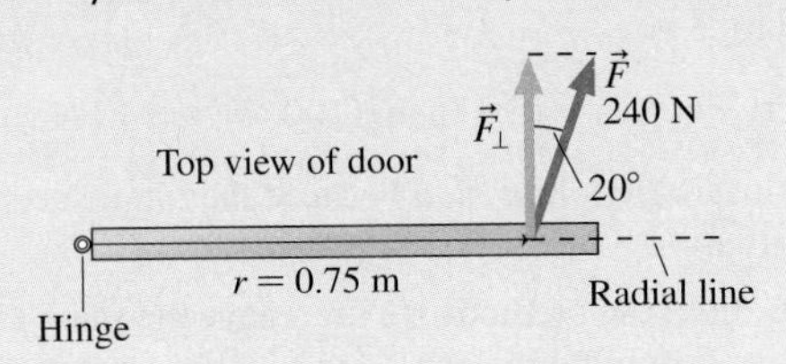

PREPARE In Figure 7.20 the radial line is shown drawn from the pivot—the hinge—through the point at which the force $\vec{F}$ is applied. We see that the component of $\vec{F}$ that is perpendicular to the radial line is $F_\perp = F\cos 20° = 226\text{ N}$. The distance from the hinge to the point at which the force is applied is $r = 0.75$ m.

SOLVE We can find the torque on the door from Equation 7.9:

$$\tau = rF_\perp = (0.75\text{ m})(226\text{ N}) = 170\text{ N}\cdot\text{m}$$

The torque depends on how hard Ryan pushes, where he pushes, and at what angle. If he wants to exert more torque, he could push at a point a bit farther out from the hinge, or he could push exactly perpendicular to the door. Or he could simply push harder!

ASSESS As you'll see by doing more problems, 170 N · m is a significant torque, but this makes sense if you are trying to free a stuck door.

Equations 7.9–7.11 give only the magnitude of the torque. But torque, like a force component, has a sign. **A torque that tends to rotate the object in a counterclockwise direction is positive, while a torque that tends to rotate the object in a clockwise direction is negative.** FIGURE 7.21 summarizes what we've said about the magnitude and the sign of a torque. Notice that a force pushing straight toward the pivot or pulling straight out from the pivot exerts *no* torque.

Video Torques and Moment Arms

NOTE ▸ When calculating a torque, you must supply the appropriate sign by observing the direction in which the torque acts. ◂

KEY CONCEPT FIGURE 7.21 Signs and strengths of the torque.

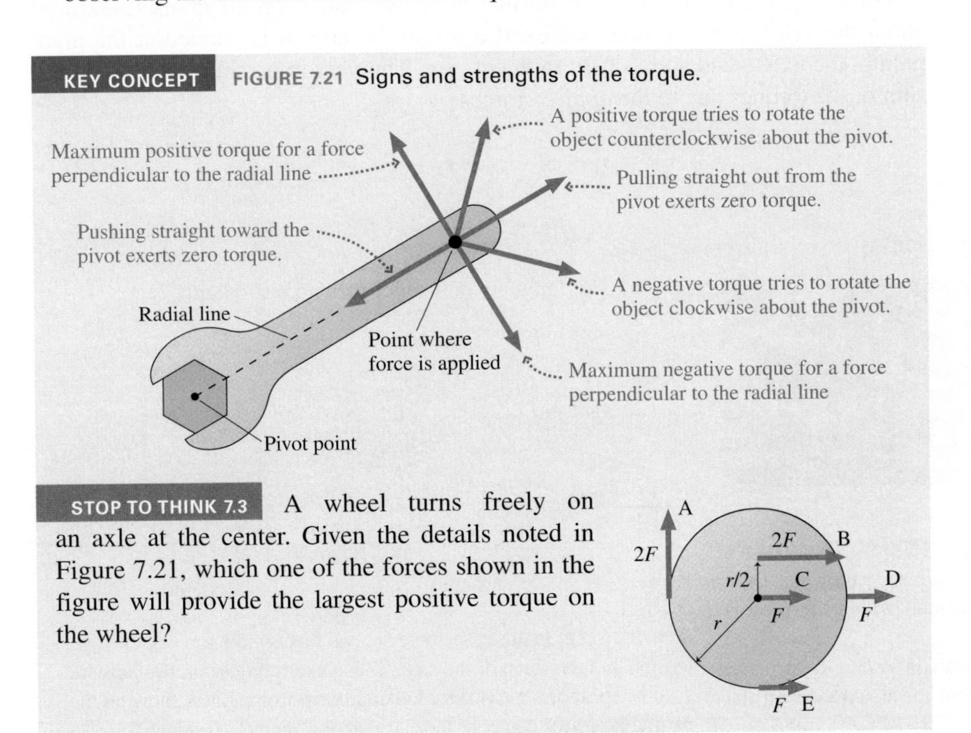

STOP TO THINK 7.3 A wheel turns freely on an axle at the center. Given the details noted in Figure 7.21, which one of the forces shown in the figure will provide the largest positive torque on the wheel?

EXAMPLE 7.10 Calculating the torque on a nut

Luis uses a 20-cm-long wrench to tighten a nut, turning it clockwise. The wrench handle is tilted 30° above the horizontal, and Luis pulls straight down on the end with a force of 100 N. How much torque does Luis exert on the nut?

STRATEGIZE We will start by drawing a visual overview that illustrates the geometry of the situation. There is a natural pivot point—the center of the nut that the wrench is turning.

PREPARE We make two diagrams in FIGURE 7.22 to illustrate the two different means of computing torque represented by Equations 7.9 and 7.10.

FIGURE 7.22 A wrench being used to turn a nut.

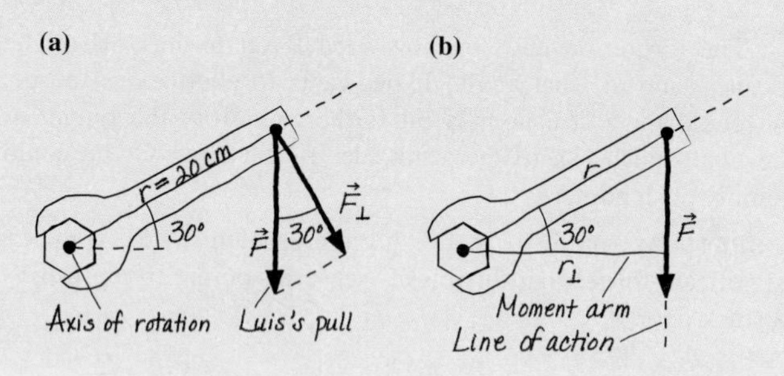
(a)
(b)
r = 20 cm
30°
30°
Axis of rotation
Luis's pull
30°
Moment arm
Line of action

SOLVE According to Equation 7.9, the torque can be calculated as $\tau = rF_\perp$. From Figure 7.22a, we see that the perpendicular component of the force is

$$F_\perp = F\cos 30° = (100\text{ N})(\cos 30°) = 86.6\text{ N}$$

This gives a torque of

$$\tau = -rF_\perp = -(0.20\text{ m})(86.6\text{ N}) = -17\text{ N}\cdot\text{m}$$

We put in the minus sign because the torque is negative—it tries to rotate the nut in a *clockwise* direction.

Alternatively, we can use Equation 7.10 to find the torque. Figure 7.22b shows the moment arm $r_\perp$, the perpendicular distance from the pivot to the line of action. From the figure we see that

$$r_\perp = r\cos 30° = (0.20\text{ m})(\cos 30°) = 0.173\text{ m}$$

Then the torque is

$$\tau = -r_\perp F = -(0.173\text{ m})(100\text{ N}) = -17\text{ N}\cdot\text{m}$$

Again, we insert the minus sign because the torque acts to give a clockwise rotation.

ASSESS Both of the methods we used gave the same answer for the torque, as should be the case, and this gives us confidence in our results.

Net Torque

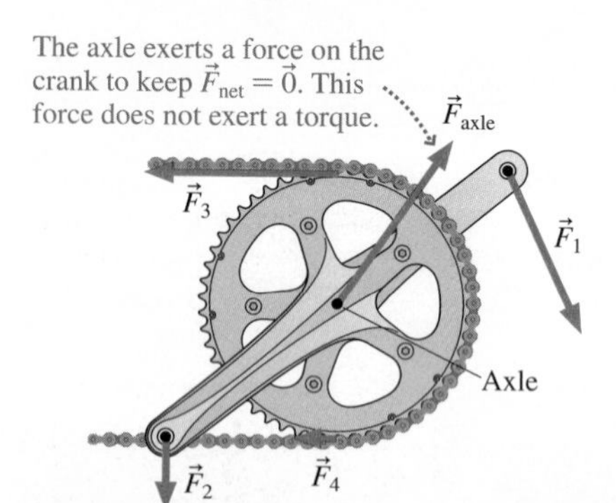
The axle exerts a force on the crank to keep $\vec{F}_{net} = \vec{0}$. This force does not exert a torque.
$\vec{F}_{axle}$
$\vec{F}_3$
$\vec{F}_1$
Axle
$\vec{F}_2$
$\vec{F}_4$

FIGURE 7.23 The forces exert a net torque about the pivot point.

FIGURE 7.23 shows the forces acting on the crankset of a bicycle. Forces $\vec{F}_1$ and $\vec{F}_2$ are due to the rider pushing on the pedals, and $\vec{F}_3$ and $\vec{F}_4$ are tension forces from the chain. The crankset is free to rotate about a fixed axle, but the axle prevents it from having any translational motion with respect to the bike frame. It does so by exerting force $\vec{F}_{axle}$ on the object to balance the other forces and keep $\vec{F}_{net} = \vec{0}$.

Forces $\vec{F}_1$, $\vec{F}_2$, $\vec{F}_3$, and $\vec{F}_4$ exert torques τ_1, τ_2, τ_3, and τ_4 on the crank (measured about the axle), but $\vec{F}_{axle}$ does *not* exert a torque because it is applied at the pivot point—the axle—and so has zero moment arm. The *net* torque about the axle is the sum of the torques due to the *applied* forces:

$$\tau_{net} = \tau_1 + \tau_2 + \tau_3 + \tau_4 + \cdots = \sum \tau \qquad (7.12)$$

EXAMPLE 7.11 Force in turning a capstan

A capstan is a device used on old sailing ships to raise the anchor. A sailor pushes the long lever, turning the capstan and winding up the anchor rope. If the capstan turns at a constant speed, the net torque on it, as we'll learn later in the chapter, is zero. The rope tension due to the weight of the anchor is 1500 N. The distance from the axis to the point on the lever where the sailor pushes is seven times the radius of the capstan around which the rope is wound.

With what force must the sailor push in order to turn the capstan at a constant speed, meaning that the net torque on the capstan is zero?

FIGURE 7.24 Top view of a sailor turning a capstan.

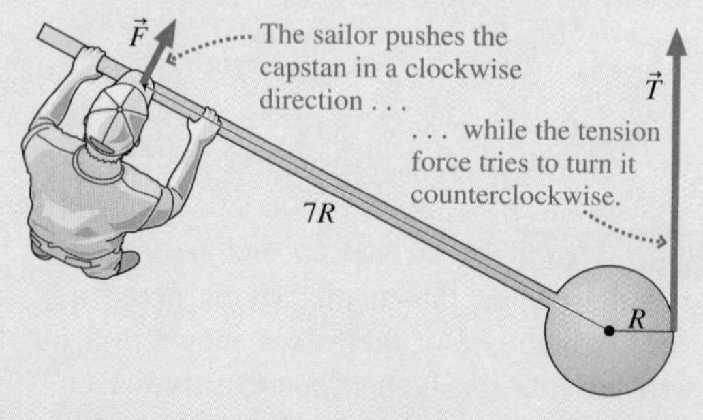
$\vec{F}$
The sailor pushes the capstan in a clockwise direction . . .
$\vec{T}$
. . . while the tension force tries to turn it counterclockwise.
7R
R

STRATEGIZE In this case, there are two forces—the force of the rope tension and the force of the sailor—that exert torques on the capstan. FIGURE 7.24 shows a top view looking down from above, showing the natural pivot (the center of the capstan), the geometry, and the forces.

PREPARE Figure 7.24 illustrates the situation. The rope pulls with a tension force $\vec{T}$ at distance R from the axis of rotation. The sailor pushes with a force $\vec{F}$ at distance $7R$ from the axis. Both forces are perpendicular to their radial lines, so ϕ in Equation 7.11 is 90°.

SOLVE The torque due to the tension in the rope is

$$\tau_T = RT\sin 90° = RT$$

We don't know the capstan radius, so we'll just leave it as R for now. This torque is positive because it tries to turn the capstan counterclockwise. The torque due to the sailor is

$$\tau_S = -(7R)F\sin 90° = -7RF$$

We put the minus sign in because this torque acts in the clockwise (negative) direction. The net torque is zero, so we have $\tau_T + \tau_S = 0$, or

$$RT - 7RF = 0$$

Note that the radius R cancels, leaving

$$F = \frac{T}{7} = \frac{1500\text{ N}}{7} = 210\text{ N}$$

ASSESS 210 N is about 50 lb, a reasonable number to expect for the pushing force of a sailor. The force the sailor must exert is one-seventh the force the rope exerts: The long lever helps him lift the heavy anchor. In the HMS *Warrior*, built in 1860, it took 200 men turning the capstan to lift the huge anchor that weighed close to 55,000 N!

Note that forces $\vec{F}$ and $\vec{T}$ point in different directions. Their torques depend only on their directions with respect to their own radial lines, not on the directions of the forces with respect to each other. The force the sailor needs to apply remains unchanged as he circles the capstan.

STOP TO THINK 7.4 Two forces act on the wheel shown. What third force, acting at point P, will make the net torque on the wheel zero?

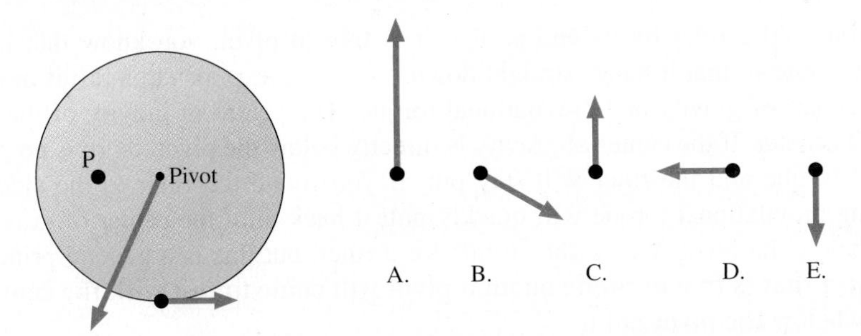

7.4 Gravitational Torque and the Center of Gravity

As the gymnast in FIGURE 7.25 pivots around the bar, a torque due to the force of gravity causes her to rotate toward a vertical position. A falling tree and a car hood slamming shut are other examples where gravity exerts a torque on an object. Stationary objects can also experience a torque due to gravity. A diving board experiences a gravitational torque about its fixed end. It doesn't rotate because of a counteracting torque provided by forces from the base at its fixed end.

We've learned how to calculate the torque due to a single force acting on an object. But gravity doesn't act at a single point on an object. It pulls downward on *every particle* that makes up the object, as shown for the gymnast in Figure 7.25a, and so each particle experiences a small torque due to the force of gravity that acts on it. The gravitational torque on the object as a whole is then the *net* torque exerted on all the particles. We won't prove it, but the gravitational torque can be calculated by assuming that the net force of gravity—that is, the object's weight $\vec{w}$—acts at a single special point on the object called its **center of gravity** (symbol ⊙). Then we can calculate the torque due to gravity by the methods we learned earlier for a single force ($\vec{w}$) acting at a single point (the center of gravity). Figure 7.25b shows how we can consider the gymnast's weight as acting at her center of gravity.

FIGURE 7.25 The center of gravity is the point where the weight appears to act.

(a) Gravity exerts a force and a torque on each particle that makes up the gymnast.

(b) The weight force provides a torque about the rotation axis.

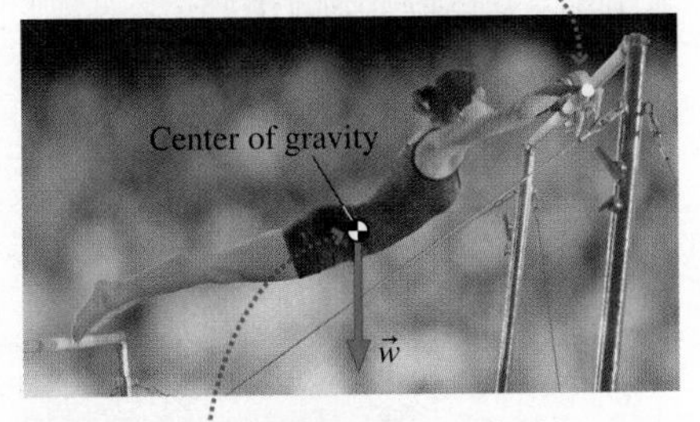

The gymnast responds *as if* her entire weight acts at her center of gravity.

EXAMPLE 7.12 The gravitational torque on a flagpole

A 3.2 kg flagpole extends from a wall at an angle of 25° from the horizontal. Its center of gravity is 1.6 m from the point where the pole is attached to the wall. What is the gravitational torque on the flagpole about the point of attachment?

STRATEGIZE In all the problems in this chapter, a visual overview is crucial to help you understand the details, so we will start with the diagram in FIGURE 7.26. We'll note dimensions, draw the weight vector at the proper point, and choose one of the approaches we've seen for calculating the torque.

PREPARE For the purpose of calculating gravitational torque, we can consider the entire weight of the pole as acting at the center of gravity. Because the moment arm $r_\perp$ is simple to visualize here, we'll use Equation 7.10 for the torque.

FIGURE 7.26 Visual overview of the flagpole.

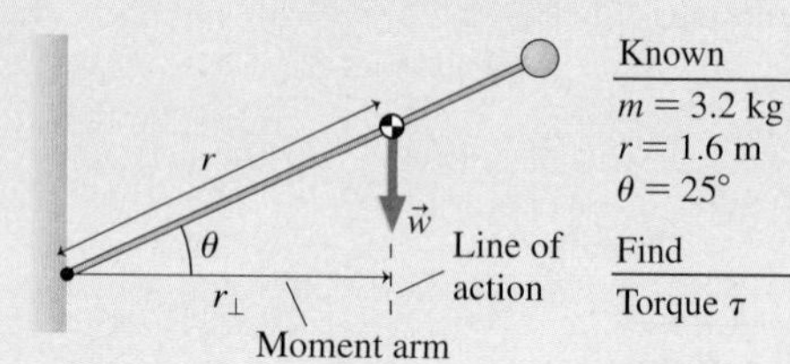

Known
$m = 3.2$ kg
$r = 1.6$ m
$\theta = 25°$
Find
Torque τ
r
θ
$\vec{w}$
Line of action
$r_\perp$
Moment arm

SOLVE From Figure 7.26, we see that the moment arm is $r_\perp = (1.6\text{ m})\cos 25° = 1.45\text{ m}$. Thus the gravitational torque on the flagpole, about the point where it attaches to the wall, is

$$\tau = -r_\perp w = -r_\perp mg = -(1.45\text{ m})(3.2\text{ kg})(9.8\text{ m/s}^2)$$
$$= -45\text{ N}\cdot\text{m}$$

We inserted the minus sign because the torque tries to rotate the pole in a clockwise direction. If the pole were attached to the wall by a hinge, the gravitational torque would cause the pole to fall. However, the actual rigid connection provides a counteracting (positive) torque to the pole that prevents this. The net torque is zero.

ASSESS A torque is a product of a force and a distance. The force is the weight of the pole, just over 30 N; the distance is the moment arm, just shy of 1.5 m. So 45 N · m is a reasonable result.

FIGURE 7.27 Suspending a ruler.

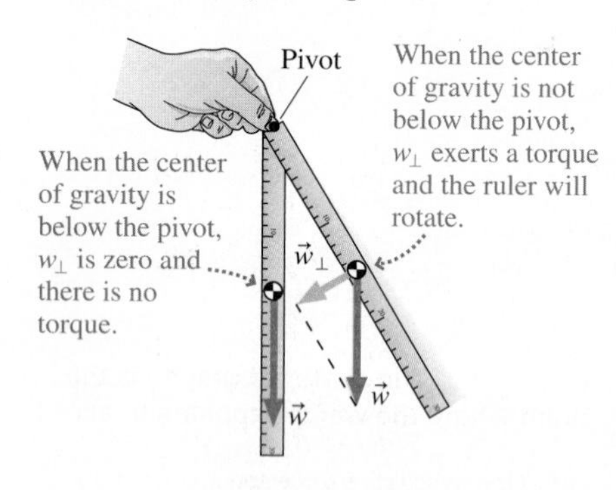

Pivot
When the center of gravity is not below the pivot, $w_\perp$ exerts a torque and the ruler will rotate.
When the center of gravity is below the pivot, $w_\perp$ is zero and there is no torque.
$\vec{w}_\perp$
$\vec{w}$
$\vec{w}$

If you hold a ruler by its end so that it is free to pivot, you know that it will quickly rotate so that it hangs straight down. FIGURE 7.27 explains this result in terms of the center of gravity and gravitational torque. The center of gravity of the ruler lies at its center. If the center of gravity is directly below the pivot, there is no gravitational torque and the ruler will stay put. If you rotate the ruler to the side, the resulting gravitational torque will quickly pull it back until the center of gravity is again below the pivot. We've shown this for a ruler, but this is a general principle: **An object that is free to rotate about a pivot will come to rest with the center of gravity below the pivot point.**

If the center of gravity lies directly *above* the pivot, as in FIGURE 7.28, there is no torque due to the object's weight and it can remain balanced. However, if the object is even slightly displaced to either side, the gravitational torque will no longer be zero and the object will begin to rotate. This question of *balance*—the behavior of an object whose center of gravity lies above the pivot—will be explored in depth in Chapter 8.

FIGURE 7.28 Balancing a ruler.

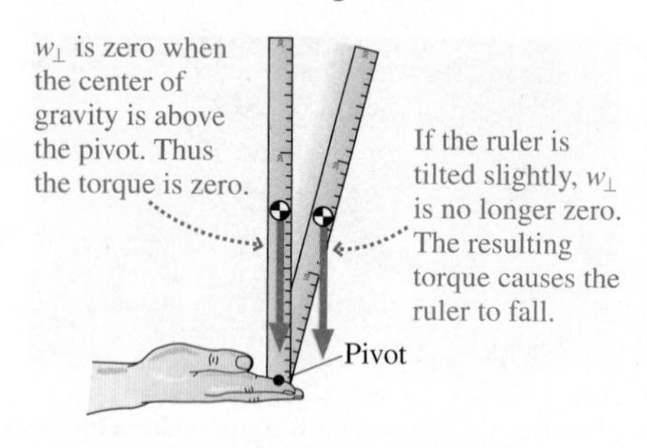

$w_\perp$ is zero when the center of gravity is above the pivot. Thus the torque is zero.
If the ruler is tilted slightly, $w_\perp$ is no longer zero. The resulting torque causes the ruler to fall.
Pivot

Calculating the Position of the Center of Gravity

Because there's no gravitational torque when the center of gravity lies either directly above or directly below the pivot, it must be the case that **the torque due to gravity when the pivot is *at* the center of gravity is zero.** We can use this fact to find a general expression for the position of the center of gravity.

Consider the dumbbell shown in FIGURE 7.29. If we slide the triangular pivot back and forth until the dumbbell balances, the pivot must then be at the center of gravity (at position x_{cg}) and the torque due to gravity must therefore be zero. But we can calculate the gravitational torque directly by calculating and summing the torques about this point due to the two individual weights. Gravity acts on weight 1 with moment arm r_1, so the torque about the pivot at position x_{cg} is

$$\tau_1 = r_1 w_1 = (x_{cg} - x_1)m_1 g$$

Similarly, the torque due to weight 2 is

$$\tau_2 = -r_2 w_2 = -(x_2 - x_{cg})m_2 g$$

FIGURE 7.29 Finding the center of gravity of a dumbbell.

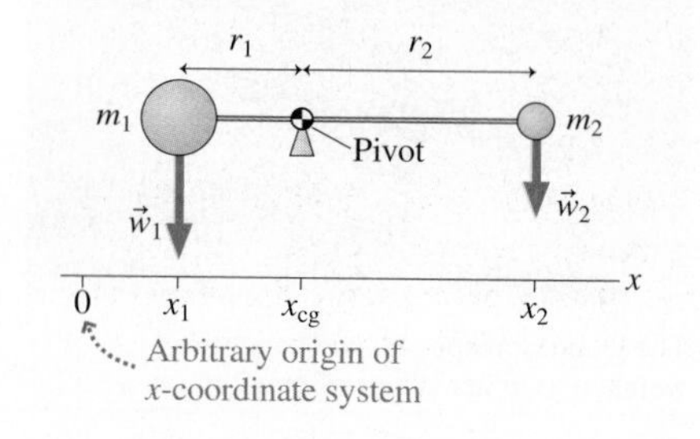

r_1
r_2
m_1
m_2
Pivot
$\vec{w}_1$
$\vec{w}_2$
0
x_1
x_{cg}
x_2
x
Arbitrary origin of x-coordinate system

This torque is negative because it tends to rotate the dumbbell in a clockwise direction. We've just argued that the net torque must be zero because the pivot is directly under the center of gravity, so

$$\tau_{net} = 0 = \tau_1 + \tau_2 = (x_{cg} - x_1)m_1 g - (x_2 - x_{cg})m_2 g$$

We can solve this equation for the position of the center of gravity x_{cg}:

$$x_{cg} = \frac{x_1 m_1 + x_2 m_2}{m_1 + m_2} \quad (7.13)$$

The following Tactics Box shows how Equation 7.13 can be generalized to find the center of gravity of *any* number of particles. If the particles don't all lie along the x-axis, then we'll also need to find the y-coordinate of the center of gravity.

TACTICS BOX 7.1 Finding the center of gravity

1. Choose an origin for your coordinate system. You can choose any convenient point as the origin.
2. Determine the coordinates (x_1, y_1), (x_2, y_2), (x_3, y_3), . . . for the particles of masses $m_1, m_2, m_3, \ldots$, respectively.
3. The x-coordinate of the center of gravity is

$$x_{cg} = \frac{x_1 m_1 + x_2 m_2 + x_3 m_3 + \cdots}{m_1 + m_2 + m_3 + \cdots} \quad (7.14)$$

4. Similarly, the y-coordinate of the center of gravity is

$$y_{cg} = \frac{y_1 m_1 + y_2 m_2 + y_3 m_3 + \cdots}{m_1 + m_2 + m_3 + \cdots} \quad (7.15)$$

Exercises 12–14

Video Balancing a Meter Stick

Because the center of gravity depends on products such as $x_1 m_1$, objects with large masses count more heavily than objects with small masses. Consequently, **the center of gravity tends to lie closer to the heavier objects or particles** that make up the entire object.

EXAMPLE 7.13 Balancing a seesaw

Emma and Noah are playing on a seesaw. Noah, mass 22 kg, is seated 1.6 m away from the pivot point. Emma, mass 25 kg, wants to sit at a point that will make the seesaw exactly balance on its pivot. How far away from the pivot must she sit? Assume that the mass of the seesaw itself is small enough to ignore.

STRATEGIZE We'll interpret the statement of Noah's position to mean that his center of gravity is 1.6 m from the pivot point; we'll mark the position of his center of gravity at that point. We are looking for the position of Emma's center of gravity, which should be placed so that the combined center of gravity for the children is right at the pivot.

PREPARE In FIGURE 7.30, we've defined coordinates so that the pivot point is at $x = 0$. Noah is 1.6 m from the pivot, so $x_N = -1.6$ m. We are looking for x_E, Emma's position, given that the combined center of gravity of the two children, with masses noted, is at the pivot.

FIGURE 7.30 Children balance on a seesaw.

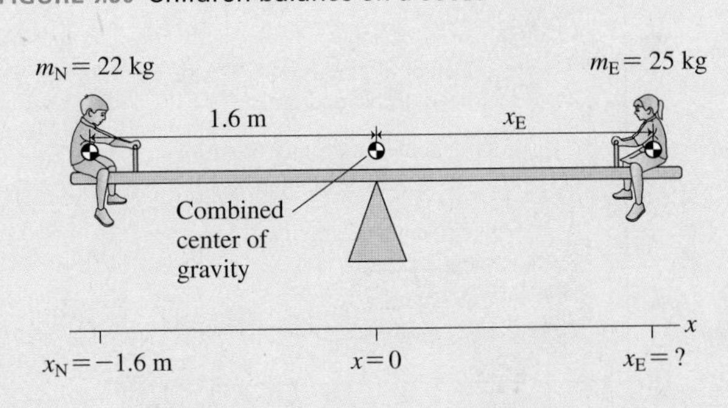

SOLVE There are two masses, the masses of Noah and Emma, and the position of the center of gravity is at the pivot, so we can simplify Equation 7.14 for the position of the center of gravity:

$$x_{cg} = 0 = \frac{x_N m_N + x_E m_E}{m_N + m_E}$$

If the fraction on the right is equal to zero, the numerator must be zero, so

$$x_N m_N + x_E m_E = 0$$

$$x_E m_E = -x_N m_N$$

$$x_E = -x_N\left(\frac{m_N}{m_E}\right) = -(-1.6\text{ m})\frac{22\text{ kg}}{25\text{ kg}} = 1.4\text{ m}$$

Emma should sit 1.4 m from the pivot.

ASSESS If you've ever played on a seesaw, you know that to achieve balance the heavier person needs to sit closer to the pivot, so our answer makes sense. It also makes sense that the combined center of gravity is closer to the heavier child.

FIGURE 7.31 Body segment masses and centers of gravity for an 80 kg man.

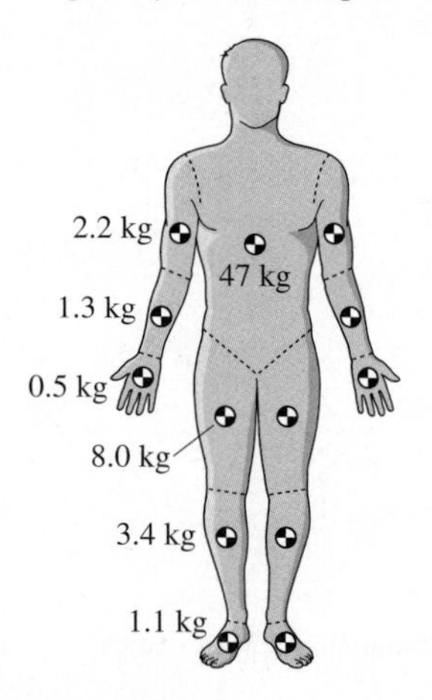

The center of gravity of an extended object can often be found by considering the object as made up of pieces, each with mass and center of gravity that are known or can be found. Then the coordinates of the entire object's center of gravity are given by Equations 7.14 and 7.15, with (x_1, y_1), (x_2, y_2), (x_3, y_3), . . . the coordinates of the center of gravity of each piece and m_1, m_2, m_3, . . . their masses.

This method is used in biomechanics and kinesiology to calculate the center of gravity of the human body. FIGURE 7.31 shows how the body can be considered to be made up of segments, each of whose mass and center of gravity have been measured. The numbers shown are appropriate for a man with a total mass of 80 kg. For a given posture the positions of the segments and their centers of gravity can be found, and thus the whole-body center of gravity from Equations 7.14 and 7.15 (and a third equation for the z-coordinate). The next example explores a simplified version of this method.

EXAMPLE 7.14 Finding the center of gravity of a gymnast

A gymnast performing on the rings holds himself in the pike position. FIGURE 7.32 shows how we can consider his body to be made up of two segments whose masses and center-of-gravity positions are shown. The upper segment includes his head, trunk, and arms, while the lower segment consists of his legs. Locate the overall center of gravity of the gymnast.

FIGURE 7.32 Centers of gravity of two segments of a gymnast.

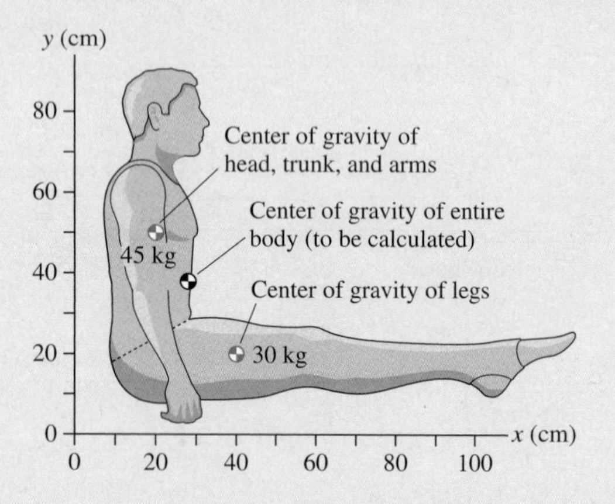

STRATEGIZE We will find the center of gravity of the gymnast by computing the combined center of gravity for the segments of his body. We will use Equation 7.14 for the x-coordinate and Equation 7.15 for the y-coordinate.

PREPARE From Figure 7.32 we can find the x- and y-coordinates of the segment centers of gravity:

$$x_{\text{trunk}} = 20\text{ cm} \qquad y_{\text{trunk}} = 50\text{ cm}$$

$$x_{\text{legs}} = 40\text{ cm} \qquad y_{\text{legs}} = 20\text{ cm}$$

SOLVE The x- and y-coordinates of the center of gravity are given by Equations 7.14 and 7.15:

$$x_{\text{cg}} = \frac{x_{\text{trunk}}m_{\text{trunk}} + x_{\text{legs}}m_{\text{legs}}}{m_{\text{trunk}} + m_{\text{legs}}}$$

$$= \frac{(20\text{ cm})(45\text{ kg}) + (40\text{ cm})(30\text{ kg})}{45\text{ kg} + 30\text{ kg}} = 28\text{ cm}$$

and

$$y_{\text{cg}} = \frac{y_{\text{trunk}}m_{\text{trunk}} + y_{\text{legs}}m_{\text{legs}}}{m_{\text{trunk}} + m_{\text{legs}}}$$

$$= \frac{(50\text{ cm})(45\text{ kg}) + (20\text{ cm})(30\text{ kg})}{45\text{ kg} + 30\text{ kg}} = 38\text{ cm}$$

ASSESS The center-of-gravity position of the entire body, shown in Figure 7.32, is closer to that of the heavier trunk segment than to that of the lighter legs. It also lies along a line connecting the two segment centers of gravity, just as it would for the center of gravity of two point particles. Note also that the gymnast's hands—the pivot point—are below his center of gravity, as we would expect for stability.

STOP TO THINK 7.5 The balls are connected by very lightweight rods pivoted at the point indicated by a dot. The rod lengths are all equal except for A, which is twice as long. Rank in order, from least to greatest, the magnitudes of the net gravitational torques about the pivots for arrangements A to D.

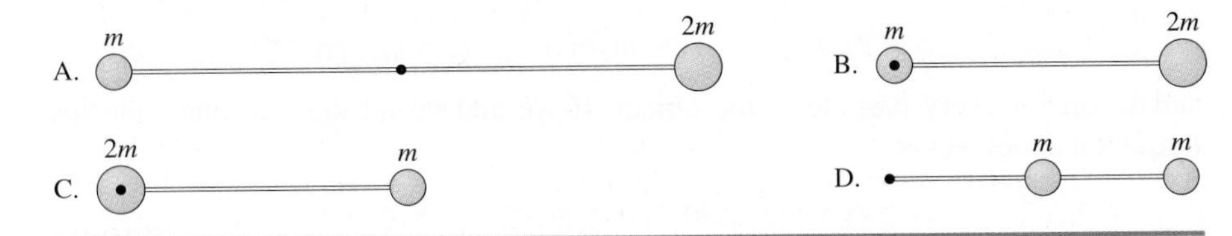

7.5 Rotational Dynamics and Moment of Inertia

In Section 7.3 we asked: What do Newton's laws tell us about rotational motion? We can now answer that question: **A torque causes an angular acceleration.** This is the rotational equivalent of our earlier assertion, for motion along a line, that a force causes an acceleration.

To see where this connection between torque and angular acceleration comes from, let's start by examining a *single particle* subject to a torque. FIGURE 7.33 shows a particle of mass m attached to a lightweight, rigid rod of length r that constrains the particle to move in a circle. The particle is subject to two forces. Because it's moving in a circle, there must be a force—here, the tension $\vec{T}$ from the rod—directed toward the center of the circle. As we learned in Chapter 6, this is the force responsible for changing the *direction* of the particle's velocity. The acceleration associated with this change in the particle's velocity is the centripetal acceleration $\vec{a}_c$.

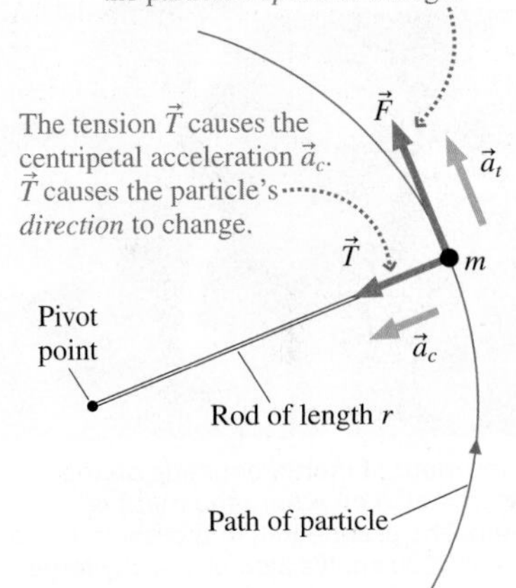

FIGURE 7.33 A tangential force $\vec{F}$ exerts a torque on the particle and causes an angular acceleration.

But the particle in Figure 7.33 is also subject to the force $\vec{F}$ that changes the *speed* of the particle. This force causes a tangential acceleration $\vec{a}_t$. Applying Newton's second law in the direction tangent to the circle gives

$$a_t = \frac{F}{m} \tag{7.16}$$

Now the tangential and angular accelerations are related by $a_t = \alpha r$, so we can rewrite Equation 7.16 as $\alpha r = F/m$, or

$$\alpha = \frac{F}{mr} \tag{7.17}$$

We can now connect this angular acceleration to the torque because force $\vec{F}$, which is perpendicular to the radial line, exerts torque

$$\tau = rF$$

With this relationship between F and τ, we can write Equation 7.17 as

$$\alpha = \frac{\tau}{mr^2} \tag{7.18}$$

Equation 7.18 gives a relationship between the torque on a single particle and its angular acceleration. Now all that remains is to expand this idea from a single particle to an extended object.

Newton's Second Law for Rotation

FIGURE 7.34 shows a rigid body that undergoes rotation about a fixed and unmoving axis. According to the rigid-body model, we can think of the object as consisting of particles with masses $m_1, m_2, m_3, \ldots$ at fixed distances $r_1, r_2, r_3, \ldots$ from the axis.

FIGURE 7.34 The forces on a rigid body exert a torque about the rotation axis.

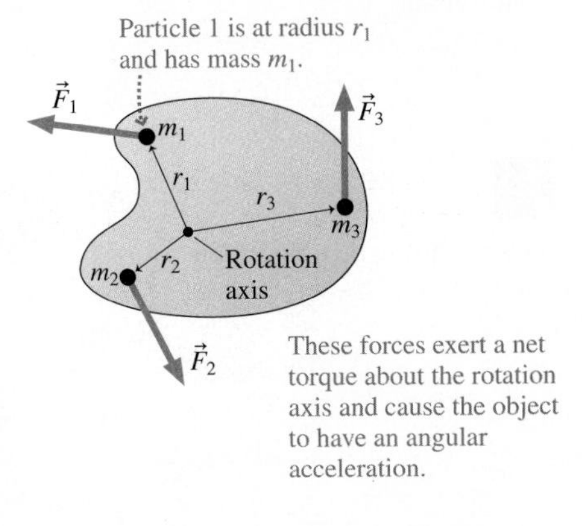

Suppose forces $\vec{F}_1, \vec{F}_2, \vec{F}_3, \ldots$ act on these particles. These forces exert torques around the rotation axis, so the object will undergo an angular acceleration α. Because all the particles that make up the object rotate together, each particle has this *same* angular acceleration α. Rearranging Equation 7.18 slightly, we can write the torques on the particles as

$$\tau_1 = m_1 r_1^2 \alpha \qquad \tau_2 = m_2 r_2^2 \alpha \qquad \tau_3 = m_3 r_3^2 \alpha$$

and so on for every particle in the object. If we add up all these torques, the *net* torque on the object is

$$\begin{aligned} \tau_{\text{net}} &= \tau_1 + \tau_2 + \tau_3 + \cdots = m_1 r_1^2 \alpha + m_2 r_2^2 \alpha + m_3 r_3^2 \alpha + \cdots \\ &= \alpha(m_1 r_1^2 + m_2 r_2^2 + m_3 r_3^2 + \cdots) = \alpha \sum m_i r_i^2 \end{aligned} \tag{7.19}$$

In factoring α out of the sum, we're making explicit use of the fact that every particle in a rotating rigid body has the *same* angular acceleration α.

The quantity Σmr^2 in Equation 7.19, which is the proportionality constant between angular acceleration and net torque, is called the object's **moment of inertia** I:

$$I = m_1 r_1^2 + m_2 r_2^2 + m_3 r_3^2 + \cdots = \sum m_i r_i^2 \tag{7.20}$$

Moment of inertia of a collection of particles

The moment of inertia depends on the mass and also on where the mass is located. The granite ball in the photo has a mass of 8200 kg. It's also physically large, with much of the mass far from the center. For both of these reasons, the moment of inertia is extremely large. The sphere can freely spin on a thin layer of pressurized water. Even though the girl exerts a large torque, the extremely large moment of inertia means that the angular acceleration is very small.

The unit of moment of inertia is mass times distance squared, or kg · m^2. An object's moment of inertia, like torque, *depends on the axis of rotation.* Once the axis is specified, allowing the values of $r_1, r_2, r_3, \ldots$ to be determined, the moment of inertia *about that axis* can be calculated from Equation 7.20.

NOTE ▶ The word "moment" in "moment of inertia" and "moment arm" has nothing to do with time. It stems from the Latin *momentum,* meaning "motion." ◀

Substituting the moment of inertia into Equation 7.19 puts the final piece of the puzzle into place, giving us the fundamental equation for rigid-body dynamics:

Newton's second law for rotation An object that experiences a net torque τ_{net} about the axis of rotation undergoes an angular acceleration

$$\alpha = \frac{\tau_{\text{net}}}{I} \tag{7.21}$$

where I is the moment of inertia of the object *about the rotation axis.*

In practice we often write $\tau_{\text{net}} = I\alpha$, but Equation 7.21 better conveys the idea that **a net torque is the cause of angular acceleration.** In the absence of a net torque ($\tau_{\text{net}} = 0$), the object has zero angular acceleration α, so it either does not rotate ($\omega = 0$) or rotates with *constant* angular velocity ($\omega = \text{constant}$).

Interpreting the Moment of Inertia

Video Walking on a Tightrope

Before rushing to calculate moments of inertia, let's get a better understanding of its meaning. First, notice that **moment of inertia is the rotational equivalent of mass.** It plays the same role in Equation 7.21 as does mass m in the now-familiar $\vec{a} = \vec{F}_{\text{net}}/m$. Recall that objects with larger mass have a larger *inertia,* meaning that they're harder to accelerate. Similarly, an object with a larger moment of inertia is harder to get rotating: It takes a larger torque to spin up an object that has a larger moment of inertia than an object with a smaller moment of inertia. The fact that "moment of inertia" retains the word "inertia" reminds us of this.

But why does the moment of inertia depend on the distances r from the rotation axis? Think about trying to start a merry-go-round from rest, as shown in FIGURE 7.35. By pushing on the rim of the merry-go-round, you exert a torque on it, and its angular velocity begins to increase. If your friends sit at the rim of the merry-go-round, as in Figure 7.35a, their distances r from the axle are large. Once you've spun up the merry-go-round to a reasonable angular velocity ω, their speeds, given by $v = \omega r$, will be large as well. If your friends sit near the axle, however, as in Figure 7.35b, then r is small and their speeds $v = \omega r$ will be small. In the first case, you've gotten your friends up to a higher speed, and it's no surprise that you will need to push harder to make that happen.

FIGURE 7.35 Moment of inertia depends on both the mass and how the mass is distributed.

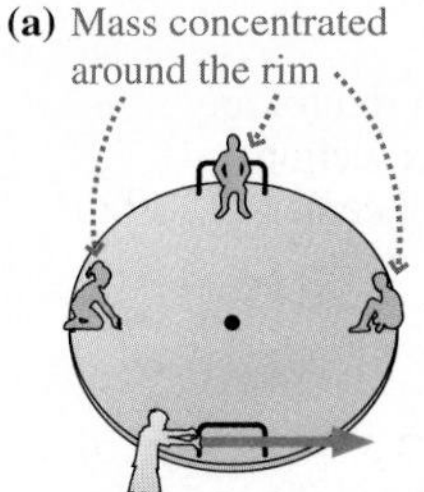
(a) Mass concentrated around the rim
Larger moment of inertia, harder to get rotating

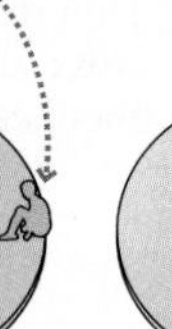
(b) Mass concentrated at the center
Smaller moment of inertia, easier to get rotating

Video Figure 7.35

We can express this result using the concept of moment of inertia. In the first case, with large values of r, Equation 7.20 says that the moment of inertia will be large. In the second case, with small values of r, Equation 7.20 gives a small value of moment of inertia. And the larger moment of inertia means you need to apply a larger torque to produce a certain angular acceleration: You'll need to push harder.

Thus an object's moment of inertia depends not only on the object's mass but also on *how the mass is distributed* around the rotation axis. This is well known to bicycle racers. Every time a cyclist accelerates, she has to "spin up" the wheels and tires. The larger the moment of inertia, the more effort it takes and the smaller her acceleration. For this reason, racers use the lightest possible tires, and they put those tires on wheels that have been designed to keep the mass as close as possible to the center without sacrificing the necessary strength and rigidity.

Synthesis 7.1 connected quantities and equations for linear and circular motion. We can now do the same thing for linear and rotational dynamics.

SYNTHESIS 7.2 Linear and rotational dynamics

The variables for linear dynamics have analogs for rotational dynamics. Newton's second law for rotational dynamics is expressed in terms of these variables.

	Linear dynamics	Rotational dynamics
Variables	Net force (N) → $\vec{F}_{net}$	τ_{net} ← Net torque (N · m)
	Mass (kg) → m	I ← Moment of inertia (kg · m^2)
	Acceleration (m/s^2) → $\vec{a}$	α ← Angular acceleration (rad/s^2)
Newton's second law	$\vec{a} = \dfrac{\vec{F}_{net}}{m}$ Acceleration is caused by forces. The larger the mass, the smaller the acceleration.	$\alpha = \dfrac{\tau_{net}}{I}$ Angular acceleration is caused by torques. The larger the moment of inertia, the smaller the angular acceleration.

▶ **Novel golf clubs** In recent years, manufacturers have introduced golf putters with heads that have high moments of inertia. When the putter hits the ball, the ball—by Newton's third law—exerts a force on the putter and thus exerts a torque that causes the head of the putter to rotate around the shaft. A large moment of inertia of the head will keep the resulting angular acceleration small, thus reducing unwanted rotation and allowing a truer putt.

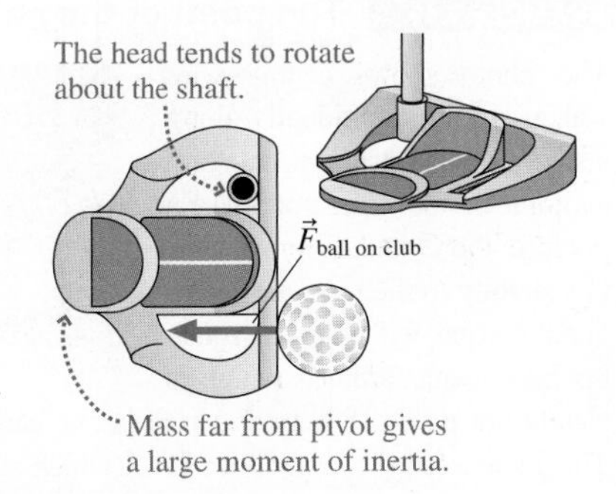
The head tends to rotate about the shaft.
$\vec{F}_{ball\ on\ club}$
Mass far from pivot gives a large moment of inertia.

The Moments of Inertia of Common Shapes

Newton's second law for rotational motion is easy to write, but we can't make use of it without knowing an object's moment of inertia. Unlike mass, we can't measure moment of inertia by putting an object on a scale. And although we can guess that

TABLE 7.1 Moments of inertia of objects with uniform density and total mass M

Object and axis	Picture	I	Object and axis	Picture	I
Thin rod (of any cross section), about center	L	$\frac{1}{12}ML^2$	Cylinder or disk, about center	R	$\frac{1}{2}MR^2$
Thin rod (of any cross section), about end	L	$\frac{1}{3}ML^2$	Cylindrical hoop, about center	R	MR^2
Plane or slab, about center	a, b	$\frac{1}{12}Ma^2$	Solid sphere, about diameter	R	$\frac{2}{5}MR^2$
Plane or slab, about edge	a, b	$\frac{1}{3}Ma^2$	Spherical shell, about diameter	R	$\frac{2}{3}MR^2$

the center of gravity of a symmetrical object is at the physical center of the object, we can *not* guess the moment of inertia of even a simple object. A short list of common moments of inertia is given in TABLE 7.1. We use a capital M for the total mass of an extended object.

We can make some general observations about the moments of inertia in Table 7.1. For instance, the cylindrical hoop is composed of particles that are all the same distance R from the axis. Thus each particle of mass m makes the *same* contribution mR^2 to the hoop's moment of inertia. Adding up all these contributions gives

$$I = m_1R^2 + m_2R^2 + m_3R^2 + \cdots = (m_1 + m_2 + m_3 + \cdots)R^2 = MR^2$$

as given in the table. The solid cylinder of the same mass and radius has a *lower* moment of inertia than the hoop because much of the cylinder's mass is nearer its center. In the same way we can see why a slab rotated about its center has a lower moment of inertia than the same slab rotated about its edge: In the latter case, some of the mass is twice as far from the axis as the farthest mass in the former case. Those particles contribute *four times* as much to the moment of inertia, leading to an overall larger moment of inertia for the slab rotated about its edge.

EXAMPLE 7.15 The point of the pole

The photo shows a high-wire walker striding confidently along a steel cable stretched nearly half a kilometer above the bottom of a gorge in the Grand Canyon. If he tips slightly to the side, a gravitational torque will start to rotate his body even farther—which is clearly not good. To help him balance, he carries a long, heavy pole. This increases the moment of inertia, which results in a smaller angular acceleration, giving him more time to recover his balance.

Model the man's body as a 1.8-m-long, 88 kg, uniform rod. The pole is 9.1 m long, with a mass of 20 kg. What is the moment of inertia of his body? Of the pole?

STRATEGIZE We will model the man's body as a rod pivoted about the end. We will treat the pole as a rod as well, with a pivot at the center, where it is supported. We'll use the equations in Table 7.1 to find the moments of inertia of these shapes. The pole is intended to provide stability, so we expect the pole's moment of inertia to be significant compared to that of the man's body.

PREPARE We model both the man's body and the pole as rods. The difference is the location of the pivot. The man's body pivots about the end of the rod, his feet, so the moment of inertia is $\frac{1}{3}ML^2$, where M is his mass and L is his height. The pole pivots about the center, so the moment of inertia is $\frac{1}{12}ML^2$, where M is the mass of the pole and L its length.

SOLVE The moment of inertia of the man's body is

$$I_{\text{man}} = \tfrac{1}{3}ML^2 = \tfrac{1}{3}(88\text{ kg})(1.8\text{ m})^2 = 95\text{ kg}\cdot\text{m}^2$$

The moment of inertia of the pole is

$$I_{\text{pole}} = \tfrac{1}{12}ML^2 = \tfrac{1}{12}(20\text{ kg})(9.1\text{ m})^2 = 140\text{ kg}\cdot\text{m}^2$$

ASSESS Although the pole's mass is less than one-quarter of the man's mass, its moment of inertia is larger. The formula for moment of inertia includes the square of a length, so the pole's greater length more than makes up for its smaller mass. Holding the pole significantly increases the man's moment of inertia; the moment of inertia of the man holding the pole is 2.5 times that of the man alone. This results in a significant decrease in angular acceleration, providing much greater stability. The pole is flexible, and this provides an additional benefit. The downward sag of the pole lowers the position of the overall center of gravity, thus decreasing gravitational torque; this improves stability as well.

STOP TO THINK 7.6 Four very lightweight disks of equal radii each have three identical heavy marbles glued to them as shown. Rank in order, from largest to smallest, the moments of inertia of the disks about the indicated axis.

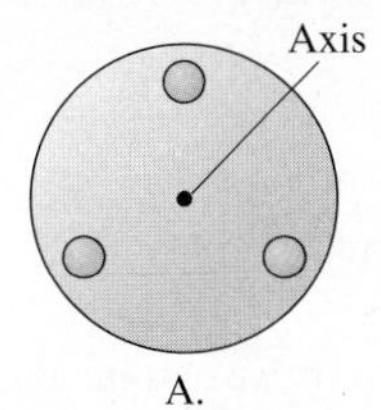

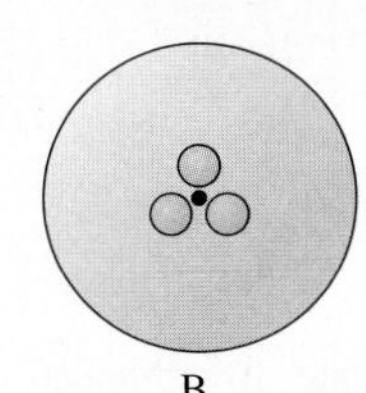

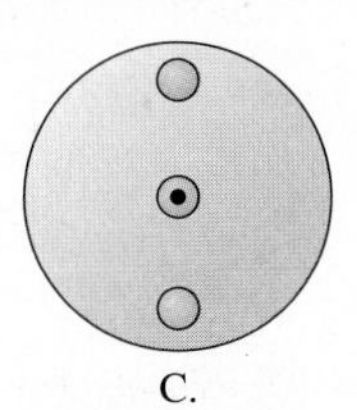

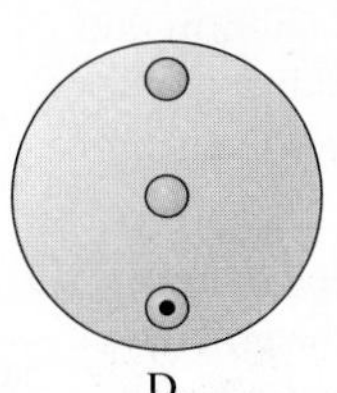

7.6 Using Newton's Second Law for Rotation

In this section we'll look at several examples of rotational dynamics for rigid bodies that rotate about a *fixed axis*. The restriction to a fixed axis avoids complications that arise for an object undergoing a combination of rotational and translational motion.

PROBLEM-SOLVING APPROACH 7.1 Rotational dynamics problems

We can use a problem-solving approach for rotational dynamics that is very similar to the approach for linear dynamics in Chapter 5.

STRATEGIZE Model the object as a simple shape. What is rotating, and what point does it rotate around? Next, identify the forces that cause the rotation, and draw a visual overview to clarify the situation.

PREPARE Develop the visual overview by defining coordinates and symbols and listing known information. Specifically:

- Identify the axis about which the object rotates.
- Identify the forces and determine their distances from the axis.
- Calculate the torques caused by the forces, and find the signs of the torques.

SOLVE The mathematical representation is based on Newton's second law for rotational motion:

$$\tau_{\text{net}} = I\alpha \quad \text{or} \quad \alpha = \frac{\tau_{\text{net}}}{I}$$

- Find the moment of inertia either by direct calculation using Equation 7.20 or from Table 7.1 for common shapes of objects.
- Use rotational kinematics to find angular positions and velocities.

ASSESS Check that your result has the correct units, is reasonable, and answers the question.

Exercise 27

EXAMPLE 7.16 Angular acceleration of a falling pole

In the caber toss, a contest of strength and skill that is part of Scottish games, contestants toss a heavy uniform pole, landing it on its end. A 5.9-m-tall pole with a mass of 79 kg has just landed on its end. It is tipped by 25° from the vertical and is starting to rotate about the end that touches the ground. Determine the angular acceleration. If we assume that the angular acceleration stays at this same value (an approximation, but it will give us a good sense of scale for this problem), how long will it take the pole to rotate by 5°?

STRATEGIZE We will follow the steps in Problem-Solving Approach 7.1. We'll start by modeling the pole as a uniform thin rod that rotates about one end under the influence of the gravitational torque. Our visual overview in FIGURE 7.36 notes the geometry, the forces that act, and where the forces act.

PREPARE Two forces are acting on the pole: the pole's weight $\vec{w}$, which acts at the center of gravity, and the force of the ground on the pole (not shown). This second force exerts no torque because it acts at the axis of rotation. The torque on the pole is thus due only to gravity. From the figure we see that this torque tends to rotate the pole in a counterclockwise direction, so the torque is positive. We can use this force to find the angular acceleration, which we can use to find the time it takes for the pole to rotate by 5°, which is 0.087 rad.

FIGURE 7.36 A falling pole undergoes an angular acceleration due to a gravitational torque.

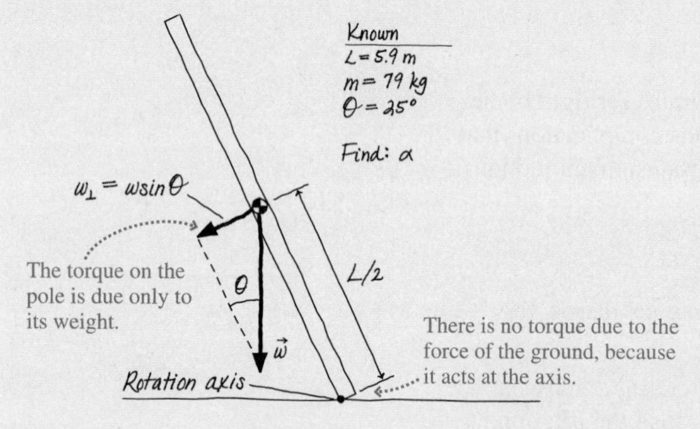
Known
L = 5.9 m
m = 79 kg
θ = 25°
Find: α
w⊥ = w sin θ
The torque on the pole is due only to its weight.
θ
w
L/2
There is no torque due to the force of the ground, because it acts at the axis.
Rotation axis

SOLVE The pole's center of gravity is at its center, a distance $L/2$ from the axis. You can see from the figure that the perpendicular component of $\vec{w}$ is $w_\perp = w\sin\theta$. Thus the torque due to gravity is

$$\tau_{\text{net}} = \left(\frac{L}{2}\right)w_\perp = \left(\frac{L}{2}\right)w\sin\theta = \frac{mgL}{2}\sin\theta$$

From Table 7.1, the moment of inertia of a thin rod rotated about its end is $I = \frac{1}{3}mL^2$. Thus, from Newton's second law for rotational motion, the angular acceleration is

$$\alpha = \frac{\tau_{\text{net}}}{I} = \frac{\frac{1}{2}mgL\sin\theta}{\frac{1}{3}mL^2} = \frac{3g\sin\theta}{2L}$$

$$= \frac{3(9.8\text{ m/s}^2)\sin 25°}{2(5.9\text{ m})} = 1.1\text{ rad/s}^2$$

Once the pole starts to rotate, it will speed up. If we assume that the angular acceleration is constant, we can use the final equation in Synthesis 7.1 to calculate the time it takes for the pole to rotate by 0.087 rad. This is $\Delta\theta$; the pole starts at rest, $\omega_i = 0$, and we find

$$0.087\text{ rad} = (0)\Delta t + \tfrac{1}{2}\alpha(\Delta t)^2$$

$$\Delta t = 0.40\text{ s}$$

ASSESS The final result for the angular acceleration did not depend on the mass, as we might expect given the analogy with free-fall problems. And the final value for the angular acceleration is quite modest; the motion is fairly gentle. For this angular acceleration, it takes just under half a second to rotate by 5°. This is reasonable: You can see that the angular acceleration is inversely proportional to the length of the pole, and it's a long pole. The modest value of angular acceleration is fortunate—the caber is pretty heavy, and folks need some time to get out of the way when it topples!

CONCEPTUAL EXAMPLE 7.17 Balancing a meter stick

You can easily balance a meter stick or a baseball bat on your palm. (Try it!) But it's nearly impossible to balance a pencil this way. Why?

REASON Suppose you've managed to balance a vertical stick on your palm, but then it starts to fall. You'll need to quickly adjust your hand to bring the stick back into balance. As Example 7.16 showed, the angular acceleration α of a thin rod is *inversely proportional* to L. Thus a long object like a meter stick topples much more slowly than a short one like a pencil. Your reaction time is fast enough to correct for a slowly tipping meter stick but not for a rapidly tipping pencil.

ASSESS If we double the length of a rod, its mass doubles and its center of gravity is twice as high, so the gravitational torque τ on it is four times as much. But because a rod's moment of inertia is $I = \frac{1}{3}ML^2$, the longer rod's moment of inertia will be *eight* times greater, so the angular acceleration will be only half as large. This gives you more time to move to put the support back under the center of gravity.

EXAMPLE 7.18 **Spinning up a flywheel**

When the electricity goes out on your campus, a backup source powers the servers that handle communications and the Internet. Rather than use batteries, many installations use a flywheel, a heavy rotating disk that spins, very rapidly, with nearly zero friction. An electric motor spins up the flywheel, which continues to spin with very little energy input. When the system needs to provide power, the flywheel's motion is used to turn an electric generator. A typical system has a 540 kg cylinder with a radius of 0.30 m. A small electric motor provides a constant 2.8 N · m torque to spin up the cylinder. If the cylinder starts at rest, how long does it take to reach its final spin rate of 13,000 rpm?

STRATEGIZE This is an angular dynamics problem. We will find the moment of inertia of the cylinder and then use the given torque to find the angular acceleration. We'll use this to find the time required to reach the final spin rate. Given the size and mass of the cylinder (half the mass of a small car), the relatively modest torque (one-sixth of the wrench torque in Example 7.10), and the very rapid spin (more than 200 rotations each second), we expect the motor to take quite a while to spin up the flywheel.

PREPARE We are given the spin rate in rpm; to do calculations, we need the angular speed in rad/s. The frequency in rev/s is

$$f = \left(13{,}000\ \frac{\text{rev}}{\text{min}}\right)\left(\frac{1\ \text{min}}{60\ \text{s}}\right) = 217\ \text{rev/s}$$

so the angular speed in rad/s is

$$\omega = (2\pi\ \text{rad})(217\ \text{rev/s}) = 1360\ \text{rad/s}$$

We'll keep an extra significant figure for the intermediate stages in the calculation.

SOLVE The flywheel is a cylinder rotated around the center. The moment of inertia is

$$I = \tfrac{1}{2}MR^2 = \tfrac{1}{2}(540\ \text{kg})(0.30\ \text{m})^2 = 24.3\ \text{kg}\cdot\text{m}^2$$

The angular acceleration is

$$\alpha = \frac{\tau}{I} = \frac{2.8\ \text{N}\cdot\text{m}}{24.3\ \text{kg}\cdot\text{m}^2} = 0.115\ \text{rad/s}^2$$

We can rearrange the definition of angular acceleration, Equation 7.7, to find the time to increase the angular speed to 1360 rad/s:

$$\Delta t = \frac{\Delta\omega}{\alpha} = \frac{1360\ \text{rad/s}}{0.115\ \text{rad/s}^2} = 11{,}800\ \text{s} = 3.3\ \text{hours}$$

ASSESS Given the modest torque, large flywheel, and rapid spin, we expected a long time, so our answer makes sense. The numbers given in the problem are real numbers for an actual system. The rated time for the initial spin-up is quite close to the calculated value.

Constraints Due to Ropes and Pulleys

FIGURE 7.37 shows a rope passing over a pulley and connected to an object in linear motion. If the pulley turns *without the rope slipping on it,* then the rope's speed v_{rope} must exactly match the speed of the rim of the pulley, which is $v_{\text{rim}} = \omega R$. If the pulley has an angular acceleration, the rope's acceleration a_{rope} must match the *tangential* acceleration of the rim of the pulley, $a_t = \alpha R$.

The object attached to the other end of the rope has the same speed and acceleration as the rope. Consequently, the object must obey the constraints

$$v_{\text{obj}} = \omega R$$
$$a_{\text{obj}} = \alpha R \qquad (7.22)$$

Motion constraints for an object connected to a pulley of radius R by a nonslipping rope

FIGURE 7.37 The rope's motion must match the motion of the rim of the pulley.

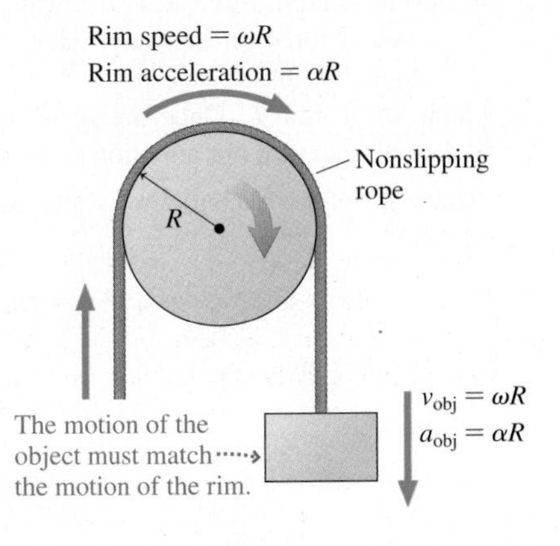

These constraints are similar to the acceleration constraints introduced in Chapter 5 for two objects connected by a string or rope.

NOTE ▶ The constraints are given as magnitudes. Specific problems will require you to specify signs that depend on the direction of motion and on the choice of coordinate system. ◀

EXAMPLE 7.19 Time for a bucket to fall

Josh has just raised a 2.5 kg bucket of water using a well's winch when he accidentally lets go of the handle. The winch consists of a rope wrapped around a 3.0 kg, 4.0-cm-diameter cylinder, which rotates on an axle through the center. The bucket is released from rest 4.0 m above the water level of the well. How long does it take to reach the water?

STRATEGIZE There's a downward force on the bucket—the weight force. There is an upward force on the bucket, too—the rope tension. The tension force also pulls on the cylinder, so as the bucket falls, the cylinder speeds up. We can assume that the rope is massless and does not slip. No slipping means that the downward acceleration of the bucket matches the tangential acceleration of the rim of the cylinder, so we'll treat the fall of the bucket and the spin of the cylinder as two connected problems. FIGURE 7.38a gives a visual overview of the system, FIGURE 7.38b shows the free-body diagram for the bucket, and FIGURE 7.38c shows the tension force acting on the cylinder.

FIGURE 7.38 Visual overview of a falling bucket.

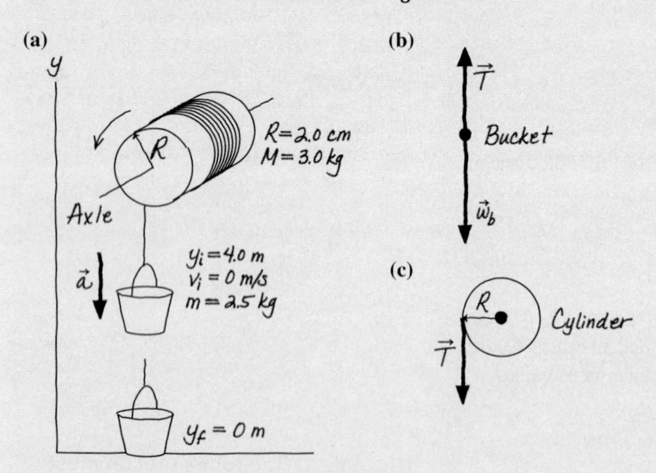

PREPARE The rope tension exerts an upward force on the bucket; the weight force acts downward. The rope tension exerts a downward force on the outer edge of the cylinder. The rope is massless, so these two tension forces have equal magnitudes, which we'll call T. Once we've analyzed the motion of the bucket, we can turn our attention to the cylinder.

SOLVE Newton's second law applied to the linear motion of the bucket is

$$ma_y = T - mg$$

where, as usual, the y-axis points upward. What about the cylinder? There are forces that act at the axle, but these forces don't exert a torque because they act at the rotation axis. The only torque comes from the rope tension. The moment arm for the tension is $r_\perp = R$, and the torque is positive because the rope turns the cylinder counterclockwise. Thus $\tau_{\text{rope}} = TR$, and Newton's second law for the rotational motion is

$$\alpha = \frac{\tau_{\text{net}}}{I} = \frac{TR}{\frac{1}{2}MR^2} = \frac{2T}{MR}$$

The moment of inertia of a cylinder rotating about a center axis was taken from Table 7.1.

The last piece of information we need is the constraint due to the fact that the rope doesn't slip. Equation 7.22 relates only the magnitudes of the linear and angular accelerations, but in this problem α is positive (counterclockwise acceleration), while a_y is negative (downward acceleration). Hence

$$a_y = -\alpha R$$

Using α from the cylinder's equation in the constraint, we find

$$a_y = -\alpha R = -\frac{2T}{MR}R = -\frac{2T}{M}$$

Thus the tension is $T = -\frac{1}{2}Ma_y$. If we use this value of the tension in the bucket's equation, we can solve for the acceleration:

$$ma_y = -\tfrac{1}{2}Ma_y - mg$$

$$a_y = -\frac{g}{(1 + M/2m)} = -6.13\ \text{m/s}^2$$

The time to fall through $\Delta y = y_f - y_i = -4.0$ m is found from kinematics:

$$\Delta y = \tfrac{1}{2}a_y(\Delta t)^2$$

$$\Delta t = \sqrt{\frac{2\Delta y}{a_y}} = \sqrt{\frac{2(-4.0\ \text{m})}{-6.13\ \text{m/s}^2}} = 1.1\ \text{s}$$

ASSESS The expression for the acceleration gives $a_y = -g$ if $M = 0$. This makes sense because the bucket would be in free fall if there were no cylinder. When the cylinder has mass, the downward force of gravity on the bucket has to accelerate the bucket and spin the cylinder. Consequently, the acceleration is reduced and the bucket takes longer to fall.

7.7 Rolling Motion

Rolling is a *combination motion* in which an object rotates about an axis that is moving along a straight-line trajectory. For example, FIGURE 7.39 is a time-exposure photo of a rolling wheel with one lightbulb on the axis and a second lightbulb at the edge. The axis light moves straight ahead, but the edge light follows a curve called a *cycloid.*

FIGURE 7.39 The trajectories of the center of a wheel and of a point on the rim are seen in a time-exposure photograph.

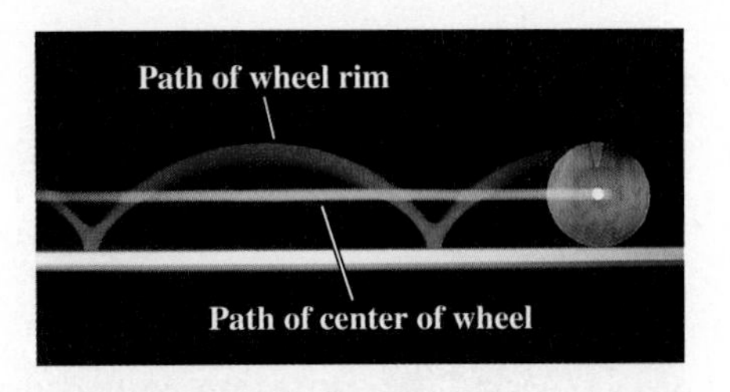

FIGURE 7.40 An object rolling through one revolution.

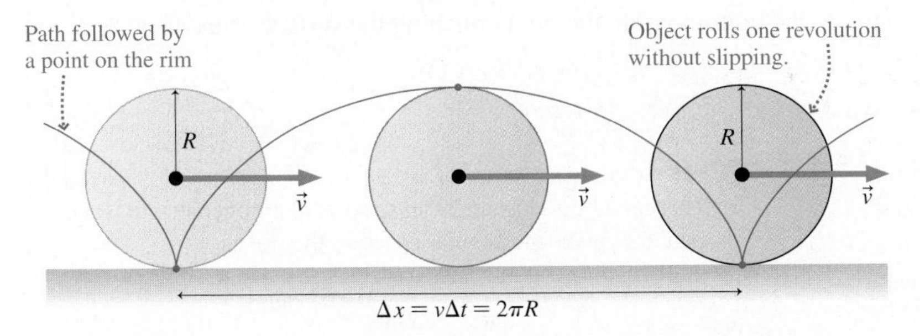

To understand rolling motion, consider FIGURE 7.40, which shows a round object—a wheel or a sphere—that rolls forward, *without slipping,* exactly one revolution. The point initially at the bottom follows the blue curve to the top and then back to the bottom. The overall position of the object is measured by the position x of the object's center. Because the object doesn't slip, in one revolution the center moves forward exactly one circumference, so that $\Delta x = 2\pi R$. The time for the object to turn one revolution is its period T, so we can compute the speed of the object's center as

$$v = \frac{\Delta x}{T} = \frac{2\pi R}{T} \tag{7.23}$$

But $2\pi/T$ is the angular velocity ω, as you learned in Chapter 6, which leads to

$$v = \omega R \tag{7.24}$$

Equation 7.24 is the **rolling constraint,** the basic link between translation and rotation for objects that roll without slipping.

We can find the velocity for any point on a rolling object by adding the velocity of that point when the object is in pure translation, without rolling, to the velocity of the point when the object is in pure rotation, without translating. FIGURE 7.41 shows how the velocity vectors at the top, center, and bottom of a rotating wheel are found in this way.

Ancient movers The great stone *moai* of Easter Island were moved as far as 16 km from a quarry to their final positions. Archeologists believe that one possible method of moving these 14 ton statues was to place them on rollers. One disadvantage of this method is that the statues, placed on top of the rollers, move twice as fast as the rollers themselves. Thus rollers are continuously left behind and have to be carried back to the front and reinserted. Sadly, the indiscriminate cutting of trees for moving *moai* may have hastened the demise of this island civilization.

FIGURE 7.41 Rolling is a combination of translation and rotation.

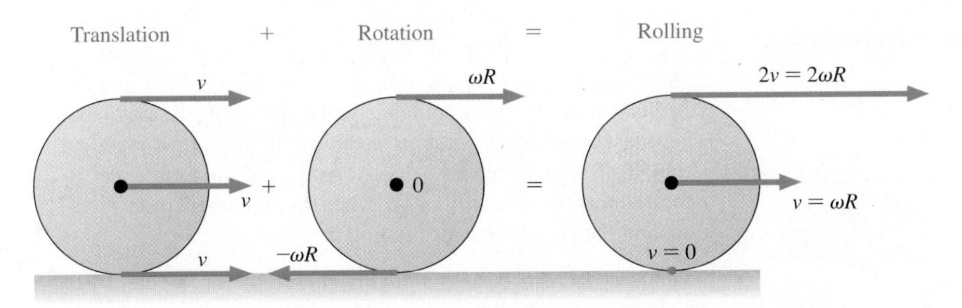

Thus the point at the top of the wheel has a forward speed of v due to its translational motion plus a forward speed of $\omega R = v$ due to its rotational motion. The speed of a point at the top of a wheel is then $2v = 2\omega R$, or *twice* the speed of its center. On the other hand, the point at the bottom of the wheel, where it touches the ground, still has a forward speed of v due to its translational motion. But its velocity due to rotation points *backward* with a magnitude of $\omega R = v$. Adding these, we find that the velocity of this lowest point is *zero*. In other words, **the point on the bottom of a rolling object is instantaneously at rest.**

Although this seems surprising, it is really what we mean by "rolling without slipping." If the bottom point had a velocity, it would be moving horizontally relative to the surface. In other words, it would be slipping or sliding across the surface. To roll without slipping, the bottom point, the point touching the surface, must be at rest.

EXAMPLE 7.20 Rotating your tires

You are driving down the highway at 65 mph, which is 29 m/s. Your tires have a radius of 0.30 m.

a. How many times per second does each tire rotate?
b. What is the speed of a point at the top of a tire, relative to the ground?

STRATEGIZE The questions concern your tires, which are in rolling motion. You've certainly noticed the tires on nearby cars as you drive down the highway; they are spinning at a pretty good clip, many times per second, so we will expect a large number for the tire's rotation frequency, the number of times per second the wheel goes around.

PREPARE We can use Equation 7.24 to find the angular speed of the wheel given the speed of the car. If we know this, we can find the frequency of the rotation. Figure 7.41 shows us how to find the speed of a point at the top of a tire.

SOLVE

a. The car is moving at 29 m/s, so a rearrangement of Equation 7.24 gives the angular speed of the tire as

$$\omega = \frac{v}{R} = \frac{29\text{ m/s}}{0.30\text{ m}} = 97\text{ rad/s}$$

The frequency is

$$f = \frac{\omega}{2\pi\text{ rad}} = \frac{97\text{ rad/s}}{2\pi\text{ rad}} = 15\text{ rev/s}$$

b. The velocity of a point on the top of the tire is the sum of the velocity of the car and the velocity of the rotational motion of the wheel. Figure 7.41 shows us that this is twice the speed of the car, or 58 m/s.

ASSESS The tire is spinning 15 times per second, a large number, as we expected. If you drive for an hour at highway speeds, your tires rotate more than 50,000 times!

STOP TO THINK 7.7 A wheel rolls without slipping. Which is the correct velocity vector for point P on the wheel?

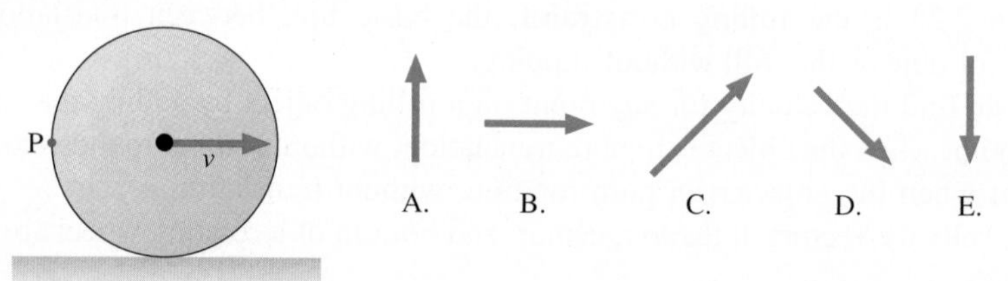

INTEGRATED EXAMPLE 7.21 Spinning a gyroscope

A gyroscope is a top-like toy consisting of a heavy ring attached by light spokes to a central axle. The axle and ring are free to turn on bearings. To get the gyroscope spinning, a 30-cm-long string is wrapped around the 2.0-mm-diameter axle, then pulled with a constant force of 5.0 N. If the ring's diameter is 5.0 cm and its mass is 30 g, at what rate is it spinning, in rpm, once the string is completely unwound?

STRATEGIZE This is a question about the ring's final angular velocity—how fast it is spinning when the string is unwound. The ring's angular acceleration is caused by the torque from the string. To find this angular velocity, we'll find the torque, calculate the moment of inertia of the ring, find the angular acceleration, and then use kinematics to find the final value.

PREPARE FIGURE 7.42 is a visual overview of the problem that illustrates the geometry and provides the known information in the correct SI units.

FIGURE 7.42 Visual overview of a gyroscope being spun.

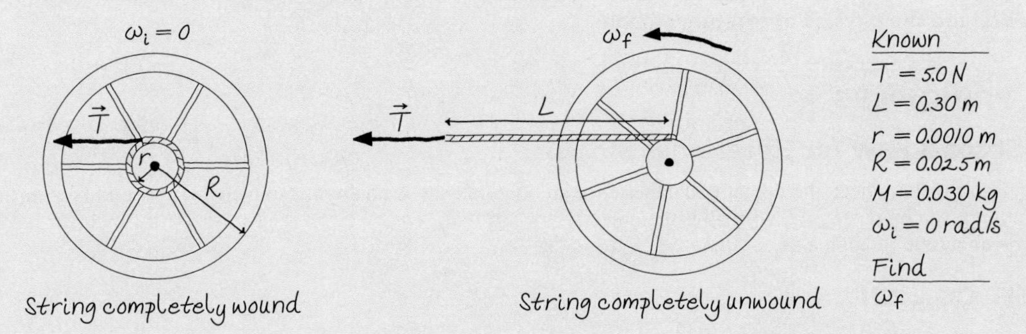

The string is light, so the tension in the string has the same magnitude as the force that pulls on the string, 5.0 N. Because the ring is heavy compared to the spokes and the axle, we model it as a cylindrical hoop, taking its moment of inertia from Table 7.1 to be $I = MR^2$. The radius R is half the 5.0 cm ring diameter. The tension force acts at a distance r from the axle, where r is half the 2.0 mm axle diameter. The ring starts at rest, so the initial angular velocity is $\omega_i = 0$ rad/s. We are looking for the ring's final angular velocity, ω_f.

SOLVE The torque on the ring is due to the tension in the string. Because the string—and the line of action of the tension—is tangent to the axle, the moment arm of the tension force is the radius r of the axle. Thus $\tau = r_\perp T = rT$. Now we can apply Newton's second law for rotational motion, Equation 7.21, to find the angular acceleration:

$$\alpha = \frac{\tau_{net}}{I} = \frac{rT}{MR^2} = \frac{(0.0010\text{ m})(5.0\text{ N})}{(0.030\text{ kg})(0.025\text{ m})^2} = 267\text{ rad/s}^2$$

We next use constant-angular-acceleration kinematics to find the final angular velocity. For the equation $\Delta\theta = \omega_i\,\Delta t + \frac{1}{2}\alpha\,\Delta t^2$ in Synthesis 7.1, we know α and ω_i, and we should be able to find $\Delta\theta$ from the length of string unwound, but we don't know Δt. For the equation $\Delta\omega = \omega_f - \omega_i = \alpha\,\Delta t$, we know α and ω_i, and ω_f is what we want to find, but again we don't know Δt. To find an equation that doesn't contain Δt, we first write

$$\Delta t = \frac{\omega_f - \omega_i}{\alpha}$$

from the second kinematic equation. Inserting this value for Δt into the first equation gives

$$\Delta\theta = \omega_i\frac{\omega_f - \omega_i}{\alpha} + \frac{1}{2}\alpha\left(\frac{\omega_f - \omega_i}{\alpha}\right)^2$$

which can be simplified to

$$\omega_f^2 = \omega_i^2 + 2\alpha\,\Delta\theta$$

FIGURE 7.43 Relating the angle turned to the length of string unwound.

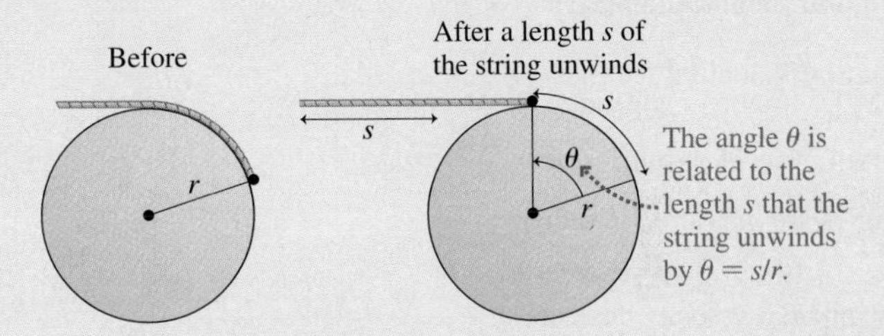

This equation, which is the rotational analog of the linear motion Equation 2.13, will allow us to find ω_f once $\Delta\theta$ is known.

FIGURE 7.43 shows how to find $\Delta\theta$. As a segment of string of length s unwinds, the axle turns through an angle (based on the definition of radian measure) $\theta = s/r$. Thus as the whole string, of length L, unwinds, the axle (and the ring) turns through an angular displacement

$$\Delta\theta = \frac{L}{r} = \frac{0.30\text{ m}}{0.0010\text{ m}} = 300\text{ rad}$$

Now we can use our kinematic equation to find that

$$\omega_f^2 = \omega_i^2 + 2\alpha\,\Delta\theta = (0\text{ rad/s})^2 + 2(267\text{ rad/s}^2)(300\text{ rad})$$
$$= 160{,}000\ (\text{rad/s})^2$$

from which we find that $\omega_f = 400$ rad/s. Converting rad/s to rpm, we find that the gyroscope ring is spinning at

$$400\text{ rad/s} = \left(\frac{400\text{ rad}}{\text{s}}\right)\left(\frac{60\text{ s}}{1\text{ min}}\right)\left(\frac{1\text{ rev}}{2\pi\text{ rad}}\right) = 3800\text{ rpm}$$

ASSESS This is fast, about the same as your car engine when it's on the highway, but if you've ever played with a gyroscope or a string-wound top, you know you can really get it spinning fast.

SUMMARY

GOAL To understand the physics of rotating objects.

GENERAL PRINCIPLES

Newton's Second Law for Rotational Motion

If a net torque τ_{net} acts on an object, the object will experience an angular acceleration given by $\alpha = \tau_{net}/I$, where I is the object's moment of inertia about the rotation axis.

This law is analogous to Newton's second law for linear motion, $\vec{a} = \vec{F}_{net}/m$.

IMPORTANT CONCEPTS

Describing circular motion

We define new variables for circular motion. By convention, counterclockwise is positive.

Angular displacement: $\Delta\theta = \theta_f - \theta_i$

Angular velocity: $\omega = \dfrac{\Delta\theta}{\Delta t}$

Angular acceleration: $\alpha = \dfrac{\Delta\omega}{\Delta t}$

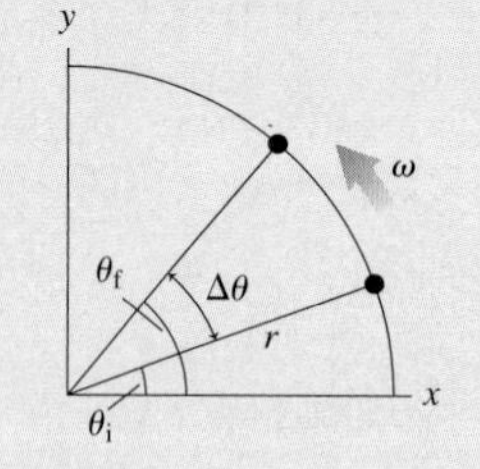

Angles are measured in radians:

$$1 \text{ rev} = 360^\circ = 2\pi \text{ rad}$$

The angular velocity depends on the frequency and period: $\omega = \dfrac{2\pi}{T} = 2\pi f$

Torque

A force causes an object to undergo a linear acceleration; a torque causes an object to undergo an angular acceleration.

There are two interpretations of torque:

Interpretation 1: $\tau = rF_\perp$ Interpretation 2: $\tau = r_\perp F$

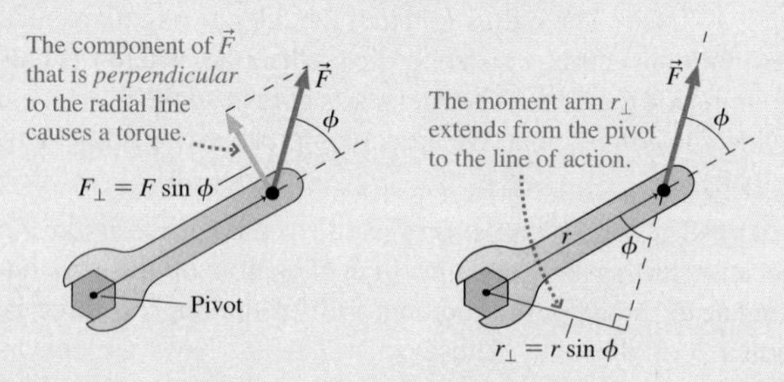

Both interpretations give the same expression for the magnitude of the torque: $\tau = rF\sin\phi$

Relating linear and circular motion quantities

Linear and angular speeds are related by: $v = \omega r$

If the particle's speed is increasing, it will also have a tangential acceleration $\vec{a}_t$ directed tangent to the circle and an angular acceleration α.

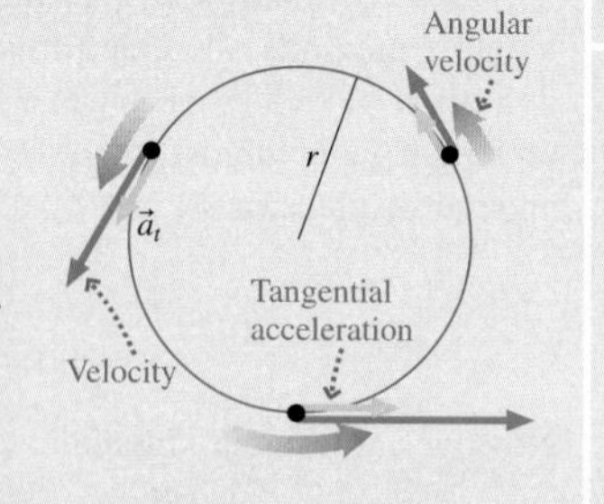

Angular and tangential accelerations are related by: $a_t = \alpha r$

Center of gravity

The **center of gravity** of an object is the point at which gravity can be considered to be acting.

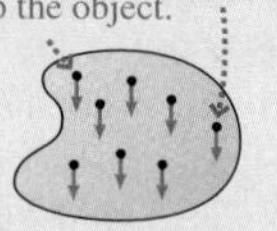

Gravity acts on each particle that makes up the object.

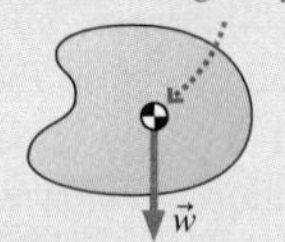

The object responds *as if* its entire weight acts at the center of gravity.
w

The **moment of inertia** is the rotational equivalent of mass. For an object made up of particles of masses $m_1, m_2, \ldots$ at distances $r_1, r_2, \ldots$ from the axis, the moment of inertia is

$$I = m_1 r_1^2 + m_2 r_2^2 + m_3 r_3^2 + \cdots$$

The **position of the center of gravity** depends on the distance $x_1, x_2, \ldots$ of each particle of mass $m_1, m_2, \ldots$ from the origin:

$$x_{cg} = \frac{x_1 m_1 + x_2 m_2 + x_3 m_3 + \cdots}{m_1 + m_2 + m_3 + \cdots}$$

APPLICATIONS

Moments of inertia of common shapes

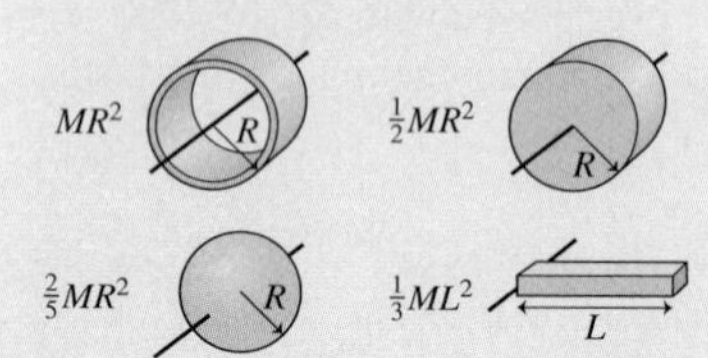

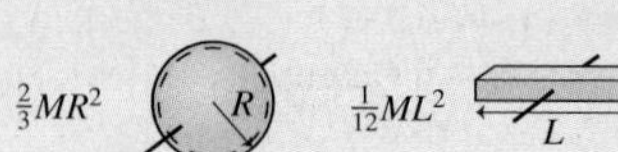

Rotation about a fixed axis

When a net torque is applied to an object that rotates about a fixed axis, the object will undergo an **angular acceleration** given by

$$\alpha = \frac{\tau_{net}}{I}$$

If a rope unwinds from a pulley of radius R, the linear motion of an object tied to the rope is related to the angular motion of the pulley by

$$a_{obj} = \alpha R \qquad v_{obj} = \omega R$$

Rolling motion

For an object that rolls without slipping,

$$v = \omega R$$

The velocity of a point at the top of the object is twice that of the center.

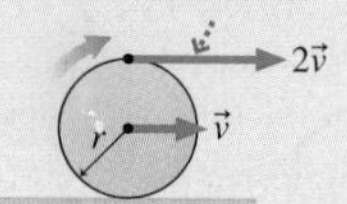

Learning Objectives After studying this chapter, you should be able to:

- Calculate angular velocity and interpret motion graphs for rotational motion. *Conceptual Question 7.2; Problems 7.3, 7.4, 7.5, 7.6, 7.9*
- Calculate angular acceleration and tangential acceleration. *Conceptual Question 7.10; Problems 7.13, 7.14, 7.15, 7.16, 7.17*
- Calculate the torque exerted on an extended object. *Conceptual Questions 7.4, 7.7; Problems 7.21, 7.22, 7.23, 7.24, 7.26*
- Determine an object's center of gravity and gravitational torque. *Conceptual Question 7.11; Problems 7.31, 7.33, 7.35, 7.37, 7.38*
- Calculate an object's moment of inertia. *Conceptual Questions 7.14, 7.18; Problems 7.39, 7.40, 7.42, 7.43, 7.44*
- Use Newton's second law for rotational motion to solve problems about rotational dynamics. *Conceptual Questions 7.12, 7.13; Problems 7.45, 7.47, 7.48, 7.49, 7.53*
- Analyze rolling motion. *Conceptual Question 7.17; Problems 7.57, 7.58, 7.59, 7.60*

STOP TO THINK ANSWERS

Chapter Preview Stop to Think: B. The speed is proportional to the frequency. Doubling the frequency means doubling the speed.

Stop to Think 7.1: A. Because $5\pi/2$ rad $= 2\pi$ rad $+ \pi/2$ rad, the particle's position is one complete revolution (2π rad) plus an extra $\pi/2$ rad. This extra $\pi/2$ rad puts the particle at position A.

Stop to Think 7.2: a. constant (but not zero), b. constant (but not zero), c. constant (but not zero), d. zero, e. zero. The angular velocity ω is constant. Thus the magnitudes of the velocity $v = \omega r$ and the centripetal acceleration $a_c = \omega^2 r$ are constant. This also means that the ball's angular acceleration α and tangential acceleration $a_t = \alpha r$ are both zero.

Stop to Think 7.3: E. Forces D and C act at or in line with the pivot and provide no torque. Of the others, only E tries to rotate the wheel counterclockwise, so it is the only choice that gives a positive nonzero torque.

Stop to Think 7.4: A. The force acting at the axis exerts no torque. Thus the third force needs to exert an equal but opposite torque to that exerted by the force acting at the rim. Force A, which has twice the magnitude but acts at half the distance from the axis, does so.

Stop to Think 7.5: $\tau_B > \tau_D > \tau_A = \tau_C$. The torques are $\tau_B = 2mgL$, $\tau_D = \frac{3}{2}mgL$, and $\tau_A = \tau_C = mgL$, where L is the length of the rod in B.

Stop to Think 7.6: $I_D > I_A > I_C > I_B$. The moments of inertia are $I_B \approx 0$, $I_C = 2mr^2$, $I_A = 3mr^2$, and $I_D = mr^2 + m(2r)^2 = 5mr^2$.

Stop to Think 7.7: C. The velocity of P is the vector sum of $\vec{v}$ directed to the right and an upward velocity of the same magnitude due to the rotation of the wheel.

Video Tutor Solution Chapter 7

QUESTIONS

Conceptual Questions

1. The batter in a baseball game hits a home run. As he circles the bases, is his angular velocity positive or negative?
2. Viewed from somewhere in space above the north pole, would a point on the earth's equator have a positive or negative angular velocity due to the earth's rotation?
3. Figure Q7.3 shows four pulleys, each with a heavy and a light block strung over it. The blocks' velocities are shown. What are the signs (+ or −) of the angular velocity and angular acceleration of the pulley in each case?

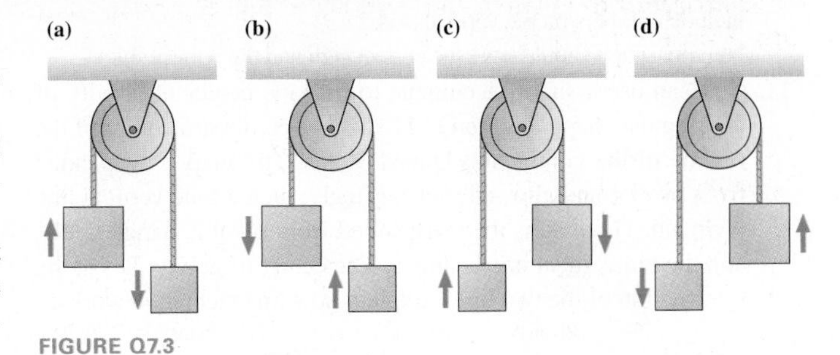

FIGURE Q7.3

Problem difficulty is labeled as | (straightforward) to ||||| (challenging). INT Problems labeled integrate significant material from earlier chapters; BIO are of biological or medical interest.

The eText icon indicates when there is a video tutor solution available for the chapter or for a specific problem. To launch these videos, log into your eText through Mastering™ Physics or log into the Study Area.

4. If you are using a wrench to loosen a very stubborn nut, you can make the job easier by using a "cheater pipe." This is a piece of pipe that slides over the handle of the wrench, as shown in Figure Q7.4, making it effectively much longer. Explain why this would help you loosen the nut.

FIGURE Q7.4

5. The torque needed to open a factory-sealed jar is about the same as the torque required to turn the wheel on a passenger car. You know that the force necessary to turn your car's steering wheel is less than the force needed to open a jar. How can the torques be the same?
6. The chain on a bicycle turns the rear wheel via a gear cluster, as shown in Figure Q7.6. The gears vary in size; the smallest gear has a radius of 2.3 cm, the largest a radius of 6.0 cm. If we assume a constant tension in the chain, which gear—the smallest or the largest—is the better choice for a rapid start?

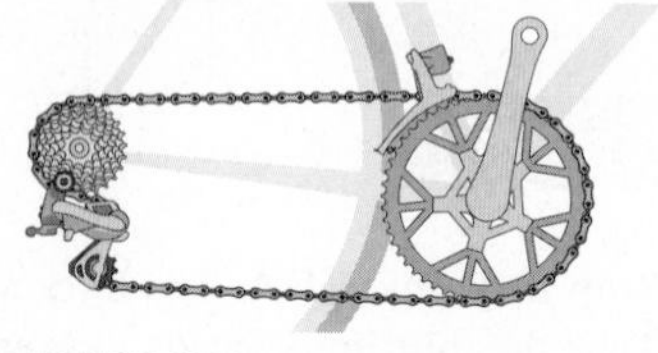

FIGURE Q7.6

7. Five forces are applied to a door, as seen from above in Figure Q7.7. For each force, is the torque about the hinge positive, negative, or zero?

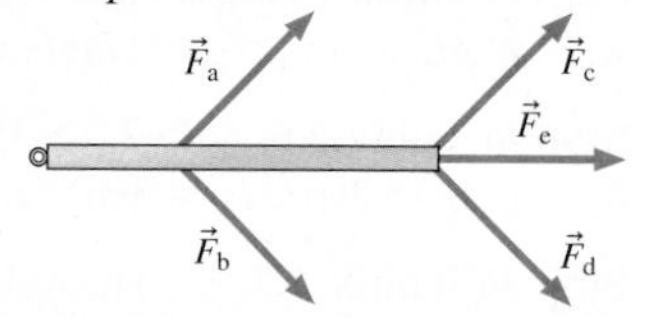

FIGURE Q7.7

8. A screwdriver with a very thick handle requires less force to operate than one with a very skinny handle. Explain why this is so.
9. If you have ever driven a truck, you likely found that it had a steering wheel with a larger diameter than that of a passenger car. Why is this?
10. A student gives a steady push to a ball at the end of a massless, rigid rod for 1 s, causing the ball to rotate clockwise in a horizontal circle as shown in Figure Q7.10. The rod's pivot is frictionless. Sketch a graph of the ball's angular velocity as a function of time for the first 3 s of the ball's motion. You won't be able to include numbers on the vertical axis, but your graph should have the correct sign and the correct shape.

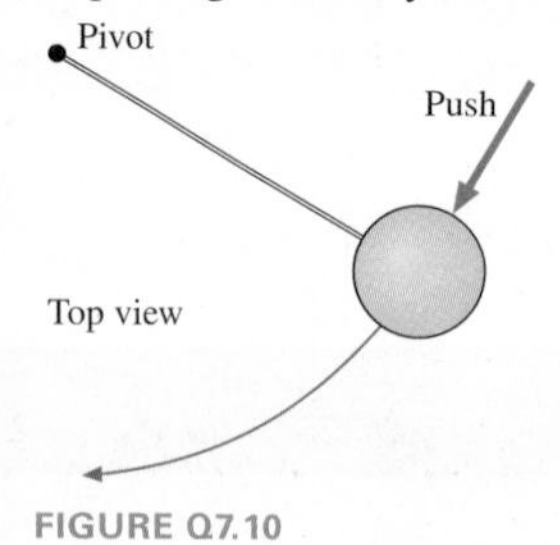

FIGURE Q7.10

11. You can use a simple technique to find the center of gravity of an irregular shape. Figure Q7.11 shows a cardboard cutout of the outline of the continental United States. The map is suspended from pivot 1 and allowed to hang freely; then a blue vertical line is drawn. The map is then suspended from pivot 2, hangs freely, and the red vertical line is drawn. The center of gravity lies at the intersection of the two lines. Explain how this technique works.

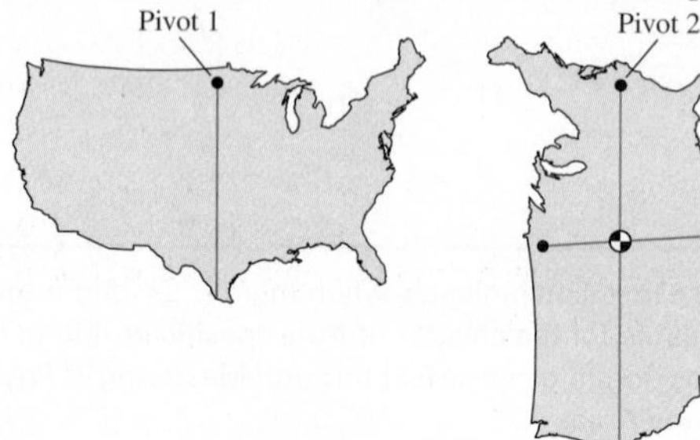

FIGURE Q7.11

12. If you grasp a hammer by its lightweight handle and wave it back and forth, and then grasp it by its much heavier head and wave it back and forth, as in the figure, you'll find that you can wave the hammer much more rapidly in the second case, when you grasp it by the head. Explain why this is so.

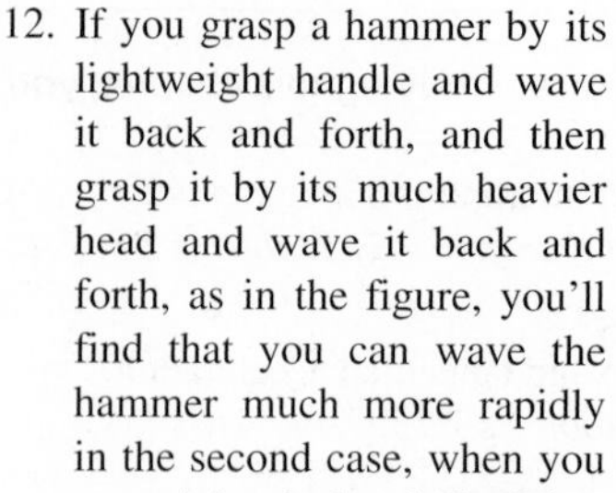

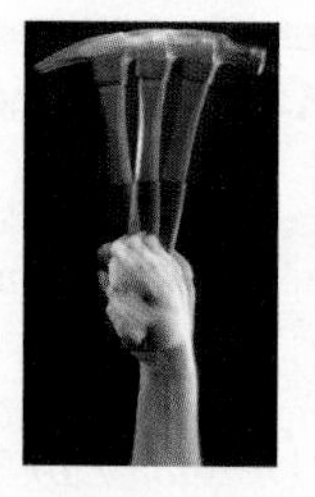

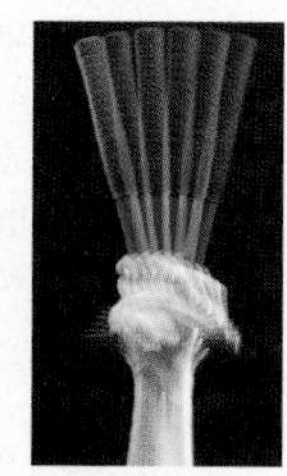

13. You have two identical-looking metal spheres of the same size and the same mass. One is solid; the other is hollow. If you place them on a ramp, the gravitational torques that make them roll down the slope are the same. But one sphere has a greater angular acceleration, so it reaches the bottom of the ramp first. Which one gets to the bottom of the ramp first, and why?
14. Getting on a so-called "tall bike" is harder than mounting a regular bike. But once a rider is in place, it is easier to balance on the tall bike, and the bike can be safely ridden at very slow speeds. Outline the physical principles that explain why you would expect balance to be easier on a taller bike.

15. Water parks often include a log rolling area, in which participants try to stay upright while balancing on top of floating logs that are free to rotate in the water. If you've tried this, you know that the larger the diameter of the log, the easier it is to balance on top. Explain why this is so.

16. The moment of inertia of a uniform rod about an axis through its center is $ML^2/12$. The moment of inertia about an axis at one end is $ML^2/3$. Explain *why* the moment of inertia is larger about the end than about the center.
17. The wheel in Figure Q7.17 is rolling to the right without slipping. Rank in order, from fastest to slowest, the *speeds* of the points labeled 1 through 5. Explain your reasoning.

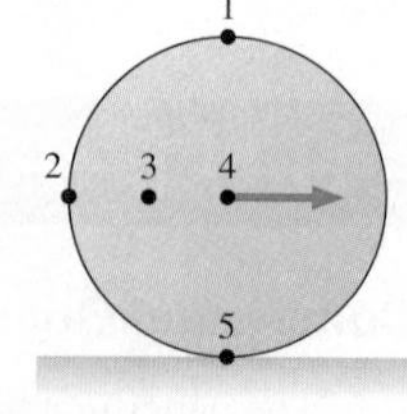

FIGURE Q7.17

18. With care, it's possible to walk on top of a barrel as it rolls. It is much easier to do this if the barrel is full than if it is empty. Explain why this is so.

Multiple-Choice Questions

19. | A nut needs to be tightened with a wrench. Which force shown in Figure Q7.19 will apply the greatest torque to the nut?

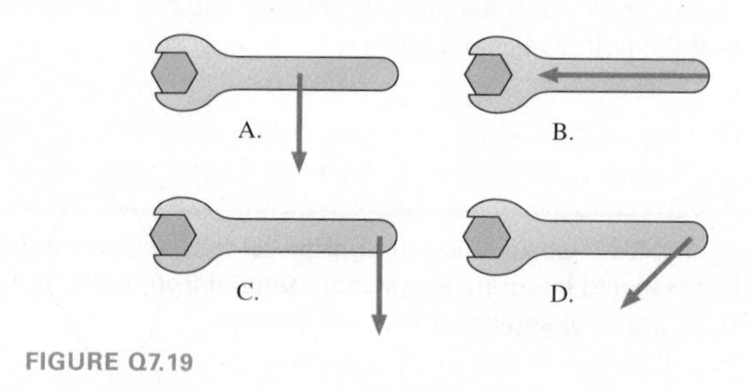

FIGURE Q7.19

20. | Specifications require the oil filter for your car to be tightened with a torque of 30 N·m. You are using a 15-cm-long oil filter wrench, and you apply a force at the very end of the wrench in the direction that produces maximum torque. How much force should you apply?
A. 2000 N B. 400 N C. 200 N D. 30 N

21. | A machine part is made up of two pieces, with centers of gravity shown in Figure Q7.21. Which point could be the center of gravity of the entire part?

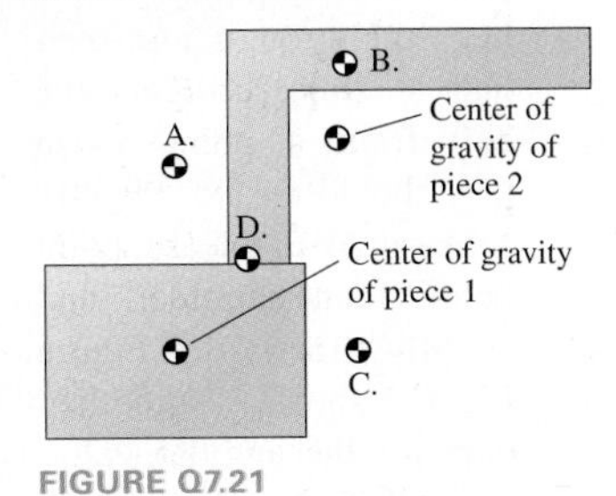

FIGURE Q7.21

22. | The two ends of the barbell shown in Figure Q7.22 are made of the same material. Which of the points shown is at the barbell's center of gravity?

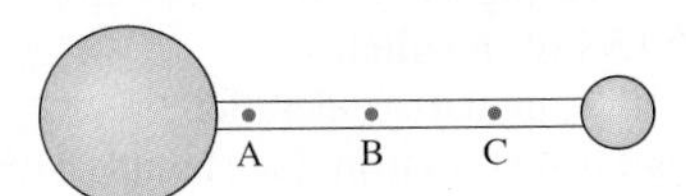

FIGURE Q7.22

23. | A man is holding the yoga BIO position shown in Figure Q7.23. The center of gravity of the combination of his head, arms, and torso is shown as a red symbol on his torso; the center of mass of his legs and feet is a red symbol on his legs. Which of positions A through D corresponds to the center of mass of his entire body?

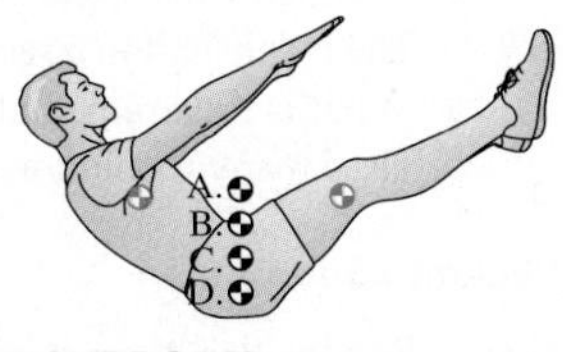

FIGURE Q7.23

24. || A typical compact disk has a mass of 15 g and a diameter of 120 mm. What is its moment of inertia about an axis through its center, perpendicular to the disk?
A. 2.7×10^{-5} kg·m^2 B. 5.4×10^{-5} kg·m^2
C. 1.1×10^{-4} kg·m^2 D. 2.2×10^{-4} kg·m^2

25. | Two horizontal rods are each held up by vertical strings tied to their ends. Rod 1 has length L and mass M; rod 2 has length $2L$ and mass $2M$. Each rod then has one of its supporting strings cut, causing the rod to begin pivoting about the end that is still tied up. Which rod has a larger initial angular acceleration?
A. Rod 1 B. Rod 2
C. The initial angular acceleration is the same for both.

26. | A baseball bat has a thick, heavy barrel and a thin, light handle. If you want to hold a baseball bat on your palm so that it balances vertically, you should
A. Put the end of the handle in your palm, with the barrel up.
B. Put the end of the barrel in your palm, with the handle up.
C. The bat will be equally easy to balance in either configuration.

Questions 27 through 29 concern a classic figure-skating jump called the axel. A skater starts the jump moving forward as shown in Figure Q7.27, leaps into the air, and turns one-and-a-half revolutions before landing. The typical skater is in the air for about 0.5 s, and the skater's hands are located about 0.8 m from the rotation axis.

FIGURE Q7.27

27. || What is the approximate angular speed of the skater during the leap?
A. 2 rad/s B. 6 rad/s C. 9 rad/s D. 20 rad/s

28. | The skater's arms are fully extended during the jump. What is the approximate centripetal acceleration of the skater's hand?
A. 10 m/s^2 B. 30 m/s^2 C. 300 m/s^2 D. 450 m/s^2

29. | What is the approximate speed of the skater's hand?
A. 1 m/s B. 3 m/s C. 9 m/s D. 15 m/s

PROBLEMS

Section 7.1 Describing Circular and Rotational Motion

1. || What is the angular position in radians of the minute hand of a clock at (a) 5:00, (b) 7:15, and (c) 3:35?
2. | A child on a merry-go-round takes 3.0 s to go around once. What is his angular displacement during a 1.0 s time interval?
3. ||| What is the angular speed of the tip of the minute hand on a clock, in rad/s?
4. || An old-fashioned vinyl record rotates on a turntable at 45 rpm. What are (a) the angular speed in rad/s and (b) the period of the motion?
5. ||| The earth's radius is about 4000 miles. Kampala, the capital of Uganda, and Singapore are both nearly on the equator. The distance between them is 5000 miles.
   a. Through what angle do you turn, relative to the earth, if you fly from Kampala to Singapore? Give your answer in both radians and degrees.
   b. The flight from Kampala to Singapore takes 9 hours. What is the plane's angular speed relative to the earth?
6. | Two children are playing tetherball, in which a ball at the end of a cord spins around a pole. After a really good hit, the ball makes three complete revolutions in 2.0 s. What is the angular speed of the ball?
7. ||||| A turntable rotates counterclockwise at 78 rpm. A speck of dust on the turntable is at $\theta = 0.45$ rad at $t = 0$ s. What is the angle of the speck at $t = 8.0$ s? Your answer should be between 0 and 2π rad.
8. || A Ferris wheel on a California pier is 27 m high and rotates once every 32 seconds. When the wheel starts turning, you are at the very top.
   a. What is your angular position 75 seconds after the wheel starts turning, measured counterclockwise from the top? Express your answer as an angle between 0° and 360°.
   b. What is your speed v?

9. ‖ Figure P7.9 shows the angular position of a potter's wheel.

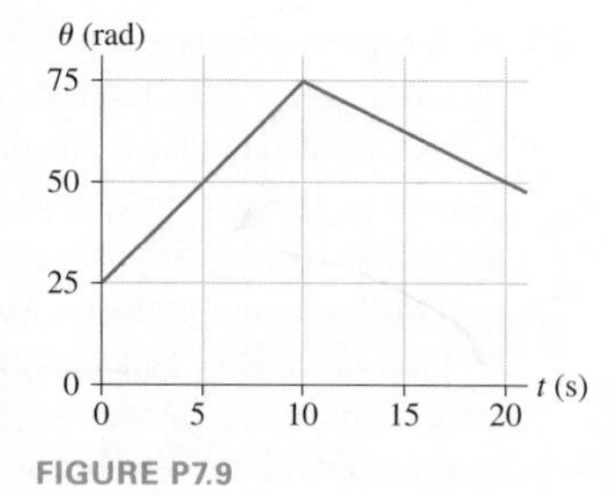

FIGURE P7.9

a. What is the angular displacement of the wheel between $t = 5$ s and $t = 15$ s?
b. What is the angular velocity of the wheel at $t = 15$ s?
c. What is the maximum speed of a point on the outside of the wheel, 15 cm from the axle?

10. ‖ Figure P7.10 shows a graph of the motion of the blades of a blender that is starts empty and is initially turned off. Then the power is turned on at the highest speed setting, some ice cubes are added, and the blades slow down. The blades are 3.0 cm long.

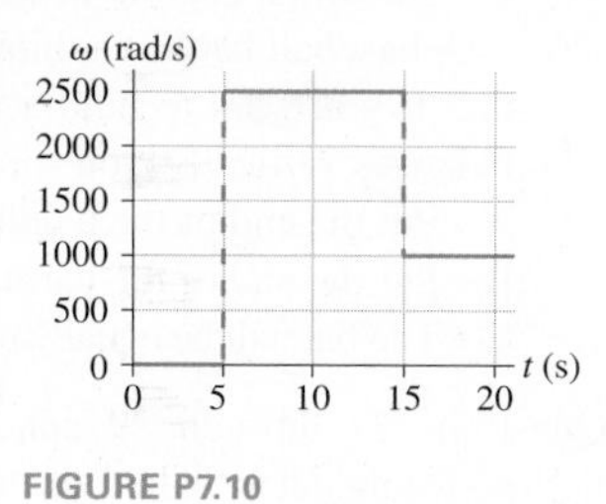

FIGURE P7.10

a. When the blender is empty and running at high speed, how much time is required for a blade to turn by 90°?
b. How many revolutions does a blade complete between 15 and 20 seconds?
c. How fast is the tip of a blade moving when the blender is running empty?

Section 7.2 The Rotation of a Rigid Body

11. | The 1.00-cm-long second hand on a watch rotates smoothly.
a. What is its angular velocity?
b. What is the speed of the tip of the hand?

12. ‖ The earth's radius is 6.37×10^6 m; it rotates once every 24 hours.
a. What is the earth's angular speed?
b. Viewed from a point above the north pole, is the angular velocity positive or negative?
c. What is the speed of a point on the equator?
d. What is the speed of a point on the earth's surface halfway between the equator and the pole? (Hint: What is the radius of the circle in which the point moves?)

13. ||| To throw a discus, the thrower holds it with a fully outstretched arm. Starting from rest, he begins to turn with a constant angular acceleration, releasing the discus after making one complete revolution. The diameter of the circle in which the discus moves is about 1.8 m. If the thrower takes 1.0 s to complete one revolution, starting from rest, what will be the speed of the discus at release?

14. ‖ BIO The bigclaw snapping shrimp shown in Figure P7.14 is aptly named—it has one big claw that snaps shut with remarkable speed. The part of the claw that moves rotates through a 90° angle in 1.3 ms. If we assume that the claw is 1.5 cm long and that it undergoes a constant angular acceleration:

FIGURE P7.14

a. What is the angular acceleration in rad/s^2?
b. What is the final angular speed of the claw?
c. What is the tangential acceleration of the tip of the claw?
d. How fast is the tip of the claw moving at the end of its motion?

15. ||| A computer hard disk starts from rest, then speeds up with an angular acceleration of 190 rad/s^2 until it reaches its final angular speed of 7200 rpm. How many revolutions has the disk made 10.0 s after it starts up?

16. ||| BIO In a softball windmill pitch, the pitcher rotates her arm through just over half a circle, bringing the ball from a point above her shoulder and slightly forward to a release point below her shoulder and slightly forward. Figure P7.16 shows smoothed data for the angular velocity of the upper arm of a college softball pitcher doing a windmill pitch; at time $t = 0$ her arm is vertical and already in motion. For the first 0.15 s there is a steady increase in speed, leading to a final push with a greater acceleration during the final 0.05 s before the release. During the first 0.15 s of the pitch:

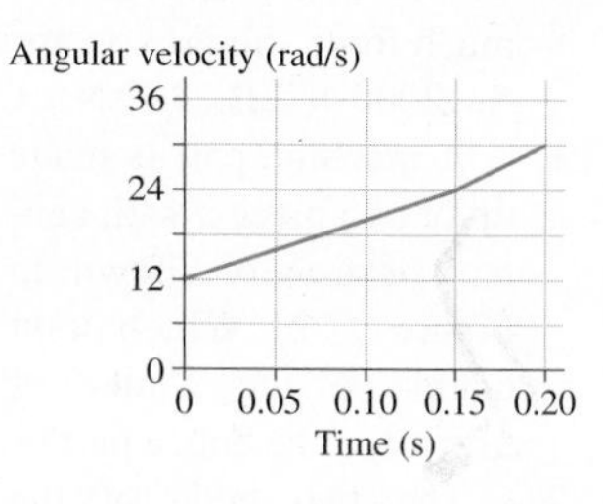

FIGURE P7.16

a. What is the angular acceleration?
b. If the ball is 0.60 m from her shoulder, what is the tangential acceleration of the ball? This is the key quantity here—it's a measure of how much the ball is speeding up. Express your answer in m/s^2 and in units of g.
c. Through what angle does her arm rotate?

17. ||| The crankshaft in a race car goes from rest to 3000 rpm in 2.0 s.
a. What is the crankshaft's angular acceleration?
b. How many revolutions does it make while reaching 3000 rpm?

Section 7.3 Torque

18. | Reconsider the situation in Example 7.10. If Luis pulls straight down on the end of a wrench that is in the same orientation but is 35 cm long, rather than 20 cm, what force must he apply to exert the same torque?

19. ||| Balls are attached to light rods and can move in horizontal circles as shown in Figure P7.19. Rank in order, from smallest to largest, the torques τ_1 to τ_4 about the centers of the circles. Explain.

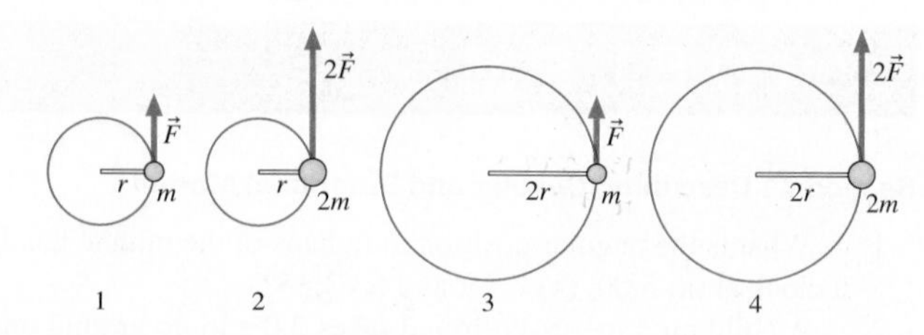

FIGURE P7.19

20. ||| BIO INT We can model a pine tree in the forest as having a compact canopy at the top of a relatively bare trunk. Wind blowing on the top of the tree exerts a horizontal force, and thus a torque that can topple the tree if there is no opposing torque. Suppose a tree's canopy presents an area of 9.0 m^2 to the wind centered at a height of 7.0 m above the ground. (These are reasonable values for forest trees.) If the wind blows at 6.5 m/s:
a. What is the magnitude of the drag force of the wind on the canopy? Assume a drag coefficient of 0.50.
b. What torque does this force exert on the tree, measured about the point where the trunk meets the ground?

21. | What is the net torque about the axle on the pulley in Figure P7.21?

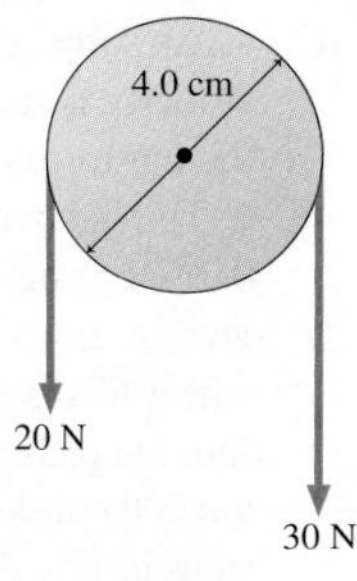

FIGURE P7.21

22. || The tune-up specifications of a car call for the spark plugs to be tightened to a torque of 38 N · m. You plan to tighten the plugs by pulling on the end of a 25-cm-long wrench. Because of the cramped space under the hood, you'll need to pull at an angle of 120° with respect to the wrench shaft. With what force must you pull?
23. || In Figure P7.23, force $\vec{F}_2$ acts half as far from the pivot as $\vec{F}_1$. What magnitude of $\vec{F}_2$ causes the net torque on the rod to be zero?

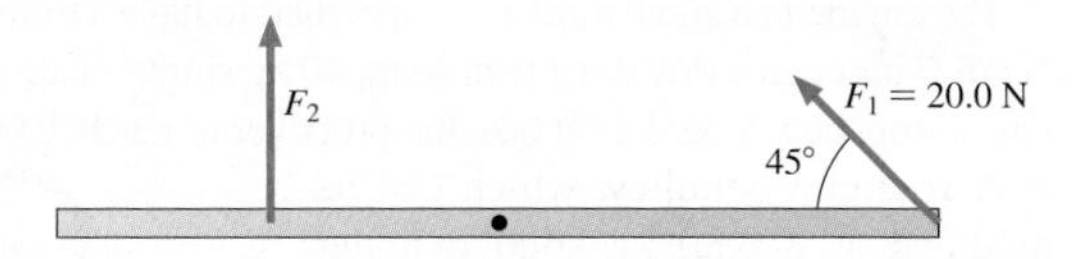

FIGURE P7.23

24. ||| A professor's office door is 0.91 m wide, 2.0 m high, and 4.0 cm thick; has a mass of 25 kg; and pivots on frictionless hinges. A "door closer" is attached to the door and the top of the door frame. When the door is open and at rest, the door closer exerts a torque of 5.2 N · m. What is the least force that you need to apply to the door to hold it open?
25. || What is the net torque on the bar shown in Figure P7.25, about the axis indicated by the dot?

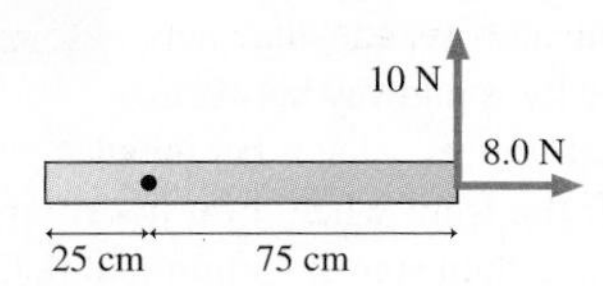

FIGURE P7.25

26. || A typical jar that has been tightened to a reasonable degree requires 2.0 N · m to open. If you grab a 7.0-cm-diameter jar lid with one hand so that your thumb and fingers exert equal magnitude forces on opposite sides of the lid, as in Figure P7.26, what is the magnitude of each of the forces?

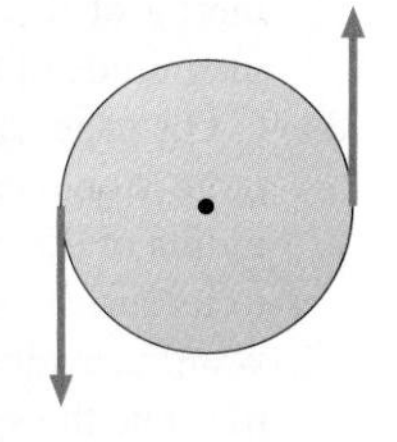
FIGURE P7.26

27. || What is the net torque on the bar shown in Figure P7.27, about the axis indicated by the dot?

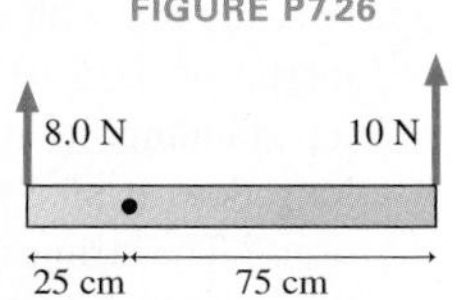

FIGURE P7.27

28. || A driver holds his hands on opposite sides of the 35-cm-diameter steering wheel in a modern sports car. A torque of 4.5 N · m is required to turn the wheel. If the driver applies an equal force on each side of the wheel, what is the minimum force each hand must supply?
29. || What is the net torque on the bar shown in Figure P7.29, about the axis indicated by the dot?

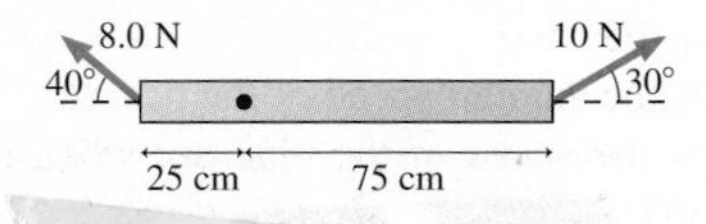

FIGURE P7.29

Section 7.4 Gravitational Torque and the Center of Gravity

30. || BIO INT When you stand quietly, you pivot back and forth a very small amount about your ankles. It's easier to maintain stability if you tip slightly forward, so that your center of gravity is slightly in front of your ankles. The 68 kg man shown in Figure P7.30 is 1.8 m tall; his center of gravity is 1.0 m above the ground and, during quiet standing, is 5.5 cm in front of his ankles. What is the magnitude of the gravitational torque about his ankles?

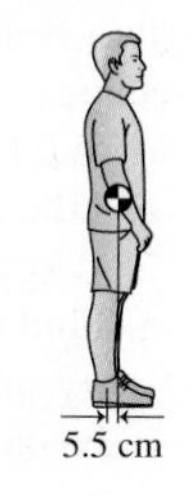

FIGURE P7.30

31. || The 2.0 kg, uniform, horizontal rod in Figure P7.31 is seen from the side. What is the gravitational torque about the point shown?

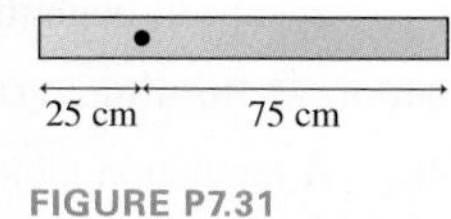

FIGURE P7.31

32. |||| A 4.00-m-long, 500 kg steel beam extends horizontally from the point where it has been bolted to the framework of a new building under construction. A 70.0 kg construction worker stands at the far end of the beam. What is the magnitude of the gravitational torque about the point where the beam is bolted into place?
33. ||| BIO An athlete at the gym holds a 3.0 kg steel ball in his hand. His arm is 70 cm long and has a mass of 4.0 kg. What is the magnitude of the gravitational torque about his shoulder if he holds his arm
   a. Straight out to his side, parallel to the floor?
   b. Straight, but 45° below horizontal?
34. ||| The 2.0-m-long, 15 kg beam in Figure P7.34 is hinged at its left end. It is "falling" (rotating clockwise, under the influence of gravity), and the figure shows its position at three different times. What is the gravitational torque on the beam about an axis through the hinged end when the beam is at the
   a. Upper position?
   b. Middle position?
   c. Lower position?

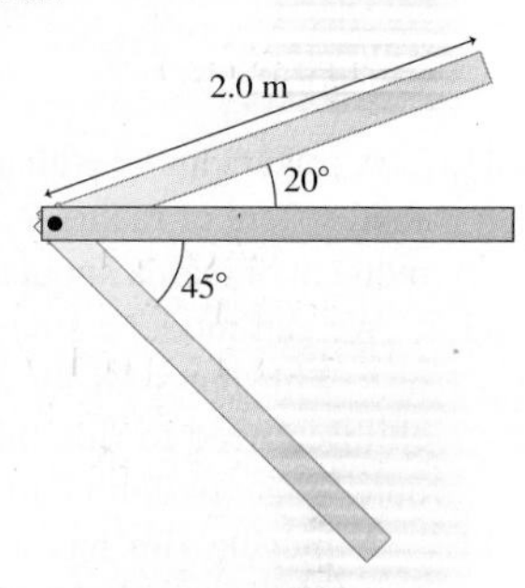

FIGURE P7.34

35. || Three identical coins lie on three corners of a square 10.0 cm on a side, as shown in Figure P7.35. Determine the x- and y-coordinates of the center of gravity of the group of three coins.

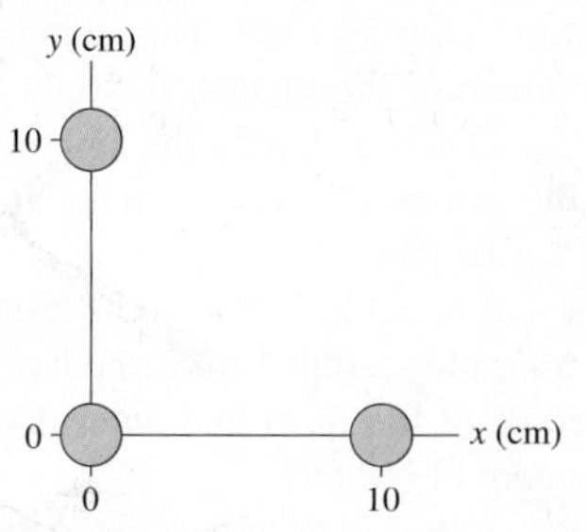

FIGURE P7.35

36. | BIO Hold your arm outstretched so that it is horizontal. Estimate the mass of your arm and the position of its center of gravity. What is the gravitational torque on your arm in this position, computed around the shoulder joint?
37. || U.S. nickels have a mass of 5.00 g and are 1.95 mm thick. If you stack 3 nickels on a table, how far above the table is their center of gravity?

38. ||| Figure P7.38 shows two thin beams joined at right angles. The vertical beam is 15.0 kg and 1.00 m long and the horizontal beam is 25.0 kg and 2.00 m long.
   a. Find the center of gravity of the two joined beams. Express your answer in the form (x, y), taking the origin at the corner where the beams join.
   b. Calculate the gravitational torque on the joined beams about an axis through the corner. The beams are seen from the side.

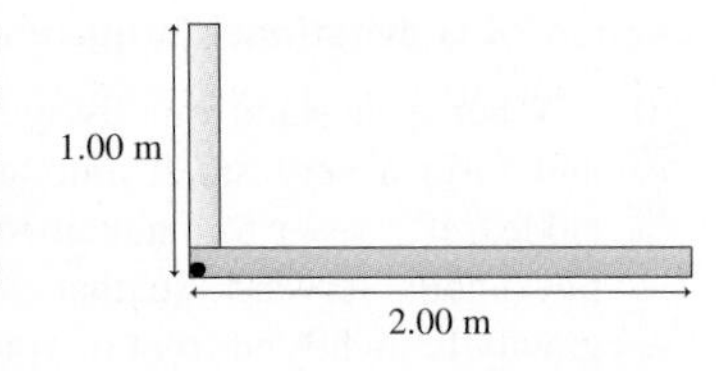

FIGURE P7.38

Section 7.5 Rotational Dynamics and Moment of Inertia

39. ||| A regulation table tennis ball is a thin spherical shell 40 mm in diameter with a mass of 2.7 g. What is its moment of inertia about an axis that passes through its center?
40. || Three pairs of balls are connected by very light rods as shown in Figure P7.40. Rank in order, from smallest to largest, the moments of inertia I_1, I_2, and I_3 about axes through the centers of the rods.

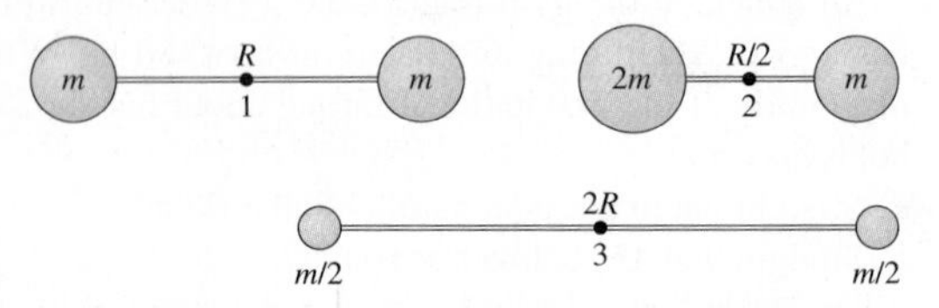

FIGURE P7.40

41. |||| A solid cylinder with a radius of 4.0 cm has the same mass as a solid sphere of radius R. If the cylinder and sphere have the same moment of inertia about their centers, what is the sphere's radius?
42. || We can model a small merry-go-round as a uniform circular disk with mass 88 kg and diameter 1.8 m. How many 22 kg children need to ride the merry-go-round, standing right at the outer edge, to double the moment of inertia of the system?
43. || A bicycle rim has a diameter of 0.65 m and a moment of inertia, measured about its center, of 0.19 kg · m². What is the mass of the rim?
44. ||| A bowling ball is far from uniform. Lightweight bowling balls are made of a relatively low-density core surrounded by a thin shell with much higher density. A 7.0 lb (3.2 kg) bowling ball has a diameter of 0.216 m; 0.196 m of this is a 1.6 kg core, surrounded by a 1.6 kg shell. This composition gives the ball a higher moment of inertia than it would have if it were made of a uniform material. Given the importance of the angular motion of the ball as it moves down the alley, this has real consequences for the game.
   a. Model a real bowling ball as a 0.196-m-diameter core with mass 1.6 kg plus a thin 1.6 kg shell with diameter 0.206 m (the average of the inner and outer diameters). What is the total moment of inertia?
   b. How does your answer in part a compare to the moment of inertia of a uniform 3.2 kg ball with diameter 0.216 m?

Section 7.6 Using Newton's Second Law for Rotation

45. | A small grinding wheel has a moment of inertia of 4.0×10^{-5} kg · m². What net torque must be applied to the wheel for its angular acceleration to be 150 rad/s²?
46. |||| a. What is the moment of inertia of the door in Problem 24?
   b. If you let go of the open door, what is its angular acceleration immediately afterward?
47. | An object's moment of inertia is 2.0 kg · m². Its angular velocity is increasing at the rate of 4.0 rad/s per second. What is the net torque on the object?
48. ||| The lightweight wheel on a road bike has a moment of inertia of 0.097 kg · m². A mechanic, checking the alignment of the wheel, gives it a quick spin; it completes 5 rotations in 2.0 s. To bring the wheel to rest, the mechanic gently applies the disk brakes, which squeeze pads against a metal disk connected to the wheel. The pads touch the disk 7.1 cm from the axle, and the wheel slows down and stops in 1.5 s. What is the magnitude of the friction force on the disk?
49. |||| A 200 g, 20-cm-diameter plastic disk is spun on an axle through its center by an electric motor. What torque must the motor supply to take the disk from 0 to 1800 rpm in 4.0 s?
50. || The engine in a small airplane is specified to have a torque of 500 N · m. This engine drives a 2.0-m-long, 40 kg single-blade propeller. On startup, how long does it take the propeller to reach 2000 rpm?
51. ||| A frictionless pulley, which can be modeled as a 0.80 kg solid cylinder with a 0.30 m radius, has a rope going over it, as shown in Figure P7.51. The tension in the rope is 10 N on one side and 12 N on the other. What is the angular acceleration of the pulley?

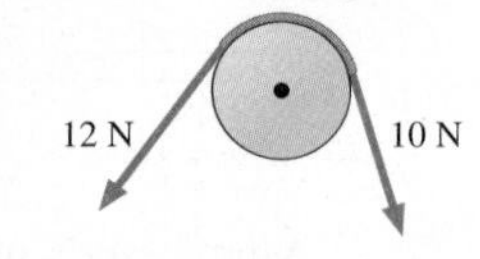

FIGURE P7.51

52. || A gymnast on the uneven parallel bars is at rest, tipped at a 45° angle from the vertical. The distance from her hands to her feet is 1.8 m. If we model her body as having a uniform cross section and assume that her center of gravity is midway between her hands and her feet, what is her initial angular acceleration?

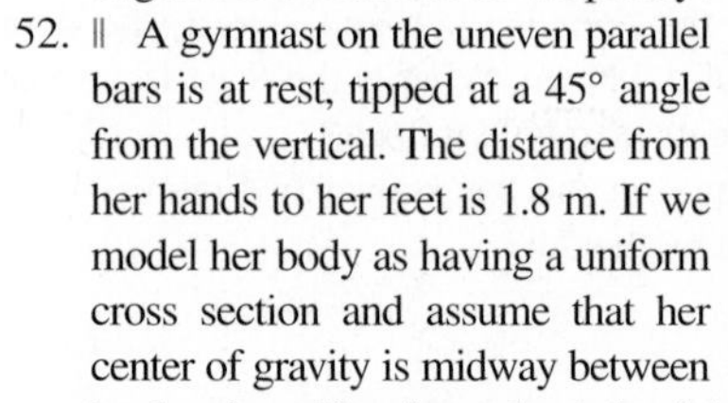

53. ||| If you lift the front wheel of a poorly maintained bicycle off the ground and then start it spinning at 0.72 rev/s, friction in the bearings causes the wheel to stop in just 12 s. If the moment of inertia of the wheel about its axle is 0.30 kg · m², what is the magnitude of the frictional torque?
54. || On page 224 there is a photograph of a girl pushing on a large stone sphere. The sphere has a mass of 8200 kg and a radius of 90 cm. Suppose that she pushes on the sphere tangent to its surface with a steady force of 50 N and that the pressured water provides a frictionless support. How long will it take her to rotate the sphere one time, starting from rest?
55. ||||| A toy top with a spool of diameter 5.0 cm has a moment of inertia of 3.0×10^{-5} kg · m² about its rotation axis. To get the top spinning, its string is pulled with a tension of 0.30 N. How long does it take for the top to complete the first five revolutions? The string is long enough that it is wrapped around the top more than five turns.
56. ||||| A 1.5 kg block and a 2.5 kg block are attached to opposite ends of a light rope. The rope hangs over a solid, frictionless pulley that is 30 cm in diameter and has a mass of 0.75 kg. When the blocks are released, what is the acceleration of the lighter block?

Section 7.7 Rolling Motion

57. ||| A bicycle with 0.80-m-diameter tires is coasting on a level road at 5.6 m/s. A small blue dot has been painted on the tread of the rear tire.
   a. What is the angular speed of the tires?
   b. What is the speed of the blue dot when it is 0.80 m above the road?
   c. What is the speed of the blue dot when it is 0.40 m above the road?

58. || A typical road bike wheel has a diameter of 70 cm including the tire. In a time trial, when a cyclist is racing along at 12 m/s:
    a. How fast is a point at the top of the tire moving?
    b. How fast, in rpm, are the wheels spinning?
59. || A 2.0-m-long slab of concrete is supported by rollers, as shown in Figure P7.59. If the slab is pushed to the right, it will move off the supporting rollers, one by one, on the left side. How far can the slab be moved before its center of gravity is to the right of the contact point with the rightmost roller—at which point the slab begins to tip?

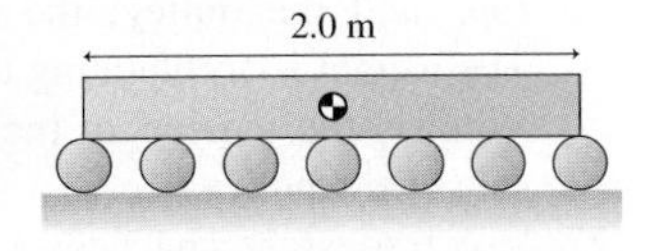

FIGURE P7.59

60. || A man in a barrel walking competition is moving along smoothly, with his barrel moving forward at 1.0 m/s.
    a. Think about how the man moves his legs. Is he walking forward or backward?
    b. From the point of view of the top of the barrel—the man's walking surface—how fast is he walking?

General Problems

61. | Figure P7.61 shows the angular position-versus-time graph for a particle moving in a circle.
INT
    a. Write a description of the particle's motion.
    b. Draw the angular velocity-versus-time graph.

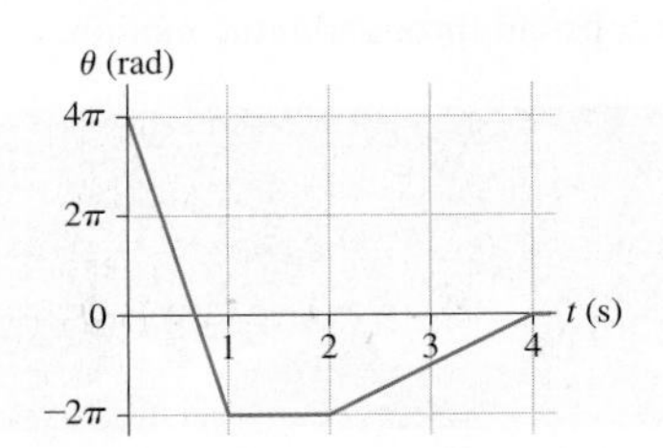

FIGURE P7.61

62. | The graph in Figure P7.62 shows the angular velocity of the crankshaft in a car. Draw a graph of the angular acceleration versus time. Include appropriate numerical scales on both axes.
INT

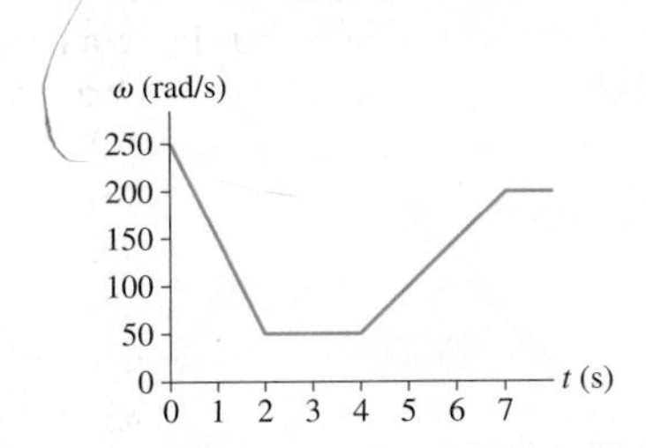

FIGURE P7.62

63. || A car with 58-cm-diameter tires accelerates uniformly from rest to 20 m/s in 10 s. How many times does each tire rotate?
INT
64. ||| The cable lifting an elevator is wrapped around a 1.0-m-diameter cylinder that is turned by the elevator's motor. The elevator is moving upward at a speed of 1.6 m/s. It then slows to a stop as the cylinder makes one complete turn at constant angular acceleration. How long does it take for the elevator to stop?
65. |||| The 20-cm-diameter disk in Figure P7.65 can rotate on an axle through its center. What is the net torque about the axle?

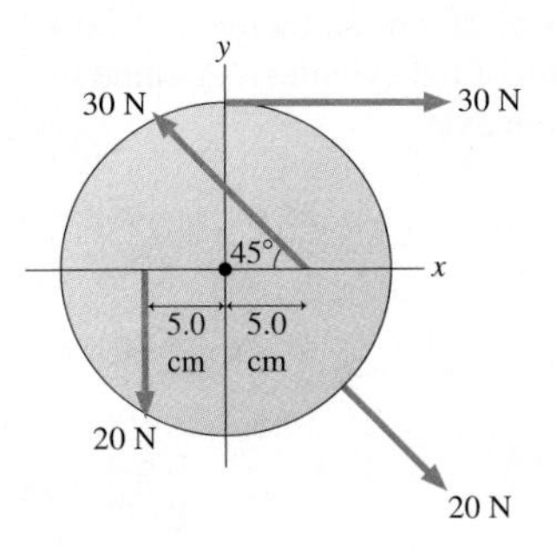

FIGURE P7.65

66. |||| A combination lock has a 1.0-cm-diameter knob that is part of the dial you turn to unlock the lock. To turn that knob, you grip it between your thumb and forefinger with a force of 0.60 N as you twist your wrist. Suppose the coefficient of static friction between the knob and your fingers is only 0.12 because some oil accidentally got onto the knob. What is the most torque you can exert on the knob without having it slip between your fingers?
INT
67. |||| A 70 kg man's arm, including the hand, can be modeled as a 75-cm-long uniform cylinder with a mass of 3.5 kg. In raising both his arms, from hanging down to straight up, by how much does he raise his center of gravity?
BIO
68. |||| The three masses shown in Figure P7.68 are connected by massless, rigid rods.
    a. Find the coordinates of the center of gravity.
    b. Find the moment of inertia about an axis that passes through mass A and is perpendicular to the page.
    c. Find the moment of inertia about an axis that passes through masses B and C.

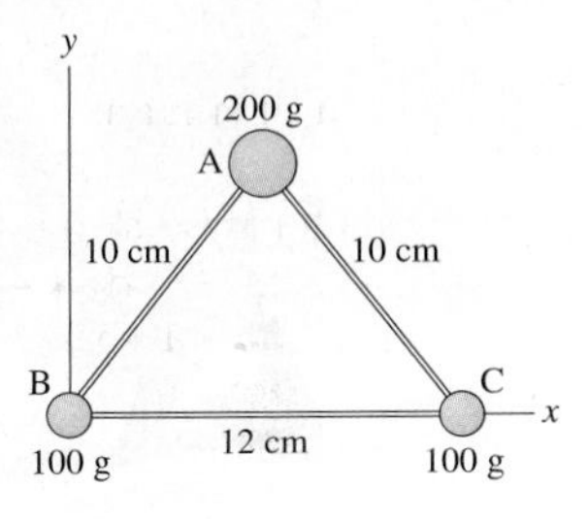

FIGURE P7.68

69. || A reasonable estimate of the moment of inertia of an ice skater spinning with her arms at her sides can be made by modeling most of her body as a uniform cylinder. Suppose the skater has a mass of 64 kg. One-eighth of that mass is in her arms, which are 60 cm long and 20 cm from the vertical axis about which she rotates. The rest of her mass is approximately in the form of a 20-cm-radius cylinder.
BIO
    a. Estimate the skater's moment of inertia to two significant figures.
    b. If she were to hold her arms outward, rather than at her sides, would her moment of inertia increase, decrease, or remain unchanged? Explain.
70. ||| A 17 g audio compact disk has a diameter of 12 cm. The disk spins under a laser that reads encoded data. The first track to be read is 2.3 cm from the axis; as the disk plays, the laser scans tracks farther and farther from the center. The part of the disk directly under the read head moves at a constant 1.2 m/s.
    a. When a disk is inserted, it takes 2.4 s to spin up from rest. What is the torque of the motor?
    b. As the disk plays, how does the angular speed change?

71. ||| The ropes in Figure P7.71 are each wrapped around a cylinder, and the two cylinders are fastened together. The smaller cylinder has a diameter of 10 cm and a mass of 5.0 kg; the larger cylinder has a diameter of 20 cm and a mass of 20 kg. What is the angular acceleration of the cylinders? Assume that the cylinders turn on a frictionless axle.

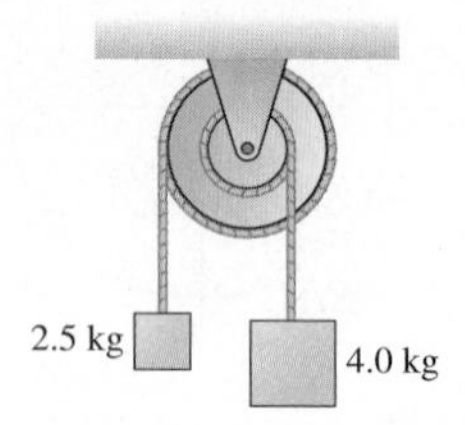

FIGURE P7.71

72. ||| A spin bike is an indoor bike that is designed to duplicate the feeling of regular road cycling. A typical spin bike has a very heavy flywheel. A friction pad or other brake provides damping that mirrors the drag and other resistive forces that a road cyclist experiences; the inertia of the wheel simulates the way a regular bike keeps moving even after you stop pedaling. A spin bike has a flywheel in two parts—a 12.5 kg disk with radius 0.23 m, and a 7.0 kg ring with mass concentrated at the outer edge of the disk. A friction pad exerts a force of 9.7 N on the outside of the disk. A cyclist is pedaling, spinning the disk at a typical 180 rpm. If she stops pedaling, how long will it take for the flywheel to come to a stop?

73. |||| Flywheels are large, massive wheels used to store energy. They can be spun up slowly, then the wheel's energy can be released quickly to accomplish a task that demands high power. An industrial flywheel has a 1.5 m diameter and a mass of 250 kg. A motor spins up the flywheel with a constant torque of 50 N · m. How long does it take the flywheel to reach top angular speed of 1200 rpm?

74. ||| A trap-jaw ant has mandibles that can snap shut with some BIO INT force, as you might expect from its name. The formidable snap is good for more than capturing prey. When an ant snaps its jaws against the ground, the resulting force can launch the ant into the air. Here are typical data: An ant rotates its mandible, of length 1.30 mm and mass 130 μg (which we can model as a uniform rod rotated about its end), at a high angular speed. As the tip strikes the ground, it undergoes an angular acceleration of 3.5×10^8 rad/s^2. If we assume that the tip of the mandible hits perpendicular to the ground, what is the force on the tip? How does this compare to the weight of a 12 mg ant?

75. |||| A 1.0 kg ball and a 2.0 kg ball are connected by a 1.0-m-long rigid, massless rod. The rod and balls are rotating clockwise about their center of gravity at 20 rpm. What torque will bring the balls to a halt in 5.0 s?

76. |||| A 1.5 kg block is connected by a rope INT across a 50-cm-diameter, 2.0 kg pulley, as shown in Figure P7.76. There is no friction in the axle, but there is friction between the rope and the pulley; the rope doesn't slip. The weight is accelerating upward at 1.2 m/s^2. What is the tension in the rope on the right side of the pulley?

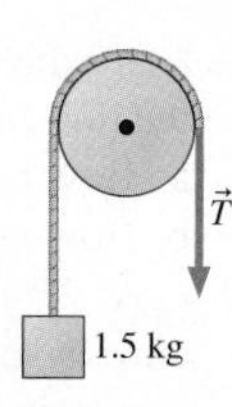

FIGURE P7.76

77. |||| A tradesman sharpens a knife by pushing it with a constant INT force against the rim of a grindstone. The 30-cm-diameter stone is spinning at 200 rpm and has a mass of 28 kg. The coefficient of kinetic friction between the knife and the stone is 0.20. If the stone slows steadily to 180 rpm in 10 s of grinding, what is the force with which the man presses the knife against the stone?

MCAT-Style Passage Problems

The Bunchberry BIO

The bunchberry flower has the fastest-moving parts ever seen in a plant. Initially, the stamens are held by the petals in a bent position, storing energy like a coiled spring. As the petals release, the tips of the stamens fly up and quickly release a burst of pollen. Figure P7.78 shows the details of the motion. The tips of the stamens act like a catapult, flipping through a 60° angle; the times on the earlier photos show that this happens in just 0.30 ms. We can model a stamen tip as a 1.0-mm-long, 10 μg rigid rod with a 10 μg anther sac at one end and a pivot point at the opposite end. Though an oversimplification, we will model the motion by assuming the angular acceleration is constant throughout the motion.

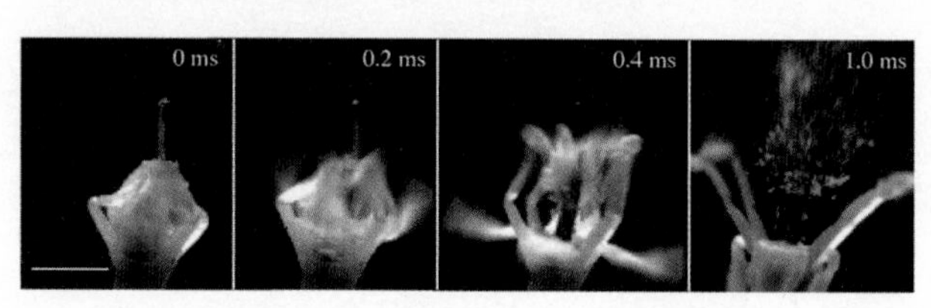

78. | What is the angular acceleration of the anther sac during the motion?
A. 3.5×10^3 rad/s^2 B. 7.0×10^3 rad/s^2
C. 1.2×10^7 rad/s^2 D. 2.3×10^7 rad/s^2

Pollen Pollen
Anther sac
1.0 mm
60°
Stamen
Initial configuration Final configuration

FIGURE P7.78

79. | What is the speed of the anther sac as it releases its pollen?
A. 3.5 m/s B. 7.0 m/s
C. 10 m/s D. 14 m/s

80. || How large is the "straightening torque"? (You can omit gravitational forces from your calculation; the gravitational torque is much less than this.)
A. 2.3×10^{-7} N · m B. 3.1×10^{-7} N · m
C. 2.3×10^{-5} N · m D. 3.1×10^{-5} N · m

The Illusion of Flight

The grand jeté is a classic ballet maneuver in which a dancer executes a horizontal leap while moving her arms and legs up and then down. At the center of the leap, the arms and legs are gracefully extended, as we see in Figure P7.81a. The goal of the leap is to create the illusion of flight. As the dancer moves through the air, he or she is in free fall. In Chapter 3, we saw that this leads to projectile motion. But what part of the dancer follows the usual parabolic path? It won't come as a surprise to learn that it's the center of gravity. But when you watch a dancer leap through the air, you don't watch her center of gravity, you watch her head. If the translational motion of her head is horizontal—not parabolic—this creates the illusion that she is flying through the air, held up by unseen forces.

Figure P7.81b illustrates how the dancer creates this illusion. While in the air, she changes the position of her center of gravity relative to her body by moving her arms and legs up, then down. Her center of gravity moves in a parabolic path, but her head moves in a straight line. It's not flight, but it will appear that way, at least for a moment.

FIGURE P7.81

81. | To perform this maneuver, the dancer relies on the fact that the position of her center of gravity
    A. Is near the center of the torso.
    B. Is determined by the positions of her arms and legs.
    C. Moves in a horizontal path.
    D. Is outside of her body.
82. | Suppose you wish to make a vertical leap with the goal of getting your head as high as possible above the ground. At the top of your leap, your arms should be
    A. Held at your sides.
    B. Raised above your head.
    C. Outstretched, away from your body.
83. | When the dancer is in the air, is there a gravitational torque on her? Take the dancer's rotation axis to be through her center of gravity.
    A. Yes, there is a gravitational torque.
    B. No, there is not a gravitational torque.
    C. It depends on the positions of her arms and legs.
84. | In addition to changing her center of gravity, a dancer may change her moment of inertia. Consider her moment of inertia about a vertical axis through the center of her body. When she raises her arms and legs, this
    A. Increases her moment of inertia.
    B. Decreases her moment of inertia.
    C. Does not change her moment of inertia.

8 Equilibrium and Elasticity

How does a dancer balance so gracefully *en pointe*? And how does her foot withstand the great stresses concentrated on her toes? In this chapter we'll find answers to both these questions.

LOOKING AHEAD ➤

Static Equilibrium

As this cyclist balances on his back tire, the net force *and* the net torque on him must be zero.

You'll learn to analyze objects that are in **static equilibrium**.

Springs

When a rider takes a seat, the spring is compressed and exerts a **restoring force,** pushing upward.

You'll learn to solve problems involving stretched and compressed springs.

Properties of Materials

All materials have some "give"; if you pull on them, they stretch, and at some point they break.

Is this spider silk as strong as steel? You'll learn to think about what the question means, and how to answer it.

GOAL **To learn about the static equilibrium of extended objects, and the basic properties of springs and elastic materials.**

LOOKING BACK ◂

Torque

In Chapter 7, you learned to calculate the torque on an object due to an applied force.

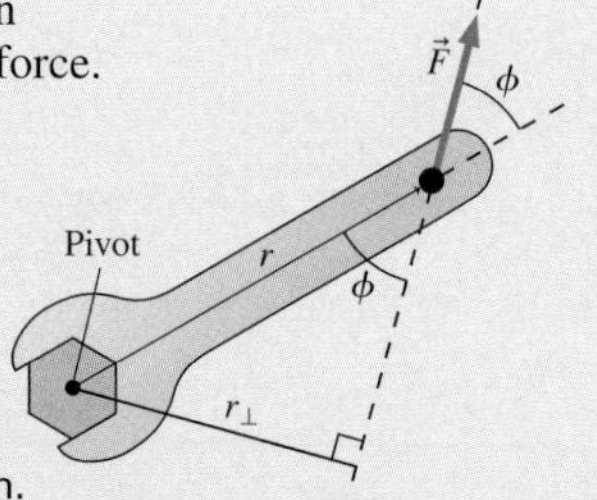

In this chapter, you'll extend your analysis to consider objects with many forces—and many torques—that act on them.

STOP TO THINK

An old-fashioned tire swing exerts a force on the branch and a torque about the point where the branch meets the trunk. If you hang the swing closer to the trunk, this will ______ the force and ______ the torque.

A. increase, increase
B. not change, increase
C. not change, not change
D. not change, decrease
E. decrease, not change
F. decrease, decrease

8.1 Torque and Static Equilibrium

We have now spent several chapters studying motion and its causes. In many disciplines, it is just as important to understand the conditions under which objects do *not* move. Buildings and dams must be designed such that they remain practically motionless, even when huge forces act on them. And joints in the body must sustain large forces when the body is supporting heavy loads, as in holding or carrying heavy objects.

Recall from ◀◀ SECTION 5.1 that an object at rest is in *static equilibrium.* As long as an object can be modeled as a *particle,* the condition necessary for static equilibrium is that the net force $\vec{F}_{\text{net}}$ on the particle is zero, as in FIGURE 8.1a, where the two forces applied to the particle balance and the particle can remain at rest.

But in Chapter 7 we moved beyond the particle model to study extended objects that can rotate. Consider, for example, the block in FIGURE 8.1b. In this case the two forces act along the same line, the net force is zero, and the block is in equilibrium. But what about the block in FIGURE 8.1c? The net force is still zero, but this time the block begins to rotate because the two forces exert a net *torque.* For an extended object, $\vec{F}_{\text{net}} = \vec{0}$ is not by itself enough to ensure static equilibrium. There is a second condition for static equilibrium of an extended object: The net torque τ_{net} on the object must also be zero.

FIGURE 8.1 A block with no net force acting on it may still be out of equilibrium.

(a) When the net force on a particle is zero, the particle is in static equilibrium.

(b) Both the net force and the net torque are zero, so the block is in static equilibrium.

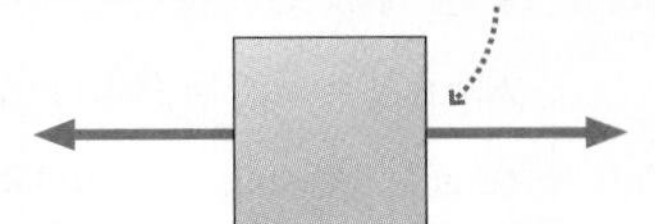

(c) The net force is still zero, but the net torque is *not* zero. The block is not in equilibrium.

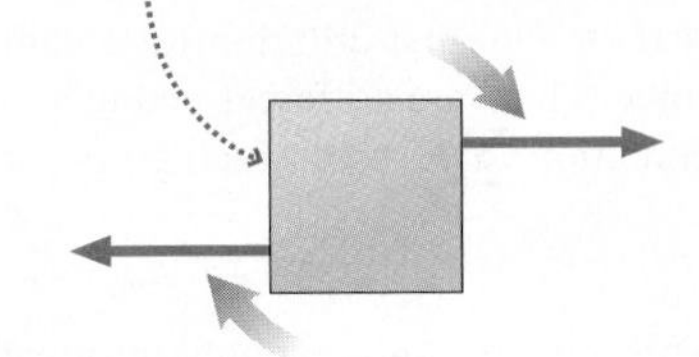

If we write the net force in component form, the conditions for static equilibrium of an extended object are

$$\left.\begin{aligned}\sum F_x &= 0\\ \sum F_y &= 0\end{aligned}\right\}\ \text{No net force}$$
$$\sum \tau = 0\ \}\ \text{No net torque} \tag{8.1}$$

Conditions for static equilibrium of an extended object

Let's look at an example to see how we can use these conditions to analyze a physical situation.

EXAMPLE 8.1 Lifting a load

A participant in a "strongman" competition uses an old-fashioned device to move a large load with only muscle power. A 1 ton (910 kg) bucket of rocks is suspended from a sturdy, lightweight beam 0.85 m from a pivot. The man lifts the beam at its end, 3.6 m from the pivot, and holds it steady. How much force must the man apply? What is the force on the beam from the pivot?

STRATEGIZE This is a static equilibrium problem—nothing is moving. The first step in our solution is to find an object to focus on. The way we've drawn the diagram in FIGURE 8.2, with the highlight color on the beam, hints at our choice. The object of interest in this case is the beam, not the man or the bucket. The beam is the element that links all of the pieces together; it touches the man, the bucket, and the pivot. The beam has different forces acting on it at different points, and it's not going anywhere. For the beam, the net force and the net torque must both be zero. Our next step is to draw the visual overview in FIGURE 8.3, in which we clearly identify both the forces that act on the beam and where they act. There is a downward force on the beam at the point where the bucket hangs, an upward force where the man supports the beam, and an upward force at the pivot. We will measure the distances from the pivot, which is the natural point about which to compute torques.

FIGURE 8.2 Using a beam to lift a load.

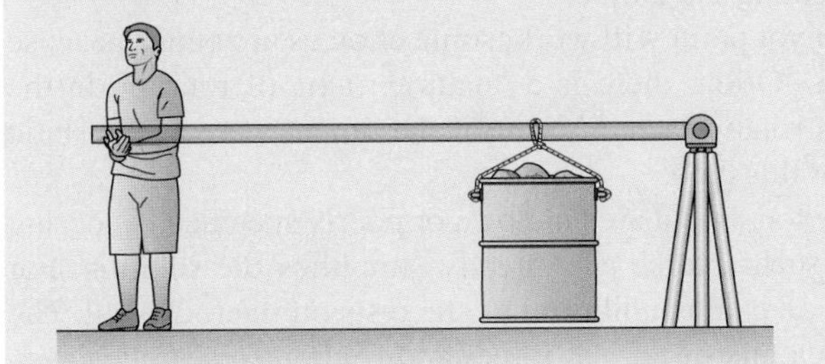

FIGURE 8.3 The forces acting on the beam.

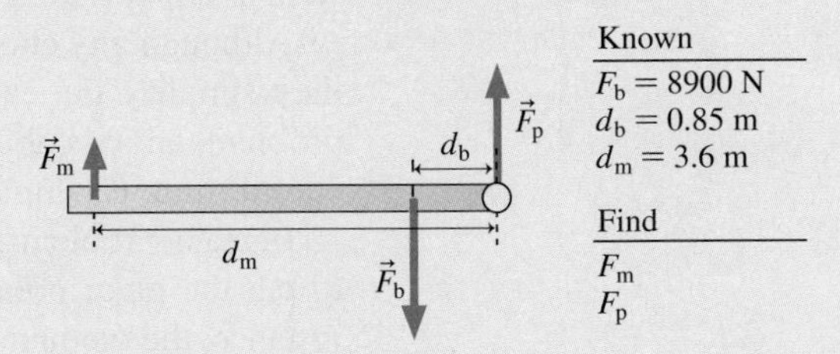

Continued

We've made some assumptions in drawing the diagram, assumptions that we will use to check our work at the end. The bucket is closer to the pivot than to the man, so we've drawn the force of the pivot as greater than the force of the man. And we expect that the total force supporting the beam—the upward force of the man plus the upward force of the pivot—will be equal to the weight of the bucket. We've drawn the arrows to match.

PREPARE We can add some details to the visual overview of Figure 8.3. The force of the bucket on the beam is equal to the weight of the bucket:

$$F_b = w_b = m_b g = (910 \text{ kg})(9.8 \text{ m/s}^2) = 8900 \text{ N}$$

This force acts a distance $d_b = 0.85$ m from the pivot. The man exerts a force 3.6 m from the pivot. We don't know the force of the man on the beam or the force of the pivot on the beam; these are the numbers we are trying to find.

SOLVE The first criterion for equilibrium is that there is no net force. There are no forces acting horizontally, so we consider the condition $\sum F_y = 0$, which gives

$$\sum F_y = F_m + F_p - F_b = 0$$

This clearly isn't enough information to solve the problem; there are two unknown forces in this equation. But we've got another condition, $\sum \tau = 0$. All the forces act perpendicular to the beam, so we can compute the torques as $\tau = rF_\perp$, where $F_\perp$ is the magnitude of each force. There is no torque from the force of the pivot because this force acts right at the pivot. The force of the bucket tries to rotate the beam counterclockwise, while the force of the man tries to rotate the beam clockwise. So our torque equation is

$$\sum \tau = F_b d_b - F_m d_m = 0$$

We know the value of every quantity in this equation except F_m, so we can calculate the force from the man:

$$F_m = F_b \frac{d_b}{d_m} = (8900 \text{ N}) \frac{0.85 \text{ N}}{3.6 \text{ m}} = 2100 \text{ N}$$

We can now revisit the equation for forces to compute the force of the pivot:

$$F_p = F_b - F_m = 8900 \text{ N} - 2100 \text{ N} = 6800 \text{ N}$$

ASSESS The force the man must apply is 2100 N, the weight of about 200 kg. That's a lot of force, but it is a strongman competition, so this makes sense. This force is less than the force of the pivot, as we expected.

STOP TO THINK 8.1 Which of these objects is in static equilibrium?

A. B. C. D.

Choosing the Pivot Point

In Chapter 7, we saw that the value of the torque depends on the choice of the pivot point. In Example 8.1, there was a natural choice for the pivot point, but we could have chosen another point. Would that choice have affected the result of our calculation?

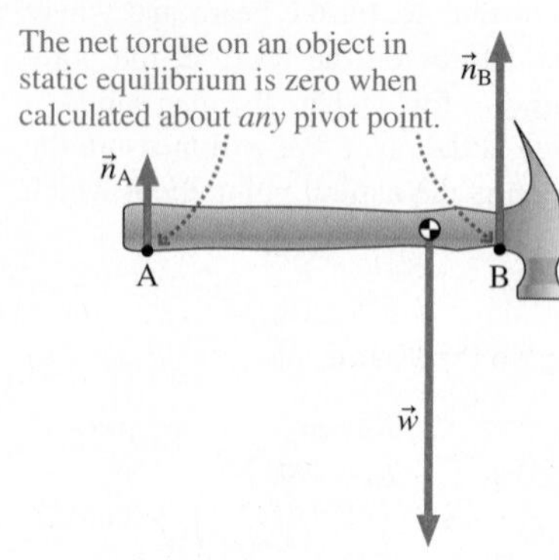

The net torque on an object in static equilibrium is zero when calculated about *any* pivot point.
$\vec{n}_B$
$\vec{n}_A$
A
B
$\vec{w}$

FIGURE 8.4 A hammer resting on two pegs.

Consider the hammer shown in FIGURE 8.4, supported on a pegboard by two pegs A and B. Because the hammer is in static equilibrium, the net torque around the pivot at peg A must be zero: The clockwise torque due to the weight $\vec{w}$ is exactly balanced by the counterclockwise torque due to the force $\vec{n}_B$ of peg B. (Recall that the torque due to $\vec{n}_A$ is zero because here $\vec{n}_A$ acts at the pivot A.) But if instead we take B as the pivot, the net torque is still zero. The counterclockwise torque due to $\vec{w}$ (with a large force but small moment arm) balances the clockwise torque due to $\vec{n}_A$ (with a small force but large moment arm). Indeed, **for an object in static equilibrium, the net torque about *every* point must be zero.** This means you can pick *any* point you wish as a pivot point for calculating the torque.

Although any choice of a pivot point will work, some choices are better because they simplify the calculations. Often, there is a "natural" axis of rotation in the problem, an axis about which rotation *would* occur if the object were not in static equilibrium. Example 8.1 is of this type.

In many problems, there are forces that are unknown or poorly specified. Choosing to put the pivot point where such a force acts greatly simplifies the solution. For instance, the woman in FIGURE 8.5 is in equilibrium as she rests on the rock wall. The force of the wall on her feet is a mix of normal forces and friction forces; the direction isn't well specified. The directions of the other two forces are well known: The tension force points along the rope, and the weight force points down. So a good choice of pivot point is where the woman's foot contacts the wall because this choice eliminates the torque due to the force of the wall on her foot, which isn't as well known as the other forces.

FIGURE 8.5 Choosing the pivot for a woman rappelling down a rock wall.

PROBLEM-SOLVING APPROACH 8.1

Static equilibrium problems

For situations involving static equilibrium, we can use the fact that there is no net force and no net torque as a basis for solving problems.

STRATEGIZE Begin by making decisions:

- Decide on an object of interest. This should be an extended object with different forces acting on it at different points.
- Decide on a pivot point about which to compute torques.

Next, model the object as a simple shape, and begin to draw a visual overview of the system.

PREPARE Add detail to your visual overview:

- Determine what forces act, and where they act. Add them to your diagram, and list values (if known) in your table.
- Determine the distances from the pivot to the points where forces act.

Think about the forces that act and the torques that they create. Think about whether the torques will be positive or negative. Remember that the torques due to any forces acting *at* the pivot are zero.

SOLVE The mathematical steps are based on the conditions:

$$\vec{F}_{\text{net}} = \vec{0} \qquad \text{and} \qquad \tau_{\text{net}} = 0$$

- Write equations for $\Sigma F_x = 0$, $\Sigma F_y = 0$, and $\Sigma \tau = 0$.
- Solve the resulting equations.

ASSESS Check that your result is reasonable and answers the question.

EXAMPLE 8.2 **Forces on a board resting on sawhorses**

A board weighing 100 N sits across two sawhorses, as shown in FIGURE 8.6. What are the magnitudes of the normal forces of the sawhorses acting on the board?

FIGURE 8.6 A board sitting on two sawhorses.

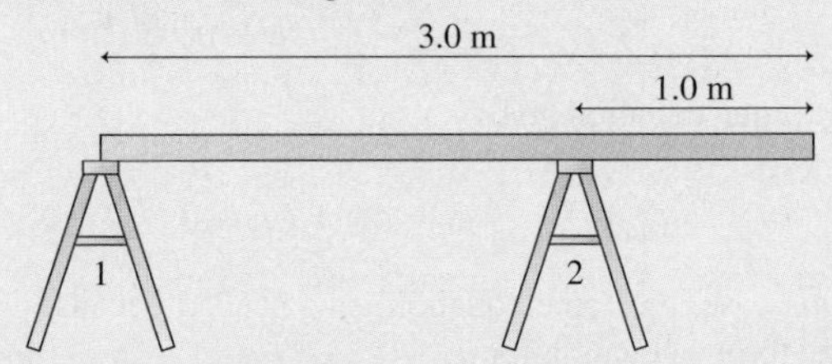

STRATEGIZE We'll choose the object of interest to be the board. There are two supporting forces acting on it at separated points, plus the weight of the board itself. As discussed previously, a good choice for the pivot is a point at which an unknown force acts because that force contributes nothing to the torque. We know the weight of the board, but we don't know the forces from either of the sawhorses, so a good choice for the pivot is at one of the sawhorses. We'll choose the left end of the board, where sawhorse 1 supports the board.

FIGURE 8.7 Visual overview of a board on two sawhorses.

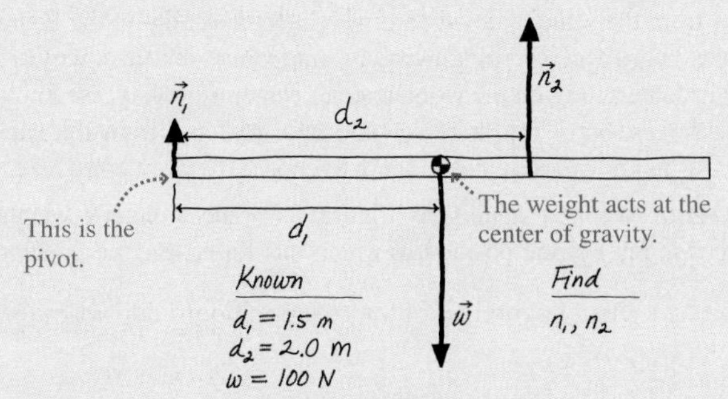

The board and the forces acting on it are shown in FIGURE 8.7. We will simplify things by having the forces from the sawhorses act at single points, the center point of each sawhorse. The center of gravity of the board is closer to sawhorse 2, so we will expect the force from this sawhorse to be greater.

PREPARE We add detail to the visual overview in Figure 8.7. $\vec{n}_1$ and $\vec{n}_2$ are the normal forces of the sawhorses supporting the

Continued

board, and $\vec{w}$ is the weight of the board, which acts at the center of gravity. The moment arm for $\vec{w}$ is $d_1 = 1.5$ m, half the board's length. Because $\vec{w}$ tends to rotate the board clockwise, its torque is negative. The moment arm for $\vec{n}_2$ is the distance of the second sawhorse from the pivot point, which is $d_2 = 2.0$ m. This force tends to rotate the board counterclockwise, so it exerts a positive torque.

SOLVE The board is in static equilibrium, so the net force $\vec{F}_{net}$ and the net torque τ_{net} must both be zero. The forces have only y-components, so the force equation is

$$\sum F_y = n_1 - w + n_2 = 0$$

As we've seen, the gravitational torque is negative and the torque from the upward force from the sawhorse is positive, so the torque equation is

$$\tau_{net} = -wd_1 + n_2 d_2 = 0$$

We now have two equations with the two unknowns n_1 and n_2. We can solve for n_2 in the torque equation and then substitute that result into the force equation. From the torque equation,

$$n_2 = \frac{d_1 w}{d_2} = \frac{(1.5 \text{ m})(100 \text{ N})}{2.0 \text{ m}} = 75 \text{ N}$$

The force equation is then $n_1 - 100 \text{ N} + 75 \text{ N} = 0$, which we can solve for n_1:

$$n_1 = w - n_2 = 100 \text{ N} - 75 \text{ N} = 25 \text{ N}$$

ASSESS $n_2 > n_1$, so sawhorse 2 supports more weight, as we expected.

A video to support a section's topic is embedded in the eText.

Video Walking the Plank

The center of gravity of a uniform board is easy to see—it's right at the midpoint—but how about the center of gravity of your body? In Chapter 7, we saw a model for the human body that gave average positions of the center of gravity of different body segments for an 80 kg man, but these were average positions. Individuals vary quite a bit around these averages, as does the exact location of the overall center of gravity.

The location of your center of gravity depends on how you are posed, but consider the simple case of standing up straight with your arms at your side. Your center of gravity is approximately at the midpoint of your body, but to determine the position of your center of gravity with any accuracy, you need to measure it. The next example illustrates the use of a reaction board and a scale to make this determination. This is a standard measurement in biomechanics.

EXAMPLE 8.3 Finding the center of gravity of the human body BIO

A woman weighing 600 N lies on a 2.5-m-long, 60 N reaction board with her feet over a fixed support. The scale on the right reads 250 N. What is the distance d from the woman's feet to her center of gravity?

STRATEGIZE The problem is about the woman and the position of her center of gravity, but it makes sense to take the woman and the board, considered together, as the object of interest. As we show in the visual overview in FIGURE 8.8, there is an upward force from the scale, a downward weight force acting on the board (at the board's center of gravity, its midpoint), and a downward weight force acting on the woman at her center of gravity. We know these forces, but a fourth force—the upward force from the support—is unknown, so it makes sense to choose the pivot point here.

PREPARE We add details to Figure 8.8. The woman's weight is known, but not the position at which this force acts, the position of her center of gravity; this is what we are looking for. The known weight force of the board acts at its midpoint, and the scale exerts a known upward normal force at the end of the board. The two weight forces exert negative torques, and the normal force of the scale exerts a positive torque. There is no torque due to the force from the support because we have taken the pivot point here.

FIGURE 8.8 Visual overview of the reaction board and woman.

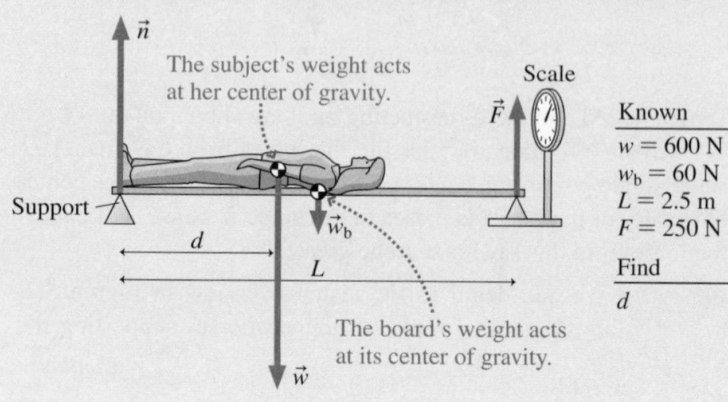

SOLVE Because the board and woman are in static equilibrium, the net force and net torque on them must be zero. The force equation reads

$$\sum F_y = n - w_b - w + F = 0$$

and the torque equation gives

$$\sum \tau = -\frac{L}{2} w_b - dw + LF = 0$$

In this case, the force equation isn't needed because we can solve the torque equation for d:

$$d = \frac{LF - \frac{1}{2}Lw_b}{w} = \frac{(2.5 \text{ m})(250 \text{ N}) - \frac{1}{2}(2.5 \text{ m})(60 \text{ N})}{600 \text{ N}}$$
$$= 0.92 \text{ m}$$

ASSESS If the woman is 5′ 6″ (1.68 m) tall, her center of gravity is $(0.92 \text{ m})/(1.68 \text{ m}) = 55\%$ of her height, or a little more than halfway up her body. This seems reasonable and this is, in fact, a typical value for women. The center of gravity for men tends to be a bit higher in the body.

EXAMPLE 8.4 Will the ladder slip?

A 3.0-m-long ladder leans against a wall at an angle of 60° with respect to the floor. What is the minimum value of μ_s, the coefficient of static friction with the ground, that will prevent the ladder from slipping? Assume that friction between the ladder and the wall is negligible.

STRATEGIZE The object of interest in this problem is straightforward to determine—it's the ladder. You know that the ladder would slide if there was no friction force; this force must be large enough to keep the ladder in static equilibrium. There is also a natural choice of pivot point in this problem. As FIGURE 8.9 shows, at the point where the ladder touches the ground, there are two forces—a normal force and a friction force. Placing the pivot point here means that there will be no torque for these forces, so the torque condition for equilibrium will be easily solved. Once we've completed the torque condition, we'll turn to the force conditions—one for the x-axis, one for the y-axis. At this point, friction will reenter the picture, as it must; the problem asks about the coefficient of friction. We are not given the weight of the ladder, so at some point in our process, the weight should cancel; this will be a good check on our work.

FIGURE 8.9 Visual overview of a ladder in static equilibrium.

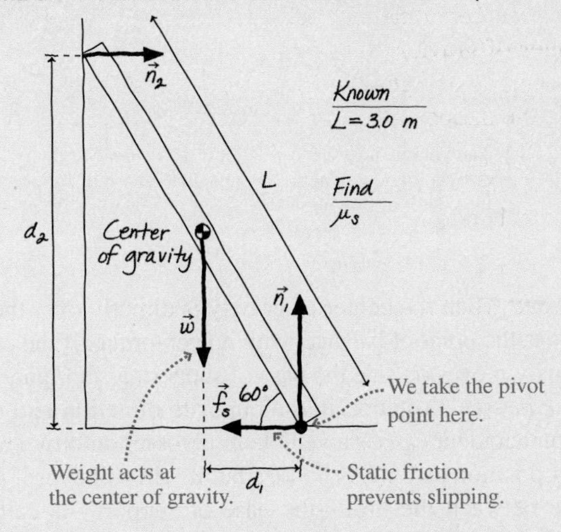

PREPARE Let's continue developing the details in the visual overview in Figure 8.9. The four vectors show where the forces act. It's easy to visualize the moment arm for the wall's normal force, d_2, and the moment arm for the weight force, d_1, so we compute torques using moment arms. Given the angles and distances noted in the figure, the two moment arms are

$$d_1 = \frac{1}{2}L\cos 60° \quad d_2 = L\sin 60°$$

We don't know the values of any of the forces, but we can write the relationship between the friction force and the normal force of the ground: $f_s \le \mu_s n_1$.

SOLVE There is no net torque about the pivot point we've chosen, and only two forces that exert torques. The weight force exerts a positive torque, $\tau_w = d_1 w$. The normal force exerts a negative torque, $\tau_n = -d_2 n_2$. Our torque condition is thus

$$\sum \tau = d_1 w - d_2 n_2 = 0$$

We can substitute the expressions we found for the distances and solve for n_2:

$$\frac{1}{2}(L\cos 60°)w = (L\sin 60°)n_2$$

$$n_2 = \frac{\frac{1}{2}(L\cos 60°)w}{L\sin 60°} = \frac{w}{2\tan 60°}$$

There is no net torque, but there is also no net force. This gives us two additional conditions:

$$\sum F_x = n_2 - f_s = 0$$
$$\sum F_y = n_1 - w = 0$$

The second force condition simplifies to

$$n_1 = w$$

Our condition for the friction force thus becomes

$$f_s \le \mu_s w$$

The first force condition simplifies to

$$f_s = n_2$$

We then substitute the expression for n_2 that we found earlier, giving

$$f_s = \frac{w}{2\tan 60°}$$

Finally, we combine our expressions for the friction force:

$$f_s = \frac{w}{2\tan 60°} \le \mu_s w$$

The weight force cancels, and we simplify to

$$\mu_s \ge \frac{1}{2\tan 60°} = 0.29$$

The minimum value of the static coefficient of friction is thus 0.29.

ASSESS The weight force did cancel out as we worked, which gives us confidence in our process. Our final result makes sense as well. You have certainly seen ladders leaning against walls at angles like this with no slipping, so we expect a final value for the coefficient of friction that's relatively modest and that most surfaces will satisfy. 0.29 is less than values we've seen for concrete and other surfaces, so our result makes sense.

STOP TO THINK 8.2 A beam with a pivot on its left end is suspended from a rope. In which direction is the force of the pivot on the beam?

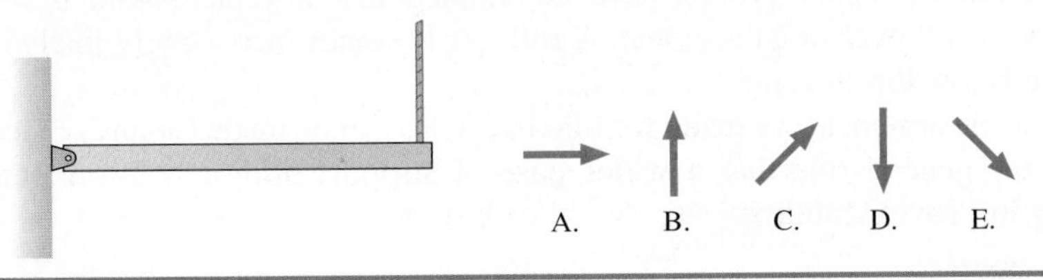

8.2 Stability and Balance

Video Center of Gravity and Stability

If you tilt a box up on one edge by a small amount and let go, it falls back down. If you tilt it too much, it falls over. And if you tilt it "just right," you can get the box to balance on its edge. What determines these three possible outcomes?

FIGURE 8.10 illustrates the idea with a car, but the results are general and apply in many situations. An extended object has a *base of support* on which it rests when in static equilibrium. If you tilt the object, one edge of the base of support becomes a pivot point. As long as the object's center of gravity remains over the base of support, torque due to gravity will rotate the object back toward its stable equilibrium position; we say that the object is **stable**. This is the situation in Figure 8.10b.

FIGURE 8.10 A car—or any object—will fall over when tilted too far.

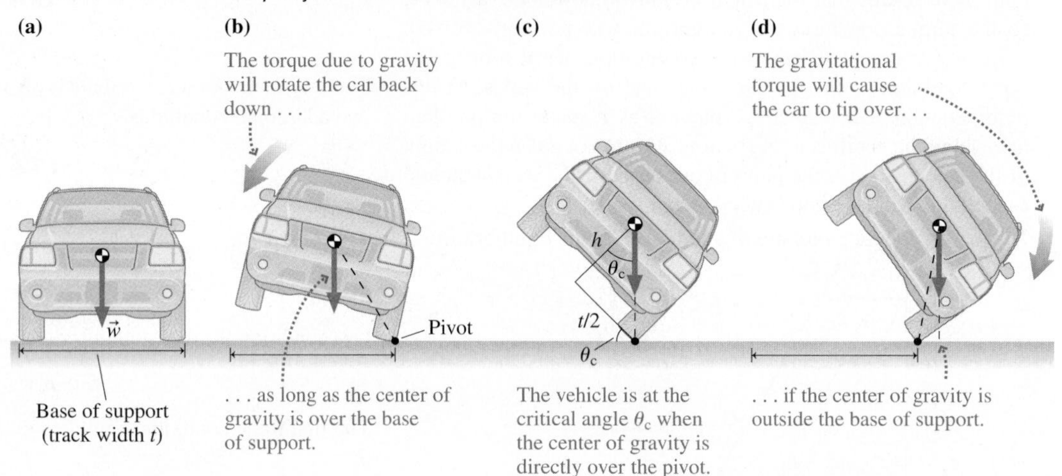
(a)
(b)
(c)
(d)
The torque due to gravity will rotate the car back down . . .
The gravitational torque will cause the car to tip over. . .
Pivot
h
θ_c
t/2
$\vec{w}$
Base of support (track width t)
. . . as long as the center of gravity is over the base of support.
The vehicle is at the critical angle θ_c when the center of gravity is directly over the pivot.
. . . if the center of gravity is outside the base of support.

Video Figure 8.10

A *critical angle* θ_c is reached when the center of gravity is directly over the pivot point, as in Figure 8.10c. This is the point of balance, with no net torque. If the car continues to tip, the center of gravity moves outside the base of support, as in Figure 8.10d. Now, the gravitational torque causes a rotation in the opposite direction and the car rolls over; it is **unstable**. If an accident causes a vehicle to pivot up onto two wheels, it will roll back to an upright position as long as $\theta < \theta_c$, but it will roll over if $\theta > \theta_c$.

For vehicles, the distance between the tires—the base of support—is called the track width t. The height of the center of gravity is h. You can see from Figure 8.10c that the ratio of the two, the height-to-width ratio, determines when the critical angle is reached. An analysis of the triangle in the third image in Figure 8.10 shows how the critical angle depends on the track width t and the height of the center of gravity h:

$$\theta_c = \tan^{-1}\left(\frac{\frac{1}{2}t}{h}\right)$$

Decreasing the height of the center of gravity or increasing the width of the base of support will increase the critical angle, leading to a more stable situation.

FIGURE 8.11 compares a passenger car and a sport utility vehicle (SUV). For the passenger car, with $h/t = 0.33$, the critical angle is $\theta_c = 57°$. But for the SUV, with $h/t = 0.47$, the critical angle is $\theta_c = 47°$. Loading an SUV with cargo further raises the center of gravity, especially if the roof rack is used, reducing θ_c even more. Various automobile safety groups have determined that a vehicle with $\theta_c > 50°$ is unlikely to roll over in an accident. A rollover becomes increasingly likely as θ_c is reduced below this threshold.

The same argument we made for tilted vehicles can be made for any object, leading to the general rule that **a wider base of support and/or a lower center of gravity improves stability.**

FIGURE 8.11 The passenger car has a lower center of gravity relative to its track width than does an SUV.

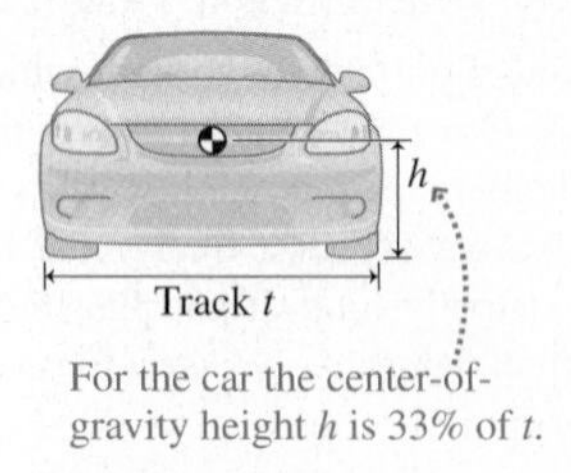
h
Track t
For the car the center-of-gravity height h is 33% of t.

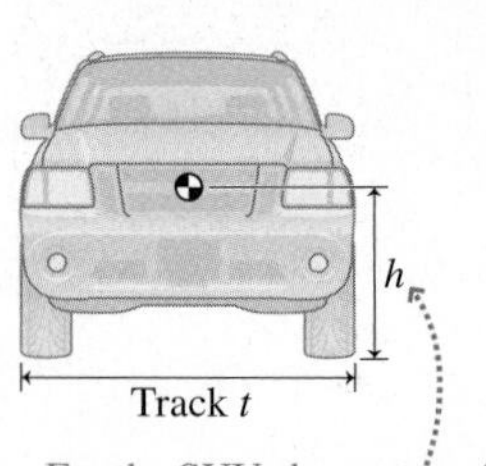
h
Track t
For the SUV, the center-of-gravity height h is 47% of t.

CONCEPTUAL EXAMPLE 8.5 Understanding ladder safety

A man is standing on a ladder to paint a wall. He (unwisely) leans out from the side of the ladder, as in FIGURE 8.12. If he leans too far, the ladder will tip to the side. Explain why and when the ladder will tip.

REASON For stability, the combined center of gravity of the man and the ladder must be over the ladder's base. When the man leans to the side, his center of gravity shifts. This changes the position of the combined center of gravity quite a bit; the man has more mass than the ladder, so the combined center of gravity is closer to his center of gravity than that of the ladder. FIGURE 8.13 shows that the situation is at a literal tipping point—if the man moves any farther to the left, the center of gravity will be to the left of the ladder's supports, and the ladder will tip.

ASSESS This result makes sense to anyone who has used a ladder for such work; there is a very narrow lean angle for safe use. The design of the ladder, with a wider base, enhances safety, but only up to a point.

FIGURE 8.12 Unsafe lean on a ladder.

FIGURE 8.13 Analyzing the unsafe lean.

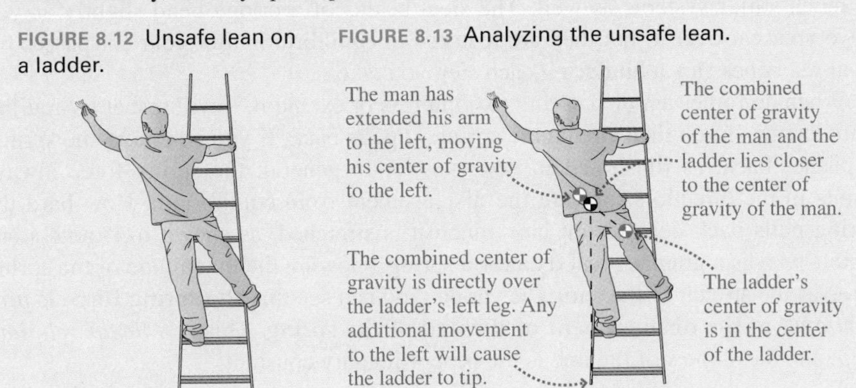

The man has extended his arm to the left, moving his center of gravity to the left.
The combined center of gravity of the man and the ladder lies closer to the center of gravity of the man.
The combined center of gravity is directly over the ladder's left leg. Any additional movement to the left will cause the ladder to tip.
The ladder's center of gravity is in the center of the ladder.

Balancing a soda can Try to balance a soda can—full or empty—on the narrow bevel at the bottom. It can't be done because, either full or empty, the center of gravity is near the center of the can. If the can is tilted enough to sit on the bevel, the center of gravity lies far outside this small base of support. But if you put about 2 ounces (60 ml) of water in an empty can, the center of gravity will be right over the bevel and the can will balance.

Stability and Balance of the Human Body BIO

The human body is remarkable for its ability to constantly adjust its stance to remain stable on just two points of support. In walking, running, or even the simple act of rising from a chair, the position of the body's center of gravity is constantly changing. To maintain stability, we unconsciously adjust the positions of our arms and legs to keep our center of gravity over our base of support.

FIGURE 8.14a shows the body in its normal standing position. Notice that the center of gravity is well centered over the base of support (the feet), ensuring stability. If the person were now to stand on tiptoes *without* otherwise adjusting the body position, her center of gravity would fall behind the base of support, which is now the balls of the feet, and she would fall backward. To prevent this, as shown in FIGURE 8.14b, the body naturally leans forward, regaining stability by moving the center of gravity over the balls of the feet. Try this: Stand facing a wall with your toes touching the base of the wall. Now try standing on your toes. Your body can't move forward to keep your center of gravity over your toes, so you can't do it!

The chapter opened with a photo of a dancer who is performing a very delicate balancing act, adjusting the positions of her torso, head, arms, and legs so that her center of gravity is above a very small point of support. Now that you understand the basics of balance, you can see what a remarkable feat this is!

FIGURE 8.14 Standing on tiptoes.

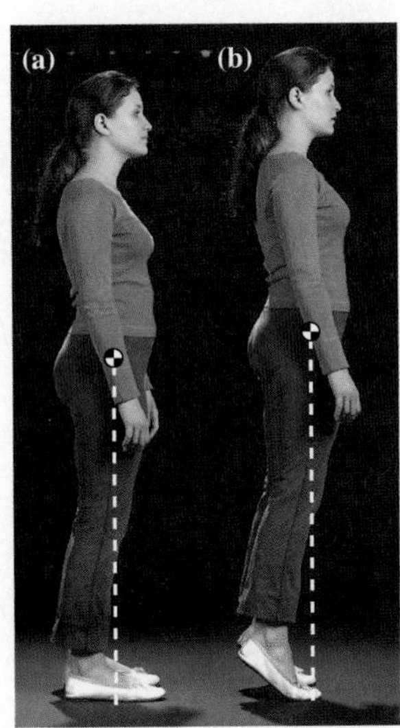

STOP TO THINK 8.3 Rank in order, from least stable to most stable, the three objects shown in the figure. The positions of their centers of gravity are marked. (For the centers of gravity to be positioned like this, the objects must have a nonuniform composition.)

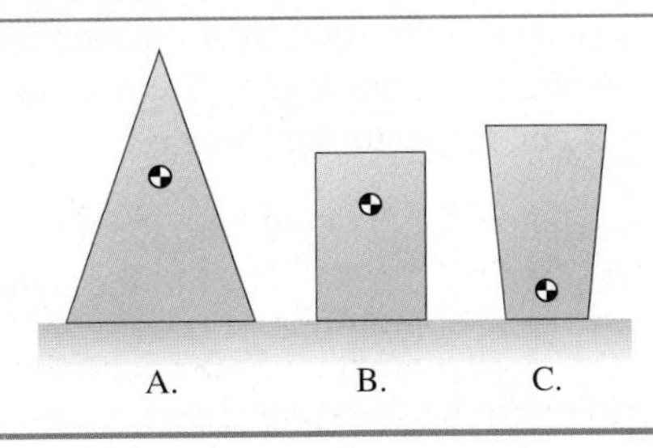

8.3 Springs and Hooke's Law

Elasticity in action A golf ball compresses quite a bit when struck. The restoring force that pushes the ball back into its original shape helps launch the ball off the face of the club, making for a longer drive.

We have assumed that objects in equilibrium maintain their shape as forces and torques are applied to them. In reality this is an oversimplification. Every solid object stretches, compresses, or deforms when a force acts on it. This change is easy to see when you press on a green twig on a tree, but even the largest branch on the tree will bend slightly under your weight.

If you stretch a rubber band, there is a force that tries to pull the rubber band back to its equilibrium, or unstretched, length. A force that restores a system to an equilibrium position is called a **restoring force.** Systems that exhibit such restoring forces are called **elastic.** The most basic examples of **elasticity** are things like springs and rubber bands. We introduced the spring force in ◀◀ SECTION 4.2. As we saw, if you stretch a spring, a tension-like force pulls back. Similarly, a compressed spring tries to re-expand to its equilibrium length. Elasticity and restoring forces are properties of much stiffer systems as well. The steel beams of a bridge bend slightly as you drive your car over it, but they are restored to equilibrium after your car passes by. Your leg bones flex a bit during each step you take.

When no forces act on a spring to compress or extend it, it will rest at its equilibrium length. If you then stretch the spring, it pulls back. If you compress the spring, it pushes back, as illustrated in FIGURE 8.15a. In general, the spring force always points in the direction opposite the displacement from equilibrium. How hard the spring pulls back depends on how much it is stretched, as shown in FIGURE 8.15b. FIGURE 8.15c is a graph of real data for a spring, showing the magnitude of the spring force as the stretch of the spring is varied. You can see that **the spring force is *proportional* to the displacement of the end of the spring.** This is a *linear relationship,* and the slope k of the line is the proportionality constant:

$$F_{sp} = k\,\Delta x \tag{8.2}$$

Compressing or stretching the spring twice as far results in a restoring force that is twice as large.

KEY CONCEPT FIGURE 8.15 The restoring force for a spring.

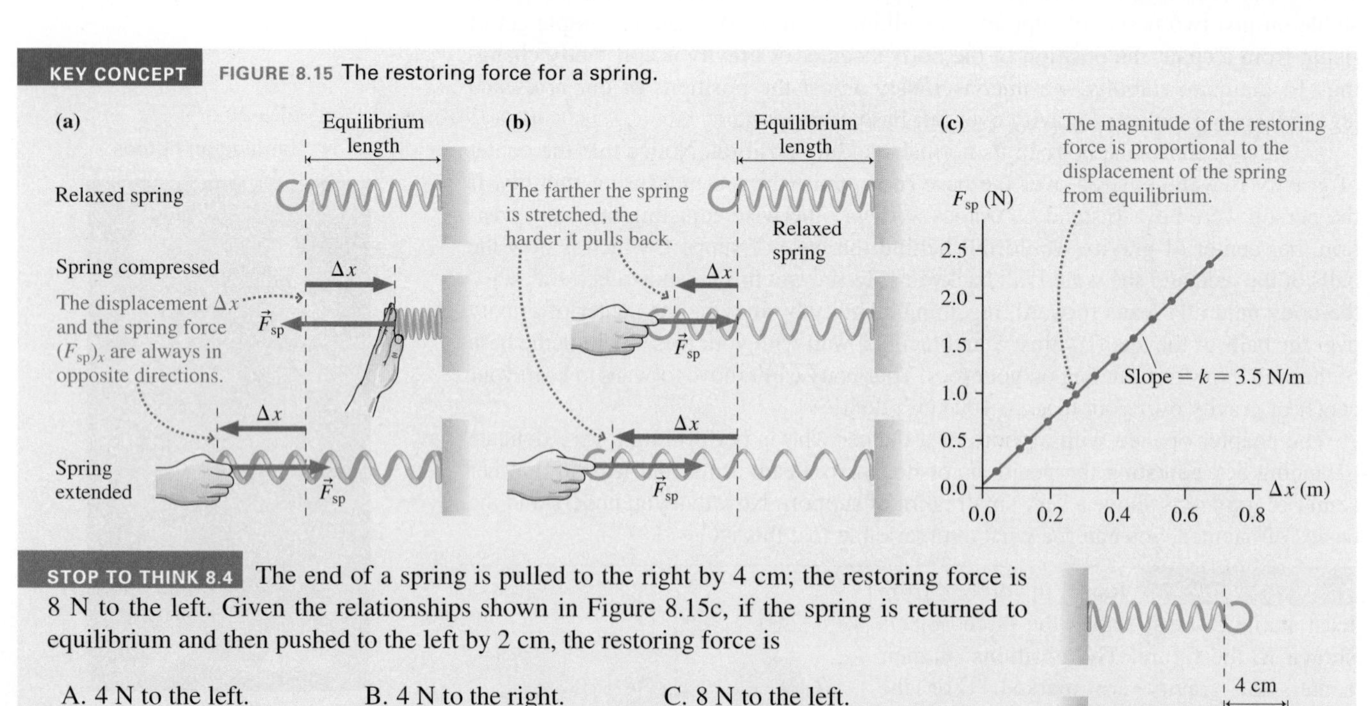

STOP TO THINK 8.4 The end of a spring is pulled to the right by 4 cm; the restoring force is 8 N to the left. Given the relationships shown in Figure 8.15c, if the spring is returned to equilibrium and then pushed to the left by 2 cm, the restoring force is

A. 4 N to the left.
B. 4 N to the right.
C. 8 N to the left.
D. 8 N to the right.
E. 16 N to the left.
F. 16 N to the right.

4 cm
8 N

The value of the constant k in Equation 8.2 depends on the spring. We call k the **spring constant**; it has units of N/m. The spring constant k is a property that characterizes a spring, just as the mass m characterizes a particle. If k is large, it takes a large pull to cause a significant stretch, and we call the spring a "stiff" spring. If k is small, we can stretch the spring with very little force, and we call it a "soft" spring. Every spring has its own unique value of k. The spring constant for the data in Figure 8.15c can be determined from the slope of the straight line to be $k = 3.5$ N/m.

As Figure 8.15a shows, if the spring is compressed, Δx is positive and, because $\vec{F}_{sp}$ points to the left, its component $(F_{sp})_x$ is negative. If the spring is stretched, however, Δx is negative and, because $\vec{F}_{sp}$ points to the right, its component $(F_{sp})_x$ is positive. If we rewrite Equation 8.2 in terms of the *component* of the spring force, we get the most general form of the relationship between the restoring force and the displacement of the end of a spring, which is known as **Hooke's law:**

$$(F_{sp})_x = -k\Delta x \quad (8.3)$$

x-component of the restoring force of the spring (N)

Displacement of the end of the spring (m)

Spring constant (N/m)

The negative sign says the restoring force and the displacement are in *opposite* directions.

$(F_{sp})_x$ — Δx — PROPORTIONAL p. 39

For motion in the vertical (y) direction, Hooke's law is $(F_{sp})_y = -k\,\Delta y$.

Hooke's law is not a true "law of nature" in the sense that Newton's laws are. It is actually just a model of a restoring force. It works extremely well for some springs, such as the one in Figure 8.15c, but less well for others. Hooke's law will fail for any spring that is compressed or stretched too far, as we'll see in the next section.

NOTE ▶ Just as we used massless strings, we will adopt the idealization of a *massless spring*. Though not a perfect description, it is a good approximation if the mass attached to a spring is much greater than the mass of the spring itself. ◀

EXAMPLE 8.6 A spring in your step

Several companies manufacture running shoes that have thick heel pads containing steel springs to launch a runner back into the air after a heel strike. When a 70 kg runner rocks back on his heels so that all of his weight is supported by the heel springs, the springs each compress by 1.2 mm. When he is running hard and hits the ground on one heel, the force compressing the spring is 5.0 times his weight.

a. What is the spring constant k of a heel spring?
b. By how much do the springs compress during a heel strike?

STRATEGIZE When the springs in the heels are compressed, they produce a restoring force that pushes upward on the runner's foot. When the runner rocks back on his heels, we can assume that half of the runner's weight is on each heel. We know the force that is compressing the spring and how much it compresses, so we will use those values to find the spring constant. With the spring constant in hand, we will find the compression in part b.

PREPARE The visual overview in FIGURE 8.16 shows the situation at one heel. The runner is in static equilibrium, so

$$\sum F_y = F_{sp} - \frac{1}{2}w = 0$$

$$F_{sp} = \frac{1}{2}mg$$

FIGURE 8.16 Visual overview of the heel spring.

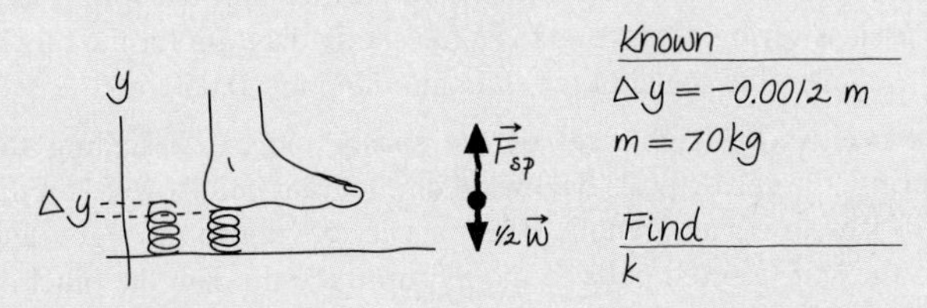

The spring's restoring force is determined by the compression, according to Equation 8.2:

$$F_{sp} = k\Delta y = k(0.0012\text{ m})$$

SOLVE

a. We know the restoring force of the spring, and we know how this is related to the compression. Combining these, we get

$$\frac{1}{2}(70\text{ kg})g = k(0.0012\text{ m})$$

$$k = 2.9 \times 10^5\text{ N/m}$$

Continued

b. In the second case, the restoring force is 5.0 times the weight. We use this force to find the new compression of the spring:

$$(5.0)mg = (29 \times 10^5 \text{ N/m})\Delta y$$
$$\Delta y = 0.012 \text{ m} = 1.2 \text{ cm}$$

During the greater force of a heel strike, the spring compresses by 1.2 cm.

ASSESS The spring is quite stiff, approximately 300,000 N/m. This is reasonable; the spring compresses only a little under the large force of the runner's weight. The force of the heel strike is 5 times the runner's weight, 10 times the original force, half the runner's weight. This gives a compression that is 10 times the original compression, 1.2 cm instead of 1.2 mm, as we'd expect.

EXAMPLE 8.7 When does the block slip?

FIGURE 8.17 shows a spring attached to a 2.0 kg block. The other end of the spring is pulled by a motorized toy train that moves forward at 5.0 cm/s. The spring constant is 50 N/m, and the coefficient of static friction between the block and the surface is 0.60. The spring is at its equilibrium length at $t = 0$ s when the train starts to move. When does the block slip?

FIGURE 8.17 A toy train stretches the spring until the block slips.

STRATEGIZE Recall that the tension in a massless string pulls equally at both ends of the string. The same is true for the spring force: It pulls (or pushes) equally at both ends. Imagine holding a rubber band with your left hand and stretching it with your right hand. Your left hand feels the pulling force, even though it is the right end of the rubber band that moves. This is the key to solving the problem: As the train moves, the spring stretches and the spring force increases. This spring force pulls to the right on the block. The static friction force can increase to keep the block in place, but there is a limit to the value of this force. At some point, the spring force will exceed the maximum value of the friction force, and then the block will slide.

PREPARE We model the block as a particle and the spring as a massless spring. FIGURE 8.18 is a free-body diagram for the block. We convert the speed of the train into m/s: $v = 0.050$ m/s.

SOLVE As the right end of the spring moves, stretching the spring, the spring pulls backward on the train *and* forward on the block with equal strength. The train is moving to the right, and so the spring force pulls to the left on the train. But the block is at the other end of the spring; the spring force pulls to the right

FIGURE 8.18 Free-body diagram for the block.

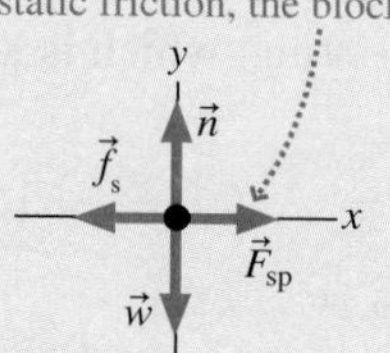

When the spring force exceeds the maximum force of static friction, the block will slip.
y
x

on the block, as shown in Figure 8.18. As the spring stretches, the static friction force on the block increases in magnitude to keep the block at rest. The block is in static equilibrium, so

$$\sum F_x = (F_{sp})_x + (f_s)_x = F_{sp} - f_s = 0$$

where F_{sp} is the magnitude of the spring force. This magnitude is $F_{sp} = k\,\Delta x$, where $\Delta x = vt$ is the distance the train has moved. Thus

$$f_s = F_{sp} = k\,\Delta x$$

The block slips when the static friction force reaches its maximum value $f_{s\,max} = \mu_s n = \mu_s mg$. This occurs when the train has moved a distance

$$\Delta x = \frac{f_{s\,max}}{k} = \frac{\mu_s mg}{k} = \frac{(0.60)(2.0 \text{ kg})(9.8 \text{ m/s}^2)}{50 \text{ N/m}} = 0.235 \text{ m}$$

The time at which the block slips is

$$t = \frac{\Delta x}{v} = \frac{0.235 \text{ m}}{0.050 \text{ m/s}} = 4.7 \text{ s}$$

ASSESS The result of about 5 s seems reasonable for a slowly moving toy train to stretch the spring enough for the block to slip.

Example 8.7 is a bit artificial, but it is also instructive. The train moves, stretching the spring, until the restoring force is larger than the friction force. At this point, the block slides forward. Once the block is sliding, the friction is kinetic, with a lower coefficient of friction, so the sliding continues for some time. When the block finally comes to rest, the friction is back to static, and the process begins again. This *stick-slip* process is important in applications ranging from the motion of arthritic joints to the gradual buildup of stress in the earth's crust before an earthquake.

STOP TO THINK 8.5 A 1.0 kg weight is suspended from a spring, stretching it by 5.0 cm. How much does the spring stretch if the 1.0 kg weight is replaced by a 3.0 kg weight?

A. 5.0 cm B. 10.0 cm C. 15.0 cm D. 20.0 cm

8.4 Stretching and Compressing Materials

In Chapter 4 we noted that we can model most solid materials as being made of particle-like atoms connected by spring-like bonds. We can model a steel rod this way, as illustrated in FIGURE 8.19a. The spring-like bonds between the atoms in steel are quite stiff, but they can be stretched or compressed, meaning that even a steel rod is elastic. If you pull on the end of a steel rod, as in Figure 8.19a, you will slightly stretch the bonds between the particles that make it up, and the rod itself will stretch. The stretched bonds pull back on your hand with a restoring force that causes the rod to return to its original length when released. In this sense, the entire rod acts like a very stiff spring. As is the case for a spring, a restoring force is also produced by compressing the rod.

In FIGURE 8.19b, real data for a 1.0-m-long, 1.0-cm-diameter steel rod show that, just as for a spring, the restoring force is proportional to the change in length. However, the *scale* of the stretch of the rod and the restoring force is much different from that for a spring. It would take a force of 16,000 N to stretch the rod by only 1 mm, corresponding to a spring constant of 1.6×10^7 N/m! Steel is elastic, but under normal forces, it experiences only very small changes in dimension. Materials of this sort are called **rigid**.

The behavior of other materials, such as the rubber in a rubber band, can be different. A rubber band can be stretched quite far with a very small force, and then snaps back to its original shape when released. Materials that show large deformations with small forces are called **pliant.**

A rod's spring constant depends on several factors, as shown in FIGURE 8.20. First, we expect that a thick rod, with a large cross-section area A, will be more difficult to stretch than a thinner rod. Second, a rod with a long length L will be easier to stretch by a given amount than a short rod (think of trying to stretch a rope by 1 cm—this would be easy to do for a 10-m-long rope, but it would be pretty hard for a piece of rope only 10 cm long). Finally, the stiffness of the rod will depend on the material that it's made of. Experiments bear out these observations, and it is found that the spring constant of the rod can be written as

$$k = \frac{YA}{L} \tag{8.4}$$

where the constant Y is called **Young's modulus.** Young's modulus is a property of the *material* from which the rod is made—it does not depend on shape or size.

From Equation 8.2, the magnitude of the restoring force for a spring is related to the change in its length as $F_{sp} = k\,\Delta x$. Writing the change in the length of a rod as ΔL, as shown in Figure 8.20, we can use Equation 8.4 to write the restoring force F of a rod as

$$F = \frac{YA}{L}\,\Delta L \tag{8.5}$$

Equation 8.5 applies both to elongation (stretching) and to compression.

It's useful to rearrange Equation 8.5 in terms of two new ratios, the *stress* and the *strain*:

$$\frac{F}{A} = Y\left(\frac{\Delta L}{L}\right) \tag{8.6}$$

The ratio of force to cross-section area is called **stress.** The ratio of the change in length to the original length is called **strain.**

The strain is a measure of how much the rod stretches. The dimensionless ratio $\Delta L/L$ is the fractional change in the rod's length. If the rod's length changes by 1%, the strain is 0.01. The unit of stress is N/m^2. If the stress is due to stretching, we call it a **tensile stress.** Because strain is dimensionless, Young's modulus Y has the same units as stress. TABLE 8.1 lists values of Young's modulus for several rigid materials. Large values of Y characterize materials that are stiff. "Softer" materials have smaller values of Y.

If the rod is in static equilibrium, then the magnitude of the restoring force is equal to the force that is stretching the rod, and we can reframe this discussion with

FIGURE 8.19 Stretching a steel rod.

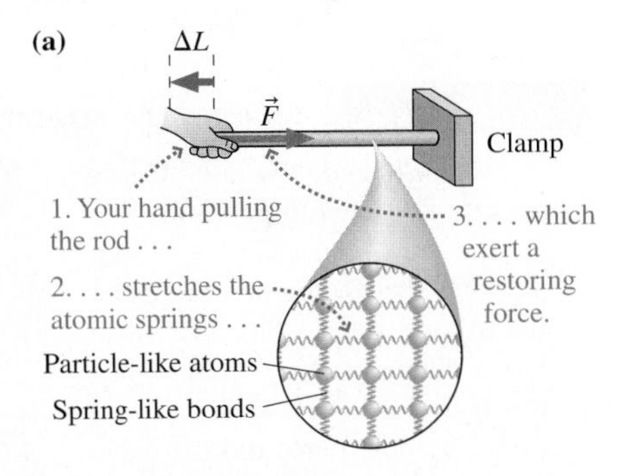

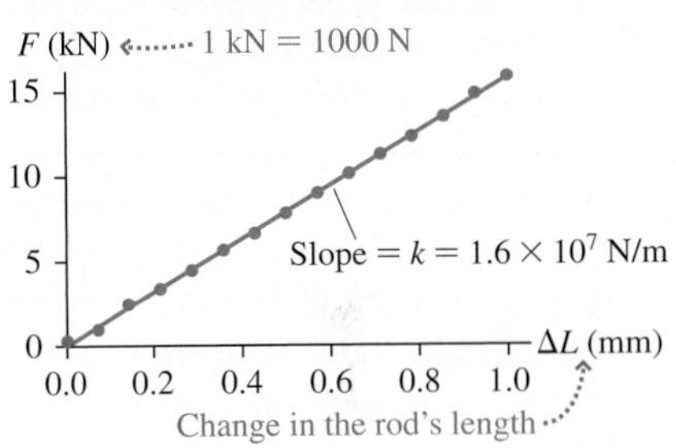

FIGURE 8.20 A rod stretched by length ΔL.

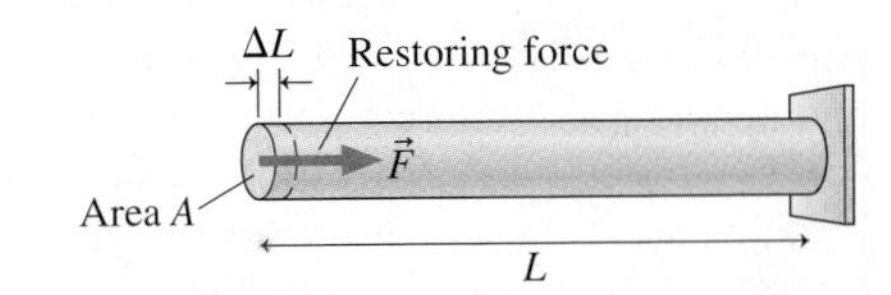

TABLE 8.1 Young's modulus for rigid materials

Material	Young's modulus (10^{10} N/m^2)
Cast iron	20
Steel	20
Silicon	13
Copper	11
Aluminum	7
Glass	7
Concrete	3
Wood (Douglas Fir)	1

the strain—the deformation of the rod—resulting from an applied stress. The mathematics is the same, but this way of thinking is more intuitive in some cases. If we apply a force to a material, the deformation depends on the force we apply as well as the area over which the force is applied.

EXAMPLE 8.8 Finding the stretch of a cable

A *Foucault pendulum* in a physics department (used to prove that the earth rotates) consists of a 120 kg steel ball that swings at the end of a 6.0-m-long steel cable. The cable has a diameter of 2.5 mm. When the ball was first hung from the cable, by how much did the cable stretch?

STRATEGIZE When the ball is suspended from the cable, it stretches—there is a strain. This results in a restoring force—there is a stress. The stress and the strain are related by Equation 8.6.

PREPARE The amount by which the cable stretches depends on the elasticity of the steel cable. Young's modulus for steel is given in Table 8.1 as $Y = 20 \times 10^{10}$ N/m^2. Equation 8.6 uses the cross-section area of the cable, which is

$$A = \pi r^2 = \pi(0.00125 \text{ m})^2 = 4.91 \times 10^{-6} \text{ m}^2$$

SOLVE Equation 8.6 relates the stretch of the cable ΔL to the restoring force F and to the properties of the cable. Rearranging terms, we find that the cable stretches by

$$\Delta L = \frac{LF}{AY}$$

The restoring force of the cable is equal to the ball's weight:

$$F = w = mg = (120 \text{ kg})(9.8 \text{ m/s}^2) = 1180 \text{ N}$$

The change in length is thus

$$\Delta L = \frac{(6.0 \text{ m})(1180 \text{ N})}{(4.91 \times 10^{-6} \text{ m}^2)(20 \times 10^{10} \text{ N/m}^2)} = 0.0072 \text{ m} = 7.2 \text{ mm}$$

ASSESS If you've ever strung a guitar with steel strings, you know that the strings stretch several millimeters with the force you can apply by turning the tuning pegs. So a stretch of 7 mm under a 120 kg load seems reasonable.

Beyond the Elastic Limit

FIGURE 8.21 Stretch data for a steel rod.

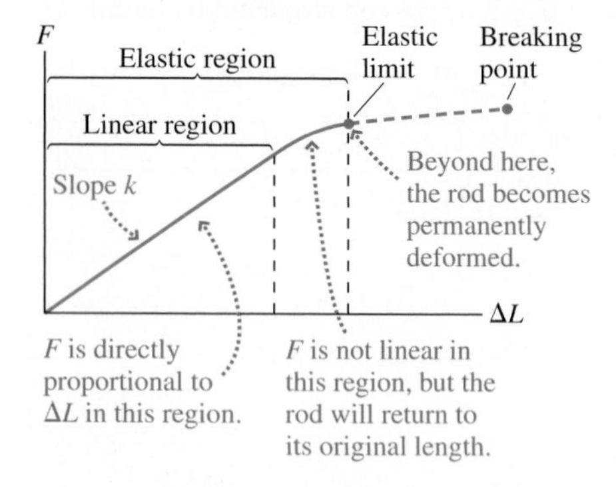

In the previous discussion, we found that if we stretch a rod by a small amount ΔL, it will pull back with a restoring force F, according to Equation 8.5. But if we continue to stretch the rod, this simple linear relationship between ΔL and F will eventually break down. FIGURE 8.21 is a graph of the rod's restoring force from the start of the stretch until the rod finally breaks.

As you can see, the graph has a *linear region*, the region where F and ΔL are proportional to each other, obeying Hooke's law: $F = k\,\Delta L$. **As long as the stretch stays within the linear region, a solid rod acts like a spring and obeys Hooke's law.**

How far can you stretch the rod before damaging it? As long as the stretch is less than the **elastic limit,** the rod will return to its initial length L when the force is removed. The elastic limit is the end of the **elastic region.** Stretching the rod beyond the elastic limit will permanently deform it, and the rod won't return to its original length. Finally, at a certain point the rod will reach a breaking point, where it will snap in two. The force that causes the rod to break depends on the area: A thicker rod can sustain a larger force. For a rod or cable of a particular material, we can determine an *ultimate stress*, also known as the **tensile strength,** the largest stress that the material can sustain before breaking:

$$\text{Tensile strength} = \frac{F_{max}}{A} \tag{8.7}$$

(Tensile strength: Largest stress that can be sustained (N/m^2); F_{max}: Largest force that can be sustained (N); A: Cross-section area (m^2))

TABLE 8.2 lists values of tensile strength for rigid materials. When we speak of the *strength* of a material, we are referring to its tensile strength.

Objects can withstand a certain stress before deforming permanently. It's not the *force* that matters, it's the *stress*—the force divided by the area. Will a heavy weight dent a wood floor? Not if the weight is spread out over a large enough area, to minimize the stress.

We can rewrite Equation 8.7 to solve for the maximum force that can be sustained:

$$F_{max} = (\text{tensile strength})A \tag{8.8}$$

If a cable can't provide enough force to support a certain weight, you can either get a cable made of stronger stuff (higher tensile strength) or get a thicker cable (larger area).

TABLE 8.2 Tensile strengths of rigid materials

Material	Tensile strength (N/m^2)
Polypropylene	20×10^6
Glass	60×10^6
Cast iron	150×10^6
Aluminum	400×10^6
Steel	1000×10^6

EXAMPLE 8.9 Finding the weight that a cable can support

After a late night of studying physics, several 80 kg students decide it would be fun to swing on the Foucault pendulum of Example 8.8. What's the maximum number of students that the pendulum cable could support?

STRATEGIZE The weight that can be supported depends on the area and the tensile strength of the cable, both of which we know or can compute.

PREPARE The tensile strength for steel is given in Table 8.2 as $1000 \times 10^6\ \text{N/m}^2$, or $1.0 \times 10^9\ \text{N/m}^2$. The area of the cable is

$$A = \pi r^2 = \pi(0.00125\ \text{m})^2 = 4.91 \times 10^{-6}\ \text{m}^2$$

SOLVE We can use Equation 8.8 to find the maximum force the cable can support:

$$F_{\text{max}} = (1.0 \times 10^9\ \text{N/m}^2)(4.91 \times 10^{-6}\ \text{m}^2) = 4.9 \times 10^3\ \text{N}$$

This force is the weight of the heaviest mass the cable can support: $w = m_{\text{max}}g$. The maximum mass that can be supported is

$$m_{\text{max}} = \frac{F_{\text{max}}}{g} = 500\ \text{kg}$$

The ball has a mass of 120 kg, leaving 380 kg for the students. Four students have a mass of 320 kg, which is less than this value. But five students, totaling 400 kg, would cause the cable to break.

ASSESS Steel has a very large tensile strength, so it's reasonable that this very narrow wire can still support 4900 N $\approx$ 1100 lb.

Elasticity of Biological Materials BIO

Suppose we take equal lengths of spider silk and steel wire, stretch each, and measure the restoring force of each until it breaks. The graph of stress versus strain might appear as in FIGURE 8.22.

FIGURE 8.22 Stress-versus-strain graphs for steel and spider silk.

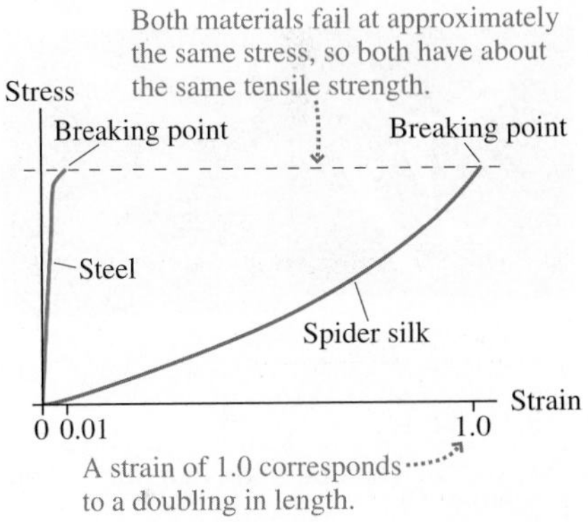

Both materials fail at approximately the same stress, so both have about the same tensile strength.
Stress
Breaking point
Breaking point
Steel
Spider silk
Strain
0 0.01
1.0
A strain of 1.0 corresponds to a doubling in length.

The spider silk is certainly less stiff: For a given stress, the silk will stretch about 100 times farther than steel. Interestingly, though, spider silk and steel eventually fail at approximately the same stress. In this sense, spider silk is "as strong as steel." Many pliant biological materials share this combination of low stiffness and large tensile strength. These materials can undergo significant deformations without failing. Tendons, the walls of arteries, and the web of a spider are all quite strong but nonetheless capable of significant stretch.

Most bones in your body are made of two different kinds of bony material: dense and rigid compact bone on the outside, and porous, flexible spongy bone on the inside. FIGURE 8.23 shows a cross section of a typical bone. Compact bone and spongy bone have very different values of Young's modulus. Young's modulus for compact bone approaches that of concrete, so it is very rigid, with little ability to stretch or compress. In contrast, spongy bone has a much lower Young's modulus. Consequently, the elastic properties of bones can be well modeled as those of a hollow cylinder.

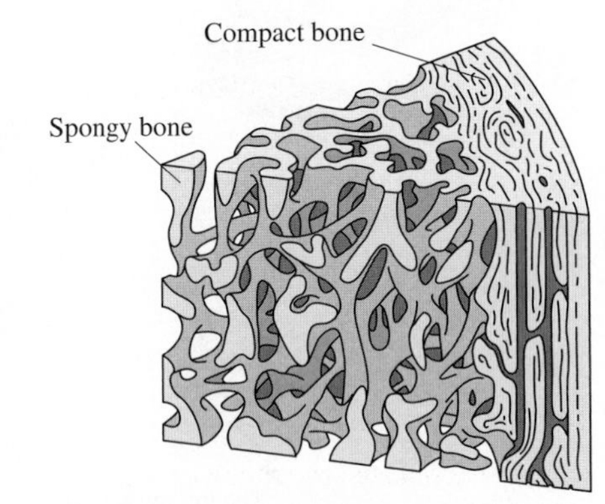

FIGURE 8.23 Cross section of a long bone.

The structure of bones in birds actually approximates a hollow cylinder quite well. FIGURE 8.24 shows that a typical bone is a thin-walled tube of compact bone with a tenuous structure of spongy bone inside. Most of a cylinder's rigidity comes from the material near its surface. A hollow cylinder retains most of the rigidity of a solid one, but it is much lighter. Bird bones carry this idea to its extreme.

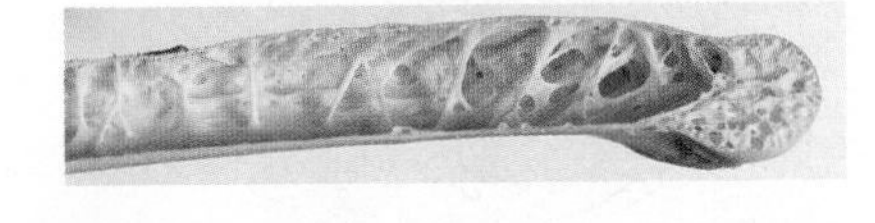
FIGURE 8.24 Section of a bone from a bird.

TABLE 8.3 gives values of Young's modulus for biological materials. Note the large difference between pliant and rigid materials. TABLE 8.4 shows the tensile strengths for biological materials. The values in Table 8.4 are for static forces—forces applied for a long time in a testing machine. Bone can withstand significantly greater stresses if the forces are applied for only a very short period of time.

TABLE 8.3 Young's modulus for biological materials

Material	Young's modulus ($10^{10}\ \text{N/m}^2$)
Tooth enamel	6
Compact bone	1.6
Spongy bone	0.02–0.3
Spider silk	0.2
Tendon	0.15
Cartilage	0.0001
Blood vessel (aorta)	0.00005

TABLE 8.4 Tensile strengths of biological materials

Material	Tensile strength (N/m^2)
Spongy bone	5×10^6
Compact bone	100×10^6
Tendon	100×10^6
Spider silk	1000×10^6

Video What the Physics? Elastic Ligament

EXAMPLE 8.10 Finding the compression of a bone BIO

The femur, the long bone in the thigh, can be modeled as a tube of compact bone for most of its length. The cross-section area of the compact bone in the femur of a 70 kg person is $4.8 \times 10^{-4}\ \text{m}^2$, a typical value.

a. If this person supports his entire weight on one leg, what fraction of the tensile strength of the bone does this stress represent?
b. By what fraction of its length does the femur shorten?

PREPARE The stress on the femur is F/A. Here F, the force compressing the femur, is the person's weight, so $F = mg$. The fractional change $\Delta L/L$ in the femur is the strain, which we can find using Equation 8.6, taking the value of Young's modulus for compact bone from Table 8.3.

SOLVE

a. The person's weight is $mg = (70\ \text{kg})(9.8\ \text{m/s}^2) = 690\ \text{N}$. The resulting stress on the femur is

$$\frac{F}{A} = \frac{690\ \text{N}}{4.8 \times 10^{-4}\ \text{m}^2} = 1.4 \times 10^6\ \text{N/m}^2$$

A stress of $1.4 \times 10^6\ \text{N/m}^2$ is 1.4% of the tensile strength of compact bone given in Table 8.4.

b. We can compute the strain as

$$\frac{\Delta L}{L} = \left(\frac{1}{Y}\right)\frac{F}{A} = \left(\frac{1}{1.6 \times 10^{10}\ \text{N/m}^2}\right)(1.4 \times 10^6\ \text{N/m}^2)$$
$$= 8.8 \times 10^{-5} \approx 0.0001$$

The femur compression is $\Delta L \approx 0.0001L$, or $\approx 0.01\%$ of its length. (The femur is far from a uniform structure, so we've expressed the answer as an approximate result to one significant figure.)

ASSESS It makes sense that, under ordinary standing conditions, the stress on the femur is only a percent or so of the maximum value it can sustain.

FIGURE 8.25 Rear legs of model skeletons of a cat (left) and an armadillo (right).

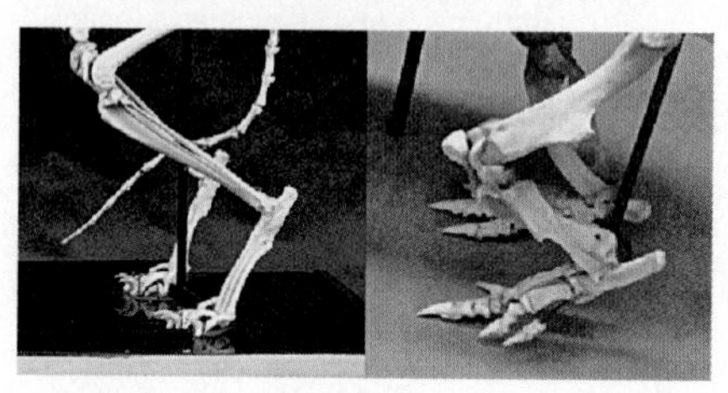

The dancer in the chapter-opening photo stands *en pointe,* balanced delicately on the tip of her shoe with her entire weight supported on a very small area. The stress on the bones in her toes is very large, but it is still much less than the tensile strength of bone.

STOP TO THINK 8.6 A 10 kg mass is hung from a 1-m-long cable, causing the cable to stretch by 2 mm. Suppose a 10 kg mass is hung from a 2 m length of the same cable. By how much does the cable stretch?

A. 0.5 mm B. 1 mm C. 2 mm D. 3 mm E. 4 mm

8.5 Forces and Torques in the Body BIO

FIGURE 8.26 The operation of a nutcracker.

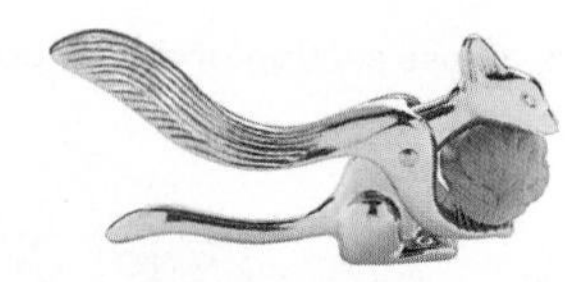

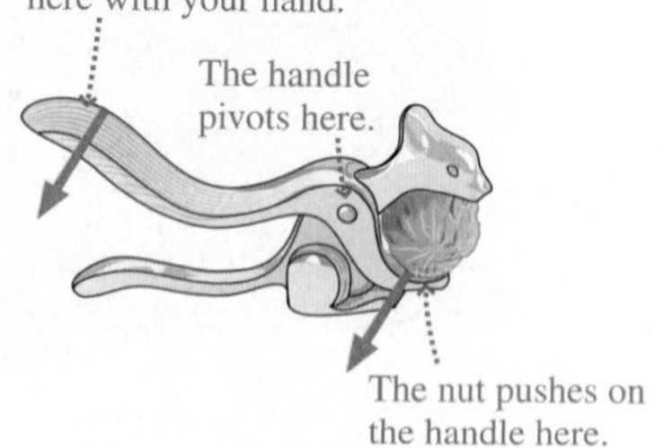
(b) You push the handle here with your hand.
The handle pivots here.
The nut pushes on the handle here.

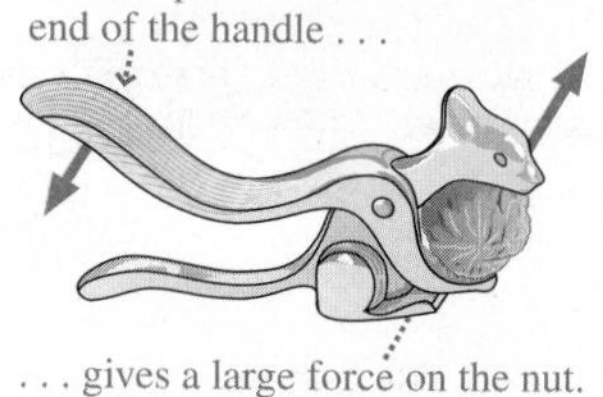
(c) A small push on the end of the handle . . .
. . . gives a large force on the nut.

FIGURE 8.25 shows two mounted skeletons, one of a cat and one of an armadillo. Which one is a jumper and which is a digger? It's easy to spot differences between the two skeletons. One obvious difference is the thicker bones of the armadillo, with clear protrusions for muscle attachment. This is a creature that is built for strength. The cat, however, is built for speed, with lighter bones and longer limbs.

It's not just the *thickness* of the bones that speaks to the strength-versus-speed issue—there are also details of the *shapes* of the bones. Look at the ankle joints on the two animals. In particular, note the long extension of the bones of the foot on the armadillo skeleton. This is the attachment for the tendon that pushes the toes downward. In the armadillo, this attachment is quite far back from the joint, which gives a large torque for a given muscle force. This leads to greater force at the toes, a good thing for an animal that gets its food by digging. But there's a cost: Having the attachment point so close to the joint permits large forces but limits the possible speed. In the skeleton of the cat, an agile jumper, we see that the attachment point is much closer to the joint. This permits a greater range of motion as well as more rapid motion. The armadillo is built for strength; the cat is built for speed.

Mechanical Advantage

FIGURE 8.26a shows a nutcracker. Why do you need a tool to crack a nut? In a word: force. It takes a good deal of force to crack a nut, more than you can easily provide with your muscles alone.

How does using a nutcracker give more force? This is a static equilibrium problem that we can solve. Let's look not at the nut, but at the handle, which pivots about the rivet. The handle is in static equilibrium, so we can use the techniques of this chapter to understand the nutcracker. As FIGURE 8.26b shows, when you push down on the handle, the handle experiences a force from the nut as well. If we take the pivot as the point about which we compute torques, there is a counterclockwise torque from the force of your hand, and a clockwise torque from the force of the nut. The magnitudes of these two torques must be the same, so the force from the nut (which is applied much closer to the pivot) must be much larger than the force of your hand. This is the point of the nutcracker, as FIGURE 8.26c shows. The nut pushes on the handle with a large force, so the handle exerts an equally large force on the nut. A small force on the handle leads to a large force on the nut. For the particular situation illustrated in Figure 8.26, the force on the nut is 3 times the force from your hand. In this way, the tool has amplified the force that you are able to bring to bear on the nut.

There are many hand tools that have a similar action, for which modest forces far from the pivot lead to large forces near the pivot. Can and bottle openers, pliers, shovels, pry bars—all of these rely on this principle, which we might call *leverage*, or **mechanical advantage.** If the nutcracker enables you apply 3 times the force you can apply with your hand, we say that it has a mechanical advantage of 3.

Strength Versus Range of Motion and Speed

Now let's look at the operation of your ankle joint. As we'll see, the geometry is the same as that of the nutcracker, with one crucial difference—the muscle force is applied close to the pivot, not far away.

EXAMPLE 8.11 Forces in the ankle joint BIO

For the purposes of this problem, we'll assume a simple model of the operation of your foot and ankle. We'll assume that your foot pivots as a single structure about a single pivot in the ankle. This simple model will help us explore the operation of the ankle joint and draw some interesting conclusions.

Let's take your foot as the object of interest. When you stand on tiptoe, your foot pivots about your ankle. As shown in FIGURE 8.27, the forces on one foot are an upward force on your toes from the floor, a downward force on your ankle from the lower leg bone, and an upward force on the heel of your foot from your Achilles tendon. Suppose a 61 kg woman stands on one foot, on tiptoe, with the sole of her foot making a 25° angle with the floor; the distances are as shown in Figure 8.27. What is the magnitude of the tension force in the tendon? By what fraction does this force exceed the woman's weight? What is the magnitude of the force in the ankle joint?

FIGURE 8.27 Forces on the foot when standing on tiptoe.

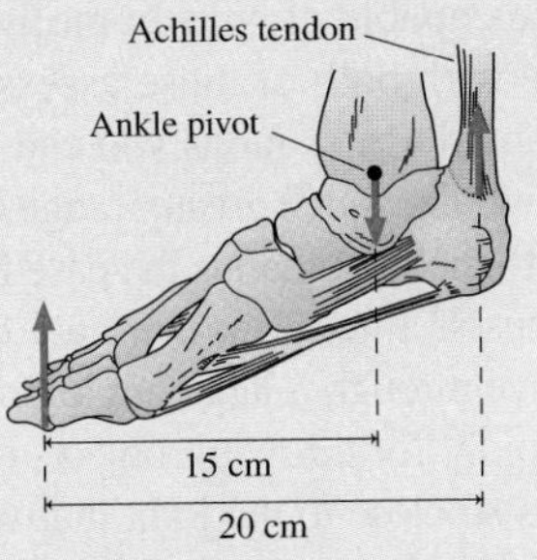

STRATEGIZE As we've noted, the object of interest is the foot—this is the object that pivots, that has forces applied at different points. We'll look at the static equilibrium of the foot given the forces that act on it. We'll take as our pivot point the ankle pivot. We are given the perpendicular distance to the pivot—the moment arm—for each force, so we'll compute torques using $\tau = r_\perp F$.

PREPARE There are three forces that act on the foot, as shown in the simplified view of the situation in FIGURE 8.28. The force of the floor has only a *y*-component; the same is true of the tendon force. There must be no net force along either the *x*- or *y*-axes, so we've drawn the force at the ankle as vertical as well.

The magnitude of the upward force from the floor is equal to the woman's weight, which we round to two significant figures:

$$F_{\text{floor}} = w = (61 \text{ kg})(9.8 \text{ m/s}^2) = 600 \text{ N}$$

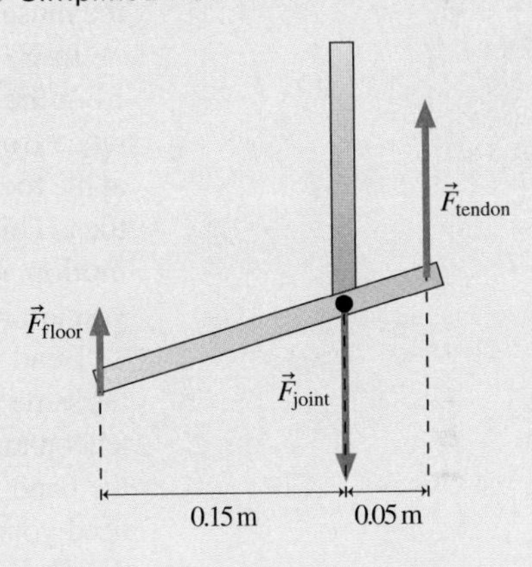

FIGURE 8.28 Simplified view of the forces on the foot.

This force acts at a perpendicular distance of 15 cm, or 0.15 m, from the pivot and is directed upward. The force from the tendon is simply the tension in the tendon, which is what we are solving for:

$$F_{\text{tendon}} = T$$

This force acts at a perpendicular distance of 5 cm, or 0.05 m, from the pivot; this force is directed upward as well. The final force is the force at the pivot, the force in the joint, which we call F_{joint}. This is the force with which the bones of the upper leg press on the bones of the foot. This force acts at the pivot and so contributes no torque; it must be directed downward to balance the two upward forces.

SOLVE The foot is in static equilibrium, so it experiences no net force and no net torque. The force at the pivot contributes no torque, so the torque condition reduces to

$$\sum \tau = 0 = (r_\perp)_{\text{tendon}} F_{\text{tendon}} - (r_\perp)_{\text{floor}} F_{\text{floor}}$$
$$= (0.05 \text{ m})T - (0.15 \text{ m})(600 \text{ N})$$

Continued

We can solve for the tension in the tendon:

$$T = 1800\ \text{N}$$

This is 3 times the woman's weight. Now, let's look at the force condition $\sum F_y = 0$ for static equilibrium:

$$\sum F_y = 0 = F_{\text{floor}} - F_{\text{joint}} + F_{\text{tendon}} = 600\ \text{N} - F_{\text{joint}} + 1800\ \text{N}$$

We can solve this equation for the force in the joint:

$$F_{\text{joint}} = 2400\ \text{N}$$

This force is 4 times her weight.

ASSESS In terms of basic physics, our results make sense. The force of the tendon acts closer to the pivot than the force of the floor, so we expect this force to be larger than her weight.

In the last step of the example, we noted that the result made sense in terms of basic physics. But it does raise certain questions. For the woman to lift her body on her toes, her tendon—and the muscles that pull on it—must pull with a force that is triple her body weight. This is the opposite of the situation we saw with the nutcracker, where a small applied force resulted in a large force on the nut. Lifting your weight by standing on your toes requires your muscles to provide a force that is 3 times your weight, which seems a bit surprising. If the attachment point of the tendon was farther from the joint, it would take less muscle force for her to raise her body. Isn't that desirable?

There are animals—like the armadillo—that do have the tendon connected farther from the ankle pivot. For such an animal, a given muscle force results in a larger force at the toes. This is very useful for the armadillo, who digs as it searches for food. But there's a cost. A muscle can contract by no more than 1/3 of its resting length. The farther from the pivot the muscle or attached tendon pulls, the smaller the range of motion in the limb. Muscles also have a maximum contraction speed. The farther from the pivot the muscle or attached tendon pulls, the slower the resulting motion of the limb.

In Example 8.11, the distance to the tendon attachment was 1/3 of the distance from the ankle to the toes. For the armadillo, the tendon attaches at about 1/2 the distance from the ankle to the toes. This makes sense for a digger that needs large forces at the toes. For a cat, the tendon attaches at about 1/5 the distance from the ankle to the toes. This means a cat isn't likely to be a great digger, but it has a bigger range of motion and is apt to be capable of speedy motion of its paws. If you are a cat owner, you are likely to agree with both of these assessments.

Bend your foot up and down as far as you can, and note the range of motion. Now do the same with your elbow. The much greater range of motion in the elbow joint requires the tendon to attach closer to the joint than is the case for your ankle. This reduces the force at the hand for a given muscle force. But there are considerations other than strength: You need your elbow to move through a large angle for normal use, and our ancestors needed their hands to move rapidly for throwing objects or throwing a punch. In the case of the elbow, the attachment is so close to the joint that very large muscle and tendon forces are required, in some cases near the limits of what muscle and tendon can support.

EXAMPLE 8.12 Analyzing forces and torques in the elbow BIO

In weightlifting, where the applied forces are large, holding the body in static equilibrium requires muscle forces that are quite large indeed. In the strict curl event, a standing athlete lifts a barbell by moving only his forearms, which pivot at the elbow. Record lifts are in the range of 110 kg. FIGURE 8.29 shows the arm bones and the main lifting muscle when the forearm is horizontal. The distance from the tendon to the elbow joint is 4.0 cm, and from the barbell to the elbow 35 cm. For a typical tendon cross-section area of $1.3 \times 10^{-4}\ \text{m}^2$, what percentage of the maximum tension the tendon can support is required to perform a 110 kg strict curl?

STRATEGIZE We'll solve the problem for one arm, which we assume is lifting half the weight. The forearm is in motion, but the speed and changes in speed are quite modest given the forces at work, so we will treat this as a static equilibrium problem. The object of interest is the forearm. We know the forces and where they act, and the elbow is a natural pivot. All of the forces are perpendicular to the arm, so there is no need to take components.

Solving the static equilibrium problem will allow us to find the tension in the tendon. We will then compare this with the maximum possible tension, computed with the known area of the tendon and the tensile strength of tendon given in Table 8.4.

PREPARE FIGURE 8.30 shows a simplified model of the situation. There are three forces at work. The magnitude of the force of the tendon is just the tension in the tendon, which we are trying to find:

$$F_{\text{tendon}} = T$$

This force acts a perpendicular distance of 0.040 m from the pivot. The force of the barbell is the weight of half the mass:

$$F_{\text{barbell}} = \left(\frac{110\ \text{kg}}{2}\right)g = 540\ \text{N}$$

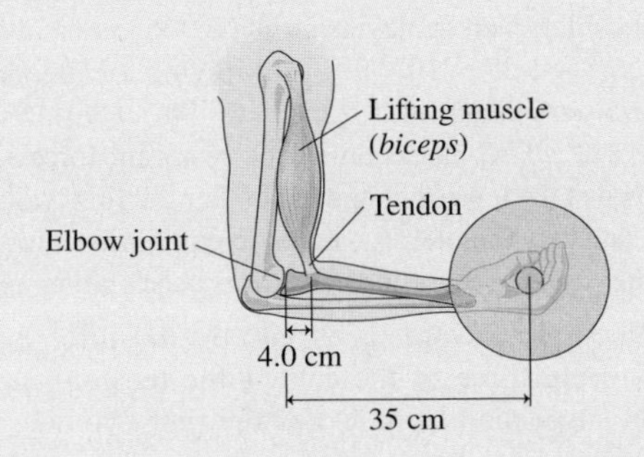

FIGURE 8.29 The arm lifting a barbell.

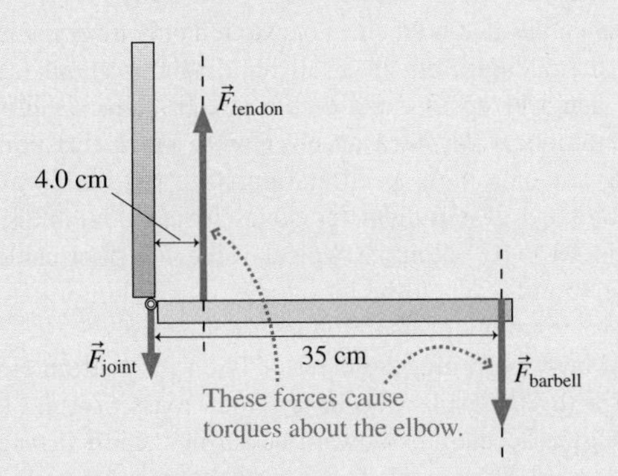

FIGURE 8.30 A simplified model of the arm and weight.

This force acts 35 cm from the pivot. The force in the joint acts at the pivot, so it does not contribute a torque.

SOLVE The tension in the tendon tries to rotate the arm counterclockwise, so it produces a positive torque. The torque due to the barbell, which tries to rotate the arm in a clockwise direction, is negative. For static equilibrium, the magnitudes of the two torques must be equal. Given the forces and distances we identified, we can write

$$T(4.0 \text{ cm}) = (540 \text{ N})(35 \text{ cm})$$

We can solve this equation for the tension in the tendon. The distances appear in a ratio, so the units cancel; there is no need for unit conversion:

$$T = (540 \text{ N})\frac{35 \text{ cm}}{4.0 \text{ cm}} = 4700 \text{ N}$$

The tendon tension comes from the muscles, which must provide a force nearly 9 times the weight lifted! For this lift, the biceps in the arms are pulling with a combined force of about 1 ton, which makes this impressive lift seem even more amazing.

The tendon sustains a very large tension force. The maximum possible tendon tension is fixed by the cross-section area of the tendon and the tensile strength given in Table 8.4:

$$T_{max} = (100 \times 10^6 \text{ N/m}^2)(1.3 \times 10^{-4} \text{ m}^2) = 13{,}000 \text{ N}$$

The required tension for the lift is 36% of the maximum possible tension.

ASSESS The large value for the tendon tension makes sense, given the problem statement, as does the fact that the tension is a significant fraction of what the tendon can support. The lift is possible, but it's nearing the limit of what the tissues of the body can do.

Let's look at one more example, the motion of the jaw. A typical person can generate a bite force of 1200 N at the second molars, a force that is probably greater than the person's weight. The masseter muscle that provides most of the force to close your jaw isn't a particularly large muscle, but its attachment is quite favorable for providing large forces, as FIGURE 8.31a shows. The force vector shows the approximate line of force of the masseter muscle. The line of the force is about 5 cm from the pivot, compared to about 7 cm from the molars. This means that the force at the molars is nearly equal to the full force of the muscle. This is a dramatic difference from the previous examples. There's a trade-off, though—you may be capable of great bite strength, but your jaw has a very limited range of motion and you have limited bite speed. FIGURE 8.31b shows a dog jaw along with the pivot and the approximate line of force of the masseter. The prominent canine teeth at the front of the jaw are much farther from the pivot than the muscle, so the canine teeth are well adapted for rapid, slashing bites. Cats have much shorter jaws than dogs. What does this imply about their bite speed and force?

Making light work of moving BIO
The tendon force is so large in Example 8.12 because the weight is supported much farther from the elbow than the point where the tendon attaches. If the weight is supported closer to the elbow, the downward torque of the weight is much less, reducing the necessary tendon and muscle force. In the picture, two people are using lifting straps to carry a heavy appliance. The straps hang very close to the elbow, so the required muscle force to support the weight is much less than it would otherwise be.

FIGURE 8.31 Jaw motion in a human and a dog.

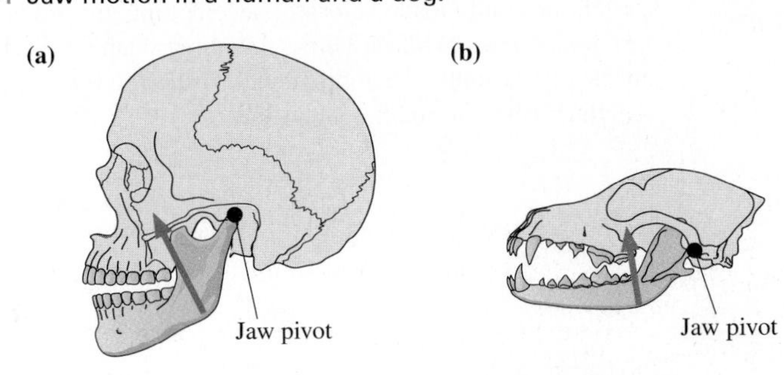

INTEGRATED EXAMPLE 8.13 Elevator cable stretch

The steel cables that hold elevators stretch only a very small fraction of their length, but in a tall building this small fractional change can add up to a noticeable stretch. This example uses realistic numbers for such an elevator to make this point. The 2300 kg car of a high-speed elevator in a tall building is supported by six 1.27-cm-diameter cables. Young's modulus for the cables is $10 \times 10^{10}\ \text{N/m}^2$, a typical value for steel cables with multiple strands. When the elevator is on the bottom floor, the cables rise 90 m up the shaft to the motor above.

On a busy morning, the elevator is on the bottom floor and fills up with 20 people who have a total mass of 1500 kg. The elevator then accelerates upward at $2.3\ \text{m/s}^2$ until it reaches its cruising speed. How much do the cables stretch due to the weight of the car alone? How much additional stretch occurs when the passengers are in the car? And, what is the total stretch of the cables while the elevator is accelerating? In all cases, you can ignore the mass of the cables.

STRATEGIZE The restoring force of the cable is the force that supports the elevator. We will determine the force that supports the elevator, and then use this force to determine the stretch of the cable.

PREPARE We can compute the stretch of the cables by rewriting Equation 8.6, the equation relating stress and strain, to get

$$\Delta L = \frac{LF}{YA}$$

Here L is the length of the cables, F is the restoring force exerted by the cables, and Y is Young's modulus, which we're given. Although there are six cables, we can imagine them combined into one cable with a cross-section area A six times that of each individual cable. Each cable has radius 0.00635 m and cross-section area $\pi r^2 = 1.27 \times 10^{-4}\ \text{m}^2$. Multiplying the area by 6 gives a total cross-section area $A = 7.62 \times 10^{-4}\ \text{m}^2$.

FIGURE 8.32 shows the details. The restoring force exerted by the cable is just the tension in the cable. For the first two questions, the cable supports the elevator in static equilibrium; for the third, there is a net force because the elevator is accelerating upward.

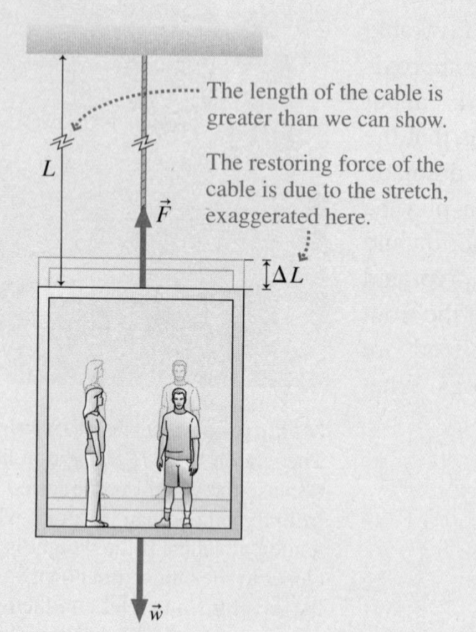
The length of the cable is greater than we can show.
The restoring force of the cable is due to the stretch, exaggerated here.
L
$\vec{F}$
ΔL
$\vec{w}$

FIGURE 8.32 Details of the cable stretch and the forces acting on the elevator.

SOLVE When the elevator is at rest, the net force is zero and so the restoring force of the cable—the tension—is equal to the weight suspended from it. For the first two questions, the forces are

$$F_1 = m_{\text{car}}g = (2300\ \text{kg})(9.8\ \text{m/s}^2) = 22{,}500\ \text{N}$$

$$F_2 = m_{\text{car+passengers}}g = (2300\text{kg} + 1500\text{kg})(9.8\text{m/s}^2) = 37{,}200\text{N}$$

These forces stretch the cable by

$$\Delta L_1 = \frac{(90\ \text{m})(22{,}500\ \text{N})}{(10 \times 10^{10}\ \text{N/m}^2)(7.62 \times 10^{-4}\ \text{m}^2)} = 0.027\ \text{m} = 2.7\ \text{cm}$$

$$\Delta L_2 = \frac{(90\ \text{m})(37{,}200\ \text{N})}{(10 \times 10^{10}\ \text{N/m}^2)(7.62 \times 10^{-4}\ \text{m}^2)} = 0.044\ \text{m} = 4.4\ \text{cm}$$

The additional stretch when the passengers board is

$$4.4\ \text{cm} - 2.7\ \text{cm} = 1.7\ \text{cm}$$

When the elevator is accelerating upward, the tension in the cables must increase. Newton's second law for the vertical motion is

$$\Sigma F_y = F - w = ma_y$$

The restoring force, or tension, is thus

$$F = w + ma_y = mg + ma_y = (3800\ \text{kg})(9.8\ \text{m/s}^2 + 2.3\ \text{m/s}^2)$$
$$= 46{,}000\ \text{N}$$

Thus right at the start of the motion, when the full 90 m of cable is still deployed, we find that the cables are stretched by

$$\Delta L = \frac{(90\ \text{m})(46{,}000\ \text{N})}{(10 \times 10^{10}\ \text{N/m}^2)(7.62 \times 10^{-4}\ \text{m}^2)} = 0.054\ \text{m} = 5.4\ \text{cm}$$

ASSESS When the passengers enter the car, the cable stretches by 1.7 cm, or about two-thirds of an inch. This is large enough to notice (as we might expect given the problem statement) but not large enough to cause concern to the passengers. The total stretch for a fully loaded elevator accelerating upward is 5.4 cm, just greater than 2 inches. This is not unreasonable for cables that are nearly as long as a football field—the fractional change in length is still quite small.

SUMMARY

GOAL To learn about the static equilibrium of extended objects, and the basic properties of springs and elastic materials.

GENERAL PRINCIPLES

Static Equilibrium

An object in **static equilibrium** must have no net force on it and no net torque. Mathematically, we express this as

$$\sum F_x = 0$$
$$\sum F_y = 0$$
$$\sum \tau = 0$$

Since the net torque is zero about *any* point, the pivot point for calculating the torque can be chosen at any convenient location.

Springs and Hooke's Law

When a spring is stretched or compressed, it exerts a force proportional to the change Δx in its length but in the opposite direction. This is known as **Hooke's law:**

$$(F_{sp})_x = -k\,\Delta x$$

The constant of proportionality k is called the **spring constant.** It is larger for a "stiff" spring.

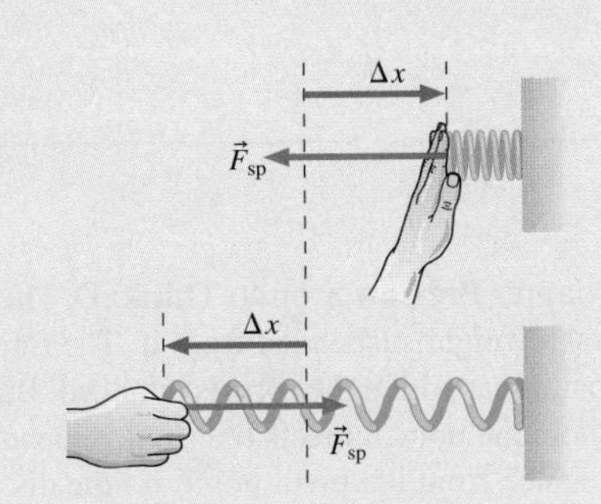

IMPORTANT CONCEPTS

Stability

An object is **stable** if its center of gravity is over its base of support; otherwise, it is **unstable.**

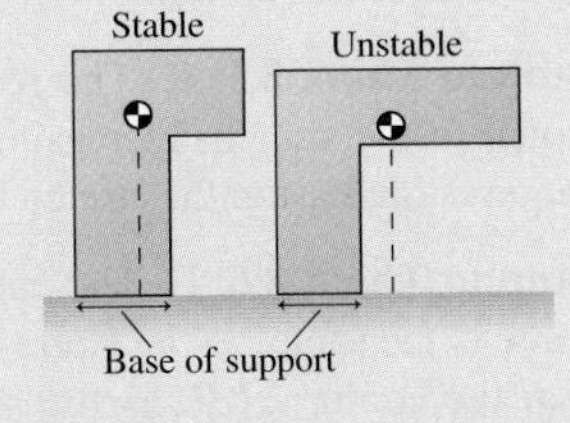

If an object is tipped, it will reach the limit of its stability when its center of gravity is over the edge of the base. This defines the **critical angle** θ_c.

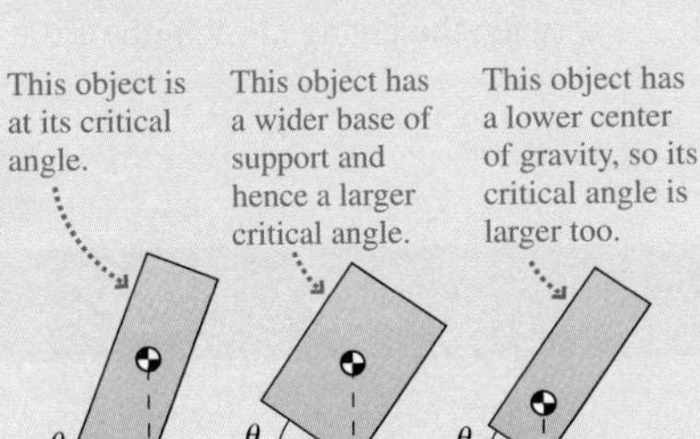

Greater stability is possible with a lower center of gravity or a broader base of support.

Elastic materials and Young's modulus

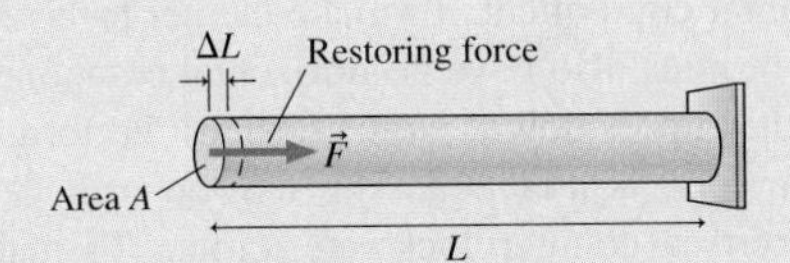

A solid rod illustrates how materials respond when stretched or compressed.

$$\left(\frac{F}{A}\right) = Y\left(\frac{\Delta L}{L}\right)$$

Stress is the restoring force of the rod divided by its cross-section area. **Strain** is the fractional change in the rod's length. Y: Young's modulus.

This equation can also be written as

$$F = \left(\frac{YA}{L}\right)\Delta L$$

This is the "spring constant" k for the rod.

showing that a rod obeys Hooke's law and acts like a very stiff spring.

APPLICATIONS

Forces and torques in the body

Muscles and tendons apply the forces and torques needed to maintain static equilibrium. These forces may be quite large.

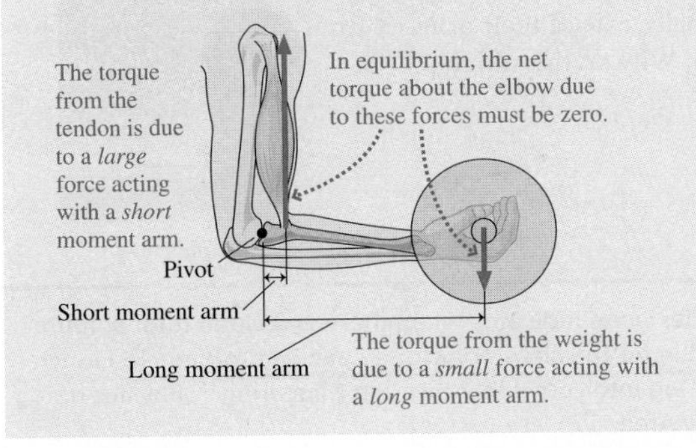

The torque from the tendon is due to a *large* force acting with a *short* moment arm.
In equilibrium, the net torque about the elbow due to these forces must be zero.
Pivot
Short moment arm
Long moment arm
The torque from the weight is due to a *small* force acting with a *long* moment arm.

The elastic limit and beyond

If a rod or other object is not stretched too far, when released it will return to its original shape.

If stretched too far, an object will permanently deform and finally break. The stress at which an object breaks is its **tensile stress.**

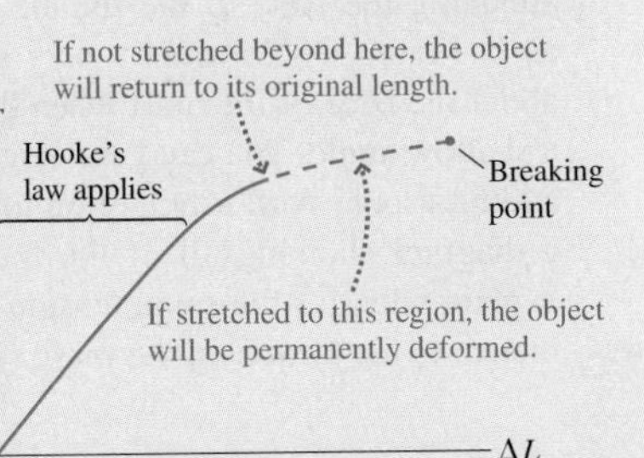

Learning Objectives After studying this chapter, you should be able to:

- Use force and torque to solve static equilibrium problems. *Conceptual Questions 8.2, 8.15; Problems 8.3, 8.4, 8.7, 8.10, 8.47*
- Determine an object's stability. *Conceptual Questions 8.6, 8.7; Problems 8.15, 8.16, 8.17, 8.18, 8.19*
- Use Hooke's law to calculate the force exerted by a spring. *Conceptual Questions 8.10, 8.11; Problems 8.23, 8.24, 8.25, 8.27, 8.28*
- Use stress, strain, and Young's modulus to calculate the elastic properties of materials. *Conceptual Questions 8.12, 8.17; Problems 8.30, 8.31, 8.32, 8.33, 8.36, 8.41*

STOP TO THINK ANSWERS

Chapter Preview Stop to Think: D. The tension in the rope is equal to the weight suspended from it. The force does not change with the position at which the rope is attached. But the torque about the point where the branch meets the trunk depends on both this force and the distance from the pivot point. As the distance decreases, so does the torque.

Stop to Think 8.1: D. Only object D has both zero net force and zero net torque.

Stop to Think 8.2: B. The tension in the rope and the weight have no horizontal component. To make the net force zero, the force due to the pivot must also have no horizontal component, so we know it points either up or down. Now consider the torque about the point where the rope is attached. The tension provides no torque. The weight exerts a counterclockwise torque. To make the net torque zero, the pivot force must exert a *clockwise* torque, which it can do only if it points *up*.

Stop to Think 8.3: B, A, C. The critical angle θ_c, shown in the figure, measures how far the object can be tipped before falling. B has the smallest critical angle, followed by A, then C.

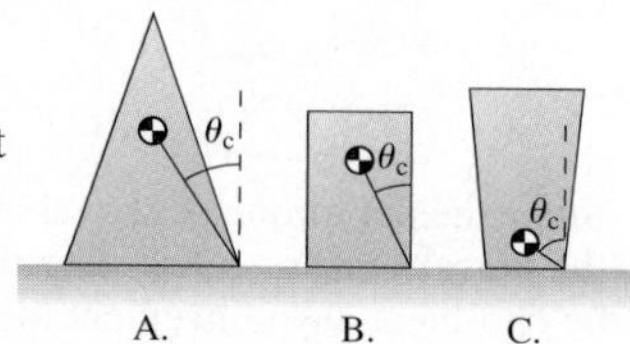

Stop to Think 8.4: B. The spring is now compressed, so the restoring force will be to the right. The end of the spring is displaced by half of what it was in the earlier case, so the restoring force will be half what it was as well.

Stop to Think 8.5: C. The restoring force of the spring is proportional to the stretch. Increasing the restoring force by a factor of 3 requires increasing the stretch by a factor of 3.

Stop to Think 8.6: E. The cables have the same diameter, and the force is the same, so the stress is the same in both cases. This means that the strain, $\Delta L/L$, is the same. The 2 m cable will experience twice the change in length of the 1 m cable.

 **Video Tutor Solution** Chapter 8

QUESTIONS

Conceptual Questions

1. Sketch a force acting at point P in Figure Q8.1 that would make the rod be in static equilibrium. Is there only one such force?

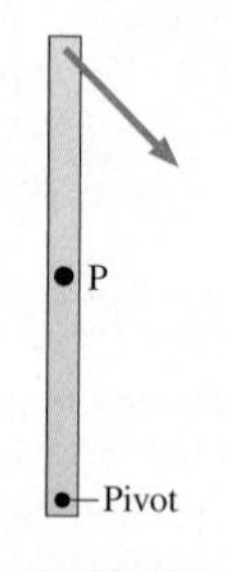

FIGURE Q8.1

2. Could a ladder on a level floor lean against a wall in static equilibrium if there were no friction forces? Explain.
3. If you are using a rope to raise a tall mast, attaching the rope to the middle of the mast as in Figure Q8.3a gives a very small torque about the base of the mast when the mast is at a shallow angle. You can get a larger torque by adding a pole with a pulley on top, as in Figure Q8.3b. Draw a diagram showing all of the forces acting on the mast and explain why, for the same tension in the rope, adding this pole increases the torque on the mast.

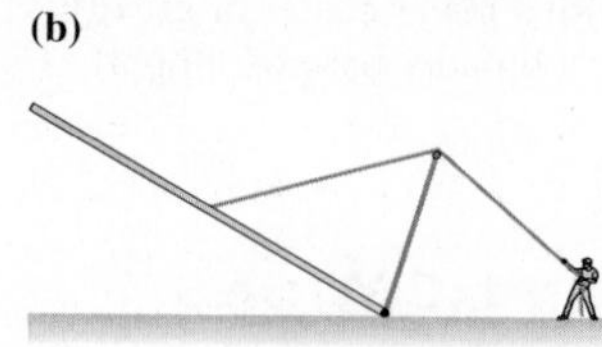

FIGURE Q8.3

4. As divers stand on tiptoes on the edge of a diving platform, in preparation for a high dive, as shown in Figure Q8.4, they usually extend their arms in front of them. Why do they do this?

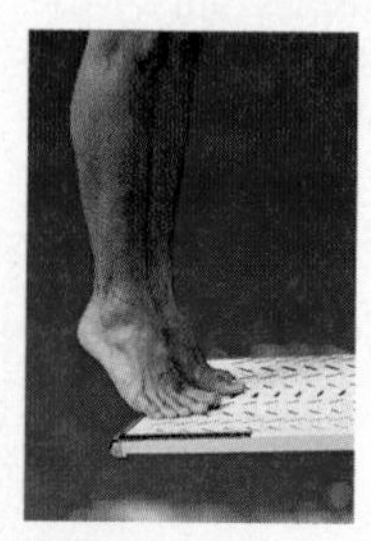

FIGURE Q8.4

Problem difficulty is labeled as | (straightforward) to ||||| (challenging). INT Problems labeled integrate significant material from earlier chapters; BIO are of biological or medical interest.

The eText icon indicates when there is a video tutor solution available for the chapter or for a specific problem. To launch these videos, log into your eText through Mastering^{TM} Physics or log into the Study Area.

5. Where are the centers of gravity of the two people doing the classic yoga poses shown in Figure Q8.5?

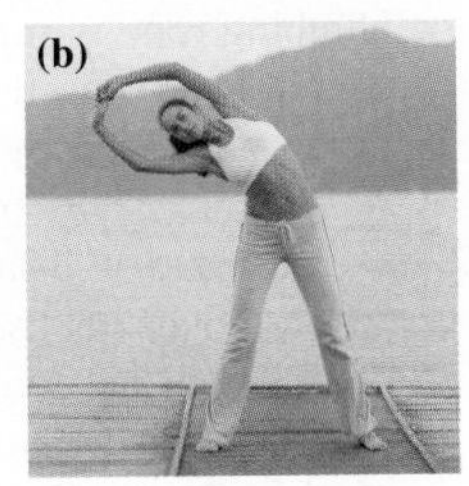

FIGURE Q8.5

6. You must lean quite far forward as you rise from a chair (try it!). Explain why.
7. If you stand with your back and heels against a wall, and bend forward at the waist, you will topple forward. Explain why.
8. A spring exerts a 10 N force after being stretched by 1 cm from its equilibrium length. By how much will the spring force *increase* if the spring is stretched from 4 cm away from equilibrium to 5 cm from equilibrium?
9. A spring is attached to the floor and pulled straight up by a string. The string's tension is measured. The graph in Figure Q8.9 shows the tension in the spring as a function of the spring's length L. What is the spring constant?

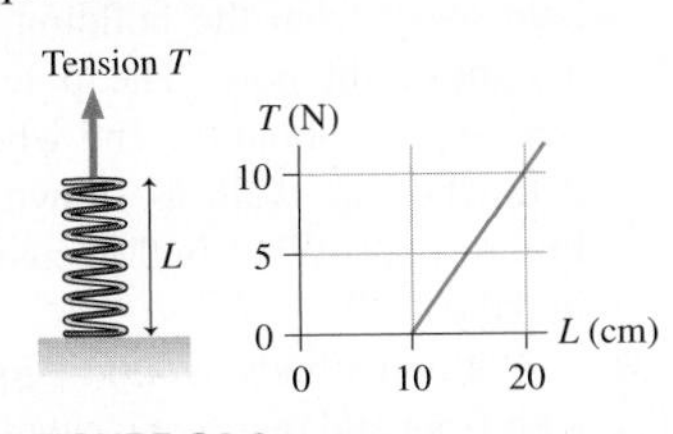

FIGURE Q8.9

10. A typical mattress has a network of springs that provide support. If you sit on a mattress, the springs compress. A heavier person compresses the springs more than a lighter person. Use the properties of springs and spring forces to explain why.
11. Take a spring and cut it in half to make two springs. How does the spring constant of the smaller springs relate to that of the original spring?
12. A wire is stretched to its breaking point by a 5000 N force. A longer wire made of the same material has the same diameter. Is the force that will stretch it right to its breaking point larger than, smaller than, or equal to 5000 N? Explain.
13. If you put a heavy load in a wheelbarrow, the force you apply to the handles is much less than the weight of the load and the wheelbarrow. To lift with the least force, you put the load toward the front of the wheelbarrow, near to the front wheel. Explain how this reduces the force to support the handles.

14. When you carry shopping bags, rather than grasp the handles with your hand as in Figure Q8.14a, you might choose to put them over your arm and slide the handle toward your elbow as in Figure Q8.14b. Explain why this leads to less muscle effort to carry the bags and less force in your elbow joint. BIO

FIGURE Q8.14

15. You can apply a much larger bite force with your molars, in the back of your mouth, than with your incisors, at the front of your mouth. Explain why this is so. BIO
16. An isometric exercise is one in which the joint angle does not change during the application of muscle force. For instance, you can join your hands together, pushing down with one hand and pushing up with the other. In each arm, the muscles that provide the force are connected to the bones of the arm by tendons. If you increase the forces but keep the angles the same, the muscles will still contract. Explain how this is possible. BIO
17. Steel nails are rigid and unbending. Steel wool is soft and squishy. How would you account for this difference?

Multiple-Choice Questions

18. || Two children hold opposite ends of a lightweight, 1.8-m-long horizontal pole with a water bucket hanging from it. The older child supports twice as much weight as the younger child. How far is the bucket from the older child?
A. 0.3 m B. 0.6 m
C. 0.9 m D. 1.2 m
19. || The uniform rod in Figure Q8.19 has a weight of 14.0 N. What is the magnitude of the normal force on the surface?
A. 7 N B. 14 N
C. 20 N D. 28 N

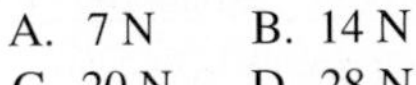

Frictionless surface

FIGURE Q8.19

20. | A student lies on a very light, rigid board with a scale under each end. Her feet are directly over one scale, and her body is positioned as shown in Figure Q8.20. The two scales read the values shown in the figure. What is the student's weight?

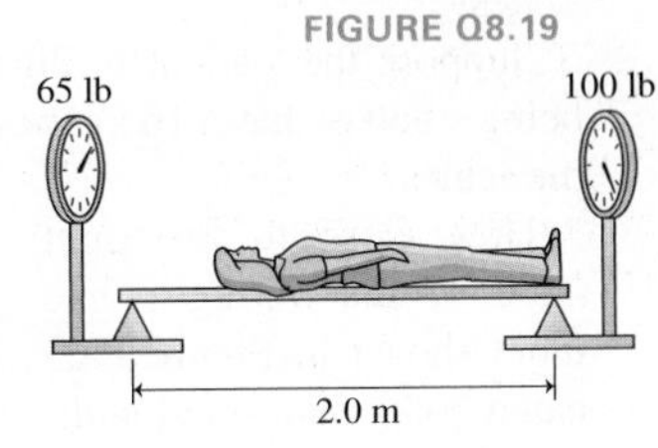

FIGURE Q8.20

A. 65 lb B. 75 lb C. 100 lb D. 165 lb
21. | For the student in Figure Q8.20, approximately how far from her feet is her center of gravity?
A. 0.6 m B. 0.8 m C. 1.0 m D. 1.2 m

Questions 22 through 24 use the information in the following paragraph and figure.

When you stand on a trampoline, the surface depresses below equilibrium, and the surface pushes up on you, as the data for a real trampoline in Figure Q8.22 show. The linear variation of the force as a function of distance means that we can model the restoring force as that of a spring.

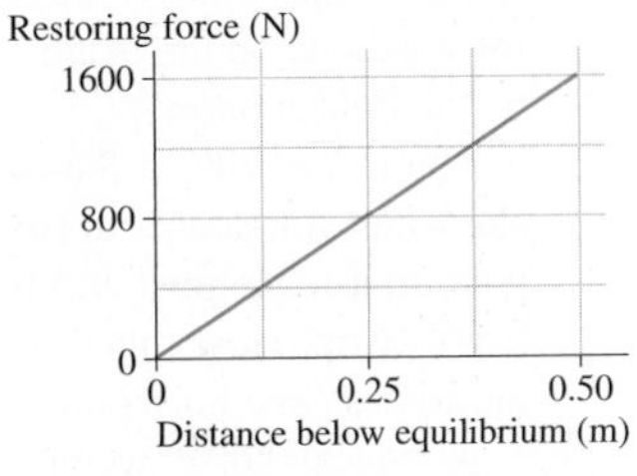

FIGURE Q8.22

22. || What is the spring constant of the trampoline?
A. 800 N/m B. 1600 N/m C. 2400 N/m D. 3200 N/m
23. || A 65 kg gymnast stands on the trampoline. How much does the center of the trampoline compress?
A. 5.0 cm B. 10 cm C. 15 cm D. 20 cm
24. || A 65 kg gymnast jumps on the trampoline. At the lowest point of her motion, the trampoline is depressed 0.50 m below equilibrium. What is her approximate upward acceleration at this instant?
A. 10 m/s^2 B. 15 m/s^2 C. 20 m/s^2 D. 25 m/s^2

Watch Video Solution Problem 8.9

25. || Some climbing ropes are designed to noticeably stretch under a load to lessen the forces in a fall. A weight is attached to 10 m length of climbing rope, which then stretches by 20 cm. Now, this single rope is replaced by a doubled rope—two pieces of rope next to each other. How much does the doubled rope stretch?
A. 5.0 cm
B. 10 cm
C. 20 cm
D. 40 cm

26. || Some climbing ropes are designed to noticeably stretch under a load to lessen the forces in a fall. A 10 m length of climbing rope is supporting a climber; this causes the rope to stretch by 60 cm. If the climber is supported by a 20 m length of the same rope, by how much does the rope stretch?
A. 30 cm B. 60 cm C. 90 cm D. 120 cm

27. || You have a heavy piece of equipment hanging from a 1.0-mm-diameter wire. Your supervisor asks that the length of the wire be doubled without changing how far the wire stretches. What diameter must the new wire have?
A. 1.0 mm B. 1.4 mm C. 2.0 mm D. 4.0 mm

PROBLEMS

Section 8.1 Torque and Static Equilibrium

1. || A 64 kg student stands on a very light, rigid board that rests on a bathroom scale at each end, as shown in Figure P8.1. What is the reading on each of the scales?

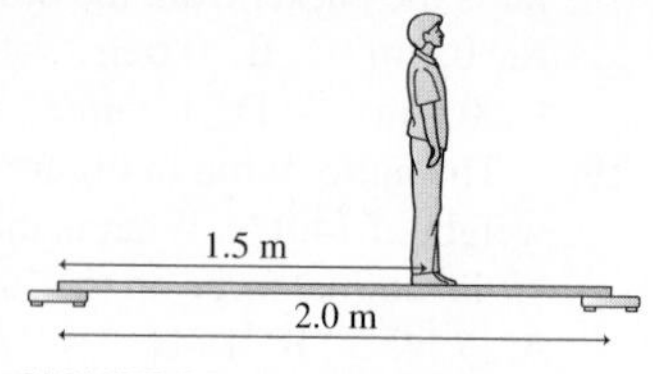

FIGURE P8.1

2. ||| Suppose the student in Figure P8.1 is 54 kg, and the board being stood on has a 10 kg mass. What is the reading on each of the scales?

3. || How close to the right edge of the 56 kg picnic table shown in Figure P8.3 can a 70 kg man stand without the table tipping over? **Hint:** When the table is just about to tip, what is the force of the ground on the table's left leg?

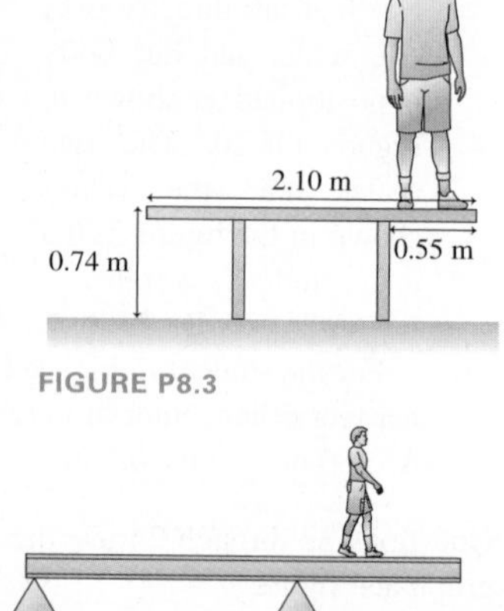

FIGURE P8.3

4. || In Figure P8.4, a 70 kg man walks out on a 10 kg beam that rests on, but is not attached to, two supports. When the beam just starts to tip, what is the force exerted on the beam by the right support?

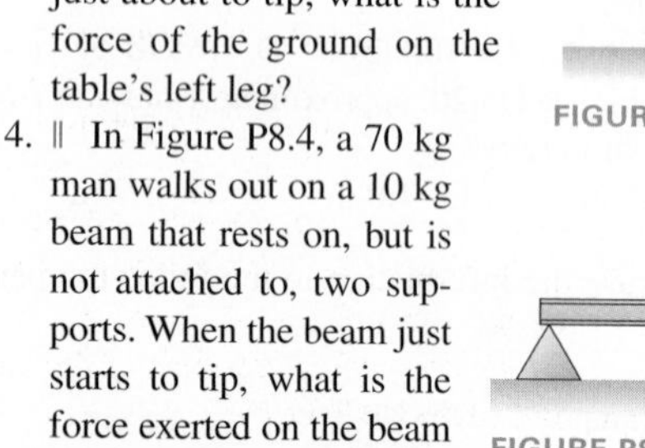
FIGURE P8.4

5. ||| You're carrying a 3.6-m-long, 25 kg pole to a construction site when you decide to stop for a rest. You place one end of the pole on a fence post and hold the other end of the pole 35 cm from its tip. How much force must you exert to keep the pole motionless in a horizontal position?

6. | A typical horse weighs 5000 N. The distance between the front and rear hooves and the distance from the rear hooves to the center of mass for a typical horse are shown in Figure P8.6. What fraction of the horse's weight is borne by the front hooves?
BIO

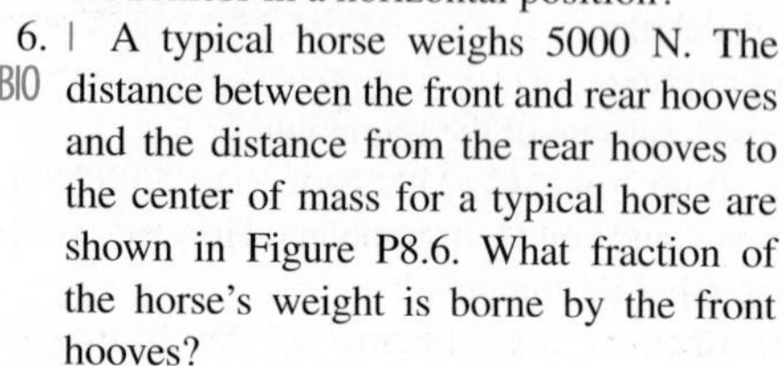
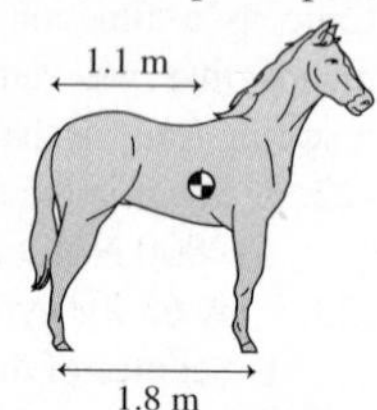

FIGURE P8.6

7. | A vendor hangs an 8.0 kg sign in front of his shop with a cable held away from the building by a lightweight pole. The pole is free to pivot about the end where it touches the wall, as shown in Figure P8.7. What is the tension in the cable?

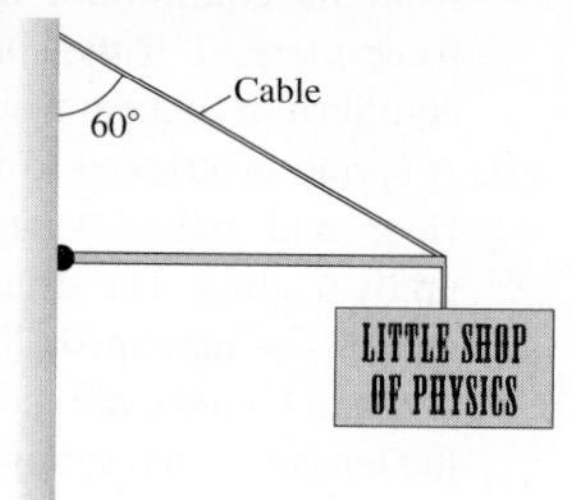

FIGURE P8.7

8. || A hippo's body is 4.0 m long with front and rear feet located as in Figure P8.8. The hippo carries 60% of its weight on its front feet. How far from its tail is the hippo's center of gravity?
BIO

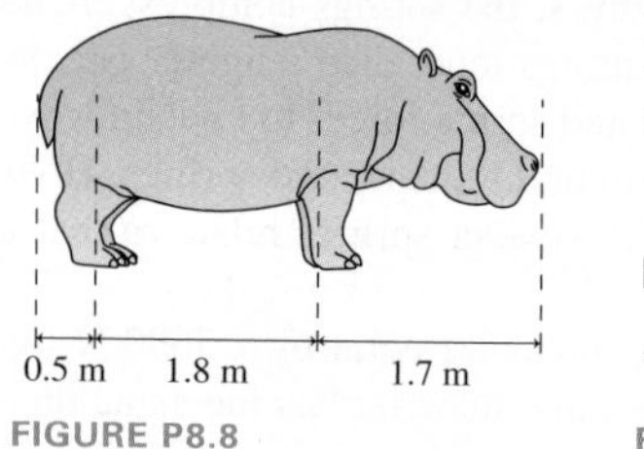

FIGURE P8.8

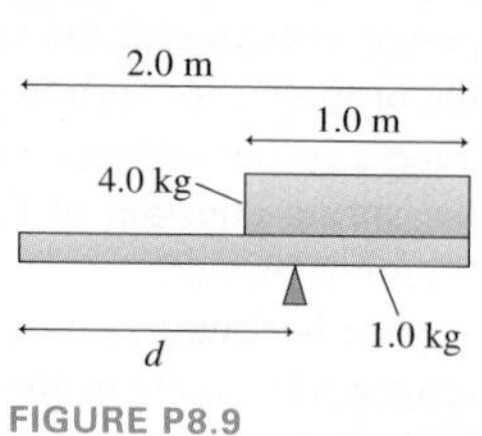

FIGURE P8.9

9. ||| The two objects in Figure P8.9 are balanced on the pivot. What is distance d?

10. || A bicycle mechanic is checking a road bike's chain. He applies a 45 N force to a pedal at the angle shown in Figure P8.10 while keeping the wheel from rotating. The pedal is 17 cm from the center of the crank; the gear has a diameter of 16 cm. What is the tension in the chain?

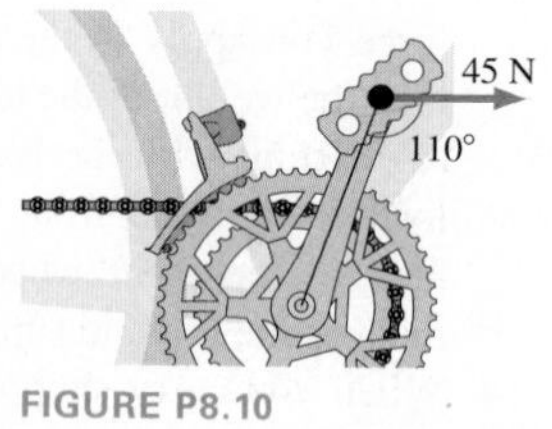

FIGURE P8.10

11. ||| A 60 kg diver stands at the end of a 30 kg springboard, as shown in Figure P8.11. The board is attached to a hinge at the left end but simply rests on the right support. What is the magnitude of the vertical force exerted by the hinge on the board?

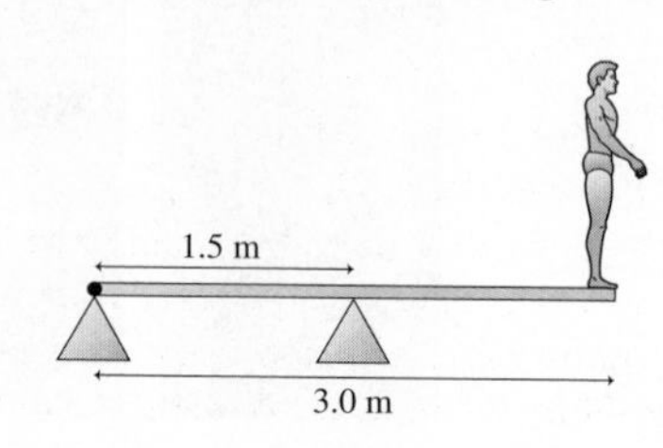

FIGURE P8.11

12. ||| A bike chain can support a tension of no more than 9800 N. The pedal connects to a crank 17 cm from the axle, and the gear pulling the chain has a 9.1 cm radius. When riding at a constant speed, with the crank and pedal horizontal, as in Figure P8.12, what is the maximum force that can be applied to the pedal before the chain breaks?

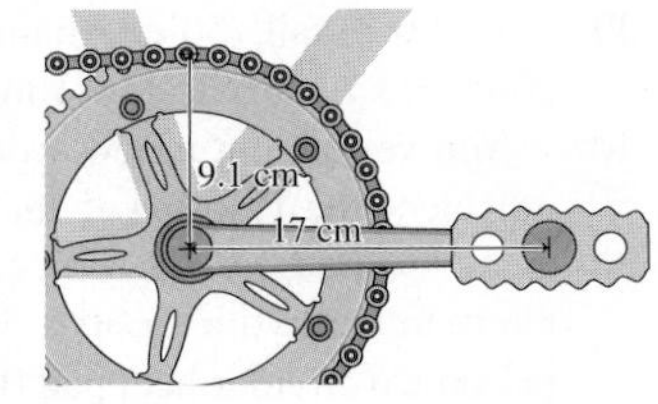

FIGURE P8.12

13. || A uniform beam of length 1.0 m and mass 10 kg is attached to a wall by a cable, as shown in Figure P8.13. The beam is free to pivot at the point where it attaches to the wall. What is the tension in the cable?

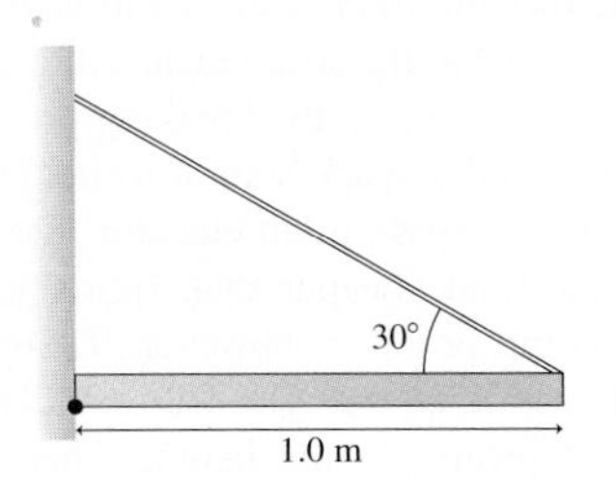

FIGURE P8.13

14. ||| The towers holding small wind turbines are often raised and lowered for easy servicing of the turbine. Figure P8.14 shows a 1000 kg wind turbine mounted on the end of a 24-m-long, 700 kg tower that connects to a support column at a pivot. A piston connected 3.0 m from the pivot applies the force needed to raise or lower the tower. At the instant shown, the wind turbine is being raised at a very slow, constant speed. What magnitude force is the piston applying?

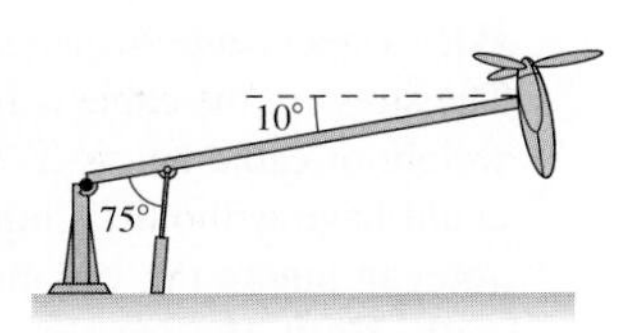

FIGURE P8.14

Section 8.2 Stability and Balance

15. | A standard four-drawer filing cabinet is 52 inches high and 15 inches wide. If it is evenly loaded, the center of gravity is at the center of the cabinet. A worker is tilting a filing cabinet to the side to clean under it. To what angle can he tilt the cabinet before it tips over?

16. | A double-decker London bus might be in danger of rolling over in a highway accident, but at the low speeds of its urban environment, it's plenty stable. The track width is 2.05 m. With no passengers, the height of the center of gravity is 1.45 m, rising to 1.73 m when the bus is loaded to capacity. What are the critical angles for both the unloaded and loaded bus?

17. | The stability of a vehicle is often rated by the *static stability factor*, which is one-half the track width divided by the height of the center of gravity above the road. A typical SUV has a static stability factor of 1.2. What is the critical angle?

18. | A magazine rack has a center of gravity 16 cm above the floor, as shown in Figure P8.18. Through what maximum angle, in degrees, can the rack be tilted without falling over?

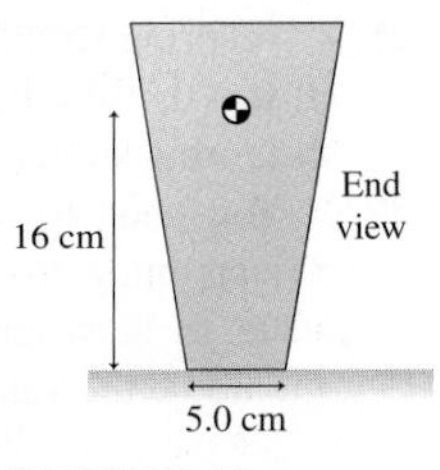

FIGURE P8.18

19. || A car manufacturer claims that you can drive its new vehicle across a hill with a 47° slope before the vehicle starts to tip. If the vehicle is 2.0 m wide, how high is its center of gravity?

20. || A thin 2.00 kg box rests on a 6.00 kg board that hangs over the end of a table, as shown in Figure P8.20. How far can the center of the box be from the end of the table before the board begins to tilt?

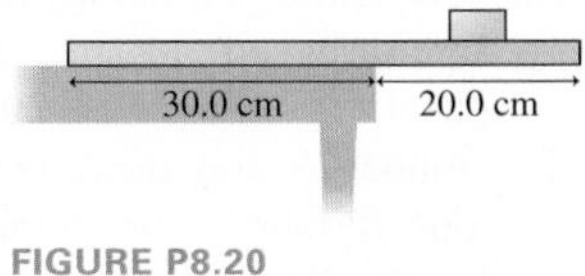

FIGURE P8.20

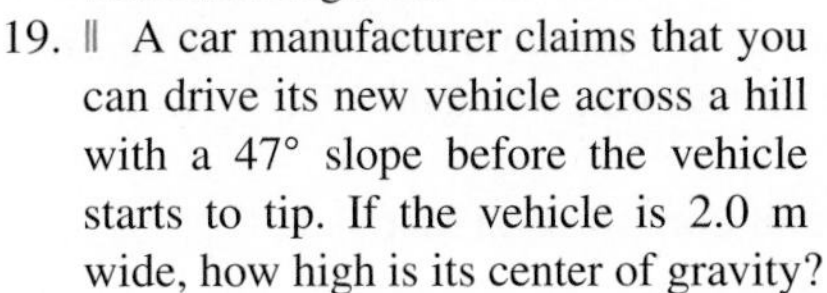

Section 8.3 Springs and Hooke's Law

21. | An orthodontic spring, connected between the upper and BIO lower jaws, is adjusted to provide no force with the mouth open. When the patient closes her mouth, however, the spring compresses by 6.0 mm. What force is exerted if the spring constant is 160 N/m?

22. || The four wheels of a car are connected to the car's body by spring assemblies that let the wheels move up and down over bumps and dips in the road. When a 68 kg (about 150 lb) person sits on the left front fender of a small car, this corner of the car dips by about 1.2 cm (about ½ in). If we treat the spring assembly as a single spring, what is the approximate spring constant?

23. | Experiments using "optical tweezers" measure the elasticity BIO of individual DNA molecules. For small enough changes in length, the elasticity has the same form as that of a spring. A DNA molecule is anchored at one end, then a force of 1.5 nN $(1.5 \times 10^{-9}\text{ N})$ pulls on the other end, causing the molecule to stretch by 5.0 nm $(5.0 \times 10^{-9}\text{ m})$. What is the spring constant of that DNA molecule?

24. || A passenger railroad car has a total of 8 wheels. Springs on each wheel compress—slightly—when the car is loaded. Ratings for the car give the stiffness per wheel (the spring constant, treating the entire spring assembly as a single spring) as 2.8×10^7 N/m. When 30 passengers, each with average mass 80 kg, board the car, how much does the car move down on its spring suspension? Assume that each wheel supports 1/8 the weight of the car.

25. || One end of a 10-cm-long spring is attached to the ceiling. When a 2.0 kg mass is hung from the other end, the spring stretches to a length of 15 cm.
    a. What is the spring constant?
    b. How long is the spring when a 3.0 kg mass is suspended from it?

26. | A scale used to weigh fish consists of a spring hung from a support. The spring's equilibrium length is 10.0 cm. When a 4.0 kg fish is suspended from the end of the spring, it stretches to a length of 12.4 cm.
    a. What is the spring constant k for this spring?
    b. If an 8.0 kg fish is suspended from the spring, what will be the length of the spring?

27. || A spring has an unstretched length of 10 cm. It exerts a restoring force F when stretched to a length of 11 cm.
    a. For what total stretched length of the spring is its restoring force $3F$?
    b. At what compressed length is the restoring force $2F$?

Watch Video Solution Problem 8.35

28. || A spring stretches 5.0 cm when a 0.20 kg block is hung from it. If a 0.70 kg block replaces the 0.20 kg block, how far does the spring stretch?
29. | You need to make a spring scale to measure the mass of objects hung from it. You want each 1.0 cm length along the scale to correspond to a mass difference of 0.10 kg. What should be the value of the spring constant?

Section 8.4 Stretching and Compressing Materials

30. || Dynamic climbing ropes are designed to be quite pliant, allowing a falling climber to slow down over a long distance. The graph in Figure P8.30 shows force-versus-strain data for an 11-mm-diameter climbing rope. What is the Young's modulus for this rope?

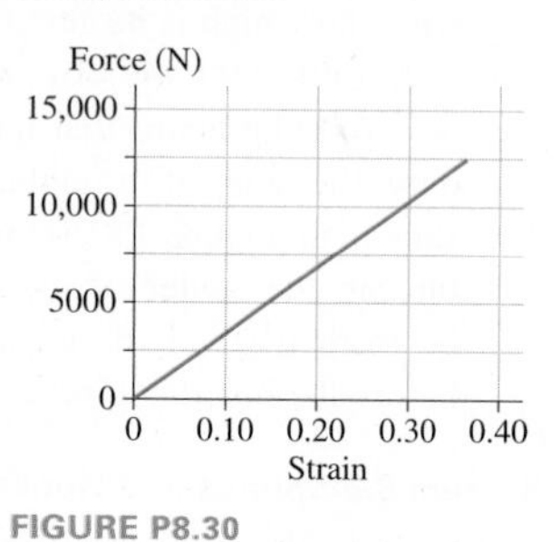

FIGURE P8.30

31. || A force stretches a wire by 1.0 mm.
    a. A second wire of the same material has the same cross section and twice the length. How far will it be stretched by the same force?
    b. A third wire of the same material has the same length and twice the diameter as the first. How far will it be stretched by the same force?
32. || Static climbing ropes are designed to be relatively stiff so that they stretch less than dynamic ropes. To meet a certain specification, an 11-mm-diameter rope must experience a maximum elongation of 5.0% when supporting a 150 kg load. What is the minimum Young's modulus?
33. ||| What hanging mass will stretch a 2.0-m-long, 0.50-mm-diameter steel wire by 1.0 mm?
34. ||| An 80-cm-long, 1.0-mm-diameter steel guitar string must be tightened to a tension of 2.0 kN by turning the tuning screws. By how much is the string stretched?
35. ||| A mineshaft has an ore elevator hung from a single braided cable of diameter 2.5 cm. Young's modulus of the cable is 10×10^{10} N/m^2. When the cable is fully extended, the end of the cable is 800 m below the support. How much does the fully extended cable stretch when 1000 kg of ore is loaded?
36. || The normal force of the ground on the foot can reach three times a runner's body weight when the foot strikes the pavement. By what amount does the 52-cm-long femur of an 80 kg runner compress at this moment? The cross-section area of the bone of the femur can be taken as 5.2×10^{-4} m^2. BIO
37. ||| A three-legged wooden bar stool made out of solid Douglas fir has legs that are 2.0 cm in diameter. When a 75 kg man sits on the stool, by what percent does the length of the legs decrease? Assume, for simplicity, that the stool's legs are vertical and that each bears the same load.
38. || To penetrate armor, a projectile's point concentrates force in a small area, creating a stress large enough that the armor fails. A species of jellyfish launches a pointed needle that can penetrate the hard shell of a crustacean. The rapid deceleration on impact creates a 32 μN force on the tip, which has a very small 15 nm radius. What is the resulting stress? How does this compare to the ultimate stress of steel? BIO INT
39. ||| A 3.0-m-tall, 50-cm-diameter concrete column supports a 200,000 kg load. By how much is the column compressed?
40. || You've just put a new wood floor in your house. An object will dent the flooring if the stress—the force divided by the area—exerted by the object is great enough. Who is more likely to dent your floor: a 50 kg woman in high-heeled shoes (assume a circular heel pad 0.50 cm in diameter) with all of her weight on one heel or a 5000 kg African elephant (assume a circular contact area of 40 cm in diameter for one foot) standing on all four feet? BIO INT
41. || A glass optical fiber in a communications system has a diameter of 9.0 μm.
    a. What maximum tension could this fiber support without breaking?
    b. Assume that the fiber stretches in a linear fashion until the instant it breaks. By how much will a 10-m-long fiber have stretched when it is at the breaking point?
42. ||| If you tethered a space station to the earth by a long cable, you could get to space in an elevator that rides up the cable—much simpler and cheaper than riding to space on a rocket. There's one big problem, however: There is no way to create a cable that is long enough. The cable would need to reach 36,000 km upward, to the height where a satellite orbits at the same speed as the earth rotates; a cable this long made of ordinary materials couldn't even support its own weight. Consider a steel cable suspended from a point high above the earth. The stress in the cable is highest at the top; it must support the weight of cable below it. What is the greatest length the cable could have without failing? For the purposes of this problem, you can ignore the variation in gravity near the surface of the earth. **Hint:** The mass of the cable is the volume of the cable multiplied by the density. The density of steel is 7900 kg/m^3. INT
43. || The Achilles tendon connects the muscles in your calf to the back of your foot. When you are sprinting, your Achilles tendon alternately stretches, as you bring your weight down onto your forward foot, and contracts to push you off the ground. A 70 kg runner has an Achilles tendon that is 15 cm long with a typical 1.1×10^{-4} m^2 area. BIO
    a. By how much will the runner's Achilles tendon stretch if the maximum force on it is 8.0 times his weight, a typical value while running?
    b. What fraction of the tendon's length does this correspond to?

Section 8.5 Forces and Torques in the Body

44. || A woman is pushing a load in a wheelbarrow, as in Figure P8.44. The combined mass of the wheelbarrow and the load is 110 kg, with a center of gravity 0.25 m behind the axle. The woman supports the wheelbarrow at the handles, 1.1 m behind the axle.
    a. What is the force required to support the wheelbarrow?
    b. What fraction of the weight of the wheelbarrow and the load does this force represent?

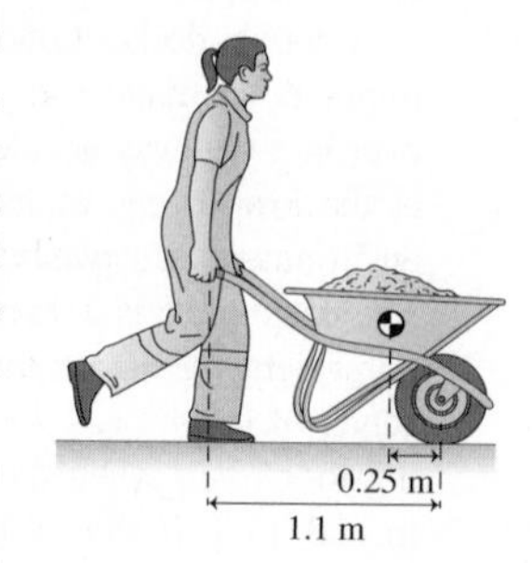

FIGURE P8.44

45. ‖ Figure P8.45 shows the operation of a garlic press. The lower part of the press is held steady, and the upper handle is pushed down, thereby crushing a garlic clove through a screen. Approximate distances are shown in the figure. If the user exerts a 12 N force on the upper handle, estimate the force on the clove.

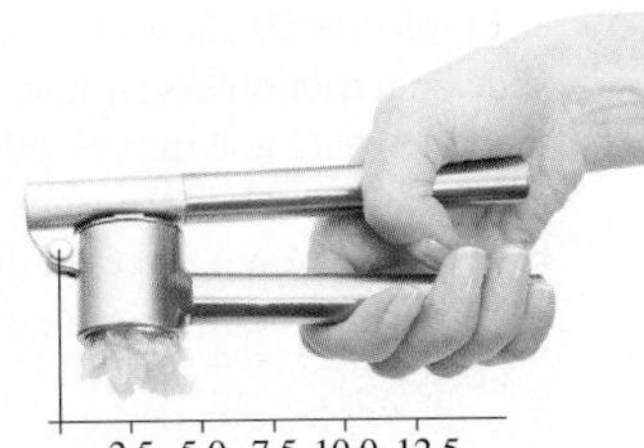

FIGURE P8.45

46. ||| Consider a rower in a scull as in Figure P8.46. The oars aren't accelerating, and they are rotating at a constant speed, so the net force and net torque on the oars are zero. An oar is 2.8 m long, and the rower pulls with a 250 N force on the handle, which is 0.92 m from the pivot.
    a. Assume that the oar touches the water at its very end. What is the drag force from the water on the oar? Assume that the oar is perpendicular to the boat, and that the force of the rower and the drag force are both perpendicular to the oar.
    b. Given that both oars are the same, what is the total force propelling the boat forward?

FIGURE P8.46

47. ‖ Hold your upper arm vertical and your lower arm horizontal BIO with your hand palm-down on a table, as shown in Figure P8.47. If you now push down on the table, you'll feel that your triceps muscle has contracted and is trying to pivot your lower arm about the elbow joint. If a person with the arm dimensions shown pushes down hard with a 90 N force (about 20 lb), what force must the triceps muscle provide? You can ignore the mass of the arm and hand in your calculation.

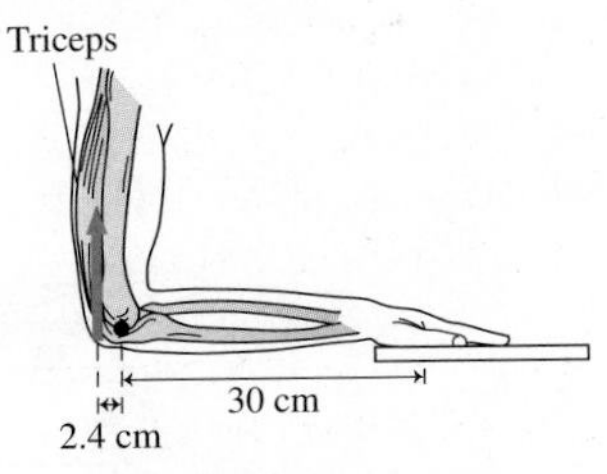

FIGURE P8.47

48. ‖ If you stand on one foot while holding your other leg up BIO behind you, your muscles apply a force to hold your leg in this raised position. We can model this situation as in Figure P8.48. The leg pivots at the knee joint, and the force that holds the leg up is provided by a tendon attached to the lower leg as shown. Assume that the lower leg and the foot have a combined mass of 4.0 kg, and that their combined center of gravity is at the center of the lower leg.
    a. How much force must the tendon exert to keep the leg in this position?
    b. As you hold your leg in this position, the upper leg exerts a force on the lower leg at the knee joint. What are the magnitude and direction of this force?

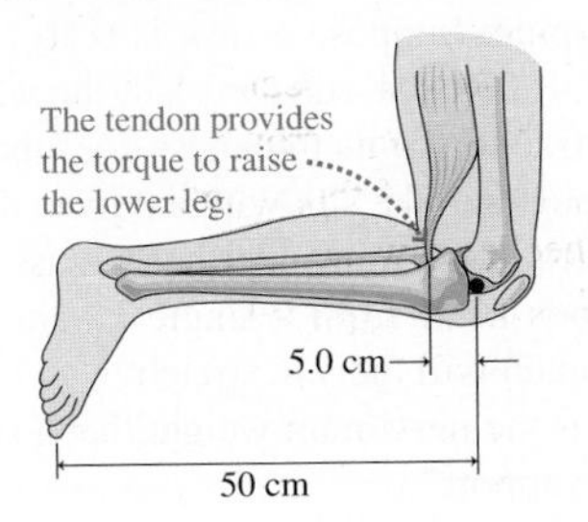

The tendon provides
the torque to raise
the lower leg.
5.0 cm
50 cm

FIGURE P8.48

49. ‖ If you hold your arm outstretched with palm upward, as in BIO Figure P8.49, the force to keep your arm from falling comes from your deltoid muscle. The arm of a typical person has mass 4.0 kg and the distances and angles shown in the figure.
    a. What force must the deltoid muscle provide to keep the arm in this position?
    b. By what factor does this force exceed the weight of the arm?

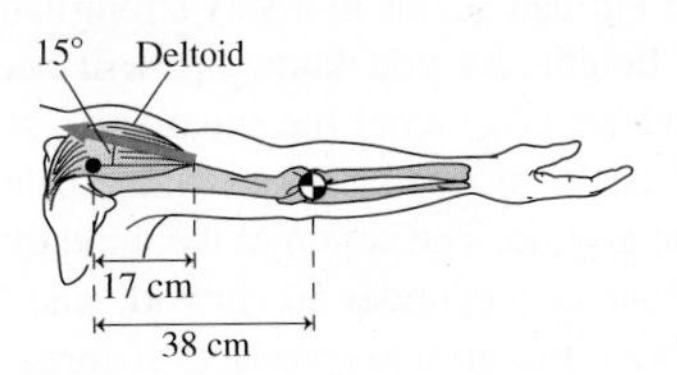

FIGURE P8.49

50. ||| Dogs—like many animals—stand and walk on their toes. A BIO photo of the rear foot of a dog is shown in Figure P8.50a; Figure P8.50b shows the bones of the leg and foot along with relevant distances. The colored element corresponds to your foot, and the connection with the leg corresponds to your ankle. The Achilles tendon pulls on the end of the foot, along a line 4.0 cm from the ankle. What is the tension in the tendon if a 20 kg dog is supporting ¼ of its weight on one rear foot?

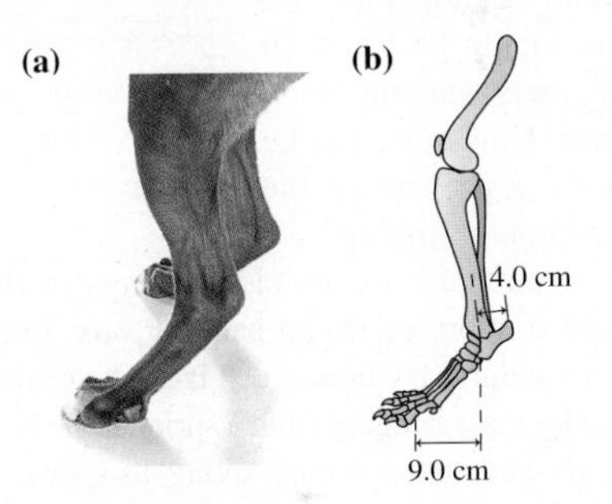

FIGURE P8.50

General Problems

51. |||| A 3.0-m-long rigid beam with a mass of 100 kg is supported at each end, as shown in Figure P8.51. An 80 kg student stands 2.0 m from support 1. How much upward force does each support exert on the beam?

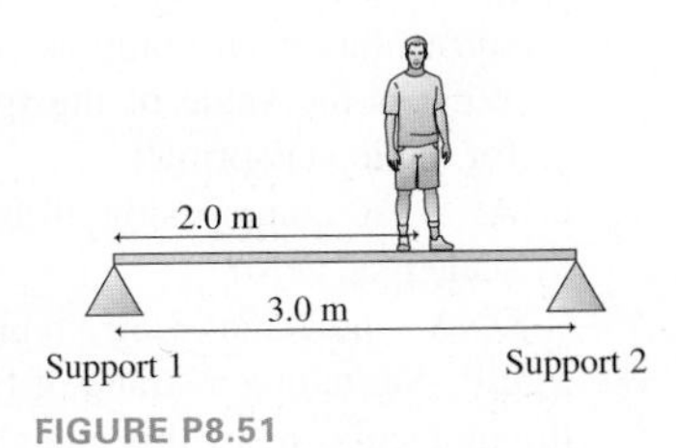

FIGURE P8.51

52. ||| An 80 kg construction worker sits down 2.0 m from the end of a 1450 kg steel beam to eat his lunch, as shown in Figure P8.52. The cable supporting the beam is rated at 15,000 N. Should the worker be worried?

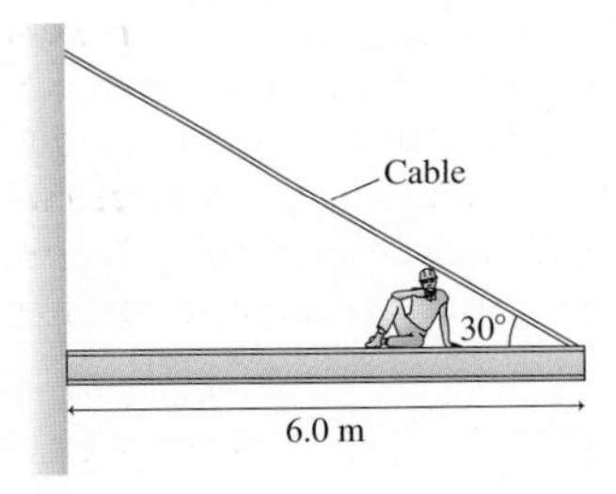

FIGURE P8.52

Watch Video Solution Problems 8.57 and 8.61

53. ||| A man is attempting to raise a 7.5-m-long, 28 kg flagpole that has a hinge at the base by pulling on a rope attached to the top of the pole, as shown in Figure P8.53. With what force does the man have to pull on the rope to hold the pole motionless in this position?

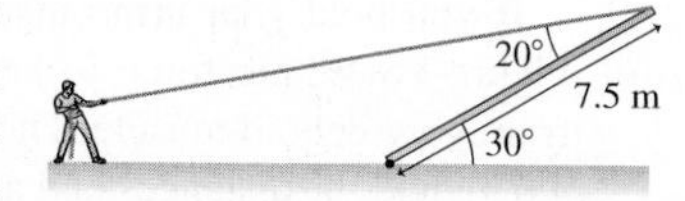

FIGURE P8.53

54. ||| An 85 kg man stands in a very strong wind moving at 14 m/s at torso height. As you know, he will need to lean in to the wind, and we can model the situation to see why. Assume that the man has a mass of 85 kg, with a center of gravity 1.0 m above the ground. The action of the wind on his torso, which we approximate as a cylinder 50 cm wide and 90 cm long centered 1.2 m above the ground, produces a force that tries to tip him over backward. To keep from falling over, he must lean forward.
BIO INT
   a. What is the magnitude of the torque provided by the wind force? Take the pivot point at his feet. Assume that he is standing vertically.
   b. At what angle to the vertical must the man lean to provide a gravitational torque that is equal to this torque due to the wind force?

55. |||| A 40 kg, 5.0-m-long beam is supported by, but not attached to, the two posts in Figure P8.55. A 20 kg boy starts walking along the beam. How close can he get to the right end of the beam without it tipping?

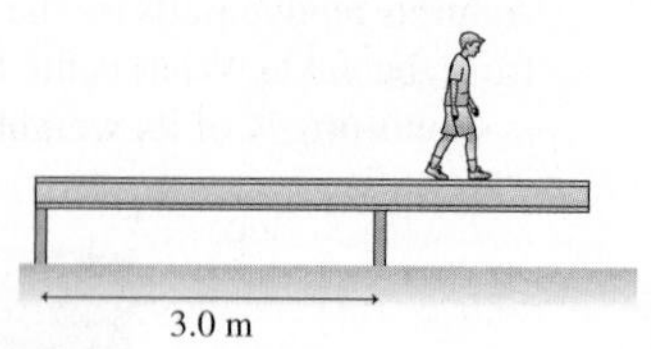

FIGURE P8.55

56. ||| Two identical, side-by-side springs with spring constant 240 N/m support a 2.00 kg hanging box. Each spring supports the same weight. By how much is each spring stretched?

57. | A 5.0 kg mass hanging from a spring scale is slowly lowered onto a vertical spring, as shown in Figure P8.57. The scale reads in newtons.
   a. What does the spring scale read just before the mass touches the lower spring?
   b. The scale reads 20 N when the lower spring has been compressed by 2.0 cm. What is the value of the spring constant for the lower spring?
   c. At what compression distance will the scale read zero?

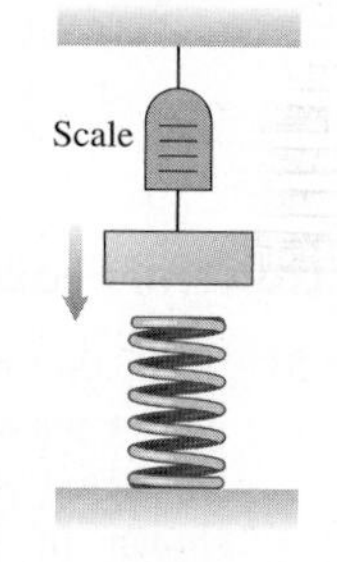

FIGURE P8.57

58. | DNA molecules are typically folded tightly. Stretching a strand of DNA means straightening it, and the molecules resist this straightening. Investigators can attach beads to the ends of a strand of DNA and, using "optical tweezers," measure the force required to produce a certain extension. Data for the stretch of a 3500 base pair strand of DNA approximately follow the line in the graph in Figure P8.58. What is the spring constant for this strand of DNA?
BIO

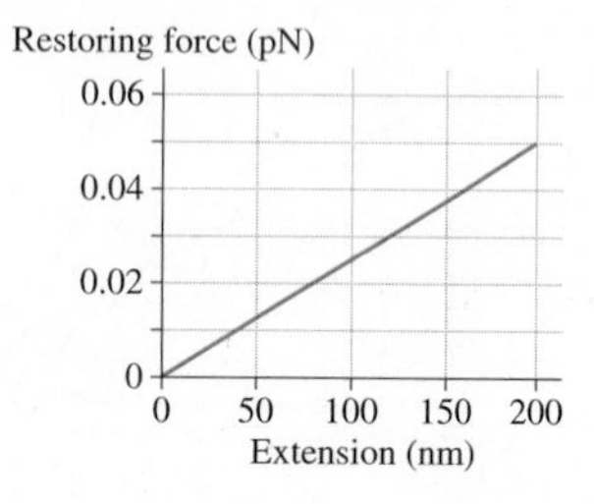

FIGURE P8.58

59. || Figure P8.59 shows two springs attached to a block that can slide on a frictionless surface. In the block's equilibrium position, the left spring is compressed by 2.0 cm.
   a. By how much is the right spring compressed?
   b. What is the net force on the block if it is moved 15 cm to the right of its equilibrium position?

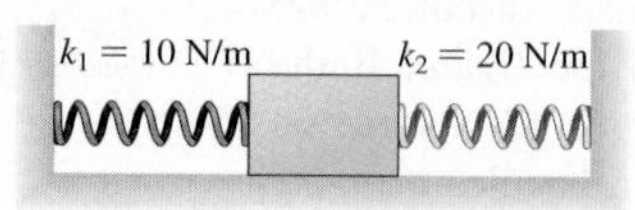

FIGURE P8.59

60. ||| A 25 kg child bounces on a pogo stick. The pogo stick has a spring with spring constant 2.0×10^4 N/m. When the child makes a nice big bounce, she finds that at the bottom of the bounce she is accelerating *upward* at 9.8 m/s^2. How much is the spring compressed?
INT

61. ||| Figure P8.61 shows a lightweight plank supported at its right end by a 7.0-mm-diameter rope with a tensile strength of 6.0×10^7 N/m^2.
   a. What is the maximum force that the rope can support?
   b. What is the greatest distance, measured from the pivot, that the center of gravity of an 800 kg piece of heavy machinery can be placed without snapping the rope?

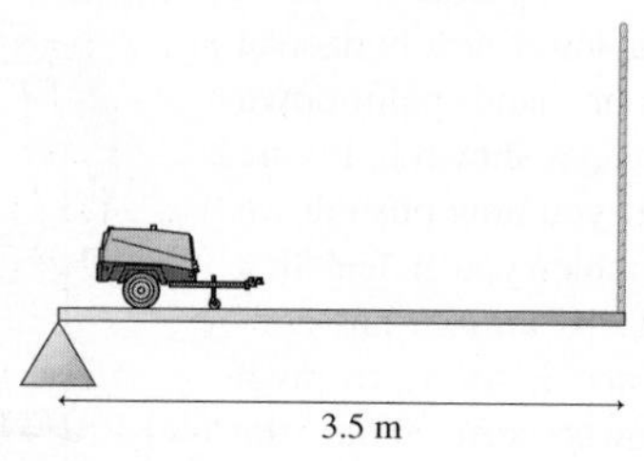

FIGURE P8.61

62. || In the hammer throw, an athlete spins a heavy mass in a circle at the end of a cable before releasing it for distance. For male athletes, the "hammer" is a mass of 7.3 kg at the end of a 1.2 m cable, which is typically a 3.0-mm-diameter steel cable. A world-class thrower can get the hammer up to a speed of 29 m/s. If an athlete swings the mass in a horizontal circle centered on the handle he uses to hold the cable
INT

   a. What is the tension in the cable?
   b. How much does the cable stretch?

63. ||| There is a disk of cartilage between each pair of vertebrae in your spine. Suppose a disk is 0.50 cm thick and 4.0 cm in diameter. If this disk supports half the weight of a 65 kg person, by what fraction of its thickness does the disk compress?
BIO

64. || Orb spiders make silk with a typical diameter of 0.15 mm.
BIO
   a. A typical large orb spider has a mass of 0.50 g. If this spider suspends itself from a single 12-cm-long strand of silk, by how much will the silk stretch?
   b. What is the maximum weight that a single thread of this silk could support?

65. ||| Larger animals have sturdier bones than smaller animals. BIO A mouse's skeleton is only a few percent of its body weight, compared to 16% for an elephant. To see why this must be so, recall, from Example 8.10, that the stress on the femur for a man standing on one leg is 1.4% of the bone's tensile strength. Suppose we scale this man up by a factor of 10 in all dimensions, keeping the same body proportions. Use the data for Example 8.10 to compute the following.
   a. Both the inside and outside diameter of the femur, the region of compact bone, will increase by a factor of 10. What will be the new cross-section area?
   b. The man's body will increase by a factor of 10 in each dimension. What will be his new mass?
   c. If the scaled-up man now stands on one leg, what fraction of the tensile strength is the stress on the femur?

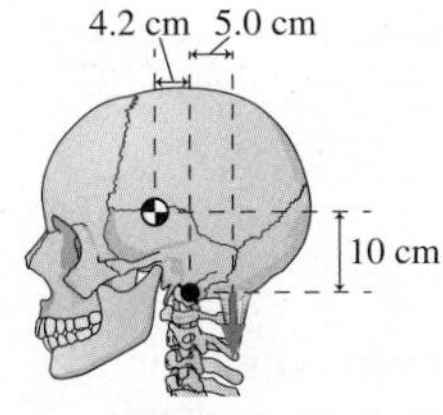

FIGURE P8.66

66. ||| The main muscles that hold your BIO head upright attach to your spine in back of the point where your head pivots on your neck. Figure P8.66 shows typical numbers for the distance from the pivot to the muscle attachment point and the distance from the pivot to the center of gravity of the head. The muscles pull down to keep your head upright. If the muscle relaxes—if, for instance, you doze in one of your classes besides Physics—your head tips forward. In the questions that follow, assume that your head has a mass of 4.8 kg, and that you maintain the relative angle between your head and your spine.
   a. With the head held level, as in Figure P8.66, what muscle force is needed to keep a 4.8 kg head upright?
   b. If you tip your body forward so that your spine is level with the ground, what muscle force is needed to keep your head in the same orientation relative to the spine?
   c. If you tip your body backward, you will reach a point where no muscle force is needed to keep your head upright. For the distances given in Figure P8.66, at what angle does this balance occur?

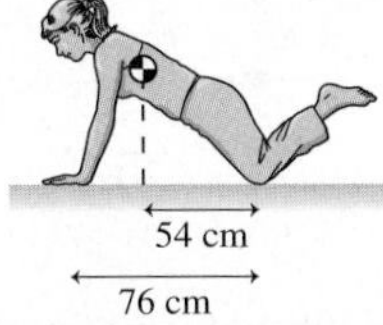

FIGURE P8.67

67. ||| A woman weighing 580 N does a BIO pushup from her knees, as shown in Figure P8.67. What are the normal forces of the floor on (a) each of her hands and (b) each of her knees?

68. |||| When you bend over, a series of large muscles, the erector BIO spinae, pull on your spine to hold you up. Figure P8.68 shows a simplified model of the spine as a rod of length L that pivots at its lower end. In this model, the center of gravity of the 320 N weight of the upper torso is at the center of the spine. The 160 N weight of the head and arms acts at the top of the spine. The erector spinae muscles are modeled as a single muscle that acts at an 12° angle to the spine. Suppose the person in Figure P8.68 bends over to an angle of 30° from the horizontal.
   a. What is the tension in the erector muscle?
      Hint: Align your x-axis with the axis of the spine.
   b. A force from the pelvic girdle acts on the base of the spine. What is the component of this force in the direction of the spine? (This large force is the cause of many back injuries).

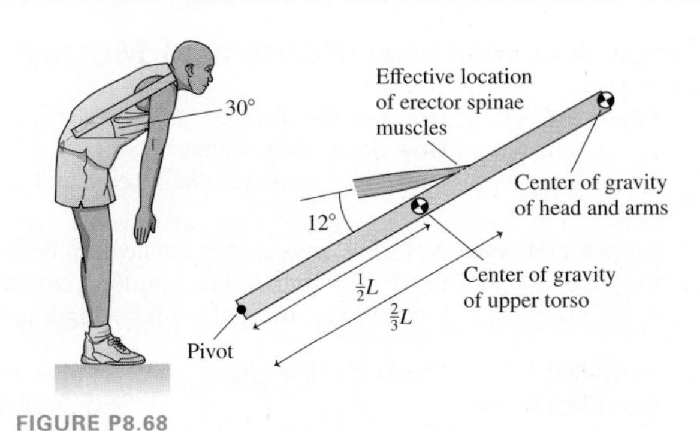

FIGURE P8.68

MCAT-Style Passage Problems

Elasticity of a Ligament BIO

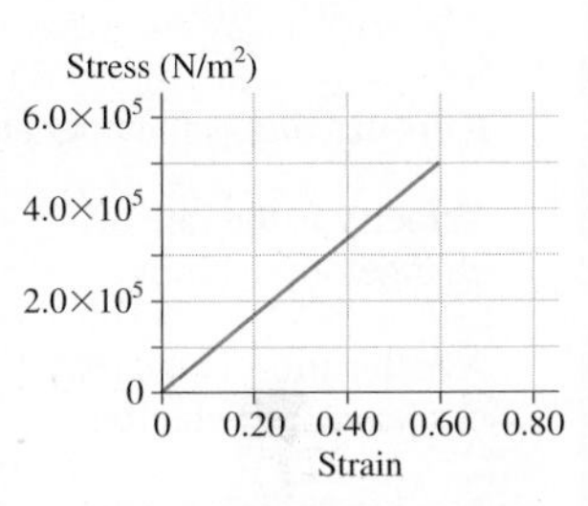

FIGURE P8.69

The nuchal ligament in a horse supports the weight of the horse's head. This ligament is much more elastic than a typical ligament, stretching from 15% to 45% longer than its resting length as a horse's head moves up and down while it runs. This stretch of the ligament stores energy, making locomotion more efficient. Measurements on a segment of ligament show a linear stress-versus-strain relationship until the stress approaches 0.80. Smoothed data for the stretch are shown in Figure P8.69.

69. || What is the approximate Young's modulus for the ligament?
   A. 4×10^5 N/m^2
   B. 8×10^5 N/m^2
   C. 3×10^6 N/m^2
   D. 6×10^6 N/m^2

70. || The segment of ligament tested has a resting length of 40 mm. How long is the ligament at a strain of 0.60?
   A. 46 mm
   B. 52 mm
   C. 58 mm
   D. 64 mm

71. || Suppose the ligament has a circular cross section. For a certain ligament, an investigator measures the restoring force at a strain of 0.40. If the ligament is replaced with one that has twice the diameter, by what factor does the restoring force increase?
   A. 1.4
   B. 2
   C. 4
   D. 8

72. || The volume of the ligament stays the same as it stretches, so the cross-section area decreases as the length increases. Given this, how would a force F versus change in length ΔL curve appear?

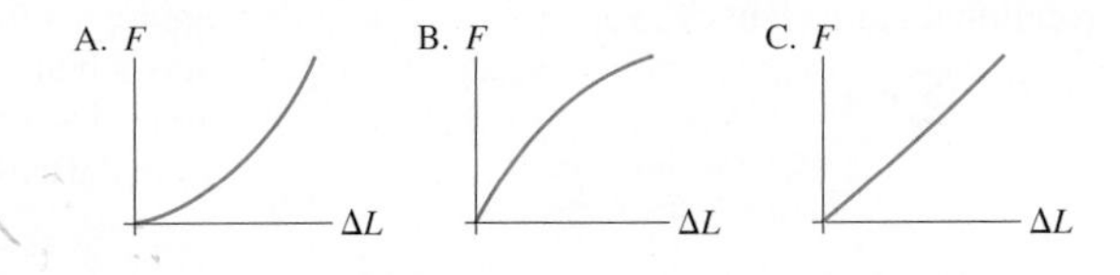

KNOWLEDGE STRUCTURE I Force and Motion

BASIC GOALS How can we describe motion? How does an object respond to a force? How do systems interact? What is the nature of the force of gravity? How can we analyze the motion and deformation of extended objects?

CROSS-CUTTING CONCEPTS Newton's laws are the connection between forces and motion. Using these laws, we can understand how particles move on a straight line, or along curved paths like that of a rider on a Ferris wheel or a baseball in flight. Newton's laws are also the key to understanding the motion of rotating objects.

GENERAL PRINCIPLES

Newton's first law An object with no forces acting on it will remain at rest or move in a straight line at a constant speed.

Newton's second law $\vec{F}_{net} = m\vec{a}$

Newton's third law $\vec{F}_{A\text{ on }B} = -\vec{F}_{B\text{ on }A}$

Newton's law of gravity $F_{1\text{ on }2} = F_{2\text{ on }1} = \dfrac{Gm_1m_2}{r^2}$

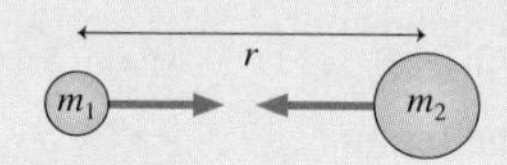

Kinematics is the mathematical description of motion.

Velocity is the rate of change of position: $v_x = \dfrac{\Delta x}{\Delta t}$

Acceleration is the rate of change of velocity: $a_x = \dfrac{\Delta v_x}{\Delta t}$

The **kinematic equations** for motion with constant acceleration are

$$(v_x)_f = (v_x)_i + a_x\Delta t$$

$$x_f = x_i + (v_x)_i\Delta t + \frac{1}{2}a_x(\Delta t)^2$$

$$(v_x)_f^2 = (v_x)_i^2 + 2a_x\Delta x$$

Projectile motion consists of vertical motion with $a_y = -g$ and horizontal motion with constant velocity $(v_x)_i$.

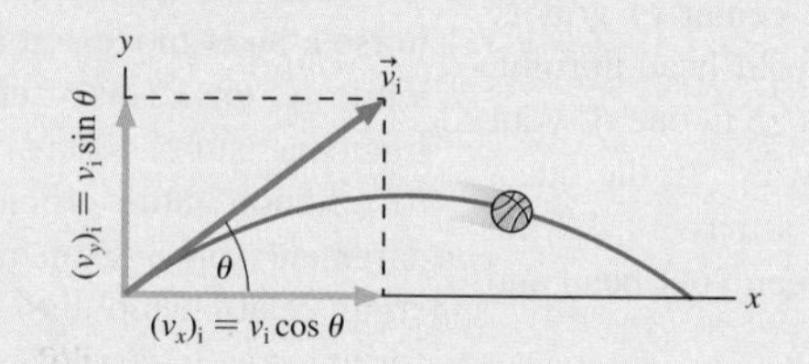

$$x_f = x_i + (v_x)_i\Delta t \qquad y_f = y_i + (v_y)_i\Delta t - \frac{1}{2}g(\Delta t)^2$$

$$(v_x)_f = (v_x)_i = \text{constant} \qquad (v_y)_f = (v_y)_i - g\Delta t$$

In **uniform circular motion** the acceleration points toward the center of the circle.

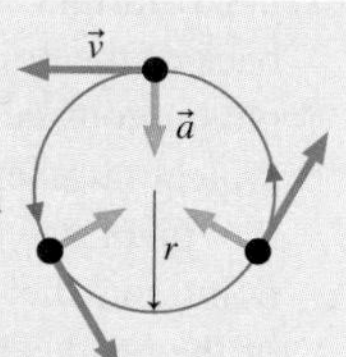

The frequency f, period T, angular velocity ω, speed v, and acceleration a are related by

$$f = \frac{1}{T} \qquad \omega = 2\pi f$$

$$v = \omega r = \frac{2\pi r}{T} \qquad a = \frac{v^2}{r} = \omega^2 r$$

Dynamics is the study of **forces** and how they cause motion.

Common forces:

Weight $\vec{w} = (mg, \text{downward})$

Friction $\vec{f}_k = (\mu_k n, \text{opposite motion})$

$\vec{f}_s = (0 \text{ to } \mu_s n, \text{direction to prevent motion})$

Drag $\vec{D} = (\frac{1}{2}C_D\rho Av^2, \text{opposite motion})$

Spring $\vec{F}_{sp} = (-k\Delta x, \text{opposite displacement})$

A **free-body diagram** shows all the forces acting on an object.

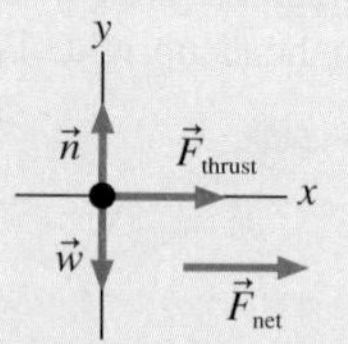

Dynamics problems are solved by writing Newton's second law in component form:

$$\sum F_x = F_{thrust} = ma_x$$

$$\sum F_y = n - mg = ma_y$$

Circular-motion dynamics: To provide the centripetal acceleration, there must be a net force directed toward the center of the circle.

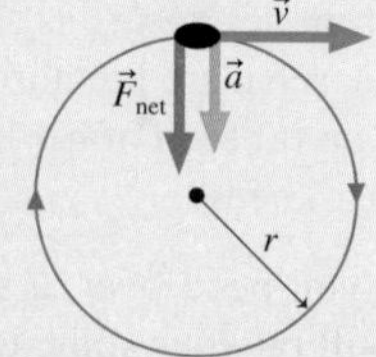

$$\vec{F}_{net} = \left(\frac{mv^2}{r}, \text{toward center of circle}\right)$$

Rigid bodies

An object's **moment of inertia** is the rotational equivalent of mass:

$$I = \sum m_i r_i^2$$

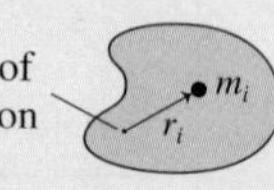

Its **center of gravity** is the point at which gravity acts:

$$x_{cg} = \frac{x_1m_1 + x_2m_2 + \cdots}{m_1 + m_2 + \cdots}$$

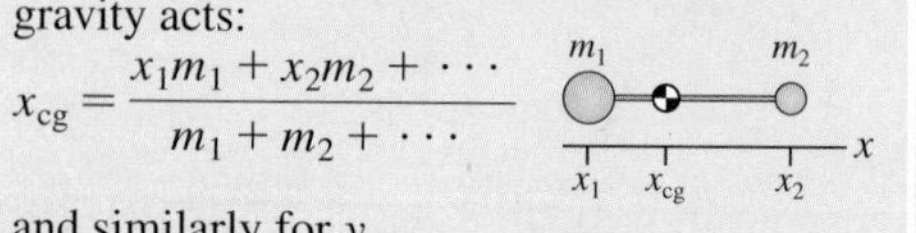

and similarly for y_{cg}.

Torque and rotational dynamics

Torque is the rotational analog of force. A torque acts when a force is applied at a point away from the axis of rotation:

$$\tau = rF\sin\phi$$

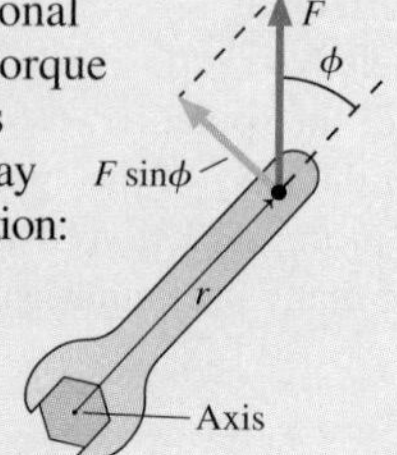

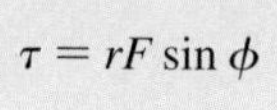

Rotation about a fixed axis

When a net torque is applied to an object with moment of inertia I that can rotate about a fixed axis, the object undergoes an **angular acceleration** given by

$$\alpha = \frac{\tau_{net}}{I}$$

Dark Matter and the Structure of the Universe

The idea that the earth exerts a gravitational force on us is something we now accept without questioning. But when Isaac Newton developed this idea to show that the gravitational force also holds the moon in its orbit, it was a remarkable, ground-breaking insight. It changed the way that we look at the universe we live in.

Newton's laws of motion and gravity are tools that allow us to continue Newton's quest to better understand our place in the cosmos. But it sometimes seems that the more we learn, the more we realize how little we actually know and understand.

Here's an example. Advances in astronomy over the past 100 years have given us great insight into the structure of the universe. But everything our telescopes can see appears to be only a small fraction of what is out there. Approximately 80% of the mass in the universe is *dark matter*—matter that gives off no light or other radiation that we can detect. Everything that we have ever seen through a telescope is merely the tip of the cosmic iceberg.

What is this dark matter? Black holes? Neutrinos? Some form of exotic particle? We simply aren't sure. It could be any of these, or all of them—or something entirely different that no one has yet dreamed of. You might wonder how we know that such matter exists if no one has seen it. Even though we can't directly observe dark matter, we see its effects. And you now know enough physics to understand why.

Whatever dark matter is, it has mass, and so it has gravity. This picture of the Andromeda galaxy shows a typical spiral galaxy structure: a dense collection of stars in the center surrounded by a disk of stars and other matter. This is the shape of our own Milky Way galaxy.

The spiral Andromeda galaxy.

This structure is reminiscent of the structure of the solar system: a dense mass (the sun) in the center surrounded by a disk of other matter (the planets, asteroids, and comets). The sun's gravity keeps the planets in their orbits, but the planets would fall into the sun if they were not in constant motion around it. The same is true of a spiral galaxy; everything in the galaxy orbits its center. Our solar system orbits the center of our galaxy with a period of about 200 million years.

The orbital speed of an object depends on the mass that pulls on it. If you analyze our sun's motion about the center of the Milky Way, or the motion of stars in the Andromeda galaxy about its center, you find that the orbits are much faster than they should be, based on how many stars we see. There must be some other mass present.

There's another problem with the orbital motion of stars around the center of their galaxies. We know that the orbital speeds of planets decrease with distance from the sun; Neptune orbits at a much slower speed than the earth. We might expect something similar for galaxies: Stars farther from the center should orbit at reduced speeds. But they don't. As we measure outward from the center of the galaxy, the orbital speed stays about the same—even as we get to the edge of the visible disk. There must be some other mass—the invisible dark matter—exerting a gravitational force on the stars. This dark matter, which far outweighs the matter we can see, seems to form a halo around the centers of galaxies, providing the gravitational force necessary to produce the observed rotation. Other observations of the motions of galaxies with respect to each other verify this basic idea.

On a cosmic scale, the picture is even stranger. The universe is currently expanding. The mutual gravitational attraction of all matter—regular and dark—in the universe should slow this expansion. But recent observations of the speeds of distant galaxies imply that the expansion of the universe is accelerating, so there must be yet another component to the universe, something that "pushes out." The best explanation at present is that the acceleration is caused by *dark energy*. The nature of dark matter isn't known, but the nature of dark energy is even more mysterious. If current theories hold, it's the most abundant stuff in the universe. And we don't know what it is.

This sort of mystery is what drives scientific investigation. It's what drove Newton to wonder about the connection between the fall of an apple and the motion of the moon, and what drove investigators to develop all of the techniques and theories you will learn about in the coming chapters.

The following questions are related to the passage "Dark Matter and the Structure of the Universe" on the previous page.

1. As noted in the passage, our solar system orbits the center of the Milky Way galaxy in about 200 million years. If there were no dark matter in our galaxy, this period would be
   A. Longer.
   B. The same.
   C. Shorter.
2. Saturn is approximately 10 times as far away from the sun as the earth. This means that its orbital acceleration is ________ that of the earth.
   A. 1/10
   B. 1/100
   C. 1/1000
   D. 1/10,000
3. Saturn is approximately 10 times as far away from the sun as the earth. If dark matter changed the orbital properties of the planets so that Saturn had the same orbital speed as the earth, Saturn's orbital acceleration would be ________ that of the earth.
   A. 1/10
   B. 1/100
   C. 1/1000
   D. 1/10,000
4. Which of the following might you expect to be an additional consequence of the fact that galaxies contain more mass than expected?
   A. The gravitational force between galaxies is greater than expected.
   B. Galaxies appear less bright than expected.
   C. Galaxies are farther away than expected.
   D. There are more galaxies than expected.

The following passages and associated questions are based on the material of Part I.

Animal Athletes BIO

Different animals have very different capacities for running. A horse can maintain a top speed of 20 m/s for a long distance but has a maximum acceleration of only 6.0 m/s^2, half what a good human sprinter can achieve with a block to push against. Greyhounds, dogs especially bred for feats of running, have a top speed of 17 m/s, but their acceleration is much greater than that of the horse. Greyhounds are particularly adept at turning corners at a run.

FIGURE I.1

5. If a horse starts from rest and accelerates at the maximum value until reaching its top speed, how much time elapses, to the nearest second?
   A. 1 s B. 2 s
   C. 3 s D. 4 s
6. If a horse starts from rest and accelerates at the maximum value until reaching its top speed, how far does it run, to the nearest 10 m?
   A. 40 m B. 30 m
   C. 20 m D. 10 m
7. A greyhound on a racetrack turns a corner at a constant speed of 15 m/s with an acceleration of 7.1 m/s^2. What is the radius of the turn?
   A. 40 m B. 30 m
   C. 20 m D. 10 m
8. A human sprinter of mass 70 kg starts a run at the maximum possible acceleration, pushing backward against a block set in the track. What is the force of his foot on the block?
   A. 1500 N B. 840 N
   C. 690 N D. 420 N
9. In the photograph of the greyhounds in Figure I.1, what is the direction of the net force on each dog?
   A. Up
   B. Down
   C. Left, toward the outside of the turn
   D. Right, toward the inside of the turn

Sticky Liquids BIO

At small scales, viscous drag becomes very important. To a paramecium (Figure I.2), a single-celled animal that can propel itself through water with fine hairs on its body, swimming through water feels like swimming through honey would to you. We can model a paramecium as a sphere of diameter 50 μm, with a mass of 6.5×10^{-11} kg.

FIGURE I.2

10. A paramecium swimming at a constant speed of 0.25 mm/s ceases propelling itself and slows to a stop. At the instant it stops swimming, what is the magnitude of its acceleration?
    A. $0.2g$ B. $0.5g$
    C. $2g$ D. $5g$
11. If the acceleration of the paramecium in Problem 10 were to stay constant as it came to rest, approximately how far would it travel before stopping?
    A. 0.02 μm B. 0.2 μm
    C. 2 μm D. 20 μm
12. If the paramecium doubles its swimming speed, how does this change the drag force?
    A. The drag force decreases by a factor of 2.
    B. The drag force is unaffected.
    C. The drag force increases by a factor of 2.
    D. The drag force increases by a factor of 4.
13. You can test the viscosity of a liquid by dropping a steel sphere into it and measuring the speed at which it sinks. For viscous fluids, the sphere will rapidly reach a terminal speed. At this terminal speed, the net force on the sphere is
    A. Directed downward.
    B. Zero.
    C. Directed upward.

Pulling Out of a Dive BIO

Falcons are excellent fliers that can reach very high speeds by diving nearly straight down. To pull out of such a dive, a falcon extends its wings and flies through a circular arc that redirects its motion. The forces on the falcon that control its motion are its weight and an upward lift force—like an airplane—due to the air flowing over its wings. At the bottom of the arc, as in Figure I.3, a falcon can easily achieve an acceleration of 15 m/s^2.

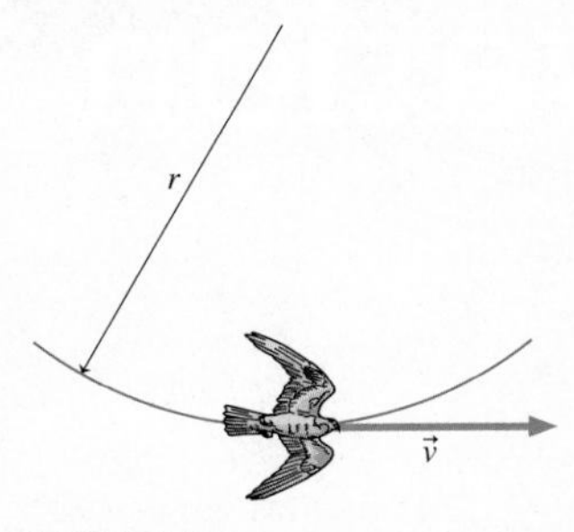

FIGURE I.3

14. At the bottom of the arc, as in Figure I.3, what is the direction of the net force on the falcon?
    A. To the left, opposite the motion
    B. To the right, in the direction of the motion
    C. Up
    D. Down
    E. The net force is zero.
15. Suppose the falcon weighs 8.0 N and is turning with an acceleration of 15 m/s^2 at the lowest point of the arc. What is the magnitude of the upward lift force at this instant?
    A. 8.0 N B. 12 N
    C. 16 N D. 20 N
16. A falcon starts from rest, does a free-fall dive from a height of 30 m, and then pulls out by flying in a circular arc of radius 50 m. Which segment of the motion has a higher acceleration?
    A. The free-fall dive
    B. The circular arc
    C. The two accelerations are equal.

Bending Beams

If you bend a rod down, it compresses the lower side of the rod and stretches the top, resulting in a restoring force. Figure I.4 shows a beam of length L, width w, and thickness t fixed at one end and free to move at the other. Deflecting the end of the beam causes a restoring force F at the end of the beam. The magnitude of the restoring force F depends on the dimensions of the beam, the Young's modulus Y for the material, and the deflection d. For small values of the deflection, the restoring force is

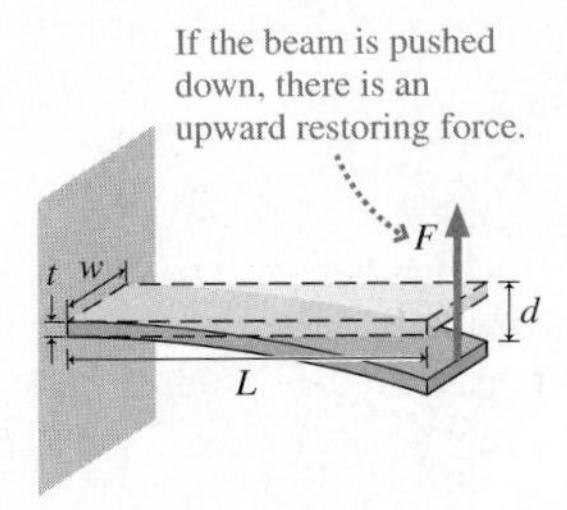

If the beam is pushed down, there is an upward restoring force.
F
t
w
d
L

FIGURE I.4

$$F = \left[\frac{Ywt^3}{4L^3}\right]d$$

This is similar to the formula for the restoring force of a spring, with the quantity in brackets playing the role of the spring constant k.

When a 70 kg man stands on the end of a springboard (a type of diving board), the board deflects by 4.0 cm.

17. If a 35 kg child stands at the end of the board, the deflection is
    A. 1.0 cm B. 2.0 cm
    C. 3.0 cm D. 4.0 cm
18. A 70 kg man jumps up and lands on the end of the board, deflecting it by 12 cm. At this instant, what is the approximate magnitude of the upward force the board exerts on his feet?
    A. 700 N B. 1400 N
    C. 2100 N D. 2800 N
19. If the board is replaced by one that is half the length but otherwise identical, how much will it deflect when a 70 kg man stands on the end?
    A. 0.50 cm B. 1.0 cm
    C. 2.0 cm D. 4.0 cm

Additional Integrated Problems

20. You go to the playground and slide down the slide, a 3.0-m-long ramp at an angle of 40° with respect to horizontal. The pants that you've worn aren't very slippery; the coefficient of kinetic friction between your pants and the slide is $\mu_k = 0.45$. A friend gives you a very slight push to get you started. How long does it take you to reach the bottom of the slide?
21. If you stand on a scale at the equator, the scale will read slightly less than your true weight due to your circular motion with the rotation of the earth.
    a. Draw a free-body diagram to show why this is so.
    b. By how much is the scale reading reduced for a person with a true weight of 800 N?
22. Dolphins and other sea creatures can leap to great heights by swimming straight up and exiting the water at a high speed. A 210 kg dolphin leaps straight up to a height of 7.0 m. When the dolphin reenters the water, drag from the water brings it to a stop in 1.5 m. Assuming that the force of the water on the dolphin stays constant as it slows down,
    a. How much time does it take for the dolphin to come to rest?
    b. What is the force of the water on the dolphin as it is coming to rest?

PART II

Conservation Laws

A peregrine falcon pulls in its wings and begins a nearly vertical dive. Speeding up as it falls, the falcon can reach a speed of 80 m/s or higher, moving faster than any other animal. When the falcon strikes a slower-moving bird at this high speed, the force of the resulting impact may exceed 100 times the falcon's weight, making for a very effective strike. In the next section, we'll learn how analyze the falcon's fall and the forces in an impact using the principles of conservation of energy and momentum.

OVERVIEW

Why Some Things Stay the Same

Part I of this textbook was about *change*. Simple observations show us that most things in the world around us are changing. Even so, there are some things that *don't* change even as everything else is changing around them. Our emphasis in Part II will be on things that stay the same.

Consider, for example, a strong, sealed box in which you have replaced all the air with a mixture of hydrogen and oxygen. The mass of the box plus the gases inside is 600.0 g. Now, suppose you use a spark to ignite the hydrogen and oxygen. As you know, this is an explosive reaction, with the hydrogen and oxygen combining to create water—and quite a bang. But the strong box contains the explosion and all of its products.

What is the mass of the box after the reaction? The gas inside the box is different now, but a careful measurement would reveal that the mass hasn't changed—it's still 600.0 g! We say that the mass is *conserved*. Of course, this is true only if the box has stayed sealed. For conservation of mass to apply, the system must be *closed*.

Conservation Laws

A closed system of interacting objects has another remarkable property. Each system is characterized by a certain number, and no matter how complex the interactions, the value of this number never changes. This number is called the *energy* of the system, and the fact that it never changes is called the *law of conservation of energy*. It is, perhaps, the single most important physical law ever discovered.

The law of conservation of energy is much more general than Newton's laws. Energy can be converted to many different forms, and, in all cases, the total energy stays the same:

- Gasoline, diesel, and jet engines convert the energy of a fuel into the mechanical energy of moving pistons, wheels, and gears.
- A solar cell converts the electromagnetic energy of light into electric energy.
- An organism converts the chemical energy of food into a variety of other forms of energy, including kinetic energy, sound energy, and thermal energy.

Energy will be *the* most important concept throughout the remainder of this text and much of Part II will focus on understanding what energy is and how it is used.

But energy is not the only conserved quantity. We will begin Part II with the study of two other quantities that are conserved in a closed system: *momentum* and *angular momentum*. Their conservation will help us understand a wide range of physical processes, from the forces when two rams butt heads to the graceful spins of ice skaters.

Conservation laws will give us a new and different *perspective* on motion. Some situations are most easily analyzed from the perspective of Newton's laws, but others make much more sense when analyzed from a conservation-law perspective. An important goal of Part II is to learn which perspective is best for a given problem.

9 Momentum

Male rams butt heads at high speeds in a ritual to assert their dominance. How can the force of this collision be minimized so as to avoid damage to their brains?

LOOKING AHEAD ▸

Impulse

This golf club delivers an **impulse** to the ball as the club strikes it.

You'll learn that a longer-lasting, stronger force delivers a greater impulse to an object.

Momentum and Impulse

The impulse delivered by the player's head *changes* the ball's **momentum.**

You'll learn how to calculate this momentum change using the **impulse-momentum theorem**.

Conservation of Momentum

As the octopus sends out a jet of water, the momentum of the octopus plus the water stays the same. The momentum is **conserved.**

You'll learn a powerful new *before-and-after* problem-solving approach using this **law of conservation of momentum.**

GOAL To learn about impulse, momentum, and a new problem-solving approach based on conservation laws.

LOOKING BACK ◂

Newton's Third Law

In Section 4.7 you learned about Newton's third law. In this chapter, you'll apply this law in order to understand the conservation of momentum.

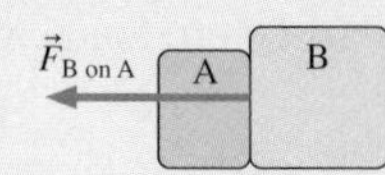

Newton's third law states that the force that object B exerts on A has *equal magnitude* but is *directed opposite to* the force that A exerts on B.

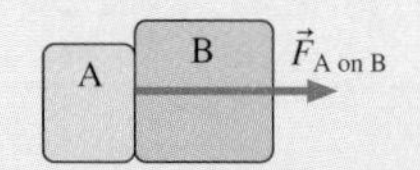

STOP TO THINK

A hammer hits a nail. The force of the nail on the hammer is

A. Greater than the force of the hammer on the nail.
B. Less than the force of the hammer on the nail.
C. Equal to the force of the hammer on the nail.
D. Zero.

9.1 Impulse

A **collision** is a short-duration interaction between two objects. The collision between a tennis ball and a racket, or your foot and a soccer ball, may seem instantaneous to your eye, but that is a limitation of your perception. The sequence of high-speed photos of a soccer kick shown in FIGURE 9.1 reveals that the ball is compressed as the foot begins its contact. It takes time to compress the ball, and more time for the ball to re-expand as it leaves the foot.

FIGURE 9.1 A sequence of high-speed photos of a soccer ball being kicked.

Contact begins

Maximum compression

Contact ends

The duration of a collision depends on the materials from which the objects are made. The harder the objects, the shorter the contact time. A collision between two pool balls lasts less than 1 ms; the collision between the foot and the soccer ball in Figure 9.1 lasted more than 20 ms.

Let's look at the soccer ball kick in Figure 9.1 again. As the foot and the ball first come into contact, as shown in the left frame, the ball is just beginning to compress. By the middle frame of the figure, the ball has sped up and become greatly compressed. Finally, as shown in the right frame, the ball, now moving very fast, is again only slightly compressed. The ball is elastic, as you know; when you compress it, it exerts a restoring force equal to the force of the foot on the ball. Greater compression implies a greater force, just as for a spring.

FIGURE 9.2 is a graph of force versus time for a typical soccer ball kick. The force is zero until the foot first contacts the ball, rises quickly to a maximum value, and then falls back to zero as the ball leaves the foot. Thus there is a well-defined duration Δt of the force. A large force like this exerted during a short interval of time is called an **impulsive force.** The forces of a hammer on a nail and of a bat on a baseball are other examples of impulsive forces.

FIGURE 9.2 The force on a soccer ball changes rapidly.

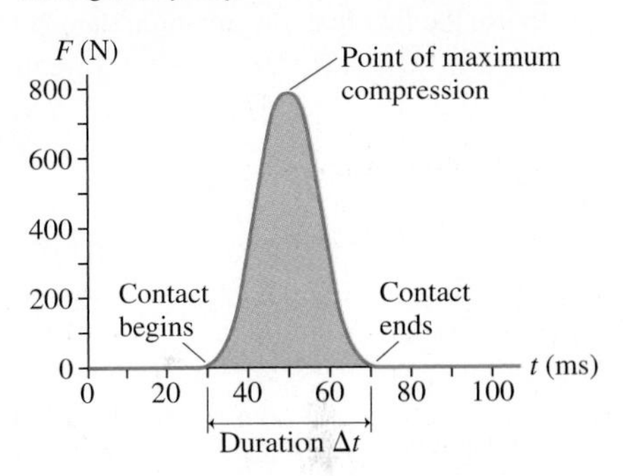
F (N)
800
600
400
200
0
Point of maximum compression
Contact begins
Contact ends
0
20
40
60
80
100
t (ms)
Duration Δt

A harder kick (i.e., a taller force curve) or a kick of longer duration (a wider force curve) causes the ball to leave the kicker's foot with a higher speed; that is, the *effect* of the kick is larger. A taller or wider force-versus-time curve has a larger *area* between the curve and the axis (i.e., the area "under" the force curve is larger), so we can say that **the effect of an impulsive force is proportional to the area under the force-versus-time curve.** This area, shown in FIGURE 9.3a, is the **impulse** J of the force.

Impulsive forces can be complex, and the shape of the force-versus-time graph often changes in a complicated way. Consequently, it is often useful to think of the collision in terms of an *average* force F_{avg}. As FIGURE 9.3b shows, F_{avg} is defined to be the constant force that has the same duration Δt and the same area under the force curve as the real force. You can see from the figure that the area under the force curve can be written simply as $F_{avg}\,\Delta t$. Thus

$$\text{impulse } J = \text{area under the force curve} = F_{avg}\,\Delta t \qquad (9.1)$$

Impulse due to a force acting for a duration Δt

From Equation 9.1 we see that impulse has units of N · s, but N · s are equivalent to kg · m/s. We'll see shortly why the latter are the preferred units for impulse.

FIGURE 9.3 Looking at the impulse graphically.

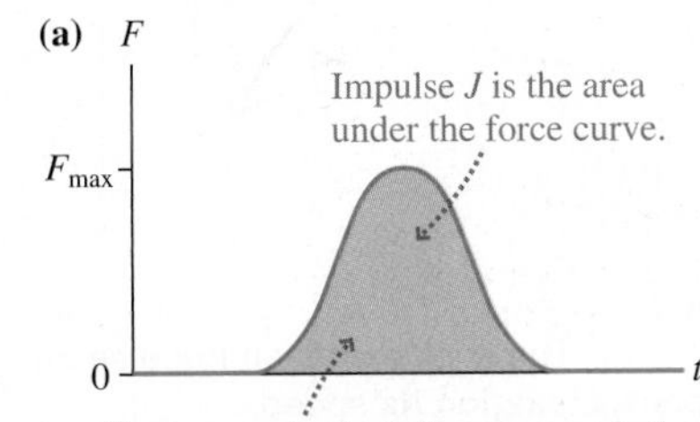
(a) F
Impulse J is the area under the force curve.
F_max
0
t
The area under both force curves is the same; thus, both forces deliver the same impulse. This means that they have the same effect on the object.

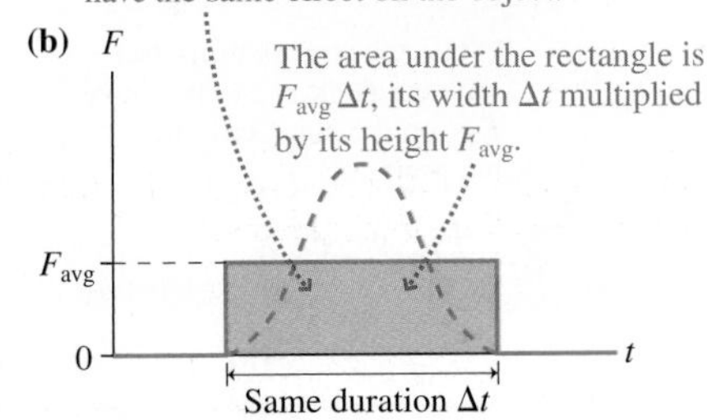
(b) F
The area under the rectangle is F_avg Δt, its width Δt multiplied by its height F_avg.
F_avg
0
t
Same duration Δt

So far, we've been assuming the force is directed along a coordinate axis, such as the x-axis. In this case impulse is a *signed* quantity—it can be positive or negative. A positive impulse results from an average force directed in the positive x-direction (that is, F_{avg} is positive), while a negative impulse is due to a force directed in the negative x-direction (F_{avg} is negative). More generally, the impulse is a *vector* quantity pointing in the direction of the average force vector:

$$\vec{J} = \vec{F}_{avg}\,\Delta t \tag{9.2}$$

EXAMPLE 9.1 Finding the impulse on a bouncing ball

FIGURE 9.4 shows smoothed force-versus-time data for the bounce of a rubber ball.

a. What is the impulse on the ball?
b. What is the average force on the ball?

FIGURE 9.4 The force of the floor on a bouncing ball.

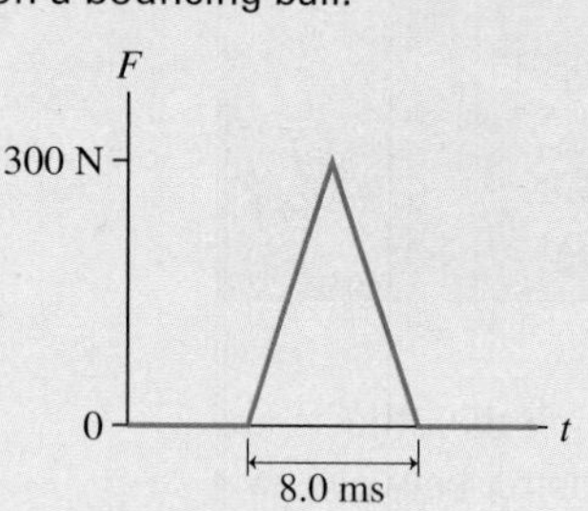

STRATEGIZE The impulse is the area under the force curve. After we compute the area, and thus the impulse, we will calculate the average force using Equation 9.1.

PREPARE Here the shape of the graph is triangular, so we'll need to use the fact that the area of a triangle is $\frac{1}{2} \times$ height $\times$ base.

SOLVE

a. The impulse is

$$J = \frac{1}{2}(300\ \text{N})(0.0080\ \text{s}) = 1.2\ \text{N}\cdot\text{s} = 1.2\ \text{kg}\cdot\text{m/s}$$

b. We can rearrange Equation 9.1 to find the average force that would give this same impulse:

$$F_{avg} = \frac{J}{\Delta t} = \frac{1.2\ \text{N}\cdot\text{s}}{0.0080\ \text{s}} = 150\ \text{N}$$

ASSESS In this example, the average value of the force is half the maximum value. This is not surprising for a triangular force because the area of a triangle is half the base times the height.

STOP TO THINK 9.1 The two graphs show smoothed force-versus-time data for two collisions. Which force delivers the greater impulse?

A. Force A
B. Force B
C. Both forces deliver the same impulse.

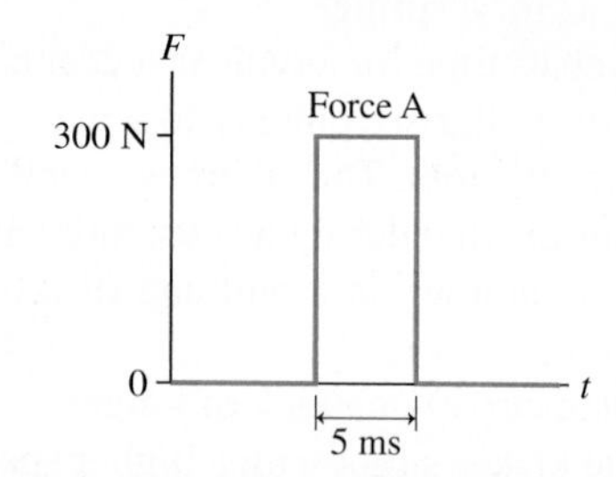

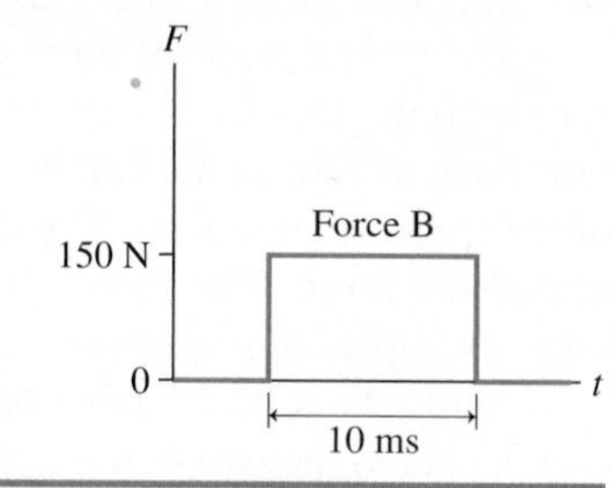

9.2 Momentum and the Impulse-Momentum Theorem

We've noted that the effect of an impulsive force depends on the impulse delivered to the object. The effect also depends on the object's mass. Our experience tells us that giving a kick to a heavy object will change its velocity much less than giving the same kick to a light object. We want now to find a quantitative relationship for impulse, mass, and velocity change.

FIGURE 9.5 The stick exerts an impulse on the puck, changing its speed.

While the stick is in contact with the puck, it exerts a force $\vec{F}_{avg}$ on the puck, causing it to accelerate.

Momentum

Consider the puck of mass m in FIGURE 9.5, sliding with an initial velocity $\vec{v}_i$. It is struck by a hockey stick that delivers an impulse $\vec{J} = \vec{F}_{avg}\,\Delta t$ to the puck. After the impulse, the puck leaves the stick with a final velocity $\vec{v}_f$. How is this final velocity related to the initial velocity?

From Newton's second law, the average acceleration of the puck during the time the stick is in contact with it is

$$\vec{a}_{avg} = \frac{\vec{F}_{avg}}{m} \tag{9.3}$$

The average acceleration is related to the change in the velocity by

$$\vec{a}_{avg} = \frac{\Delta\vec{v}}{\Delta t} = \frac{\vec{v}_f - \vec{v}_i}{\Delta t} \quad (9.4)$$

Combining Equations 9.3 and 9.4, we have

$$\frac{\vec{F}_{avg}}{m} = \vec{a}_{avg} = \frac{\vec{v}_f - \vec{v}_i}{\Delta t}$$

or, rearranging,

$$\vec{F}_{avg}\,\Delta t = m\vec{v}_f - m\vec{v}_i \quad (9.5)$$

A video to support a section's topic is embedded in the eText.

Video Force and Momentum Change

We recognize the left side of this equation as the impulse $\vec{J}$. The right side is the *change* in the quantity $m\vec{v}$. This quantity, the product of the object's mass and velocity, is called the **momentum** of the object. The symbol for momentum is $\vec{p}$:

$$\vec{p} = m\vec{v} \quad (9.6)$$

Momentum of an object of mass m and velocity $\vec{v}$

From Equation 9.6, the units of momentum are those of mass times velocity, or kg · m/s. We noted previously that kg · m/s are the preferred units of impulse. Now we see that the reason for that preference is to match the units of momentum.

FIGURE 9.6 shows that the momentum $\vec{p}$ is a *vector* quantity that points in the same direction as the velocity vector $\vec{v}$. Like any vector, $\vec{p}$ can be decomposed into x- and y-components. Equation 9.6, which is a vector equation, is a shorthand way to write the two equations

$$p_x = mv_x$$
$$p_y = mv_y \quad (9.7)$$

FIGURE 9.6 An object's momentum vector $\vec{p}$ can be decomposed into x- and y-components.

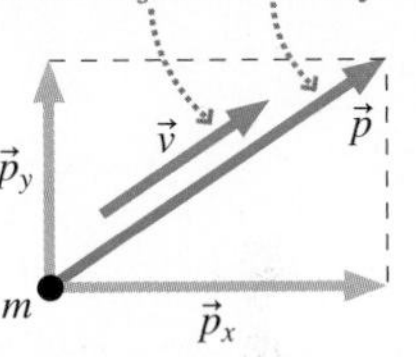
Momentum is a vector that points in the same direction as the object's velocity.
$\vec{v}$
$\vec{p}$
$\vec{p}_y$
m
$\vec{p}_x$

NOTE ▶ One of the most common errors in momentum problems is failure to use the correct signs. The momentum component p_x has the same sign as v_x. Just like velocity, momentum is positive for an object moving to the right (on the x-axis) or up (on the y-axis), but *negative* for an object moving to the left or down. ◀

The *magnitude* of an object's momentum is simply the product of the object's mass and speed, or $p = mv$. A heavy, fast-moving object will have a great deal of momentum, while a light, slow-moving object will have very little. Two objects with very different masses can have similar values of momentum if their speeds are very different as well. TABLE 9.1 gives some typical values of the momentum of various moving objects.

TABLE 9.1 Some typical values of momentum (approximate)

Object	Mass (kg)	Speed (m/s)	Momentum (kg · m/s)
Falling raindrop	2×10^{-5}	5	10^{-4}
Bullet	0.004	500	2
Pitched baseball	0.15	40	6
Running person	70	3	200
Car on highway	1000	30	3×10^4

The Impulse-Momentum Theorem

We can now write Equation 9.5 in terms of impulse and momentum:

$$\vec{J} = \vec{p}_f - \vec{p}_i = \Delta\vec{p} \quad (9.8)$$

Impulse-momentum theorem

where $\vec{p}_i = m\vec{v}_i$ is the object's initial momentum, $\vec{p}_f = m\vec{v}_f$ is its final momentum after the impulse, and $\Delta\vec{p} = \vec{p}_f - \vec{p}_i$ is the *change* in its momentum. This expression is known as the **impulse-momentum theorem.** It states that **an impulse delivered to an object causes the object's momentum to change.** That is, the *effect* of an impulsive force is to change the object's momentum from $\vec{p}_i$ to

$$\vec{p}_f = \vec{p}_i + \vec{J} \quad (9.9)$$

Legging it BIO A frog making a jump wants to gain as much momentum as possible before leaving the ground. This means that he wants the greatest impulse $J = F_{avg}\,\Delta t$ delivered to him by the ground. There is a maximum force that muscles can exert, limiting F_{avg}. But the time interval Δt over which the force is exerted can be greatly increased by having long legs. Many animals that are good jumpers have particularly long legs.

Equation 9.8 can also be written in terms of its x- and y-components as

$$
\begin{aligned}
J_x &= \Delta p_x = (p_x)_\mathrm{f} - (p_x)_\mathrm{i} = m(v_x)_\mathrm{f} - m(v_x)_\mathrm{i} \\
J_y &= \Delta p_y = (p_y)_\mathrm{f} - (p_y)_\mathrm{i} = m(v_y)_\mathrm{f} - m(v_y)_\mathrm{i}
\end{aligned}
\tag{9.10}
$$

The impulse-momentum theorem is illustrated by two examples in FIGURE 9.7. In Figure 9.7a, the putter strikes the ball, exerting a force on it and delivering an impulse $\vec{J} = \vec{F}_\mathrm{avg}\,\Delta t$. Notice that the direction of the impulse is the same as that of the force. Because $\vec{p}_\mathrm{i} = \vec{0}$ in this situation, we can use the impulse-momentum theorem to find that the ball leaves the putter with momentum $\vec{p}_\mathrm{f} = \vec{p}_\mathrm{i} + \vec{J} = \vec{J}$.

NOTE ▶ You can think of the putter as changing the ball's momentum by transferring momentum to it as an impulse. Thus we say the putter *delivers* an impulse to the ball, and the ball *receives* an impulse from the putter. ◀

The soccer player in Figure 9.7b presents a more complicated case. Here, the initial momentum of the ball is directed downward to the left. The impulse delivered to it by the player's head, upward to the right, is strong enough to reverse the ball's motion and send it off in a new direction. The graphical addition of vectors in Figure 9.7b again shows that $\vec{p}_\mathrm{f} = \vec{p}_\mathrm{i} + \vec{J}$.

KEY CONCEPT **FIGURE 9.7** Impulse causes a *change* in momentum.

(a) A putter delivers an impulse to a golf ball, changing its momentum.

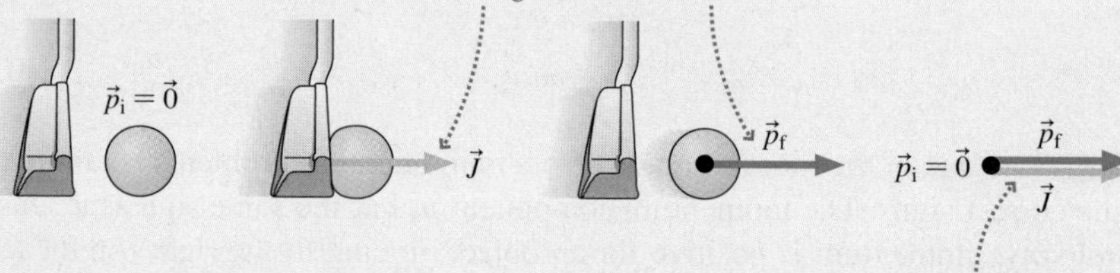
The club delivers an impulse $\vec{J}$ to the ball, increasing the ball's momentum.
$\vec{p}_\mathrm{i} = \vec{0}$
$\vec{J}$
$\vec{p}_\mathrm{f}$
$\vec{p}_\mathrm{i} = \vec{0}$
$\vec{p}_\mathrm{f}$
$\vec{J}$

(b) A soccer player delivers an impulse to a soccer ball, changing its momentum.

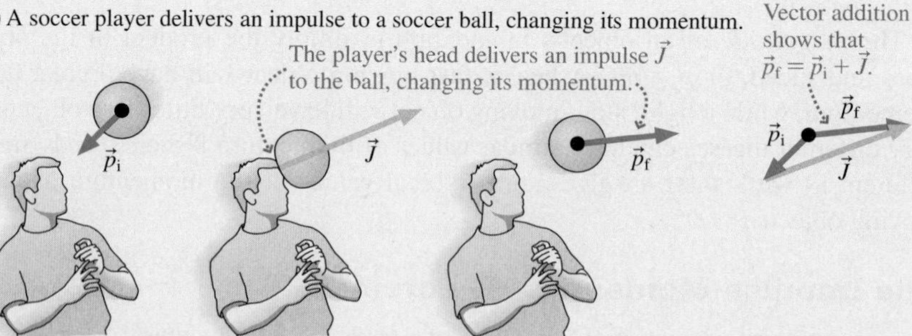
Vector addition shows that $\vec{p}_\mathrm{f} = \vec{p}_\mathrm{i} + \vec{J}$.
The player's head delivers an impulse $\vec{J}$ to the ball, changing its momentum.
$\vec{p}_\mathrm{i}$
$\vec{J}$
$\vec{p}_\mathrm{f}$
$\vec{p}_\mathrm{f}$
$\vec{p}_\mathrm{i}$
$\vec{J}$

STOP TO THINK 9.2 A puck, seen from above, was moving with an initial momentum when it received an impulse $\vec{J}$ from a hockey stick, giving it the final momentum shown. Using the ideas of Figure 9.7, which arrow best represents the puck's initial momentum?

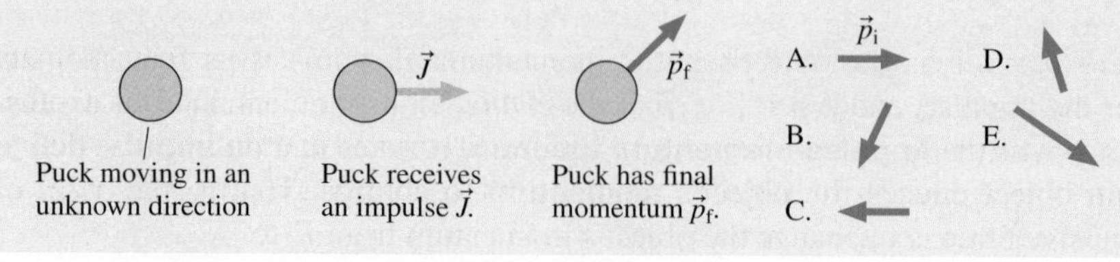
$\vec{J}$
$\vec{p}_\mathrm{f}$
Puck moving in an unknown direction
Puck receives an impulse $\vec{J}$.
Puck has final momentum $\vec{p}_\mathrm{f}$.
$\vec{p}_\mathrm{i}$
A.
B.
C.
D.
E.

EXAMPLE 9.2 Calculating the change in momentum

A ball of mass $m = 0.25$ kg rolling to the right at 1.3 m/s strikes a wall and rebounds to the left at 1.1 m/s. What is the change in the ball's momentum? What is the impulse delivered to it by the wall?

STRATEGIZE The impulse-momentum theorem deals with the change in momentum due to an impulse. The details of how the impulse is delivered don't matter. In problems like this one, we don't need to worry about the details of the collision, only the change from the initial state to the final state. To solve such problems, we'll introduce a new type of visual overview, one in which we consider the situation before the interaction and then after the interaction. This before-and-after visual overview is shown in FIGURE 9.8. In this case, we'll also include a diagram showing the force during the collision with the wall. We won't include such diagrams in all cases, but it's useful to visualize the force that causes the momentum change. The force is directed to the left, as we know it must be.

FIGURE 9.8 Visual overview for a ball bouncing off a wall.

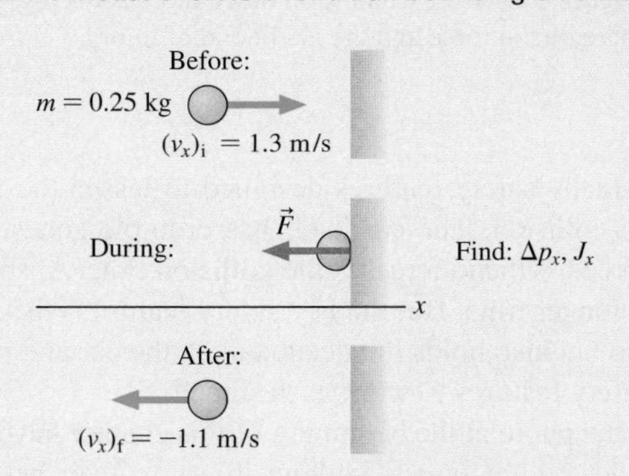
Before:
$m = 0.25$ kg
$(v_x)_i = 1.3$ m/s
During:
$\vec{F}$
Find: Δp_x, J_x
x
After:
$(v_x)_f = -1.1$ m/s

PREPARE The ball is moving along the x-axis, so we write the momentum in component form, as in Equation 9.7. The change in momentum is then the difference between the final and initial values of the momentum. By the impulse-momentum theorem, the impulse is equal to this change in momentum.

SOLVE The x-component of the initial momentum is

$$(p_x)_i = m(v_x)_i = (0.25 \text{ kg})(1.3 \text{ m/s}) = 0.325 \text{ kg} \cdot \text{m/s}$$

The y-component of the momentum is zero both before and after the bounce. After the ball rebounds, the x-component is

$$(p_x)_f = m(v_x)_f = (0.25 \text{ kg})(-1.1 \text{ m/s}) = -0.275 \text{ kg} \cdot \text{m/s}$$

It is particularly important to notice that the x-component of the momentum, like that of the velocity, is negative. This indicates that the ball is moving to the *left*. The change in momentum is

$$\Delta p_x = (p_x)_f - (p_x)_i = (-0.275 \text{ kg} \cdot \text{m/s}) - (0.325 \text{ kg} \cdot \text{m/s})$$
$$= -0.60 \text{ kg} \cdot \text{m/s}$$

By the impulse-momentum theorem, the impulse delivered to the ball by the wall is equal to this change, so

$$J_x = \Delta p_x = -0.60 \text{ kg} \cdot \text{m/s}$$

ASSESS The impulse is negative, indicating that the force that causes the impulse is pointing to the left, which we noted must be the case.

SYNTHESIS 9.1 Momentum and impulse

A moving object has momentum. A force acting on an object delivers an *impulse* that changes the object's momentum.

The **momentum** of an object is the product of its mass and its velocity.

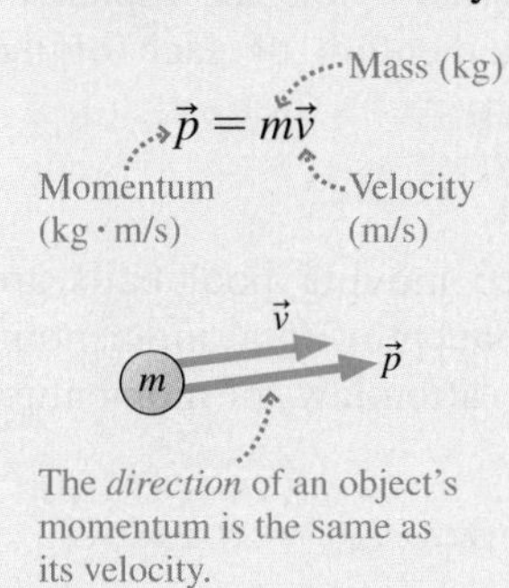
Mass (kg)
$\vec{p} = m\vec{v}$
Momentum (kg · m/s)
Velocity (m/s)
$\vec{v}$
m
$\vec{p}$
The *direction* of an object's momentum is the same as its velocity.

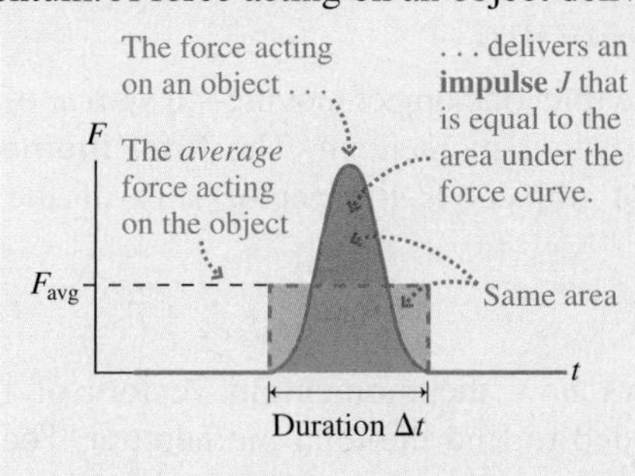
The force acting on an object . . .
. . . delivers an **impulse** J that is equal to the area under the force curve.
F
The *average* force acting on the object
F_{avg}
Same area
t
Duration Δt

Impulse (kg · m/s) $\vec{J} = \vec{F}_{avg}\,\Delta t$ Duration (s)

Average force (N)

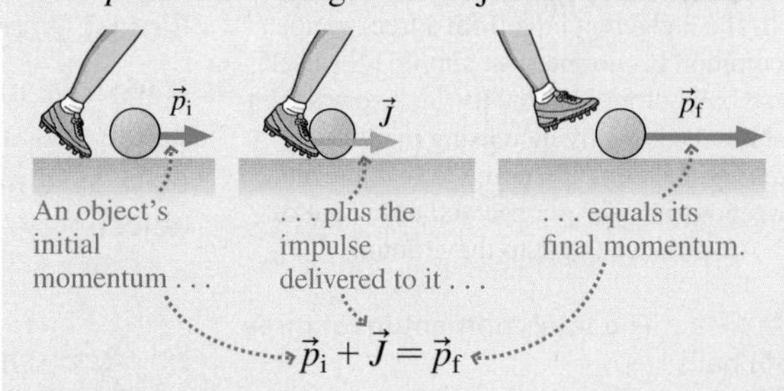
$\vec{p}_i$
$\vec{J}$
$\vec{p}_f$
An object's initial momentum . . .
. . . plus the impulse delivered to it . . .
. . . equals its final momentum.
$\vec{p}_i + \vec{J} = \vec{p}_f$

This relationship can also be written in terms of the *change* in the object's momentum:

The impulse delivered . . . $\vec{J} = \vec{p}_f - \vec{p}_i = \Delta\vec{p}$ . . . equals the change in the momentum.

The rams in the chapter-opening photo butt heads—and their heads go from moving to standing still. A change in momentum occurs, which means an impulse. An impulse means a force, and a large force on the head can cause injury. How do the rams keep from getting hurt? A similar situation is at work in a car crash—the car goes from moving to standing still, as do the driver and the passengers in the car. There is a change in momentum; this is unavoidable. What strategies reduce the chances of injury to the occupants of the car?

In these examples, the object has momentum $\vec{p}_i$ just before impact and zero momentum after (i.e., $\vec{p}_f = \vec{0}$). The impulse-momentum theorem tells us that

$$\vec{J} = \vec{F}_{avg}\,\Delta t = \Delta\vec{p} = \vec{p}_f - \vec{p}_i = -\vec{p}_i$$

or

$$\vec{F}_{avg} = -\frac{\vec{p}_i}{\Delta t} \tag{9.11}$$

INVERSE p. 116

That is, the average force needed to stop an object is *inversely proportional* to the duration Δt of the collision. **If the duration of the collision can be increased, the force of the impact will be decreased.**

CONCEPTUAL EXAMPLE 9.3 A stretch in time

The seatbelts in your car are made of webbing that is designed to have a significant stretch when supporting large forces. Why is this useful?

REASON Suppose a car crashes into a fixed barrier and comes to rest. The driver will come to rest as well. Increasing the time over which the driver comes to rest will decrease the force on the driver. If the seatbelt can stretch, the driver can still be moving after the car has stopped—the driver can come to rest over a longer time. Increasing the time reduces the force, and thus the chance for injury.

ASSESS This makes sense. If the seatbelt could not stretch at all, a driver or passenger would stop much more suddenly, which would increase the force and the likelihood of injury.

A spiny cushion BIO The spines of a hedgehog obviously help protect it from predators. But they serve another function as well. If a hedgehog falls from a tree—a not uncommon occurrence—it simply rolls itself into a ball before it lands. Its thick spines then cushion the blow by increasing the time it takes for the animal to come to rest. Indeed, hedgehogs have been observed to fall out of trees on purpose to get to the ground!

Modern automobiles have many safety features designed to lessen the force on occupants of the car during a collision. The car itself has crumple zones, regions designed to fail under large forces, which increases the collision time. Air bags also bring occupants to rest over a longer time. But the key safety feature is the seatbelt, which stretches to reduce force but also holds the occupants in the car and positions them correctly for the other safety features to work as designed.

The butting rams shown in the photo at the beginning of this chapter have adaptations that allow them to collide at high speeds without injury to their brains. The cranium has a double wall to prevent skull injuries, and there is a thick spongy mass that increases the time it takes for the brain to come to rest upon impact, reducing the magnitude of the force on the brain.

Total Momentum

If we have more than one object moving—a *system* of objects—then the system as a whole has an overall momentum. The **total momentum** $\vec{P}$ (note the capital P) of a system of objects is the vector sum of the momentum of each of the objects:

$$\vec{P} = \vec{p}_1 + \vec{p}_2 + \vec{p}_3 + \cdots$$

FIGURE 9.9 shows how the momentum vectors of three moving pool balls are graphically added to find the total momentum. The concept of total momentum will be of key importance when we discuss the conservation law for momentum in Section 9.4.

FIGURE 9.9 The total momentum of three pool balls.

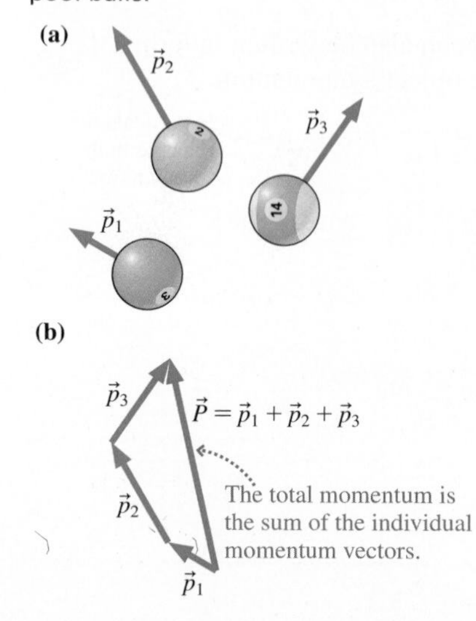

STOP TO THINK 9.3 The cart's change of momentum is

A. $-30\ \text{kg}\cdot\text{m/s}$
B. $-20\ \text{kg}\cdot\text{m/s}$
C. $-10\ \text{kg}\cdot\text{m/s}$
D. $10\ \text{kg}\cdot\text{m/s}$
E. $20\ \text{kg}\cdot\text{m/s}$
F. $30\ \text{kg}\cdot\text{m/s}$

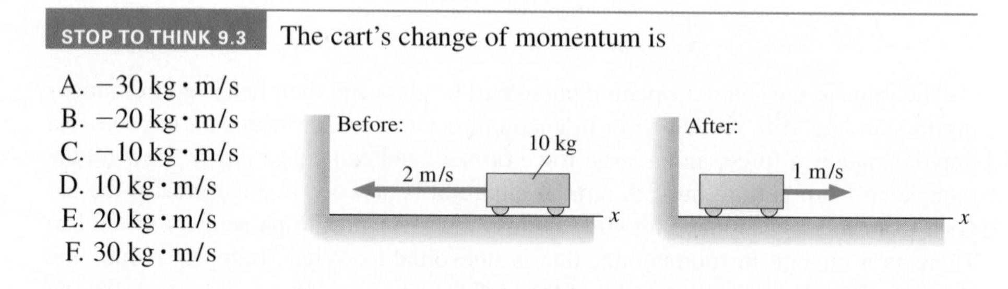

9.3 Solving Impulse and Momentum Problems

The visual overviews and free-body diagrams that you learned to draw in Chapters 1–8 were oriented toward the use of Newton's laws and a subsequent kinematic analysis. Now that we are working with momentum, a conserved quantity, we want to consider the situation before and after an impact or other event. We'll draw a before-and-after visual overview, an idea we introduced in Example 9.2.

TACTICS BOX 9.1 Drawing a before-and-after visual overview

1. **Sketch the situation.** Use two drawings, labeled "Before" and "After," to show the objects *immediately before* they interact and again *immediately after* they interact.
2. **Establish a coordinate system.** Select your axes to match the motion.
3. **Define symbols.** Define symbols for the masses and for the velocities before and after the interaction. Position and time are not needed.
4. **List known information.** List the values of quantities known from the problem statement or that can be found quickly with simple geometry or unit conversions. Before-and-after pictures are usually simpler than the pictures you used for dynamics problems, so listing known information on the sketch is often adequate.
5. **Identify the desired unknowns.** What quantity or quantities will allow you to answer the question? These should have been defined as symbols in step 3.

Exercises 9–11

EXAMPLE 9.4 Force in hitting a baseball

A 150 g baseball is thrown with a speed of 20 m/s. It is hit straight back toward the pitcher at a speed of 40 m/s. The impulsive force of the bat on the ball has the shape shown in FIGURE 9.10. What is the *maximum* force F_{max} that the bat exerts on the ball? What is the *average* force that the bat exerts on the ball?

FIGURE 9.10 The interaction force between the baseball and the bat.

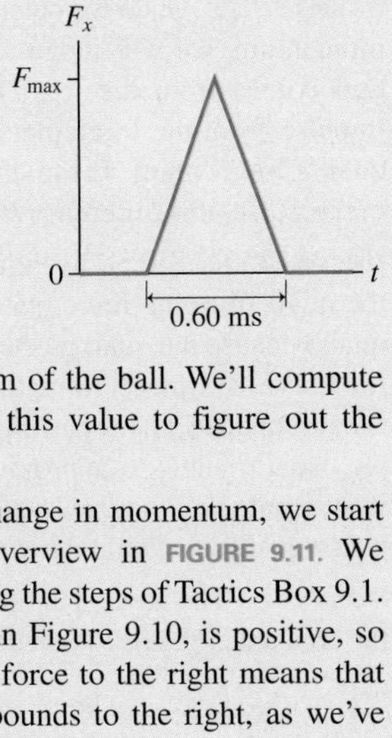

STRATEGIZE When a baseball collides with the bat, the impulse from the bat changes the momentum of the ball. We'll compute the change in momentum and use this value to figure out the details of the impulse.

PREPARE To help determine the change in momentum, we start with the before-and-after visual overview in FIGURE 9.11. We develop the visual overview following the steps of Tactics Box 9.1. F_x, the force of the bat on the ball in Figure 9.10, is positive, so the force is directed to the right. A force to the right means that the ball came from the left and rebounds to the right, as we've drawn.

SOLVE In the last several chapters we've started the mathematical solution with Newton's second law. Now we want to use the impulse-momentum theorem:

$$\Delta p_x = J_x = \text{area under the force curve}$$

FIGURE 9.11 A before-and-after visual overview.

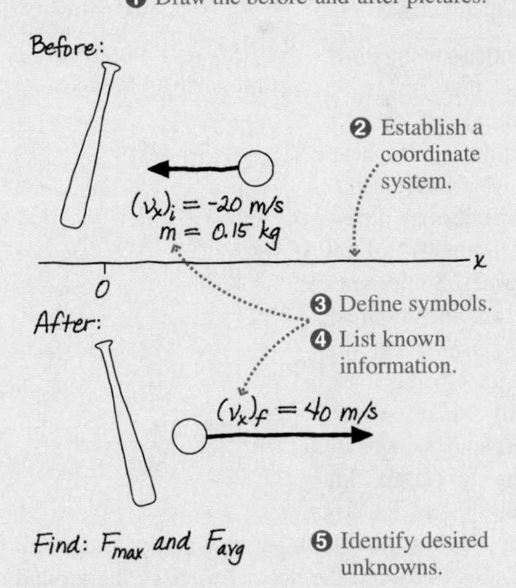

We know the velocities before and after the collision, so we can find the change in the ball's momentum:

$$\Delta p_x = m(v_x)_f - m(v_x)_i = (0.15 \text{ kg})(40 \text{ m/s} - (-20 \text{ m/s}))$$
$$= 9.0 \text{ kg} \cdot \text{m/s}$$

Continued

The force curve is a triangle with height F_{max} and width 0.60 ms. As in Example 9.1, the area under the curve is

$$J_x = \text{area} = \frac{1}{2} \times F_{max} \times (6.0 \times 10^{-4}\ \text{s})$$
$$= (F_{max})(3.0 \times 10^{-4})$$

According to the impulse-momentum theorem, $\Delta p_x = J_x$, so we have

$$9.0\ \text{kg} \cdot \text{m/s} = (F_{max})(3.0 \times 10^{-4}\ \text{s})$$

Thus the *maximum* force is

$$F_{max} = \frac{9.0\ \text{kg} \cdot \text{m/s}}{3.0 \times 10^{-4}\ \text{s}} = 30{,}000\ \text{N}$$

Using Equation 9.1, we find that the *average* force, which depends on the collision duration $\Delta t = 6.0 \times 10^{-4}$ s, has a smaller value:

$$F_{avg} = \frac{J_x}{\Delta t} = \frac{\Delta p_x}{\Delta t} = \frac{9.0\ \text{kg} \cdot \text{m/s}}{6.0 \times 10^{-4}\ \text{s}} = 15{,}000\ \text{N}$$

ASSESS F_{max} is a large force, but we expect a large force for such a short collision time and a large change in momentum.

The Impulse Approximation

When two objects interact during a collision or other brief interaction, such as that between the bat and ball of Example 9.4, the forces *between* them are generally quite large. Other forces may also act on the interacting objects, but usually these forces are *much* smaller than the interaction forces. In Example 9.4, for example, the 1.5 N weight of the ball is vastly less than the 30,000 N force of the bat on the ball. We can reasonably ignore these small forces *during* the brief time of the impulsive force. Doing so is called the **impulse approximation.**

When we use the impulse approximation, $(p_x)_i$ and $(p_x)_f$ —and $(v_x)_i$ and $(v_x)_f$ — are the values of the momentum (and velocity) *immediately* before and *immediately* after the collision. For example, the velocities in Example 9.4 are those of the ball just before and after it collides with the bat. The values will subsequently change as other forces, such as weight and drag, exert their influence.

EXAMPLE 9.5 How does the bird take flight?

Scientists who study birds that make a rapid takeoff from a standing start have discovered that different species use different strategies to get airborne. Some species use mostly their legs, pushing off the ground; some use mostly their wings, with rapid wingbeats giving a large lift force. In one experiment, a 190 g quail took off essentially straight up from a force plate that measured the normal force of the ground on the quail. The net upward force on the quail is plotted in FIGURE 9.12. The quail left the ground at 3.5 m/s.

FIGURE 9.12 Force versus time for a quail taking off.

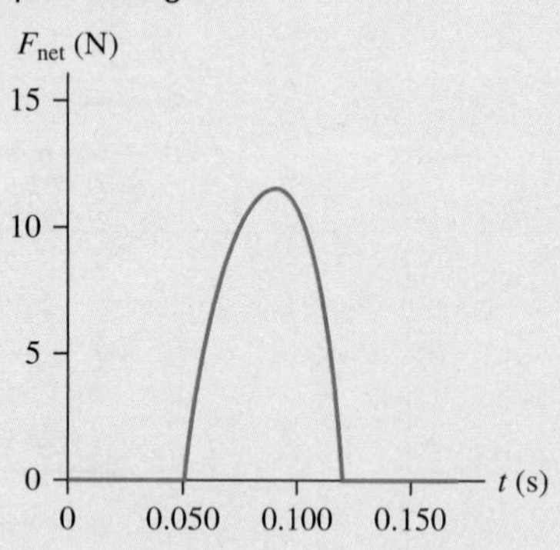

a. Approximate the impulse by estimating the area under the force-versus-time curve.
b. What is the change in momentum of the quail as it leaves the ground?
c. Given the values you found, as the quail executes a rapid takeoff, does it use mostly its legs or mostly its wings?

STRATEGIZE Once we determine the impulse and the change in momentum, we will judge what fraction of the takeoff momentum comes from the legs. If the quail uses only its legs, the impulse from the force plate will be equal to the change in the quail's momentum. If the change in momentum is greater than the impulse, this difference is due to the lift force from the wings.

PREPARE The graph in Figure 9.12 is the net upward force on the quail from the force plate. The force plate pushes up on the quail because the quail pushes down with its legs. The area under the curve is a measure of the impulse delivered to the quail as a consequence of its pushing off with its legs. In FIGURE 9.13a, we use a triangle to approximate the area under the curve to let us compute the impulse. FIGURE 9.13b is a before-and-after visual overview of the quail taking off.

SOLVE a. The area of the triangle in Figure 9.13a approximates the area under the force-versus-time curve, which is the impulse J_y:

$$J_y = \text{area under curve} = \frac{1}{2} \times \text{base} \times \text{height}$$
$$= \frac{1}{2}(0.125\ \text{s} - 0.045\ \text{s})(13\ \text{N}) = 0.52\ \text{kg} \cdot \text{m/s}$$

b. The quail starts at rest, so the change in momentum is the final momentum:

$$\Delta p_y = (p_y)_f = (0.19\ \text{kg})(3.5\ \text{m/s}) = 0.67\ \text{kg} \cdot \text{m/s}$$

FIGURE 9.13 Estimating the impulse and determining the change in momentum of the quail.

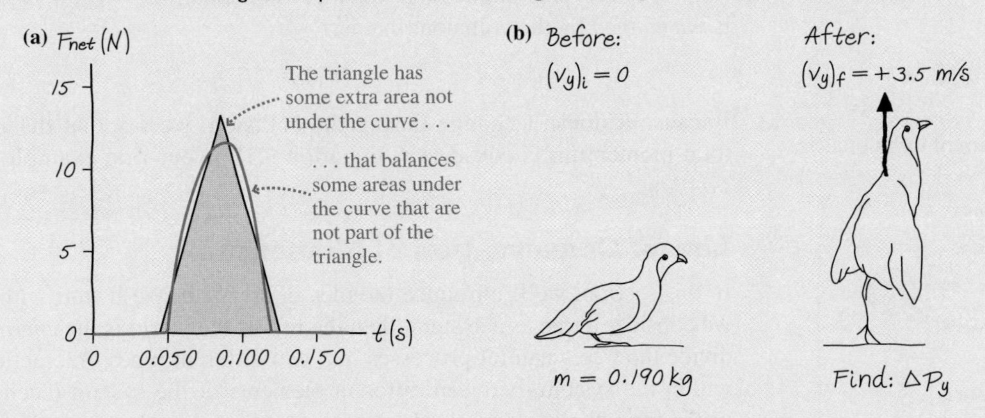

c. The impulse is nearly 80% of the change in momentum, so most of the change in momentum is provided by the impulse from the legs, with only a small amount coming from the lift force provided by the wings.

ASSESS Quail tend to make short flights, they spend most of their time on the ground, and they are good runners, so it's reasonable to expect them to rely mostly on their legs to get airborne. Our results seem reasonable.

STOP TO THINK 9.4 A 10 g rubber ball and a 10 g clay ball are each thrown at a wall with equal speeds. The rubber ball bounces; the clay ball sticks. Which ball receives the greater impulse from the wall?

A. The clay ball receives a greater impulse because it sticks.
B. The rubber ball receives a greater impulse because it bounces.
C. They receive equal impulses because they have equal values of momentum.
D. Neither receives an impulse because the wall doesn't move.

9.4 Conservation of Momentum

Let's consider two objects, such as the rams shown in the opening photo of this chapter, that interact during the brief moment of a collision. During a collision, two objects exert forces on each other that vary in a complex way. We usually don't even know the magnitudes of these forces. Using Newton's second law to predict the outcome of such a collision would thus be a daunting challenge. However, by using Newton's *third* law in the language of impulse and momentum, we'll find that it's possible to describe the *outcome* of a collision—the final speeds and directions of the colliding objects—in a simple way.

FIGURE 9.14 A collision between two balls.

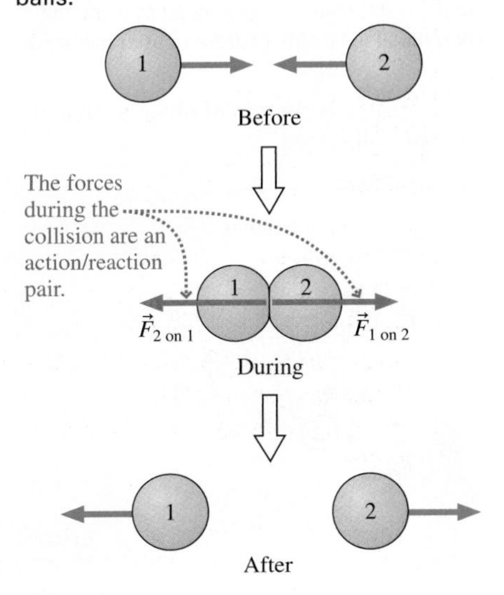

FIGURE 9.14 shows two balls initially headed toward each other. The balls collide, then bounce apart. The forces during the collision, when the balls are interacting, form an action/reaction pair $\vec{F}_{1\text{ on }2}$ and $\vec{F}_{2\text{ on }1}$. For now, we'll continue to assume that the motion is one dimensional along the x-axis.

During the collision, the impulse J_{2x} delivered to ball 2 by ball 1 is the average value of $\vec{F}_{1\text{ on }2}$ multiplied by the collision time Δt. Likewise, the impulse J_{1x} delivered to ball 1 by ball 2 is the average value of $\vec{F}_{2\text{ on }1}$ multiplied by Δt. Because $\vec{F}_{1\text{ on }2}$ and $\vec{F}_{2\text{ on }1}$ form an action/reaction pair, they have equal magnitudes but opposite directions. As a result, the two impulses J_{1x} and J_{2x} are also equal in magnitude but opposite in sign, so that $J_{1x} = -J_{2x}$.

According to the impulse-momentum theorem, the change in the momentum of ball 1 is $\Delta p_{1x} = J_{1x}$ and the change in the momentum of ball 2 is $\Delta p_{2x} = J_{2x}$. Because $J_{1x} = -J_{2x}$, the change in the momentum of ball 1 is equal in magnitude but opposite in sign to the change in momentum of ball 2. If ball 1's momentum increases by a certain amount during the collision, ball 2's momentum will *decrease* by exactly the same amount. This implies that the total momentum $P_x = p_{1x} + p_{2x}$ of the two balls is *unchanged* by the collision; that is,

$$(P_x)_f = (P_x)_i \tag{9.12}$$

Because it doesn't change during the collision, we say that the x-component of total momentum is *conserved*. Equation 9.12 is our first example of a *conservation law*.

FIGURE 9.15 A system of three objects.

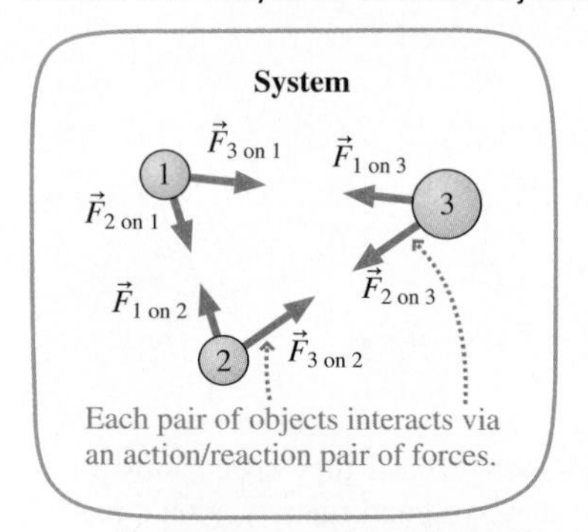

Those are the breaks At the start of a game of pool, a player hits the cue ball into the triangle of stationary numbered balls. The forces between the balls are much larger than friction or the other forces at work, so we can use the impulse approximation. If we define the system as the collection of all the balls on the table, then the only forces that matter are internal forces. This means that the total momentum of the system is the same before and after the break. In the photo you can see the balance between motion to the left and to the right; there is no momentum along the horizontal axis, either before or after the break.

Law of Conservation of Momentum

In this section, we'll introduce the idea of a *system*. We'll start with an example: If we consider a tree as a system, then the rest of the world is the *environment*. We can divide the tree's natural processes into two different types: interactions that happen within the system, between different elements of the system (such as transport of water from the roots to the leaves), and interactions between the system and the environment (such as light from the sun falling on the leaves).

This approach of dividing the world into system and environment will be very important to us. As we'll see in future chapters, how we define the system can dramatically change the way that we approach the solution to a problem. And the boundaries aren't always clear, as we can see by considering the example of the tree. Bacteria live on the roots of the tree, facilitating the uptake of nutrients. Are these bacteria part of the system or part of the environment? We could define the system either way depending on the questions we are asking.

Let's continue our discussion with a simple system, three objects that interact with each other—you might imagine three balls on a pool table. We've drawn the system in FIGURE 9.15, including a red line around the three objects to denote the boundary of the system. Arrows show the forces between the three objects. Forces that act only between objects within the system are called **internal forces.** Forces that come from agents outside the system are called **external forces.** When one of the balls smacks into a second ball, this is an internal force; it's a force between two objects of the system. When someone uses a cue to strike one of the balls, this is an external force; it's from an agent outside the system.

Each pair of objects in the system in Figure 9.15 interacts via forces that are an action/reaction pair. Exactly as for the two-object collision, the change in momentum of object 2 due to the force from object 3 is equal in magnitude, but opposite in direction, to the change in object 3's momentum due to object 2. The net change in the momentum of these two objects due to their interaction forces is thus zero. The same argument holds for every pair, with the result that, no matter how complicated the forces between the objects, **there is no change in the total momentum $\vec{P}$ of the system.** The total momentum of the system remains constant: It is *conserved*. We can generalize this principle as follows: **The total momentum of a system subject to only internal forces is conserved.**

FIGURE 9.16 A system of objects subject to external forces.

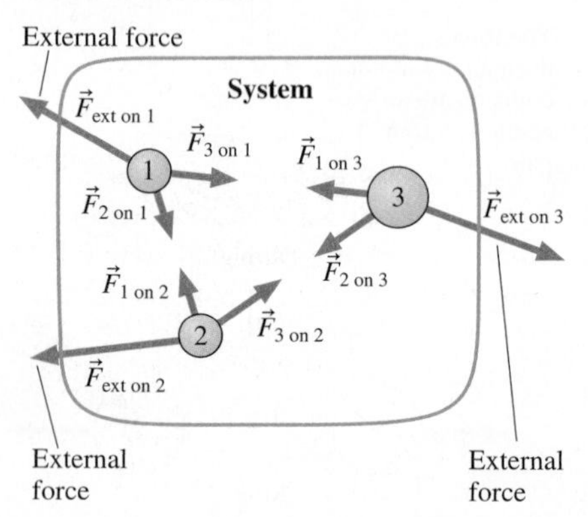

In FIGURE 9.16 we show the same three-object system of Figure 9.15, but now with *external* forces acting on the three objects. These external forces *can* change the momentum of the system. During a time interval Δt, for instance, the external force $\vec{F}_{\text{ext on }1}$ acting on object 1 changes its momentum, according to the impulse-momentum theorem, by $\Delta\vec{p}_1 = (\vec{F}_{\text{ext on }1})\Delta t$. The momentum of each of the other two objects changes similarly. Thus the change in the total momentum is

$$\begin{aligned}\Delta\vec{P} &= \Delta\vec{p}_1 + \Delta\vec{p}_2 + \Delta\vec{p}_3 \\ &= (\vec{F}_{\text{ext on }1}\Delta t) + (\vec{F}_{\text{ext on }2}\Delta t) + (\vec{F}_{\text{ext on }3}\Delta t) \\ &= (\vec{F}_{\text{ext on }1} + \vec{F}_{\text{ext on }2} + \vec{F}_{\text{ext on }3})\,\Delta t \\ &= \vec{F}_{\text{net}}\Delta t\end{aligned} \tag{9.13}$$

where $\vec{F}_{\text{net}}$ is the net force due to *external forces*.

Equation 9.13 has a very important implication in the case where the net force on a system is zero: **If $\vec{F}_{net} = \vec{0}$, the *total* momentum $\vec{P}$ of the system does not change.** The total momentum remains constant, *regardless* of whatever interactions are going on *inside* the system. With no net external force acting that can change its momentum, we call a system with $\vec{F}_{net} = \vec{0}$ an **isolated system.**

The importance of these results is sufficient to elevate them to a law of nature, alongside Newton's laws.

Law of conservation of momentum The total momentum $\vec{P}$ of an isolated system is a constant. Interactions within the system do not change the system's total momentum.

NOTE ▶ It is worth emphasizing the critical role of Newton's third law in the derivation of Equation 9.13. The law of conservation of momentum is a direct consequence of the fact that interactions within an isolated system are action/reaction pairs. ◀

Mathematically, the law of conservation of momentum for an isolated system is

$$\vec{P}_f = \vec{P}_i \tag{9.14}$$

Law of conservation of momentum for an isolated system

The total momentum after an interaction is equal to the total momentum before the interaction. Because Equation 9.14 is a vector equation, the equality is true for each of the components of the momentum vector; that is,

$$\text{x-component: } \underbrace{(p_{1x})_f + (p_{2x})_f + (p_{3x})_f + \cdots}_{\text{Final momentum}} = \underbrace{(p_{1x})_i + (p_{2x})_i + (p_{3x})_i + \cdots}_{\text{Initial momentum}}$$

Object 1 Object 2 Object 3 (9.15)

$$\text{y-component: } (p_{1y})_f + (p_{2y})_f + (p_{3y})_f + \cdots = (p_{1y})_i + (p_{2y})_i + (p_{3y})_i + \cdots$$

EXAMPLE 9.6 **Speed of ice skaters pushing off**

Two ice skaters, Sandra and David, stand facing each other on frictionless ice. Sandra has a mass of 45 kg, David a mass of 80 kg. They then push off from each other. After the push, Sandra moves off at a speed of 2.2 m/s. What is David's speed?

STRATEGIZE We will consider the system to be the two skaters. For each skater, the upward normal force of the ice balances the downward weight force to make $F_{net} = \vec{0}$. We will thus consider the two skaters to be an isolated system. The momentum of the system is therefore conserved—the total momentum of the skaters before and after the push will be the same. Even though we know nothing about the details of the forces between the two skaters—how hard they push or for how long—we will solve this problem by looking at the "before" and "after" cases and considering conservation of momentum.

PREPARE We start with FIGURE 9.17, which shows a before-and-after visual overview for the two skaters. The total momentum before they push off is $\vec{P}_i = \vec{0}$ because both skaters are at rest. Consequently, the total momentum will still be $\vec{0}$ *after* they push off.

FIGURE 9.17 Before-and-after visual overview for two skaters pushing off from each other.

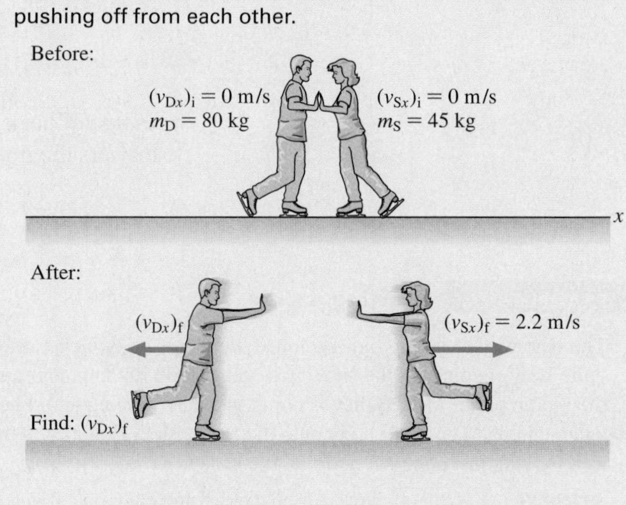

Continued

SOLVE Since the motion is only in the x-direction, we need to consider only x-components of momentum. We write Sandra's initial momentum as $(p_{Sx})_i = m_S(v_{Sx})_i$, where m_S is her mass and $(v_{Sx})_i$ her initial velocity. Similarly, we write David's initial momentum as $(p_{Dx})_i = m_D(v_{Dx})_i$. Both of the skaters have zero initial momentum because both are initially at rest.

We can now apply the mathematical statement of momentum conservation, Equation 9.15. Writing the final momentum of Sandra as $m_S(v_{Sx})_f$ and that of David as $m_D(v_{Dx})_f$, we have

$$\underbrace{m_S(v_{Sx})_f + m_D(v_{Dx})_f}_{\text{The skaters' final momentum . . .}} = \underbrace{m_S(v_{Sx})_i + m_D(v_{Dx})_i}_{\text{. . . equals their initial momentum . . .}} = \underbrace{0}_{\text{. . . which was zero.}}$$

Solving for $(v_{Dx})_f$, we find

$$(v_{Dx})_f = -\frac{m_S}{m_D}(v_{Sx})_f = -\frac{45\text{ kg}}{80\text{ kg}} \times 2.2\text{ m/s} = -1.2\text{ m/s}$$

David moves backward with a *speed* of 1.2 m/s.

Notice that we didn't need to know any details about the force between David and Sandra in order to find David's final speed. Conservation of momentum *mandates* this result.

ASSESS It seems reasonable that Sandra, whose mass is less than David's, has the greater final speed.

PROBLEM-SOLVING APPROACH 9.1 Conservation of momentum problems

We can use the law of conservation of momentum to relate the values of velocity and momentum of objects *after* an interaction to their values *before* the interaction.

STRATEGIZE Clearly define the *system*.

- If possible, choose a system that is isolated ($\vec{F}_{net} = \vec{0}$) or within which the interactions are sufficiently short and intense that you can ignore external forces for the duration of the interaction (the impulse approximation). Momentum is then conserved.
- If it's not possible to choose an isolated system, try to divide the problem into parts such that momentum is conserved during one segment of the motion. Other segments of the motion can be analyzed using Newton's laws or, as you'll learn in Chapter 10, conservation of energy.

PREPARE Following Tactics Box 9.1, draw a before-and-after visual overview. Define symbols that will be used in the problem, list known values, and identify what you're trying to find.

SOLVE The mathematical representation is based on the law of conservation of momentum, Equations 9.15. Because we generally want to solve for the velocities of objects, we usually use Equations 9.15 in the equivalent form

$$m_1(v_{1x})_f + m_2(v_{2x})_f + \cdots = m_1(v_{1x})_i + m_2(v_{2x})_i + \cdots$$
$$m_1(v_{1y})_f + m_2(v_{2y})_f + \cdots = m_1(v_{1y})_i + m_2(v_{2y})_i + \cdots$$

ASSESS Check that your result has the correct units, is reasonable, and answers the question.

Exercise 17

EXAMPLE 9.7 Starting a sled

The best way to get an old-fashioned wooden sled (which can be quite heavy) going on the snow is to get a good running start and then jump on the sled. Bailey is doing just that, running at 4.0 m/s and launching her 26 kg body onto the 5.9 kg sled. How fast is the sled moving just after she lands on it?

STRATEGIZE When Bailey lands on the sled and grabs onto it, many forces are at work. But if we choose the system as Bailey + sled, the forces involved in this collision—friction forces between Bailey and the sled, the force of her holding the handles—are *internal* forces. The normal force balances the weight of both Bailey and the sled, so there is no net vertical force on the system. There is a small friction force from the runners of the sled on the snow, but if we consider the motion just after Bailey lands on the sled, we can ignore this force; it is much smaller than the internal forces. We will

FIGURE 9.18 Before-and-after visual overview of Bailey and the sled.

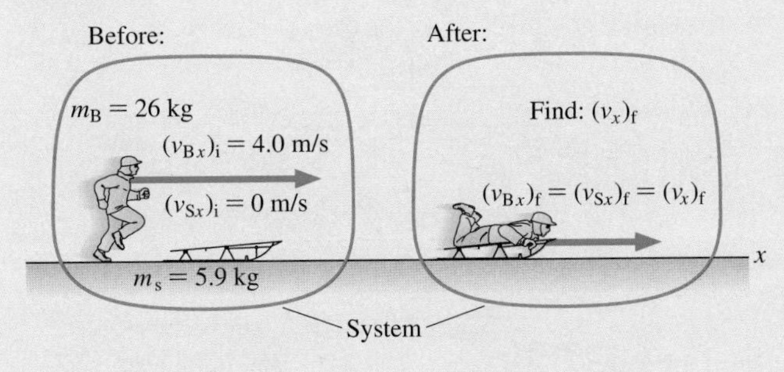

consider Bailey + sled as an isolated system, and the total momentum of the system will be conserved.

PREPARE The visual overview in FIGURE 9.18 has two parts, before and after, with the system clearly labeled. Before Bailey lands on the sled, she is moving with velocity $(v_{Bx})_i$ but the sled is stationary. After she lands on the sled, she and the sled move at a common final velocity, $(v_x)_f$.

SOLVE To determine the final velocity, we use conservation of momentum. The total momentum of the system before and after is the same: $(P_x)_f = (P_x)_i$. The initial momentum is the sum of the momentum of each of the individual objects. Since the sled is initially at rest, we can simplify:

$$(P_x)_i = m_B(v_{Bx})_i + m_S(v_{Sx})_i = m_B(v_{Bx})_i$$

The final momentum is the combined mass of Bailey and the sled moving at the common speed:

$$(P_x)_f = (m_B + m_S)(v_x)_f$$

The final momentum is equal to the initial momentum, $(P_x)_f = (P_x)_i$, so we can write

$$(m_B + m_S)(v_x)_f = m_B(v_{Bx})_i$$

Using known values, we can solve this equation for the final speed:

$$(v_x)_f = \left(\frac{m_B}{m_B + m_S}\right)(v_{Bx})_i = \left(\frac{26\text{ kg}}{26\text{ kg} + 5.9\text{ kg}}\right)(4.0\text{ m/s}) = 3.3\text{ m/s}$$

The speed of the sled is 3.3 m/s after Bailey jumps on—plenty fast for a good start to an epic run down the hill!

ASSESS The final speed is just a bit less than Bailey's initial speed, which makes sense. Bailey's mass is much greater than that of the sled.

Notice how straightforward this example was! No forces, no kinematic equations, no simultaneous equations. Why didn't we think of this before? Although conservation laws are indeed powerful, they can answer only certain questions. If we had wanted to know whether Bailey slid across the sled, or the magnitude of the force on her hands, or the acceleration of the sled, or how long the collision took, we would not have been able to find answers on the basis of the conservation law. There is a price to pay for finding a simple connection between before and after, and that price is the loss of information about the details of the interaction. If we are satisfied with knowing only about before and after, then conservation laws are a straightforward way to proceed.

It Depends on the System

The first step in Problem-Solving Approach 9.1 is to clearly define the *system*. This is worth emphasizing because many problem-solving errors arise from trying to apply momentum conservation to an inappropriate system. **The goal is to choose a system whose momentum will be conserved.** Even then, it is the *total* momentum of the system that is conserved, not the momentum of any individual object within the system.

Video Conservation of Linear Momentum

In Example 9.7, the question asked for the speed of the sled after Bailey lands on it. Suppose we had chosen to consider the sled alone as the system. The sled experiences a force from Bailey. This is an external force, so we can't consider the sled as an isolated system and conservation of momentum isn't a useful tool. (Indeed, we know that the sled can't be an isolated system; it speeds up, so its momentum changes.) That's why, in our solution, we added Bailey to the mix, taking the system as Bailey plus the sled. We can ignore the external forces, so momentum is conserved. The sled's momentum increases and Bailey's momentum decreases, but the increase in the sled's momentum is exactly equal to the decrease in Bailey's momentum. We were able to quickly calculate the final speed of Bailey and the sled.

Explosions

An **explosion,** where the particles of the system move apart after a brief, intense interaction, is the opposite of a collision. The explosive forces, which could be from an expanding spring or from expanding hot gases, are *internal* forces. If the system is isolated, its total momentum during the explosion will be conserved.

Video What the Physics? Rocket Chair

EXAMPLE 9.8 Recoil speed of a toy rifle

A 30 g ball is fired from a 1.2 kg spring-loaded toy rifle with a speed of 15 m/s. What is the recoil speed of the rifle?

STRATEGIZE As the ball moves down the barrel, there are complicated forces exerted on the ball and on the rifle. However, if we take the system to be ball + rifle, these are *internal* forces that do not change the total momentum.

The *external* forces of the rifle's and ball's weights are balanced by the external force exerted by the person holding the rifle, so $\vec{F}_{net} = \vec{0}$. This is an isolated system and the law of conservation of momentum applies.

PREPARE FIGURE 9.19 shows a visual overview before and after the ball is fired. We assume the ball is fired in the $+x$-direction.

SOLVE The x-component of the total momentum is $P_x = p_{Bx} + p_{Rx}$. Everything is at rest before the trigger is pulled, so the initial momentum is zero. After the trigger is pulled, the internal force of the spring pushes the ball down the barrel *and* pushes the rifle backward. Conservation of momentum gives

$$(P_x)_f = m_B(v_{Bx})_f + m_R(v_{Rx})_f = (P_x)_i = 0$$

Solving for the rifle's velocity, we find

$$(v_{Rx})_f = -\frac{m_B}{m_R}(v_{Bx})_f = -\frac{0.030\ \text{kg}}{1.2\ \text{kg}} \times 15\ \text{m/s} = -0.38\ \text{m/s}$$

FIGURE 9.19 Before-and-after visual overview for a toy rifle.

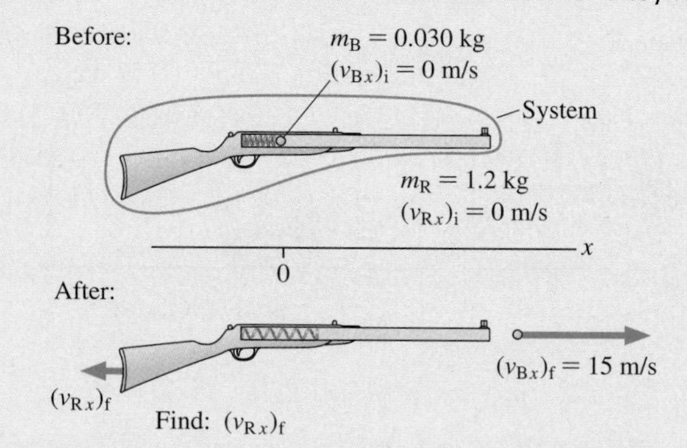

The minus sign indicates that the rifle's recoil is to the left. The recoil *speed* is 0.38 m/s.

ASSESS The bullet is much lighter than the rifle, so it makes sense that the recoil speed of the rifle is much less than the speed of the bullet. Real rifles fire their bullets at much higher velocities, and their recoil is correspondingly higher. Shooters need to brace themselves against the "kick" of the rifle back against their shoulder.

This would be a very difficult problem to solve using Newton's laws. But Example 9.8 is straightforward when approached from the before-and-after perspective of a conservation law. The selection of ball + rifle as the "system" was the critical step. For momentum conservation to be useful, we had to select a system in which the complicated forces due to the spring and to friction were all internal forces. The rifle by itself is *not* an isolated system, so its momentum is *not* conserved.

Much the same reasoning explains how a rocket or jet aircraft accelerates. FIGURE 9.20 shows a rocket with a parcel of fuel on board. Burning converts the fuel to hot gases that are expelled from the rocket motor. If we choose rocket + gases to be the system, then the burning and expulsion are internal forces. In deep space there are no other forces, so the total momentum of the rocket + gases system must be conserved. The rocket gains forward velocity and momentum as the exhaust gases are shot out the back, but the *total* momentum of the system remains zero.

FIGURE 9.20 Rocket propulsion is an example of conservation of momentum.

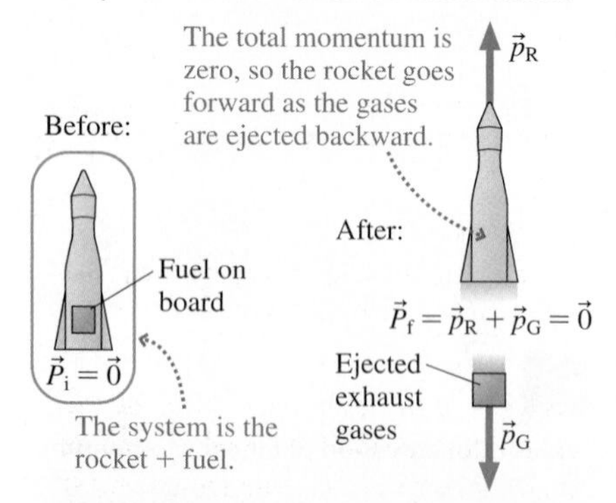

Many people find it hard to understand how a rocket can accelerate in the vacuum of space because there is nothing to "push against." Thinking in terms of momentum, you can see that the rocket does not push against anything *external,* but only against the gases that it pushes out the back. In return, in accordance with Newton's third law, the gases push forward on the rocket.

Squid propulsion BIO Squids use a form of jet propulsion to make quick movements to escape enemies or catch prey. The squid draws in water through a pair of valves in its outer sheath, or mantle, and then quickly expels the water through a funnel, propelling the squid backward.

STOP TO THINK 9.5 An explosion in a rigid pipe shoots three balls out of its ends. A 6 g ball comes out the right end. A 4 g ball comes out the left end with twice the speed of the 6 g ball. From which end, left or right, does the third ball emerge?

9.5 Inelastic Collisions

A rubber ball dropped on the floor bounces—it's *elastic*—but a ball of clay sticks to the floor without bouncing; we call such a collision *inelastic*. A golf club hitting a golf ball causes the ball to rebound away from the club (elastic), but a bullet striking a block of wood becomes embedded in the block (inelastic).

A collision in which the two objects stick together and move with a common final velocity is called a **perfectly inelastic collision.** The clay hitting the floor and the bullet embedding itself in the wood are examples of perfectly inelastic collisions. Other examples include railroad cars coupling together upon impact and darts hitting a dart board. FIGURE 9.21 emphasizes the fact that the two objects have a common final velocity after they collide. (We have drawn the combined object moving to the right, but it could have ended up moving to the left, or not moving at all, depending on the objects' masses and initial velocities.)

In other collisions, the two objects bounce apart. We've looked at some examples of these kinds of collisions, but a full analysis requires some ideas about energy. We will learn more about collisions and energy in Chapter 10.

FIGURE 9.21 A perfectly inelastic collision.

Two objects approach and collide.

Before: $(v_{1x})_i$ $(v_{2x})_i$ m_1 m_2

They stick and move together.

After:

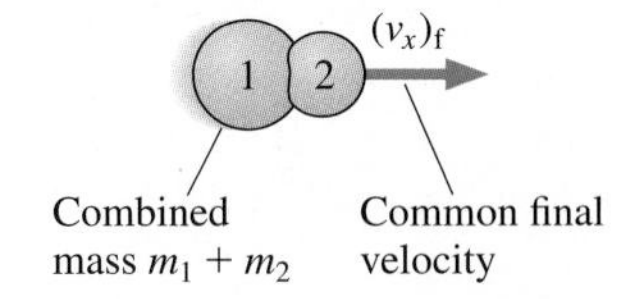

EXAMPLE 9.9 A perfectly inelastic collision of railroad cars

In assembling a train from several railroad cars, two of the cars, with masses 2.0×10^4 kg and 4.0×10^4 kg, are rolled toward each other. When they meet, they couple and stick together. The lighter car has an initial speed of 1.5 m/s; the collision causes it to reverse direction at 0.25 m/s. What was the initial speed of the heavier car?

STRATEGIZE We will define the system to be the two railroad cars. The large forces between the cars when they collide are internal forces. Friction and other external forces are much smaller. We will use the impulse approximation and consider the system of the two cars as isolated, so total momentum is conserved in the collision. The cars stick together, so this is a perfectly inelastic collision.

PREPARE FIGURE 9.22 shows a visual overview. We've chosen to let the 2.0×10^4 kg car (car 1) start out moving to the right, so

FIGURE 9.22 Before-and-after visual overview for two train cars colliding on a track.

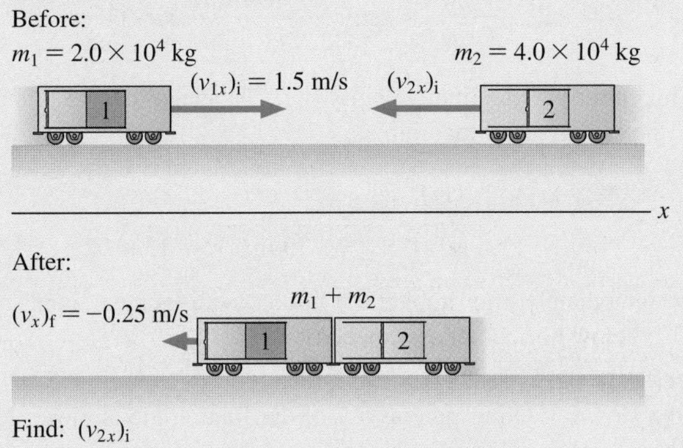

Before: $m_1 = 2.0 \times 10^4$ kg, $m_2 = 4.0 \times 10^4$ kg, $(v_{1x})_i = 1.5$ m/s, $(v_{2x})_i$

After: $(v_x)_f = -0.25$ m/s, $m_1 + m_2$

Find: $(v_{2x})_i$

$(v_{1x})_i$ is a positive 1.5 m/s. The cars move to the left after the collision, so their common final velocity is $(v_x)_f = -0.25$ m/s. You can see that velocity $(v_{2x})_i$ must be negative in order to "turn around" both cars.

SOLVE The law of conservation of momentum, $(P_x)_f = (P_x)_i$, is

$$(m_1 + m_2)(v_x)_f = m_1(v_{1x})_i + m_2(v_{2x})_i$$

where we make use of the fact that the combined mass $m_1 + m_2$ moves together after the collision. We can solve for the initial velocity of the 4.0×10^4 kg car:

$$(v_{2x})_i = \frac{(m_1 + m_2)(v_x)_f - m_1(v_{1x})_i}{m_2}$$

$$= \frac{(6.0 \times 10^4\ \text{kg})(-0.25\ \text{m/s}) - (2.0 \times 10^4\ \text{kg})(1.5\ \text{m/s})}{4.0 \times 10^4\ \text{kg}}$$

$$= -1.1\ \text{m/s}$$

The negative sign, which we anticipated, indicates that the heavier car started out moving to the left. The initial *speed* of the car, which we were asked to find, is 1.1 m/s.

ASSESS Our final result makes sense. The final speed of the two cars is only one-sixth the initial speed of the lighter car. The small final momentum means that the initial momentum must also be small, so the magnitude of the initial momentum of the heavier car should be about the same as that of the lighter car. The heavier car has twice the mass, so its speed should be about half that of the lighter car, just as we found.

STOP TO THINK 9.6 The two objects shown collide and stick together. After the collision, the combined objects

A. Move to the right as shown.
B. Move to the left.
C. Are at rest.

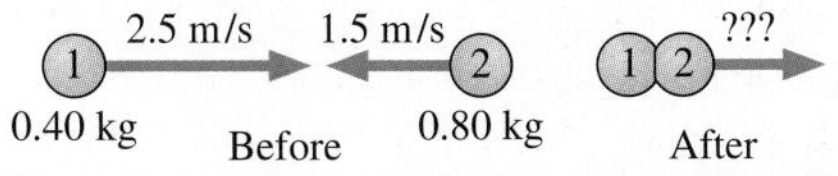

Before: 2.5 m/s, 1.5 m/s, 0.40 kg, 0.80 kg — After: ???

9.6 Momentum and Collisions in Two Dimensions

Our examples thus far have been confined to motion along a one-dimensional axis; now we'll consider motion in a plane. The total momentum $\vec{P}$ is the *vector* sum of the momentum $\vec{p} = m\vec{v}$ of each of the individual objects. Consequently, as Equations 9.15 showed, momentum is conserved only if each component of $\vec{P}$ is conserved:

$$(p_{1x})_f + (p_{2x})_f + (p_{3x})_f + \cdots = (p_{1x})_i + (p_{2x})_i + (p_{3x})_i + \cdots$$
$$(p_{1y})_f + (p_{2y})_f + (p_{3y})_f + \cdots = (p_{1y})_i + (p_{2y})_i + (p_{3y})_i + \cdots$$

The steps of Problem-Solving Approach 9.1 still apply.

EXAMPLE 9.10 **Analyzing a peregrine falcon strike** BIO

Peregrine falcons often grab their prey in flight. A falcon, flying at 18 m/s, swoops down at a 45° angle from behind a pigeon flying horizontally at 9.0 m/s. The falcon has a mass of 0.80 kg and the pigeon a mass of 0.36 kg. What are the speed and direction of the falcon (now holding the pigeon) immediately after impact?

STRATEGIZE We'll take the system to be falcon + pigeon. The external forces at work are negligible compared to the internal forces of the impact, so the total momentum of the falcon + pigeon system is conserved. This is a perfectly inelastic collision because after the collision the birds move at a common velocity.

PREPARE For a two-dimensional collision, conservation of momentum means that the x-component of the total momentum before the collision must equal the x-component of the total momentum after the collision, and similarly for the y-components. FIGURE 9.23 is a before-and-after visual overview.

FIGURE 9.23 Before-and-after visual overview for a falcon catching a pigeon.

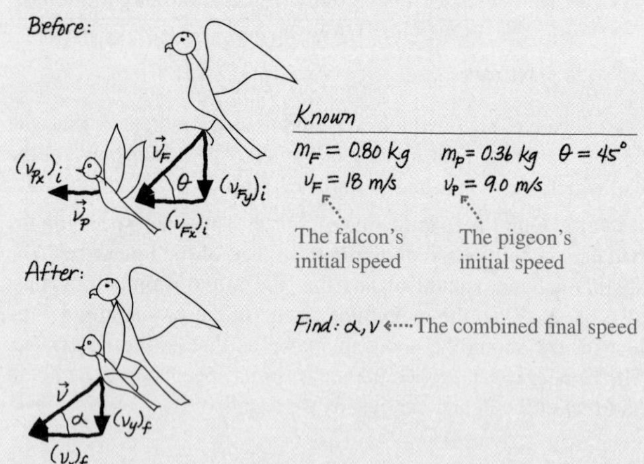

SOLVE We start by finding the x- and y-components of the momentum before the collision. For the x-component we have

The x-component of the initial momentum . . .

(Both velocity components are negative, since they point to the left.)

$$(P_x)_i = m_F(v_{Fx})_i + m_P(v_{Px})_i = m_F(-v_F\cos\theta) + m_P(-v_P)$$

. . . equals the x-component of the initial momentum of the falcon . . .

. . . plus the x-component of the initial momentum of the pigeon.

$$= (0.80\text{ kg})(-18\text{ m/s})(\cos 45°) + (0.36\text{ kg})(-9.0\text{ m/s})$$
$$= -13.4\text{ kg}\cdot\text{m/s}$$

Similarly, for the y-component of the initial momentum we have

$$(P_y)_i = m_F(v_{Fy})_i + m_P(v_{Py})_i = m_F(-v_F\sin\theta) + 0$$
$$= (0.80\text{ kg})(-18.0\text{ m/s})(\sin 45°) = -10.2\text{ kg}\cdot\text{m/s}$$

After the collision, the two birds move with a common velocity $\vec{v}$ that is directed at an angle α from the horizontal. The x-component of the final momentum is then

$$(P_x)_f = (m_F + m_P)(v_x)_f$$

Momentum conservation requires $(P_x)_f = (P_x)_i$, so

$$(v_x)_f = \frac{(P_x)_i}{m_F + m_P} = \frac{-13.4\text{ kg}\cdot\text{m/s}}{(0.80\text{ kg}) + (0.36\text{ kg})} = -11.6\text{ m/s}$$

Similarly, $(P_y)_f = (P_y)_i$ gives

$$(v_y)_f = \frac{(P_y)_i}{m_F + m_P} = \frac{-10.2\text{ kg}\cdot\text{m/s}}{(0.80\text{ kg}) + (0.36\text{ kg})} = -8.79\text{ m/s}$$

From the figure we see that $\tan\alpha = (v_y)_f/(v_x)_f$, so that

$$\alpha = \tan^{-1}\left(\frac{(v_y)_f}{(v_x)_f}\right) = \tan^{-1}\left(\frac{-8.79\text{ m/s}}{-11.6\text{ m/s}}\right) = 37°$$

The magnitude of the final velocity (i.e., the speed) can be found from the Pythagorean theorem as

$$v = \sqrt{(v_x)_f^2 + (v_y)_f^2}$$
$$= \sqrt{(-11.6\text{ m/s})^2 + (-8.79\text{ m/s})^2} = 15\text{ m/s}$$

Thus immediately after impact the falcon, with its meal, is moving 37° below horizontal at a speed of 15 m/s.

ASSESS It makes sense that the falcon slows down after catching the slower-moving pigeon. Also, the final angle is closer to the horizontal than the falcon's initial angle. This seems reasonable because the pigeon was initially flying horizontally, making the total momentum vector more horizontal than the direction of the falcon's initial momentum.

It's instructive to examine the collision in this example with a picture of the momentum vectors. The vectors $\vec{p}_F$ and $\vec{p}_P$ before the collision, and their sum $\vec{P} = \vec{p}_F + \vec{p}_P$, are shown in FIGURE 9.24. You can see that the total momentum vector makes a 37° angle with the negative x-axis. The individual values of momentum change in the collision, *but the total momentum does not.*

FIGURE 9.24 The momentum vectors of the falcon strike.

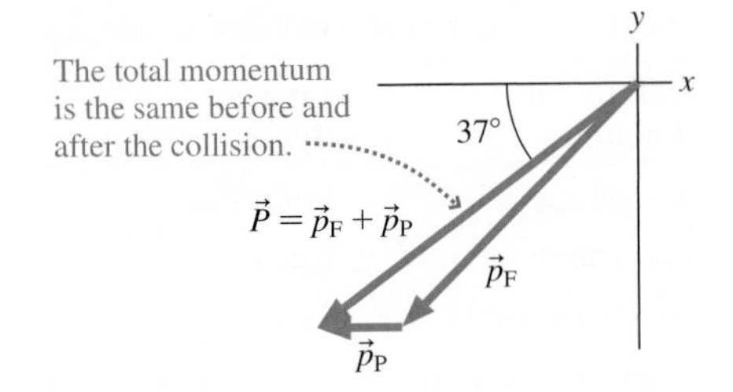
y
x
The total momentum
is the same before and
after the collision.
37°
$\vec{P} = \vec{p}_F + \vec{p}_P$
$\vec{p}_F$
$\vec{p}_P$

9.7 Angular Momentum

For a single object, we can think of the law of conservation of momentum as an alternative way of stating Newton's first law. Rather than saying that an object will continue to move in a straight line at constant velocity unless acted on by a net force, we can say that the momentum of an isolated object is conserved. Both express the idea that an object moving in a straight line tends to "keep going" unless something acts on it to change its motion.

Next, we want to come up with a rotational analog for the conservation of momentum, just as we found rotational analogs for motion (angular velocity, angular acceleration) and force (torque). Momentum isn't conserved for an object that is moving in a circle. Momentum is a vector, and the momentum of an object in circular motion changes as the direction of motion changes. Nonetheless, a ball that is moving in a circle at the end of a string tends to "keep going" in a circular path. And a spinning bicycle wheel would keep turning if it were not for friction. The quantity that expresses this idea for circular motion is called *angular momentum.*

Video Angular Momentum

Let's look at the example of pushing a merry-go-round, as in FIGURE 9.25. If you push tangentially to the rim, you are applying a torque to the merry-go-round. As you learned in ◀◀ SECTION 7.5, the merry-go-round's angular speed will continue to increase for as long as you apply this torque. The harder (greater torque) or longer (greater time) you push, the greater the increase in the angular velocity.

FIGURE 9.25 By applying a torque to the merry-go-round, the girl is increasing its angular momentum.

Suppose you apply a constant torque τ_{net} to the merry-go-round for a time Δt. By how much will the merry-go-round's angular speed increase? In Section 7.5 we found that the angular acceleration α is given by the rotational equivalent of Newton's second law, or

$$\alpha = \frac{\tau_{\text{net}}}{I} \tag{9.16}$$

where I is the merry-go-round's moment of inertia.

The angular acceleration is the rate of change of the angular velocity, so

$$\alpha = \frac{\Delta\omega}{\Delta t} \tag{9.17}$$

Setting Equations 9.16 and 9.17 equal to each other gives

$$\frac{\Delta\omega}{\Delta t} = \frac{\tau_{\text{net}}}{I}$$

or, rearranging,

$$\tau_{\text{net}}\,\Delta t = I\,\Delta\omega \tag{9.18}$$

If you recall the impulse-momentum theorem for *linear* motion, which is

$$\vec{F}_{\text{net}}\,\Delta t = m\,\Delta\vec{v} = \Delta\vec{p} \tag{9.19}$$

you can see that Equation 9.18 is an analogous statement about rotational motion. Because the quantity $I\omega$ is evidently the rotational equivalent of $m\vec{v}$, the linear momentum $\vec{p}$, it seems reasonable to define the **angular momentum** L to be

$$L = I\omega \tag{9.20}$$

Angular momentum of an object with moment of inertia I rotating at angular velocity ω

The SI units of angular momentum are those of moment of inertia times angular velocity, or $kg \cdot m^2/s$.

TABLE 9.2 Rotational and linear dynamics

Rotational dynamics	Linear dynamics
Torque τ_{net}	Force $\vec{F}_{net}$
Moment of inertia I	Mass m
Angular velocity ω	Velocity $\vec{v}$
Angular momentum $L = I\omega$	Linear momentum $\vec{p} = m\vec{v}$

Just as an object in linear motion can have a large momentum by having either a large mass or a high speed, a rotating object can have a large angular momentum by having a large moment of inertia or a large angular velocity. The merry-go-round in Figure 9.25 has a larger angular momentum if it's spinning fast than if it's spinning slowly. Also, the merry-go-round (large I) has a much larger angular momentum than a toy top (small I) spinning with the same angular velocity.

TABLE 9.2 summarizes the analogies between linear and rotational quantities that you learned in Chapter 7 and adds the analogy between linear momentum and angular momentum.

Conservation of Angular Momentum

Having now defined angular momentum, we can write Equation 9.18 as

$$\tau_{net}\,\Delta t = \Delta L \tag{9.21}$$

in exact analogy with its linear dynamics equivalent, Equation 9.19. This equation implies that the change in the angular momentum of an object is proportional to the net torque applied to the object. If the net external torque on an object is *zero,* then the change in the angular momentum is zero as well. That is, a rotating object will continue to rotate with *constant* angular momentum—to "keep going"—unless acted upon by an external torque. We can state this conclusion as the *law of conservation of angular momentum*:

Law of conservation of angular momentum The angular momentum of a rotating object subject to no net external torque ($\tau_{net} = 0$) is a constant. The final angular momentum L_f is equal to the initial angular momentum L_i.

For a system that consists of more than one rotating object we define the **total angular momentum** as the sum of the angular momentum of each of the objects in the system. If no net external torque acts on the system, the mathematical statement of the law of conservation of angular momentum is then

Final (f) and initial (i) moment of inertia and angular velocity of object 1

$$(I_1)_f(\omega_1)_f + (I_2)_f(\omega_2)_f + \cdots = (I_1)_i(\omega_1)_i + (I_2)_i(\omega_2)_i + \cdots \tag{9.22}$$

Final (f) and initial (i) moment of inertia and angular velocity of object 2

We will look at problems involving multiple objects in which the angular momentum before and after an interaction is the same. The solution is similar to what we've done for conservation of momentum in one and two dimensions.

Varying Moment of Inertia

There is one aspect of angular momentum that is not analogous to linear momentum. As we've noted, for an isolated object, the *linear* momentum $\vec{p} = m\vec{v}$ is constant. The object's mass can't change, so the velocity can't change either. The object keeps moving at a constant speed. For an isolated object, the *angular* momentum $L = I\omega$ is constant as well. The object's mass can't change, but it can experience a change in its moment of inertia because the *distribution* of mass can change. This means that **an isolated, rotating object can experience a change in angular velocity.**

The classic example is a spinning figure skater. A skater in a spin on the ice experiences very little friction. The normal force balances her weight force, so she is, to a good approximation, an isolated system. If she starts the spin with her arms and legs far out from the axis of her spin, she'll start with a large moment of inertia. As she pulls her arms and legs in, her moment of inertia decreases, so her angular velocity increases. FIGURE 9.26 illustrates this idea. By dramatically reducing their moment of inertia, world-class skaters can increase the angular speed of a spin by a factor of 5 or more.

FIGURE 9.26 A spinning figure skater.

Video Figure 9.26

EXAMPLE 9.11 Period of a merry-go-round

Joey, whose mass is 36 kg, stands at the center of a 200 kg merry-go-round that is rotating once every 2.5 s. While it is rotating, Joey walks out to the edge of the merry-go-round, 2.0 m from its center. What is the rotational period of the merry-go-round when Joey gets to the edge?

STRATEGIZE Take the system to be Joey + merry-go-round and assume frictionless bearings. There is no external torque on this system, so the angular momentum of the system will be conserved.

PREPARE As shown in the visual overview of FIGURE 9.27, we model the merry-go-round as a uniform disk of radius $R = 2.0$ m. From Table 7.1, the moment of inertia of a disk is $I_{\text{disk}} = \frac{1}{2}MR^2$. If we model Joey as an object of mass m, his moment of inertia is zero when he is at the center, but it increases to mR^2 when he reaches the edge.

FIGURE 9.27 Visual overview of the merry-go-round.

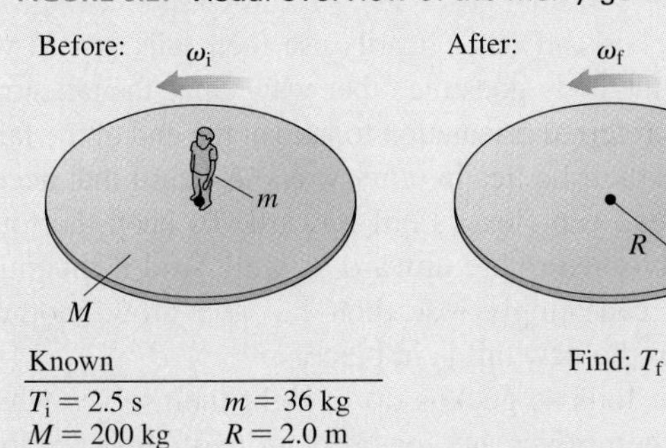

SOLVE The mathematical statement of the law of conservation of angular momentum is Equation 9.22. The initial angular momentum is

$$L_i = (I_{\text{Joey}})_i(\omega_{\text{Joey}})_i + (I_{\text{disk}})_i(\omega_{\text{disk}})_i = 0 \cdot \omega_i + \frac{1}{2}MR^2\omega_i = \frac{1}{2}MR^2\omega_i$$

Here we have used the fact that both Joey and the disk have the same initial angular velocity, which we have called ω_i. Similarly, the final angular momentum is

$$L_f = (I_{\text{Joey}})_f\omega_f + (I_{\text{disk}})_f\omega_f = mR^2\omega_f + \frac{1}{2}MR^2\omega_f = \left(mR^2 + \frac{1}{2}MR^2\right)\omega_f$$

where ω_f is the common final angular velocity of both Joey and the disk.

The law of conservation of angular momentum states that $L_f = L_i$, so that

$$\left(mR^2 + \frac{1}{2}MR^2\right)\omega_f = \frac{1}{2}MR^2\omega_i$$

Canceling the R^2 terms from both sides and solving for ω_f give

$$\omega_f = \left(\frac{M}{M + 2m}\right)\omega_i$$

The initial angular velocity is related to the initial period of rotation T_i by

$$\omega_i = \frac{2\pi}{T_i} = \frac{2\pi}{2.5\text{ s}} = 2.51\text{ rad/s}$$

Thus the final angular velocity is

$$\omega_f = \left(\frac{200\text{ kg}}{200\text{ kg} + 2(36\text{ kg})}\right)(2.51\text{ rad/s}) = 1.85\text{ rad/s}$$

When Joey reaches the edge, the period of the merry-go-round has increased to

$$T_f = \frac{2\pi}{\omega_f} = \frac{2\pi}{1.85\text{ rad/s}} = 3.4\text{ s}$$

ASSESS The merry-go-round rotates *more slowly* after Joey moves out to the edge. This makes sense because if the system's moment of inertia increases, as it does when Joey moves out, the angular velocity must decrease to keep the angular momentum constant.

◀ **The eye of a hurricane** As air masses from the slowly rotating outer zones are drawn toward the low-pressure center, their moment of inertia decreases. Because the angular momentum of these air masses is conserved, their speed must *increase* as they approach the center, leading to the high wind speeds near the center of the storm.

Solving Example 9.11 using Newton's laws would be quite difficult. We would have to deal with internal forces, such as Joey's feet against the merry-go-round, and other complications. Problems like this show the true power of conservation laws to solve problems. As long as we are worried only about the before and after states, we can use conservation laws to make a connection between the two without being concerned about the details of the interactions.

CONCEPTUAL EXAMPLE 9.12 Free falling, with a twist

Geckos have feet that stick to surfaces, which gives them a remarkable ability to climb, including the ability to scale smooth vertical surfaces. Sometimes, though, their feet fail them, and they fall. No matter their orientation when this happens, geckos can use their stout tails to flip themselves over and land on their feet. Suppose a gecko is falling, upside down, and is not rotating. How can it possibly right itself before landing?

REASON When the gecko begins to fall, it is not rotating—its angular momentum is zero. If we treat the gecko as the system, its angular momentum can't change. The gecko can't start rotating. Instead, let's consider the system as being made up of two parts, the body of the gecko and the stout tail. The total angular momentum of the body + tail system is zero. If the gecko spins its tail in one direction, the body will spin in the opposite direction. When the gecko's body reaches the correct orientation, the gecko simply stops rotating its tail. The gecko's body will then stop rotating as well, and the gecko will land on its feet. FIGURE 9.28 outlines how the gecko might do this.

FIGURE 9.28 Motion of a falling gecko.

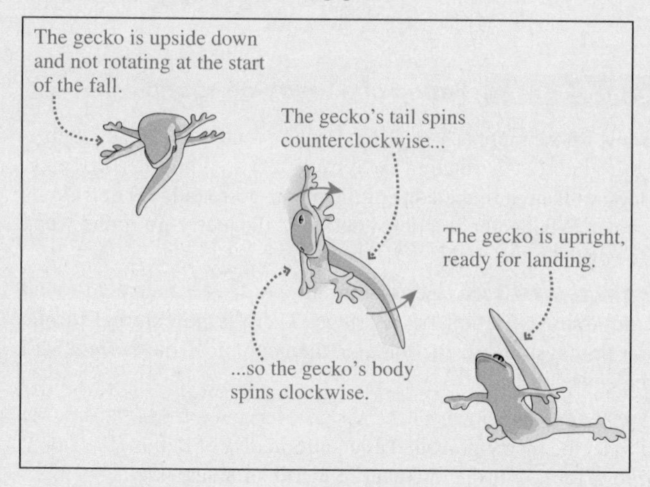

The gecko is upside down and not rotating at the start of the fall.
The gecko's tail spins counterclockwise...
The gecko is upright, ready for landing.
...so the gecko's body spins clockwise.

ASSESS This description makes sense. Lizards certainly use their tails for balance, and the flip described would be a natural extension of this ability. We were told that the gecko has a stout tail; having a large mass—and therefore a large moment of inertia—in the tail would be necessary for the gecko to right itself as described.

In fact, video studies show that geckos and other lizards use their tails just as we described. The tail spins one way, and the body goes the other way. Once the tail stops spinning, so does the body, now in the preferred orientation to land at the end of the fall.

Lizards also use their tails to rotate their bodies in other ways. A lizard that needs to raise its body in the middle of a leap can flex its tail upward. To keep the total angular momentum constant, the body must rotate upward as well. And if this puts the tail in harm's way, some lizards can simply lose their tail and grow another, keeping acrobatics to a minimum until the new tail is in place.

You might think that cats use their tails as geckos do to right themselves when they fall. In fact, falling cats do right themselves, but they rely more on their flexible bodies than their fluffy tails. Cats can bend in the middle and complete a complicated twisting motion that leads to a rotation of the body. Tailless cats can right themselves just fine, but the same is not true for geckos.

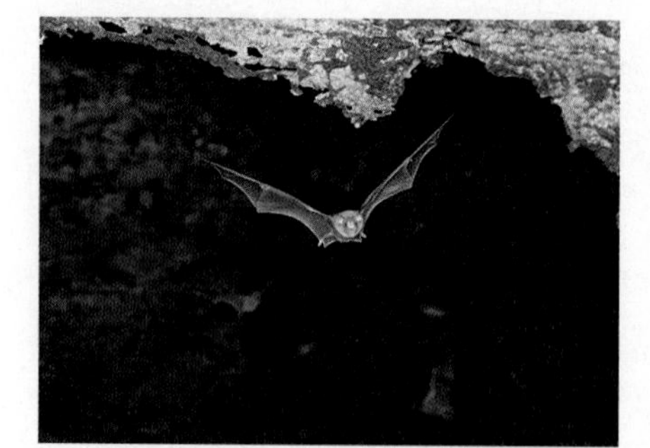

Acrobatic bats Bats have heavier wings than other fliers. This makes flight more energetically expensive for bats than for birds, but it enhances maneuverability. Twisting their heavy wings this way and that allows the bats to reorient their bodies in flight, much as the gecko spins its substantial tail to reposition itself. This is a real asset to an animal that must fly in tight spaces and complete complex maneuvers such as landing on a ceiling perch.

STOP TO THINK 9.7 The left figure shows two boys of equal mass standing halfway to the edge on a turntable that is freely rotating at angular speed ω_i. They then walk to the positions shown in the right figure. The final angular speed ω_f is

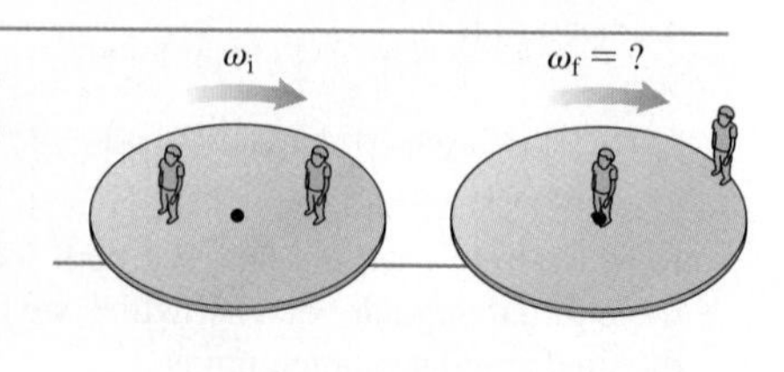

A. Greater than ω_i. B. Less than ω_i. C. Equal to ω_i.

INTEGRATED EXAMPLE 9.13 Aerial firefighting

A forest fire is easiest to attack when it's just getting started. In remote locations, this often means using airplanes to rapidly deliver large quantities of water and fire suppressant to the blaze.

The "Superscooper" is an amphibious aircraft that can pick up a 6000 kg load of water by skimming over the surface of a river or lake and scooping water directly into its storage tanks. As it approaches the water's surface at a speed of 35 m/s, an empty Superscooper has a mass of 13,000 kg.

a. It takes the plane 12 s to pick up a full load of water. If we ignore the force on the plane due to the thrust of its propellers, what is its speed immediately after picking up the water?
b. What is the impulse delivered to the plane by the water?
c. What is the average force of the water on the plane?
d. The plane then flies over the fire zone at 40 m/s. It releases water by opening doors in the belly of the plane, allowing the water to fall straight down with respect to the plane. What is the plane's speed after dropping the water if it takes 5.0 s to do so?

STRATEGIZE We can solve part a, and later part d, using conservation of momentum, following Problem-Solving Approach 9.1. We'll need to choose the system with care, so that $\vec{F}_{net} = \vec{0}$. The plane alone is not an appropriate system for using conservation of momentum: As the plane scoops up the water, the water exerts a large external drag force on the plane, so $\vec{F}_{net}$ is definitely not zero. Instead, we should choose the plane *and* the water it is going to scoop up as the system. Then there are no external forces in the x-direction, and the net force in the y-direction is zero, since neither plane nor water accelerates appreciably in this direction during the scooping process. The complicated forces between plane and water are now *internal* forces that do not change the total momentum of the plane + water system.

PREPARE With the system chosen, we follow the steps of Tactics Box 9.1 to prepare the before-and-after visual overview shown in FIGURE 9.29.

Parts b and c are impulse-and-momentum problems, so to solve them we'll use the impulse-momentum theorem, Equation 9.8. The impulse-momentum theorem considers the dynamics of a *single* object—here, the plane—subject to external forces—in this case, from the water.

SOLVE a. The x-component of the law of conservation of momentum is

$$(P_x)_f = (P_x)_i$$

FIGURE 9.29 Visual overview of the plane and water.

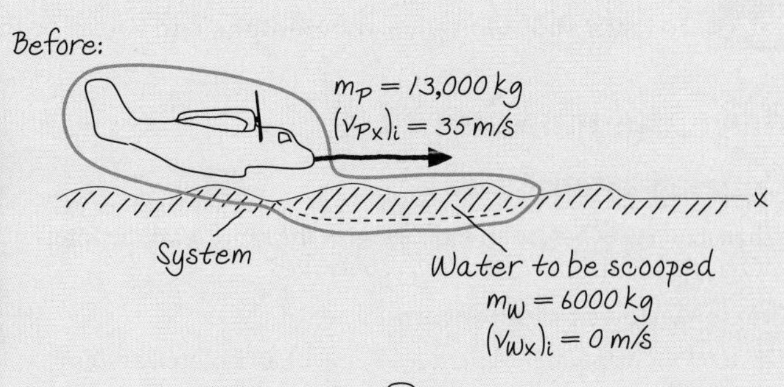

or

$$(m_P + m_W)(v_x)_f = m_P(v_{Px})_i + m_W(v_{Wx})_i = m_P(v_{Px})_i + 0$$

Here we've used the facts that the initial velocity of the water is zero and that the final situation, as in an inelastic collision, has the combined mass of the plane and water moving with the same velocity $(v_x)_f$. Solving for $(v_x)_f$, we find

$$(v_x)_f = \frac{m_P(v_{Px})_i}{m_P + m_W} = \frac{(13{,}000 \text{ kg})(35 \text{ m/s})}{(13{,}000 \text{ kg}) + (6000 \text{ kg})} = 24 \text{ m/s}$$

b. The impulse-momentum theorem is $J_x = \Delta p_x$, where $\Delta p_x = m_P \Delta v_x$ is the change in the plane's momentum. Thus

$$J_x = m_P \Delta v_x = m_P[(v_x)_f - (v_{Px})_i]$$
$$= (13{,}000 \text{ kg})(24 \text{ m/s} - 35 \text{ m/s}) = -1.4 \times 10^5 \text{ kg} \cdot \text{m/s}$$

c. From Equation 9.1, the definition of impulse, we have

$$(F_{avg})_x = \frac{J_x}{\Delta t} = \frac{-1.4 \times 10^5 \text{ kg} \cdot \text{m/s}}{12 \text{ s}} = -12{,}000 \text{ N}$$

d. Because the water drops straight down *relative to the plane,* it has the same x-component of velocity immediately after being dropped as before being dropped. That is, simply opening the doors doesn't cause the water to speed up or slow down horizontally, so the water's horizontal momentum doesn't change upon being dropped. Because the total momentum of the plane + water system is conserved, the momentum of the plane doesn't change either. The plane's speed after the drop is still 40 m/s.

ASSESS The mass of the water is nearly half that of the plane, so the significant decrease in the plane's velocity as it scoops up the water is reasonable. The force of the water on the plane is large, but is still only about 10% of the plane's weight, $mg = 130{,}000$ N, so the answer seems to be reasonable.

SUMMARY

GOAL To learn about impulse, momentum, and a new problem-solving approach based on conservation laws.

GENERAL PRINCIPLES

Conservation Laws

When a quantity *before* an interaction is the same *after* the interaction, we say that the quantity is **conserved.**

Conservation of momentum

The total momentum $\vec{P} = \vec{p}_1 + \vec{p}_2 + \cdots$ of an **isolated system**—one on which no net force acts—is a constant. Thus

$$\vec{P}_f = \vec{P}_i$$

Conservation of angular momentum

The angular momentum L of a rotating object or system of objects subject to zero net external torque is a constant. Thus

$$L_f = L_i$$

This can be written in terms of the initial and final moments of inertia I and angular velocities ω as

$$(I_1)_f(\omega_1)_f + (I_2)_f(\omega_2)_f + \cdots = (I_1)_i(\omega_1)_i + (I_2)_i(\omega_2)_i + \cdots$$

Solving Momentum Conservation Problems

STRATEGIZE Choose an isolated system or a system that is isolated during at least part of the problem.

PREPARE Draw a visual overview of the system before and after the interaction.

SOLVE Write the law of conservation of momentum in terms of vector components:

$$(p_{1x})_f + (p_{2x})_f + \cdots = (p_{1x})_i + (p_{2x})_i + \cdots$$
$$(p_{1y})_f + (p_{2y})_f + \cdots = (p_{1y})_i + (p_{2y})_i + \cdots$$

In terms of masses and velocities, this is

$$m_1(v_{1x})_f + m_2(v_{2x})_f + \cdots = m_1(v_{1x})_i + m_2(v_{2x})_i + \cdots$$
$$m_1(v_{1y})_f + m_2(v_{2y})_f + \cdots = m_1(v_{1y})_i + m_2(v_{2y})_i + \cdots$$

ASSESS Is the result reasonable?

IMPORTANT CONCEPTS

Momentum $\vec{p} = m\vec{v}$

$\vec{v}$ m $\vec{p}$

Impulse J_x = area under force curve

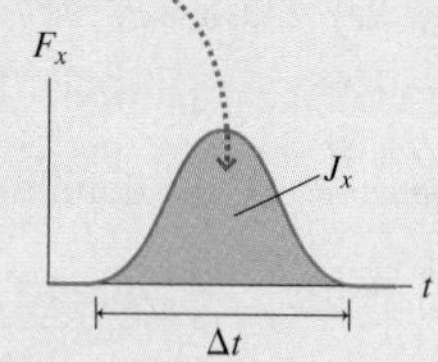

Impulse and momentum are related by the **impulse-momentum theorem**

$$\Delta p_x = J_x$$

This is an alternative statement of Newton's second law.

Angular momentum $L = I\omega$ is the rotational analog of linear momentum $\vec{p} = m\vec{v}$.

System A group of interacting objects

Isolated system A system on which the net external force is zero

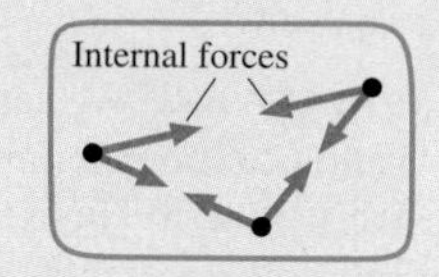

Before-and-after visual overview

- Define the system.
- Use two drawings to show the system *before* and *after* the interaction.
- List known information and identify what you are trying to find.

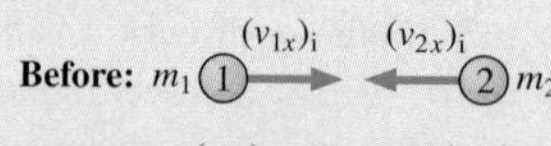

APPLICATIONS

Collisions Two or more objects come together. In a perfectly inelastic collision, they stick together and move with a common final velocity.

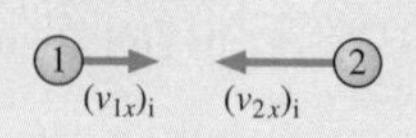

Explosions Two or more objects move away from each other.

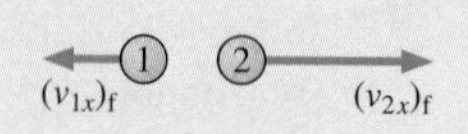

Two dimensions Both the x- and y-components of the total momentum $\vec{P}$ must be conserved, giving two simultaneous equations.

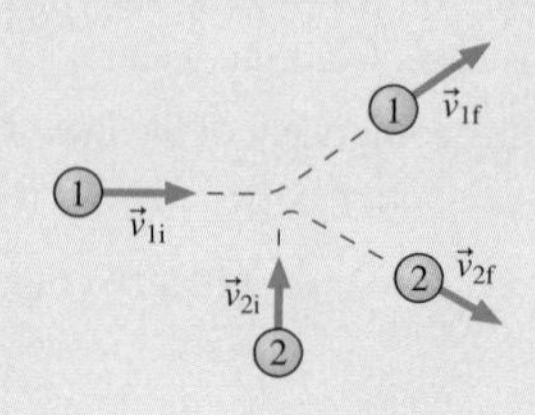

Learning Objectives After studying this chapter, you should be able to:

- Calculate the momentum of and impulse on an object. *Conceptual Questions 9.1, 9.3; Problems 9.1, 9.2, 9.5, 9.7*
- Use the impulse-momentum theorem to solve impulse and momentum problems. *Conceptual Questions 9.4, 9.10; Problems 9.3, 9.4, 9.6, 9.9, 9.16*
- Use Problem-Solving Approach 9.1 to solve conservation of momentum problems. *Conceptual Questions 9.9, 9.12; Problems 9.10, 9.13, 9.15, 9.20, 9.23*
- Apply conservation of momentum to collisions and explosions. *Conceptual Questions 9.7, 9.17; Problems 9.19, 9.26, 9.28, 9.30, 9.33*
- Use momentum in two dimensions. *Problems 9.35, 9.36, 9.37, 9.38, 9.39*
- Understand and use angular momentum and its conservation. *Conceptual Questions 9.18, 9.19; Problems 9.40, 9.41, 9.42, 9.43, 9.45*

STOP TO THINK ANSWERS

Chapter Preview Stop to Think: C. The force of the hammer on the nail and the force of the nail on the hammer are the two members of an action/reaction pair and thus, according to Newton's third law, must have the same magnitude.

Stop to Think 9.1: C. Impulse equals the area under the force-versus-time curve or, for these rectangular graphs, the force times the duration Δt. Force A has twice the magnitude of force B, but the collision lasts for half the time. Hence both curves have the same area and the same impulse.

Stop to Think 9.2: D. We know that $\vec{p}_f = \vec{p}_i + \vec{J}$. As shown in the figure at right, the only initial momentum vector that satisfies this relationship is vector D.

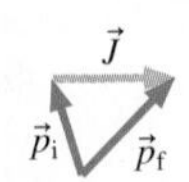

Stop to Think 9.3: F. The cart is initially moving in the negative x-direction, so $(p_x)_i = -20 \text{ kg} \cdot \text{m/s}$. After it bounces, $(p_x)_f = 10 \text{ kg} \cdot \text{m/s}$. Thus $\Delta p_x = (10 \text{ kg} \cdot \text{m/s}) - (-20 \text{ kg} \cdot \text{m/s}) = 30 \text{ kg} \cdot \text{m/s}$.

Stop to Think 9.4: B. The clay ball goes from $(v_x)_i = v$ to $(v_x)_f = 0$, so $J_{clay} = \Delta p_x = -mv$. The rubber ball rebounds, going from $(v_x)_i = v$ to $(v_x)_f = -v$ (same speed, opposite direction). Thus $J_{rubber} = \Delta p_x = -2mv$. The rubber ball has a greater momentum change, and this requires a greater impulse.

Stop to Think 9.5: Right end. The balls started at rest, so the total momentum of the system is zero. It's an isolated system, so the total momentum after the explosion is still zero. The 6 g ball has momentum $6v$. The 4 g ball, with velocity $-2v$, has momentum $-8v$. The combined momentum of these two balls is $-2v$. In order for P to be zero, the third ball must have a *positive* momentum $(+2v)$ and thus a positive velocity.

Stop to Think 9.6: B. The momentum of object 1 is $(0.40 \text{ kg})(2.5 \text{ m/s}) = 1.0 \text{ kg} \cdot \text{m/s}$, while that of object 2 is $(0.80 \text{ kg})(-1.5 \text{ m/s}) = -1.2 \text{ kg} \cdot \text{m/s}$. The total momentum is then $1.0 \text{ kg} \cdot \text{m/s} - 1.2 \text{ kg} \cdot \text{m/s} = -0.2 \text{ kg} \cdot \text{m/s}$. Because it's negative, the total momentum, and hence the final velocity of the objects, is directed to the left.

Stop to Think 9.7: B. Angular momentum $L = I\omega$ is conserved. Both boys have mass m and initially stand distance $R/2$ from the axis. Thus the initial moment of inertia is $I_i = I_{disk} + 2 \times m(R/2)^2 = I_{disk} + \frac{1}{2}mR^2$. The final moment of inertia is $I_f = I_{disk} + 0 + mR^2$, because the boy standing at the axis contributes nothing to the moment of inertia. Because $I_f > I_i$ we must have $\omega_f < \omega_i$.

Video Tutor Solution Chapter 9

QUESTIONS

Conceptual Questions

1. You are bowling with a 10-pound ball that you throw at 10 mph, and you've decided that you need to hit the pins with a ball that has more momentum. What are two possible changes you could make?
2. Rank in order, from largest to smallest, the magnitudes of the momentum p_{1x} through p_{5x} of the objects presented in Figure Q9.2. Explain.

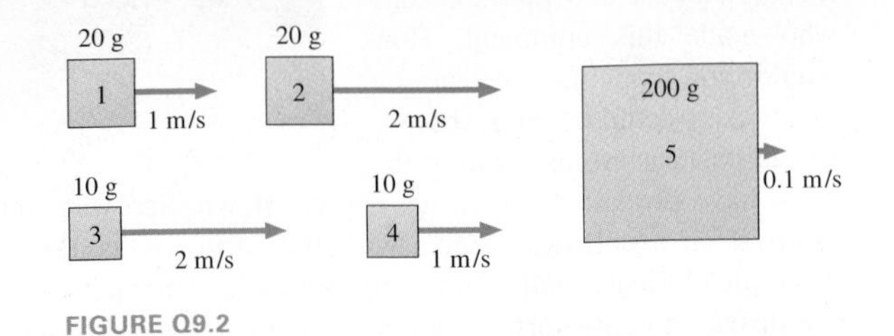

FIGURE Q9.2

Problem difficulty is labeled as | (straightforward) to ||||| (challenging). INT Problems labeled integrate significant material from earlier chapters; BIO are of biological or medical interest.

The eText icon indicates when there is a video tutor solution available for the chapter or for a specific problem. To launch these videos, log into your eText through Mastering™ Physics or log into the Study Area.

3. Two pucks, of mass m and $4m$, lie on a frictionless table. Equal forces are used to push both pucks forward a distance of 1 m.
   a. Which puck takes longer to travel the distance? Explain.
   b. Which puck has the greater momentum upon completing the distance? Explain.
4. When you catch a water balloon, it's best to start with your hand in motion, moving with the balloon, and then gradually slow it to rest. Why is this approach desirable?
5. When you leap down from a high perch, you have a gentler landing with less force if you bend your knees as you land. Use the principles of impulse and momentum to explain why this approach reduces the force on you.
6. Students in a technology class are racing cars propelled by carbon dioxide cartridges. A burst of gas is released when the seal of the cartridge is broken, and this propels the car down the track. The cartridges are designed to provide a certain impulse. Use the concepts of impulse and momentum to explain why such a cartridge will propel a lighter car to a higher speed.
7. A stationary firecracker explodes into three pieces. One piece travels off to the east; a second travels to the north. Which of the vectors of Figure Q9.7 could be the velocity of the third piece? Explain.

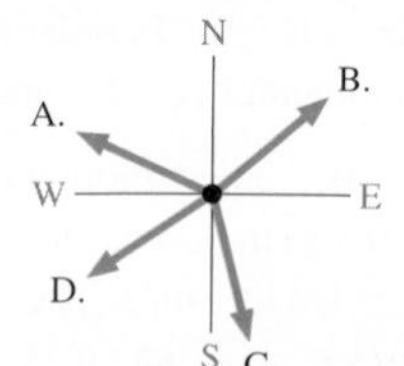

FIGURE Q9.7

8. Two students stand at rest, facing each other on frictionless skates. They then start tossing a heavy ball back and forth between them. Describe their subsequent motion.
9. A 2 kg cart rolling to the right at 3 m/s runs into a 3 kg cart rolling to the left. After the collision, both carts are stationary. What was the original speed of the 3 kg cart?
10. Automobiles are designed with "crumple zones" intended to collapse in a collision. Why would a manufacturer design part of a car so that it collapses in a collision?
11. You probably know that it feels better to catch a baseball if you are wearing a padded glove. Explain why this is so, using the ideas of momentum and impulse.
12. In the early days of rocketry, some people claimed that rockets couldn't fly in outer space as there was no air for the rockets to push against. Suppose you were an early rocket scientist and met someone who made this argument. How would you convince the person that rockets could travel in space?

13. In the past, asteroids striking the earth have produced disastrous results. If we discovered an asteroid on a collision course with the earth, we could, in principle, deflect it and avoid an impact by focusing a laser on the surface. Intense surface heating from the laser could cause surface material to be ejected into space at high speed. How would this deflect the asteroid?
14. Two ice skaters, Megan and Jason, push off from each other on frictionless ice. Jason's mass is twice that of Megan.
   a. Which skater, if either, experiences the greater impulse during the push? Explain.
   b. Which skater, if either, has the greater speed after the push-off? Explain.
15. Suppose a rubber ball and a steel ball collide. Which, if either, receives the larger impulse? Explain.
16. While standing still on a basketball court, you throw the ball to a teammate. Why do you not move backward as a result? Is the law of conservation of momentum violated?
17. To win a prize at the county fair, you're trying to knock down a bowling pin by hitting it with a thrown object. Should you choose to throw a rubber ball, which will bounce off the pin, or a beanbag, which will strike the pin and not bounce? Assume the ball and beanbag have equal size and weight. Explain.
18. In the playground game of tetherball, a ball tied to a rope circles a pole, wrapping the rope around the pole as it goes. Opponents compete to wrap the ball around the pole in one direction or the other; the game ends when the rope is entirely wrapped around the pole and the ball touches the pole. If you get the ball going around in one direction, as the rope shortens and the ball goes in smaller and smaller circles, the ball goes around with a shorter and shorter period. Explain why this happens.

19. Monica stands at the edge of a circular platform that is slowly rotating on a frictionless axle. She then walks toward the opposite edge, passing through the platform's center. Describe the motion of the platform as Monica makes her trip.
20. If the earth warms significantly, the polar ice caps will melt. Water will move from the poles, near the earth's rotation axis, and will spread out around the globe. In principle, this will change the length of the day. Why? Will the length of the day increase or decrease?
21. If you jump from a diving board and find that your body is rotating forward, you can spin your arms to stop the rotation and make your body land feet first. Explain in which direction you would rotate your arms and how this would help.
22. A dancer leaping into the air and trying for the maximum possible vertical height will move her arms upward as she is pushing off with her legs, putting her arms in rapid upward motion before she leaves the ground. Explain how this approach can give greater height.

Multiple-Choice Questions

23. || Model rocket engines are rated by their thrust force and by the impulse they provide. You can use this information to determine the time interval at which the engines fire. Two rocket engines provide the same impulse. The first engine provides 6 N of thrust for 2 s; the second provides 4 N of thrust. For how long does this second engine fire?
   A. 1 s B. 2 s C. 3 s D. 4 s
24. | Curling is a sport played with 20 kg stones that slide across an ice surface. Suppose a curling stone sliding at 1 m/s strikes another, stationary stone and comes to rest in 2 ms. Approximately how much force is there on the stone during the impact?
   A. 200 N B. 1000 N C. 2000 N D. 10,000 N
25. | Two balls are hung from cords. The first ball, of mass 1.0 kg, is pulled to the side and released, reaching a speed of 2.0 m/s at the bottom of its arc. Then, as shown in Figure Q9.25, it hits and sticks to another ball. The speed of the pair just after the collision is 1.2 m/s. What is the mass of the second ball?
   A. 0.67 kg B. 2.0 kg C. 1.7 kg D. 1.0 kg

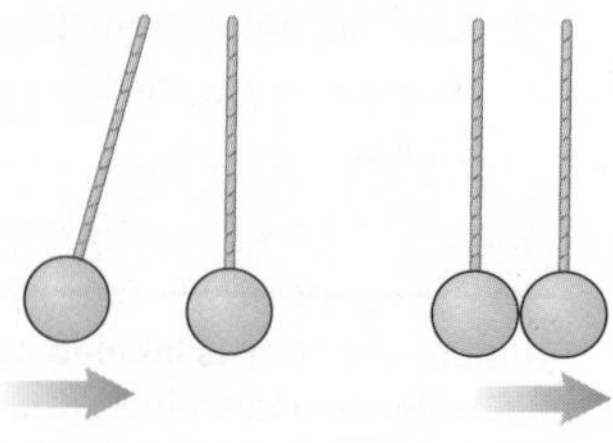

FIGURE Q9.25

26. | Figure Q9.26 shows two blocks sliding on a frictionless surface. Eventually the smaller block catches up with the larger one, collides with it, and sticks. What is the speed of the two blocks after the collision?
A. $v_i/2$ B. $4v_i/5$ C. v_i D. $5v_i/4$ E. $2v_i$

FIGURE Q9.26

27. | Two friends are sitting in a stationary canoe. At $t = 3.0$ s the person at the front tosses a sack to the person in the rear, who catches the sack 0.2 s later. Which plot in Figure Q9.27 shows the velocity of the boat as a function of time? Positive velocity is forward, negative velocity is backward. Neglect any drag force on the canoe from the water.

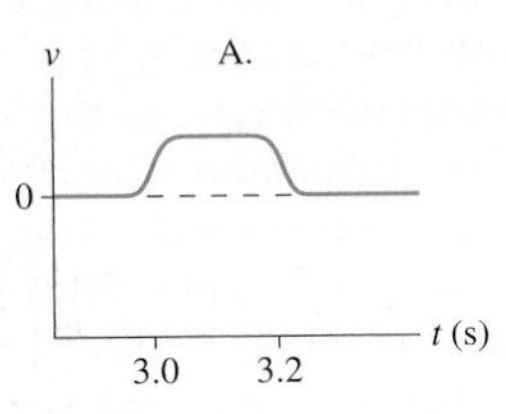

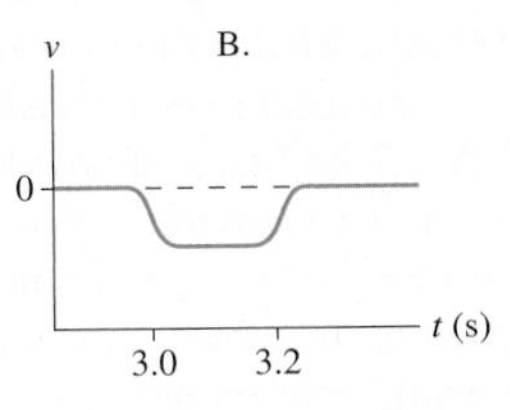

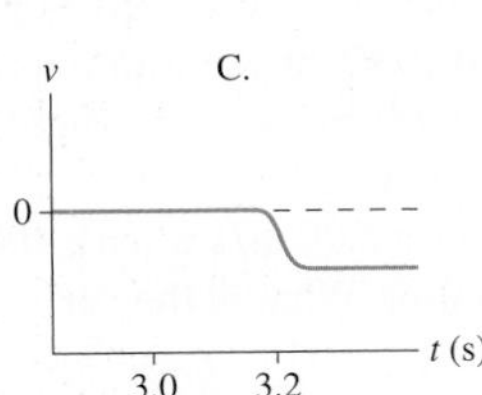

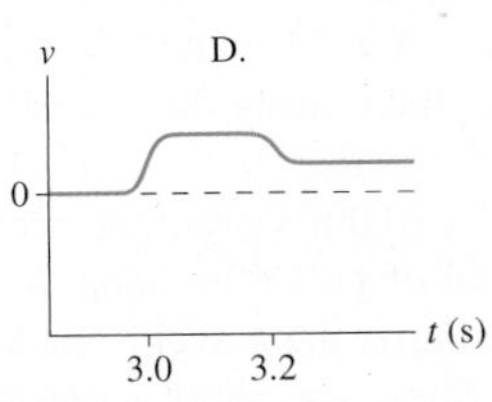

FIGURE Q9.27

28. || A 4.0-m-diameter playground merry-go-round, with a moment of inertia of 400 kg · m^2, is freely rotating with an angular velocity of 2.0 rad/s. Ryan, whose mass is 80 kg, runs on the ground around the outer edge of the merry-go-round in the opposite direction to its rotation. Still moving, he jumps directly onto the rim of the merry-go-round, bringing it (and himself) to a halt. How fast was Ryan running when he jumped on?
A. 2.0 m/s
B. 4.0 m/s
C. 5.0 m/s
D. 7.5 m/s
E. 10 m/s

29. || A disk rotates freely on a vertical axis with an angular velocity of 30 rpm. An identical disk rotates above it in the same direction about the same axis, but without touching the lower disk, at 20 rpm. The upper disk then drops onto the lower disk. After a short time, because of friction, they rotate together. The final angular velocity of the disks is
A. 50 rpm
B. 40 rpm
C. 25 rpm
D. 20 rpm
E. 10 rpm

PROBLEMS

Section 9.1 Impulse

Section 9.2 Momentum and the Impulse-Momentum Theorem

1. | At what speed do a bicycle and its rider, with a combined mass of 100 kg, have the same momentum as a 1500 kg car traveling at 1.0 m/s?
2. | Two physics students are doing a side competition during a game of bowling, seeing who can toss a ball with the larger momentum. The first bowler throws a 4.5 kg ball at 7.2 m/s. A second bowler throws a 6.4 kg ball. What speed must she beat to win the competition?
3. | A 57 g tennis ball is served at 45 m/s. If the ball started from rest, what impulse was applied to the ball by the racket?
4. || A large raindrop—the type that lands with a definite splat—has a mass of 0.014 g and hits your roof at a speed of 8.1 m/s.
   a. What is the magnitude of the impulse delivered to your roof?
   b. If the raindrop comes to rest in 0.37 ms, what is the magnitude of the average force of the impact?
5. || A student throws a 120 g snowball at 7.5 m/s at the side of the schoolhouse, where it hits and sticks. What is the magnitude of the average force on the wall if the duration of the collision is 0.15 s?
6. || Model rocket engines are rated by the impulse that they deliver when they fire. A particular engine is rated to deliver an impulse of 3.5 kg · m/s. The engine powers a 120 g rocket, including the mass of the engine. What is the final speed of the rocket once the engine has fired? (Ignore the change in mass as the engine fires and ignore the weight force during the short duration firing of the engine.)
7. |||| In Figure P9.7, what value of F_{max} gives an impulse of 6.0 N · s?

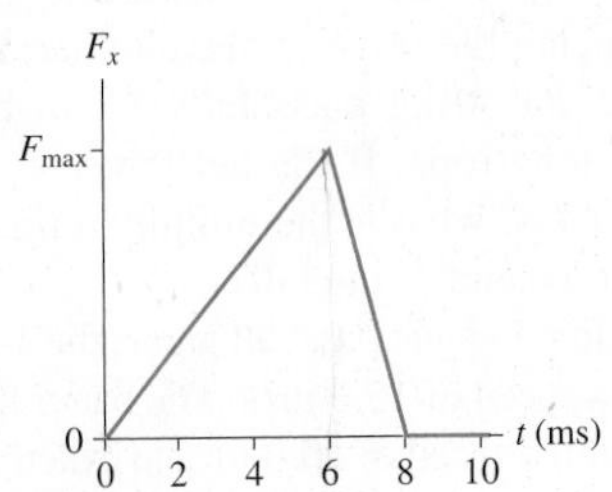

FIGURE P9.7

Section 9.3 Solving Impulse and Momentum Problems

8. || A billiard ball of mass 0.28 kg hits a second, identical ball at a speed of 7.2 m/s and comes to rest as the second ball flies off. The collision takes 250 μs. What is the average force on the first ball? The second ball?

Watch Video Solution Problem 9.25

9. || Use the impulse-momentum theorem to find how long a stone falling straight down takes to increase its speed from 5.5 m/s to 10.4 m/s.

10. || BIO A trap-jaw ant snaps its mandibles shut at very high speed, a good trait for catching small prey. But an ant can also slam its mandibles into the ground; the resulting force can launch the ant into the air for a quick escape. A 12 mg ant hits the ground with an average force of 47 mN for a time of 0.13 ms; these are all typical values. At what speed does it leave the ground?

11. |||| A 60 g tennis ball with an initial speed of 32 m/s hits a wall and rebounds with the same speed. Figure P9.11 shows the force of the wall on the ball during the collision. What is the value of F_{max}, the maximum value of the contact force during the collision?

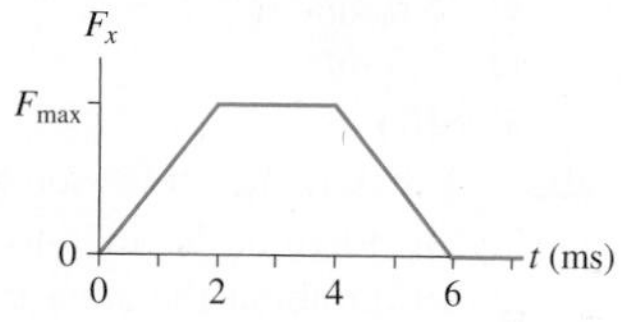

FIGURE P9.11

12. || BIO Investigators studying the effect of hitting a soccer ball with the head are using a force plate to look at the forces in ball collisions; the force when the ball hits a player's head will be similar. A 0.43 kg ball is launched at a force plate at 16 m/s. Smoothed data for the force versus time for the collision are shown in Figure P9.12. At what speed does the ball rebound from the plate?

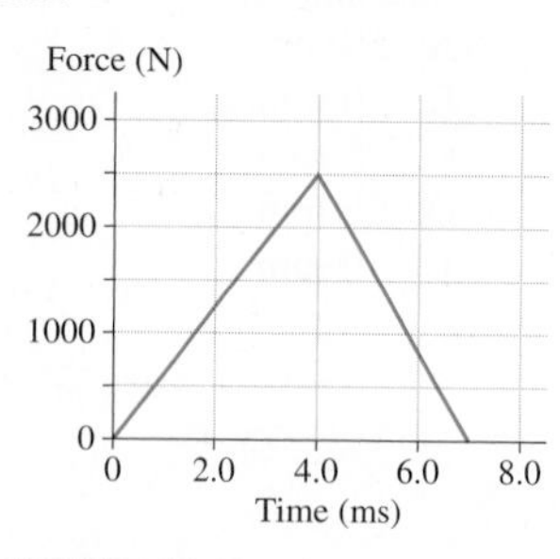

FIGURE P9.12

13. || A child is sliding on a sled at 1.5 m/s to the right. You stop the sled by pushing on it for 0.50 s in a direction opposite to its motion. If the mass of the child and sled is 35 kg, what average force do you need to apply to stop the sled? Use the concepts of impulse and momentum.

14. || BIO Ferns spread spores instead of seeds, and some ferns eject the spores at surprisingly high speeds. One species accelerates 1.4 μg spores to a 4.5 m/s ejection speed in a time of 1.0 ms. What impulse is provided to the spores? What is the average force on a spore?

15. | As part of a safety investigation, two 1400 kg cars traveling at 20 m/s are crashed into different barriers. Find the average forces exerted on (a) the car that hits a line of water barrels and takes 1.5 s to stop, and (b) the car that hits a concrete barrier and takes 0.10 s to stop.

16. || INT Climbing ropes stretch when they catch a falling climber, thus increasing the time it takes the climber to come to rest and reducing the force on the climber. In one standardized test of ropes, an 80 kg mass falls 4.8 m before being caught by a 2.5-m-long rope. If the net force on the mass must be kept below 11 kN, what is the minimum time for the mass to come to rest at the end of the fall?

17. || In a Little League baseball game, the 145 g ball reaches the batter with a speed of 15.0 m/s. The batter hits the ball, and it leaves his bat with a speed of 20.0 m/s in exactly the opposite direction.
   a. What is the magnitude of the impulse delivered by the bat to the ball?
   b. If the bat is in contact with the ball for 1.5 ms, what is the magnitude of the average force exerted by the bat on the ball?

18. || BIO The flowers of the bunchberry plant open with astonishing force and speed, with pollen grains accelerating from rest to 7.5 m/s.
   a. What impulse is delivered to a 1.0×10^{-10} kg pollen grain?
   b. What is the average force if this impulse is delivered in 0.30 ms?
   c. How does this force compare to the weight of the pollen grain?

Section 9.4 Conservation of Momentum

19. |||| A small, 100 g cart is moving at 1.20 m/s on a frictionless track when it collides with a larger, 1.00 kg cart at rest. After the collision, the small cart recoils at 0.850 m/s. What is the speed of the large cart after the collision?

20. || In principle, when you fire a rifle, the recoil should push you backward. How big a push will it give? Let's find out by doing a calculation in a very artificial situation. Suppose a man standing on frictionless ice fires a rifle horizontally. The mass of the man together with the rifle is 70 kg, and the mass of the bullet is 10 g. If the bullet leaves the muzzle at a speed of 500 m/s, what is the final speed of the man? Given your result, do you expect to be pushed backward when you fire a rifle?

21. ||| A 2.7 kg block of wood sits on a frictionless table. A 3.0 g bullet, fired horizontally at a speed of 500 m/s, goes completely through the block, emerging at a speed of 220 m/s. What is the speed of the block immediately after the bullet exits?

22. || BIO Squid rely on jet propulsion when a rapid escape is necessary. A 1.5 kg squid at rest pulls 0.10 kg of water into its mantle, then ejects this water at a remarkable 45 m/s. Right after this ejection, how fast is the squid moving?

23. || A 10,000 kg railroad car is rolling at 2.00 m/s when a 4000 kg load of gravel is suddenly dropped in. What is the car's speed just after the gravel is loaded?

24. || BIO Some species of jellyfish use jet propulsion to get around, a much gentler form of jet propulsion than squid use. A small jellyfish takes water into its bell; the total mass of jellyfish and water is 3.2 g. The jellyfish then rapidly ejects 1.3 g of water, achieving a speed of 0.070 m/s. What is the speed of the ejected water?

25. | A 55 kg hunter, standing on frictionless ice, shoots a 42 g bullet at a speed of 620 m/s. What is the recoil speed of the hunter?

26. || If you free the cork in a highly pressurized champagne bottle, the resulting launch of the cork will, in principle, cause the bottle to recoil. A filled champagne bottle has a mass of 1.8 kg. The cork has a mass of 7.5 g and is launched at 20 m/s. If the bottle could move freely, with what speed would it recoil? Is this something you are likely to notice?

Section 9.5 Inelastic Collisions

27. || A 300 g bird flying along at 6.0 m/s sees a 10 g insect heading straight toward it with a speed of 30 m/s. The bird opens its mouth wide and enjoys a nice lunch. What is the bird's speed immediately after swallowing?

28. || BIO Peregrine falcons frequently grab prey birds from the air, as in Example 9.10. Sometimes they strike at high enough speeds that the force of the impact disables prey birds. A 480 g peregrine falcon high in the sky spies a 240 g pigeon some distance below. The falcon slows to a near stop, then goes into a dive—called a stoop—and picks up speed as she falls. The falcon reaches a vertical speed of 45 m/s before striking the pigeon, which we can assume is stationary. The falcon strikes the pigeon and grabs it in her talons. The collision between the birds lasts 0.015 s.
   a. What is the final speed of the falcon and pigeon?
   b. What is the average force on the pigeon during the impact?

29. || A 71 kg baseball player jumps straight up to catch a hard-hit ball. If the 140 g ball is moving horizontally at 28 m/s, and the catch is made when the ballplayer is at the highest point of his leap, what is his speed immediately after stopping the ball?
30. || In a football game, a 90 kg receiver leaps straight up in the air to catch the 0.42 kg ball the quarterback threw to him at a vigorous 21 m/s, catching the ball at the highest point in his jump. Right after catching the ball, how fast is the receiver moving? Is he likely to notice this recoil?
31. ||| A kid at the junior high cafeteria wants to propel an empty milk carton along a lunch table by hitting it with a 3.0 g spit ball. If he wants the speed of the 20 g carton just after the spit ball hits it to be 0.30 m/s, at what speed should his spit ball hit the carton?
32. | The parking brake on a 2000 kg Cadillac has failed, and it is rolling slowly, at 1 mph, toward a group of small children. Seeing the situation, you realize you have just enough time to drive your 1000 kg Volkswagen head-on into the Cadillac and save the children. With what speed should you impact the Cadillac to bring it to a halt?
33. | A 2.0 kg block slides along a frictionless surface at 1.0 m/s. A second block, sliding at 4.0 m/s, collides with the first from behind and sticks to it. The final velocity of the combined blocks is 2.0 m/s. What was the mass of the second block?
34. || Erica (36 kg) and Danny (47 kg) are bouncing on a trampoline. Just as Erica reaches the high point of her bounce, Danny is moving upward past her at 4.1 m/s. At that instant he grabs hold of her. What is their speed just after he grabs her?

Section 9.6 Momentum and Collisions in Two Dimensions

35. |||| A 20 g ball of clay traveling east at 3.0 m/s collides with a 30 g ball of clay traveling north at 2.0 m/s. What are the speed and the direction of the resulting 50 g blob of clay?
36. ||| Casey is driving a 1600 kg car toward the east. She goes through an intersection at a speed of 16 m/s (approximately 35 mph), the speed limit on both roads of the intersection. Kerry is driving a car of mass 1200 kg into the intersection, going north, and doesn't see or doesn't heed a red light, and slams into Casey's car. The cars lock together and skid to a stop. Later, the two review the scene with the police. Skid marks from the instant after the collision reveal that the two cars were moving exactly northeast. Kerry claims to have been driving at the speed limit, but Casey says that Kerry seemed to be going over the speed limit before the collision. Who is correct? Use the concept of conservation of momentum to make your case.
37. || Two objects collide and bounce apart. Figure P9.37 shows the initial momentum of each object and the final momentum of object 2. What is the final momentum of object 1? Show your answer by copying the figure and drawing the final momentum vector on the figure.

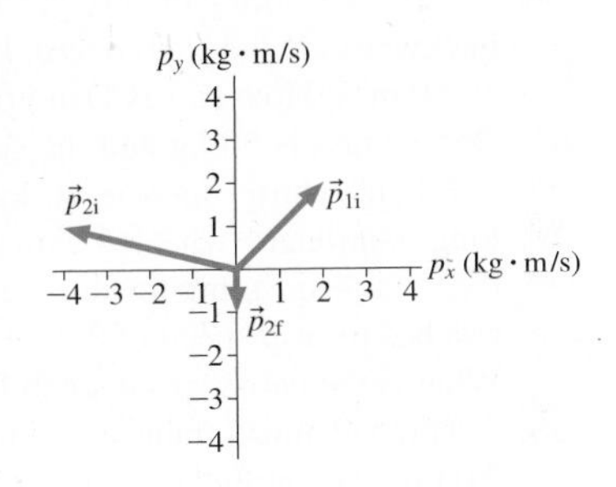

FIGURE P9.37

38. || A 20 g ball of clay traveling east at 2.0 m/s collides with a 30 g ball of clay traveling 30° south of west at 1.0 m/s. What are the speed and direction of the resulting 50 g blob of clay?
39. || A firecracker in a coconut blows the coconut into three pieces. Two pieces of equal mass fly off south and west, perpendicular to each other, at 20 m/s. The third piece has twice the mass of the other two. What are the speed and direction of the third piece?

Section 9.7 Angular Momentum

40. || What is the angular momentum about the axle of the 500 g rotating bar in Figure P9.40?

FIGURE P9.40

41. |||| What is the angular momentum about the axle of the 2.0 kg, 4.0-cm-diameter rotating disk in Figure P9.41?

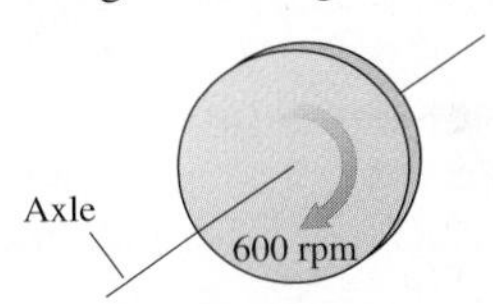

FIGURE P9.41

42. | Divers change their body position in midair while rotating
BIO about their center of mass. In one dive, the diver leaves the board with her body nearly straight, then tucks into a somersault position. If the moment of inertia of the diver in a straight position is 14 kg · m^2 and in a tucked position is 4.0 kg · m^2, by what factor does her angular speed increase?
43. || Ice skaters often end their performances with spin turns,
BIO where they spin very fast about their center of mass with their arms folded in and legs together. Upon ending, their arms extend outward, proclaiming their finish. Not quite as noticeably, one leg goes out as well. Suppose that the moment of inertia of a skater with arms out and one leg extended is 3.2 kg · m^2 and for arms and legs in is 0.80 kg · m^2. If she starts out spinning at 5.0 rev/s, what is her angular speed (in rev/s) when her arms and one leg open outward?
44. ||| A diver leaves the platform with her body straight.
BIO
INT Her body is in a relatively slow rotation, with an angular speed of 4.0 rad/s. She then tucks into a pike position, with her body essentially folded in half. We can use a simple model to understand what happens next. First, model her 50 kg, 1.8 m body as uniform. Next, assume that when she goes into a pike position, she really does fold her body exactly in half. In terms of this model,

   a. What is her initial moment of inertia?
   b. What is her moment of inertia in the pike position?
   c. What is her angular speed in the pike position?
   d. How many rotations does she complete in the 1.3 s that she holds the pike position?

Watch Video Solution Problems 9.51 and 9.57

45. ||| What is the angular momentum of the moon around INT the earth? The moon's mass is 7.4×10^{22} kg and it orbits 3.8×10^{8} m from the earth.

General Problems

46. ||| What is the impulse on a 3.0 kg object that experiences the force described by the graph in Figure P9.46?

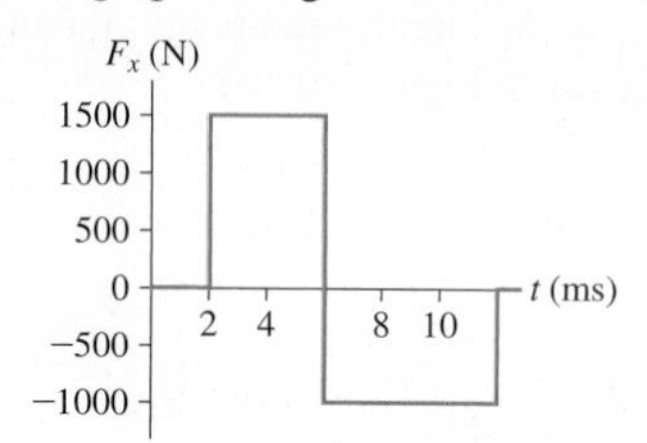

FIGURE P9.46

47. ||| A 600 g air-track glider collides with a spring at one end of the track. Figure P9.47 shows the glider's velocity and the force exerted on the glider by the spring. How long is the glider in contact with the spring?

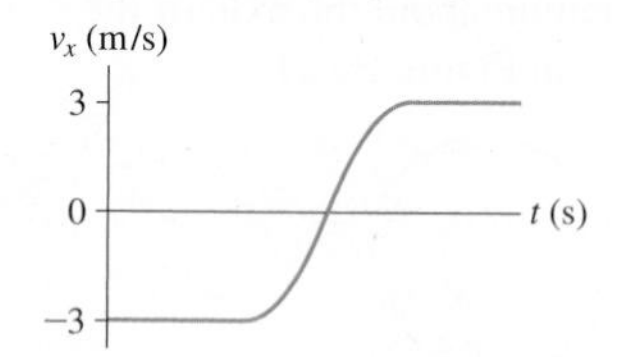

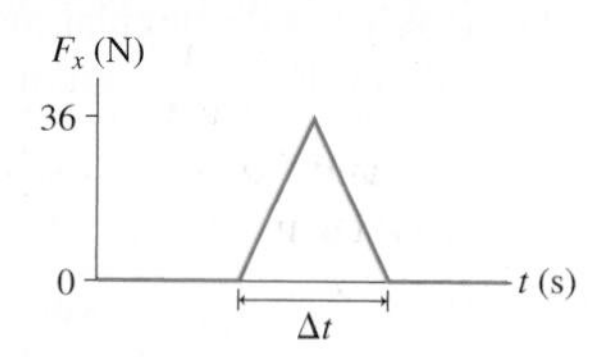

FIGURE P9.47

48. || Far in space, where gravity is negligible, a 425 kg rocket traveling at 75.0 m/s in the positive x-direction fires its engines. Figure P9.48 shows the thrust force as a function of time. The mass lost by the rocket during these 30.0 s is negligible.
   a. What impulse does the engine impart to the rocket?
   b. At what time does the rocket reach its maximum speed? What is the maximum speed?

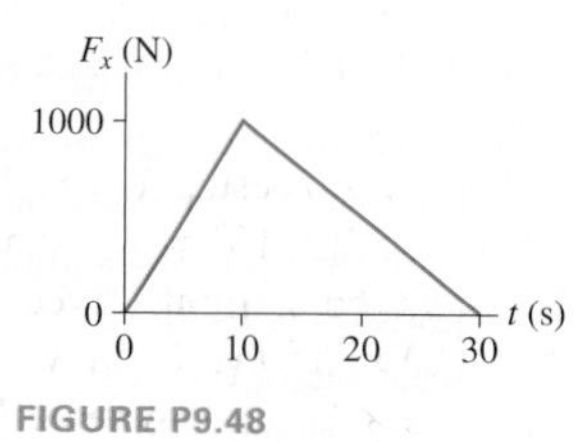

FIGURE P9.48

49. |||| A 200 g ball is INT dropped from a height of 2.0 m, bounces on a hard floor, and rebounds to a height of 1.5 m. Figure P9.49 shows the impulse received from the floor. What maximum force does the floor exert on the ball?

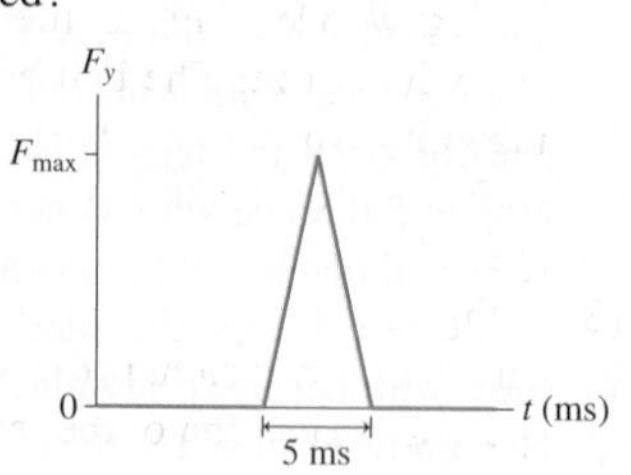

FIGURE P9.49

50. |||| A 200 g ball is dropped INT from a height of 2.0 m and bounces on a hard floor. The force on the ball from the floor is shown in Figure P9.50. How high does the ball rebound?

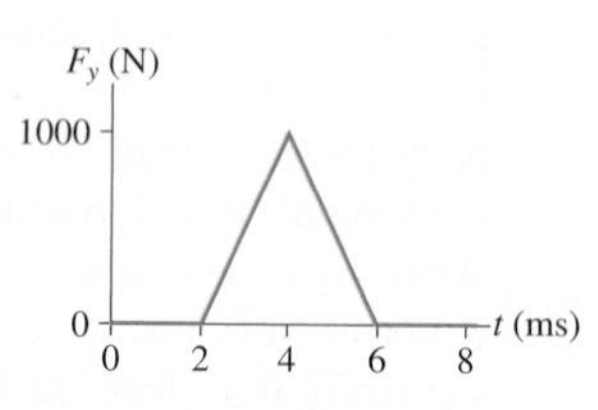

FIGURE P9.50

51. ||| Figure P9.51 is a graph of INT the force exerted by the floor on a woman making a vertical jump. At what speed does she leave the ground?
Hint: The force of the floor is not the only force acting on the woman.

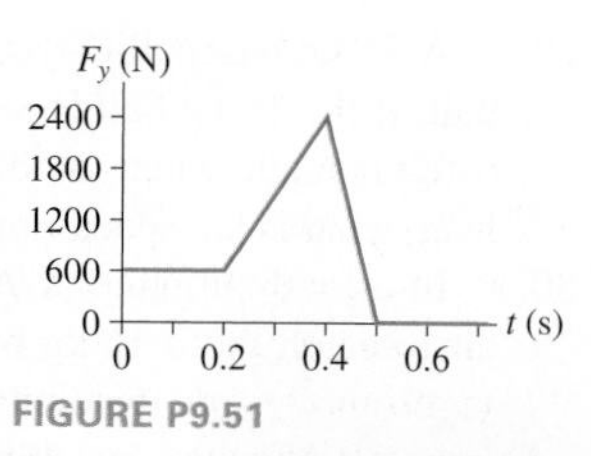

FIGURE P9.51

52. | We've seen that squid can escape BIO INT from predators by ejecting water. Some squid do this at the surface of the ocean, thus launching themselves into the air—a particularly effective escape strategy. Suppose a 36 kg squid (not including water) at rest at the surface of the water brings in and quickly ejects 3.0 kg of water to achieve a takeoff speed of 3.5 m/s; these are typical numbers.
   a. At what speed does the squid eject the water?
   b. If we ignore lift and drag forces as the squid flies through the air, what is the maximum horizontal range that the squid can achieve before splashing down?

53. ||| A 140 g baseball is moving horizontally to the right at 35 m/s when it is hit by the bat. The ball flies off to the left at 55 m/s, at an angle of 25° above the horizontal. What are the magnitude and direction of the impulse that the bat delivers to the ball?
54. ||| A tennis player swings her 1000 g racket with a speed of 10 m/s. She hits a 60 g tennis ball that was approaching her at a speed of 20 m/s. The ball rebounds at 40 m/s.
   a. How fast is her racket moving immediately after the impact? You can ignore the interaction of the racket with her hand for the brief duration of the collision.
   b. If the tennis ball and racket are in contact for 10 ms, what is the average force that the racket exerts on the ball?
55. || A 20 g ball of clay is thrown horizontally at 30 m/s toward a 1.0 kg block sitting at rest on a frictionless surface. The clay hits and sticks to the block.
   a. What is the speed of the block and clay right after the collision?
   b. Use the block's initial and final speeds to calculate the impulse the clay exerts on the block.
   c. Use the clay's initial and final speeds to calculate the impulse the block exerts on the clay.
   d. Does $\vec{J}_{\text{block on clay}} = -\vec{J}_{\text{clay on block}}$?
56. || Dan is gliding on his skateboard at 4.0 m/s. He jumps backward off the skateboard, kicking the skateboard forward at 8.0 m/s. How fast is Dan going as his feet hit the ground? Dan's mass is 50 kg and the skateboard's mass is 5.0 kg.
57. ||| Ethan, whose mass is 80 kg, stands at one end of a very INT long, stationary wheeled cart that has a mass of 500 kg. He then starts sprinting toward the other end of the cart. He soon reaches his top speed of 8.0 m/s, measured relative to the cart. What is the cart's speed when Ethan reaches top speed?
58. || Three identical train cars, coupled together, are rolling east at 2.0 m/s. A fourth car traveling east at 4.0 m/s catches up with the three and couples to make a four-car train. A moment later, the train cars hit a fifth car that was at rest on the tracks, and it couples to make a five-car train. What is the speed of the five-car train?

59. | A 110 kg linebacker running at 2.0 m/s and an 82 kg quarterback running at 3.0 m/s have a head-on collision in midair. The linebacker grabs and holds onto the quarterback. Who ends up moving forward after they hit?

60. || Most geologists believe that the dinosaurs became extinct INT 65 million years ago when a large comet or asteroid struck the earth, throwing up so much dust that the sun was blocked out for a period of many months. Suppose an asteroid with a diameter of 2.0 km and a mass of 1.0×10^{13} kg hits the earth with an impact speed of 4.0×10^4 m/s.
   a. What is the earth's recoil speed after such a collision? (Use a reference frame in which the earth was initially at rest.)
   b. What percentage is this of the earth's speed around the sun? (Use the astronomical data inside the back cover.)

61. |||| Two ice skaters, with masses of 75 kg and 55 kg, stand facing INT each other on a 15-m-wide frozen river. The skaters push off against each other, glide backward straight toward the river's edges, and reach the edges at exactly the same time. How far did the 75 kg skater glide?

62. || Two ice skaters, with masses of 50 kg and 75 kg, are at the INT center of a 60-m-diameter circular rink. The skaters push off against each other and glide to opposite edges of the rink. If the heavier skater reaches the edge in 20 s, how long does the lighter skater take to reach the edge?

63. ||| One billiard ball is shot east at 2.00 m/s. A second, identical billiard ball is shot west at 1.00 m/s. The balls have a glancing collision, not a head-on collision, deflecting the second ball by 90° and sending it north at 1.41 m/s. What are the speed and direction of the first ball after the collision?

64. |||| A 10 g bullet is fired into a 10 kg wood block that is at INT rest on a wood table. The block, with the bullet embedded, slides 5.0 cm across the table. What was the speed of the bullet?

65. ||| A typical raindrop is much more massive than a mosquito and BIO much faster than a mosquito flies. How does a mosquito survive the impact? Recent research has found that the collision of a falling raindrop with a mosquito is a perfectly inelastic collision. That is, the mosquito is "swept up" by the raindrop and ends up traveling along with the raindrop. Once the relative speed between the mosquito and the raindrop is zero, the mosquito is able to detach itself from the drop and fly away.
   a. A hovering mosquito is hit by a raindrop that is 40 times as massive and falling at 8.2 m/s, a typical raindrop speed. How fast is the raindrop, with the attached mosquito, falling immediately afterward if the collision is perfectly inelastic?
   b. Because a raindrop is "soft" and deformable, the collision duration is a relatively long 8.0 ms. What is the mosquito's average acceleration, in g's, during the collision? The peak acceleration is roughly twice the value you found, but the mosquito's rigid exoskeleton allows it to survive accelerations of this magnitude. In contrast, humans cannot survive an acceleration of more than about $10g$.

66. |||| A 15 g bullet is fired at 610 m/s into a 4.0 kg block that sits INT at the edge of a 75-cm-high table. The bullet embeds itself in the block and carries it off the table. How far from the point directly below the table's edge does the block land?

67. |||| A 1500 kg weather rocket accelerates upward at 10.0 m/s^2. INT It explodes 2.00 s after liftoff and breaks into two fragments, one twice as massive as the other. Photos reveal that the lighter fragment traveled straight up and reached a maximum height of 530 m. What were the speed and direction of the heavier fragment just after the explosion?

68. ||| A canoe is designed to have very little drag when it moves INT along its length. Riley, mass 52 kg, sits in a 21 kg canoe in the middle of a lake. She dives into the water off the front of the canoe, along the axis of the canoe. She dives forward at 1.6 m/s relative to the boat. Just after her leap,
   a. How fast is she moving relative to the water?
   b. How fast is the canoe moving relative to the water?

69. || Two 500 g blocks of wood are 2.0 m apart on a frictionless table. A 10 g bullet is fired at 400 m/s toward the blocks. It passes all the way through the first block, then embeds itself in the second block. The speed of the first block immediately afterward is 6.0 m/s. What is the speed of the second block after the bullet stops?

70. ||| A water pipe in a building delivers 1000 liters (with mass 1000 kg) of water per second. The water is moving through the pipe at 1.4 m/s. The pipe has a 90° bend, and the pipe will require a supporting structure, called a *thrust block,* at the bend, as in Figure P9.70. We can use the ideas of momentum and impulse to understand why. Each second, 1000 kg of water moving at $v_x = 1.4$ m/s changes direction to move at $v_y = 1.4$ m/s.
   a. What are the magnitude and direction of the change in momentum of the 1000 kg of water?
   b. What are the magnitude and direction of the necessary impulse?
   c. This impulse takes place over 1.0 s. What is the necessary force?

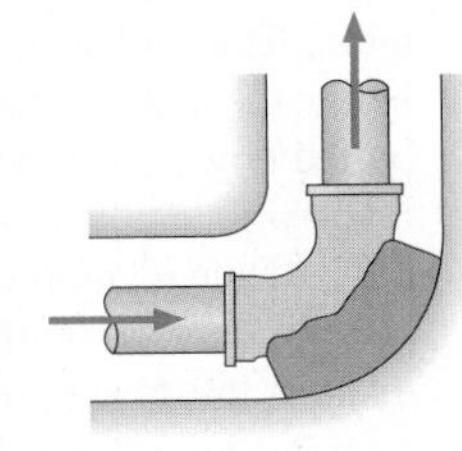
FIGURE P9.70

71. || A spaceship of mass 2.0×10^6 kg is cruising at a speed of 5.0×10^6 m/s when the antimatter reactor fails, blowing the ship into three pieces. One section, having a mass of 5.0×10^5 kg, is blown straight backward with a speed of 2.0×10^6 m/s. A second piece, with mass 8.0×10^5 kg, continues forward at 1.0×10^6 m/s. What are the direction and speed of the third piece?

72. ||| At the county fair, Chris throws a 0.15 kg baseball at a 2.0 kg wooden milk bottle, hoping to knock it off its stand and win a prize. The ball bounces straight back at 20% of its incoming speed, knocking the bottle straight forward. What is the bottle's speed, as a percentage of the ball's incoming speed?

73. ||| Figure P9.73 shows a collision between three balls of clay. The three hit simultaneously and stick together. What are the speed and direction of the resulting blob of clay?

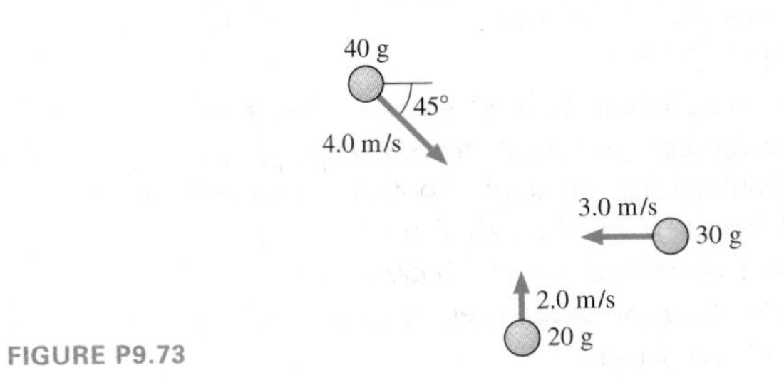

FIGURE P9.73

Watch Video Solution Problem 9.78

74. ||| The carbon isotope ^{14}C is used for carbon dating of archeological artifacts. ^{14}C (mass 2.34×10^{-26} kg) decays by the process known as *beta decay* in which the nucleus emits an electron (the beta particle) and a subatomic particle called a neutrino. In one such decay, the electron and the neutrino are emitted at right angles to each other. The electron (mass 9.11×10^{-31} kg) has a speed of 5.00×10^{7} m/s and the neutrino has a momentum of 8.00×10^{-24} kg·m/s. What is the recoil speed of the nucleus?

75. ||| INT Figure P9.75 shows a 100 g puck revolving at 100 rpm on a 20-cm-radius circle on a frictionless table. A string attached to the puck passes through a hole in the middle of the table. The end of the string below the table is then slowly pulled down until the puck is revolving in a 10-cm-radius circle. How many revolutions per minute does the puck make at this new radius?

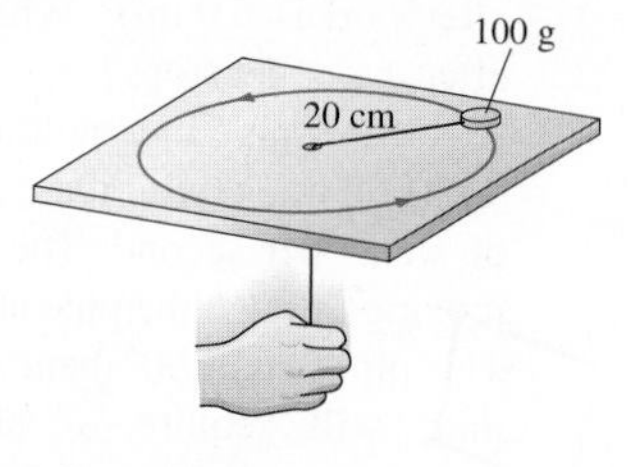

FIGURE P9.75

76. || A 2.0 kg, 20-cm-diameter turntable rotates at 100 rpm on frictionless bearings. Two 500 g blocks fall from above, hit the turntable simultaneously at opposite ends of a diagonal, and stick. What is the turntable's angular speed, in rpm, just after this event?

77. || Joey, from Example 9.11, stands at rest at the outer edge of the frictionless merry-go-round of Figure 9.27. The merry-go-round is also at rest. Joey then begins to run around the perimeter of the merry-go-round, finally reaching a constant speed, measured relative to the ground, of 5.0 m/s. What is the final angular speed of the merry-go-round?

78. ||| A 3.0-m-diameter merry-go-round with a mass of 250 kg is spinning at 20 rpm. John runs around the merry-go-round at 5.0 m/s, in the same direction that it is turning, and jumps onto the outer edge. John's mass is 30 kg. What is the merry-go-round's angular speed, in rpm, after John jumps on?

79. ||| Disk A, with a mass of 2.0 kg and a radius of 40 cm, rotates clockwise about a frictionless vertical axle at 30 rev/s. Disk B, also 2.0 kg but with a radius of 20 cm, rotates counterclockwise about that same axle, but at a greater height than disk A, at 30 rev/s. Disk B slides down the axle until it lands on top of disk A, after which they rotate together. After the collision, what is their common angular speed (in rev/s) and in which direction do they rotate?

MCAT-Style Passage Problems

Hitting a Golf Ball

Consider a golf club hitting a golf ball. To a good approximation, we can model this as a collision between the rapidly moving head of the golf club and the stationary golf ball, ignoring the shaft of the club and the golfer.

A golf ball has a mass of 46 g. Suppose a 200 g club head is moving at a speed of 40 m/s just before striking the golf ball. After the collision, the golf ball's speed is 60 m/s.

80. | What is the momentum of the club + ball system right before the collision?
A. 1.8 kg·m/s B. 8.0 kg·m/s
C. 3220 kg·m/s D. 8000 kg·m/s

81. | Immediately after the collision, the momentum of the club + ball system will be
A. Less than before the collision.
B. The same as before the collision.
C. Greater than before the collision.

82. | A manufacturer makes a golf ball that compresses more than a traditional golf ball when struck by a club. How will this affect the average force during the collision?
A. The force will decrease.
B. The force will not be affected.
C. The force will increase.

83. | By approximately how much does the club head slow down as a result of hitting the ball?
A. 4 m/s B. 6 m/s C. 14 m/s D. 26 m/s

Analyzing a Dance Move BIO INT

Understanding the details of timing and forces in motion can improve the performance of athletes, including dancers. For Problems 84–87, consider the forces involved in a ballet jump called a sauté demi plié. Figure P9.84a shows the sequence of moves in the jump. The dancer starts upright, then quickly bends her knees, moving downward. After she reaches the bottom of this dip, she extends her legs, pushing herself upward. After this upward push, she leaves the ground, beginning a short period of time in the air. Figure P9.84b is a slightly idealized graph of the net force on a 42 kg dancer executing this move.

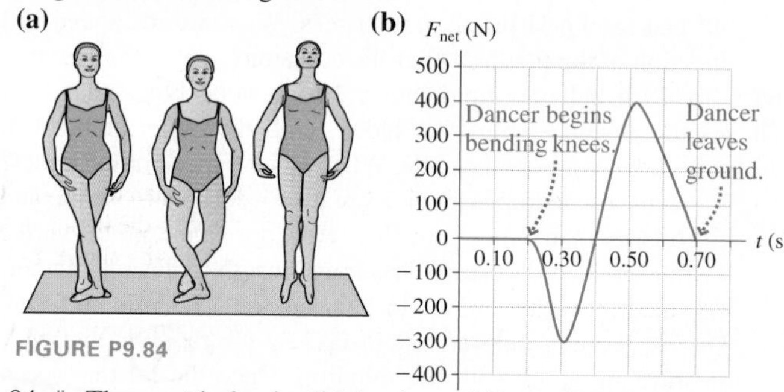

FIGURE P9.84

84. || The sauté demi plié begins with a phase in which the net force on the dancer is negative. During this phase of the jump,
A. The normal force of the floor on her is zero.
B. The normal force of the floor on her is less than her weight but greater than zero.
C. The normal force of the floor on her is equal to her weight.
D. The normal force of the floor on her is greater than her weight.

85. || At what time does the dancer reach the lowest point of her motion, when her speed is zero?
A. 0.20 s
B. 0.40 s
C. Between 0.40 s and 0.70 s
D. After 0.70 s

86. || What is the approximate net impulse on the dancer, from the moment she begins bending her knees to the instant she leaves the floor?
A. 15 kg·m/s^2 B. 30 kg·m/s^2
C. 60 kg·m/s^2 D. 90 kg·m/s^2

87. || To the nearest m/s, how fast is the dancer moving when she leaves the floor?
A. 1 m/s B. 2 m/s C. 3 m/s D. 4 m/s

10 Energy and Work

A greyhound can rapidly go from a standstill to a very speedy run—meaning a very rapid increase in kinetic energy, the energy of motion. How does the greyhound's ability to convert energy from one form to another compare to that of other animals?

LOOKING AHEAD ►

Forms of Energy

This dolphin has lots of **kinetic energy** as it leaves the water. At its highest point its energy is mostly **potential energy.**

You'll learn about several of the most important forms of energy—kinetic, potential, and thermal.

Work and Energy

The woman does **work** on the jack, applying a force to the handle and pushing it down. This is a *transfer* of energy into the system, increasing the potential energy as the car is lifted.

You'll learn how to calculate the work done by a force, and how this work is related to the *change* in a system's energy.

Conservation of Energy

As they slide, their potential energy decreases and their kinetic energy increases, but their total energy is unchanged: It is **conserved.**

How fast will they be moving when they reach the bottom? You'll use a new before-and-after analysis to find out.

GOAL To introduce the concept of energy and to learn a new problem-solving strategy based on conservation of energy.

LOOKING BACK ◄

Motion with Constant Acceleration

In Chapter 2 you learned how to describe the motion of a particle that has a constant acceleration. In this chapter, you'll use the constant-acceleration equations to connect work and energy.

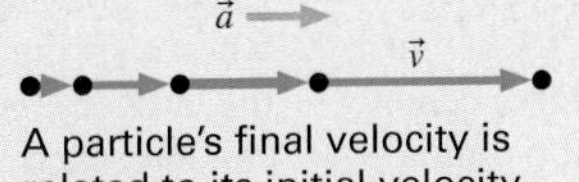

A particle's final velocity is related to its initial velocity, its acceleration, and its displacement by

$$(v_x)_f^2 = (v_x)_i^2 + 2a_x\,\Delta x$$

STOP TO THINK

A car pulls away from a stop sign with a constant acceleration. After traveling 10 m, its speed is 5 m/s. What will its speed be after traveling 40 m?

A. 10 m/s
B. 20 m/s
C. 30 m/s
D. 40 m/s

10.1 The Basic Energy Model

Energy. It's a word you hear all the time. We use chemical energy to heat our homes and bodies, electric energy to run our lights and computers, and solar energy to grow our crops and forests. We're told to use energy wisely and not to waste it. Athletes and weary students consume "energy bars" and "energy drinks."

But just what is energy? The concept of energy has grown and changed over time, and it is not easy to define in a general way just what energy is. Rather than starting with a formal definition, we'll let the concept of energy expand slowly over the course of several chapters. In this chapter we introduce several fundamental forms of energy, including kinetic energy, potential energy, and thermal energy. Our goal is to understand the characteristics of energy, how energy is used, and, especially important, how energy is transformed from one form into another. Understanding these transformations will allow us to understand and explore a wide variety of physical phenomena. Anything that happens involves a transformation of energy from one form to another, so the range of topics we'll consider in this chapter is extensive.

In solving problems, we'll use a key fact about energy: **Energy is neither created nor destroyed: If one form of energy in a system decreases, it must appear in an equal amount in another form.** Many scientists consider this law of conservation of energy to be the most important of all the laws of nature.

Systems and Forms of Energy

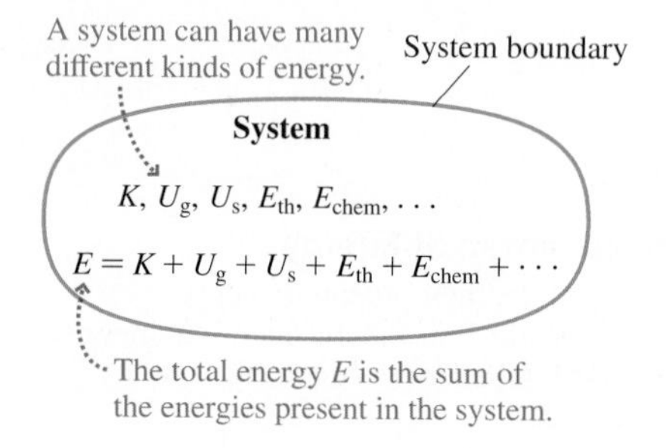
A system can have many different kinds of energy.
System boundary
System
$K, U_g, U_s, E_{th}, E_{chem}, \ldots$
$E = K + U_g + U_s + E_{th} + E_{chem} + \cdots$
The total energy E is the sum of the energies present in the system.

FIGURE 10.1 A system and its energies.

In Chapter 9 we introduced the idea of a *system* of interacting objects. A system can be as simple as a falling acorn or as complex as a city. But whether simple or complex, every system in nature has associated with it a quantity we call its **total energy** E. The total energy is the sum of the different kinds of energies present in the system. In the table below, we give a brief overview of some of the more important forms of energy; in the rest of the chapter, we'll look at several of these forms of energy in greater detail.

A system may have many of these kinds of energy at one time. For instance, a moving car has kinetic energy of motion, chemical energy stored in its gasoline, thermal energy in its hot engine, and many other forms of energy. FIGURE 10.1 illustrates the idea that the total energy of the system, E, is the *sum* of all the different energies present in the system:

$$E = K + U_g + U_s + E_{th} + E_{chem} + \cdots \quad (10.1)$$

The energies shown in this sum are the forms of energy in which we'll be most interested in this and the next chapter. The ellipses ($\cdots$) stand for other forms of energy, such as nuclear or electric, that also might be present. We'll treat these and others in later chapters.

Some important forms of energy

Kinetic energy K

Kinetic energy is the energy of *motion*. All moving objects have kinetic energy. The heavier an object and the faster it moves, the more kinetic energy it has. The wrecking ball in this picture is effective in part because of its large kinetic energy.

Gravitational potential energy U_g

Gravitational potential energy is *stored* energy associated with an object's *height above the ground*. As this coaster ascends, energy is stored as gravitational potential energy. As it descends, this stored energy is converted into kinetic energy.

Elastic or spring potential energy U_s

Elastic potential energy is energy stored when a spring or other elastic object, such as this archer's bow, is *stretched*. This energy can later be transformed into the kinetic energy of the arrow.

Continued

Thermal energy E_{th}

Hot objects have more *thermal energy* than cold ones because the molecules in a hot object jiggle around more than those in a cold object. Thermal energy is the sum of the microscopic kinetic and potential energies of all the molecules in an object.

Chemical energy E_{chem}

Electric forces cause atoms to bind together to make molecules. Energy can be stored in these bonds, energy that can later be released as the bonds are rearranged during chemical reactions. All animals eat, taking in chemical energy to provide energy to move muscles and fuel processes of the body.

Nuclear energy $E_{nuclear}$

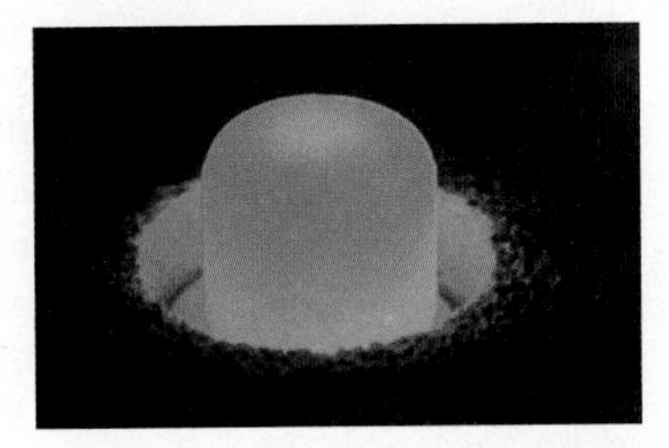

The forces that hold together the particles in the nucleus of the atom are much stronger than the electric forces that hold together molecules, so they store a great deal more energy. Certain nuclei break apart into smaller fragments, releasing some of this *nuclear energy*. The energy is transformed into the kinetic energy of the fragments and then into thermal energy.

Energy Transformations

If the amounts of each form of energy never changed, the world would be a very dull place. What makes the world interesting is that **energy of one kind can be transformed into energy of another kind.** The following table illustrates a few common energy transformations. In this table, we use an arrow $\rightarrow$ as a shorthand way of representing an energy transformation.

Some energy transformations

A weightlifter lifts a barbell over her head

The barbell has much more gravitational potential energy when high above her head than when on the floor. To lift the barbell, she transforms chemical energy in her body into gravitational potential energy of the barbell.

$$E_{chem} \rightarrow U_g$$

A base runner slides into the base

When running, he has lots of kinetic energy. After sliding, he has none. His kinetic energy is transformed mainly into thermal energy: The ground and his legs are slightly warmer.

$$K \rightarrow E_{th}$$

A burning campfire

The wood contains considerable chemical energy. When the carbon in the wood combines chemically with oxygen in the air, this chemical energy is transformed largely into thermal energy of the hot gases and embers.

$$E_{chem} \rightarrow E_{th}$$

A springboard diver

Here's a two-step energy transformation. At the instant shown, the board is flexed to its maximum extent, so that elastic potential energy is stored in the board. Soon this energy will begin to be transformed into kinetic energy; then, as the diver rises into the air and slows, this kinetic energy will be transformed into gravitational potential energy.

$$U_s \rightarrow K \rightarrow U_g$$

FIGURE 10.2 Energy transformations within the system.

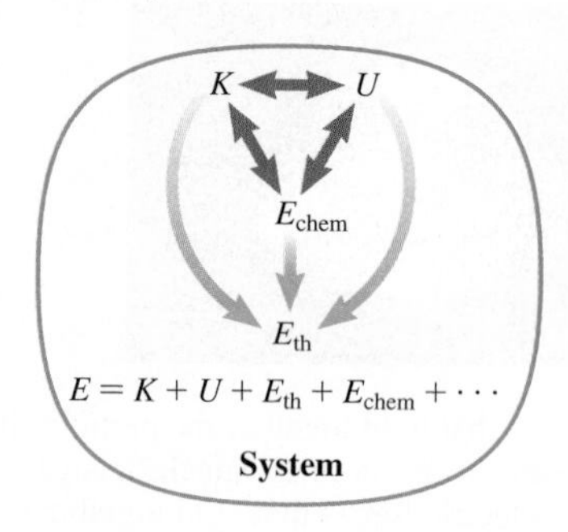
K
U
E_{chem}
E_{th}
$E = K + U + E_{th} + E_{chem} + \cdots$
System

FIGURE 10.2 reinforces the idea that **energy transformations are changes of energy *within* the system from one form to another.** (The U in this figure is a generic potential energy; it could be gravitational potential energy U_g, spring potential energy U_s, or some other form of potential energy.) There are two types of arrows in the figure. The arrow between K and U is a two-way arrow; it's easy to transform energy back and forth between these forms. When the springboard diver goes up in the air, his kinetic energy is transformed into gravitational potential energy; when he comes back down, this process is reversed. But the arrow between K and E_{th} is a one-way arrow pointing toward E_{th}. When the runner slides into the base, his kinetic energy is transformed into thermal energy. This process doesn't spontaneously reverse, although this would certainly make baseball a more exciting game. In Chapter 11, we'll see that it is possible to transform thermal energy into other forms, but it's not easy, and there are real limitations.

A video to support a section's topic is embedded in the eText.

Video The Basic Energy Model

Energy Transfers and Work

We've just seen that energy *transformations* occur between forms of energy *within* a system. But every physical system also interacts with the world around it—that is, with its *environment.* In the course of these interactions, the system can exchange energy with the environment. **An exchange of energy between system and environment is called an energy *transfer.*** There are two primary energy-transfer processes: **work,** the *mechanical* transfer of energy to or from a system by pushing or pulling on it, and **heat,** the *nonmechanical* transfer of energy from the environment to the system (or vice versa) because of a temperature difference between the two.

FIGURE 10.3 Work and heat are energy transfers into and out of the system.

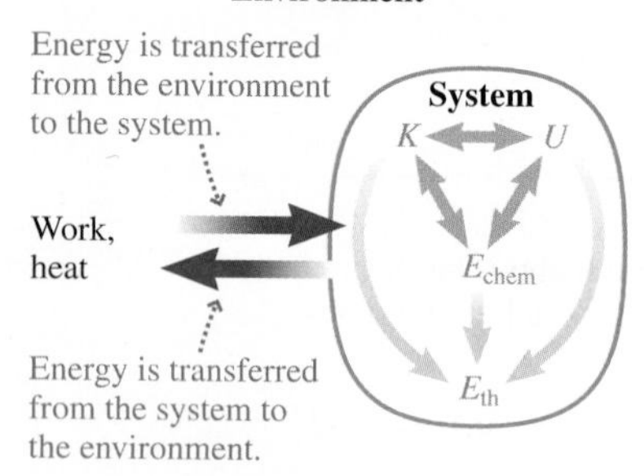
The *environment* is everything that is *not* part of the system.
Environment
Energy is transferred from the environment to the system.
System
K
U
Work, heat
E_{chem}
E_{th}
Energy is transferred from the system to the environment.

FIGURE 10.3, which we call the **basic energy model,** shows how our energy model is modified to include energy transfers into and out of the system as well as energy transformations within the system. In this chapter we'll consider energy transfers by means of work; the concept of heat will be developed in Chapters 11 and 12.

"Work" is a common word in the English language, with many meanings. When you first think of work, you probably think of physical effort or the job you do to make a living. In physics, "work" is the process of *transferring* energy from the environment to a system, or from a system to the environment, by the application of mechanical forces—pushes and pulls—to the system. Once the energy has been transferred to the system, it can appear in many forms. Exactly what form it takes depends on the details of the system and how the forces are applied. The table below gives three examples of energy transfers due to work. We use W as the symbol for work.

Energy transfers: work

Putting a shot

The system: The shot

The environment: The athlete

As the athlete pushes on the shot to get it moving, he is doing work on the system; that is, he is transferring energy from himself to the shot. The energy transferred to the system appears as kinetic energy.

The transfer: $W \rightarrow K$

Striking a match

The system: The match and matchbox

The environment: The hand

As the hand quickly pulls the match across the box, the hand does work on the system, increasing its thermal energy. The match head becomes hot enough to ignite.

The transfer: $W \rightarrow E_{th}$

Firing a slingshot

The system: The slingshot

The environment: The boy

As the boy pulls back on the elastic bands, he does work on the system, increasing its elastic potential energy.

The transfer: $W \rightarrow U_s$

Notice that in each example on the preceding page, the environment applies a force while the system undergoes a *displacement.* Energy is transferred as work only when the system *moves* while the force acts. A force applied to a stationary object, such as when you push against a wall, transfers no energy to the object and thus does no work.

NOTE ▸ In the table on the preceding page, energy is being transferred *from* the athlete *to* the shot by the force of his hand. We say he "does work" on the shot. We speak similarly for the other examples. The hand does work on the match and matchbox, and the boy does work on the slingshot. ◂

The Law of Conservation of Energy

Work done on a system represents energy that is transferred into or out of the system. This transferred energy *changes* the system's energy by exactly the amount of work W that was done. Writing the change in the system's energy as ΔE, we can represent this idea mathematically as

$$\Delta E = W \tag{10.2}$$

Now the total energy E of a system is, according to Equation 10.1, the sum of the different energies present in the system. Thus the change in E is the sum of the *changes* in the different energies present. Then Equation 10.2 gives what is called the *work-energy equation:*

The work-energy equation The total energy of a system changes by the amount of work done on it:

$$\Delta E = \Delta K + \Delta U_g + \Delta U_s + \Delta E_{th} + \Delta E_{chem} + \cdots = W \tag{10.3}$$

NOTE ▸ Equation 10.3, the work-energy equation, is the mathematical representation of the basic energy model of Figure 10.3. Together, they are the heart of what the subject of energy is all about. ◂

Suppose we have an **isolated system,** one that is separated from its surrounding environment in such a way that no energy is transferred into or out of the system. This means that *no work is done on the system.* The energy within the system may be transformed from one form into another, but it is a deep and remarkable fact of nature that, during these transformations, the total energy of an isolated system—the *sum* of all the individual kinds of energy—remains *constant,* as shown in FIGURE 10.4. We say that **the total energy of an isolated system is *conserved*.**

For an isolated system, we must set $W = 0$ in Equation 10.3, leading to the following statement of the *law of conservation of energy:*

Law of conservation of energy The total energy of an isolated system remains constant:

$$\Delta E = \Delta K + \Delta U_g + \Delta U_s + \Delta E_{th} + \Delta E_{chem} + \cdots = 0 \tag{10.4}$$

FIGURE 10.4 An isolated system.

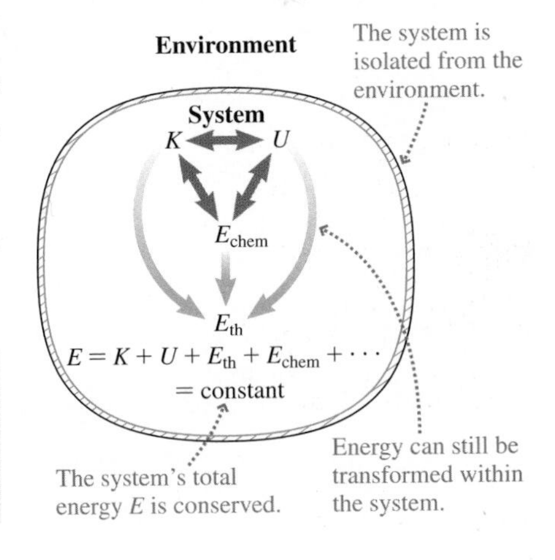
Environment
The system is isolated from the environment.
System
K
U
E_chem
E_th
E = K + U + E_th + E_chem + · · ·
= constant
The system's total energy E is conserved.
Energy can still be transformed within the system.

The law of conservation of energy is similar to the law of conservation of momentum. A system's momentum changes when an external force acts on it, but the total momentum of an *isolated* system doesn't change. Similarly, a system's energy changes when external forces do work on it, but the total energy of an *isolated* system doesn't change.

In solving momentum problems, we adopted a new before-and-after perspective: The momentum *after* an interaction was the same as the momentum *before* the interaction. We will introduce a similar before-and-after perspective for energy that will lead to an extremely powerful problem-solving strategy.

STOP TO THINK 10.1 A roller coaster slows as it goes up a hill. The energy transformation is

A. $U_g \rightarrow K$ B. $U_g \rightarrow E_{th}$ C. $K \rightarrow U_g$ D. $K \rightarrow E_{th}$

10.2 Work

We've seen that work is the transfer of energy to or from a system by the application of forces exerted on the system by the environment. Thus work is done on a system by forces *outside* the system; we call such forces *external forces*. Only external forces can change the energy of a system. *Internal forces*—forces between objects *within* the system—cause energy transformations within the system but don't change the system's total energy. In order for energy to be transferred as work, the system must undergo a displacement—it must *move*—during the time that the force is applied.

FIGURE 10.5 The force of the wind does work on the system, increasing its kinetic energy K.

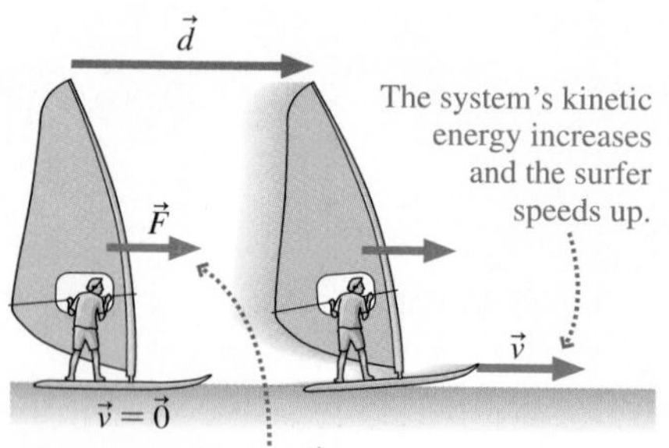
$\vec{d}$
The system's kinetic energy increases and the surfer speeds up.
$\vec{F}$
$\vec{v}$
$\vec{v} = \vec{0}$
The force of the wind $\vec{F}$ does work on the system.

Consider a system consisting of a windsurfer at rest, as shown on the left in FIGURE 10.5. Let's assume that there is no friction or drag force acting on the board. Initially the system has no kinetic energy. But if a force from outside the system, such as the force due to the wind, begins to act on the system, the surfer will begin to speed up, and his kinetic energy will increase. In terms of energy transfers, we would say that the energy of the system has increased because of the work done on the system by the force of the wind.

What determines how much work is done by the force of the wind? First, we note that the greater the distance over which the wind pushes the surfer, the faster the surfer goes, and the more his kinetic energy increases. This implies a greater transfer of energy. So, **the larger the displacement, the greater the work done.** Second, if the wind pushes with a stronger force, the surfer speeds up more rapidly, and the change in his kinetic energy is greater than with a weaker force. **The stronger the force, the greater the work done.**

This suggests that the amount of energy transferred to a system by a force $\vec{F}$—that is, the amount of work done by $\vec{F}$—depends on both the magnitude F of the force *and* the displacement d of the system. Many experiments of this kind have established that the amount of work done by $\vec{F}$ is *proportional* to both F and d. For the simplest case described above, where the force $\vec{F}$ is constant and points in the direction of the object's displacement, the expression for the work done is found to be

$$W = Fd \tag{10.5}$$

Work done by a constant force $\vec{F}$ in the direction of a displacement $\vec{d}$

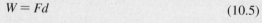

The unit of work, that of force multiplied by distance, is N · m. This unit is so important that it has been given its own name, the **joule** (rhymes with *cool*). We define:

$$1 \text{ joule} = 1 \text{ J} = 1 \text{ N} \cdot \text{m}$$

Because work is simply energy being transferred, **the joule is the unit of *all* forms of energy.** Note that work, unlike momentum, is a *scalar* quantity—it has a magnitude but not a direction.

EXAMPLE 10.1 Working like a dog

A dog in a weight-pulling competition tugs a sled 4.9 m across a snowy track at a constant speed. The force needed to keep the sled moving is 350 N. How much work does the dog do? Where does this energy go?

STRATEGIZE Let's take the system to be the sled + snow. The friction force between the runners and the snow is thus an internal force, so it won't change the total energy of the system, just the form of the energy. But the rope extends outside the system; this is an external interaction, so the tension force of the rope does work on the sled as it moves. Since the dog pulls on the end of the rope, we can say, informally, that the dog does work on the system.

PREPARE We'll continue, as we did with momentum problems in Chapter 9, with a before-and-after visual overview, shown in FIGURE 10.6. The tension force is in the direction of the sled's motion, so we can use Equation 10.5 to calculate the work that the dog does on the sled.

SOLVE The work done is

$$W = Fd = (350 \text{ N})(4.9 \text{ m}) = 1700 \text{ J}$$

The dog does work on the system, but the kinetic energy doesn't increase (the sled doesn't speed up) and the gravitational potential energy doesn't increase (the track is level). The energy the dog puts into the system goes to increasing the system's thermal energy as friction warms up the runners and the snow.

ASSESS 1700 J is a decent amount of energy, as we'll see, but pulling with a 350 N force (about 80 pounds) for a distance of 4.9 m (about 16 feet) sounds like a lot of work, so our result makes sense.

FIGURE 10.6 A dog pulling a loaded sled.

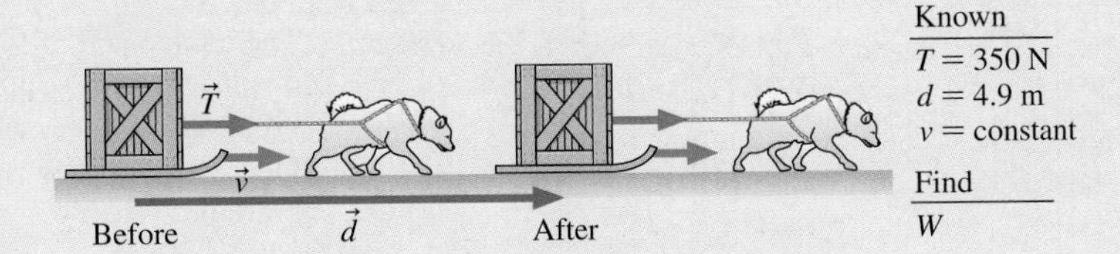

Work by Forces at an Angle to the Displacement

A force does the greatest possible amount of work on an object when the force points in the same direction as the object's displacement. Less work is done when the force acts at an angle to the displacement. To see this, consider the kite buggy of FIGURE 10.7a, pulled along a horizontal path by the angled force of the kite string $\vec{F}$. As shown in FIGURE 10.7b, we can divide $\vec{F}$ into a component $F_\perp$ perpendicular to the motion, and a component $F_\parallel$ parallel to the motion. Only the parallel component acts to accelerate the rider and increase her kinetic energy, so only the parallel component does work on the rider. From Figure 10.7b, we see that if the angle between $\vec{F}$ and the displacement is θ, then the parallel component is $F_\parallel = F\cos\theta$. So, when the force acts at an angle θ to the direction of the displacement, we have

$$W = F_\parallel d = Fd\cos\theta \quad (10.6)$$

Work done by a constant force $\vec{F}$ at an angle θ to the displacement $\vec{d}$

FIGURE 10.7 The force on the kite buggy is at an angle to the displacement.

(a)

(b) The rider undergoes a displacement $\vec{d}$.

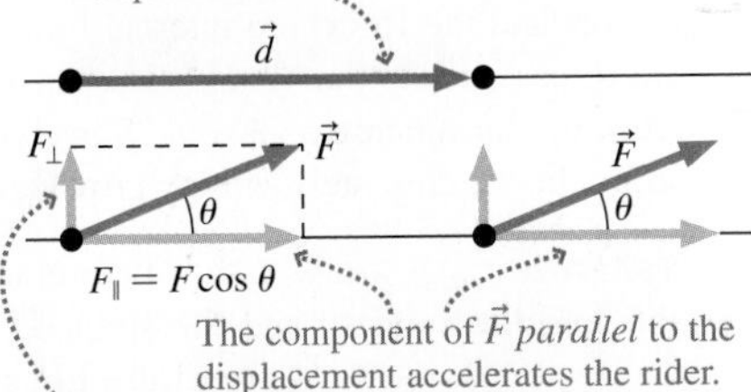

The component of $\vec{F}$ *parallel* to the displacement accelerates the rider.

The component of $\vec{F}$ *perpendicular* to the displacement only pulls up on the rider. It doesn't accelerate her.

Notice that this more general definition of work agrees with Equation 10.5 if $\theta = 0°$.

Tactics Box 10.1 shows how to calculate the work done by a force at any angle to the direction of motion. The system illustrated is a block sliding on a frictionless, horizontal surface, so that only the kinetic energy is changing. However, the same relationships hold for any object undergoing a displacement.

The quantities F and d are always positive, so **the sign of W is determined entirely by the angle θ between the force and the displacement.** Note that Equation 10.6, $W = Fd\cos\theta$, is valid for any angle θ. In three special cases, $\theta = 0°$, $\theta = 90°$, and $\theta = 180°$, however, there are simple versions of Equation 10.6 that you can use. These are noted in Tactics Box 10.1.

TACTICS BOX 10.1 Calculating the work done by a constant force

Direction of force relative to displacement	Angles and work done	Sign of W	Energy transfer
Before: $\vec{v}_i$ After: $\vec{v}_f$ $\vec{d}$ $\vec{F}$ $\theta = 0°$	$\theta = 0°$ $\cos\theta = 1$ $W = Fd$	+	The force is in the direction of motion. The block has its greatest positive acceleration. K increases the most: **Maximum energy transfer *into* the system.**
$\theta < 90°$ $\vec{F}$ $\vec{d}$	$\theta < 90°$ $W = Fd\cos\theta$	+	The component of force parallel to the displacement is less than F. The block has a smaller positive acceleration. K increases less: **Decreased energy transfer *into* the system.**
$\theta = 90°$ $\vec{F}$ $\vec{d}$	$\theta = 90°$ $\cos\theta = 0$ $W = 0$	0	There is no component of force in the direction of motion. The block moves at constant speed. No change in K: **No energy transferred.**
$\theta > 90°$ $\vec{F}$ $\vec{d}$	$\theta > 90°$ $W = Fd\cos\theta$	–	The component of force parallel to the displacement is opposite the motion. The block slows down, and K decreases: **Decreased energy transfer *out of* the system.**
$\theta = 180°$ $\vec{F}$ $\vec{d}$	$\theta = 180°$ $\cos\theta = -1$ $W = -Fd$	–	The force is directly opposite the motion. The block has its greatest deceleration. K decreases the most: **Maximum energy transfer *out of* the system.**

Exercises 5–6

EXAMPLE 10.2 Work done in pulling a suitcase

It's 120 m from one gate to another in the airport. You use a strap inclined upward at a 45° angle to pull your suitcase through the airport. The tension in the strap is 20 N. How much work do you do?

STRATEGIZE Let's take the system to be the suitcase + floor. As with the dog sled, friction forces (in the wheels or between the wheels and the floor) are internal forces. Both the strap and you are forces outside the system. The tension force of the strap does work on the suitcase as it rolls. Since you are the one pulling the strap, this is, ultimately, energy provided by you.

PREPARE FIGURE 10.8 is a before-and-after visual overview showing the suitcase and the strap. The force is at an angle to the displacement, so we must use Equation 10.6 to calculate the work.

SOLVE The tension force does work

$$W = Td\cos\theta = (20\text{ N})(120\text{ m})\cos(45°) = 1700\text{ J}$$

FIGURE 10.8 A suitcase pulled by a strap.

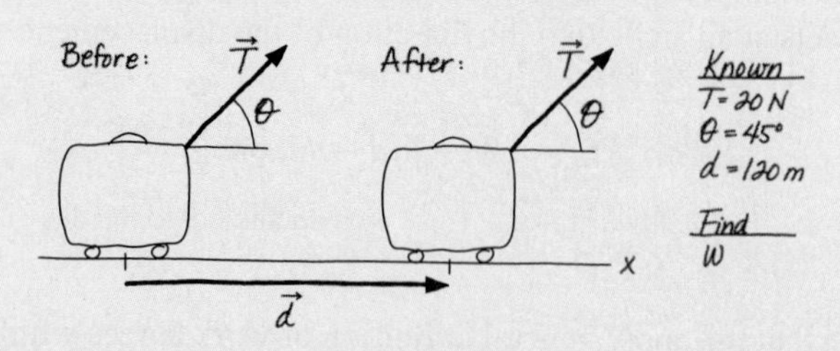

ASSESS This is the same amount of work that the dog did pulling the sled. The force is much less, but the distance is much greater, so this result makes sense.

CONCEPTUAL EXAMPLE 10.3 **Work done by a parachute**

A drag racer is slowed by a parachute. What is the sign of the work done?

REASON The drag force on the drag racer is shown in FIGURE 10.9, along with the dragster's displacement as it slows. The force points in the direction opposite the displacement, so the angle θ in Equation 10.6 is 180°. Then $\cos\theta = \cos(180°) = -1$. Because F and d in Equation 10.6 are magnitudes, and hence positive, the work $W = Fd\cos\theta = -Fd$ done by the drag force is *negative.*

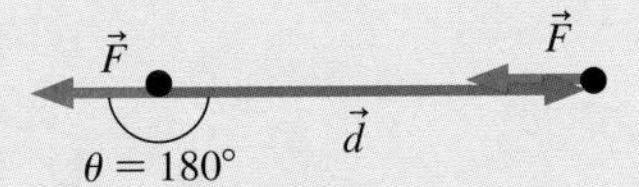

FIGURE 10.9 The force acting on a drag racer.

ASSESS Applying Equation 10.3 to this situation, we have

$$\Delta K = W$$

because the only system energy that changes is the racer's kinetic energy K. Because the kinetic energy is decreasing, its change ΔK is negative. This agrees with the sign of W. This example illustrates the general principle that **negative work represents a transfer of energy out of the system.**

If several forces act on an object that undergoes a displacement, each does work on the object. The **total** (or **net**) **work** W_{total} is the sum of the work done by each force. The total work represents the total energy transfer *to* the system from the environment (if $W_{total} > 0$) or *from* the system to the environment (if $W_{total} < 0$).

Forces That Do No Work

The fact that a force acts on an object doesn't mean that the force will do work on the object. The table below shows three common cases where a force does no work.

Forces that do no work

If the object undergoes no displacement while the force acts, no work is done.

This can sometimes seem counterintuitive. The weightlifter struggles mightily to hold the barbell over his head. But during the time the barbell remains stationary, he does no work on it because its displacement is zero. Why then is it so hard for him to hold it there? Your muscles use energy to apply a force even if there is no displacement and thus no work. We'll talk about the energy that you use to perform a task in Chapter 11.

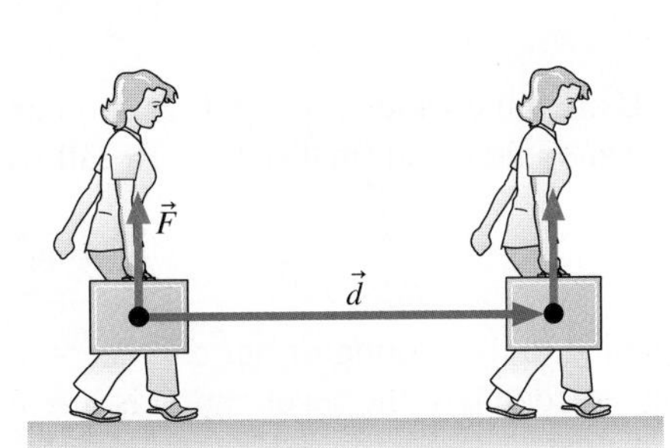

A force perpendicular to the displacement does no work.

The woman exerts only a vertical force on the briefcase she's carrying. This force has no component in the direction of the displacement, so the briefcase moves at a constant velocity and its kinetic energy remains constant. Since the energy of the briefcase doesn't change, it must be that no energy is being transferred to it as work.

(This is the case where $\theta = 90°$ in Tactics Box 10.1.)

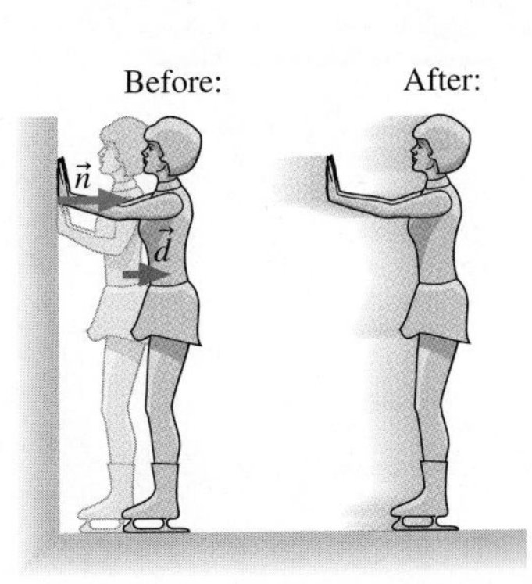

If the part of the object on which the force acts undergoes no displacement, no work is done.

Even though the wall pushes on the skater with a normal force $\vec{n}$ and she undergoes a displacement $\vec{d}$, the wall does no work on her, because the point of her body on which $\vec{n}$ acts—her hands—undergoes no displacement. This makes sense: How could energy be transferred as work from an inert, stationary object? The energy to get the skater moving comes, as you know, from her muscles. This is an internal transformation; chemical energy in her muscles is converted to kinetic energy of her motion.

STOP TO THINK 10.2 Which force does the most work?

A. The 10 N force
B. The 8 N force
C. The 6 N force
D. They all do the same amount of work.

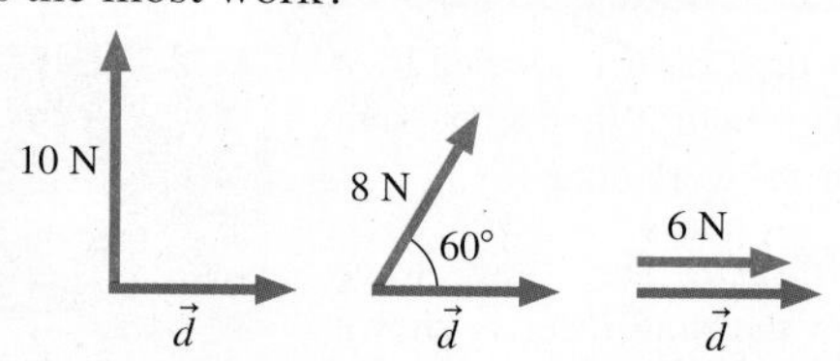

10.3 Kinetic Energy

Kinetic energy is an object's energy of motion. We can use what we've learned about work, and some simple kinematics, to find quantitative expressions for kinetic energy.

We'll start with the case of an object in motion along a line. Such an object has **translational kinetic energy.** In Chapter 7, we introduced the idea of rotational motion: Objects can be in motion even if they aren't going anywhere. An object, like the blade of a wind turbine, rotating about a fixed axis has **rotational kinetic energy,** the kinetic energy of the rotational motion.

Translational Kinetic Energy

FIGURE 10.10 The work done by the tow rope increases the car's kinetic energy.

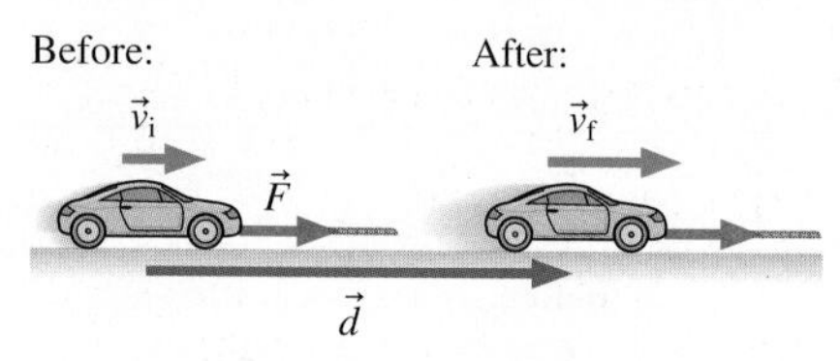

Consider a car being pulled by a tow rope, as in FIGURE 10.10. The rope pulls with a constant force $\vec{F}$ while the car undergoes a displacement $\vec{d}$, so the force does work $W = Fd$ on the car. If we ignore friction and drag, the work done by $\vec{F}$ is transferred entirely into the car's energy of motion—its kinetic energy. In this case, the change in the car's kinetic energy is given by the work-energy equation, Equation 10.3, as

$$W = \Delta K = K_f - K_i \tag{10.7}$$

Using kinematics, we can find another expression for the work done, in terms of the car's initial and final speeds. Recall from ◀◀ SECTION 2.5 the kinematic equation

$$v_f^2 = v_i^2 + 2a\,\Delta x$$

Applied to the motion of our car, $\Delta x = d$ is the car's displacement and, from Newton's second law, the acceleration is $a = F/m$. Thus we can write

$$v_f^2 = v_i^2 + \frac{2Fd}{m} = v_i^2 + \frac{2W}{m}$$

where we have replaced Fd with the work W. If we now solve for the work, we find

$$W = \frac{1}{2}m(v_f^2 - v_i^2) = \frac{1}{2}mv_f^2 - \frac{1}{2}mv_i^2$$

If we compare this result with Equation 10.7, we see that

$$K_f = \frac{1}{2}mv_f^2 \quad \text{and} \quad K_i = \frac{1}{2}mv_i^2$$

In general, then, an object of mass m moving with speed v has kinetic energy

$$K = \frac{1}{2}mv^2 \quad (10.8)$$

Kinetic energy of an object of mass m moving with speed v

p.50 QUADRATIC

TABLE 10.1 Some approximate kinetic energies

Object	Kinetic energy
Ant walking	1×10^{-8} J
Coin dropped 1 m	5×10^{-3} J
Person walking	70 J
Fastball, 100 mph	150 J
Bullet	5000 J
Car, 60 mph	5×10^{5} J
Supertanker, 20 mph	2×10^{10} J

From Equation 10.8, the units of kinetic energy are those of mass times speed squared, or $\text{kg} \cdot (\text{m/s})^2$. But

$$1\ \text{kg} \cdot (\text{m/s})^2 = \underbrace{1\ \text{kg} \cdot (\text{m/s}^2)}_{1\ \text{N}} \cdot \text{m} = 1\ \text{N} \cdot \text{m} = 1\ \text{J}$$

We see that the units of kinetic energy are the same as those of work, as they must be. TABLE 10.1 gives some approximate kinetic energies.

EXAMPLE 10.4 Finding the work to set a boat in motion

At a history center, an old canal boat is pulled by two draft horses. It doesn't take much force to keep the boat moving; the drag force is quite small. But it takes some work to get the 55,000 kg boat up to speed! The horses can pull with a steady force and put a 1400 N tension in the rope that connects to the boat. The rope is straight and level. The boat starts from rest, and the horses pull steadily as they begin their walk down the tow-path. How much distance do the horses cover as they bring the boat up to its final speed of 0.70 m/s?

STRATEGIZE Let's take the system to be the boat. We could include the water, but since we can ignore the drag force (we're told that it's small), it's not important to do so. The rope is not part of the system, so the tension force does work on the boat. It's this work, which comes from energy provided by the horses, that increases the kinetic energy, and thus the speed, of the boat. We'll consider the initial state to be the boat at rest, the final state to be the boat in motion at its final speed.

PREPARE FIGURE 10.11 is a before-and-after visual overview of the situation. The work that is done by the rope will change the energy of the system, so we can use Equation 10.3, the work-energy equation. Because the only thing that changes is the speed, the only form of energy that changes is the kinetic energy, so we can simplify the equation to

$$\Delta K = W$$

This makes sense—the work done changes the kinetic energy of the boat. The tension force is in the direction of the motion, so the work done is $W = Td$. The boat starts at rest, with kinetic energy equal to zero, so the change in kinetic energy is just the final kinetic energy: $\Delta K = \frac{1}{2}mv_f^2$.

SOLVE With the details noted, the work-energy equation reduces to

$$\frac{1}{2}mv_f^2 = Td$$

We are looking for the distance the horses pull the boat:

$$d = \frac{mv_f^2}{2T} = \frac{(55{,}000\ \text{kg})(0.70\ \text{m/s})^2}{2(1400\ \text{N})} = 9.6\ \text{m}$$

ASSESS This distance is about 30 feet. This seems a reasonable distance; the horses would be pulling for several strides as they get the boat up to speed.

FIGURE 10.11 Getting the canal boat up to speed.

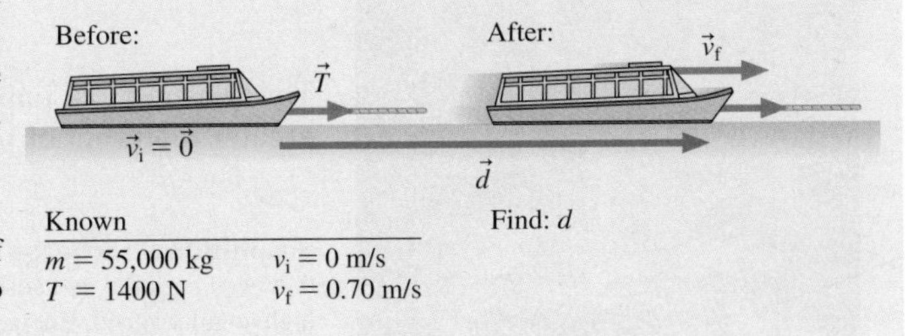

STOP TO THINK 10.3 Rank in order, from greatest to least, the kinetic energies of the sliding pucks.

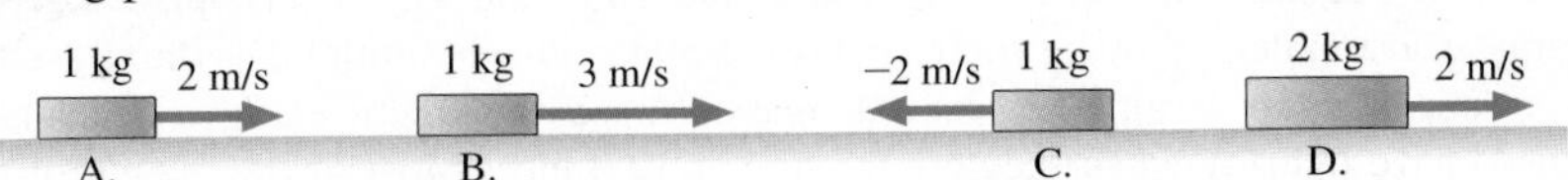

Rotational Kinetic Energy

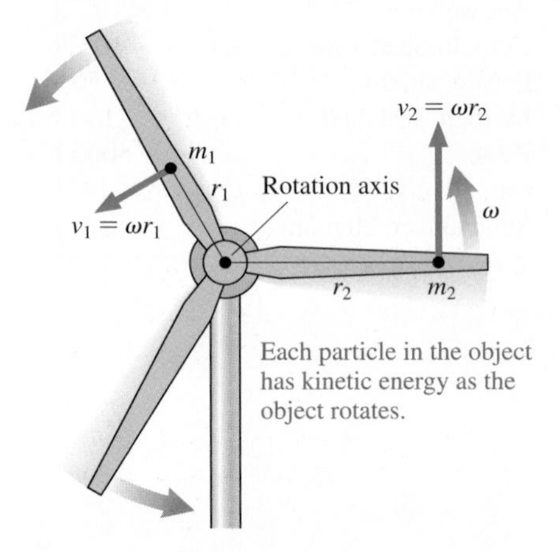

FIGURE 10.12 Rotational kinetic energy of a spinning wind turbine.

FIGURE 10.12 shows the rotating blades of a wind turbine. Although the blades have no overall translational motion, each particle in the blades is moving and hence has kinetic energy. Adding up the kinetic energy of all the particles that make up the blades, we find that the blades have rotational kinetic energy, the kinetic energy due to rotation.

In Figure 10.12, we focus on the motion of two particles in the wind turbine blades. The blade assembly rotates with angular velocity ω. Recall from ◀◀ SECTION 7.1 that a particle moving with angular velocity ω in a circle of radius r has a speed $v = \omega r$. Thus particle 1, which rotates in a circle of radius r_1, moves with speed $v_1 = r_1\omega$ and so has kinetic energy $\frac{1}{2}m_1v_1^2 = \frac{1}{2}m_1r_1^2\omega^2$. Similarly, particle 2, which rotates in a circle with a larger radius r_2, has kinetic energy $\frac{1}{2}m_2r_2^2\omega^2$. The object's rotational kinetic energy is the sum of the kinetic energies of *all* the particles:

$$K_{\text{rot}} = \frac{1}{2}m_1r_1^2\omega^2 + \frac{1}{2}m_2r_2^2\omega^2 + \cdots = \frac{1}{2}\left(\sum mr^2\right)\omega^2$$

You will recognize the term in parentheses as our old friend, the moment of inertia I. Thus the rotational kinetic energy is

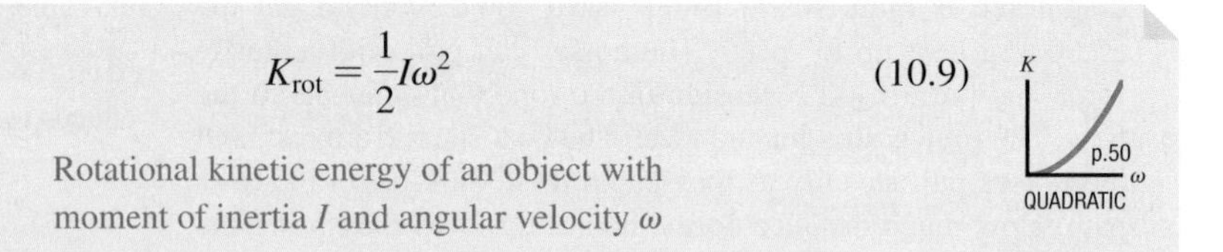

$$K_{\text{rot}} = \frac{1}{2}I\omega^2 \qquad (10.9)$$

Rotational kinetic energy of an object with moment of inertia I and angular velocity ω

NOTE ▶ Rotational kinetic energy is *not* a new form of energy. It is the ordinary kinetic energy of motion, only now expressed in a form that is especially convenient for rotational motion. Comparison with the familiar $\frac{1}{2}mv^2$ shows again that the moment of inertia I is the rotational equivalent of mass. ◀

A rolling object, such as a wheel, is undergoing both rotational *and* translational motions. Consequently, its total kinetic energy is the sum of its rotational and translational kinetic energies:

$$K = K_{\text{trans}} + K_{\text{rot}} = \frac{1}{2}mv^2 + \frac{1}{2}I\omega^2 \qquad (10.10)$$

This illustrates an important fact: **The kinetic energy of a rolling object is always greater than that of a nonrotating object moving at the same speed.**

◀ **Rotational recharge** A promising new technology would replace spacecraft batteries that need periodic and costly replacement with a *flywheel*—a cylinder rotating at a very high angular speed. Energy from solar panels is used to speed up the flywheel, which stores energy as rotational kinetic energy that can then be converted back into electric energy as needed.

EXAMPLE 10.5 Where should you trim the weight?

Any time a cyclist stops, it will take energy to get moving again. Using less energy to get going means more energy is available to go farther or go faster, so racing cyclists want their bikes to be as light as possible. It's particularly important to have lightweight wheels, as this example will show. Consider two bikes that have the same total mass but different mass wheels. Bike 1 has a 10.0 kg frame and two 1.00 kg wheels; bike 2 has a 9.00 kg frame and two 1.50 kg wheels. Both bikes thus have the same 12.0 kg total mass. What is the kinetic energy of each bike when they are moving at 12.0 m/s? Most of the weight of the tire and wheel is at the rim, so we can model each wheel as a hoop.

STRATEGIZE As the bike moves, the wheels rotate. The bike has translational kinetic energy, but the wheels have both translational and rotational kinetic energy. If the bike is moving at speed v, we know from Chapter 7 that the wheels rotate at $\omega = v/R$, where R is the radius of a wheel.

PREPARE Each bike's frame has only translational kinetic energy $K_{\text{frame}} = \frac{1}{2}mv^2$, where m is the mass of the frame. The kinetic energy of each rolling wheel is given by Equation 10.10. From Table 7.1, we find that I for a hoop is MR^2, where M is the mass of one wheel.

SOLVE From Equation 10.10 the kinetic energy of each rolling wheel is

$$K_{\text{wheel}} = \frac{1}{2}Mv^2 + \frac{1}{2}I\omega^2 = \frac{1}{2}Mv^2 + \frac{1}{2}\underbrace{(MR^2)}_{I}\underbrace{\left(\frac{v}{R}\right)^2}_{\omega^2} = Mv^2$$

Then the total kinetic energy of a bike is

$$K = K_{\text{frame}} + 2K_{\text{wheel}} = \frac{1}{2}mv^2 + 2Mv^2$$

The factor of 2 in the second term occurs because each bike has two wheels. Thus the kinetic energies of the two bikes are

$$K_1 = \frac{1}{2}(10.0\text{ kg})(12.0\text{ m/s})^2 + 2(1.00\text{ kg})(12.0\text{ m/s})^2$$
$$= 1010\text{ J}$$
$$K_2 = \frac{1}{2}(9.00\text{ kg})(12.0\text{ m/s})^2 + 2(1.50\text{ kg})(12.0\text{ m/s})^2$$
$$= 1080\text{ J}$$

The kinetic energy of bike 2 is about 7% higher than that of bike 1. Note that the radius of the wheels was not needed in this calculation.

ASSESS We were told that it's particularly important for cyclists to have lightweight wheels, so this result makes sense. Both of the bikes in the example have the same total mass, but the one with lighter wheels takes less energy to get moving. Shaving a little extra weight off your bike's wheels is more useful than taking that same weight off the bike's frame.

10.4 Potential Energy

When two or more objects in a system interact, it is sometimes possible to *store* energy in the system in a way that the energy can be easily recovered. For instance, the earth and a ball interact by the gravitational force between them. If the ball is lifted up into the air, energy is stored in the ball + earth system, energy that can later be recovered as kinetic energy when the ball is released and falls. Similarly, a spring is a system made up of countless atoms that interact via their atomic "springs." If we push a box against a spring, energy is stored that can be recovered when the spring later pushes the box across the table. This sort of stored energy is called **potential energy,** since it has the *potential* to be converted into other forms of energy, such as kinetic or thermal energy.

NOTE ▸ Potential energy is really a property of a *system*, but we often speak informally of the potential energy of an *object*. We might say, for instance, that raising a ball increases its potential energy. This is fine as long as we remember that this energy is really stored in the ball + earth system. ◂

The forces due to gravity and springs are special in that they allow for the storage of energy. Other interaction forces do not. When a dog pulls a sled, the sled interacts with the ground via the force of friction, and the work that the dog does on the sled is converted into thermal energy. The energy is *not* stored up for later recovery—it slowly diffuses into the environment and cannot be recovered.

FIGURE 10.13 Lifting a book increases the system's gravitational potential energy.

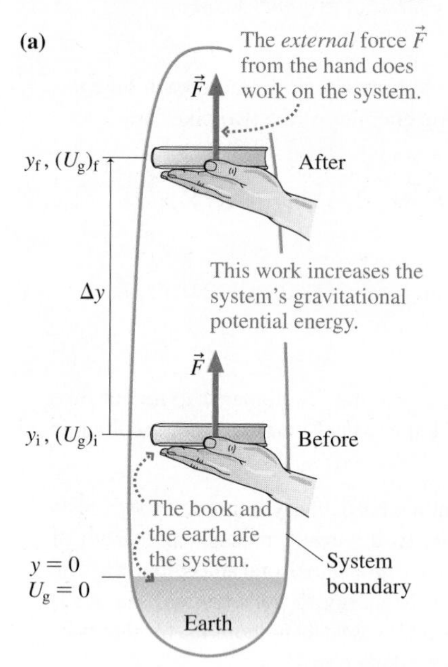
(a)
The *external* force $\vec{F}$ from the hand does work on the system.
$\vec{F}$
y_f, $(U_g)_f$
After
Δy
This work increases the system's gravitational potential energy.
$\vec{F}$
y_i, $(U_g)_i$
Before
The book and the earth are the system.
$y = 0$
$U_g = 0$
System boundary
Earth

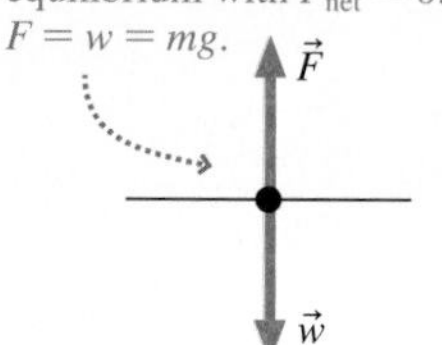
(b) Because the book is being lifted at a constant speed, it is in dynamic equilibrium with $\vec{F}_{net} = \vec{0}$. Thus $F = w = mg$.
$\vec{F}$
$\vec{w}$

Gravitational Potential Energy

To find an expression for **gravitational potential energy U_g**, let's consider the system of the book and the earth shown in FIGURE 10.13a. The book is lifted at a constant speed from its initial position at y_i to a final height y_f. The lifting force of the hand is external to the system and so does work W on the system, increasing its energy. The book is lifted at a constant speed, so its kinetic energy doesn't change. Because there's no friction, the book's thermal energy doesn't change either. Thus the work done goes entirely into increasing the gravitational potential energy of the system. According to Equation 10.3, the work-energy equation, this can be written as $\Delta U_g = W$. Because $\Delta U_g = (U_g)_f - (U_g)_i$, Equation 10.3 can be written

$$(U_g)_f = (U_g)_i + W \tag{10.11}$$

The work done is $W = Fd$, where $d = \Delta y = y_f - y_i$ is the vertical distance that the book is lifted. From the free-body diagram of FIGURE 10.13b, we see that $F = mg$. Thus $W = mg\,\Delta y$, and so

$$(U_g)_f = (U_g)_i + mg\Delta y \tag{10.12}$$

Because our final height was greater than our initial height, Δy is positive and $(U_g)_f > (U_g)_i$. **The higher the object is lifted, the greater the gravitational potential energy in the object + earth system.**

We can express Equation 10.12 in terms of the change in potential energy, $\Delta U_g = (U_g)_f - (U_g)_i$:

$$\Delta U_g = mg\Delta y \tag{10.13}$$

If we lift a 1.5 kg book up by $\Delta y = 2.0$ m, we increase the system's gravitational potential energy by $\Delta U_g = (1.5\text{ kg})(9.8\text{ m/s}^2)(2.0\text{ m}) = 29.4$ J. This increase is *independent* of the book's starting height: The gravitational potential energy increases by 29.4 J whether we lift the book 2.0 m starting at sea level or starting at the top of the Washington Monument. This illustrates an important general fact about *every* form of potential energy: **Only *changes* in potential energy are significant.**

Because of this fact, we are free to choose a *reference level* where we define U_g to be zero. Our expression for U_g is particularly simple if we choose this reference level to be at $y = 0$. We then have

$$U_g = mgy \tag{10.14}$$

Gravitational potential energy of an object of mass m at height y (assuming $U_g = 0$ when the object is at $y = 0$)

EXAMPLE 10.6 Racing up a skyscraper

In the Empire State Building Run-Up, competitors race up the 1576 steps of the Empire State Building, climbing a total vertical distance of 320 m. How much gravitational potential energy does a 70 kg racer gain during this race?

Racers head up the staircase in the Empire State Building Run-Up.

STRATEGIZE We'll take the system to be the racer + earth so that we can consider gravitational potential energy.

PREPARE We are asked for the change in gravitational potential energy as the racer goes up the stairs, so we need only consider the change in height, which is given. We can use Equation 10.13 to compute the change in potential energy during the run.

SOLVE As the racer goes up the stairs, her change in gravitational potential energy is

$$\Delta U_g = mg\Delta y = (70\text{ kg})(9.8\text{ m/s}^2)(320\text{ m}) = 2.2 \times 10^5\text{ J}$$

ASSESS This is a lot of energy. According to Table 10.1, it's comparable to the energy of a speeding car. But the difference in height is pretty great, so this seems reasonable. In Chapter 11, we'll consider how much food energy you'd need to consume to fuel this climb.

An important conclusion from Equation 10.14 is that gravitational potential energy depends only on the height of the object above the reference level $y = 0$, not on the object's horizontal position. To understand why, consider carrying a briefcase while walking on level ground at a constant speed. As shown in the table on page 317, the vertical force of your hand on the briefcase is *perpendicular* to the displacement. No work is done on the briefcase, so its gravitational potential energy remains constant as long as its height above the ground doesn't change.

This idea can be applied to more complicated cases, such as the 82 kg hiker in FIGURE 10.14. His gravitational potential energy depends *only* on his height y above the reference level. Along path A, it's the same value $U_g = mgy = 80$ kJ at any point where he is at height $y = 100$ m above the reference level. If he had instead taken path B, his gravitational potential energy at $y = 100$ m would be the same 80 kJ. It doesn't matter *how* he gets to the 100 m elevation; his potential energy at that height is always the same. **Gravitational potential energy depends only on the *height* of an object and not on the path the object took to get to that position.** This fact will allow us to use the law of conservation of energy to easily solve a variety of problems that would be very difficult to solve using Newton's laws alone.

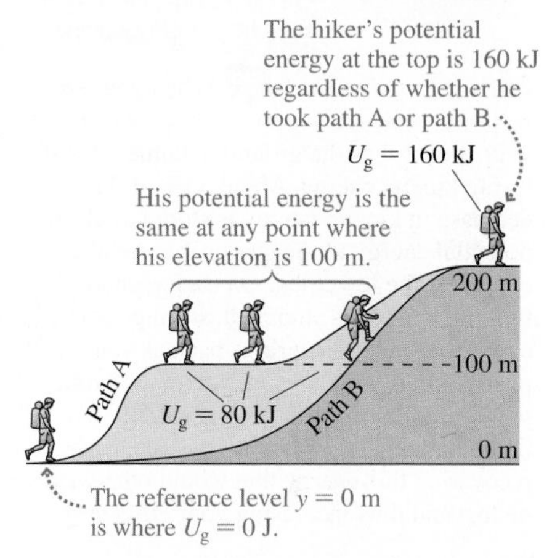
The hiker's potential energy at the top is 160 kJ regardless of whether he took path A or path B.
$U_g = 160$ kJ
His potential energy is the same at any point where his elevation is 100 m.
200 m
100 m
0 m
Path A
Path B
$U_g = 80$ kJ
The reference level $y = 0$ m is where $U_g = 0$ J.

FIGURE 10.14 The hiker's gravitational potential energy depends only on his height above the $y = 0$ m reference level.

STOP TO THINK 10.4 Rank in order, from largest to smallest, the gravitational potential energies of identical balls 1 through 4.

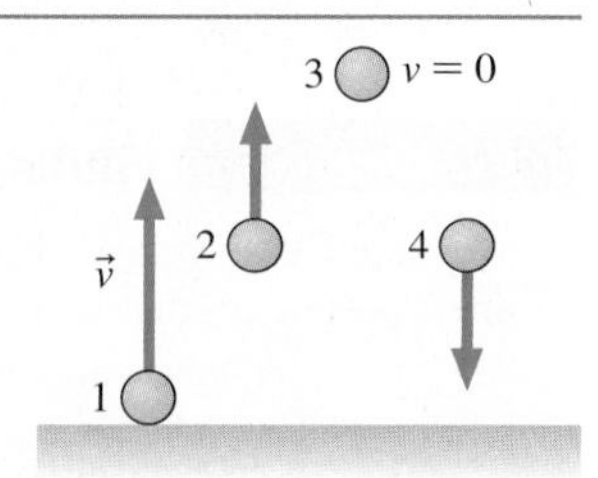
3
$v = 0$
2
4
$\vec{v}$
1

Elastic Potential Energy

Energy can also be stored in a compressed or extended spring as **elastic** (or **spring**) **potential energy** U_s. We can find out how much energy is stored in a spring by using an external force to slowly compress the spring. This external force does work on the spring, transferring energy to the spring. Since only the elastic potential energy of the spring is changing, Equation 10.3 becomes

$$\Delta U_s = W \tag{10.15}$$

That is, we can find out how much elastic potential energy is stored in the spring by calculating the amount of work needed to compress the spring.

FIGURE 10.15 shows a spring being compressed by a hand. In ◀◀ SECTION 8.3 we found that the force the spring exerts on the hand is $F_s = -k\,\Delta x$ (Hooke's law), where Δx is the displacement of the end of the spring from its equilibrium position and k is the spring constant. In Figure 10.15 we have set the origin of our coordinate system at the equilibrium position. The displacement from equilibrium Δx is therefore equal to x, and the spring force is then $-kx$. By Newton's third law, the force that the hand exerts on the spring is thus $F = +kx$.

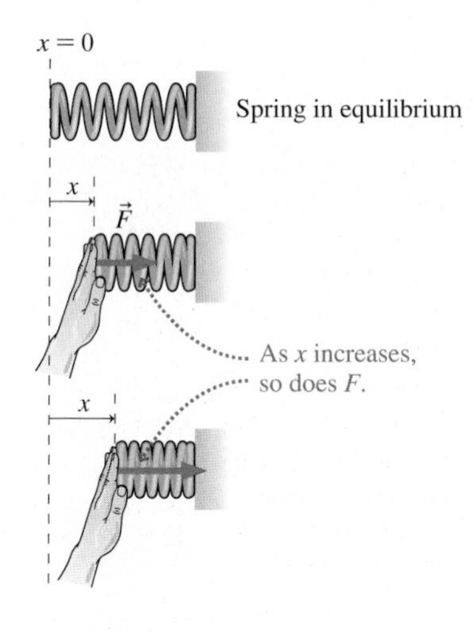
$x = 0$
Spring in equilibrium
x
$\vec{F}$
As x increases, so does F.
x

FIGURE 10.15 The force required to compress a spring is not constant.

As the hand pushes the end of the spring from its equilibrium position to a final position x, the applied force increases from 0 to kx. This is not a constant force, so we can't use Equation 10.5, $W = Fd$, to find the work done. However, it seems reasonable to calculate the work by using the *average* force in Equation 10.5. Because the force varies from $F_i = 0$ to $F_f = kx$, the average force used to compress the spring is $F_{avg} = \frac{1}{2}kx$. Thus the work done by the hand is

$$W = F_{avg}d = F_{avg}x = \left(\frac{1}{2}kx\right)x = \frac{1}{2}kx^2$$

This work is stored as potential energy in the spring, so we can use Equation 10.15 to find that as the spring is compressed, the elastic potential energy increases by

$$\Delta U_s = \frac{1}{2}kx^2$$

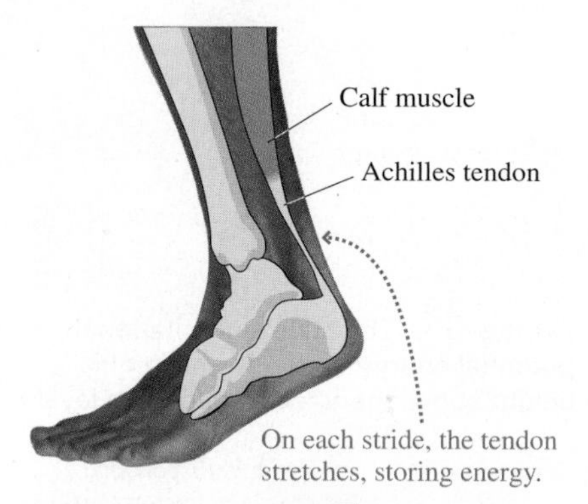

Spring in your step BIO When you run, your feet repeatedly stop and start; when your foot strikes the ground, it comes to rest, losing kinetic energy. About 35% of this decrease in kinetic energy is stored as elastic potential energy in the stretchable Achilles tendon of the lower leg. On each plant of the foot, the tendon is stretched, storing some energy. The tendon springs back as you push off the ground again, helping to propel you forward and returning the stored elastic potential energy back to kinetic energy, recovering this energy that would otherwise be lost and thus increasing your efficiency.

Just as in the case of gravitational potential energy, we have found an expression for the *change* in U_s, not U_s itself. Again, we are free to set $U_s = 0$ at any convenient spring extension. An obvious choice is to set $U_s = 0$ at the point where the spring is in equilibrium, neither compressed nor stretched—that is, at $x = 0$. With this choice we have

$$U_s = \frac{1}{2}kx^2 \tag{10.16}$$

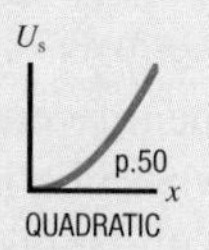

Elastic potential energy of a spring displaced a distance x from equilibrium (assuming $U_s = 0$ when the end of the spring is at $x = 0$)

NOTE ▶ Because U_s depends on the *square* of the displacement x, U_s is the same whether x is positive (the spring is compressed as in Figure 10.15) or negative (the spring is stretched). ◀

EXAMPLE 10.7 Finding the energy stored in a stretched tendon BIO

We noted that your Achilles tendon stretches when you run, and this stores some energy—energy that is returned to you when you push off the ground with your foot. FIGURE 10.16 shows smoothed data for restoring force versus extension for the Achilles tendon and attached muscle in a female subject. When she runs, at one point in her stride the stretch reaches a maximum of 0.50 cm. What energy is stored for this stretch?

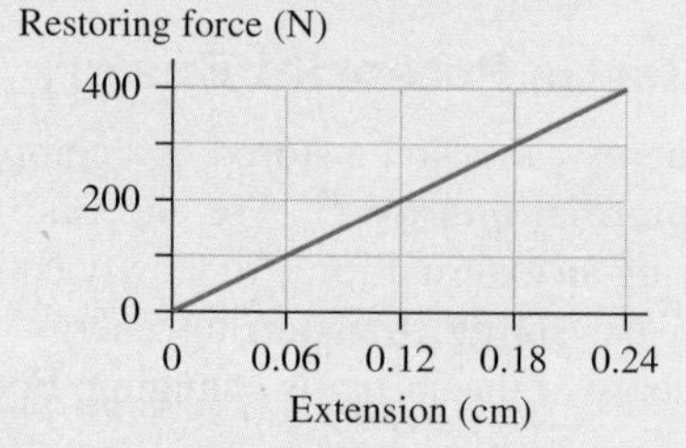

FIGURE 10.16 Force data for the stretch of the Achilles tendon.

STRATEGIZE The force increases linearly with extension, so we can model the tendon as a spring. If we find the spring constant, we can compute the stored energy using Equation 10.16.

PREPARE The spring constant k is the slope of the graph in Figure 10.16. At the top right, the line goes through a point that is easy to read off the axes. Using this point, we determine the slope to be

$$k = \text{slope} = \frac{400 \text{ N}}{0.0024 \text{ m}} = 1.67 \times 10^5 \text{ N/m}$$

SOLVE With this spring constant, the energy stored for a 0.50 cm (0.0050 m) stretch is

$$U_s = \frac{1}{2}kx^2 = \frac{1}{2}(1.67 \times 10^5 \text{ N/m})(0.0050 \text{ m})^2 = 2.1 \text{ J}$$

ASSESS Table 10.1 gives 70 J as the kinetic energy for a person walking; 2.1 J is a few percent of that value. This is reasonable: When you run, the only part of your body that stops and starts is your feet; it's a fraction of this energy that is recovered, and we expect this to be a small fraction of the kinetic energy of your body. However, saving even this small amount of energy is useful; over a long run of many steps, it will add up!

STOP TO THINK 10.5 When a spring is stretched by 5 cm, its elastic potential energy is 1 J. What will its elastic potential energy be if it is *compressed* by 10 cm?

A. −4 J B. −2 J C. 2 J D. 4 J

10.5 Thermal Energy

We noted earlier that thermal energy is related to the microscopic motion of the atoms of an object. As FIGURE 10.17 shows, the atoms in a hot object jiggle around their average positions more than the atoms in a cold object. This has two consequences. First, each atom is on average moving faster in the hot object. This means that each atom has a higher *kinetic energy*. Second, each atom in the hot object tends to stray farther from its equilibrium position, leading to a greater stretching or compressing of the spring-like molecular bonds. This means that each atom has on average a higher *potential energy*. The potential energy stored in any one bond and the kinetic energy of any one atom are both exceedingly small, but there are incredibly many bonds and atoms. The sum of all these microscopic potential and kinetic energies is what we call **thermal energy E_{th}**. Increasing an object's thermal energy corresponds to increasing its temperature.

FIGURE 10.18 A thermograph of a box that's been dragged across the floor.

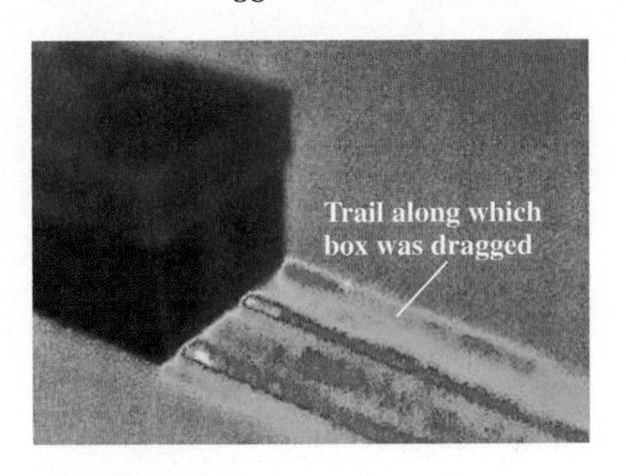

Trail along which box was dragged

FIGURE 10.17 An atomic view of thermal energy.

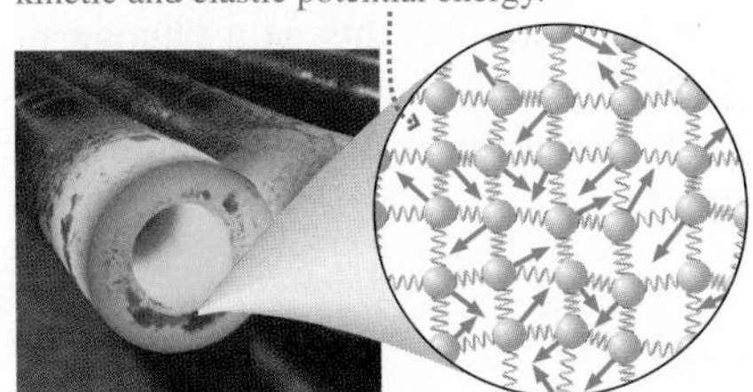

Hot object: Fast-moving atoms have lots of kinetic and elastic potential energy.

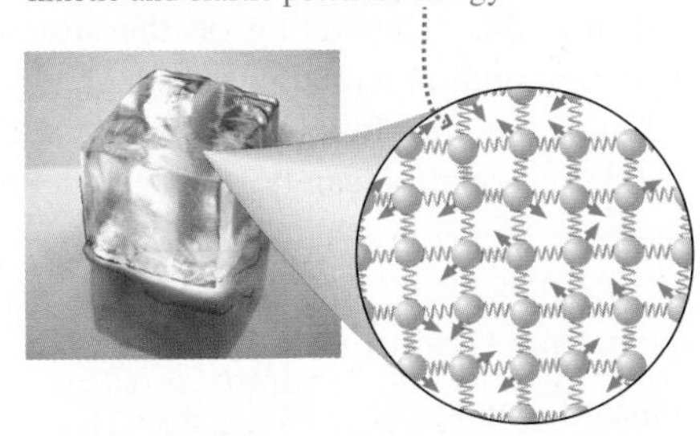

Cold object: Slow-moving atoms have little kinetic and elastic potential energy.

FIGURE 10.18 shows a thermograph of a heavy box and the floor across which it has just been dragged. In this image, cool areas appear in shades of blue and green, warm areas in shades of red. You can clearly see that the patch of floor that the box has been dragged across is much warmer than the box or the rest of the floor. Dragging the box across the floor caused the thermal energy of the system to increase.

This increase in thermal energy is a general feature of any system in which there is friction between sliding objects. An atomic-level explanation is illustrated in FIGURE 10.19. The interaction between the surfaces that leads to the force of friction also leads to increased thermal energy of the sliding surfaces.

FIGURE 10.19 How friction causes an increase in thermal energy.

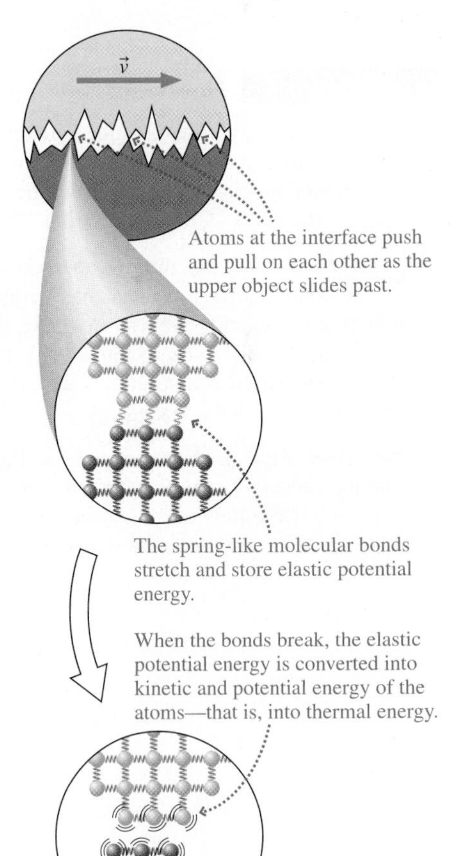

Atoms at the interface push and pull on each other as the upper object slides past.

We can find a quantitative expression for the change in thermal energy by considering the case of the box pulled by a rope at a constant speed. Let's consider the system to be the box + floor. As the box is pulled across the floor, the rope exerts a constant forward force $\vec{F}$ on the box, while the friction force $\vec{f}_k$ exerts a constant force on the box that is directed backward. Because the box moves at a constant speed, the magnitudes of these two forces are equal: $F = f_k$. As the box moves through a displacement $d = \Delta x$, the rope does work $W = F\Delta x = f_k\Delta x$ on the box. This work represents energy transferred into the system, so the system's energy must increase. The box's kinetic energy and gravitational potential energy don't change, so the increased energy must be in the form of thermal energy E_{th}. Because all of the work into the system shows up as increased thermal energy, we can say that

$$\Delta E_{th} = f_k \Delta x \qquad (10.17)$$

The increased thermal energy is distributed between the two surfaces, the box and the floor, the two elements of the system. Although we arrived at Equation 10.17 by considering energy transferred into the system via work done by an external force, the equation is equally valid for the transformation of energy into thermal energy when, for instance, an object slides to a halt on a rough surface.

Video Figure 10.19

If an object moves through air, water, or another fluid at a constant speed, there is a constant drag force opposite the motion, so we can do an analysis similar to that for the friction force. If we consider the object + fluid to be the system, as the object moves at a constant speed, the drag force D transforms energy into thermal energy:

$$\Delta E_{th} = D\Delta x \tag{10.18}$$

The change in thermal energy is mostly a change in the air or the water: Collisions between the object and particles of the fluid cause the particles to move more quickly, thus increasing thermal energy.

The work-energy equation, Equation 10.3, states that the change in the total energy of a system equals the energy transferred to or from the system as work. If we consider only those forms of energy that are typically transformed during the motion of ordinary objects—kinetic energy K, gravitational and elastic potential energies U_g and U_s, and thermal energy E_{th}—then the work-energy equation can be written as

$$\Delta K + \Delta U + \Delta E_{th} = W \tag{10.19}$$

NOTE ▶ We've written the change in potential energy as a single potential energy change ΔU. Depending on the situation, we can interpret this as a change in gravitational potential energy, a change in elastic potential energy, or a combination of the two. In later chapters, we'll add additional forms of potential energy, and this equation can be adapted accordingly. ◀

EXAMPLE 10.8 How much energy does it take to swim a kilometer?

How much energy is required for a 70 kg swimmer to complete a 1.0 km swim at a steady 1.4 m/s? We can assume typical data for a swimmer moving through the water: frontal area 0.080 m^2, drag coefficient 0.45, density of water 1000 kg/m^3.

STRATEGIZE Moving through the water at a constant speed means continuously replacing the energy that drag transforms into thermal energy. We can compute the change in thermal energy during the swim to find energy that the swimmer must supply. The speed is constant, so we can use Equation 10.18 to find the change in thermal energy.

PREPARE We'll need to compute the drag force before we can find the necessary energy. A swimmer moving through the water has a very large Reynold's number, so we can use Equation 5.12 to compute the drag force, as shown in the next column:

$$D = \frac{1}{2}C_D\rho A v^2$$
$$= \frac{1}{2}(0.45)(1000\ \text{kg/m}^3)(0.080\ \text{m}^2)(1.4\ \text{m/s})^2 = 35\ \text{N}$$

SOLVE With the drag force in hand, we can find the energy converted to thermal energy:

$$\Delta E_{th} = D\Delta x = (35\ \text{N})(1000\ \text{m}) = 3.5 \times 10^4\ \text{J}$$

This is the energy that the swimmer must supply, which is all converted to thermal energy. The net effect of swimming laps is to warm up the water in the pool!

ASSESS This is a lot of energy, about one-sixth of the energy needed to climb the Empire State Building, which seems reasonable.

We've done a few calculations of the energy required for certain tasks. But there's another factor to consider—efficiency. How much energy would your body actually *use* in completing these tasks? The swimmer needs to supply 3.5×10^4 J to move through the water. But how much metabolic energy will it cost the swimmer to provide this energy? Swimming is, for humans, a reasonably inefficient form of locomotion, so the energy used by the swimmer is quite a bit greater than the value we found. We'll return to this issue in Chapter 11.

STOP TO THINK 10.6 A block with an initial kinetic energy of 4.0 J comes to rest after sliding 1.0 m. How far would the block slide if it had 8.0 J of initial kinetic energy?

A. 1.4 m B. 2.0 m C. 3.0 m D. 4.0 m

10.6 Conservation of Energy

Video Breaking Boards

Just as for momentum conservation, we will develop a before-and-after perspective for energy conservation. We'll limit our consideration for now to kinetic energy, potential energy (both gravitational and elastic), thermal energy, and work. We then note that $\Delta K = K_f - K_i$ and $\Delta U = U_f - U_i$. Now we can rewrite Equation 10.19 as a rule that we can use to solve problems:

$$K_f + U_f + \Delta E_{th} = K_i + U_i + W \quad (10.20)$$

Before-and-after work-energy equation

Video Chin Basher?

Equation 10.20 states that a system's final energy, including any change in the system's thermal energy, equals its initial energy plus any energy added to the system as work. This equation is the basis for a powerful problem-solving approach.

NOTE ▶ We don't write ΔE_{th} as $(E_{th})_f - (E_{th})_i$ in Equation 10.20 because the initial and final values of the thermal energy are typically unknown; only their difference ΔE_{th} can be measured. ◀

In Section 10.1 we introduced the idea of an isolated system—one in which no work is done on the system and no energy is transferred into or out of the system. In that case, $W = 0$ in Equation 10.20, so the final energy, including any change in thermal energy, equals the initial energy:

$$K_f + U_f + \Delta E_{th} = K_i + U_i$$

The following table shows how to choose an isolated system for common situations.

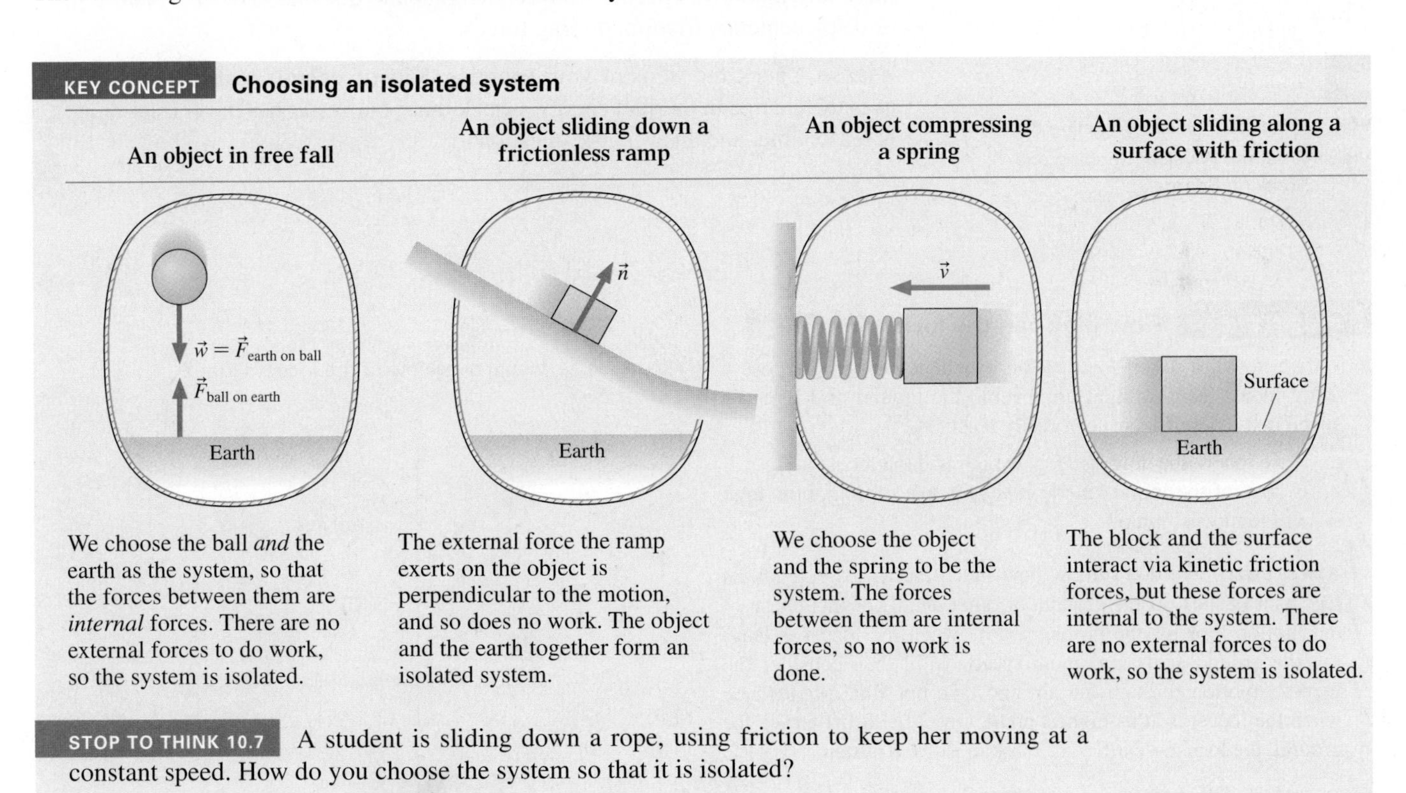

KEY CONCEPT Choosing an isolated system
An object in free fall
An object sliding down a frictionless ramp
An object compressing a spring
An object sliding along a surface with friction
$\vec{w} = \vec{F}_{\text{earth on ball}}$
$\vec{F}_{\text{ball on earth}}$
Earth
$\vec{n}$
Earth
$\vec{v}$
Surface
Earth
We choose the ball *and* the earth as the system, so that the forces between them are *internal* forces. There are no external forces to do work, so the system is isolated.
The external force the ramp exerts on the object is perpendicular to the motion, and so does no work. The object and the earth together form an isolated system.
We choose the object and the spring to be the system. The forces between them are internal forces, so no work is done.
The block and the surface interact via kinetic friction forces, but these forces are internal to the system. There are no external forces to do work, so the system is isolated.

STOP TO THINK 10.7 A student is sliding down a rope, using friction to keep her moving at a constant speed. How do you choose the system so that it is isolated?

Using the Law of Conservation of Energy

Now that we have mathematical expressions for different forms of energy and a general before-and-after equation expressing the law of conservation of energy, we have all the tools we need to formulate a problem-solving approach. We'll sketch out the details and then use it to solve a range of problems.

Spring into action BIO A locust can jump as far as 1 meter, an impressive distance for such a small animal. To make such a jump, its legs must extend much more rapidly than muscles can ordinarily contract. Thus, instead of using its muscles to make the jump directly, the locust uses them to more slowly stretch an internal "spring" near its knee joint. This stores elastic potential energy in the spring. When the muscles relax, the spring is suddenly released, and its energy is rapidly converted into kinetic energy of the insect.

Video What the Physics? Bouncing Balloons

Video What the Physics? Bending and Breaking Boards

PROBLEM-SOLVING APPROACH 10.1 Conservation of energy problems

The work-energy equation and the law of conservation of energy relate a system's *final* energy to its *initial* energy. We can solve for initial and final heights, speeds, and displacements from these energies.

STRATEGIZE The first step in a conservation of energy problem is to choose the system. This means thinking about the forces and the energies involved. We'll consider the situation before and after a process or an interaction, so we must also decide on the initial and final states.

PREPARE As we did for momentum problems, we'll start with a before-and-after visual overview, as outlined in Tactics Box 9.1. Note the known quantities, and identify what you're trying to find.

SOLVE Apply Equation 10.20:

$$K_f + U_f + \Delta E_{th} = K_i + U_i + W$$

Start with this general equation, then specialize to the case at hand:

- Use the appropriate form or forms of potential energy.
- If the system is isolated, no work is done. Set $W = 0$.
- If there is no friction, drag, or similar force, set $\Delta E_{th} = 0$.

Depending on the problem, you'll need to calculate the initial and/or final values of these energies. You can then solve for the unknown energies, and from these any unknown speeds (from K), heights and distances (from U_g and U_s), or displacements, friction, or drag forces.

ASSESS Check the signs of your energies. Kinetic energy is always positive, as is the change in thermal energy. Check that your result has the correct units, is reasonable, and answers the question.

Exercise 23 

EXAMPLE 10.9 How high can the locust jump? BIO

As we noted, a desert locust is an excellent jumper. Suppose a 2.0 g locust leaps straight up, leaving the ground at 3.1 m/s, a speed that a desert locust can easily reach.

a. If we ignore the drag force, how high will the locust jump?
b. If 20% of the initial kinetic energy is lost to drag, how high will the locust jump?

STRATEGIZE We don't know how much the locust extends its legs as it pushes off the ground, or other details of this phase of the motion. For us, the problem starts when the locust is leaving the ground at its maximum speed. We'll then consider the locust's motion through the air, and take our final point to be when the locust is at its highest point. Once the locust leaves the ground, the locust + earth + air form an isolated system.

PREPARE The before-and-after visual overview is shown in FIGURE 10.20. We'll describe the situation using Equation 10.20 with $W = 0$, because we've identified an isolated system. For the initial and final points we chose our working equation becomes

$$K_f + (U_g)_f + \Delta E_{th} = K_i + (U_g)_i$$

FIGURE 10.20 Visual overview of the locust's jump.

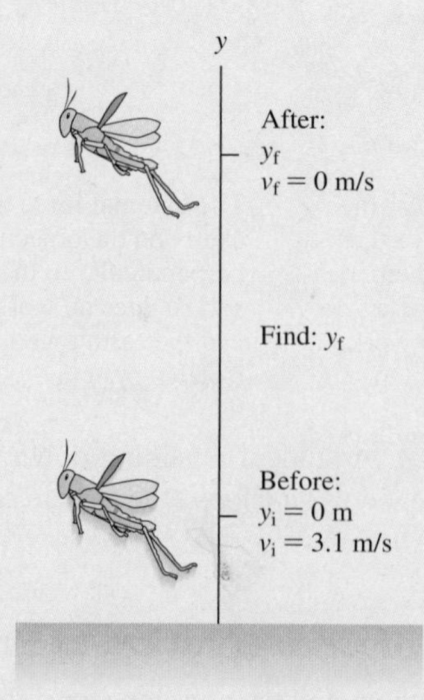

SOLVE a. For this part, we ignore drag, so $\Delta E_{th} = 0$. We then substitute expressions for the various forms of energy to find

$$\frac{1}{2}mv_f^2 + mgy_f = \frac{1}{2}mv_i^2 + mgy_i$$

The mass appears in every term, so we can cancel out this factor. We'll take our starting position to be $y_i = 0$ and note that the final velocity $v_f = 0$ to simplify further:

$$gy_f = \frac{1}{2}v_i^2$$

$$y_f = \frac{v_i^2}{2g} = \frac{(3.1\ \text{m/s})^2}{2(9.8\ \text{m/s}^2)} = 0.49\ \text{m}$$

b. For this part of the problem, we assume that 20% of the initial kinetic energy is lost to drag. The "lost" energy is transformed into thermal energy, so $\Delta E_{th} = (0.20)K_i$. Our working equation then becomes

$$K_f + (U_g)_f + (0.20)K_i = K_i + (U_g)_i$$

$$K_f + (U_g)_f = (0.80)K_i + (U_g)_i$$

Simplifying as we did before, we find

$$gy_f = (0.80)\frac{1}{2}v_i^2$$

$$y_f = (0.80)\frac{v_i^2}{2g} = (0.80)\frac{(3.1\ \text{m/s})^2}{2(9.8\ \text{m/s}^2)} = 0.39\ \text{m}$$

ASSESS We noted that the desert locust can jump a horizontal distance of 1 m, so a vertical leap of half a meter seems reasonable. Notice that the locust's mass didn't enter into our calculation; this isn't a surprise, given that the motion is free fall. This also gives us confidence in our solution.

EXAMPLE 10.10 To push or not to push?

The Summit Plummet is an extreme water slide—one of the steepest and fastest in the world. Riders drop 36 m from the start until they hit a run-out at the bottom. If you give yourself a good push at the start, so that you begin your plunge moving at 2.0 m/s, how fast are you moving when you get to the bottom? How fast would you be moving if you skipped the push? The slide is steep and slippery, so assume that you can ignore friction and drag forces.

STRATEGIZE We'll take the system to be the rider + earth. The initial state has the rider moving at 2.0 m/s at a height of 36 m above the bottom of the slide; the final state is at the bottom.

PREPARE The visual overview in FIGURE 10.21 shows the initial and final states and the slide in between. The exact shape of the slide doesn't matter; we care only about the difference in height. We'll describe the situation using Equation 10.20 with $W = 0$, because this is an isolated system—once you start the ride, no one is giving you a push! There is no elastic potential energy, only gravitational, and we can ignore friction, so our working equation becomes

$$K_f + (U_g)_f = K_i + (U_g)_i$$

FIGURE 10.21 Visual overview of the trip down the water slide.

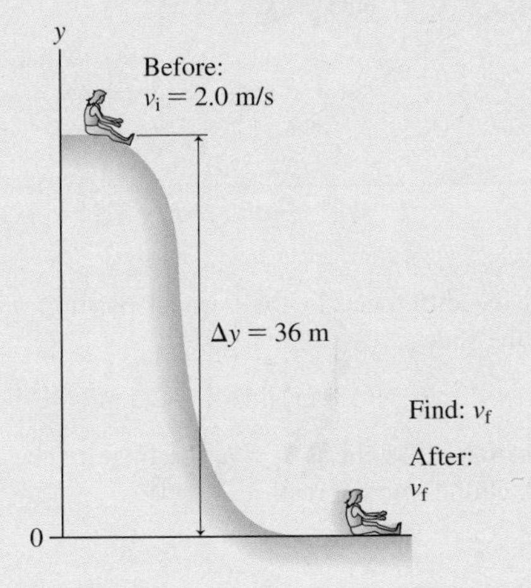

SOLVE If we write our working equation in terms of the change in potential energy, we can express this in terms of the change in height using Equation 10.13:

$$K_f = K_i + ((U_g)_i - (U_g)_f) = K_i + \Delta U_g = K_i + mg\Delta y$$

We rewrite the kinetic energy in terms of speed and then solve for the final speed:

$$\frac{1}{2}mv_f^2 = \frac{1}{2}mv_i^2 + mg\Delta y$$

$$v_f = \sqrt{v_i^2 + 2g\Delta y} = \sqrt{(2.0\ \text{m/s})^2 + 2(9.8\ \text{m/s}^2)(36\ \text{m})}$$
$$= 27\ \text{m/s}$$

This is pretty speedy! Now, suppose you skip the initial push. How much does this change the final result? The calculation is the same but with $v_i = 0$:

$$v_f = \sqrt{2g\Delta y} = \sqrt{2(9.8\ \text{m/s}^2)(36\ \text{m})} = 27\ \text{m/s}$$

We get exactly the same result to 2 significant figures: greater precision is not warranted given the approximations we've made. Push or not—the final result is about the same! Most of your energy at the end of the ride comes from the change in potential energy, not your initial push.

ASSESS We weren't given the mass of the rider, but the mass canceled along the way, which gives us confidence in the process. The final result, 27 m/s (about 60 mph), is pretty fast. But this is an extreme slide, and a website for this slide claims that you can expect to reach 60 mph, so the result of the calculation is reasonable, even if actually riding the slide isn't.

EXAMPLE 10.11 Speed of a spring-launched ball

A spring-loaded toy gun is used to launch a 10 g plastic ball. The spring, which has a spring constant of 10 N/m, is compressed by 10 cm as the ball is pushed into the barrel. When the trigger is pulled, the spring is released and shoots the ball back out horizontally. What is the ball's speed as it leaves the barrel? Assume that friction is negligible.

STRATEGIZE Let's take the system to be the ball + spring. The initial state has the compressed spring touching the stationary ball; the final state is the expanded spring and the ball in motion.

PREPARE The visual overview is shown in FIGURE 10.22. We have chosen the origin of the coordinate system to be the equilibrium position of the free end of the spring, making $x_i = -10$ cm and $x_f = 0$ cm. Work is done on the spring during the compression, but during the time the spring is expanding, the ball + spring is an isolated system, so $W = 0$. We are ignoring friction, so $\Delta E_{th} = 0$. Because the launch is horizontal, we can ignore changes in gravitational potential energy. With these assumptions, the work-energy equation becomes

FIGURE 10.22 Before-and-after visual overview of a ball being shot out of a spring-loaded toy gun.

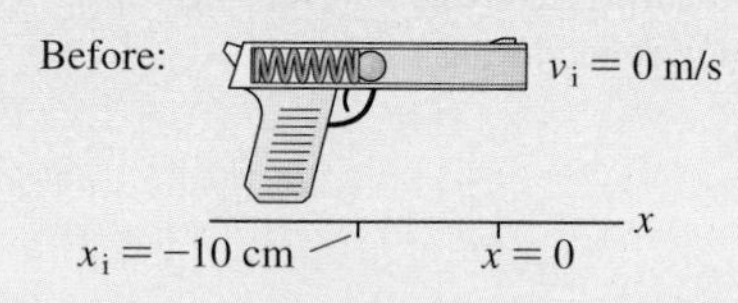

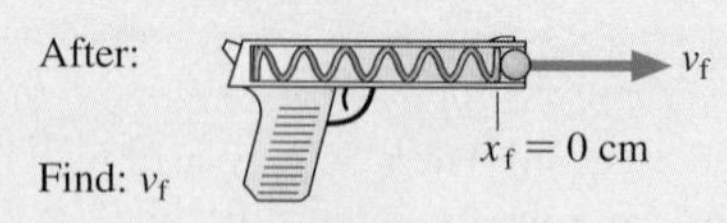

$$K_f + (U_s)_f = K_i + (U_s)_i$$

SOLVE We can use expressions for kinetic energy and elastic potential energy to rewrite this equation as

$$\tfrac{1}{2}mv_f^2 + \tfrac{1}{2}kx_f^2 = \tfrac{1}{2}mv_i^2 + \tfrac{1}{2}kx_i^2$$

We know that $x_f = 0$ m and $v_i = 0$ m/s, so this simplifies to

$$\tfrac{1}{2}mv_f^2 = \tfrac{1}{2}kx_i^2$$

It is now straightforward to solve for the ball's speed:

$$v_f = \sqrt{\frac{kx_i^2}{m}} = \sqrt{\frac{(10\ \text{N/m})(-0.10\ \text{m})^2}{0.010\ \text{kg}}} = 3.2\ \text{m/s}$$

ASSESS The ball moves pretty slowly, which we expect for a toy gun. Our result seems reasonable.

This is not a problem that we could have easily solved with Newton's laws. The acceleration is not constant, and we have not learned how to handle the kinematics of variable acceleration. But with conservation of energy, this was a straightforward problem.

EXAMPLE 10.12 The thermal energy of a trip down a slide

Quinn is at rest at the top of a playground slide. The main part of the slide is 5.0 m long, and it is tipped at a 30° angle. Quinn starts sliding, moving down the tipped section, and then the slide levels out so that he leaves the slide at 2.0 m/s, moving horizontally. If Quinn's mass is 24 kg, how much thermal energy is deposited in his trousers and in the slide?

STRATEGIZE We'll choose Quinn + earth + slide to be the system. The initial point will be when Quinn is motionless at the top of the slide; the final point will be when Quinn has descended the slide and is moving off the end. As Quinn goes down the slide, potential energy decreases and kinetic energy increases. But there is clearly friction, so that his speed is slower than it would be otherwise. Some of the energy is transformed into thermal energy of his trousers and the slide, internal to the system. It's this change in thermal energy that we'll solve for.

PREPARE The visual overview in FIGURE 10.23 shows the initial and final points of the motion. We've chosen an isolated system, so $W = 0$. The only form of potential energy is gravitational potential energy, so the work-energy equation reduces to

$$K_f + (U_g)_f + \Delta E_{th} = K_i + (U_g)_i$$

FIGURE 10.23 Visual overview for motion down the slide.

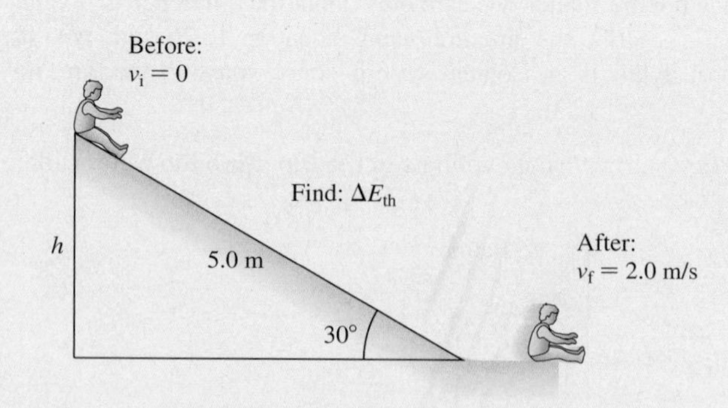

We can find the difference in the vertical position h from the geometry of the slide:

$$h = (5.0\ \text{m})\sin 30° = 2.5\ \text{m}$$

We'll take the initial height as $y_i = h$, the final height as $y_f = 0$, so $(U_g)_f = 0$. Quinn starts at rest, so $K_i = 0$.

SOLVE With all of these parts in hand we can simplify the work-energy equation further and then solve for ΔE_{th}:

$$\Delta E_{th} = (U_g)_i - K_f$$
$$= mgy - \frac{1}{2}mv_f^2$$
$$= 588\text{ J} - 48\text{ J} = 540\text{ J}$$

We've found values for the change in potential energy and the final kinetic energy so that you can see the relative magnitudes. The values in the problem are typical values for a slide—about an 8 foot drop, kids launched off the end at a slow jogging pace. During the slide, the most important energy transformation that takes place is the increase in thermal energy. The slide is mostly about warming things up rather than getting kids up to speed!

ASSESS If you remember going down the slide as a child, you no doubt remember the appreciable warming during the motion, so our result makes sense.

SYNTHESIS 10.1 Energy and its conservation

The energies present in an isolated system can transform from one kind into another, but the total energy is *conserved*. The unit of all types of energy is the **joule** (J).

Kinetic energy is the energy of motion.

$$K = \frac{1}{2}mv^2$$

Mass (kg)
Velocity (m/s)

Gravitational potential energy is stored energy associated with an object's height above the ground.

$$U_g = mgy$$

Free-fall acceleration
Mass (kg)
Height (m) above a reference level $y = 0$

Elastic potential energy is stored energy associated with a stretched or compressed spring.

$$U_s = \frac{1}{2}kx^2$$

Spring constant (N/m)
Displacement of end of spring from equilibrium (m)

Work is the transfer of energy into or out of a system by an external force:

$$W = F_{\parallel}d$$

Work into (+) or out of (−) a system
Force parallel to motion
Displacement

The before-and-after work-energy equation captures the **law of conservation of energy**:

$$K_f + U_f + \Delta E_{th} = K_i + U_i + W$$

Final kinetic and potential energy plus change in thermal energy
Initial kinetic and potential energy plus energy transferred by work

STOP TO THINK 10.8 At the water park, Katie slides down each of the frictionless slides shown. At the top, she is given a push so that she has the same initial speed each time. At the bottom of which slide is she moving the fastest?

A. Slide A
B. Slide B
C. Slide C
D. Her speed is the same at the bottom of all three slides.

Video Loop-the-Loop

10.7 Energy Diagrams

Energy is a central concept in physics, but it's also crucial for understanding chemistry, biology, and other sciences. In this section and the following one, we'll develop different means of describing energy that connect to these other subjects that you are likely studying or have studied. It's not that energy is different in chemistry, but the way it is treated, the language used to describe it, may be.

In this section, we'll consider isolated systems for which there is no friction or drag. In this case, there is no work and there is no change in thermal energy, so the work-energy equation becomes

$$K_f + U_f = K_i + U_i$$

In other words, the sum of the kinetic and potential energy is constant. We'll define the total energy E as the sum of these two quantities. For the systems we consider in this section,

$$E = K + U = \text{constant}$$

Kinetic energy depends on an object's speed, but potential energy depends on its *position*. A tossed ball's gravitational potential energy depends on its height y, while the elastic potential energy of a compressed spring depends on the displacement x. Other potential energies also depend in some way on position. A graph showing a system's potential energy and total energy as a function of position is called an **energy diagram.** We'll spend some time learning about energy diagrams so that we can use them to think about bonds and chemical reactions in the next section.

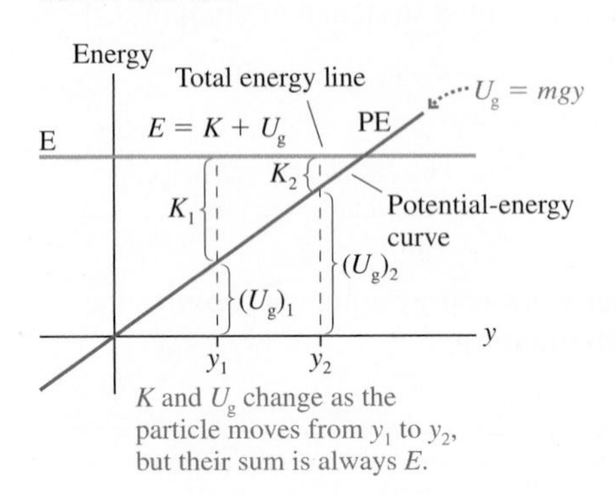

FIGURE 10.24 The energy diagram of a ball in free fall.

FIGURE 10.24 is the energy diagram of a ball in free fall. This is a bit different from most graphs we've seen. It doesn't include time; the horizontal axis is the vertical position y, and the vertical axis represents energy. The lines on the graph show different energies as a function of the vertical position. The gravitational potential energy increases with the vertical position; the mathematical relationship is $U_g = mgy$, and a graph of mgy versus y is a straight line through the origin with slope mg. The resulting blue *potential-energy curve* is labeled PE. The tan line labeled E is the system's total energy. This line is always horizontal because the sum of kinetic and potential energy is the same at every point.

Suppose we consider a ball that is at a vertical position y_1 and is moving upward. When the ball is at height y_1, the distance from the axis up to the potential-energy curve is the potential energy $(U_g)_1$ at that position. Because $K_1 = E - (U_g)_1$, the kinetic energy is represented graphically as the distance between the potential-energy curve and the total energy line. Now, the ball continues to rise. Some time later it is at height y_2. The energy diagram shows that the potential energy $(U_g)_2$ has increased while the ball's kinetic energy K_2 has decreased, as we know must be the case. Kinetic energy has been transformed into potential energy, but their sum has not changed.

NOTE ▸ In graphs like this, the potential-energy curve PE is determined by the physical properties of the system—for example, the mass or the spring constant. But the total energy line E is under your control. If you change the *initial conditions*, such as throwing the ball upward with a different speed or compressing a spring by a different amount, the total energy line will appear at a different position. We can thus use an energy diagram to see how changing the initial conditions affects the subsequent motion. ◂

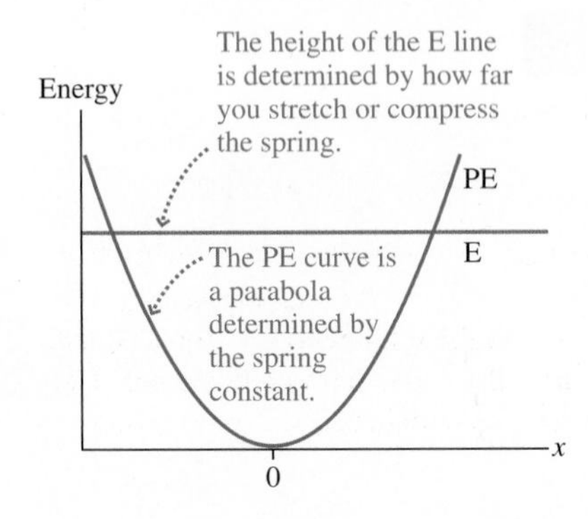

FIGURE 10.25 The energy diagram of a mass on a horizontal spring.

FIGURE 10.25 is the energy diagram of a mass on a horizontal spring. In this case, the blue potential-energy curve $U_s = \frac{1}{2}kx^2$ is a parabola centered at $x = 0$, the equilibrium position of the end of the spring. The blue PE curve is determined by the spring constant; we can't change it. But we can set the tan E line to any height we wish by stretching or compressing the spring to different lengths. The figure shows one possible E line.

Suppose you pull the mass out to position x_R and release it from rest. FIGURE 10.26 shows a five-frame "movie" of the subsequent motion. Initially, in frame a, the energy is entirely potential—the energy of a stretched spring—so the E line has been drawn to cross the PE line at $x_a = x_R$. This is the graphical statement that initially $E = U_s$ and $K = 0$.

The restoring force pulls the mass toward the origin. In frame b, where the mass has reached x_b, the potential energy has decreased while the kinetic energy—the

FIGURE 10.26 A five-frame movie of a mass oscillating on a spring.

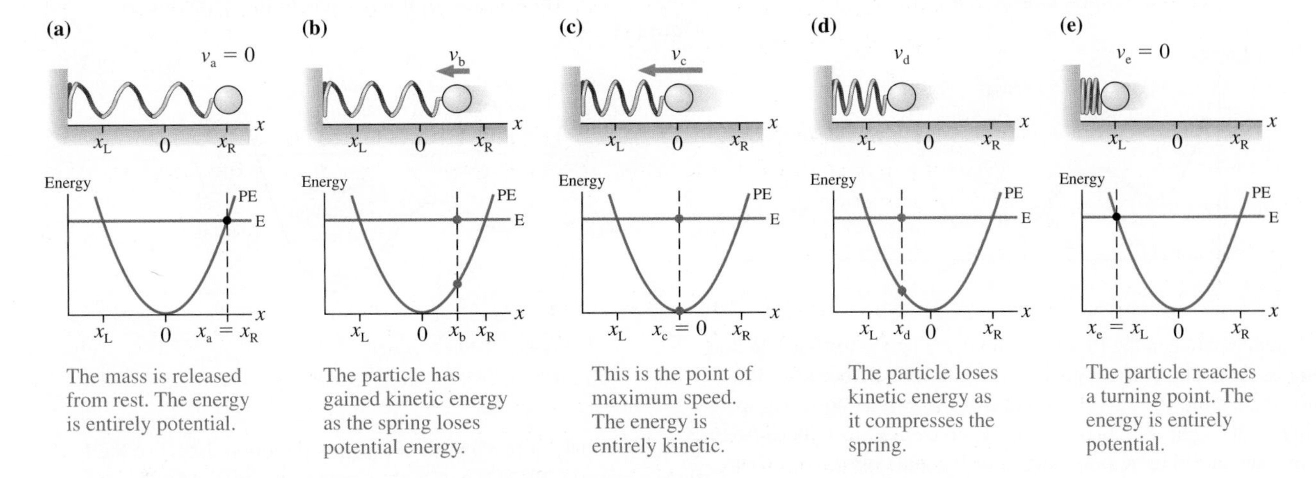
(a)
$v_a = 0$
Energy
PE
E
x_L 0 $x_a = x_R$
The mass is released from rest. The energy is entirely potential.
(b)
v_b
Energy
PE
E
x_L 0 x_b x_R
The particle has gained kinetic energy as the spring loses potential energy.
(c)
v_c
Energy
PE
E
x_L $x_c = 0$ x_R
This is the point of maximum speed. The energy is entirely kinetic.
(d)
v_d
Energy
PE
E
x_L x_d 0 x_R
The particle loses kinetic energy as it compresses the spring.
(e)
$v_e = 0$
Energy
PE
E
$x_e = x_L$ 0 x_R
The particle reaches a turning point. The energy is entirely potential.

distance *above* the PE curve—has increased. Notice that the total energy—the brown dot—hasn't changed. The mass continues to speed up until it reaches maximum speed at $x_c = 0$, where the PE curve is at a minimum and the distance above the PE curve is maximum. At position x_d, the mass has started to slow down as it begins to transform kinetic energy back into elastic potential energy.

The mass continues moving to the left until, in frame e, it reaches position x_L, where the total energy line crosses the potential-energy curve. This point, where $K = 0$ and the energy is entirely potential, is a *turning point* where the mass reverses direction. A mass would need negative kinetic energy to be to the left of x_L, and that's not physically possible. You should be able to see, from the energy diagram, that the mass will *oscillate* back and forth between positions x_L and x_R, having maximum kinetic energy (and thus maximum speed) each time it passes through $x = 0$.

Now, let's consider a different initial condition. Suppose you pull the mass out to a greater initial distance. You've increased the potential energy in the system, and thus the total energy. The tan E line is now at a greater height, and it will intersect the PE graph at two points that are farther from the equilibrium point. With this new initial condition, the mass will oscillate back and forth between two points at a greater distance from equilibrium.

Interpreting Energy Diagrams

The lessons we learn from Figure 10.26 are true for any energy diagram:

- At any position, the distance from the axis to the PE curve is the object's potential energy. The distance from the PE curve to the E line is its kinetic energy.
- The object cannot be at a position where the PE curve is above the E line.
- A position where the E line crosses the PE curve is a turning point where the object reverses direction.
- If the E line crosses the PE curve at two positions, the object will oscillate between those two positions. Its speed will be maximum at the position where the PE curve is a minimum.

CONCEPTUAL EXAMPLE 10.13 Interpreting an energy diagram

FIGURE 10.27 is a more general energy diagram. We don't know how this potential energy was created, but we can still use the energy diagram to understand how a particle with this potential energy will move. Suppose a particle begins at rest at the position shown in the figure and is then released. Describe its subsequent motion.

REASON We've added details to the graph and sketched out details of the motion in FIGURE 10.28. The particle is at rest at the starting point, so $K = 0$ and the total energy is equal to the potential energy. We can draw the E line through this point. The PE curve tells us the particle's potential energy at each position.

Continued

FIGURE 10.27 A more general energy diagram.

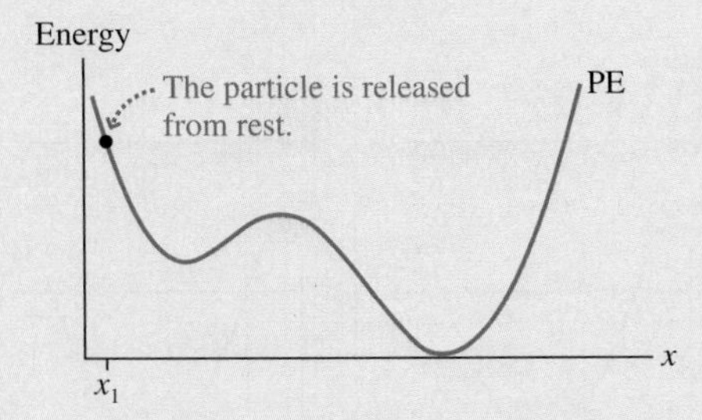

FIGURE 10.28 The motion of the particle in the potential of Figure 10.27.

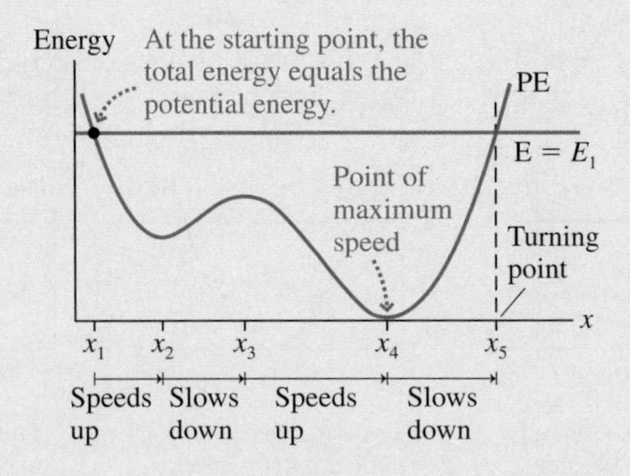

The distance between the PE curve and the E line is the particle's kinetic energy. The particle cannot move to the left because that would require the PE curve to go above the E line, so it begins moving to the right. The particle speeds up from x_1 to x_2 because U decreases and thus K must increase. It then slows down (but doesn't stop) from x_2 to x_3 as it goes over the "potential-energy hill." It speeds up after x_3 until it reaches maximum speed at x_4, where the PE curve is a minimum. The particle then steadily slows from x_4 to x_5 as kinetic energy is transformed into an increasing potential energy. Position x_5 is a turning point, a position where the E line crosses the PE curve. The particle is instantaneously at rest and then reverses direction. Because the E line crosses the PE curve at both x_1 and x_5 the particle will oscillate back and forth between these two points, speeding up and slowing down as described.

ASSESS Our results make sense. The particle is moving fastest where the PE line is lowest, as it must, and it turns around where the E and PE lines cross, meaning $K = 0$ and the particle is at rest.

Equilibrium Positions

Positions x_2, x_3, and x_4 in Figure 10.28, where the potential energy has a local minimum or maximum, are special positions. Consider the particle at position x_3 with energy E_3 in FIGURE 10.29a. Its energy is entirely potential energy and its kinetic energy is zero. It must be *at rest*—in equilibrium.

But suppose this particle is slightly disturbed—a tiny push to the right or left—giving it a very small amount of kinetic energy. The particle will begin to move away from x_3, moving away faster and faster because the potential energy decreases—and thus kinetic energy increases—on both sides of x_3. The situation is analogous to trying to balance a marble on the top of a hill; we can do so if the positioning is absolutely perfect, but any small displacement or disturbance will cause the marble to roll down the hill. An equilibrium position for which any small disturbance drives the particle away from equilibrium is called a point of **unstable equilibrium. Any local maximum in the PE curve is a point of unstable equilibrium.**

In contrast, consider a particle at position x_2 with energy E_2 in FIGURE 10.29b. Its kinetic energy is zero and, as we just discussed, the particle must be at rest. Position x_2 is also an equilibrium position, this time for a particle with energy E_2. What happens if this particle is slightly disturbed, raising the E line by a very small amount? Now the E line will intersect the PE curve just slightly to either side of x_2. These

FIGURE 10.29 Positions of (a) unstable and (b) stable equilibrium.

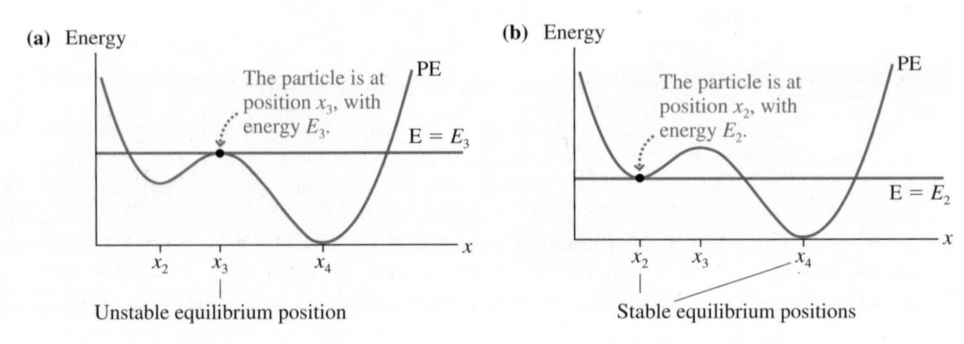

intersections are turning points, so the particle will undergo a very small oscillation centered on x_2, rather like a marble in the bottom of a bowl. An equilibrium for which a small disturbance causes only a small oscillation around the equilibrium position is called a point of **stable equilibrium.** You should recognize that **any local minimum in the PE curve is a point of stable equilibrium.** Position x_4 is also a point of stable equilibrium—in this case for a particle with $E_4 = 0$.

In the next section, we'll see how ideas about stable and unstable equilibrium help us understand molecular bonds and chemical reactions.

STOP TO THINK 10.9 The figures below show blue PE curves and tan E lines for four identical particles. Which particle has the highest maximum speed?

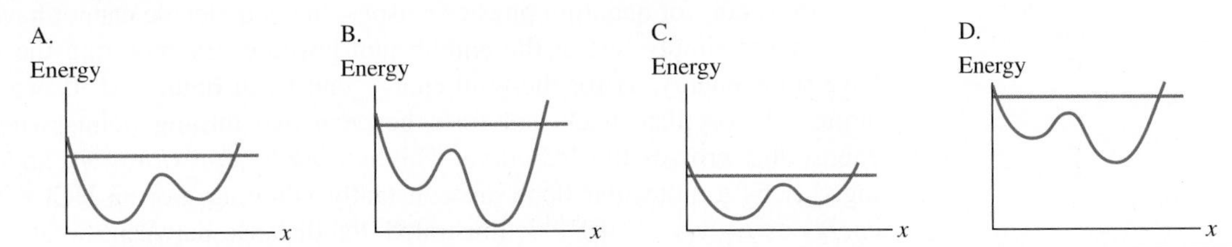

10.8 Molecular Bonds and Chemical Energy

With few exceptions, the materials of everyday life are made of atoms bound together into larger molecules. The *molecular bond* that holds two atoms together is an electric interaction between the atoms' negative electrons and positive nuclei. The electric force, like the gravitational force, is a force that can store energy. Fortunately, we don't need to know any details about electric potential energy—a topic we'll take up in Chapter 21—to deduce the energy diagram of a molecular bond.

We've noted that molecular bonds are somewhat analogous to springs: The normal force when an object rests on a table arises from the compression of spring-like bonds, and thermal energy is due, in part, to spring-like vibrations of atoms around an equilibrium position. This suggests that the energy diagram of two atoms connected by a molecular bond should look similar to the Figure 10.25 energy diagram of a mass on a spring.

FIGURE 10.30 shows the experimentally determined energy diagram of the diatomic molecule HCl (hydrogen chloride). Distance x is the *atomic separation*, the distance between the hydrogen and chlorine atoms. Note the very small distances: 1 nm $= 10^{-9}$ m. The left side of the PE curve looks very much like the PE curve of a spring; the right side starts out similar to the PE curve of a spring and then levels off. We can interpret and understand this potential-energy diagram by using what we learned in Section 10.7.

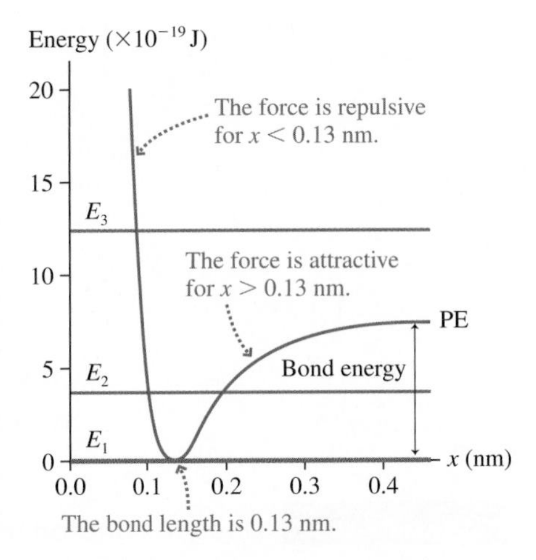

FIGURE 10.30 The energy diagram of the diatomic molecule HCl.

There is a clear minimum of the potential energy curve, a position of stable equilibrium. We've set the potential energy equal to zero at this position. If the total energy is zero as well, as is the case for the total energy line E_1, the atoms will rest at this separation. They will have no kinetic energy—no molecular vibration—and will form a molecule with an atomic separation of 0.13 nm. This is the **bond length** of HCl.

If we try to push the atoms closer together, the potential energy rises very rapidly. Physically, this is an electric repulsion between the negative electrons orbiting each atom, but it's analogous to the increasingly strong repulsive force we get when we compress a spring. Thus the PE curve to the left of the equilibrium position looks very much like the PE curve of a spring.

There are also attractive forces between two atoms. These can be the attractive force between two oppositely charged ions, as is the case for HCl; the attractive forces of covalent bonds when electrons are shared; or even weak *polarization forces* that are related to the static electricity force by which a comb that has been brushed through your hair attracts small pieces of paper. For any of these, the attractive force resists if we try to pull the atoms apart—analogous to stretching a spring—and thus potential energy increases to the right. The equilibrium position, with minimum potential energy, is the separation at which the repulsive force between electrons and the attractive force are exactly balanced.

The repulsive force gets stronger as we push the atoms closer together, but the attractive force gets *weaker* as we pull them farther apart. If we pull too hard, the bond breaks and the atoms come apart. Consequently, the PE curve becomes *less steep* as x increases, eventually leveling off when the atoms are so far apart that they cease interacting with each other. This difference between the attractive and repulsive forces explains the asymmetric PE curve in Figure 10.30.

It turns out, for quantum physics reasons, that a molecule cannot have $E = 0$ and thus cannot simply rest at the equilibrium position. By requiring the molecule to have some energy, as for the total energy line E_2 in Figure 10.30, we see that the atoms will oscillate back and forth between two turning points where the total energy line crosses the PE curve. This is a *molecular vibration*, and atoms held together by a molecular bond are constantly vibrating. For an HCl molecule with energy as $E_2 = 3.5 \times 10^{-19}$ J, illustrated, the distance between the atoms oscillates between roughly 0.10 nm and 0.18 nm.

As we've seen, an object's thermal energy is the sum of the energies of all the moving and vibrating atoms and molecules. Increasing a system's thermal energy increases the energy of each molecule. If we imagine the line E_2 in Figure 10.30 being raised, we can see that increased thermal energy, and thus increased temperature, corresponds to molecules vibrating more vigorously, with larger amplitude and more kinetic energy.

Suppose the molecule's energy is increased to $E_3 = 12.5 \times 10^{-19}$ J. This could happen, for example, if the molecule absorbs some light. We can see from the energy diagram that the molecules will keep moving apart. By raising the molecule's energy to E_3, we've broken the molecular bond. The breaking of molecular bonds by the absorption of light is called **photodissociation.** Light-mediated reactions, from sun tanning to photosynthesis to vision, are very similar to photodissociation but involve conformational changes in macromolecules—which require energy—rather than the actual breaking of bonds.

The **bond energy** is the minimum energy required to break a bond when the molecule's energy corresponds to a "room temperature" of 25°C. Bond energy is shown on an energy diagram as the vertical distance from the total energy line to the potential-energy "plateau" on the right side of the diagram. HCl molecules at room temperature have energy $E \simeq 0.04 \times 10^{-19}$ J, barely distinguishable from zero energy. We can see in Figure 10.30 that the bond energy of HCl is approximately 7.5×10^{-19} J.

NOTE ▸ Chemists and biologists usually quote molecular energies in kJ/mol or kcal/mol. Physicists prefer to work directly with the energy per molecule or energy per bond in J. To find energies in kJ/mol, simply multiply the bond energy by Avogadro's number. For example, the 7.5×10^{-19} J bond energy of HCl becomes 450 kJ/mol. ◂

EXAMPLE 10.14 Does the photon have enough energy?

An energy diagram for molecular oxygen, O_2, is shown in FIGURE 10.31. A germicidal lamp for sterilizing equipment uses short-wavelength ultraviolet radiation at 185 nm. At this wavelength, each photon, or quantum, of ultraviolet light has 10.7×10^{-19} J of energy. If a molecule of O_2 at room temperature absorbs one photon of light from the lamp, does this provide enough energy to split the molecule? If so, what will be the kinetic energy of the atoms after they have separated?

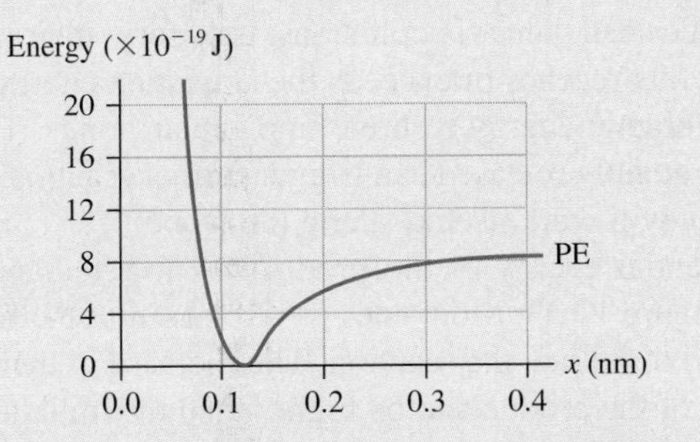

FIGURE 10.31 The energy diagram for molecular oxygen, O_2.

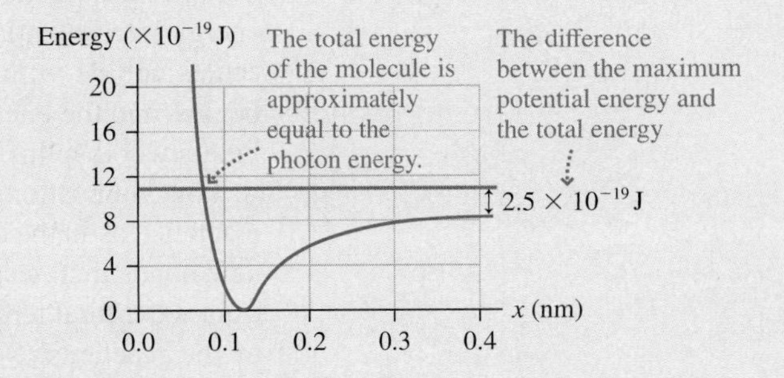

FIGURE 10.32 Comparing the photon energy to the potential energy for O_2.

STRATEGIZE We can use the molecular energy diagram in Figure 10.31 to determine what happens at different energies.

PREPARE We'll assume that the room-temperature energy is similar to that for HCl, very close to zero. After the molecule absorbs the photon, the energy of the molecule will be nearly equal to the photon energy.

SOLVE FIGURE 10.32 is the O_2 energy diagram with a total energy line added that corresponds to the photon energy. The total energy is much greater than the maximum of the potential energy on the right side of the graph. After absorbing the photon, the two oxygen molecules will separate. The residual kinetic energy will be equal to the difference between the total energy and the maximum of the potential energy, which we estimate to be 2.5×10^{-19} J.

ASSESS Germicidal lamps at this wavelength are known to produce O_3, ozone, a very reactive form of oxygen—so it's clear that the photons have enough energy to break apart the normally stable O_2 molecules. Therefore, our answer makes sense.

Chemical Reactions

Energy ideas are key to understanding what happens during a chemical reaction. The basic idea of any chemical reaction—involving simple diatomic molecules or large biological macromolecules—is that some molecular bonds are broken and new molecular bonds are formed. For example, a simple reaction that we can symbolize as $AB + CD \rightarrow AC + BD$ requires the bonds of molecules AB and CD to be broken and then new bonds to form between atoms A and C and between atoms B and D. And, as we've just seen, it takes energy to break molecular bonds.

FIGURE 10.33 is the energy diagram of a chemical reaction. It's very much like the energy diagrams we've been using, but with one important difference: The position coordinate of the horizontal axis has been replaced with an abstract **reaction coordinate.** The reaction coordinate is not a physical quantity that could be measured; instead, it shows in a general sense the progress of bond breaking and bond formation as a reaction moves from reactants, on the left, to products, on the right.

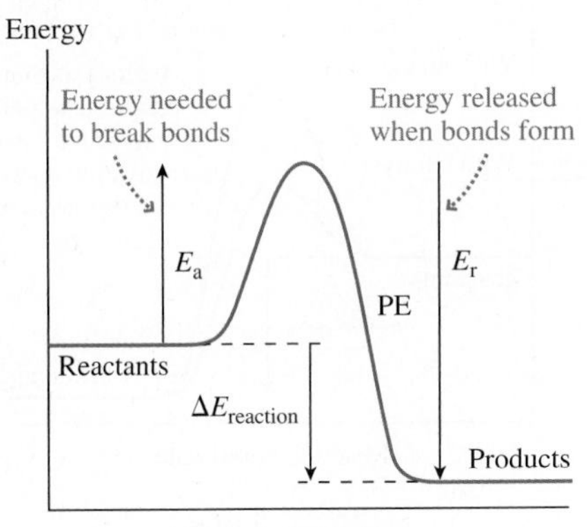

FIGURE 10.33 A reaction energy diagram.

All reaction energy diagrams have a large hump, or *energy barrier*, in the middle. This represents the energy required to break the bonds of the reactant molecules. For the reaction to take place, the reactants must increase their potential energy by the amount E_a, called the **activation energy.** Graphically, the activation energy is the height of the energy barrier above the initial potential energy of the reactants.

How does this happen? The reactant molecules have thermal energy, which means that the individual molecules are moving around and vibrating. When molecules collide, this energy can be transformed into the increased potential energy of stretched bonds. If the thermal energy is too low, the increased potential energy is less than the activation energy, meaning that the bonds don't break and the reaction doesn't occur. Wood and oxygen don't react at room temperature, even though the reaction—combustion—is energetically favorable, because the reactants don't have enough thermal energy to allow bond breaking during collisions. In essence, the total energy line is lower than the energy barrier, so there's a turning point in the reaction coordinate.

To burn wood, you must substantially increase the thermal energy of at least a portion of the fuel. You can do this with the high-temperature flame from a match. Some of the reactant molecules can then, via collisions, transform their large kinetic energy into potential energy that reaches or exceeds the activation energy—the molecules collide with enough kinetic energy to break molecular bonds. The reaction begins, and the energy subsequently released can trigger further reactions. Once you light a wood splint, the flame will work its way along the wood.

For combustion, the potential energy of the products is lower than that of the reactants, as is the case in Figure 10.33. More energy is released in the formation of new bonds than was required to initiate the reaction. The increase in thermal energy causes the final temperature of the products to be higher than the initial temperature of the reactants. Such reactions are called **exothermic reactions.** In contrast, an **endothermic reaction** releases less energy than was required to initiate it. Such reactions require a continuous input of energy to keep going.

NOTE ► Chemists and biologists often describe reactions in terms of what is called *free energy*. Free energy is a more appropriate description when some or all of the energy released in a reaction is used to do work rather than increasing the thermal energy. This is often the case in biology, where the energy released by a reaction does work by moving molecules around or changing the configuration of macromolecules instead of simply heating up the cell. We'll leave the definition and use of free energy to your chemistry and biology classes, simply noting that the analysis of a reaction in terms of free energy is exactly the same as the analysis presented here. ◄

Reaction Rates and Catalysts

The reaction energy diagram tells us nothing about the *rate of reaction*—how fast a reaction proceeds. A more detailed theory, which you will study in chemistry, finds that the rate of reaction increases *exponentially* as the activation energy decreases. An exponential change means that even a small decrease in activation energy can produce a large increase in the rate of reaction.

The role of a catalyst is to provide an alternate reaction pathway with a lower activation energy, thus dramatically speeding up the rate of reaction. FIGURE 10.34 shows an exothermic reaction, one that is energetically favorable but where the energy barrier is so high that this reaction will not happen at room temperature because the reaction rate is essentially zero. In Figure 10.34, we see that a catalyst offers an alternate pathway whose activation energy is easily exceeded by room-temperature molecules. A catalyst can dramatically increase reaction rates.

FIGURE 10.34 A reaction energy diagram for a chemical reaction with and without a catalyst.

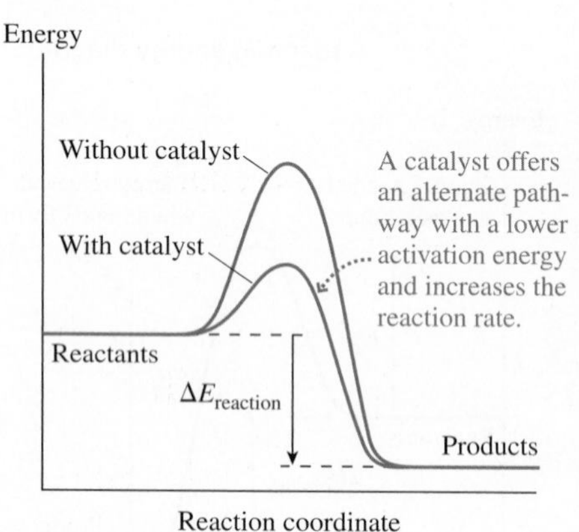
Energy
Without catalyst
With catalyst
A catalyst offers an alternate pathway with a lower activation energy and increases the reaction rate.
Reactants
$\Delta E_{\text{reaction}}$
Products
Reaction coordinate

Most of biochemistry is mediated by catalysts in the form of *enzymes*. Processes such as respiration, photosynthesis, and protein synthesis involve energetically favorable exothermic (also called *exergonic*) reactions, but the activation energy is so high that the reactants, on their own, would react barely, if at all, at normal temperatures. Enzymes catalyze these reactions, allowing them to proceed at a rate sufficient for cellular functions.

Chemical Energy

The law of conservation of energy includes the term ΔE_{chem}, the change in chemical energy. Physics usually focuses on systems in which chemical energy is not important, but energy conservation also has to apply to chemistry and biology. Chemical energy is simply a name for the total electric potential energy stored in all the molecular bonds of a system. If there are no reactions, the chemical energy doesn't change and we can ignore it, because only energy *changes* enter into the law of energy conservation.

If there are chemical reactions, then the breaking and creation of molecular bonds change the system's chemical energy. In Figure 10.33 the energy of the products is lower than the energy of the reactants. To make the reaction go, potential energy must be increased by the activation energy E_a—the energy of breaking bonds. Once the reaction is over the energy barrier, the formation of new bonds releases energy E_r; that is, potential energy is transformed into thermal energy. The difference, $\Delta E_{reaction} = E_r - E_a$, is the net energy released in *one* reaction due to the change in bonds.

Any realistic system has a vast number of chemical reactions taking place. The change in chemical energy is simply the total energy released by all these reactions. If N reactions take place, then

$$\Delta E_{chem} = N \Delta E_{reaction} \tag{10.21}$$

In biological systems, the production of chemical energy via the catalyzed reactions of respiration powers the cellular machinery and maintains body temperature.This change in energy will be an important part of the story in Chapter 11.

10.9 Energy in Collisions

In Chapter 9 we studied collisions between two objects. We found that if no external forces are acting on the objects, the total *momentum* of the objects will be conserved. Now we wish to study what happens to *energy* in collisions.

Let's first re-examine a perfectly inelastic collision. We studied just such a collision in Example 9.8. Recall that in such a collision the two objects stick together and then move with a common final velocity. What happens to the energy?

EXAMPLE 10.15 How much energy is transformed in a collision between railroad cars?

FIGURE 10.35 shows two train cars that move toward each other, collide, and couple together. In Example 9.8, we used conservation of momentum to find the final velocity shown in Figure 10.35 from the given initial velocities. How much thermal energy is created in this collision?

FIGURE 10.35 Before-and-after visual overview of a collision between two train cars.

Before:

$(v_{1x})_i = 1.5$ m/s $\quad (v_{2x})_i = -1.1$ m/s

$m_1 = 2.0 \times 10^4$ kg $\quad m_2 = 4.0 \times 10^4$ kg

x

After:

$m_1 + m_2$

$(v_x)_f = -0.25$ m/s

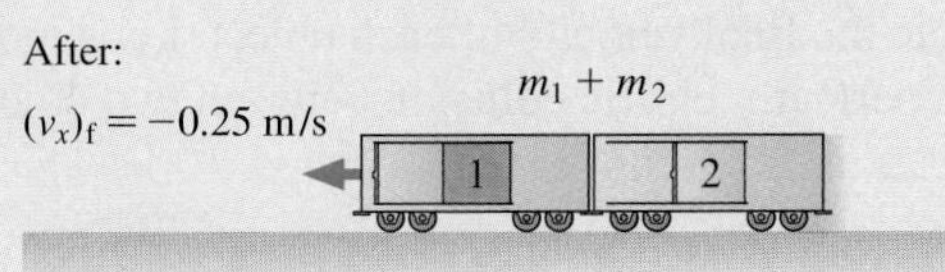

STRATEGIZE We'll choose our system to be the two cars. The initial state is the instant before the collision; the final state is the instant just after.

PREPARE This is an isolated system, so $W = 0$. Because the track is horizontal, there is no change in potential energy. Thus the work-energy equation reduces to

$$K_f + \Delta E_{th} = K_i$$

Energy is conserved, but *kinetic* energy is not; it will be lower after the collision than before.

SOLVE The initial kinetic energy is

$$K_i = \frac{1}{2}m_1(v_{1x})_i^2 + \frac{1}{2}m_2(v_{2x})_i^2$$
$$= \frac{1}{2}(2.0 \times 10^4\ \text{kg})(1.5\ \text{m/s})^2 + \frac{1}{2}(4.0 \times 10^4\ \text{kg})(-1.1\ \text{m/s})^2$$
$$= 4.7 \times 10^4\ \text{J}$$

Because the cars stick together and move as a single object with mass $m_1 + m_2$, the final kinetic energy is

$$K_f = \frac{1}{2}(m_1 + m_2)(v_x)_f^2$$
$$= \frac{1}{2}(6.0 \times 10^4\ \text{kg})(-0.25\ \text{m/s})^2 = 1900\ \text{J}$$

From the conservation of energy equation above, we find that the thermal energy increases by

$$\Delta E_{th} = K_i - K_f = 4.7 \times 10^4\ \text{J} - 1900\ \text{J} = 4.5 \times 10^4\ \text{J}$$

This amount of the initial kinetic energy is transformed into thermal energy during the impact of the collision.

ASSESS The cars are moving much more slowly after the collision than before, so we expect that most of the kinetic energy is transformed into thermal energy, just as we observed.

Elastic Collisions

FIGURE 10.36 A tennis ball collides with a racket. Notice that the ball is compressed and the strings are stretched.

FIGURE 10.36 shows a collision of a tennis ball with a racket. The ball is compressed and the racket strings stretch as the two collide, then the ball expands and the strings rebound as the two are pushed apart. In the language of energy, the kinetic energy of the objects is transformed into the elastic potential energy of the ball and strings, then back into kinetic energy as the two objects spring apart. If *all* of the kinetic energy is stored as elastic potential energy, and *all* of the elastic potential energy is transformed back into the post-collision kinetic energy of the objects, then mechanical energy is conserved. A collision in which mechanical energy is conserved is called a **perfectly elastic collision.**

Needless to say, most real collisions fall somewhere between perfectly elastic and perfectly inelastic. A rubber ball bouncing on the floor might "lose" 20% of its kinetic energy on each bounce and return to only 80% of the height of the preceding bounce. But collisions between two very hard objects, such as two pool balls or two steel balls, come close to being perfectly elastic. And collisions between microscopic particles, such as atoms or electrons, can be perfectly elastic.

FIGURE 10.37 A perfectly elastic collision.

Before: $\vec{v}_{1i}$ K_i

During:

Energy is stored in compressed molecular bonds, then released as the bonds re-expand.

After:

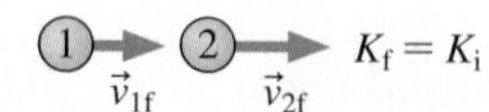

$\vec{v}_{1f}$ $\vec{v}_{2f}$ $K_f = K_i$

FIGURE 10.37 shows a head-on, perfectly elastic collision of a ball of mass m_1, having initial velocity $(v_{1x})_i$, with a ball of mass m_2 that is initially at rest. The balls' velocities after the collision are $(v_{1x})_f$ and $(v_{2x})_f$. These are velocities, not speeds, and have signs. Ball 1, in particular, might bounce backward and have a negative value for $(v_{1x})_f$.

The collision must obey two conservation laws: conservation of momentum (obeyed in any collision) and conservation of mechanical energy (because the collision is perfectly elastic). Although the energy is transformed into potential energy during the collision, the mechanical energy before and after the collision is purely kinetic energy. Thus,

$$\text{momentum conservation:} \quad m_1(v_{1x})_i = m_1(v_{1x})_f + m_2(v_{2x})_f$$

$$\text{energy conservation:} \quad \frac{1}{2}m_1(v_{1x})_i^2 = \frac{1}{2}m_1(v_{1x})_f^2 + \frac{1}{2}m_2(v_{2x})_f^2$$

Momentum conservation alone is not sufficient to analyze the collision because there are two unknowns: the two final velocities. That is why we did not consider perfectly elastic collisions in Chapter 9. Energy conservation gives us another condition. The complete solution of these two equations involves straightforward but rather lengthy algebra. We'll just give the solution here:

$$(v_{1x})_f = \frac{m_1 - m_2}{m_1 + m_2}(v_{1x})_i \qquad (v_{2x})_f = \frac{2m_1}{m_1 + m_2}(v_{1x})_i \tag{10.22}$$

Perfectly elastic collision with object 2 initially at rest

Equations 10.22 allow us to compute the final velocity of each object. Let's look at a common and important example: a perfectly elastic collision between two objects of equal mass.

EXAMPLE 10.16 Finding the aftermath of a collision between air hockey pucks

On an air hockey table, a moving puck, traveling to the right at 2.3 m/s, makes a head-on collision with an identical puck at rest. What is the final velocity of each puck?

PREPARE The before-and-after visual overview is shown in FIGURE 10.38. We've shown the final velocities in the picture, but we don't really know yet which way the pucks will move. Because one puck was initially at rest, we can use Equations 10.22 to find the final velocities of the pucks. The pucks are identical, so we have $m_1 = m_2 = m$.

SOLVE We use Equations 10.22 with $m_1 = m_2 = m$ to get

$$(v_{1x})_f = \frac{m - m}{m + m}(v_{1x})_i = 0 \text{ m/s}$$

$$(v_{2x})_f = \frac{2m}{m + m}(v_{1x})_i = (v_{1x})_i = 2.3 \text{ m/s}$$

FIGURE 10.38 A moving puck collides with a stationary puck.

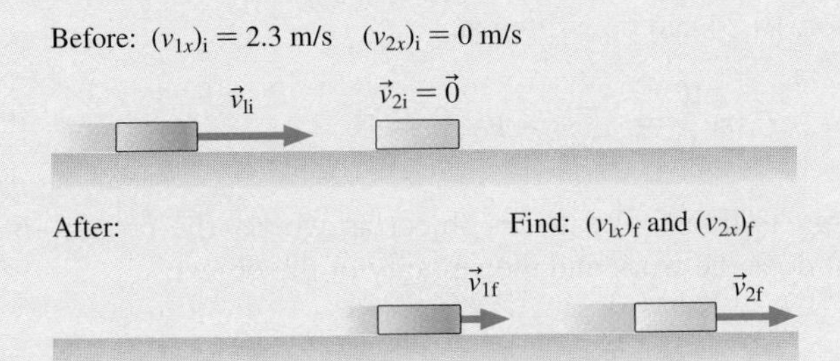

The incoming puck stops dead, and the initially stationary puck goes off with the same velocity that the incoming one had.

ASSESS You can see that momentum and energy are conserved: The incoming puck's momentum and energy are completely transferred to the outgoing puck. If you've ever played pool, you've probably seen this sort of collision when you hit a ball head-on with the cue ball. The cue ball stops and the other ball picks up the cue ball's velocity.

STOP TO THINK 10.10 A small ball with mass M is at rest. It is then struck by a ball with twice the mass, moving at speed v_0. The situation after the collision is shown in the figure. Is this possible?

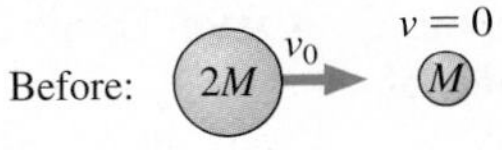

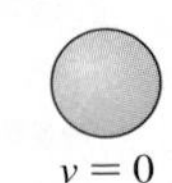

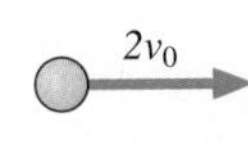

A. Yes
B. No, because momentum is not conserved
C. No, because energy is not conserved
D. No, because neither momentum nor energy is conserved

10.10 Power

We've now studied how energy can be transformed from one kind into another and how it can be transferred between the environment and the system as work. In many situations we would like to know *how quickly* the energy is transformed or transferred. Is a transfer of energy very rapid, or does it take place over a long time? In passing a truck, your car needs to transform a certain amount of the chemical energy in its fuel into kinetic energy. It makes a *big* difference whether your engine can do this in 20 s or 60 s!

The question How quickly? implies that we are talking about a *rate*. For example, the velocity of an object—how fast it is going—is the *rate of change* of position. So, when we raise the issue of how fast the energy is transformed, we are talking about the *rate of transformation* of energy. Suppose in a time interval Δt an amount of energy ΔE is transformed from one form into another. The rate at which this energy is transformed is called the **power** P and is defined as

$$P = \frac{\Delta E}{\Delta t} \tag{10.23}$$

Power when an amount of energy ΔE is transformed in a time interval Δt

Both these cars take about the same energy to reach 60 mph, but the race car gets there in a much shorter time, so its *power* is much greater.

The unit of power is the **watt,** which is defined as 1 watt = 1 W = 1 J/s.

Power also measures the rate at which energy is transferred into or out of a system as work W. If work W is done in time interval Δt, the rate of energy *transfer* is

$$P = \frac{W}{\Delta t} \tag{10.24}$$

Power when an amount of work W is done in a time interval Δt

If a person, animal, vehicle, or device is transforming or transferring energy at a rate of 3 J/s, we say that it has an **output power** of 3 W.

The English unit of power is the *horsepower*. The conversion factor to watts is

1 horsepower = 1 hp = 746 W

Many common appliances, such as motors, are rated in hp.

We can express Equation 10.24 in a different form. If in the time interval Δt an object undergoes a displacement Δx, the work done by a force acting on the object is $W = F\Delta x$. Then Equation 10.24 can be written as

$$P = \frac{W}{\Delta t} = \frac{F\Delta x}{\Delta t} = F\frac{\Delta x}{\Delta t} = Fv$$

The rate at which energy is transferred to an object as work—the power—is the product of the force that does the work and the velocity of the object:

$$P = Fv \tag{10.25}$$

Rate of energy transfer due to a force F acting on an object moving at velocity v

EXAMPLE 10.17 Finding the output power for a weightlifter BIO

A 100 kg weightlifter performs a lift called a clean and jerk, raising a 190 kg bar from the ground to a height of 1.9 m in a time of 1.8 s. What is his output power?

STRATEGIZE We'll take the system to be the weightlifter + bar + earth. We'll assume that the bar is stationary before and after the lift, so the relevant energy change is the increase in gravitational potential energy of the bar. This is an isolated system, so the change is an internal transformation. We'll use Equation 10.23 to compute the power.

PREPARE The change in potential energy depends on the change in height:

$$\Delta U_g = mg\Delta y = (190\text{ kg})(9.8\text{ m/s}^2)(1.9\text{ m}) = 3540\text{ J}$$

SOLVE The power of the transformation is

$$P = \frac{\Delta E}{\Delta t} = \frac{3540\text{ J}}{1.8\text{ s}} = 2000\text{ W}$$

ASSESS This is a lot of power—about 2.7 horsepower! But we'd expect such a large output power for a large weightlifter.

The 100 kg weightlifter has a very large output power, which isn't surprising. How about someone smaller?

EXAMPLE 10.18 Finding the output power for a sprinter BIO

A 50 kg sprinter accelerates from 0 to 11 m/s in 3.0 s. What is the output power for this rapid start?

STRATEGIZE We can take the system to be the runner + earth. Let's assume that the track is level, so there is no change in potential energy, only a change in kinetic energy. We can safely ignore drag and other forces, so this is isolated system and the change is an internal transformation.

PREPARE The initial kinetic energy is zero, so the change in kinetic energy is equal to the final kinetic energy:

$$\Delta K = K_f = \frac{1}{2}mv_f^2 = \frac{1}{2}(50\text{ kg})(11\text{ m/s})^2 = 3000\text{ J}$$

SOLVE The power of the transformation is

$$P = \frac{\Delta E}{\Delta t} = \frac{3000\text{ J}}{3.0\text{s}} = 1000\text{ W}$$

ASSESS This is a lot of power—about 1.3 horsepower—but less than for the weightlifter, which makes sense. In this case, as for the weightlifter, the power came from the athlete's muscles.

A 100 kg weightlifter can produce more output power than a 50 kg sprinter, which makes sense. It's worthwhile to consider what we'll call the **specific power**—the output power divided by the mass of the person (or animal, device, or machine) doing the transformation or the work:

$$\text{specific power} = \frac{\text{power of a transformation or a transfer}}{\text{mass of agent causing the transformation or transfer}}$$

Let's compute the specific power for the weightlifter and the sprinter:

$$\text{sprinter:} \qquad \text{specific power} = \frac{1000\ \text{W}}{50\ \text{kg}} = 20\ \text{W/kg}$$

$$\text{weightlifter:} \qquad \text{specific power} = \frac{2000\ \text{W}}{100\ \text{kg}} = 20\ \text{W/kg}$$

The numbers for the sprinter and the weightlifter are at the extreme end of what humans are capable of; these are numbers typical of world-class athletes. It's interesting to note that the specific power for both cases is about the same. Humans in peak condition who are skilled at athletic pursuits are capable of short bursts of about 20 W/kg. Larger athletes can produce more output power, but the power per kilogram is about the same. As we'll see in the next chapter, humans can't sustain this level of power output; this number applies for only short bursts that use the large muscles of the body. Sustained activities such as cycling or swimming correspond to specific powers of perhaps 5 W/kg for elite athletes.

Smaller animals are generally capable of higher specific powers. A bushbaby, a 200 g primate that gets around by executing rapid leaps in the trees it calls home, is able to push off with its legs with sufficient force to accelerate to 6.7 m/s in 0.16 s, corresponding to a specific power of 140 W/kg. This is the upper end of what can be accomplished with muscle power alone. The leap of the 2.0 g desert locust that we considered earlier has an even higher specific power, but this power comes from springs in the legs; muscles alone could not get the locust to such a high speed in such a short time. Many of the impressive jumpers of the insect world, from fleas to springtails, use energy storage systems corresponding to springs to power their leaps.

Of course, the notion of specific power can be applied to other systems as well. We can do a similar calculation for a passenger car, either starting from rest or climbing a hill, to find a specific power in the range of 90 W/kg. This is an interesting measure of a vehicle.

STOP TO THINK 10.11 Four students run up the stairs in the times shown. Rank in order, from largest to smallest, their power outputs P_A through P_D.

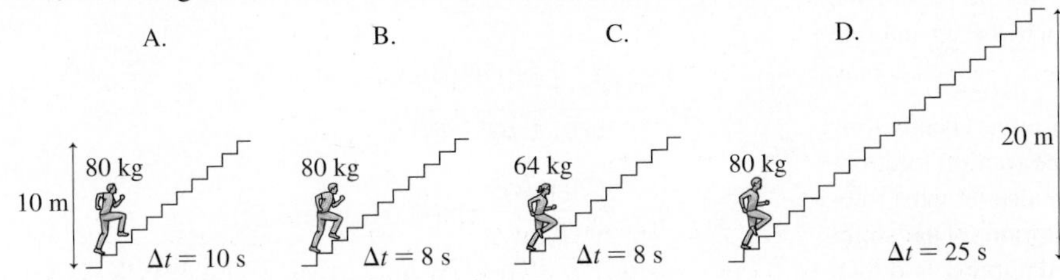

INTEGRATED EXAMPLE 10.19 Stopping a runaway truck

A truck's brakes can overheat and fail while descending mountain highways, leading to an extremely dangerous runaway truck. Some highways have *runaway-truck ramps* to safely bring out-of-control trucks to a stop. These uphill ramps are covered with a deep bed of gravel. The uphill slope and the large coefficient of rolling friction as the tires sink into the gravel bring the truck to a safe halt.

A 22,000 kg truck heading down a 3.5° slope at 20 m/s ($\approx$45 mph) suddenly has its brakes fail. Fortunately, there's a runaway-truck ramp 600 m ahead. The ramp slopes upward at an angle of 10°, and the coefficient of rolling friction between the truck's tires and the loose gravel is $\mu_r = 0.40$. Ignore air resistance and rolling friction as the truck rolls down the highway.

a. Use conservation of energy to find how far along the ramp the truck travels before stopping.
b. By how much does the thermal energy of the truck and ramp increase as the truck stops?

STRATEGIZE We'll follow the steps of Problem-Solving Approach 10.1. We start by defining the system as the truck + ramp + earth. Gravitational potential energy will be part of the solution, and the change in thermal energy will be an internal transformation.

PREPARE FIGURE 10.39 shows a before-and-after visual overview. Because we're going to need to determine friction forces to calculate the increase in thermal energy, we've also drawn a free-body diagram for the truck as it moves up the ramp. One slight complication is that the y-axis of free-body diagrams is drawn perpendicular to the slope, whereas the calculation of gravitational potential energy needs a vertical y-axis to measure height.

FIGURE 10.39 Visual overview of the runaway truck.

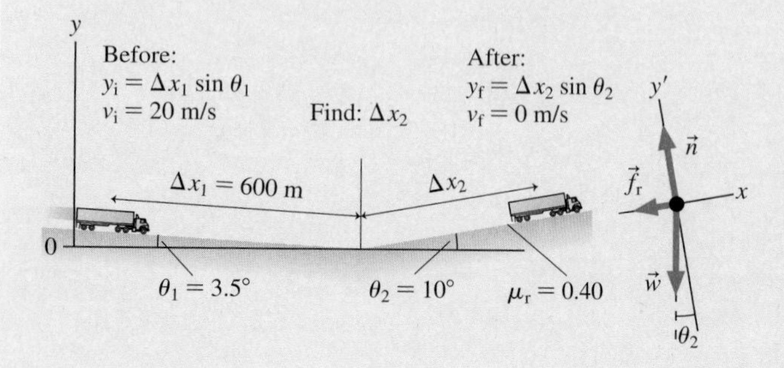

We've dealt with this by labeling the free-body diagram axis the y'-axis.

SOLVE a. The work-energy equation for the motion of the truck, from the moment its brakes fail to when it finally stops, is

$$K_f + (U_g)_f + \Delta E_{th} = K_i + (U_g)_i$$

Because friction is present only along the ramp, thermal energy will increase only as the truck moves up the ramp. This thermal energy is then given by $\Delta E_{th} = f_r \, \Delta x_2$, because Δx_2 is the length of the ramp. The conservation of energy equation then is

$$\frac{1}{2} m v_f^2 + mgy_f + f_r \, \Delta x_2 = \frac{1}{2} m v_i^2 + mgy_i$$

From Figure 10.39 we have $y_i = \Delta x_1 \sin \theta_1$, $y_f = \Delta x_2 \sin \theta_2$, and $v_f = 0$, so the equation becomes

$$mg \, \Delta x_2 \sin \theta_2 + f_r \, \Delta x_2 = \frac{1}{2} m v_i^2 + mg \, \Delta x_1 \sin \theta_1$$

To find $f_r = \mu_r n$ we need to find the normal force n. The free-body diagram shows that

$$\sum F_{y'} = n - mg \cos \theta_2 = a_{y'} = 0$$

from which $f_r = \mu_r n = \mu_r mg \cos \theta_2$. With this result for f_r, our conservation of energy equation is

$$mg \, \Delta x_2 \sin \theta_2 + \mu_r mg \cos \theta_2 \, \Delta x_2 = \frac{1}{2} m v_i^2 + mg \, \Delta x_1 \sin \theta_1$$

which, after we divide both sides by mg, simplifies to

$$\Delta x_2 \sin \theta_2 + \mu_r \cos \theta_2 \Delta x_2 = \frac{v_i^2}{2g} + \Delta x_1 \sin \theta_1$$

Solving this for Δx_2 gives

$$\Delta x_2 = \frac{\dfrac{v_i^2}{2g} + \Delta x_1 \sin \theta_1}{\sin \theta_2 + \mu_r \cos \theta_2}$$

$$= \frac{\dfrac{(20 \text{ m/s})^2}{2(9.8 \text{ m/s}^2)} + (600 \text{ m})(\sin 3.5°)}{\sin 10° + 0.40(\cos 10°)} = 100 \text{ m}$$

b. We know that $\Delta E_{th} = f_r \, \Delta x_2 = (\mu_r mg \cos \theta_2) \Delta x_2$, so that

$$\Delta E_{th} = (0.40)(22{,}000 \text{ kg})(9.8 \text{ m/s}^2)(\cos 10°)(100 \text{ m})$$

$$= 8.5 \times 10^6 \text{ J}$$

ASSESS It seems reasonable that a truck that speeds up as it rolls 600 m downhill takes only 100 m to stop on a steeper, high-friction ramp. At the top of the hill the truck's kinetic energy is $K_i = \frac{1}{2} m v_i^2 = \frac{1}{2}(22{,}000 \text{ kg})(20 \text{ m/s})^2 = 4.4 \times 10^6$ J, which is of the same order of magnitude as ΔE_{th}. Our answer is reasonable.

SUMMARY

GOAL To introduce the concept of energy and to learn a new problem-solving strategy based on conservation of energy.

GENERAL PRINCIPLES

Basic Energy Model

Within a system, energy can be **transformed** between various forms.

Energy can be **transferred** into or out of a system in two basic ways:

- **Work:** The transfer of energy by mechanical forces
- **Heat:** The nonmechanical transfer of energy from a hotter to a colder object

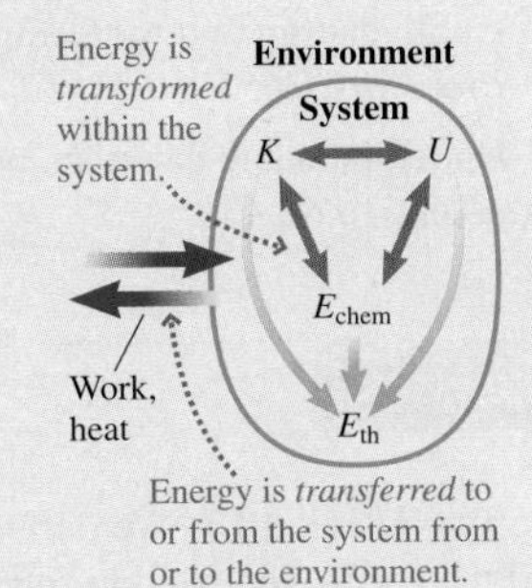

Conservation of Energy

When work W is done on a system, the system's total energy changes by the amount of work done. In mathematical form, this is the **work-energy equation:**

$$\Delta E = \Delta K + \Delta U_g + \Delta U_s + \Delta E_{th} + \Delta E_{chem} + \cdots = W$$

A system is isolated when no energy is transferred into or out of the system. This means the work is zero, giving the **law of conservation of energy:**

$$\Delta K + \Delta U_g + \Delta U_s + \Delta E_{th} + \Delta E_{chem} + \cdots = 0$$

Solving Energy Transfer and Energy Conservation Problems

STRATEGIZE Choose the system. Determine the initial and final states.

PREPARE Draw a before-and-after visual overview.

SOLVE Use the before-and-after version of the work-energy equation:

$$K_f + U_f + \Delta E_{th} = K_i + U_i + W$$

Start with this general equation, then specialize to the case at hand:

- Use the appropriate form or forms of potential energy.
- If the system is isolated, set $W = 0$.
- If there is no friction or drag, set $\Delta E_{th} = 0$.

ASSESS See if the numbers make sense—and if the numbers add up. Energy is conserved, and kinetic energy and the change in thermal energy are always positive.

IMPORTANT CONCEPTS

Kinetic energy is an energy of motion:

$$K = \underbrace{\frac{1}{2}mv^2}_{\text{Translational}} + \underbrace{\frac{1}{2}I\omega^2}_{\text{Rotational}}$$

Potential energy is energy stored in a system of interacting objects.

- **Gravitational potential energy:** $U_g = mgy$
- **Elastic potential energy:** $U_s = \frac{1}{2}kx^2$

Thermal energy is the sum of the microscopic kinetic and potential energies of all the molecules in an object. The hotter an object, the more thermal energy it has. When kinetic (sliding) friction is present, the increase in the thermal energy is $\Delta E_{th} = f_k \Delta x$. When the drag force is present, the increase in the thermal energy is $\Delta E_{th} = D\Delta x$.

Work is the process by which energy is transferred to or from a system by the application of mechanical forces.

If a particle moves through a displacement $\vec{d}$ while acted upon by a constant force $\vec{F}$, the force does work

$$W = F_{\parallel}d = Fd\cos\theta$$

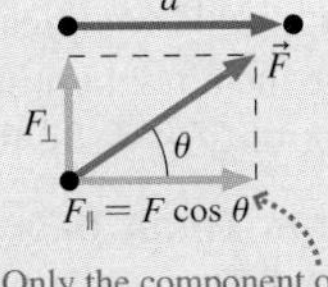

Only the component of the force parallel to the displacement does work.

APPLICATIONS

Energy diagrams are a useful way to analyze physical systems.

This curve shows the potential energy of a two-atom molecule as a function of the atomic separation.

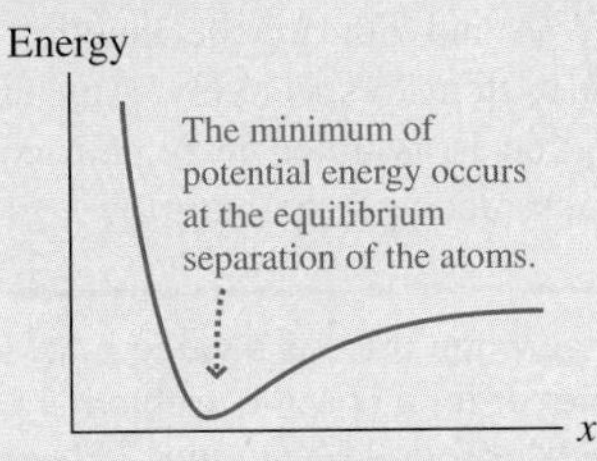

Energy
The minimum of potential energy occurs at the equilibrium separation of the atoms.
x

Perfectly elastic collisions Both mechanical energy and momentum are conserved.

Object 2 initially at rest

Before: (1) $(v_{1x})_i$ → (2) K_i

After: $K_f = K_i$ (1) $(v_{1x})_f$ (2) $(v_{2x})_f$

$$(v_{1x})_f = \frac{m_1 - m_2}{m_1 + m_2}(v_{1x})_i$$

$$(v_{2x})_f = \frac{2m_1}{m_1 + m_2}(v_{1x})_i$$

Power is the rate at which energy is transformed . . .

$$P = \frac{\Delta E}{\Delta t}$$

ΔE: Amount of energy transformed; Δt: Time required to transform it

. . . or at which work is done.

$$P = \frac{W}{\Delta t}$$

W: Amount of work done; Δt: Time required to do work

Learning Objectives After studying this chapter, you should be able to:

- Calculate the work done on an object. *Problems 10.1, 10.2, 10.3, 10.4, 10.5, 10.6, 10.7*
- Calculate an object's kinetic and potential energy. *Conceptual Questions 10.17, 10.18; Problems 10.14, 10.17, 10.20, 10.23, 10.25*
- Understand and calculate the change in thermal energy. *Conceptual Question 10.12; Problems 10.29, 10.30, 10.31, 10.32, 10.34*
- Use the problem-solving approach to solve conservation of energy problems. *Conceptual Questions 10.20, 10.22; Problems 10.36, 10.37, 10.39, 10.41, 10.48*
- Draw and interpret energy diagrams. *Conceptual Question 10.24; Problems 10.51, 10.52*
- Interpret and use molecular bond energies. *Problems 10.53, 10.54*
- Apply energy and momentum conservation to elastic collisions. *Problems 10.55, 10.56, 10.58*
- Understand and calculate power. *Problems 10.60, 10.61, 10.62, 10.65, 10.68*

STOP TO THINK ANSWERS

Chapter Preview Stop to Think: A. Because the car starts from rest, $v_i = 0$ and the kinematic equation is $(v_x)_f^2 = 2a_x\Delta x$, so that $(v_x)_f = \sqrt{2a_x\Delta x}$. Thus the speed is proportional to the square root of the displacement. If, as in this question, the displacement increases by a factor of 4, the speed only doubles. So the speed will increase from 5 m/s to 10 m/s.

Stop to Think 10.1: C. The coaster slows. Its kinetic energy is decreasing because kinetic energy is transformed into gravitational potential energy as the coaster climbs the hill.

Stop to Think 10.2: C. $W = Fd\cos\theta$. The 10 N force at 90° does no work at all. $\cos 60° = \frac{1}{2}$, so the 8 N force does less work than the 6 N force.

Stop to Think 10.3: B > D > A = C. $K = \frac{1}{2}mv^2$. Using the given masses and velocities, we find $K_A = 2.0$ J, $K_B = 4.5$ J, $K_C = 2.0$ J, $K_D = 4.0$ J.

Stop to Think 10.4: $(U_g)_3 > (U_g)_2 = (U_g)_4 > (U_g)_1$. Gravitational potential energy depends only on height, not speed.

Stop to Think 10.5: D. The potential energy of a spring depends on the *square* of the displacement x, so the energy is positive whether the spring is compressed or extended. If the spring is compressed by twice the amount it had been stretched, the energy will increase by a factor of $2^2 = 4$. So the energy will be $4 \times 1\text{ J} = 4\text{ J}$.

Stop to Think 10.6: B. We can use conservation of energy to write $\Delta K + \Delta E_{th} = 0$. Now if the initial kinetic energy doubles, so does ΔK, so ΔE_{th} must double as well. But $\Delta E_{th} = f_k\,\Delta x$, so if ΔE_{th} doubles, then Δx doubles to 2.0 m.

Stop to Think 10.7: We define the system as the student, the rope, and the earth. The friction force and the weight force are internal to the system, so it is an isolated system.

Stop to Think 10.8: D. In all three cases, Katie has the same initial kinetic energy and potential energy. Thus her energy must be the same at the bottom of the slide in all three cases. Because she has only kinetic energy at the bottom, her speed there must be the same in all three cases as well.

Stop to Think 10.9: B. The kinetic energy is the difference between the total energy and the potential energy. This is highest at the bottom of the right well in B.

Stop to Think 10.10: C. The initial momentum is $(2M)v_0 + 0$, and the final momentum is $0 + M(2v_0)$. These are equal, so momentum is conserved. The initial kinetic energy is $\frac{1}{2}(2M)v_0^2 = Mv_0^2$, and the final kinetic energy is $\frac{1}{2}M(2v_0)^2 = 2Mv_0^2$. The final kinetic energy is *greater* than the initial kinetic energy, so this collision is not possible. (If the final kinetic energy had been less than the initial kinetic energy, the collision could be possible because the difference in energy could be converted into thermal energy.)

Stop to Think 10.11: $P_B > P_A = P_C > P_D$. The power here is the rate at which each runner's internal chemical energy is converted into gravitational potential energy. The change in gravitational potential energy is $mg\Delta y$, so the power is $mg\Delta y/\Delta t$. For runner A, the ratio $m\Delta y/\Delta t$ equals $(80\text{ kg})(10\text{ m})/(10\text{ s}) = 80\text{ kg}\cdot\text{m/s}$. For C, the ratio is also $80\text{ kg}\cdot\text{m/s}$. For B, it's $100\text{ kg}\cdot\text{m/s}$, while for D the ratio is $64\text{ kg}\cdot\text{m/s}$.

Video Tutor Solution Chapter 10

QUESTIONS

Conceptual Questions

1. The brake shoes of your car are made of a material that can tolerate very high temperatures without being damaged. Why is this so?

For Questions 2 through 9, give a specific example of a system with the energy transformation shown. In these questions, W is the work done on the system, and K, U, and Eth are the kinetic, potential, and thermal energies of the system, respectively. Any energy not mentioned in the transformation is assumed to remain constant; if work is not mentioned, it is assumed to be zero.

Problem difficulty is labeled as | (straightforward) to ||||| (challenging). INT Problems labeled integrate significant material from earlier chapters; BIO are of biological or medical interest.

The eText icon indicates when there is a video tutor solution available for the chapter or for a specific problem. To launch these videos, log into your eText through Mastering™ Physics or log into the Study Area.

2. $W \rightarrow K$
3. $W \rightarrow U$
4. $K \rightarrow U$
5. $K \rightarrow W$
6. $U \rightarrow K$
7. $W \rightarrow \Delta E_{th}$
8. $U \rightarrow \Delta E_{th}$
9. $K \rightarrow \Delta E_{th}$
10. A ball of putty is dropped from a height of 2 m onto a hard floor, where it sticks. What object or objects need to be included within the system if the system is to be isolated during this process?
11. A diver leaps from a high platform, speeds up as she falls, and then slows to a stop in the water. How do you define the system so that the energy changes are all transformations internal to an isolated system?
12. When your hands are cold, you can rub them together to warm them. Explain the energy transformations that make this possible.
13. Puck B has twice the mass of puck A. Starting from rest, both pucks are pulled the same distance across frictionless ice by strings with the same tension.
    a. Compare the final kinetic energies of pucks A and B.
    b. Compare the final speeds of pucks A and B.
14. To change a tire, you need to use a jack to raise one corner of your car. While doing so, you happen to notice that pushing the jack handle down 20 cm raises the car only 0.2 cm. Use energy concepts to explain why the handle must be moved so far to raise the car by such a small amount.
15. You drop two balls from a tower, one of mass m and the other of mass $2m$. Just before they hit the ground, which ball, if either, has the larger kinetic energy? Explain.
16. If you fall and skid to a stop on a carpeted floor, you can get a rug burn. Much of this discomfort comes from abrasion, but there can also be a real burn where the skin was too hot. How does this happen?
17. A roller coaster car rolls down a frictionless track, reaching speed v at the bottom.
    a. If you want the car to go twice as fast at the bottom, by what factor must you increase the height of the track?
    b. Does your answer to part a depend on whether the track is straight or not? Explain.
18. A spring gun shoots out a plastic ball at speed v. The spring is then compressed twice the distance it was on the first shot.
    a. By what factor is the spring's potential energy increased?
    b. By what factor is the ball's speed increased? Explain.
19. A baseball pitcher can throw a baseball (mass 0.14 kg) much faster than a football quarterback can throw a football (mass 0.42 kg). Use energy concepts to explain why you would expect this to be true.
20. Sandy and Chris stand on the edge of a cliff and throw identical mass rocks at the same speed. Sandy throws her rock horizontally while Chris throws his upward at an angle of 45° to the horizontal. Are the rocks moving at the same speed when they hit the ground, or is one moving faster than the other? If one is moving faster, which one? Explain.
21. A solid cylinder and a hollow cylinder have the same mass, same radius, and turn on frictionless, horizontal axles. (The hollow cylinder has lightweight spokes connecting it to the axle.) A rope is wrapped around each cylinder and tied to a block. The blocks have the same mass and are held the same height above the ground as shown in Figure Q10.21. Both blocks are released simultaneously. The ropes do not slip. Which block hits the ground first? Or is it a tie? Explain.

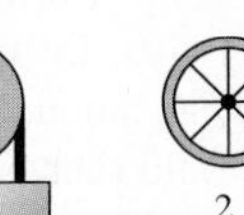
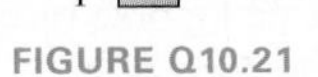

FIGURE Q10.21

22. A bowler tosses a ball without spin. The ball slides down the alley. At some point, friction with the alley makes the ball start to roll; eventually, it rolls without sliding. When the ball reaches this point, it is moving at a lower speed than the original toss. Use energy concepts to give two reasons for this change.
23. BIO Ferns that eject spores generally do so in pairs, with two spores flying off in opposite directions. The structure from which the spores are launched is quite lightweight. If it takes a certain amount of energy to eject each spore, explain how launching the spores in pairs provides for the greatest initial launch speed for each spore.
24. Figure Q10.24 shows a potential-energy diagram for a particle. The particle is at rest at point A and is then given a slight nudge to the right. Describe the subsequent motion.

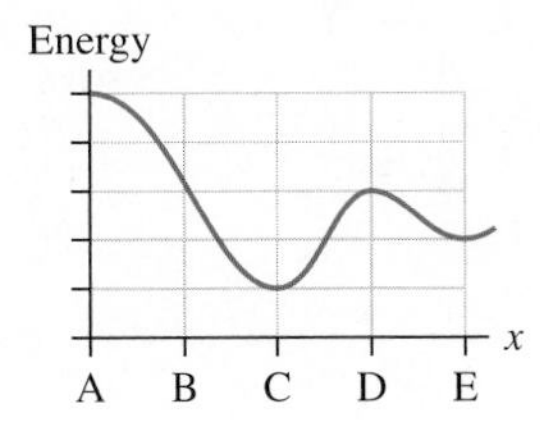

FIGURE Q10.24

Multiple-Choice Questions

25. || A roller coaster starts from rest at its highest point and then descends on its (frictionless) track. Its speed is 30 m/s when it reaches ground level. What was its speed when its height was half that of its starting point?
    A. 11 m/s B. 15 m/s C. 21 m/s D. 25 m/s
26. | A woman uses a pulley and a rope to raise a 20 kg weight to a height of 2 m. If it takes 4 s to do this, about how much power is she supplying?
    A. 100 W B. 200 W C. 300 W D. 400 W
27. | A hockey puck sliding along frictionless ice with speed v to the right collides with a horizontal spring and compresses it by 2.0 cm before coming to a momentary stop. What will be the spring's maximum compression if the same puck hits it at a speed of $2v$?
    A. 2.0 cm B. 2.8 cm C. 4.0 cm
    D. 5.6 cm E. 8.0 cm
28. || A block slides down a smooth ramp, starting from rest at a height h. When it reaches the bottom it's moving at speed v. It then continues to slide up a second smooth ramp. At what height is its speed equal to $v/2$?
    A. $h/4$ B. $h/2$ C. $3h/4$ D. $2h$
29. | A wrecking ball is suspended from a 5.0-m-long cable that makes a 30° angle with the vertical. The ball is released and swings down. What is the ball's speed at the lowest point?
    A. 7.7 m/s B. 4.4 m/s C. 3.6 m/s D. 3.1 m/s
30. || BIO A dog can provide sufficient power to pull a sled with a 60 N force at a steady 2.0 m/s. Suppose the dog is hitched to a different sled that requires 120 N to move at a constant speed. How fast can the dog pull this second sled?
    A. 0.50 m/s B. 1.0 m/s C. 1.5 m/s D. 2.0 m/s
31. ||| Most of the energy you expend in cycling is dissipated by the drag force. If you double your speed, you increase the drag force by a factor of 4. This increases the power to cycle at this greater speed by what factor?
    A. 2 B. 4 C. 8 D. 16

Watch Video Solution Problem 10.5

PROBLEMS

Section 10.2 Work

1. || A 2.0 kg book is lying on a 0.75-m-high table. You pick it up and place it on a bookshelf 2.3 m above the floor. During this process,
   a. How much work does gravity do on the book?
   b. How much work does your hand do on the book?
2. || The two ropes seen in Figure P10.2 are used to lower a 255 kg piano exactly 5 m from a second-story window to the ground. How much work is done by each of the three forces?

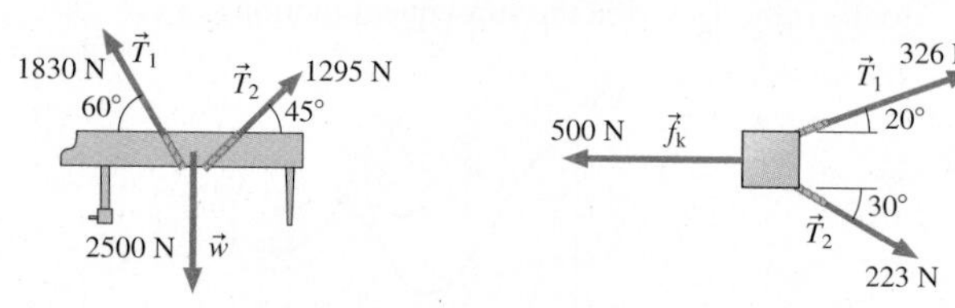

FIGURE P10.2 FIGURE P10.3

3. | The two ropes shown in the bird's-eye view of Figure P10.3 are used to drag a crate exactly 3 m across the floor. How much work is done by each of the ropes on the crate?
4. | You are pulling a child in a wagon. The rope handle is inclined upward at a 60° angle. The tension in the handle is 20 N. How much work do you do if you pull the wagon 100 m at a constant speed?
5. | A boy flies a kite with the string at a 30° angle to the horizontal. The tension in the string is 4.5 N. How much work does the string do on the boy if the boy
   a. Stands still?
   b. Walks a horizontal distance of 11 m away from the kite?
   c. Walks a horizontal distance of 11 m toward the kite?
6. || BIO A typical muscle fiber is 2.0 cm long and has a cross-section area of 3.1×10^{-9} m^2. When the muscle fiber is stimulated, it pulls with a force of 1.2 mN. What is the work done by the muscle fiber as it contracts to a length of 1.6 cm?
7. || A crate slides down a ramp that makes a 20° angle with the ground. To keep the crate moving at a steady speed, Paige pushes back on it with a 68 N horizontal force. How much work does Paige do on the crate as it slides 3.5 m down the ramp?

Section 10.3 Kinetic Energy

8. || A wind turbine works by slowing the air that passes its blades and converting much of the extracted kinetic energy to electric energy. A large wind turbine has 45-m-radius blades. In typical conditions, 92,000 kg of air moves past the blades every second. If the air is moving at 12 m/s before it passes the blades and the wind turbine extracts 40% of this kinetic energy, how much energy is extracted every second?
9. || At what speed does a 1000 kg compact car have the same kinetic energy as a 20,000 kg truck going 25 km/h?
10. | A 60 kg runner in a sprint moves at 11 m/s. A 60 kg cheetah in a sprint moves at 33 m/s. By what factor does the kinetic energy of the cheetah exceed that of the human runner?
11. | A car is traveling at 10 m/s.
    a. How fast would the car need to go to double its kinetic energy?
    b. By what factor does the car's kinetic energy increase if its speed is doubled to 20 m/s?
12. || The opposite of a wind turbine is an electric fan: The electric energy that powers the fan is converted to the kinetic energy of moving air. A fan is putting 1.0 J of kinetic energy into the air every second. Then the fan speed is increased by a factor of 2. Air moves through the fan faster, so the fan moves twice as much air at twice the speed. How much kinetic energy goes into the air every second?
13. | How fast would an 80 kg man need to run in order to have the same kinetic energy as an 8.0 g bullet fired at 400 m/s?
14. ||| A fielder tosses a 0.15 kg baseball at 32 m/s at a 30° angle to the horizontal. What is the ball's kinetic energy at the start of its motion? What is the kinetic energy at the highest point of its arc?
15. ||| Sam's job at the amusement park is to slow down and bring to a stop the boats in the log ride. If a boat and its riders have a mass of 1200 kg and the boat drifts in at 1.2 m/s, how much work does Sam do to stop it?
16. || A school has installed a modestly-sized wind turbine. The three blades are 4.6 m long; each blade has a mass of 45 kg. You can assume that the blades are uniform along their lengths. When the blades spin at 240 rpm, what is the kinetic energy of the blade assembly?
17. ||| The turntable in a microwave oven has a moment of inertia of 0.040 kg · m^2 and rotates continuously, making a complete revolution every 4.0 s. What is its kinetic energy?
18. || A typical meteor that hits the earth's upper atmosphere has a mass of only 2.5 g, about the same as a penny, but it is moving at an impressive 40 km/s. As the meteor slows, the resulting thermal energy makes a glowing streak across the sky, a shooting star. The small mass packs a surprising punch. At what speed would a 900 kg compact car need to move to have the same kinetic energy?

19. |||| An energy storage system based on a flywheel (a rotating disk) can store a maximum of 4.0 MJ when the flywheel is rotating at 20,000 revolutions per minute. What is the moment of inertia of the flywheel?

Section 10.4 Potential Energy

20. || The lowest point in Death Valley is 85.0 m below sea level. The summit of nearby Mt. Whitney has an elevation of 4420 m. What is the change in gravitational potential energy of an energetic 65.0 kg hiker who makes it from the floor of Death Valley to the top of Mt. Whitney?
21. | The world's fastest humans can reach speeds of about 11 m/s. In order to increase his gravitational potential energy by an amount equal to his kinetic energy at full speed, how high would such a sprinter need to climb?
22. | A 72 kg bike racer climbs a 1200-m-long section of road that has a slope of 4.3°. By how much does his gravitational potential energy change during this climb?
23. || A 1000 kg wrecking ball hangs from a 15-m-long cable. The ball is pulled back until the cable makes an angle of 25° with the vertical. By how much has the gravitational potential energy of the ball changed?

24. | How far must you stretch a spring with $k = 1000$ N/m to store 200 J of energy?

25. | How much energy can be stored in a spring with a spring constant of 500 N/m if its maximum possible stretch is 20 cm?

26. | The spring in a retractable ballpoint pen is 1.8 cm long, with a 300 N/m spring constant. When the pen is retracted, the spring is compressed by 1.0 mm. When you click the button to extend the pen, you compress the spring by an additional 6.0 mm. How much energy is required to extend the pen?

27. |||| The elastic energy stored in your tendons can contribute up BIO to 35% of your energy needs when running. Sports scientists have studied the change in length of the knee extensor tendon in sprinters and nonathletes. They find (on average) that the sprinters' tendons stretch 41 mm, while nonathletes' stretch only 33 mm. The spring constant for the tendon is the same for both groups, 33 N/mm. What is the difference in maximum stored energy between the sprinters and the nonathletes?

28. | Scallops use muscles BIO INT to close their shells. Opening the shell is another story—muscles can only pull, they can't push. Instead of muscles, the shell is opened by a spring, a pad of a very elastic biological material called abductin. When the shell closes, the pad compresses; a restoring force then pushes the shell back open. The energy to open the shell comes from the elastic energy that was stored when the shell was closed. Figure P10.28 shows smoothed data for the restoring force of an abductin pad versus the compression. When the shell closes, the pad compresses by 0.15 mm. How much elastic potential energy is stored?

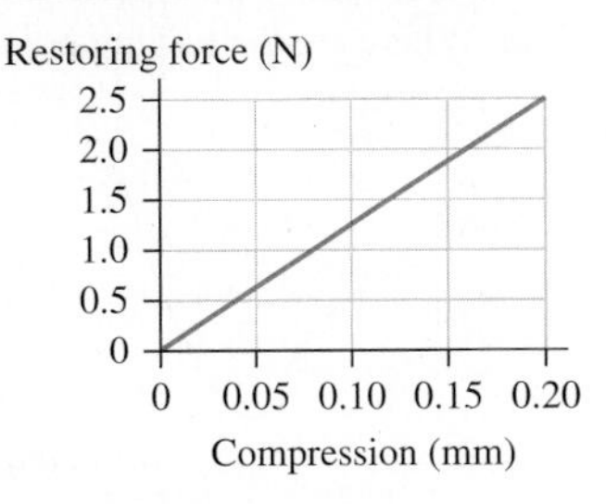

FIGURE P10.28

Section 10.5 Thermal Energy

29. | Mark pushes his broken car 150 m down the block to his friend's house. He has to exert a 110 N horizontal force to push the car at a constant speed. How much thermal energy is created in the tires and road during this short trip?

30. | When you skid to a stop on your bike, you can significantly INT heat the small patch of tire that rubs against the road surface. Suppose a person skids to a stop by hitting the brake on his back tire, which supports half the 80 kg combined mass of the bike and rider, leaving a skid mark that is 40 cm long. Assume a coefficient of kinetic friction of 0.80. How much thermal energy is deposited in the tire and the road surface?

31. ||| A 900 N crate slides 12 m down a ramp that makes an angle of 35° with the horizontal. If the crate slides at a constant speed, how much thermal energy is created?

32. | If you slide down a rope, it's possible to create enough ther- INT mal energy to burn your hands or your legs where they grip the rope. Suppose a 40 kg child slides down a rope at a playground, descending 2.0 m at a constant speed. How much thermal energy is created as she slides down the rope?

33. ||| A 25 kg child slides down a playground slide at a *constant speed*. The slide has a height of 3.0 m and is 7.0 m long. Using the law of conservation of energy, find the magnitude of the kinetic friction force acting on the child.

34. | Some runners train with parachutes that trail behind them BIO INT to provide a large drag force. These parachutes are designed to have a large drag coefficient. One model expands to a square 1.8 m on a side, with a drag coefficient of 1.4. A runner completes a 200 m run at 5.0 m/s with this chute trailing behind. How much thermal energy is added to the air by the drag force?

Section 10.6 Conservation of Energy

35. | A boy reaches out of a window and tosses a ball straight up with a speed of 10 m/s. The ball is 20 m above the ground as he releases it. Use conservation of energy to find
    a. The ball's maximum height above the ground.
    b. The ball's speed as it passes the window on its way down.
    c. The speed of impact on the ground.

36. | The famous cliff divers of Acapulco leap from a perch 35 m INT above the ocean. How fast are they moving when they reach the water surface? What happens to their kinetic energy as they slow to a stop in the water?

37. ||| What minimum speed does a 100 g puck need to make it to the top of a frictionless ramp that is 3.0 m long and inclined at 20°?

38. | You can, in an emergency, start a manual transmission car by putting it in neutral, letting the car roll down a hill to pick up speed, then putting it in gear and quickly letting out the clutch. If the car needs to be moving at 3.5 m/s for this to work, how high a hill do you need? (You can ignore friction and drag.)

39. ||| A 1500 kg car is approaching the hill shown in Figure P10.39 at 10 m/s when it suddenly runs out of gas.
    a. Can the car make it to the top of the hill by coasting?
    b. If your answer to part a is yes, what is the car's speed after coasting down the other side?

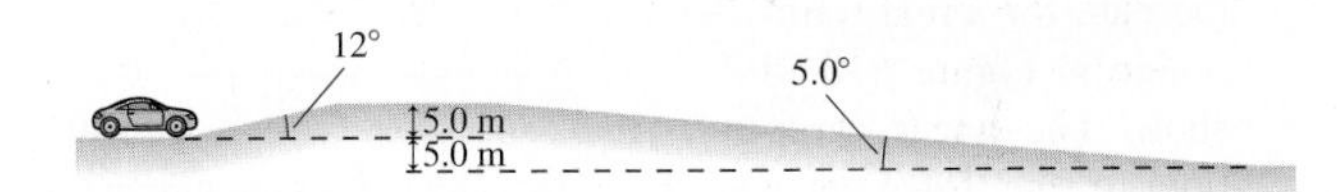

FIGURE P10.39

40. | A 480 g peregrine falcon reaches a BIO speed of 75 m/s in a vertical dive called a stoop. If we assume that the falcon speeds up under the influence of gravity only, what is the minimum height of the dive needed to achieve this speed?

41. | A fireman of mass 80 kg slides down a pole. When he reaches the bottom, 4.2 m below his starting point, his speed is 2.2 m/s. By how much has thermal energy increased during his slide?

42. | A 20 kg child slides down a 3.0-m-high playground slide. She starts from rest, and her speed at the bottom is 2.0 m/s.
    a. What energy transfers and transformations occur during the slide?
    b. What is the total change in the thermal energy of the slide and the seat of her pants?

43. | A hockey puck is given an initial speed of 5.0 m/s. If the coefficient of kinetic friction between the puck and the ice is 0.05, how far does the puck slide before coming to rest? Solve this problem using conservation of energy.

44. ||| Monica pulls her daughter Jessie in a bike trailer. The trailer and Jessie together have a mass of 25 kg. Monica starts up a 100-m-long slope that's 4.0 m high. On the slope, Monica's bike pulls on the trailer with a constant force of 8.0 N. They start out at the bottom of the slope with a speed of 5.3 m/s. What is their speed at the top of the slope?

45. || In the winter activity of tubing, riders slide down snow-covered slopes while sitting on large inflated rubber tubes. To get to the top of the slope, a rider and his tube, with a total mass of 80 kg, are pulled at a constant speed by a tow rope that maintains a constant tension of 340 N. How much thermal energy is created in the slope and the tube during the ascent of a 30-m-high, 120-m-long slope?

46. || Mosses don't spread by dispersing seeds; they disperse tiny spores. The spores are so small that they will stay aloft and move with the wind, but getting them to be windborne requires the moss to shoot the spores upward. Some species do this by using a spore-containing capsule that dries out and shrinks. The pressure of the air trapped inside the capsule increases. At a certain point, the capsule pops, and a stream of spores is ejected upward at 3.6 m/s, reaching an ultimate height of 20 cm. What fraction of the initial kinetic energy is converted to the final potential energy? What happens to the "lost" energy?
BIO INT

47. || A cyclist is coasting at 12 m/s when she starts down a 450-m-long slope that is 30 m high. The cyclist and her bicycle have a combined mass of 70 kg. A steady 12 N drag force due to air resistance acts on her as she coasts all the way to the bottom. What is her speed at the bottom of the slope?

48. || When you stand on a trampoline, the surface depresses below equilibrium, and the surface pushes up on you, as the data for a real trampoline in Figure P10.48 show. The linear variation of the force as a function of distance means that we can model the restoring force as that of a spring. A 72 kg gymnast jumps on the trampoline. At the lowest point of his motion, he is 0.80 m below equilibrium. If we assume that all of the energy stored in the trampoline goes into his motion, how high above this lowest point will he rise?
INT

Restoring force (N)
1600
800
0
0 0.25 0.50
Distance below equilibrium (m)

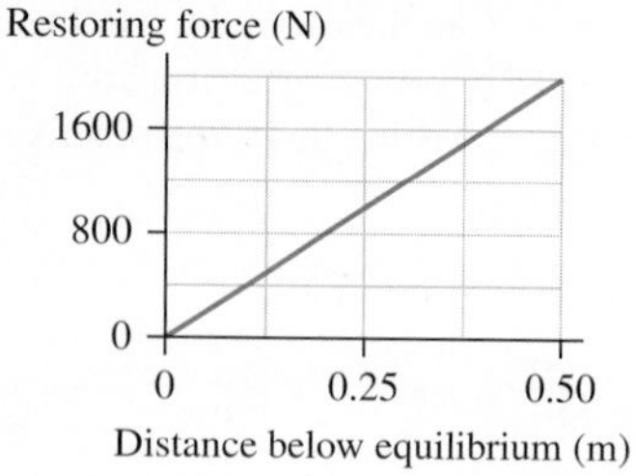

FIGURE P10.48

49. || The 5.0-m-long rope in Figure P10.49 hangs vertically from a tree right at the edge of a ravine. A woman wants to use the rope to swing to the other side of the ravine. She runs as fast as she can, grabs the rope, and swings out over the ravine.

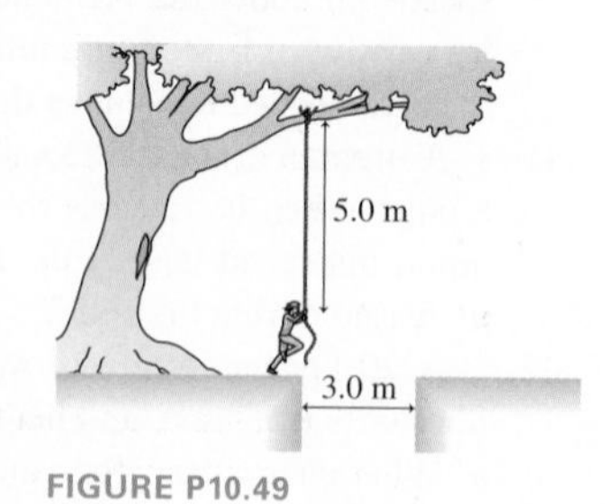

FIGURE P10.49

a. As she swings, what energy conversion is taking place?
b. When she's directly over the far edge of the ravine, how much higher is she than when she started?
c. Given your answers to parts a and b, how fast must she be running when she grabs the rope in order to swing all the way across the ravine?

50. ||| The Special Olympics raises money through "plane pull" events in which teams of 25 people compete to see who can pull a 74,000 kg airplane 3.7 m across the tarmac. The inertia of the plane is an issue—but so is the 14,000 N rolling friction force that works against the teams. If a team pulls with a constant force and moves the plane 3.7 m in 6.1 s (an excellent time), what fraction of the team's work goes to kinetic energy and what fraction goes to thermal energy?
INT

Section 10.7 Energy Diagrams

51. || Figure P10.51 is the potential-energy diagram for a 20 g particle that is released from rest at $x = 1.0$ m.
a. Will the particle move to the right or to the left? How can you tell?
b. What is the particle's maximum speed? At what position does it have this speed?
c. Where are the turning points of the motion?

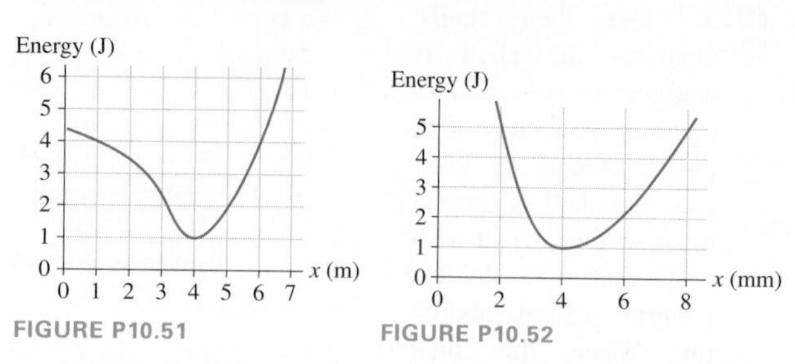

FIGURE P10.51

FIGURE P10.52

52. || For the potential-energy diagram in Figure P10.52, what is the maximum speed of a 2.0 g particle that oscillates between $x = 2.0$ mm and $x = 8.0$ mm?

Section 10.8 Molecular Bonds and Chemical Energy

At normal temperatures and pressures, hydrogen gas is composed of H_2 molecules. An energy diagram for a hydrogen molecule appears in Figure P10.53. Use this information to answer Problems 10.53 and 10.54.

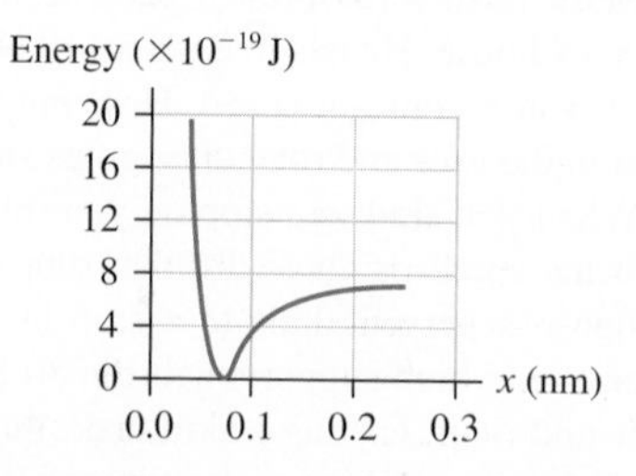

FIGURE P10.53

53. | How far apart are the individual atoms in a molecule of H_2?

54. || What energy photon is needed to dissociate a molecule of H_2?

Section 10.9 Energy in Collisions

55. || A 50 g marble moving at 2.0 m/s strikes a 20 g marble at rest. What is the speed of each marble immediately after the collision? Assume the collision is perfectly elastic and the marbles collide head-on.

56. || Ball 1, with a mass of 100 g and traveling at 10 m/s, collides head-on with ball 2, which has a mass of 300 g and is initially at rest. What are the final velocities of each ball if the collision is (a) perfectly elastic? (b) perfectly inelastic?

57. | An air-track glider undergoes a perfectly inelastic collision with an identical glider that is initially at rest. What fraction of the first glider's initial kinetic energy is transformed into thermal energy in this collision?
58. | Two balls undergo a perfectly elastic head-on collision, with one ball initially at rest. If the incoming ball has a speed of 200 m/s, what are the final speed and direction of each ball if
    a. The incoming ball is *much* more massive than the stationary ball?
    b. The stationary ball is *much* more massive than the incoming ball?

Section 10.10 Power

59. || a. How much work must you do to push a 10 kg block of steel across a steel table at a steady speed of 1.0 m/s for 3.0 s? The coefficient of kinetic friction for steel on steel is 0.60.
    b. What is your power output while doing so?
60. || A shooting star is actually the track of a meteor, typically a small chunk of debris from a comet that has entered the earth's atmosphere. As the drag force slows the meteor down, its kinetic energy is converted to thermal energy, leaving a glowing trail across the sky. A typical meteor has a surprisingly small mass, but what it lacks in size it makes up for in speed. Assume that a meteor has a mass of 1.5 g and is moving at an impressive 50 km/s, both typical values. What power is generated if the meteor slows down over a typical 2.1 s? Can you see how this tiny object can make a glowing trail that can be seen hundreds of kilometers away?
61. | a. How much work does an elevator motor do to lift a 1000 kg elevator a height of 100 m at a constant speed?
    b. How much power must the motor supply to do this in 50 s at constant speed?
62. || A 500 kg horse can provide a steady output power of 750 W (that is, 1 horsepower) when pulling a load. How about a 38 kg sled dog? Data show that a 38 kg dog can pull a sled that requires a pulling force of 60 N at a steady 2.2 m/s. What are the specific power values for the dog and the horse? What is the minimum number of dogs needed to provide the same power as one horse? BIO
63. ||| A 1000 kg sports car accelerates from 0 to 30 m/s in 10 s. What is the average power of the engine?
64. || A world-class sprinter running a 100 m dash was clocked at 5.4 m/s 1.0 s after starting running and at 9.8 m/s 1.5 s later. In which of these time intervals, 0 to 1.0 s or 1.0 s to 2.5 s, was his output power greater? BIO
65. || An elite Tour de France cyclist can maintain an output power of 450 W during a sustained climb. At this output power, how long would it take an 85 kg cyclist (including the mass of his bike) to climb the famed 1100-m-high Alpe d'Huez mountain stage? BIO
66. || A 70 kg human sprinter can accelerate from rest to 10 m/s in 3.0 s. During the same time interval, a 30 kg greyhound can accelerate from rest to 20 m/s. What is the specific power for each of these athletes? BIO
67. || A 710 kg car drives at a constant speed of 23 m/s. It is subject to a drag force of 500 N. What power is required from the car's engine to drive the car
    a. On level ground?
    b. Up a hill with a slope of 2.0°?
68. || A 95 kg quarterback accelerates a 0.42 kg ball from rest to 24 m/s in 0.083 s. What is the specific power for this toss? BIO
69. ||| An elevator weighing 2500 N ascends at a constant speed of 8.0 m/s. How much power must the motor supply to do this?

70. || Humans can produce an output power as great as 20 W/kg during extreme exercise. Sloths are not so energetic. At its maximum speed, a 4.0 kg sloth can climb a height of 6.0 m in 2.0 min. What's the specific power for this climb? BIO

General Problems

71. || A 550 kg elevator accelerates upward at 1.2 m/s^2 for the first 15 m of its motion. How much work is done during this part of its motion by the cable that lifts the elevator? INT
72. || The energy yield of a nuclear weapon is often defined in terms of the equivalent mass of a conventional explosive. 1 ton of a conventional explosive releases 4.2 GJ. A typical nuclear warhead releases 250,000 times more, so the yield is expressed as 250 kilotons. That is a staggering explosion, but the asteroid impact that wiped out the dinosaurs was significantly greater. Assume that the asteroid was a sphere 10 km in diameter, with a density of 2500 kg/m^3 and moving at 30 km/s. What energy was released at impact, in joules and in kilotons?
73. || A 2.3 kg box, starting from rest, is pushed up a ramp by a 10 N force parallel to the ramp. The ramp is 2.0 m long and tilted at 17°. The speed of the box at the top of the ramp is 0.80 m/s. Consider the system to be the box + ramp + earth.
    a. How much work W does the force do on the system?
    b. What is the change ΔK in the kinetic energy of the system?
    c. What is the change ΔU_g in the gravitational potential energy of the system?
    d. What is the change ΔE_{th} in the thermal energy of the system?
74. ||| A 55 kg skateboarder wants to just make it to the upper edge of a "half-pipe" with a radius of 3.0 m, as shown in Figure P10.74. What speed does he need at the bottom if he will coast all the way up? The skateboarder isn't a simple particle: Assume that his mass in a deep crouch is concentrated 0.75 m from the half-pipe. If he remains in that position all the way up, what initial speed does he need to reach the upper edge?

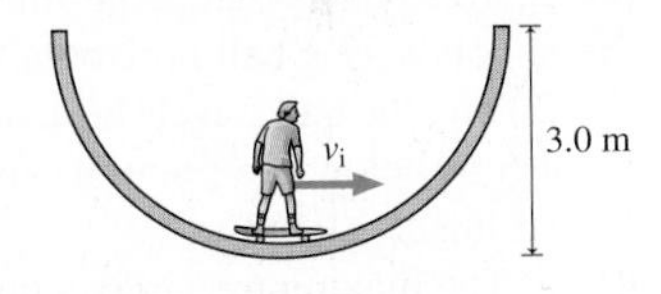

FIGURE P10.74

75. ||| Fleas have remarkable jumping ability. A 0.50 mg flea, jumping straight up, would reach a height of 40 cm if there were no air resistance. In reality, air resistance limits the height to 20 cm. BIO
    a. What is the flea's kinetic energy as it leaves the ground?
    b. At its highest point, what fraction of the initial kinetic energy has been converted to potential energy?
76. ||| You are driving your 1500 kg car at 20 m/s down a hill with a 5.0° slope when a deer suddenly jumps out onto the roadway. You slam on your brakes, skidding to a stop. How far do you skid before stopping if the kinetic friction force between your tires and the road is 1.2 × 10^4 N? Solve this problem using conservation of energy.

Watch Video Solution Problem 10.79

77. || A 20 kg child is on a swing that hangs from 3.0-m-long chains, as shown in Figure P10.77. What is her speed v_i at the bottom of the arc if she swings out to a 45° angle before reversing direction?

FIGURE P10.77 FIGURE P10.78

78. || Suppose you lift a 20 kg box by a height of 1.0 m.
    a. How much work do you do in lifting the box?
    Instead of lifting the box straight up, suppose you push it up a 1.0-m-high ramp that makes a 30° degree angle with the horizontal, as shown in Figure P10.78. Being clever, you choose a ramp with no friction.
    b. How much force F is required to push the box straight up the slope at a constant speed?
    c. How long is the ramp?
    d. Use your force and distance results to calculate the work you do in pushing the box up the ramp. How does this compare to your answer to part a?
79. | The sledder shown in Figure P10.79 starts from the top of a frictionless hill and slides down into the valley. What initial speed v_i does the sledder need to just make it over the next hill?

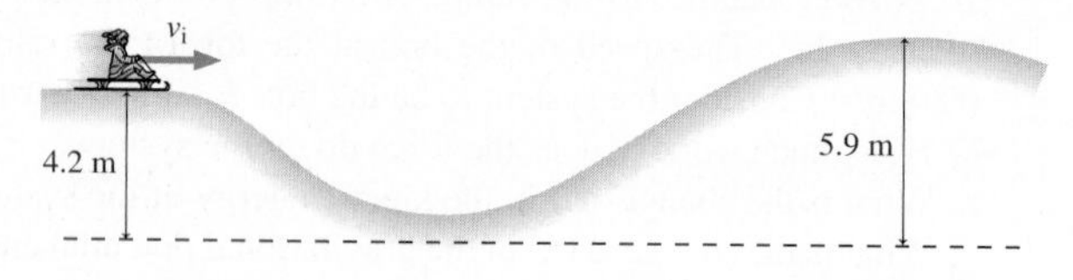

FIGURE P10.79

80. ||||| In a physics lab experiment, a spring clamped to the table
INT shoots a 20 g ball horizontally. When the spring is compressed 20 cm, the ball travels horizontally 5.0 m and lands on the floor 1.5 m below the point at which it left the spring. What is the spring constant?
81. ||||| The maximum energy a bone can absorb without breaking is
BIO surprisingly small. For a healthy human of mass 60 kg, experimental data show that the leg bones of both legs can absorb about 200 J.
    a. From what maximum height could a person jump and land rigidly upright on both feet without breaking his legs? Assume that all the energy is absorbed in the leg bones in a rigid landing.
    b. People jump from much greater heights than this; explain how this is possible.
    Hint: Think about how people land when they jump from greater heights.
82. || In an amusement park water slide, people slide down an
INT essentially frictionless tube. The top of the slide is 3.0 m above the bottom where they exit the slide, moving horizontally, 1.2 m above a swimming pool. What horizontal distance do they travel from the exit point before hitting the water? Does the mass of the person make any difference?
83. ||| You have been asked to design a "ballistic spring system" to measure the speed of bullets. A bullet of mass m is fired into a block of mass M. The block, with the embedded bullet, then slides across a frictionless table and collides with a horizontal spring whose spring constant is k. The opposite end of the spring is anchored to a wall. The spring's maximum compression d is measured.
    a. Find an expression for the bullet's initial speed v_B in terms of m, M, k, and d.
    Hint: This is a two-part problem. The bullet's collision with the block is an inelastic collision. What quantity is conserved in an inelastic collision? Subsequently the block hits a spring on a frictionless surface. What quantity is conserved in this collision?
    b. What was the speed of a 5.0 g bullet if the block's mass is 2.0 kg and if the spring, with $k = 50$ N/m, was compressed by 10 cm?
    c. What fraction of the bullet's initial kinetic energy is "lost"? Where did it go?
84. || Boxes A and B in Figure P10.84 have masses of 12.0 kg and 4.0 kg, respectively. The two boxes are released from rest. Use conservation of energy to find the boxes' speed when box B has fallen a distance of 0.50 m. Assume a frictionless upper surface.

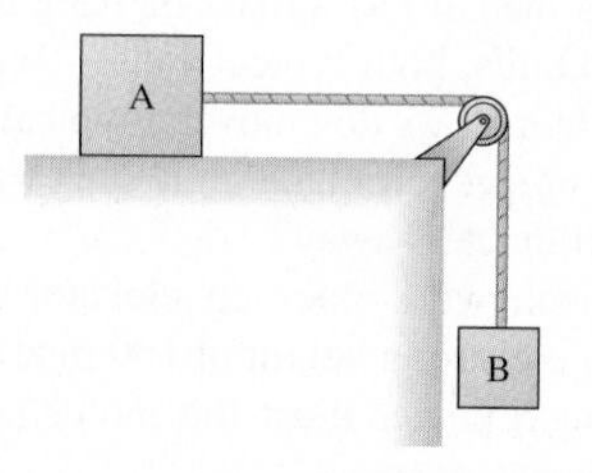

FIGURE P10.84

85. || Two coupled boxcars are rolling along at 2.5 m/s when they
INT collide with and couple to a third, stationary boxcar.
    a. What is the final speed of the three coupled boxcars?
    b. What fraction of the cars' initial kinetic energy is transformed into thermal energy?
86. ||| A 50 g ball of clay traveling at 6.5 m/s hits and sticks to a
INT 1.0 kg block sitting at rest on a frictionless surface.
    a. What is the speed of the block after the collision?
    b. Show that the mechanical energy is *not* conserved in this collision. What percentage of the ball's initial kinetic energy is "lost"? Where did this kinetic energy go?
87. || A package of mass m is
INT released from rest at a warehouse loading dock and slides down a 3.0-m-high frictionless chute to a waiting truck. Unfortunately, the truck driver went on a break without having removed the previous package, of mass $2m$, from the bottom of the chute as shown in Figure P10.87.

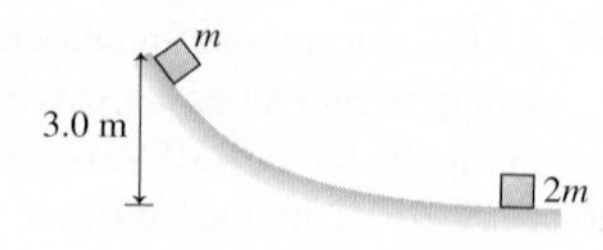

FIGURE P10.87

    a. Suppose the packages stick together. What is their common speed after the collision?
    b. Suppose the collision between the packages is perfectly elastic. To what height does the package of mass m rebound?

88. ||| Swordfish are capable of stunning output power for short bursts. A 650 kg swordfish has a cross-section area of $0.92\ m^2$ and a drag coefficient of 0.0091—exceptionally low due to a number of adaptations. Such a fish can sustain a speed of 30 m/s for a few seconds. Assume seawater has a density of $1026\ kg/m^3$. What is the specific power for motion at this high speed?
BIO
INT

89. || The mass of an elevator and its occupants is 1200 kg. The electric motor that lifts the elevator can provide a maximum power of 15 kW. What is the maximum constant speed at which this motor can lift the elevator?

MCAT-Style Passage Problems

Tennis Ball Testing

A tennis ball bouncing on a hard surface compresses and then rebounds. The details of the rebound are specified in tennis regulations. Tennis balls, to be acceptable for tournament play, must have a mass of 57.5 g. When dropped from a height of 2.5 m onto a concrete surface, a ball must rebound to a height of 1.4 m. During impact, the ball compresses by approximately 6 mm.

90. | How fast is the ball moving when it hits the concrete surface? (Ignore air resistance.)
A. 5 m/s B. 7 m/s
C. 25 m/s D. 50 m/s
91. | If the ball accelerates uniformly when it hits the floor, what is its approximate acceleration as it comes to rest before rebounding?
A. $1000\ m/s^2$ B. $2000\ m/s^2$
C. $3000\ m/s^2$ D. $4000\ m/s^2$
92. | The ball's kinetic energy just after the bounce is less than just before the bounce. In what form does this lost energy end up?
A. Elastic potential energy
B. Gravitational potential energy
C. Thermal energy
D. Rotational kinetic energy
93. | By approximately what percent does the kinetic energy decrease?
A. 35% B. 45% C. 55% D. 65%
94. | When a tennis ball bounces from a racket, the ball loses approximately 30% of its kinetic energy to thermal energy. A ball that hits a racket at a speed of 10 m/s will rebound with approximately what speed?
A. 8.5 m/s B. 7.0 m/s C. 4.5 m/s D. 3.0 m/s

Work and Power in Cycling

When you ride a bicycle at constant speed, almost all of the energy you expend goes into the work you do against the drag force of the air. In this problem, assume that *all* of the energy expended goes into working against drag. As we saw in Section 5.6, the drag force on an object is approximately proportional to the square of its speed with respect to the air. For this problem, assume that $F \propto v^2$ exactly and that the air is motionless with respect to the ground unless noted otherwise. Suppose a cyclist and her bicycle have a combined mass of 60 kg and she is cycling along at a speed of 5 m/s.

95. | If the drag force on the cyclist is 10 N, how much energy does she use in cycling 1 km?
A. 6 kJ B. 10 kJ C. 50 kJ D. 100 kJ
96. | Under these conditions, how much power does she expend as she cycles?
A. 10 W B. 50 W C. 100 W D. 200 W
97. | If she doubles her speed to 10 m/s, how much energy does she use in cycling 1 km?
A. 20 kJ B. 40 kJ C. 200 kJ D. 400 kJ
98. | How much power does she expend when cycling at that speed?
A. 100 W B. 200 W C. 400 W D. 1000 W
99. | Upon reducing her speed back down to 5 m/s, she hits a headwind of 5 m/s. How much power is she expending now?
A. 100 W B. 200 W C. 500 W D. 1000 W

11 Using Energy

The odd hopping gait of a kangaroo has a very practical purpose: It lets the kangaroo cover great distances with minimal energy input. How is this efficiency possible?

LOOKING AHEAD ▸▸

Energy Use in the Body

The energy to run, to move, and to stay warm comes from the chemical energy in the food you eat.

You'll learn to calculate how much energy your body requires to complete a range of tasks.

Temperature and Heat

This tea kettle gets hotter because of the transfer of energy from the burner as **heat.**

You'll learn that *heat* is energy transferred due to a temperature difference between objects.

Heat Engines

A **heat engine** is a device, such as this Icelandic geothermal power plant, that transforms thermal energy into useful work.

You'll learn how to calculate the maximum efficiency for converting thermal energy into work.

GOAL To learn about practical energy transformations and transfers, and the limits on how efficiently energy can be used.

LOOKING BACK ◂◂

The Basic Energy Model

The basic energy model you learned about in Chapter 10 emphasized work and mechanical energy. In this chapter, we'll focus on thermal energy, chemical energy, and energy transfers in the form of heat.

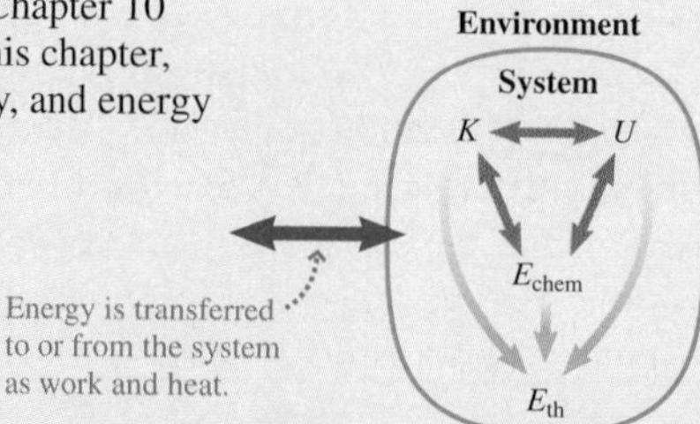

Environment
System
K
U
E_{chem}
E_{th}
Energy is transferred to or from the system as work and heat.

Work and heat are energy transfers that change the system's total energy. If the system is isolated, the total energy is *conserved.*

STOP TO THINK

Christina throws a javelin into the air. As she propels it forward from rest, she does 270 J of work on it. At its highest point, its gravitational potential energy has increased by 70 J. What is the javelin's kinetic energy at this point?

A. 270 J B. 340 J C. 200 J
D. −200 J E. −340 J

11.1 Transforming Energy

As we saw in Chapter 10, energy can't be created or destroyed; it can only be converted from one form to another. When we say we are *using energy,* we mean that we are transforming it, such as transforming the chemical energy of food into the kinetic energy of your body. The following table revisits the idea of energy transformations, considering some realistic situations that have interesting theoretical and practical limitations.

Energy transformations

Light energy hitting a solar cell on top of this walkway light is converted to electric energy and then stored as chemical energy in a battery.

At night, the battery's chemical energy is converted to electric energy that is then converted to light energy in a light- emitting diode.

Light energy is absorbed by photosynthetic pigments in soybean plants, which use this energy to create concentrated chemical energy.

The soybeans are harvested and their oil is used to make candles. When the candle burns, the stored chemical energy is transformed into light energy and thermal energy.

A wind turbine converts the translational kinetic energy of moving air into electric energy.

Miles away, this energy can be used by a fan, which transforms the electric energy back into the kinetic energy of moving air.

In each of the processes in the table, energy is transformed from one form into another and another, ending up in the same form as it began. But some energy appears to have been "lost" along the way. The garden light certainly shines with a glow that is much less bright than the sun that shined on it. When sunlight falls on the soybean plants, only a small fraction of the light energy ends up as chemical energy in the plants. No energy is lost, of course; we know that total energy is conserved. But at each stage in the processes in the table, much—or most—of the energy is converted into forms that are less useful to us.

The table also illustrates another key point: We are broadening our scope to include forms of energy beyond those we considered in Chapter 10, including radiant energy (the light that hits the solar cell and the plants) and electric energy (energy that is transferred from the wind turbine to the fan).

Recall the work-energy equation from Chapter 10:

$$\Delta E = \Delta K + \Delta U + \Delta E_{\text{th}} + \Delta E_{\text{chem}} + \cdots = W \tag{11.1}$$

This equation includes work, an energy transfer. Work can be positive or negative; it's positive when energy is transferred into a system, negative when energy is transferred out of a system. In Chapter 10 we defined work as a mechanical transfer of energy; in this chapter we'll broaden our definition of work to include electric energy into and out of motors and generators.

Let's look at a practical example of a chain of energy transformations, focusing on the energy that is "lost" and the energy that remains. Electric power companies sometimes use excess electric energy to pump water uphill to a reservoir. They reclaim some of this energy when demand increases by using the water to generate electricity as it flows back down. FIGURE 11.1 shows the energy transformations in a simple case. We assume 100 J of electric energy at the start and define the system as the water and the network of pipes, pumps, and generators that make up the plant. The electric energy that is used to run the pump is a work input; the electric energy that is extracted at the end is a work output. Pumping the water uphill requires 100 J of work, but this

FIGURE 11.1 Energy transformations for a pumped storage system.

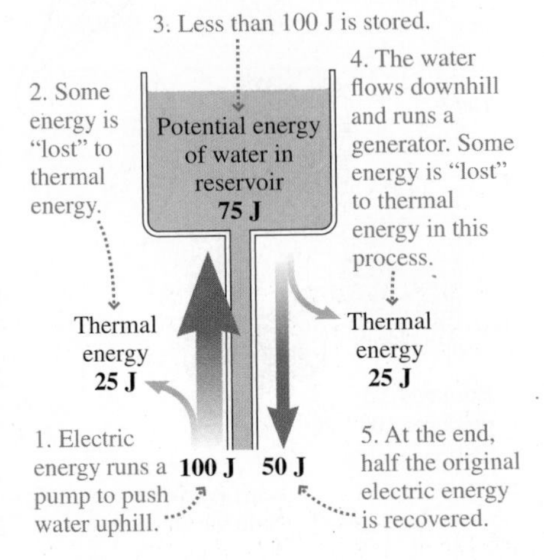

increases the potential energy of the water by only 75 J, with 25 J transformed into thermal energy by friction in the pump and other causes. When the water flows back downhill, a generator converts some—but not all—of the potential energy to electric energy. Ultimately, a 100 J input results in a 50 J output, with a 50 J increase in thermal energy of the system. This transformation into thermal energy is an irreversible change. There is no practical way to recover this thermal energy and convert it back to electric energy. **The energy isn't lost, but it is lost to our use.**

In Chapter 10, we considered cases in which the losses were small or could be ignored. In this chapter, we'll focus on more realistic situations in which this is not the case, and we'll explore and explain the reasons why.

Efficiency

To *get* 50 J of energy out of your pumped storage plant, you would actually need to *pay* to put in 100 J of energy, as we saw in Figure 11.1. Because only 50% of the energy input is returned as useful energy output, we can say that this plant has an efficiency of 50%. Generally, we can define efficiency as

$$e = \frac{\text{what you get}}{\text{what you had to pay}} \tag{11.2}$$

General definition of efficiency

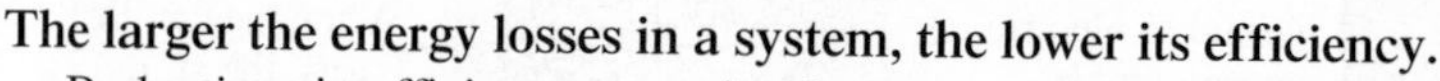

These cooling towers release thermal energy from a coal-fired power plant.

The larger the energy losses in a system, the lower its efficiency.

Reductions in efficiency can arise from two different sources:

- **Process limitations.** In some cases, energy losses may be due to practical details of an energy transformation process. You could, in principle, design a process that entailed smaller losses.
- **Fundamental limitations.** In other cases, energy losses are due to physical laws that cannot be circumvented. The best process that could theoretically be designed will have less than 100% efficiency. As we will see, these limitations result from the difficulty of transforming thermal energy into other forms of energy.

As we'll see, if you are of average weight, walking up a typical flight of stairs increases your body's potential energy by about 1800 J. But a measurement of the energy used by your body to climb the stairs would show that your body uses about 7200 J of chemical energy to complete this task. The 1800 J increase in potential energy is "what you get"; the 7200 J your body uses is "what you had to pay." We can use Equation 11.2 to compute the efficiency for this action:

$$e = \frac{1800\ \text{J}}{7200\ \text{J}} = 0.25 = 25\%$$

This relatively low efficiency is due to *process limitations*. The efficiency is less than 100% because of the biochemistry of how your food is metabolized and the biomechanics of how you move.

The electric energy you use daily must be generated from other sources. In a coal-fired power plant, chemical energy is converted to thermal energy by burning, and much of the resulting thermal energy is converted to electric energy. In fact, as we'll see, less than half of the energy is transformed into electric energy; *most* of the thermal energy from the burning coal is simply exhausted into the environment.

A typical power-plant cycle is shown in FIGURE 11.2. "What you get" is the energy output, the 40 J of electric energy. "What you had to pay" is the energy input, the 100 J of chemical energy, giving an efficiency of

$$e = \frac{40\ \text{J}}{100\ \text{J}} = 0.40 = 40\%$$

Only 40% of the energy in the coal is used to generate electricity; 60% goes to warming the area around the plant. Why would you design a plant to waste more than half of the energy input?

FIGURE 11.2 Energy transformations in a coal-fired power plant.

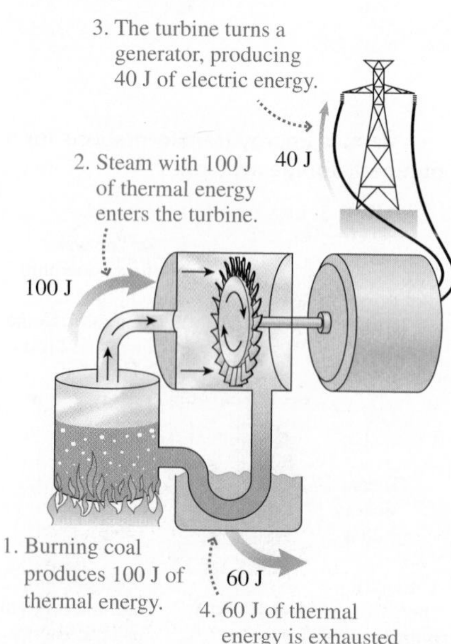
3. The turbine turns a generator, producing 40 J of electric energy.
40 J
2. Steam with 100 J of thermal energy enters the turbine.
100 J
1. Burning coal produces 100 J of thermal energy.
60 J
4. 60 J of thermal energy is exhausted into the environment.

Unlike your stair-climbing efficiency, the rather modest power-plant efficiency turns out to be largely due to a *fundamental limitation:* Thermal energy cannot be transformed into other forms of energy with 100% efficiency. In later sections, we will explore the fundamental properties of thermal energy that make this so.

PROBLEM-SOLVING APPROACH 11.1 Energy efficiency problems

The efficiency of a process is the ratio of the useful energy output (what you get) to the original energy input (what you had to pay).

STRATEGIZE To start, you need to decide what to identify as the input and the output:

1. Choose what energy is "what you get." This could be the useful energy output of an engine or process or the work that is done in completing a process. For example, when you climb a flight of stairs, "what you get" is your change in potential energy.
2. Decide what energy is "what you had to pay." This will generally be the total energy input needed for an engine, task, or process. For example, when you climb a flight of stairs, "what you had to pay" is the chemical energy used by your body.

PREPARE You may need to do additional calculations:

- Compute values for "what you get" and "what you had to pay."
- Be certain that all energy values are in the same units.

SOLVE Compute the efficiency using $e = \frac{\text{what you get}}{\text{what you had to pay}}$.

ASSESS Check your answer to see if it is reasonable, given what you know about typical efficiencies for the process under consideration.

EXAMPLE 11.1 Determining the efficiency of lightbulbs

A 75 W incandescent bulb and a 15 W compact fluorescent bulb, as shown in FIGURE 11.3, each produce 3.0 W of visible-light energy. What are the efficiencies of these two types of bulbs for converting electric energy into light?

FIGURE 11.3 Incandescent and compact fluorescent bulbs.

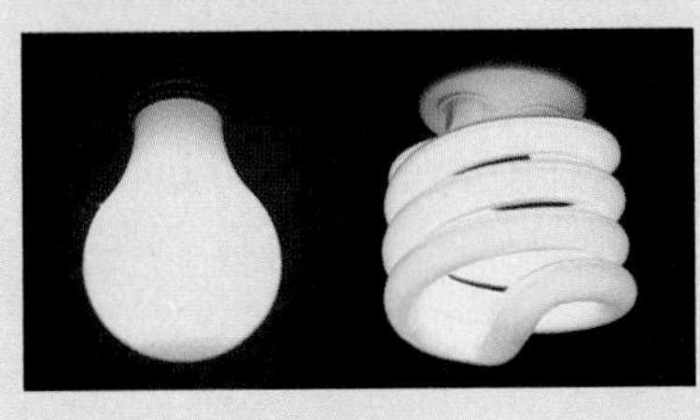

STRATEGIZE In this case, what you have to pay is the energy to light the bulb, the electric energy input. (If you purchase your electricity from the local utility, this is, literally, what you need to pay.) The energy input is the rating of the bulb: 15 W for the compact fluorescent bulb and 75 W for the incandescent bulb. The purpose of a lightbulb is to produce light, so what you get is light output, the 3.0 W of visible-light energy.

PREPARE The problem statement doesn't give values for energy; we are given values for power. But 15 W is 15 J/s, so we can consider the energy in or out in 1 second. In 1 second, the visible-light output for each bulb is 3.0 J. In 1 second, the energy input for the compact fluorescent bulb is 15 J; for the incandescent bulb, it is 75 J.

SOLVE The efficiencies of the two bulbs are computed using the energy in 1 second:

$$e(\text{compact fluorescent bulb}) = \frac{3.0\text{ J}}{15\text{ J}} = 0.20 = 20\%$$

$$e(\text{incandescent bulb}) = \frac{3.0\text{ J}}{75\text{ J}} = 0.040 = 4\%$$

ASSESS Both bulbs produce the same visible-light output, but the compact fluorescent bulb does so with a significantly lower energy input, so it is more efficient, as we know to be the case.

EXAMPLE 11.2 How much water flows through the dam?

A large hydroelectric power plant uses water from a reservoir to turn a generator. The water leaves the generator below the dam, 60 m below the level of the reservoir. The plant uses an 85% efficient generator to supply 1.0 GW of electricity. What mass of water flows through the plant each second?

STRATEGIZE What you get is clear: electric energy. We are given the plant's power output, so we'll consider the electric-energy output in 1 second. The plant's power output is 1.0 GW, so in 1 second, the plant puts out 1.0×10^9 J of electric energy. The change in potential energy of the water is what you had to pay. The ratio of these two values is the plant's efficiency, which is given.

PREPARE If water of mass m goes through the plant, the change in potential energy is

$$\Delta U_g = mg\Delta y = mg(-60\text{ m})$$

The water loses potential energy, so this is a negative number. The energy used by the generator is $mg(60\text{ m})$.

SOLVE The efficiency of the plant is 0.85; this is the ratio of what you get to what you had to pay, so we can write

$$\frac{1.0 \times 10^9\text{ J}}{mg(60\text{ m})} = 0.85$$

We can solve this equation for the mass of water:

$$m = \frac{1.0 \times 10^9\text{ J}}{g(60\text{ m})(0.85)} = 2.0 \times 10^6\text{ kg}$$

The density of water is 1000 kg/m^3, so this corresponds to 2000 m^3 of water.

ASSESS In 1 second, enough water flows through the plant to fill two Olympic-size swimming pools. But 1.0 GW is a lot of power, enough for a city of a million people, so this result makes sense. These are actual numbers for a real plant, and this much water really passes through the dam's giant generators.

STOP TO THINK 11.1 Crane 1 uses 10 kJ of energy to lift a 50 kg box to the roof of a building. Crane 2 uses 20 kJ to lift a 100 kg box the same distance. Which crane is more efficient?

A. Crane 1
B. Crane 2
C. Both cranes have the same efficiency.

11.2 Energy in the Body BIO

FIGURE 11.4 Energy of the body, considered as the system.

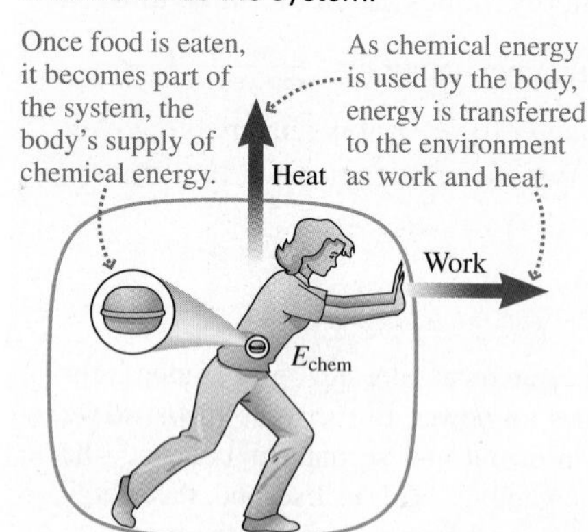

Once food is eaten, it becomes part of the system, the body's supply of chemical energy.
As chemical energy is used by the body, energy is transferred to the environment as work and heat.
Heat
Work
E_{chem}

In this section, we will look at energy in the body, which will give us the opportunity to explore a number of different energy transformations and transfers in a practical context. FIGURE 11.4 shows the body considered as the system for energy analysis. The chemical energy in food provides the necessary energy input for your body to function. It is this energy that is used for energy transfers with the environment.

Getting Energy from Food: Energy Inputs

When you walk up a flight of stairs, where does the energy come from to increase your body's potential energy? At some point, the energy came from the food you ate, but what were the intermediate steps? The chemical energy in food is made available to the cells in the body by a two-step process. First, the digestive system breaks down food into simpler molecules such as glucose, a simple sugar. These molecules are delivered via the bloodstream to cells in the body, where they are metabolized by combining with oxygen, as in Equation 11.3:

Glucose from the digestion of food combines with oxygen that is breathed in to produce . . .
. . . carbon dioxide, which is exhaled; water, which can be used by the body; and energy.

$$\underset{\text{Glucose}}{C_6H_{12}O_6} + \underset{\text{Oxygen}}{6O_2} \longrightarrow \underset{\text{Carbon dioxide}}{6CO_2} + \underset{\text{Water}}{6H_2O} + \text{energy} \tag{11.3}$$

This metabolism releases energy, much of which is stored in a molecule called adenosine triphosphate, or ATP. Cells in the body use this ATP to do all the work of life: Muscle cells use it to contract, nerve cells use it to produce electrical signals.

EXAMPLE 11.3 Computing basic biochemical efficiency BIO

Power for cellular processes comes from the conversion of ATP to ADP, adenosine diphosphate. The cell then recycles the ADP molecules, using chemical energy from glucose to convert the ADP back to ATP. Converting one molecule of ATP to ADP releases 5.1×10^{-20} J. Metabolizing one molecule of glucose releases 4.7×10^{-18} J. If everything goes smoothly in the biochemical pathways that use this energy to produce ATP, metabolizing one glucose molecule produces 38 molecules of ATP. What is the overall efficiency of this energy conversion process?

STRATEGIZE What you get is the energy of 38 molecules of ATP; what you have to pay is the energy from the metabolism of one molecule of glucose.

PREPARE The metabolism of one molecule of glucose releases 4.7×10^{-18} J; each molecule of ATP can release 5.1×10^{-20} J.

SOLVE The efficiency is

$$e = \frac{\text{what you get}}{\text{what you had to pay}} = \frac{(38)(5.1 \times 10^{-20}\ \text{J})}{4.7 \times 10^{-18}\ \text{J}} = 0.41$$

ASSESS We know that the efficiency of this process must be less than 1.0; we also know that it must be greater than 0.25, which is the efficiency of the body in using energy to climb stairs. Our result falls between these two extremes, which gives us confidence in our result.

Metabolizing 1 g of glucose (or any other carbohydrate) from food or body stores releases approximately 17 kJ of energy; 1 g of fat from food or body stores provides 37 kJ. TABLE 11.1 compares the energy contents of carbohydrates and other foods with other common sources of chemical energy.

It is possible to measure the chemical energy content of food by burning it. Burning food may seem quite different from metabolizing it, but if glucose is burned, the chemical formula for the reaction is again Equation 11.3—the two reactions are the same. Burning food transforms its chemical energy into thermal energy, which we can measure. Thermal energy is often measured in units of calories (cal) rather than joules; 1.00 calorie is equivalent to 4.19 joules. The energy content of food is usually given in Calories (Cal, with a capital "C"); 1 Calorie (also called a "food calorie") is equal to 1000 calories. We usually express energy in joules, so it's worthwhile to sum up the relationships among these different units:

$$1.00 \text{ Cal} = 1000 \text{ cal} = 4190 \text{ J} = 4.19 \text{ kJ}$$
$$1.00 \text{ kJ} = 1000 \text{ J} = 239 \text{ cal} = 0.239 \text{ Cal}$$

TABLE 11.1 Energy in fuels

Fuel	Energy in 1 g of fuel (in kJ)
Hydrogen	121
Gasoline	44
Fat (in food)	37
Coal	27
Carbohydrates (in food)	17
Wood chips	15

EXAMPLE 11.4 How much energy does a soda provide? BIO

A 12 oz can of soda contains 40 g (or a bit less than 1/4 cup) of sugar, a simple carbohydrate. What is the chemical energy in joules? How many Calories is this?

SOLVE From Table 11.1, 1 g of sugar contains 17 kJ of energy, so 40 g contains

$$40 \text{ g} \times \frac{17 \times 10^3 \text{ J}}{1 \text{ g}} = 68 \times 10^4 \text{ J} = 680 \text{ kJ}$$

Converting to Calories, we get

$$680 \text{ kJ} = (680 \text{ kJ})\frac{1.00 \text{ Cal}}{4.19 \text{ kJ}} = 160 \text{ Cal}$$

ASSESS 160 Calories is a typical value for the energy content of a 12 oz can of soda (check the nutrition label on one to see), so this result seems reasonable.

Counting calories BIO Most dry foods burn quite well, as this photo of corn chips illustrates. You could set food on fire to measure its energy content, but this isn't really necessary. The chemical energies of the basic components of food (carbohydrates, proteins, fats) have been carefully measured—by burning—in a device called a *calorimeter.* Foods are analyzed to determine their composition, and their chemical energy can then be calculated.

The first item on the nutrition label on packaged foods is Calories—a measure of the chemical energy in the food. (In Europe, where SI units are standard, you will find the energy contents listed in kJ.) The energy contents of some common foods are given in TABLE 11.2.

TABLE 11.2 Energy content of foods

Food	Energy content in Cal	Energy content in kJ	Food	Energy content in Cal	Energy content in kJ
Carrot (large)	30	125	Slice of pizza	300	1260
Fried egg	100	420	Frozen burrito	350	1470
Apple (large)	125	525	Apple pie slice	400	1680
Beer (can)	150	630	Fast-food meal:		
BBQ chicken wing	180	750	burger, fries,		
Latte (whole milk)	260	1090	drink (large)	1350	5660

Using Energy in the Body: Energy Outputs

You know that your body uses energy when you exercise. But even at rest, your body uses energy to build and repair tissue, digest food, and keep warm. TABLE 11.3 lists the amounts of power used by different tissues in the resting body. All told, your body uses 100 W, or 100 J per second, when at rest, and all of this energy is

TABLE 11.3 Energy use at rest

Organ	Resting power (W) of 68 kg individual
Liver	26
Brain	19
Kidneys	11
Heart	7
Skeletal muscle	18
Remainder of body	19
Total	**100**

FIGURE 11.5 The mask measures the oxygen used by the athlete.

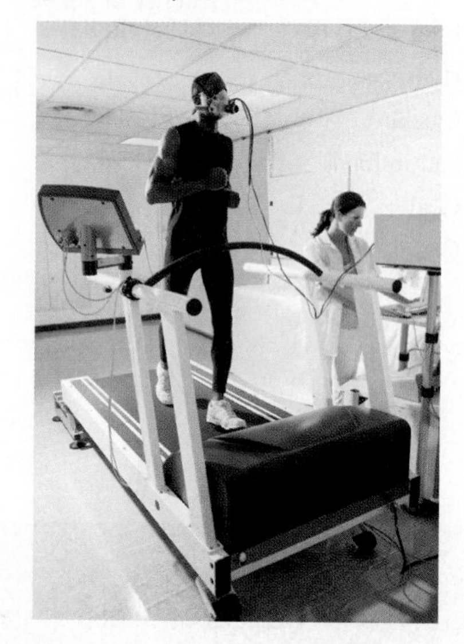

ultimately converted to thermal energy. Your body stays at a constant temperature, so this energy is dissipated as heat into the environment. This process has practical consequences. If you take physics in a class of 100 people in a lecture hall, the class, as a whole, dissipates 10,000 W into the room—the equivalent of seven electric space heaters running at full blast. In the winter, this extra heat is welcome; in the summer, the air conditioning system must work harder to keep the room cool.

Your cells use ATP to power cellular processes, so the cells of your body must continually replenish their ATP stores with metabolic reactions such as the one shown in Equation 11.3. This requires oxygen, so it is possible to measure the energy used by the body, at rest or engaged in exercise, by measuring the oxygen used. Using a respiratory apparatus, as shown in FIGURE 11.5, physiologists can precisely measure the body's energy use by measuring how much oxygen the body is taking up. The device determines the body's *total metabolic energy use*—all of the energy used by the body while performing an activity. This total includes all of the body's basic processes such as breathing plus whatever additional energy is needed to perform the activity. This corresponds to measuring "what you had to pay."

The metabolic energy used in an activity depends on an individual's size, level of fitness, and other variables. But we can make reasonable estimates for the power used in various activities for a typical individual. Some values are given in TABLE 11.4.

TABLE 11.4 Metabolic power use during activities

Activity	Metabolic power (W) of 68 kg individual
Typing	125
Ballroom dancing	250
Walking at 5 km/h	380
Cycling at 15 km/h	480
Swimming at a fast crawl	800
Running at 15 km/h	1150

FIGURE 11.6 Climbing a set of stairs.

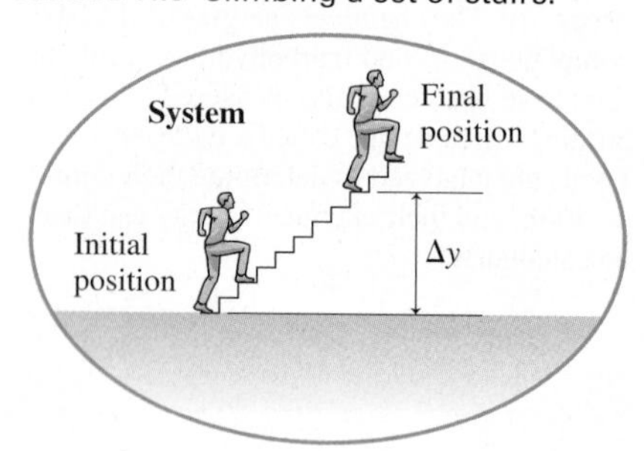

Efficiency of the Human Body

Suppose you climb a set of stairs at a constant speed, as in FIGURE 11.6. What is your body's efficiency for this process? To find out, let's apply the work-energy equation. We note first that no work is done on you: There is no external input of energy, as there would be if you took the elevator. Furthermore, if you climb at a constant speed, there's no change in your kinetic energy. However, your gravitational potential energy clearly increases as you climb, as does the overall thermal energy of you and perhaps the surrounding air. This latter fact is something you know well: If you climb several sets of stairs, you certainly warm up in the process! And, finally, your body must use chemical energy to power your muscles for the climb.

For this case of climbing stairs, then, the work-energy equation reduces to

$$\Delta E_{\text{chem}} + \Delta U_{\text{g}} + \Delta E_{\text{th}} = 0 \tag{11.4}$$

Thermal energy and gravitational potential energy are increasing, so ΔE_{th} and ΔU_{g} are positive; chemical energy is being used, so ΔE_{chem} is a negative number. We can get a better feeling about what is happening by rewriting Equation 11.4 as

$$|\Delta E_{\text{chem}}| = \Delta U_{\text{g}} + \Delta E_{\text{th}} \tag{11.5}$$

The *magnitude* of the change in the chemical energy is equal to the sum of the changes in the gravitational potential and thermal energies. Chemical energy from your body is converted into potential energy and thermal energy; in the final position, you are at a greater height and your body is slightly warmer.

Earlier in the chapter, we noted that the efficiency for stair climbing is about 25%. Let's see where that number comes from.

1. *What you get.* What you get is the change in potential energy: You have raised your body to the top of the stairs. If you climb a flight of stairs of vertical height Δy, the increase in potential energy is $\Delta U_{\text{g}} = mg\,\Delta y$. Assuming a mass of 68 kg and a change in height of 2.7 m (about 9 ft, a reasonable value for a flight of stairs), we compute

$$\Delta U_{\text{g}} = (68\text{ kg})(9.8\text{ m/s}^2)(2.7\text{ m}) = 1800\text{ J}$$

2. *What you had to pay*. The cost is the metabolic energy your body used in completing the task. As we've seen, physiologists can measure directly how much energy $|\Delta E_{\text{chem}}|$ your body uses to perform a task. A typical value for climbing a flight of stairs is

$$|\Delta E_{\text{chem}}| = 7200 \text{ J}$$

Given the definition of efficiency in Equation 11.2, we can compute an efficiency for climbing the stairs:

$$e = \frac{\Delta U_{\text{g}}}{|\Delta E_{\text{chem}}|} = \frac{1800 \text{ J}}{7200 \text{ J}} = 0.25 = 25\%$$

For the types of activities we will consider in this chapter, such as running, walking, and cycling, the body's efficiency is typically in the range of 20–30%. **We will generally use a value of 25% for the body's efficiency for our calculations.** Efficiency varies from individual to individual and from activity to activity, but this rough approximation will be sufficient for our purposes.

The metabolic power values given in Table 11.4 represent the energy *used by the body* while these activities are being performed. Given that we assume an efficiency of 25%, the body's actual *useful power output* is then quite a bit less than this. The table's value for cycling at 15 km/h (a bit less than 10 mph) is 480 W. If we assume that the efficiency for cycling is 25%, the actual power going to forward propulsion will be only 120 W. An elite racing cyclist whizzing along at 35 km/h is using about 300 W for forward propulsion, so his metabolic power is about 1200 W.

The energy you use per second while running is proportional to your speed; running twice as fast takes approximately twice as much power. But running twice as fast takes you twice as far in the same time, so the energy you use to run a certain *distance* doesn't depend on how fast you run! Running a marathon takes approximately the same amount of energy whether you complete it in 2 hours, 3 hours, or 4 hours; it is only the power that varies.

High heating costs? BIO The daily energy use of mammals is much higher than that of reptiles, largely because mammals use energy to maintain a constant body temperature. A 40 kg timber wolf uses approximately 19,000 kJ during the course of a day. A Komodo dragon, a reptilian predator of the same size, uses only 2100 kJ.

NOTE ▸ It is important to remember the distinction between the metabolic energy used to perform a task and the work done in a physics sense; these values can be quite different. Your muscles use power when applying a force, even when there is no motion. Holding a weight above your head involves no external work, but it clearly takes metabolic power to keep the weight in place! ◂

This distinction—between the work done (what you get) and the energy used by the body (what you had to pay)—is important to keep in mind when you do calculations on energy used by the body. We'll consider two different types of problems:

- **Type 1: "What you get" can be quantified.** For some tasks, such as climbing stairs, we can easily compute the energy change that is the outcome of the task; that is, we can compute what you get. We assume an efficiency of 25%, so the energy used by the body ("what you had to pay") is 4 times this amount. There is no need to use table data; you can do the calculation directly.
- **Type 2: "What you get" can't be easily quantified.** For other tasks, it is not so easy to find the energy that corresponds to "what you get." If you ride your bike at a constant speed on a level course, you don't change your potential energy or your kinetic energy. In energy terms, there is no good way to compute "what you get." In such cases, we consider data from metabolic studies, such as the data in Table 11.4. This is the actual power used by the body to complete a task—in other words, "what you had to pay." We assume an efficiency of 25%, so the useful power output ("what you get") is ¼ of this value.

CONCEPTUAL EXAMPLE 11.5 Determining energy transformations in weightlifting BIO

A weightlifter lifts a 50 kg bar from the floor to a position over his head and back to the floor again 10 times in succession. At the end of this exercise, what energy transformations have taken place?

REASON We will take the system to be the weightlifter plus the bar. The environment does no work on the system, and we will assume that no heat is transferred from the system to the environment. The bar has returned to its starting position and is not moving, so there has been no change in potential or kinetic energy. The equation for energy conservation is thus $\Delta E_{chem} + \Delta E_{th} = 0$. This equation tells us that ΔE_{chem} must be negative. This makes sense because the muscles *use* chemical energy—depleting the body's store—each time the bar is raised or lowered. Ultimately, all of this energy is transformed into thermal energy.

ASSESS You know that you warm up when you are working out, so this result makes sense. Most exercises in the gym—lifting weights, running on a treadmill—involve only the transformation of chemical energy into thermal energy.

EXAMPLE 11.6 Determining the energy usage for a cyclist BIO

A cyclist pedals for 20 min at a speed of 15 km/h. How much metabolic energy is required? How much energy is used for forward propulsion?

STRATEGIZE This is a Type 2 problem. There is no easy way to compute what you get—there is no change in potential or kinetic energy. As the cyclist moves, all of the chemical energy the body uses ends up as thermal energy in the cyclist's body or in the air. We'll use the value in Table 11.4 for the energy used. This is the energy the body uses, so it is the input—what you had to pay. If we assume an efficiency of 25%, the energy output is ¼ of this.

PREPARE Table 11.4 gives a value of 480 W for the metabolic power used in cycling at 15 km/h. 480 W is the power used by the body; the power going into forward propulsion is much less than this. The cyclist uses energy at this rate for 20 min, or 1200 s.

SOLVE We know the power and the time, so we can compute the energy needed by the body as follows:

$$\Delta E = P\Delta t = (480\ \text{J/s})(1200\ \text{s}) = 580\ \text{kJ}$$

If we assume an efficiency of 25%, then only 25%, or 140 kJ, of this energy is used for forward propulsion. The remainder goes into thermal energy.

ASSESS This result is reasonable. 15 km/h isn't that speedy; most recreational cyclists could keep up this pace. So we don't expect the energy expenditure to be all that great—a glance at Table 11.2 shows that 580 kJ is slightly more than the energy available in a large apple.

Your body is really quite efficient. The typical 25% efficiency that we've used might not sound that great, but it's better than the efficiency of a typical automobile or that of other devices that use chemical energy as a power source. This efficiency is something to be aware of, though, if you have overindulged and plan to "work off" the calories. Based on the previous examples, the energy in a typical fast-food meal would power the cyclist for more than 3 hours!

EXAMPLE 11.7 How high can you climb? BIO

How many flights of stairs could you climb on the energy contained in a 12 oz can of soda? During the climb, how much energy is transformed into thermal energy? Assume that your mass is 68 kg and that a flight of stairs has a vertical height of 2.7 m.

STRATEGIZE This is a Type 1 problem. It's easy to quantify what you get—a change in potential energy. This is the energy output. What you had to pay is the energy input, the energy in the soda. We will determine the energy in the soda. Then, if we assume a typical efficiency, 25% of this energy will go to the increase in potential energy, which we will use to determine the height. The rest will go to thermal energy.

PREPARE In Example 11.4, we determined that the energy content of a can of soda is 680 kJ. This is the chemical energy input. We considered the work-energy equation for this case earlier in the section; we can repeat Equation 11.5, noting the relative sizes of the different energies:

$$|\Delta E_{chem}| = \Delta U_g + \Delta E_{th}$$

This is "what you get," the change in potential energy. It is 25% of the total.

The balance, 75% of the total, is transformed into thermal energy.

SOLVE At 25% efficiency, the amount of chemical energy transformed into increased potential energy is

$$\Delta U_g = (0.25)(680 \times 10^3\ \text{J}) = 1.7 \times 10^5\ \text{J}$$

Because $\Delta U_g = mg\Delta y$, the height gained is

$$\Delta y = \frac{\Delta U_g}{mg} = \frac{1.7 \times 10^5\ \text{J}}{(68\ \text{kg})(9.8\ \text{m/s}^2)} = 255\ \text{m}$$

With each flight of stairs having a height of 2.7 m, the number of flights climbed is

$$\frac{255\ \text{m}}{2.7\ \text{m}} \approx 94\ \text{flights}$$

The balance of the energy is transformed into thermal energy:

$$\Delta E_{th} = |\Delta E_{chem}| - \Delta U_g = 6.8 \times 10^5\ \text{J} - 1.7 \times 10^5\ \text{J}$$
$$= 5.1 \times 10^5\ \text{J} = 510\ \text{kJ}$$

ASSESS 94 flights is almost enough to get to the top of the Empire State Building! But this makes sense; there is a lot of energy in a can of soda. And if you climb several sets of stairs, you know that your body warms up, so the large increase in thermal energy makes sense as well.

Energy Storage

If the energy that the body gets from food is not used, it is stored. A small amount of energy needed for immediate use is stored as ATP. A larger amount of energy is stored as chemical energy of simple carbohydrates in muscle tissue and the liver. A healthy adult might store 400 g of these carbohydrates, which is a little more carbohydrate than is typically consumed in one day.

If the energy input from food continuously exceeds the energy outputs of the body, this energy will be stored in the form of fat under the skin and around the organs. From an energy point of view, gaining weight is simply explained!

EXAMPLE 11.8 How far can you run? BIO

The body stores about 400 g of carbohydrates. Approximately how far could a 68 kg runner travel on this stored energy?

STRATEGIZE This is a Type 2 problem; we can't easily quantify what you get, so we'll use the value from Table 11.4 to determine the power. We will use the data in Table 11.1 to determine the energy in the carbohydrate stored in the body. From the energy available and the rate at which the body uses this energy, we can determine how long the energy will last, and then we will calculate the distance the runner can travel.

PREPARE Table 11.1 gives a value of 17 kJ per g of carbohydrate. The 400 g of carbohydrates in the body contain an energy of

$$E_{chem} = (400\text{ g})(17 \times 10^3\text{ J/g}) = 6.8 \times 10^6\text{ J}$$

SOLVE Table 11.4 gives the power used in running at 15 km/h as 1150 W. The time that the stored chemical energy will last at this rate is

$$\Delta t = \frac{\Delta E_{chem}}{P} = \frac{6.8 \times 10^6\text{ J}}{1150\text{ W}} = 5.91 \times 10^3\text{ s} = 1.64\text{ h}$$

And the distance that can be covered during this time at 15 km/h is

$$\Delta x = v\Delta t = (15\text{ km/h})(1.64\text{ h}) = 25\text{ km}$$

to two significant figures.

ASSESS A marathon is longer than this—just over 42 km. Even with "carbo loading" before the event (eating high-carbohydrate meals), many marathon runners "hit the wall" before the end of the race as they reach the point where they have exhausted their store of carbohydrates. Given that this is a problem for long-distance runners, our answer makes sense. But if runners deplete their carbohydrate reserves, how can they finish the race? The body has other energy stores (in fats, for instance), but the rate that they can be drawn on is much lower.

Energy and Locomotion

When you walk at a constant speed on level ground, your kinetic energy is constant. Your potential energy is also constant. So why does your body need energy to walk? Where does this energy go?

We use energy to walk because of mechanical inefficiencies in our gait. FIGURE 11.7 shows how the speed of your foot typically changes during each stride. The kinetic energy of your leg and foot increases, only to go to zero at the end of the stride. The kinetic energy is mostly transformed into thermal energy in your muscles and in your shoes. This thermal energy is lost; it can't be used for making more strides.

This inefficiency is a process limitation. It's possible to do better. Footwear can be designed to minimize the loss of kinetic energy to thermal energy. A spring in the sole of the shoe can store potential energy, which can be returned to kinetic energy during the next stride. Such a spring will make the collision with the ground more elastic. We saw in Chapter 10 that the tendons in the ankle store a certain amount of energy during a stride; very stout tendons in the legs of kangaroos store energy even more efficiently. Their peculiar hopping gait is quite efficient at high speeds.

FIGURE 11.7 Human locomotion analysis.

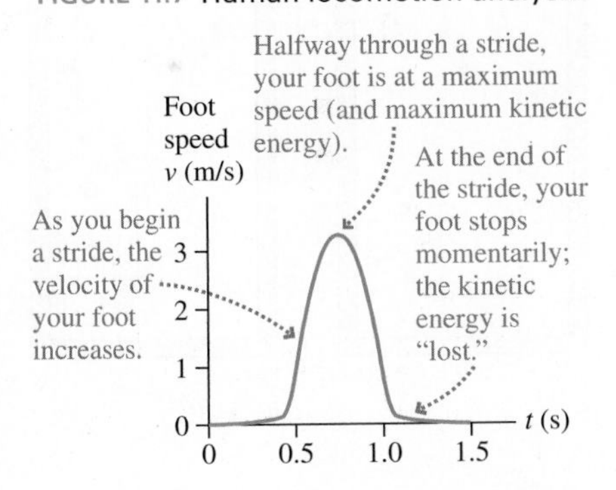
Halfway through a stride, your foot is at a maximum speed (and maximum kinetic energy).
Foot speed v (m/s)
At the end of the stride, your foot stops momentarily; the kinetic energy is "lost."
As you begin a stride, the velocity of your foot increases.
3
2
1
0
0
0.5
1.0
1.5
t (s)

▶ **Where do you wear the weights? BIO** If you wear a backpack with a mass equal to 1% of your body mass, your energy expenditure for walking will increase by 1%. But if you wear ankle weights with a combined mass of 1% of your body mass, the increase in energy expenditure is 6%, because you must repeatedly accelerate this extra mass. If you want to "burn more fat," wear the weights on your ankles, not on your back! If you are a runner who wants to shave seconds off your time in the mile, you might try lighter shoes.

STOP TO THINK 11.2 A runner is moving at a constant speed on level ground. Chemical energy in the runner's body is being transformed into other forms of energy. Most of the chemical energy is transformed into

A. Kinetic energy. B. Potential energy. C. Thermal energy.

11.3 Temperature, Thermal Energy, and Heat

What do you mean when you say something is "hot"? Do you mean that it has a high temperature? Or do you mean that it has a lot of thermal energy? Or maybe that the object in some way contains "heat"? Are all of these definitions the same? Let's give some thought to the meanings of temperature, thermal energy, and heat, and the relationship between them.

Relating Thermal Energy and Temperature

FIGURE 11.8 The brake disks on a race car can reach very high temperatures during rapid braking.

As we saw in ◀◀ SECTION 10.5, the thermal energy of an object is the sum of the kinetic and elastic potential energies of the object's atoms. When we add thermal energy, the atoms move more quickly and the bonds between atoms stretch. This change in thermal energy also leads to a higher temperature.

When a race car driver hits the brakes, the kinetic energy of the moving vehicle is transformed into thermal energy of the brake disks. This increases the temperature of the brake disks, as the photo in FIGURE 11.8 shows; the disks are at a high enough temperature to glow like the burner on an electric stove. The thermal energy of the disks must increase as the car comes to rest, but it's useful to limit the resulting change in temperature. In practice, race car designers can do this by making one simple change: using more massive brake disks. How does increasing the disks' mass keep them at a lower temperature? At this point, we'll need to carefully consider the meanings of thermal energy and temperature. We'll do this by considering a gas, but the results are general and apply to all phases of matter.

FIGURE 11.9 Atoms of a gas at low temperature and high temperature.

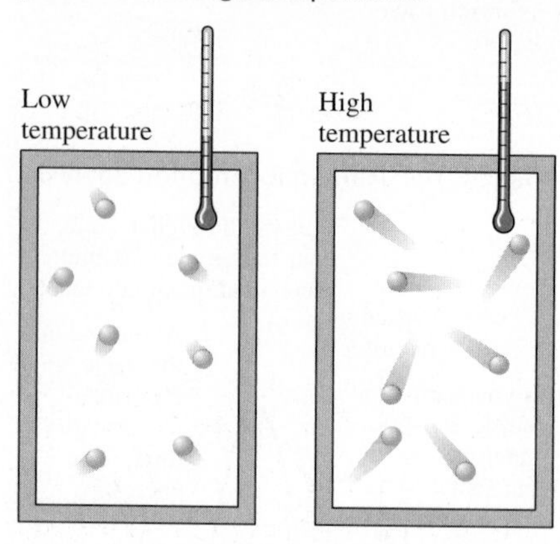

FIGURE 11.9 shows two samples of gas, one at a low temperature and one at a high temperature. You know what this means: The atoms in the high-temperature gas move at a higher speed than the atoms in the low-temperature gas. You can also see that the gas sample on the right has a higher thermal energy than the gas sample on the left. The two samples have the same number of atoms, but the atoms in the sample on the right have higher kinetic energy.

Now let's do a thought experiment: Add more and more atoms to the sample on the left, with each new atom having the same speed as the atoms already there. This clearly increases the thermal energy of the system—it increases the total energy of the atoms—but it won't increase the temperature.

The temperature of a system does not depend on the size of the system. If you mix together two glasses of water, each at a temperature of 20°C, you will have a larger volume of water at the same temperature of 20°C. The combined volume has more atoms, and therefore more total thermal energy, but each atom is moving just about as it was before, and so the average kinetic energy per atom is unchanged. It is this average kinetic energy of the atoms that is related to the temperature. **The temperature of an object is a measure of the *average* energy of the atoms that make up the object.**

Thermal energy is the *total* energy of the atoms; temperature is the *average* energy. This is an important distinction.

CONCEPTUAL EXAMPLE 11.9 How does increased mass lead to decreased temperature?

A race car driver notices that during periods of aggressive driving, the brake disks on her car reach temperatures that are too high. She switches to new disks that are made of the same material but have twice the mass, and the temperature of the disks now stays in a reasonable range. How did increasing the mass of the disks solve the problem?

REASON During braking, kinetic energy of the car is transferred to thermal energy in the brake disks. If the mass of the disks is increased, this energy is shared over a larger number of atoms, so the average energy of each atom increases by less. The addition of thermal energy therefore causes a smaller temperature rise.

ASSESS This explanation makes sense. You know, from experience, that it takes more energy to raise the temperature of larger objects, so a given energy input will cause a smaller temperature change in a larger object.

Temperature Scales

A common glass-tube thermometer works by allowing a small volume of mercury or alcohol to expand or contract when placed in contact with a "hot" or "cold" object. Other thermometers work in different ways, but all—at a microscopic level—depend on the speed of the object's atoms as they collide with the atoms in the thermometer. All thermometers are sampling the average kinetic energy of the atoms in the object.

The temperature scale used in scientific work (and in almost every country in the world) is the *Celsius scale*. As you likely know, the Celsius scale is defined so that the freezing point of water is 0°C and the boiling point is 100°C. The units of the Celsius temperature scale are "degrees Celsius," which we abbreviate as °C. The Fahrenheit scale, still widely used in the United States, is related to the Celsius scale by

$$T(°C) = \frac{5}{9}(T(°F) - 32°) \qquad T(°F) = \frac{9}{5}T(°C) + 32° \qquad (11.6)$$

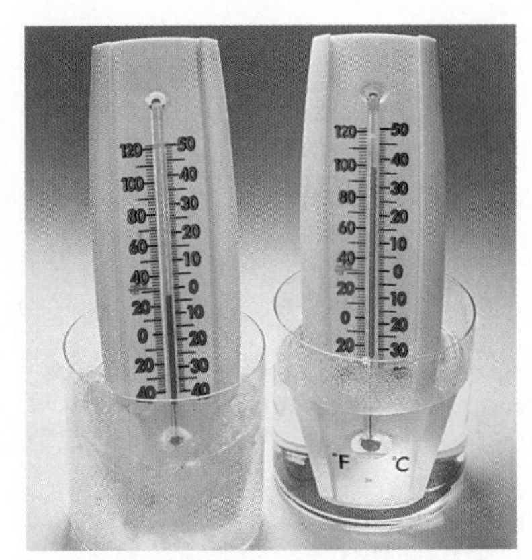

Thermal expansion of the liquid in the thermometer pushes it higher when immersed in hot water than in ice water.

Both the Celsius and the Fahrenheit scales have a zero point that is arbitrary—simply an agreed-upon convention—and both allow negative temperatures. But we can define a temperature scale for which the temperature is always positive. Consider the gas samples in Figure 11.9. If we slow the atoms down, this reduces the average kinetic energy of the atoms and so reduces the temperature. But there's a limit. If the atoms slow to a stop, the kinetic energy is zero; no lower temperature is possible. This is the temperature we call **absolute zero.** If we take this temperature as the zero of our temperature scale, temperatures below zero are not possible.

This is how zero is defined on the temperature scale called the *Kelvin scale:* **Zero degrees is the point at which the kinetic energy of atoms is zero.** All temperatures on the Kelvin scale are positive, so it is often called an *absolute temperature scale.* The units are "kelvin" (not degrees kelvin!), abbreviated K.

The spacing between divisions on the Kelvin scale is the same as that on the Celsius scale; the only difference is the position of the zero point. Absolute zero—the temperature at which atoms would cease moving—is −273°C. The conversion between Celsius and Kelvin temperatures is therefore quite straightforward:

$$T(K) = T(°C) + 273 \qquad T(°C) = T(K) - 273 \qquad (11.7)$$

One Celsius degree and one kelvin are the same. This means that a temperature *difference* is the same on both scales:

$$\Delta T(K) = \Delta T(°C)$$

On the Kelvin scale, the freezing point of water at 0°C is $T = 0 + 273 = 273$ K. A 30°C warm summer day is $T = 303$ K on the Kelvin scale. FIGURE 11.10 gives a side-by-side comparison of these scales.

FIGURE 11.10 Celsius and Kelvin temperature scales.

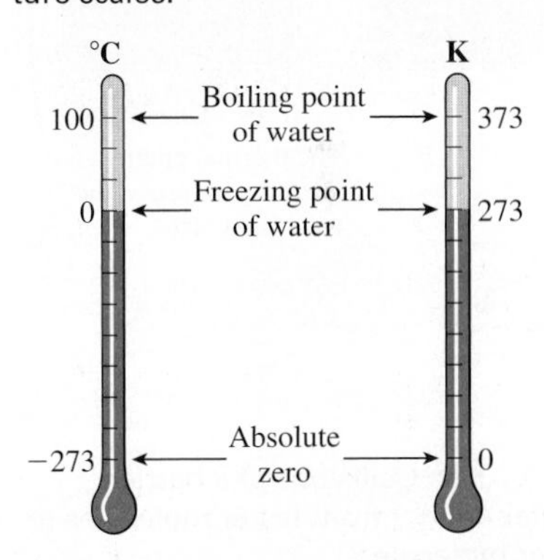

Temperature *differences* are the same on the Celsius and Kelvin scales. The temperature difference between the freezing point and boiling point of water is 100°C or 100 K.

NOTE ▸From now on, we will use the symbol T for temperature in kelvin. We will denote other scales by showing the units in parentheses. In the equations in this chapter and the rest of the text, ***T* should be interpreted as a temperature in kelvin.** ◂

EXAMPLE 11.10 Finding the temperature on different scales

The coldest temperature ever measured on earth was −129°F, in Antarctica. What is this in °C and K?

SOLVE We first use Equation 11.6 to convert the temperature to the Celsius scale:

$$T(°C) = \frac{5}{9}(-129° - 32°) = -89°C$$

We can then use Equation 11.7 to convert this to kelvin:

$$T = -89 + 273 = 184 \text{ K}$$

ASSESS This is cold, but still far warmer than the coldest temperatures achieved in the laboratory, which are very close to 0 K.

Optical molasses It isn't possible to reach absolute zero, where the atoms would be still, but it is possible to get quite close by slowing the atoms down directly. These crossed laser beams produce what is known as "optical molasses." As we will see in Chapter 28, light is made of photons, which carry energy and momentum. Interactions of the atoms of a diffuse gas with the photons cause the atoms to slow down. In this manner atoms can be slowed to speeds that correspond to a temperature as cold as 5×10^{-10} K!

STOP TO THINK 11.3 Two samples of a gas, sample 1 and sample 2, have the same thermal energy. Sample 1 has twice as many atoms as sample 2. What can we say about the temperatures of the two samples?

A. $T_1 > T_2$ B. $T_1 = T_2$ C. $T_1 < T_2$

What Is Heat?

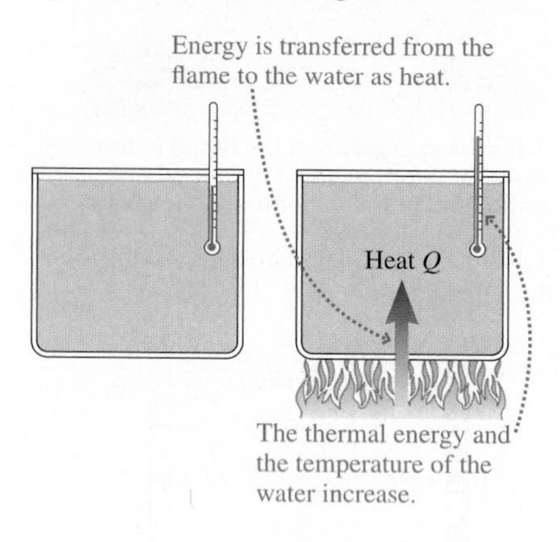

Energy is transferred from the flame to the water as heat.
Heat Q
The thermal energy and the temperature of the water increase.

FIGURE 11.11 Raising the temperature of a pan of water by heating it.

In Chapter 10, we saw that a system could exchange energy with the environment through two different means: work and heat. Work was treated in detail in Chapter 10; now it is time to look at the transfer of energy by heat.

Heat is a more elusive concept than work. We use the word "heat" very loosely in the English language, often as synonymous with "hot." We might say, on a very hot day, "This heat is oppressive." If your apartment is cold, you may say, "Turn up the heat." It's time to develop more precise language to discuss these concepts.

Suppose you put a pan of cold water on the stove. If you light the burner so that there is a hot flame under the pan, as in FIGURE 11.11, the temperature of the water increases. We know that this increased temperature means that the water has more thermal energy, so this process must have *transferred* energy to the system. This energy transferred from the hotter flame to the cooler water is what we call heat. In general, **heat is energy transferred between two objects because of a temperature difference between them.** Heat always flows from the hotter object to the cooler one, never in the opposite direction. If there is no temperature difference, no energy is transferred as heat. We use Q as the symbol for heat.

An Atomic Model of Heat

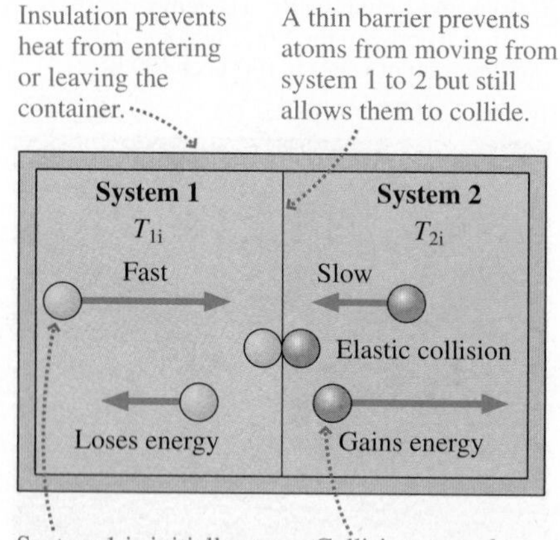

Insulation prevents heat from entering or leaving the container.
A thin barrier prevents atoms from moving from system 1 to 2 but still allows them to collide.
System 1
T_{1i}
Fast
System 2
T_{2i}
Slow
Elastic collision
Loses energy
Gains energy
System 1 is initially at a higher temperature, and so has atoms with a higher average speed.
Collisions transfer energy to the atoms in system 2.

FIGURE 11.12 Collisions at a barrier transfer energy from faster molecules to slower molecules.

Let's consider an atomic model to explain why thermal energy is transferred from higher temperatures to lower. FIGURE 11.12 shows a rigid, insulated container that is divided into two sections by a very thin barrier. Each side is filled with a gas of the same kind of atoms. The left side, which we'll call system 1, is at an initial temperature T_{1i}. System 2 on the right is at an initial temperature T_{2i}. We imagine the barrier to be so thin that atoms can collide at the boundary as if the barrier were not there, yet it prevents atoms from moving from one side to the other.

Suppose that system 1 is initially at a higher temperature: $T_{1i} > T_{2i}$. This means that the atoms in system 1 have a higher average kinetic energy. Figure 11.12 shows a fast atom and a slow atom approaching the barrier from opposite sides. They undergo a perfectly elastic collision at the barrier. Although no net energy is lost in a perfectly elastic collision, the faster atom loses energy while the slower one gains energy. There is an energy *transfer* from the faster atom's side to the slower atom's side.

Because the atoms in system 1 are, on average, more energetic than the atoms in system 2, the collisions transfer energy from system 1 to system 2. The net energy transfer, from all collisions, is from the warmer system 1 to the cooler system 2. This transfer of energy is heat; **thermal energy is transferred from the faster atoms on the warmer side to the slower atoms on the cooler side.**

This transfer will continue until a stable situation is reached. This is a situation we call **thermal equilibrium.** How do the systems "know" when they've reached thermal equilibrium? Energy transfer continues until the atoms on both sides of the barrier have the *same average kinetic energy.* Once the average kinetic energies are the same, individual collisions will still transfer energy from one side to the other. But since both sides have atoms with the same average kinetic energies, the amount of energy transferred from 1 to 2 will equal the amount transferred from 2 to 1. Once the average kinetic energies are the same, there is no net energy transfer.

As we've seen, the average kinetic energy of the atoms in a system is a measure of the system's temperature. If two systems exchange energy until their atoms have the same average kinetic energy, we can say that

$$T_{1f} = T_{2f} = T_f$$

That is, heat is transferred until the two systems reach a common final temperature; this is the state of thermal equilibrium. We considered a rather artificial system in this case, but the result is quite general: **Two systems placed in thermal contact will transfer thermal energy from hot to cold until their final temperatures are the same.** This process is illustrated in FIGURE 11.13.

FIGURE 11.13 Two systems in thermal contact exchange thermal energy.

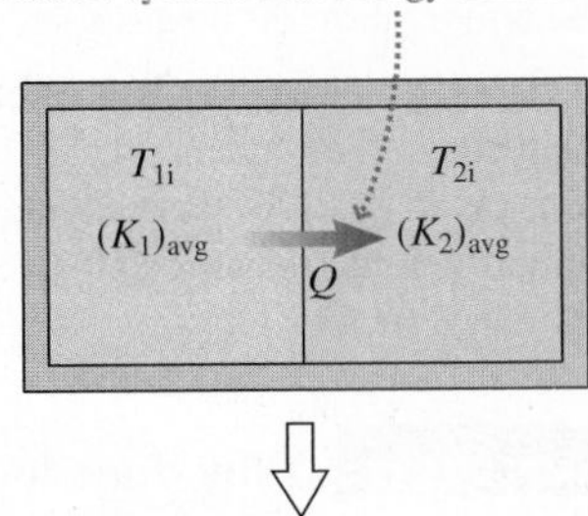

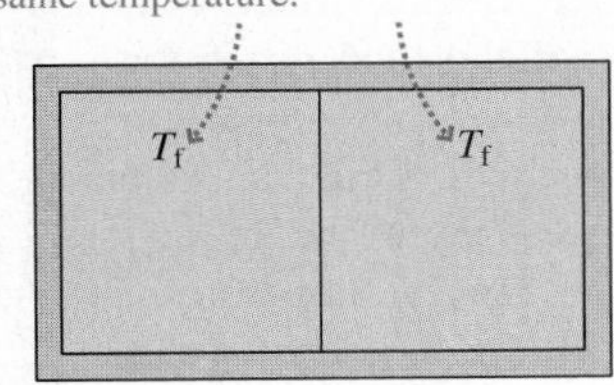

Heat is a transfer of energy. The sign that we use for transfers is defined in FIGURE 11.14. In the process of Figure 11.13, Q_1 is negative because system 1 loses energy; Q_2 is positive because system 2 gains energy. No energy escapes from the container, so all of the energy that was lost by system 1 was gained by system 2. We can write that as

$$Q_2 = -Q_1$$

The heat energy lost by one system is gained by the other.

FIGURE 11.14 The sign of Q.

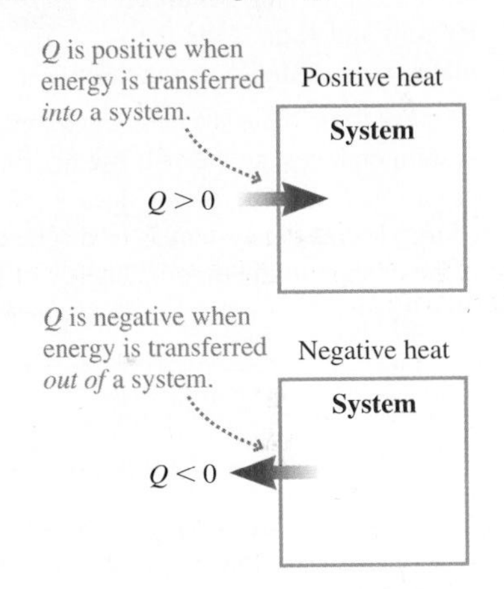

11.4 The First Law of Thermodynamics

We need to broaden our work-energy equation, Equation 11.1, to include energy transfers in the form of heat. The sign conventions for heat, as shown in Figure 11.14, have the same form as those for work—a positive value means a transfer into the system, a negative value means a transfer out of the system. Thus, when we include heat Q in the work-energy equation, it appears on the right side of the equation along with work W:

$$\Delta K + \Delta U + \Delta E_{th} + \Delta E_{chem} + \cdots = W + Q \quad (11.8)$$

This equation is our most general statement to date about the conservation of energy; it includes all of the energy transfers and transformations we have discussed.

In Chapter 10, we focused on systems where the potential and kinetic energies could change, such as a sled moving down a hill. Earlier in this chapter, we looked at the body, where the chemical energy changes. Now lets consider systems in which only the thermal energy changes, systems that aren't moving, that aren't changing chemically, but whose temperatures can change. Such systems are the province of **thermodynamics,** the study of thermal energy and heat and their relationships to other forms of energy and energy transfer. The question of how to keep your house cool in the summer is a question of thermodynamics. If energy is transferred into your house, the thermal energy increases and the temperature rises. To reduce the temperature, you must transfer energy out of the house. This is the purpose of an air conditioner.

If we consider cases in which only thermal energy changes, Equation 11.8 can be simplified. This simpler version is a statement of conservation of energy for systems in which only the thermal energy changes; we call it the **first law of thermodynamics:**

A video to support a section's topic is embedded in the eText.

Video Work and Thermal Energy in Gases

First law of thermodynamics For systems in which only the thermal energy changes, the change in thermal energy is equal to the energy transferred into or out of the system as work W, heat Q, or both:

$$\Delta E_{th} = W + Q \quad (11.9)$$

Work and heat are the two ways by which energy can be transferred between a system and its environment, thereby changing the system's energy. And, in thermodynamic systems, the only energy that changes is the thermal energy. Whether this energy increases or decreases depends on the signs of W and Q. The possible energy transfers between a system and the environment are illustrated in FIGURE 11.15.

FIGURE 11.15 Energy transfers in a thermodynamic system.

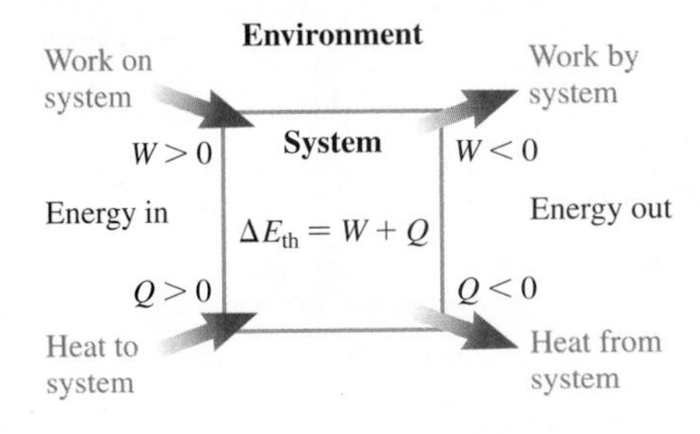

CONCEPTUAL EXAMPLE 11.11 What happens when you compress a gas?

Suppose a gas is in an insulated container, so that no heat energy can escape. If a piston is used to compress the gas, what happens to the temperature of the gas?

REASON The piston applies a force to the gas, and there is a displacement. This means that work is done on the gas by the piston ($W > 0$). No thermal energy can be exchanged with the environment, meaning $Q = 0$. Since energy is transferred into the system, the thermal energy of the gas must increase. This means that the temperature must increase as well.

ASSESS This result makes sense in terms of everyday observations you may have made. When you use a bike pump to inflate a tire, the pump and the tire get warm. This temperature increase is largely due to the warming of the air by the work you do compressing the gas.

EXAMPLE 11.12 Why does food heat up in a blender?

If you mix food in a blender, the electric motor does work on the system, which consists of the food inside the container. This work can noticeably warm up the food. Suppose the blender motor runs at a power of 250 W for 40 s. During this time, 2000 J of heat flow from the now-warmer food to its cooler surroundings. By how much does the thermal energy of the food increase?

STRATEGIZE This is a case where only the thermal energy of the system changes, so we will use the first law of thermodynamics, Equation 11.9. The motor does work on the system, and some energy leaves the system as heat. The difference between the two is the change in the thermal energy of the system.

PREPARE We can find the work done by the motor from the power it uses and the time it runs. Equation 10.24 relates work to power and time. The work done on the system is

$$W = P\Delta t = (250\ \text{W})(40\ \text{s}) = 10{,}000\ \text{J}$$

Because heat *leaves* the system, its sign is negative, so $Q = -2000$ J.

SOLVE With these values in hand, the first law of thermodynamics gives

$$\Delta E_{\text{th}} = W + Q = 10{,}000\ \text{J} - 2000\ \text{J} = 8000\ \text{J}$$

ASSESS It seems reasonable that the work done by the powerful motor rapidly increases the thermal energy, while thermal energy only slowly leaks out as heat. The increased thermal energy of the food implies an increased temperature. If you run a blender long enough, the food can actually start to steam, as the photo shows.

FIGURE 11.16 Energy-transfer diagrams.

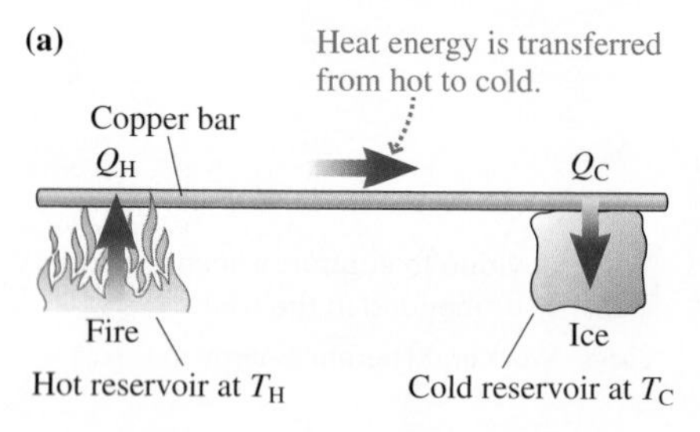
(a)
Heat energy is transferred from hot to cold.
Copper bar
Q_H
Q_C
Fire
Ice
Hot reservoir at T_H
Cold reservoir at T_C

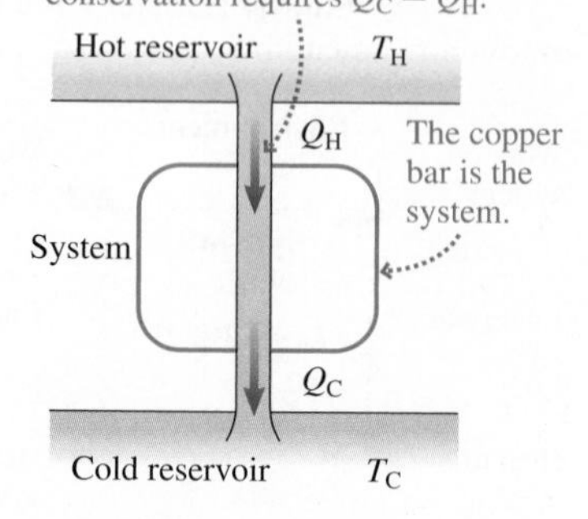
(b)
Heat energy is transferred from a hot reservoir to a cold reservoir. Energy conservation requires $Q_C = Q_H$.
Hot reservoir
T_H
Q_H
The copper bar is the system.
System
Q_C
Cold reservoir
T_C

Energy-Transfer Diagrams

Suppose you drop a hot rock into the ocean. Heat is transferred from the rock to the ocean until the rock and ocean are the same temperature. Although the ocean warms up ever so slightly, ΔT_{ocean} is so small as to be completely insignificant.

An **energy reservoir** is an object or a part of the environment so large that, like the ocean, its temperature does not noticeably change when heat is transferred between the system and the reservoir. A reservoir at a higher temperature than the system is called a *hot reservoir.* A vigorously burning flame is a hot reservoir for small objects placed in the flame. A reservoir at a lower temperature than the system is called a *cold reservoir.* The ocean is a cold reservoir for the hot rock. We will use T_H and T_C to designate the temperatures of the hot and cold reservoirs.

Heat energy is transferred between a system and a reservoir if they have different temperatures. We will define

$$Q_H = \text{amount of heat transferred to or from a hot reservoir}$$

$$Q_C = \text{amount of heat transferred to or from a cold reservoir}$$

By definition, Q_H and Q_C are *positive* quantities.

FIGURE 11.16a shows a heavy copper bar placed between a hot reservoir (at temperature T_H) and a cold reservoir (at temperature T_C). Heat Q_H is transferred from the hot reservoir into the copper, and heat Q_C is transferred from the copper into the cold reservoir. FIGURE 11.16b is an **energy-transfer diagram** for this process. The hot reservoir is generally drawn at the top, the cold reservoir at the bottom, and the system—the copper bar in this case—between them. The reservoirs and the system are connected by "pipes" that show the energy transfers. Figure 11.16b shows heat Q_H being transferred into the system and Q_C being transferred out.

FIGURE 11.17 illustrates an important fact about heat transfers that we have discussed: Spontaneous transfers go in one direction only, from hot to cold. This is an important result that has significant practical implications.

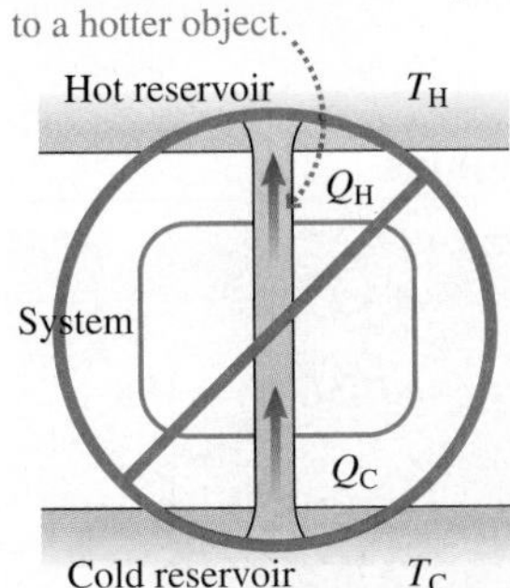

Heat is never spontaneously transferred from a colder object to a hotter object.
Hot reservoir
T_H
Q_H
System
Q_C
Cold reservoir
T_C

FIGURE 11.17 An impossible energy transfer.

CONCEPTUAL EXAMPLE 11.13 Keeping your cool BIO

Why—in physics terms—is it more taxing on the body to exercise in very hot weather?

REASON Your body continuously converts chemical energy to thermal energy, as we have seen. In order to maintain a constant body temperature, your body must continuously transfer heat to the environment. This is a simple matter in cool weather when heat is spontaneously transferred to the environment, but when the air temperature is higher than your body temperature, your body cannot cool itself this way and must use other mechanisms to transfer this energy, such as perspiring. These mechanisms require additional energy expenditure.

ASSESS Strenuous exercise in hot weather can easily lead to a rise in body temperature if the body cannot exhaust heat quickly enough.

STOP TO THINK 11.4 You have driven your car for a while and now turn off the engine. Your car's radiator is at a higher temperature than the air around it. Considering the radiator as the system, as the radiator cools down we can say that

A. $Q > 0$ B. $Q = 0$ C. $Q < 0$

11.5 Heat Engines

In the early stages of the industrial revolution, most of the energy needed to run mills and factories came from water power. Water in a high reservoir will naturally flow downhill. A waterwheel can be used to harness this natural flow of water to produce some useful energy because some of the potential energy lost by the water as it flows downhill can be converted into other forms.

It is possible to do something similar with heat. Thermal energy is naturally transferred from a hot reservoir to a cold reservoir; it is possible to take some of this energy as it is transferred and convert it to other forms. This is the job of a device known as a **heat engine.**

The energy-transfer diagram of FIGURE 11.18a illustrates the basic physics of a heat engine. It takes in energy as heat from the hot reservoir, turns some of it into useful work, and exhausts the balance as waste heat into the cold reservoir. Any heat engine has exactly the same schematic.

FIGURE 11.18 The operation of a heat engine.

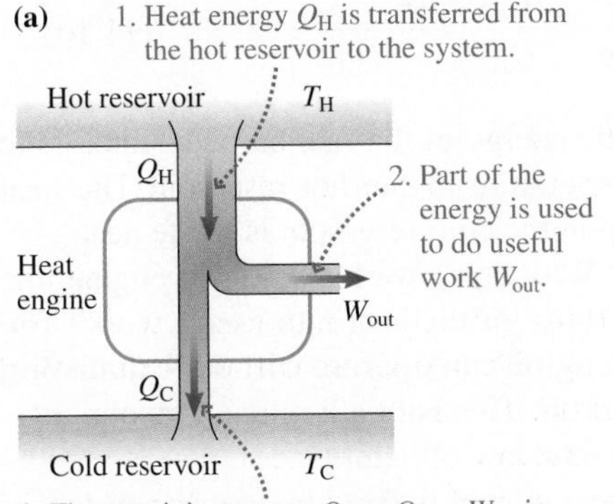

(a)
1. Heat energy Q_H is transferred from the hot reservoir to the system.
Hot reservoir
T_H
Q_H
2. Part of the energy is used to do useful work W_{out}.
Heat engine
W_{out}
Q_C
Cold reservoir
T_C
3. The remaining energy $Q_C = Q_H - W_{out}$ is exhausted to the cold reservoir as waste heat.

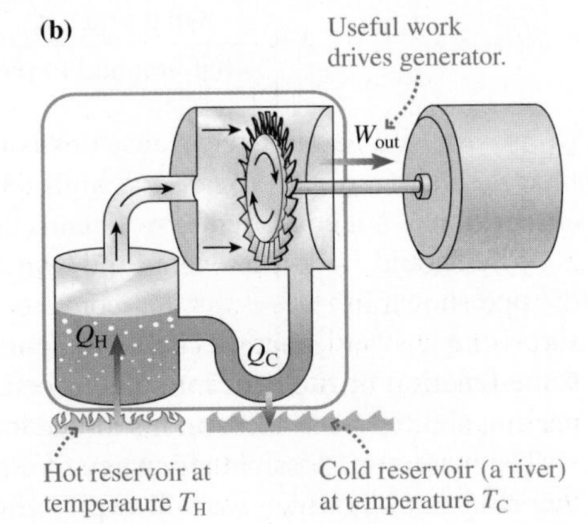

(b)
Useful work drives generator.
W_{out}
Q_H
Q_C
Hot reservoir at temperature T_H
Cold reservoir (a river) at temperature T_C

Video Figure 11.18

FIGURE 11.18b shows the directions of the energy flows and the locations of the reservoirs for a real heat engine, the power plant discussed earlier in this chapter. The cold reservoir for a power plant can be a river or lake, as shown, or simply the atmosphere, to which heat Q_C is transferred by cooling towers.

Most of the energy that you use daily comes from the conversion of chemical energy into thermal energy and the subsequent conversion of that energy into other forms. The following table shows some common examples of heat engines.

Heat engines

Much of the electricity that you use was generated by heat engines. Natural gas or other fossil fuels are burned to produce high-temperature, high-pressure steam. The steam does work by spinning a turbine attached to a generator, which produces electricity. Some of the energy of the steam is extracted in this way, but more than half simply flows "downhill" and is deposited in a cold reservoir, often a lake or a river.

Your car gets the energy it needs to run from the chemical energy in gasoline. The gasoline is burned; the resulting hot gases are the hot reservoir. Some of the thermal energy is converted into the kinetic energy of the moving vehicle, but more than 90% is lost as heat to the surrounding air via the radiator and the exhaust, as shown in this thermogram.

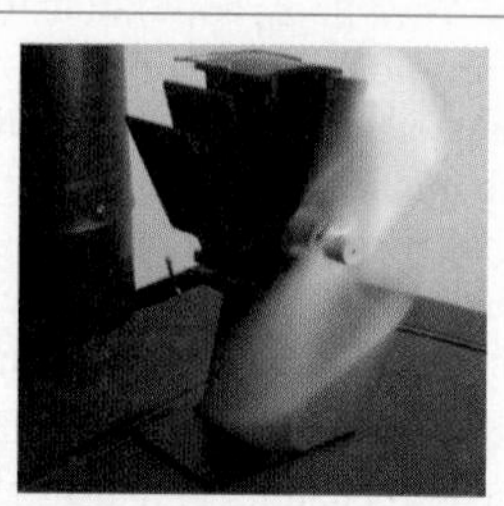

There are many small, simple heat engines that are part of things you use daily. This fan, which can be put on top of a wood stove, uses the thermal energy of the stove to provide power to drive air around the room. Where are the hot and cold reservoirs in this device?

We assume, for a heat engine, that the engine's thermal energy doesn't change. This means that there is no net energy transfer into or out of the heat engine. Because energy is conserved, we can say that the useful work extracted is equal to the difference between the heat energy transferred from the hot reservoir and the heat exhausted into the cold reservoir:

$$W_{out} = Q_H - Q_C$$

The energy input to the engine is Q_H and the energy output is W_{out}.

NOTE ▸ We earlier defined Q_H and Q_C to be positive quantities. We also define the energy output of a heat engine W_{out} to be a positive quantity. For heat engines, the directions of the transfers will always be clear, and we will take all of these basic quantities to be positive. ◂

We can use the definition of efficiency, from earlier in the chapter, to compute the heat engine's efficiency:

$$e = \frac{\text{what you get}}{\text{what you had to pay}} = \frac{W_{out}}{Q_H} = \frac{Q_H - Q_C}{Q_H} \quad (11.10)$$

Q_H is what you had to pay because this is the energy of the fuel burned—and, quite literally, paid for—to provide the high temperature of the hot reservoir. The heat energy that is not converted to work ends up in the cold reservoir as waste heat.

Why should we waste energy this way? Why don't we make a heat engine like the one shown in FIGURE 11.19 that converts 100% of the heat into useful work? The surprising answer is that we can't. **No heat engine can operate without exhausting some fraction of the heat into a cold reservoir.** This isn't a limitation on our engineering abilities. As we'll see, it's a fundamental law of nature.

The maximum possible efficiency of a heat engine is fixed by the *second law of thermodynamics,* which we will explore in Section 11.7. We will not do a detailed

FIGURE 11.19 A perfect (and impossible!) heat engine.

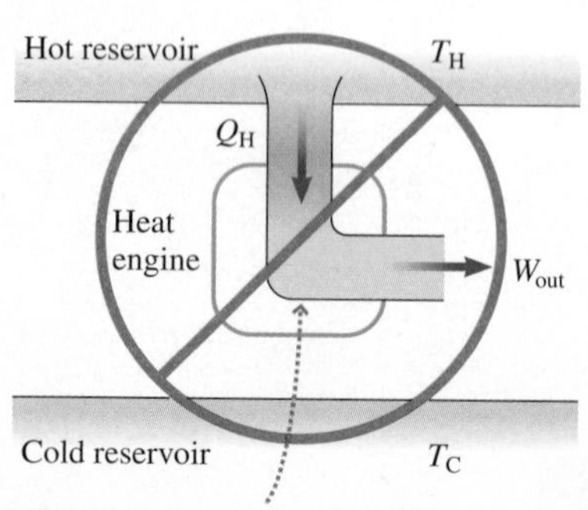
Hot reservoir
T_H
Q_H
Heat engine
W_{out}
Cold reservoir
T_C
This is impossible! No heat engine can convert 100% of heat into useful work.

derivation, but simply note that the second law gives the theoretical maximum efficiency of an ideal heat engine as

$$e_{max} = 1 - \frac{T_C}{T_H} \quad (11.11)$$

Maximum possible efficiency of a heat engine; T_C: Temperature of cold reservoir; T_H: Temperature of hot reservoir. Both T_C and T_H must be in *kelvin*.

The maximum possible efficiency of any heat engine is fixed by the ratio of the temperatures of the hot and cold reservoirs. We can increase the efficiency of a heat engine by increasing the temperature of the hot reservoir or by decreasing the temperature of the cold reservoir.

NOTE ▶ The actual efficiency of real heat engines is usually noticeably less than the theoretical maximum. Equation 11.10 is a statement of energy conservation; it applies to *any* heat engine. Equation 11.11 applies only to an *ideal* heat engine; it's the upper limit of what is possible. ◀

The work done by a heat engine is always less than the heat input: $W < Q_H$. Consequently, there *must* be heat Q_C exhausted to the cold reservoir. We can increase the efficiency and reduce the quantity of heat that is "wasted" this way, but we can't eliminate it. This is a *fundamental* limit on efficiency due to a law of nature. No matter how well you design your power plant, its efficiency can't exceed the limit of Equation 11.11.

EXAMPLE 11.14 Determining the theoretical and actual efficiencies for a heat engine

Iceland has both high geothermal activity, with high temperatures near the surface, and abundant cold surface water. Iceland has many power plants that take advantage of the proximity of these natural hot and cold reservoirs. One plant uses an underground source at 122°C as the hot reservoir and a nearby lake at 5°C as the cold reservoir. The plant draws 16 MW from the hot reservoir to produce 1.8 MW of electricity. How does the actual efficiency of the plant compare to the theoretical maximum efficiency?

STRATEGIZE Computing the actual efficiency will be straightforward; it's just the ratio of what you get (the electricity output) to what you had to pay (the heat input). We will determine the theoretical efficiency from the temperatures of the hot and cold reservoirs, which are given.

PREPARE The plant's operation is specified in terms of power, but Equation 11.10 uses energy, so we can consider the energy transferred or transformed in 1 second. In 1 second, the plant draws 16 MJ from the hot reservoir and produces 1.8 MJ of electric energy.

To compute the maximum possible efficiency, we need the absolute temperatures of the hot and cold reservoirs:

$$T_H = 122°C = 395 \text{ K} \qquad T_H = 5°C = 278 \text{ K}$$

SOLVE The actual efficiency of the plant is

$$e = \frac{W_{out}}{Q_H} = \frac{1.8 \text{ MJ}}{16 \text{ MJ}} = 0.11$$

The maximum possible efficiency for the plant is given by Equation 11.11:

$$e_{max} = 1 - \frac{T_C}{T_H} = 1 - \frac{278 \text{ K}}{395 \text{ K}} = 0.30$$

The theoretical maximum efficiency is nearly triple the actual efficiency.

ASSESS The actual efficiency is less than the maximum efficiency, as it must be, and this gives us confidence in our result. It's perhaps not surprising that this particular plant's efficiency is so much less than the theoretical maximum, because there is no cost for the energy input. There might not be a huge incentive to increase efficiency if you don't literally have to pay for "what you had to pay."

There are fundamental limits that specify that some of the energy in a heat engine must be deposited in a cold reservoir, but there is no law of nature that says this energy must be wasted. It's common practice in Europe for power plants to use the heat that they would otherwise exhaust into the environment to heat nearby homes. Another option is shown in the photo of the Icelandic geothermal power plant in the chapter preview. The lagoon shown is the cold reservoir of the power plant in the background; the excess heat from the plant's operation warms the water, making it a very popular bathing spot. The cold reservoir of this particular heat engine is the country's largest tourist attraction!

STOP TO THINK 11.5 Which of the following changes (there may be more than one) would increase the maximum theoretical efficiency of a heat engine?

A. Increase T_H B. Increase T_C C. Decrease T_H D. Decrease T_C

11.6 Heat Pumps

The inside of your refrigerator is colder than the air in your kitchen, so heat will always flow from the room to the inside of the refrigerator, warming it. This happens every time you open the door and as heat "leaks" through the refrigerator's walls. To keep your refrigerator cool, you need some way to move this heat back out to the warmer room. Transferring heat energy from a cold reservoir to a hot reservoir—the opposite of the natural direction—is the job of a **heat pump.**

FIGURE 11.20 A heat pump provides hot and cold water.

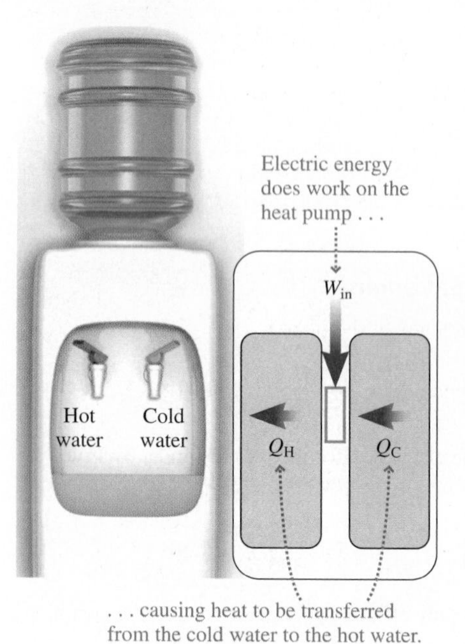
Electric energy does work on the heat pump . . .
W_{in}
Hot water
Cold water
Q_H
Q_C
. . . causing heat to be transferred from the cold water to the hot water.

If you have a container of hot water next to a container of cold water, heat will flow from the hot water to the cold. But a heat pump can cause the heat to move the other way. The water dispenser in FIGURE 11.20 has containers for hot and cold water. A heat pump between them moves heat from the cold water to the hot water. The energy used to chill the cold water results in increased thermal energy in the hot water. Paying for cold water gets you hot water for free.

Heat pumps are most often used for cooling, by transferring heat to another location. The heat pump in a refrigerator transfers heat from the cold air inside the refrigerator to the warmer air in the room. An air conditioner is a heat pump that performs a similar task, transferring heat from the cool air inside a house (or a car) to the warmer air outside. But a heat pump can also be used for heating; on a cold winter day, an electric heat pump can extract heat from the cold outside air to warm the inside of a house. A reversible heat pump can do both—move heat out of the house in the summer and into the house in the winter.

In all of these cases, we are moving energy against the natural direction it would flow. This requires an energy input—work must be done—as shown in the energy-transfer diagram of a heat pump in FIGURE 11.21. Energy must be conserved, so the heat deposited in the hot side must equal the sum of the heat removed from the cold side and the work input:

$$Q_H = Q_C + W_{in}$$

For heat pumps, rather than compute efficiency, we compute an analogous quantity called the **coefficient of performance (COP).** A refrigerator uses a heat pump for cooling, removing heat from a cold reservoir to keep it cold. As Figure 11.21 shows, we must do work to make this happen. If we use a heat pump for cooling, we define the coefficient of performance as

FIGURE 11.21 The operation of a heat pump.

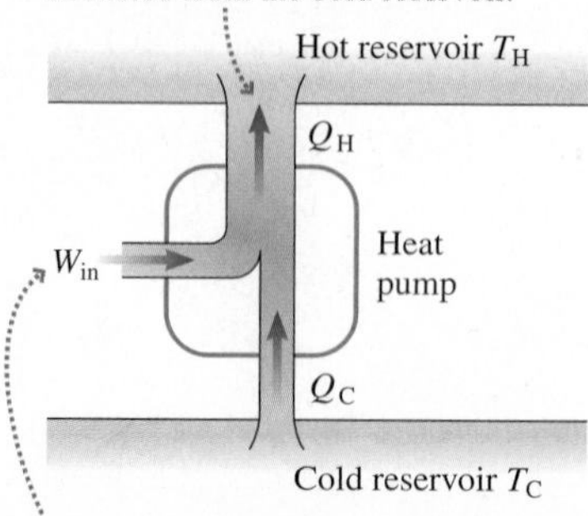
The amount of heat exhausted to the hot reservoir is larger than the amount of heat extracted from the cold reservoir.
Hot reservoir T_H
Q_H
W_{in}
Heat pump
Q_C
Cold reservoir T_C
External work is used to remove heat from a cold reservoir and exhaust heat to a hot reservoir.

$$\text{COP} = \frac{\text{what you get}}{\text{what you had to pay}} = \frac{\text{energy removed from cold reservoir}}{\text{work required to perform the transfer}} = \frac{Q_C}{W_{in}} \quad (11.12)$$

The second law of thermodynamics limits the efficiency of a heat pump just as it limits the efficiency of a heat engine. The maximum possible coefficient of performance is related to the temperatures of the hot and cold reservoirs:

$$\text{COP}_{max} = \frac{T_C}{T_H - T_C} \quad (11.13)$$

Theoretical maximum coefficient of performance of a heat pump used for cooling

We can also use a heat pump for heating, moving heat from a cold reservoir to a hot reservoir to keep it warm. In that case, we define the coefficient of performance as

$$\text{COP} = \frac{\text{what you get}}{\text{what you had to pay}} = \frac{\text{energy added to hot reservoir}}{\text{work required to perform the transfer}} = \frac{Q_H}{W_{in}} \quad (11.14)$$

In this case, the maximum possible coefficient of performance is

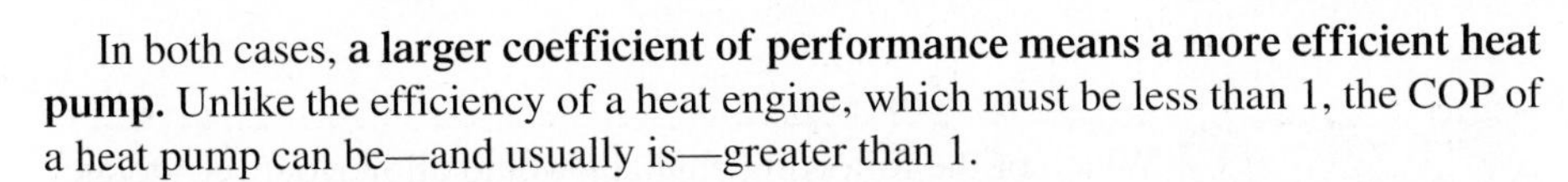

$$\text{COP}_{max} = \frac{T_H}{T_H - T_C} \quad (11.15)$$

Theoretical maximum coefficient of performance of a heat pump used for heating

Hot or cold lunch? Small coolers like this use *Peltier devices* that run off the 12 V electrical system of your car. They can transfer heat from the interior of the cooler to the outside, keeping food or drinks cool. But Peltier devices, like some other heat pumps, are reversible; switching the direction of current reverses the direction of heat transfer, causing heat to be transferred into the interior and keeping your lunch warm!

In both cases, **a larger coefficient of performance means a more efficient heat pump.** Unlike the efficiency of a heat engine, which must be less than 1, the COP of a heat pump can be—and usually is—greater than 1.

NOTE ▶ As for heat engines, the actual efficiency of real heat pumps is usually noticeably less than the theoretical maximum. Equations 11.12 and 11.14 are statements of energy conservation; they apply to *any* heat pump. Equations 11.13 and 11.15 apply only to *ideal* heat pumps; they give the upper limits of what is possible. ◀

EXAMPLE 11.15 How much energy does it take to heat the house?

A heat pump moves heat from the chilly 8°C air outside a house to the toastier 21°C air inside. The actual coefficient of performance of the system is 3.2, a typical value.

a. If the system adds 11 kJ to the house each second, how much electric energy is required?

b. How much energy would it take to add 11 kJ each second if the heat pump ran at the maximum theoretical efficiency?

STRATEGIZE This heat pump is used for heating, so we will use Equation 11.14 to compute the energy required for the system. We will use Equation 11.15 to find the maximum possible COP.

PREPARE To compute the maximum possible COP, we need to know the absolute temperatures of the cold and hot reservoirs:

$$T_H = 21°\text{C} = 294 \text{ K} \qquad T_H = 8°\text{C} = 281 \text{ K}$$

SOLVE We rearrange Equation 11.14 to find W_{in}, the electric energy required, for the given COP of 3.2:

$$W_{in} = \frac{Q_H}{\text{COP}} = \frac{11 \text{ kJ}}{3.2} = 3.4 \text{ kJ}$$

The maximum possible COP for the heat pump is given by Equation 11.15:

$$\text{COP}_{max} = \frac{T_H}{T_H - T_C} = \frac{294 \text{ K}}{294 \text{ K} - 281 \text{ K}} = 23$$

With this greater efficiency, it would take much less energy to add 11 kJ to the house:

$$W_{in} = \frac{Q_H}{\text{COP}_{max}} = \frac{11 \text{ kJ}}{23} = 0.48 \text{ kJ}$$

ASSESS The 3.4 kJ input of electric energy is much less than the 11 kJ of heat added to the house, so we have confidence in our results—that's the way a heat pump is supposed to work. The actual COP of 3.2 is much less than the theoretical maximum, so it's no surprise that a heat pump running at its theoretical maximum efficiency can achieve the same result with much less energy input.

A heat pump like the one in the example is a great idea if you live in an area where electric heat is common. An electric space heater uses 3.4 kJ of electricity to add 3.4 kJ of heat to your house. The heat pump uses this 3.4 kJ of electric energy to add 11 kJ of heat to your house—a clear improvement. But a heat pump working at the maximum possible efficiency could add 78 kJ to your house for this 3.4 kJ input. There are practical limitations of design, but the big difference between the theoretical and actual efficiencies makes it clear that we could be doing better.

CONCEPTUAL EXAMPLE 11.16 Keeping your cool?

It's a hot day, and your apartment is rather warm. Your roommate suggests cooling off the apartment by keeping the door of the refrigerator open. Will this help the situation?

REASON The energy-transfer diagram for this process is shown in FIGURE 11.22. The heat pump in your refrigerator moves heat from its interior to the air in the apartment as usual. But, with the door open, heat flows spontaneously from the apartment back into the refrigerator. Note, however, that more heat enters the apartment than leaves it—the open refrigerator is actually warming up the apartment!

ASSESS Even a closed refrigerator slowly warms the room it's in. The work done by the motor is always transformed into extra heat exhausted to the hot reservoir.

FIGURE 11.22 Energy-transfer diagram for an open refrigerator.

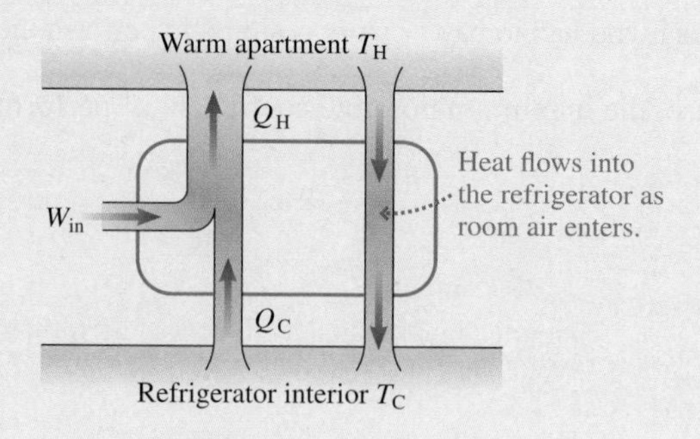

STOP TO THINK 11.6 Which of the following changes would allow your refrigerator to use less energy to run? (There may be more than one correct answer.)

A. Increasing the temperature inside the refrigerator
B. Increasing the temperature of the kitchen
C. Decreasing the temperature inside the refrigerator
D. Decreasing the temperature of the kitchen

FIGURE 11.23 Molecular collisions are reversible.

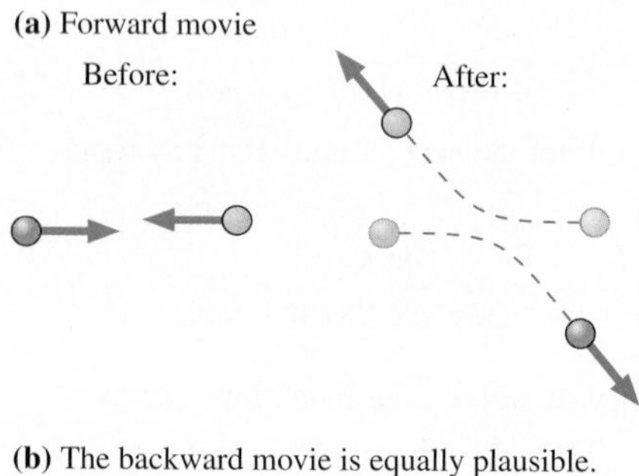

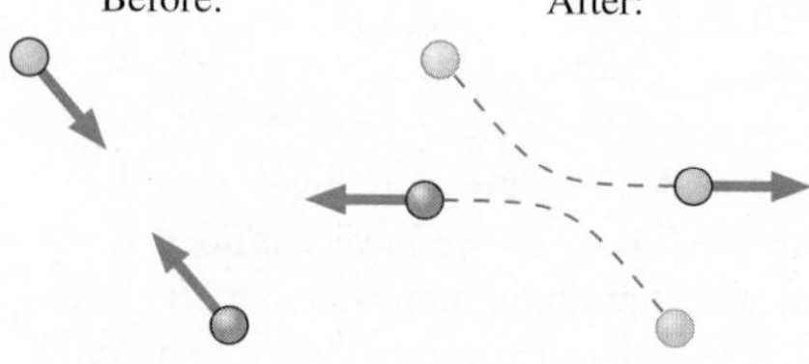

FIGURE 11.24 Macroscopic collisions are not reversible.

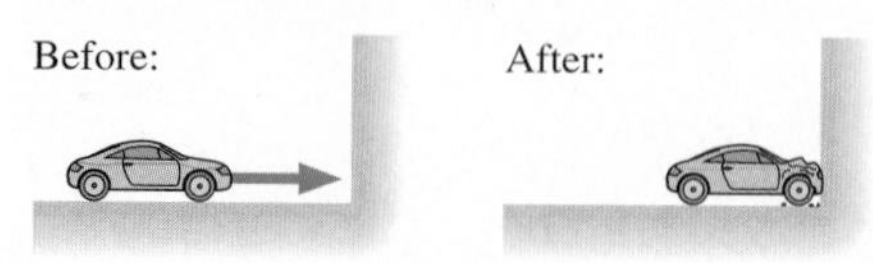

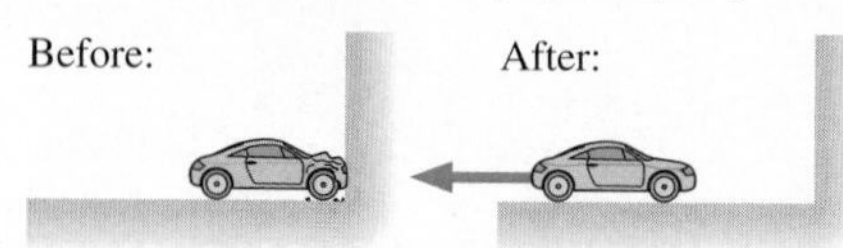

11.7 Entropy and the Second Law of Thermodynamics

Throughout the chapter, we have noticed certain trends and certain limitations in energy transformations and transfers. Heat is transferred spontaneously from hot to cold, not from cold to hot. The spontaneous transfer of heat from hot to cold is an example of an **irreversible** process, a process that can happen in only one direction. Why are some processes irreversible? The spontaneous transfer of heat from cold to hot would not violate any law of physics that we have seen to this point, but it is never observed. There must be another law of physics that prevents it.

Reversible and Irreversible Processes

At a microscopic level, collisions between molecules are completely reversible. In FIGURE 11.23 we see two possible movies of a collision between two gas molecules, one forward and the other backward. You can't tell by looking which is really going forward and which is being played backward. Nothing in either collision looks wrong, and no measurements you might make on either would reveal any violations of Newton's laws. Interactions at the molecular level are **reversible** processes.

At a macroscopic level, it's a different story. FIGURE 11.24 shows two possible movies of the collision of a car with a barrier. One movie is being run forward, the other backward. The backward movie of Figure 11.24b is obviously wrong. But what has been violated in the backward movie? To have the car return to its original shape and spring away from the wall would not violate any laws of physics we have so far discussed.

If microscopic motions are all reversible, how can macroscopic phenomena such as the car crash end up being irreversible? If reversible collisions can cause heat to be transferred from hot to cold, why do they never cause heat to be transferred from cold to hot?

Thermal Energy Spreads

A fundamental difference between microscopic and macroscopic objects is that macroscopic objects have thermal energy, the kinetic and potential energies of all the molecules that make up the object. A macroscopic object can store energy internally, but a point-like particle cannot. Thermal energy didn't appear in our study of the mechanics of particles; it became an issue only when we considered macroscopic objects.

A macroscopic object contains a vast number of atoms and molecules. The thermal energy is the total energy of all these particles, but knowing the thermal energy tells us nothing about how the energy is distributed. FIGURE 11.25a shows a container of gas in which one molecule is moving while all the others are at rest. The thermal energy of the gas is simply the kinetic energy of the one moving molecule. This is one possible way that the energy could be distributed, but this is not a stable, equilibrium situation. The moving molecule will quickly collide with other molecules, giving them kinetic energy while the original fast molecule loses energy. Eventually, we expect that every atom in the gas will be moving, something like the situation shown in FIGURE 11.25b. The thermal energy is the same in both cases; it's just distributed differently.

FIGURE 11.25 Two different distributions of the thermal energy of a gas.

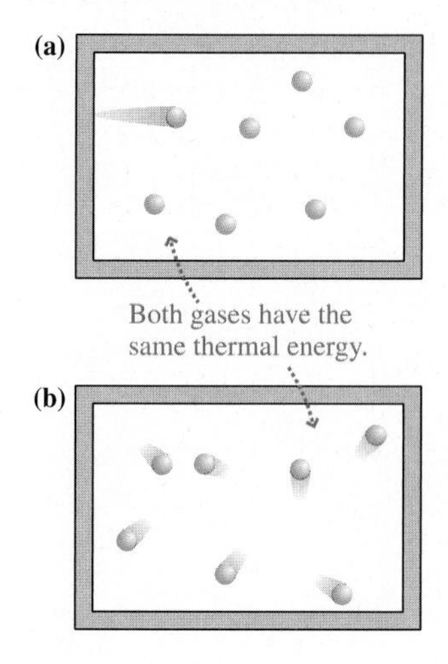

Let's think about what happened here. Initially, the system's thermal energy was concentrated in a single molecule. As time went on, collisions and interactions caused the system's energy to spread out, or disperse, until it was shared among all the constituents. And this spreading will not spontaneously reverse. Ordinary molecular collisions cause a concentrated energy to spread out, but we never see a system spontaneously concentrating dispersed energy into a single constituent. **The spreading of energy is an irreversible process.**

All thermal interactions involve the spreading of energy. Suppose a hot object and a cold object are brought into thermal contact, as in FIGURE 11.26. The thermal energy is initially concentrated in the hot object; it is not shared equally. As we've seen, collisions at the boundary transfer energy from the more energetic molecules on the warmer side to the less energetic molecules on the cooler side until the thermal energy is shared equally among the constituents of the combined system. After some time, thermal equilibrium is reached, which means that no additional sharing or spreading is possible.

FIGURE 11.26 A thermal interaction causes energy to spread until it is shared equally.

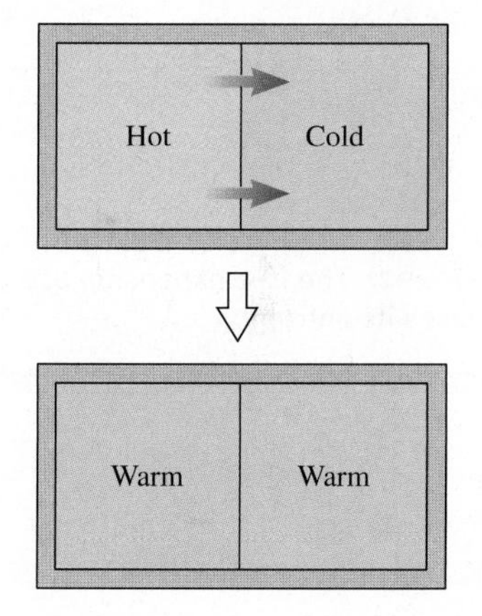

The *spreading* of thermal energy happens spontaneously—it's simply what happens when macroscopic systems interact. The *concentration* of thermal energy does not happen spontaneously. You could imagine that energy continues to be transferred from left to right in Figure 11.26, making the right side hotter than the left by concentrating the energy in the right side, but this never happens in practice.

Entropy

Scientists use the term **entropy** to quantify the spread or dispersal of thermal energy. The formal definition is fairly technical, but a qualitative discussion will capture the key elements. A system in which energy is concentrated—not spread out—has low entropy. The initial situations in Figures 11.25 and 11.26 are states of low entropy. In each case, the system spontaneously evolved to a state in which the energy is more spread out. These are states of higher entropy; the entropy increases as thermal energy becomes more spread out. Because equilibrium is a state in which thermal energy is maximally dispersed, or shared among all the constituents, **the entropy of a system is maximum when the system is in equilibrium.**

This observation about how thermal energy spreads and how macroscopic systems evolve irreversibly toward equilibrium is a new law of physics, the **second law of thermodynamics,** that is usually stated in terms of entropy:

> **Second law of thermodynamics** The entropy of an isolated system never decreases. The entropy either increases until the system reaches equilibrium or, if the system began in equilibrium, stays the same. Entropy is maximum when the system is in equilibrium.

NOTE ► The qualifier "isolated" is crucial. We could reduce a system's entropy by reaching in from the outside, perhaps using tiny tweezers to give most of the thermal energy to a few fast molecules while freezing the others in place. Similarly, we can transfer heat from cold to hot by using a refrigerator. The second law is about what a system can or cannot do spontaneously, on its own, without outside intervention. ◄

Thermodynamics is a science of energy, and we need two laws of energy to understand how macroscopic systems behave. We can restate the two laws of energy as follows:

- **First law of thermodynamics:** Energy is conserved. It can be transferred or transformed, but the total amount does not change.
- **Second law of thermodynamics:** Energy spreads out. An isolated system evolves until the thermal energy is maximally dispersed among all the system's constituents.

It is the second law that creates the "arrow of time" that we observe in irreversible processes. Systems can evolve over time only in ways that continue to spread out thermal energy, thus increasing entropy, not in ways that concentrate thermal energy, thus decreasing entropy.

Understanding the thermal interaction of Figure 11.26 requires not only the first law—energy is transferred from the hotter to the colder object without change—but also the second law. Energy would also be conserved if it were transferred from the colder object to the hotter one, thus increasing their temperature difference, but that cannot happen because it would lower the overall entropy by reducing the energy spread. The evolution toward thermal equilibrium, with equal temperatures, is the route of increasing entropy—increasing energy spread—and that is the only evolution allowed by the second law.

Reasoning with Entropy

If we add heat to a system, we increase its thermal energy. By increasing the thermal energy, we also increase the ability of the system to disperse thermal energy, so we increase the entropy. Similarly, if we take heat out of a system, we reduce the ability of the system to disperse thermal energy, so we reduce the entropy.

FIGURE 11.27 The free expansion of a gas increases its entropy.

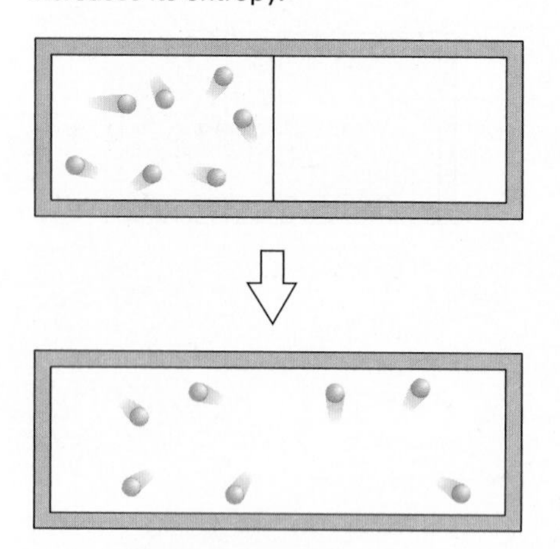

The thermal energy hasn't changed, but the entropy has.

The entropy can also change if there is no change in thermal energy. FIGURE 11.27 shows a container of a gas with the molecules kept on one side by a partition. The container is an isolated system; no heat is transferred to or from the environment. If the partition is removed, the gas expands to fill the cylinder—what we call a *free expansion*. The volume has changed and the pressure will change. But if we think about energy, we know that this expansion does not change the gas temperature. No work is done because the expanding gas isn't pushing against anything, and no heat is transferred from the environment, so the first law of thermodynamics tells us there's no change in thermal energy and hence no change in temperature. But the system's entropy has increased! The thermal energy may not have changed in value, but it's more spatially dispersed because each molecule has more room to move around. And a larger dispersal of energy corresponds to greater entropy. The fact that entropy increases makes the free expansion an irreversible process; the reverse process of all the gas molecules moving into the left half of the cylinder never occurs spontaneously.

Something else that changes is the ability of the gas to do work. Instead of simply removing the partition, we could have allowed the compressed gas to do work by pushing on a piston. This would have transferred some of the thermal energy to the environment by doing mechanical work. The thermal energy is unchanged by the free expansion, but the gas now has less ability to do work. In general, our ability to extract useful work from thermal energy depends not only on the quantity of thermal energy but also on the system's entropy. Increased entropy, with the energy more dispersed, means less ability to do work.

This is why biochemical reactions are analyzed using free energy, which, loosely speaking, is the amount of work that can be done. Any increase in entropy lessens the work that can be done and thus lessens the free energy even if the total energy is unchanged.

CONCEPTUAL EXAMPLE 11.17 **What is the entropy change when ice melts?**

A phase change, such as melting or freezing, occurs at a fixed temperature. What happens to the entropy of the water if a block of frozen water at 0°C becomes liquid water at 0°C?

REASON The entropy can differ, even at the same temperature, if one of the phases of matter—solid or liquid in this case—is able to spread out the energy more. Ice is a solid, so water molecules are frozen into a lattice. Molecules jiggle around—that's what thermal energy is—but their energy is confined to small volumes around the lattice positions. In contrast, the molecules in liquid water can move freely to all points in space. This allows the thermal energy to be more spatially dispersed, just as it was for the free expansion of a gas. Thus the entropy increases as ice melts.

ASSESS This result makes sense. You know that you need to add heat to melt ice, so you expect this change to increase the entropy.

Of all the forms of energy we've seen, only thermal energy is associated with entropy. As long as thermal energy isn't involved, you can freely and reversibly convert between different forms of energy, as we've seen. When you toss a ball into the air, the kinetic energy is converted to potential energy as the ball rises, then back to kinetic energy as the ball falls. But when some form of energy is transformed into thermal energy, entropy has increased, and the second law of thermodynamics tells us that this change is irreversible. When you drop a ball on the floor, it bounces several times and then stops moving. The kinetic energy of the ball in motion has become thermal energy of the slightly warmer ball at rest. This process won't reverse—the ball won't suddenly cool and jump into the air! This process would conserve energy, but it would reduce entropy, and so would violate this new law of physics.

CONCEPTUAL EXAMPLE 11.18 **Why are hybrid vehicles more efficient?**

Hybrid vehicles are powered by a gasoline engine paired with an electric motor and batteries. Hybrids get much better mileage in the stop and go of city driving than conventional vehicles do. When you brake to a stop in a conventional car, friction converts the kinetic energy of the car's motion into thermal energy in the brakes. In a typical hybrid car, depressing the brake pedal will connect a generator that converts much of the car's kinetic energy into chemical energy in a battery. Explain how this makes a hybrid vehicle more efficient.

REASON When energy is transformed into thermal energy, the increase in entropy makes this change irreversible. When you brake a conventional car, the kinetic energy is transformed into the thermal energy of hot brakes and is lost to your use. In a hybrid vehicle, the kinetic energy is converted into chemical energy in a battery. This change is reversible; when the car starts again, the energy can be transformed back into kinetic energy.

ASSESS This makes sense. Whenever energy is converted to thermal energy, it is in some sense "lost," which reduces efficiency. The hybrid vehicle avoids this transformation, and so is more efficient.

STOP TO THINK 11.7 Which of the following processes does not involve a change in entropy?

A. An electric heater raises the temperature of a cup of water by 20°C.
B. A ball rolls up a ramp, decreasing in speed as it rolls higher.
C. A basketball is dropped from 2 m and bounces until it comes to rest.
D. The sun shines on a black surface and warms it.

11.8 Systems, Energy, and Entropy

We commented earlier that no heat engine can be 100% efficient, transforming all of the input heat into useful work. Doing so wouldn't violate energy conservation, but it would violate the second law of thermodynamics. Now we are in a position to explain why.

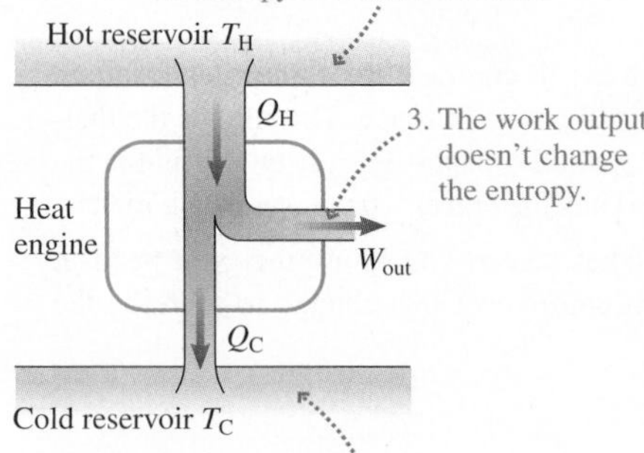
1. Heat leaving the hot reservoir decreases the entropy of the hot reservoir.
Hot reservoir T_H
Q_H
3. The work output doesn't change the entropy.
Heat engine
W_{out}
Q_C
Cold reservoir T_C
2. Heat entering the cold reservoir increases the entropy of the cold reservoir.

FIGURE 11.28 Entropy considerations for a heat engine.

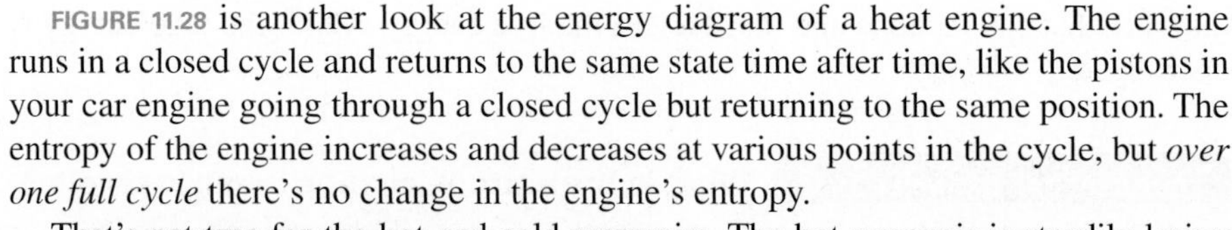

FIGURE 11.28 is another look at the energy diagram of a heat engine. The engine runs in a closed cycle and returns to the same state time after time, like the pistons in your car engine going through a closed cycle but returning to the same position. The entropy of the engine increases and decreases at various points in the cycle, but *over one full cycle* there's no change in the engine's entropy.

That's not true for the hot and cold reservoirs. The hot reservoir is steadily losing energy, the energy that it transfers to the engine. Less energy means less ability to disperse energy, so the entropy of the hot reservoir is steadily decreasing. Conversely, the entropy of the cold reservoir is steadily increasing as it receives the exhaust energy from the engine, increasing the cold reservoir's ability to disperse energy. The work output of the engine doesn't involve an entropy change because it doesn't involve thermal energy.

A 100% efficient heat engine would transform all the input heat Q_H into work and have no exhaust ($Q_C = 0$). With no exhaust, and thus no increase of entropy in the cold reservoir, the net operation of the heat engine would involve a decreasing entropy of the hot reservoir with no corresponding entropy increase. The second law forbids an overall decrease in entropy, so this is not possible.

The most efficient possible heat engine is one that has zero change in entropy. This is the efficiency specified in Equation 11.11 for an ideal heat engine, the maximum efficiency allowed by the second law of thermodynamics.

The Conservation of Energy and Energy Conservation

We have all heard that it is important to "conserve energy." We turn off lights when we leave rooms, drive our cars less, turn down our thermostats. But this brings up an interesting question: If we have a law of conservation of energy, which states that energy can't be created or destroyed, what do we really mean by "conserving energy"?

We started this chapter looking at energy transformations. We saw that whenever energy is transformed, some of it is "lost." And now we know what this means: The energy isn't really lost, but it is converted into thermal energy. This change is irreversible; thermal energy can't be efficiently converted back into other forms of energy.

We aren't, as a society or as a planet, running out of energy. We can't! What we can run out of is high-quality sources of energy. Oil is a good example. A gallon of gasoline contains a great deal of chemical energy. It is easily transported, and it is easily burned to generate heat, electricity, or motion. When you burn gasoline in your car, you don't use up its energy—you simply convert its chemical energy into thermal energy. You thus decrease the amount of high-quality chemical energy in the world and increase the supply of thermal energy. The amount of energy is still the same; it's just in a less useful form.

Perhaps the best way to "conserve energy" is to concentrate on efficiency, to reduce "what you had to pay." More efficient lightbulbs, more efficient cars—all of these use less energy to produce the same final result.

Entropy and Life BIO

The second law of thermodynamics is sometimes interpreted as saying that "things run down"—that gradually other forms of energy are converted to thermal energy and that entropy increases to the point where no useful work can be done. But living organisms seem to violate this rule:

- Plants grow from simple seeds to complex entities.
- Single-celled fertilized eggs grow into complex adult organisms.
- Over the last billion years or so, life has evolved from simple unicellular organisms to the very complex forms we see today.

Rather than running down, many living systems seem to be ramping up and becoming more complex. How can this be?

There is an important qualification to the second law of thermodynamics: It applies to isolated systems, systems that do not exchange energy with their environment. The situation is different if energy is transferred into or out of the system.

Sealed, but not isolated BIO This glass container is a completely sealed system containing living organisms, shrimp and algae. But the organisms will live and grow for many years. The reason this is possible is that the glass sphere, though sealed, is not an *isolated* system. Energy can be transferred in and out as light and heat. If the container were placed in a darkened room, the organisms would quickly perish.

Consider your body. Every day, you take in chemical energy in the food you eat. As you use this energy, most of it ends up as thermal energy that you exhaust as heat into the environment, thereby increasing the entropy of the environment. An energy diagram of this situation is shown in FIGURE 11.29. The entropy of your body stays approximately constant, but the entropy of the environment is steadily increasing. It's the energy flow *through* your body that allows you to maintain your body in an organized, low-entropy state. There's no violation of the second law because the total entropy—that of you plus that of the environment—is increasing.

To develop, grow, or evolve, organisms take in high-quality, low-entropy energy in the form of food or sunlight, then exhaust thermal energy to the environment. This continuous exchange of energy with the environment, which increases environmental entropy, makes your life—and all life—possible without violating physical laws.

FIGURE 11.29 A thermodynamic view of the body.

Chemical energy comes into your body in the food you eat.

Energy leaves your body mostly as heat, meaning the entropy of the environment increases.

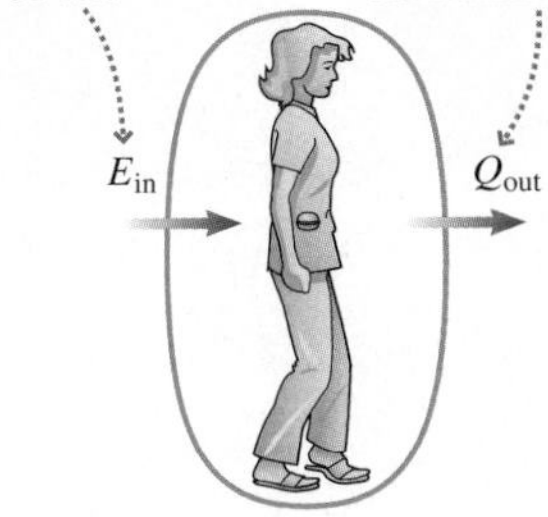

E_{in}

Q_{out}

Entropy, Free Energy, and Spontaneity

We'll conclude with a more technical section for those who have learned in chemistry to use free energy to analyze chemical reactions. We'll assume some background knowledge from your other courses in this section; we'll assume that you are familiar with the concepts of free energy and enthalpy.

A reaction can occur spontaneously only if the free energy decreases in going from reactants to products. Technically, the free energy used in reaction analysis is the *Gibbs free energy*, hence the symbol G. The requirement for a spontaneous reaction is $\Delta G < 0$. The meaning of this requirement is this: It's the second law of thermodynamics telling us that spontaneous processes require an increase in entropy.

But the entropy of what? In this case, the entropy of the universe! We can't apply the second law to the reactants and products only because they are not an isolated system; they transfer heat to the environment in an exothermic reaction or absorb heat from the environment in an endothermic reaction. So let's expand our system to be the reaction plus the surrounding environment that is affected by the reaction. We can call this the universe, although in practice nothing beyond the immediate surroundings is likely to change.

By definition, the universe is an isolated system that does not interact with anything outside it. The second law *does* apply to the universe, so any spontaneous process that takes place in the universe must obey

$$\Delta S_{univ} > 0$$

where S is the symbol for entropy. For our purposes, the universe consists of the chemicals involved in the reaction and the environment surrounding the reaction. If we make the reasonable assumption that entropies add (an assumption that can be proven), then $S_{univ} = S_{chem} + S_{environ}$. In that case, the second law requires

$$\Delta S_{chem} + \Delta S_{environ} > 0 \qquad (11.16)$$

A reaction may temporarily change the temperature of part of the system, but the end state of the reaction is when the reaction products and the environment have re-equilibrated to a common final temperature. Let's assume that the environment is so large that its temperature is not affected by the reactions, so overall the reaction takes place at constant temperature T, the temperature of the environment. Although the environment's temperature doesn't change, its entropy changes because heat is transferred to (or from) the environment by the reaction. If the environment receives heat $Q_{environ}$ at temperature T, its entropy changes by $\Delta S_{environ} = Q_{environ}/T$, a result you learned in chemistry.

Because the universe is isolated, any heat gained by the environment is heat lost by the chemicals (e.g., heat released in the reaction), so $Q_{environ} = -Q_{chem}$. Thus Equation 11.16 can be written as

$$\Delta S_{chem} - \frac{Q_{chem}}{T} > 0$$

Now we have the second-law requirement for spontaneity written in terms of only the chemicals involved in the reaction and the temperature.

Chemical reactions are also assumed to occur at constant pressure. At constant pressure, the heat of the reaction is given by the change in enthalpy H between the reactants and the products: $Q_{\text{chem}} = \Delta H_{\text{chem}}$. Thus spontaneity requires

$$\Delta S_{\text{chem}} - \frac{\Delta H_{\text{chem}}}{T} > 0$$

We can now drop the subscript, knowing that both the entropy and the enthalpy are those of the chemicals. If we first multiply through by T and then by -1 (which changes $>$ to $<$), we have

$$\Delta H - T\Delta S = \Delta(H - TS) < 0$$

But $H - TS$ is the definition of the free energy: $G = H - TS$. Thus the second-law requirement for a spontaneous process—namely, that the entropy must increase and cannot decrease—is

$$\Delta G < 0$$

The free-energy requirement for a spontaneous reaction is about entropy and the second law. A reaction can occur spontaneously if there is an overall increase in entropy.

INTEGRATED EXAMPLE 11.19 Determining the efficiency of an automobile

In the absence of external forces, Newton's first law tells us that a car would continue at a constant speed once it was moving. So why can't you get your car up to speed and then take your foot off the gas? Because there *are* external forces. At highway speeds, nearly all of the force opposing your car's motion is the drag force of the air. (Rolling friction is much smaller than drag at highway speeds, so we'll ignore it.) FIGURE 11.30 shows that as your car moves through the air it pushes the air aside. Doing so takes energy, and your car's engine must keep running to replace this lost energy.

Sports cars, with their aerodynamic shape, are often those that lose the least energy to drag. A typical sports car has a 350 hp engine, a drag coefficient of 0.30, and a low profile with the area of the front of the car being a modest 1.8 m^2. Such a car gets about 25 miles per gallon of gasoline at a highway speed of 30 m/s (just over 65 mph).

Suppose this car is driven 25 miles on a level stretch of road at 30 m/s. It will consume 1 gallon of gasoline, which contains 1.4×10^8 J of chemical energy. What is the car's efficiency for this trip?

FIGURE 11.30 A wind-tunnel test shows airflow around a car.

STRATEGIZE We'll consider the car as the system. As the car moves through the air, the drag force is taking energy away. It's this energy that must be replenished with chemical energy from the fuel.

PREPARE We can use what we learned about the drag force in Chapter 5 to compute the amount of energy needed to move the car forward through the air. We can use this value in Equation 11.2 as "what you get"; it is the minimum amount of energy that *could* be used to move the car forward at this speed. We can then use the mileage data to calculate "what you had to pay"; this is the energy actually used by the car's engine. Once we have these two pieces of information, we can calculate efficiency.

SOLVE Heat and work both play a role in this process, so we need to use our most general equation about energy and energy conservation, Equation 11.8:

$$\Delta K + \Delta U + \Delta E_{\text{th}} + \Delta E_{\text{chem}} + \cdots = W + Q$$

The car is moving at a constant speed, so its kinetic energy is not changing. The road is assumed to be level, so there is no change in potential energy. Once the car is warmed up, its temperature will be constant, so its thermal energy isn't changing. Equation 11.8 reduces to

$$\Delta E_{\text{chem}} = W + Q$$

The car's engine is using chemical energy as it burns fuel. The engine transforms some of this energy into work—the propulsion force pushing the car forward against the opposing force of air drag. But the engine also transforms much of the chemical energy into "waste heat" that is transferred to the environment through the radiator and the exhaust. ΔE_{chem} is negative because the amount of energy stored in the gas tank is decreasing as gasoline is burned. W and Q are also negative, according to the sign convention of Figure 11.15, because these energies are being transferred *out* of the system to the environment.

In Chapter 5 we saw that the drag force on an object moving at a speed v is given by

$$\vec{D} = \left(\frac{1}{2}C_{\text{D}}\rho A v^2, \text{ direction opposite the motion}\right)$$

where C_{D} is the drag coefficient, ρ is the density of air (approximately 1.2 kg/m^3), and A is the area of the front of the car. The drag force on the car moving at 30 m/s is

$$D = \frac{1}{2}(0.30)(1.2 \text{ kg/m}^3)(1.8 \text{ m}^2)(30 \text{ m/s})^2 = 292 \text{ N}$$

The drag force opposes the motion. To move the car forward at constant speed, and thus with no *net* force, requires a propulsion force $F = 292$ N in the forward direction. In Chapter 10 we found that the power—the rate of energy expenditure—to move an object at speed v with force F is $P = Fv$. Thus the power the car must supply—at the wheels—to keep the car moving down the highway is

$$P = Fv = (292 \text{ N})(30 \text{ m/s}) = 8760 \text{ W}$$

Let's compare this to the engine power by converting to horsepower:

$$P = 8760 \text{ W}\left(\frac{1 \text{ hp}}{746 \text{ W}}\right) = 12 \text{ hp}$$

Only a small fraction of the engine's 350 horsepower is needed to keep the car moving at highway speeds.

The distance traveled is

$$\Delta x = 25 \text{ mi} \times \frac{1.6 \text{ km}}{1 \text{ mi}} \times \frac{1000 \text{ m}}{1 \text{ km}} = 40{,}000 \text{ m}$$

and the time required to travel it is $\Delta t = (40{,}000 \text{ m})/(30 \text{ m/s}) = 1333 \text{ s}$. Thus the minimum energy needed to travel 40 km—the energy needed simply to push the air aside—is

$$E_{\text{min}} = P\Delta t = (8760 \text{ W})(1333 \text{ s}) = 1.17 \times 10^7 \text{ J}$$

In traveling this distance, the car uses 1 gallon of gas, with 1.4×10^8 J of chemical energy. This is, quite literally, what you had to pay to drive this distance. Thus the car's efficiency is

$$e = \frac{\text{what you get}}{\text{what you had to pay}} = \frac{1.17 \times 10^7 \text{ J}}{1.4 \times 10^8 \text{ J}} = 0.084 = 8.4\%$$

ASSESS The efficiency of the car is quite low, even compared to other engines that we've seen. Nonetheless, our calculation agrees reasonably well with actual measurements. Gasoline-powered vehicles simply are inefficient, which is one factor favoring more efficient alternative vehicles. Smaller mass and better aerodynamic design improve the efficiency of vehicles, but a large part of the inefficiency of a gasoline-powered vehicle is inherent in the thermodynamics of the engine itself and in the complex drive train needed to transfer the engine's power to the wheels.

SUMMARY

GOAL To learn about practical energy transformations and transfers, and the limits on how efficiently energy can be used.

GENERAL PRINCIPLES

Energy and Efficiency

When energy is transformed from one form into another, some may be "lost," usually to thermal energy, due to practical or theoretical constraints. This limits the efficiency of processes. We define **efficiency** as

$$e = \frac{\text{what you get}}{\text{what you had to pay}} = \text{COP}$$

Efficiency: Used for heat engines → e; COP: Used for heat pumps → COP

Entropy and Irreversibility

Thermal energy tends to spread out. This spreading is irreversible; once the energy has spread, it doesn't go back. This means that heat is transferred from a hot object to a cold object until the two are at equilibrium; once a system is at equilibrium, it will stay there.

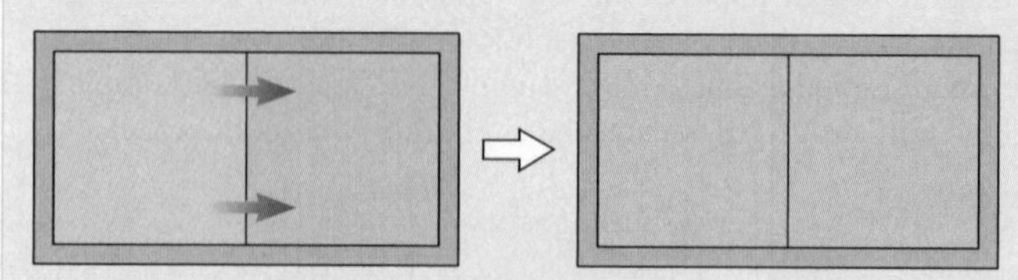

The Laws of Thermodynamics

The **first law of thermodynamics** is a statement of conservation of energy for systems in which only thermal energy changes:

$$\Delta E_{\text{th}} = W + Q$$

The **second law of thermodynamics** specifies the way that isolated systems can evolve:

The entropy of an isolated system never decreases.

This law has practical consequences:

- Heat energy spontaneously flows only from hot to cold.
- A transformation of energy into thermal energy is irreversible.
- No heat engine can be 100% efficient.

Heat is energy transferred between two objects because they are at different temperatures. Energy will be transferred until thermal equilibrium is reached.

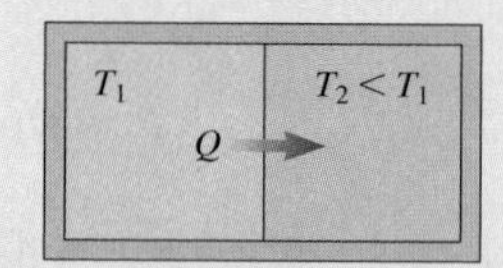

IMPORTANT CONCEPTS

Thermal energy

- For a gas, the thermal energy is the **total kinetic energy** of motion of the atoms.
- Thermal energy is random kinetic energy and so is associated with entropy.

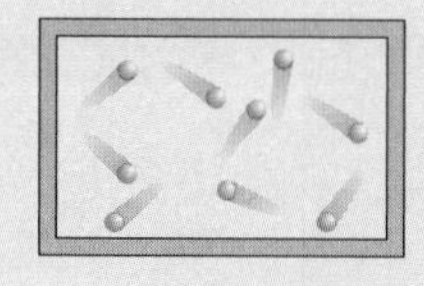

Temperature

- For a gas, temperature is related to the **average kinetic energy** of the motion of the atoms.
- Two systems are in **thermal equilibrium** if they are at the same temperature. No heat energy is transferred.

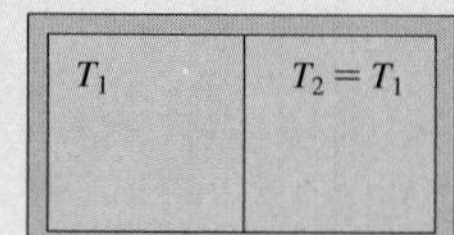

A **heat engine** converts thermal energy from a hot reservoir into useful work. Some heat is exhausted into a cold reservoir, limiting efficiency.

$$e_{\text{max}} = 1 - \frac{T_C}{T_H}$$

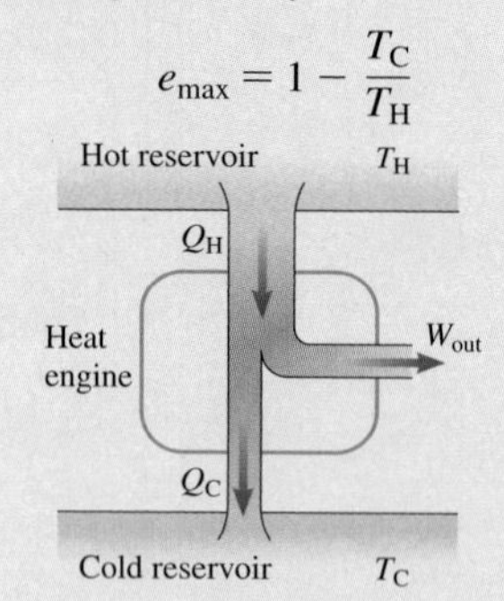

A **heat pump** uses an energy input to transfer heat from a cold side to a hot side. The **coefficient of performance** is analogous to efficiency. For cooling, its limit is

$$\text{COP}_{\text{max}} = \frac{T_C}{T_H - T_C}$$

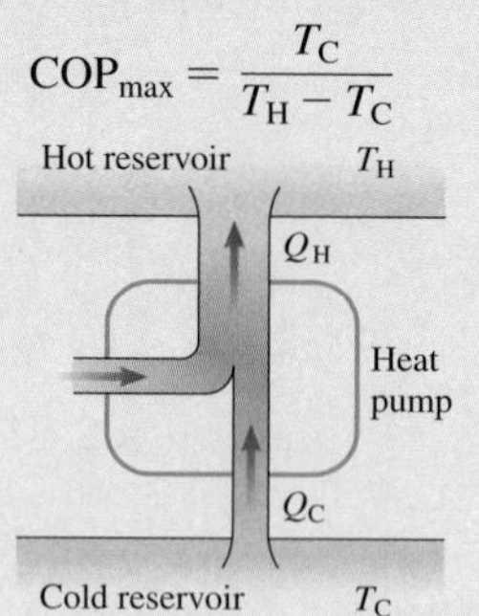

APPLICATIONS

Efficiencies

Energy in the body Cells in the body metabolize chemical energy in food. Efficiency for most actions is about 25%.

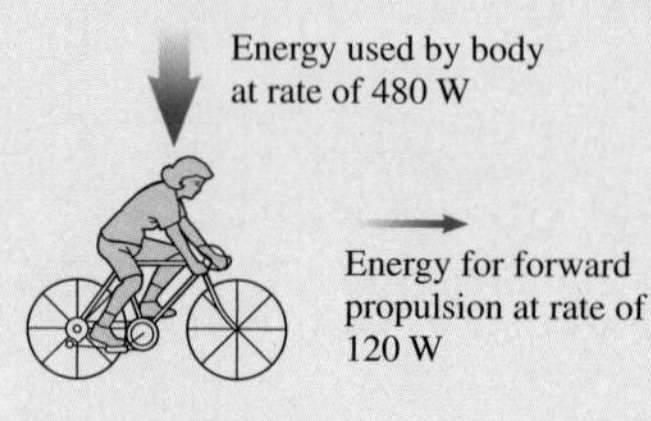

Power plants A typical power plant converts about 1/3 of the energy input into useful work. The rest is exhausted as waste heat.

Temperature scales

Zero on the **Kelvin temperature scale** is the temperature at which the kinetic energy of atoms is zero. This is **absolute zero**. The conversion from °C to K is

$$T(\text{K}) = T(°\text{C}) + 273$$

▶ All temperatures in equations must be in kelvin. ◀

Learning Objectives After studying this chapter, you should be able to:

- Identify transformations of energy and calculate their efficiencies. *Conceptual Questions 11.1, 11.7; Problems 11.2 , 11.3, 11.4, 11.5, 11.6*
- Understand how energy is used and stored in the body. *Conceptual Questions 11.2, 11.4; Problems 11.7, 11.10 , 11.11, 11.13, 11.16*
- Distinguish between work, heat, temperature, and thermal energy. *Conceptual Questions 11.12, 11.19; Problems 11.28, 11.31, 11.32, 11.34, 11.35*
- Apply the first law of thermodynamics to heat engines, heat pumps, and refrigerators. *Conceptual Questions 11.24, 11.25; Problems 11.36, 11.42, 11.47, 11.48, 11.51*
- Understand the concept of entropy and the limitations that the second law of thermodynamics places on energy transformations. *Conceptual Questions 11.28, 11.30; Problems 11.52, 11.53*

STOP TO THINK ANSWERS

Chapter Preview Stop to Think: C. The work Christina does on the javelin is transferred into kinetic energy, so the javelin has 270 J of kinetic energy as it leaves her hand. As the javelin rises, some of its kinetic energy is transformed into gravitational potential energy, but the total energy remains constant at 270 J. Thus at its highest point, the javelin's kinetic energy is 270 J − 70 J = 200 J.

Stop to Think 11.1: C. In each case, what you get is the potential-energy change of the box. Crane 2 lifts a box with twice the mass the same distance as crane 1, so you get twice as much energy with crane 2. How about what you have to pay? Crane 2 uses 20 kJ, crane 1 only 10 kJ. Comparing crane 1 and crane 2, we find crane 2 has twice the energy out for twice the energy in, so the efficiencies are the same.

Stop to Think 11.2: C. As the body uses chemical energy from food, approximately 75% is transformed into thermal energy. Also, kinetic energy of motion of the legs and feet is transformed into thermal energy with each stride. Most of the chemical energy is transformed into thermal energy.

Stop to Think 11.3: C. Samples 1 and 2 have the same thermal energy, which is the total kinetic energy of all the atoms. Sample 1 has twice as many atoms, so the average energy per atom, and thus the temperature, must be less.

Stop to Think 11.4: C. The radiator is at a higher temperature than the surrounding air. Thermal energy is transferred out of the system to the environment, so $Q < 0$.

Stop to Think 11.5: A, D. The efficiency is fixed by the ratio of T_C to T_H. Decreasing this ratio increases efficiency; the heat engine will be more efficient with a hotter hot reservoir or a colder cold reservoir.

Stop to Think 11.6: A, D. The closer the temperatures of the hot and cold reservoirs, the more efficient the heat pump can be. (It is also true that having the two temperatures be closer will cause less thermal energy to "leak" out.) Any change that makes the two temperatures closer will allow the refrigerator to use less energy to run.

Stop to Think 11.7: B. In this case, kinetic energy is transformed into potential energy; there is no entropy change. In the other cases, energy is transformed into thermal energy, meaning entropy increases.

 **Video Tutor Solution** Chapter 11

QUESTIONS

Conceptual Questions

1. Rub your hands together vigorously. What happens? Discuss the energy transfers and transformations that take place.
2. BIO Describe the energy transfers and transformations that occur from the time you sit down to breakfast until you've completed a fast bicycle ride.
3. BIO According to Table 11.4, cycling at 15 km/h requires less metabolic energy than running at 15 km/h. Suggest reasons why this is the case.
4. BIO Walking up a set of stairs and running up the stairs require approximately the same metabolic energy. Explain why you'd expect this to be the case.
5. BIO A pronghorn, the fastest North American animal, is capable of running at 18 m/s (40 mph) for 10 minutes, after which it must slow down. The time limit isn't because the pronghorn runs out of energy; it's because the pronghorn's temperature rises, and it must stop to cool down. Why does the pronghorn's temperature rise during the run?

FIGURE Q11.5

Problem difficulty is labeled as | (straightforward) to ||||| (challenging). INT Problems labeled integrate significant material from earlier chapters; BIO are of biological or medical interest.

The eText icon indicates when there is a video tutor solution available for the chapter or for a specific problem. To launch these videos, log into your eText through Mastering™ Physics or log into the Study Area.

6. When the hoof of a galloping horse hits the ground, the digital flexor tendon in its lower leg may stretch by 5% in length, a significant stretch for this 45 cm tendon. The tendon is elastic; most—but not all—of the energy stored in the stretch is returned, which allows for efficient locomotion. In an exercising horse, the tendon may reach a temperature of 46°C, much higher than the horse's 38°C body temperature. How can you explain this temperature rise?
BIO
7. For most automobiles, the number of miles per gallon decreases as highway speed increases. Fuel economy drops as speeds increase from 55 to 65 mph, then decreases further as speeds increase to 75 mph. Explain why this is the case.
8. If you want to jump as high as possible, it's best to move downward quickly to a deep crouch, stretching tendons and muscles, before pushing off and leaving the ground rather than simply pushing off from a stationary crouch. Explain how this provides additional energy for the jump.
BIO
9. When a spacecraft returns to earth, its surfaces reach a very high temperature during its high-speed reentry into the atmosphere. Is this temperature rise due to the transfer of energy as heat? Explain.
10. When you grind wheat to make flour, the flour comes out very warm—you can actually scorch the flour if you grind too quickly. Explain how this temperature rise comes about.
11. One end of a short aluminum rod is in a campfire and the other end is in a block of ice, as shown in Figure Q11.11. The temperature at every point in the rod has reached a steady value. If 100 J of energy are transferred from the fire to the rod, how much energy goes from the rod into the ice?

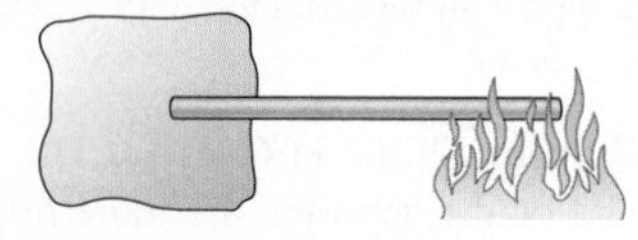
FIGURE Q11.11

12. Two blocks of copper, one of mass 1 kg and the second of mass 3 kg, are at the same temperature. Which block has more thermal energy? If the blocks are placed in thermal contact, will the thermal energy of the blocks change? If so, how?

For Questions 13 through 18, give a specific example of a process that has the energy changes and transfers described. (For example, if the question states "$\Delta E_{th} > 0$, $W = 0$," you are to describe a process that has an increase in thermal energy and no transfer of energy by work. You could write "Heating a pan of water on the stove.")

13. $\Delta E_{th} < 0$, $W = 0$
14. $\Delta E_{th} > 0$, $Q = 0$
15. $\Delta E_{th} < 0$, $Q = 0$
16. $\Delta E_{th} > 0$, $W \neq 0$, $Q \neq 0$
17. $\Delta E_{th} < 0$, $W \neq 0$, $Q \neq 0$
18. $\Delta E_{th} = 0$, $W \neq 0$, $Q \neq 0$
19. If you leave a fan running, it actually warms up the air in the room, a fact that surprises many people. Explain why a fan would be expected to warm the air that passes through it.
20. A fire piston—an impressive physics demonstration—ignites a fire without matches. The operation is shown in Figure Q11.20. A wad of cotton is placed at the bottom of a sealed syringe with a tight-fitting plunger. When the plunger is rapidly depressed, the air temperature in the syringe rises enough to ignite the cotton. Explain why the air temperature rises, and why the plunger must be pushed in very quickly.

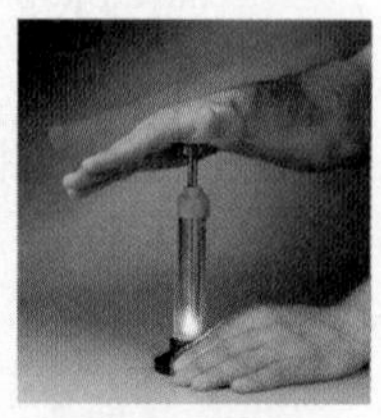
FIGURE Q11.20

21. In a gasoline engine, fuel vapors are ignited by a spark. In a diesel engine, a fuel–air mixture is drawn in, then rapidly compressed to as little as 1/20 the original volume, in the process increasing the temperature enough to ignite the fuel–air mixture. Explain why the temperature rises during the compression.
22. If you hold a rubber band loosely between two fingers and then stretch it, you can tell by touching it to the sensitive skin of your forehead that stretching the rubber band has increased its temperature. If you then let the rubber band rest against your forehead, it soon returns to its original temperature. What are the signs of W and Q for the entire process?
23. Your car's engine is a heat engine; it converts the thermal energy from burning fuel into energy to move your car and power its systems. On a cold winter day, you needn't feel guilty about cranking up the heat in your car; running the heater doesn't cost any additional energy beyond the small amount needed to run the fan. Explain why this is so.
24. The geothermal power plant at Brady Hot Springs in Nevada uses the outside air as the cold reservoir. The plant has a higher electricity output in the morning than it does in the evening. Why would you expect this to be the case?
25. You can save money on electricity if you put your refrigerator in the basement, which is usually cooler than the rest of your house. Explain.
26. The ground temperature a few meters below the surface is fairly constant throughout the year and is near the average value of the air temperature. In areas in which the air temperature drops very low in the winter, the exterior unit of a heat pump designed for heating is sometimes buried underground in order to use the earth as a thermal reservoir. Why is it worthwhile to bury the heat exchanger, even if the underground unit costs more to purchase and install than one above ground?
27. Electric vehicles increase speed by using an electric motor that draws energy from a battery. When the vehicle slows, the motor runs as a generator, recharging the battery. Explain why this means that an electric vehicle can be more efficient than a gasoline-fueled vehicle.
28. When the sun's light hits the earth, the temperature rises. Is there an entropy change to accompany this transformation? Explain.
29. A company markets an electric heater that is described as 100% efficient at converting electric energy to thermal energy. Does this violate the second law of thermodynamics?
30. Living creatures generally have a higher temperature than their environment. Explain why this makes sense.
31. A reaction happens in a perfectly sealed and insulated container. Does the total energy of the system decrease? The free energy?

Multiple-Choice Questions

32. || On clear day, sunlight delivers approximately 1000 J each second to a 1 m^2 surface; smaller areas receive proportionally less—half the area receives half the energy. A 12% efficient solar cell, a square 15 cm on a side, is in bright sunlight. How much electric power does it produce?
A. 0.9 W B. 1.8 W C. 2.7 W D. 3.6 W
33. || For locations that have no electric service, companies are designing bicycle-powered generators that are nearly 100% efficient that can be used to charge cell phones or power lights. If you are the person chosen to pedal and your friends need a total of 150 W of electric power, how much metabolic power will you use as you pedal?
A. 150 W B. 300 W C. 450 W D. 600 W

34. ||| A person is walking on level ground at constant speed. What energy transformation is taking place?
    A. Chemical energy is being transformed to thermal energy.
    B. Chemical energy is being transformed to kinetic energy.
    C. Chemical energy is being transformed to kinetic energy and thermal energy.
    D. Chemical energy and thermal energy are being transformed to kinetic energy.
35. | A person walks 1 km, turns around, and runs back to where he started. Compare the energy used and the power during the two segments.
    A. The energy used and the power are the same for both.
    B. The energy used while walking is greater, the power while running is greater.
    C. The energy used while running is greater, the power while running is greater.
    D. The energy used is the same for both segments, the power while running is greater.
36. || 200 J of heat is added to two gases, each in a sealed container. Gas 1 is in a rigid container that does not change volume. Gas 2 expands as it is heated, pushing out a piston that lifts a small weight. Which gas has the greater increase in its thermal energy?
    A. Gas 1 B. Gas 2
    C. Both gases have the same increase.
37. || An electric power plant uses energy from burning coal to generate steam at 450°C. The plant is cooled by 20°C water from a nearby river. If burning coal provides 100 MJ of heat, what is the theoretical minimum amount of heat that must be transferred to the river during the conversion of heat to electric energy?
    A. 100 MJ B. 90 MJ C. 60 MJ D. 40 MJ
38. || A refrigerator's freezer compartment is set at –10°C; the kitchen is 24°C. What is the theoretical minimum amount of electric energy necessary to pump 1.0 J of energy out of the freezer compartment?
    A. 0.89 J B. 0.87 J C. 0.13 J D. 0.11 J

PROBLEMS

Section 11.1 Transforming Energy

1. || A 10% efficient engine accelerates a 1500 kg car from rest to 15 m/s. How much energy is transferred to the engine by burning gasoline?
2. || Diesel engines give more miles per gallon than gasoline engines, but some of this is due to the higher energy content of diesel fuel. At highway speeds, it takes 0.20 MJ to move an aerodynamic car 1.0 km. At highway speeds, with a gasoline engine, a car gets 16 km per liter of fuel; with a diesel engine under the hood, the car gets 19 km per liter of fuel. One liter of diesel contains 36 MJ; 1 liter of gasoline contains only 32 MJ. What is the efficiency of each of these two engines?
3. | A typical photovoltaic cell delivers 4.0×10^{-3} W of electric energy when illuminated with 1.2×10^{-1} W of light energy. What is the efficiency of the cell?
4. ||| When the Glen Canyon hydroelectric power plant in Arizona is running at capacity, 690 m^3 of water flows through the dam each second. The water is released 220 m below the top of the reservoir. If the generators that the dam employs are 90% efficient, what is the maximum possible electric power output?
5. || An individual white LED (light-emitting diode) has an efficiency of 20% and uses 1.0 W of electric power. How many LEDs must be combined into one light source to give a total of 1.6 W of visible-light output (comparable to the light output of a 40 W incandescent bulb)? What total power is necessary to run this LED light source?
6. || A typical wind turbine extracts 40% of the kinetic energy of the wind that blows through the area swept by the blades. For a large turbine, 110,000 kg of air moves past the blades at 15 m/s every second. If the wind turbine extracts 40% of this kinetic energy, and if 80% of this energy is converted to electric energy, what is the power output of the generator?

Section 11.2 Energy in the Body

7. ||| BIO In an average human, basic life processes require energy to be supplied at a steady rate of 100 W. What daily energy intake, in Calories, is required to maintain these basic processes?
8. | BIO We noted that, under ideal conditions, metabolizing one molecule of glucose can form 38 molecules of ATP. A more realistic value is 30 molecules of ATP. For this lower number, what is the efficiency of the process of converting the chemical energy of glucose to that of ATP?
9. | BIO An "energy bar" contains 22 g of carbohydrates. How much energy is this in joules? In calories? In Calories?
10. | BIO Jessie and Jaime complete a 5.0 km race. Each has a mass of 68 kg. Jessie runs the race at 15 km/h; Jaime walks it at 5 km/h. How much metabolic energy does each use to complete the course?
11. | BIO A sleeping 68 kg man has a metabolic power of 71 W. How many Calories does he burn during an 8.0 hour sleep?
12. | BIO Tessa and Jody, each of mass 68 kg, go out for some exercise together. Tessa runs at 15 km/h; Jody cycles alongside at the same speed. After 20 minutes, how much metabolic energy has each used?
13. ||| BIO An "energy bar" contains 22 g of carbohydrates. If the energy bar was his only fuel, how far could a 68 kg person walk at 5.0 km/h?
14. || BIO Each time he does one pushup, Jose, who has a mass of 75 kg, raises his center of mass by 25 cm. He completes an impressive 150 pushups in 5 minutes, exercising at a steady rate.
    a. If we assume that lowering his body has no energetic cost, what is his metabolic power during this workout?
    b. In fact, it costs Jose a certain amount of energy to lower his body—about half of what it costs to raise it. If you include this in your calculation, what is his metabolic power?
15. |||| BIO For how long would a 68 kg athlete have to swim at a fast crawl to use all the energy available in a typical fast-food meal of burger, fries, and a drink?
16. || BIO INT The basis of muscle action is the power stroke of the myosin protein pulling on an actin filament. It takes the energy of one molecule of ATP, 5.1×10^{-20} J, to produce a displacement of 10 nm against a force of 1.0 pN. What is the efficiency?
17. |||| BIO The label on a candy bar says 400 Calories. Assuming a typical efficiency for energy use by the body, if a 60 kg person were to use the energy in this candy bar to climb stairs, how high could she go?

18. ||| Some jellyfish use jet propulsion to get around; the efficiency of this process can be quite low, as a calculation will show. A 1.9 g jellyfish is at rest. It takes water into its bell, then rapidly ejects 1.3 g of water, achieving a speed of 0.070 m/s. If we define "what you get" as the kinetic energy of the jellyfish, and "what you had to pay" as the total kinetic energy of the jellyfish and the water, what is the efficiency? If we then add in a 25% metabolic efficiency, what's the overall efficiency of this form of locomotion?
BIO INT

19. ||| A weightlifter curls a 30 kg bar, raising it each time a distance of 0.60 m. How many times must he repeat this exercise to burn off the energy in one slice of pizza?
BIO

20. || In an extreme marathon, participants run a total of 100 km; world-class athletes maintain a pace of 15 km/h. How many 230 Calorie energy bars would be required to fuel such a run for a 68 kg athlete?
BIO

21. ||| A weightlifter works out at the gym each day. Part of her routine is to lie on her back and lift a 40 kg barbell straight up from chest height to full arm extension, a distance of 0.50 m.
BIO
    a. How much work does the weightlifter do to lift the barbell one time?
    b. If the weightlifter does 20 repetitions a day, what total energy does she expend on lifting, assuming a typical efficiency for energy use by the body.
    c. How many 400 Calorie donuts can she eat a day to supply that energy?

22. || A 10 kg migratory swan cruises at 20 m/s. A calculation that takes into account the necessary forces shows that this motion requires 200 W of mechanical power. If we assume an efficiency similar to humans, a reasonable assumption, then the metabolic power of the swan is significantly higher than this. The swan doesn't stop to eat during a long day of flying; it gets the energy it needs from fat stores. Assuming an efficiency similar to humans, after 12 hours of flight,
BIO INT
    a. How far has the swan traveled?
    b. How much metabolic energy has it used?
    c. What fraction of its body mass does it lose?

23. ||| Suppose your body was able to use the chemical energy in gasoline. How far could you pedal a bicycle at 15 km/h on the energy in 1 gal of gas? (1 gal of gas has a mass of 3.2 kg.)
BIO

Questions 24 through 26 use the following information.

Flapping flight is very energy intensive. A wind tunnel test on an 89 g starling showed that the bird used 12 W of metabolic power to fly at 11 m/s.

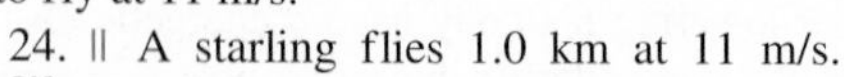

24. || A starling flies 1.0 km at 11 m/s. How many grams of carbohydrates does the bird need to consume to fuel this flight?
BIO

25. || A migrating starling flies steadily at 11 m/s for 1.0 h, using energy from its fat stores. How many grams of fat does it burn?
BIO

26. || What is the specific metabolic power for starling flight? How does this compare to the specific metabolic power for a 68 kg human running at 15 km/h?
BIO INT

Section 11.3 Temperature, Thermal Energy, and Heat

27. | Helium has the lowest boiling point of any substance, at 4.2 K. What is this temperature in °C and °F?

28. | The planet Mercury's surface temperature varies from 700 K during the day to 90 K at night. What are these values in °C and °F?

29. | A piece of metal at 100°C has its Celsius temperature doubled. By what factor does its kelvin temperature increase?

30. | A normal human body temperature can range from 33.2°C to 38.2°C. This range includes variations with gender, time of day, and where the measurement is taken. What is this range in °F?
BIO

Section 11.4 The First Law of Thermodynamics

31. || 500 J of work are done on a system in a process that decreases the system's thermal energy by 200 J. How much energy is transferred to or from the system as heat?

32. || Your roommate leaves a 120 W fan running in your apartment. Over the course of an hour, how much thermal energy does the fan add to the air?

33. | 600 J of heat energy are transferred to a system that does 400 J of work. By how much does the system's thermal energy change?

34. || Compressing air to fill a scuba tank warms it up. A dive shop compensates by putting the tank in a tub of water, keeping the tank and the gas inside it at a constant temperature as it is filled. If the system is the tank, what are the signs of W and Q for this process? What can you say about the relative magnitudes of W and Q?

35. | 10 J of heat are removed from a gas sample while it is being compressed by a piston that does 20 J of work. What is the change in the thermal energy of the gas? Does the temperature of the gas increase or decrease?

Section 11.5 Heat Engines

36. | A heat engine extracts 55 kJ from the hot reservoir and exhausts 40 kJ into the cold reservoir. What are (a) the work done and (b) the efficiency?

37. || A heat engine does 20 J of work while exhausting 30 J of waste heat. What is the engine's efficiency?

38. | A power plant running at 35% efficiency generates 300 MW of electric power. At what rate (in MW) is heat energy exhausted to the river that cools the plant?

39. ||| A heat engine operating between energy reservoirs at 20°C and 600°C has 30% of the maximum possible efficiency. How much energy does this engine extract from the hot reservoir to do 1000 J of work?

40. | A newly proposed device for generating electricity from the sun is a heat engine in which the hot reservoir is created by focusing sunlight on a small spot on one side of the engine. The cold reservoir is ambient air at 20°C. The designer claims that the efficiency will be 60%. What minimum hot-reservoir temperature, in °C, would be required to produce this efficiency?

41. | Converting sunlight to electricity with solar cells has an efficiency of ≈15%. It's possible to achieve a higher efficiency (though currently at higher cost) by using concentrated sunlight as the hot reservoir of a heat engine. Each dish in Figure P11.41 concentrates sunlight on one side of a heat engine, producing a hot-reservoir temperature of 650°C. The cold reservoir, ambient air, is approximately 30°C. The actual working efficiency of this device is ≈30%. What is the theoretical maximum efficiency?

FIGURE P11.41

42. || A California geothermal power plant generates 4.3 MW by extracting 24 MW of heat from an underground source at 108°C and then discharging waste heat into the 17°C air.
    a. What is the actual efficiency of the plant?
    b. What is the theoretical efficiency?
43. ||| Each second, a nuclear power plant generates 2000 MJ of thermal energy from nuclear reactions in the reactor's core. This energy is used to boil water and produce high-pressure steam at 300°C. The steam spins a turbine, which produces 700 MJ of electric power, then the steam is condensed and the water is cooled to 30°C before starting the cycle again.
    a. What is the maximum possible efficiency of the plant?
    b. What is the plant's actual efficiency?
44. ||| A 32% efficient electric power plant produces 900 MJ of electric energy per second and discharges waste heat into 20°C ocean water. Suppose the waste heat could be used to heat homes during the winter instead of being discharged into the ocean. A typical American house requires an average 20 kW for heating. How many homes could be heated with the waste heat of this one power plant?

Section 11.6 Heat Pumps

45. || A refrigerator takes in 20 J of work and exhausts 50 J of heat. What is the refrigerator's coefficient of performance?
46. || Air conditioners are rated by their coefficient of performance at 80°F inside temperature and 95°F outside temperature. An efficient but realistic air conditioner has a coefficient of performance of 3.2. What is the maximum possible coefficient of performance?
47. || 50 J of work are done on a refrigerator with a coefficient of performance of 4.0. How much heat is (a) extracted from the cold reservoir and (b) exhausted to the hot reservoir?
48. || A NATO base in northern Norway is warmed with a heat INT pump that uses 7.0°C ocean water as the cold reservoir. Heat extracted from the ocean water warms fluid to 80°C; this warmed fluid is used to heat the building. When the system is working at full capacity, 2000 kW of heat are delivered to the building at the cost of 600 kW of electric energy.
    a. What is the actual coefficient of performance of the system?
    b. What is the theoretical maximum coefficient of performance of the system?
49. || Find the maximum possible coefficient of performance for a heat pump used to heat a house in a northerly climate in winter. The inside is kept at 20°C while the outside is −20°C.
50. | The inside of your refrigerator is approximately 0°C. Heat from the inside of your refrigerator is deposited into the air in your kitchen, which has a temperature of approximately 20°C. At these operating temperatures, what is the maximum possible coefficient of performance of your refrigerator?
51. ||| 250 students sit in an auditorium listening to a physics lecture. Because they are thinking hard, each is using 125 W of metabolic power, slightly more than they would use at rest. An air conditioner with a COP of 5.0 is being used to keep the room at a constant temperature. What minimum electric power must be used to operate the air conditioner?

Section 11.7 Entropy and the Second Law of Thermodynamics

Section 11.8 Systems, Energy, and Entropy

52. | Which, if any, of the heat engines in Figure P11.52 below violate (a) the first law of thermodynamics or (b) the second law of thermodynamics? Explain.

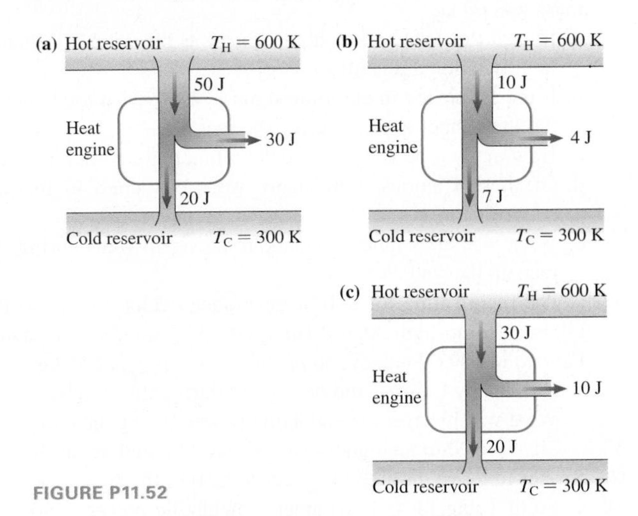

FIGURE P11.52

53. | Which, if any, of the refrigerators in Figure P11.53 below violate (a) the first law of thermodynamics or (b) the second law of thermodynamics? Explain.

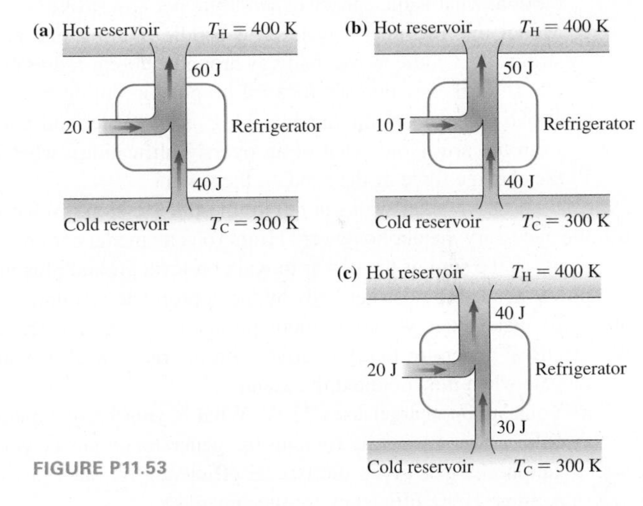

FIGURE P11.53

54. || Entropy can be used as a measure of the overall health of BIO an ecosystem—higher entropy correlates with better health. A thermal image of a landscape, where color indicates temperature, can be used for a quick assessment. What would be the thermal signature of a healthy ecosystem?

Watch Video Solution Problem 11.66

General Problems

55. ||||| How many slices of pizza must you eat to walk for 1.0 h at a
BIO speed of 5.0 km/h? (Assume your mass is 68 kg.)

56. | The winning time for the 2005 annual race up 86 floors of
BIO the Empire State Building was 10 min and 49 s. The winner's mass was 60 kg.
    a. If each floor was 3.7 m high, what was the winner's change in gravitational potential energy?
    b. If the efficiency in climbing stairs is 25%, what total energy did the winner expend during the race?
    c. How many food Calories did the winner "burn" in the race?
    d. Of those Calories, how many were converted to thermal energy?
    e. What was the winner's metabolic power in watts during the race up the stairs?

57. || The record time for a Tour de France cyclist to ascend the famed 1100-m-high Alpe d'Huez was 37.5 min, set by Marco Pantani in 1997. Pantani and his bike had a mass of 65 kg.
    a. How many Calories did he expend during this climb?
    b. What was his average metabolic power during the climb?

58. ||| Championship swimmers take about 22 s and about 30 arm
BIO strokes to move through the water in a 50 m freestyle race.
INT a. From Table 11.4, a swimmer's metabolic power is 800 W. If the efficiency for swimming is 25%, how much energy is expended moving through the water in a 50 m race?
    b. If half the energy is used in arm motion and half in leg motion, what is the energy expenditure per arm stroke?
    c. Model the swimmer's hand as a paddle. During one arm stroke, the paddle moves halfway around a 90-cm-radius circle. If all the swimmer's forward propulsion during an arm stroke comes from the hand pushing on the water and none from the arm (somewhat of an oversimplification), what is the average force of the hand on the water?

59. ||||| A 68 kg hiker walks at 5.0 km/h up a 7% slope. What is
BIO the necessary metabolic power? **Hint:** You can model her power needs as the sum of the power to walk on level ground plus the power needed to raise her body by the appropriate amount.

60. || To make your workouts more productive, you can get an
BIO electrical generator that you drive with the rear wheel of your bicycle when it is mounted in a stand.
    a. Your laptop charger uses 75 W. What is your body's metabolic power use while running the generator to power your laptop charger, given the typical efficiency for such tasks? Assume 100% efficiency for the generator.
    b. Your laptop takes 1 hour to recharge. If you run the generator for 1 hour, how much energy does your body use? Express your result in joules and in Calories.

61. || The resistance of an exercise bike is often provided by a gen-
BIO erator; that is, the energy that you expend is used to generate electric energy, which is then dissipated. Rather than dissipate the energy, it could be used for practical purposes.
    a. A typical person can maintain a steady energy expenditure of 400 W on a bicycle. Assuming a typical efficiency for the body, and a generator that is 80% efficient, what useful electric power could you produce with a bicycle-powered generator?
    b. How many people would need to ride bicycle generators simultaneously to power a 400 W TV in the gym?

62. || Smaller mammals use proportionately more energy than larger
BIO mammals; that is, it takes more energy per gram to power a mouse than a human. A typical mouse has a mass of 20 g and, at rest, needs to consume 3.0 Cal each day for basic body processes.
    a. If a 68 kg human used the same energy per kg of body mass as a mouse, how much energy would be needed each day?
    b. What resting power does this correspond to? How much greater is this than the resting power noted in the chapter?

63. || Larger animals use pro-
BIO portionately less energy than smaller animals; that is, it takes less energy per kg to power an elephant than to power a human. A 5000 kg African elephant requires about 70,000 Cal for basic needs for one day.

    a. If a 68 kg human required the same energy per kg of body mass as an elephant, how much energy would be required each day?
    b. What resting power does this correspond to? How much less is this than the resting power noted in the chapter?

64. ||| An engine does 10 J of work and exhausts 15 J of waste heat.
    a. What is the engine's efficiency?
    b. If the cold-reservoir temperature is 20°C, what is the minimum possible temperature in °C of the hot reservoir?

65. ||| An engine operating at maximum theoretical efficiency whose cold-reservoir temperature is 7°C is 40% efficient. By how much should the temperature of the hot reservoir be increased to raise the efficiency to 60%?

66. ||| An engineer claims to have measured the characteristics of a heat engine that takes in 100 J of thermal energy and produces 50 J of useful work. Is this engine possible? If so, what is the smallest possible ratio of the temperatures (in kelvin) of the hot and cold reservoirs?

67. |||| A typical coal-fired power plant burns 300 metric tons of coal *every hour* to generate 2.7×10^6 MJ of electric energy. 1 metric ton = 1000 kg; 1 metric ton of coal has a volume of 1.5 m^3. The heat of combustion of coal is 28 MJ/kg. Assume that *all* heat is transferred from the fuel to the boiler and that *all* the work done in spinning the turbine is transformed into electric energy.
    a. Suppose the coal is piled up in a 10 m $\times$ 10 m room. How tall must the pile be to operate the plant for one day?
    b. What is the power plant's efficiency?

68. || Driving on asphalt roads entails very little rolling resistance,
INT so most of the energy of the engine goes to overcoming air resistance. But driving slowly in dry sand is another story. If a 1500 kg car is driven in sand at 5.0 m/s, the coefficient of rolling friction is 0.06. In this case, nearly all of the energy that the car uses to move goes to overcoming rolling friction, so you can ignore air drag in this problem.
    a. What propulsion force is needed to keep the car moving forward at a constant speed?
    b. What power is required for propulsion at 5.0 m/s?
    c. If the car gets 15 mpg when driving on sand, what is the car's efficiency? One gallon of gasoline contains 1.4×10^8 J of chemical energy.

69. ||| Air conditioners sold in the United States are given a seasonal energy-efficiency ratio (SEER) rating that consumers can use to compare different models. A SEER rating is the ratio of heat pumped to energy input, similar to a COP but using English units, so a higher SEER rating means a more efficient model. You can determine the COP of an air conditioner by dividing the SEER rating by 3.4. For typical inside and outside temperatures when you'd be using air conditioning, estimate the theoretical maximum SEER rating of an air conditioner. (New air conditioners must have a SEER rating that exceeds 13, quite a bit less than the theoretical maximum, but there are practical issues that reduce efficiency.)
70. || The surface waters of tropical oceans are at a temperature of 27°C while water at a depth of 1200 m is at 3°C. It has been suggested these warm and cold waters could be the energy reservoirs for a heat engine, allowing us to do work or generate electricity from the thermal energy of the ocean. What is the maximum efficiency possible of such a heat engine?
71. || The light energy that falls on a square meter of ground over the course of a typical sunny day is about 20 MJ. The average rate of electric energy consumption in one house is 1.0 kW.
    a. On average, how much energy does one house use during each 24 h day?
    b. If light energy to electric energy conversion using solar cells is 15% efficient, how many square miles of land must be covered with solar cells to supply the electric energy for 250,000 houses? Assume there is no cloud cover.

MCAT-Style Passage Problems

Kangaroo Locomotion BIO

Kangaroos have very stout tendons in their legs that can be used to store energy. When a kangaroo lands on its feet, the tendons stretch, transforming kinetic energy of motion to elastic potential energy. Much of this energy can be transformed back into kinetic energy as the kangaroo takes another hop. The kangaroo's peculiar hopping gait is not very efficient at low speeds but is quite efficient at high speeds.

Figure P11.72 shows the energy cost of human and kangaroo locomotion. The graph shows oxygen uptake (in mL/s) per kg of body mass, allowing a direct comparison between the two species.

For humans, the energy used per second (i.e., power) is proportional to the speed. That is, the human curve nearly passes through the origin, so running twice as fast takes approximately twice as much power. For a hopping kangaroo, the graph of energy use has only a very small slope. In other words, the energy used per second changes very little with speed. Going faster requires very little additional power. Treadmill tests on kangaroos and observations in the wild have shown that they do not become winded at any speed at which they are able to hop. No matter how fast they hop, the necessary power is approximately the same.

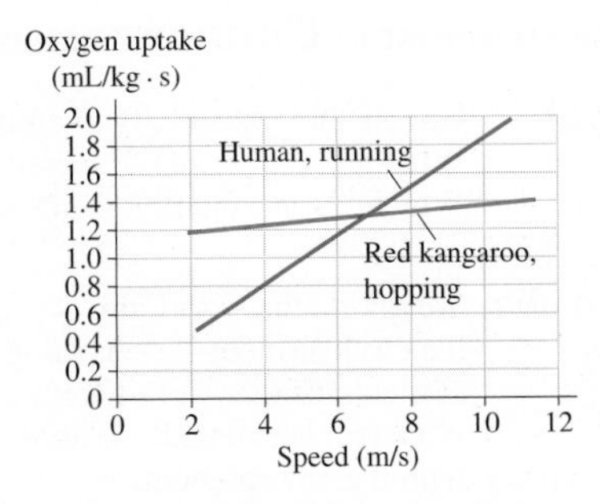

FIGURE P11.72 Oxygen uptake (a measure of energy use per second) for a running human and a hopping kangaroo.

72. | A person runs 1 km. How does his speed affect the total energy needed to cover this distance?
    A. A faster speed requires less total energy.
    B. A faster speed requires more total energy.
    C. The total energy is about the same for a fast speed and a slow speed.
73. | A kangaroo hops 1 km. How does its speed affect the total energy needed to cover this distance?
    A. A faster speed requires less total energy.
    B. A faster speed requires more total energy.
    C. The total energy is about the same for a fast speed and a slow speed.
74. | At a speed of 4 m/s,
    A. A running human is more efficient than an equal-mass hopping kangaroo.
    B. A running human is less efficient than an equal-mass hopping kangaroo.
    C. A running human and an equal-mass hopping kangaroo have about the same efficiency.
75. | At approximately what speed would a human use half the power of an equal-mass kangaroo moving at the same speed?
    A. 3 m/s B. 4 m/s C. 5 m/s D. 6 m/s
76. | At what speed does the hopping motion of the kangaroo become more efficient than the running gait of a human?
    A. 3 m/s B. 5 m/s C. 7 m/s D. 9 m/s

Animal Power

Measurements in the 1800s showed that a draft horse could provide a maximum steady output power over the course of a working day; that's how the "horsepower," about 750 W, came to be defined. Another animal that's used for pulling loads for which there are good data is the sled dog. A 38 kg dog can provide power to pull a sled with a force of 60 N at a steady 2.2 m/s. Dogs and horses have approximately the same 25% metabolic efficiency as humans.

77. || A horse is working steadily, providing 750 W of output power. How much food energy does the horse need to work for 1 hour, to the nearest MJ? BIO INT
    A. 3 MJ B. 5 MJ C. 9 MJ D. 11 MJ
78. || How many sled dogs would be needed to provide 1 horsepower? BIO INT
    A. 6 B. 12 C. 14 D. 20
79. || A dog pulls a sled that requires a 120 N pulling force. How fast can the dog pull the sled, to the nearest m/s? BIO INT
    A. 1 m/s B. 2 m/s C. 3 m/s D. 4 m/s

KNOWLEDGE STRUCTURE II Conservation Laws

BASIC GOALS

How is the system "after" an interaction related to the system "before"?
What quantities are conserved, and under what conditions?
Why are some energy changes more efficient than others?

CROSS-CUTTING CONCEPTS

We divide the world into two parts: the **system** and the **environment**—everything else. The system is a collection of elements that work together, that share energy and momentum. A system is isolated if it does not exchange energy or momentum with the environment.

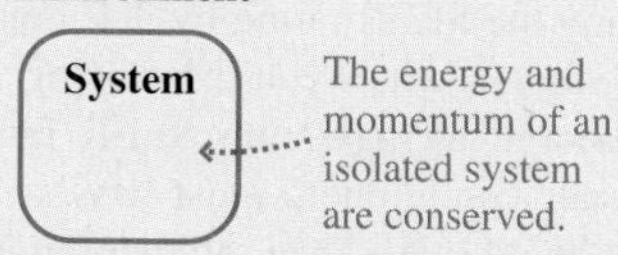

GENERAL PRINCIPLES

Law of conservation of momentum For an isolated system, $\vec{P}_f = \vec{P}_i$.
Law of conservation of energy For an isolated system, there is no change in the system's energy:

$$\Delta K + \Delta U_g + \Delta U_s + \Delta E_{th} + \Delta E_{chem} + \cdots = 0$$

Energy can be exchanged with the environment as work or heat:

$$\Delta K + \Delta U_g + \Delta U_s + \Delta E_{th} + \Delta E_{chem} + \cdots = W + Q$$

Laws of thermodynamics First law: If only thermal energy changes, $\Delta E_{th} = W + Q$.
Second law: The entropy of an isolated system always increases.

BASIC PROBLEM-SOLVING APPROACH

Draw a visual overview for the system "before" and "after"; then use the conservation of momentum or energy equation to relate the two. If necessary, calculate impulse, work, or both.

Momentum and impulse

In a collision, the total momentum

$$\vec{P} = \vec{p}_1 + \vec{p}_2 = m_1\vec{v}_1 + m_2\vec{v}_2$$

is the same before and after.

Before: m_1 ① $(v_{1x})_i$ → ← $(v_{2x})_i$ ② m_2

After: ← $(v_{1x})_f$ ① ② $(v_{2x})_f$ →

A force can change the momentum of an object. The change is the **impulse:** $\Delta p_x = J_x$.

$J_x =$ area under force curve

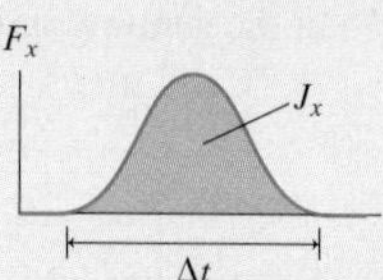

Basic model of energy

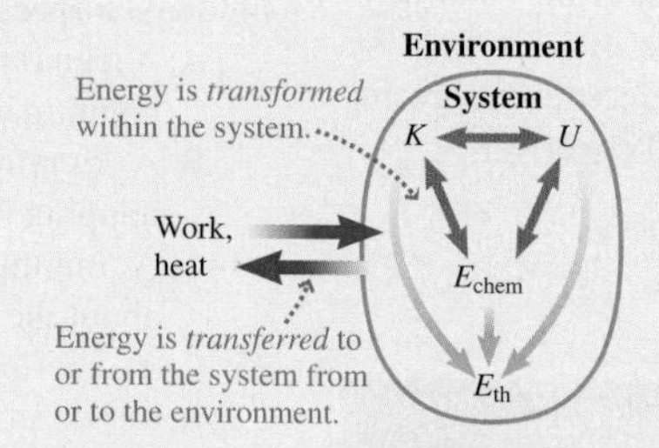

Environment
System
Energy is transformed within the system.
K
U
Work, heat
E_chem
Energy is transferred to or from the system from or to the environment.
E_th

Work $W = F_\parallel d$ is done by the component of a force parallel to a displacement.

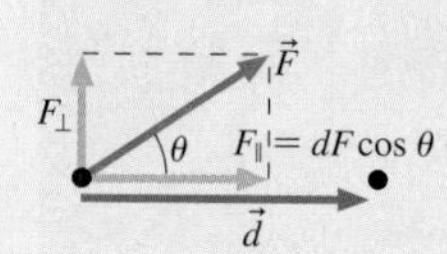

Limitations on energy transfers and transformations

Thermal energy is random kinetic energy. Changing other forms of energy to thermal energy is **irreversible.**

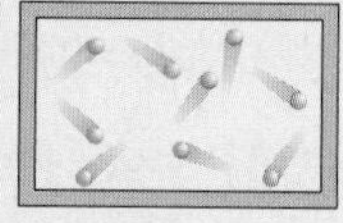

When energy is transformed from one form into another, some may be "lost" as thermal energy. This limits efficiency:

$$\text{efficiency: } e = \frac{\text{what you get}}{\text{what you had to pay}}$$

A heat engine can convert thermal energy to useful work. The efficiency must be less than 100%.

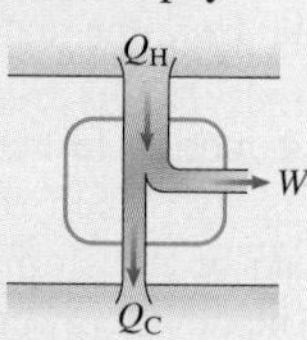

Momentum

Momentum is the product of the mass and the velocity of an object. Momentum is a vector quantity.

$$\vec{p} = m\vec{v}$$

$\vec{v}$ → ; m • $\vec{p}$ →

The units of momentum are kg · m/s.

Energy

Energy comes in a variety of different forms. Some can be easily quantified:

Kinetic energy	$K = \frac{1}{2}mv^2$
Gravitational potential energy	$U_g = mgy$
Elastic potential energy	$U_s = \frac{1}{2}kx^2$

Other important forms are:

Thermal energy	E_{th}
Chemical energy	E_{chem}

Energy is measured in joules, with 1 J = 1 N · m.

Relating Force and Momentum

The change in momentum, the impulse, is related to the average force in a collision:

$$\Delta p_x = J_x = F_{avg}\Delta t$$

Work-Energy Equation

For a system before and after a transformation or a transfer by work:

$$K_f + U_f + \Delta E_{th} = K_i + U_i + W$$

Power

Power is the rate at which energy is transformed or work is done:

$$P = \frac{\Delta E}{\Delta t} \qquad P = \frac{W}{\Delta t}$$

Power is measured in watts, with 1 W = 1 J/s.

Order Out of Chaos

The second law of thermodynamics specifies that "the future" is the direction of entropy increase. But, as we have seen, this doesn't mean that systems must invariably run down or decay. You don't need to look far to find examples of systems that spontaneously evolve to a state of greater order.

A snowflake is a perfect example. As water freezes, the random motion of water molecules is transformed into the orderly arrangement of a crystal. The entropy of the snowflake is less than that of the water vapor from which it formed. Has the second law of thermodynamics been turned on its head?

The entropy of the water molecules in the snowflake certainly decreases, but the water doesn't freeze as an isolated system. For it to freeze, heat energy must be transferred from the water to the surrounding air. The entropy of the air increases by *more* than the entropy of the water decreases. Thus the *total* entropy of the water + air system increases when a snowflake is formed, just as the second law predicts. If the system isn't isolated, its entropy can decrease without violating the second law as long as the entropy increases somewhere else.

Systems that become *more* ordered as time passes, and in which the entropy decreases, are called *self-organizing systems.* These systems can't be isolated. It is common in self-organizing systems to find a substantial flow of energy *through* the system. Your body takes in chemical energy from food, makes use of that energy, and then gives waste heat back to the environment. It is this energy flow that allows systems to develop a high degree of order and a very low entropy. The entropy of the environment undergoes a significant *increase* so as to let selected subsystems decrease their entropy and become more organized.

Self-organizing systems don't violate the second law of thermodynamics, but this fact doesn't really explain their existence. If you toss a coin, no law of physics says that you can't get heads 100 times in a row—but you don't expect this to happen. Can we show that self-organization isn't just possible, but likely?

Let's look at a simple example. Suppose you heat a shallow dish of oil at the bottom, while holding the temperature of the top constant. When the temperature difference between the top and the bottom of the dish is small, heat is transferred from the bottom to the top by conduction. But convection begins when the temperature difference becomes large enough. The pattern of convection needn't be random, though; it can develop in a stable, highly ordered pattern, as we see in the figure. Convection is a much more efficient means of transferring energy than conduction, so the spread of energy is *increased* as a result of the development of these ordered *convection cells.*

The development of the convection cells is an example of self-organization. The roughly 10^{23} molecules in the fluid had been moving randomly but now have begun behaving in a very orderly fashion. But there is more to the story. The convection cells transfer energy from the hot lower side of the dish to the cold upper side. This hot-to-cold energy transfer increases the entropy of the surrounding environment, as we have seen. In becoming more organized, the system has become more effective at transferring heat, resulting in a greater rate of entropy increase! Order has arisen out of disorder in the system, but the net result is a more rapid increase of the entropy of the universe.

Convection cells are thus a thermodynamically favorable form of order. We should expect this, because convection cells aren't confined to the laboratory. We see them in the sun, where they transfer energy from lower levels to the surface, and in the atmosphere of the earth, where they give rise to some of our most dramatic weather.

Self-organizing systems are a very active field of research in physical and biological sciences. The 1977 Nobel Prize in chemistry was awarded to the Belgian scientist Ilya Prigogine for his studies of *nonequilibrium thermodynamics,* the basic science underlying self-organizing systems. Prigogine and others have shown how energy flow through a system can, when the conditions are right, "bring order out of chaos." And this spontaneous ordering is not just possible—it can be probable. The existence and evolution of self-organizing systems, from thunderstorms to life on earth, might just be nature's preferred way of increasing entropy in the universe.

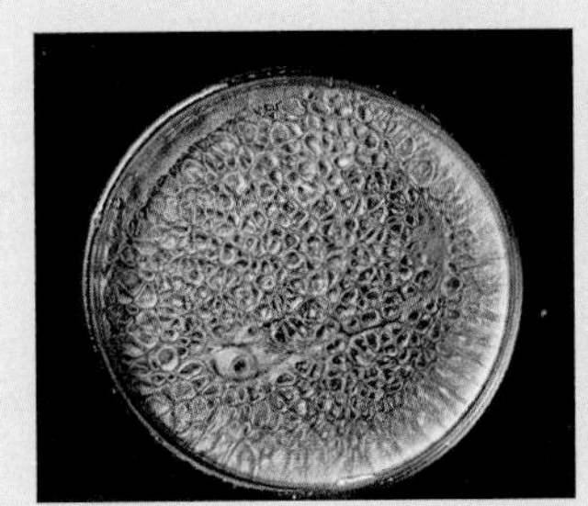
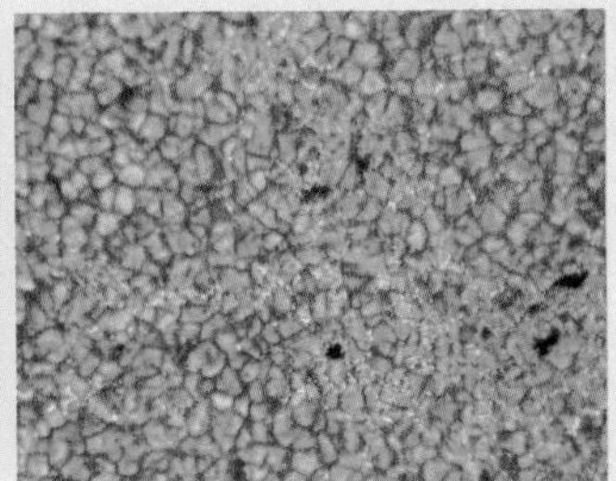

Convection cells in a shallow dish of oil heated from below (left) and in the sun (right). In both, warmer fluid is rising (lighter color) and cooler fluid is sinking (darker color).

The following questions are related to the passage "Order Out of Chaos" on the previous page.

1. When water freezes to make a snowflake crystal, the entropy of the water
   A. Decreases.
   B. Increases.
   C. Does not change.
2. When thermal energy is transferred from a hot object to a cold object, the overall entropy
   A. Decreases.
   B. Increases.
   C. Does not change.
3. Do convection cells represent a reversible process?
   A. Yes, because they are orderly.
   B. No, because they transfer thermal energy from hot to cold.
   C. It depends on the type of convection cell.
4. In an isolated system far from thermal equilibrium, as time passes,
   A. The total energy stays the same; the total entropy stays the same.
   B. The total energy decreases; the total entropy increases.
   C. The total energy stays the same; the total entropy increases.
   D. The total energy decreases; the total entropy stays the same.

The following passages and associated questions are based on the material of Part II.

Big Air

A new generation of pogo sticks lets a rider bounce more than 2 meters off the ground by using elastic bands to store energy. When the pogo's plunger hits the ground, the elastic bands stretch as the pogo and rider come to rest. At the low point of the bounce, the stretched bands start to contract, pushing out the plunger and launching the rider into the air. For a total mass of 80 kg (rider plus pogo), a stretch of 0.40 m launches a rider 2.0 m above the starting point.

5. If you were to jump to the ground from a height of 2 meters, you'd likely injure yourself. But a pogo rider can do this repeatedly, bounce after bounce. How does the pogo stick make this possible?
   A. The elastic bands absorb the energy of the bounce, keeping it from hurting the rider.
   B. The elastic bands warm up as the rider bounces, absorbing dangerous thermal energy.
   C. The elastic bands simply convert the rider's kinetic energy to potential energy.
   D. The elastic bands let the rider come to rest over a longer time, meaning less force.
6. Assuming that the elastic bands stretch and store energy like a spring, how high would the 80 kg pogo and rider go for a stretch of 0.20 m?
   A. 2.0 m B. 1.5 m C. 1.0 m D. 0.50 m
7. Suppose a much smaller rider (total mass of rider plus pogo of 40 kg) mechanically stretched the elastic bands of the pogo by 0.40 m, then got on the pogo and released the bands. How high would this unwise rider go?
   A. 8.0 m B. 6.0 m C. 4.0 m D. 3.0 m
8. A pogo and rider of 80 kg total mass at the high point of a 2.0 m jump will drop 1.6 m before the pogo plunger touches the ground, slowing to a stop over an additional 0.40 m as the elastic bands stretch. What approximate average force does the pogo stick exert on the ground during the landing?
   A. 4000 N B. 3200 N C. 1600 N D. 800 N
9. Riders can use fewer elastic bands, reducing the effective spring constant of the pogo. The maximum stretch of the bands is still 0.40 m. Reducing the number of bands will
   A. Reduce the force on the rider and give a lower jump height.
   B. Not change the force on the rider but give a lower jump height.
   C. Reduce the force on the rider but give the same jump height.
   D. Make no difference to the force on the rider or the jump height.

Testing Tennis Balls

Tennis balls are tested by being dropped from a height of 2.5 m onto a concrete floor. The 57 g ball hits the ground, compresses, then rebounds. A ball will be accepted for play if it rebounds to a height of about 1.4 m; it will be rejected if the bounce height is much more or much less than this.

10. Consider the sequence of energy transformations in the bounce. When the dropped ball is motionless on the floor, compressed, and ready to rebound, most of the energy is in the form of
    A. Kinetic energy.
    B. Gravitational potential energy.
    C. Thermal energy.
    D. Elastic potential energy.
11. If a ball is "soft," it will spend more time in contact with the floor and won't rebound as high as it is supposed to. The force on the floor of the "soft" ball is ________ the force on the floor of a "normal" ball.
    A. Greater than
    B. The same as
    C. Less than
12. Suppose a ball is dropped from 2.5 m and rebounds to 1.4 m.
    a. How fast is the ball moving just before it hits the floor?
    b. What is the ball's speed just after leaving the floor?
    c. What happens to the "lost" energy?
    d. If the time of the collision with the floor is 6.0 ms, what is the average force on the ball during the impact?

Squid Propulsion BIO

Squid usually move by using their fins, but they can utilize a form of jet propulsion, ejecting water at high speed to rocket them backward, as shown in Figure II.1. A 4.0 kg squid can slowly draw in and then quickly eject 0.30 kg of water. The water is ejected in 0.10 s at a speed of 10 m/s. This gives the squid a quick burst of speed to evade predators or catch prey.

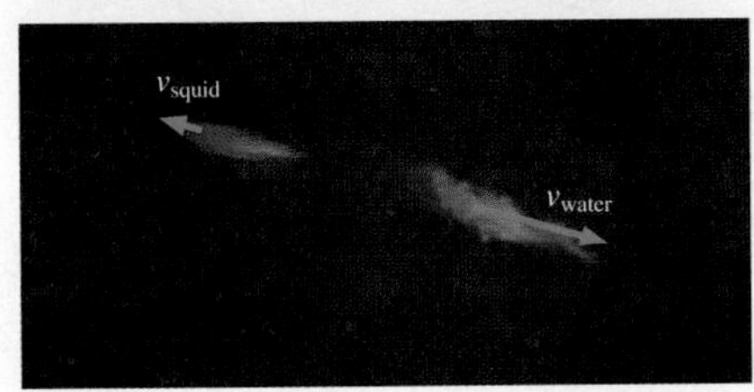

FIGURE II.1

13. What is the speed of the squid immediately after the water is ejected?
    A. 10 m/s B. 7.5 m/s C. 1.3 m/s D. 0.75 m/s
14. What is the squid's approximate acceleration, in g's?
    A. $10g$ B. $7.5g$ C. $1.0g$ D. $0.75g$
15. What is the average force on the water during the jet?
    A. 100 N B. 30 N C. 10 N D. 3.0 N
16. This form of locomotion is speedy, but is it efficient? The energy that the squid expends goes two places: the kinetic energy of the squid and the kinetic energy of the water. Think about how to define "what you get" and "what you had to pay"; then calculate an efficiency for this particular form of locomotion. (You can ignore biomechanical efficiency for this problem.)

Teeing Off

A golf club has a lightweight flexible shaft with a heavy block of wood or metal (called the head of the club) at the end. A golfer making a long shot off the tee uses a driver, a club whose 300 g head is much more massive than the 46 g ball it will hit. The golfer swings the driver so that the club head is moving at 40 m/s just before it collides with the ball. The collision is so rapid that it can be treated as the collision of a moving 300 g mass (the club head) with a stationary 46 g mass (the ball); the shaft of the club and the golfer can be ignored. The collision takes 5.0 ms, and the ball leaves the tee with a speed of 63 m/s.

17. What is the change in momentum of the ball during the collision?
    A. 1.4 kg · m/s B. 1.8 kg · m/s
    C. 2.9 kg · m/s D. 5.1 kg · m/s
18. What is the speed of the club head immediately after the collision?
    A. 30 m/s B. 25 m/s C. 19 m/s D. 11 m/s
19. Is this a perfectly elastic collision?
    A. Yes
    B. No
    C. There is insufficient information to make this determination.
20. If we define the kinetic energy of the club head before the collision as "what you had to pay" and the kinetic energy of the ball immediately after as "what you get," what is the efficiency of this energy transfer?
    A. 0.54 B. 0.46 C. 0.38 D. 0.27

Additional Integrated Problems

21. Football players measure their acceleration by seeing how fast they can sprint 40 yards (37 m). A zippy player can, from a standing start, run 40 yards in 4.1 s, reaching a top speed of about 11 m/s. For an 80 kg player, what is the average power output for this sprint?
    A. 300 W B. 600 W C. 900 W D. 1200 W
22. The unit of horsepower was defined by considering the power output of a typical horse. Working-horse guidelines in the 1900s called for them to pull with a force equal to 10% of their body weight at a speed of 3.0 mph. For a typical working horse of 1200 lb, what power does this represent in W and in hp?
23. A 100 kg football player is moving at 6.0 m/s to the east; a 130 kg player is moving at 5.0 m/s to the west. They meet, each jumping into the air and grabbing the other player. While they are still in the air, which way is the pair moving, and how fast?
24. A swift blow with the hand can break a pine board. As the hand hits the board, the kinetic energy of the hand is transformed into elastic potential energy of the bending board; if the board bends far enough, it breaks. Applying a force to the center of a particular pine board deflects the center of the board by a distance that increases in proportion to the force. Ultimately the board breaks at an applied force of 800 N and a deflection of 1.2 cm.
    a. To break the board with a blow from the hand, how fast must the hand be moving? Use 0.50 kg for the mass of the hand.
    b. If the hand is moving this fast and comes to rest in a distance of 1.2 cm, what is the average force on the hand?
25. A child's sled has rails that slide with little friction across the snow. Logan has an old wooden sled with heavy iron rails that has a mass of 10 kg—quite a bit for a 30 kg child! Logan runs at 4.0 m/s and leaps onto the stationary sled and holds on tight as it slides forward. The impact time with the sled is 0.25 s.
    a. Immediately after Logan jumps on the sled, how fast is it moving?
    b. What was the force on the sled during the impact?
    c. How much energy was "lost" in the impact? Where did this energy go?

PART III

Properties of Matter

Male emperor penguins overwinter on the Antarctic ice, going long periods without eating in bitterly cold surroundings. Minimizing heat loss to the environment is crucial because replacing lost energy draws down precious reserves. In this section, we'll consider properties of matter, including thermal energy and its transfer as heat. We'll learn to understand and appreciate some of the adaptations these remarkable birds use to keep warm during the winter.

OVERVIEW

Beyond the Particle Model

The first 11 chapters of this text have made extensive use of the *particle model* in which we represent objects as point masses. The particle model is especially useful for describing how discrete objects move through space and how they interact with each other. Whether a ball is made of metal or wood is irrelevant to calculating its trajectory.

But there are many situations where the distinction between metal and wood is crucial. If you toss a metal ball and a wood ball into a pond, one sinks and the other floats. If you stir a pan on the stove with a metal spoon, it can quickly get too hot to hold unless it has a wooden handle.

Wood and metal have different physical properties. So do air and water. Our goal in Part III is to describe and understand the similarities and differences of different materials. To do so, we must go beyond the particle model and dig deeper into the nature of matter.

Macroscopic Physics

In Part III, we will be concerned with systems that are solids, liquids, or gases. Properties such as pressure, temperature, specific heat, and viscosity are characteristics of the system as a whole, not of the individual particles. Solids, liquids, and gases are often called *macroscopic* systems—the prefix "macro" (the opposite of "micro") meaning "large." We'll make sense of the behavior of these macroscopic properties by considering a microscopic view in which we think of these systems as collections of particle-like atoms. This "micro-to-macro" development will be a key piece of the following chapters.

In the coming chapters, we'll consider a wide range of practical questions, such as:

- How do the temperature and pressure of a system change when you heat it? Why do some materials respond quickly, others slowly?
- What are the mechanisms by which a system exchanges heat energy with its environment? Why does blowing on a cup of hot coffee cause it to cool off?
- Why are there three phases of matter—solids, liquids, gases? What happens during a phase change?
- Why do some objects float while others, with the same mass, sink? What keeps a massive steel ship afloat?
- What are the laws of motion of a flowing liquid? How do they differ from the laws governing the motion of a particle?

Both Newton's laws and the law of conservation of energy will remain important tools—they are, after all, the basic laws of physics—but we'll have to learn how they apply to macroscopic systems.

It should come as no surprise that an understanding of macroscopic systems and their properties is essential for understanding the world around us. Biological systems, from cells to ecosystems, are macroscopic systems exchanging energy with their environment. On a larger scale, energy transport on earth gives us weather, and the exchange of energy between the earth and space determines our climate.

12 Thermal Properties of Matter

This thermal image shows a person with warm hands holding a much cooler tarantula. Why do the hands radiate energy? And why does the tarantula radiate less?

LOOKING AHEAD ▸

The Ideal Gas

The high pressure in a car tire is due to the countless collisions between the air molecules inside and the tire's walls.

You'll learn how gas properties are related to the microscopic motion of the gas molecules.

Heat and Temperature

Adding ice cools your drink as heat is transferred from the warm drink to the cold ice; even more heat is used to melt the ice.

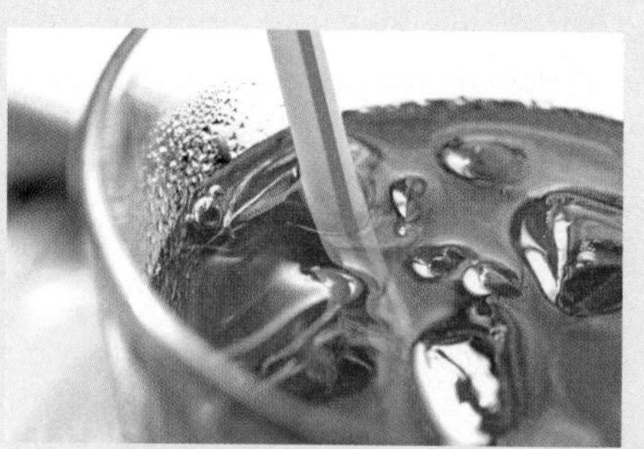

You'll learn how to compute the temperature changes that occur when heat is transferred or during a **phase change** such as melting.

Heat Transfer

Its large ears keep an elephant cool. Blood flowing through large vessels loses heat to the environment, returning to the body at a lower temperature.

You'll learn about the heat-transfer mechanisms of **conduction, convection,** and **radiation.**

GOAL To use the atomic model of matter to explain many properties of matter associated with heat and temperature.

LOOKING BACK ◂

Heat

In Section 11.4 you learned about heat and the first law of thermodynamics. In this chapter we will explore some of the consequences of transferring heat to or from a system, and doing work on the system.

You learned that a system's energy can be changed by doing work on it *or* by transferring heat to it.

STOP TO THINK

A blender does 5000 J of work on the food in its bowl. During the time the blender runs, 2000 J of heat is transferred from the warm food to the cooler environment. What is the change in the thermal energy of the food?

A. +2000 J
B. +3000 J
C. +7000 J
D. −2000 J
E. −3000 J

12.1 The Atomic Model of Matter

We began exploring the concepts of thermal energy, temperature, and heat in Chapter 11, but many unanswered questions remain. How do the properties of matter depend on temperature? When you add heat to a system, by how much does its temperature change? And how is heat transferred to or from a system?

These are questions about the *macroscopic* state of systems, but we'll start our exploration by looking at a *microscopic* view, the atomic model that we've used to explain friction, elastic forces, and the nature of thermal energy. In this chapter, we'll use the atomic model to understand and explain the thermal properties of matter.

As you know, each element and most compounds can exist as a gas, liquid, or solid. You've certainly learned about these three **phases** of matter earlier in your education. An atomic view of the three phases is shown in FIGURE 12.1.

Our atomic model makes a simplification that is worth noting. The basic particles in Figure 12.1 are drawn as simple spheres; no mention is made of the nature of the particles. In a real gas, the basic particles might be helium atoms or nitrogen molecules. In a solid, the basic particles might be the gold atoms that make up a bar of gold or the water molecules that make up ice. But, because many of the properties of gases, liquids, and solids do not depend on the exact nature of the particles that make them up, it's often a reasonable assumption to ignore these details and just consider the gas, liquid, or solid as being made up of simple spherical *particles*.

FIGURE 12.1 Atomic models of the three phases of matter: gas, liquid, and solid.

In a gas, the particles are in random motion and interact only through elastic collisions.
In a liquid, the particles have weak bonds that keep them close together. The particles can slide around each other, so the liquid can flow.

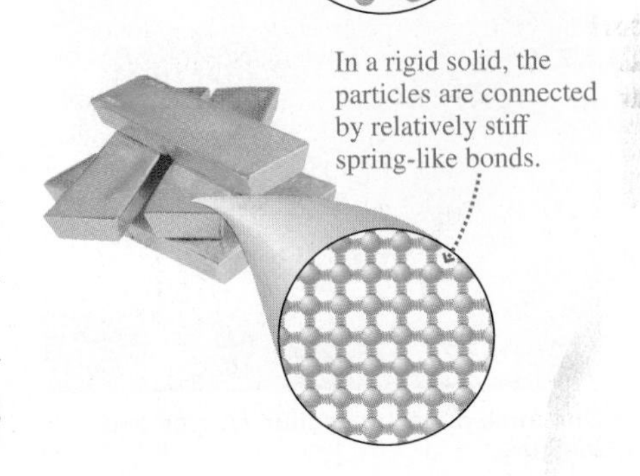

In a rigid solid, the particles are connected by relatively stiff spring-like bonds.

Atomic Mass and Atomic Mass Number

Before we see how the atomic model explains the thermal properties of matter, we need to remind you of some "atomic accounting." Recall that atoms of different elements have different masses. The mass of an atom is determined primarily by its most massive constituents: the protons and neutrons in its nucleus. The *sum* of the number of protons and the number of neutrons is the **atomic mass number** A:

$$A = \text{number of protons} + \text{number of neutrons}$$

A, which by definition is an integer, is written as a leading superscript on the atomic symbol. For example, the primary isotope of carbon, with six protons (which is what makes it carbon) and six neutrons, has $A = 12$ and is written ^{12}C. The radioactive isotope ^{14}C, used for carbon dating of archeological finds, contains six protons and eight neutrons.

The **atomic mass** scale is established by defining the mass of ^{12}C to be exactly 12 u, where u is the symbol for the *atomic mass unit.* That is, $m(^{12}C) = 12$ u. In kg, the atomic mass unit is

$$1\text{ u} = 1.66 \times 10^{-27}\text{ kg}$$

Atomic masses are all very nearly equal to the integer atomic mass number A. For example, the mass of ^{1}H, with $A = 1$, is $m = 1.0078$ u. For our present purposes, it will be sufficient to use the integer atomic mass numbers as the values of the atomic mass. That is, we'll use $m(^{1}H) = 1$ u, $m(^{4}He) = 4$ u, and $m(^{16}O) = 16$ u. For molecules, the **molecular mass** is the sum of the atomic masses of the atoms that form the molecule. Thus the molecular mass of the molecule O_2, the constituent of oxygen gas, is $m(O_2) = 2m(^{16}O) = 32$ u.

NOTE ▶ An element's atomic mass number is *not* the same as its atomic number. The *atomic number,* which gives the element's position in the periodic table, is the number of protons. ◀

TABLE 12.1 lists the atomic mass numbers of some of the elements that we'll use for examples and homework problems. A complete periodic table, including atomic masses, is found in Appendix B.

TABLE 12.1 Some atomic mass numbers

Element	Symbol	A
Hydrogen	^{1}H	1
Helium	^{4}He	4
Carbon	^{12}C	12
Nitrogen	^{14}N	14
Oxygen	^{16}O	16
Neon	^{20}Ne	20
Aluminum	^{27}Al	27
Argon	^{40}Ar	40
Lead	^{207}Pb	207

The Definition of the Mole

One way to specify the amount of substance in a system is to give its mass. Another way, one connected to the number of atoms, is to measure the amount of substance in *moles*. **1 mole of substance, abbreviated 1 mol, is 6.02×10^{23} basic particles.**

The basic particle depends on the substance. Helium is a **monatomic gas,** meaning that the basic particle is the helium atom. Thus 6.02×10^{23} helium atoms are 1 mol of helium. But oxygen gas is a **diatomic gas** because the basic particle is the two-atom diatomic molecule O_2. 1 mol of oxygen gas contains 6.02×10^{23} *molecules* of O_2 and thus $2 \times 6.02 \times 10^{23}$ oxygen atoms. TABLE 12.2 lists the monatomic and diatomic gases that we will use for examples and problems.

TABLE 12.2 Monatomic and diatomic gases

Monatomic	Diatomic
Helium (He)	Hydrogen (H_2)
Neon (Ne)	Nitrogen (N_2)
Argon (Ar)	Oxygen (O_2)

The number of basic particles per mole of substance is called **Avogadro's number** N_A. The value of Avogadro's number is thus

$$N_A = 6.02 \times 10^{23} \text{ mol}^{-1}$$

The number n of moles in a substance containing N basic particles is

$$n = \frac{N}{N_A} \tag{12.1}$$

Moles of a substance in terms of the number of basic particles

One mole of helium, sulfur, copper, and mercury.

The **molar mass** of a substance, M_{mol}, is the mass *in grams* of 1 mol of substance. To a good approximation, the numerical value of the molar mass equals the numerical value of the atomic or molecular mass. That is, the molar mass of He, with $m = 4$ u, is M_{mol} (He) $= 4$ g/mol, and the molar mass of diatomic O_2 is $M_{mol}(O_2) = 32$ g/mol.

You can use the molar mass to determine the number of moles. In one of the few instances where the proper units are *grams* rather than kilograms, the number of moles contained in a system of mass M consisting of atoms or molecules with molar mass M_{mol} is

$$n = \frac{M \text{ (in grams)}}{M_{mol}} \tag{12.2}$$

Moles of a substance in terms of its mass

EXAMPLE 12.1 Determining quantities of oxygen

A system contains 100 g of oxygen. How many moles does it contain? How many molecules?

SOLVE The diatomic oxygen molecule O_2 has molar mass $M_{mol} = 32$ g/mol. From Equation 12.2,

$$n = \frac{100 \text{ g}}{32 \text{ g/mol}} = 3.1 \text{ mol}$$

Each mole contains N_A molecules, so the total number is $N = nN_A = 1.9 \times 10^{24}$ molecules.

Volume

The volume V of a macroscopic system is the amount of space the system occupies. The SI unit of volume is m^3. Nonetheless, both cm^3 and, to some extent, liters (L) are widely used metric units of volume. In most cases, you *must* convert these to m^3

before doing calculations. Some important conversions for volumes are listed here. Although it is true that $1\ \text{m} = 100\ \text{cm}$, it is not true that $1\ \text{m}^3 = 100\ \text{cm}^3$. FIGURE 12.2 shows that the volume conversion factor is $1\ \text{m}^3 = 10^6\ \text{cm}^3$.

$$1\ \text{m}^3 = 1000\ \text{liters} = 10^6\ \text{cm}^3 \qquad 1\ \text{liter} = 1000\ \text{cm}^3 \qquad 1\ \text{cm}^3 = 1\ \text{mL}$$

FIGURE 12.2 There are $10^6\ \text{cm}^3$ in $1\ \text{m}^3$.

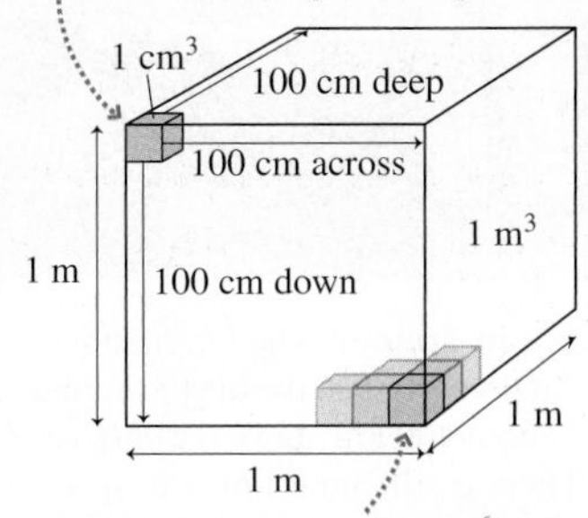

Subdivide the 1 m × 1 m × 1 m cube into little cubes 1 cm on a side. You will get 100 subdivisions along each edge.

STOP TO THINK 12.1 Which system contains more atoms: 5 mol of helium ($A = 4$) or 1 mol of neon ($A = 20$)?

A. Helium B. Neon C. They have the same number of atoms.

12.2 The Atomic Model of an Ideal Gas

Solids and liquids are nearly incompressible because the atomic particles are in close contact with each other. Gases, in contrast, are highly compressible because the atomic particles are far apart. When we work with gases, we'll assume a set of simplifications that we call the **ideal-gas model,** summed up in FIGURE 12.3. It turns out that most real gases at the temperatures and pressures we work with can be explained quite well using this model; the simplifications are not far from reality. We discussed thermal energy in the context of a gas in ◀◀ SECTION 11.3. The thermal energy of a gas is simply the total kinetic energy of the atoms of the gas; the temperature is a measure of the average kinetic energy of the atoms. Adding heat to a gas makes its atoms move faster, increasing the thermal energy and the temperature.

For an ideal gas, there is a straightforward relationship between the temperature and the average kinetic energy per atom K_{avg}:

$$T = \frac{2}{3}\frac{K_{\text{avg}}}{k_{\text{B}}} \tag{12.3}$$

where k_{B} is a constant known as **Boltzmann's constant.** Its value is

$$k_{\text{B}} = 1.38 \times 10^{-23}\ \text{J/K}$$

We can rearrange Equation 12.3 to give the average kinetic energy in terms of the temperature:

$$K_{\text{avg}} = \frac{3}{2}k_{\text{B}}T \tag{12.4}$$

The thermal energy of an ideal gas consisting of N atoms is the sum of the kinetic energies of the individual atoms:

$$E_{\text{th}} = NK_{\text{avg}} = \frac{3}{2}Nk_{\text{B}}T \tag{12.5}$$

Thermal energy of an ideal gas of N atoms

E_{th} vs N: PROPORTIONAL p. 39; E_{th} vs T: PROPORTIONAL p. 39

FIGURE 12.3 The ideal-gas model.

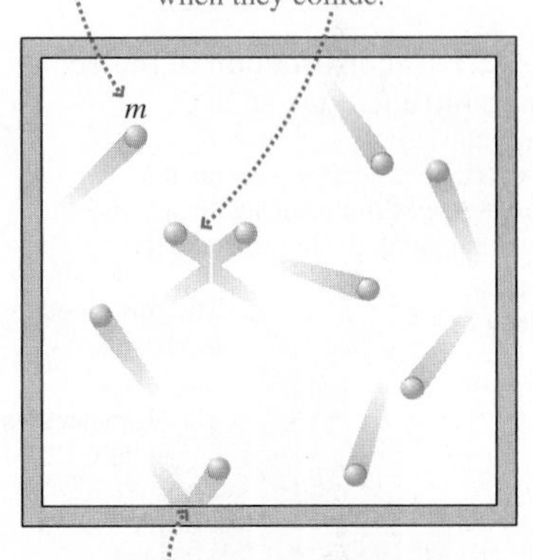

The gas is made up of a large number N of particles of mass m, each moving randomly.
The particles are quite far from each other and interact only rarely when they collide.
m
The collisions of the particles with each other (and with walls of the container) are elastic; no energy is lost in these collisions.

For an ideal gas, **thermal energy is directly proportional to temperature.** Consequently, a change in the thermal energy of an ideal gas is proportional to a change in temperature:

$$\Delta E_{\text{th}} = \frac{3}{2}Nk_{\text{B}}\Delta T \tag{12.6}$$

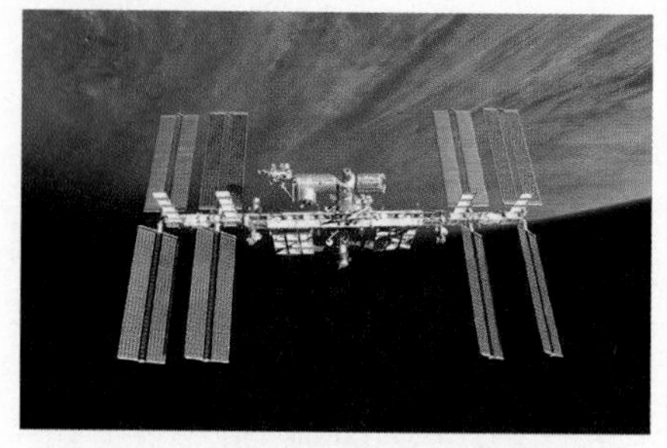

Is it cold in space? The International Space Station orbits in the upper thermosphere, about 400 km above the surface of the earth. There is still a trace of atmosphere left at this altitude, and it has quite a high temperature—over 1000°C. Although the average speed of the air molecules here is high, there are so few air molecules present that the thermal energy is extremely low.

EXAMPLE 12.2 Finding the energy needed to warm up a room

A large bedroom contains about 1×10^{27} molecules of air. If we model the air as an ideal gas, how much energy is required to raise the temperature of the air in the room by 5°C.

STRATEGIZE We can use Equation 12.6 to determine the change in thermal energy corresponding to the noted change in temperature.

PREPARE The actual temperature of the gas doesn't matter—only the change. The temperature increase is given as 5°C, implying a change in the absolute temperature by the same amount: $\Delta T = 5\text{ K}$.

SOLVE We can use Equation 12.6 to calculate the amount by which the room's thermal energy must be increased:

$$\Delta E_{th} = \frac{3}{2}Nk_B\Delta T = \frac{3}{2}(1 \times 10^{27})(1.38 \times 10^{-23}\text{ J/K})(5\text{ K}) = 1 \times 10^5\text{ J} = 100\text{ kJ}$$

This is the energy we would have to supply—probably in the form of heat from a furnace—to raise the temperature.

ASSESS 100 kJ isn't that much energy. Table 11.2 showed it to be less than the food energy in a carrot! This seems reasonable because you know that your furnace can quickly warm up the air in a room. Heating up the walls and furnishings is another story.

Molecular Speeds and Temperature

The atomic model of an ideal gas is based on random motion, so it's no surprise that the individual atoms in a gas are moving at different speeds. FIGURE 12.4 shows data from an experiment to measure the molecular speeds in nitrogen gas at 20°C. The results are presented as a histogram, a bar chart in which the height of the bar indicates what percentage of the molecules have a speed in the range of speeds shown below the bar. For example, 16% of the molecules have speeds in the range from 600 m/s to 700 m/s. The most probable speed, as judged from the tallest bar, is $\sim 500\text{ m/s}$. This is quite fast: ~1200 mph!

FIGURE 12.4 The distribution of molecular speeds in nitrogen gas at 20°C.

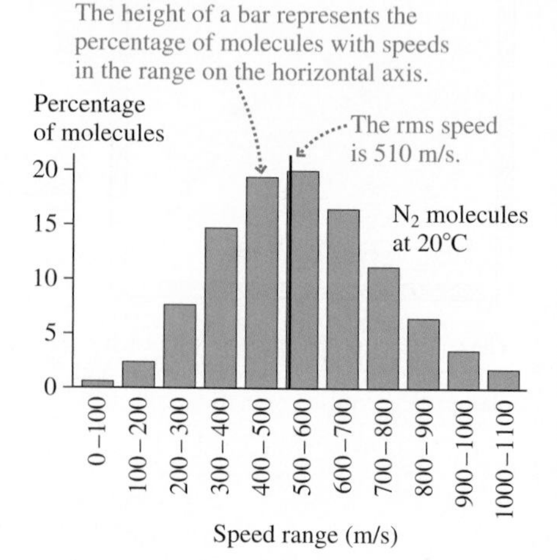

Because temperature is proportional to the average kinetic energy of the atoms, it will be useful to calculate the average kinetic energy for this distribution. An individual atom of mass m and velocity v has kinetic energy $K = \frac{1}{2}mv^2$. Recall that the average of a series of measurements is found by adding all the values and then dividing the total by the number of data points. Thus, we can find the average kinetic energy by adding up all the kinetic energies of all the atoms and then dividing the total by the number of atoms:

$$K_{avg} = \frac{\sum \frac{1}{2}mv^2}{N} = \frac{1}{2}m\frac{\sum v^2}{N} = \frac{1}{2}m(v^2)_{avg} \tag{12.7}$$

The quantity $\sum v^2/N$ is the sum of the values of v^2 for all the atoms divided by the number of atoms. By definition, this is the average of the *squares* of all the individual speeds, which we've written $(v^2)_{avg}$.

The square root of this average is about how fast a typical atom in the gas is moving. Because we'll be taking the square root of the average, or mean, of the square of the speeds, we define the **root-mean-square speed** as

$$v_{rms} = \sqrt{(v^2)_{avg}} = \text{speed of a typical atom} \tag{12.8}$$

The root-mean-square speed is often referred to as the *rms speed*. The rms speed isn't the average speed of atoms in the gas; it's the speed of an atom with the average kinetic energy. But the average speed and the rms speed are very nearly equal, so we'll interpret an rms speed as telling us the speed of a typical atom in the gas.

Rewriting Equation 12.7 in terms of v_{rms} gives the average kinetic energy per atom:

$$K_{avg} = \frac{1}{2}mv_{rms}^2 \tag{12.9}$$

We can relate the temperature to the speeds of the atoms by substituting Equation 12.9 into Equation 12.4:

$$T = \frac{1}{3}\frac{mv_{rms}^2}{k_B} \quad (12.10)$$

Solving Equation 12.10 for the rms speed of the atoms, we find that

$$v_{rms} = \sqrt{\frac{3k_BT}{m}} \quad (12.11)$$

rms speed of an atom of mass m in an ideal gas at temperature T

NOTE ► You must use absolute temperature, in kelvin, to compute rms speeds. ◄

We have been considering ideal gases made of atoms, but our results are equally valid for real gases made of either atoms (such as helium, He) or molecules (such as oxygen, O_2). At a given temperature, Equation 12.11 shows that the speed of atoms or molecules in a gas varies with the atomic or molecular mass. A gas with lighter atoms will have faster atoms, on average, than a gas with heavier atoms. The equation also shows that higher temperatures correspond to faster atomic or molecular speeds. The rms speed is proportional to the *square root* of the temperature. This is a new mathematical form that we will see again, so we will take a look at its properties.

Martian airsicles The atmosphere of Mars is mostly carbon dioxide. At night, the temperature may drop so low that the molecules in the atmosphere will slow down enough to stick together—the atmosphere actually freezes. The frost on the surface in this image from the Viking 2 lander is composed partially of frozen carbon dioxide.

Square-root relationships

Two quantities are said to have a **square-root relationship** if y is proportional to the square root of x. We write the mathematical relationship as

$$y = A\sqrt{x}$$

y is proportional to the square root of x

The graph of a square-root relationship is a parabola that has been rotated by 90°.

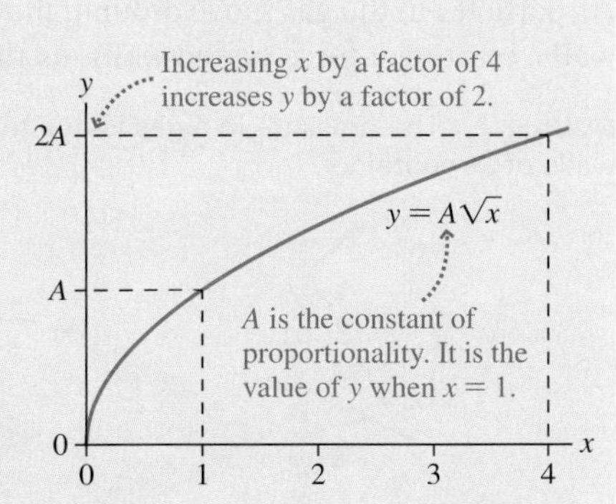
Increasing x by a factor of 4 increases y by a factor of 2.
y = A√x
A is the constant of proportionality. It is the value of y when x = 1.
y
x
2A
A
0
0
1
2
3
4

SCALING If x has the initial value x_1, then y has the initial value y_1. Changing x from x_1 to x_2 changes y from y_1 to y_2. The ratio of y_2 to y_1 is

$$\frac{y_2}{y_1} = \frac{A\sqrt{x_2}}{A\sqrt{x_1}} = \sqrt{\frac{x_2}{x_1}}$$

which is the square root of the ratio of x_2 to x_1.

- If you increase x by a factor of 4, you increase y by a factor of $\sqrt{4} = 2$.
- If you decrease x by a factor of 9, you decrease y by a factor of $\sqrt{9} = 3$.

These examples illustrate a general rule:

Changing x by a factor of c changes y by a factor of $\sqrt{c}$.

Exercises 7, 8

EXAMPLE 12.3 Finding the speeds of air molecules

Most of the earth's atmosphere is the gas nitrogen, which consists of molecules, N_2. At the coldest temperature ever recorded on earth, –95°C, what is the root-mean-square speed of the nitrogen molecules? Does the temperature at the earth's surface ever get high enough that a typical molecule is moving at twice this speed? (The highest temperature ever recorded on earth was 54°C.)

STRATEGIZE The first part of the problem is a straightforward calculation. The second part is best done by considering ratios. The rms speed varies as the square root of the temperature, so we will determine the factor by which we'd need to increase the temperature to double the rms speed.

Continued

PREPARE We can use the periodic table to determine that the mass of a nitrogen atom is 14 u. A molecule consists of two atoms, so its mass is 28 u. Thus the molecular mass in SI units (i.e., kg) is

$$m = 28\text{ u} \times \frac{1.66 \times 10^{-27}\text{ kg}}{1\text{ u}} = 4.6 \times 10^{-26}\text{ kg}$$

The problem statement gives two temperatures we'll call T_1 and T_2; we need to express these in kelvin. The lowest temperature ever observed on earth is $T_1 = -95 + 273 = 178$ K; the highest temperature is $T_2 = 54 + 273 = 327$ K.

SOLVE We use Equation 12.11 to find v_{rms} for the nitrogen molecules at T_1:

$$v_{rms} = \sqrt{\frac{3k_BT_1}{m}} = \sqrt{\frac{3(1.38 \times 10^{-23}\text{ J/K})(178\text{ K})}{4.6 \times 10^{-26}\text{ kg}}} = 400\text{ m/s}$$

Because the rms speed is proportional to the square root of the temperature, doubling the rms speed would require increasing the temperature by a factor of 4. The ratio of the highest temperature ever recorded to the lowest temperature ever recorded is less than this:

$$\frac{T_2}{T_1} = \frac{327\text{ K}}{178\text{ K}} = 1.8$$

The temperature at the earth's surface is never high enough that a typical nitrogen molecule would move at twice the computed speed.

ASSESS We can use the square-root relationship to assess our computed result for the molecular speed. Figure 12.4 shows an rms speed of 510 m/s for nitrogen molecules at 20°C, or 293 K. Temperature T_1 is about 0.6 of this, so we'd expect to compute a speed that is lower by the square root of 0.6 about 0.8, which is what we found.

Pressure

Everyone has some sense of the concept of *pressure.* If you get a hole in your bicycle tire, the higher-pressure air inside comes squirting out. It's hard to get the lid off a vacuum-sealed jar because of the low pressure inside. But just what is pressure?

Let's take an atomic-scale view of pressure, defining it in terms of the motion of particles of a gas. Suppose we have a sample of gas in a container with rigid walls. As particles in the gas move around, they sometimes collide with and bounce off the walls, creating a force on the walls, as illustrated in FIGURE 12.5a.

FIGURE 12.5 The pressure in a gas is due to the force of the particles colliding with the walls of its container.

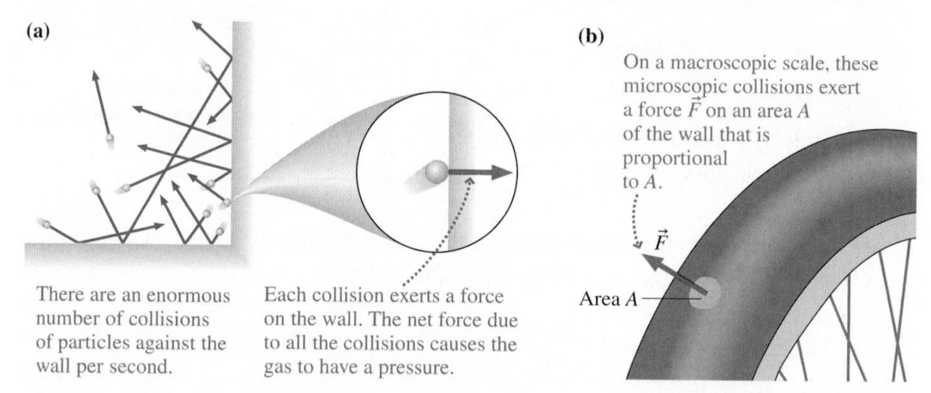

(a)
There are an enormous number of collisions of particles against the wall per second.
Each collision exerts a force on the wall. The net force due to all the collisions causes the gas to have a pressure.
(b)
On a macroscopic scale, these microscopic collisions exert a force $\vec{F}$ on an area A of the wall that is proportional to A.
$\vec{F}$
Area A

These countless microscopic collisions are what lead to the pressure in the gas. On a small patch of the container wall with surface area A, these collisions result in a continuous macroscopic force of magnitude F directed perpendicular to the wall, as shown for the bicycle tire in FIGURE 12.5b. If the size of the patch is doubled, then twice as many particles will hit it every second, leading to a doubling of the force. This implies that the force is proportional to the area of the patch, and so the *ratio* F/A is constant. We define this ratio to be the **pressure** p of the gas:

$$p = \frac{F}{A} \tag{12.12}$$

Definition of pressure in a gas

You can see from Equation 12.12 that a gas exerts a force of magnitude

$$F = pA \tag{12.13}$$

on a surface of area A.

NOTE ▶ We sometimes talk informally about "the force exerted by the pressure," but it's really the gas that exerts the force, not the pressure. ◀

A video to support a section's topic is embedded in the eText.

Video Force and Pressure

From its definition, you can see that pressure has units of N/m^2. The SI unit of pressure is the **pascal,** defined as

$$1 \text{ pascal} = 1 \text{ Pa} = 1 \frac{\text{N}}{\text{m}^2}$$

A pascal is a very small pressure, so we usually see pressures in kilopascals, where 1 kPa = 1000 Pa.

The pressure with the most significance for our daily lives is the pressure of the atmosphere, caused by the microscopic collisions of the air molecules that surround us. The pressure of the atmosphere varies with altitude and the weather, but the global average pressure at sea level, called the *standard atmosphere,* is

$$1 \text{ standard atmosphere} = 1 \text{ atm} = 101{,}300 \text{ Pa} = 101.3 \text{ kPa}$$

Just as we measured acceleration in units of g, we will often measure pressure in units of atm.

In the United States, pressure is often expressed in pounds per square inch, or psi. When you measure pressure in the tires on your car or bike, you probably use a gauge that reads in psi. The conversion factor is

$$1 \text{ atm} = 14.7 \text{ psi}$$

The total force on the surface of your body due to the pressure of the atmosphere is over 40,000 pounds. Why doesn't this enormous force pushing in simply crush you? The key is that there is also a force pushing out. FIGURE 12.6a shows an empty plastic soda bottle. The bottle isn't a very sturdy structure, but the inward force due to the pressure of the atmosphere outside the bottle doesn't crush it because there is an equal pressure due to air inside the bottle that pushes out. The forces pushing on both sides are quite large, but they exactly balance and so there is no net force.

A net pressure force is exerted only where there's a pressure *difference* between the two sides of a surface. FIGURE 12.6b shows a surface of area A with a pressure difference $\Delta p = p_2 - p_1$ between the two sides. The net pressure force is

$$F_{\text{net}} = F_2 - F_1 = p_2 A - p_1 A = A(p_2 - p_1) = A\,\Delta p$$

This is the force that holds the lid on a vacuum-sealed jar, where the pressure inside is less than the pressure outside.

Decreasing the number of molecules in a container decreases the pressure because there are fewer collisions with the walls. The pressure in a completely empty container would be $p = 0$ Pa. This is called a *perfect vacuum.* A perfect vacuum cannot be achieved because it's impossible to remove every molecule from a region of space. In practice, a **vacuum** is an enclosed space in which $p \ll 1$ atm. Using $p = 0$ Pa is then a good approximation.

The basic principle behind pressure measurements is expressed by Equation 12.12. A pressure gauge measures the force exerted by the gas on a known area. The actual or **absolute pressure** p is directly proportional to this force. Because the effects of pressure depend on pressure differences, most gauges measure not the absolute pressure but what is called the **gauge pressure** p_g, the *difference* between the absolute pressure and atmospheric pressure. Most gauges measure this difference directly because they measure a force on a surface. FIGURE 12.7 shows how this works for a tire gauge. The gauge measures the difference between the pressure inside the tire and the pressure outside the tire, which is just the pressure of the atmosphere, p_{atmos}.

The calculations we'll do later in the chapter involve absolute pressure, but tire pressures and other common pressures are reported as gauge pressures, so we'll often need to convert between the two. The gauge pressure is, as in the case of the

FIGURE 12.6 The net force depends on the pressure difference.

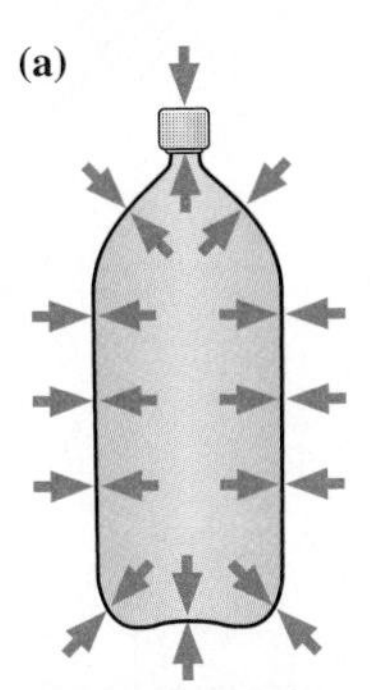
(a)
The pressure inside the bottle equals the pressure outside, so there is no net force.

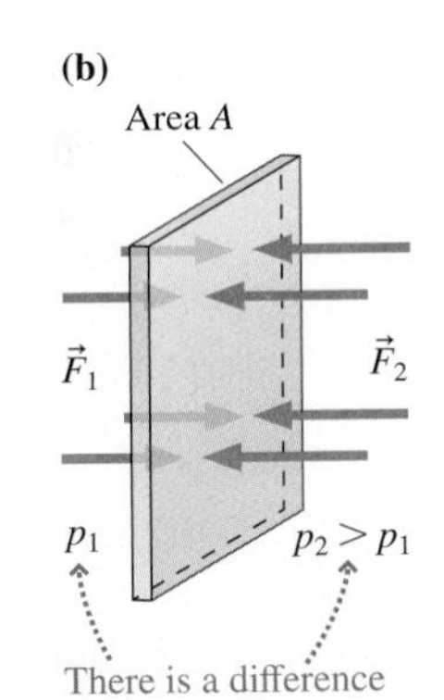
(b)
Area A
$\vec{F}_1$
$\vec{F}_2$
p_1
$p_2 > p_1$
There is a difference in pressure between the two sides.

Too little pressure too fast BIO The rockfish is a popular game fish that can be caught at ocean depths up to 1000 ft. The great pressure at these depths is balanced by an equally large pressure inside the fish's gas-filled *swim bladder.* If the fish is hooked and rapidly raised to the lower pressure of the surface, the pressure inside the swim bladder is suddenly much greater than the pressure outside, causing the swim bladder to expand dramatically. When reeled in from a great depth, these fish seldom survive.

FIGURE 12.7 A tire gauge measures the difference between the tire's pressure and atmospheric pressure.

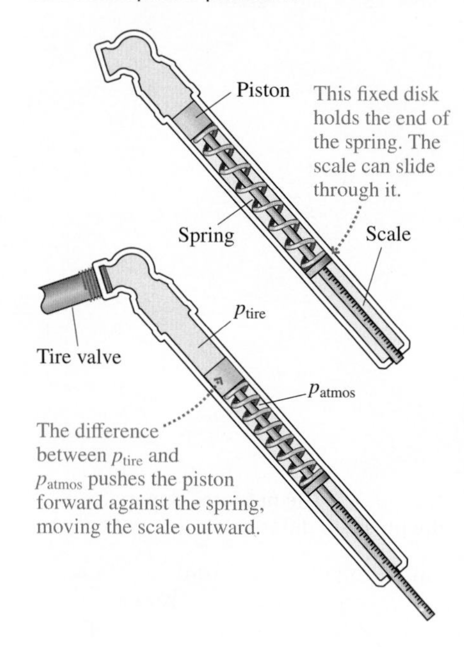
Piston
This fixed disk holds the end of the spring. The scale can slide through it.
Spring
Scale
p_{tire}
Tire valve
p_{atmos}
The difference between p_{tire} and p_{atmos} pushes the piston forward against the spring, moving the scale outward.

tire gauge, the difference between an absolute pressure and atmospheric pressure, so we can write

$$p_g = p - p_{atmos} \qquad p = p_g + p_{atmos} \tag{12.14}$$

If you are at sea level, you can assume that $p_{atmos} = 1.0$ atm; at other elevations, you'll need to use the local pressure.

EXAMPLE 12.4 Finding the force due to a pressure difference

Patients suffering from decompression sickness may be treated in a hyperbaric oxygen chamber filled with oxygen at greater than atmospheric pressure. A cylindrical chamber with flat end plates of diameter 0.75 m is filled with oxygen to a gauge pressure of 27 kPa. What is the resulting force on the end plate of the cylinder?

STRATEGIZE The net force $F_{net} = A\Delta p$ on the end plate depends on Δp, the difference between the pressure inside and the pressure outside the chamber. This is the gauge pressure, 27 kPa.

PREPARE The end plates are circular, with area $A = \pi r^2 = \pi(0.375\text{ m})^2 = 0.442\text{ m}^2$.

SOLVE The pressure difference results in a net force

$$F_{net} = A\,\Delta p = (0.442\text{ m}^2)(27{,}000\text{ Pa}) = 12\text{ kN}$$

ASSESS The area of the end plate is large, so we expect a large force, which is exactly what we get—the force is equivalent to the weight of about 15 people. If you've seen chambers like this, you know that the end plate is fastened in place with stout bolts. It is remarkable to think that this large force results from the collisions of tiny air molecules with the plate!

The Ideal-Gas Law

We can use the fact that the pressure in a gas is due to the collisions of particles with the walls to make some qualitative predictions. FIGURE 12.8 presents a few such predictions.

FIGURE 12.8 Relating gas pressure to other variables.

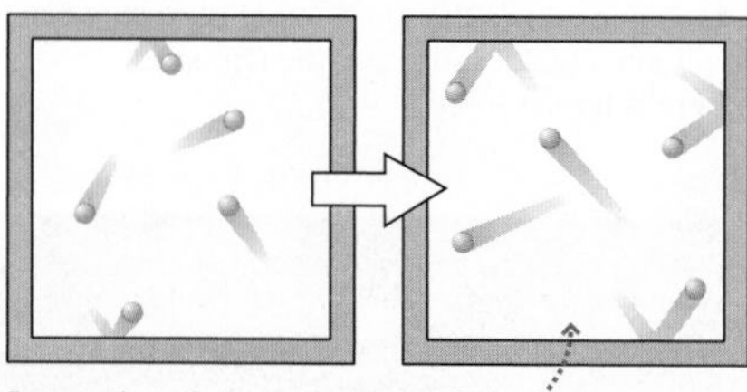

Increasing the temperature of the gas means the particles move at higher speeds. They hit the walls more often and with more force, so there is more pressure.

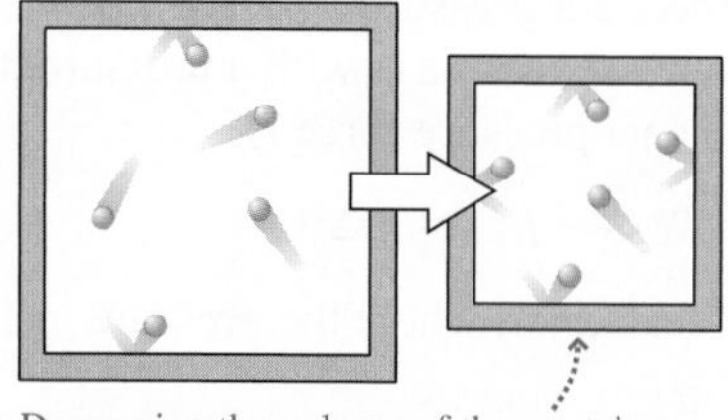

Decreasing the volume of the container means more frequent collisions with the walls of the container, and thus more pressure.

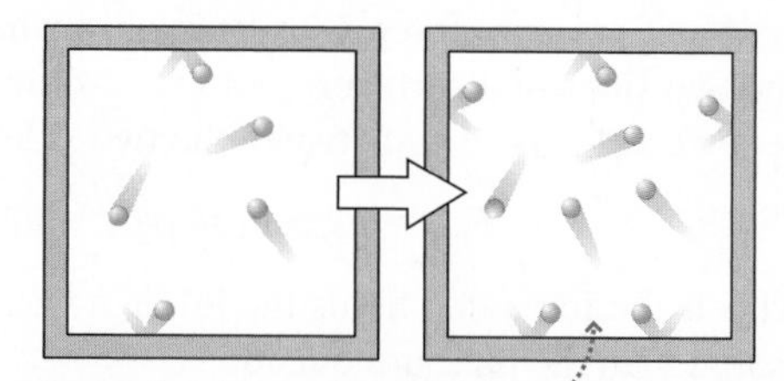

Increasing the number of particles in the container means more frequent collisions with the walls of the container, and thus more pressure.

Pumping up a bicycle tire adds more particles to the fixed volume of the tire, increasing the pressure.

Based on the reasoning in Figure 12.8, we expect the following proportionalities:

- Pressure should be proportional to the temperature of the gas: $p \propto T$.
- Pressure should be inversely proportional to the volume of the container: $p \propto 1/V$.
- Pressure should be proportional to the number of gas particles: $p \propto N$.

In fact, careful experiments back up each of these predictions, leading to a single equation that expresses these proportionalities:

$$p = C\frac{NT}{V}$$

The proportionality constant C turns out to be none other than Boltzmann's constant k_B, which allows us to write

$$pV = Nk_B T \quad (12.15)$$

Ideal-gas law, version 1

Equation 12.15 is known as the **ideal-gas law.**

Equation 12.15 is written in terms of the number N of particles in the gas, whereas the ideal-gas law is stated in chemistry in terms of the number n of moles. But the change is straightforward to make. The number of particles is $N = nN_A$, so we can rewrite Equation 12.15 as

$$pV = nN_A k_B T = nRT \quad (12.16)$$

Ideal-gas law, version 2

In this version of the equation, the proportionality constant—known as the *gas constant*—is

$$R = N_A k_B = 8.31 \text{ J/mol} \cdot \text{K}$$

The units may seem unusual, but the product of Pa and m^3, the units of pV, is equivalent to J.

Let's review the meanings and the units of the various quantities in the ideal-gas law:

$$pV = nRT$$

EXAMPLE 12.5 **Finding the volume of a mole of a gas**

What volume is occupied by 1 mole of an ideal gas at a pressure of 1.00 atm and a temperature of 0°C?

STRATEGIZE We are given the pressure, the temperature, and the number of moles. We will use version 2 of the ideal-gas law to find the volume.

PREPARE The first step in ideal-gas law calculations is to convert all quantities to SI units:

$$p = 1.00 \text{ atm} = 101.3 \times 10^3 \text{ Pa}$$

$$T = 0 + 273 = 273 \text{ K}$$

SOLVE We rearrange version 2 of the ideal-gas law equation to compute

$$V = \frac{nRT}{p} = \frac{(1.00 \text{ mol})(8.31 \text{ J/mol} \cdot \text{K})(273 \text{ K})}{101.3 \times 10^3 \text{ Pa}} = 0.0224 \text{ m}^3$$

We recall from earlier in the chapter that 1.00 m^3 = 1000 L, so we can write

$$V = 22.4 \text{ L}$$

ASSESS At this temperature and pressure, we find that the volume of 1 mole of a gas is 22.4 L, a result you might recall from chemistry. When we do calculations using gases, it will be useful to keep this volume in mind to see if our answers make physical sense.

STOP TO THINK 12.2 A sample of ideal gas is in a sealed container. The temperature of the gas and the volume of the container are both increased. What other properties of the gas necessarily change? (More than one answer may be correct.)

A. The rms speed of the gas atoms
B. The thermal energy of the gas
C. The pressure of the gas
D. The number of molecules of gas

12.3 Ideal-Gas Processes

Suppose you measure the pressure in the tires on your car on a cold morning. How much will the tire pressure increase after the tires warm up in the sun and the temperature of the air in the tires has increased? We will solve this problem later in the chapter, but, for now, note these properties of this process:

- The quantity of gas is fixed. No air is added to or removed from the tire.
- There is a well-defined initial state. The initial values of pressure, volume, and temperature will be designated p_i, V_i, and T_i.
- There is a well-defined final state in which the pressure, volume, and temperature have values p_f, V_f, and T_f.

Processes with these properties are called **ideal-gas processes.**

For gases in sealed containers, the number of moles (and the number of molecules) does not change. In that case, the ideal-gas law can be written as

$$\frac{pV}{T} = nR = \text{constant}$$

The values of the variables in the initial and final states are then related by

$$\frac{p_f V_f}{T_f} = \frac{p_i V_i}{T_i} \tag{12.17}$$

Initial and final states for an ideal gas in a sealed container

This before-and-after relationship between the two states, reminiscent of a conservation law, will be valuable for solving many problems.

NOTE ▶ Because pressure and volume appear on both sides of the equation, pressure and volume can be in any units; we don't necessarily need to convert to SI units. However, temperature *must* be in K. Unit-conversion factors for pressure and volume are multiplicative factors, and the same factor on both sides of the equation cancels. But the conversion from K to °C is an *additive* factor, and additive factors in the denominator don't cancel. ◀

FIGURE 12.9 The state of the gas and ideal-gas processes can be shown on a pV diagram.

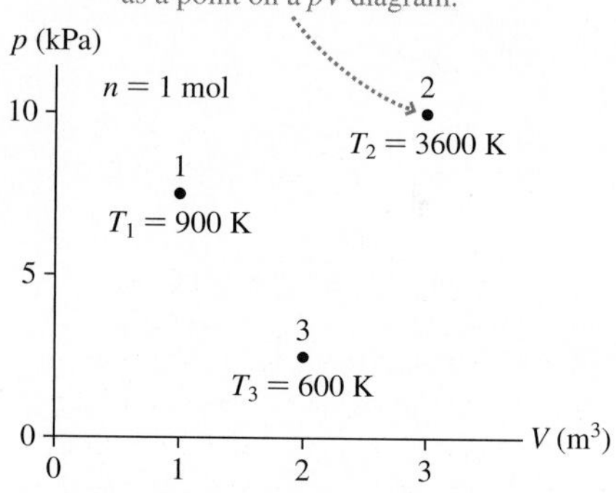

(a)
Each state of an ideal gas is represented as a point on a pV diagram.
p (kPa)
n = 1 mol
10
5
0
1
2
3
T1 = 900 K
T2 = 3600 K
T3 = 600 K
0
V (m3)

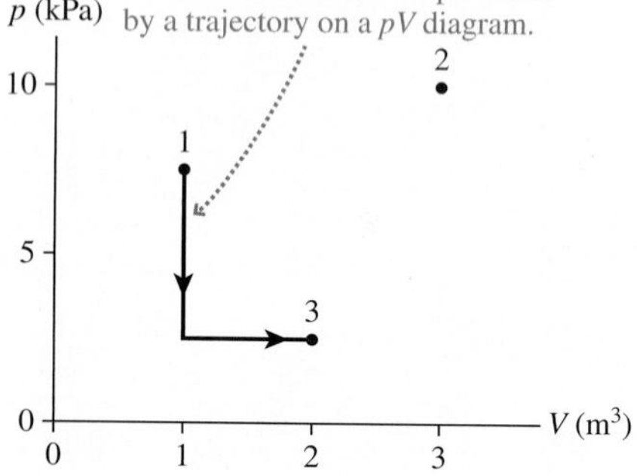

(b)
A process that changes the gas from one state to another is represented by a trajectory on a pV diagram.
p (kPa)
10
5
0
1
2
3
V (m3)

pV Diagrams

It's useful to represent ideal-gas processes on a graph called a ***pV* diagram.** The important idea behind a pV diagram is that each point on the graph represents a single, unique state of the gas. This may seem surprising because a point on a graph specifies only the values of pressure and volume. But knowing p and V, and assuming that n is known for a sealed container, we can find the temperature from the ideal-gas law. Thus each point on a pV diagram actually represents a triplet of values (p, V, T) specifying the state of the gas.

For example, FIGURE 12.9a is a pV diagram showing three states of a system consisting of 1 mol of gas. The values of p and V can be read from the axes, then the temperature at that point calculated from the ideal-gas law. An ideal-gas process—a process that changes the state of the gas by, for example, heating it or compressing it—can be represented as a "trajectory" in the pV diagram. FIGURE 12.9b shows one possible process by which the gas of Figure 12.9a is changed from state 1 to state 3.

FIGURE 12.10 A constant-volume process.

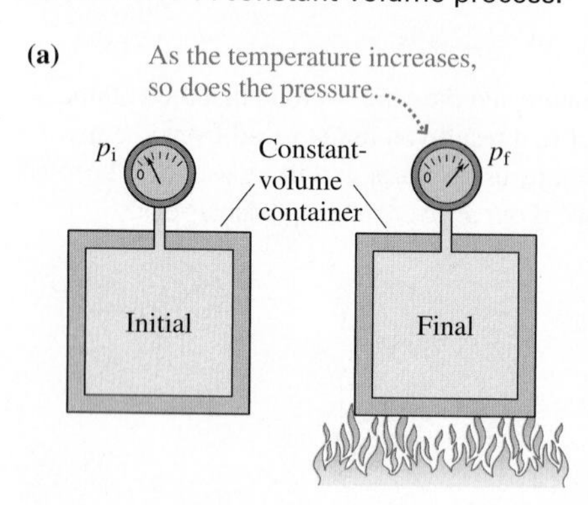

(a)
As the temperature increases, so does the pressure.
pi
Constant-volume container
pf
Initial
Final

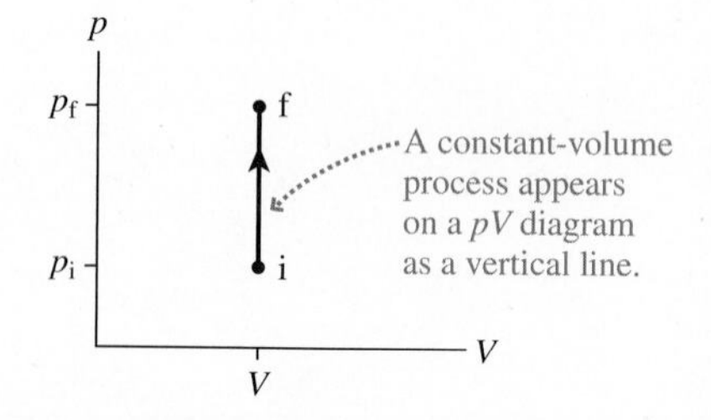

(b)
p
pf
pi
f
i
A constant-volume process appears on a pV diagram as a vertical line.
V
V

Constant-Volume Processes

Suppose you have a gas in the closed, rigid container shown in FIGURE 12.10a. Warming the gas will raise its pressure without changing its volume. This is an example of a **constant-volume process.** $V_f = V_i$ for a constant-volume process.

Because the value of V doesn't change, this process is shown as the vertical line $i \rightarrow f$ on the pV diagram of FIGURE 12.10b. **A constant-volume process appears on a pV diagram as a vertical line.**

EXAMPLE 12.6 Computing tire pressure on a hot day

The pressure in a car tire measures 30.0 psi on a cool morning when the air temperature is 0°C. After the day warms up and bright sun shines on the black tire, the temperature of the air inside the tire reaches 30°C. What is the tire pressure at this temperature?

STRATEGIZE A tire is (to a good approximation) a sealed container with constant volume, so this is a constant-volume process.

PREPARE We weren't told anything about elevation, so we assume that the atmospheric pressure is 1.0 atm. When we solve this problem, we don't need to convert units, but we do need to be sure that we are using the correct form of the pressure. The measured tire pressure is a gauge pressure, but the ideal-gas law requires an absolute pressure. We can use Equation 12.14 to convert to absolute pressure. The initial pressure is

$$p_i = (p_g)_i + 1.00 \text{ atm} = 30.0 \text{ psi} + 14.7 \text{ psi} = 44.7 \text{ psi}$$

Temperatures must be in kelvin, so we convert:

$$T_i = 0°\text{C} + 273 = 273 \text{ K}$$
$$T_f = 30°\text{C} + 273 = 303 \text{ K}$$

SOLVE We can use Equation 12.17 to solve for the final pressure. In this equation, we divide both sides by V_f and then cancel the ratio of the two volumes, which is equal to 1 for this constant-volume process:

$$p_f = p_i \frac{V_i}{V_f}\frac{T_f}{T_i} = p_i \frac{T_f}{T_i}$$

The units for p_f will be the same as those for p_i, so we can keep the initial pressure in psi. The pressure at the higher temperature is

$$p_f = 44.7 \text{ psi} \times \frac{303 \text{ K}}{273 \text{ K}} = 49.6 \text{ psi}$$

This is an absolute pressure, but the problem asks for the measured pressure in the tire—a gauge pressure. Converting to gauge pressure gives

$$(p_g)_f = p_f - 1.00 \text{ atm} = 49.6 \text{ psi} - 14.7 \text{ psi} = 34.9 \text{ psi}$$

ASSESS The temperature has changed by 30 K, which is a bit more than 10% of the initial temperature, so we expect a large change in pressure. Our result seems reasonable, and it has practical implications: If you check the pressure in your tires when they are at a particular temperature, don't expect the pressure to be the same when conditions change!

Constant-Pressure Processes

Many gas processes take place at a constant, unchanging pressure. A constant-pressure process is also called an **isobaric process.** For a constant-pressure process $p_f = p_i$.

One way to produce a constant-pressure process is shown in FIGURE 12.11a, where a gas is sealed in a cylinder by a lightweight, tight-fitting cap—a *piston*—that is free to slide up and down. In fact, the piston *will* slide up or down, compressing or expanding the gas inside, until it reaches the position at which $p_{gas} = p_{ext}$. That's the equilibrium position for the piston, the position at which the upward force $F_{gas} = p_{gas}A$, where A is the area of the face of the piston, exactly balances the downward force $F_{ext} = p_{ext}A$ due to the external pressure p_{ext}. The gas pressure in this situation is thus equal to the external pressure. As long as the external pressure doesn't change, neither can the gas pressure inside the cylinder.

FIGURE 12.11 A constant-pressure process.

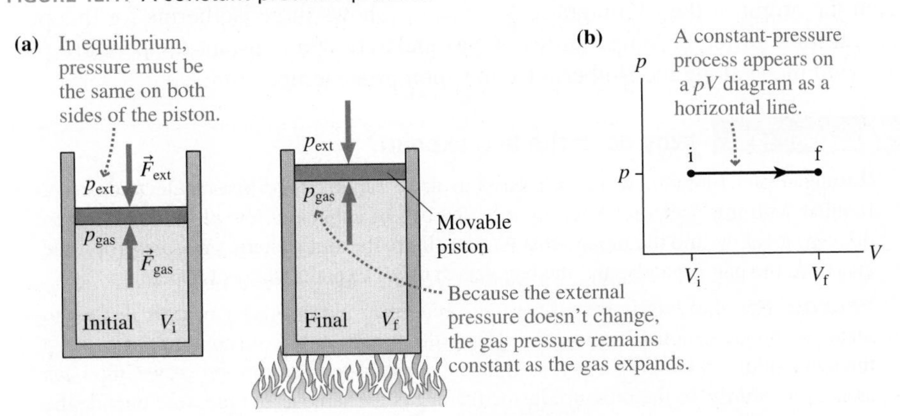
(a) In equilibrium, pressure must be the same on both sides of the piston.
$\vec{F}_{ext}$
p_{ext}
p_{gas}
$\vec{F}_{gas}$
Initial V_i
p_{ext}
p_{gas}
Final V_f
Movable piston
Because the external pressure doesn't change, the gas pressure remains constant as the gas expands.
(b)
A constant-pressure process appears on a pV diagram as a horizontal line.
p
i
f
V
V_i
V_f

Suppose we heat the gas in the cylinder. The gas pressure doesn't change because the pressure is controlled by the unchanging external pressure, not by the temperature. But as the temperature rises, the faster-moving atoms cause the gas to expand, pushing the piston outward. Because the pressure is always the same, this is a constant-pressure process, with a trajectory as shown in FIGURE 12.11b. **A constant-pressure process appears on a pV diagram as a horizontal line.**

EXAMPLE 12.7 Finding the final volume of a compressed gas

A gas in a cylinder with a movable piston occupies 50.0 cm^3 at 50°C. The gas is cooled at constant pressure until the temperature is 10°C. What is the final volume?

STRATEGIZE The pressure of the gas doesn't change, so this is a constant-pressure process, with $p_i/p_f = 1$.

PREPARE The temperatures must be in kelvin, so we convert:

$$T_i = 50°\text{C} + 273 = 323\text{ K}$$
$$T_f = 10°\text{C} + 273 = 283\text{ K}$$

SOLVE We can use Equation 12.17 to solve for V_f:

$$V_f = V_i \frac{p_i}{p_f}\frac{T_f}{T_i} = 50.0\text{ cm}^3 \times 1 \times \frac{283\text{ K}}{323\text{ K}} = 43.8\text{ cm}^3$$

ASSESS Our answer makes sense. The gas cools, so we expect a smaller volume. The cooling is—in absolute temperature terms—relatively modest, so we expect a modest change in volume.

In this example and in Example 12.6, we have not converted pressure and volume units because these multiplicative factors cancel. But we did convert temperature to kelvin because this *additive* factor does *not* cancel.

Constant-Temperature Processes

A constant-temperature process is also called an **isothermal process.** For a constant-temperature process $T_f = T_i$. One possible constant-temperature process is illustrated in FIGURE 12.12a. A piston is being pushed down to compress a gas, but the gas cylinder is submerged in a large container of liquid that is held at a constant temperature. If the piston is pushed *slowly,* then heat-energy transfer through the walls of the cylinder will keep the gas at the same temperature as the surrounding liquid. This is an *isothermal compression.* The reverse process, with the piston slowly pulled out, is an *isothermal expansion.*

FIGURE 12.12 A constant-temperature process.

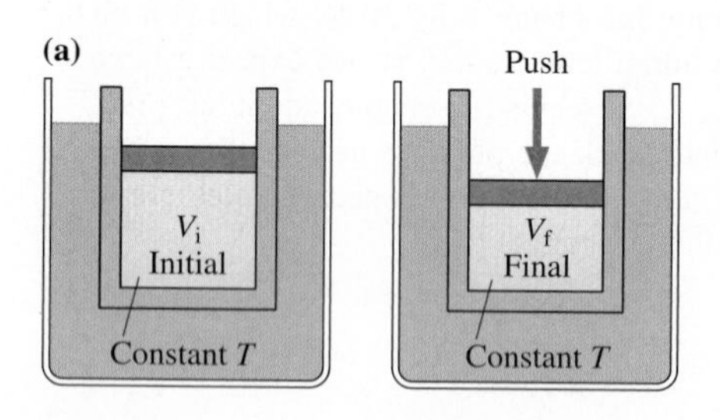

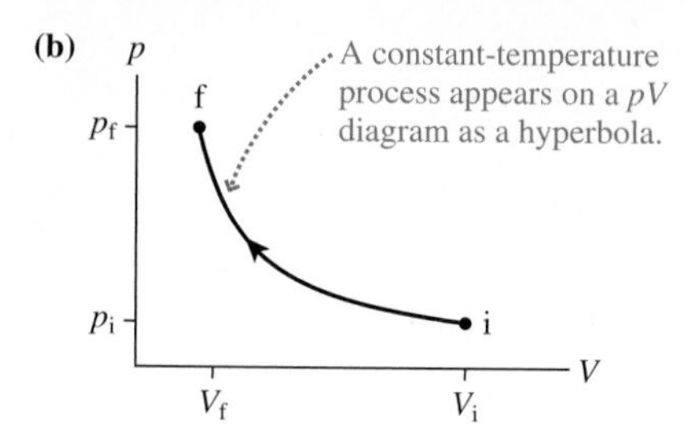

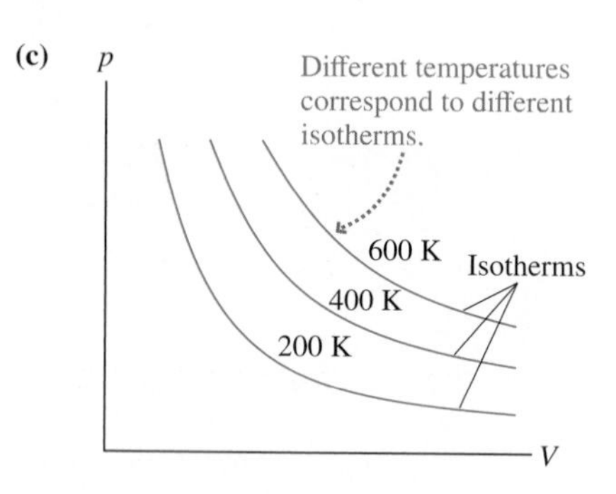

Representing a constant-temperature process on the pV diagram is a little more complicated than the two preceding processes because both p and V change. As long as T remains fixed, we have the relationship

$$p = \frac{nRT}{V} = \frac{\text{constant}}{V} \tag{12.18}$$

INVERSE p.116

Because there is an inverse relationship between p and V, the graph of a constant-temperature process is a *hyperbola*.

Figure 12.12a shows an isothermal compression, which is represented graphically as the hyperbola i → f in FIGURE 12.12b. An isothermal expansion would move in the opposite direction along the hyperbola. The graph of an isothermal process is known as an **isotherm.**

The location of the hyperbola depends on the value of T. If we use a higher constant temperature for the process in Figure 12.12a, the isotherm will move farther from the origin of the pV diagram. FIGURE 12.12c shows three isotherms for this process at three different temperatures. A gas undergoing a constant-temperature process will move along the isotherm for the appropriate temperature.

EXAMPLE 12.8 Why does the bag expand?

Residents of Colorado, where it's easy to drive rapidly to different elevations, are familiar with this scenario: You buy a sealed bag of chips at a low elevation and then drive your car up into the mountains. As you climb, the atmospheric pressure drops, the air inside the bag expands, and the bag gets "puffy." Explain what is happening.

REASON The chip bag is sealed—no gas gets in or out—so we can treat this as an ideal-gas process. Folks in the car will use the heater or the air conditioner to keep the temperature in the car steady, so this is a constant-temperature process. Chip bags aren't very sturdy, so the pressure inside the bag is the same as the pressure outside the bag. As the pressure outside the bag drops, so does the pressure inside the bag. The volume of air in the bag is inversely proportional to the pressure, so as the pressure drops, the volume of the air increases and the bag expands.

ASSESS The result makes sense. A change in pressure also occurs on an airplane, so you may have seen something similar during a flight, or you may have had bottles with caps that pop off and contents pushed out as the gas inside expands.

EXAMPLE 12.9 Finding the volume of air in the lungs BIO

A snorkeler takes a deep breath at the surface, filling his lungs with 4.0 L of air. He then descends to a depth of 5.0 m, where the pressure is 0.50 atm higher than at the surface. At this depth, what is the volume of air in the snorkeler's lungs?

STRATEGIZE Most of the snorkeler's air is nitrogen, which his body doesn't use. Oxygen is used by the body, but for every oxygen molecule taken from the lungs, a molecule of carbon dioxide is returned. The ideal-gas law doesn't care what the molecules are, just the total number. The total number of gas molecules stays approximately constant, and we can treat this as an ideal-gas process. The air stays at body temperature, so this is a constant-temperature process with $T_f = T_i$, so $T_f/T_i = 1$. As the snorkeler descends, the pressure inside his lungs rises to match the pressure of the surrounding water because the body can't sustain large pressure differences between inside and out.

PREPARE We need absolute pressures for our calculation. At the surface, the pressure of the air inside the snorkeler's lungs is 1.0 atm—it's the pressure at sea level. The additional pressure at the noted depth is 0.50 atm, so the absolute pressure at this depth is 1.5 atm.

SOLVE The ideal-gas law for a sealed container (the lungs) gives

$$V_f = V_i \frac{p_i}{p_f} \frac{T_f}{T_i} = 4.0\text{ L} \times \frac{1.0\text{ atm}}{1.5\text{ atm}} \times 1 = 2.7\text{ L}$$

Notice that we didn't need to convert pressure to SI units. As long as the units are the same in the numerator and the denominator, they cancel.

ASSESS The air has a smaller volume at the higher pressure, as we would expect. The air inside your lungs does compress—significantly!—when you dive below the surface.

Thermodynamics of Ideal-Gas Processes

Chapter 11 introduced the first law of thermodynamics, and we saw that heat and work are just two different ways to add energy to a system. We've been considering the changes when we heat gases, but now we want to consider the other form of energy transfer—work.

When gases expand, they can do work by pushing against a piston. This is how the engine under the hood of your car works: When the spark plug fires in a cylinder in the engine, it ignites the gaseous fuel-air mixture inside. The hot gas expands, pushing the piston out and, through various mechanical linkages, turning the wheels of your car. Energy is transferred out of the gas as work; we say that the gas does work on the piston. Similarly, the gas in Figure 12.11a does work as it pushes on and moves the piston.

You learned in ◀◀ SECTION 10.2 that the work done by a constant force F in pushing an object a distance d is $W = Fd$. Let's apply this idea to a gas. FIGURE 12.13a shows a gas cylinder sealed at one end by a movable piston. Force $\vec{F}_{gas}$ is due to the gas pressure and has magnitude $F_{gas} = pA$. Force $\vec{F}_{ext}$, perhaps a force applied by a piston rod, is equal in magnitude and opposite in direction to $\vec{F}_{gas}$. The gas pressure would blow the piston out if the external force weren't there!

FIGURE 12.13 The expanding gas does work on the piston.

(a)

The gas pushes on the piston with force $\vec{F}_{gas}$.

To keep the piston in place, an external force must be equal and opposite to $\vec{F}_{gas}$.

Pressure p

$\vec{F}_{gas}$ $\vec{F}_{ext}$

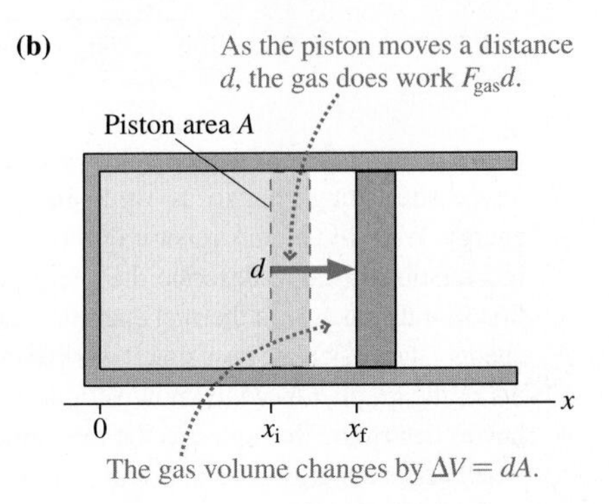
(b)
As the piston moves a distance d, the gas does work $F_{gas}d$.
Piston area A
d
0
x_i
x_f
x
The gas volume changes by $\Delta V = dA$.

Suppose the gas expands at constant pressure, pushing the piston outward from x_i to x_f, a distance $d = x_f - x_i$, as shown in FIGURE 12.13b. As it does, the force due to the gas pressure does work

$$W_{gas} = F_{gas} d = (pA)(x_f - x_i) = p(x_f A - x_i A)$$

But $x_i A$ is the cylinder's initial volume V_i (recall that the volume of a cylinder is the length times the area of the base) and $x_f A$ is the final volume V_f. Thus the work done is

$$W_{gas} = p(V_f - V_i) = p\,\Delta V \tag{12.19}$$

Work done by a gas in a constant-pressure process

where ΔV is the *change* in volume.

Equation 12.19 has a particularly simple interpretation on a pV diagram. As FIGURE 12.14a on the next page shows, $p\Delta V$ is the "area under the pV graph" between V_i and V_f. Although we've shown this result for only a constant-pressure process, it turns out to be true for all ideal-gas processes. That is, as FIGURE 12.14b shows,

$$W_{gas} = \text{area under the } pV \text{ graph between } V_i \text{ and } V_f$$

FIGURE 12.14 Calculating the work done in an ideal-gas process.

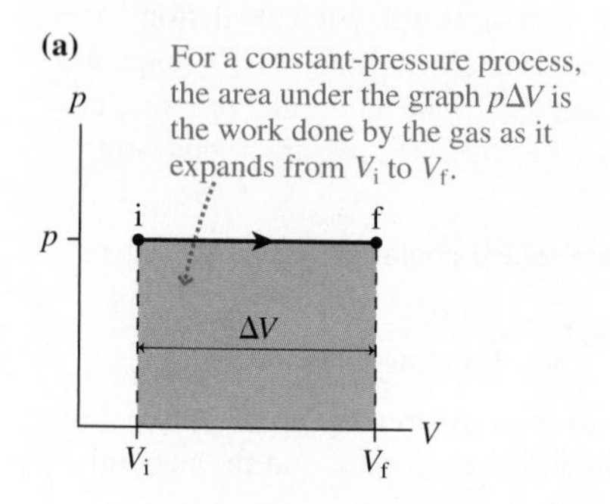
(a)
For a constant-pressure process, the area under the graph $p\Delta V$ is the work done by the gas as it expands from V_i to V_f.
p
i
f
ΔV
V
V_i
V_f

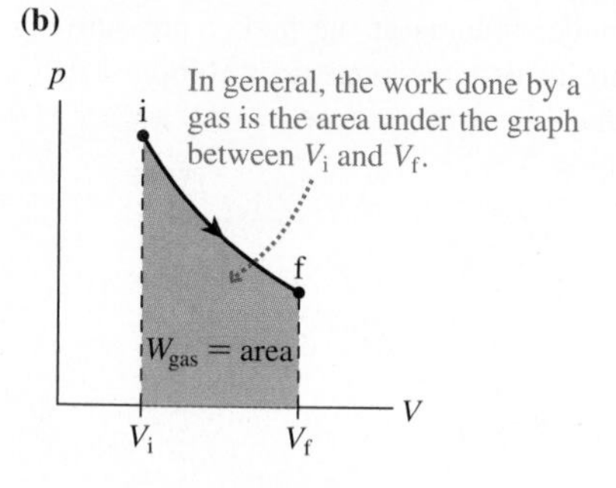
(b)
p
In general, the work done by a gas is the area under the graph between V_i and V_f.
i
f
W_{gas} = area
V
V_i
V_f

There are a few things to clarify:

- In order for the gas to do work, its volume must change. No work is done in a constant-volume process.
- The simple relationship of Equation 12.19 applies only to constant-pressure processes. For any other ideal-gas process, you must use the geometry of the pV diagram to calculate the area under the graph.
- To calculate work, pressure must be in Pa and volume in m^3. The product of Pa (which is N/m^2) and m^3 is $N \cdot m$. But $1\ N \cdot m$ is 1 J—the unit of work and energy.
- W_{gas} is positive if the gas expands ($\Delta V > 0$). The gas does work by pushing against the piston. In this case, the work done is energy transferred out of the system, and the energy of the gas decreases. W_{gas} is negative if the piston compresses the gas ($\Delta V < 0$) because the force $\vec{F}_{gas}$ is opposite the displacement of the piston. Energy is transferred into the system as work, and the energy of the gas increases. We often say "work is done *on* the gas," but this just means that W_{gas} is negative.

In the first law of thermodynamics, $\Delta E_{th} = Q + W$, W is the work done by the environment—that is, by force $\vec{F}_{ext}$ acting on the system. But $\vec{F}_{ext}$ and $\vec{F}_{gas}$ are equal and opposite forces, as we noted previously, so the work done by the environment is the negative of the work done by the gas: $W = -W_{gas}$. Consequently, the first law of thermodynamics can be written as

$$\Delta E_{th} = Q - W_{gas} \tag{12.20}$$

The thermal energy of an ideal gas depends only on its temperature as $E_{th} = \frac{3}{2}Nk_B T$. Comparing Equations 12.15 and 12.16 shows that Nk_B is equal to nR, so we can write the change in thermal energy of an ideal gas as

$$\Delta E_{th} = \frac{3}{2}Nk_B\,\Delta T = \frac{3}{2}nR\,\Delta T \tag{12.21}$$

EXAMPLE 12.10 Finding the heat for an ideal-gas process

A cylinder with a movable piston contains 0.016 mol of helium. A researcher expands the gas via the process illustrated in FIGURE 12.15. To achieve this, does she need to heat the gas? If so, how much heat energy must be added or removed?

FIGURE 12.15 pV diagram for Example 12.10.

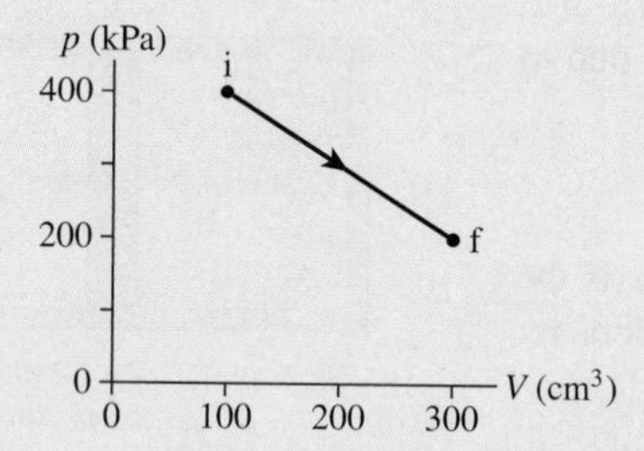
p (kPa)
400
200
0
i
f
0
100
200
300
V (cm³)

STRATEGIZE As the gas expands, it does work on the piston. Its temperature may change as well, implying a change in thermal energy. We will use the version of the first law of thermodynamics in Equation 12.20 to describe the energy changes in the gas. We'll first find the change in thermal energy by computing the temperature change, then calculate how much work is done by looking at the area under the graph. Once we know W_{gas} and ΔE_{th}, we will use the first law to determine the sign and the magnitude of the heat—telling us whether heat energy goes in or out, and how much.

PREPARE The graph tells us the pressure and the volume, so we can use the ideal-gas law to compute the temperature at the initial and final points. To do this we need the volumes in SI units. Reading the initial and final volumes on the graph and converting, we find

$$V_i = 100\ \text{cm}^3 \times \frac{1\ \text{m}^3}{10^6\ \text{cm}^3} = 1.0 \times 10^{-4}\ \text{m}^3$$

$$V_f = 300\ \text{cm}^3 \times \frac{1\ \text{m}^3}{10^6\ \text{cm}^3} = 3.0 \times 10^{-4}\ \text{m}^3$$

SOLVE The initial and final temperatures are found using the ideal-gas law:

$$T_i = \frac{p_i V_i}{nR} = \frac{(4.0 \times 10^5\ \text{Pa})(1.0 \times 10^{-4}\ \text{m}^3)}{(0.016\ \text{mol})(8.31\ \text{J/mol} \cdot \text{K})} = 300\ \text{K}$$

$$T_f = \frac{p_f V_f}{nR} = \frac{(2.0 \times 10^5\ \text{Pa})(3.0 \times 10^{-4}\ \text{m}^3)}{(0.016\ \text{mol})(8.31\ \text{J/mol} \cdot \text{K})} = 450\ \text{K}$$

The temperature increases, and so must the thermal energy. We can use Equation 12.21 to compute this change:

$$\Delta E_{th} = \frac{3}{2}(0.016\ \text{mol})(8.31\ \text{J/mol} \cdot \text{K})(450\ \text{K} - 300\ \text{K}) = 30\ \text{J}$$

The other piece of the puzzle is to compute the work done. We do this by finding the area under the graph for the process. FIGURE 12.16 shows that we can do this calculation by viewing the area as a triangle on top of a rectangle. Notice that the areas are in joules because they are the product of Pa and m^3.

The total work is

$$W_{gas} = \text{area of triangle} + \text{area of rectangle}$$
$$= 20\text{ J} + 40\text{ J} = 60\text{ J}$$

Now we can use the first law as written in Equation 12.20 to find the heat:

$$Q = \Delta E_{th} + W_{gas} = 30\text{ J} + 60\text{ J} = 90\text{ J}$$

This is a positive number, so—using the conventions introduced in Chapter 11—we see that 90 J of heat energy must be added to the gas.

FIGURE 12.16 The work done by the expanding gas is the total area under the graph.

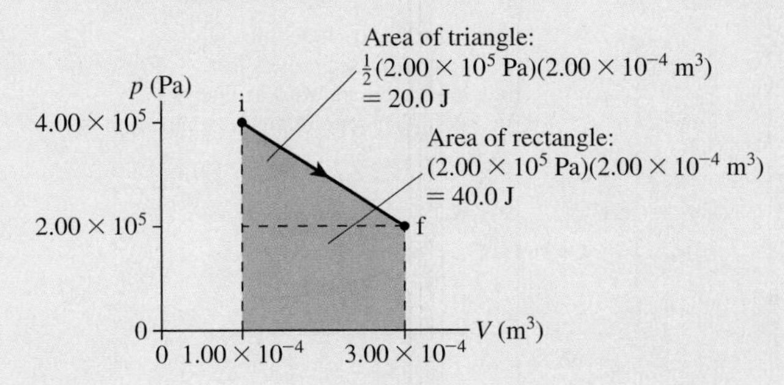

ASSESS The gas does work—a loss of energy—but its temperature increases, so it makes sense that heat energy must be added.

Adiabatic Processes

You may have noticed that when you pump up a bicycle tire with a hand pump, the pump gets warm. We noted the reason for this in Chapter 11: When you press down on the handle of the pump, a piston in the pump's chamber compresses the gas, doing work on it. According to the first law of thermodynamics, doing work on the gas increases its thermal energy. So the gas temperature goes up, and heat is then transferred through the walls of the pump to your hand.

Now suppose you compress a gas in an insulated container, so that no heat is exchanged with the environment, or you compress a gas so quickly that there is no time for heat to be transferred. In either case, $Q = 0$. If a gas process has $Q = 0$, for either a compression or an expansion, we call this an **adiabatic process.**

An expanding gas does work, so $W_{gas} > 0$. If the expansion is adiabatic, meaning $Q = 0$, then the first law of thermodynamics as written in Equation 12.20 tells us that $\Delta E_{th} < 0$. Temperature is proportional to thermal energy, so the temperature will decrease as well. **An adiabatic expansion lowers the temperature of a gas.** If the gas is compressed, work is done on the gas ($W_{gas} < 0$). If the compression is adiabatic, the first law of thermodynamics implies that $\Delta E_{th} > 0$ and thus that the temperature increases. **An adiabatic compression raises the temperature of a gas.** Adiabatic processes allow you to use work, rather than heat, to change the temperature of a gas.

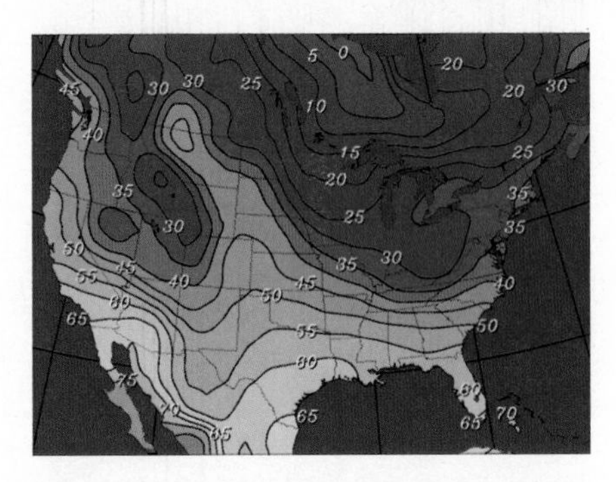

Warm mountain winds This image shows surface temperatures (in °F) in North America on a winter day. Notice the bright green area of unseasonably warm temperatures extending north and west from the center of the continent. On this day, a strong westerly wind, known as a Chinook wind, was blowing down off the Rocky Mountains, rapidly moving from high elevations (and low pressures) to low elevations (and higher pressures). The air was rapidly compressed as it descended. The compression was so rapid that no heat was exchanged with the environment, so this was an adiabatic process that significantly increased the air temperature.

CONCEPTUAL EXAMPLE 12.11 What is the shape of the curve?

FIGURE 12.17 shows the pV diagram of a gas undergoing an isothermal compression from point 1 to point 2. Sketch how the pV diagram would look if the gas were compressed from point 1 to the same final pressure by a rapid adiabatic compression.

REASON An adiabatic compression increases the temperature of the gas as the work done on the gas is transformed into thermal energy. Consequently, as seen in FIGURE 12.18 on the next page, the curve of the adiabatic compression cuts across the isotherms to end on a higher-temperature isotherm when the gas pressure reaches p_2.

FIGURE 12.17 pV diagram for an isothermal compression.

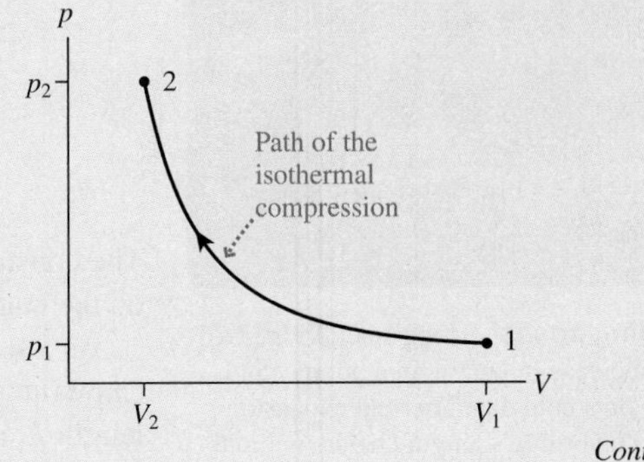

Continued

FIGURE 12.18 pV diagram for an adiabatic compression.

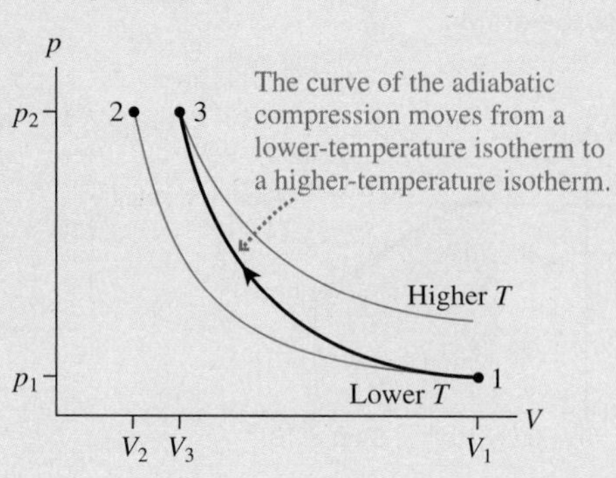

ASSESS In an isothermal compression, heat energy is transferred out of the gas so that the gas temperature stays the same. This heat transfer doesn't happen in an adiabatic compression, so we'd expect the gas to have a higher final temperature. In general, the temperature at the final point of an adiabatic compression is higher than at the starting point. Similarly, an adiabatic expansion ends on a lower-temperature isotherm.

STOP TO THINK 12.3 What is the ratio T_f/T_i for this process?

A. 1/4
B. 1/2
C. 1 (no change)
D. 2
E. 4
F. There is not enough information to decide.

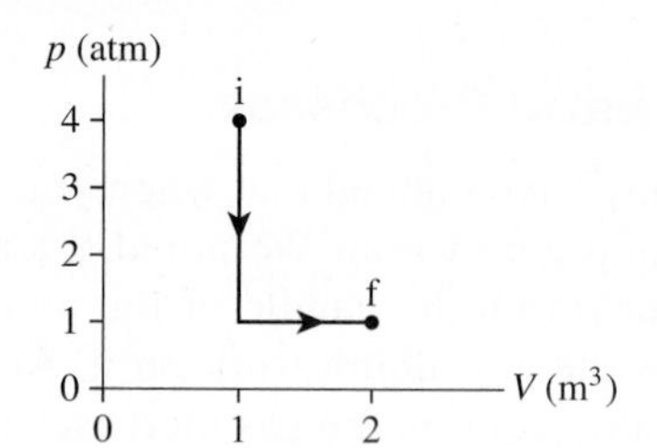

12.4 Thermal Expansion

FIGURE 12.19 Increasing the temperature of a solid causes it to expand.

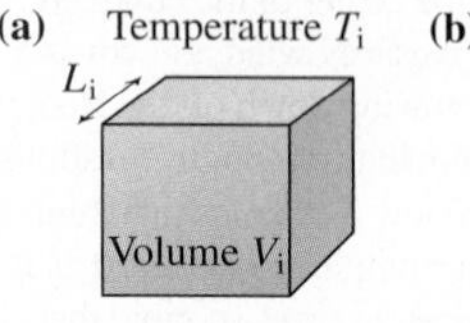

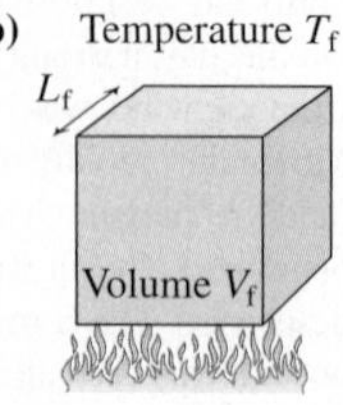

The bonds between atoms in solids and liquids mean that solids and liquids are much less compressible than gases, as we've noted. But raising the temperature of a solid or a liquid does produce a small, measurable change in volume. This **thermal expansion** underlies many practical phenomena.

FIGURE 12.19a shows a cube of material, initially at temperature T_i. The edge of the cube has length L_i, and the cube's volume is V_i. We then heat the cube, increasing its temperature to T_f, as shown in FIGURE 12.19b. The cube's edge length increases to L_f, and its volume increases to V_f.

For most substances, the change in volume $\Delta V = V_f - V_i$ is linearly related to the change in temperature $\Delta T = T_f - T_i$ by

$$\Delta V = \beta V_i \Delta T \quad (12.22)$$

Volume thermal expansion

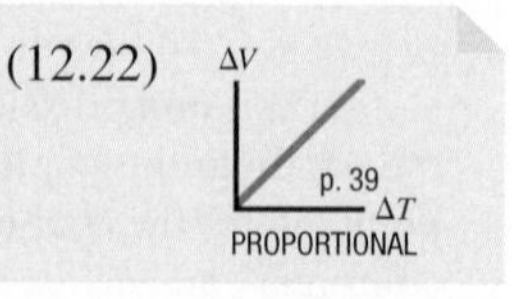

Expanding spans A long steel bridge will slightly increase in length on a hot day and decrease on a cold day. Thermal expansion joints let the bridge's length change without causing the roadway to buckle.

The constant β is known as the **coefficient of volume expansion.** Its value depends on the material the object is made of. ΔT is measured in K, so the units of β are K^{-1}.

As a solid's volume increases, each of its linear dimensions increases as well. We can write a similar expression for this linear thermal expansion. If an object of initial length L_i undergoes a temperature change ΔT, its length changes to L_f. The change in length, $\Delta L = L_f - L_i$, is given by

$$\Delta L = \alpha L_i \Delta T \quad (12.23)$$

Linear thermal expansion

The constant α is the **coefficient of linear expansion.** Note that Equations 12.22 and 12.23 apply equally well to thermal *contractions,* in which case both ΔT and ΔV (or ΔL) are negative.

Values of α and β for common materials are listed in TABLE 12.3. The volume expansion of a liquid can be measured, but, because a liquid can change shape, we don't assign a coefficient of linear expansion to a liquid. The values of α and β are very small, so the changes in length and volume, ΔL and ΔV, are always a small fraction of their original values.

TABLE 12.3 Coefficients of linear and volume thermal expansion at 20°C

Substance	Linear α (K^{-1})	Volume β (K^{-1})
Glass	9×10^{-6}	27×10^{-6}
Iron or steel	12×10^{-6}	36×10^{-6}
Concrete	12×10^{-6}	36×10^{-6}
Aluminum	23×10^{-6}	69×10^{-6}
Water		210×10^{-6}
Ethyl alcohol		1100×10^{-6}
Air (and other gases)		3400×10^{-6}

NOTE ▶ The expressions for thermal expansion are only approximate expressions that apply over a limited range of temperatures. They are what we call **empirical formulas;** they are a good fit to measured data, but they do not represent any underlying fundamental law. ◀

EXAMPLE 12.12 How much closer to space?

The height of the Space Needle, a steel observation tower in Seattle, is 180 meters on a 0°C winter day. How much taller is it on a hot summer day when the temperature is 30°C?

STRATEGIZE The steel expands because of an increase in temperature. We are interested in the change in height, so we'll consider the linear expansion.

PREPARE The expansion depends on the difference in temperature, which is

$$\Delta T = T_f - T_i = 30°C - 0°C = 30°C = 30 \text{ K}$$

SOLVE The coefficient of linear expansion is given in Table 12.3; we can use this value in Equation 12.23 to compute the increase in height:

$$\Delta L = \alpha L_i \Delta T = (12 \times 10^{-6} \text{ K}^{-1})(180 \text{ m})(30 \text{ K}) = 0.065 \text{ m}$$

ASSESS You don't notice buildings getting taller on hot days, so we expect the final answer to be small. The change is a small fraction of the height of the tower, as we expect. Compared to 180 m, an expansion of 6.5 cm is not something you would easily notice—but it isn't negligible. The thermal expansion of structural elements in towers and bridges must be accounted for in the design to avoid damaging stresses. When designers failed to properly account for thermal stresses in the marble panels cladding the Amoco Building in Chicago, all 43,000 panels had to be replaced, at great cost.

CONCEPTUAL EXAMPLE 12.13 What happens to the hole?

A metal plate has a circular hole in it. As the plate is heated, does the hole get larger or smaller?

REASON As the plate expands, you might think the hole would shrink (with the metal expanding into the hole). But suppose we took a metal plate and simply drew a circle where a hole could be cut. On heating, the plate and the marked area both expand, as we see in FIGURE 12.20. We could cut a hole on the marked line before or after heating; the size of the hole would be larger in the latter case. Therefore, the size of the hole must expand as the plate expands.

ASSESS When an object undergoes thermal expansion (or contraction), all dimensions increase (or decrease) by the *same percentage*.

FIGURE 12.20 The thermal expansion of a hole.

Cut
hole
The marked area increases in size as the temperature increases.
Increase temperature
A hole can be cut before or after heating.
Cut
hole

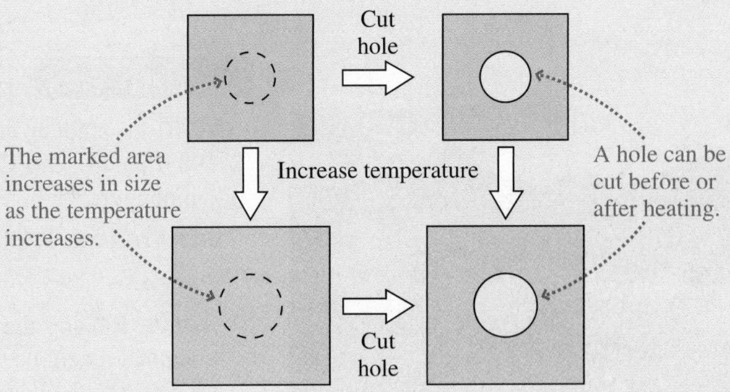

STOP TO THINK 12.4 An aluminum ring is tight around a solid iron rod. If we wish to loosen the ring to remove it from the rod, we should

A. Increase the temperature of the ring and rod.
B. Decrease the temperature of the ring and rod.

12.5 Specific Heat and Heat of Transformation

If you hold a glass of cold water in your hand, the heat from your hand will raise the temperature of the water. The heat from your hand will also melt an ice cube; melting is an example of what we will call a **phase change,** a change of state from one phase to another. In this section, we'll consider these two types of changes.

Specific Heat

Adding 4190 J of heat energy to 1 kg of water raises its temperature by 1 K. If you are fortunate enough to have 1 kg of gold, you need only 129 J of heat to raise its temperature by 1 K. The amount of heat that raises the temperature of 1 kg of a substance by 1 K is called the **specific heat** of that substance. The symbol for specific heat is c. Water has specific heat $c_{water} = 4190$ J/kg · K, and the specific heat of gold is $c_{gold} = 129$ J/kg · K. Specific heat depends only on the material from which an object is made. TABLE 12.4 lists the specific heats of some common liquids and solids.

TABLE 12.4 Specific heats of solids and liquids

Substance	c (J/kg · K)
Solids	
Lead	128
Gold	129
Copper	385
Iron	449
Aluminum	900
Water ice	2090
Mammalian body	3400
Liquids	
Mercury	140
Ethyl alcohol	2400
Water	4190

If heat c is required to raise the temperature of 1 kg of a substance by 1 K, then heat Mc is needed to raise the temperature of mass M by 1 K and $Mc\,\Delta T$ is needed to raise the temperature of mass M by ΔT. In general, the heat needed to bring about a temperature change ΔT is

$$Q = Mc\,\Delta T \tag{12.24}$$

Heat needed to produce a temperature change ΔT for mass M with specific heat c

Q can be either positive (temperature goes up) or negative (temperature goes down).

It takes more heat energy to change the temperature of a substance with a large specific heat than to change the temperature of a substance with a small specific heat. Water, with a very large specific heat, is slow to warm up and slow to cool down. This large "thermal inertia" of water is essential for the biological processes of life.

Video Heating Water and Aluminum

Temperate lakes At night, the large specific heat of water prevents the temperature of a body of water from dropping nearly as much as that of the surrounding air. Early in the morning, water vapor evaporating from a warm lake quickly condenses in the colder air above, forming mist. During the day, the opposite happens: The air becomes much warmer than the water.

EXAMPLE 12.14 How much energy is needed to run a fever?

A 70 kg student catches the flu, and his body temperature increases from 37.0°C (98.6°F) to 39.0°C (102.2°F). How much energy is required to raise his body's temperature?

STRATEGIZE The increase in temperature requires the addition of energy.

PREPARE The change in temperature ΔT is 2.0°C, or 2.0 K.

SOLVE Raising the temperature of the body uses energy supplied internally from the chemical reactions of the body's metabolism, which transfer heat to the body. The specific heat of the body is given in Table 12.4 as 3400 J/kg · K. We can use Equation 12.24 to find the necessary heat energy:

$$Q = Mc\,\Delta T = (70\text{ kg})(3400\text{ J/kg}\cdot\text{K})(2.0\text{ K}) = 4.8\times 10^5\text{ J}$$

ASSESS The body is mostly water, with a large specific heat, and the mass of the body is large, so we'd expect a large amount of energy to be necessary. Looking back to Chapter 11, we see that this is approximately the energy in a large apple, or the amount of energy required to walk 1 mile.

Phase Changes

Suppose you remove a few ice cubes from the freezer and place them in a sealed container with a thermometer. Then, as in FIGURE 12.21a, you put a steady flame under the container. We'll assume that the heating is done slowly so that the inside of the container always has a single, uniform temperature.

FIGURE 12.21b shows a graph of the temperature as a function of time. In stage 1, the ice steadily warms without melting until it reaches 0°C. During stage 2, the temperature remains fixed at 0°C for an extended period of time during which the ice melts. As the ice is melting, both the ice temperature and the liquid water temperature remain at 0°C. Even though the system is being heated, the temperature doesn't begin to rise again until all the ice has melted.

KEY CONCEPT FIGURE 12.21 The temperature as a function of time as water is transformed from solid to liquid to gas.

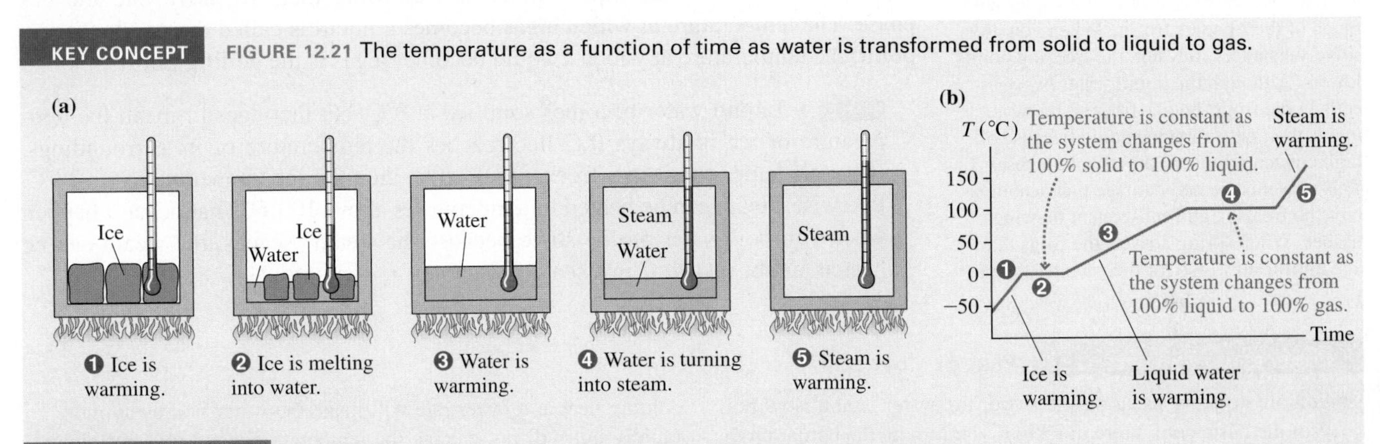

STOP TO THINK 12.5 In Figure 12.21, by comparing the slope of the graph during the time the liquid water is warming to the slope as steam is warming, we can say that

A. The specific heat of water is larger than that of steam.
B. The specific heat of water is smaller than that of steam.
C. The specific heat of water is equal to that of steam.
D. The slope of the graph is not related to the specific heat.

At the end of stage 2, the ice has completely melted into liquid water. During stage 3, the flame slowly warms this liquid water, raising its temperature until it reaches 100°C. During stage 4, the liquid water remains at 100°C as it turns into water vapor—steam—which also remains at 100°C. Even though the system is being heated, the temperature doesn't begin to rise again until all the water has been converted to steam. Finally, stage 5 begins, in which the system is pure steam whose temperature rises continuously from 100°C.

NOTE ► In everyday language, the three phases of water are called *ice, water,* and *steam.* The term "water" implies the liquid phase. Scientifically, these are the solid, liquid, and gas phases of the compound called *water.* When we are working with different phases of water, we'll use the term "water" in the scientific sense of a collection of H_2O molecules. We'll say either *liquid* or *liquid water* to denote the liquid phase. ◄

CONCEPTUAL EXAMPLE 12.15 **Strategy for cooling a drink**

If you have a warm soda that you wish to cool, is it more effective to add 25 g of liquid water at 0°C or 25 g of water ice at 0°C?

REASON If you add liquid water at 0°C, heat will be transferred from the soda to the water, raising the temperature of the water and lowering that of the soda. If you add water ice at 0°C, heat will first be transferred from the soda to the ice to melt it, transforming the 0°C ice to 0°C liquid water, then will be transferred to the liquid water to raise its temperature. Thus more thermal energy will be removed from the soda, giving it a lower final temperature, if ice is used rather than liquid water.

ASSESS This makes sense because you know that this is what you do in practice. To cool a drink, you drop in an ice cube.

Frozen frogs BIO It seems impossible, but common wood frogs survive the winter with much of their bodies frozen. When you dissolve substances in water, the freezing point lowers. Although the liquid water *between* cells in the frogs' bodies freezes, the water *inside* their cells remains liquid because of high concentrations of dissolved glucose. This prevents the cell damage that accompanies the freezing and subsequent thawing of tissues. When spring arrives, the frogs thaw and appear no worse for their winter freeze.

The temperature at which—if the thermal energy is increased—a solid becomes a liquid is called the **melting point;** if the thermal energy is instead decreased, a liquid becomes a solid at the **freezing point.** Melting and freezing are *phase changes.* A system at the melting point is in **phase equilibrium,** meaning that any amount of solid can coexist with any amount of liquid. Raise the temperature ever so slightly and the entire system soon becomes liquid. Lower it slightly and it all becomes solid.

You can see another region of phase equilibrium of water in Figure 12.21b at 100°C. This is a phase equilibrium between the liquid phase and the gas phase, and any amount of liquid can coexist with any amount of gas at this temperature. As heat is added to the system, the temperature stays the same. The added energy is used to break bonds between the liquid molecules, allowing them to move into the gas phase. The temperature at which a gas becomes a liquid is called the **condensation point;** the temperature at which a liquid becomes a gas is the **boiling point.**

NOTE ▶ Liquid water becomes solid ice at 0°C, but that doesn't mean the temperature of ice is always 0°C. Ice reaches the temperature of its surroundings. If the air temperature in a freezer is −20°C, then the ice temperature is −20°C. Likewise, steam can be heated to temperatures above 100°C. That doesn't happen when you boil water on the stove because the steam escapes, but steam can be heated far above 100°C in a sealed container. ◀

CONCEPTUAL EXAMPLE 12.16 Fast or slow boil?

You are cooking pasta on the stove; the water is at a slow boil. Will the pasta cook more quickly if you turn up the burner on the stove so that the water is at a fast boil?

REASON Water boils at 100°C; no matter how vigorously the water is boiling, the temperature is the same. It is the temperature of the water that determines how fast the cooking takes place. Adding heat at a faster rate will make the water boil away more rapidly but will not change the temperature—and will not alter the cooking time.

ASSESS This result may seem counterintuitive, but you can try the experiment next time you cook pasta!

Heat of Transformation

Lava—molten rock—undergoes a phase change from liquid to solid when it contacts liquid water; the transfer of heat to the water causes the water to undergo a phase change from liquid to gas.

In Figure 12.21b, the phase changes appeared as horizontal line segments on the graph. During these segments, heat is being transferred to the system but the temperature isn't changing. The thermal energy continues to increase during a phase change, but, as noted, the additional energy goes into breaking molecular bonds rather than speeding up the molecules. **A phase change is characterized by a change in thermal energy without a change in temperature.**

The amount of heat energy that causes 1 kg of a substance to undergo a phase change is called the **heat of transformation** of that substance. For example, laboratory experiments show that 333,000 J of heat are needed to melt 1 kg of ice at 0°C. The symbol for heat of transformation is L. The heat required for the entire system of mass M to undergo a phase change is

$$Q = ML \tag{12.25}$$

"Heat of transformation" is a generic term that refers to any phase change. Two particular heats of transformation are the **heat of fusion** L_f, the heat of transformation between a solid and a liquid, and the **heat of vaporization** L_v, the heat of transformation between a liquid and a gas. The heat needed for these phase changes is

$$Q = \begin{cases} \pm ML_f & \text{Heat needed to melt/freeze mass } M \\ \pm ML_v & \text{Heat needed to boil/condense mass } M \end{cases} \tag{12.26}$$

The ± indicates that heat must be added to the system during melting or boiling but removed from the system during freezing or condensing. **You must explicitly include the minus sign when it is needed.** Carefully consider the nature of the phase change and whether heat is added or removed. When an ice cube melts in your hand, heat goes from your hand to the ice cube. In your freezer, where the temperature is below the freezing point of water, liquid water freezes as heat goes from the liquid water to the surrounding air. If you put a pan of water on the stove, you must add heat to make the water boil. When you breathe on a mirror, water vapor from your lungs condenses on the glass, and heat goes from the water vapor into the cooler glass.

TABLE 12.5 lists some heats of transformation. Notice that the heat of vaporization is always much higher than the heat of fusion.

TABLE 12.5 Melting and boiling temperatures and heats of transformation at standard atmospheric pressure

	Solid to Liquid		Liquid to Gas	
Substance	Melting Point T_m (°C)	Heat of Fusion L_f (J/kg)	Boiling Point T_b (°C)	Heat of Vaporization L_v (J/kg)
Nitrogen (N_2)	−210	0.26×10^5	−196	1.99×10^5
Ethyl alcohol	−114	1.09×10^5	78	8.79×10^5
Mercury	−39	0.11×10^5	357	2.96×10^5
Water	0	3.33×10^5	100	22.6×10^5
Lead	328	0.25×10^5	1750	8.58×10^5

EXAMPLE 12.17 How much energy is needed to melt a popsicle?

A girl eats a 45 g frozen popsicle that was taken out of a −10°C freezer. How much energy does her body use to bring the popsicle up to body temperature?

STRATEGIZE We will assume that the popsicle is pure water. There are three parts to the problem, corresponding to stages 1–3 in Figure 12.21: The popsicle must be warmed to 0°C, the popsicle must melt, and then the resulting water must be warmed to body temperature. We will figure out how much heat is needed for each stage; the sum is the total energy required.

PREPARE Normal body temperature is 37°C. The specific heats of ice and liquid water are given in Table 12.4; the heat of fusion of water is given in Table 12.5.

SOLVE The heat needed to warm the frozen water by $\Delta T = 10°\text{C} = 10\text{ K}$ to the melting point is

$$Q_1 = Mc_{\text{ice}}\,\Delta T = (0.045\text{ kg})(2090\text{ J/kg}\cdot\text{K})(10\text{ K}) = 940\text{ J}$$

Note that we use the specific heat of water ice, not liquid water, in this equation. Melting 45 g of ice requires heat

$$Q_2 = ML_f = (0.045\text{ kg})(3.33 \times 10^5\text{ J/kg}) = 15{,}000\text{ J}$$

The liquid water must now be warmed to body temperature; this requires heat

$$Q_3 = Mc_{\text{water}}\,\Delta T = (0.045\text{ kg})(4190\text{ J/kg}\cdot\text{K})(37\text{ K})$$
$$= 7000\text{ J}$$

The total energy is the sum of these three values: $Q_{\text{total}} = 23{,}000$ J.

ASSESS More energy is needed to melt the ice than to warm the water, as we would expect. A commercial popsicle has 40 Calories, which is about 170 kJ. Roughly 15% of the chemical energy in this frozen treat is used to bring it up to body temperature!

Evaporation

Water boils at 100°C. But individual molecules of water can move from the liquid phase to the gas phase at lower temperatures. This process is known as **evaporation.** Water evaporates as sweat from your skin at a temperature well below 100°C. Different particles in a liquid move at different speeds, as they do in a gas. At any temperature, some molecules will be moving fast enough to go into the gas phase. And they will do so, carrying away thermal energy as they go. The molecules that leave the liquid are the ones that have the highest kinetic energy, so evaporation reduces the average kinetic energy (and thus the temperature) of the liquid left behind.

Keeping your cool BIO Humans (and cattle and horses) have sweat glands, so we can perspire to moisten our skin, allowing evaporation to cool our bodies. Animals that do not perspire can also use evaporation to keep cool. Dogs, goats, rabbits, and even birds pant, evaporating water from their respiratory passages. Elephants spray water on their skin; other animals may lick their fur.

The opposite of sweating You are sitting outside, enjoying a cold beverage and you notice that your glass "sweats"—drops of water collect on the outside. Your glass isn't really sweating, of course. The water doesn't come from inside the glass but from outside, as water vapor from the air condenses onto the sides of the glass. Heat leaves the water vapor when it condenses—and so heat is added to your glass. Sweating may keep you cool due to evaporation, but a "sweating" glass means condensation, and this means a warming drink.

Video What the Physics? Fighting Fire with Evaporation

The evaporation of water, both from sweat and in moisture you exhale, is one of the body's methods of exhausting the excess heat of metabolism to the environment, allowing you to maintain a steady body temperature. The heat to evaporate a mass M of water is $Q = ML_v$, and this amount of heat is removed from your body. However, the heat of vaporization L_v is a little larger than the value for boiling. At a skin temperature of 30°C, the heat of vaporization of water is

$$L_v(\text{at } 30°) = 24 \times 10^5 \text{ J/kg}$$

This is 6% higher than the value at 100°C in Table 12.5. You should use this value for L_v when you consider the heat transfer due to evaporation of perspiration from the skin and similar situations.

EXAMPLE 12.18 Computing heat loss by perspiration BIO

The human body can produce approximately 30 g of perspiration per minute. At what rate is it possible to exhaust heat by the evaporation of perspiration?

STRATEGIZE We are looking for a rate of exhausting heat, so we are looking for a power in watts, an energy divided by a time. We'll assume that perspiration is pure water, which is a good approximation. The energy is the heat required to evaporate 30 g of water from the skin; the time is 1 minute.

PREPARE The water is evaporated from the skin, which we assume is at 30°C. The value for the heat of transformation is therefore $L_v = 24 \times 10^5$ J/kg.

SOLVE The evaporation of 30 g of perspiration at normal body temperature requires heat energy

$$Q = ML_v = (0.030 \text{ kg})(24 \times 10^5 \text{ J/kg}) = 7.2 \times 10^4 \text{ J}$$

This is the heat lost per minute; the rate of heat loss is

$$\frac{Q}{\Delta t} = \frac{7.2 \times 10^4 \text{ J}}{60 \text{ s}} = 1200 \text{ W}$$

ASSESS Given the metabolic power required for different activities, as listed in Chapter 11, this is sufficient to keep the body cool even when exercising in hot weather—as long as the person drinks enough water to keep up this rate of perspiration.

STOP TO THINK 12.6 1 kg of barely molten lead, at 328°C, is poured into a large beaker holding liquid water right at the boiling point, 100°C. What is the mass of the water that will be boiled away as the lead solidifies?

A. 0 kg B. <1 kg C. 1 kg D. >1 kg

12.6 Calorimetry

If you've put an ice cube into a hot drink to cool it quickly, you engaged, in a trial-and-error way, in a practical aspect of heat transfer known as **calorimetry,** the quantitative measurement of the heat transferred between systems or evolved in reactions. You know that heat energy will be transferred from the hot drink into the cold ice cube, reducing the temperature of the drink, as shown in FIGURE 12.22.

FIGURE 12.22 Cooling hot coffee with ice.

Let's make this qualitative picture more precise. FIGURE 12.23 shows two systems that can exchange heat with each other but that are isolated from everything else. Suppose they start at different temperatures T_1 and T_2. As you know, heat energy will be transferred from the hotter to the colder system until they reach a common final temperature T_f. In the cooling coffee example of Figure 12.22, the coffee is system 1, the ice is system 2, and the insulating barrier is the mug.

The insulation prevents any heat energy from being transferred to or from the environment, so energy conservation tells us that any energy leaving the hotter system must enter the colder system. The concept is straightforward, but to state the idea mathematically we need to be careful with signs.

Let Q_1 be the energy transferred to system 1 as heat. Q_1 is positive if energy *enters* system 1, negative if energy *leaves* system 1. Similarly, Q_2 is the energy transferred to system 2. The fact that the systems are merely exchanging energy can be written as $|Q_1| = |Q_2|$. But Q_1 and Q_2 have opposite signs, so $Q_1 = -Q_2$. No energy is exchanged with the environment, so it makes more sense to write this relationship as

$$Q_{\text{net}} = Q_1 + Q_2 = 0 \tag{12.27}$$

FIGURE 12.23 Heat is being transferred from system 1 to system 2, so $Q_2 > 0$ and $Q_1 < 0$.

The magnitude $|Q_1|$ of the heat leaving system 1 equals the magnitude $|Q_2|$ of the heat entering system 2.

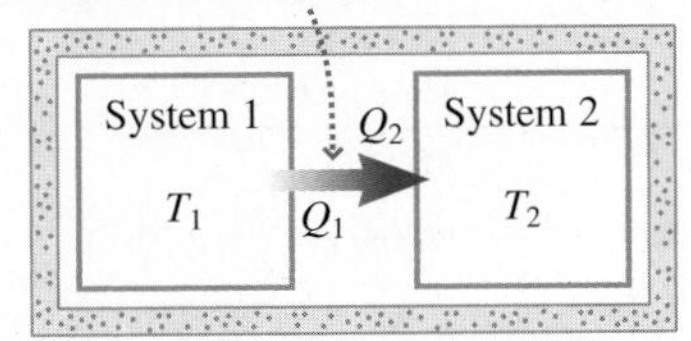

Opposite signs mean that $Q_{\text{net}} = Q_1 + Q_2 = 0$

NOTE ▸ The signs are very important in calorimetry problems. ΔT is always $T_f - T_i$, so ΔT and Q are negative for any system whose temperature decreases. The proper sign of Q for any phase change must be supplied *by you,* depending on the direction of the phase change. ◂

PROBLEM-SOLVING APPROACH 12.1 Calorimetry problems

When two systems are brought into thermal contact, we use calorimetry to find the heat transferred between them and their final equilibrium temperature.

STRATEGIZE Identify the individual interacting systems. Assume that they are isolated from the environment.

PREPARE List the known information and identify what you need to find. Convert all quantities to SI units.

SOLVE The statement of energy conservation is

$$Q_{\text{net}} = Q_1 + Q_2 + \cdots = 0$$

- For systems that undergo a temperature change, $Q_{\Delta T} = Mc(T_f - T_i)$. Be sure to have the temperatures T_i and T_f in the correct order.
- For systems that undergo a phase change, $Q_{\text{phase}} = \pm ML$. Supply the correct sign by observing whether energy enters or leaves the system during the transition.
- Some systems may undergo a temperature change *and* a phase change. Treat the changes separately. The heat energy is $Q = Q_{\Delta T} + Q_{\text{phase}}$.

ASSESS The final temperature should be between the initial temperatures. A T_f that is higher or lower than all initial temperatures is an indication that something is wrong, usually a sign error.

EXAMPLE 12.19 Using calorimetry to identify a metal

200 g of an unknown metal is heated to 90.0°C, then dropped into 50.0 g of water at 20.0°C in an insulated container. The water temperature rises within a few seconds to 27.7°C, then changes no further. Identify the metal.

STRATEGIZE The interacting systems are the metal and the water. We'll assume that the two systems quickly come to a final temperature, so there is no time for heat to be exchanged with the environment.

PREPARE We label the temperatures as follows: The initial temperature of the metal is T_m; the initial temperature of the water is T_w. The common final temperature is T_f. For water, $c_w = 4190$ J/kg·K is known from Table 12.4. Only the specific heat c_m of the metal is unknown.

Continued

SOLVE Energy conservation requires that $Q_w + Q_m = 0$. Using $Q = Mc(T_f - T_i)$ for each, we have

$$Q_w + Q_m = M_w c_w (T_f - T_w) + M_m c_m (T_f - T_m) = 0$$

This is solved for the unknown specific heat:

$$c_m = \frac{-M_w c_w (T_f - T_w)}{M_m (T_f - T_m)}$$

$$= \frac{-(0.0500\text{ kg})(4190\text{ J/kg}\cdot\text{K})(27.7°\text{C} - 20.0°\text{C})}{(0.200\text{ kg})(27.7°\text{C} - 90.0°\text{C})}$$

$$= 129\text{ J/kg}\cdot\text{K}$$

Referring to Table 12.4, we find we have either 200 g of gold or, if we made an ever-so-slight experimental error, 200 g of lead!

ASSESS The temperature of the unknown metal changed much more than the temperature of the water. This means that the specific heat of the metal must be much less than that of water, which is exactly what we found.

EXAMPLE 12.20 Calorimetry with a phase change

Your 500 mL diet soda, with a mass of 500 g, is at 20°C, room temperature, so you cool it by adding 100 g of ice from the −20°C freezer. Does all the ice melt? If so, what is the final temperature? If not, what fraction of the ice melts? Assume that you have a well-insulated cup.

STRATEGIZE The interacting systems are the soda and the ice; heat will be transferred from the soda to the ice. You have a well-insulated cup, so we'll assume that the systems are isolated. We will also assume that the diet soda has the same specific heat as water; this is a good approximation.

There are two possible outcomes. If all the ice melts as the soda cools, then $T_f > 0°\text{C}$. It's also possible that the soda will cool to 0°C before all the ice has melted, leaving the ice and liquid in equilibrium at 0°C. We will need to distinguish between these outcomes before we can complete the solution.

PREPARE The soda has mass $M_{soda} = 0.50$ kg and an initial temperature of 20°C. The ice has mass $M_{ice} = 0.10$ kg and an initial temperature of −20°C. Since we are looking for a temperature difference, we can keep the temperatures in Celsius. We are looking for the final temperature of the combined system.

SOLVE Let's first calculate the heat needed to melt all the ice and leave it as liquid water at 0°C. To do so, we must warm the ice by 20 K to 0°C, then change it to water. The heat input for this two-stage process is

$$Q_{melt} = \underbrace{M_{ice}c_{ice}(20\text{ K})}_{\substack{\text{This is the energy to raise the}\\ \text{temperature of the ice from}\\ -20°\text{C to }0°\text{C. }\Delta T = 20\text{ K.}}} + \underbrace{M_{ice}L_f}_{\substack{\text{This is the energy}\\ \text{to melt the ice once}\\ \text{it reaches }0°\text{C.}}} = 37{,}500\text{ J}$$

where L_f is the heat of fusion of water.

Q_{melt} is a *positive* quantity because we must *add* heat to melt the ice. Next, let's calculate how much heat energy will leave the 500 g soda if it cools all the way to 0°C:

$$Q_{cool} = M_{soda} c_{water} (-20\text{ K}) = -41{,}900\text{ J}$$

where $\Delta T = -20$ K because the temperature decreases. Because $|Q_{cool}| > Q_{melt}$, the soda has sufficient energy to melt all the ice. Hence the final state will be all liquid at $T_f > 0$. (Had we found $|Q_{cool}| < Q_{melt}$, then the final state would have been an ice-liquid mixture at 0°C.)

Energy conservation requires $Q_{ice} + Q_{soda} = 0$. The heat Q_{ice} consists of three terms: warming the ice to 0°C, melting the ice to water at 0°C, then warming the 0°C water to T_f. The mass will still be M_{ice} in the last of these steps because it is the "ice system," but we need to use the specific heat of *liquid water*. Thus

$$Q_{ice} + Q_{soda} = [M_{ice} c_{ice} (20\text{ K}) + M_{ice} L_f + M_{ice} c_{water} (T_f - 0°\text{C})] + M_{soda} c_{water} (T_f - 20°\text{C}) = 0$$

We've already done part of the calculation, allowing us to write

$$37{,}500\text{ J} + M_{ice} c_{water} (T_f - 0°\text{C}) + M_{soda} c_{water} (T_f - 20°\text{C}) = 0$$

Solving for T_f gives

$$T_f = \frac{20 M_{soda} c_{water} - 37{,}500\text{ J}}{M_{ice} c_{water} + M_{soda} c_{water}} = 1.8°\text{C}$$

ASSESS A good deal of ice has been put in the soda, so it ends up being cooled nearly to the freezing point, as we might expect.

STOP TO THINK 12.7 1 kg of lead at 100°C is dropped into a container holding 1 kg of water at 0°C. Once the lead and water reach thermal equilibrium, the final temperature is

A. <50°C B. 50°C C. >50°C

12.7 Specific Heats of Gases

Just as for solids and liquids, heating a gas changes its temperature. But by how much? FIGURE 12.24 shows two isotherms on the pV diagram for a gas. Processes A and B, which start on the T_i isotherm and end on the T_f isotherm, have the *same* temperature change $\Delta T = T_f - T_i$, so we might expect them both to require the same amount of heat. But it turns out that process A, which takes place at constant volume, requires *less* heat than does process B, which occurs at constant pressure. The reason is that work is done in process B but not in process A.

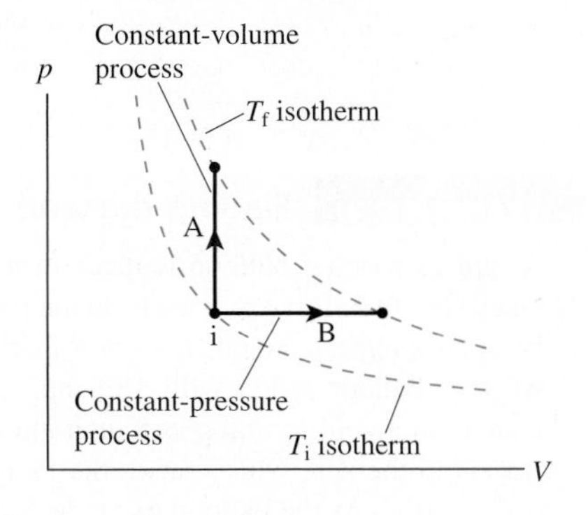

FIGURE 12.24 Processes A and B have the same ΔT and the same ΔE_{th}, but they require different amounts of heat.

It is useful to define two different versions of the specific heat of gases: one for constant-volume processes and one for constant-pressure processes. We will define these as *molar* specific heats because we usually do gas calculations using moles instead of mass. The quantity of heat needed to change the temperature of n moles of gas by ΔT is, for a constant-volume process,

$$Q = nC_V \Delta T \tag{12.28}$$

and for a constant-pressure process,

$$Q = nC_P \Delta T \tag{12.29}$$

C_V is the **molar specific heat at constant volume,** and C_P is the **molar specific heat at constant pressure.** TABLE 12.6 gives the values of C_V and C_P for some common gases. The units are J/mol · K. The values for air are essentially equal to those for N_2.

TABLE 12.6 Molar specific heats of gases (J/mol · K) at 20°C

Gas	C_P	C_V
Monatomic Gases		
He	20.8	12.5
Ne	20.8	12.5
Ar	20.8	12.5
Diatomic Gases		
H_2	28.7	20.4
N_2	29.1	20.8
O_2	29.2	20.9

It's interesting that all the monatomic gases in Table 12.6 have the same values for C_P and for C_V. Why should this be? Monatomic gases are really close to ideal, so let's go back to our atomic model of the ideal gas. We know that the thermal energy of an ideal gas of N atoms is $E_{th} = \frac{3}{2}Nk_BT = \frac{3}{2}nRT$. If the temperature of an ideal gas changes by ΔT, its thermal energy changes by

$$\Delta E_{th} = \frac{3}{2}nR\,\Delta T \tag{12.30}$$

If we keep the volume of the gas constant, so that no work is done, this energy change can come only from heat, so

$$Q = \frac{3}{2}nR\,\Delta T \tag{12.31}$$

Comparing Equation 12.31 with the definition of molar specific heat in Equation 12.28, we see that the molar specific heat at constant volume must be

$$C_V\,(\text{monatomic gas}) = \frac{3}{2}R = 12.5\text{ J/mol}\cdot\text{K} \tag{12.32}$$

This *predicted* value from the ideal-gas model is exactly the *measured* value of C_V for the monatomic gases in Table 12.6, a good check that the model we have been using is correct.

If you heat a gas in a sealed container so that there is no change in volume, then no work is done. But if you heat a sample of gas in a cylinder with a piston to keep it at constant pressure, the gas must expand, and it will do work as it expands. The expression for ΔE_{th} in Equation 12.30 is still valid, but now, according to the first law of thermodynamics, $Q = \Delta E_{th} + W_{gas}$. The work done by the gas in a constant-pressure process is $W_{gas} = p\,\Delta V$, so the heat required is

$$Q = \Delta E_{th} + W_{gas} = \frac{3}{2}nR\,\Delta T + p\,\Delta V \tag{12.33}$$

The ideal-gas law, $pV = nRT$, implies that if p is constant and only V and T change, then $p\Delta V = nR\,\Delta T$. Using this result in Equation 12.33, we find that the heat needed to change the temperature by ΔT in a constant-pressure process is

$$Q = \frac{3}{2}nR\,\Delta T + nR\,\Delta T = \frac{5}{2}nR\,\Delta T$$

A comparison with the definition of molar specific heat shows that

$$C_P = \frac{5}{2}R = 20.8\ \text{J/mol}\cdot\text{K} \tag{12.34}$$

This is larger than C_V, as expected, and in perfect agreement for all the monatomic gases in Table 12.6.

EXAMPLE 12.21 Finding the work done by an expanding gas

A typical weather balloon is made of a thin latex envelope that takes very little force to stretch, so the pressure inside the balloon is approximately equal to atmospheric pressure. Suppose a weather balloon filled with 180 mol of helium is waiting for launch on a cold morning at a high-altitude station. The balloon warms in the sun, which raises the temperature of the gas from 0°C to 30°C. As the balloon expands, how much work is done by the expanding gas?

STRATEGIZE The work done is equal to $p\Delta V$, but we don't know the pressure (it's not sea level and we don't know the altitude) and we don't know the volume of the balloon. Instead, we'll use the first law of thermodynamics. We can compute how much heat energy is transferred to the balloon as it warms because this is a temperature change at constant pressure, and we can compute how much the thermal energy of the gas increases because we know ΔT.

PREPARE We can rewrite Equation 12.20 as

$$W_{gas} = Q - \Delta E_{th}$$

The change in temperature of the gas is 30°C, so $\Delta T = 30$ K.

SOLVE The heat required to increase the temperature of the gas is given by Equation 12.29:

$$Q = nC_P\Delta T = (180\,\text{mol})(20.8\ \text{J/mol}\cdot\text{K})(30\,\text{K}) = 112\ \text{kJ}$$

The change in thermal energy depends on the change in temperature according to Equation 12.21:

$$\Delta E_{th} = \frac{3}{2}nR\Delta T = \frac{3}{2}(180\ \text{mol})(8.31\ \text{J/mol}\cdot\text{K})(30\ \text{K})$$

$$= 67.3\ \text{kJ}$$

The work done by the expanding balloon is just the difference between these two values:

$$W_{gas} = Q - \Delta E_{th} = 112\ \text{kJ} - 67.3\ \text{kJ} = 45\ \text{kJ}$$

ASSESS The numbers are large—it's a lot of heat and a large change in thermal energy—but it's a big balloon with a lot of gas, so this seems reasonable.

Now let's turn to the diatomic gases in Table 12.6. The molar specific heats are higher than for monatomic gases, and our atomic model explains why. The thermal energy of a monatomic gas consists exclusively of the translational kinetic energy of the atoms; heating a monatomic gas simply means that the atoms move faster. The thermal energy of a diatomic gas is more than just the translational energy, as shown in FIGURE 12.25. Heating a diatomic gas makes the molecules move faster, but it also causes them to rotate more rapidly. Energy goes into the translational kinetic energy of the molecules (thus increasing the temperature), but some goes into rotational kinetic energy. Because some of the added heat goes into rotation and not into translation, the specific heat of a diatomic gas is higher than that of a monatomic gas, as we see in Table 12.6.

FIGURE 12.25 Thermal energy of a diatomic gas.

The thermal energy of a diatomic gas is the sum of the translational kinetic energy of the molecules . . .

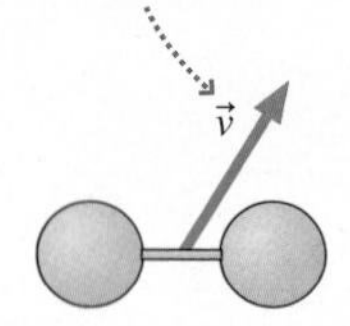

. . . and the rotational kinetic energy of the molecules.

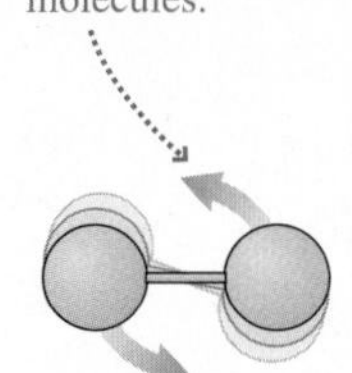

12.8 Heat Transfer

You feel warmer when the sun is shining on you, colder when you are sitting on a cold concrete bench or when a stiff wind is blowing. This is due to the transfer of heat. Although we've talked about heat and heat transfers a lot in these last two chapters, we haven't said much about *how* heat is transferred from a hotter object to a colder object. There are four basic mechanisms, described in the following table, by which objects exchange heat with other objects or their surroundings. Evaporation was treated in an earlier section; in this section we will consider the other mechanisms.

Heat-transfer mechanisms

When two objects are in direct physical contact, such as the soldering iron and the circuit board, heat is transferred by *conduction.* **Energy is transferred by direct contact.**

This special photograph shows air currents near a flame. Hot gases from the combustion rise, carrying thermal energy in a process known as *convection.* **Energy is transferred by the bulk motion of molecules with high thermal energy.**

The lamp shines on the lambs huddled below, warming them. The energy is transferred by infrared *radiation,* a form of electromagnetic waves. **Energy is transferred by electromagnetic waves.**

When you blow on a cup of cocoa, this increases the rate of *evaporation,* rapidly cooling it. **Energy is transferred by the removal of molecules with high thermal energy.**

Conduction

If you hold a metal spoon in a cup of hot coffee, the handle of the spoon soon gets warm. Thermal energy is transferred along the spoon from the coffee to your hand. The difference in temperature between the two ends drives this heat transfer by a process known as **conduction.** Conduction is the transfer of thermal energy directly through a physical material.

FIGURE 12.26 shows a copper rod placed between a hot reservoir (a fire) and a cold reservoir (a block of ice). We can use our atomic model to see how thermal energy is transferred along the rod by the interaction between atoms in the rod; fast-moving atoms at the hot end transfer energy to slower-moving atoms at the cold end.

Suppose we set up a series of experiments to measure the heat Q transferred through various rods. We would find the following trends in our data:

- Q increases if the temperature difference ΔT between the hot end and the cold end is increased.
- Q increases if the cross-section area A of the rod is increased.
- Q decreases if the length L of the rod is increased.
- Some materials (such as metals) transfer heat quite readily. Other materials (such as wood) transfer very little heat.

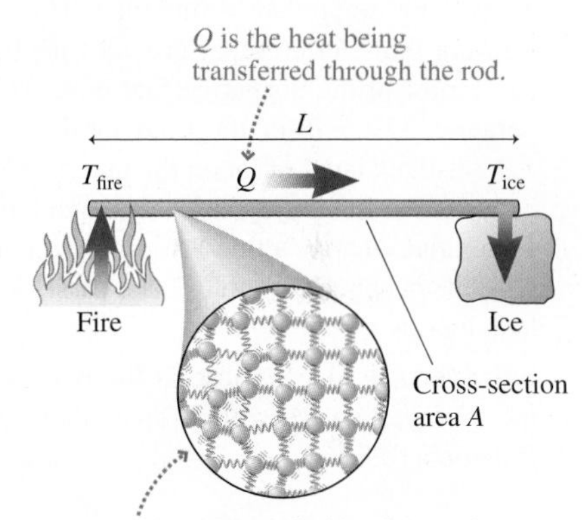

Q is the heat being transferred through the rod.
L
T_{fire}
Q
T_{ice}
Fire
Ice
Cross-section area A
The particles on the left side of the rod are vibrating more vigorously than the particles on the right. The particles on the left transfer energy to the particles on the right via the bonds connecting them.

FIGURE 12.26 Conduction of heat in a solid rod.

The final observation is one that is familiar to you: If you are stirring a pot of hot soup on the stove, you generally use a wood or plastic spoon rather than a metal one.

These experimental observations about heat conduction can be summarized in a single formula. If heat Q is transferred in a time interval Δt, the *rate* of heat transfer (joules per second, or watts) is $Q/\Delta t$. For a material of cross-section area A and length L, spanning a temperature difference ΔT, the rate of heat transfer is

$$\frac{Q}{\Delta t} = \left(\frac{kA}{L}\right)\Delta T \tag{12.35}$$

Rate of conduction of heat across a temperature difference

The quantity k, which characterizes whether the material is a good or a poor conductor of heat, is called the **thermal conductivity** of the material. Because the heat transfer rate J/s is a *power,* measured in watts, the units of k are W/m · K. Values of k

Cold feet, warm heart BIO A penguin standing on the cold Antarctic ice loses very little heat through its feet. Its thick skin with limited thermal conductivity helps, but more important, the penguin's feet are very cold, which minimizes heat loss by conduction. The penguin can have cold feet so close to the warmth of the body because of an adaptation called *countercurrent heat exchange*. Arteries carrying warm blood to the feet run next to veins carrying cold blood back from the feet. Heat goes from the arteries to the veins, cooling the blood going to the feet and warming the blood going to the body—so the feet stay cool while the body stays warm.

for some common materials are listed in TABLE 12.7; a larger number for k means a material is a better conductor of heat.

TABLE 12.7 Thermal conductivity values (measured at 20°C)

Material	k (W/m · K)	Material	k (W/m · K)
Diamond	1000	Skin	0.50
Silver	420	Muscle	0.46
Copper	400	Fat	0.21
Iron	72	Wood	0.2
Stainless steel	14	Carpet	0.04
Ice	1.7	Fur, feathers	0.02–0.06
Concrete	0.8	Air (27°C, 100 kPa)	0.026
Plate glass	0.75		

The weak bonds between molecules make most biological materials poor conductors of heat. Fat is a worse conductor than muscle, so sea mammals have thick layers of fat for insulation. The bodies of land mammals are insulated with fur, and those of birds with feathers. Both fur and feathers trap a good deal of air, so their conductivity is similar to that of air, as Table 12.7 shows.

EXAMPLE 12.22 Warming the bench

At the start of the section we noted that you "feel cold" when you sit on a cold concrete bench. "Feeling cold" really means that your body is losing a significant amount of heat. How significant? Suppose you are sitting on a 10°C concrete bench. You are wearing thin clothing that provides negligible insulation. In this case, most of the insulation that protects your body's core (temperature 37°C) from the cold of the bench is provided by a 1.0-cm-thick layer of fat on the part of your body that touches the bench. (The thickness varies from person to person, but this is a reasonable average value.) A good estimate of the area of contact with the bench is 0.10 m^2. Given these details, what is the rate of heat loss by conduction?

STRATEGIZE Heat is lost to the bench by conduction through the fat layer, so we will compute the rate of heat loss by using Equation 12.35.

PREPARE The thickness of the conducting layer is 0.010 m, the area is 0.10 m^2, and the thermal conductivity of fat is given in Table 12.7. The temperature difference is the difference between your body's core temperature (37°C) and the temperature of the bench (10°C), a difference of 27°C, or 27 K.

SOLVE We have all of the data we need to use Equation 12.35 to compute the rate of heat loss:

$$\frac{Q}{\Delta t} = \left(\frac{(0.21\ \text{W/m}\cdot\text{K})(0.10\ \text{m}^2)}{0.010\ \text{m}}\right)(27\ \text{K}) = 57\ \text{W}$$

ASSESS 57 W is more than half your body's resting power, which we learned in Chapter 11 is approximately 100 W. That's a significant loss, so your body will feel cold, a result that seems reasonable if you've ever sat on a cold bench for any length of time.

A feather coat BIO A penguin's short, dense feathers serve a different role than the flight feathers of other birds: They trap air to provide thermal insulation. The fluffy feathers of the juvenile penguin makes this trapping of air very clear, but the smooth covering of the adults works in a similar manner.

Convection

In conduction, faster-moving atoms transfer thermal energy to adjacent atoms. But in fluids such as water or air, there is a more efficient means to move energy: by transferring the faster-moving atoms themselves. When you place a pan of cold water on a burner on the stove, it's heated on the bottom. This heated water expands and becomes less dense than the water above it, so it rises to the surface while cooler, denser water sinks to take its place. This transfer of thermal energy by the motion of a fluid is known as **convection**.

Convection is usually the main mechanism for heat transfer in fluid systems. On a small scale, convection mixes the pan of water that you heat on the stove; on a large scale, convection is responsible for making the wind blow and ocean currents circulate. Air is a very poor thermal conductor, but it is very effective at transferring energy by convection. To use air for thermal insulation, it is necessary to trap the air in small pockets to limit convection. And that's exactly what feathers, fur, double-paned windows, and fiberglass insulation do.

Radiation

You *feel* the warmth from the glowing red coals in a fireplace. On a cool day, you prefer to sit in the sun rather than the shade so that the sunlight keeps you warm. In both cases heat energy is being transferred to your body in the form of **radiation.**

Radiation consists of electromagnetic waves—a topic we will explore further in later chapters—that transfer energy from the object that emits the radiation to the object that absorbs it. All warm objects emit radiation in this way. Objects near room temperature emit radiation in the invisible infrared part of the electromagnetic spectrum. Thermal images, like the one of the teapot in FIGURE 12.27, are made with cameras having special infrared-sensitive detectors that allow them to "see" radiation that our eyes don't respond to. Hotter objects, such as glowing embers, still emit most of their radiant energy in the infrared, but some is emitted as visible red light.

FIGURE 12.27 A thermal image of a teapot.

NOTE ▶ The word "radiation" comes from "radiate," which means "to beam." You have likely heard the word used to refer to x rays and radioactive materials. This is not the sense we use in this chapter. Here, we use "radiation" to mean electromagnetic waves that "beam" from an object. ◀

Radiation is a significant part of the energy balance that keeps your body at the proper temperature. You can change your body temperature by absorbing more radiation (sitting next to a fire) or by emitting more radiation (taking off your hat and scarf to expose more skin).

Video Figure 12.27

The energy radiated by an object shows a strong dependence on temperature. In the photo that opened the chapter, the person's warm hands emit quite a bit more energy than the cooler spider. We can quantify the dependence of radiated energy on temperature. If heat energy Q is radiated in a time interval Δt by an object with surface area A and absolute temperature T, the *rate* of heat transfer $Q/\Delta t$ (joules per second) is found to be

$$\frac{Q}{\Delta t} = e\sigma AT^4 \tag{12.36}$$

Rate of heat transfer by radiation at temperature T (Stefan's law)

Video Energy Transfer by Radiation

Quantities in this equation are defined as follows:

- e is the **emissivity** of a surface, a measure of the effectiveness of radiation. The value of e ranges from 0 to 1. Human skin of any color is a very effective radiator at body temperature, with $e = 0.97$.
- T is the absolute temperature in kelvin.
- A is the surface area in m^2.
- σ is a constant with the value $\sigma = 5.67 \times 10^{-8}\ \text{W/m}^2 \cdot \text{K}^4$.

Notice the very strong fourth-power dependence on temperature. Doubling the absolute temperature of an object increases the radiant heat transfer by a factor of 16!

An adult human with bare skin in a room at a comfortable temperature has a skin temperature of approximately 33°C, or 306 K. A typical value for the skin's surface area is $1.8\ \text{m}^2$. With these values and the emissivity of skin noted previously, we can calculate the rate of heat transfer via radiation from the skin:

$$\frac{Q}{\Delta t} = e\sigma AT^4 = (0.97)\left(5.67 \times 10^{-8}\ \frac{\text{W}}{\text{m}^2 \cdot \text{K}^4}\right)(1.8\ \text{m}^2)(306\ \text{K})^4 = 870\ \text{W}$$

As we learned in Chapter 11, the body at rest generates approximately 100 W of thermal energy. If the body radiated energy at 870 W, it would quickly cool. At this rate of emission, the body temperature would drop by 1°C every 7 minutes! Clearly, there must be some mechanism to balance this emitted radiation, and there is: the radiation *absorbed* by the body.

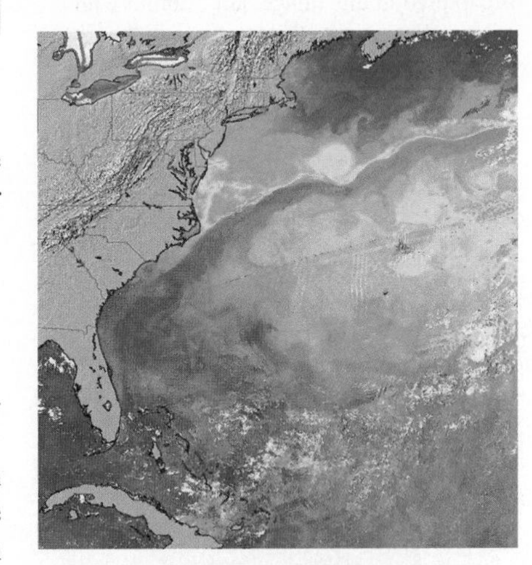

Global heat transfer This satellite image shows radiation emitted by the waters of the ocean off the east coast of the United States. You can clearly see the warm waters of the Gulf Stream. The satellite can see this radiation from the earth because it readily passes through the atmosphere into space. This radiation is also the only way the earth can cool. This has consequences for the energy balance of the earth, as we will discuss in One Step Beyond at the end of Part VI.

When you sit in the sun, your skin warms due to the radiation you absorb. Even if you are not in the sun, you are absorbing the radiation emitted by the objects surrounding you. Suppose an object at temperature T is surrounded by an environment at temperature T_0. The *net* rate at which the object radiates heat energy—that is, radiation emitted minus radiation absorbed—is

$$\frac{Q_{\text{net}}}{\Delta t} = e\sigma A(T^4 - T_0^4) \qquad (12.37)$$

This makes sense. An object should have no net energy transfer by radiation if it's in thermal equilibrium ($T = T_0$) with its surroundings. Note that the emissivity e appears for absorption as well; objects that are good emitters are also good absorbers.

BIO The infrared radiation emitted by a dog is captured in this image. His cool nose and paws radiate much less energy than the rest of his body.

EXAMPLE 12.23 Determining energy loss by radiation for the body BIO

A person with a skin temperature of 33°C is in a room at 24°C. What is the net rate of heat transfer by radiation?

STRATEGIZE We know the temperature of the skin and the environment, so we will compute the net rate of transfer using Equation 12.37.

PREPARE Body temperature is $T = 33 + 273 = 306$ K; the temperature of the room is $T_0 = 24 + 273 = 297$ K.

SOLVE The net radiation rate is

$$\frac{Q_{\text{net}}}{\Delta t} = e\sigma A(T^4 - T_0^4)$$

$$= (0.97)\left(5.67 \times 10^{-8} \frac{\text{W}}{\text{m}^2 \cdot \text{K}^4}\right)(1.8\ \text{m}^2)[(306\ \text{K})^4 - (297\ \text{K})^4] = 98\ \text{W}$$

ASSESS This is a reasonable value, roughly matching your resting metabolic power. When you are dressed (little convection) and sitting on wood or plastic (little conduction), radiation is your body's primary mechanism for dissipating the excess thermal energy of metabolism.

STOP TO THINK 12.8 Suppose you are an astronaut in the vacuum of space, hard at work in your sealed spacesuit. The only way that you can transfer heat to the environment is by

A. Conduction. B. Convection. C. Radiation. D. Evaporation.

12.9 Diffusion

FIGURE 12.28 The random motion of a pollen grain undergoing Brownian motion.

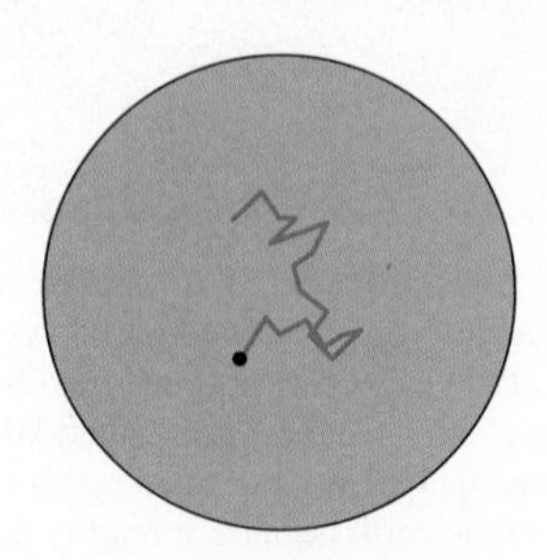

In 1827, the botanist Robert Brown was observing pollen grains through a microscope. He noticed that pollen grains suspended in water were not stationary, as expected, but moved about continuously in a jittery, erratic fashion rather like that shown in FIGURE 12.28. This apparently random motion, today called **Brownian motion,** is caused by the ceaseless bombardment of the pollen grains by individual water molecules.

Brownian motion is an example of a **random walk.** Suppose you stood at the origin of an *xy*-coordinate system, turned through a randomly chosen angle, and then took a step with a randomly chosen length. Wherever you end up, you repeat the process again. And again. You would be engaging in a random walk, and your path would look something like the path of the particle in Figure 12.28.

We can't see or track individual molecules, but their motion—the motion of thermal energy—is also a random walk. At atmospheric pressure, a molecule in air can move only about 80 nm before it collides with another molecule. Each collision sends the molecule off in a new direction, and the trajectory of an individual molecule also looks similar to Figure 12.28—a sequence of straight lines of varying lengths connected by random turns.

Imagine using a pipette to put a very tiny droplet of dye molecules in a beaker of water. The droplet will gradually disperse into the water. This net movement of molecules from a region of higher concentration (the initial droplet) to a region of lower concentration (the water) is called **diffusion.** Microscopically, diffusion is simply the random walks of the dye molecules as they collide both with each other and with the more numerous water molecules.

FIGURE 12.29 shows 50 molecules initially grouped so close together that we can't distinguish them. Then each begins a random walk, and their positions are shown at $t = 1$ s and $t = 9$ s. We can see that the molecules are diffusing outward. A random step is just as likely to go right as to go left, or to go up as to go down, so the *average position* of all the molecules is $(0, 0)$. However, the *average distance* from the origin is not zero. And it increases with time.

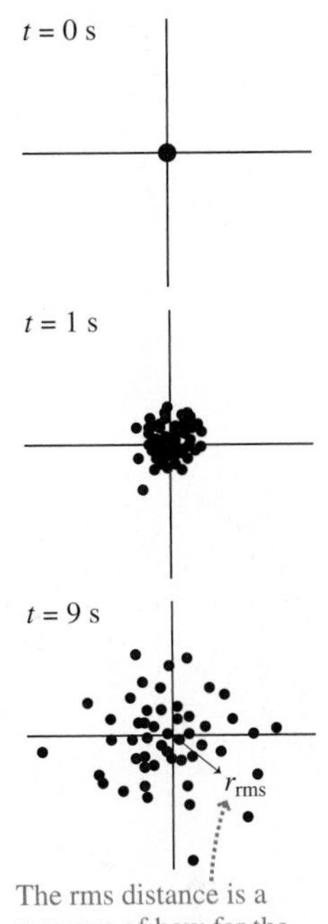

t = 0 s
t = 1 s
t = 9 s
r_rms
The rms distance is a measure of how far the molecules have diffused.

FIGURE 12.29 Random walks cause a group of molecules to diffuse outward.

We earlier defined the root-mean-square speed v_{rms}. Similarly, we can define a **root-mean-square distance** r_{rms} as

$$r_{rms} = \sqrt{(r^2)_{avg}} = \text{distance of a typical molecule} \tag{12.38}$$

The rms distance is a rough measure of how far the molecules have diffused. Figure 12.29 shows the rms distance at $t = 9$ s. A mathematical analysis of a random walk in three dimensions finds

$$r_{rms} = \sqrt{6Dt} \tag{12.39}$$

where D is called the **diffusion constant.** Its value, with units of m^2/s, depends on the type of molecules that are diffusing, the type of molecules that they are diffusing through, and the temperature and pressure. TABLE 12.8 lists some typical values. We notice two trends in the values. First, larger molecules have smaller diffusion constants and thus diffuse more slowly. Second, the constants for diffusion in air are roughly 10^4 times larger than are those for diffusion in water. This shouldn't be surprising: Water is much denser than air, so the distances between molecular collisions—the lengths of the steps of the random walk—are much shorter.

TABLE 12.8 Diffusion constants

Substance	$D(m^2/s)$
Diffusion in air, 1 atm, 25°C	
O_2	1.8×10^{-5}
CO_2	1.9×10^{-5}
H_2O	2.5×10^{-5}
Diffusion in water, 25°C	
Acetylcholine	4.0×10^{-10}
Glucose	7.1×10^{-10}
Urea	1.3×10^{-9}
CO_2	2.0×10^{-9}
O_2	2.3×10^{-9}

If a group of molecules is released close together, how long will it take them to diffuse a distance L? There's no definitive answer to this question because the diffusing molecules don't all arrive at once. However, we've seen that r_{rms} is about how far the molecules have diffused, so we can estimate the time needed to diffuse a distance L by setting $r_{rms} = L$ in Equation 12.39 and solving for time:

$$t = \frac{L^2}{6D} \tag{12.40}$$

Estimate of time to diffuse a distance L with diffusion constant D

The time depends on the square of the distance, so an increase in the distance results in a much larger increase in time, as Example 12.24 illustrates.

EXAMPLE 12.24 Diffusion of oxygen through water BIO

A cell doesn't need a circulatory system, but your body does. Let's do a quick calculation to see why. A typical cell has a diameter of $10\,\mu m$. The smallest mammals in the world, shrews, are about 10 mm across. Compute the diffusion time for oxygen molecules through water at 25°C for these two distances.

STRATEGIZE We will use Equation 12.40 to find the diffusion times for the two distances.

Continued

PREPARE Table 12.8 gives the diffusion constant of oxygen through water at 25°C as $D = 2.3 \times 10^{-9}\,\text{m}^2/\text{s}$.

SOLVE For the two distances we're considering, we find

$$\text{For } L = 10\,\mu\text{m},\ t = \frac{(10 \times 10^{-6}\,\text{m}^2)^2}{6(2.3 \times 10^{-9}\,\text{m}^2/\text{s})} = 0.0072\,\text{s}$$

$$\text{For } L = 10\,\text{mm},\ t = \frac{(10 \times 10^{-3}\,\text{m}^2)^2}{6(2.3 \times 10^{-9}\,\text{m}^2/\text{s})} = 7200\,\text{s}$$

The time for oxygen to diffuse across the full diameter of a cell is about 7 ms; this is fast enough that there is no need for active transport. For a size typical of the smallest mammals, the diffusion time is much longer—about 2 hours! Clearly, mammals need a circulatory system to deliver oxygen to the cells of the body.

ASSESS Diffusion works for a cell, but not for a shrew. This result makes sense. We know that all but the smallest animals have circulatory systems. The diffusion time increases as the *square* of the distance, so it takes 1,000,000 times longer to diffuse 1000 times as far!

Your circulatory system delivers oxygen to different parts of your body, but once the oxygen reaches the cells, diffusion can take over. The cellular transport of oxygen, nutrients, and larger macromolecules is often by passive diffusion:

- Air in the alveoli of the lungs, the small gas-filled sacs through which gas exchange with the blood takes place, is separated from blood in the pulmonary capillaries by a membrane that is about 1 μm thick. Oxygen in the lungs rapidly diffuses through this membrane into the capillaries, while excess carbon dioxide in the capillaries diffuses into the lungs and is exhaled.
- Oxygen, glucose, and other molecules diffuse across the cell wall from capillaries into cells. Waste products, such as carbon dioxide or urea, diffuse from the cell into the capillaries for removal.
- A nerve impulse moves across a synapse when a neurotransmitter, such as acetylcholine, is released from the pre-synaptic membrane and diffuses across the synaptic cleft to the post-synaptic membrane. The transmission takes less than a microsecond because the synaptic cleft is so narrow.

The body needs active transport systems, such as the circulatory system, to move molecules through distances of more than a few cell diameters because diffusion is not an effective mechanism at these scales.

FIGURE 12.30 Diffusion of molecules across a concentration difference.

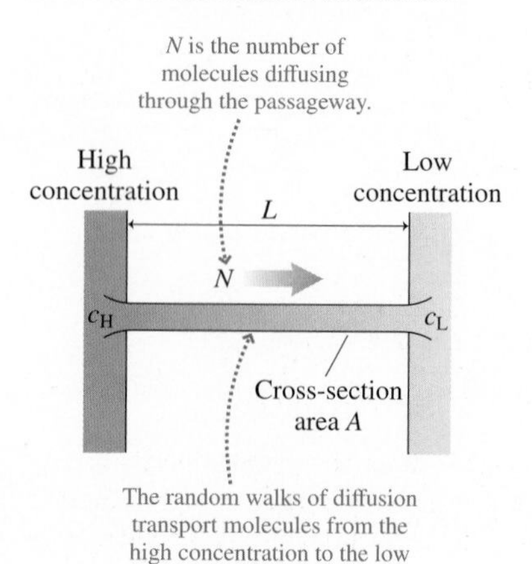
N is the number of molecules diffusing through the passageway.
High concentration
Low concentration
L
N
c_H
c_L
Cross-section area A
The random walks of diffusion transport molecules from the high concentration to the low concentration.

Fick's Law

Heat conduction *transfers energy* between reservoirs with different temperatures. Similarly, diffusion *transfers matter* between reservoirs with different concentrations.

FIGURE 12.30 shows two reservoirs, a distance L apart, holding different concentrations of some substance: a high concentration c_H and a low concentration c_L. When N molecules occupy volume V, their **concentration** is

$$c = \frac{N}{V} \tag{12.41}$$

The concentration is the number of molecules per cubic meter, with units m^{-3}.

NOTE ▶ Chemists and biologists measure concentration in other units, such as molecules per cubic centimeter, moles per liter, or parts per million. The concepts in this section still apply, but the diffusion constant may be in different units. ◀

Suppose the reservoirs are connected by a passageway with cross-section area A through which molecules diffuse. Diffusion is a random walk, so at any instant there are molecules moving in both directions. However, more molecules move from the high concentration toward the low concentration than in the other direction, so diffusion causes a net transfer of molecules when there is a *concentration difference* $\Delta c = c_H - c_L$. If N molecules are transferred in a time interval Δt, the *rate* of transfer (molecules per second) is $N/\Delta t$.

According to **Fick's law,** which can be derived from an analysis of random walks, the rate of molecule transfer is

$$\frac{N}{\Delta t} = \left(\frac{DA}{L}\right)\Delta c \qquad (12.42)$$

Rate of transfer of molecules across a concentration difference

Wrinkly respiration BIO The hellbender salamander, native to the eastern United States, can be over half a meter long, with a mass of 2 kilograms. Remarkably, this large amphibian doesn't have lungs or gills. Instead, its extracts oxygen from water by diffusion through its skin. The wrinkles on the sides of the salamander aren't just for looks; they increase the surface area of its skin so that diffusion can deliver sufficient oxygen.

where D is the diffusion constant. Fick's law is exactly analogous to the heat-conduction equation, Equation 12.35, with the diffusion constant D playing the same role as the thermal conductivity constant k.

EXAMPLE 12.25 Carbon dioxide diffusion in photosynthesis BIO

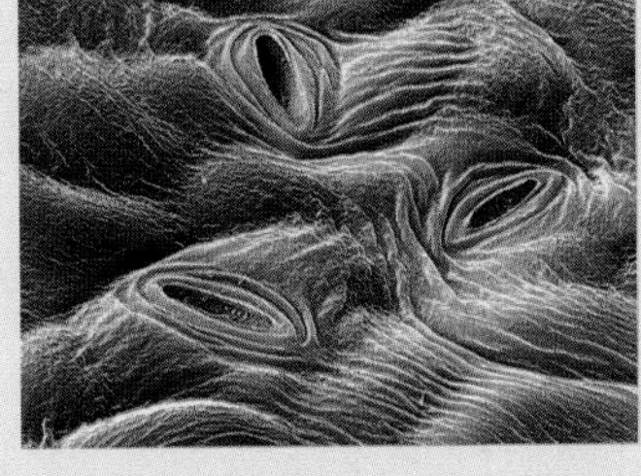

Carbon dioxide makes up about 0.040% of the atmosphere. Carbon dioxide enters a green leaf, where photosynthesis occurs, by diffusing through small pores called *stomata,* pores with a depth of about 10 μm. When the stomata are open, we can assume that their total area is 2.0% of the leaf area. Photosynthesis consumes CO_2, and the concentration within the leaf is about half the atmospheric value. Estimate, in both molecules/s and μmol/s, the rate of CO_2 uptake by a leaf with an area of $7.0 \times 10^{-4}\,\text{m}^2$.

STRATEGIZE The atmosphere and the leaf are reservoirs with different concentrations of CO_2 connected by pores through which matter transfer can occur via diffusion. Fick's law describes the rate of CO_2 uptake.

PREPARE From Table 12.8, the diffusion constant of carbon dioxide through air at 25°C is $D = 1.9 \times 10^{-5}\,\text{m}^2/\text{s}$. The stomata depth is the length across which diffusion occurs, so $L = 10 \times 10^{-6}\,\text{m}$, and the total area of the stomata is 2.0% of the leaf area, or $1.4 \times 10^{-5}\,\text{m}^2$.

Now, let's consider the concentration. We can use the ideal-gas law to determine the number of molecules in a cubic meter of air at 1 atm (1.013×10^6 Pa) pressure and a temperature of 25°:

$$N = \frac{pV}{k_B T} = \frac{(1.013 \times 10^5\,\text{Pa})(1.0\,\text{m}^3)}{(1.38 \times 10^{-23}\,\text{J/K})(298\,\text{K})} = 2.5 \times 10^{25}$$

CO_2 represents 0.040% of the molecules, so the concentration of CO_2 in the high concentration reservoir is

$$c_H = (0.00040)\frac{N}{V} = (0.00040)\frac{2.5 \times 10^{25}}{1.0\,\text{m}^3} = 1.0 \times 10^{22}\,\text{m}^{-3}$$

We're told that $c_L \approx \frac{1}{2}c_H$, so the concentration difference is $\Delta c \approx \frac{1}{2}c_H$.

SOLVE We use Fick's law to calculate the rate at which molecules diffuse from the air into the leaf. The area is the total area of the stomata; the length across which diffusion occurs is the length of the stomata:

$$\frac{N}{\Delta t} = \left(\frac{DA}{L}\right)\Delta c$$
$$= \left(\frac{(1.9 \times 10^{-5}\,\text{m}^2/\text{s})(1.4 \times 10^{-5}\,\text{m}^2)}{10 \times 10^{-6}\,\text{m}}\right)(0.50 \times 10^{22}\,\text{m}^{-3})$$
$$= 1.3 \times 10^{17}\ \text{molecules/s}$$

We can convert to moles by dividing by Avogadro's number:

$$\frac{N}{\Delta t} = \frac{1.3 \times 10^{17}\ \text{molecules/s}}{6.02 \times 10^{23}\ \text{molecules/mol}}$$
$$= 2.2 \times 10^{-7}\,\text{mol/s} = 0.22\,\mu\text{mol/s}$$

ASSESS CO_2 uptake by plants is more complicated than we've assumed, so it's no surprise that our final number is a bit lower than typical values, but it is of the correct order of magnitude.

INTEGRATED EXAMPLE 12.26 Breathing in cold air BIO

On a cold day, breathing costs your body energy; as the cold air comes into contact with the warm tissues of your lungs, the air warms due to heat transferred from your body. The thermal image of the person inflating a balloon in FIGURE 12.31 shows that exhaled air is quite a bit warmer than the surroundings.

FIGURE 12.31 A thermal image of a person blowing up a balloon on a cold day.

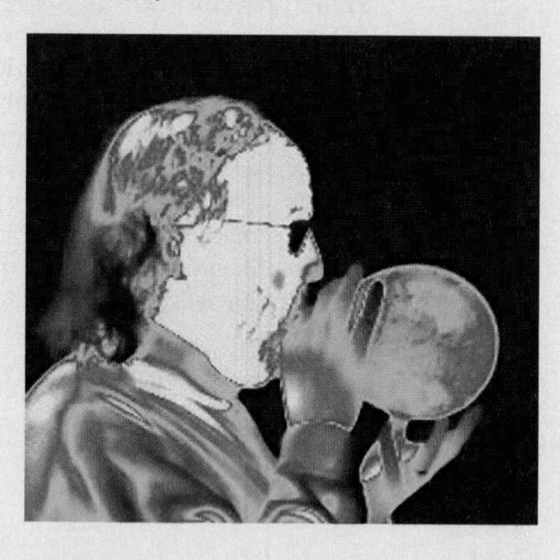

FIGURE 12.32 shows this process for a frosty $-10°C$ (14°F) day. The inhaled air warms to nearly the temperature of the interior of your body, 37°C. When you exhale, some heat is retained by the body, but most is lost; the exhaled air is still about 30°C.

FIGURE 12.32 Breathing warms the air.

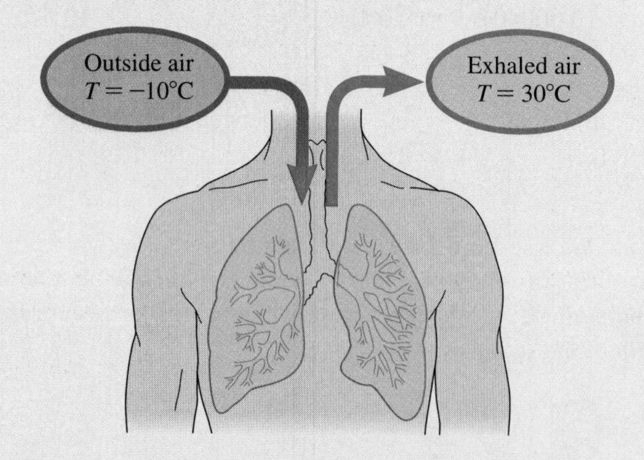

Your lungs hold several liters of air, but only a small part of the air is exchanged during each breath. A typical person takes 12 breaths each minute, with each breath drawing in 0.50 L of outside air. If the air warms up from $-10°C$ to 30°C, what is the volume of the air exhaled with each breath? What fraction of your body's resting power goes to warming the air? (Note that gases are exchanged as you breathe—oxygen to carbon dioxide—but to a good approximation the number of atoms, and thus the number of moles, stays the same.) Consider only the energy required to warm the air, not the energy lost to evaporation from the tissues of the lungs.

STRATEGIZE There are two parts to the problem. First, we'll figure out the increase in volume of the air as it is warmed. Next, we'll figure out the amount of energy needed to warm the air.

PREPARE When your body warms the air, its temperature increases, so its volume will increase as well. The initial and final states represented in Figure 12.32 are at atmospheric pressure, so we will treat this change as a constant-pressure process. We need the absolute temperatures of the initial and final states:

$$T_i = -10°C + 273 = 263 \text{ K}$$
$$T_f = 30°C + 273 = 303 \text{ K}$$

Because the initial and final pressures are the same, we can find the energy needed to warm the air by computing the heat needed to raise the temperature of a gas at constant pressure. The change in temperature is +40°C, so we can use $\Delta T = 40$ K. Air is a mix of nitrogen and oxygen with a small amount of other gases. C_P for nitrogen and oxygen is the same to two significant figures, so we assume the gas has $C_P = 29 \text{ J/mol} \cdot \text{K}$.

SOLVE The change in volume of the air in your lungs is a constant-pressure process. 0.50 L of air is breathed in; this is V_i. When the temperature increases, so does the volume. The gas isn't in a sealed container, but we are considering the same "parcel" of gas before and after, so we can use the ideal-gas law to find the volume after the temperature increase:

$$V_f = V_i \frac{p_i}{p_f} \frac{T_f}{T_i} = (0.50 \text{ L}) \times 1 \times \frac{303 \text{ K}}{263 \text{ K}} = 0.58 \text{ L}$$

The volume increases by a little over 10%, from 0.50 L to 0.58 L.

Now we can move on to the second part: determining the energy required. We need the number of moles of gas, which we can compute from the ideal-gas law:

$$n = \frac{pV}{RT} = \frac{(101.3 \times 10^3 \text{ Pa})(0.50 \times 10^{-3} \text{ m}^3)}{(8.31 \text{ J/mol} \cdot \text{K})(263 \text{ K})} = 0.023 \text{ mol}$$

In doing this calculation, we used $1 \text{ m}^3 = 1000$ L to convert the 0.50 L volume to m^3. Now we can compute the heat needed to warm one breath, using Equation 12.29:

$$Q(\text{one breath}) = nC_P\Delta T = (0.023 \text{ mol})(29 \text{ J/mol} \cdot \text{K})(40 \text{ K})$$
$$= 27 \text{ J}$$

If you take 12 breaths a minute, a single breath takes 1/12 of a minute, or 5.0 s. Thus the heat power your body supplies to warm the incoming air is

$$P = \frac{Q}{\Delta t} = \frac{27 \text{ J}}{5.0 \text{ s}} = 5.4 \text{ W}$$

At rest, your body typically uses 100 W, so this is just over 5% of your body's resting power. You breathe in a good deal of air each minute and warm it by quite a bit, but the specific heat of air is small enough that the energy required is reasonably modest.

ASSESS The air expands slightly and a small amount of energy goes into heating it. If you've been outside on a cold day, you know that neither change is dramatic, so this final result seems reasonable. The energy loss is noticeable but reasonably modest; other forms of energy loss are more important when you are outside on a cold day.

SUMMARY

GOAL To use the atomic model of matter to explain many properties of matter associated with heat and temperature.

GENERAL PRINCIPLES

Atomic Model

We model matter as being made of simple basic particles. The relationship of these particles to each other defines the phase.

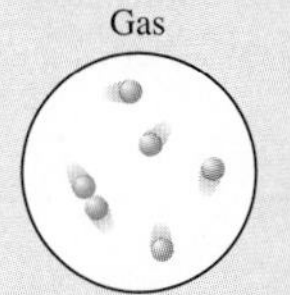

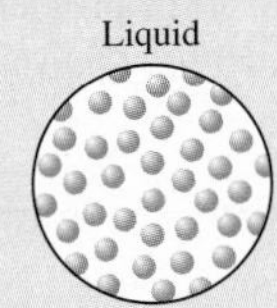

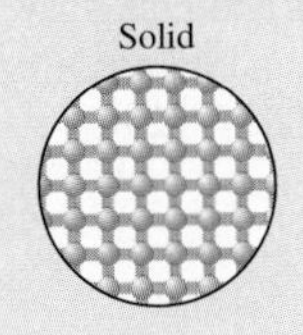

The atomic model explains thermal expansion, specific heat, and heat transfer.

Atomic Model of a Gas

Macroscopic properties of gases can be explained in terms of the atomic model of the gas. The speed of the particles is related to the temperature:

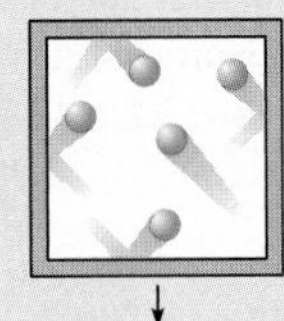

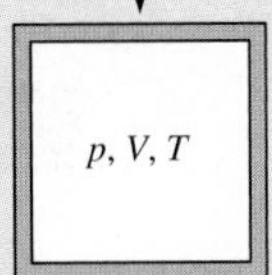

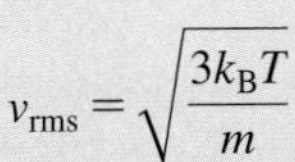

$$v_{rms} = \sqrt{\frac{3k_B T}{m}}$$

The collisions of particles with each other and with the walls of the container determine the pressure.

Ideal-Gas Law

The ideal-gas law relates the pressure, volume, and temperature in a sample of gas. We can express the law in terms of the number of atoms or the number of moles in the sample:

$$pV = Nk_B T$$
$$pV = nRT$$

For a gas process in a sealed container,

$$\frac{p_i V_i}{T_i} = \frac{p_f V_f}{T_f}$$

IMPORTANT CONCEPTS

Effects of heat transfer

A system that is heated can either change temperature or change phase.

The **specific heat** c of a material is the heat required to raise 1 kg by 1 K.

$$Q = Mc\,\Delta T$$

The **heat of transformation** is the energy necessary to change the phase of 1 kg of a substance. Heat is added to change a solid to a liquid or a liquid to a gas; heat is removed to reverse these changes.

$$Q = \begin{cases} \pm ML_f \text{ (melt/freeze)} \\ \pm ML_v \text{ (boil/condense)} \end{cases}$$

The **molar specific heat** of a gas depends on the process.

For a constant-volume process: $Q = nC_V\,\Delta T$

For a constant-pressure process: $Q = nC_P\,\Delta T$

Mechanisms of heat transfer

An object can transfer heat to other objects or to its environment:

Conduction is the transfer of heat by direct physical contact.

$$\frac{Q}{\Delta t} = \left(\frac{kA}{L}\right)\Delta T$$

Convection is the transfer of heat by the motion of a fluid.

Radiation is the transfer of heat by electromagnetic waves.

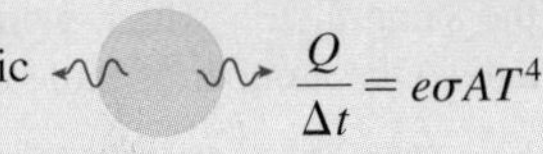

$$\frac{Q}{\Delta t} = e\sigma AT^4$$

A ***pV*** **(pressure-volume) diagram** is a useful means of looking at a process involving a gas.

A **constant-volume** process follows a vertical line between initial and final states.

A **constant-pressure** process follows a horizontal line between initial and final states.

A **constant-temperature** process follows a curved path between initial and final states.

The work done by a gas is the area under the graph.

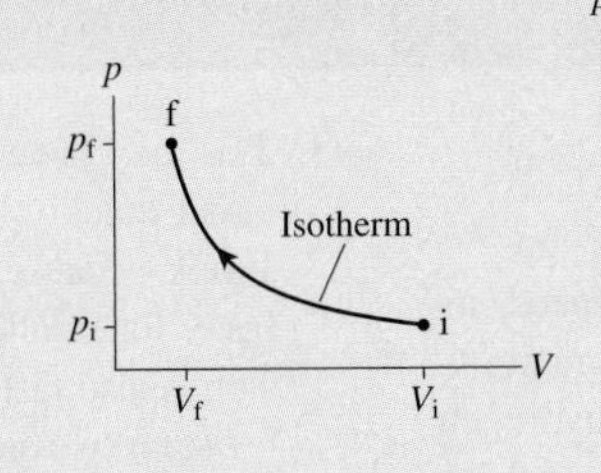

Diffusion

The random motion of molecules causes diffusion from an area of high concentration to an area of low concentration.

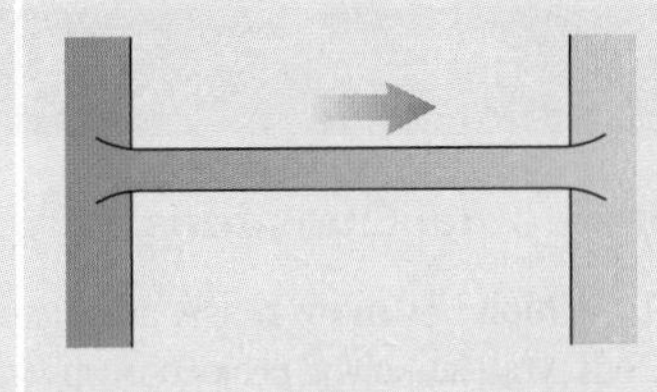

The time for diffusion varies with the square of the distance over which the diffusion occurs:

$$t = \frac{L^2}{6D}$$

APPLICATIONS

Thermal expansion Objects experience an increase in volume and an increase in length when their temperature changes:

$$\Delta V = \beta V_i\,\Delta T \qquad \Delta L = \alpha L_i\,\Delta T$$

Calorimetry When two systems interact thermally, they come to a common final temperature determined by

$$Q_{net} = Q_1 + Q_2 = 0$$

The number of **moles** is

$$n = \frac{M \text{ (in grams)}}{M_{mol}}$$

Learning Objectives After studying this chapter, you should be able to:

- Work with moles and atomic masses. *Conceptual Question 12.1; Problems 12.1, 12.2, 12.3, 12.4, 12.5*
- Understand the atomic model of gas temperature and pressure. *Conceptual Questions 12.2, 12.5; Problems 12.8, 12.14, 12.17, 12.19, 12.22*
- Use the ideal-gas law, pV diagrams, and the first law of thermodynamics to solve problems about ideal-gas processes. *Conceptual Questions 12.9, 12.20; Problems 12.29, 12.31, 12.32, 12.36, 12.38*
- Calculate the thermal expansion of solids and liquids. *Conceptual Questions 12.11, 12.12; Problems 12.42, 12.43, 12.44, 12.45*
- Solve calorimetry and phase-change problems for solids, liquids, and gases. *Conceptual Questions 12.14, 12.16; Problems 12.49, 12.60, 12.61, 12.65, 12.68*
- Calculate the rate of heat transfer via conduction and radiation. *Conceptual Questions 12.23, 12.24; Problems 12.73, 12.74, 12.75, 12.76, 12.77*
- Use random walks and Fick's law to solve problems about the diffusion of gases and liquids. *Conceptual Questions 12.28, 12.29; Problems 12.80, 12.81, 12.82*

STOP TO THINK ANSWERS

Chapter Preview Stop to Think: B. We invoke the first law of thermodynamics, $\Delta E_{th} = W + Q$. The work done by the blender is positive because it adds energy to the food. But the heat transfer is negative because energy is leaving the food as heat. Thus $\Delta E_{th} = 5000\text{ J} - 2000\text{ J} = 3000\text{ J}$.

Stop to Think 12.1: A. Both helium and neon are monatomic gases, where the basic particles are atoms. 5 mol of helium contain 5 times as many atoms as 1 mol of neon, though both samples have the same mass.

Stop to Think 12.2: A, B. An increase in temperature means that the atoms have a larger average kinetic energy and will thus have a larger rms speed. Because the thermal energy of the gas is simply the total kinetic energy of the atoms, this must increase as well. The pressure *could* change, but it's not required; T and V could increase by the same factor, which would keep p constant. The container is sealed, so the number of molecules does not change.

Stop to Think 12.3: B. The product pV/T is constant. During the process, pV decreases by a factor of 2, so T must decrease by a factor of 2 as well.

Stop to Think 12.4: A. The thermal expansion coefficients of aluminum are greater than those of iron. Heating the rod and the ring will expand the outer diameter of the rod and the inner diameter of the ring, but the ring's expansion will be greater.

Stop to Think 12.5: A. Because heat is being added at a constant rate, the slope tells us the ratio of the temperature change ΔT to the heat added Q; that is, the slope is proportional to $\Delta T/Q$. But by the definition of specific heat, this ratio is equal to $1/Mc$. Thus the segment with the *lesser* slope (water) has the *larger* specific heat c.

Stop to Think 12.6: B. To solidify the lead, heat must be removed; this heat boils the water. The heat of vaporization of water is 10 times the heat of fusion of lead, so much less than 1 kg of water vaporizes as 1 kg of lead solidifies.

Stop to Think 12.7: A. The lead cools and the water warms as heat is transferred from the lead to the water. The specific heat of water is much larger than that of lead, so the temperature change of the water is much less than that of the lead.

Stop to Think 12.8: C. With a sealed suit and no matter around you, there is no way to transfer heat to the environment except by radiation.

Video Tutor Solution Chapter 12

QUESTIONS

Conceptual Questions

1. Which has more mass, a mole of Ne gas or a mole of N_2 gas?
2. If you launch a projectile upward with a high enough speed, its kinetic energy is sufficient to allow it to escape the earth's gravity—it will go up and not come back down. Given enough time, hydrogen and helium gas atoms in the earth's atmosphere will escape, so these elements are not present in our atmosphere. Explain why hydrogen and helium atoms have the necessary speed to escape but why other elements, such as oxygen and nitrogen, do not.
3. You may have noticed that latex helium balloons tend to shrink rather quickly; a balloon filled with air lasts a lot longer. Balloons shrink because gas diffuses out of them. The rate of diffusion is faster for smaller particles and for particles of higher speed. Diffusion is also faster when there is a large difference in concentration between two sides of a membrane. Given these facts, explain why an air-filled balloon lasts longer than a helium balloon.
4. If you double the typical speed of the molecules in a gas, by what factor does the pressure change? Give a simple explanation why the pressure changes by this factor.

Problem difficulty is labeled as | (straightforward) to ||||| (challenging). INT Problems labeled integrate significant material from earlier chapters; BIO are of biological or medical interest.

The eText icon indicates when there is a video tutor solution available for the chapter or for a specific problem. To launch these videos, log into your eText through Mastering™ Physics or log into the Study Area.

5. Two gases have the same number of molecules per cubic meter (N/V) and the same rms speed. The molecules of gas 2 are more massive than the molecules of gas 1.
   a. Do the two gases have the same pressure? If not, which is larger?
   b. Do the two gases have the same temperature? If not, which is larger?
6. If the temperature T of an ideal gas doubles, by what factor does the average kinetic energy of the atoms change?
7. A bottle of helium gas and a bottle of argon gas contain equal numbers of atoms at the same temperature. Which bottle, if either, has the greater total thermal energy?
8. A gas cylinder contains 1.0 mol of helium at a temperature of 20°C. A second identical cylinder contains 1.0 mol of neon at 20°C. The helium atoms are moving with a higher average speed, but the gas pressure in the two containers is the same. Explain how this is possible.
9. A gas is in a sealed container. By what factor does the gas pressure change if
   a. The volume is doubled and the temperature is tripled?
   b. The volume is halved and the temperature is tripled?
10. A gas is in a sealed container. By what factor does the gas temperature change if
    a. The volume is doubled and the pressure is tripled?
    b. The volume is halved and the pressure is tripled?
11. You need to precisely measure the dimensions of a large wood panel for a construction project. Your metal tape measure was left outside for hours in the sun on a hot summer day, and now the tape is so hot it's painful to pick up. How will your measurements differ from those taken by your coworker, whose tape stayed in the shade? Explain.
12. A common trick for opening a stubborn lid on a jar is to run very hot water over the lid for a short time. Explain how this helps to loosen the lid.

13. You fill up a steel container with cool water to the very top. Now you place the container in the sun. As the container warms up, the water expands, but so does the container. Does the water overflow? Explain.
14. Materials A and B have equal densities, but A has a larger specific heat than B. You have 100 g cubes of each material. Cube A, initially at 0°C, is placed in good thermal contact with cube B, initially at 200°C. The cubes are inside a well-insulated container where they don't interact with their surroundings. Is their final temperature greater than, less than, or equal to 100°C? Explain.
15. Two containers hold equal masses of nitrogen gas at equal temperatures. You supply 10 J of heat to container A while not allowing its volume to change, and you supply 10 J of heat to container B while not allowing its pressure to change. Afterward, is temperature T_A greater than, less than, or equal to T_B? Explain.
16. You need to raise the temperature of a gas by 10°C. To use the smallest amount of heat energy, should you heat the gas at constant pressure or at constant volume? Explain.
17. The interior of the earth is at a higher temperature than it would be if it were cooling steadily; there must be internal sources of heat. Current models of the earth's interior suggest that one source of heat is the crystallization of liquid iron. Explain how this process provides heat to the earth's interior.
18. If you are exposed to water vapor at 100°C, you are likely to experience a worse burn than if you are exposed to liquid water at 100°C. Why is water vapor more damaging than liquid water at the same temperature?
19. When night falls, the temperature of the earth's surface starts to drop. On a cool night, dew starts to form on the grass as water vapor condenses. Once dew starts to form, the rate of temperature decrease slows. Explain why this change occurs.
20. A sample of ideal gas is in a cylinder with a movable piston. 600 J of heat is added to the gas in an isothermal process. As the gas expands, pushing against the piston, how much work does it do?
21. A student is heating chocolate in a pan on the stove. He uses a cooking thermometer to measure the temperature of the chocolate and sees it varies as shown in Figure Q12.21. Describe what is happening to the chocolate in each of the three portions of the graph.

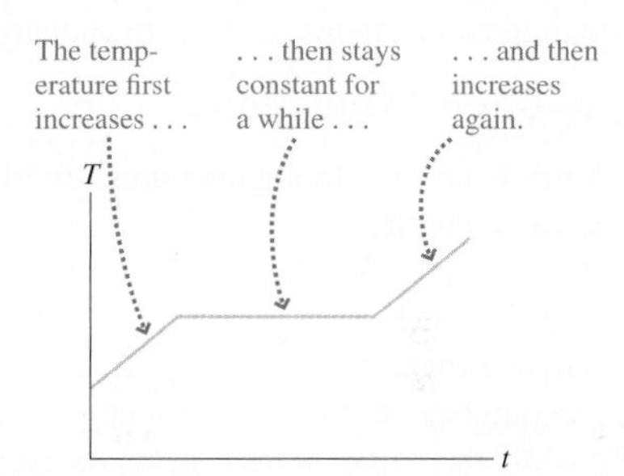

The temperature first increases . . .
. . . then stays constant for a while . . .
. . . and then increases again.
T
t

FIGURE Q12.21

22. If you bake a cake at high elevation, where atmospheric pressure is lower than at sea level, you will need to adjust the recipe. You will need to cook the cake for a longer time, and you will need to add less baking powder. (Baking powder is a leavening agent. As it heats, it releases gas bubbles that cause the cake to rise.) Explain why those adjustments are necessary.
23. Snakes and other reptiles that live on land warm up in the sun. Their bodies can reach temperatures well above the ambient air temperature. But fish don't do this. Most fish are at nearly the same temperature as the water in which they swim. How do you explain the difference?
24. Male emperor penguins spend the long Antarctic night on the ice. They keep from freezing with exquisite thermal insulation. Their insulation is so effective that the surface temperature of their feathers can actually be *lower* than that of the surrounding air. Explain how this is possible.
25. A student is asked to sketch a pV diagram for a gas that goes through a cycle consisting of (a) an isobaric expansion, (b) a constant-volume reduction in temperature, and (c) an isothermal process that returns the gas to its initial state. The student draws the diagram shown in Figure Q12.25. What, if anything, is wrong with the student's diagram?

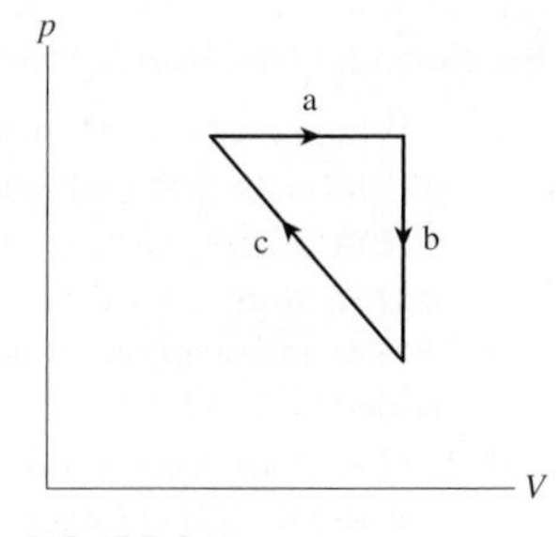

p
a
b
c
V

FIGURE Q12.25

26. You can buy an expensive frying pan that is made of copper but has a handle made of stainless steel. Why are different metals used for the part of the pan that touches the stove and for the part that touches your hand?

27. At night, the surface of the earth cools, mostly by radiation. The surface radiates energy upward; the much cooler sky above radiates much less. Patches of ground under trees can stay warmer, which is why, if you live somewhere that has cold, clear nights, you may have noticed some mornings when there is frost on open patches of ground but not under trees. Why does being under a tree keep the ground warmer?

28. The diffusion constants in Table 12.8 are given at 25°C. In fact, diffusion constants vary significantly with temperature. Explain why you would expect this to be true. Would you expect the values to increase or decrease as the temperature rises?
29. If a person's lungs are compromised, they can use supplemental oxygen to increase the concentration of oxygen in their lungs. Explain how this helps increase oxygen uptake.

Multiple-Choice Questions

30. | A tire is inflated to a gauge pressure of 35 psi. The absolute pressure in the tire is
    A. Less than 35 psi.
    B. Equal to 35 psi.
    C. Greater than 35 psi.
31. | The number of atoms in a container is increased by a factor of 2 while the temperature is held constant. The pressure
    A. Decreases by a factor of 4.
    B. Decreases by a factor of 2.
    C. Stays the same.
    D. Increases by a factor of 2.
    E. Increases by a factor of 4.
32. ||| A gas is compressed by an isothermal process that decreases its volume by a factor of 2. In this process, the pressure
    A. Does not change.
    B. Increases by a factor of less than 2.
    C. Increases by a factor of 2.
    D. Increases by a factor of more than 2.
33. | The thermal energy of a container of helium gas is halved. What happens to the temperature, in kelvin?
    A. It decreases to one-fourth its initial value.
    B. It decreases to one-half its initial value.
    C. It stays the same.
    D. It increases to twice its initial value.
34. ||| A gas is compressed by an adiabatic process that decreases its volume by a factor of 2. In this process, the pressure
    A. Does not change.
    B. Increases by a factor of less than 2.
    C. Increases by a factor of 2.
    D. Increases by a factor of more than 2.
35. | A cup of water is heated with a heating coil that delivers 100 W of heat. In one minute, the temperature of the water rises by 20°C. What is the mass of the water?
    A. 72 g B. 140 g
    C. 720 g D. 1.4 kg
36. | Three identical beakers each hold 1000 g of water at 20°C. 100 g of liquid water at 0°C is added to the first beaker, 100 g of ice at 0°C is added to the second beaker, and the third beaker gets 100 g of aluminum at 0°C. The contents of which container end up at the lowest final temperature?
    A. The first beaker.
    B. The second beaker.
    C. The third beaker.
    D. All end up at the same temperature.
37. || Steam at 100°C causes worse burns than liquid water at 100°C. This is because (BIO)
    A. The steam is hotter than the water.
    B. Heat is transferred to the skin as steam condenses.
    C. Steam has a higher specific heat than water.
    D. Evaporation of liquid water on the skin causes cooling.
38. || Oxygen diffuses across the membranes that separate the air in the alveoli in the lungs from the blood in the capillaries. If you double the thickness of the separating membrane, by what factor will the diffusion time increase? (BIO)
    A. $\sqrt{2}$ B. 2
    C. $2\sqrt{2}$ D. 4

PROBLEMS

Section 12.1 The Atomic Model of Matter

1. | Which contains the most moles: 10 g of hydrogen gas, 100 g of carbon, or 500 g of lead?
2. |||| How many grams of water (H_2O) have the same number of oxygen atoms as 1.0 mol of oxygen gas?
3. |||| How many atoms of hydrogen are in 100 g of hydrogen peroxide (H_2O_2)?
4. || How many cubic millimeters (mm^3) are in 1 L?
5. || A box is 200 cm wide, 40 cm deep, and 3.0 cm high. What is its volume in m^3?

Section 12.2 The Atomic Model of an Ideal Gas

6. | What is the typical speed of a helium atom in a child's balloon at a room temperature of 20°C?
7. | An ideal gas is at 20°C. If we double the average kinetic energy of the gas atoms, what is the new temperature in °C?
8. || A typical helium balloon contains 1.1 g of helium gas. At 20°C room temperature, what is the total kinetic energy of the helium in the balloon?
9. || An ideal gas is at 20°C. The gas is cooled, reducing the thermal energy by 10%. What is the new temperature in °C?
10. || An ideal gas at 0°C consists of 1.0×10^{23} atoms. 10 J of thermal energy are added to the gas. What is the new temperature in °C?
11. || An ideal gas at 20°C consists of 2.2×10^{22} atoms. 4.3 J of thermal energy are removed from the gas. What is the new temperature in °C?
12. |||| Total lung capacity of a typical adult is approximately 5.0 L. Approximately 20% of the air is oxygen. At sea level and at an average body temperature of 37°C, how many moles of oxygen do the lungs contain at the end of an inhalation? (BIO)

13. |||| Many cultures around the world still use a simple weapon BIO called a blowgun, a tube with a dart that fits tightly inside. A sharp breath into the end of the tube launches the dart. When exhaling forcefully, a healthy person can supply air at a gauge pressure of 6.0 kPa. What force does this pressure exert on a dart in a 1.5-cm-diameter tube?

14. ||| When you stifle a sneeze, you can damage delicate tissues BIO because the pressure of the air that is not allowed to escape may rise by up to 45 kPa. If this extra pressure acts on the inside of your 8.4-mm-diameter eardrum, what is the outward force?

15. | What is the absolute pressure of the air in your car's tires, in psi, when your pressure gauge indicates they are inflated to 35.0 psi? Assume you are at sea level.

16. || For a normal car riding on tires with relatively flexible sidewalls, the weight of the car is held up, in large measure, by the pressure of the air in the tires. If you look at one of your car's tires, you'll note that the tire is flattened slightly to make a rectangle where it touches the ground. The area of the resulting "contact patch" depends on the pressure in the tires. To a good approximation, the upward normal force of the ground (which we can assume is equal to ¼ of the car's weight) on this patch of the tire is equal to the downward pressure force on the patch.
   a. Suppose you inflate your 2000 kg car's tires to the recommended pressure, as measured by a gauge. The resulting contact patch is 18 cm wide and 12 cm long. What does the gauge read?
   b. If you let a bit of air out of your tire, what happens to the area of the contact patch?

17. || Mars has an atmosphere composed almost entirely of carbon dioxide, with an average temperature of −63°C. What is the rms speed of a molecule in Mars's atmosphere?

18. || We've discussed elevators that are raised with cables, but many INT elevators use pistons. A pump provides fluid at very high pressure to the piston; the resulting force on the piston raises the elevator and passengers. For a typical piston elevator, the mass of the elevator and passengers is 2000 kg and the radius of the piston is a surprisingly small 4.2 cm. The elevator begins to move upward, accelerating at 0.80 m/s^2. What is the pressure in the piston?

19. | The lowest pressure ever obtained in a laboratory setting is 4.0×10^{-11} Pa. At this pressure, how many molecules of air would there be in a 20°C experimental chamber with a volume of 0.090 m^3?

20. ||| The pressure inside a champagne bottle can be quite high INT and can launch a cork explosively. Suppose you open a bottle at sea level. The absolute pressure inside a champagne bottle is 6 times atmospheric pressure; the cork has a mass of 7.5 g and a diameter of 18 mm. Assume that once the cork starts to move, the only force that matters is the pressure force. What is the acceleration of the cork?

21. || Helium has the lowest condensation point of any substance; the gas liquefies at 4.2 K. 1.0 L of liquid helium has a mass of 125 g. What is the volume of this amount of helium in gaseous form at STP (1 atm and 0°C)?

22. || Some jellyfish can eject stingers with remarkable force and BIO INT speed. One species does this by building up a 15 MPa gauge pressure that pushes against the base of a 2.0-μm-diameter stylet, forcing it outward.
   a. What force does the excess pressure exert on the stylet?
   b. The mass set into motion is a tiny 1.0×10^{-12} kg. What is the resulting acceleration?

23. || A college student is working on her physics homework in her dorm room. Her room contains a total of 6.0×10^{26} gas molecules. As she works, her body is converting chemical energy into thermal energy at a rate of 125 W. If her dorm room were an isolated system (dorm rooms can certainly feel like that) and if all of this thermal energy were transferred to the air in the room, by how much would the temperature increase in 10 min?

24. || Party stores sell small tanks containing 30 g of helium gas. If you use such a tank to fill 0.010 m^3 foil balloons (which don't stretch, and so have an internal pressure that is very close to atmospheric pressure), how many balloons can you expect to fill?

25. || On average, each person in the industrialized world is responsible for the emission of 10,000 kg of carbon dioxide (CO_2) every year. This includes the CO_2 that you generate directly, by burning fossil fuels to operate your car or your furnace, as well as the CO_2 generated on your behalf by electric generating stations and manufacturing plants. CO_2 is a greenhouse gas that contributes to climate change. What is the volume, in m^3, of 10,000 kg of CO_2 at 20°C and 1.0 atm pressure? To put this in perspective, the volume of a typical hot-air balloon is 3000 m^3.

Section 12.3 Ideal-Gas Processes

26. ||| Suppose you start your day in Denver, on a cool 10°C spring day. The local atmospheric pressure is 85 kPa. You fill your car's tires until the gauge shows 210 kPa (about 30 psi). You then drive up to Fairplay, Colorado, where the atmospheric pressure is lower—70 kPa—and the temperature drops to 0°C. When you finish lunch, you check your tire pressure. What does the gauge read?

27. || A cylinder contains 3.0 L of oxygen at 300 K and 2.4 atm. The gas is heated, causing a piston in the cylinder to move outward. The heating causes the temperature to rise to 600 K and the volume of the cylinder to increase to 9.0 L. What is the final gas pressure?

28. || On a chilly 10°C day, you quickly take a deep breath—all BIO your lungs can hold, 4.0 L. The air warms to your body temperature of 37°C. If the air starts at a pressure of 1.0 atm, and you hold the volume of your lungs constant (a good approximation) and the number of molecules in your lungs stays constant as well (also a good approximation), what is the increase in pressure inside your lungs?

29. ||| 0.10 mol of argon gas is admitted to an evacuated 50 cm^3 container at 20°C. The gas then undergoes heating at constant volume to a temperature of 300°C.
   a. What is the final pressure of the gas?
   b. Show the process on a pV diagram. Include a proper scale on both axes.

30. || A football is inflated in the locker room before the game. The air warms as it is pumped, so it enters the ball at a temperature of 27°C. The ball is inflated to a gauge pressure of 13 psi. The ball is used for play at 10°C. Once the ball cools, what is the pressure in the ball? Assume that atmospheric pressure is 14.7 psi.

31. | 0.10 mol of argon gas is admitted to an evacuated 50 cm^3 container at 20°C. The gas then undergoes an isobaric heating to a temperature of 300°C.
   a. What is the final volume of the gas?
   b. Show the process on a pV diagram. Include a proper scale on both axes.

32. ||| 0.10 mol of argon gas is admitted to an evacuated 50 cm^3 container at 20°C. The gas then undergoes an isothermal expansion to a volume of 200 cm^3.
    a. What is the final pressure of the gas?
    b. Show the process on a pV diagram. Include a proper scale on both axes.
33. || 0.0040 mol of gas undergoes the process shown in Figure P12.33.
    a. What type of process is this?
    b. What are the initial and final temperatures?

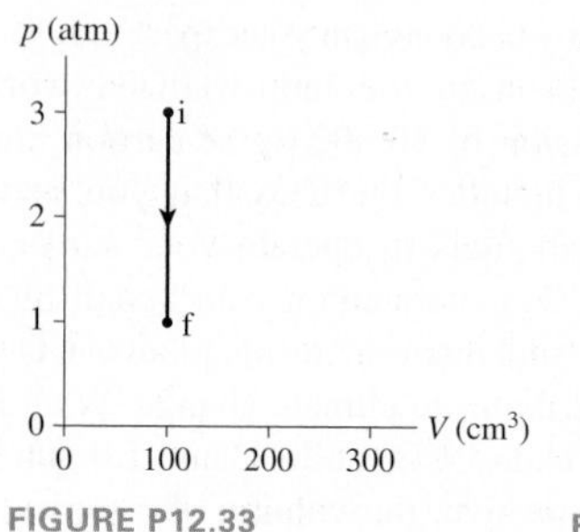

FIGURE P12.33

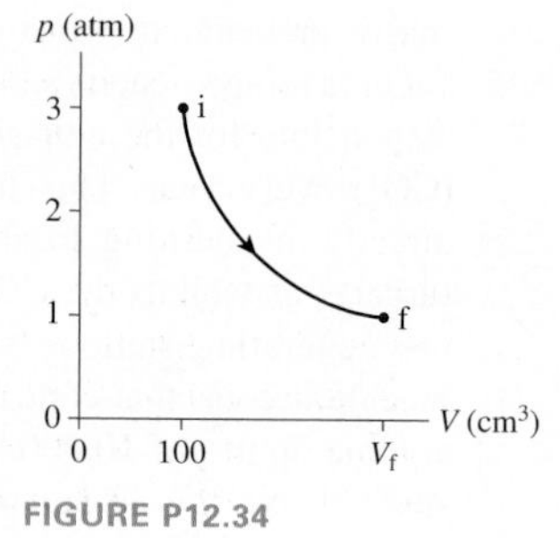

FIGURE P12.34

34. ||| 0.0040 mol of gas follows the hyperbolic trajectory shown in Figure P12.34.
    a. What type of process is this?
    b. What are the initial and final temperatures?
    c. What is the final volume V_f?
35. || A gas with an initial temperature of 900°C undergoes the process shown in Figure P12.35.
    a. What type of process is this?
    b. What is the final temperature?
    c. How many moles of gas are there?

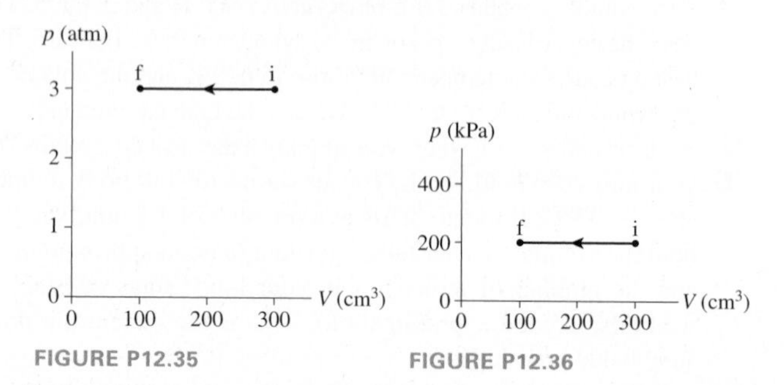

FIGURE P12.35

FIGURE P12.36

36. ||| How much work is done on the gas in the process shown in Figure P12.36?
37. || A 1.0 cm^3 air bubble is released from the sandy bottom of a warm, shallow sea, where the gauge pressure is 1.5 atm. The bubble rises slowly enough that the air inside remains at the same constant temperature as the water.
    a. What is the volume of the bubble as it reaches the surface?
    b. As the bubble rises, is heat energy transferred from the water to the bubble or from the bubble to the water? Explain.
38. || A weather balloon rises through the atmosphere, its volume expanding from 4.0 m^3 to 12 m^3 as the temperature drops from 20°C to −10°C. If the initial gas pressure inside the balloon is 1.0 atm, what is the final pressure?
39. ||| On a cool morning, when the temperature is 15°C, you measure the pressure in your car tires to be 30 psi. After driving 20 mi on the freeway, the temperature of your tires is 45°C. What pressure will your tire gauge now show?

Watch Video Solution Problem 12.38

40. || Many fish maintain buoyancy with a gas-filled swim bladder. The pressure inside the swim bladder is the same as the outside water pressure, so when a fish descends to a greater depth, the gas compresses. Adding gas to restore the original volume requires energy. A fish at a depth where the absolute pressure is 3.0 atm has a swim bladder with the desired volume of 5.0×10^{-4} m^3. The fish now descends to a depth where the absolute pressure is 5.0 atm.
BIO
    a. The gas in the swim bladder is always the same temperature as the fish's body. What is the volume of the swim bladder at the greater depth?
    b. The fish remains at the greater depth, slowly adding gas to the swim bladder to return it to its desired volume. How much work is required?
41. ||| Suppose you inflate your car tires to 35 psi on a 20°C day. Later, the temperature drops to 0°C. What is the pressure in your tires now?

Section 12.4 Thermal Expansion

42. || When you apply the brakes on your car, the kinetic energy of your vehicle is transformed into thermal energy in your brake disks. During a mountain descent, a 28.00-cm-diameter iron brake disk heats up from 30°C to 180°C. What is the diameter of the disk after it heats up?
43. || The length of a steel beam increases by 0.73 mm when its temperature is raised from 22°C to 35°C. What is the length of the beam at 22°C?
44. | In the chapter, you saw that bridges often have expansion joints to account for the changes in size due to temperature changes. Suppose a bridge is supported by steel girders; the central span has a girder that is 30 m long. If the girder warms from 0°C to 30°C, by how much does its length increase?
45. || Older railroad tracks in the U.S. are made of 12-m-long pieces of steel. When the tracks are laid, gaps are left between the sections to prevent buckling when the steel thermally expands. If a track is laid at 16°C, how large should the gaps be if the track is not to buckle when the temperature is as high as 50°C?
46. ||| At 20°C, the hole in an aluminum ring is 2.500 cm in diameter. You need to slip this ring over a steel shaft that has a room-temperature diameter of 2.506 cm. To what common temperature should the ring and the shaft be heated so that the ring will just fit onto the shaft?
47. |||| The temperature of an aluminum disk is increased by 120°C. By what percentage does its volume increase?

Section 12.5 Specific Heat and Heat of Transformation

48. ||| When people exercise, they report feeling exhaustion when their body temperature rises to 39.7°C. A 68 kg man is running at 15 km/h on a hot, muggy day—so hot and muggy that he can't lose any heat to the environment. His body temperature starts at 37.0°C. How long can he run before he feels exhausted?
BIO INT
49. ||| How much energy must be removed from a 200 g block of ice to cool it from 0°C to −30°C?
50. || Suppose that the 800 W of radiation in a microwave oven is absorbed by 250 g of water in a very lightweight cup. How long will it take for the water to warm up from 20°C to 80°C?
51. || a. 100 J of heat energy are transferred to 20 g of mercury initially at 20°C. By how much does the temperature increase?
    b. How much heat is needed to raise the temperature of 20 g of water by the same amount?

52. II Dogs keep themselves cool by panting, rapidly breathing air BIO in and out. Panting results in evaporation from moist tissues of the airway and lungs, which cools the animal. Measurements show that, on a 35°C day with a relative humidity of 50%, a 12 kg dog loses 1.0 g of water per minute if it is panting vigorously. What rate of heat loss, in watts, does this achieve?
53. II The maximum amount of water an adult in temperate cli- BIO mates can perspire in one hour is typically 1.8 L. However, after several weeks in a tropical climate the body can adapt, increasing the maximum perspiration rate to 3.5 L/h. At what rate, in watts, is energy being removed when perspiring that rapidly? Assume all of the perspired water evaporates. At body temperature, the heat of vaporization of water is $L_v = 24 \times 10^5$ J/kg.
54. II A pronghorn antelope can run at a remarkable 18 m/s for BIO INT up to 10 minutes, almost triple the speed that an elite human runner can maintain. For a 32 kg pronghorn, this requires an astonishing 3.4 kW of metabolic power, which leads to a significant increase in body temperature. If the pronghorn had no way to exhaust heat to the environment, by how much would its body temperature increase during this run? (In fact, it will lose some heat, so the rise won't be this dramatic, but it will be quite noticeable, requiring adaptations that keep the pronghorn's brain cooler than its body in such circumstances.)
55. II Alligators and other reptiles BIO don't use enough metabolic energy to keep their body temperatures constant. They cool off at night and must warm up in the sun in the morning. Suppose a 300 kg alligator with an early-morning body temperature of 25°C is absorbing radiation from the sun at a rate of 1200 W. How long will the alligator need to warm up to a more favorable 30°C? (Assume that the specific heat of the reptilian body is the same as that of the mammalian body.)

56. III The Bodø NATO base in northern Norway uses a heat pump INT to extract heat from ocean water. 7.0°C ocean water is continuously drawn into the system, heat is extracted, and the water is returned to the ocean at 3.0°C. When the system is working at full capacity, 2000 kW of heat is delivered to the building at the cost of 600 kW of electric energy.
    a. How much energy is removed from the water (the cold reservoir) each second?
    b. How much water moves through the system each second?
57. II When air is inhaled, it quickly becomes saturated with water vapor as it passes through the moist upper airways. When a person breathes dry air, about 25 mg of water are exhaled with each breath. At 12 breaths/min, what is the rate of energy loss due to evaporation? Express your answer in both watts and Calories per day. At body temperature, the heat of vaporization of water is $L_v = 24 \times 10^5$ J/kg.
58. II Over the course of a day, 4.0 kg of water evaporates from the BIO leaves of a corn plant.
    a. How much energy is required to evaporate the water? (Assume that the temperature of the leaves is 30°C.)
    b. If the plant is active for 12 hours, how much power does this correspond to? You can think of this as the necessary power to drive transport in the plant.
59. III It is important for the body to have mechanisms to effectively cool itself; if not, moderate exercise could easily increase body temperatures to dangerous levels. Suppose a 70 kg man runs on a treadmill for 30 min, using a metabolic power of 1000 W. Assume that all of this power goes to thermal energy in the body. If he couldn't perspire or otherwise cool his body, by how much would his body temperature rise during this exercise?
60. III What minimum heat is needed to bring 100 g of water at 20°C to the boiling point and completely boil it away?

Section 12.6 Calorimetry

61. II Brewed coffee is often too hot to drink right away. You can cool it with an ice cube, but this dilutes it. Or you can buy a device that will cool your coffee without dilution—a 200 g aluminum cylinder that you take from your freezer and place in a mug of hot coffee. If the cylinder is cooled to –20°C, a typical freezer temperature, and then dropped into a large cup of coffee (essentially water, with a mass of 500 g) at 85°C, what is the final temperature of the coffee?
62. II If you pour whisky over ice, the ice will cool the drink, but it INT will also dilute it. A solution is to use whisky stones. Suppose Ernest pours 55 g of whisky at 22°C room temperature, and then adds three whisky stones to cool it. Each stone is a 32 g soapstone cube that is stored in the freezer at –15°C. The specific heat of soapstone is 980 J/kg · K; the specific heat of whisky is 3400 J/kg · K. What is the final temperature of the whisky?
63. III 30 g of copper pellets are removed from a 300°C oven and immediately dropped into 100 mL of water at 20°C in an insulated cup. What will the new water temperature be?
64. III It's possible to boil water by adding hot rocks to it, a technique that has been used in many societies over time. If you heat a rock in the fire, you can easily get it to a temperature of 500°C. If you use granite or other similar stones, the specific heat is about 800 J/kg · K. If 5.0 kg of water at 10°C is in a leak-proof vessel, what minimum number of 1.0 kg stones must be added to bring the water to a boil?
65. III A copper block is removed from a 300°C oven and dropped into 1.00 kg of water at 20.0°C. The water quickly reaches 25.5°C and then remains at that temperature. What is the mass of the copper block?
66. II If you ask for an iced coffee, the barista will add ice cubes to hot coffee to cool it down. To cool 350 mL of coffee at 90°C to 0°C, what is the minimum mass of 0°C ice required?
67. II If a person has a dangerously high fever, submerging her in BIO ice water is a bad idea, but an ice pack can help to quickly bring her body temperature down. How many grams of ice at 0°C will be melted in bringing down a 60 kg patient's fever from 40°C to 39°C?

Section 12.7 Specific Heats of Gases

68. I A container holds 1.0 g of argon at a pressure of 8.0 atm.
    a. How much heat is required to increase the temperature by 100°C at constant volume?
    b. How much will the temperature increase if this amount of heat energy is transferred to the gas at constant pressure?
69. II A container holds 1.0 g of oxygen at a pressure of 8.0 atm.
    a. How much heat is required to increase the temperature by 100°C at constant pressure?
    b. How much will the temperature increase if this amount of heat energy is transferred to the gas at constant volume?

70. | What is the temperature change of 1.0 mol of a monatomic gas if its thermal energy is increased by 1.0 J?

71. | Heating 2.5 mol of neon in a rigid container causes the temperature to increase by 15°C. By how much would the temperature increase if the same amount of heat was added at a constant pressure?

72. || You draw in a deep breath on a chilly day, inhaling 3.0 L of 0°C air. If the pressure in your lungs is a constant 1.0 atm, how much heat must your body supply to warm the air to your 37°C internal body temperature? BIO

Section 12.8 Heat Transfer

73. ||| A 1.8-cm-thick wood floor covers a 4.0 m × 5.5 m room. The subfloor on which the flooring sits is at a temperature of 16.2°C, while the air in the room is at 19.6°C. What is the rate of heat conduction through the floor?

74. ||||| A stainless-steel-bottomed kettle, its bottom 24 cm in diameter and 1.6 mm thick, sits on a burner. The kettle holds boiling water, and energy flows into the water from the kettle bottom at 800 W. What is the temperature of the bottom surface of the kettle?

75. ||| Seals may cool themselves by using *thermal windows,* patches on their bodies with much higher than average surface temperature. Suppose a seal has a 0.030 m^2 thermal window at a temperature of 30°C. If the seal's surroundings are a frosty −10°C, what is the net rate of energy loss by radiation? Assume an emissivity equal to that of a human. BIO

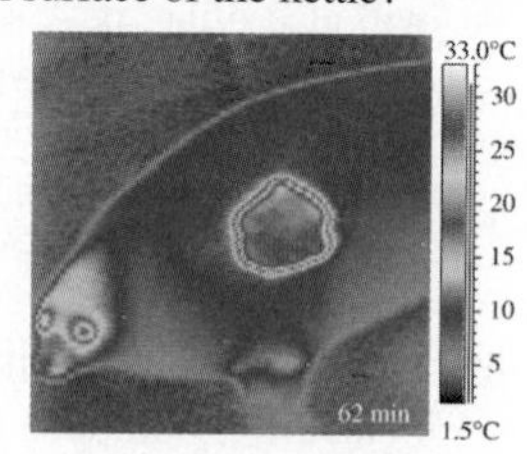

76. || Electronics and inhabitants of the International Space Station generate a significant amount of thermal energy that the station must get rid of. The only way that the station can exhaust thermal energy is by radiation, which it does using thin, 1.8-m-by-3.6-m panels that have a working temperature of about 6°C. How much power is radiated from each panel? Assume that the panels are in the shade so that the absorbed radiation will be negligible. Assume that the emissivity of the panels is 1.0. **Hint:** Don't forget that the panels have two sides!

77. || The glowing filament in a lamp is radiating energy at a rate of 60 W. At the filament's temperature of 1500°C, the emissivity is 0.23. What is the surface area of the filament?

78. ||| A 30 kg male emperor penguin under a clear sky in the Antarctic winter loses very little heat to the environment by convection; its feathers provide very good insulation. It does lose some heat through its feet to the ice, and some heat due to evaporation as it breathes; the combined power is about 12 W. The outside of the penguin's body is a chilly −22°C, but its surroundings are an even chillier −38°C. The penguin's surface area is 0.56 m^2, and its emissivity is 0.97. BIO
   a. What is the net rate of energy loss by radiation?
   b. If the penguin has a 45 W basal metabolic rate, will it feel warm or cold under these circumstances?

79. ||| If you lie on the ground at night with no cover, you get cold rather quickly. Much of this is due to energy loss by radiation. At night in a dry climate, the temperature of the sky can drop to −40°C. If you are lying on the ground with thin clothing that provides little insulation, the surface temperature of your skin and clothes will be about 30°C. Estimate the net rate at which your body loses energy by radiation to the night sky under these conditions. **Hint:** What area should you use? BIO

Section 12.9 Diffusion

80. || A nerve impulse is propagated across a synapse when the neurotransmitter acetylcholine is released from the pre-synaptic membrane and diffuses across the 20-nm-wide synaptic cleft to the post-synaptic membrane. BIO
   a. How long does it take a nerve signal to cross a synapse? You can assume that the synaptic fluid is essentially water.
   b. What is the speed of transmission, in m/s?

81. || The small capillaries in the lungs are in close contact with the alveoli. A red blood cell takes up oxygen during the 0.5 s that it squeezes through a capillary at the surface of an alveolus. What is the diffusion time for oxygen across the 1-μm-thick membrane separating air from blood? Assume that the diffusion coefficient for oxygen in tissue is 2×10^{-11} m^2/s. Give your answer to 1 significant figure. BIO

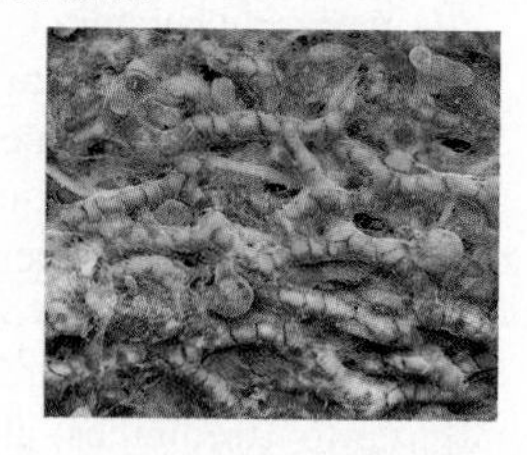

82. ||| The partial pressure of oxygen in the lungs is about 150 mm of Hg. (The partial pressure is the pressure of the oxygen alone, if all other gases were removed.) This corresponds to a concentration of 5.3×10^{24} molecules per m^3. In the oxygen-depleted blood entering the pulmonary capillaries, the concentration is 1.4×10^{24} molecules per m^3. The blood is separated from air in the alveoli of the lungs by a 1-μm-thick membrane. What is the rate of transfer of oxygen to the blood through the 5×10^{-9} m^2 surface area of one alveolus? Give your answer in both molecules/s and μmol/s. Assume The diffusion coefficient for oxygen in tissue is 2×10^{-11} m^2/s. Give your answer to 1 significant figure. BIO

General Problems

83. || A compressed-air cylinder is known to fail if the pressure exceeds 110 atm. A cylinder that was filled to 25 atm at 20°C is stored in a warehouse. Unfortunately, the warehouse catches fire and the temperature reaches 950°C. Does the cylinder explode?

84. ||| 80 J of work are done on the gas in the process shown in Figure P12.84. What is V_f in cm^3?

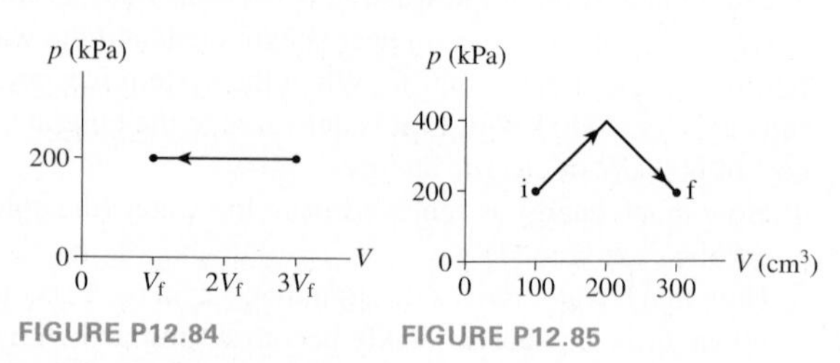

FIGURE P12.84 FIGURE P12.85

85. ||| How much work is done by the gas in the process shown in Figure P12.85?

86. | 0.10 mol of gas undergoes the process 1 → 2 shown in Figure P12.86.
   a. What are temperatures T_1 and T_2?
   b. What type of process is this?
   c. The gas undergoes constant-volume heating from point 2 until the pressure is restored to the value it had at point 1. What is the final temperature of the gas?

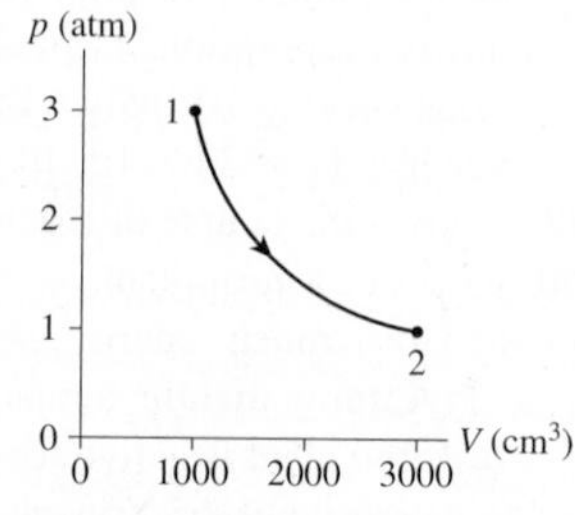

FIGURE P12.86

87. || Your heart does work to pump blood through your body.
BIO Each minute, 5.0 L of blood travels through your circulatory
INT system. The pressure drops from 16 kPa as it exits the heart to approximately zero as it returns to the heart. We can calculate the work required to move a fluid volume V through a pressure difference with a formula very similar to the one you saw for work in gases:
   a. What is the work done to pump the blood?
   b. What is the output power of the heart?

88. ||| A 5.0-m-diameter garden pond holds 5.9×10^3 kg of water. 400 W/m^2 of solar energy strikes the pond's surface, meaning each square meter of the pond's surface is absorbing 400 W. How many hours will it take for the pond to warm from 15°C to 25°C?

89. || James Joule (after whom the unit of energy is named) claimed that the water at the bottom of Niagara Falls should be warmer than the water at the top, 51 m above the bottom. He reasoned that the falling water would transform its gravitational potential energy at the top into thermal energy at the bottom, where turbulence brings the water almost to a halt. If this transformation is the only process occurring, how much warmer will the water at the bottom be?

90. |||| Susan, whose mass is 68 kg, climbs 59 m to the top of the
BIO Cape Hatteras lighthouse.
INT a. During the climb, by how much does her potential energy increase?
   b. For a typical efficiency of 25%, what metabolic energy does she require to complete the climb?
   c. When exercising, the body must perspire and use other mechanisms to cool itself to avoid potentially dangerous increases on body temperature. If we assume that Susan doesn't perspire or otherwise cool herself and that all of the "lost" energy goes into increasing her body temperature, by how much would her body temperature increase during this climb?

91. ||| A typical nuclear reactor generates 1000 MW of electric energy. In doing so, it produces "waste heat" at a rate of 2000 MW, and this heat must be removed from the reactor. Many reactors are sited next to large bodies of water so that they can use the water for cooling. Consider a reactor where the intake water is at 18°C. State regulations limit the temperature of the output water to 30°C so as not to harm aquatic organisms. How many kilograms of cooling water have to be pumped through the reactor each minute?

92. ||| An often-overlooked consequence of power generation is
INT "thermal pollution" from excess heat deposited in the environment. The heating can be considerable. Here are some typical numbers: A nuclear power plant generates 1.0 GW of electric power with an operating efficiency of 35%. The reactor has a single-pass cooling system; water from a nearby river is brought in, warmed, and returned to the river. When operating at full power, the plant takes in 90 m^3 of water per second, comparable to the flow of a small river. If the river water comes in at 10°C, at what temperature does it emerge?

93. |||| A 68 kg woman cycles at a constant 15 km/h. All of the met-
BIO abolic energy that does not go to forward propulsion is con-
INT verted to thermal energy in her body. If the only way her body has to keep cool is by evaporation, how many kilograms of water must she lose to perspiration each hour to keep her body temperature constant?

94. ||| A 1200 kg car traveling at 60 mph quickly brakes to a halt.
INT The kinetic energy of the car is converted to thermal energy of the disk brakes. The brake disks (one per wheel) are iron disks with a mass of 4.0 kg. Estimate the temperature rise in each disk as the car stops.

95. ||| A 5000 kg African elephant has a resting metabolic rate of
BIO 2500 W. On a hot day, the elephant's environment is likely to be
INT nearly the same temperature as the animal itself, so cooling by radiation is not effective. The only plausible way to keep cool is by evaporation, and elephants spray water on their body to accomplish this.

If this is the only possible means of cooling, how many kilograms of water per hour must be evaporated from an elephant's skin to keep it at a constant temperature?

96. ||| When you put on the brakes on your bicycle, friction heats
INT the steel rims of your wheels. Could this heating be a problem? Suppose a 65 kg cyclist with a 15 kg bike is descending Trail Ridge Road in Rocky Mountain National Park, going down a 7.0% grade and thus losing 7.0 m in height for every 100 m of travel along the road. If the cyclist keeps a constant speed of 6.0 m/s, and we assume that all of the "lost" energy ends up as thermal energy in the two steel rims, each of mass 0.80 kg, by how much does the temperature of each rim rise in 1.0 minute?

97. || An experiment measures the temperature of a 200 g substance while steadily supplying heat to it. Figure P12.97 shows the results of the experiment. What are (a) the specific heat of the liquid phase and (b) the heat of vaporization?

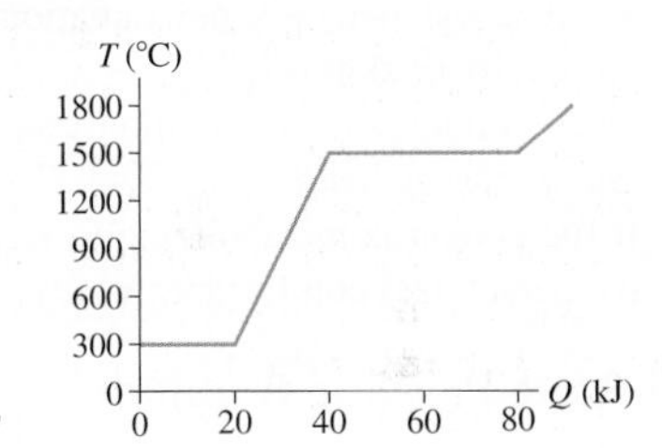

FIGURE P12.97

98. |||| Your 300 mL cup of coffee is too hot to drink when served at 90°C. What is the mass of an ice cube, taken from a −20°C freezer, that will cool your coffee to a pleasant 60°C?

99. || A gas is compressed from 600 cm^3 to 200 cm^3 at a constant pressure of 400 kPa. At the same time, 100 J of heat energy is transferred out of the gas. What is the change in thermal energy of the gas during this process?

100. |||| An expandable cube, initially 20 cm on each side, contains 3.0 g of helium at 20°C. 1000 J of heat energy are transferred to this gas. What are (a) the final pressure if the process is at constant volume and (b) the final volume if the process is at constant pressure?

101. ||| 0.10 mol of a monatomic gas follows the process shown in Figure P12.101.
   a. How much heat energy is transferred to or from the gas during process 1 → 2?
   b. How much heat energy is transferred to or from the gas during process 2 → 3?
   c. What is the total change in thermal energy of the gas?

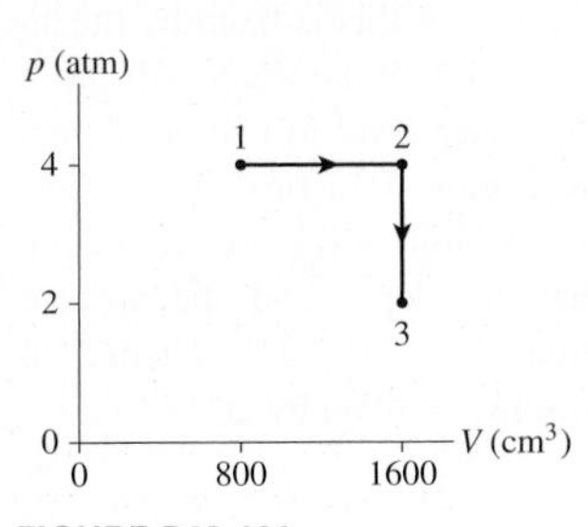

FIGURE P12.101

102. || A monatomic gas follows the process $1 \rightarrow 2 \rightarrow 3$ shown in Figure P12.102. How much heat is needed for (a) process $1 \rightarrow 2$ and (b) process $2 \rightarrow 3$?

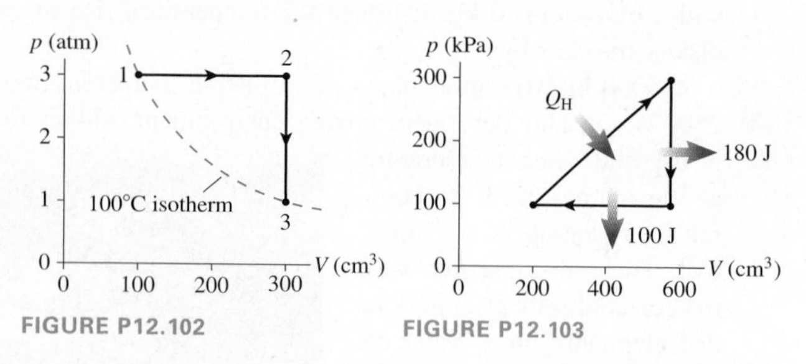

FIGURE P12.102

FIGURE P12.103

103. ||| What are (a) the heat Q_H extracted from the hot reservoir and
INT (b) the efficiency for a heat engine described by the pV diagram of Figure P12.103?

104. | Homes are often insulated with fiberglass insulation in their walls and ceiling. The thermal conductivity of fiberglass is 0.040 W/m·K. Suppose that the total surface area of the walls and roof of a windowless house is 370 m^2 and that the thickness of the insulation is 10 cm. At what rate does heat leave the house on a day when the outside temperature is 30°C colder than the inside temperature?

105. || The surface area of an adult human is about 1.8 m^2. Suppose
BIO a person with a skin temperature of 34°C is standing with bare skin in a room where the air is 25°C but the walls are 17°C.

a. There is a "dead-air" layer next to your skin that acts as insulation. If the dead-air layer is 5.0 mm thick, what is the person's rate of heat loss by conduction?
b. What is the person's net radiation loss to the walls? The emissivity of skin is 0.97.
c. Does conduction or radiation contribute more to the total rate of energy loss?
d. If the person is metabolizing food at a rate of 155 W, does the person feel comfortable, chilly, or too warm?

MCAT-Style Passage Problems

Thermal Properties of the Oceans

Seasonal temperature changes in the ocean only affect the top layer of water, to a depth of 500 m or so. This "mixed" layer is thermally isolated from the cold, deep water below. The average temperature of this top layer of the world's oceans, which has area 3.6×10^8 km^2, is approximately 17°C.

In addition to seasonal temperature changes, the oceans have experienced an overall warming trend over the last century that is expected to continue as the earth's climate changes. A warmer ocean means a larger volume of water; the oceans will rise. Suppose the average temperature of the top layer of the world's oceans were to increase from a temperature T_i to a temperature T_f. The area of the oceans will not change, as this is fixed by the size of the ocean basin, so any thermal expansion of the water will cause the water level to rise, as shown in Figure P12.106. The original volume is the product of the original depth and the surface area, $V_i = Ad_i$. The change in volume is given by $\Delta V = A\,\Delta d$.

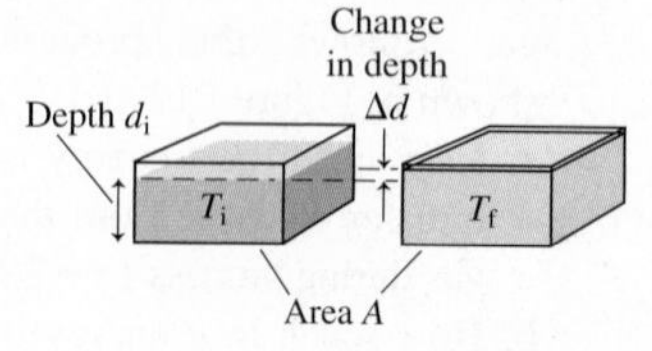

FIGURE P12.106

106. | If the top 500 m of ocean water increased in temperature from 17°C to 18°C, what would be the resulting rise in ocean height?
A. 0.11 m
B. 0.22 m
C. 0.44 m
D. 0.88 m

107. | Approximately how much energy would be required to raise the temperature of the top layer of the oceans by 1°C? (1 m^3 of water has a mass of 1000 kg.)
A. 1×10^{24} J
B. 1×10^{21} J
C. 1×10^{18} J
D. 1×10^{15} J

108. | An increase in temperature of the water may cause other changes. An increase in surface water temperature is likely to ________ the rate of evaporation.
A. Increase
B. Not affect
C. Decrease

109. | The ocean is mostly heated from the top, by light from the sun. The warmer surface water doesn't mix much with the colder deep ocean water. This lack of mixing can be ascribed to a lack of
A. Conduction.
B. Convection.
C. Radiation.
D. Evaporation.

The Work of Breathing

We've seen that the area *under* a pV graph is the work done in an ideal-gas process. If the process follows a closed curve, the work is the area *inside* the curve. The graph in Figure P12.110 shows the gauge pressure in the lungs versus the volume of gas in the lungs for a person who is taking rapid, deep breaths. During one complete breath, the pressure-versus-volume data trace out the curve shown, in the direction of the arrows. The energy expended in one complete breath is represented by the shaded area inside the curve. If you graph pressure-versus-volume data for normal breathing, the upper and lower lines are much closer together and the volume range is much smaller.

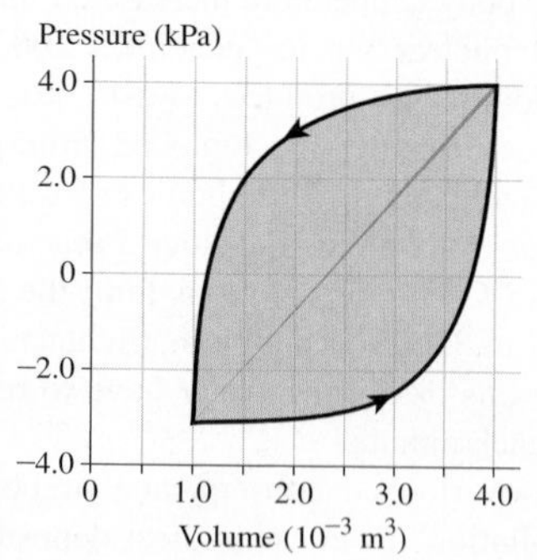

FIGURE P12.110

110. || In the graph in Figure P12.110, the ________ line corresponds to inhalation; the ________ line corresponds to
INT BIO exhalation.
A. upper, lower
B. lower, upper

111. || For the graph in the figure, approximately how much energy
INT is required for one complete breath?
BIO A. 5 J
B. 15 J
C. 25 J
D. 35 J

112. || For one cycle of normal breathing, we expect the energy
INT expended to be ________ for the cycle illustrated in the figure.
BIO A. more than
B. the same as
C. less than

13 Fluids

The albatross must build up speed before it can take off. Why does this graceful flier require such an awkward start to get airborne?

LOOKING AHEAD ▸▸

Pressure in Liquids

A liquid's pressure increases with depth. The high pressure at the base of this water tower pushes water throughout the city.

You'll learn about **hydrostatics**—how liquids behave when they're in equilibrium.

Buoyancy

These students are competing in a concrete canoe contest. How can such heavy, dense objects stay afloat?

You'll learn how to find the **buoyant force** on an object in a fluid using **Archimedes' principle**.

The Circulatory System

This woman is measuring her blood pressure. Why does she wear the cuff where she does?

You'll explore how the principles of **fluid dynamics** apply to blood flow in arteries, capillaries, and veins.

GOAL To understand the static and dynamic properties of fluids.

LOOKING BACK ◂◂

Equilibrium

In Section 5.1, you learned that for an object to be at rest—in *static equilibrium*—the net force on it must be zero. We'll use the principle of equilibrium in this chapter to understand how an object floats.

This mountain goat is in equilibrium: Its weight is balanced by the normal force of the rock.

STOP TO THINK

Three identical books are stacked vertically. The normal force of book 1 on book 2 is

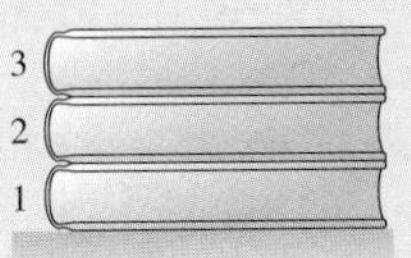

A. Equal to the weight of one book.
B. Less than the weight of one book.
C. Greater than the weight of one book.

13.1 Fluids and Density

A **fluid** is simply a substance that flows. Because they flow, fluids take the shape of their containers rather than retaining a shape of their own. Gases and liquids are both fluids, and their similarities are often more important than their differences.

As you learned in ◀◀ SECTION 12.2, a gas, as shown in FIGURE 13.1a, is a system in which each molecule moves freely through space until, on occasion, it collides with another molecule or with the wall of the container. Gases are *compressible;* that is, the volume of a gas is easily increased or decreased, a consequence of the "empty space" between the molecules in a gas.

Liquids are more complicated than either gases or solids. Liquids, like solids, are essentially *incompressible.* This property tells us that the molecules in a liquid, as in a solid, are about as close together as they can get. At the same time, a liquid flows and deforms to fit the shape of its container. The fluid nature of a liquid tells us that the molecules are free to move around. Together, these observations suggest the model of a liquid shown in FIGURE 13.1b.

FIGURE 13.1 Simple atomic-level models of gases and liquids.

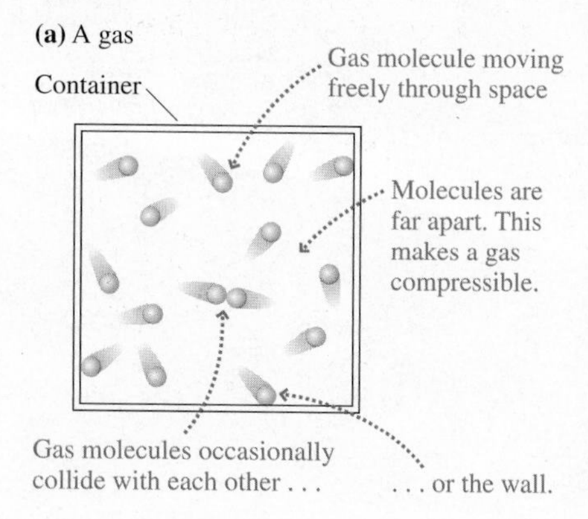
(a) A gas
Container
Gas molecule moving freely through space
Molecules are far apart. This makes a gas compressible.
Gas molecules occasionally collide with each other . . .
. . . or the wall.

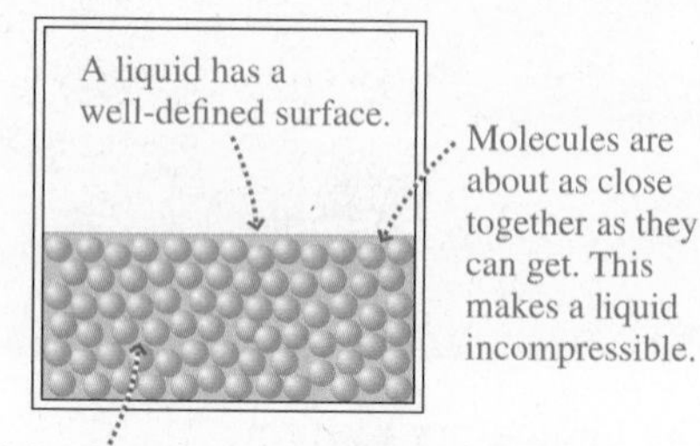
(b) A liquid
A liquid has a well-defined surface.
Molecules are about as close together as they can get. This makes a liquid incompressible.
Molecules make weak bonds with each other that keep them close together. But the molecules can slide around each other, allowing the liquid to flow and conform to the shape of its container.

Density

Suppose we have several blocks of copper, each of different size. Each block has a different mass m and a different volume V. Nonetheless, all the blocks are copper, so there should be some quantity that has the *same* value for all the blocks, telling us, "This is copper, not some other material." The most important such parameter is the *ratio* of mass to volume, which we call the **mass density** ρ (lowercase Greek rho):

$$\rho = \frac{m}{V} \tag{13.1}$$

Mass density of an object of mass m and volume V

Therefore, an object of mass density ρ and volume V has mass

$$m = \rho V \tag{13.2}$$

The SI units of mass density are kg/m^3. Nonetheless, units of g/cm^3 are widely used. You need to convert these to SI units before doing most calculations:

$$1\ \text{g/cm}^3 = 1000\ \text{kg/m}^3$$

The mass density is usually called simply "the density" if there is no danger of confusion. Density is independent of the object's size. That is, mass and volume are parameters that characterize a *specific piece* of some substance—say, copper—whereas density characterizes the substance itself. All pieces of copper have the same density, which differs from the density of almost any other substance.

TABLE 13.1 provides a short list of the densities of various fluids. Notice the enormous difference between the densities of gases and liquids. Gases have lower densities because the molecules in gases are farther apart than in liquids. Also, the density of a liquid varies only slightly with temperature because its molecules are always nearly in contact. The density of a gas, such as air, has a larger variation with temperature because it's easy to change the already large distance between the molecules.

TABLE 13.1 Densities of fluids at 1 atm pressure

Substance	ρ (kg/m^3)
Hydrogen gas (20°C)	0.083
Helium gas (20°C)	0.166
Air (20°C)	1.20
Air (0°C)	1.28
Gasoline	680
Ethyl alcohol	790
Oil (typical)	900
Water	1000
Seawater	1030
Blood (whole)	1060
Glycerin	1260
Mercury	13,600

What does it *mean* to say that the density of gasoline is $680\ \text{kg/m}^3$? Recall in Chapter 1 we discussed the meaning of the word "per." We found that it meant "for each," so that 2 miles per hour means you travel 2 miles *for each* hour that passes. In the same way, saying that the density of gasoline is 680 kg per cubic meter means that there are 680 kg of gasoline *for each* 1 cubic meter of the liquid. If we have $2\ \text{m}^3$ of gasoline, each will have a mass of 680 kg, so the total mass will be $2 \times 680\ \text{kg} = 1360\ \text{kg}$.

EXAMPLE 13.1 **Weighing the air in a living room**

What is the mass of air in a living room with dimensions $4.0\text{ m}\times 6.0\text{ m}\times 2.5\text{ m}$?

STRATEGIZE We will compute the volume of air in the room, and with this figure, use the density to compute the mass.

PREPARE Table 13.1 gives air density at a temperature of 20°C, which is about room temperature.

SOLVE The room's volume is

$$V = (4.0\text{ m}) \times (6.0\text{ m}) \times (2.5\text{ m}) = 60\text{ m}^3$$

The mass of the air is

$$m = \rho V = (1.20\text{ kg/m}^3)(60\text{ m}^3) = 72\text{ kg}$$

ASSESS This is perhaps more mass—about that of an adult person—than you might have expected from a substance that hardly seems to be there. For comparison, a swimming pool this size would contain 60,000 kg of water.

STOP TO THINK 13.1 A piece of glass is broken into two pieces of different size. Rank in order, from largest to smallest, the densities of pieces 1, 2, and 3.

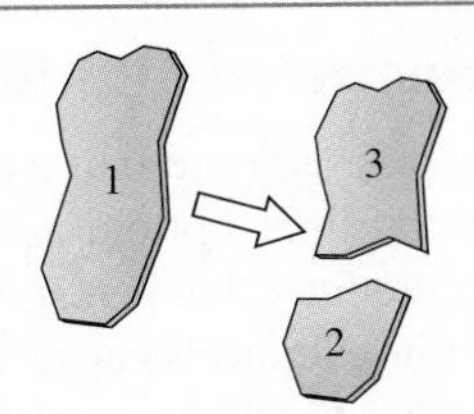
1
3
2

13.2 Pressure

In ◀◀ SECTION 12.2, you learned how a gas exerts a force on the walls of its container. Liquids also exert forces on the walls of their containers, as shown in FIGURE 13.2, where a force $\vec{F}$ due to the liquid pushes against a small area A of the wall. Just as for a gas, we define the pressure at this point in the fluid to be the ratio of the force to the area on which the force is exerted:

$$p = \frac{F}{A} \tag{13.3}$$

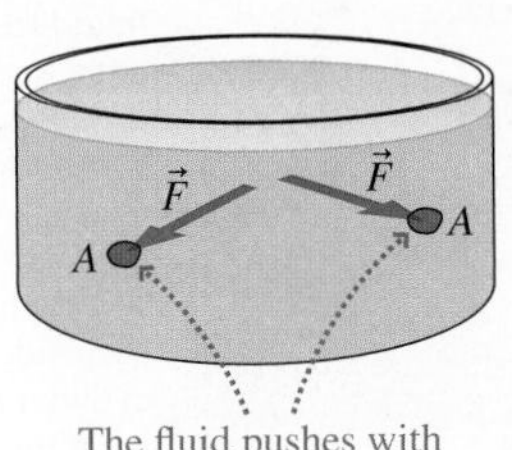
$\vec{F}$
$\vec{F}$
A
A
The fluid pushes with force $\vec{F}$ against area A.

FIGURE 13.2 The fluid presses against area A with force $\vec{F}$.

This is the same as Equation 12.12 of Chapter 12. Recall also from Chapter 12 that the SI unit of pressure, the pascal, is defined as

$$1\text{ pascal} = 1\text{ Pa} = 1\,\frac{\text{N}}{\text{m}^2}$$

The force due to a fluid's pressure pushes not only on the walls of its container, but on *all* parts of the fluid itself. If we punch holes in a container of water, the water spurts out from the holes, as in FIGURE 13.3. It is the force due to the pressure of the water behind each hole that pushes the water forward through the holes.

FIGURE 13.3 Pressure pushes the water *sideways,* out of the holes.

Let's imagine measuring the pressure at different points within a fluid using the simple pressure-measuring device shown in FIGURE 13.4a. Because we know the spring constant k and the area A, we can determine the pressure by measuring the compression of the spring. FIGURE 13.4b shows the results of some simple experiments.

FIGURE 13.4 Learning about pressure.

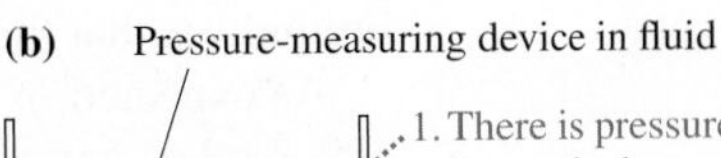
(a) Piston attached to spring
Vacuum; no fluid force is exerted on the piston from this side.
1. The fluid exerts force $\vec{F}$ on a piston with surface area A.
2. The force compresses the spring. Because the spring constant k is known, we can use the spring's compression to find F.
3. Because A is known, we can find the pressure from $p = F/A$.
(b) Pressure-measuring device in fluid
1. There is pressure *everywhere* in a fluid, not just at the bottom or at the walls of the container.
2. The pressure at a given depth in the fluid is the same whether you point the pressure-measuring device up, down, or sideways. The fluid pushes up, down, and sideways with equal strength.
3. In a *liquid*, the pressure increases rapidly with depth below the surface. In a *gas*, the pressure changes much less rapidly.

The first statement in Figure 13.4b emphasizes again that pressure exists at *all* points within a fluid, not just at the walls of the container. You may recall that tension exists at *all* points in a string, not only at its ends where it is tied to an object. We understood tension as the different parts of the string *pulling* against each other. Pressure is an analogous idea, except that the different parts of a fluid are *pushing* against each other.

The spring-and-piston gauge shown in Figure 13.4 measures the actual or *absolute pressure* p, an idea introduced in Chapter 12. But you also learned in ◀◀ SECTION 12.2 that most practical pressure gauges, such as tire gauges, measure the *gauge pressure* p_g, which is the amount the actual pressure is above atmospheric pressure; that is, the gauge pressure is $p_g = p - p_{atmos}$.

Pressure in Liquids

If we pour a liquid into a container, the force of gravity pulls the liquid down, causing it to fill the bottom of the container. It is this force of gravity—that is, the weight of the liquid—that is responsible for the pressure in a liquid. Pressure increases with depth in a liquid because the liquid below is being squeezed by all the liquid above, as well as the pressure of the air above the liquid.

We'd like to determine the pressure at a depth d below the surface of the liquid. We will assume that the liquid is at rest; flowing liquids will be considered later in this chapter. The darker shaded cylinder of liquid in FIGURE 13.5 extends from the surface to depth d. This cylinder, like the rest of the liquid, is in static equilibrium with $\vec{F}_{net} = \vec{0}$. Several forces act on this cylinder: its weight mg, a downward force p_0A due to the pressure p_0 at the surface of the liquid, an upward force pA due to the liquid beneath the cylinder pushing up on the bottom of the cylinder, and inward-directed forces due to the liquid pushing in on the sides of the cylinder. The forces due to the liquid pushing on the cylinder are a consequence of our earlier observation that different parts of a fluid push against each other. Pressure p, the pressure at the bottom of the cylinder, is what we're trying to find.

FIGURE 13.5 Measuring the pressure at depth d in a liquid.

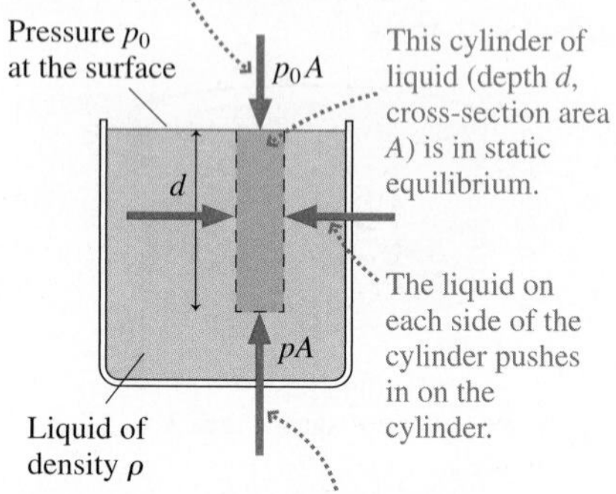
Whatever is above the liquid pushes down on the top of the cylinder.
Pressure p_0 at the surface
p_0A
This cylinder of liquid (depth d, cross-section area A) is in static equilibrium.
d
The liquid on each side of the cylinder pushes in on the cylinder.
pA
Liquid of density ρ
The liquid beneath the cylinder pushes up on the cylinder. The pressure at depth d is p.

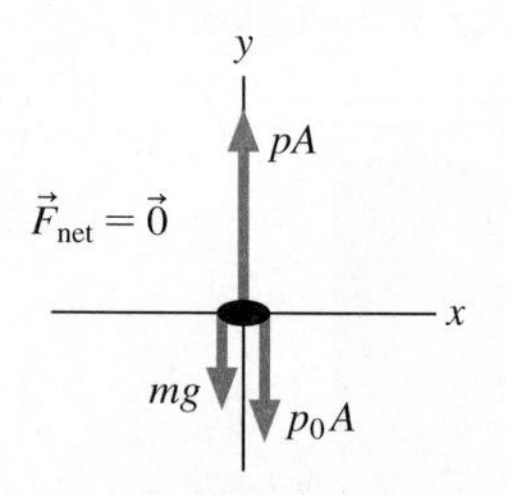
y
pA
$\vec{F}_{net} = \vec{0}$
x
mg
p_0A

Free-body diagram of the column of liquid. The horizontal forces cancel and are not shown.

The horizontal forces cancel each other. The upward force balances the two downward forces, so

$$pA = p_0A + mg \tag{13.4}$$

The liquid is a cylinder of cross-section area A and height d. Its volume is $V = Ad$ and its mass is $m = \rho V = \rho Ad$. Substituting this expression for the mass of the liquid into Equation 13.4, we find that the area A cancels from all terms. The pressure at depth d in a liquid is then

$$p = p_0 + \rho g d \tag{13.5}$$

Pressure of a liquid with density ρ at depth d

Because of our assumption that the liquid is at rest, the pressure given by Equation 13.5 is called the **hydrostatic pressure.** The fact that g appears in Equation 13.5 reminds us that the origin of this pressure is the gravitational force on the fluid.

As expected, $p = p_0$ at the surface, where $d = 0$. Pressure p_0 is usually due to the air or other gas above the liquid. For a liquid that is open to the air at sea level, $p_0 = 1$ atm $= 101$ kPa, as we learned in ◀◀ SECTION 12.2. In other situations, p_0 might be the pressure due to a piston or a closed surface pushing down on the top of the liquid.

NOTE ▶ Equation 13.5 assumes that the fluid is *incompressible;* that is, its density ρ doesn't increase with depth. This is an excellent assumption for liquids, but not a good one for a gas. Equation 13.5 should not be used for calculating the pressure of a gas. ◀

A video to support a section's topic is embedded in the eText.
Video Pressures in Water and Alcohol

EXAMPLE 13.2 **The pressure on a submarine**

A submarine cruises at a depth of 300 m. What is the pressure at this depth? Give the answer in both pascals and atmospheres.

STRATEGIZE Pressure varies with depth below the surface. We will use Equation 13.5 to find the pressure at the depth of the submarine.

PREPARE p_0 is the pressure at the surface, which is 101 kPa. The density of seawater is given in Table 13.1 as 1030 kg/m^3.

SOLVE The pressure at depth $d = 300$ m is

$$\begin{aligned} p &= p_0 + \rho g d \\ &= (1.01 \times 10^5 \text{ Pa}) + (1030 \text{ kg/m}^3)(9.80 \text{ m/s}^2)(300 \text{ m}) \\ &= 3.13 \times 10^6 \text{ Pa} \end{aligned}$$

Converting the answer to atmospheres gives

$$p = (3.13 \times 10^6 \text{ Pa}) \times \frac{1 \text{ atm}}{1.013 \times 10^5 \text{ Pa}} = 30.9 \text{ atm}$$

ASSESS At this great depth, we expect a large pressure, so our result makes sense.

Thick windows The deeper you dive in the ocean, the greater the pressure. At the 4500 m depths the research submarine *Alvin* reaches, the pressure is greater than 450 atm! The pilots certainly want to look outside, but putting a window in a craft at this depth is no mean feat. In the photo, a technician inspects the window that will fill a viewport. The tapered shape will have its larger face toward the sea. The water pressure will push the window firmly into its conical seat, making a tight seal.

FIGURE 13.6a shows two connected tubes. It's certainly true that the larger volume of liquid in the wide tube weighs more than the liquid in the narrow tube. You might think that this extra weight would push the liquid in the narrow tube higher than in the wide tube. But it doesn't. If d_1 were larger than d_2, then, according to the hydrostatic pressure equation, the pressure at the bottom of the narrow tube would be higher than the pressure at the bottom of the wide tube. This *pressure difference* would cause the liquid to *flow* from right to left until the heights were equal. Thus a first conclusion: **A connected liquid in hydrostatic equilibrium rises to the same height in all open regions of the container.**

FIGURE 13.6b shows two connected tubes of different shape. The conical tube holds more liquid above the dashed line, so you might think that $p_1 > p_2$. But it isn't. Both points are at the same depth; thus $p_1 = p_2$. If p_1 were greater than p_2, the pressure at the bottom of the left tube would be greater than the pressure at the bottom of the right tube. This would cause the liquid to flow until the pressures were equal. Thus a second conclusion: **In hydrostatic equilibrium, the pressure is the same at all points on a horizontal line through a connected liquid of a single kind.**

NOTE ► Both of these conclusions are restricted to liquids in hydrostatic equilibrium. The situation is entirely different for flowing fluids, as we'll see later on. ◄

FIGURE 13.6 Liquid in connected tubes.

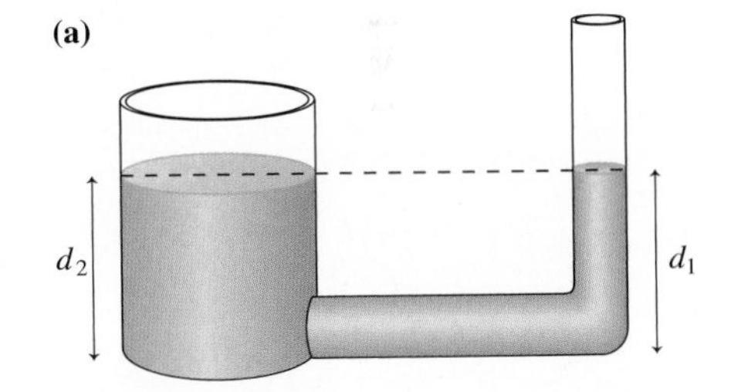

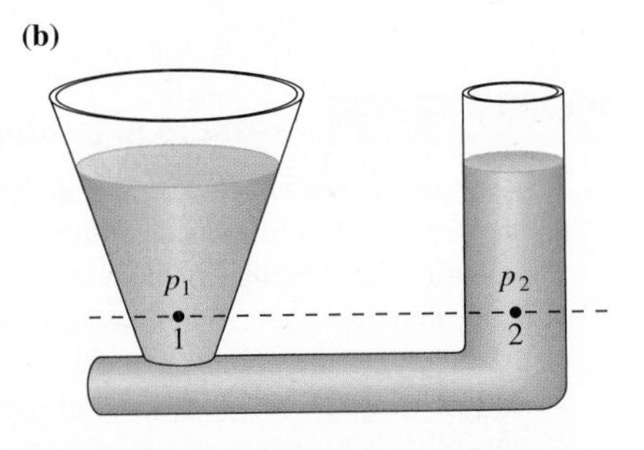

EXAMPLE 13.3 **Pressure in a closed tube**

Water fills the tube shown in FIGURE 13.7. What is the pressure at the top of the closed tube?

FIGURE 13.7 A bent tube closed at one end.

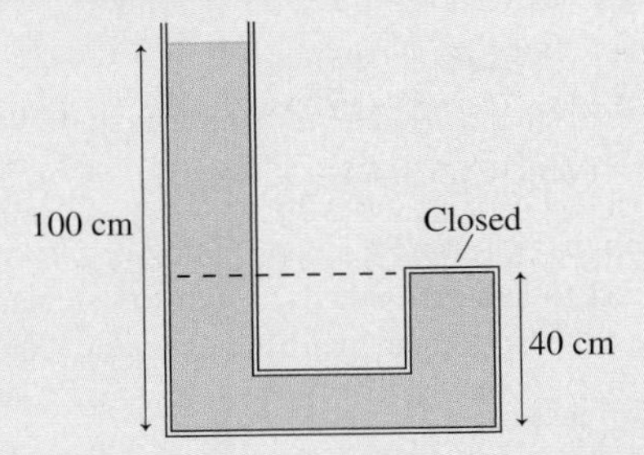

STRATEGIZE This is a liquid in hydrostatic equilibrium. The closed tube is not an open region of the container, so the water cannot rise to an equal height. Nevertheless, the pressure is still the same at all points on a horizontal line. In particular, the pressure at the top of the closed tube equals the pressure in the open tube at the height of the dashed line.

PREPARE The top of the container is open to the atmosphere. We assume that $p_0 = 1$ atm $= 101$ kPa.

SOLVE A point 40 cm above the bottom of the open tube is at a depth of 60 cm. The pressure at this depth is

$$\begin{aligned} p &= p_0 + \rho g d \\ &= (1.01 \times 10^5 \text{ Pa}) + (1000 \text{ kg/m}^3)(9.80 \text{ m/s}^2)(0.60 \text{ m}) \\ &= 1.07 \times 10^5 \text{ Pa} = 1.06 \text{ atm} \end{aligned}$$

ASSESS The water column that creates this pressure is not very tall, so it makes sense that the pressure is only a little higher than atmospheric pressure.

We can draw one more conclusion from the hydrostatic pressure equation $p = p_0 + \rho g d$. If we change the pressure at the surface to $p_1 = p_0 + \Delta p$, so that Δp is the *change* in pressure, then the pressure at a point at a depth d becomes

$$p' = p_1 + \rho g d = (p_0 + \Delta p) + \rho g d = (p_0 + \rho g d) + \Delta p = p + \Delta p$$

That is, the pressure at depth d changes by the same amount as it did at the surface. This idea is called *Pascal's principle:*

Video Water Level in Pascal's Vases

Pascal's principle If the pressure at one point in an incompressible fluid is changed, the pressure at every other point in the fluid changes by the same amount.

We now have enough information to formulate a set of rules for working with hydrostatic problems.

TACTICS BOX 13.1 Hydrostatics

1. **Draw a picture.** Show open surfaces, pistons, boundaries, and other features that affect pressure. Include height and area measurements and fluid densities. Identify the points at which you need to find the pressure.
2. **Determine the pressure p_0 at surfaces.**
   - **Surface open to the air:** $p_0 = p_{atmos}$, usually 1 atm.
   - **Surface in contact with a gas:** $p_0 = p_{gas}$.
   - **Closed surface:** $p_0 = F/A$, where F is the force that the surface, such as a piston, exerts on the fluid.
3. **Use horizontal lines.** The pressure in a connected fluid (of one kind) is the same at any point along a horizontal line.
4. **Allow for gauge pressure.** Pressure gauges read $p_g = p - p_{atmos}$.
5. **Use the hydrostatic pressure equation:** $p = p_0 + \rho g d$.

Exercises 5–7

EXAMPLE 13.4 Pressure in a tube with two liquids

A U-shaped tube is closed at one end; the other end is open to the atmosphere. Water fills the side of the tube that includes the closed end, while oil, floating on the water, fills the side of the tube open to the atmosphere. The two liquids do not mix. The height of the oil above the point where the two liquids touch is 75 cm, while the height of the closed end of the tube above this point is 25 cm. What is the gauge pressure at the closed end?

STRATEGIZE Following the steps in Tactics Box 13.1, we will start by drawing the picture shown in FIGURE 13.8. We assume that the pressure at the open surface of the oil is $p_0 = 1$ atm. Pressures p_1 and p_2 are the same because they are on a horizontal line that connects two points in the *same* fluid. (The pressure at point A is *not* equal to p_3, even though point A and the closed end are on the same horizontal line, because the two points are in *different* fluids.)

We will apply the hydrostatic pressure equation twice: once to find the pressure p_1 by its known depth below the open end at pressure p_0, and again to find the pressure p_3 at the closed end once we know p_2 a distance d below it.

FIGURE 13.8 A tube containing two different liquids.

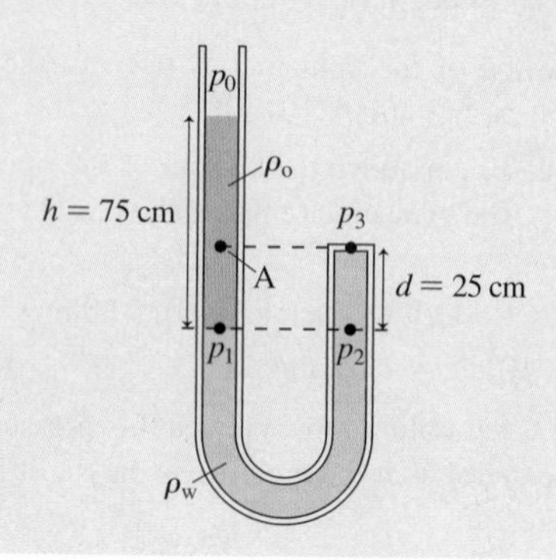

PREPARE We need the densities of water and oil, which are found in Table 13.1 to be $\rho_w = 1000\ \text{kg/m}^3$ and $\rho_o = 900\ \text{kg/m}^3$.

SOLVE The pressure at point 1, 75 cm below the open end, is

$$\begin{aligned} p_1 &= p_0 + \rho_o g h \\ &= 1\ \text{atm} + (900\ \text{kg/m}^3)(9.8\ \text{m/s}^2)(0.75\ \text{m}) \\ &= 1\ \text{atm} + 6620\ \text{Pa} \end{aligned}$$

(We will keep $p_0 = 1$ atm separate in this result because we'll eventually need to subtract exactly 1 atm to calculate the gauge pressure.) We can also use the hydrostatic pressure equation to find

$$\begin{aligned} p_2 &= p_3 + \rho_w g d \\ &= p_3 + (1000\ \text{kg/m}^3)(9.8\ \text{m/s}^2)(0.25\ \text{m}) \\ &= p_3 + 2450\ \text{Pa} \end{aligned}$$

But we know that $p_2 = p_1$, so

$$p_3 = p_2 - 2450\ \text{Pa} = p_1 - 2450\ \text{Pa}$$
$$= 1\ \text{atm} + 6620\ \text{Pa} - 2450\ \text{Pa}$$
$$= 1\ \text{atm} + 4200\ \text{Pa}$$

The gauge pressure at point 3, the closed end of the tube, is $p_3 - 1$ atm or 4200 Pa.

ASSESS The oil's open surface is 50 cm higher than the water's closed surface. Their densities are not too different, so we expect a pressure difference of roughly $\rho g(0.50\ \text{m}) = 5000$ Pa. This is not too far from our answer, giving us confidence that it's correct.

Atmospheric Pressure

We live at the bottom of a "sea" of air that extends up many kilometers. As FIGURE 13.9 shows, there is no well-defined top to the atmosphere; it just gets less and less dense with increasing height until reaching zero in the vacuum of space. Nonetheless, 99% of the air in the atmosphere is below about 30 km.

If we recall that a gas like air is compressible, we can see why the atmosphere becomes less dense with increasing altitude. In a liquid, pressure increases with depth because of the weight of the liquid above. The same holds true for the air in the atmosphere, but because the air is compressible, the weight of the air above compresses the air below, increasing its density. At high altitudes there is very little air above to push down, so the density is less.

We learned in Section 12.2 that the global average sea-level pressure, the *standard atmosphere,* is 1 atm = 101,300 Pa. The standard atmosphere, usually referred to simply as "atmospheres," is a commonly used unit of pressure. But it is not an SI unit, so you must convert atmospheres to pascals before doing most calculations with pressure.

FIGURE 13.9 Atmospheric pressure and density.

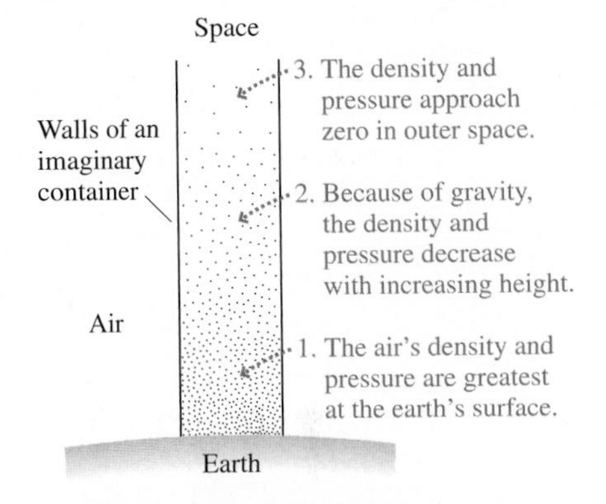
Space
3. The density and pressure approach zero in outer space.
Walls of an imaginary container
2. Because of gravity, the density and pressure decrease with increasing height.
Air
1. The air's density and pressure are greatest at the earth's surface.
Earth

NOTE ▶ Unless you happen to live right at sea level, the atmospheric pressure around you is not exactly 1 atm. A pressure gauge must be used to determine the actual atmospheric pressure. For simplicity, this text will assume that the pressure of the air is $p_{atmos} = 1$ atm unless stated otherwise. ◀

Atmospheric pressure varies not only with altitude, but also with changes in the weather. Local winds and weather are largely determined by the presence and movement of air masses of differing pressure. You may have seen weather maps like the one shown in FIGURE 13.10 on the evening news. The letters H and L denote regions of high and low atmospheric pressure.

FIGURE 13.10 High- and low-pressure zones on a weather map.

Measuring Atmospheric Pressure

FIGURE 13.11a shows a glass tube, sealed at the bottom, that has been completely filled with a liquid. If we temporarily seal the top end, we can invert the tube and place it in a beaker of the same liquid. When the temporary seal is removed, some, but not all, of the liquid runs out, leaving a liquid column in the tube that is a height h above the surface of the liquid in the beaker. This device, shown in FIGURE 13.11b, is a **barometer.** What does it measure? And why doesn't *all* the liquid in the tube run out?

We can analyze the barometer using the steps of Tactics Box 13.1. Point 1 in Figure 13.11b is open to the atmosphere, so $p_1 = p_{atmos}$. The pressure at point 2 is the pressure due to the weight of the liquid in the tube plus the pressure of the gas above the liquid. But in this case there is no gas above the liquid! Because the tube had been completely full of liquid when it was inverted, the space left behind when the liquid ran out is essentially a vacuum, with $p_0 = 0$. Thus pressure p_2 is simply ρgh.

Because points 1 and 2 are on a horizontal line, and the liquid is in hydrostatic equilibrium, the pressures at these two points must be equal. Equating these two pressures gives

$$p_{atmos} = \rho gh \qquad (13.6)$$

Thus we can measure the atmosphere's pressure by measuring the height of the liquid column in a barometer.

FIGURE 13.11 A barometer.

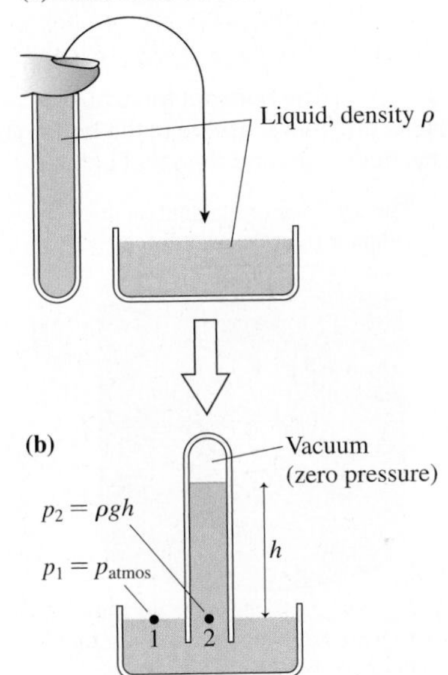
(a) Seal and invert tube.
Liquid, density ρ
(b)
Vacuum (zero pressure)
$p_2 = \rho gh$
h
$p_1 = p_{atmos}$
1
2

Equation 13.6 shows that the liquid height is $h = p_{atmos}/\rho g$. If a barometer were made using water, with $\rho = 1000\ \text{kg/m}^3$, the liquid column would be

$$h = \frac{101{,}300\ \text{Pa}}{(1000\ \text{kg/m}^3)(9.8\ \text{m/s}^2)} \approx 10\ \text{m}$$

which is impractical. Instead, mercury, with its high density of $13{,}600\ \text{kg/m}^3$, is usually used. The average air pressure at sea level causes a column of mercury in a mercury barometer to stand 760 mm above the surface.

Because of the importance of mercury-filled barometers in measuring pressure, the height of a column of mercury in millimeters is a common unit of pressure. From our discussion of barometers, 760 millimeters of mercury (abbreviated "mm Hg") corresponds to a pressure of 1 atm.

Pressure Units

In practice, pressure is measured in a number of different units. This plethora of units and abbreviations has arisen historically as scientists and engineers working on different subjects (liquids, high-pressure gases, low-pressure gases, weather, etc.) developed what seemed to them the most convenient units. These units continue in use through tradition, so it is necessary to convert back and forth among them. TABLE 13.2 gives the basic conversions.

TABLE 13.2 Pressure units

Unit	Abbreviation	Conversion to Pa	Uses
pascal	Pa		SI unit: $1\ \text{Pa} = 1\ \text{N/m}^2$ used in most calculations
atmosphere	atm	$1\ \text{atm} = 101.3\ \text{kPa}$	general
millimeters of mercury	mm Hg	$1\ \text{mm Hg} = 133\ \text{Pa}$	gases and barometric pressure
inches of mercury	in	$1\ \text{in} = 3.39\ \text{kPa}$	barometric pressure in U.S. weather forecasting
pounds per square inch	psi	$1\ \text{psi} = 6.89\ \text{kPa}$	U.S. engineering and industry

STOP TO THINK 13.2 A U-shaped tube is open to the atmosphere on both ends. Water is poured into the tube, followed by oil, which floats on the water because it is less dense than water. Which figure shows the correct equilibrium configuration?

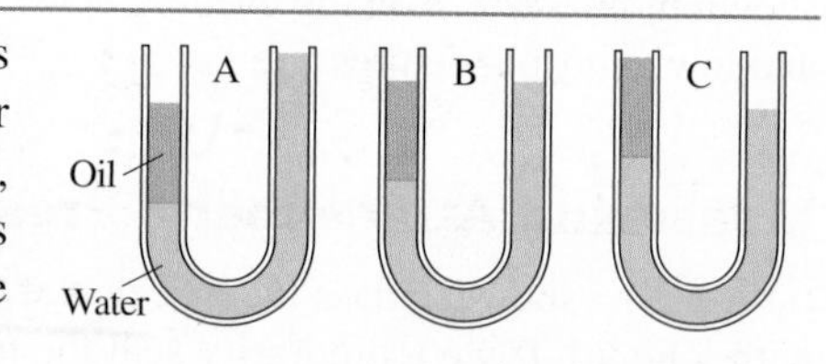
A
B
C
Oil
Water

FIGURE 13.12 The buoyant force arises because the fluid pressure at the bottom of the cylinder is greater than that at the top.

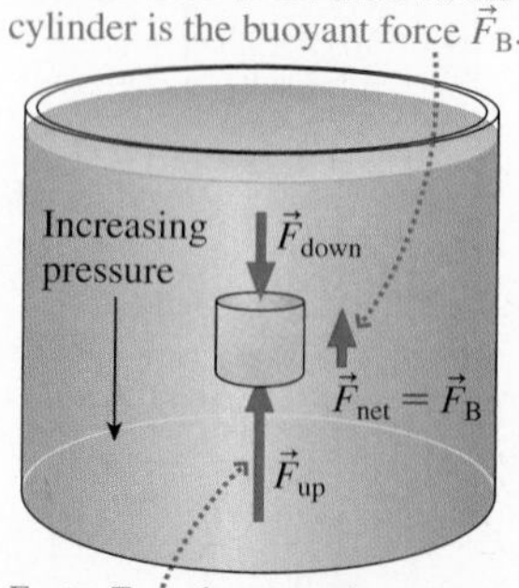
The net force of the fluid on the cylinder is the buoyant force $\vec{F}_B$.
Increasing pressure
$\vec{F}_{down}$
$\vec{F}_{net} = \vec{F}_B$
$\vec{F}_{up}$
$F_{up} > F_{down}$ because the pressure is greater at the bottom. Hence the fluid exerts a net upward force.

13.3 Buoyancy

Wood floats on the surface of a lake. A penny with a mass of a few grams sinks, but a massive steel barge floats. How can we understand these diverse phenomena?

An air mattress floats effortlessly on the surface of a swimming pool. But if you've ever tried to push an air mattress underwater, you know it is nearly impossible. As you push down, the water pushes up. This upward force of a fluid is called the **buoyant force.**

The basic reason for the buoyant force is straightforward. FIGURE 13.12 shows a cylinder submerged in a liquid. The pressure in the liquid increases with depth, so the pressure at the bottom of the cylinder is greater than at the top. Both cylinder ends have equal area, so force $\vec{F}_{up}$ is greater than force $\vec{F}_{down}$. (Remember that pressure forces push in *all* directions.) Consequently, the pressure in the liquid exerts a *net upward force* on the cylinder of magnitude $F_{net} = F_{up} - F_{down}$. This is the buoyant force.

The submerged cylinder illustrates the idea in a simple way, but the result is not limited to cylinders or to liquids. Suppose we isolate a parcel of fluid of arbitrary shape and volume by drawing an imaginary boundary around it, as shown in FIGURE 13.13a. This parcel is in static equilibrium. Consequently, the parcel's weight force pulling it down must be balanced by an upward force. The upward force, which is exerted on this parcel of fluid by the surrounding fluid, is the buoyant force $\vec{F}_B$. The buoyant force matches the weight of the fluid: $F_B = w$.

Now imagine that we remove this parcel of fluid and instantaneously replace it with an object having exactly the same shape and size, as shown in FIGURE 13.13b. Because the buoyant force is exerted by the *surrounding* fluid, and the surrounding fluid hasn't changed, the buoyant force on this new object is *exactly the same* as the buoyant force on the parcel of fluid that we removed.

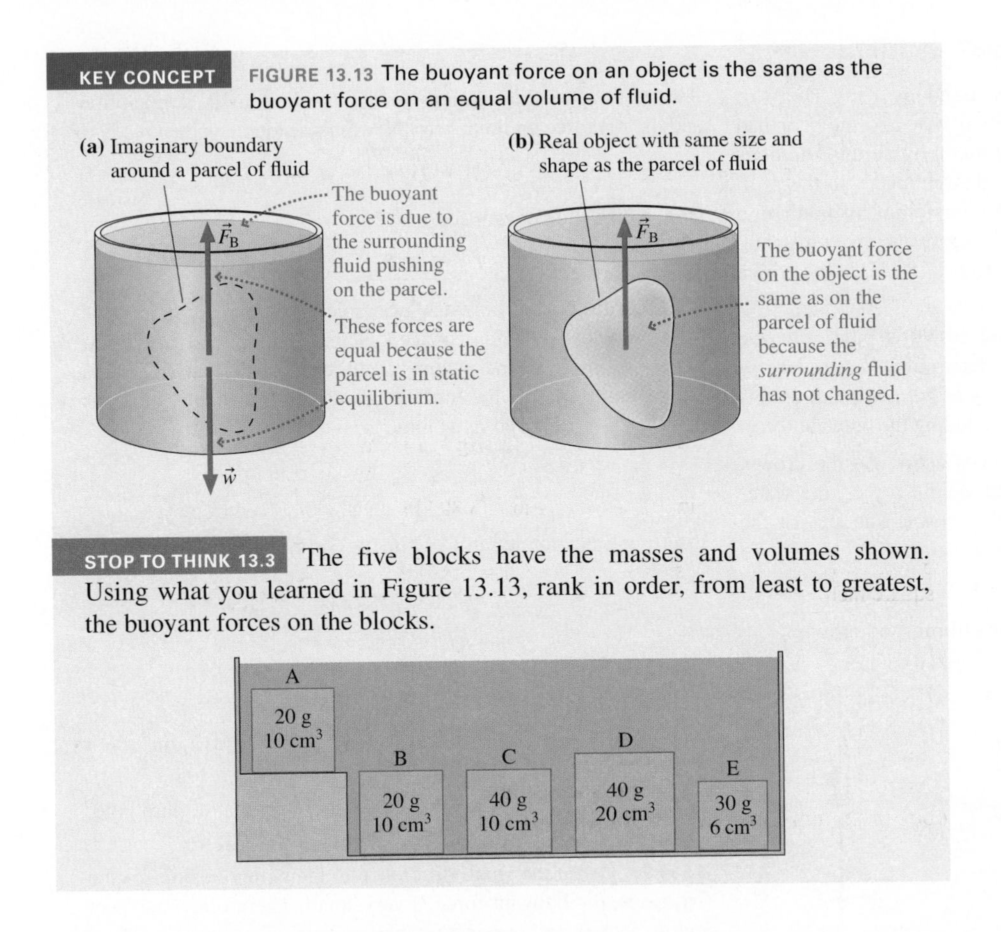
KEY CONCEPT
(a) Imaginary boundary around a parcel of fluid
$\vec{F}_B$
The buoyant force is due to the surrounding fluid pushing on the parcel.
These forces are equal because the parcel is in static equilibrium.
$\vec{w}$
(b) Real object with same size and shape as the parcel of fluid
$\vec{F}_B$
The buoyant force on the object is the same as on the parcel of fluid because the *surrounding* fluid has not changed.
A
20 g
10 cm³
B
20 g
10 cm³
C
40 g
10 cm³
D
40 g
20 cm³
E
30 g
6 cm³

FIGURE 13.13 The buoyant force on an object is the same as the buoyant force on an equal volume of fluid.

STOP TO THINK 13.3 The five blocks have the masses and volumes shown. Using what you learned in Figure 13.13, rank in order, from least to greatest, the buoyant forces on the blocks.

When an object (or a portion of an object) is immersed in a fluid, it *displaces* fluid that would otherwise fill that region of space. The displaced fluid's volume is exactly the volume of the portion of the object that is immersed in the fluid. Figure 13.13 leads us to conclude that the magnitude of the upward buoyant force matches the weight of this displaced fluid.

This idea was first recognized by the ancient Greek mathematician and scientist Archimedes, and today we know it as *Archimedes' principle:*

Archimedes' principle A fluid exerts an upward buoyant force $\vec{F}_B$ on an object immersed in or floating on the fluid. The magnitude of the buoyant force equals the weight of the fluid displaced by the object.

Suppose the fluid has density ρ_f and the object displaces a volume V_f of fluid. The mass of the displaced fluid is then $m_f = \rho_f V_f$ and so its weight is $w_f = \rho_f V_f g$. Thus Archimedes' principle in equation form is

$$F_B = \rho_f V_f g \tag{13.7}$$

Buoyant force on an object displacing volume V_f of fluid of density ρ_f

Video Weighing Weights in Water

NOTE ► It is important to distinguish the density and volume of the displaced fluid from the density and volume of the object. To do so, we use subscript f for the fluid and o for the object. ◄

If the object is completely submerged, the volume of fluid displaced is the same as the volume of the object, so $V_f = V_o$.

EXAMPLE 13.5 Is the crown gold?

Legend has it that Archimedes was asked by King Hiero of Syracuse to determine whether a crown was pure gold or had been adulterated with a less valuable metal by an unscrupulous goldsmith. It was this problem that led Archimedes to the principle that bears his name. In a modern version of his method, a crown weighing 8.30 N is suspended underwater from a string. The tension in the string is measured to be 7.81 N. Is the crown pure gold?

STRATEGIZE To discover whether the crown is pure gold, we will proceed as Archimedes did—by determining its density ρ_o and comparing that to the known density of gold. We will do this by analyzing the forces on the crown, including the buoyant force.

PREPARE FIGURE 13.14 shows the forces acting on the crown. In addition to the familiar tension and weight forces, the water exerts an upward buoyant force on the crown. The size of the buoyant force is given by Archimedes' principle.

FIGURE 13.14 The forces acting on the submerged crown.

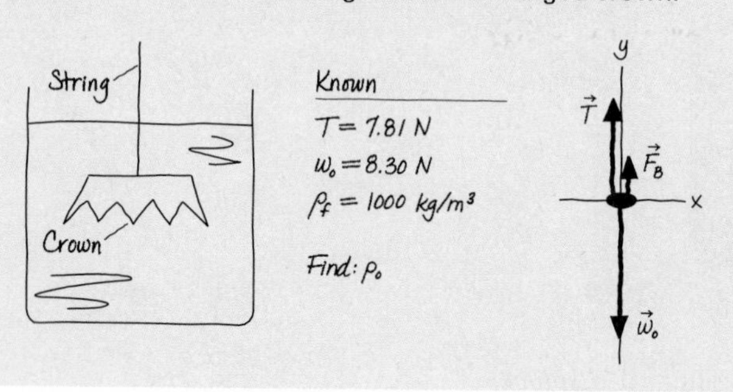

SOLVE Because the crown is in static equilibrium, its acceleration and the net force on it are zero. Newton's second law then reads

$$\sum F_y = F_B + T - w_o = 0$$

from which the buoyant force is

$$F_B = w_o - T = 8.30\text{ N} - 7.81\text{ N} = 0.49\text{ N}$$

According to Archimedes' principle, $F_B = \rho_f V_f g$, where V_f is the volume of the fluid displaced. Here, where the crown is completely submerged, the volume of the fluid displaced is equal to the volume V_o of the crown. Now the crown's weight is $w_o = m_o g = \rho_o V_o g$, so its volume is

$$V_o = \frac{w_o}{\rho_o g}$$

Inserting this volume into Archimedes' principle gives

$$F_B = \rho_f V_o g = \rho_f\left(\frac{w_o}{\rho_o g}\right)g = \frac{\rho_f}{\rho_o}w_o$$

or, solving for ρ_o,

$$\rho_o = \frac{\rho_f w_o}{F_B} = \frac{(1000\text{ kg/m}^3)(8.30\text{ N})}{0.49\text{ N}} = 17{,}000\text{ kg/m}^3$$

The crown's density is considerably lower than that of pure gold, which is 19,300 kg/m^3. The crown is not pure gold.

ASSESS Given the small difference in the string tension for the two cases, the buoyant force is very small. Therefore, we expect a large density and our answer makes sense.

Float or Sink?

If we *hold* an object underwater and then release it, it rises to the surface, sinks, or remains "hanging" in the water. How can we predict which it will do? Whether it heads for the surface or the bottom depends on whether the upward buoyant force F_B on the object is larger or smaller than the downward weight force w_o.

The magnitude of the buoyant force is $\rho_f V_f g$. The weight of a uniform object, such as a block of steel, is simply $\rho_o V_o g$. But a compound object, such as a scuba diver, may have pieces of varying density. If we define the **average density** to be $\rho_{avg} = m_o/V_o$, the weight of a compound object can be written as $w_o = \rho_{avg} V_o g$.

Video What the Physics? Buoyant Balloons

Comparing $\rho_f V_f g$ to $\rho_{avg} V_o g$, and noting that $V_f = V_o$ for an object that is fully submerged, we see that an object floats or sinks depending on whether the fluid density ρ_f is larger or smaller than the object's average density ρ_{avg}. If the densities are equal, the object is in static equilibrium and hangs motionless. This is called **neutral buoyancy.** These conditions are summarized in Tactics Box 13.2.

TACTICS BOX 13.2 Finding whether an object floats or sinks

❶ Object sinks

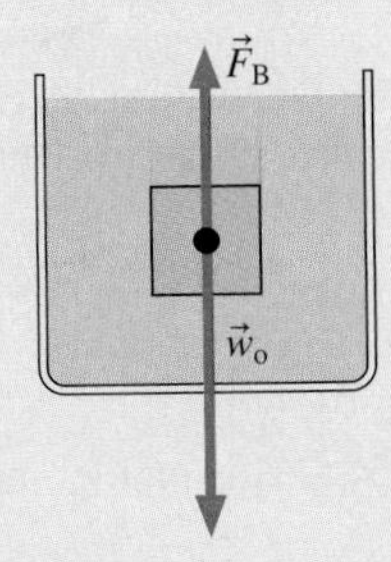

An object sinks if it weighs more than the fluid it displaces—that is, if its average density is greater than the density of the fluid:

$$\rho_{avg} > \rho_f$$

❷ Object floats

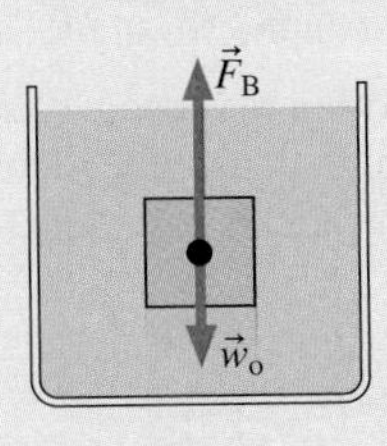

An object rises to the surface if it weighs less than the fluid it displaces—that is, if its average density is less than the density of the fluid:

$$\rho_{avg} < \rho_f$$

❸ Object has neutral buoyancy

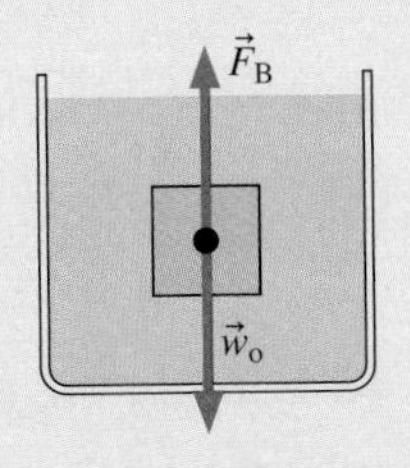

An object hangs motionless if it weighs exactly the same as the fluid it displaces—that is, if its average density equals the density of the fluid:

$$\rho_{avg} = \rho_f$$

Exercises 10–12 

Steel is denser than water, so a chunk of steel sinks. Oil is less dense than water, so oil floats on water. Scuba divers use weighted belts to adjust their average density to match the density of the water, so they maintain their depth in the water without exerting effort.

If we release a block of wood underwater, the net upward force causes the block to shoot to the surface. Then what? To understand floating, let's begin with a *uniform* object such as the block shown in FIGURE 13.15. This object contains nothing tricky, like indentations or voids. Because it's floating, it must be the case that $\rho_o < \rho_f$.

Now that the object is floating, it's in static equilibrium. Thus, the upward buoyant force, given by Archimedes' principle, exactly balances the downward weight of the object; that is,

$$F_B = \rho_f V_f g = w_o = \rho_o V_o g \tag{13.8}$$

For a floating object, the volume of the displaced fluid is *not* the same as the volume of the object. In fact, we can see from Equation 13.8 that the volume of fluid displaced by a floating object of uniform density is

$$V_f = \frac{\rho_o}{\rho_f} V_o \tag{13.9}$$

which is *less* than V_o because $\rho_o < \rho_f$.

NOTE ► Equation 13.9 applies only to *uniform* objects. It does not apply to boats, hollow spheres, or other objects of nonuniform composition. ◄

Video Buoyancy and Density Part 1

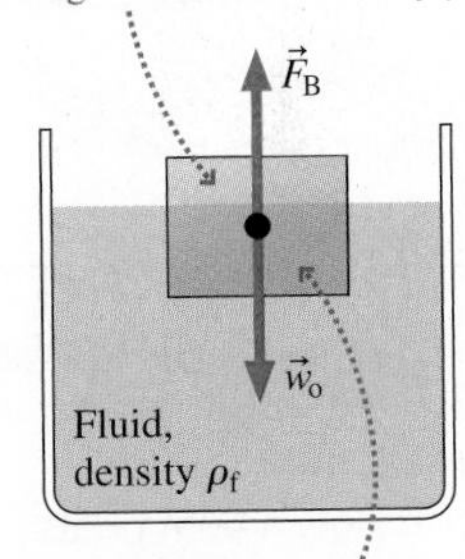

FIGURE 13.15 A floating object is in static equilibrium.

Video Figure 13.15

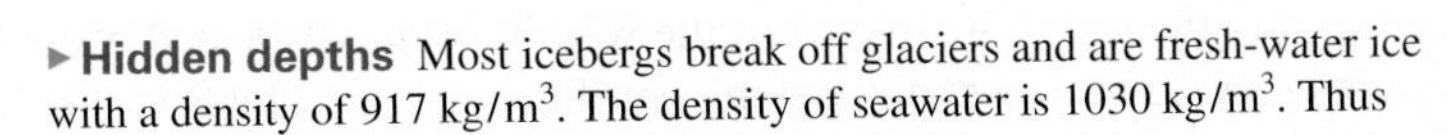

► **Hidden depths** Most icebergs break off glaciers and are fresh-water ice with a density of 917 kg/m^3. The density of seawater is 1030 kg/m^3. Thus

$$V_f = \frac{917\ \text{kg/m}^3}{1030\ \text{kg/m}^3} V_o = 0.89 V_o$$

V_f, the volume of the displaced water, is also the volume of the iceberg that is underwater. The saying "90% of an iceberg is underwater" is correct!

CONCEPTUAL EXAMPLE 13.6 Which has the greater buoyant force?

A block of iron sinks to the bottom of a vessel of water while a block of wood of the *same size* floats. On which is the buoyant force greater?

REASON The buoyant force is equal to the volume of water displaced. The iron block is completely submerged, so it displaces a volume of water equal to its own volume. The wood block floats, so it displaces only the fraction of its volume that is under water, which is *less* than its own volume. The buoyant force on the iron block is therefore greater than on the wood block.

ASSESS This result may seem counterintuitive, but remember that the iron block sinks because of its high density, while the wood block floats because of its low density. A smaller buoyant force is sufficient to keep it floating.

EXAMPLE 13.7 Measuring the density of an unknown liquid

You need to determine the density of an unknown liquid. You notice that a wooden block floats in this liquid with 4.6 cm of the side of the block submerged. When the block is placed in water, it also floats but with 5.8 cm submerged. What is the density of the unknown liquid?

STRATEGIZE We will write expressions for the volume submerged—the volume of fluid displaced—for the two cases. The block floats in both liquids, so we write Equation 13.9 for the two cases. We will then combine the two equations to solve for the density of the fluid.

PREPARE It's useful to draw a picture to visualize the situation. We don't know the area of the block, so we simply label it A. FIGURE 13.16 shows the block and submerged lengths h_u in the unknown liquid and h_w in water.

FIGURE 13.16 A wooden block floating in two liquids.

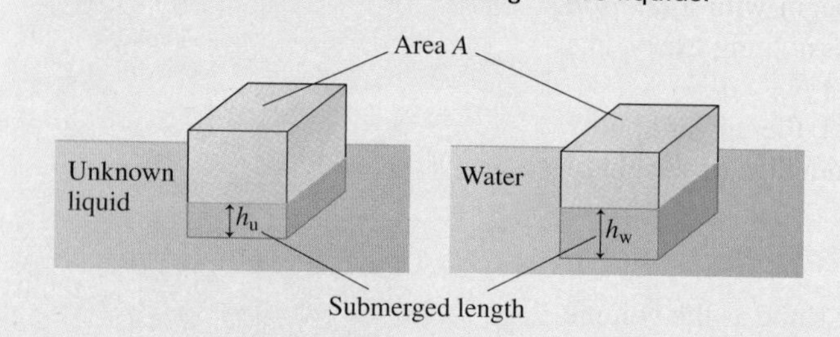

SOLVE The block displaces volume $V_u = Ah_u$ of the unknown liquid. Thus

$$V_u = Ah_u = \frac{\rho_o}{\rho_u} V_o$$

Similarly, the block displaces volume $V_w = Ah_w$ of the water, leading to

$$V_w = Ah_w = \frac{\rho_o}{\rho_w} V_o$$

Because there are two fluids, we've used subscripts w for water and u for the unknown in place of the fluid subscript f. The product $\rho_o V_o$ appears in both equations. In the first $\rho_o V_o = \rho_u A h_u$, and in the second $\rho_o V_o = \rho_w A h_w$. Equating the right-hand sides gives

$$\rho_u A h_u = \rho_w A h_w$$

The area A cancels, and the density of the unknown liquid is

$$\rho_u = \frac{h_w}{h_u}\rho_w = \frac{5.8 \text{ cm}}{4.6 \text{ cm}} 1000 \text{ kg/m}^3 = 1300 \text{ kg/m}^3$$

Comparison with Table 13.1 shows that the unknown liquid is likely to be glycerin.

ASSESS The object floats slightly higher in the unknown liquid than in water, so we expect a density slightly greater than that of water, as we found.

FIGURE 13.17 How a boat floats.

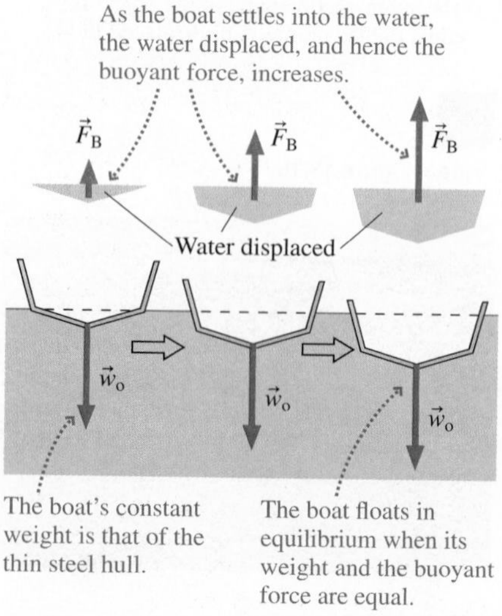

As the boat settles into the water, the water displaced, and hence the buoyant force, increases.

$\vec{F}_B$

Water displaced

$\vec{w}_o$

The boat's constant weight is that of the thin steel hull.

The boat floats in equilibrium when its weight and the buoyant force are equal.

Boats and Balloons

A chunk of steel sinks, so how does a steel-hulled boat float? As we've seen, an object floats if the upward buoyant force—the weight of the displaced water—balances the weight of the object. A boat is really a large hollow shell whose weight is determined by the volume of steel in the hull. As FIGURE 13.17 shows, the volume of water displaced by a shell is *much* larger than the volume of the hull itself. As a boat settles into the water, it sinks until the weight of the displaced water exactly matches the boat's weight. It is then in static equilibrium, so it floats at that level.

The concept of buoyancy and flotation applies to all fluids, not just liquids. An object immersed in a gas such as air feels a buoyant force as well. Because the density of air is so low, this buoyant force is generally negligible. Nonetheless, even though the buoyant force due to air is small, an object will float in air if it weighs less than the air that it displaces. For a balloon to float, it must be filled with a gas that has a *lower* density than that of air. The following example illustrates how this works.

EXAMPLE 13.8 **How fast will the balloon rise?**

A 2.0 g latex balloon is filled with helium. When it is completely inflated, its shape is approximately spherical with a diameter of 24 cm. If the balloon is released, what is its initial upward acceleration?

STRATEGIZE The balloon plus the helium inside it has a certain mass; this leads to a downward weight force. But the balloon displaces a certain amount of air, and this leads to an upward buoyant force. We're told that the balloon accelerates upward, so we know that the upward buoyant force exceeds the downward weight force, as in FIGURE 13.18. This net force will produce an upward acceleration.

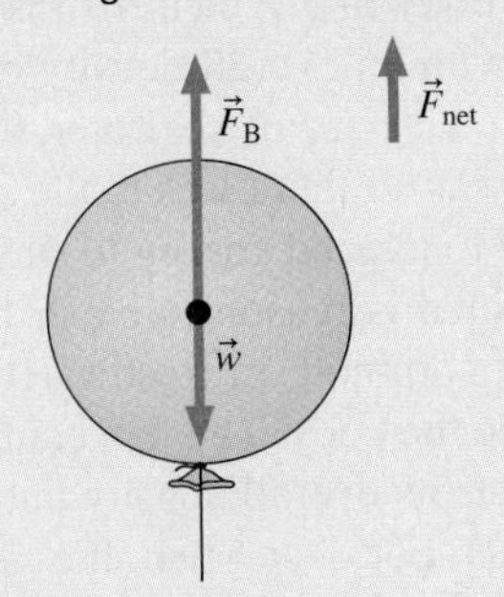

FIGURE 13.18 The forces acting on a helium balloon.

PREPARE We assume that the atmospheric pressure is 1 atm; we can also assume that the pressure inside the balloon is the same. Table 13.1 lists the densities for helium and for air at 1 atm pressure: $\rho_{He} = 0.166\ kg/m^3$ and $\rho_{air} = 1.20\ kg/m^3$. The volume of the balloon in m^3 is

$$V = \frac{4}{3}\pi r^3 = \frac{4}{3}\pi(0.12\ m)^3 = 0.00724\ m^3$$

The thickness of the latex is negligible, so this is the volume of helium in the balloon as well as V_f, the volume of fluid displaced.

SOLVE The total mass is the sum of the mass of the latex balloon (0.0020 kg) and the mass of the helium:

$$m = 0.0020\ kg + \rho_{He}V = 0.0020\ kg + (0.166\ kg/m^3)(0.00724\ m^3) = 0.0032\ kg$$

The weight force is

$$w = (0.0032\ kg)(9.8\ m/s^2) = 0.031\ N$$

The upward buoyant force is

$$F_B = \rho_f V g = (1.20\ kg/m^3)(0.00724\ m^3)(9.8\ m/s^2) = 0.085\ N$$

The net force is the difference of these two forces:

$$F_{net} = F_B - w = 0.085\ N - 0.031\ N = 0.054\ N$$

The initial acceleration of the balloon is

$$a = \frac{F_{net}}{m} = \frac{0.054\ N}{0.0032\ kg} = 17\ m/s^2$$

ASSESS The buoyant force is much greater than the weight force, so we expect a very large initial acceleration. If you've ever been holding a helium balloon and loosened your grip for just a second, you've no doubt seen this in action!

Hot air rising A hot-air balloon is filled with a low-density gas: hot air! You learned in Chapter 12 that gases expand upon heating, which lowers their density. The air at the top of a hot-air balloon is surprisingly toasty—about 100°C. The density of the heated air is about 80% that of room-temperature air. The weight of the air displaced significantly exceeds the weight of the air inside the balloon, allowing the balloon to lift the weight of the passengers and the basket.

Video Buoyancy and Density Part 2

TABLE 13.3 Densities of body components

Body Component	Density (kg/m^3)
Fat	0.90
Water	1.00
Blood	1.05
Muscle	1.06
Bone	1.28

Masters of density BIO A manatee may look chubby, but these mammals live in fresh water and must dive for their food. Too much fat would make them positively buoyant, which would cost them energy to forage. A typical body fat percentage for a manatee is about 7%, comparable to elite human athletes, and this reduces the energy cost for diving. Manatees have other adaptations to reduce the energy cost of moving in the water. They carefully adjust the amount of residual air in their lungs to achieve nearly neutral buoyancy. They can also move air from one lung to the other to roll, and from the top of the lungs to the bottom to tip forward for a dive.

Buoyancy and Bodies BIO

Different components of your body have different densities, as shown in TABLE 13.3. Of the different body components, only fat has a density lower than that of water. An average college student's body is about 20% fat, with the balance a mix of water, muscle, blood, and bone. With this typical fat percentage, the overall density of the body is about 1050 kg/m^3. Submerged in water, a person with this density would sink. But if the person takes a deep breath, she reduces her overall density to approximately 990 kg/m^3 and will float. This is a common experience for most folks: After taking a deep breath, you float; if you exhale as completely as possible, you sink.

In Example 13.5, we saw that the density of an object could be determined by weighing it both underwater and in air. The same thing can be done with a person. Since fat has a lower density than muscle or bone, a lower overall body density implies a greater proportion of body fat. To determine a person's density, he is first weighed in air and then lowered completely into water, asked to exhale as completely as possible, and weighed again. It's then a reasonably straightforward calculation to determine the percentage of body fat. Empirical formulas take into account the air space that remains in the body and expected percentages of the other elements to give an estimate of the amount of fat in a person's body.

If an animal lives in water, then buoyancy is a fact of life that it must deal with. The body of a typical fish is similar to yours. It is made of muscle, skin, bone, and fat, with an overall density of about 1050 kg/m^3. This is greater than the density of fresh water or of seawater, and fish have no lungs filled with air to compensate. If this was all there was to the story, fish would have a tendency to sink, so they would need to expend energy to stay at the same depth.

For this reason, all fish have structures that provide additional buoyancy. The most common, in freshwater and saltwater fish, is a *swim bladder,* a gas-filled sac inside the body. The fish controls the volume of gas in the swim bladder, and thereby adjusts its overall density until it is a reasonably close match to the water in which it swims. For ocean fish, the swim bladder might take up 1–3% of the volume of the body; for freshwater fish, 5–7% is more typical. A fish must continually adjust the amount of gas in the swim bladder as it moves to greater and lesser depths; changes in pressure change the volume of the gas.

EXAMPLE 13.9 How big a bladder?

With its swim bladder deflated, a 9.5 kg yellowfin tuna has a density of 1050 kg/m^3. It is swimming near the surface of the ocean. What volume of gas must the fish have in its swim bladder to achieve neutral buoyancy?

STRATEGIZE For neutral buoyancy, the density of the fish must equal the density of seawater.

PREPARE The density of seawater is given in Table 13.1 as 1030 kg/m^3. We can use the fish's density and its mass to find its volume V_f:

$$V_f = \frac{9.5 \text{ kg}}{1050 \text{ kg/m}^3} = 0.00905 \text{ m}^3$$

SOLVE The density of the fish including the swim bladder is the ratio of the mass of the fish (we can ignore the mass of the air in the swim bladder) to the volume of the fish plus the volume of the swim bladder V_b. We can set this equal to the density of seawater:

$$\frac{9.5 \text{ kg}}{0.00905 \text{ m}^3 + V_b} = 1030 \text{ kg/m}^3$$

Solving for V_b, we find

$$V_b = 0.00017 \text{ m}^3$$

ASSESS This is about 2% of the fish's volume, which is typical for an ocean fish, so our result seems reasonable. A fish would typically maintain a slight negative buoyancy, so real data for tuna show a volume slightly less than this.

Sharks don't have swim bladders; neither do marine mammals. Sharks and many marine mammals have areas of low-density fats that reduce their average density. Sperm whales have a great deal of low-density oil in an organ in their heads. There is evidence that these deep-diving whales adjust the temperature of the oil to change its density. They cool the oil to make them negatively buoyant at the start of a dive; they then warm the oil during the dive so they achieve positive buoyancy when it is time to resurface.

STOP TO THINK 13.4 An ice cube is floating in a glass of water that is filled to the brim. When the ice cube melts, the water level will

A. Fall. B. Stay the same. C. Rise, causing the water to spill.

13.4 Fluids in Motion

We've focused thus far on fluid statics, but it's time to turn our attention to fluids in motion—water coming out of your tap, the wind blowing over the land. Fluid flow is a complex subject. Many aspects of fluid flow, especially turbulence and the formation of eddies, are still not well understood and are areas of current research. We can avoid most of the complexity by making certain assumptions. In our discussion, we'll assume:

- **Fluids are incompressible.** This is a very good assumption for liquids, but it also holds reasonably well for a moving gas, such as air. For instance, even when a 100 mph wind slams into a wall, its density changes by only about 1%.
- **The flow is steady.** That is, the fluid velocity at each point in the fluid is constant; it does not fluctuate or change with time. Flow under these conditions is called *laminar flow*, and it is distinguished from *turbulent flow*.

The rising smoke in FIGURE 13.19 begins as laminar flow, recognizable by its smooth contours, but at some point it undergoes a transition to turbulent flow. Our model of fluids can be applied to the laminar flow, but not to the turbulent flow. This limitation isn't as limiting as it might seem. The motion of blood in your circulatory system, for instance, is always laminar if things are working as they should—turbulence spells trouble. Turbulence in blood vessels makes a very particular sound in a stethoscope, which allows a physician listening to your heart and lungs to detect potentially serious problems in a routine physical exam.

FIGURE 13.19 Rising smoke changes from laminar flow to turbulent flow.

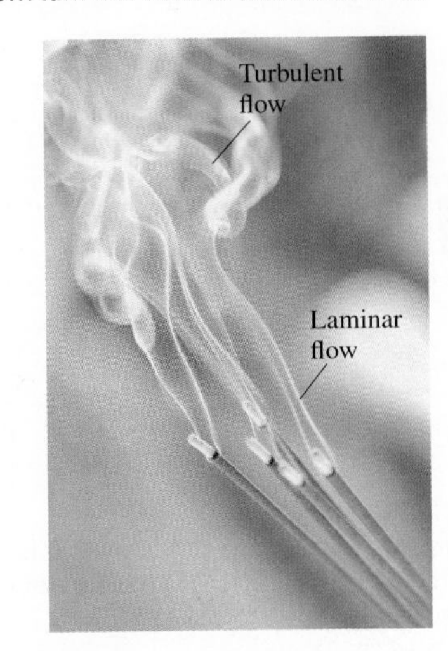

The Role of Viscosity

Water flows much more easily than honey, which has a much higher *viscosity*, a quality you learned about in ◀◀ SECTION 5.6. For some flows—such as the flow of water in a river—we can ignore viscosity. Viscosity is resistance to flow, and assuming that we can ignore viscosity is analogous to assuming that we can ignore friction, or that we can ignore drag. For other flows—such as the motion of blood in the small vessels—viscosity is very important.

The different situations are distinguished by the value of the *Reynolds number*, which you also learned about in ◀◀ SECTION 5.6. The Reynolds number is the ratio of inertial forces to viscous forces for an object moving through a fluid, but it can be applied to the motion of fluids as well. For computing the Reynolds number for a fluid flow, the length scale is the depth of the flow in a channel or the diameter of a pipe carrying a flow. If the Reynolds number is large, the viscous forces are small compared to the other forces at work and can be ignored. For water flowing at 1 m/s in a river that is 2 m deep, the Reynolds number is

$$Re = \frac{\rho v L}{\eta} = \frac{(1000\ \text{kg/m}^3)(1\ \text{m/s})(2\ \text{m})}{1 \times 10^{-3}\ \text{Pa} \cdot \text{s}} = 2{,}000{,}000$$

This is clearly large enough that we needn't worry about viscosity. However, for blood (viscosity 3.5×10^{-3} Pa · s) moving in an arteriole with a diameter of 0.00020 m at a speed of 0.0030 m/s, the Reynolds number is

$$Re = \frac{\rho v L}{\eta} = \frac{(1050\ \text{kg/m}^3)(0.0030\ \text{m/s})(0.00020\ \text{m})}{3.5 \times 10^{-3}\ \text{Pa} \cdot \text{s}} = 0.18$$

In this case, the motion is dominated by viscosity.

The Reynolds number also determines the transition between laminar and turbulent flow. We'll see an example when we treat fluid flow in the circulatory system later in the chapter.

We'll start our treatment of fluid flow by considering properties that apply regardless of viscosity, regardless of the value of the Reynolds number for the flow.

The Equation of Continuity

Consider a fluid flowing through a tube—oil through a pipe or blood through an artery. If the tube's diameter changes, as happens in FIGURE 13.20, what happens to the speed of the fluid?

When you squeeze a toothpaste tube, the volume of toothpaste that emerges matches the amount by which you reduce the volume of the tube. An *incompressible* fluid flowing through a rigid tube or pipe acts the same way. Fluid is neither created

FIGURE 13.20 Flow speed changes through a tapered tube.

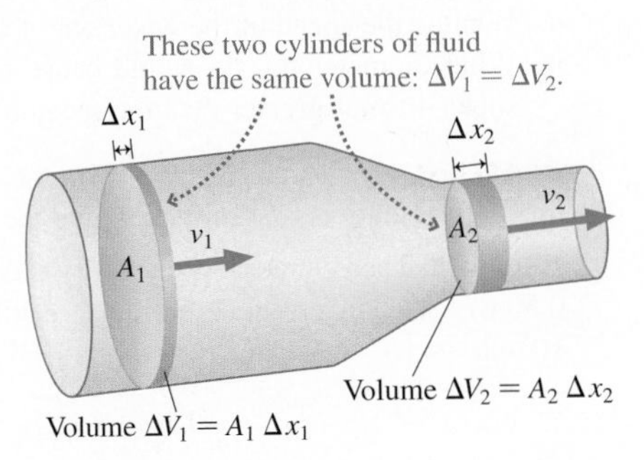

nor destroyed within the tube, and there's no place to store any extra fluid introduced into the tube. If volume V enters the tube during some interval of time Δt, then an equal volume of fluid must leave the tube.

To see the implications of this idea, suppose all the molecules of the fluid in Figure 13.20 are moving to the right with speed v_1 at a point where the cross-section area is A_1. Farther along the tube, where the cross-section area is A_2, their speed is v_2. During an interval of time Δt, the molecules in the wider section move forward a distance $\Delta x_1 = v_1 \Delta t$ and those in the narrower section move $\Delta x_2 = v_2 \Delta t$. Because the fluid is incompressible, the volumes ΔV_1 and ΔV_2 must be equal; that is,

$$\Delta V_1 = A_1 \Delta x_1 = A_1 v_1 \Delta t = \Delta V_2 = A_2 \Delta x_2 = A_2 v_2 \Delta t \qquad (13.10)$$

Dividing both sides of the equation by Δt gives the **equation of continuity:**

Video Continuity

$$v_1 A_1 = v_2 A_2 \qquad (13.11)$$

The equation of continuity relating the speed v of an incompressible fluid to the cross-section area A of the tube in which it flows

Equations 13.10 and 13.11 say that **the volume of an incompressible fluid entering one part of a tube or pipe must be matched by an equal volume leaving downstream.**

An important consequence of the equation of continuity is that **flow is faster in narrower parts of a tube, slower in wider parts.** You're familiar with this conclusion from many everyday observations. The garden hose shown in FIGURE 13.21a squirts water farther after you put a nozzle on it. This is because the narrower opening of the nozzle gives the water a higher exit speed. Water flowing from the faucet shown in FIGURE 13.21b picks up speed as it falls. As a result, the flow tube "necks down" to a smaller diameter.

FIGURE 13.21 The speed of the water is inversely proportional to the diameter of the stream.

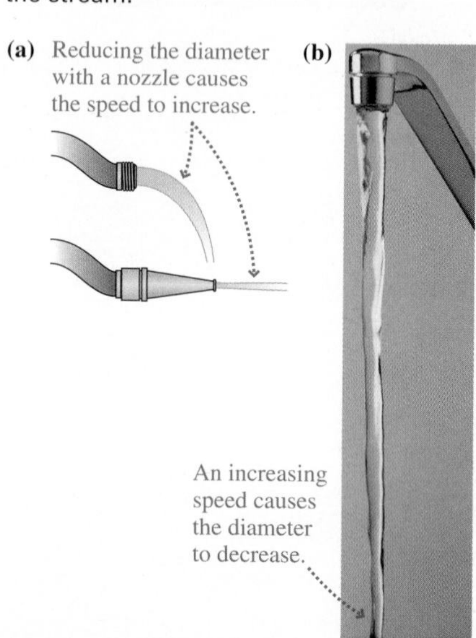
(a) Reducing the diameter with a nozzle causes the speed to increase.
(b)
An increasing speed causes the diameter to decrease.

The *rate* at which fluid flows through the tube—volume per second—is $\Delta V/\Delta t$. This is called the **volume flow rate** Q. We can see from Equation 13.10 that

$$Q = \frac{\Delta V}{\Delta t} = vA \qquad (13.12)$$

Volume flow rate for liquid moving at speed v through a tube of cross-section area A

The SI units of Q are m^3/s, although in practice Q may be measured in cm^3/s, liters per minute, or, in the United States, gallons per minute and cubic feet per minute. Another way to express the meaning of the equation of continuity is to say that **the volume flow rate is constant at all points in a tube.**

EXAMPLE 13.10 Speed of water through a hose

A garden hose has an inside diameter of 16 mm. The hose can fill a 10 L bucket in 20 s.

a. What is the speed of the water out of the end of the hose?
b. What diameter nozzle would cause the water to exit with a speed 4 times greater than the speed inside the hose?

STRATEGIZE We are given the volume flow rate; we will use this with Equation 13.12 to determine the speed of the flow.

PREPARE The volume flow rate is $Q = \Delta V/\Delta t = (10\text{ L})/(20\text{ s}) = 0.50\text{ L/s}$. To convert this to SI units, recall that $1\text{ L} = 1000\text{ mL} = 10^3\text{ cm}^3 = 10^{-3}\text{ m}^3$. Thus $Q = 5.0 \times 10^{-4}\text{ m}^3/\text{s}$.

SOLVE a. The speed of the water is

$$v = \frac{Q}{A} = \frac{Q}{\pi r^2} = \frac{5.0 \times 10^{-4}\text{ m}^3/\text{s}}{\pi(0.0080\text{ m})^2} = 2.5\text{ m/s}$$

b. The quantity $Q = vA$ remains constant as the water flows through the hose and then the nozzle. To increase v by a factor of 4, A must be reduced by a factor of 4. The cross-section area depends on the square of the diameter, so the area is reduced by a factor of 4 if the diameter is reduced by a factor of 2. Thus the necessary nozzle diameter is 8 mm.

ASSESS This seems like a reasonable speed for water through a garden hose.

Representing Fluid Flow: Streamlines and Fluid Elements

Representing the flow of fluid is more complicated than representing the motion of a point particle because fluid flow is the collective motion of a vast number of particles. FIGURE 13.22 gives us an idea of one possible fluid-flow representation. Here smoke is being used to help engineers visualize the airflow around a car in a wind tunnel. The smoothness of the flow tells us this is laminar flow. But notice also how the individual smoke trails retain their identity. They don't cross or get mixed together. Each smoke trail represents a *streamline* in the fluid.

Imagine that we could inject a tiny colored drop of water into a stream of water undergoing laminar flow. Because the flow is steady and the water is incompressible, this colored drop would maintain its identity as it flowed along. The path or trajectory followed by this "particle of fluid" is called a **streamline.** Smoke particles mixed with the air allow us to see the streamlines in the wind-tunnel photograph of Figure 13.22. FIGURE 13.23 illustrates three important properties of streamlines.

FIGURE 13.22 Streamlines in the laminar airflow around a car in a wind tunnel.

FIGURE 13.23 Particles in a fluid move along streamlines.

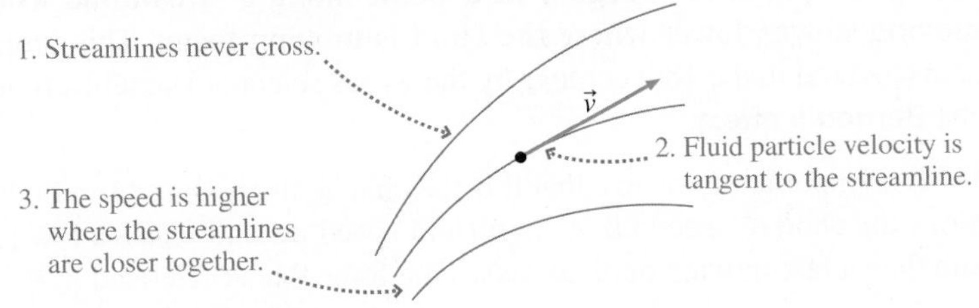

As we study the motion of a fluid, it is often also useful to consider a small *volume* of fluid, a volume containing many particles of fluid. Such a volume is called a **fluid element.** FIGURE 13.24 shows two important properties of a fluid element. Unlike a particle, a fluid element has an actual shape and volume. Although the shape of a fluid element may change as it moves, the equation of continuity requires that its volume remain constant.

FIGURE 13.24 Motion of a fluid element.

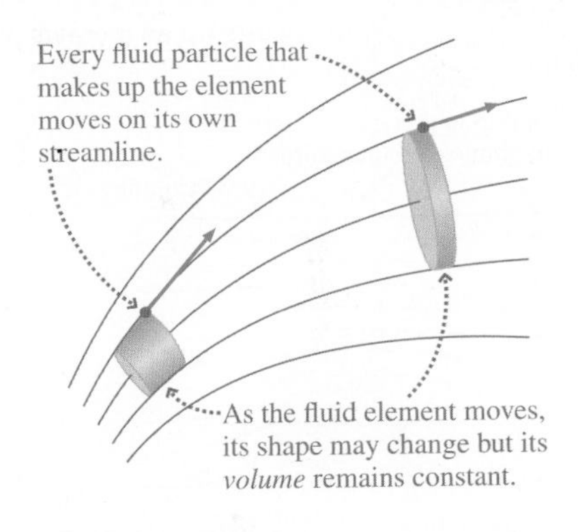

Every fluid particle that makes up the element moves on its own streamline.
As the fluid element moves, its shape may change but its *volume* remains constant.

STOP TO THINK 13.5

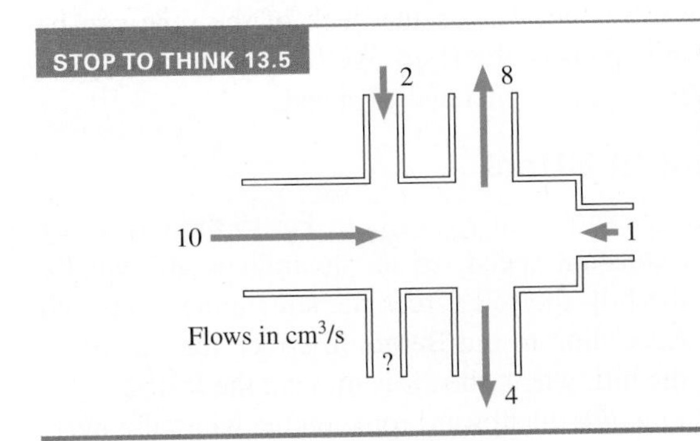

The figure shows volume flow rates (in cm^3/s) for all but one tube. What is the volume flow rate through the unmarked tube? Is the flow direction in or out?

13.5 Fluid Dynamics

The equation of continuity describes a moving fluid but doesn't tell us anything about *why* the fluid is in motion. To understand the dynamics, consider the fluid moving from left to right through the tube shown in FIGURE 13.25. The fluid moves at a steady speed v_1 in the wider part of the tube. In accordance with the equation of continuity, its speed is a higher, but steady, v_2 in the narrower part of the tube. If we follow a fluid element through the tube, we see that it undergoes an *acceleration* from v_1 to v_2 in the tapered section of the tube. According to Newton's second law, there must be a net force acting on this fluid element to accelerate it.

FIGURE 13.25 A motion diagram of a fluid element moving through a narrowing tube.

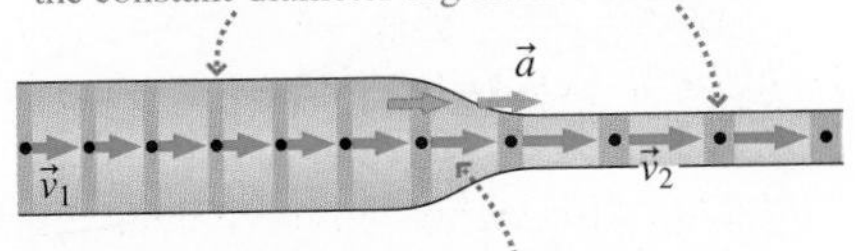

FIGURE 13.26 The net force on a fluid element due to pressure points from high to low pressure.

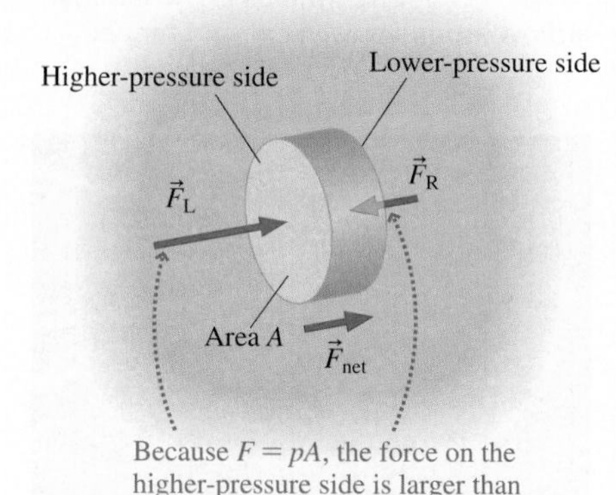

Because $F = pA$, the force on the higher-pressure side is larger than that on the lower-pressure side.

What is the origin of this force? There are no external forces, and the horizontal motion rules out gravity. Instead, the fluid element is pushed from both ends by the *surrounding fluid*—that is, by *pressure forces*. The fluid element with cross-section area A in FIGURE 13.26 has a higher pressure on its left side than on its right. Thus the force $F_L = p_L A$ of the fluid pushing on its left side—a force pushing to the right—is greater than the force $F_R = p_R A$ of the fluid on its right side. The net force, which points from the higher-pressure side of the element to the lower-pressure side, is

$$F_{net} = F_L - F_R = (p_L - p_R)A = A\,\Delta p$$

In other words, there's a net force on the fluid element, causing it to change speed, if and only if there's a pressure *difference* Δp between the two faces.

Thus, in order to accelerate the fluid elements in Figure 13.25 through the neck, the pressure p_1 in the wider section of the tube must be higher than the pressure p_2 in the narrower section. When the pressure is changing from one point in a fluid to another, we say that there is a **pressure gradient** in that region. We can also say that pressure forces are caused by pressure gradients, so if we can ignore viscosity, **a fluid accelerates wherever there is a pressure gradient.**

As a result, **the pressure is higher at a point along a streamline where the fluid is moving slower, lower where the fluid is moving faster.** This property of fluids was discovered in the 18th century by the Swiss scientist Daniel Bernoulli and is called the **Bernoulli effect.**

NOTE ▶ It is important to realize that it is the change in pressure from high to low that *causes* the fluid to speed up. A high fluid speed doesn't *cause* a low pressure any more than a fast-moving particle causes the force that accelerated it. ◀

FIGURE 13.27 A Venturi tube measures flow speeds of a fluid.

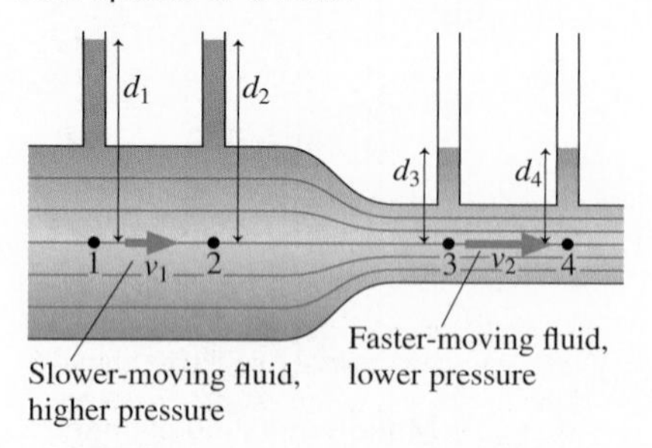

This relationship between pressure and fluid speed can be used to measure the speed of a fluid with a device called a *Venturi tube*. FIGURE 13.27 shows a simple Venturi tube suitable for a flowing liquid. The high pressure at point 1, where the fluid is moving slowly, causes the fluid to rise in the vertical pipe to a total height d_1. Because there's no vertical motion of the fluid, we can use the hydrostatic pressure equation to find that the pressure at point 1 is $p_1 = p_0 + \rho g d_1$. At point 3, on the same streamline, the fluid is moving faster and the pressure is lower; thus the fluid in the vertical pipe rises to a lower height d_3. The pressure *difference* is $\Delta p = \rho g(d_1 - d_3)$, so the pressure difference across the neck of the pipe can be found by measuring the difference in heights of the fluid. We'll see later in this section how to relate this pressure difference to the increase of speed.

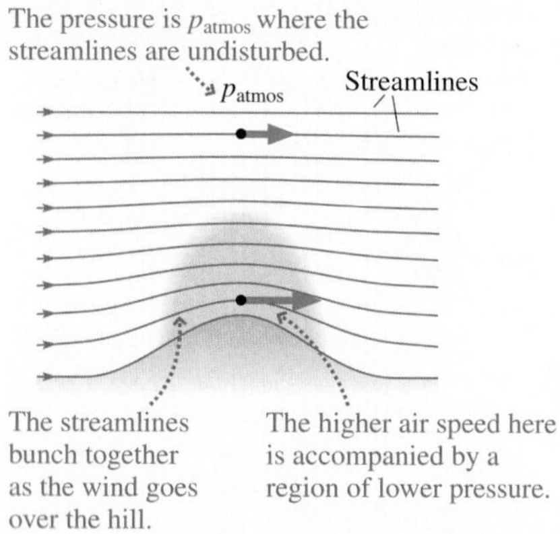

FIGURE 13.28 Wind speeds up as it crests a hill.

The pressure is p_{atmos} where the streamlines are undisturbed.

p_{atmos}

Streamlines

The streamlines bunch together as the wind goes over the hill.

The higher air speed here is accompanied by a region of lower pressure.

Applications of the Bernoulli Effect

Consider the flow of air over a hill, as shown in FIGURE 13.28. Far to the left, away from the hill, the wind blows at a constant speed, so its streamlines are equally spaced. But as the air moves over the hill, the hill forces the streamlines to bunch together so that the air speeds up. According to the Bernoulli effect, there exists a zone of *low pressure* at the crest of the hill, where the air is moving the fastest.

Using these ideas, we can understand *lift*, the upward force on the wing of a moving airplane that makes flight possible. FIGURE 13.29 shows an airplane wing, seen in cross section, for an airplane flying to the left. The figure is drawn in the reference frame of the airplane, so that the wing appears stationary and the air flows past it to the right. The shape of the wing is such that, just as for the hill in Figure 13.28, the streamlines of the air must squeeze together as they pass over the wing. This increased speed is, by the Bernoulli effect, accompanied by a region of low pressure above the wing. The air speed below the wing is actually slowed, so that there's a region of high pressure below. This high pressure pushes up on the wing, while the low-pressure air above it presses down, but less strongly. The result is a net upward force—the lift. The same principles are at work on the wings of soaring birds. The lift force results from the motion of air over the wings, so the albatross in the chapter-opening photo must build up some speed before it generates enough lift to fly.

FIGURE 13.29 Airflow over a wing generates lift by creating unequal pressures above and below.

As the air squeezes over the top of the wing, it speeds up. There is a region of low pressure associated with this faster-moving air.

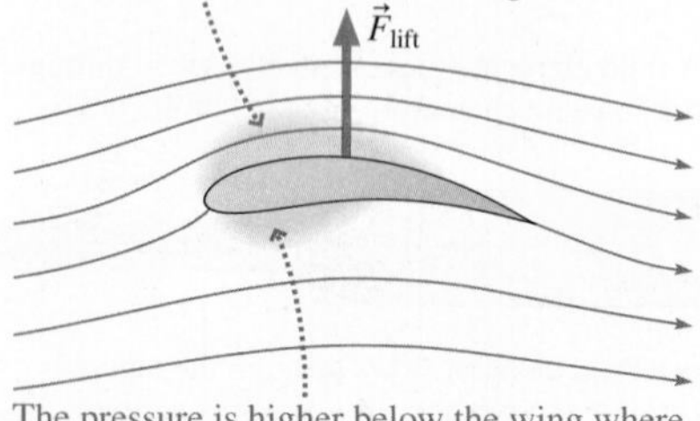

The pressure is higher below the wing where the air is moving more slowly. The high pressure below and the low pressure above result in a net upward lift force.

Bernoulli's Equation

We've seen that a pressure gradient causes a fluid to accelerate horizontally. Not surprisingly, gravity can also cause a fluid to speed up or slow down if the fluid changes elevation. These are the key ideas of fluid dynamics. Now we would like to make these ideas quantitative by finding a numerical relationship for pressure, height, and the speed of a fluid. We can do so by applying a version of the work-energy equation you learned in ◀◀ SECTION 10.6,

$$\Delta K + \Delta U = W$$

where U is the gravitational potential energy and W is the work done by other forces—in this case, pressure forces. Recall that we're still ignoring viscosity, so there's no dissipation of energy to thermal energy.

FIGURE 13.30a shows fluid flowing through a tube. The tube narrows from cross-section area A_1 to area A_2 as it bends uphill. Let's concentrate on the large volume of the fluid that is shaded in Figure 13.30a. **This moving fluid element will be our system for the purpose of applying conservation of energy.**

To use conservation of energy, we need to draw a before-and-after overview. The "before" situation is shown in FIGURE 13.30b. A short time Δt later, the system has moved along the tube a bit, as shown in the "after" drawing of FIGURE 13.30c. Because the tube is not of uniform diameter, the two ends of the fluid system do not move the same distance during Δt: The lower end moves distance Δx_1, while the upper end moves Δx_2. Thus the system moves *out of* a cylindrical volume $\Delta V_1 = A_1 \Delta x_1$ at the lower end and *into* a volume $\Delta V_2 = A_2 \Delta x_2$ at the upper end. The equation of continuity tells us that these two volumes must be the same, so $A_1 \Delta x_1 = A_2 \Delta x_2 = \Delta V$.

From the "before" situation to the "after" situation, the system *loses* the kinetic and potential energy it originally had in the volume ΔV_1 but *gains* the kinetic and potential energy in the volume ΔV_2 that it later occupies. (The energy it has in the region between these two small volumes is unchanged.) Let's find the kinetic energy in each of these small volumes. The mass of fluid in each cylinder is $m = \rho \Delta V$, where ρ is the density of the fluid. The kinetic energies of the two small volumes 1 and 2 are then

$$K_1 = \frac{1}{2}\underbrace{\rho \Delta V}_{m} v_1^2 \quad \text{and} \quad K_2 = \frac{1}{2}\underbrace{\rho \Delta V}_{m} v_2^2$$

Thus the net *change* in kinetic energy is

$$\Delta K = K_2 - K_1 = \frac{1}{2}\rho \Delta V v_2^2 - \frac{1}{2}\rho \Delta V v_1^2$$

Similarly, the net change in the gravitational potential energy of our fluid system is

$$\Delta U = U_2 - U_1 = \rho \Delta V g y_2 - \rho \Delta V g y_1$$

The final piece of our conservation of energy treatment is the work done on the system by the rest of the fluid. As the fluid system moves, positive work is done on it by the force $\vec{F}_1$ due to the pressure p_1 of the fluid to the left of the system, while negative work is done on the system by the force $\vec{F}_2$ due to the pressure p_2 of the fluid to the right of the system. In ◀◀ SECTION 10.2 you learned that the work done by a force F acting on a system as it moves through a displacement Δx is $F \Delta x$. Thus the positive work is

$$W_1 = F_1 \Delta x_1 = (p_1 A_1)\Delta x_1 = p_1(A_1 \Delta x_1) = p_1 \Delta V$$

Similarly the negative work is

$$W_2 = -F_2 \Delta x_2 = -(p_2 A_2)\Delta x_2 = -p_2(A_2 \Delta x_2) = -p_2 \Delta V$$

Thus the *net* work done on the system is

$$W = W_1 + W_2 = p_1 \Delta V - p_2 \Delta V = (p_1 - p_2)\Delta V$$

We can now use these expressions for ΔK, ΔU, and W to write the energy equation as

$$\underbrace{\frac{1}{2}\rho \Delta V v_2^2 - \frac{1}{2}\rho \Delta V v_1^2}_{\Delta K} + \underbrace{\rho \Delta V g y_2 - \rho \Delta V g y_1}_{\Delta U} = \underbrace{(p_1 - p_2)\Delta V}_{W}$$

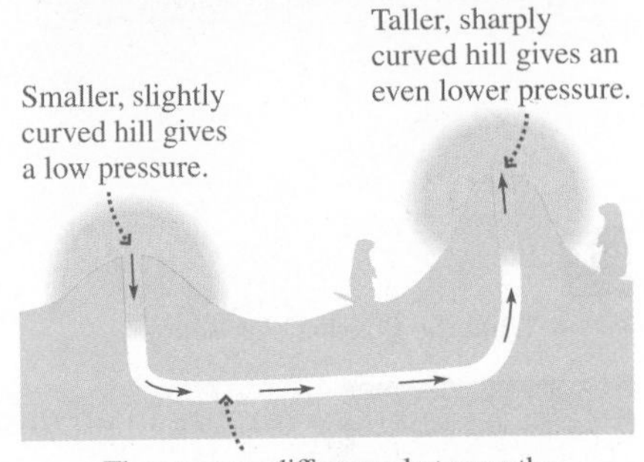

Smaller, slightly curved hill gives a low pressure.
Taller, sharply curved hill gives an even lower pressure.
The pressure difference between the two ends pushes air through the burrow.

Nature's air conditioning BIO Prairie dogs ventilate their underground burrows with the same aerodynamic forces and pressures that give airplanes lift. The two entrances to their burrows are surrounded by mounds, one higher than the other. When the wind blows across these mounds, the pressure is reduced at the top. The taller mound, with its greater curvature, has the lower pressure of the two entrances. Air then is pushed through the burrow toward this lower-pressure side.

FIGURE 13.30 A fluid flowing through a tube.

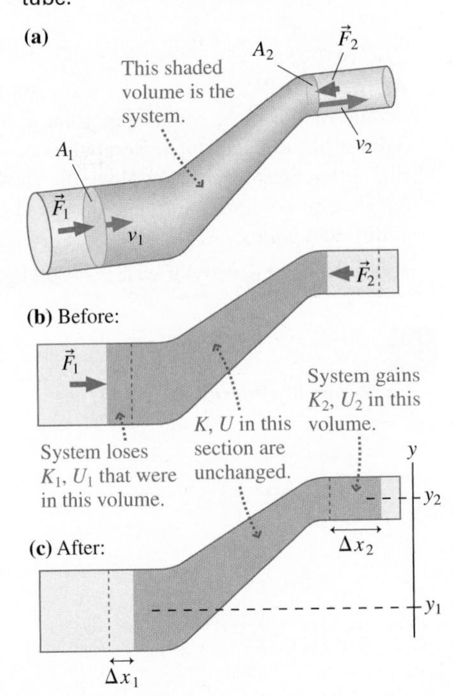

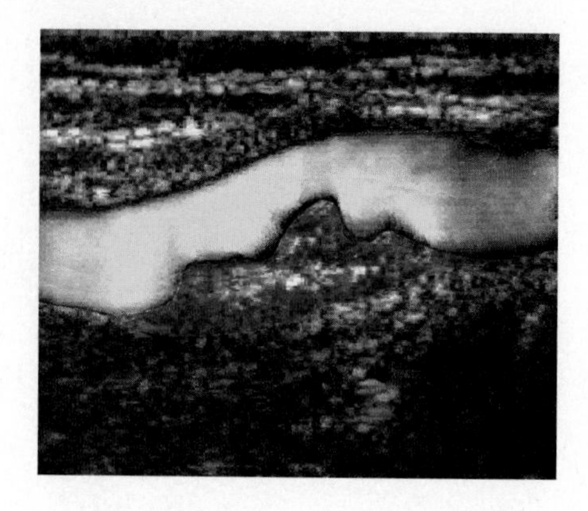

◀ **Living under pressure** BIO When plaque builds up in major arteries, dangerous drops in blood pressure can result. *Doppler ultrasound* uses sound waves to detect the velocity of flowing blood. The image shows the blood flow through a carotid artery with significant plaque buildup; yellow indicates a higher blood velocity than red. Once the velocities are known at two points along the flow, Bernoulli's equation can be used to deduce the corresponding pressure drop.

The ΔV's cancel, and we can rearrange the remaining terms to get **Bernoulli's equation,** which relates fluid quantities at two points along a streamline:

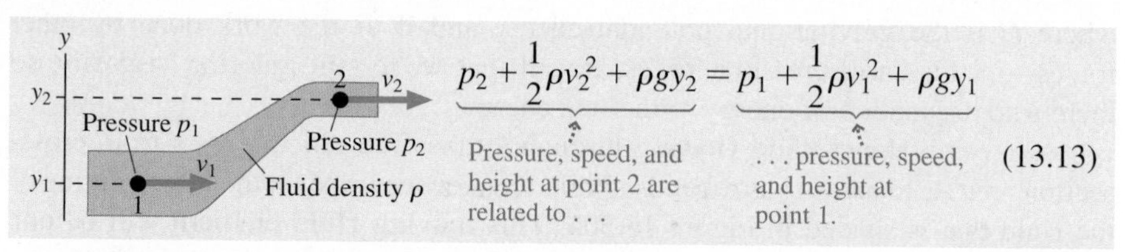

$$p_2 + \frac{1}{2}\rho v_2^2 + \rho g y_2 = p_1 + \frac{1}{2}\rho v_1^2 + \rho g y_1 \quad (13.13)$$

Video What the Physics? Air Bender

Equation 13.13, a quantitative statement of the ideas we developed earlier in this section, is really nothing more than a statement about work and energy. Using Bernoulli's equation is very much like using the law of conservation of energy. Rather than identifying a "before" and "after," we want to identify two points on a streamline.

EXAMPLE 13.11 Pressure in an irrigation system

Water flows through the pipes shown in FIGURE 13.31. The water's speed through the lower pipe is 5.0 m/s, and a pressure gauge reads 75 kPa. What is the reading of the pressure gauge on the upper pipe?

FIGURE 13.31 The water pipes of an irrigation system.

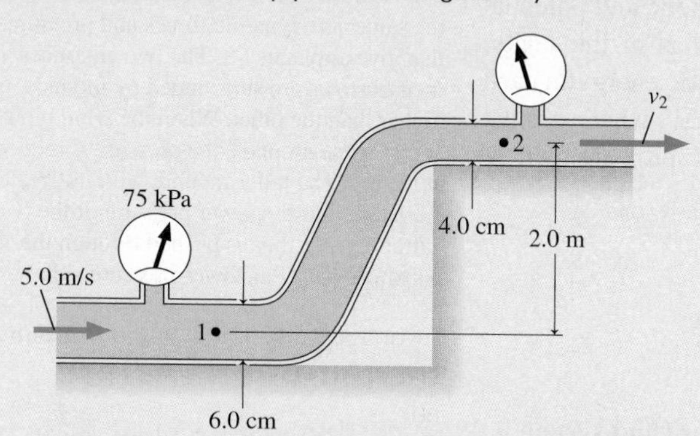

STRATEGIZE We will consider a streamline that goes along the middle of the pipe, connecting point 1 in the lower pipe with point 2 in the upper pipe. Bernoulli's equation will allow us to find differences in pressure along the flow. We can also use the equation of continuity to relate the speed of the flow to the area at different points.

PREPARE The density of water is 1000 kg/m^3.

SOLVE Bernoulli's equation, Equation 13.13, relates the pressures, fluid speeds, and heights at points 1 and 2. It can be used to solve for the pressure p_2 at point 2:

$$p_2 = p_1 + \frac{1}{2}\rho v_1^2 - \frac{1}{2}\rho v_2^2 + \rho g y_1 - \rho g y_2$$
$$= p_1 + \frac{1}{2}\rho(v_1^2 - v_2^2) + \rho g(y_1 - y_2)$$

All quantities on the right are known except v_2, and that is where the equation of continuity will be useful. The cross-section areas and water speeds at points 1 and 2 are related by

$$v_1 A_1 = v_2 A_2$$

from which we find

$$v_2 = \frac{A_1}{A_2}v_1 = \frac{r_1^2}{r_2^2}v_1 = \frac{(0.030\text{ m})^2}{(0.020\text{ m})^2}(5.0\text{ m/s}) = 11.25\text{ m/s}$$

The pressure at point 1 is $p_1 = 75\text{ kPa} + 1\text{ atm} = 176{,}300\text{ Pa}$. We can now use the above expression for p_2 to calculate $p_2 = 105{,}900\text{ Pa}$. This is the absolute pressure; the pressure gauge on the upper pipe is

$$p_2 = 105{,}900\text{ Pa} - 1\text{ atm} = 4.6\text{ kPa}$$

ASSESS We find a lower pressure at point 2; this makes sense. Reducing the pipe size decreases the pressure because it makes $v_2 > v_1$. Gaining elevation also reduces the pressure.

Video Bernoulli's Principle: Venturi Tubes

STOP TO THINK 13.6 Rank in order, from highest to lowest, the liquid heights h_1 to h_4 in tubes 1 to 4. The airflow is from left to right.

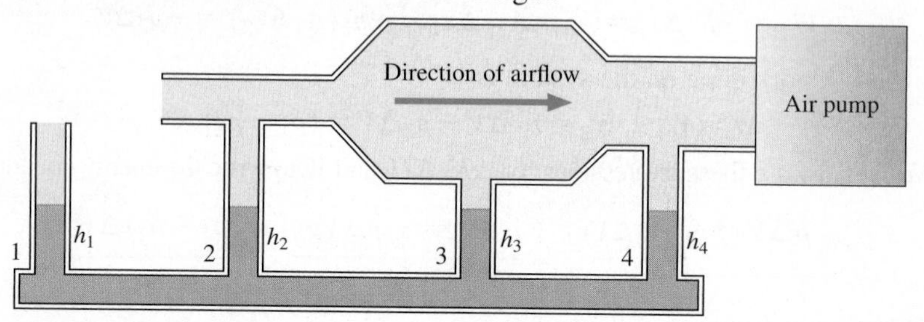

13.6 Viscosity and Poiseuille's Equation

In this section, we'll begin to include the effects of viscosity. Viscosity is important to understanding many real-world situations, from blood flow to the flight of birds to throwing a curveball.

If there is no viscosity, a fluid will "coast" at constant speed through a constant-diameter tube with no change in pressure. That's why the fluid heights are the same in the first and second pressure-measuring columns on the Venturi tube of Figure 13.27. But as FIGURE 13.32 shows, a viscous fluid requires a *pressure difference* between the ends of a tube to keep the fluid moving at constant speed. As the fluid moves, viscosity works like friction, converting some of the energy of the flow into thermal energy. The pressure difference does work to keep the energy of the flow constant. A greater viscosity means that more energy is lost and a greater pressure difference is required to maintain the flow. Think about how much harder you have to suck on a straw to drink a thick milkshake than to drink water or to pull air through the straw.

FIGURE 13.32 A viscous fluid needs a pressure difference to keep it moving.

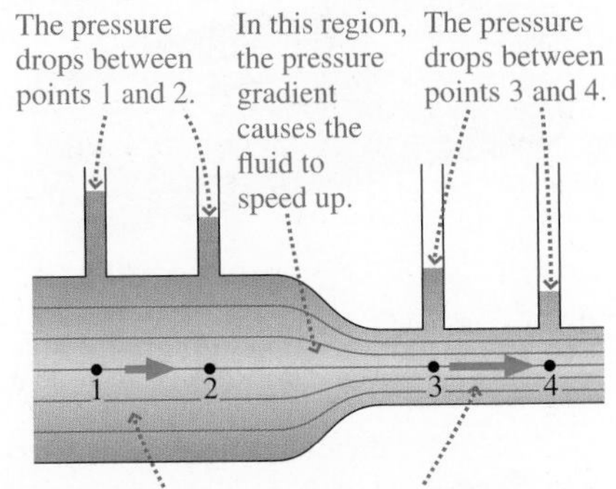
The pressure drops between points 1 and 2.
In this region, the pressure gradient causes the fluid to speed up.
The pressure drops between points 3 and 4.
1
2
3
4
In these regions, a pressure gradient is needed simply to keep the fluid moving with constant speed.

FIGURE 13.33 shows a viscous fluid flowing with a constant average speed v_{avg} through a tube of length L and cross-section area A. Experiments show that the pressure difference needed to keep the fluid moving is proportional to v_{avg} and to L and inversely proportional to A. We can write

$$\Delta p = 8\pi\eta \frac{Lv_{avg}}{A} \tag{13.14}$$

where η is the viscosity.

FIGURE 13.33 The pressure difference needed to keep the fluid flowing is proportional to the fluid's viscosity.

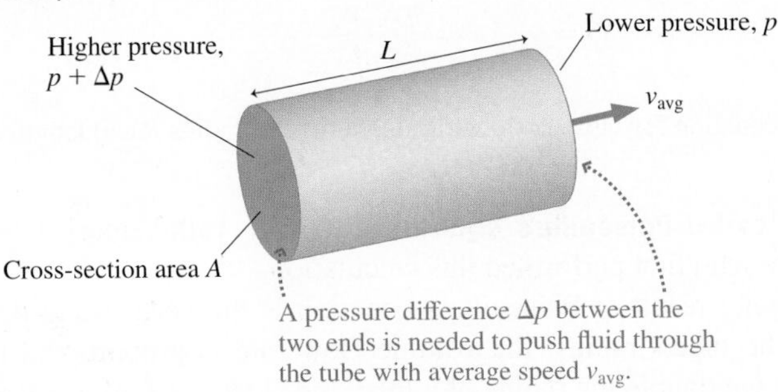
Higher pressure, $p + \Delta p$
L
Lower pressure, p
v_{avg}
Cross-section area A
A pressure difference Δp between the two ends is needed to push fluid through the tube with average speed v_{avg}.

Equation 13.14 makes sense. A more viscous fluid needs a larger pressure difference to push it through the tube; if there is no viscosity, $\eta = 0$, and the fluid will keep flowing without any pressure difference. We can also see from Equation 13.14 that the units of viscosity are Pa · s. TABLE 13.4 gives values of η for some common fluids, a wider selection than we saw in Chapter 5. Note that the viscosity of many liquids decreases *very* rapidly with temperature. Cold oil hardly flows at all, but hot oil pours almost like water.

TABLE 13.4 Viscosities of fluids

Fluid	η (Pa · s)	Fluid	η (Pa · s)
Air 20°C	1.8×10^{-5}	Motor oil:	
Water:		−30°C	3×10^{5}
20°C	1.0×10^{-3}	40°C	0.07
40°C	0.7×10^{-3}	100°C	0.01
60°C	0.5×10^{-3}	Molasses:	
		20°C	50
Whole blood 37°C	3.5×10^{-3}	50°C	2

FIGURE 13.34 Viscosity alters the velocities of the fluid particles.

(a) Ideal fluid

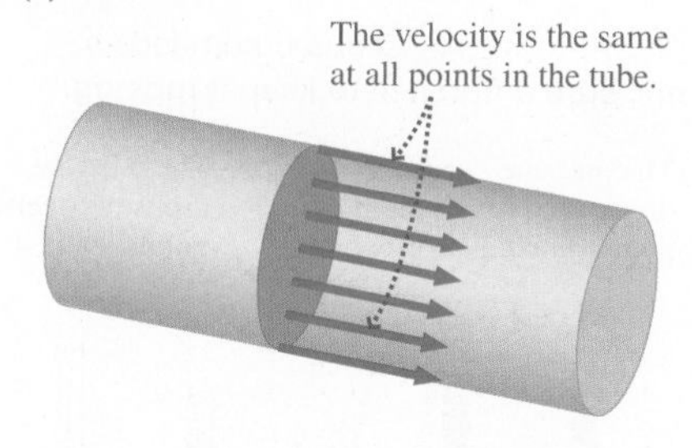
The velocity is the same at all points in the tube.

(b) Viscous fluid

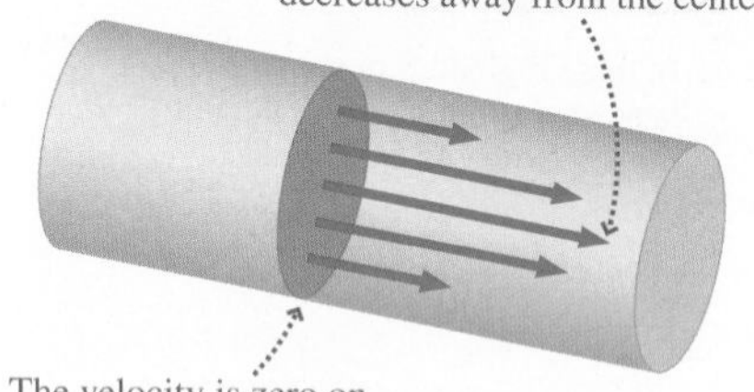
The velocity is maximum at the center of the tube. It decreases away from the center.
The velocity is zero on the walls of the tube.

Poiseuille's Equation

Viscosity has a profound effect on how a fluid moves through a tube. FIGURE 13.34a shows that in an ideal fluid, all fluid particles move with the same speed v, the speed that appears in the equation of continuity. For a viscous fluid, FIGURE 13.34b shows that the fluid moves fastest in the center of the tube. The speed decreases as we move away from the center of the tube until it reaches zero on the walls of the tube. That is, the layer of fluid in contact with the tube doesn't move at all. Whether it is water through pipes or blood through arteries, the fact that the fluid at the outer edges "lingers" and barely moves allows deposits to build on the inside walls of a tube.

Although we can't characterize the flow of a viscous liquid by a single flow speed v, we can still define the *average* flow speed. Suppose a fluid with viscosity η flows through a circular pipe with radius R and cross-section area $A = \pi R^2$. From Equation 13.14, a pressure difference Δp between the ends of the pipe causes the fluid to flow with average speed

$$v_{avg} = \frac{R^2}{8\eta L}\Delta p \qquad (13.15)$$

The average flow speed is directly proportional to the pressure difference; for the fluid to flow twice as fast, we would need to double the pressure difference between the ends of the pipe.

Equation 13.12 defined the volume flow rate $Q = \Delta V/\Delta t$ and found that $Q = vA$ for an ideal fluid. For viscous flow, where v isn't constant throughout the fluid, we simply need to replace v with the average speed v_{avg} found in Equation 13.15. Using $A = \pi R^2$ for a circular tube, we see that a pressure difference Δp causes a volume flow rate

$$Q = v_{avg}A = \frac{\pi R^4 \Delta p}{8\eta L} \qquad (13.16)$$

Poiseuille's equation for viscous flow through a tube of radius R and length L

This result is called **Poiseuille's equation** after the 19th-century French scientist Jean Poiseuille who first performed this calculation.

One surprising result of Poiseuille's equation is the very strong dependence of the flow on the tube's radius; the volume flow rate is proportional to the *fourth* power of R. If we double the radius of a tube, the flow rate will increase by a factor of $2^4 = 16$.

EXAMPLE 13.12 What causes the pressure drop?

When water goes through the small holes in a showerhead, the total area for the flow decreases, so the speed of the flow increases. The gauge pressure drops from the value in the water line behind the showerhead to zero after it exits; a greater difference in pressure permits a greater increase in speed and a more vigorous shower.

The pressure at the showerhead is less than where the supply line connects to the city's water supply; a difference in pressure is necessary to move the water through the pipe. A greater flow requires a greater drop in pressure. If someone else starts to use water when you are taking a shower, there can be a small but noticeable pressure drop. You'll notice this by the reduction in flow of your shower.

Let's look at some typical numbers. Suppose you are taking a shower after school. Your home's water travels 40 m from the city's main line to your house, through a pipe of radius 0.0072 m. The water is a chilly 10°C, so the viscosity is 1.3×10^{-3} Pa · s, slightly greater than the value given for 20°C in Table 13.4. The gauge pressure at the showerhead is 200 kPa, enough to provide a vigorous shower. You are the only person in the house using water, and the flow rate of your shower is 2.0×10^{-4} m^3/s. But then your roommate, in another bathroom, flushes a toilet. As the toilet's tank refills, the supply line is supplying a total flow of 6.0×10^{-4} m^3/s. What is the new pressure at the shower head? Assume that there is no appreciable change in height throughout the system.

STRATEGIZE We will assume that the pressure at the start of the supply line, where it meets the city water supply, is constant. We will use the flow rate and the characteristics of the pipe to find the pressure difference necessary to drive the flow. This will give us the pressure at the start of the supply line. The higher flow will require a greater pressure difference, leading to a lower pressure at the showerhead.

PREPARE All of the quantities in the problem are given in the proper SI units, so no unit conversions are necessary.

SOLVE We rearrange Equation 13.16 to give the pressure difference in terms of the flow rate:

$$\Delta p = \frac{8\eta LQ}{\pi R^4} = \frac{8(1.3 \times 10^{-3}\ \text{Pa}\cdot\text{s})(40\ \text{m})(2.0 \times 10^{-4}\ \text{m}^3/\text{s})}{\pi(0.0072\ \text{m})^4}$$
$$= 9.9\ \text{kPa} \approx 10\ \text{kPa}$$

This is the pressure difference across the supply line, so the city water supply has a pressure of 210 kPa. At the higher flow rate, the difference in pressure is

$$\Delta p = \frac{8\eta LQ}{\pi R^4} = \frac{8(1.3 \times 10^{-3}\ \text{Pa}\cdot\text{s})(40\ \text{m})(6.0 \times 10^{-4}\ \text{m}^3/\text{s})}{\pi(0.0072\ \text{m})^4}$$
$$= 30\ \text{kPa}$$

With the toilet and the shower both using water, the pressure drops by 30 kPa, so the pressure at the showerhead is

$$210\ \text{kPa} - 30\ \text{kPa} = 180\ \text{kPa}$$

ASSESS This is a small but noticeable change in pressure from the original situation, exactly as expected.

STOP TO THINK 13.7 A viscous fluid flows through the pipe shown. The three marked segments are of equal length. Across which segment is the pressure difference the greatest?

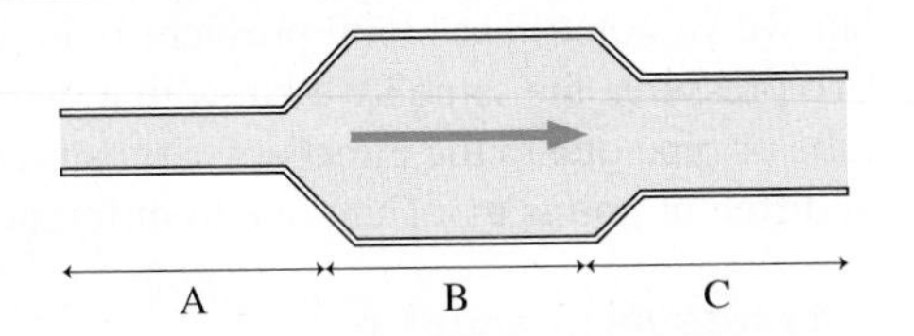

13.7 The Circulatory System BIO

Once every second, more or less, your heart beats, and the contraction of the chambers of your heart sends blood flowing through your body's arteries, capillaries, and veins. In this section, we'll use the principles that we've learned in earlier sections to explain and explore the motion of blood through your circulatory system.

We'll focus on the motion of blood through your body, from your heart's left ventricle to the right atrium; we won't examine the pulmonary circulation through your lungs. As usual, we'll make a simplified model of the system. We'll assume, as we've done before, that the fluid is incompressible (a very good approximation) and that the flow is laminar. The relationships seen earlier in the chapter will therefore apply. In the larger vessels, we'll be able to assume that the effect of viscosity is negligible; in the smaller vessels, it isn't.

FIGURE 13.35 Blood pressure during one cycle of a heartbeat.

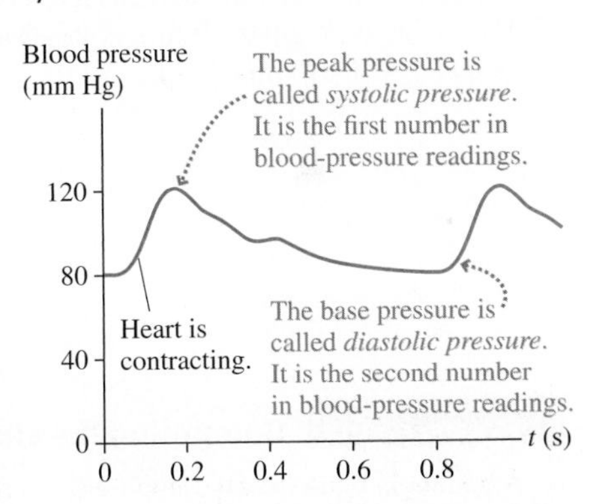

Blood Pressure

When your heart muscle contracts, the pressure of the blood increases, and blood leaves the left ventricle and moves into the large artery called the aorta. FIGURE 13.35 is a pressure graph showing how the blood pressure at the heart changes during one cycle of the heartbeat. Blood pressure is a gauge pressure, the pressure in excess of the pressure of the atmosphere, and it is traditionally measured in mm Hg. The graph shows a fairly typical blood pressure for a healthy young adult, with a *systolic* (peak) pressure of 120 mm Hg and a *diastolic* (base) pressure of 80 mm Hg—that is, a blood pressure of "120 over 80."

FIGURE 13.36 Measuring blood pressure.

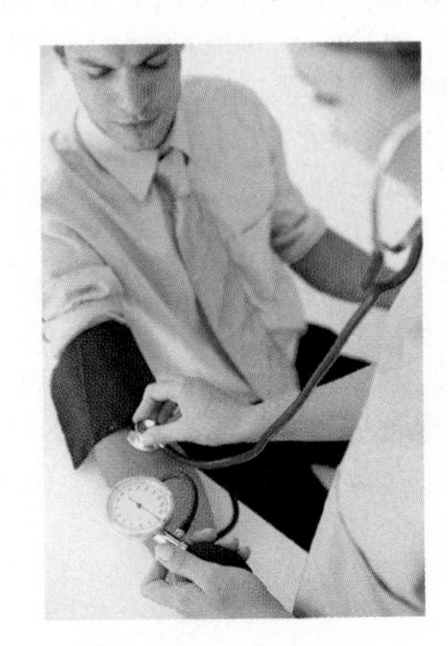

If your blood pressure is too high, it can cause many negative health effects. If your blood pressure is too low, it may indicate other health problems. For this reason, a visit to the doctor usually starts with a measurement of blood pressure. As shown in FIGURE 13.36, blood pressure is measured with a cuff that goes around the upper arm. A nurse pumps air into the cuff while using a stethoscope to listen to the flow of blood in a large artery and reading a gauge that shows the pressure in the cuff. Your arteries are flexible; if the pressure of the air in the cuff is greater than the pressure of the blood in the artery, the greater pressure outside the artery squeezes it shut, stopping blood flow. The nurse pumps air into the cuff until the pressure is well above the systolic pressure so that the cuff squeezes the artery shut and cuts off the blood flow. He or she then slowly lets air leak out of the cuff, listening for telltale sounds of blood flow in your artery. When the cuff pressure drops below the systolic pressure, the pressure pulse during each beat of your heart forces the artery to open briefly and a squirt of blood goes through. The nurse records the pressure when he or she hears the blood start to flow. This is your systolic pressure. This pulsing of the blood through your artery lasts until the cuff pressure reaches the diastolic pressure, when the artery remains open continuously and the blood flows smoothly. This transition is easily heard in the stethoscope, and the nurse records your diastolic pressure.

If we need only an average value, we'll take 100 mm Hg, the average of typical systolic and diastolic values, as a typical blood pressure. Our equations require SI units, so we'll remind you of the conversion of mm Hg to Pa:

$$1 \text{ mm Hg} = 133 \text{ Pa}$$

The clinically interesting measure for blood pressure is the pressure at the heart, but we've noted that blood pressure is measured in an artery in the arm. Are these two pressures the same? We know that there can be a difference in pressure across a tube or pipe due to the effects of viscosity; there can also be a difference in pressure at different points in a fluid due to differences in elevation.

EXAMPLE 13.13 How much does blood pressure change? BIO

A woman is lying down with her arms horizontal. Along her upper arm, her brachial artery stretches 20 cm with a diameter of 4.0 mm. Blood moves through the artery at an average speed of 9.0 cm/s. What is the pressure drop along the 20 cm length?

STRATEGIZE Any pressure difference along the artery isn't due to a height difference because the artery is horizontal. Therefore, any pressure difference must be due to the viscosity of the blood, so we will compute the pressure difference using Equation 13.14.

PREPARE The viscosity of blood is given in Table 13.4 as 3.5×10^{-3} Pa·s. We are looking for the pressure drop along 20 cm of artery, so the pressure drop along a tube of length $L = 0.20$ m. The radius is $R = 0.0020$ m and the fluid moves at an average speed $v_{avg} = 0.090$ m/s.

SOLVE Using the relevant values in Equation 13.14 gives the pressure drop as

$$\Delta p = \frac{8\pi(3.5 \times 10^{-3} \text{ Pa} \cdot \text{s})(0.20 \text{ m})(0.090 \text{ m/s})}{\pi(0.0020 \text{ m})^2}$$
$$= 130 \text{ Pa} = 0.95 \text{ mm Hg}$$

ASSESS We know that, clinically, the blood pressure in the arm is taken to be the same as the blood pressure at the heart, so we'd expect a very small difference in pressure along large arteries such as the brachial artery. And, indeed, the number that we found is a small fraction of the average blood pressure—less than 1%—and certainly small compared to the rapid fluctuations in blood pressure that occur naturally.

There is only a minimal change in pressure due to viscous losses along your arteries. But there can certainly be differences in pressure due to differences in elevation. The blood in your arteries is a connected fluid, so the pressures are lower at greater heights. In the blood pressure measurement shown in Figure 13.36, the cuff is placed on the arm at the level of the heart so that there is no difference in pressure. How much error is introduced if this is not the case?

EXAMPLE 13.14 Computing the change in blood pressure with height BIO

A patient is lying in bed. Suppose she raises her arm for a blood pressure measurement, with the cuff placed 15 cm (about 6 inches) above her heart. How much of an error does this introduce in the measurement? Ignore any pressure drop due to the viscosity of the blood.

STRATEGIZE Blood flows from the heart through the aorta and other arteries until it reaches the artery compressed by the cuff. We will use Equation 13.5 to determine the pressure at different depths in this connected fluid.

PREPARE In Equation 13.5, we can take p_0 as the pressure at the cuff; the pressure p is the pressure at the heart, $d = 0.15$ m below the cuff.

SOLVE The pressure at the cuff is lower than the pressure at the heart by

$$p - p_0 = \rho g d = (1050 \text{ kg/m}^3)(9.8 \text{ m/s}^2)(0.15 \text{ m})$$
$$= 1540 \text{ Pa} \approx 12 \text{ mm Hg}$$

ASSESS This is a significant difference—more than 10 times the difference in pressure due to viscous losses. To avoid introducing such an error, the blood pressure cuff must be placed at the level of the heart.

Blood pressure decreases with increasing height above the heart. When you are upright, the blood pressure in your brain is noticeably lower than the pressure at your heart. For animals with long necks, this difference in pressure is correspondingly greater.

▸ **Extreme blood pressure** BIO A giraffe's head can be more than 2 meters above its heart. If the blood pressure in the brain is greater than zero (as it must be), the pressure at the heart must be quite high. A typical blood pressure for a resting giraffe is 240/160 mm Hg—a number that would raise serious alarms in a human!

The Arteries and Capillaries

In the human body, blood pumped from the heart to the body starts its journey in a single large artery, the aorta. The flow then branches into smaller blood vessels, the large arteries that feed the head, the trunk, and the limbs. These branch into still smaller arteries, which then branch into a network of much smaller arterioles, which branch further into the capillaries. FIGURE 13.37 shows a schematic outline of the circulation, with average values for the diameters of the individual vessels, the total cross-section area of all of each type of vessel considered together, and the pressure in these vessels, assuming that the person is lying down so that there is no pressure change due to differences in elevation.

FIGURE 13.37 Schematic overview of blood flow in the circulatory system.

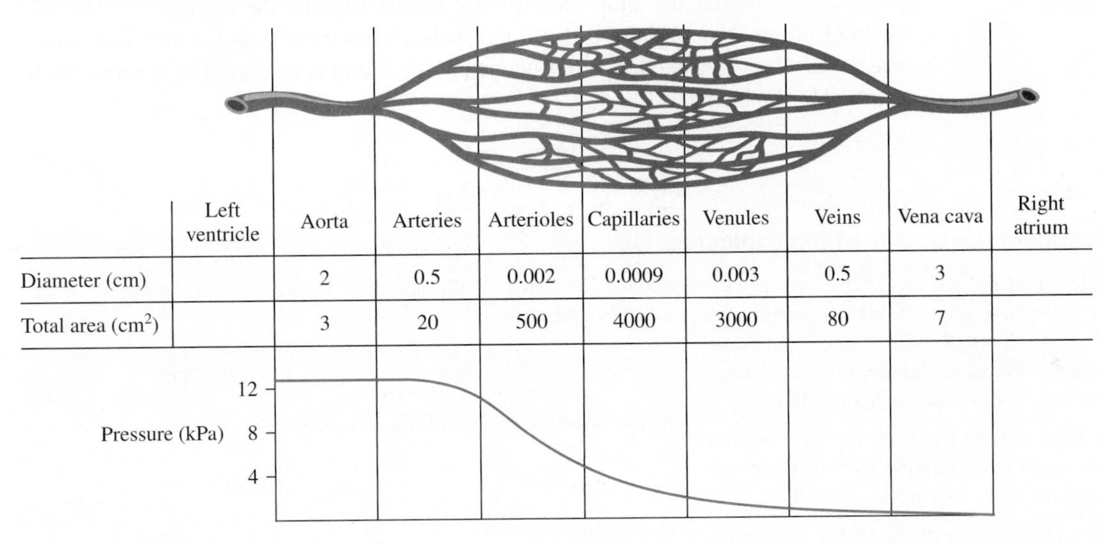

	Left ventricle	Aorta	Arteries	Arterioles	Capillaries	Venules	Veins	Vena cava	Right atrium
Diameter (cm)		2	0.5	0.002	0.0009	0.003	0.5	3	
Total area (cm^2)		3	20	500	4000	3000	80	7	

The numbers in the table are averages from different sources, representing variations among individuals of different sizes, ages, and levels of physical conditioning. The numbers are given to one significant figure; we really can't claim more precision. Nonetheless, a few broad trends emerge:

- As we've seen, there is only a very small change in pressure across the larger arteries. The pressure begins to drop only when the blood enters the smaller arteries and, even more so, the arterioles, where viscosity starts to have a significant effect. Most of the pressure drop occurs in these smaller vessels. Changes in the size of the small arteries and arterioles have significant effects on blood flow and on blood pressure.
- As blood moves from the aorta to the arteries and then to the arterioles, the diameter of the individual vessels decreases but the total area of all of the vessels increases. There is a high degree of branching; going from the aorta to the arterioles, the area of an individual vessel decreases by a factor of 1,000,000, but the area of all of the vessels considered together increases by a factor of 200. This implies that the single aorta eventually branches to 200,000,000 arterioles!

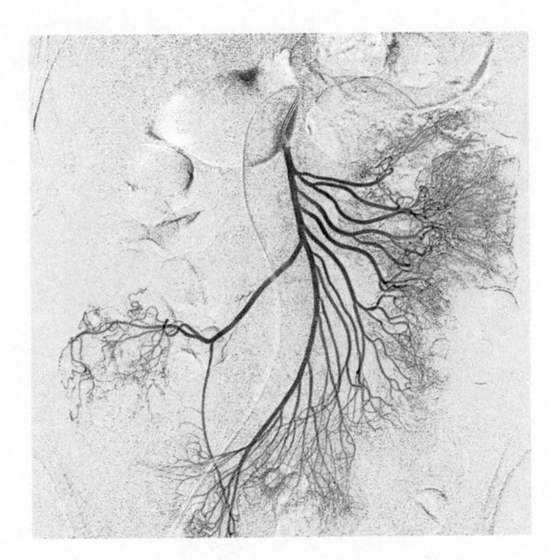

This preserved section of blood vessels shows the tremendous increase in number and in total area as blood vessels branch from large arteries to arterioles. One large artery gives rise to thousands of smaller vessels.

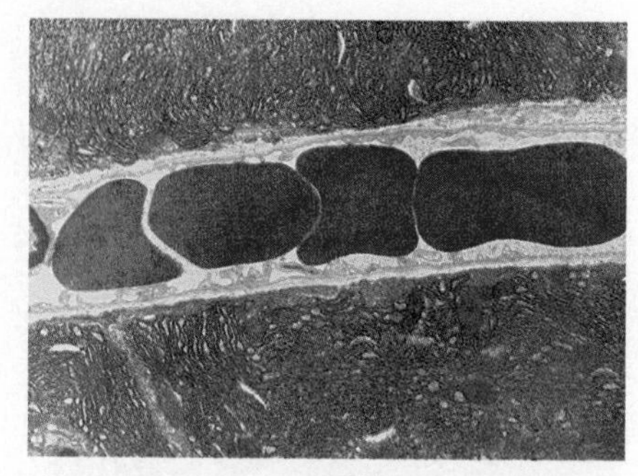

Single file, please BIO The capillaries are about as small as they can be because red blood cells can barely fit through. The red blood cells move through one after another as shown in the photo; they even deform to fit the contours of the capillary. This puts the cells close to the walls of the capillary, which enhances diffusion. This ordering actually reduces the viscosity of the blood and so reduces the pressure needed to move blood through these vessels.

- As the total cross-section area for the flow increases, the continuity equation indicates that the flow speed must decrease. As the flow branches from a large vessel to a number of smaller vessels, the area of the branches exceeds that of the initial vessel, so the speed decreases.

If we assume that 5 liters of blood move through the circulatory system in 1 minute, the volume flow rate is

$$Q = \frac{5\text{ liters}}{1\text{ minute}} = \frac{0.005\text{ m}^3}{60\text{ s}} = 8.3 \times 10^{-5}\text{ m}^3/\text{s}$$

All of this blood goes through the aorta, for which Figure 13.37 shows a diameter of 2 cm. The speed of the flow can be computed using Equation 13.12:

$$v = \frac{Q}{A} = \frac{8.3 \times 10^{-5}\text{ m}^3/\text{s}}{\pi(0.010\text{ m})^2} = 0.26\text{ m/s}$$

This is a reasonable number for the average flow speed. As the vessels branch and the total area increases, the velocity decreases, as we've discussed.

This decrease in velocity is important for the primary purpose of blood—delivering oxygen to and removing carbon dioxide from cells in the body. These gases diffuse across the thin walls of the tiniest vessels, the capillaries. The low speed of blood flow in these tiny, thin vessels allows more time for this diffusion to take place. Once gas exchange has occurred, the blood is collected by a venule to be returned to the heart.

EXAMPLE 13.15 Determining the details of the capillaries BIO

Figure 13.37 gives typical values for the diameter of a capillary as well as the total cross-section area of all the capillaries together. A typical capillary has a length of 1.0 mm. If the heart is pumping blood at a typical 5 liters per minute, how much time does it take for a red blood cell to move through a capillary?

STRATEGIZE Given the volume flow rate and the cross-section area, we can find the flow speed in the capillaries, which we will use to find the time a blood cell takes to traverse a capillary.

PREPARE The total cross-section area of all of the capillaries is given in Figure 13.37 as $4000\text{ cm}^2 = 0.40\text{ m}^2$. We showed previously that a flow rate of 5 liters per minute corresponds to $8.3 \times 10^{-5}\text{ m}^3/\text{s}$.

SOLVE We can find the speed of blood through the capillaries using Equation 13.12:

$$v = \frac{Q}{A} = \frac{8.3 \times 10^{-5}\text{ m}^3/\text{s}}{0.40\text{ m}^2} = 2.1 \times 10^{-4}\text{ m/s}$$

At this speed, it takes a blood cell a time

$$\Delta t = \frac{\Delta x}{v} = \frac{0.0010\text{ m}}{2.1 \times 10^{-4}\text{ m/s}} = 4.8\text{ s} \approx 5\text{ s}$$

to traverse a capillary.

ASSESS The relatively long time for the blood to traverse the capillary makes sense; there must be sufficient time for diffusion to occur.

We've assumed that the flow in the circulatory system is laminar. For flow in tubes, the transition between laminar and turbulent flow occurs at a Reynolds number of about 2000. For tubes, the appropriate length parameter is the diameter of the tube, so for the flow in the aorta just described, the Reynolds number is

$$Re = \frac{\rho v L}{\eta} = \frac{(1050\text{ kg/m}^3)(0.26\text{ m/s})(0.020\text{ m})}{3.5 \times 10^{-3}\text{ Pa}\cdot\text{s}} = 1600$$

This is just below the threshold for turbulence to develop. If there is narrowing in the large arteries, speeds increase, and the combination of these two factors can put *Re* above the critical value, leading to turbulent flow. A physician can detect turbulent flow from its characteristic sound. This is one thing a physician is screening for when listening to your heart; the telltale sound of turbulent flow is a sign of trouble.

EXAMPLE 13.16 **Using the difference in blood speeds to compute a blood pressure drop** BIO

If the aortic valve, through which blood exits the left ventricle, becomes narrowed, it can lead to significant health problems. One of the diagnostic criteria is a drop in pressure as the blood traverses the valve. If this pressure drop exceeds 40 mm Hg, a diagnosis of severe aortic stenosis (a stenosis is a narrowing of a blood vessel) is warranted. Clinicians can use Doppler echocardiography, a noninvasive test using ultrasound, to measure the speeds of the blood upstream and downstream of the valve. The speeds can then be used to compute a corresponding pressure drop. In one patient, as the left ventricle is ejecting blood, measurements show speeds of 1.1 m/s upstream of the valve and 3.5 m/s just past the valve. Does the patient meet the criterion for severe aortic stenosis?

STRATEGIZE In the aorta and other large blood vessels, we can ignore losses due to viscosity, so we will treat this problem using Bernoulli's equation, Equation 13.13.

PREPARE We can ignore any change in vertical position. We are looking for the difference in pressure before (p_1) and after (p_2) the valve, so Equation 13.13 reduces to

$$\Delta p = p_1 - p_2 = \frac{1}{2}\rho v_2^2 - \frac{1}{2}\rho v_1^2$$

SOLVE The drop in pressure is

$$\Delta p = p_1 - p_2 = \frac{1}{2}\rho(v_2^2 - v_1^2)$$
$$= \frac{1}{2}(1050\ \text{kg/m}^3)\Big((3.5\ \text{m/s})^2 - (1.1\ \text{m/s})^2\Big)$$
$$= 5.8\ \text{kPa} = 44\ \text{mm Hg}$$

This is above the threshold, so a positive diagnosis is warranted.

ASSESS The clinical data show a significant increase in speed and a corresponding drop in pressure as the blood exits the valve, so it's no surprise that this person has a serious condition. In a normal heart, there is some increase in speed as the blood exits the aortic valve. This causes a reduction in pressure, but normally the pressure increases again as the flow slows down on entering the aorta. If the increase in speed is too great, as it is for this patient, the flow may cross the boundary between laminar and turbulent flow, with negative consequences.

The Veins

The pressure difference along the arteries keeps the blood moving. But this doesn't work for the veins. Figure 13.37 shows that, when the blood reaches the veins, the pressure is quite low. How then does the blood make its way back to the heart? It can't be a pressure difference along the veins that keeps the blood moving; the pressures are simply too low. The veins must have other structures that keep the blood moving.

To further prove this point, let's consider the veins in your lower legs and feet. When you are lying down, the pressure in these veins is nearly zero. When you are upright, this pressure rises. If you are walking or exercising, the venous pressure rises to about 30 mm Hg. What would the pressure be if the veins were simple tubes?

EXAMPLE 13.17 **What is the pressure in the veins?** BIO

A woman's feet are 1.3 m below her heart. If we assume her veins are simple tubes, what is the blood pressure in the veins of her feet?

STRATEGIZE When the blood returns to the right atrium, the pressure is approximately zero. If the veins from the feet to the heart are simple tubes, then the blood in the veins is a connected fluid, so the pressure will increase with distance below the heart.

PREPARE We can use Equation 13.5 to find the pressure at the feet. We are looking at the pressure at $d = 1.3$ m, the distance below the heart.

SOLVE Equation 13.5 gives the venous pressure in the feet as

$$p = p_0 + \rho g d = 0 + (1050\ \text{kg/m}^3)(9.8\ \text{m/s}^2)(1.3\ \text{m})$$
$$= 13.4\ \text{kPa} = 100\ \text{mm Hg}$$

ASSESS The necessary pressure is far above what is observed, so the veins can't be simple tubes; there must be more to the story.

The veins of the legs, and other veins in the body, have one-way valves. The contraction of muscles surrounding the veins squeezes them, thereby moving the blood. The valves ensure that this motion is toward the heart, and they prevent the blood from moving back down the veins when the muscles relax. If you sit or stand for a long time, this system can't do its job, and blood can collect in your legs. (Your veins are quite expandable; they change size to accommodate changes in blood volume and blood flow.) At some point, you'll want to get up and move around, using your leg muscles, which pushes the blood back toward the heart.

◀ **Walking improves the circulation** BIO A horse has no muscles in its lower legs or hoofs that aid in blood return to the heart. Instead, the normal force of the ground on each hoof compresses a network of veins in the hoof, pushing blood upward, where valves in the veins prevent return flow. When a horse walks, the changing forces on each hoof make this system work as a pump, efficiently returning blood to the heart.

INTEGRATED EXAMPLE 13.18 An intravenous transfusion BIO

At the hospital, a patient often receives fluids via an intravenous (IV) infusion. A bag of the fluid is held at a fixed height above the patient's body. The fluid then travels down a large-diameter, flexible tube to a catheter—a short tube with a small diameter—inserted into the patient's vein.

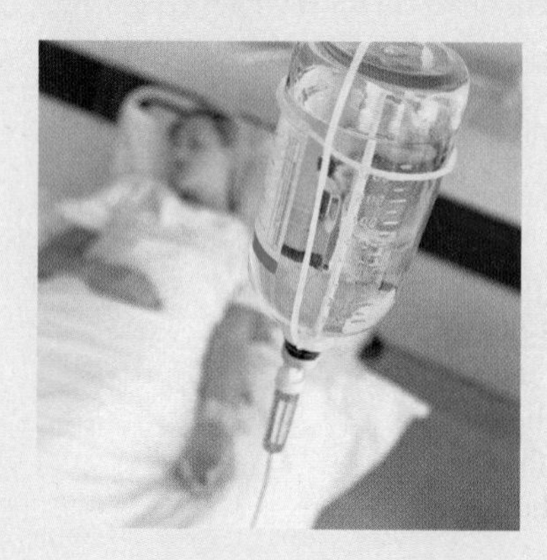

1.0 L of saline solution, with a density of 1020 kg/m^3 and a viscosity of 1.1×10^{-3} Pa·s, is to be infused into a patient in 8.0 h. The catheter is 30 mm long and has an inner diameter of 0.30 mm. The pressure in the patient's vein is 10 mm Hg. How high above the patient should the bag be positioned to get the desired flow rate?

STRATEGIZE We're concerned with the flow of a viscous fluid. According to Poiseuille's equation, the flow rate depends inversely on the fourth power of a tube's radius. The tube from the elevated bag to the catheter has a large diameter, while the diameter of the catheter is small. Thus, we expect the flow rate to be determined entirely by the flow through the narrow catheter; the wide tube has a negligible effect on the rate. We will determine the pressure difference across the catheter that is necessary to provide the needed flow. Then we will use this to determine the necessary pressure in the bag, and thus the height at which it must be placed.

PREPARE FIGURE 13.38 shows a sketch of the situation, defines variables, and lists the known information.

To use Poiseuille's equation for the catheter, we need the pressure difference Δp between the ends of the catheter. We know the pressure on the end of the catheter in the patient's vein is $p_v = 10$ mm Hg or, converting to SI units using Table 13.2,

$$p_v = 10 \text{ mm Hg} \times \frac{101 \times 10^3 \text{ Pa}}{760 \text{ mm Hg}} = 1330 \text{ Pa}$$

This is a gauge pressure, the pressure in excess of 1 atm. The pressure on the fluid side of the catheter is due to the hydrostatic pressure of the saline solution filling the bag and the flexible tube leading to the catheter. This pressure is given by the hydrostatic pressure equation $p = p_0 + \rho g d$, where d is the "depth" of the catheter below the bag. Thus we'll use Δp to find d.

SOLVE The desired volume flow rate is

$$Q = \frac{\Delta V}{\Delta t} = \frac{1.0 \text{ L}}{8.0 \text{ h}} = 0.125 \text{ L/h}$$

FIGURE 13.38 Visual overview of an IV transfusion.

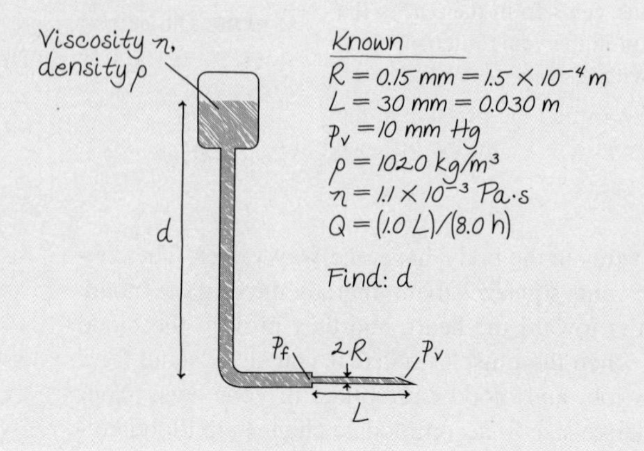

Converting to SI units using 1.0 L = 1.0×10^{-3} m^3, we have

$$Q = 0.125\,\frac{\text{L}}{\text{h}} \times \frac{1.0 \times 10^{-3}\ \text{m}^3}{\text{L}} \times \frac{1\ \text{h}}{3600\ \text{s}} = 3.47 \times 10^{-8}\ \text{m}^3/\text{s}$$

Poiseuille's equation for viscous fluid flow is

$$Q = \frac{\pi R^4 \Delta p}{8\eta L}$$

Thus the pressure difference needed between the ends of the tube to produce the desired flow rate Q is

$$\Delta p = \frac{8\eta L Q}{\pi R^4}$$

$$= \frac{8(1.1 \times 10^{-3}\ \text{Pa}\cdot\text{s})(0.030\ \text{m})(3.47 \times 10^{-8}\ \text{m}^3/\text{s})}{\pi(1.5 \times 10^{-4}\ \text{m})^4}$$

$$= 5760\ \text{Pa}$$

Now Δp is the difference between the fluid pressure p_f at one end of the catheter and the vein pressure p_v at the other end: $\Delta p = p_f - p_v$. We know p_v, so

$$p_f = p_v + \Delta p = 1330\ \text{Pa} + 5760\ \text{Pa} = 7090\ \text{Pa}$$

This pressure, like the vein pressure, is a gauge pressure. The true hydrostatic pressure at the catheter is $p = 1\ \text{atm} + 7090\ \text{Pa}$. But the hydrostatic pressure in the fluid is

$$p = p_0 + \rho g d = 1\ \text{atm} + \rho g d$$

We see that $\rho g d = 7090$ Pa. Solving for d, we find the required elevation of the bag above the patient's arm:

$$d = \frac{p_f}{\rho g} = \frac{7090\ \text{Pa}}{(1020\ \text{kg/m}^3)(9.8\ \text{m/s}^2)} = 0.71\ \text{m}$$

ASSESS This height of almost a meter seems reasonable for the height of an IV bag. In practice, the bag can be raised or lowered to adjust the fluid flow rate.

SUMMARY

GOAL To understand the static and dynamic properties of fluids.

GENERAL PRINCIPLES

Fluid Statics

Gases

- Freely moving particles
- Compressible
- Pressure mainly due to particle collisions with walls

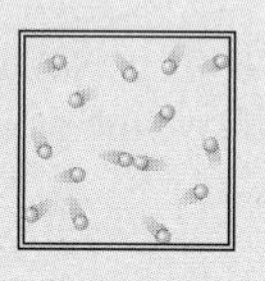

Liquids

- Loosely bound particles
- Incompressible
- Pressure due to the weight of the liquid
- Hydrostatic pressure at depth d is $p = p_0 + \rho g d$
- The pressure is the same at all points on a horizontal line through a liquid (of one kind) in hydrostatic equilibrium

Fluid Dynamics

We assume that fluids are incompressible, flow is laminar

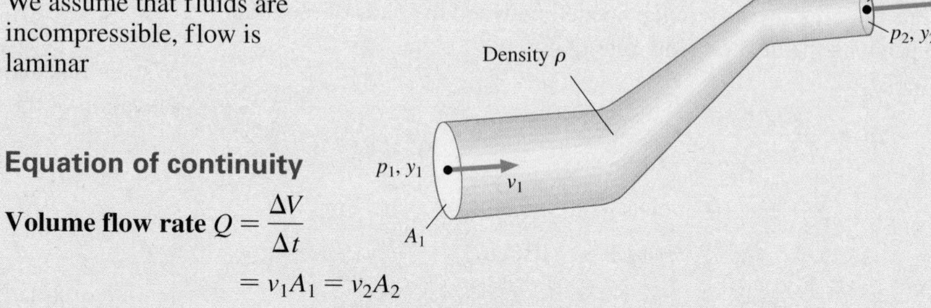

Equation of continuity

Volume flow rate $Q = \dfrac{\Delta V}{\Delta t}$

$= v_1 A_1 = v_2 A_2$

Bernoulli's equation is a statement of energy conservation:

$$p_1 + \frac{1}{2}\rho v_1{}^2 + \rho g y_1 = p_2 + \frac{1}{2}\rho v_2{}^2 + \rho g y_2$$

Poiseuille's equation governs viscous flow through a tube:

$$Q = v_{avg} A = \frac{\pi R^4 \Delta p}{8\eta L}$$

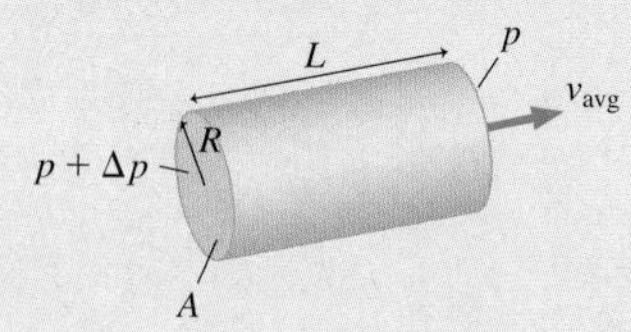

IMPORTANT CONCEPTS

Density $\rho = m/V$, where m is mass and V is volume.

Pressure $p = F/A$, where F is force magnitude and A is the area on which the force acts.

- Pressure exists at all points in a fluid.
- Pressure pushes equally in all directions.
- Gauge pressure $p_g = p - 1$ atm.

Viscosity η is the property of a fluid that makes it resist flowing.

Representing fluid flow
Streamlines are the paths of individual fluid particles.

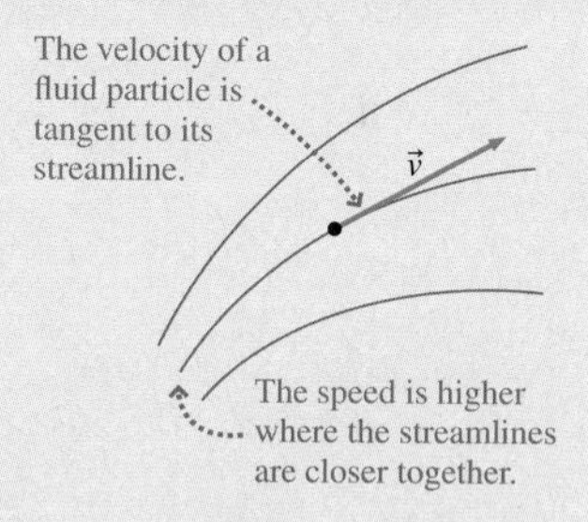

Fluid elements contain a fixed volume of fluid. Their shape may change as they move.

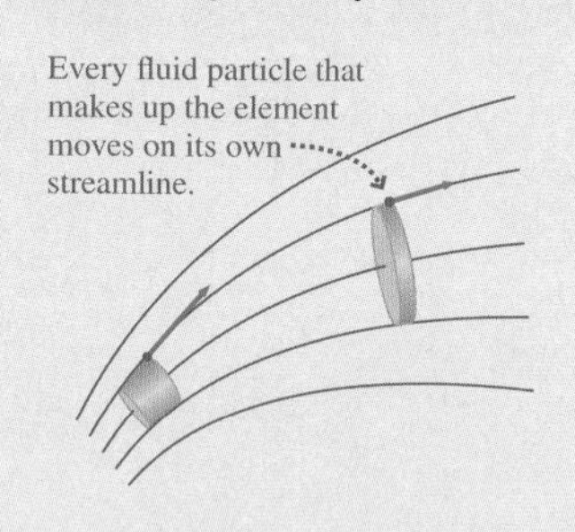

Every fluid particle that makes up the element moves on its own streamline.

APPLICATIONS

Buoyancy is the upward force of a fluid on an object immersed in the fluid.

Archimedes' principle: The magnitude of the buoyant force equals the weight of the fluid displaced by the object.

Sink:	$\rho_{avg} > \rho_f$	$F_B < w_o$
Float:	$\rho_{avg} < \rho_f$	$F_B > w_o$
Neutrally buoyant:	$\rho_{avg} = \rho_f$	$F_B = w_o$

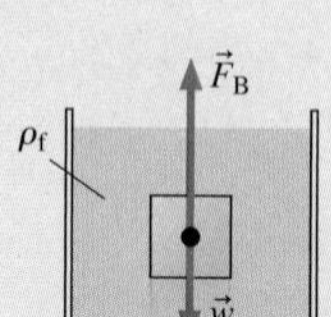

Barometers measure atmospheric pressure. Atmospheric pressure is related to the height of the liquid column by $p_{atmos} = \rho g h$.

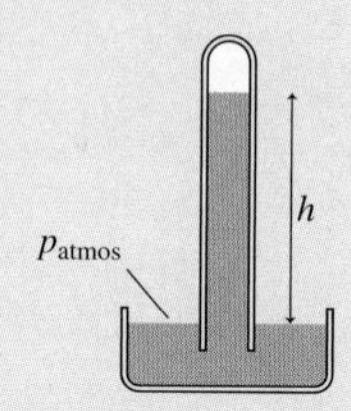

Our understanding of fluids can be applied to the motion of blood in the **circulatory system.**

- The flow is laminar under normal circumstances.
- Blood in the large arteries forms a connected fluid, so height differences matter, but viscosity can be ignored.
- In the small arteries and arterioles, viscosity is a major factor.

Learning Objectives After studying this chapter, you should be able to:

- Work with the density of gases and liquids. *Conceptual Questions 13.2, 13.3; Problems 13.2, 13.3, 13.4, 13.5, 13.6*
- Calculate and use the pressure in a liquid. *Conceptual Questions 13.8, 13.9; Problems 13.8, 13.9, 13.10, 13.17, 13.18*
- Use Archimedes' principle to understand buoyancy. *Conceptual Questions 13.19, 13.20; Problems 13.23, 13.24, 13.25, 13.26, 13.27*
- Use the equation of continuity to solve problems about fluid flow. *Conceptual Questions 13.28, 13.29; Problems 13.32, 13.33, 13.34, 13.35*
- Use Bernoulli's and Poiseuille's equations to solve problems about fluid dynamics. *Conceptual Questions 13.30, 13.31; Problems 13.35, 13.39, 13.41, 13.42, 13.45*
- Understand the motion of blood in the circulatory system. *Problems 13.47, 13.48, 13.49, 13.50, 13.51*

STOP TO THINK ANSWERS

Chapter Preview Stop to Think: C. Consider books 2 and 3 to be a single object in equilibrium. Then the downward weight force of these two books must be balanced by the upward normal force of book 1 on book 2, so that the normal force is equal to the weight of *two* books.

Stop to Think 13.1: $\rho_1 = \rho_2 = \rho_3$. Density depends only on what the object is made of, not how big the pieces are.

Stop to Think 13.2: C. Because points 1 and 2 are at the same height and are connected by the same fluid (water), they must be at the same pressure. This means that these two points can support the same weight of fluid above them. The oil is less dense than water, so this means a taller column of oil can be supported on the left side than of water on the right side.

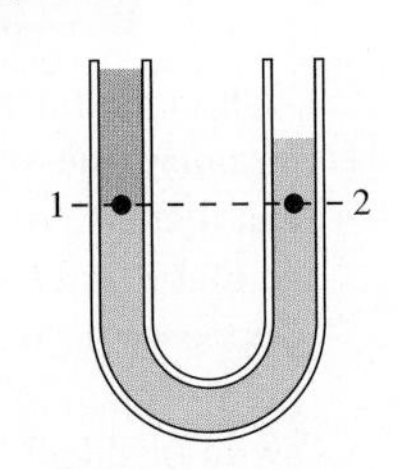

Stop to Think 13.3: $(F_B)_D > (F_B)_A = (F_B)_B = (F_B)_C > (F_B)_E$. The buoyant force on an object is equal to the weight of the fluid it displaces. Block D has the greatest volume and thus displaces the greatest amount of fluid. Blocks A, B, and C have equal volumes, so they displace equal amounts of fluids, but less than D displaces. And block E has the smallest volume and displaces the least amount of fluid. The buoyant force on an object does not depend on the object's mass, or its position in the fluid.

Stop to Think 13.4: B. The weight of the displaced water equals the weight of the ice cube. When the ice cube melts and turns into water, that amount of water will exactly fill the volume that the ice cube is now displacing.

Stop to Think 13.5: 1 cm^3/s out. The fluid is incompressible, so the sum of what flows in must match the sum of what flows out. 13 cm^3/s is known to be flowing in while 12 cm^3/s flows out. An additional 1 cm^3/s must flow out to achieve balance.

Stop to Think 13.6: $h_2 > h_4 > h_3 > h_1$. The liquid level is higher where the pressure is lower. The pressure is lower where the flow speed is higher. The flow speed is highest in the narrowest tube, zero in the open air.

Stop to Think 13.7: A. All three segments have the same volume flow rate Q. According to Poiseuille's equation, the segment with the smallest radius R has the greatest pressure difference Δp.

Video Tutor Solution Chapter 13

QUESTIONS

Conceptual Questions

1. Which has the greater density, 1 g of mercury or 1000 g of water?
2. A 1×10^{-3} m^3 chunk of material has a mass of 3 kg.
   a. What is the material's density?
   b. Would a 2×10^{-3} m^3 chunk of the same material have the same mass? Explain.
   c. Would a 2×10^{-3} m^3 chunk of the same material have the same density? Explain.
3. Object 1 has an irregular shape. Its density is 4000 kg/m^3.
   a. Object 2 has the same shape and dimensions as object 1, but it is twice as massive. What is the density of object 2?
   b. Object 3 has the same mass and the same *shape* as object 1, but its size in all three dimensions is twice that of object 1. What is the density of object 3?
4. BIO When you get a blood transfusion the bag of blood is held above your body, but when you donate blood the collection bag is held below. Why is this?
5. BIO To explore the bottom of a 10-m-deep lake, your friend Tom proposes to get a long garden hose, put one end on land and the other in his mouth for breathing underwater, and descend into the depths. Susan, who overhears the conversation, reacts with horror and warns Tom that he will not be able to inhale when he is at the lake bottom. Why is Susan so worried?
6. If you let go of a helium balloon, it quickly rises. As it rises, the balloon gets larger and larger until it pops. Why does the balloon expand as it rises?

Problem difficulty is labeled as | (straightforward) to ||||| (challenging). INT Problems labeled integrate significant material from earlier chapters; BIO are of biological or medical interest.

The eText icon indicates when there is a video tutor solution available for the chapter or for a specific problem. To launch these videos, log into your eText through Mastering™ Physics or log into the Study Area.

7. Scuba divers are warned that if they must make a rapid ascent, they should exhale on the way up. If a diver rapidly ascends to the surface with lungs full of air, his lungs could be damaged. Explain why this is so.
8. Rank in order, from largest to smallest, the pressures at A, B, and C in Figure Q13.8. Explain.
9. Refer to Figure Q13.8. Rank in order, from largest to smallest, the pressures at D, E, and F. Explain.
10. A steel cylinder at sea level contains air at a high pressure. Attached to the tank are two gauges, one that reads absolute pressure and one that reads gauge pressure. The tank is then brought to the top of a mountain. For each of the two gauges, does the pressure reading decrease, increase, or remain the same? Does the difference between the two readings decrease, increase, or remain the same? Explain.

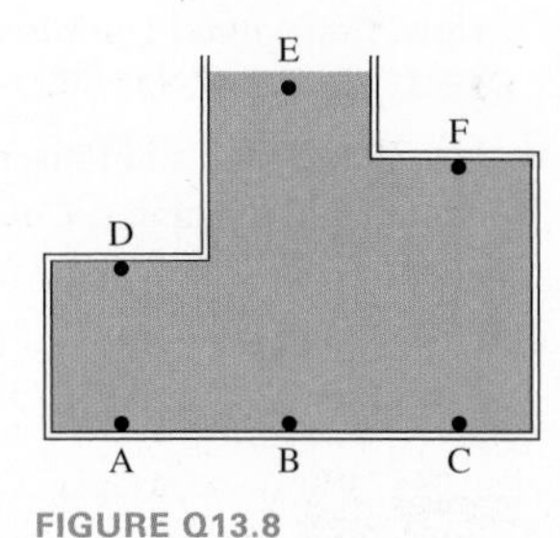

FIGURE Q13.8

11. In Figure Q13.11, A and B are rectangular tanks full of water. They have equal heights and equal side lengths (the dimension into the page), but different widths.
    a. Compare the forces the water exerts on the bottoms of the tanks. Is F_A larger, smaller, or equal to F_B? Explain.
    b. Compare the forces the water exerts on the sides of the tanks. Is F_A larger, smaller, or equal to F_B? Explain.

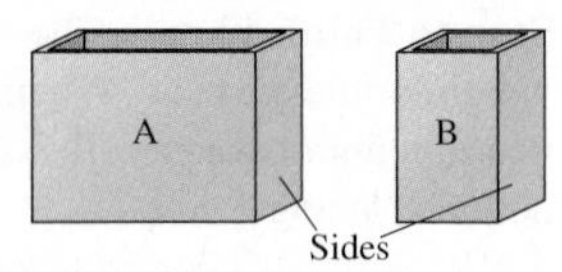

FIGURE Q13.11

12. Imagine a square column of the atmosphere, 1 m on a side, that extends all the way to the top of the atmosphere. How much does this column of air weigh in newtons?
13. Water expands when heated. Suppose a beaker of water is heated from 10°C to 90°C. Does the pressure at the bottom of the beaker increase, decrease, or stay the same? Explain.
14. In Figure Q13.14, is p_A larger, smaller, or equal to p_B? Explain.
15. A beaker of water rests on a scale. A metal ball is then lowered into the beaker using a string tied to the ball. The ball doesn't touch the sides or bottom of the beaker, and no water spills from the beaker. Does the scale reading decrease, increase, or stay the same? Explain.

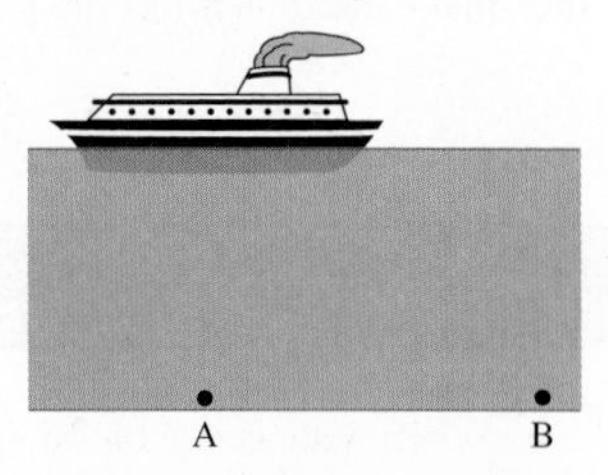

FIGURE Q13.14

16. Rank in order, from largest to smallest, the densities of objects A, B, and C in Figure Q13.16. Explain.

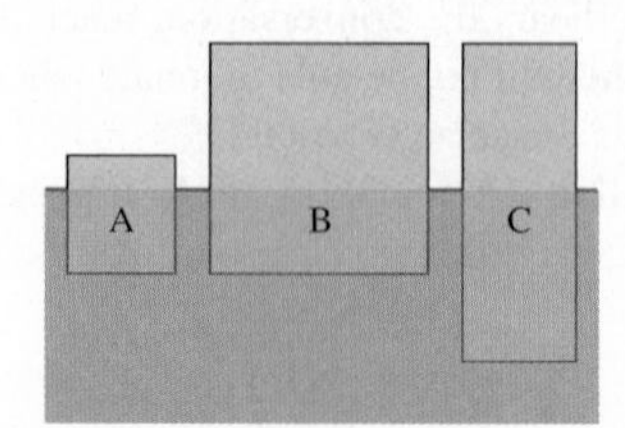

FIGURE Q13.16

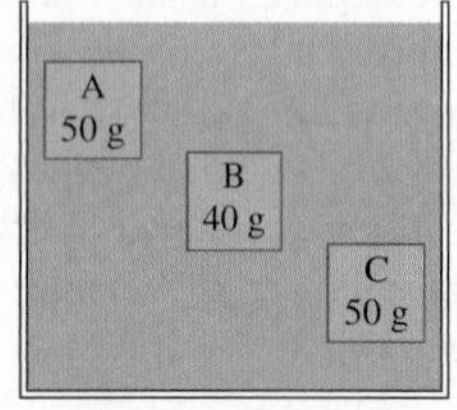

FIGURE Q13.17

17. Objects A, B, and C in Figure Q13.17 have the same volume. Rank in order, from largest to smallest, the sizes of the buoyant forces F_A, F_B, and F_C on A, B, and C. Explain.
18. Refer to Figure Q13.17. Now A, B, and C have the same density, but still have the masses given in the figure. Rank in order, from largest to smallest, the sizes of the buoyant forces on A, B, and C. Explain.
19. A heavy lead block and a light aluminum block of equal size sit at rest at the bottom of a pool of water. Is the buoyant force on the lead block greater than, less than, or equal to the buoyant force on the aluminum block? Explain.
20. When you place an egg in water, it sinks. If you start adding salt to the water, after some time the egg floats. Explain.
21. The water of the Dead Sea is extremely salty, which gives it a very high density of 1240 kg/m^3. Explain why a person floats much higher in the Dead Sea than in ordinary water.

22. BIO Fish can adjust their buoyancy with an organ called the *swim bladder.* The swim bladder is a flexible gas-filled sac; the fish can increase or decrease the amount of gas in the swim bladder so that it stays neutrally buoyant—neither sinking nor floating. Suppose the fish is neutrally buoyant at some depth and then goes deeper. What needs to happen to the volume of air in the swim bladder? Will the fish need to add or remove gas from the swim bladder to maintain its neutral buoyancy?
23. BIO Freshwater fish tend to have larger swim bladders than saltwater fish. Explain why you would expect this to be true.
24. BIO Elephant seals do deep dives to forage for food. The energy used by the seals in diving depends on their body composition; for example, fatter seals use less energy to swim to the surface at the end of a dive, allowing them to take longer, more productive dives. Explain why you'd expect this to be the case.

25. BIO A higher level of hemoglobin in the blood increases the blood's density. This is the basis for a simple test that can be used to see if a prospective blood donor has a high enough hemoglobin level to donate safely. A drop of blood is placed in a copper sulfate solution, and the time for the drop to sink to the bottom is measured. If this time is too long, the hemoglobin level is too low. Explain how this test works.
26. Early airships—craft that relied on large volumes of low-density gas to rise from the ground—were filled with hydrogen, but safety concerns necessitated a switch to helium. When an airship made the switch, how did this affect its cargo capacity? Explain.
27. The air in a hot-air balloon can be heated only so much; the temperature of the gas inside the balloon can only be so high before damaging the nylon envelope. This means that hot-air balloons can lift a larger load on a cool morning than on a warm afternoon. Explain.

28. A liquid with negligible viscosity flows through the pipe shown in Figure Q13.28. This is an overhead view.
   a. Rank in order, from largest to smallest, the flow speeds v_1 to v_4 at points 1 to 4. Explain.
   b. Rank in order, from largest to smallest, the pressures p_1 to p_4 at points 1 to 4. Explain.

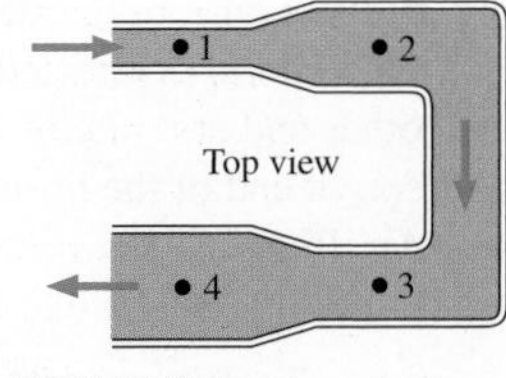

FIGURE Q13.28

29. You are looking for a deep spot to take a swim in a local river. The width of the river is about the same all along its length, but there are places where the water flows quickly and places where it flows slowly. Which is a better choice for a swim?
30. Is it possible for a fluid in a tube to flow in the direction from low pressure to high pressure? If so, give an example. If not, why not?
31. Wind blows over the house shown in Figure Q13.31. A window on the ground floor is open. Is there an airflow through the house? If so, does the air flow in the window and out the chimney, or in the chimney and out the window? Explain.

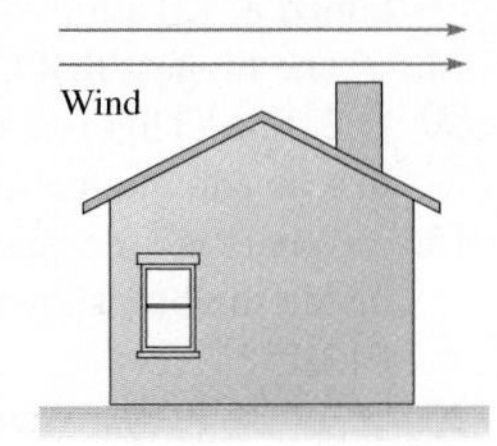

FIGURE Q13.31

Multiple-Choice Questions

32. | In Figure Q13.32, water is slowly poured into the container until the water level has risen into tubes 1, 2, and 3. The water doesn't overflow from any of the tubes. How do the water depths in the three columns compare?
   A. $d_1 > d_2 > d_3$ B. $d_1 < d_2 < d_3$
   C. $d_1 = d_2 = d_3$ D. $d_1 = d_2 > d_3$
   E. $d_1 < d_2 = d_3$

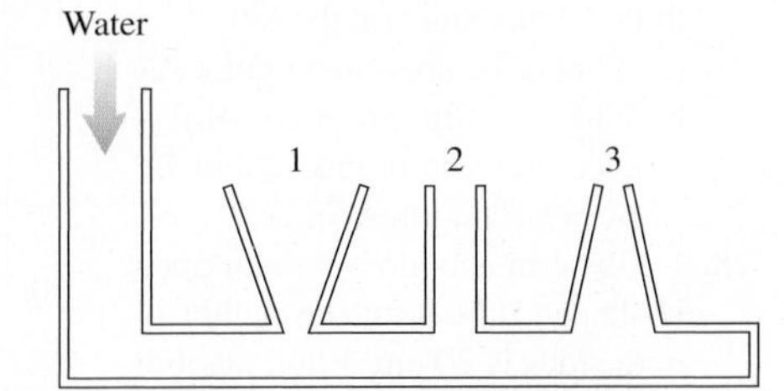

FIGURE Q13.32

33. | Figure Q13.33 shows a 100 g block of copper ($\rho = 8900\ \text{kg/m}^3$) and a 100 g block of aluminum ($\rho = 2700\ \text{kg/m}^3$) connected by a massless string that runs over two massless, frictionless pulleys. The two blocks exactly balance, since they have the same mass. Now suppose that the whole system is submerged in water. What will happen?
   A. The copper block will fall, the aluminum block will rise.
   B. The aluminum block will fall, the copper block will rise.
   C. Nothing will change.
   D. Both blocks will rise.

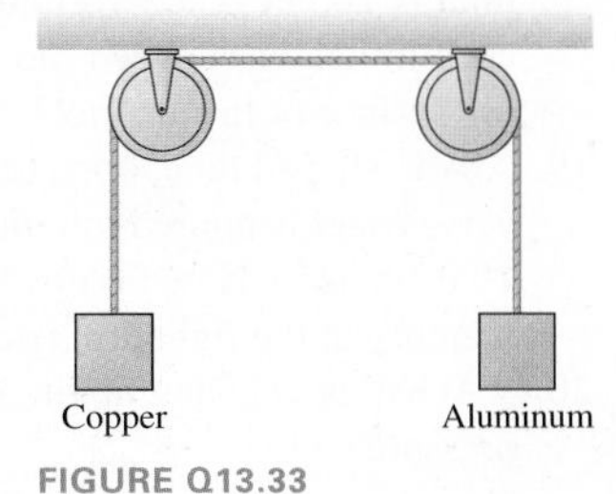

FIGURE Q13.33

34. | Masses A and B rest on very light pistons that enclose a fluid, as shown in Figure Q13.34. There is no friction between the pistons and the cylinders they fit inside. Which of the following is true?
   A. Mass A is greater. B. Mass B is greater.
   C. Mass A and mass B are the same.

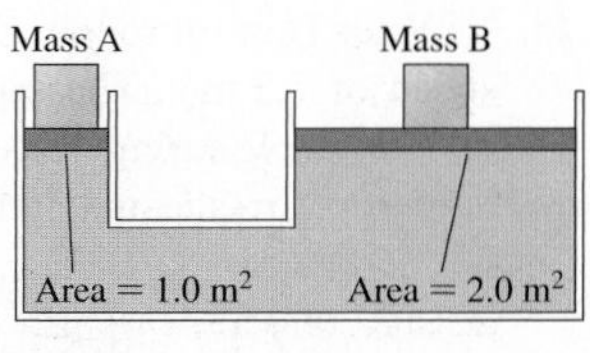

FIGURE Q13.34

35. | If you dive underwater, you notice an uncomfortable pressure on your eardrums due to the increased pressure. The human eardrum has an area of about 70 $\text{mm}^2 (7 \times 10^{-5}\ \text{m}^2)$, and it can sustain a force of about 7 N without rupturing. If your body had no means of balancing the extra pressure (which, in reality, it does), what would be the maximum depth you could dive without rupturing your eardrum?
BIO
   A. 0.3 m B. 1 m C. 3 m D. 10 m
36. || An 8.0 lb bowling ball has a diameter of 8.5 inches. When lowered into water, this ball will
   A. Float. B. Sink. C. Have neutral buoyancy.
37. | A large beaker of water is filled to its rim with water. A block of wood is then carefully lowered into the beaker until the block is floating. In this process, some water is pushed over the edge and collects in a tray. The weight of the water in the tray is
   A. Greater than the weight of the block.
   B. Less than the weight of the block.
   C. Equal to the weight of the block.
38. ||| A 55 g soapstone cube—a whisky stone—is used to chill a glass of whisky. Soapstone has a density of 3000 kg/m^3, whisky a density of 940 kg/m^3. What is the approximate normal force of the bottom of the glass on a single stone?
   A. 0.1 N B. 0.2 N C. 0.3 N D. 0.4 N
39. || At the surface of a freshwater spring, a manatee with a small amount of air in its lungs is neutrally buoyant. The manatee now dives to a depth of 10 m. At this greater depth, the buoyant force is
BIO
   A. Less than the weight force.
   B. The same as the weight force.
   C. Greater than the weight force.
40. || The density of a typical hippo's body is 1030 kg/m^3, so it will sink in fresh water. What is the buoyant force on a submerged 1500 kg hippo?
BIO
   A. 14,000 N B. 14,300 N
   C. 14,600 N D. 14,900 N
41. | An object floats in water, with 75% of its volume submerged. What is its approximate density?
   A. 250 kg/m^3 B. 750 kg/m^3
   C. 1000 kg/m^3 D. 1250 kg/m^3
42. | A syringe is being used to squirt water as shown in Figure Q13.42. The water is ejected from the nozzle at 10 m/s. At what speed is the plunger of the syringe being depressed?
   A. 0.01 m/s
   B. 0.1 m/s
   C. 1 m/s
   D. 10 m/s

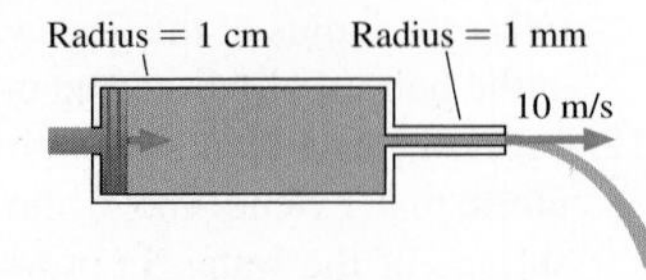

FIGURE Q13.42

43. || Water flows through a 4.0-cm-diameter horizontal pipe at a speed of 1.3 m/s. The pipe then narrows down to a diameter of 2.0 cm. Ignoring viscosity, what is the pressure difference between the wide and narrow sections of the pipe?
A. 850 Pa B. 3400 Pa C. 9300 Pa
D. 12,700 Pa E. 13,500 Pa

44. || A 15-m-long garden hose has an inner diameter of 2.5 cm. One end is connected to a spigot; 20°C water flows from the other end at a rate of 1.2 L/s. What is the gauge pressure at the spigot end of the hose?
A. 1900 Pa B. 2700 Pa C. 4200 Pa
D. 5800 Pa E. 7300 Pa

PROBLEMS

Section 13.1 Fluids and Density

1. || A 100 mL beaker holds 120 g of liquid. What is the liquid's density in SI units?
2. | A 68 kg college student's body has a typical density of 1050 kg/m^2. What is his volume? BIO
3. | A standard gold bar stored at Fort Knox, Kentucky, is 7.00 inches long, 3.63 inches wide, and 1.75 inches tall. Gold has a density of 19,300 kg/m^3. What is the mass of such a gold bar?
4. || A 5.0 kg freshwater fish at the surface of a lake is neutrally buoyant. If the density of its body with its swim bladder deflated is 1060 kg/m^3, what volume of gas must be in the swim bladder for the fish to be neutrally buoyant? BIO
5. || Air enclosed in a sphere has density $\rho = 1.4$ kg/m^3. What will the density be if the radius of the sphere is halved, compressing the air within?
6. || Fat cells in humans are composed almost entirely of pure triglycerides with an average density of about 900 kg/m^3. If 20% of the mass of a 70 kg student's body is fat (a typical value), what is the total volume of the fat in his body? BIO

Section 13.2 Pressure

7. || The deepest point in the ocean is 11 km below sea level, deeper than Mt. Everest is tall. What is the pressure in atmospheres at this depth?
8. || James Cameron piloted a submersible craft to the bottom of the Challenger Deep, the deepest point on the ocean's floor, 11,000 m below the surface. What was the total inward force on the 1.1-m-diameter pilot sphere in which Cameron sat?
9. || A tall cylinder contains 25 cm of water. Oil is carefully poured into the cylinder, where it floats on top of the water, until the total liquid depth is 40 cm. What is the gauge pressure at the bottom of the cylinder?
10. ||| A 35-cm-tall, 5.0-cm-diameter cylindrical beaker is filled to its brim with water. What is the downward force of the water on the bottom of the beaker?
11. || The gauge pressure at the bottom of a cylinder of liquid is $p_g = 0.40$ atm. The liquid is poured into another cylinder with twice the radius of the first cylinder. What is the gauge pressure at the bottom of the second cylinder?
12. || Snorkelers breathe through tubes that extend above the surface of the water. In principle, a snorkeler could go deeper with a longer tube, but the extra pressure at the greater depth would make breathing too difficult. If a snorkeler can develop a pressure in her lungs that is 10 kPa below the pressure outside her body, what is the longest snorkel she could use in the ocean?

13. ||||| A research submarine has a 20-cm-diameter window 8.0 cm thick. The manufacturer says the window can withstand forces up to 1.0×10^6 N. What is the submarine's maximum safe depth in seawater? The pressure inside the submarine is maintained at 1.0 atm.
14. ||||| The highest that George can suck water up a very long straw is 2.0 m. (This is a typical value.) What is the lowest pressure that he can maintain in his mouth? BIO
15. |||| What is the gas pressure inside the box shown in Figure P13.15?

p_{gas}
Mercury
16 cm
6.0 cm

FIGURE P13.15

16. || A city uses a water tower to store water for times of high demand. When demand is light, water is pumped into the tower. When demand is heavy, water can flow from the tower without overwhelming the pumps. To provide water at a typical 350 kPa gauge pressure, how tall must the tower be?
17. || The container shown in Figure P13.17 is filled with oil. It is open to the atmosphere on the left.
a. What is the pressure at point A?
b. What is the pressure difference between points A and B? Between points A and C?

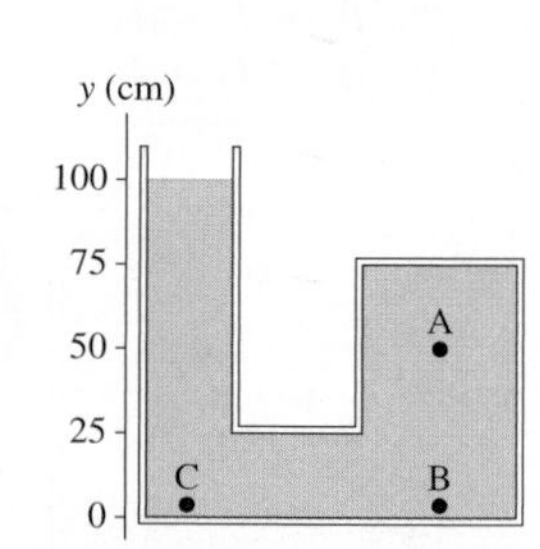

FIGURE P13.17

18. ||| Glycerin is poured into an open U-shaped tube until the height in both sides is 20 cm. Ethyl alcohol is then poured into one arm until the height of the alcohol column is 20 cm. The two liquids do not mix. What is the difference in height between the top surface of the glycerin and the top surface of the alcohol?
19. ||| A U-shaped tube, open to the air on both ends, contains mercury. Water is poured into the left arm until the water column is 10.0 cm deep. How far upward from its initial position does the mercury in the right arm rise?
20. | What is the height of a water barometer at atmospheric pressure?

Section 13.3 Buoyancy

21. ||| A cargo barge is loaded in a saltwater harbor for a trip up a freshwater river. If the rectangular barge is 3.0 m by 20.0 m and sits 0.80 m deep in the harbor, how deep will it sit in the river?

22. ‖ Hippos spend much of their lives in water, but amazingly, they don't swim. They have, like manatees, very little body fat. The density of a hippo's body is approximately 1030 kg/m^3, so it sinks to the bottom of the freshwater lakes and rivers it frequents—and then it simply walks on the bottom. A 1500 kg hippo is completely submerged, standing on the bottom of a lake. What is the approximate value of the upward normal force on the hippo?

23. ‖ A 10 cm × 10 cm × 10 cm wood block with a density of 700 kg/m^3 floats in water.
    a. What is the distance from the top of the block to the water if the water is fresh?
    b. If it's seawater?
24. ‖ The classic Goodyear blimp is essentially a helium balloon—a big one, containing 5700 m^3 of helium. If the envelope and gondola have a total mass of 4300 kg, what is the maximum cargo load when the blimp flies at a sea-level location? Assume an air temperature of 20°C.
25. ‖ What is the tension in the string in Figure P13.25?

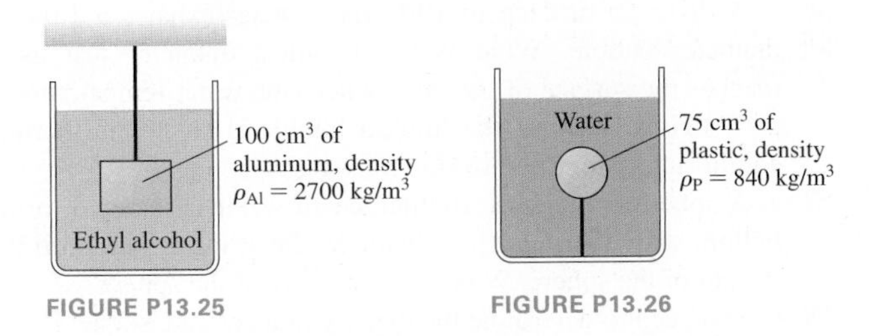

FIGURE P13.25 FIGURE P13.26

26. ‖ What is the tension in the string in Figure P13.26?
27. ‖‖ BIO To determine an athlete's body fat, she is weighed first in air and then again while she's completely underwater, as discussed on page 466. It is found that she weighs 690 N when weighed in air and 42 N when weighed underwater. What is her average density?
28. ‖ BIO Sharks are generally negatively buoyant; the upward buoyant force is less than the weight force. This is one reason sharks tend to swim continuously; water moving past their fins causes a lift force that keeps sharks from sinking. A 92 kg bull shark has a density of 1040 kg/m^3.
    a. What lift force must the shark's fins provide if the shark is swimming in seawater?
    b. Bull sharks often swim into freshwater rivers. What lift force is required in a river?
29. ‖‖ Styrofoam has a density of 32 kg/m^3. What is the maximum mass that can hang without sinking from a 50-cm-diameter Styrofoam sphere in water? Assume the volume of the mass is negligible compared to that of the sphere.
30. ‖‖ INT A 2.5 g latex balloon is filled with 2.4 g of helium. When filled, the balloon is a 30-cm-diameter sphere. When released, the balloon accelerates upward until it reaches a terminal speed. What is this speed? Assume an air density of 1.2 kg/m^3.
31. ‖‖‖ Calculate the buoyant force due to the surrounding air on a man weighing 800 N. Assume his average density is the same as that of water.

Section 13.4 Fluids in Motion

32. ‖ Your low-flow showerhead is delivering water at 1.1×10^{-4} m^3/s, about 1.8 gallons per minute. If this is the only water being used in your house, how fast is the water moving through your house's water supply line, which has a diameter of 0.021 m (about ¾ of an inch)?
33. ‖ Water flowing through a 2.0-cm-diameter pipe can fill a 300 L bathtub in 5.0 min. What is the speed of the water in the pipe?
34. ‖‖‖ A pump is used to empty a 6000 L wading pool. The water exits the 2.5-cm-diameter hose at a speed of 2.1 m/s. How long will it take to empty the pool?
35. ‖ A 1.0-cm-diameter pipe widens to 2.0 cm, then narrows to 0.50 cm. Liquid flows through the first segment at a speed of 4.0 m/s.
    a. What are the speeds in the second and third segments?
    b. What is the volume flow rate through the pipe?

Section 13.5 Fluid Dynamics

36. ‖ BIO During typical urination, a man releases about 400 mL of urine in about 30 seconds through the urethra, which we can model as a tube 4 mm in diameter and 20 cm long. Assume that urine has the same density as water, and that viscosity can be ignored for this flow.
    a. What is the flow speed in the urethra?
    b. If we assume that the fluid is released at the same height as the bladder and that the fluid is at rest in the bladder (a reasonable approximation), what bladder pressure would be necessary to produce this flow? (In fact, there are additional factors that require additional pressure; the actual pressure is higher than this.)
37. ‖ What does the top pressure gauge in Figure P13.37 read?

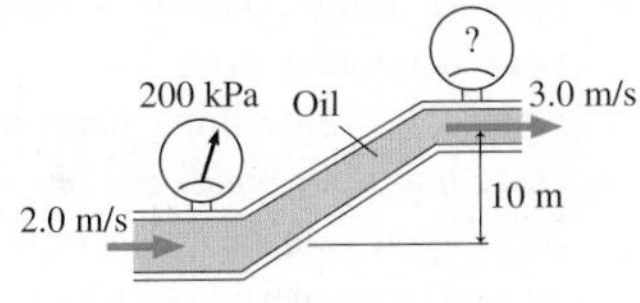

FIGURE P13.37

38. ‖‖ BIO When you urinate, you increase pressure in your bladder to produce the flow. For an elephant, gravity does the work. An elephant urinates at a remarkable rate of 0.0060 m^3 (a bit over a gallon and a half) per second. Assume that the urine exits 1.0 m below the bladder and passes through the urethra, which we can model as a tube of diameter 8.0 cm and length 1.2 m. Assume that urine has the same density as water, and that viscosity can be ignored for this flow.
    a. What is the speed of the flow?
    b. If we assume that the liquid is at rest in the bladder (a reasonable assumption) and that the pressure where the urine exits is equal to atmospheric pressure, what does Bernoulli's equation give for the pressure in the bladder? (In fact, the pressure is higher than this; other factors are at work. But you can see that no increase in bladder pressure is needed!)
39. ‖‖ A rectangular trough, 2.0 m long, 0.60 m wide, and 0.45 m deep, is completely full of water. One end of the trough has a small drain plug right at the bottom edge. When you pull the plug, at what speed does water emerge from the hole?
40. ‖‖‖ The 3.0-cm-diameter water line in Figure P13.40 splits into two 1.0-cm-diameter pipes. All pipes are circular and at the same elevation. At point A, the water speed is 2.0 m/s and the gauge pressure is 50 kPa. What is the gauge pressure at point B?

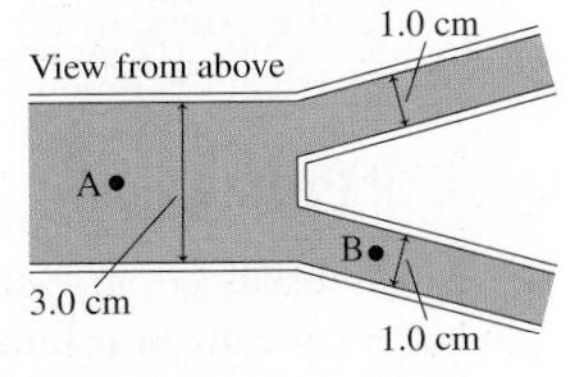

FIGURE P13.40

Section 13.6 Viscosity and Poiseuille's Equation

41. |||| What pressure difference is required between the ends of a 2.0-m-long, 1.0-mm-diameter horizontal tube for 40°C water to flow through it at an average speed of 4.0 m/s?
42. | A milkshake has a viscosity of 0.50 Pa·s. To drink this shake through a straw of diameter 0.56 cm and length 22 cm, you need to reduce the pressure at the top of the straw to less than atmospheric pressure. If you want to drain a 480 mL shake in 2.0 minutes, what pressure difference is needed? You can ignore the height difference between the top and the bottom of the straw.
43. |||| Water flows at 0.25 L/s through a 10-m-long garden hose 2.5 cm in diameter that is lying flat on the ground. The temperature of the water is 20°C. What is the gauge pressure of the water where it enters the hose?
44. | How easy is it to breathe through a straw? When you breathe BIO deeply, you pull in 4.0 L of air in about 3.0 s. This requires a pressure difference of about 4.0 kPa between the air in your lungs and the outside air. What additional pressure difference is required to pull 20°C air through a straw that is 22 cm long and 0.56 cm in diameter?
45. | Figure P13.45 shows a water-filled syringe with a 4.0-cm-long needle. What is the gauge pressure of the water at the point P, where the needle meets the wider chamber of the syringe?

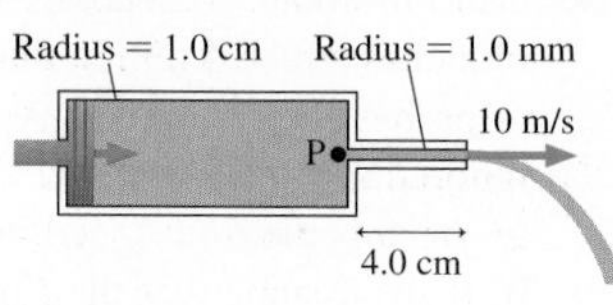

FIGURE P13.45

Section 13.7 The Circulatory System

46. | When you hold your hands at your sides, you may have noticed BIO that the veins sometimes bulge—the height difference between your heart and your hands produces increased pressure in the veins. The same thing happens in the arteries. Estimate the distance that your hands are below your heart. If the average arterial pressure at your heart is a typical 100 mm Hg, what is the average arterial pressure in your hands when they are held at your side?
47. | The top of your head is about 30 cm above your heart. What BIO is the blood pressure difference between your heart and the top of your head?
48. | To keep blood from pooling in their lower legs on plane trips, BIO some people wear compression socks. These socks are sold by the pressure they apply; a typical rating is 20 mm Hg. Over what vertical distance can this pressure move the blood?
49. | Using the data in Figure 13.37 for the venules, and assuming BIO a typical blood flow rate of 5.0 L per minute,
    a. What is the flow speed in the venules?
    b. What is the pressure difference across a 1.0 cm length of venule?
50. | Using the data in Figure 13.37 for the arterioles, and assum- BIO ing a typical blood flow rate of 5.0 L per minute,
    a. What is the flow speed in the arterioles?
    b. What is the pressure difference across a 1.0 cm length of arteriole?
51. ||| Blood in a carotid artery carrying blood to the head is mov- BIO ing at 0.15 m/s when it reaches a section where plaque has narrowed the artery to 80% of its diameter. What pressure drop occurs when the blood reaches this narrow section?

General Problems

52. ||| As discussed in Section 13.3, a person's percentage of BIO body fat can be estimated by weighing the person both in air and underwater, from which the average density ρ_{avg} can be calculated. A widely used equation that relates average body density to fat percentage is the *Siri equation,* % body fat $= (495/\rho_{avg}) - 450$, where ρ_{avg} is measured in g/cm^3. One person's weight in air is 947 N while his weight underwater is 44 N. What is his percentage of body fat?
53. ||| The density of aluminum is 2700 kg/m^3. How many atoms INT are in a 2.0 cm × 2.0 cm × 2.0 cm cube of aluminum?
54. |||| An oil layer floats on 85 cm of water in a tank. The absolute pressure at the bottom of the tank is 112.0 kPa. How thick is the oil?
55. | a. In Figure P13.55, how much force does the fluid exert on the end of the cylinder at A?
    b. How much force does the fluid exert on the end of the cylinder at B?

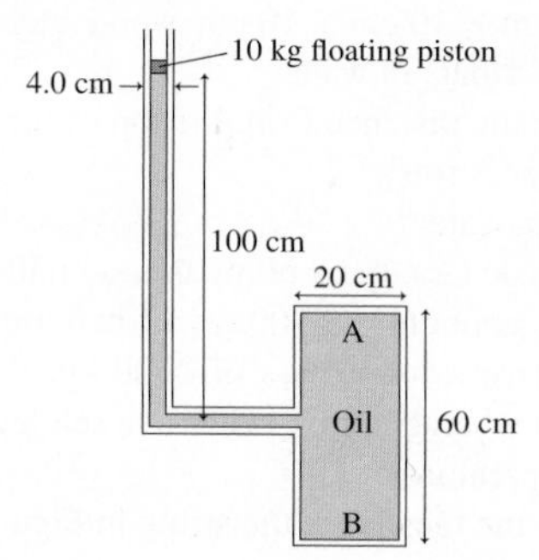

FIGURE P13.55

56. ||| A diver 50 m deep in 10°C fresh water exhales a 1.0-cm- INT diameter bubble. What is the bubble's diameter just as it reaches the surface of the lake, where the water temperature is 20°C? **Hint:** Assume that the air bubble is always in thermal equilibrium with the surrounding water.
57. | A sphere completely submerged in water is tethered to the bottom with a string. The tension in the string is one-third the weight of the sphere. What is the density of the sphere?
58. | You need to determine the density of a ceramic statue. If you suspend it from a spring scale, the scale reads 28.4 N. If you then lower the statue into a tub of water so that it is completely submerged, the scale reads 17.0 N. What is the density?
59. | A 5.0 kg rock whose density is 4800 kg/m^3 is suspended by a string such that half of the rock's volume is under water. What is the tension in the string?
60. | A flat slab of styrofoam, with a density of 32 kg/m^3, floats on a lake. What minimum volume must the slab have so that a 40 kg boy can sit on the slab without it sinking?
61. |||| A 2.0 mL syringe has an inner diameter of 6.0 mm, a needle BIO inner diameter of 0.25 mm, and a plunger pad diameter (where you place your finger) of 1.2 cm. A nurse uses the syringe to inject medicine into a patient whose blood pressure is 140/100. Assume the liquid is an ideal fluid.
    a. What is the minimum force the nurse needs to apply to the syringe?
    b. The nurse empties the syringe in 2.0 s. What is the flow speed of the medicine through the needle?
62. ||| A child's water pistol shoots water through a 1.0-mm- INT diameter hole. If the pistol is fired horizontally 70 cm above the ground, a squirt hits the ground 1.2 m away. What is the volume flow rate during the squirt? Ignore air resistance.
63. | The leaves of a tree lose water to the atmosphere via the pro- BIO cess of *transpiration*. A particular tree loses water at the rate of 3×10^{-8} m^3/s; this water is replenished by the upward flow of sap through vessels in the trunk. This tree's trunk contains about 2000 vessels, each 100 μm in diameter. What is the speed of the sap flowing in the vessels?

64. || A hurricane wind blows across a 6.00 m × 15.0 m flat roof at a speed of 130 km/h.
    a. Is the air pressure above the roof higher or lower than the pressure inside the house? Explain.
    b. What is the pressure difference?
    c. How much force is exerted on the roof? If the roof cannot withstand this much force, will it "blow in" or "blow out"?

65. ||||| Water flows from the pipe shown in Figure P13.65 with a speed of 4.0 m/s.
    a. What is the water pressure as it exits into the air?
    b. What is the height h of the standing column of water?

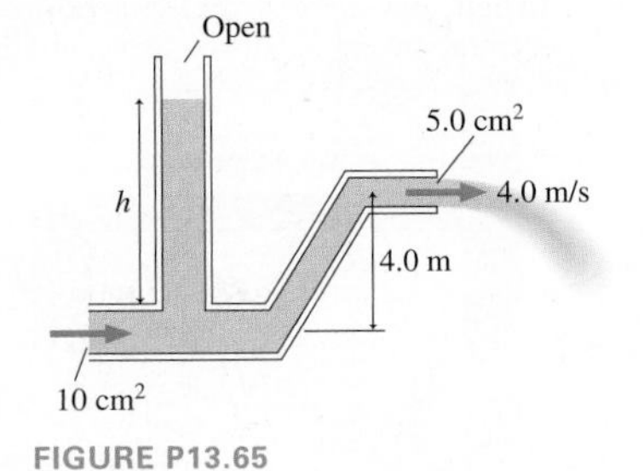

FIGURE P13.65

66. |||| Air at 20°C flows through the tube shown in Figure P13.66. Assume that air is an ideal fluid.
    a. What are the air speeds v_1 and v_2 at points 1 and 2?
    b. What is the volume flow rate?

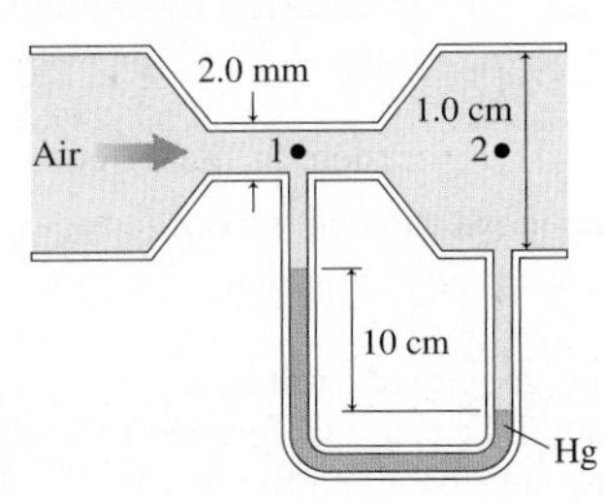

FIGURE P13.66

67. ||| Air at 20°C flows through the tube shown in Figure P13.67 INT at a rate of 1200 cm^3/s. Assume that air is an ideal fluid. What is the height h of mercury in the right side of the U-tube?

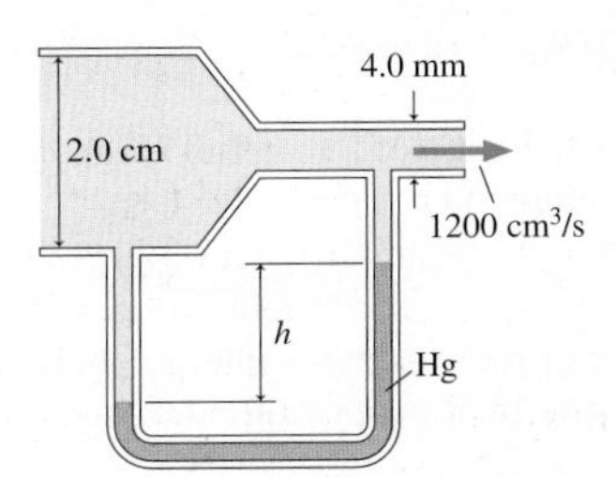

FIGURE P13.67

68. |||| There are two carotid arteries that feed blood to the brain, BIO one on each side of the neck and head. One patient's carotid arteries are each 11.2 cm long and have an inside diameter of 5.2 mm. Near the middle of the left artery, however, is a 2.0-cm-long *stenosis*, a section of the artery with a smaller diameter of 3.4 mm. For the same blood flow rate, what is the ratio of the pressure drop along the patient's left carotid artery to the drop along his right artery?

69. ||||| Figure P13.69 shows a section of a long tube that narrows near its open end to a diameter of 1.0 mm. Water at 20°C flows out of the open end at 0.020 L/s. What is the gauge pressure at point P, where the diameter is 4.0 mm?

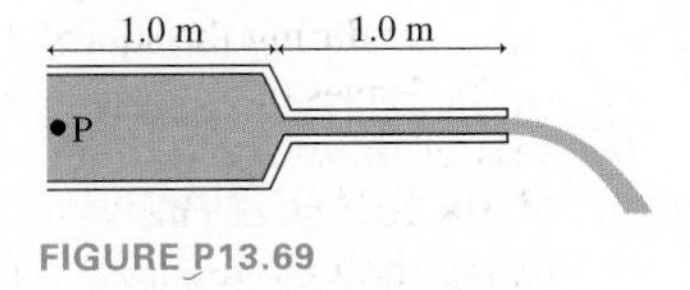

FIGURE P13.69

70. || Smoking tobacco is bad for your circulatory health. In an BIO attempt to maintain the blood's capacity to deliver oxygen, the body increases its red blood cell production, and this increases the viscosity of the blood. In addition, nicotine from tobacco causes arteries to constrict.

    For a nonsmoker, normal blood flow requires a pressure difference of 8.0 mm Hg between the two ends of an artery. Assume that smoking increases viscosity by 10% and reduces the arterial diameter to 90% of its previous value. For a smoker, what pressure difference is needed to maintain the same blood flow?

71. ||| A stiff, 10-cm-long tube with an inner diameter of 3.0 mm is attached to a small hole in the side of a tall beaker. The tube sticks out horizontally. The beaker is filled with 20°C water to a level 45 cm above the hole, and it is continually topped off to maintain that level. What is the volume flow rate through the tube?

MCAT-Style Passage Problems

Blood Pressure and Blood Flow BIO

As blood goes from the left ventricle through the arteries and veins of the human body, both its speed and pressure change. The arteries and arterioles can either constrict, reducing the area, or dilate, increasing the area, in response to certain conditions. Both of these changes can affect blood flow and blood pressure. An artery can also develop a permanent narrow area (stenosis) or a permanent wide area (aneurysm). Both of these changes can have significant health consequences.

72. | Suppose that in response to some stimulus a small blood vessel narrows to 90% of its original diameter. If there is no change in the pressure across the vessel, what is the ratio of the new volume flow rate to the original flow rate?
    A. 0.66 B. 0.73 C. 0.81 D. 0.90

73. | Sustained exercise can increase the blood flow rate of the heart by a factor of 5 with only a modest increase in blood pressure. This is a large change in flow. Although several factors come into play, which of the following physiological changes would most plausibly account for such a large increase in flow with a small change in pressure?
    A. A decrease in the viscosity of the blood
    B. Dilation of the smaller blood vessels to larger diameters
    C. Dilation of the aorta to larger diameter
    D. An increase in the oxygen carried by the blood

74. || A patient has developed a narrowing of the coronary arteries BIO that provide blood to the heart. If the rate of blood flow through the arteries is to stay the same,
    A. The length of the arteries must decrease as well.
    B. The speed of the flow must not change.
    C. The pressure difference across the arteries must stay the same, requiring a decrease in blood pressure.
    D. The pressure difference across the arteries must increase, requiring an increase in blood pressure.

75. || The nicotine in tobacco causes the arteries to constrict. A BIO rapid constriction of the arteries likely leads to
    A. A rapid decrease in blood pressure.
    B. No change in blood pressure.
    C. A rapid increase in blood pressure.

76. || A patient has developed an aneurysm in the aorta, a short BIO section where the diameter is twice the normal diameter. In the aneurysm, the speed of the blood is ______ than in the section before the aneurysm, and the pressure is ________ than in the section before the aneurysm.
    A. Greater, greater B. Greater, less
    C. Less, greater D. Less, less

KNOWLEDGE STRUCTURE III Properties of Matter

BASIC GOALS How can we describe the macroscopic flows of energy and matter in heat transfer and fluid flow?

CROSS-CUTTING CONCEPTS

Flows of energy and matter

A difference in temperature leads to transfer of heat: **conduction.**

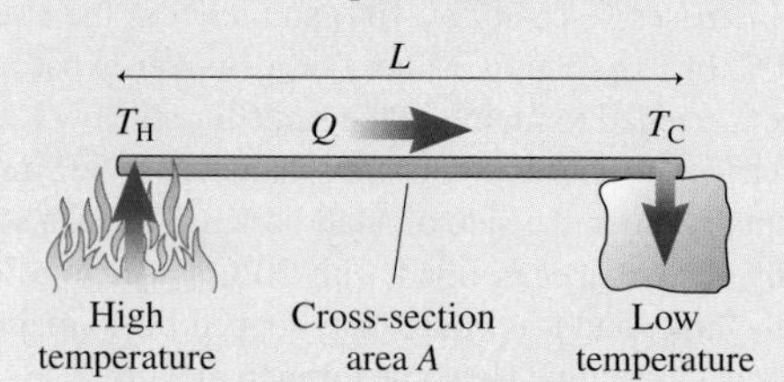

$$\frac{Q}{\Delta t} = \left(\frac{kA}{L}\right)\Delta T$$

A difference in concentration leads to a transfer of matter: **diffusion.**

N is the number of molecules diffusing through the passageway.

High concentration | Low concentration | L | N | c_H | c_L | Cross-section area A

$$\frac{N}{\Delta t} = \left(\frac{DA}{L}\right)\Delta c$$

GENERAL PRINCIPLES **Phases of matter**

Gas: Particles are freely moving. Thermal energy is kinetic energy of motion of particles. Pressure is due to collisions of particles with walls of container.

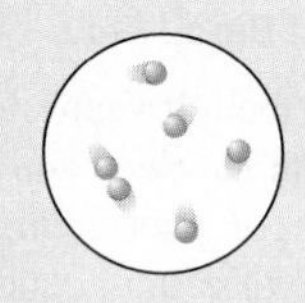

Liquid: Particles are loosely bound, but flow is possible. Heat is transferred by convection and conduction. Pressure is due to gravity.

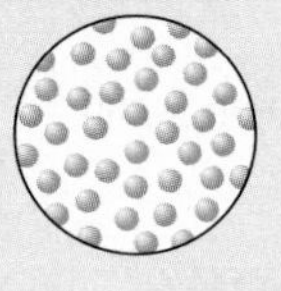

Solid: Particles are joined by spring-like bonds. Increasing temperature causes expansion. Heat is transferred by conduction only.

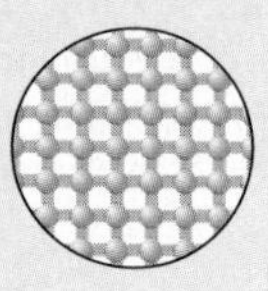

Heat must be added to change a solid to a liquid and a liquid to a gas. Heat must be removed to reverse these changes.

$$Q = \begin{cases} \pm ML_f & \text{Heat needed to melt/freeze mass } M \\ \pm ML_v & \text{Heat needed to boil/condense mass } M \end{cases}$$

Fluids and pressure

A **fluid** is a substance that can flow, a liquid or a gas. A fluid exerts **pressure** on the walls of its container.

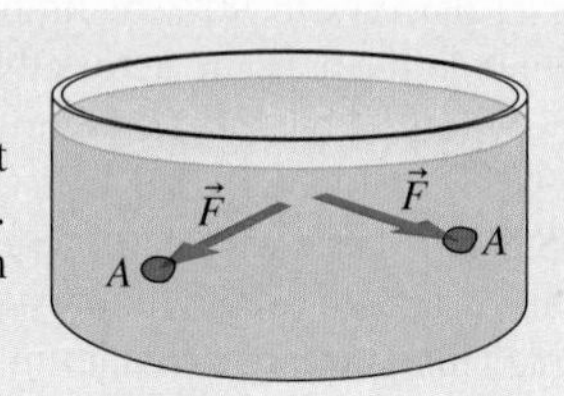

The pressure is the ratio of the force to the area over which this force acts.

$$p = \frac{F}{A}$$

The gauge pressure in a container is the difference in pressure between the inside and the outside. For containers of gases, this is usually atmospheric pressure.

$$p_g = p - p_{atmos}$$

The pressure in the gas laws is the absolute pressure, the actual pressure in a container.

Ideal-gas processes

We can graph processes on a pV diagram. Work done by the gas is the area under the graph.

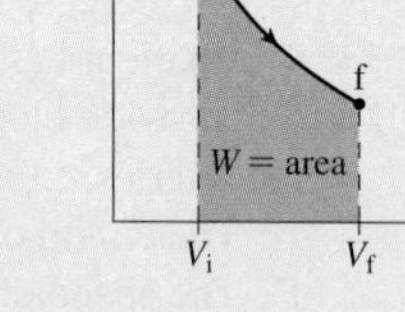

For a gas process in a sealed container, the ideal-gas law relates values of pressure, volume, and temperature:

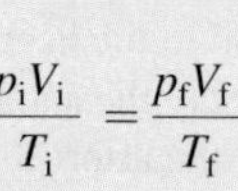

$$\frac{p_i V_i}{T_i} = \frac{p_f V_f}{T_f}$$

- Constant-volume process V = constant
- Isothermal process T = constant
- Isobaric process p = constant
- Adiabatic process $Q = 0$

Heat and heat transfer

The **specific heat** c of a material is the heat required to raise the temperature of 1 kg by 1 K:

$$Q = Mc\,\Delta T$$

When two or more systems interact thermally, they come to a common final temperature determined by

$$Q_{net} = Q_1 + Q_2 + Q_3 + \cdots = 0$$

Conduction transfers heat by direct physical contact.
Convection transfers heat by the motion of a fluid.
Radiation transfers heat by electromagnetic waves.

Fluid statics

Archimedes' principle: The magnitude of the buoyant force equals the weight of the fluid displaced by the object.

An object sinks if it is more dense than the fluid in which it is submerged; if it is less dense, it floats.

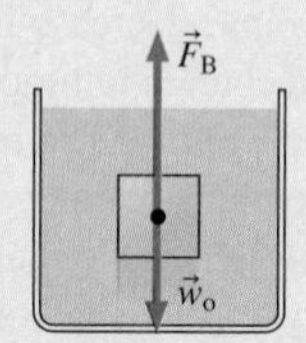

Pressure increases with depth in a liquid. The pressure at depth d is

$$p = p_0 + \rho g d$$

Fluid flow

Ideal fluid flow is laminar, incompressible, and nonviscous. The **equation of continuity** is

$$v_1 A_1 = v_2 A_2$$

Bernoulli's equation is

$$p_1 + \tfrac{1}{2}\rho v_1^2 + \rho g y_1 = p_2 + \tfrac{1}{2}\rho v_2^2 + \rho g y_2$$

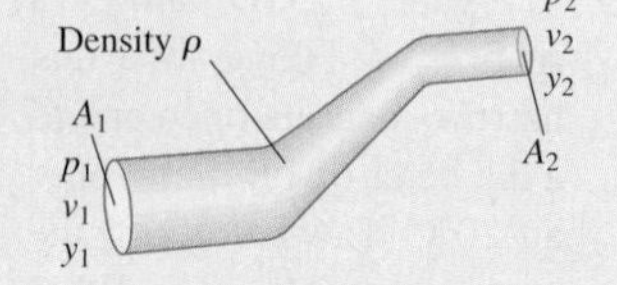

Size and Life

Physicists look for simple models and general principles that underlie and explain diverse physical phenomena. In the first 13 chapters of this textbook, you've seen that just a handful of general principles and laws can be used to solve a wide range of problems. Can this approach have any relevance to a subject like biology? It may seem surprising, but there *are* general "laws of biology" that apply, with quantitative accuracy, to organisms as diverse as elephants and mice.

Let's look at an example. An elephant uses more metabolic power than a mouse. This is not surprising, as an elephant is much bigger. But recasting the data shows an interesting trend. When we looked at the energy required to raise the temperature of different substances, we considered *specific heat*. The "specific" meant that we considered the heat required for 1 kilogram. For animals, rather than metabolic rate, we can look at the *specific metabolic rate,* the metabolic power used *per kilogram* of tissue. If we factor out the mass difference between a mouse and an elephant, are their specific metabolic powers the same?

In fact, the specific metabolic rate varies quite a bit among mammals, as the graph of specific metabolic rate versus mass shows. But there is an interesting trend: All of the data points lie on a single smooth curve. In other words, there really is a biological *law* we can use to predict a mammal's metabolic rate knowing only its mass M. In particular, the specific metabolic rate is proportional to $M^{-0.25}$. Because a 4000 kg elephant is 160,000 times more massive than a 25 g mouse, the mouse's specific metabolic power is $(160{,}000)^{0.25} = 20$ times that of the elephant. A law that shows how a property scales with the size of a system is called a *scaling law.*

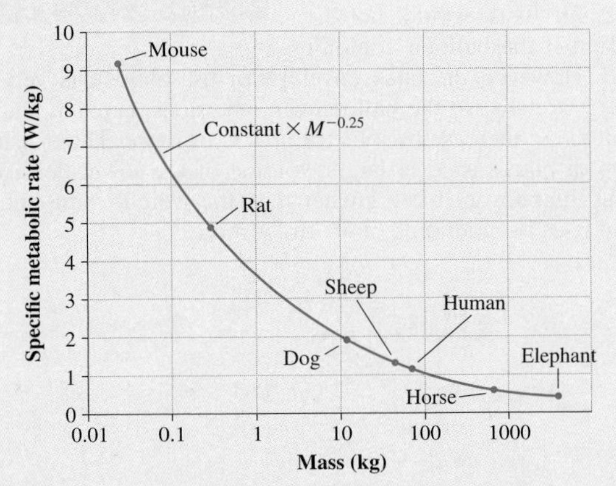

Specific metabolic rate as a function of body mass follows a simple scaling law.

A similar scaling law holds for birds, reptiles, and even bacteria. Why should a single simple relationship hold true for organisms that range in size from a single cell to a 100 ton blue whale? Interestingly, no one knows for sure. It is a matter of current research to find out just what this and other scaling laws tell us about the nature of life.

Perhaps the metabolic-power scaling law is a result of heat transfer. In Chapter 12, we noted that all metabolic energy used by an animal ends up as heat, which must be transferred to the environment. A 4000 kg elephant has 160,000 times the mass of a 25 g mouse, but it has only about 3000 times the surface area. The heat transferred to the environment depends on the surface area; the more surface area, the greater the rate of heat transfer. An elephant with a mouse-sized metabolism simply wouldn't be able to dissipate heat fast enough—it would quickly overheat and die.

If heat dissipation were the only factor limiting metabolism, we can show that the specific metabolic rate should scale as $M^{-0.33}$, quite different from the $M^{-0.25}$ scaling observed. Clearly, another factor is at work. Exactly what underlies the $M^{-0.25}$ scaling is still a matter of debate, but some recent analysis suggests the scaling is due to limitations not of heat transfer but of fluid flow. Cells in mice, elephants, and all mammals receive nutrients and oxygen for metabolism from the bloodstream. Because the minimum size of a capillary is about the same for all mammals, the structure of the circulatory system must vary from animal to animal. The human aorta has a diameter of about 1 inch; in a mouse, the diameter is approximately 1/20th of this. Thus a mouse has fewer levels of branching to smaller and smaller blood vessels as we move from the aorta to the capillaries. The smaller blood vessels in mice mean that viscosity is more of a factor throughout the circulatory system. The circulatory system of a mouse is quite different from that of an elephant.

A model of specific metabolic rate based on blood-flow limitations predicts a $M^{-0.25}$ law, exactly as observed. The model also makes other testable predictions. For example, the model predicts that the smallest possible mammal should have a body mass of about 1 gram—exactly the size of the smallest shrew. Even smaller animals have different types of circulatory systems; in the smallest animals, nutrient transport is by diffusion alone. But the model can be extended to predict that the specific metabolic rate for these animals will follow a scaling law similar to that for mammals, exactly as observed. It is too soon to know if this model will ultimately prove to be correct, but it's indisputable that there are large-scale regularities in biology that follow mathematical relationships based on the laws of physics.

PART III PROBLEMS

The following questions are related to the passage "Size and Life" on the previous page. BIO

1. A typical timber wolf has a mass of 40 kg, a typical jackrabbit a mass of 2.5 kg. Given the scaling law presented in the passage, we'd expect the specific metabolic rate of the jackrabbit to be higher by a factor of
   A. 2 B. 4 C. 8 D. 16
2. A typical timber wolf has a mass of 40 kg, a typical jackrabbit a mass of 2.5 kg. Given the scaling law presented in the passage, we'd expect the wolf to use _______ times more energy than a jackrabbit in the course of a day.
   A. 2 B. 4 C. 8 D. 16
3. Given the data of the graph, approximately how much energy, in Calories, would a 200 g rat use during the course of a day?
   A. 10 B. 20 C. 100 D. 200
4. All other things being equal, species that inhabit cold climates tend to be larger than related species that inhabit hot climates. For instance, the Alaskan hare is the largest North American hare, with a typical mass of 5.0 kg, double that of a jackrabbit. A likely explanation is that
   A. Larger animals have more blood flow, allowing for better thermoregulation.
   B. Larger animals need less food to survive than smaller animals.
   C. Larger animals have larger blood volumes than smaller animals.
   D. Larger animals lose heat less quickly than smaller animals.
5. The passage proposes that there are quantitative "laws" of biology that have their basis in physical principles, using the scaling of specific metabolic rate with body mass as an example. Which of the following regularities among animals might also be an example of such a "law"?
   A. As a group, birds have better color vision than mammals.
   B. Reptiles have a much lower specific metabolic rate than mammals.
   C. Predators tend to have very good binocular vision; prey animals tend to be able to see over a very wide angle.
   D. Jump height varies very little among animals. Nearly all animals, ranging in size from a flea to a horse, have a maximum vertical leap that is quite similar.

The following passages and associated questions are based on the material of Part III.

Keeping Your Cool BIO

A 68 kg cyclist is pedaling down the road at 15 km/h, using a total metabolic power of 480 W. A certain fraction of this energy is used to move the bicycle forward, but the balance ends up as thermal energy in his body, which he must get rid of to keep cool. On a very warm day, conduction, convection, and radiation transfer little energy, and so he does this by perspiring, with the evaporation of water taking away the excess thermal energy.

6. If the cyclist reaches his 15 km/h cruising speed by rolling down a hill, what is the approximate height of the hill?
   A. 22 m B. 11 m C. 2 m D. 1 m
7. As he cycles at a constant speed on level ground, at what rate is chemical energy being converted to thermal energy in his body, assuming a typical efficiency of 25% for the conversion of chemical energy to the mechanical energy of motion?
   A. 480 W B. 360 W C. 240 W D. 120 W
8. To keep from overheating, the cyclist must get rid of the excess thermal energy generated in his body. If he cycles at this rate for 2 hours, how many liters of water must he perspire, to the nearest 0.1 liter?
   A. 0.4 L B. 0.9 L C. 1.1 L D. 1.4 L
9. Being able to exhaust this thermal energy is very important. If he isn't able to get rid of any of the excess heat, by how much will the temperature of his body increase in 10 minutes of riding, to the nearest 0.1°C?
   A. 0.3°C B. 0.6°C C. 0.9°C D. 1.2°C

Weather Balloons

The data used to generate weather forecasts are gathered by hundreds of weather balloons launched from sites throughout the world. A typical balloon is made of latex and filled with hydrogen. A packet of sensing instruments (called a *radiosonde*) transmits information back to earth as the balloon rises into the atmosphere. At the beginning of its flight, the average density of the weather balloon package (total mass of the balloon plus cargo divided by their volume) is less than the density of the surrounding air, so the balloon rises. As it does, the density of the surrounding air decreases, as shown in Figure III.1. The balloon will rise to the point at which the buoyant force of the air exactly balances its weight. This would not be very high if the balloon couldn't expand. However, the latex envelope of the balloon is very thin and very stretchy, so the balloon can, and does, expand, allowing the volume to increase by a factor of 100 or more. The expanding balloon displaces an ever-larger volume of the lower-density air, keeping the buoyant force greater than the weight force until the balloon rises to an altitude of 40 km or more.

Density (kg/m^3) vs. Height (km)

FIGURE III.1

10. A balloon launched from sea level has a volume of approximately 4 m^3. What is the approximate buoyant force on the balloon?
A. 50 N
B. 40 N
C. 20 N
D. 10 N
11. A balloon launched from sea level with a volume of 4 m^3 will have a volume of about 12 m^3 on reaching an altitude of 10 km. What is the approximate buoyant force now?
A. 50 N
B. 40 N
C. 20 N
D. 10 N
12. The balloon expands as it rises, keeping the pressures inside and outside the balloon approximately equal. If the balloon rises slowly, heat transfers will keep the temperature inside the same as the outside air temperature. A balloon with a volume of 4.0 m^3 is launched at sea level, where the atmospheric pressure is 100 kPa and the temperature is 15°C. It then rises slowly to a height of 5500 m, where the pressure is 50 kPa and the temperature is –20°C. What is the volume of the balloon at this altitude?
A. 5.0 m^3
B. 6.0 m^3
C. 7.0 m^3
D. 8.0 m^3
13. If the balloon rises quickly, so that no heat transfer is possible, the temperature inside the balloon will drop as the gas expands. If a 4.0 m^3 balloon is launched at a pressure of 100 kPa and rapidly rises to a point where the pressure is 50 kPa, the volume of the balloon will be
A. Greater than 8.0 m^3.
B. 8.0 m^3.
C. Less than 8.0 m^3.
14. At the end of the flight, the radiosonde is dropped and falls to earth by parachute. Suppose the parachute achieves its terminal speed at a height of 30 km. As it descends into the atmosphere, how does the terminal speed change?
A. It increases.
B. It stays the same.
C. It decreases.

Passenger Balloons

Long-distance balloon flights are usually made using a hot-air-balloon/helium-balloon hybrid. The balloon has a sealed, flexible chamber of helium gas that expands or contracts to keep the helium pressure approximately equal to the air pressure outside. The helium chamber sits on top of an open (that is, air can enter or leave), constant-volume chamber of propane-heated air. Assume that the hot air and the helium are kept at a constant temperature by burning propane.

15. A balloon is launched at sea level, where the air pressure is 100 kPa. The helium has a volume of 1000 m^3 at this altitude. What is the volume of the helium when the balloon has risen to a height where the atmospheric pressure is 33 kPa?
A. 330 m^3
B. 500 m^3
C. 1000 m^3
D. 3000 m^3
16. A balloon is launched at sea level, where the air pressure is 100 kPa. The density in the hot-air chamber is 1.0 kg/m^3. What is the density of the air when the balloon has risen to a height where the atmospheric pressure is 33 kPa?
A. 3.0 kg/m^3
B. 1.0 kg/m^3
C. 0.66 kg/m^3
D. 0.33 kg/m^3
17. A balloon is at a height of 5.0 km and is descending at a constant rate. The buoyancy force is directed __________; the drag force is directed __________.
A. Up, up
B. Up, down
C. Down, up
D. Down, down

Additional Integrated Problems

18. When you exhale, all of the air in your lungs must exit through the trachea. If you exhale through your nose, this air subsequently leaves through your nostrils. The area of your nostrils is less than that of your trachea. How does the speed of the air in the trachea compare to that in the nostrils?
BIO
19. Sneezing requires an increase in pressure of the air in the lungs; a typical sneeze might result in an extra pressure of 7.0 kPa. Estimate how much force this exerts on the diaphragm, the large muscle at the bottom of the ribcage.
BIO
20. A 20 kg block of aluminum sits on the bottom of a tank of water. How much force does the block exert on the bottom of the tank?
21. We've seen that fish can control their buoyancy through the use of a swim bladder, a gas-filled organ inside the body. You can assume that the gas pressure inside the swim bladder is roughly equal to the external water pressure. A fish swimming at a particular depth adjusts the volume of its swim bladder to give it neutral buoyancy. If the fish swims upward or downward, the changing water pressure causes the bladder to expand or contract. Consequently, the fish must adjust the quantity of gas to restore the original volume and thus reestablish neutral buoyancy. Consider a large, 7.0 kg striped bass with a volume of 7.0 L. When neutrally buoyant, 7.0% of the fish's volume is taken up by the swim bladder. Assume a body temperature of 15°C.
BIO
a. How many moles of air are in the swim bladder when the fish is at a depth of 80 ft?
b. What will the volume of the swim bladder be if the fish ascends to a 50 ft depth without changing the quantity of gas?
c. To return the swim bladder to its original size, how many moles of gas must be removed?

PART IV

Oscillations and Waves

Wolves are social animals, and they howl to communicate over distances of several miles with other members of their pack. How are such sounds made? How do they travel through the air? And how are other wolves able to hear these sounds from such a great distance?

OVERVIEW

Motion That Repeats Again and Again

Up to this point in the text, we have generally considered processes that have a clear starting and ending point, such as a car accelerating from rest to a final speed, or a solid being heated from an initial to a final temperature. In Part IV, we begin to consider processes that are *periodic*—they repeat. A child on a swing, a boat bobbing on the water, and even the repetitive bass beat of a rock song are *oscillations* that happen over and over without a starting or ending point. The *period*, the time for one cycle of the motion, will be a key parameter for us to consider as we look at oscillatory motion.

Our first goal will be to develop the language and tools needed to describe oscillations, ranging from the swinging of the bob of a pendulum clock to the bouncing of a car on its springs. Once we understand oscillations, we will extend our analysis to consider oscillations that travel—*waves*.

The Wave Model

We've had great success modeling the motion of complex objects as the motion of one or more particles. We were even able to explain the macroscopic properties of matter, such as pressure and temperature, in terms of the motion of the atomic particles that comprise all matter.

Now it's time to explore another way of looking at nature, the *wave model*. Familiar examples of waves include

- Ripples on a pond.
- The sound of thunder.
- The swaying ground of an earthquake.
- A vibrating guitar string.
- The colors of a rainbow.

Despite the great diversity of types and sources of waves, there is a single, elegant physical theory that can describe them all. Our exploration of wave phenomena will call upon water waves, sound waves, and light waves for examples, but our goal will be to emphasize the unity and coherence of the ideas that are common to *all* types of waves. As was the case with the particle model, we will use the wave model to explain a wide range of phenomena.

When Waves Collide

The collision of two particles is a dramatic event. Energy and momentum are transferred as the two particles head off in different directions. Something much gentler happens when two waves come together—the two waves pass through each other unchanged. Where they overlap, we get a *superposition* of the two waves. We will finish our discussion of waves by analyzing the standing waves that result from the superposition of two waves traveling in opposite directions. The physics of standing waves will allow us to understand how your vocal tract can produce such a wide range of sounds, and how your ears are able to analyze them.

14 Oscillations

A gibbon swings below a branch, moving forward by pivoting below one handhold and then the next. How does an understanding of oscillatory motion allow us to analyze the swinging motion of a gibbon—and the walking motion of other animals?

LOOKING AHEAD ▸▸

Motion That Repeats

When the girl moves down, the springy ropes pull up. This restoring force produces an **oscillation:** one bounce after another.

You'll see many examples of systems with restoring forces that lead to oscillatory motion.

Simple Harmonic Motion

The sand records the motion of the oscillating pendulum. The **sinusoidal** shape tells us that this is **simple harmonic motion.**

All oscillations show a similar form. You'll learn to describe and analyze oscillating systems.

Resonance

When you make a system oscillate at its **natural frequency,** you can get a large amplitude. We call this **resonance.**

You'll learn how resonance of a membrane in the inner ear lets you determine the **pitch** of a musical note.

GOAL To understand systems that oscillate with simple harmonic motion.

LOOKING BACK ◂◂

Springs and Restoring Forces

In Chapter 8, you learned that a stretched spring exerts a restoring force proportional to the stretch:

$$F_{sp} = -k\,\Delta x$$

In this chapter, you'll see how this linear restoring force leads to an oscillation, with a frequency determined by the spring constant k.

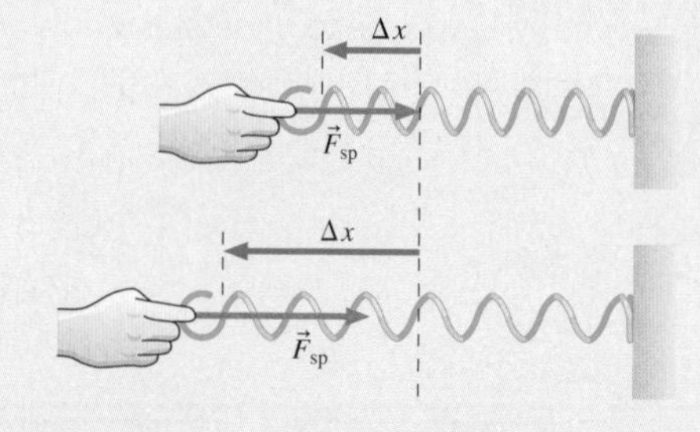

STOP TO THINK

A hanging spring has length 10 cm. A 100 g mass is hung from the spring, stretching it to 12 cm. What will be the length of the spring if this mass is replaced by a 200 g mass?

A. 14 cm B. 16 cm
C. 20 cm D. 24 cm

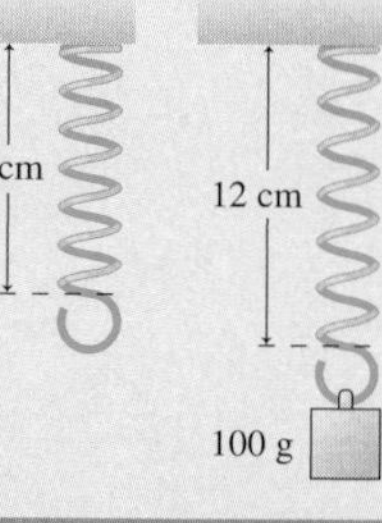

14.1 Equilibrium and Oscillation

Consider a marble that is free to roll inside a spherical bowl, as shown in FIGURE 14.1. The marble has an **equilibrium position** at the bottom of the bowl where it will rest with no net force on it. If you push the marble away from equilibrium, the marble's weight leads to a net force directed back toward the equilibrium position. We call this a **restoring force** because it acts to restore equilibrium. The magnitude of this restoring force increases if the marble is moved farther away from the equilibrium position.

If you pull the marble to the side and release it, it doesn't just roll back to the bottom of the bowl and stay put. It keeps on moving, rolling up and down each side of the bowl, repeatedly moving through its equilibrium position, as we see in FIGURE 14.2. We call such repetitive motion an **oscillation.** This oscillation is a result of an interplay between the restoring force and the marble's inertia, something we will see in all of the oscillations we consider.

We'll start our description by noting the most important fact about oscillatory motion: It repeats. Any oscillation is characterized by a *period,* the time for the motion to repeat. We met the concepts of period and frequency when we studied circular motion in Chapter 6. As a starting point then, let's review these ideas and see how they apply to oscillatory motion.

FIGURE 14.1 Equilibrium and restoring forces for a ball in a bowl.

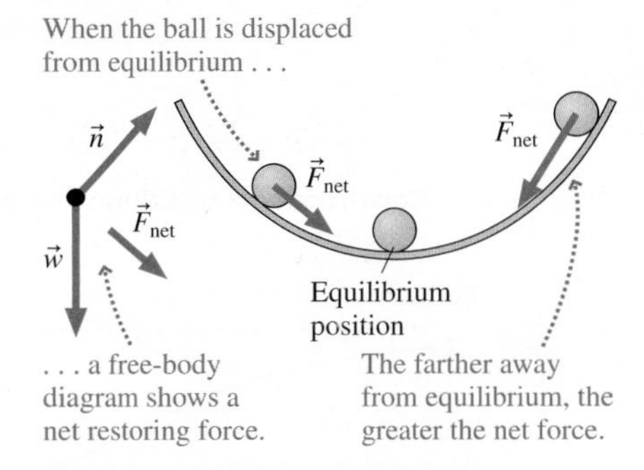
When the ball is displaced from equilibrium . . .
$\vec{n}$
$\vec{w}$
$\vec{F}_{net}$
$\vec{F}_{net}$
$\vec{F}_{net}$
Equilibrium position
. . . a free-body diagram shows a net restoring force.
The farther away from equilibrium, the greater the net force.

FIGURE 14.2 The oscillating motion of a ball rolling in a bowl.

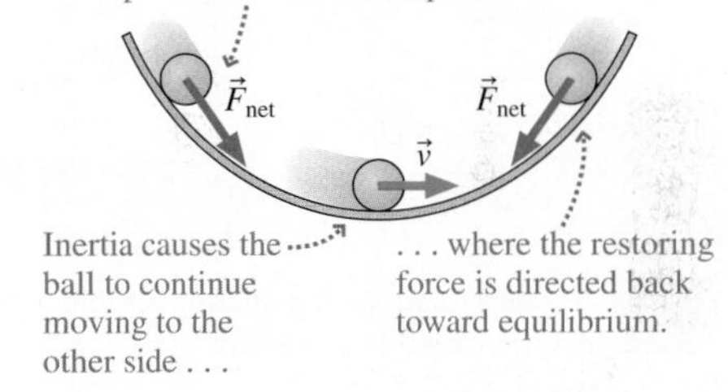
When the ball is released, a restoring force pulls it back toward equilibrium.
$\vec{F}_{net}$
$\vec{F}_{net}$
$\vec{v}$
Inertia causes the ball to continue moving to the other side . . .
. . . where the restoring force is directed back toward equilibrium.

Frequency and Period

An electrocardiogram (ECG), such as the one shown in FIGURE 14.3, is a record of the electrical signals of the heart as it beats. We will explore the ECG in some detail in Chapter 21. Although the shape of a typical ECG is rather complex, notice that it has a *repeating pattern.* For this or any oscillation, the time to complete one full cycle is called the **period** of the oscillation. Period is given the symbol T.

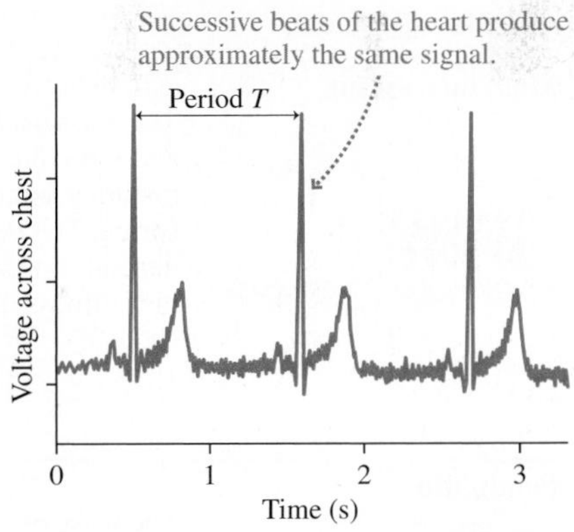

An equivalent piece of information is the number of cycles, or oscillations, completed per second. If the period is $\frac{1}{10}$ s, then the oscillator can complete 10 cycles in 1 second. Conversely, an oscillation period of 10 s allows only $\frac{1}{10}$ of a cycle to be completed per second. In general, T seconds per cycle implies that $1/T$ cycles are completed each second. The number of cycles per second is called the **frequency** f of the oscillation. The relationship between frequency and period is therefore

$$f = \frac{1}{T} \quad \text{or} \quad T = \frac{1}{f} \tag{14.1}$$

The units of frequency are **hertz,** abbreviated Hz. By definition,

$$1 \text{ Hz} = 1 \text{ cycle per second} = 1 \text{ s}^{-1}$$

We will frequently deal with very rapid oscillations and make use of the units shown in TABLE 14.1.

NOTE ▶ Uppercase and lowercase letters *are* important. 1 MHz is 1 megahertz = 10^6 Hz, but 1 mHz is 1 millihertz = 10^{-3} Hz! ◀

FIGURE 14.3 BIO An electrocardiogram has a well-defined period.

Successive beats of the heart produce approximately the same signal.
Period T
Voltage across chest
Time (s)
0 1 2 3

EXAMPLE 14.1 Finding the frequency and period of a radio station

An FM radio station broadcasts an oscillating radio wave at a frequency of 100 MHz. What is the period of the oscillation?

SOLVE The frequency f of oscillations in the radio transmitter is 100 MHz = 1.0×10^8 Hz. The period is the inverse of the frequency; hence,

$$T = \frac{1}{f} = \frac{1}{1.0 \times 10^8 \text{ Hz}} = 1.0 \times 10^{-8} \text{ s} = 10 \text{ ns}$$

TABLE 14.1 Common units of frequency

Frequency	Period
10^3 Hz = 1 kilohertz = 1 kHz	1 ms
10^6 Hz = 1 megahertz = 1 MHz	1 μs
10^9 Hz = 1 gigahertz = 1 GHz	1 ns

Simple Harmonic Motion

Let's make a graph of the motion of the marble in a bowl, with positions to the right of equilibrium positive and positions to the left of equilibrium negative. FIGURE 14.4 shows a series of "snapshots" of the motion and the corresponding points on the graph. This graph has the form of a *cosine function.* A graph or a function that has the form of a sine or cosine function is called **sinusoidal.** A sinusoidal oscillation is called **simple harmonic motion,** often abbreviated SHM.

FIGURE 14.4 Constructing a position-versus-time graph for a marble rolling in a bowl.

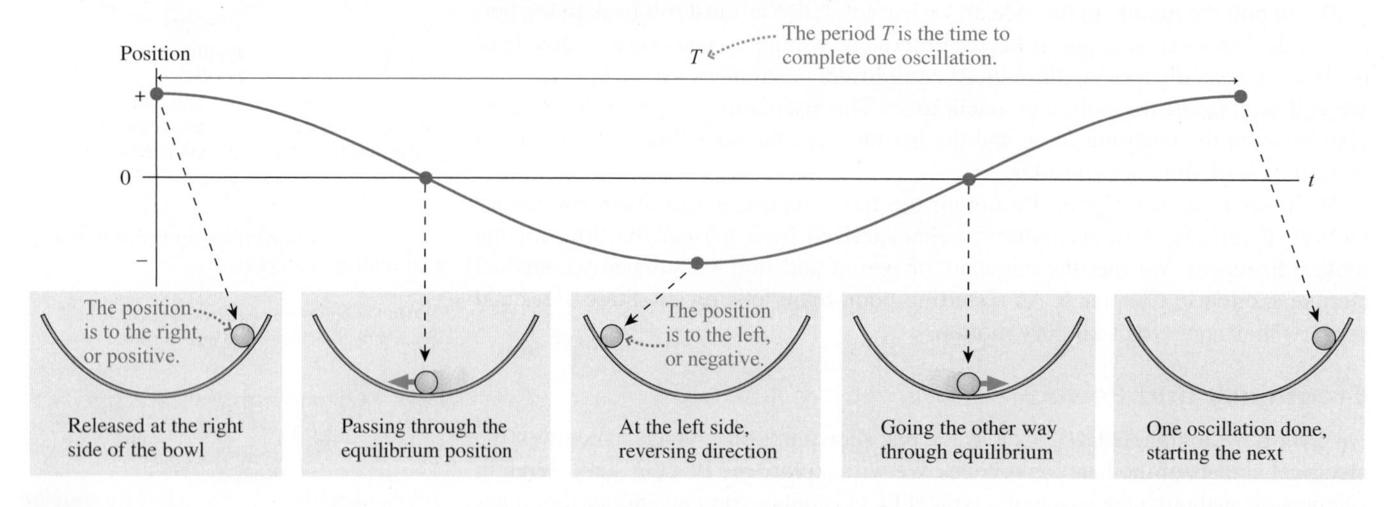

A video to support a section's topic is embedded in the eText.

Video Oscillations

A marble rolling in the bottom of a bowl undergoes simple harmonic motion, as does a car bouncing on its springs. SHM is very common, but most cases of SHM can be modeled as one of two simple systems: an object with mass (or simply a "mass") oscillating on a spring or a pendulum swinging back and forth. Examples are shown in the following table.

Examples of simple harmonic motion

Oscillating system		Related real-world example BIO	
Mass on a spring 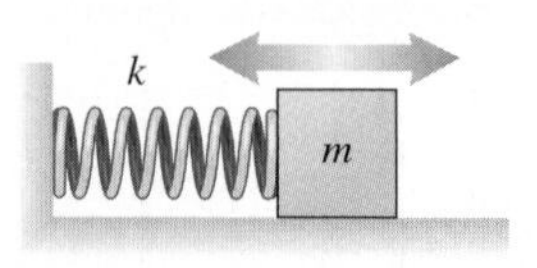	The mass oscillates back and forth due to the restoring force of the spring. The period depends on the mass and the stiffness of the spring.	**Vibrations in the ear** 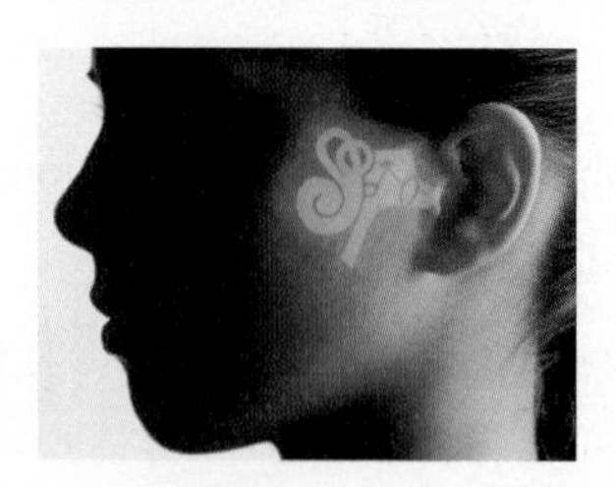	Sound waves entering the ear cause the oscillation of a membrane in the cochlea, part of the inner ear. The vibration can be modeled as a mass on a spring. The period of oscillation of a segment of the membrane depends on mass (the thickness of the membrane) and stiffness (the rigidity of the membrane).
Pendulum 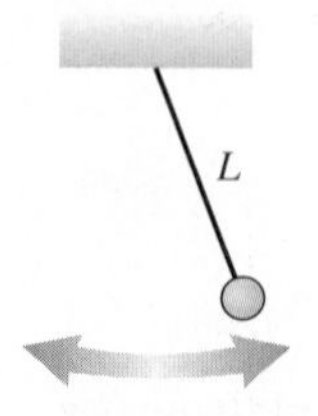	The mass oscillates back and forth due to the restoring gravitational force. The period depends on the length of the pendulum and the free-fall acceleration g.	**Motion of legs while walking** 	The motion of a walking animal's legs can be modeled as pendulum motion. The rate at which the legs swing depends on the length of the legs and the free-fall acceleration g.

STOP TO THINK 14.1 Two oscillating systems have periods T_1 and T_2, with $T_1 < T_2$. How are the frequencies of the two systems related?

A. $f_1 < f_2$ B. $f_1 = f_2$ C. $f_1 > f_2$

14.2 Linear Restoring Forces and SHM

FIGURE 14.5 shows a glider that rides with very little friction on an air track. There is a spring connecting the glider to the end of the track. When the spring is neither stretched nor compressed, the net force on the glider is zero. The glider just sits there—this is the equilibrium position.

If the glider is now displaced from this equilibrium position by Δx, the spring exerts a force back toward equilibrium—a restoring force. In ◂◂ SECTION 8.3, we found that the spring force is given by Hooke's law: $(F_{sp})_x = -k\,\Delta x$, where k is the spring constant. (Recall that a "stiffer" spring has a larger value of k.) If we set the origin of our coordinate system at the equilibrium position, the displacement from equilibrium Δx is equal to x; thus the spring force can be written as $(F_{sp})_x = -kx$.

The net force on the glider is simply the spring force, so we can write

$$(F_{net})_x = -kx \tag{14.2}$$

$(F_{net})_x$ p. 39 x PROPORTIONAL

The negative sign tells us that this is a restoring force because the force is in the direction opposite the displacement. If we pull the glider to the right (x is positive), the force is to the left (negative)—back toward equilibrium.

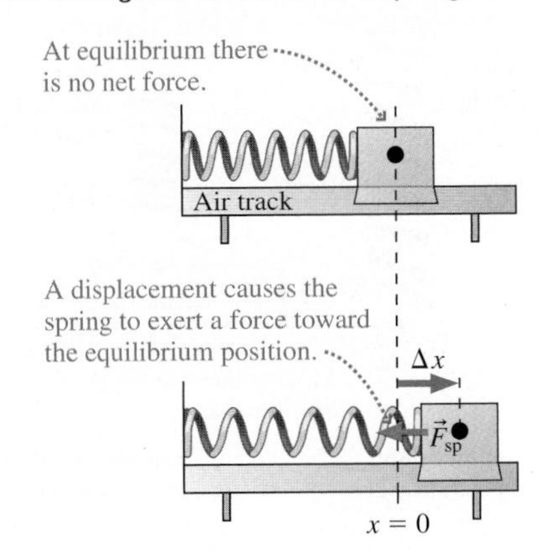
At equilibrium there is no net force.
Air track
A displacement causes the spring to exert a force toward the equilibrium position.
Δx
$\vec{F}_{sp}$
$x = 0$

FIGURE 14.5 The restoring force on an air-track glider attached to a spring.

This is a **linear restoring force;** that is, **the net force is toward the equilibrium position and is proportional to the distance from equilibrium.**

Horizontal Motion of a Mass on a Spring

If we pull the air-track glider of Figure 14.5 a short distance to the right and release it, it will oscillate back and forth. FIGURE 14.6 shows actual data from an experiment in which the position of a glider was measured 20 times every second. This is a position-versus-time graph that has been rotated 90° from its usual orientation in order for the x-axis to match the motion of the glider.

The object's maximum displacement from equilibrium is called the **amplitude** A of the motion. The object's position oscillates between $x = -A$ and $x = +A$.

NOTE ▸ The amplitude is the distance from equilibrium to the maximum, *not* the distance from the minimum to the maximum. ◂

The graph of the position is sinusoidal, so this is simple harmonic motion. **Oscillation about an equilibrium position with a linear restoring force is always simple harmonic motion.**

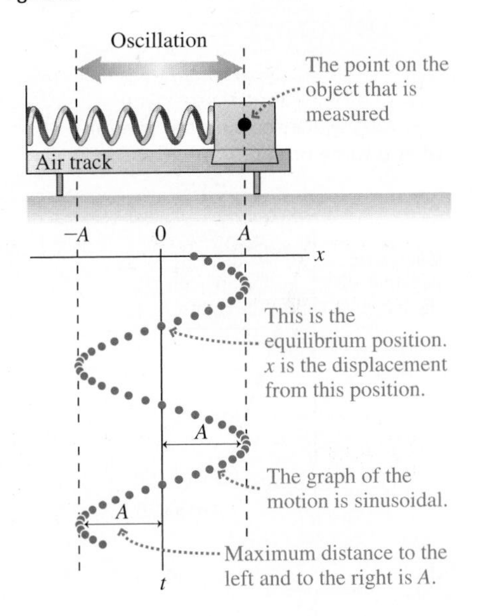
Oscillation
The point on the object that is measured
Air track
$-A$
0
A
x
This is the equilibrium position. x is the displacement from this position.
A
The graph of the motion is sinusoidal.
A
Maximum distance to the left and to the right is A.
t

FIGURE 14.6 An experiment showing the horizontal oscillation of an air-track glider.

Vertical Motion of a Mass on a Spring

FIGURE 14.7 shows a block of mass m hanging from a spring with spring constant k. Will this mass-spring system have the same simple harmonic motion as the horizontal system we just saw, or will gravity add an additional complication? An important fact to notice is that the equilibrium position of the block is *not* where the spring is at its unstretched length. At the equilibrium position of the block, where it hangs motionless, the spring has stretched by ΔL.

Finding ΔL is a static-equilibrium problem in which the upward spring force balances the downward weight force of the block. The y-component of the spring force is given by Hooke's law:

$$(F_{sp})_y = k\,\Delta L \tag{14.3}$$

Newton's first law for the block in equilibrium is

$$(F_{net})_y = (F_{sp})_y + w_y = k\,\Delta L - mg = 0 \tag{14.4}$$

from which we can find

$$\Delta L = \frac{mg}{k} \tag{14.5}$$

This is the distance the spring stretches when the block is attached to it.

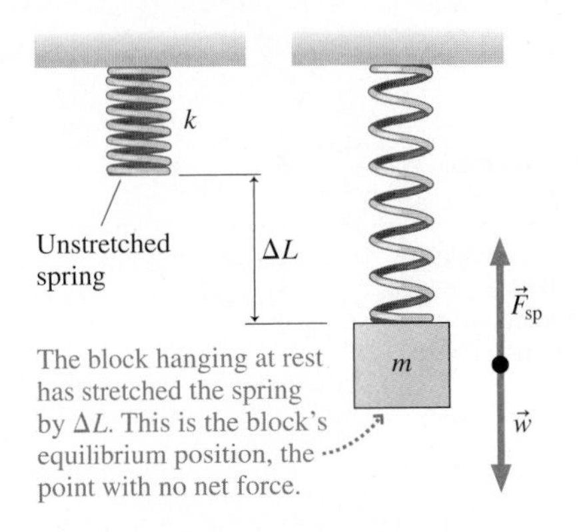
k
Unstretched spring
ΔL
m
$\vec{F}_{sp}$
$\vec{w}$
The block hanging at rest has stretched the spring by ΔL. This is the block's equilibrium position, the point with no net force.

FIGURE 14.7 The equilibrium position of a mass on a vertical spring.

FIGURE 14.8 Displacing the block from its equilibrium position produces a restoring force.

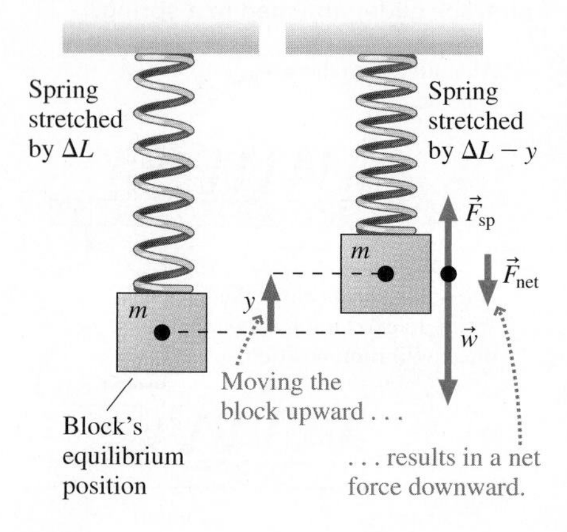
Spring
stretched
by ΔL
Spring
stretched
by ΔL − y
m
m
y
Moving the
block upward . . .
Block's
equilibrium
position
. . . results in a net
force downward.

Video Basic Oscillation Problems

Suppose we now displace the block from this equilibrium position, as shown in FIGURE 14.8. We've placed the origin of the y-axis at the block's equilibrium position in order to be consistent with our analyses of oscillations throughout this chapter. If the block moves upward, as in the figure, the spring gets shorter compared to its equilibrium length, but the spring is still *stretched* compared to its unstretched length in Figure 14.7. When the block is at position y, the spring is stretched by an amount $\Delta L - y$ and hence exerts an *upward* spring force $F_{sp} = k(\Delta L - y)$. The net force on the block at this point is

$$(F_{net})_y = (F_{sp})_y + w_y = k(\Delta L - y) - mg = (k\Delta L - mg) - ky \tag{14.6}$$

But $k\Delta L - mg = 0$, from Equation 14.4, so the net force on the block is

$$(F_{net})_y = -ky \tag{14.7}$$

Equation 14.7 for a mass hung from a spring has the same form as Equation 14.2 for the horizontal spring, where we found $(F_{net})_x = -kx$. That is, the restoring force for vertical oscillations is identical to the restoring force for horizontal oscillations. **The role of gravity is to determine where the equilibrium position is, but it doesn't affect the restoring force for displacement from the equilibrium position.** Because it has a linear restoring force, **a mass on a vertical spring oscillates with simple harmonic motion.**

The Pendulum

The chapter opened with a picture of a gibbon, whose body swings back and forth below a tree branch. The motion of the gibbon is essentially that of a **pendulum,** a mass (or *bob*) suspended from a pivot point by a light string or rod. A pendulum oscillates about its equilibrium position, but is this simple harmonic motion? To answer this question, we need to examine the restoring force on the pendulum. If the restoring force is linear, the motion will be simple harmonic.

FIGURE 14.9a shows a mass m attached to a string of length L and free to swing back and forth. The pendulum's position can be described by either the arc of length s or the angle θ, both of which are zero when the pendulum hangs straight down. Because angles are measured counterclockwise, s and θ are positive when the pendulum is to the right of center, negative when it is to the left.

FIGURE 14.9 Describing the motion of and force on a pendulum.

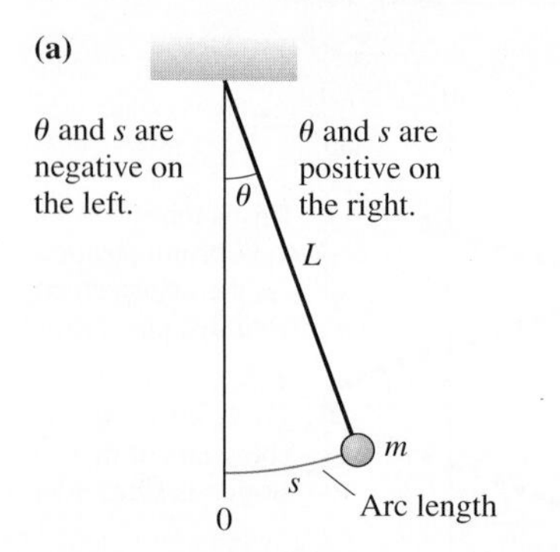
(a)
θ and s are
negative on
the left.
θ
θ and s are
positive on
the right.
L
m
s
0
Arc length

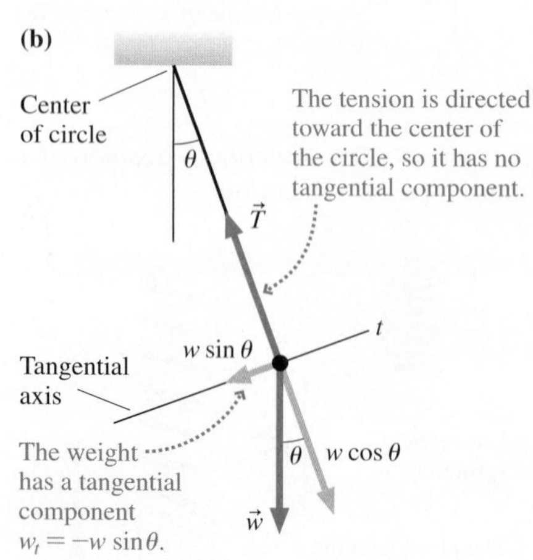
(b)
Center
of circle
θ
The tension is directed
toward the center of
the circle, so it has no
tangential component.
t
w sin θ
Tangential
axis
The weight
has a tangential
component
w_t = −w sin θ.
θ
w cos θ

As shown on the free-body diagram of FIGURE 14.9b, two forces are acting on the mass: the string tension $\vec{T}$ and the weight $\vec{w}$. The motion is along a circular arc. We choose a coordinate system centered on the mass with one axis along the radius of the circle (i.e., along the string) and the other tangent to the circle. We resolve the tension and weight forces into two components, a *tangential* component parallel to the motion (and denoted with subscript t) and a component directed along the string toward the center of the circular arc.

The mass must move along a circular arc, as noted. As Figure 14.9b shows, the net force in this direction is the tangential component of the weight:

$$(F_{net})_t = \sum F_t = w_t = -mg\sin\theta \tag{14.8}$$

This is the restoring force pulling the mass back toward the equilibrium position.

This equation becomes much simpler if we restrict the pendulum's oscillations to *small angles* of about 10° (0.17 rad) or less. For such small angles, if the angle θ is in radians, it turns out that

$$\sin\theta \approx \theta$$

This result, which applies only when the angle is in radians, is called the **small-angle approximation.**

NOTE ▶ For an angle of 10°, $\sin\theta$ differs from θ by only 0.5%. But even for a 45° angle the error is only 10%. Unless stated otherwise, you can use the small-angle approximation for all problems in this text. ◀

Recall from ◀◀ SECTION 7.1 that the angle is related to the arc length by $\theta = s/L$. Using this and the small-angle approximation, we can write the restoring force as

$$(F_{net})_t = -mg\sin\theta \approx -mg\theta = -mg\frac{s}{L} = -\left(\frac{mg}{L}\right)s \tag{14.9}$$

The net force is directed toward equilibrium, and it is linearly proportional to the displacement s from equilibrium. **The force on a pendulum is a linear restoring force for small angles, so the pendulum will undergo simple harmonic motion.**

STOP TO THINK 14.2 A ball is hung from a rope, making a pendulum. When it is pulled 5° to the side, the restoring force is 1.0 N. What will be the magnitude of the restoring force if the ball is pulled 10° to the side?

A. 0.5 N B. 1.0 N C. 1.5 N D. 2.0 N

14.3 Describing Simple Harmonic Motion

Now that we know what *causes* simple harmonic motion, we can develop graphical and mathematical descriptions. If we do this for one system, we can adapt the description to any system. The details will vary, but the form of the motion will stay the same. The following table shows one period of the oscillation of a mass on a vertical spring.

KEY CONCEPT **Details of oscillatory motion**

A mass is suspended from a vertical spring with an equilibrium position at $y = 0$. The mass is lifted by a distance A and released.

1. The mass starts at its maximum positive displacement, $y = A$. The velocity is zero, but the acceleration is negative because there is a net downward force.
2. The mass is now moving downward, so the velocity is negative. As the mass nears equilibrium, the restoring force—and thus the magnitude of the acceleration—decreases.
3. At this time the mass is moving downward with its maximum speed. It's at the equilibrium position, so the net force—and thus the acceleration—is zero.
4. The velocity is still negative but its magnitude is decreasing, so the acceleration is positive.
5. The mass has reached the lowest point of its motion, a **turning point.** The spring is at its maximum extension, so there is a net upward force and the acceleration is positive.
6. The mass has begun moving upward; the velocity and acceleration are positive.
7. The mass is passing through the equilibrium position again, in the opposite direction, so it has a positive velocity. There is no net force, so the acceleration is zero.
8. The mass continues moving upward. The velocity is positive but its magnitude is decreasing, so the acceleration is negative.
9. The mass is now back at its starting position. This is another turning point. The mass is at rest but will soon begin moving downward, and the cycle will repeat.

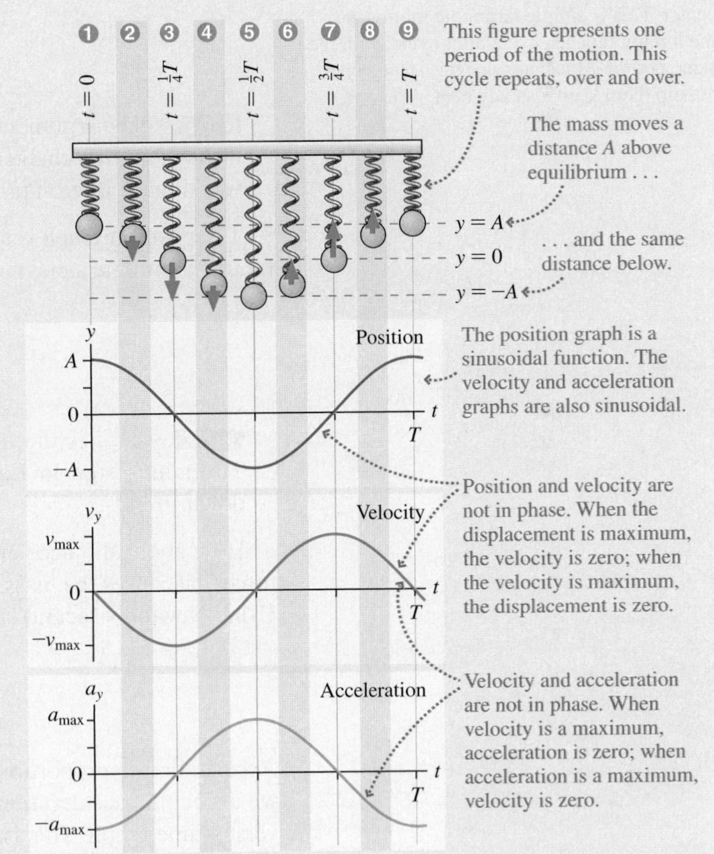
1 2 3 4 5 6 7 8 9
$t = 0$
$t = \frac{1}{4}T$
$t = \frac{1}{2}T$
$t = \frac{3}{4}T$
$t = T$
This figure represents one period of the motion. This cycle repeats, over and over.
The mass moves a distance A above equilibrium . . .
$y = A$
$y = 0$
$y = -A$
. . . and the same distance below.
Position
y
A
0
$-A$
t
T
The position graph is a sinusoidal function. The velocity and acceleration graphs are also sinusoidal.
Velocity
v_y
v_{max}
$-v_{max}$
Position and velocity are not in phase. When the displacement is maximum, the velocity is zero; when the velocity is maximum, the displacement is zero.
Acceleration
a_y
a_{max}
$-a_{max}$
Velocity and acceleration are not in phase. When velocity is a maximum, acceleration is zero; when acceleration is a maximum, velocity is zero.

STOP TO THINK 14.3 The graphs in this table apply to pendulum motion as well as the motion of a mass on a spring. A pendulum in a clock has a period of 2.0 seconds. You pull the pendulum to the right—a positive displacement—and let it go; we call this time $t = 0$ s. At what time will the pendulum (a) first be at its maximum negative displacement, (b) first have its maximum speed, (c) first have its maximum positive velocity, and (d) first have its maximum positive acceleration?

A. 0.5 s B. 1.0 s C. 1.5 s D. 2.0 s

Keeping the beat The metal rod in a metronome swings back and forth, making a loud click each time it passes through the center. This is simple harmonic motion, so the motion repeats, cycle after cycle, with the same period. Musicians use this steady click to help them keep a steady beat.

The graphs in the previous table are for an oscillation in which the object just happened to be at $y = A$ at $t = 0$. You can certainly imagine a different set of *initial conditions,* with the object at $y = -A$ or somewhere in the middle of an oscillation. This would change the starting point of the graphs but not their overall appearance. The oscillation repeats over and over, so the exact choice of starting point isn't crucial; the results we have outlined are general. Keep in mind that all the features of the motion of a mass on a spring also apply to other examples of simple harmonic motion.

The position-versus-time graph in the table on the preceding page is a cosine curve. We can write the object's position as

$$x(t) = A\cos\left(\frac{2\pi t}{T}\right) \tag{14.10}$$

where the notation $x(t)$ indicates that the position x is a *function* of time t. Because $\cos(2\pi \text{ rad}) = \cos(0 \text{ rad})$, we see that the position at time $t = T$ is the same as the position at $t = 0$. In other words, this is a cosine function with period T. We can write Equation 14.10 in an alternative form. Because the oscillation frequency is $f = 1/T$, we can write

$$x(t) = A\cos(2\pi ft) \tag{14.11}$$

NOTE ▶ The argument $2\pi ft$ of the cosine function is in *radians*. That will be true throughout this chapter. Don't forget to set your calculator to radian mode before working oscillation problems. ◀

The position graph is a cosine function; the velocity graph is an upside-down sine function with the same period T. The velocity v_x can thus be written as

$$v_x(t) = -v_{\text{max}}\sin\left(\frac{2\pi t}{T}\right) = -v_{\text{max}}\sin(2\pi ft) \tag{14.12}$$

NOTE ▶ v_{max} is the maximum *speed* and thus is inherently a *positive* number. The minus sign in Equation 14.12 is needed to turn the sine function upside down. ◀

How about the acceleration? As we saw earlier in Equation 14.2, the restoring force that causes the mass to oscillate with simple harmonic motion is $(F_{\text{net}})_x = -kx$. Using Newton's second law, we see that this force causes an acceleration

$$a_x = \frac{(F_{\text{net}})_x}{m} = -\frac{k}{m}x \tag{14.13}$$

Acceleration is proportional to the position x, but with a minus sign. Consequently we expect the acceleration-versus-time graph to be an inverted form of the position-versus-time graph. This is, indeed, what we find; the acceleration-versus-time graph presented in the earlier table is clearly an upside-down cosine function with the same period T. We can write the acceleration as

$$a_x(t) = -a_{\text{max}}\cos\left(\frac{2\pi t}{T}\right) = -a_{\text{max}}\cos(2\pi ft) \tag{14.14}$$

In this and coming chapters, we will use sine and cosine functions extensively, so we will summarize some of the key aspects of mathematical relationships using these functions.

Sinusoidal relationships

A quantity that oscillates in time and can be written

$$x = A\sin\left(\frac{2\pi t}{T}\right)$$

or

$$x = A\cos\left(\frac{2\pi t}{T}\right)$$

is called a **sinusoidal function** with **period** T. The argument of the functions, $2\pi t/T$, is in radians.

The graphs of both functions have the same shape, but they have different initial values at $t = 0$ s.

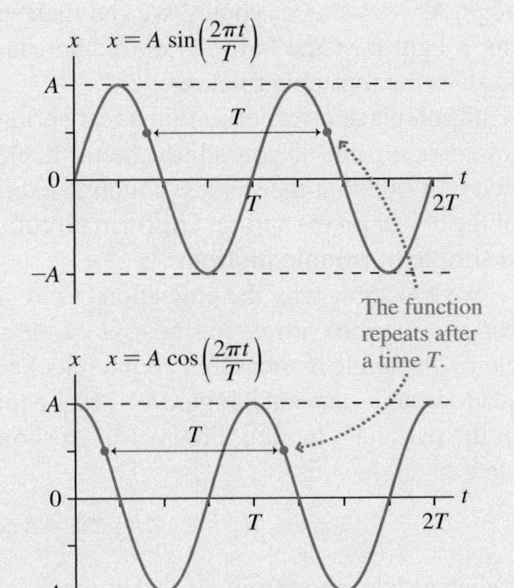

LIMITS If x is a sinusoidal function, then x is:

- *Bounded*—it can take only values between A and $-A$.
- *Periodic*—it repeats the same sequence of values over and over again. Whatever value x has at time t, it has the same value at $t + T$.

SPECIAL VALUES The function x has special values at certain times:

	$t = 0$	$t = \frac{1}{4}T$	$t = \frac{1}{2}T$	$t = \frac{3}{4}T$	$t = T$
$x = A\sin(2\pi t/T)$	0	A	0	$-A$	0
$x = A\cos(2\pi t/T)$	A	0	$-A$	0	A

Exercise 6 

EXAMPLE 14.2 **Understanding the motion of a glider on a spring**

An air-track glider oscillates horizontally on a spring at a frequency of 0.50 Hz. Suppose the glider is pulled to the right of its equilibrium position by 12 cm and then released. Where will the glider be 1.0 s after its release? What is its velocity at this point?

STRATEGIZE The motion of the glider will be sinusoidal. It is released from rest at $x = 12$ cm, so its amplitude is 12 cm.

PREPARE The frequency is 0.50 Hz, so the period is $T = 1/f = 2.0$ s. The glider is released at maximum extension from the equilibrium position, meaning that we can take this point to be $t = 0$.

SOLVE 1.0 s is exactly half the period. As the graph of the motion in FIGURE 14.10 shows, half a cycle brings the glider to its left turning point, 12 cm to the left of the equilibrium position. The velocity at this point is zero.

FIGURE 14.10 Position-versus-time graph for the glider.

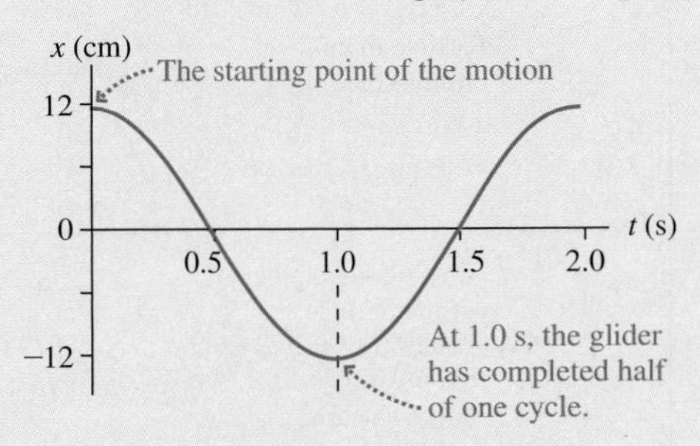

ASSESS Drawing a graph was an important step that helped us make sense of the motion.

Connecting SHM to Uniform Circular Motion

Both circular motion and simple harmonic motion are motions that repeat. Many of the concepts we have used for describing simple harmonic motion were introduced in our study of circular motion. We will use this connection to extend our knowledge of the kinematics of circular motion to simple harmonic motion.

FIGURE 14.11 Projecting the circular motion of a rotating ball.

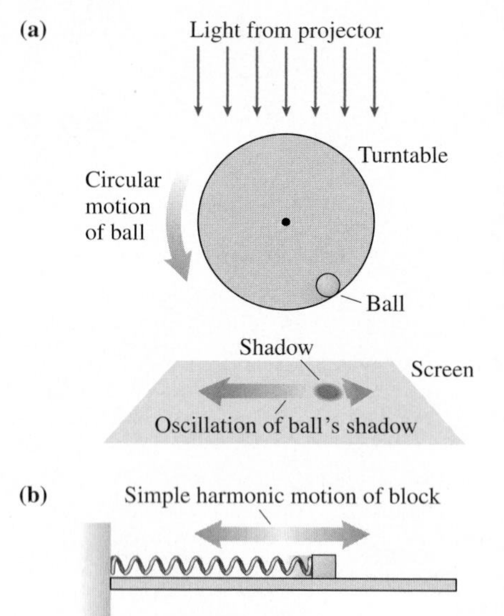
(a)
Light from projector
Turntable
Circular motion of ball
Ball
Shadow
Screen
Oscillation of ball's shadow
(b)
Simple harmonic motion of block

We can demonstrate the relationship between circular and simple harmonic motion with a simple experiment. Suppose a turntable has a small ball glued to the edge. As FIGURE 14.11a shows, we can make a "shadow movie" of the ball by projecting a light past the ball and onto a screen. The ball's shadow oscillates back and forth as the turntable rotates.

If you place a real object on a real spring directly below the shadow, as shown in FIGURE 14.11b, and if you adjust the turntable to have the same period as the spring, you will find that the shadow's motion exactly matches the simple harmonic motion of the object on the spring. **Uniform circular motion projected onto one dimension is simple harmonic motion.**

We can show why the projection of circular motion is simple harmonic motion by considering the particle in FIGURE 14.12a. As in ◀◀ SECTION 7.1, we can locate the particle by the angle ϕ measured counterclockwise from the x-axis. Projecting the ball's shadow onto a screen in Figure 14.11a is equivalent to observing just the x-component of the particle's motion. Figure 14.12a shows that the x-component, when the particle is at angle ϕ, is

$$x = A\cos\phi$$

If the particle starts from $\phi_0 = 0$ at $t = 0$, its angle at a later time t is

$$\phi = \omega t$$

where ω is the particle's *angular velocity,* as defined in Section 7.1. Recall that the angular velocity is related to the frequency f by

$$\omega = 2\pi f$$

The particle's x-component can therefore be expressed as

$$x(t) = A\cos(2\pi f t) \qquad (14.15)$$

This is identical to Equation 14.11 for the position of a mass on a spring! **The x-component of a particle in uniform circular motion is simple harmonic motion.**

FIGURE 14.12 Projections of position, velocity, and acceleration.

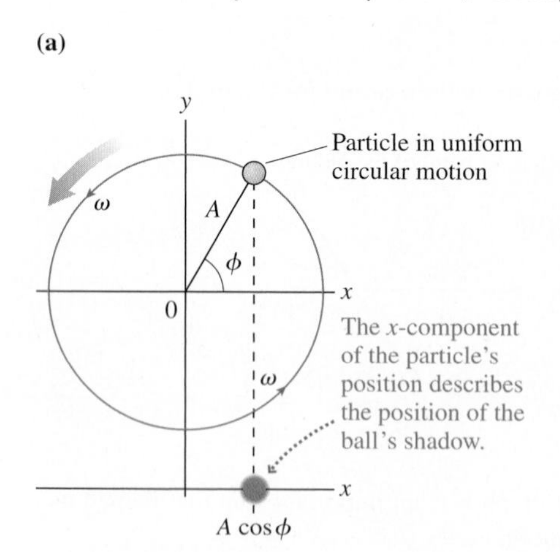
(a)
y
Particle in uniform circular motion
ω
A
φ
x
0
The x-component of the particle's position describes the position of the ball's shadow.
ω
x
A cos φ

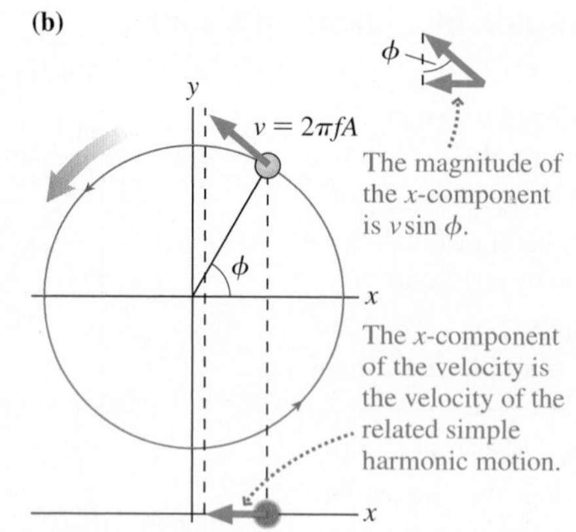
(b)
y
φ
v = 2πfA
The magnitude of the x-component is v sin φ.
φ
x
The x-component of the velocity is the velocity of the related simple harmonic motion.
x

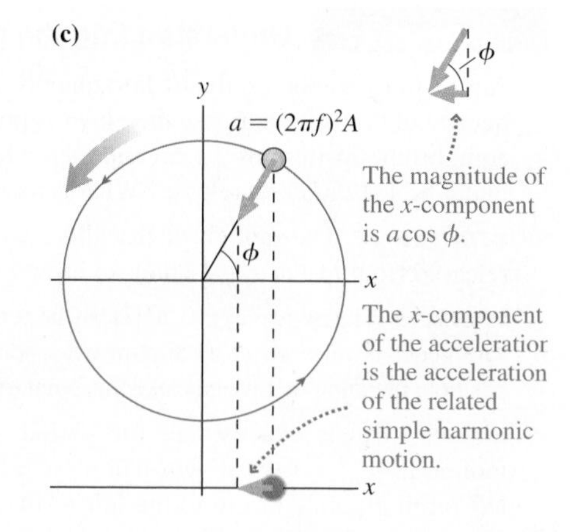
(c)
φ
y
a = (2πf)²A
The magnitude of the x-component is a cos φ.
φ
x
The x-component of the acceleration is the acceleration of the related simple harmonic motion.
x

We can use this correspondence to deduce more details. FIGURE 14.12b shows the velocity vector tangent to the circle. The magnitude of the velocity vector is the particle's speed. Recall from ◀◀ SECTION 6.1 that the speed of a particle in circular motion with radius A and frequency f is $v = 2\pi f A$. Thus the x-component of the velocity vector, which is pointing in the negative x-direction, is

$$v_x = -v\sin\phi = -(2\pi f)A\sin(2\pi f t)$$

According to the correspondence between circular motion and simple harmonic motion, this is the velocity of an object in simple harmonic motion. This is, indeed, exactly Equation 14.12, which we deduced from the graph of the motion, if we define the maximum speed to be

$$v_{max} = 2\pi fA \quad (14.16)$$

FIGURE 14.12c shows the centripetal acceleration for the circular motion. The magnitude of the centripetal acceleration is $a = v^2/A = (2\pi f)^2 A$. The x-component of the acceleration vector, which is the acceleration for simple harmonic motion, is

$$a_x = -a\cos\phi = -(2\pi f)^2 A\cos(2\pi ft)$$

The maximum acceleration is thus

$$a_{max} = (2\pi f)^2 A \quad (14.17)$$

Once you establish that motion is simple harmonic, you can describe it with the equations developed above, which we summarize in Synthesis 14.1. **If you know the amplitude and the frequency, the motion is completely specified.**

SHM in your microwave The next time you are warming a cup of water in a microwave oven, try this: As the turntable rotates, moving the cup in a circle, stand in front of the oven with your eyes level with the cup and watch it, paying attention to the side-to-side motion. You'll see something like the turntable demonstration. The cup's apparent motion is the horizontal component of the turntable's circular motion—simple harmonic motion!

SYNTHESIS 14.1 Describing simple harmonic motion

The position, velocity, and acceleration of objects undergoing simple harmonic motion are sinusoidal functions whose amplitudes and phases are related in a particular way.

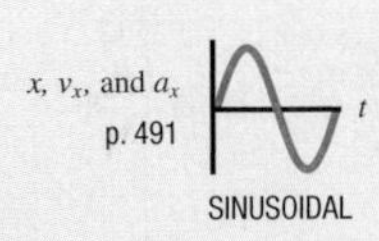

Position

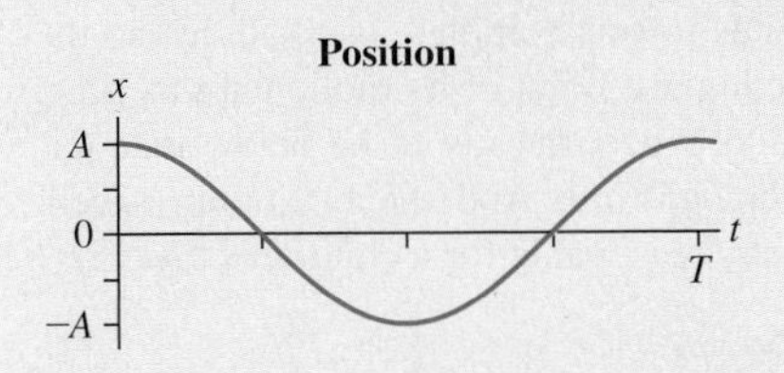

Velocity

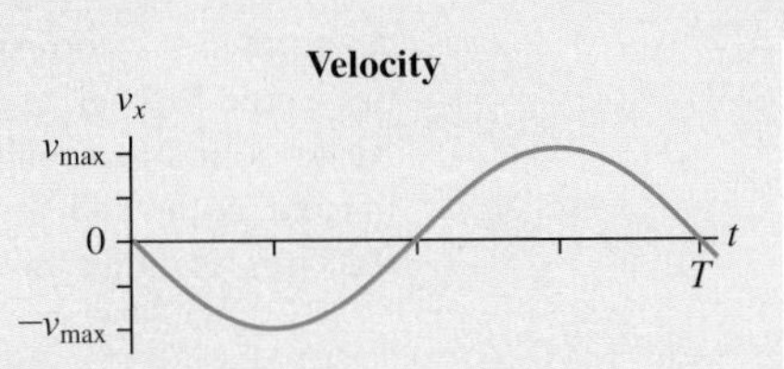

Acceleration

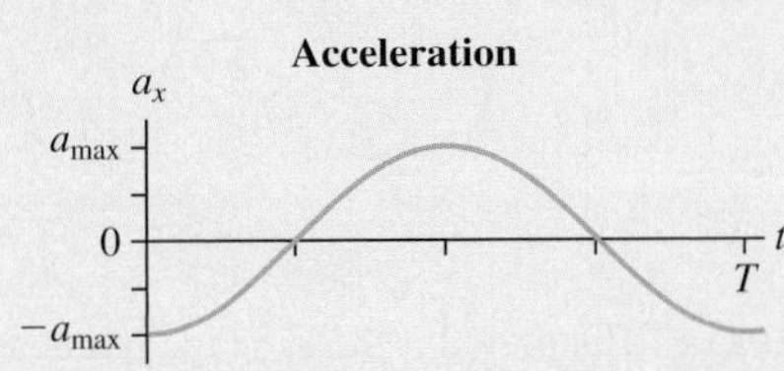

At time t, the displacement, velocity, and acceleration are given by:

$$x(t) = A\cos(2\pi ft) \qquad v_x(t) = -v_{max}\sin(2\pi ft) \qquad a_x(t) = -a_{max}\cos(2\pi ft)$$

The maximum values of the displacement, velocity, and acceleration are determined by the amplitude A and the frequency f:

$$x_{max} = A \qquad v_{max} = 2\pi fA \qquad a_{max} = (2\pi f)^2 A$$

EXAMPLE 14.3 Measuring the sway of a tall building in the wind

The John Hancock Center in Chicago is 100 stories high. Strong winds can cause the building to sway, as is the case with all tall buildings. On particularly windy days, the top of the building oscillates with an amplitude of 40 cm ($\approx$ 16 in) and a period of 7.7 s. What are the maximum speed and acceleration of the top of the building?

STRATEGIZE We will use the relationships for amplitude, velocity, and acceleration from Synthesis 14.1.

PREPARE We assume that the oscillation of the building is simple harmonic motion with amplitude $A = 0.40$ m. The frequency can be computed from the period:

$$f = \frac{1}{T} = \frac{1}{7.7\text{ s}} = 0.13\text{ Hz}$$

SOLVE We can use the equations for maximum velocity and acceleration in Synthesis 14.1 to compute:

$$v_{max} = 2\pi fA = 2\pi(0.13\text{ Hz})(0.40\text{ m}) = 0.33\text{ m/s}$$

$$a_{max} = (2\pi f)^2 A = [2\pi(0.13\text{ Hz})]^2(0.40\text{ m}) = 0.27\text{ m/s}^2$$

In terms of the free-fall acceleration, the maximum acceleration is $a_{max} = 0.027g$.

ASSESS The acceleration is quite small, as you would expect; if it were large, building occupants would certainly complain! Even if they don't notice the motion directly, office workers on high floors of tall buildings may experience a bit of nausea when the oscillations are large because the acceleration affects the equilibrium organs in the inner ear.

EXAMPLE 14.4 **Analyzing the motion of a hanging toy**

A classic children's toy consists of a wooden animal suspended from a spring. If you lift the toy up by 10 cm and let it go, it will gently bob up and down, completing 4 oscillations in 10 seconds.

a. What is the oscillation frequency?
b. When does the toy first reach its maximum speed, and what is this speed?
c. What are the position and velocity 4.0 s after you release the toy?

STRATEGIZE The toy is a mass on a vertical spring, so it undergoes simple harmonic motion. We will need to use the relationships from Synthesis 14.1, using the variable y instead of x because the motion here is vertical.

PREPARE The toy begins its motion at its maximum positive displacement of 10 cm, or 0.10 m; this is the amplitude of the motion. Because the toy is released at maximum positive displacement, we can set the moment of release as $t = 0$ s. The toy completes 4 oscillations in 10 seconds, so the period is 1/4 of 10 seconds, or $T = 2.5$ s. FIGURE 14.13 shows the first two cycles of the oscillation.

FIGURE 14.13 A position graph for the spring toy.

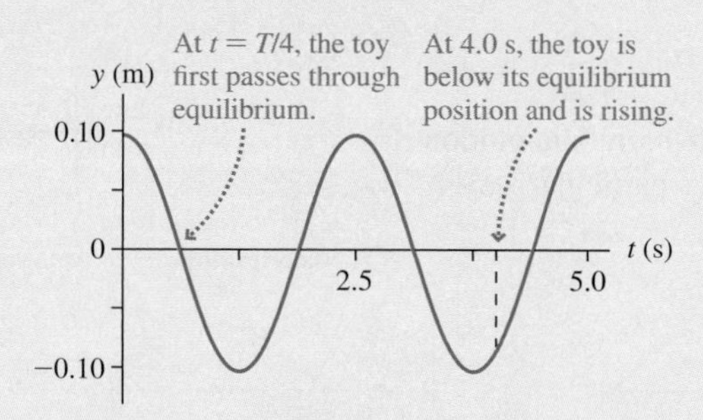

SOLVE a. The period is 2.5 s, so the frequency is

$$f = \frac{1}{T} = \frac{1}{2.5\text{ s}} = 0.40\text{ oscillation/s} = 0.40\text{ Hz}$$

b. Figure 14.13 shows that the toy first passes through the equilibrium position at $t = T/4 = (2.5\text{ s})/4 = 0.62\text{ s}$. This is a point of maximum speed, which we calculate using the maximum speed equation from Synthesis 14.1:

$$v_{max} = 2\pi f A = 2\pi(0.40\text{ Hz})(0.10\text{ m}) = 0.25\text{ m/s}$$

c. 4.0 s after release is $t = 4.0\text{ s}$. Figure 14.13 shows that the toy is below the equilibrium position and rising at this time, so we expect that the position is negative and the velocity is positive. We can use expressions from Synthesis 14.1 to find the position and velocity at this time:

$$y(t = 4.0\text{ s}) = A\cos(2\pi f t)$$
$$= (0.10\text{ m})\cos[2\pi(0.40\text{ Hz})(4.0\text{ s})]$$
$$= -0.081\text{ m}$$
$$v_y(t = 4.0\text{ s}) = -v_{max}\sin(2\pi f t)$$
$$= -(0.25\text{ m/s})\sin[2\pi(0.40\text{ Hz})(4.0\text{ s})]$$
$$= 0.15\text{ m/s}$$

Remember that your calculator must be in radian mode to do calculations like these. The toy is below its equilibrium position ($y < 0$) and moving upward ($v_y > 0$), as we expected from the graph in Figure 14.13.

ASSESS 5.0 s corresponds to two complete oscillations, so the toy would be back at the highest point of its motion at this time. At 4.0 s, it's reasonable to expect the toy to be below its equilibrium point and moving upward. And the maximum speed, 0.25 m/s, is rather modest—reasonable for a children's toy.

STOP TO THINK 14.4 The figures show four identical oscillators at different points in their motion. Which is moving fastest at the time shown?

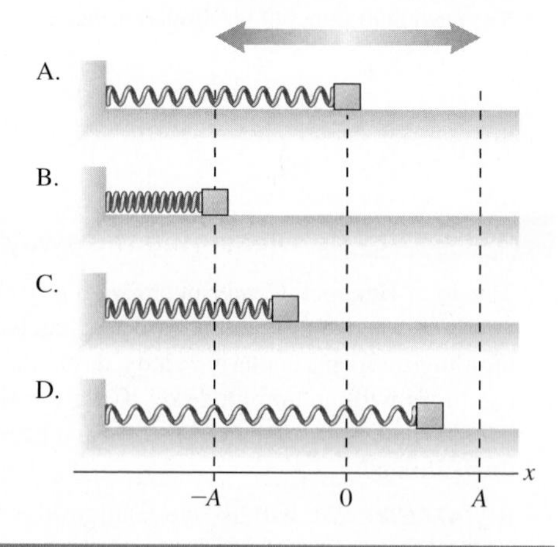

Elastic cords lead to the up-and-down motion of a bungee jump.

14.4 Energy in Simple Harmonic Motion

A bungee jumper falls, increasing in speed, until the elastic cords attached to his ankles start to stretch. The kinetic energy of his motion is transformed into the elastic potential energy of the cords. Once the cords reach their maximum stretch, his velocity reverses. He rises as the elastic potential energy of the cords is transformed back into kinetic energy. If he keeps bouncing (simple harmonic motion), this transformation happens again and again.

This interplay between kinetic and potential energy is very important for understanding simple harmonic motion. Let's explore these energy transformations by looking at the five diagrams in FIGURE 14.14a, which show the position and velocity of a mass on a spring at successive points in time.

The object begins at rest, with the spring at a maximum extension; here, the kinetic energy is zero and the potential energy is a maximum. As the spring contracts, the object speeds up until it reaches the center point of its oscillation, the equilibrium point. At this point, the potential energy is zero and the kinetic energy is a maximum. As the object continues to move, it slows down as it compresses the spring. Eventually, it reaches the turning point, where its instantaneous velocity is zero. At this point, the kinetic energy is zero and the potential energy is again a maximum.

Now suppose that the object has mass m, the spring has spring constant k, and that the motion takes place on a frictionless surface. You learned in ◀◀ SECTION 10.4 that the elastic potential energy of a spring stretched by a distance x from its equilibrium position is

$$U = \frac{1}{2}kx^2 \tag{14.18}$$

Because the surface is frictionless there is no energy loss to thermal energy, so conservation of energy for this system can be written

$$E = K + U = \frac{1}{2}mv^2 + \frac{1}{2}kx^2 = \text{constant} \tag{14.19}$$

FIGURE 14.14b shows a graph of the potential energy, kinetic energy, and total energy for the object as it moves. You can see that, as the object goes through its motion, energy is transformed from potential to kinetic and then back to potential. At maximum displacement, with $x = \pm A$ and $v_x = 0$, the energy is purely potential, so the potential energy has its maximum value:

$$E(\text{at } x = \pm A) = U_{\text{max}} = \frac{1}{2}kA^2 \tag{14.20}$$

At $x = 0$, where $v_x = \pm v_{\text{max}}$, the energy is purely kinetic, so the kinetic energy has its maximum value:

$$E(\text{at } x = 0) = K_{\text{max}} = \frac{1}{2}m(v_{\text{max}})^2 \tag{14.21}$$

FIGURE 14.14 Energy transformations for a mass on a spring.

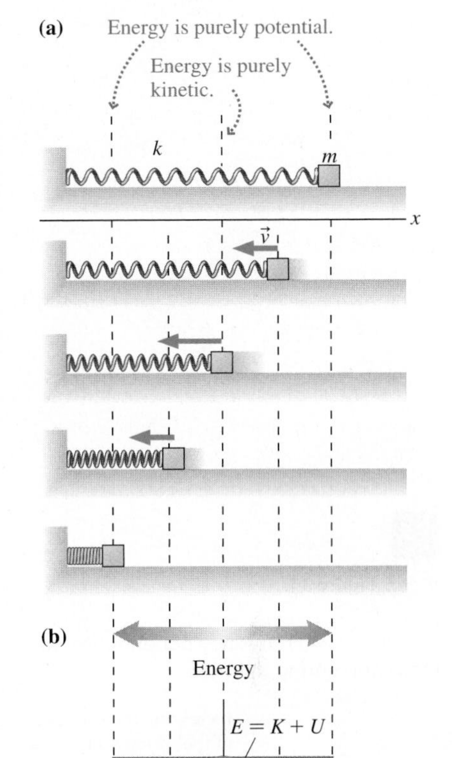

CONCEPTUAL EXAMPLE 14.5 Energy changes for a playground swing

You are at the playground, undergoing simple harmonic motion on a swing. Describe the changes in energy that occur during one cycle of the motion, starting from when you are at the farthest forward point, motionless, and just about to swing backward.

REASON The motion of the swing is that of a pendulum, with you playing the role of the pendulum bob. The energy at different points of the motion is illustrated in FIGURE 14.15. When you are motionless and at the farthest forward point, you are raised up; your energy is entirely gravitational potential energy. As you swing back, your potential energy decreases and your kinetic energy increases, reaching a maximum when the swing is at the lowest point. The swing continues to move backward; you rise up, transforming kinetic energy into potential energy. The process then reverses as you move forward.

ASSESS This description matches the experience of anyone who has been on a swing. You know that you are momentarily motionless at the highest points, and that the motion is fastest at the lowest point.

FIGURE 14.15 Energy at different points of the motion of a swing.

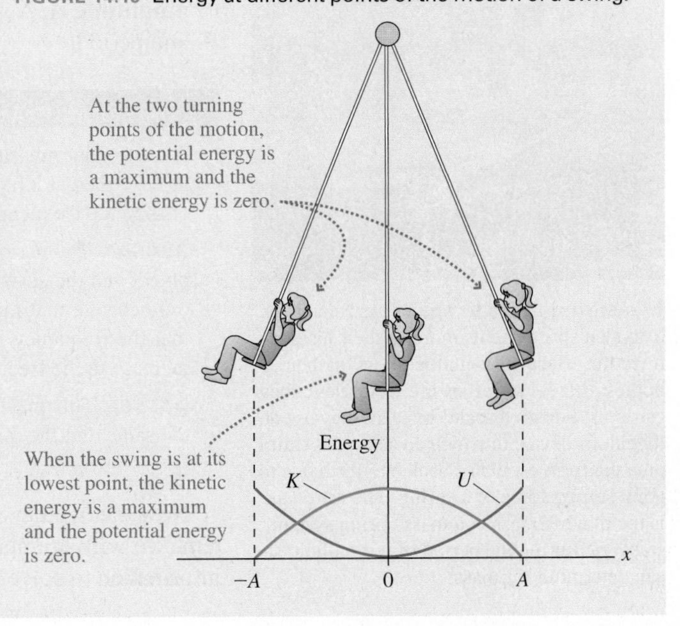

At the two turning points of the motion, the potential energy is a maximum and the kinetic energy is zero.
When the swing is at its lowest point, the kinetic energy is a maximum and the potential energy is zero.
Energy
K
U
x
−A
0
A

Finding the Frequency for Simple Harmonic Motion

We can use what we know about energy in simple harmonic motion to deduce other details of the motion. Let's return to the mass on a spring of Figure 14.14. The graph in Figure 14.14b shows energy being transformed back and forth between kinetic and potential energy. At the turning points, the energy is purely potential; at the equilibrium point, the energy is purely kinetic. Because the total energy doesn't change, the maximum kinetic energy given in Equation 14.21 must be equal to the maximum potential energy given in Equation 14.20:

$$\frac{1}{2}m(v_{\text{max}})^2 = \frac{1}{2}kA^2 \tag{14.22}$$

By solving Equation 14.22 for the maximum speed, we can see that it is related to the amplitude by

$$v_{\text{max}} = \sqrt{\frac{k}{m}}A \tag{14.23}$$

Earlier we found that

$$v_{\text{max}} = 2\pi fA \tag{14.24}$$

Comparing Equations 14.23 and 14.24, we see that the frequency, and thus the period, of an oscillating mass on a spring is determined by the spring constant k and the object's mass m:

$$f = \frac{1}{2\pi}\sqrt{\frac{k}{m}} \quad \text{and} \quad T = 2\pi\sqrt{\frac{m}{k}} \tag{14.25}$$

Frequency and period of SHM
for mass m on a spring with spring constant k

Video Figure 14.16

FIGURE 14.16 Frequency dependence on mass and spring stiffness.

High frequency
Stiff spring, low mass
Increasing the stiffness of the spring will increase the restoring force. This *increases* the frequency.
$f = \frac{1}{2\pi}\sqrt{\frac{k}{m}}$
Low frequency
Soft spring, high mass
Increasing the mass will increase the inertia of the system. This *decreases* the frequency.

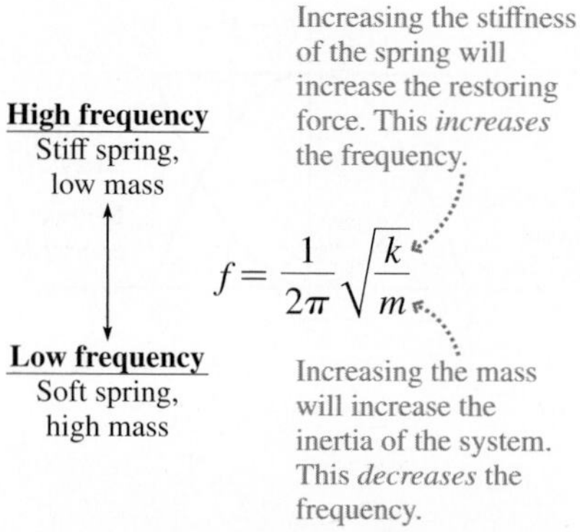

We can make two observations about these equations:

- **The frequency and period of simple harmonic motion are determined by the physical properties of the oscillator.** The frequency and period of a mass on a spring are determined by (1) the mass and (2) the stiffness of the spring, as shown in FIGURE 14.16. This dependence of frequency and period on a force term and an inertia term will also apply to other oscillators.
- **The frequency and period of simple harmonic motion do not depend on the amplitude A.** A small oscillation and a large oscillation have the same frequency and period.

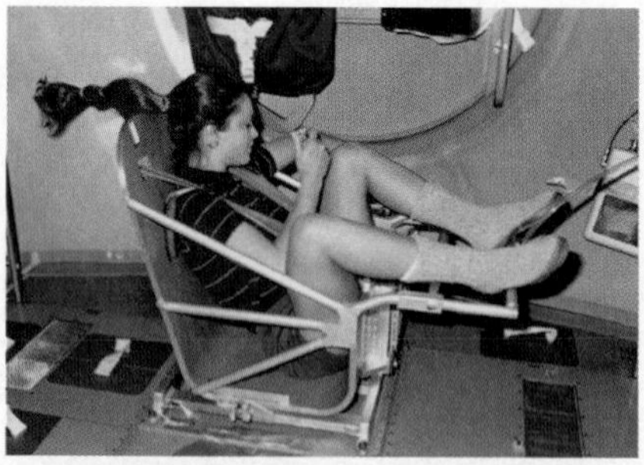

Measuring mass in space Astronauts on extended space flights monitor their mass to track the effects of weightlessness on their bodies. But because they are weightless, they can't just hop on a scale! Instead, they use an ingenious device in which an astronaut sitting on a platform oscillates back and forth due to the restoring force of a spring. The astronaut is the moving mass in a mass-spring system, so by measuring the period of her motion, she can determine her mass.

CONCEPTUAL EXAMPLE 14.6 Changing mass, changing period

An astronaut measures her mass each day using the Body Mass Measurement Device, as described at left. During an 8-day flight, her mass steadily decreases. How does this change the frequency of her oscillatory motion on the device?

REASON The period and frequency of a mass-spring system depend on the mass of the object and the spring constant. The spring constant of the device won't change, so the only change that matters is the change in the astronaut's mass. Equation 14.25 shows that the frequency is proportional to $\sqrt{k/m}$, so a decrease in her mass will cause an increase in the frequency. The oscillation will be a bit more rapid.

ASSESS This makes sense. The force of the spring—which causes the oscillation—is the same, but the mass to be accelerated is less. We expect a higher frequency.

Now that we have a complete description of simple harmonic motion in one system, we will summarize the details in Tactics Box 14.1, showing how to use this information to solve oscillation problems.

TACTICS BOX 14.1 Identifying and analyzing simple harmonic motion

1. If the net force acting on a particle is a linear restoring force, the motion is simple harmonic motion around the particle's equilibrium position.
2. The position, velocity, and acceleration as a function of time are given in Synthesis 14.1. The equations are given in terms of x, but they can be written in terms of y, θ, or some other variable if the situation calls for it.
3. The amplitude A is the maximum value of the displacement from equilibrium. The maximum speed and the maximum magnitude of the acceleration are given in Synthesis 14.1.
4. The frequency f (and hence the period $T = 1/f$) depends on the physical properties of the particular oscillator, but f does *not* depend on A.

   For a mass on a spring, the frequency is given by $f = \dfrac{1}{2\pi}\sqrt{\dfrac{k}{m}}$.
5. The sum of potential energy and kinetic energy is constant. As the oscillation proceeds, energy is transformed from kinetic into potential energy and then back again.

Exercises 10–11

EXAMPLE 14.7 Finding the frequency of a mass hanging from a spring

A spring has an unstretched length of 10.0 cm. A 25 g mass is hung from the spring, stretching it to a length of 15.0 cm. If the mass is pulled down and released so that it oscillates, what will be the frequency of the oscillation?

STRATEGIZE The spring provides a linear restoring force, so the motion will be simple harmonic, as noted in Tactics Box 14.1. The oscillation frequency depends on the spring constant, which we will determine from the stretch of the spring.

PREPARE FIGURE 14.17 gives a visual overview of the situation.

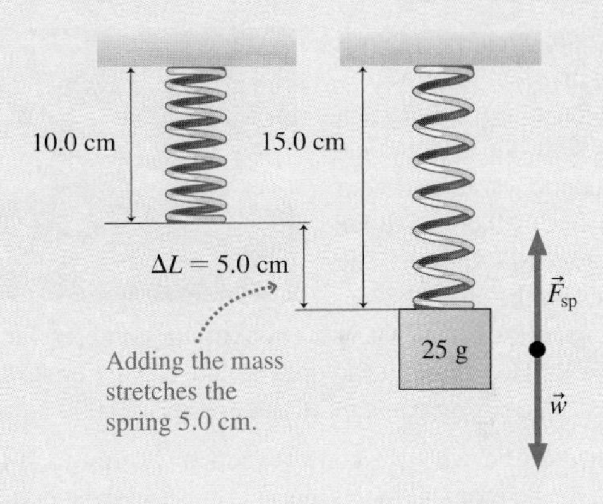

FIGURE 14.17 Visual overview of a mass suspended from a spring.

SOLVE When the mass hangs at rest, after stretching the spring to 15 cm, the net force on it must be zero. Thus the magnitude of the upward spring force equals the downward weight, giving $k\,\Delta L = mg$. The spring constant is thus

$$k = \frac{mg}{\Delta L} = \frac{(0.025\text{ kg})(9.8\text{ m/s}^2)}{0.050\text{ m}} = 4.9\text{ N/m}$$

Now that we know the spring constant, we can compute the oscillation frequency:

$$f = \frac{1}{2\pi}\sqrt{\frac{k}{m}} = \frac{1}{2\pi}\sqrt{\frac{4.9\text{ N/m}}{0.025\text{ kg}}} = 2.2\text{ Hz}$$

ASSESS 2.2 Hz is 2.2 oscillations per second. This seems like a reasonable frequency for a mass on a spring. A frequency in the kHz range (thousands of oscillations per second) would have been suspect!

EXAMPLE 14.8 Weighing DNA molecules BIO

It has recently become possible to "weigh" individual DNA molecules by measuring the influence of their mass on a nanoscale oscillator. FIGURE 14.18 shows a thin rectangular cantilever etched out of silicon. The cantilever has a mass of 3.7×10^{-16} kg. If pulled down and released, the end of the cantilever vibrates with simple harmonic motion, moving up and down like a diving board after a jump. When the end of the cantilever is bathed with DNA molecules whose ends have been modified to bind to a surface, one or more molecules may attach to the end of the cantilever. The addition of their mass causes a very slight—but measurable—decrease in the oscillation frequency.

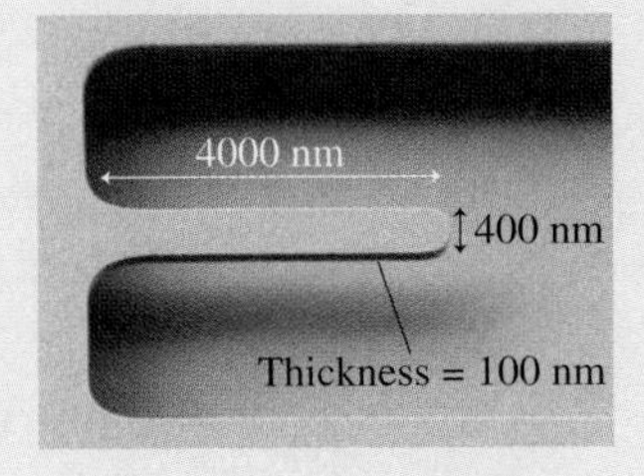

FIGURE 14.18 A nanoscale cantilever.

A vibrating cantilever of mass M can be modeled as a simple block of mass $\frac{1}{3}M$ attached to a spring. (The factor of $\frac{1}{3}$ arises from the moment of inertia of a bar pivoted at one end: $I = \frac{1}{3}ML^2$.) Neither the mass nor the spring constant can be determined very accurately—perhaps to only two significant figures—but the

Continued

oscillation frequency can be measured with very high precision simply by counting the oscillations. In one experiment, the cantilever was initially vibrating at exactly 12 MHz. Attachment of a DNA molecule caused the frequency to decrease by 50 Hz. What was the mass of the DNA molecule?

STRATEGIZE According to Equation 14.25, the oscillation frequency of a mass on a spring depends on the mass and the spring constant. When the DNA attaches, increasing the mass, the spring constant remains constant. We will use this fact to determine the change in mass.

PREPARE We model the cantilever as a block of mass $m = \frac{1}{3}M = 1.2 \times 10^{-16}$ kg oscillating on a spring with spring constant k. When the mass increases to $m + m_{\text{DNA}}$, the oscillation frequency decreases from $f_0 = 12{,}000{,}000$ Hz to $f_1 = 11{,}999{,}950$ Hz.

SOLVE We can solve Equation 14.25 for the spring constant to get $k = m(2\pi f)^2$. The spring constant k_0 before the mass is added is equal to the spring constant k_1 after it's added, so we have

$$k_0 = m(2\pi f_0)^2 = k_1 = (m + m_{\text{DNA}})(2\pi f_1)^2$$

The 2π terms cancel, and we can rearrange this equation to give

$$\frac{m + m_{\text{DNA}}}{m} = 1 + \frac{m_{\text{DNA}}}{m} = \left(\frac{f_0}{f_1}\right)^2 = \left(\frac{12{,}000{,}000\text{ Hz}}{11{,}999{,}950\text{ Hz}}\right)^2 = 1.0000083$$

Subtracting 1 from both sides gives

$$\frac{m_{\text{DNA}}}{m} = 0.0000083$$

and thus

$$m_{\text{DNA}} = 0.0000083m = (0.0000083)(1.2 \times 10^{-16}\text{ kg}) = 1.0 \times 10^{-21}\text{ kg} = 1.0 \times 10^{-18}\text{ g}$$

ASSESS This is a reasonable mass for a DNA molecule. It's a remarkable technical achievement to be able to measure a mass this small. With further improvements in sensitivity, scientists will be able to determine the number of base pairs in a strand of DNA simply by weighing it!

EXAMPLE 14.9 A gymnast jumps on a springboard

In the vault event, the gymnast begins her routine by jumping on a *springboard*, a platform with springs underneath. In one vault, a 45 kg gymnast jumps 0.60 m above the springboard's surface. She then lands on the springboard, which compresses it. What is the maximum compression of the springboard? How much time does it take for the board to fully compress? The spring constant of the board is 50,000 N/m.

STRATEGIZE We will use conservation of energy to solve this problem. At the moment the gymnast is at her highest point above the board, the system's energy is entirely her gravitational potential energy. Later, when the board is fully compressed, the energy is her (lower) gravitational potential energy plus the potential energy of the spring (we ignore the small mass of the board). By conservation of energy, these two energies are equal. We can calculate the compression time by treating the mass-spring system as an oscillator undergoing simple harmonic motion.

PREPARE Once the gymnast has landed on the platform, we can model the system as a mass (the gymnast) on a spring (the platform). FIGURE 14.19 gives a visual overview of the problem that we can use for conservation of energy. We can measure both the gymnast's position y and the platform's compression x from the platform's uncompressed position.

SOLVE The gymnast is instantaneously at rest at both the beginning and end of the motion, so she has no kinetic energy at those points. We also note that her final position is the same as the maximum compression of the platform—that is, x_f. Thus we can write conservation of energy as

$$\underbrace{mgx_f}_{(U_g)_f} + \underbrace{\tfrac{1}{2}kx_f^2}_{(U_{sp})_f} = \underbrace{mgy_i}_{(U_g)_i}$$

FIGURE 14.19 A gymnast jumps onto a springboard.

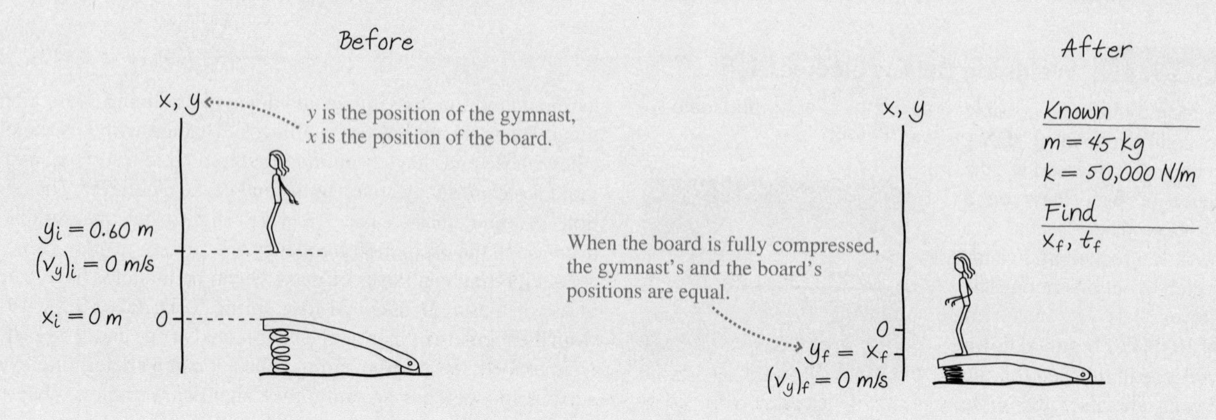

This is a quadratic equation in the variable x_f. Using the quadratic equation, the solution is

$$x_f = \frac{-mg \pm \sqrt{(mg)^2 + 4\left(\frac{1}{2}k\right)(mgy_i)}}{k}$$

Inserting the values $m = 45$ kg, $g = 9.8$ m/s^2, $k = 50{,}000$ N/m, and $y_i = 0.60$ m, we find that x_i equals either +9.4 cm (from the "+" sign in the quadratic equation) or −11.2 cm (from the "−" sign). Because the board is *compressed*, so that $x_f < 0$, we know that the correct answer is $x_f = -11.2$ cm. (The positive solution corresponds to the artificial case where the gymnast is instantaneously at rest after moving back up on the springboard; this could happen only if her feet became attached to the board so that she couldn't continue upward on her own!)

To find the time for the compression to occur, we note that the compression is $\frac{1}{4}$ of a cycle of simple harmonic motion, so the time to completely compress the board occurs at $t_f = \frac{1}{4}T$. The period is computed using Equation 14.25, so we write

$$t_f = \frac{1}{4}T = \frac{1}{4}2\pi\sqrt{\frac{m}{k}} = \frac{\pi}{2}\sqrt{\frac{45\text{ kg}}{50{,}000\text{ N/m}}} = 0.047\text{ s}$$

This is the time required for the springboard to compress.

ASSESS The compression distance of around 11 cm seems reasonable. The time to compress the board is short, but the springs in the board are very stiff so this seems plausible. It is interesting that the time to compress the springboard—that is, the time it takes for the gymnast to stop—depends on only her mass and the spring constant; it does not depend on how fast she was moving when she landed on the board, or how far the board was compressed. This is a feature of simple harmonic motion for a mass on a spring: The period depends on only the mass and the spring constant, not on other details of the motion.

We can think of the gymnast's landing on the springboard as a *collision* between her and a spring-like, or *elastic*, object. Many real-world collisions involve a moving object that is stopped by an elastic object. In other cases (such as a bouncing rubber ball) the object itself is elastic. The time of the collision is reasonably constant, independent of the speed of the collision, for the reasons noted in Example 14.9.

Automobile collision times When a car hits an obstacle, it takes approximately 0.1 s to come to rest, regardless of the initial speed. The crumpling of the front of a car is complex, but for many cars the force is approximately proportional to the displacement during the compression, so we can model the body of the car as a mass and the front of the car as a spring. In this model, the time for the car to come to rest does not depend on the initial speed.

STOP TO THINK 14.5 Four mass-spring systems have masses and spring constants shown here. Rank in order, from highest to lowest, the frequencies of the oscillations.

A. k, $4m$
B. $\frac{1}{2}k$, m
C. k, $2m$
D. $2k$, m

14.5 Pendulum Motion

As we've already seen, a simple pendulum—a mass at the end of a string or a rod that is free to pivot—is another system that exhibits simple harmonic motion. Everything we have learned about the mass on a spring can be applied to the pendulum as well.

In the first part of the chapter, we looked at the restoring force in the pendulum. For a pendulum of length L displaced by an arc length s, as in FIGURE 14.20, the tangential restoring force is

$$(F_{net})_t = -\frac{mg}{L}s \tag{14.26}$$

NOTE ► Recall that this equation holds accurately only for small angles. ◄

This linear restoring force has exactly the same form as the net force in a mass-spring system, but with the constants mg/L in place of the constant k. Given this, we can quickly deduce the essential features of pendulum motion by replacing k, wherever it occurs in the oscillating spring equations, with mg/L.

FIGURE 14.20 A simple pendulum.

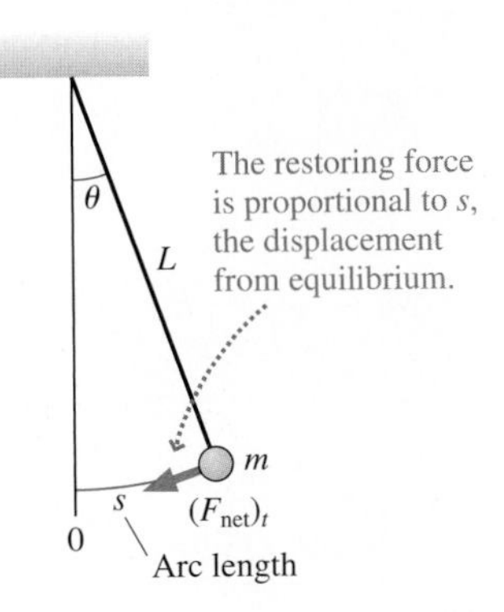

- The oscillation of a pendulum is simple harmonic motion, so the equations of motion can be written for the arc length or the angle:

$$s(t) = A\cos(2\pi ft) \qquad \text{or} \qquad \theta(t) = \theta_{\max}\cos(2\pi ft)$$

- The frequency can be obtained from the equation for the frequency of the mass on a spring by substituting mg/L in place of k:

$$f = \frac{1}{2\pi}\sqrt{\frac{g}{L}} \quad \text{and} \quad T = 2\pi\sqrt{\frac{L}{g}} \tag{14.27}$$

Frequency and period of a pendulum of length L with free-fall acceleration g

- As for a mass on a spring, the frequency does not depend on the amplitude. Note also that **the frequency, and hence the period, is independent of the mass.** It depends only on the length of the pendulum.
- Equation 14.16 showed that the maximum speed for a particle in simple harmonic motion is $v_{\max} = 2\pi fA$. Using the frequency f from Equation 14.27, we find that the maximum speed of a pendulum bob is

$$v_{\max} = \sqrt{\frac{g}{L}}A = \sqrt{gL}\,\theta_{\max} \tag{14.28}$$

The dependence of the pendulum frequency on length and gravity is summed up in FIGURE 14.21. Adjusting the period of a pendulum is a matter of adjusting the length, which is a mechanical operation that can be performed with great precision. Clocks based on the period of a pendulum were the most accurate timepieces available until well into the 20th century.

FIGURE 14.21 Frequency dependence on length and gravity for a pendulum.

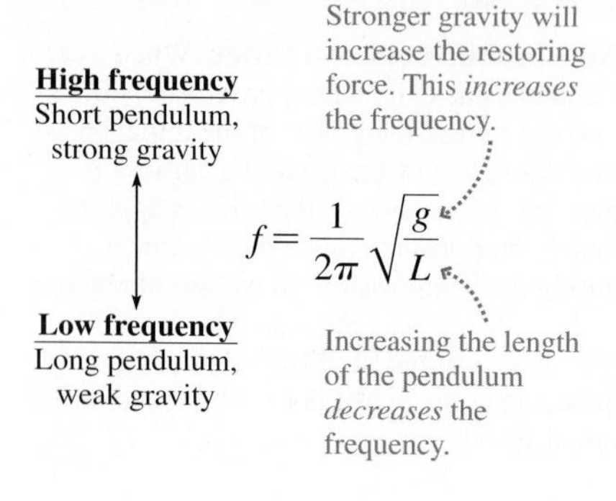

High frequency
Short pendulum, strong gravity
Stronger gravity will increase the restoring force. This increases the frequency.
f = 1/2π √(g/L)
Low frequency
Long pendulum, weak gravity
Increasing the length of the pendulum decreases the frequency.

EXAMPLE 14.10 Designing a pendulum for a clock

A grandfather clock is designed so that one swing of the pendulum in either direction takes 1.00 s. What is the length of the pendulum?

STRATEGIZE The period of a pendulum depends only on its length, according to Equation 14.27.

PREPARE One period of the pendulum is two swings, so the period is $T = 2.00$ s.

SOLVE The period is independent of the mass and depends only on the length. From Equation 14.27,

$$T = \frac{1}{f} = 2\pi\sqrt{\frac{L}{g}}$$

Solving for L, we find

$$L = g\left(\frac{T}{2\pi}\right)^2 = (9.80\ \text{m/s}^2)\left(\frac{2.00\ \text{s}}{2\pi}\right)^2 = 0.993\ \text{m}$$

ASSESS A pendulum clock with a "tick" or "tock" each second requires a long pendulum of about 1 m—which is why these clocks were originally known as "tall case clocks."

Physical Pendulums and Locomotion

In Chapter 6, we computed maximum walking speed using the ideas of circular motion. We can also model the motion of your legs during walking as pendulum motion. When you walk, you push off with your rear leg and then let it swing forward for the next stride. Your leg swings forward under the influence of gravity—like a pendulum.

Try this: Stand on one leg, and gently swing your free leg back and forth. There is a certain frequency at which it will naturally swing. This is your leg's pendulum frequency. Now, try swinging your leg at twice this frequency. You can do it, but it is very difficult. The muscles that move your leg back and forth aren't very strong because under normal circumstances they don't need to apply much force.

A pendulum, like your leg, whose mass is distributed along its length is known as a **physical pendulum.** FIGURE 14.22 shows a simple pendulum and a physical pendulum of the same length. The position of the center of gravity of the physical pendulum is at a distance d from the pivot.

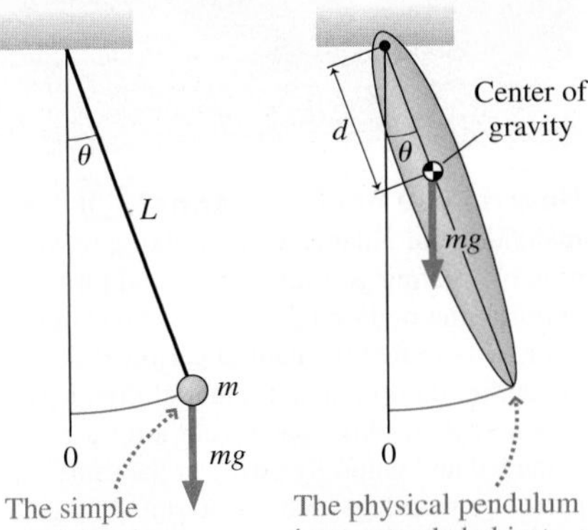

θ
L
m
0
mg
The simple pendulum is a small mass m at the end of a light rod of length L.
d
θ
Center of gravity
mg
0
The physical pendulum is an extended object with mass m, length L, and moment of inertia I.

FIGURE 14.22 A simple pendulum and a physical pendulum of equal length.

Finding the frequency is a rotational motion problem similar to those we considered in Chapter 7. We would expect the frequency to depend on the moment of inertia and the distance to the center of gravity as follows:

- The moment of inertia I is a measure of an object's resistance to rotation. Increasing the moment of inertia while keeping other variables equal should cause the frequency to decrease. In an expression for the frequency of the physical pendulum, we would expect I to appear in the denominator.
- When the pendulum is pushed to the side, a gravitational torque pulls it back. The greater the distance d of the center of gravity from the pivot point, the greater the torque. Increasing this distance while keeping the other variables constant should cause the frequency to increase. In an expression for the frequency of the physical pendulum, we would expect d to appear in the numerator.

A careful analysis of the motion of the physical pendulum produces a result for the frequency that matches these expectations:

$$f = \frac{1}{2\pi}\sqrt{\frac{mgd}{I}} \tag{14.29}$$

Frequency of a physical pendulum of mass m, moment of inertia I, with center of gravity distance d from the pivot

EXAMPLE 14.11 Finding the frequency of a swinging leg BIO

A student in a biomechanics lab measures the length of his leg, from hip to heel, to be 0.90 m. What is the frequency of the pendulum motion of the student's leg? What is the period?

STRATEGIZE Equation 14.29 for the period of a physical pendulum contains the center-of-gravity distance d and the moment of inertia I. We'll need to make some simplifying assumptions to estimate these lengths.

PREPARE We can model a human leg reasonably well as a rod of uniform cross section, pivoted at one end (the hip). Recall from Chapter 7 that the moment of inertia of a rod pivoted about its end is $\frac{1}{3}mL^2$. The center of gravity of a uniform leg is at the midpoint, so $d = L/2$.

SOLVE The frequency of a physical pendulum is given by Equation 14.29. Before we put in numbers, we will use symbolic relationships and simplify:

$$f = \frac{1}{2\pi}\sqrt{\frac{mgd}{I}} = \frac{1}{2\pi}\sqrt{\frac{mg(L/2)}{\frac{1}{3}mL^2}} = \frac{1}{2\pi}\sqrt{\frac{3}{2}\frac{g}{L}}$$

The expression for the frequency is similar to that for the simple pendulum, but with an additional numerical factor of 3/2 inside the square root. The numerical value of the frequency is

$$f = \frac{1}{2\pi}\sqrt{\left(\frac{3}{2}\right)\left(\frac{9.8\text{ m/s}^2}{0.90\text{ m}}\right)} = 0.64\text{ Hz}$$

The period is

$$T = \frac{1}{f} = 1.6\text{ s}$$

ASSESS Notice that we didn't need to know the mass of the leg to find the period. The period of a physical pendulum does not depend on the mass, just as it doesn't for the simple pendulum. The period depends only on the *distribution* of mass. When you walk, swinging your free leg forward to take another stride corresponds to half a period of this pendulum motion. For a period of 1.6 s, this is 0.80 s. For a normal walking pace, one stride in just under one second sounds about right.

As you walk, your legs swing as physical pendulums as you bring them forward. The frequency is fixed by the length of your legs and their distribution of mass; it doesn't depend on amplitude. Consequently, you don't increase your walking speed by taking more rapid steps—changing the frequency is quite difficult. You simply

How do you hold your arms? BIO You maintain your balance when walking or running by moving your arms back and forth opposite the motion of your legs. You hold your arms so that the natural period of their pendulum motion matches that of your legs. At a normal walking pace, your arms are extended and naturally swing at the same period as your legs. When you run, your gait is more rapid. To decrease the period of the pendulum motion of your arms to match, you bend them at the elbows, shortening their effective length and increasing the natural frequency of oscillation.

take longer strides, changing the amplitude but not the frequency. At some point a limit is reached, and you use another gait—you start to run.

As we saw in the photo that opens the chapter, gibbons move through the forest canopy with a hand-over-hand swinging motion called *brachiation*. The body swings under a pivot point where a hand grips a tree branch, a clear example of pendulum motion. A brachiating ape will increase its speed by taking bigger swings. But, just as for walking, at some point a limit is reached, and gibbons use a different gait, launching themselves from branch to branch through the air.

STOP TO THINK 14.6 A pendulum clock is made with a metal rod. It keeps perfect time at a temperature of 20°C. At a higher temperature, the metal rod lengthens. How will this change the clock's timekeeping?

A. The clock will run fast; the dial will be ahead of the actual time.
B. The clock will keep perfect time.
C. The clock will run slow; the dial will be behind the actual time.

14.6 Damped Oscillations

A real pendulum clock must have some energy input; otherwise, the oscillation of the pendulum would slowly decrease in amplitude due to air resistance. If you strike a bell, the oscillation will soon die away as energy is lost to sound waves in the air and dissipative forces within the metal of the bell.

Unless energy is continually added to an oscillator, its amplitude will decrease—sometimes very slowly, but other times quite quickly—as its mechanical energy is transformed into the thermal energy of the oscillator and its environment. An oscillation that runs down and stops is called a **damped oscillation.**

For a pendulum, the main energy loss is due to air resistance, which we called the *drag force* in Chapter 4. When we learned about the drag force, we noted that it depends on velocity: The faster the motion, the bigger the drag force. For this reason, the decrease in amplitude of an oscillating pendulum will be fastest at the start of the motion. For a pendulum or other oscillator with modest damping, we end up with a graph of motion like that in FIGURE 14.23a. The maximum displacement, x_{max}, decreases with time. As the oscillation decays, the *rate* of the decay decreases; the difference between successive peaks is less.

FIGURE 14.23 The motion of a damped oscillator.

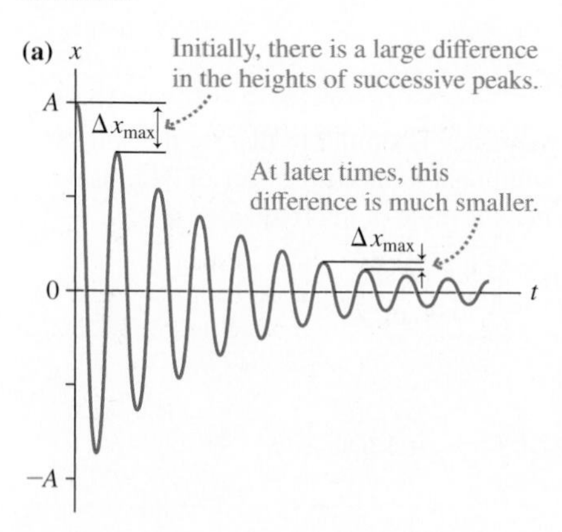

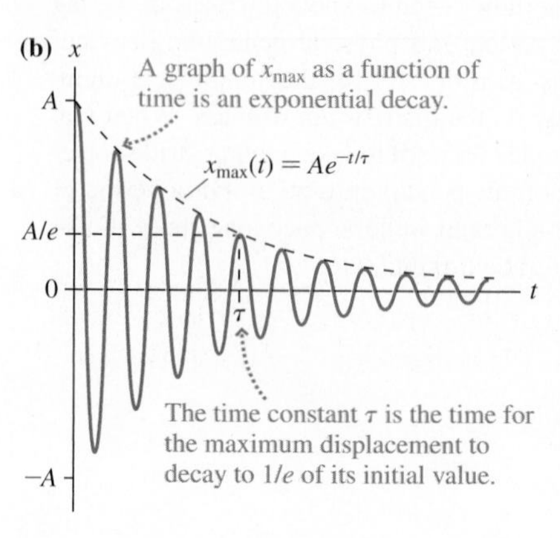

If we plot a smooth curve that connects the peaks of successive oscillations (we call such a curve an *envelope*), we get the dashed line shown in FIGURE 14.23b. It's possible, using calculus, to show that x_{max} decreases with time as

$$x_{max}(t) = Ae^{-t/\tau} \quad (14.30)$$

where $e \approx 2.718$ is the base of the natural logarithm and A is the *initial* amplitude. This steady decrease of x_{max} with time is called an **exponential decay.**

The constant τ (lowercase Greek tau) in Equation 14.30 is called the **time constant.** After one time constant has elapsed—that is, at $t = \tau$—the maximum displacement x_{max} has decreased to

$$x_{max}(\text{at } t = \tau) = Ae^{-1} = \frac{A}{e} \approx 0.37A$$

In other words, the oscillation has decreased after one time constant to about 37% of its initial value. The time constant τ measures the "characteristic time" during which damping causes the amplitude of the oscillation to decay away. An oscillation that decays quickly has a small time constant, whereas a "lightly damped" oscillator, which decays very slowly, has a large time constant.

Because we will see exponential decay again, we will look at it in more detail.

Exponential decay

Exponential decay occurs when a quantity y is proportional to the number e taken to the power $-t/\tau$. The quantity τ is known as the **time constant.** We write this mathematically as

$$y = Ae^{-t/\tau}$$

y is proportional to $e^{-t/\tau}$

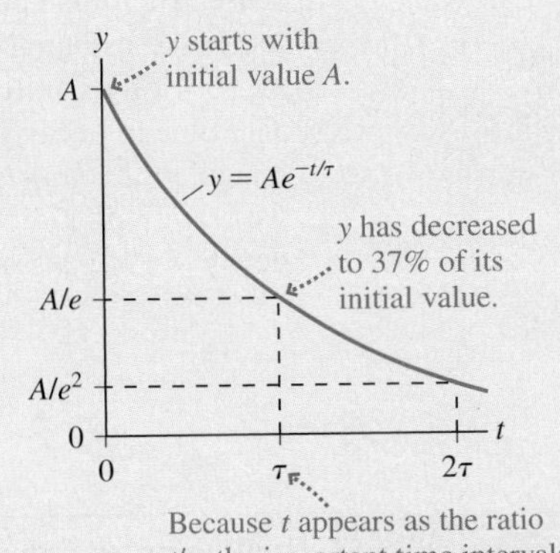

y
y starts with initial value A.
A
y = Ae^{-t/τ}
y has decreased to 37% of its initial value.
A/e
A/e²
0
0
τ
2τ
t
Because t appears as the ratio t/τ, the important time intervals are τ, 2τ, and so on.

SCALING Whenever t increases by one time constant, y decreases by a factor of $1/e$. For instance:

- At time $t = 0$, $y = A$.
- Increasing time to $t = \tau$ reduces y to A/e.
- A further increase to $t = 2\tau$ reduces y by another factor of $1/e$ to A/e^2.

Generally, we can say:

At $t = n\tau$, y has the value A/e^n.

LIMITS As t becomes large, y becomes very small and approaches zero.

Exercises 14–17

The damped oscillation shown in Figure 14.23 continues for a long time. The amplitude isn't zero after one time constant, or two, or three.... Mathematically, the oscillation never ceases, though the amplitude will eventually be so small as to be undetectable. For practical purposes, we can speak of the time constant τ as the *lifetime* of an oscillation—a measure of about how long it takes to decay. The best way to measure the relative size of the time constant is to compare it to the period. If $\tau \gg T$, the oscillation persists for many, many periods and the amplitude decrease from cycle to cycle is quite small. The oscillation of a bell after it is struck has $\tau \gg T$; the sound continues for a long time. Other oscillatory systems have very short time constants, as noted in the description of a car's suspension.

Damping smoothes the ride A car's wheels are attached to the car's body with springs so that the wheels can move up and down as the car moves over an uneven road. The car-spring system oscillates with a typical period of just under 1 second. You don't want the car to continue bouncing after hitting a bump, so a shock absorber provides damping. The time constant for the damping is about the same length as the period, so that the oscillation damps quickly.

EXAMPLE 14.12 Finding a clock's decay time

The pendulum in a grandfather clock has a period of 1.00 s. If the clock's driving spring is allowed to run down, damping due to friction will cause the pendulum to slow to a stop. If the time constant for this decay is 300 s, how long will it take for the pendulum's swing to be reduced to half its initial amplitude?

STRATEGIZE The time constant of 300 s is much greater than the 1.00 s period, so this is an example of modest damping as described by Equation 14.30. We are asked "how long," which implies that we want to find a time t. Because t is in the exponent in Equation 14.30, we'll need to use logarithms to find it.

PREPARE We need to know the time constant of 300 s, but the 1.00 s period of the pendulum is not needed.

SOLVE Equation 14.30 gives an expression for the decay of the maximum displacement of a damped harmonic oscillator:

$$x_{\max}(t) = Ae^{-t/\tau}$$

As noted in Tactics Box 14.1, we can write this equation equally well in terms of the pendulum's angle in the form

$$\theta_{\max}(t) = \theta_i e^{-t/\tau}$$

where θ_i is the initial angle of swing. We wish to find the time t at which the amplitude has decayed to half its initial value. At this time,

$$\theta_{\max}(t) = \theta_i e^{-t/\tau} = \frac{1}{2}\theta_i$$

Continued

The θ_i cancels, giving $e^{-t/\tau} = \frac{1}{2}$.

To solve this for t, we take the natural logarithm of both sides and use the logarithm property $\ln(e^a) = a$:

$$\ln(e^{-t/\tau}) = -\frac{t}{\tau} = \ln\left(\frac{1}{2}\right) = -\ln 2$$

In the last step we used the property $\ln(1/b) = -\ln b$. Now we can solve for t:

$$t = \tau \ln 2$$

The time constant was specified as $\tau = 300$ s, so

$$t = (300\text{ s})(0.693) = 208\text{ s}$$

It will take 208 s, or about 3.5 min, for the oscillations to decay by half after the spring has run down.

ASSESS The time is less than the time constant, which makes sense. The time constant is the time for the amplitude to decay to 37% of its initial value; we are looking for the time to decay to 50% of its initial value, which should be a shorter time. The time to decay to $\frac{1}{2}$ of the initial amplitude, $t = \tau \ln 2$, could be called the *half-life*. We will see this expression again when we work with radioactivity, another example of an exponential decay.

STOP TO THINK 14.7 Rank in order, from largest to smallest, the time constants τ_A to τ_D of the decays in the figures. The scales on all the graphs are the same.

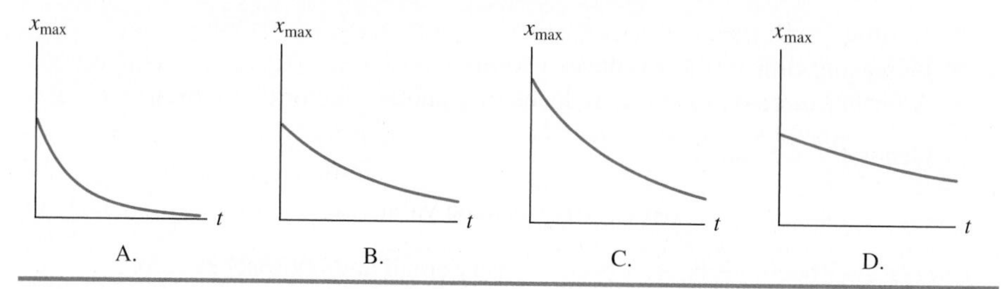

14.7 Driven Oscillations and Resonance

If you jiggle a cup of water, the water sloshes back and forth. This is an example of an oscillator (the water in the cup) subjected to a periodic external force (from your hand). This motion is called a **driven oscillation.**

We can give many examples of driven oscillations. The electromagnetic coil on the back of a loudspeaker cone provides a periodic magnetic force to drive the cone back and forth, causing it to send out sound waves. Earthquakes cause the surface of the earth to move back and forth; this motion causes buildings to oscillate, possibly producing damage or collapse.

Consider an oscillating system that, when left to itself, oscillates at a frequency f_0. We will call this the **natural frequency** of the oscillator. f_0 is simply the frequency of the system if it is displaced from equilibrium and released.

◀ **Serious sloshing** Water in Canada's Bay of Fundy oscillates back and forth with a period of 12 hours—nearly equal to the period of the moon's tidal force, which gives two daily high tides 12.5 hours apart. This *resonance*, a close match between the bay's natural frequency and the moon's driving frequency, produces a huge tidal amplitude. Low tide can be 16 m below high tide, leaving boats high and dry.

Suppose that this system is now subjected to a *periodic* external force of frequency f_{ext}. This frequency, which is called the **driving frequency,** is completely independent of the oscillator's natural frequency f_0. Somebody or something in the environment selects the frequency f_{ext} of the external force, causing the force to push on the system f_{ext} times every second. The external force causes the oscillation of the system, so it will oscillate at f_{ext}, the driving frequency, not at its natural frequency f_0.

Let's return to the example of the cup of water. If you nudge the cup, you will notice that the water sloshes back and forth at a particular frequency; this is the natural frequency f_0. Now shake the cup at some frequency; this is the driving frequency f_{ext}. As you shake the cup, the oscillation amplitude of the water depends very sensitively on the frequency f_{ext} of your hand. If the driving frequency is near the natural frequency of the system, the oscillation amplitude may become so large that water splashes out of the cup.

Any driven oscillator will show a similar dependence of amplitude on the driving frequency. Suppose a mass on a spring has a natural frequency $f_0 = 2$ Hz. We can use an external force to push and pull on the mass at frequency f_{ext}, measure the amplitude of the resulting oscillation, and then repeat this over and over for many different driving frequencies. A graph of amplitude versus driving frequency, such as the one in FIGURE 14.24, is called the oscillator's **response curve.**

In Figure 14.24, for frequencies substantially lower or higher than the oscillator's natural frequency, the system oscillates, but its amplitude is very small. The system simply does not respond well to a driving frequency that differs much from f_0. As the driving frequency gets closer and closer to the natural frequency, the amplitude of the oscillation rises dramatically. After all, f_0 is the frequency at which the system "wants" to oscillate, so it is quite happy to respond to a driving frequency near f_0. Hence the amplitude reaches a maximum when the driving frequency matches the system's natural frequency: $f_{ext} = f_0$. This large-amplitude response to a driving force whose frequency matches the natural frequency of the system is a phenomenon called **resonance.** Within the context of driven oscillations, the natural frequency f_0 is often called the **resonance frequency.**

The amplitude can become exceedingly large when the frequencies match, especially if there is very little damping. FIGURE 14.25 shows the response curve of the oscillator of Figure 14.24 with different amounts of damping. Three different graphs are plotted, each with a different time constant for damping. The three graphs have damping that ranges from $\tau = 50T$ (very little damping) to $\tau = 2T$ (significant damping).

FIGURE 14.24 The response curve shows the amplitude of a driven oscillator at frequencies near its natural frequency $f_0 = 2$ Hz.

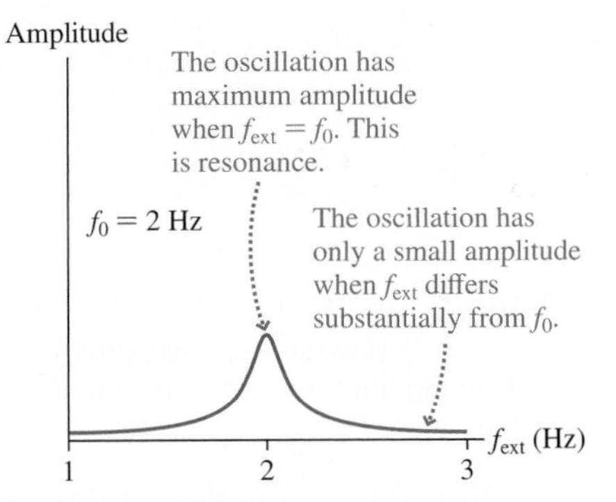

FIGURE 14.25 The response curve becomes taller and narrower as the damping is reduced.

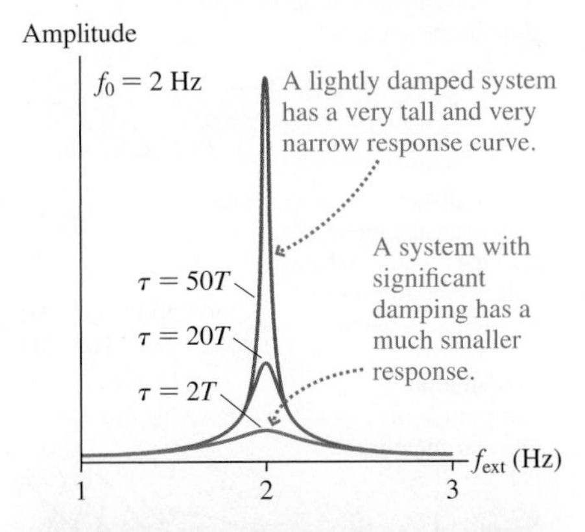

▶ **Simple harmonic music** A wine glass has a natural frequency of oscillation and a very small amount of damping. A tap on the rim of the glass causes it to "ring" like a bell. If you moisten your finger and slide it gently around the rim of the glass, it will stick and slip in quick succession. With some practice you can match the stick-slip to the frequency of oscillation of the glass. The resulting resonance can lead to a very loud sound. Adding water changes the frequency, so you can tune a set of glasses to make an unusual musical instrument.

Video Pendulum Resonance

CONCEPTUAL EXAMPLE 14.13 Fixing an unwanted resonance in a railroad car

Railroad cars have a natural frequency at which they rock side to side. This can lead to problems on certain stretches of track that have bumps where the rails join. If the joints alternate sides, with a bump on the left rail and then on the right, a train car moving down the track is bumped one way and then the other. In some cases, bumps have caused rocking with amplitude large enough to derail the train. A train moving down the track at a certain speed is experiencing a large amplitude of oscillation due to alternating joints in the track. How can the driver correct this potentially dangerous situation?

REASON The large amplitude of oscillation is produced by a resonance, a match between the frequency at which the train car rocks back and forth and the frequency at which the car hits the bumps. To eliminate this resonance, the driver must either reduce the speed of the train—decreasing the driving frequency—or increase the speed of the train—thus increasing the driving frequency.

ASSESS It's perhaps surprising that increasing the speed of the train could produce a smoother ride. But increasing the frequency at which the train hits the bumps will eliminate the match with the natural rocking frequency just as surely as decreasing the speed.

Resonance and Hearing BIO

Resonance in a system means that certain frequencies produce a large response and others do not. The phenomenon of resonance is responsible for the frequency discrimination of the ear.

As we will see in the next chapter, sound is a vibration in air. FIGURE 14.26 provides an overview of the structures by which sound waves that enter the ear produce vibrations in the cochlea, the coiled, fluid-filled, sound-sensing organ of the inner ear.

FIGURE 14.26 The structures of the ear.

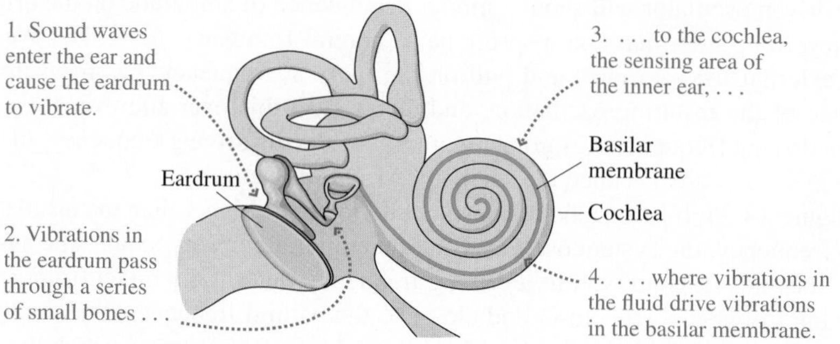

1. Sound waves enter the ear and cause the eardrum to vibrate.

Eardrum

2. Vibrations in the eardrum pass through a series of small bones . . .

3. . . . to the cochlea, the sensing area of the inner ear, . . .

Basilar membrane

Cochlea

4. . . . where vibrations in the fluid drive vibrations in the basilar membrane.

FIGURE 14.27 Resonance plays a role in determining the frequencies of sounds we hear.

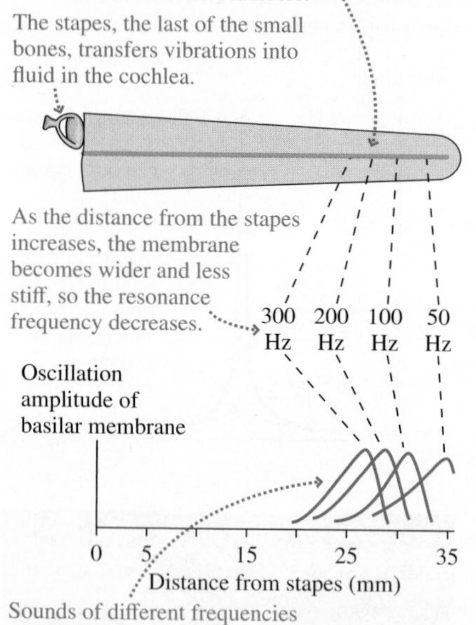
We imagine the spiral structure of the cochlea unrolled, with the basilar membrane separating two fluid-filled chambers.
The stapes, the last of the small bones, transfers vibrations into fluid in the cochlea.
As the distance from the stapes increases, the membrane becomes wider and less stiff, so the resonance frequency decreases.
300 Hz
200 Hz
100 Hz
50 Hz
Oscillation amplitude of basilar membrane
0
5
15
25
35
Distance from stapes (mm)
Sounds of different frequencies cause different responses.

FIGURE 14.27 shows a very simplified model of the cochlea. As a sound wave travels down the cochlea, it causes a large-amplitude vibration of the basilar membrane at the point where the membrane's natural oscillation frequency matches the sound frequency—a resonance. Lower-frequency sound causes a response farther from the stapes. Sensitive hair cells on the membrane sense the vibration and send nerve signals to your brain. The fact that different frequencies produce maximal response at different positions allows your brain to very accurately determine frequency because a small shift in frequency causes a detectable change in the position of the maximal response. People with no musical training can listen to two notes and easily determine which is at a higher pitch.

We now know a bit about how your ear responds to the vibration of a sound wave—but how does this vibration get from a source to your ear? This is a topic we will consider in the next chapter, when we look at waves, oscillations that travel.

INTEGRATED EXAMPLE 14.14 Analyzing the motion of a rope swing

A rope hangs down from a high tree branch at the edge of a river. Josh, who has mass 65 kg, trots at 2.5 m/s to the edge of the river, grabs the rope 7.2 m below where it is tied to the branch, and swings out over the river.

a. What is the minimum time that Josh must hang on to make it back to shore?
b. What is the maximum tension in the rope?

STRATEGIZE This is pendulum motion, with Josh as the mass at the end of a 7.2-m-long rope. The time for him to swing out and return to shore is one-half the pendulum's period, which we will find from Equation 14.27.

Once Josh starts swinging, he moves in a circular arc, so he will have a centripetal acceleration. This acceleration is caused by the tension and the weight force acting on him. We will use the ideas of circular motion dynamics to relate the tension to his speed.

PREPARE Josh grabs the rope, swings out, and comes back to shore, as shown in FIGURE 14.28. As he swings out, his speed decreases until he reaches the greatest distance from shore. When he returns to shore, he has the same speed as when he left. As we noted, the time for this motion is half the oscillation period of the pendulum.

FIGURE 14.28 Pendulum motion of a person swinging on a rope.

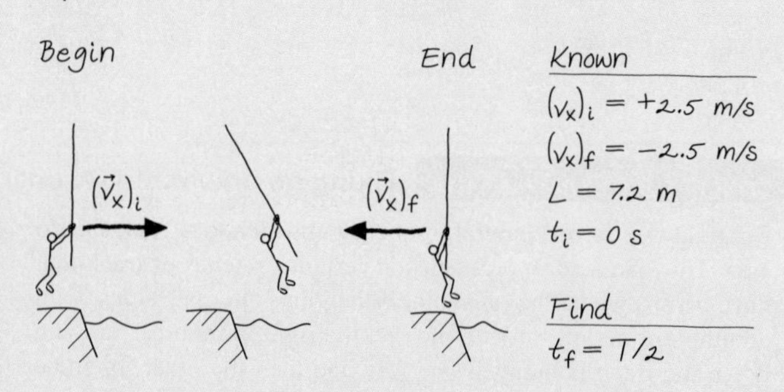
Begin
End
Known
$(v_x)_i = +2.5$ m/s
$(v_x)_f = -2.5$ m/s
$L = 7.2$ m
$t_i = 0$ s
$(\vec{v}_x)_i$
$(\vec{v}_x)_f$
Find
$t_f = T/2$

Once Josh starts swinging in a circular arc, the force required to produce his centripetal acceleration is the difference between the tension in the rope and the radial component of Josh's weight. The tension will be largest when the centripetal acceleration is largest, which is at the lowest point of the swing, where his speed is the greatest. In FIGURE 14.29 we have prepared a visual overview, including a free-body diagram, showing Josh as he returns to this lowest point. Notice that we use the symbol T_s for the tension in the rope so as not to confuse it with the symbol T for period.

FIGURE 14.29 A circular motion view gives insight into forces.

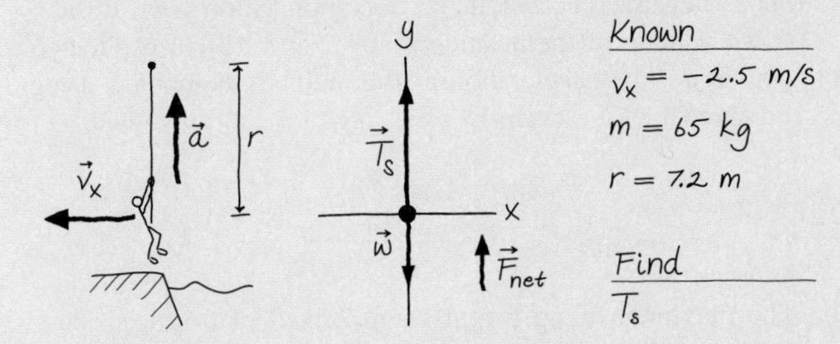

SOLVE a. The period for the pendulum motion is

$$T = 2\pi\sqrt{\frac{L}{g}} = 2\pi\sqrt{\frac{7.2\text{ m}}{9.8\text{ m/s}^2}} = 5.4\text{ s}$$

As shown in Figure 14.28, Josh must hold on until $t_f = T/2 = 2.7$ s.

b. At the lowest point of the motion, where Josh's speed is the highest, the centripetal acceleration is directed upward along the y-axis and is (from Equation 3.20)

$$a_y = \frac{v^2}{r} = \frac{(2.5\text{ m/s})^2}{7.2\text{ m}} = 0.87\text{ m/s}^2$$

As outlined in Figure 14.29, Newton's second law for motion along the y-axis is

$$\Sigma F_y = (T_s)_y + w_y = T_s - mg = ma_y$$

We know the acceleration, so we can solve for the magnitude of the tension:

$$T_s = mg + ma_y = (65\text{ kg})(9.8\text{ m/s}^2 + 0.87\text{ m/s}^2) = 690\text{ N}$$

ASSESS Both of our answers seem reasonable. A time of 2.4 s to swing out and back sounds about right. And the speed is small and the rope is long, so the magnitude of the centripetal acceleration is modest. We thus expect the tension in the rope to be only slightly larger than Josh's weight, just as we found.

INTEGRATED EXAMPLE 14.15 Springboard diving

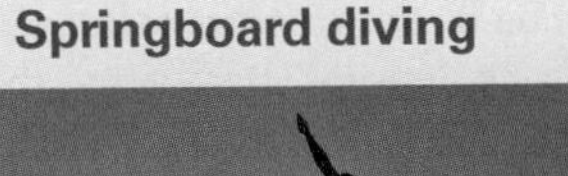

Flexible diving boards designed for large deflections are called springboards. If a diver jumps up and lands on the end of the board, the resulting deflection of the diving board produces a linear restoring force that launches him into the air. But if the diver simply bobs up and down on the end of the board, we can effectively model his motion as that of a mass oscillating on a spring.

A light and flexible springboard deflects by 15 cm when a 65 kg diver stands on its end. He then jumps and lands on the end of the board, depressing it by a total of 25 cm, after which he moves up and down with the oscillations of the end of the board.

a. What is the frequency of the oscillation?
b. What is the maximum speed of his up-and-down motion?

Suppose the diver then drives the motion of the board with his legs, gradually increasing the amplitude of the oscillation. At some point the oscillation becomes large enough that his feet leave the board.

c. What is the amplitude of the oscillation when the diver just becomes airborne at one point of the cycle? What is the acceleration at this point?
d. A diver leaving a springboard can achieve a much greater height than a diver jumping from a fixed platform. Use energy concepts to explain how the spring of the board allows a greater vertical jump.

STRATEGIZE We will model the diver on the board as a mass on a spring; we will consider the spring itself to be massless. As we've seen, the oscillation frequency is determined by the spring constant and the mass. The mass of the diver is given; we will determine the spring constant from the deflection of the springboard when the diver stands on the end.

As the diver oscillates up and down, the board will never rise above its undeflected position. To see this, imagine pushing down on a vertical spring, then slowly letting your hand rise. The spring will follow your hand until the spring reaches its undeflected position; at this point it cannot follow your hand further unless you actively pull on it.

PREPARE FIGURE 14.30 is a sketch of the oscillation that will help us visualize the motion. The equilibrium position, with the diver

FIGURE 14.30 Position-versus-time graph for the springboard.

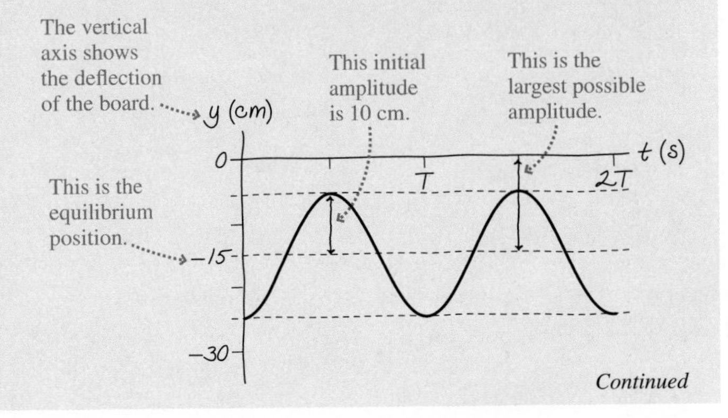

Continued

standing motionless on the end of the board, corresponds to a deflection of 15 cm. When the diver jumps on the board, the total deflection is 25 cm, which means a deflection of an additional 10 cm beyond the equilibrium position, so 10 cm is the amplitude of the subsequent oscillation. When the board rises, it won't bend beyond its undeflected position, so the maximum possible amplitude is 15 cm.

NOTE ▶ We've started the graph at the lowest point of the motion. Because we'll use only the equations for the maximum values of the speed and the acceleration, not the full equations that describe the motion, the exact starting point isn't critical. ◀

SOLVE

a. FIGURE 14.31 shows the forces on the diver as he stands motionless at the end of the board. The net force on him is zero, $\vec{F}_{net} = \vec{F}_{sp} + \vec{w} = \vec{0}$, so the two forces have equal magnitudes and we can write

$$F_{sp} = w$$

A linear restoring force means that the board obeys Hooke's law for a spring: $F_{sp} = k\,\Delta y$, where k is the spring constant. Thus the equilibrium equation is

$$k\,\Delta y = mg$$

Solving for the spring constant, we find

$$k = \frac{mg}{\Delta y} = \frac{(65\text{ kg})(9.8\text{ m/s}^2)}{0.15\text{ m}} = 4.25 \times 10^3\text{ N/m}$$

FIGURE 14.31 Forces on a springboard diver at rest at the end of the board.

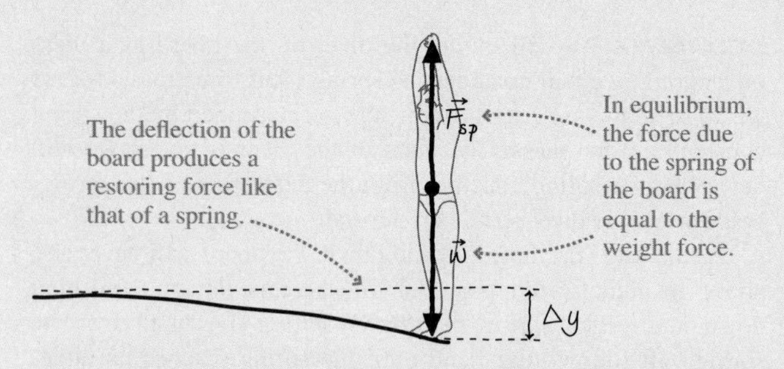
The deflection of the board produces a restoring force like that of a spring.
In equilibrium, the force due to the spring of the board is equal to the weight force.
$\vec{F}_{sp}$
$\vec{w}$
Δy

The frequency of the oscillation depends on the diver's mass and the spring constant of the board:

$$f = \frac{1}{2\pi}\sqrt{\frac{k}{m}} = \frac{1}{2\pi}\sqrt{\frac{4.25 \times 10^3\text{ N/m}}{65\text{ kg}}} = 1.29\text{ Hz}$$

b. The maximum speed of the oscillation is given by Equation 14.16:

$$v_{max} = 2\pi f A = 2\pi(1.29\text{ Hz})(0.10\text{ m}) = 0.81\text{ m/s}$$

c. We can see from Figure 14.30 that an amplitude of 15 cm returns the board to its undeflected position. At this point, the board exerts no upward force—no supporting normal force—on the diver, so the diver loses contact with the board. (His apparent weight becomes zero.) The acceleration at this point in the motion has the maximum possible magnitude but is negative because the acceleration graph is an upside-down version of the position graph. For a 15 cm oscillation amplitude, the acceleration at this point is computed using Equation 14.17:

$$a = -a_{max} = -(2\pi f)^2 A = [2\pi(1.29\text{ Hz})]^2(0.15\text{ m})$$
$$= -9.8\text{ m/s}^2$$

d. The maximum jump height from a fixed platform is determined by the maximum speed at which a jumper leaves the ground. During a jump, chemical energy in the muscles is transformed into kinetic energy. This kinetic energy is transformed into potential energy as the jumper rises, back into kinetic energy as he falls, then into thermal energy as he hits the ground. The springboard recaptures and stores this kinetic energy rather than letting it degrade to thermal energy; a diver can jump up once and land on the board, storing the energy of his initial jump as elastic potential energy of the bending board. Then, as the board rebounds, turning the stored energy back into kinetic energy, the diver can push off from this moving platform, transforming even more chemical energy and thus further increasing his kinetic energy. This allows the diver to get "two jumps worth" of chemical energy in a single jump.

ASSESS We learned in earlier chapters that an object loses contact with a surface—like a car coming off the track in a loop-the-loop—when its apparent weight becomes zero: $w_{app} = 0$. And in Chapter 5 we found that the apparent weight of an object in vertical motion becomes zero when $a = -g$, that is, when it enters free fall. This correspondence is a good check on our work. Because part c has the answer we expect, we have confidence in our earlier steps.

SUMMARY

GOAL To understand systems that oscillate with simple harmonic motion.

GENERAL PRINCIPLES

Frequency and Period

SHM occurs when a **linear restoring force** acts to return a system to an equilibrium position. The frequency and period depend on the details of the oscillator, but **the frequency and period do not depend on the amplitude.**

Mass on spring

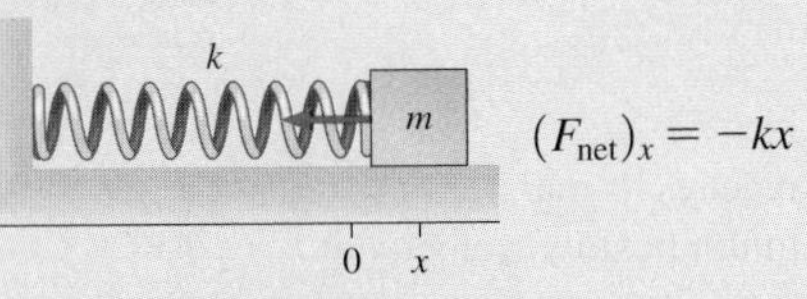

$(F_{\text{net}})_x = -kx$

The frequency and period of a mass on a spring depend on the mass and the spring constant: They are the same for horizontal and vertical systems.

$$f = \frac{1}{2\pi}\sqrt{\frac{k}{m}} \qquad T = 2\pi\sqrt{\frac{m}{k}}$$

Pendulum

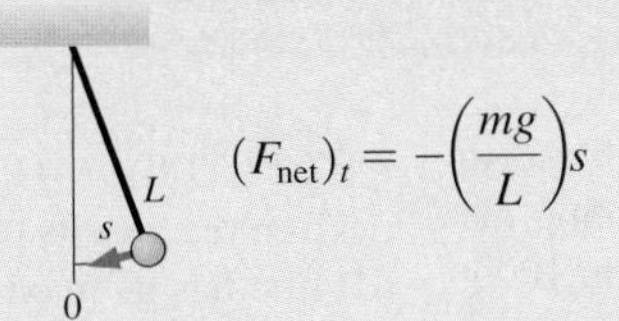

$(F_{\text{net}})_t = -\left(\frac{mg}{L}\right)s$

The frequency and period of a pendulum depend on the length and the free-fall acceleration. They do not depend on the mass.

$$f = \frac{1}{2\pi}\sqrt{\frac{g}{L}} \qquad T = 2\pi\sqrt{\frac{L}{g}}$$

Energy

If there is no friction or dissipation, kinetic and potential energies are alternately transformed into each other in SHM, with the sum of the two conserved.

$$E = \frac{1}{2}mv_x^2 + \frac{1}{2}kx^2 = \frac{1}{2}mv_{\max}^2 = \frac{1}{2}kA^2$$

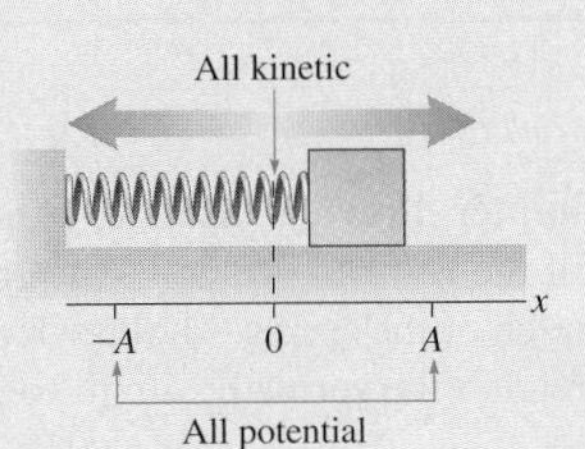

IMPORTANT CONCEPTS

Oscillation

An **oscillation** is a repetitive motion about an equilibrium position. The **amplitude** A is the maximum displacement from equilibrium. The period T is the time for one cycle. We may also characterize an oscillation by its frequency f.

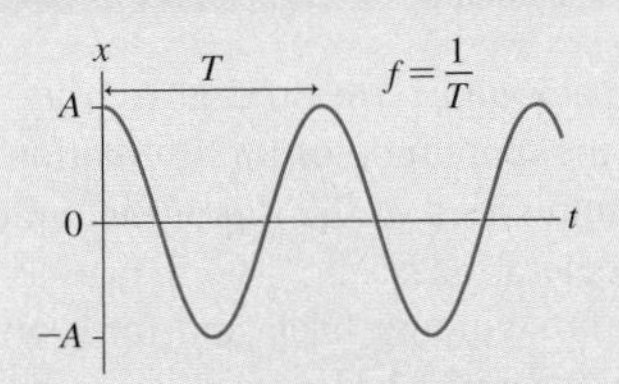

Simple Harmonic Motion (SHM)

SHM is an oscillation that is described by a sinusoidal function. All systems that undergo SHM can be described by the same functional forms.

Position-versus-time is a cosine function.

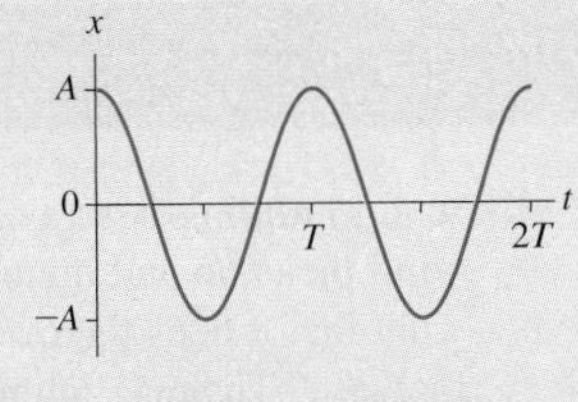

$x(t) = A\cos(2\pi ft)$

Velocity-versus-time is an inverted sine function.

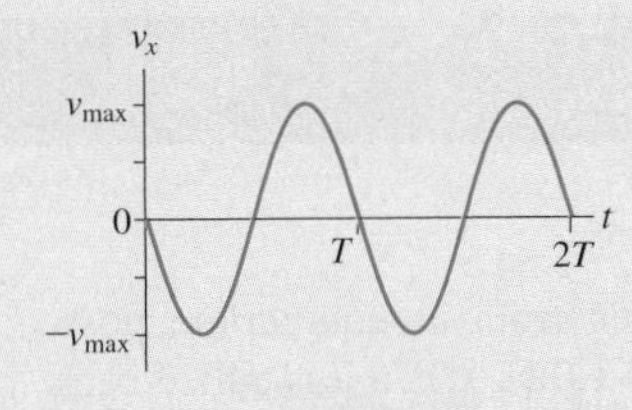

$v_x(t) = -[(2\pi f)A]\sin(2\pi ft)$

Acceleration-versus-time is an inverted cosine function.

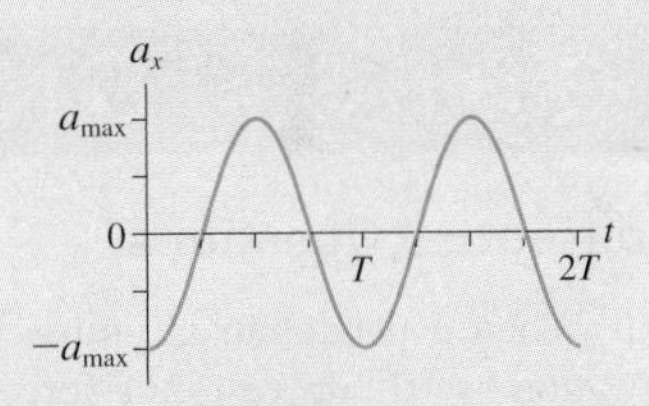

$a_x(t) = -[(2\pi f)^2A]\cos(2\pi ft)$

APPLICATIONS

Damping

Simple harmonic motion with damping (due to drag) decreases in amplitude over time. The **time constant** τ determines how quickly the amplitude decays.

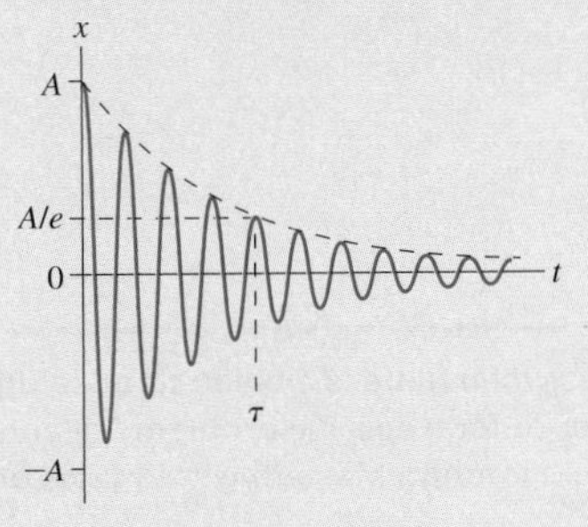

Resonance

A system that oscillates has a **natural frequency** of oscillation f_0. **Resonance** occurs if the system is driven with a frequency f_{ext} that matches this natural frequency. This may produce a large amplitude of oscillation.

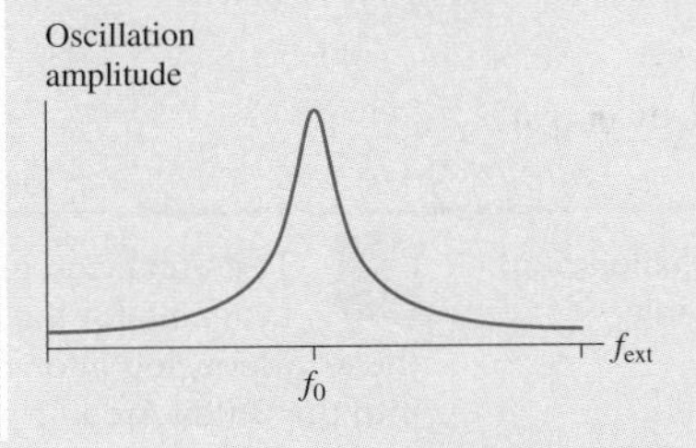

Physical pendulum

A **physical pendulum** is a pendulum with mass distributed along its length. The frequency depends on the position of the center of gravity and the moment of inertia.

The motion of legs during walking can be described using a physical pendulum model.

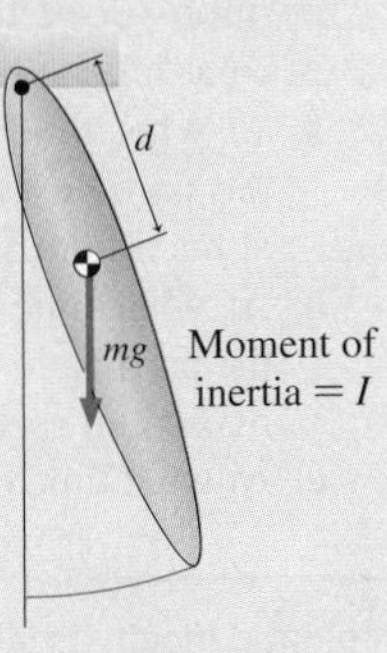

$$f = \frac{1}{2\pi}\sqrt{\frac{mgd}{I}}$$

Learning Objectives After studying this chapter, you should be able to:

- Work with the period, frequency, and amplitude of oscillatory motion. *Conceptual Questions 14.1, 14.2; Problems 14.1, 14.2, 14.3, 14.4, 14.5*
- Use the graphical and mathematical representations of simple harmonic motion. *Conceptual Questions 14.8, 14.11; Problems 14.6, 14.7, 14.8, 14.12, 14.18*
- Apply energy conservation to simple harmonic motion. *Conceptual Questions 14.6, 14.7; Problems 14.19, 14.21, 14.23, 14.24, 14.26*
- Solve problems about simple and physical pendulums. *Conceptual Questions 14.14, 14.16; Problems 14.27, 14.28, 14.30, 14.33, 14.38*
- Use the concepts of damping and resonance. *Conceptual Questions 14.20, 14.21; Problems 14.39, 14.40, 14.41, 14.44, 14.47*

STOP TO THINK ANSWERS

Chapter Preview Stop to Think: A. If the weight doubles, so will the restoring force of the spring. This also means doubling the stretch of the spring—not the length, but the *change* in length. 100 g stretches the spring by 2 cm, so 200 g will stretch it by 4 cm.

Stop to Think 14.1: C. The frequency is inversely proportional to the period, so a shorter period implies a higher frequency.

Stop to Think 14.2: D. The restoring force is proportional to the displacement. If the displacement from equilibrium is doubled, the force is doubled as well.

Stop to Think 14.3: B, A, C, B. The motion of the pendulum exactly matches the graphs in the table. The pendulum starts at rest with a maximum positive displacement and then undergoes simple harmonic motion. By looking at the graphs, we can see that the pendulum (a) first reaches its maximum negative displacement at $t = \frac{1}{2}T = 1.0$ s, (b) first reaches its maximum speed at $t = \frac{1}{4}T = 0.5$ s, (c) first reaches its maximum positive velocity at $t = \frac{3}{4}T = 1.5$ s, and (d) first has its maximum positive acceleration at $t = \frac{1}{2}T = 1.0$ s.

Stop to Think 14.4: A. The maximum speed occurs when the mass passes through its equilibrium position.

Stop to Think 14.5: $f_D > f_C = f_B > f_A$. The frequency is determined by the ratio of k to m.

Stop to Think 14.6: C. The increase in length will cause the frequency to decrease and thus the period will increase. The time between ticks will increase, so the clock will run slow.

Stop to Think 14.7: $\tau_D > \tau_B = \tau_C > \tau_A$. The time constant is the time to decay to 37% of the initial height. The time constant is independent of the initial height.

Video Tutor Solution Chapter 14

QUESTIONS

Conceptual Questions

1. When a pendulum is pulled back from its equilibrium position by 10°, the restoring force is 1.0 N. When it is pulled back to 30°, the force increases to 2.8 N. If the pendulum is then released from 30°, will its motion be periodic? Will its motion be simple harmonic? Explain.
2. A person's heart rate is given in beats per minute. Is this a period or a frequency?
3. Figure Q14.3 shows the position-versus-time graph of a particle in SHM.
   a. At what time or times is the particle moving right at maximum speed?
   b. At what time or times is the particle moving left at maximum speed?
   c. At what time or times is the particle instantaneously at rest?

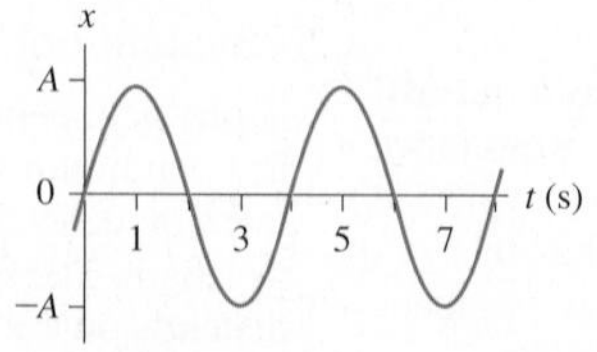

FIGURE Q14.3

4. A tall building is swaying back and forth on a gusty day. The wind picks up and doubles the amplitude of the oscillation. By what factor does the maximum speed of the top of the building increase? The maximum acceleration?
5. A child is on a swing, gently swinging back and forth with a maximum angle of 5°. A friend gives her a small push so that she now swings with a maximum angle of 10°. By what factor does this increase her maximum speed?
6. A block oscillating on a spring has an amplitude of 20 cm. What will be the amplitude if the maximum kinetic energy is doubled?
7. A block oscillating on a spring has a maximum kinetic energy of 2.0 J. What will be the maximum kinetic energy if the amplitude is doubled? Explain.

Problem difficulty is labeled as | (straightforward) to ||||| (challenging). INT Problems labeled integrate significant material from earlier chapters; BIO are of biological or medical interest.

The eText icon indicates when there is a video tutor solution available for the chapter or for a specific problem. To launch these videos, log into your eText through Mastering™ Physics or log into the Study Area.

8. A mass hanging from a spring undergoes vertical simple harmonic motion.
   a. Where in the motion is the magnitude of the net force equal to zero?
   b. Where in the motion is the velocity equal to zero?
   c. Where in the motion does the acceleration have its greatest magnitude?
   d. Where in the motion is the spring force equal to zero?
9. For the graph in Figure Q14.9, determine the frequency f and the oscillation amplitude A.

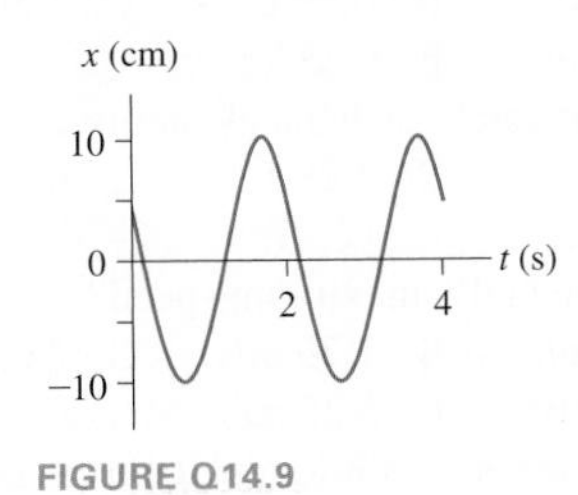

FIGURE Q14.9

10. For the graph in Figure Q14.10, determine the frequency f and the oscillation amplitude A.

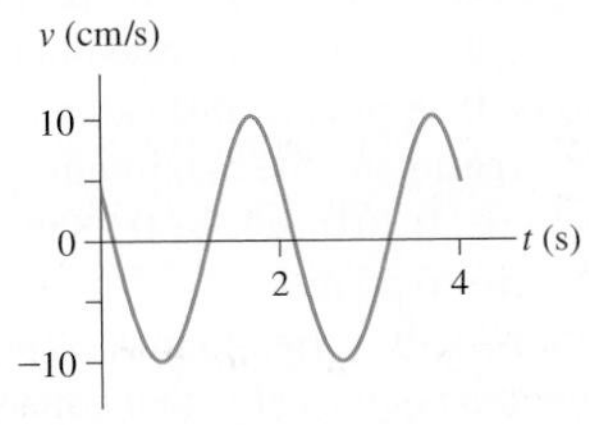

FIGURE Q14.10

11. A block oscillating on a spring has period $T = 2.0$ s.
    a. What is the period if the block's mass is doubled?
    b. What is the period if the value of the spring constant is quadrupled?
    c. What is the period if the oscillation amplitude is doubled while m and k are unchanged?

    Note: You do not know values for either m or k. Do *not* assume any particular values for them. The required analysis involves thinking about ratios.
12. If a mass on a spring moving horizontally were taken to the moon, how would its frequency change? What about a mass on a spring moving vertically? A pendulum?
13. BIO Flies flap their wings at frequencies much too high for pure muscle action. A hypothesis for how they achieve these high frequencies is that the flapping of their wings is the driven oscillation of a mass-spring system. One way to test this is to trim a fly's wings. If the oscillation of the wings can be modeled as a mass-spring system, how would this change the frequency of the wingbeats?
14. Denver is at a higher elevation than Miami; the free-fall acceleration is slightly less at this higher elevation. If a pendulum clock keeps perfect time in Miami, will it run fast or slow in Denver? Explain.
15. If you want to play a tune on wine glasses, you'll need to adjust the oscillation frequencies by adding water to the glasses. This changes the mass that oscillates (more water means more mass) but not the restoring force, which is determined by the stiffness of the glass itself. If you need to raise the frequency of a particular glass, should you add water or remove water?

16. A young girl and her mother are swinging on a swing set. Who, if either, has the longer period of oscillation? Explain.

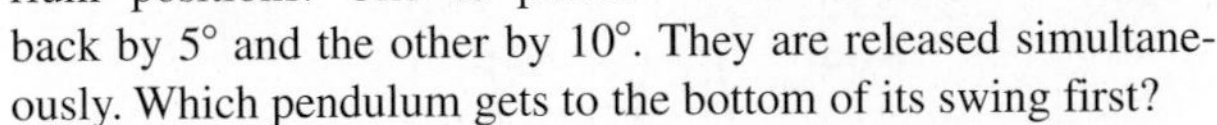

17. Two identical pendulums are pulled back from their equilibrium positions. One is pulled back by 5° and the other by 10°. They are released simultaneously. Which pendulum gets to the bottom of its swing first?
18. BIO Gibbons move through the trees by swinging from successive handholds, as we have seen. To increase their speed, gibbons may bring their legs close to their bodies. How does this help them move more quickly?

19. BIO Sprinters push off from the ball of their foot, then bend their knee to bring their foot up close to the body as they swing their leg forward for the next stride. Why is this an effective strategy for running fast?
20. BIO Humans have a range of hearing of approximately 20 Hz to 20 kHz. Mice have auditory systems similar to humans, but all of the physical elements are smaller. Given this, would you expect mice to have a higher or lower frequency range than humans? Explain.

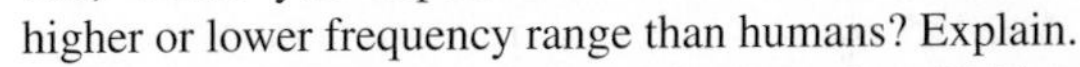

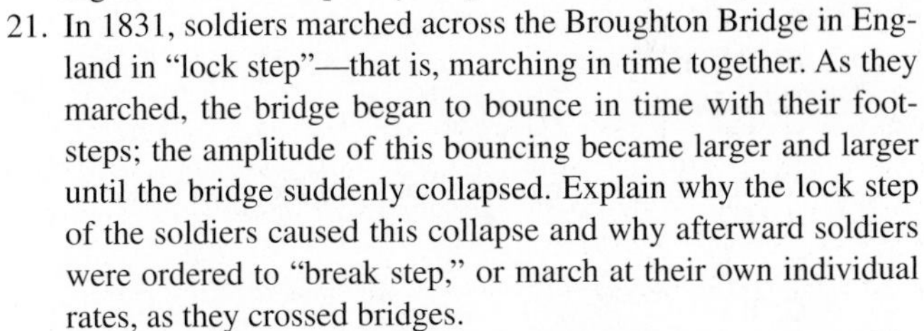

21. In 1831, soldiers marched across the Broughton Bridge in England in "lock step"—that is, marching in time together. As they marched, the bridge began to bounce in time with their footsteps; the amplitude of this bouncing became larger and larger until the bridge suddenly collapsed. Explain why the lock step of the soldiers caused this collapse and why afterward soldiers were ordered to "break step," or march at their own individual rates, as they crossed bridges.
22. BIO We've seen that stout tendons in the legs of hopping kangaroos store energy. When a kangaroo lands, much of the kinetic energy of motion is converted to elastic energy as the tendons stretch, returning to kinetic energy when the kangaroo again leaves the ground. If a hopping kangaroo increases its speed, it spends more time in the air with each bounce, but the contact time with the ground stays approximately the same. Explain why you would expect this to be the case.

Multiple-Choice Questions

23. | A spring has an unstretched length of 20 cm. A 100 g mass hanging from the spring stretches it to an equilibrium length of 30 cm.
    a. Suppose the mass is pulled down to where the spring's length is 40 cm. When it is released, it begins to oscillate. What is the amplitude of the oscillation?
       A. 5.0 cm B. 10 cm C. 20 cm D. 40 cm
    b. For the data given above, what is the frequency of the oscillation?
       A. 0.10 Hz B. 0.62 Hz C. 1.6 Hz D. 10 Hz
    c. Suppose this experiment were done on the moon, where the free-fall acceleration is approximately 1/6 of that on the earth. How would this change the frequency of the oscillation?
       A. The frequency would decrease.
       B. The frequency would increase.
       C. The frequency would stay the same.

24. || Figure Q14.24 represents the motion of a mass on a spring.
a. What is the period of this oscillation?
A. 12 s B. 24 s C. 36 s D. 48 s E. 50 s

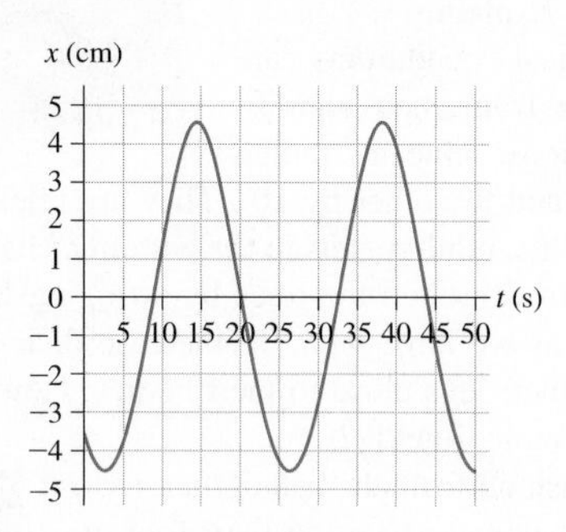

FIGURE Q14.24

b. What is the amplitude of the oscillation?
A. 1.0 cm
B. 2.5 cm
C. 4.5 cm
D. 5.0 cm
E. 9.0 cm

c. What is the position of the mass at time $t = 30$ s?
A. −4.5 cm
B. −2.5 cm
C. 0.0 cm
D. 4.5 cm
E. 30 cm

d. When is the first time the velocity of the mass is zero?
A. 0 s
B. 2 s
C. 8 s
D. 10 s
E. 13 s

e. At which of these times does the kinetic energy have its maximum value?
A. 0 s
B. 8 s
C. 13 s
D. 26 s
E. 30 s

25. | A ball of mass m oscillates on a spring with spring constant $k = 200$ N/m. The ball's position is $x = (0.350\text{ m})\cos(15.0t)$, with t measured in seconds.
a. What is the amplitude of the ball's motion?
A. 0.175 m B. 0.350 m C. 0.700 m
D. 7.50 m E. 15.0 m
b. What is the frequency of the ball's motion?
A. 0.35 Hz B. 2.39 Hz C. 5.44 Hz
D. 6.28 Hz E. 15.0 Hz
c. What is the value of the mass m?
A. 0.45 kg B. 0.89 kg C. 1.54 kg
D. 3.76 kg E. 6.33 kg
d. What is the total mechanical energy of the oscillator?
A. 1.65 J B. 3.28 J C. 6.73 J
D. 10.1 J E. 12.2 J
e. What is the ball's maximum speed?
A. 0.35 m/s B. 1.76 m/s C. 2.60 m/s
D. 3.88 m/s E. 5.25 m/s

26. | A 0.20 kg mass on a horizontal spring is pulled back 2.0 cm and released. If, instead, a 0.40 kg mass were used in this same experiment, the total energy of the system would
A. Double. B. Remain the same. C. Be half as large.

27. | A 0.20 kg mass on a horizontal spring is pulled back a certain distance and released. The maximum speed of the mass is measured to be 0.28 m/s. If, instead, a 0.40 kg mass were used in this same experiment, the maximum speed would be
A. 0.14 m/s B. 0.20 m/s C. 0.28 m/s
D. 0.40 m/s E. 0.56 m/s

28. | A heavy brass ball is used to make a pendulum with a period of 5.5 s. How long is the cable that connects the pendulum ball to the ceiling?
A. 4.7 m B. 6.2 m
C. 7.5 m D. 8.7 m

29. | Very loud sounds can damage hearing by injuring the
BIO vibration-sensing hair cells on the basilar membrane. Suppose a person has injured hair cells on a segment of the basilar membrane close to the stapes. What type of sound is most likely to have produced this particular pattern of damage?
A. Loud music with a mix of different frequencies
B. A very loud, high-frequency sound
C. A very loud, low-frequency sound

PROBLEMS

Section 14.1 Equilibrium and Oscillation
Section 14.2 Linear Restoring Forces and SHM

1. | When a guitar string plays the note "A," the string vibrates at 440 Hz. What is the period of the vibration?
2. | In the aftermath of an intense earthquake, the earth as a whole "rings" with a period of 54 minutes. What is the frequency (in Hz) of this oscillation?
3. | In taking your pulse, you count 75 heartbeats in 1 min. What
BIO are the period (in s) and frequency (in Hz) of your heart's oscillations?
4. | A spring scale hung from the ceiling stretches by 6.4 cm when a 1.0 kg mass is hung from it. The 1.0 kg mass is removed and replaced with a 1.5 kg mass. What is the stretch of the spring?
5. | A heavy steel ball is hung from a cord to make a pendulum. The ball is pulled to the side so that the cord makes a 5° angle with the vertical. Holding the ball in place takes a force of 20 N. If the ball is pulled farther to the side so that the cord makes a 10° angle, what force is required to hold the ball?

Section 14.3 Describing Simple Harmonic Motion

6. || An air-track glider attached to a spring oscillates between the 10 cm mark and the 60 cm mark on the track. The glider completes 10 oscillations in 33 s. What are the (a) period, (b) frequency, (c) amplitude, and (d) maximum speed of the glider?

7. ||| An air-track glider is attached to a spring. The glider is pulled to the right and released from rest at $t = 0$ s. It then oscillates with a period of 2.0 s and a maximum speed of 40 cm/s.
   a. What is the amplitude of the oscillation?
   b. What is the glider's position at $t = 0.25$ s?
8. | What are the (a) amplitude and (b) frequency of the oscillation shown in Figure P14.8?

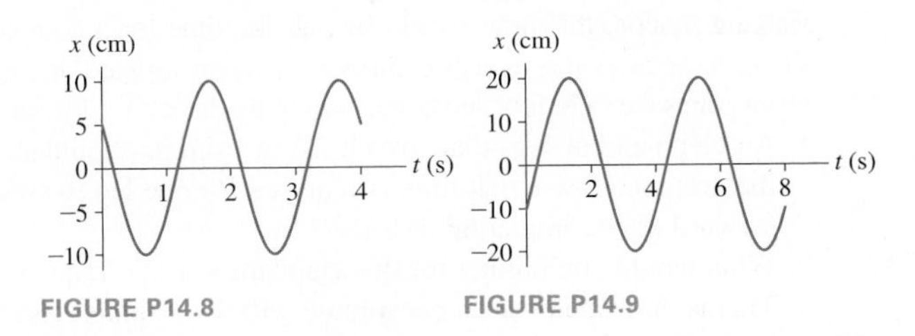

FIGURE P14.8 FIGURE P14.9

9. | What are the (a) amplitude and (b) frequency of the oscillation shown in Figure P14.9?
10. ||| A block with a mass of 0.28 kg is attached to a horizontal spring. The block is pulled back from its equilibrium position until the spring exerts a force of 1.0 N on the block. When the block is released, it oscillates with a frequency of 1.2 Hz. How far was the block pulled back before being released?
11. ||| The acceleration of an oscillator undergoing simple harmonic motion is described by the equation $a_x(t) = -(18.0 \text{ m/s}^2)\cos(30t)$, where the time t is measured in seconds. What is the amplitude of this oscillator?
12. || The end of a nanoscale cantilever used for weighing DNA BIO molecules (see Example 14.8) oscillates at 8.8 MHz with an amplitude of 5.0 nm. What are (a) the maximum speed and (b) the maximum acceleration of the end of the cantilever?
13. || Some passengers on an ocean cruise may suffer from motion sickness as the ship rocks back and forth on the waves. At one position on the ship, passengers experience a vertical motion of amplitude 1 m with a period of 15 s.
    a. To one significant figure, what is the maximum acceleration of the passengers during this motion?
    b. What fraction is this of g?
14. ||| The New England Merchants Bank Building in Boston is 152 m high. On windy days it sways with a frequency of 0.17 Hz, and the acceleration of the top of the building can reach 2.0% of the free-fall acceleration, enough to cause discomfort for occupants. What is the total distance, side to side, that the top of the building moves during such an oscillation?
15. || The motion of a nightingale's wingtips can be modeled as BIO simple harmonic motion. In one study, the tips of a bird's wings were found to move up and down with an amplitude of 8.8 cm and a period of 0.82 s. What are the wingtips' (a) maximum speed and (b) maximum acceleration?
16. | In a loudspeaker, an electromagnetic coil rapidly drives a paper cone back and forth, sending out sound waves. If the cone of a loudspeaker moves sinusoidally at 1.2 kHz with an amplitude of 3.5 μm, what are the cone's maximum speed and acceleration?
17. | We can model the motion of a bumblebee's wing as simple BIO harmonic motion. A bee beats its wings 250 times per second, and the wing tip moves at a maximum speed of 2.5 m/s. What is the amplitude of the wing tip's motion?
18. || Hummingbirds may seem BIO fragile, but their wings are capable of sustaining very large forces and accelerations. Figure P14.18 shows data for the vertical position of the wing tip of a rufous hummingbird. What is the maximum acceleration of the wing tips, in m/s^2 and in units of g?

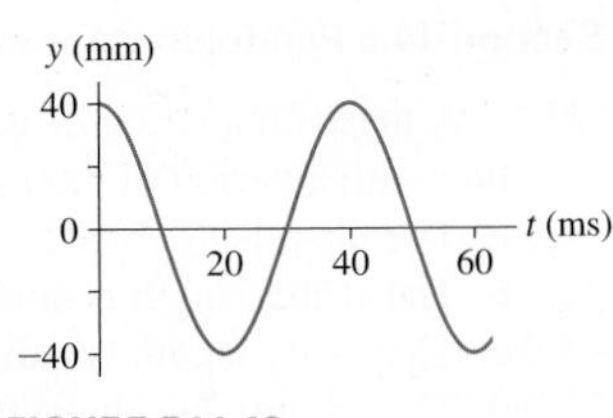

FIGURE P14.18

Section 14.4 Energy in Simple Harmonic Motion

19. ||| a. When the displacement of a mass on a spring is $\frac{1}{2}A$, what fraction of the mechanical energy is kinetic energy and what fraction is potential energy?
    b. At what displacement, as a fraction of A, is the energy half kinetic and half potential?
20. ||| A 1.0 kg block is attached to a spring with spring constant 16 N/m. While the block is sitting at rest, a student hits it with a hammer and almost instantaneously gives it a speed of 40 cm/s. What are
    a. The amplitude of the subsequent oscillations?
    b. The block's speed at the point where $x = \frac{1}{2}A$?
21. | A block attached to a spring with unknown spring constant oscillates with a period of 2.00 s. What is the period if
    a. The mass is doubled?
    b. The mass is halved?
    c. The amplitude is doubled?
    d. The spring constant is doubled?
    Parts a to d are independent questions, each referring to the initial situation.
22. || A 200 g air-track glider is attached to a spring. The glider is pushed 10.0 cm against the spring, then released. A student with a stopwatch finds that 10 oscillations take 12.0 s. What is the spring constant?
23. ||| The position of a 50 g oscillating mass is given by $x(t) = (2.0 \text{ cm})\cos(10t)$, where t is in seconds. Determine:
    a. The amplitude.
    b. The period.
    c. The spring constant.
    d. The maximum speed.
    e. The total energy.
    f. The velocity at $t = 0.40$ s.
24. ||| A 50-cm-long spring is suspended from the ceiling. A 250 g mass is connected to the end and held at rest with the spring unstretched. The mass is released and falls, stretching the spring by 20 cm before coming to rest at its lowest point. It then continues to oscillate vertically.
    a. What is the spring constant?
    b. What is the amplitude of the oscillation?
    c. What is the frequency of the oscillation?
25. || A 200 g mass attached to a horizontal spring oscillates at a frequency of 2.0 Hz. At one instant, the mass is at $x = 5.0$ cm and has $v_x = -30$ cm/s. Determine:
    a. The period. b. The amplitude.
    c. The maximum speed. d. The total energy.
26. || A 507 g mass oscillates with an amplitude of 10.0 cm on a spring whose spring constant is 20.0 N/m. Determine:
    a. The period.
    b. The maximum speed.
    c. The total energy.

Section 14.5 Pendulum Motion

27. || A mass on a string of unknown length oscillates as a pendulum with a period of 4.00 s. What is the period if
    a. The mass is doubled?
    b. The string length is doubled?
    c. The string length is halved?
    d. The amplitude is halved?

    Parts a to d are independent questions, each referring to the initial situation.
28. | The mass in a pendulum clock completes one full back-and-forth oscillation in 1.00 s. What is the length of the rod?
29. ||| A 200 g ball is tied to a string. It is pulled to an angle of 8.00° and released to swing as a pendulum. A student with a stopwatch finds that 10 oscillations take 12.0 s. How long is the string?
30. || The free-fall acceleration on the moon is 1.62 m/s^2. What is the length of a pendulum whose period on the moon matches the period of a 2.00-m-long pendulum on the earth?
31. | Astronauts on the first trip to Mars take along a pendulum that has a period on earth of 1.50 s. The period on Mars turns out to be 2.45 s. Use this data to calculate the Martian free-fall acceleration.
32. ||| A building is being knocked down with a wrecking ball, which is a big metal sphere that swings on a 10-m-long cable. You are (unwisely!) standing directly beneath the point from which the wrecking ball is hung when you notice that the ball has just been released and is swinging directly toward you. How much time do you have to move out of the way?
33. ||| BIO Interestingly, there have been several studies using cadavers to determine the moment of inertia of human body parts by letting them swing as a pendulum about a joint. In one study, the center of gravity of a 5.0 kg lower leg was found to be 18 cm from the knee. When pivoted at the knee and allowed to swing, the oscillation frequency was 1.6 Hz. What was the moment of inertia of the lower leg?
34. ||| A child on a swing set swings back and forth with a period of 3.3 s and an amplitude of 25°. What is the maximum speed of the child as she swings?
35. ||| You and your friends find a rope that hangs down 15 m from a high tree branch right at the edge of a river. You find that you can run, grab the rope, swing out over the river, and drop into the water. You run at 2.0 m/s and grab the rope, launching yourself out over the river. How long must you hang on if you want to drop into the water at the greatest possible distance from the edge?
36. ||| BIO In a study designed to better understand the dynamics of walking, a subject stood with one leg at rest, and swung the other leg at various frequencies. The metabolic power expended, in watts per kilogram of body weight, is shown in Figure P14.36. Estimate the length of this subject's legs.

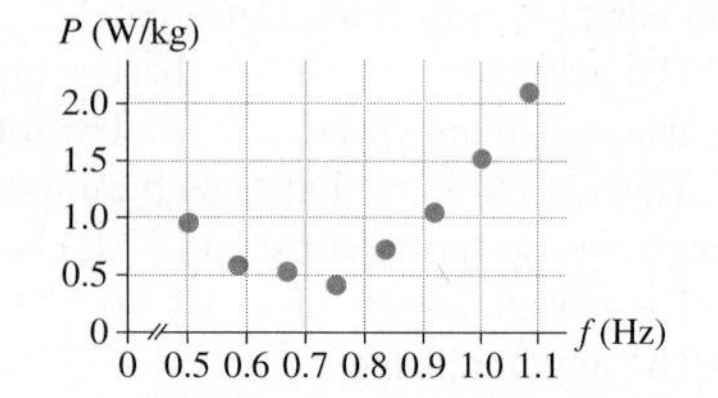

FIGURE P14.36

37. |||| BIO An elephant's legs have a reasonably uniform cross section from top to bottom, and they are quite long, pivoting high on the animal's body. When an elephant moves at a walk, it uses very little energy to bring its legs forward, simply allowing them to swing like pendulums. For fluid walking motion, this time should be half the time for a complete stride; as soon as the right leg finishes swinging forward, the elephant plants the right foot and begins swinging the left leg forward.

    a. An elephant has legs that stretch 2.3 m from its shoulders to the ground. How much time is required for one leg to swing forward after completing a stride?
    b. What would you predict for this elephant's stride frequency? That is, how many steps per minute will the elephant take?
38. ||| In a science museum, you may have seen a Foucault pendulum, which is used to demonstrate the rotation of the earth. In one museum's pendulum, the 110 kg bob swings from a 15.8-m-long cable with an amplitude of 5.0°.
    a. What is the period of this pendulum?
    b. What is the bob's maximum speed?
    c. What is the pendulum's maximum kinetic energy?
    d. When the bob is at its maximum displacement, how much higher is it than when it is at its equilibrium position?

Section 14.6 Damped Oscillations

39. | The amplitude of an oscillator decreases to 36.8% of its initial value in 10.0 s. What is the value of the time constant?
40. || A physics department has a Foucault pendulum, a long-period pendulum suspended from the ceiling. The pendulum has an electric circuit that keeps it oscillating with a constant amplitude. When the circuit is turned off, the oscillation amplitude decreases by 50% in 22 minutes. What is the pendulum's time constant? How much additional time elapses before the amplitude decreases to 25% of its initial value?
41. || Calculate and draw an accurate displacement graph from $t = 0$ s to $t = 10$ s of a damped oscillator having a frequency of 1.0 Hz and a time constant of 4.0 s.
42. || A small earthquake starts a lamppost vibrating back and forth. The amplitude of the vibration of the top of the lamppost is 6.5 cm at the moment the quake stops, and 8.0 s later it is 1.8 cm.
    a. What is the time constant for the damping of the oscillation?
    b. What was the amplitude of the oscillation 4.0 s after the quake stopped?
43. ||| BIO The common field cricket makes its characteristic loud chirping sound using a specialized vibrating structure in its wings. The motion of this structure—and the sound intensity that it produces—can be modeled as a damped oscillation. The sound intensity of such a cricket is shown in Figure P14.43.
    a. What is the frequency of the oscillations?
    b. What is the time constant for the decay of these oscillations?

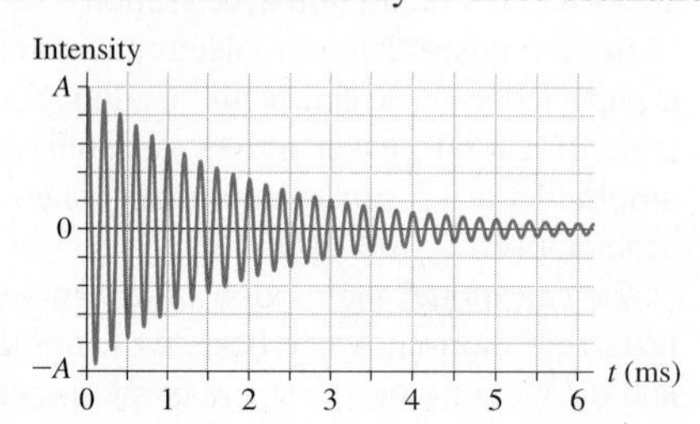

FIGURE P14.43

44. ||| A damped pendulum has a period of 0.66 s and a time constant of 4.1 s. How many oscillations will this pendulum make before its amplitude has decreased to 20% of its initial amplitude?

45. ||| When you drive your car over a bump, the springs connecting the wheels to the car compress. Your shock absorbers then damp the subsequent oscillation, keeping your car from bouncing up and down on the springs. Figure P14.45 shows real data for a car driven over a bump. We can model this as a damped oscillation, although this model is far from perfect. Estimate the frequency and time constant in this model.

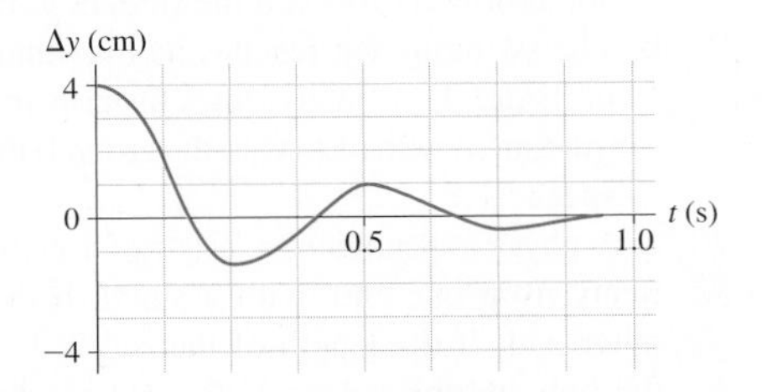

FIGURE P14.45

Section 14.7 Driven Oscillations and Resonance

46. || Taipei 101 (a 101-story building in Taiwan) is sited in an area that is prone to earthquakes and typhoons, both of which can lead to dangerous oscillations of the building. To reduce the maximum amplitude, the building has a *tuned mass damper*, a 660,000 kg mass suspended from 42-m-long cables that oscillates at the same natural frequency as the building. When the building sways, the pendulum swings, reaching an amplitude of 75 cm in strong winds or tremors. Damping the motion of the mass reduces the maximum amplitude of oscillation of the building.

    a. What is the period of oscillation of the building?
    b. During strong winds, how fast is the pendulum moving when it passes through the equilibrium position?

47. || A 25 kg child sits on a 2.0-m-long rope swing. You are going to give the child a small, brief push at regular intervals. If you want to increase the amplitude of her motion as quickly as possible, how much time should you wait between pushes?

48. ||| Your car rides on springs, so it will have a natural frequency of oscillation. Figure P14.48 shows data for the amplitude of motion of a car driven at different frequencies. The car is driven at 20 mph over a washboard road with bumps spaced 10 feet apart; the resulting ride is quite bouncy. Should the driver speed up or slow down for a smoother ride?

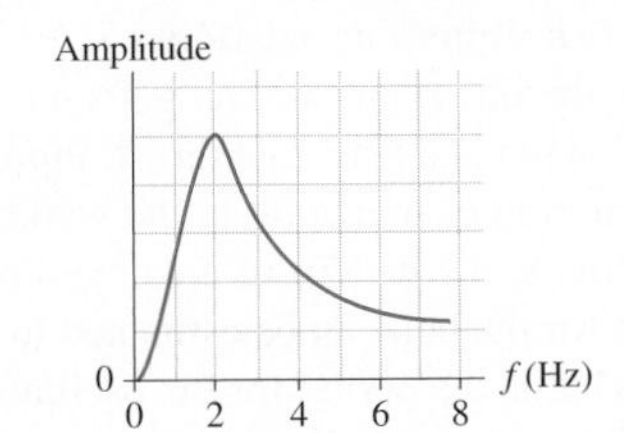

FIGURE P14.48

49. || Vision is blurred if the head is vibrated at 29 Hz because the
BIO vibrations are resonant with the natural frequency of the eyeball held by the musculature in its socket. If the mass of the eyeball is 7.5 g, a typical value, what is the effective spring constant of the musculature attached to the eyeball?

General Problems

50. || A spring has an unstretched length of 12 cm. When an 80 g ball is hung from it, the length increases by 4.0 cm. Then the ball is pulled down another 4.0 cm and released.
    a. What is the spring constant of the spring?
    b. What is the period of the oscillation?
    c. Draw a position-versus-time graph showing the motion of the ball for three cycles of the oscillation. Let the equilibrium position of the ball be $y = 0$. Be sure to include appropriate units on the axes so that the period and the amplitude of the motion can be determined from your graph.

51. || A 0.40 kg ball is suspended from a spring with spring constant 12 N/m. If the ball is pulled down 0.20 m from the equilibrium position and released, what is its maximum speed while it oscillates?

52. || When Kayla stands on her trampoline, it sags by 0.21 m. Now she starts bouncing. How much time elapses between the instant when she first lands on the trampoline's surface and when she passes the same point on the way up?

53. || A spring with spring constant 15.0 N/m hangs from the ceiling. A ball is suspended from the spring and allowed to come to rest. It is then pulled down 6.00 cm and released. If the ball makes 30 oscillations in 20.0 s, what are its (a) mass and (b) maximum speed?

54. |||| A block attached to a horizontal spring is pulled back a certain distance from equilibrium, then released from rest at $t = 0$ s. If the frequency of the block is 0.72 Hz, what is the earliest time after the block is released that its kinetic energy is exactly one-half of its potential energy?

55. |||| A circular cylinder has a diameter of 2.0 cm and a mass of 10 g. It
INT floats in water with its long axis perpendicular to the water's surface. It is pushed down into the water by a small distance and released; it then bobs up and down. What is the oscillation frequency?

56. || A spring is hung from the ceiling. When a coffee mug is attached to its end, the spring stretches 2.0 cm before reaching its new equilibrium length. The mug is then pulled down slightly and released. What is the frequency of oscillation?

57. ||| On your first trip to Planet X you happen to take along a 200 g mass, a 40.0-cm-long spring, a meter stick, and a stopwatch. You're curious about the free-fall acceleration on Planet X, where ordinary tasks seem easier than on earth, but you can't find this information in your Visitor's Guide. One night you suspend the spring from the ceiling in your room and hang the mass from it. You find that the mass stretches the spring by 31.2 cm. You then pull the mass down 10.0 cm and release it. With the stopwatch you find that 10 oscillations take 14.5 s. Can you now satisfy your curiosity?

58. ||| An object oscillating on a spring has the velocity graph shown in Figure P14.58. Draw a velocity graph if the following changes are made.
    a. The amplitude is doubled and the frequency is halved.
    b. The amplitude and spring constant are kept the same, but the mass is quadrupled.

    Parts a and b are independent questions, each starting from the graph shown.

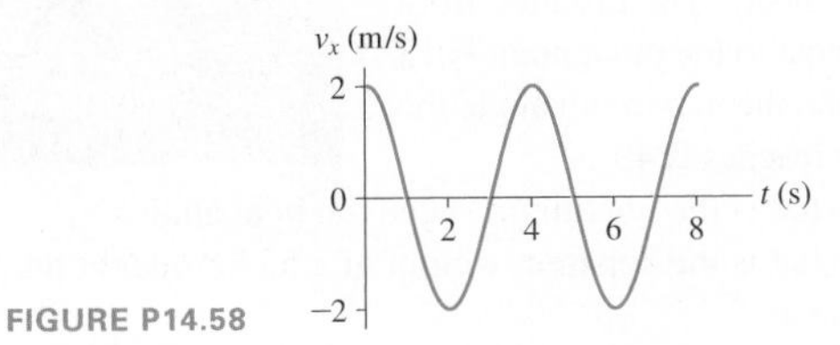

FIGURE P14.58

59. || The two graphs in Figure P14.59 are for two different vertical mass-spring systems.

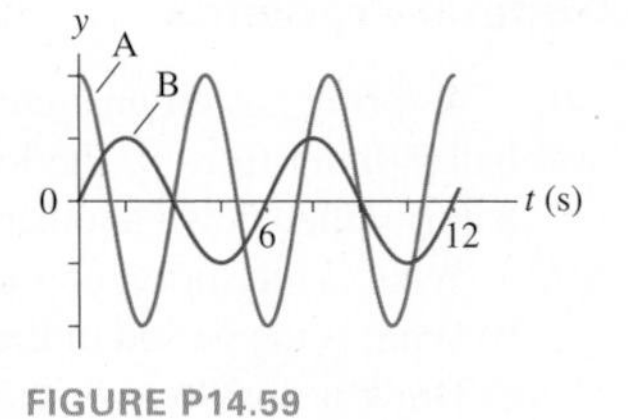

FIGURE P14.59

a. What is the frequency of system A? What is the first time at which the mass has maximum speed while traveling in the upward direction?
b. What is the period of system B? What is the first time at which the mechanical energy is all potential?
c. If both systems have the same mass, what is the ratio k_A/k_B of their spring constants?

60. ||| As we've seen, astronauts measure their mass by measuring the BIO period of oscillation when sitting in a chair connected to a spring. The Body Mass Measurement Device on Skylab, a 1970s space station, had a spring constant of 606 N/m. The empty chair oscillated with a period of 0.901 s. What is the mass of an astronaut who oscillates with a period of 2.09 s when sitting in the chair?

61. |||| A 100 g ball attached to a spring with spring constant 2.50 N/m oscillates horizontally on a frictionless table. Its velocity is 20.0 cm/s when $x = -5.00$ cm.
a. What is the amplitude of oscillation?
b. What is the speed of the ball when $x = 3.00$ cm?

62. || The ultrasonic transducer used in a medical ultrasound imag- BIO ing device is a very thin disk ($m = 0.10$ g) driven back and forth in SHM at 1.0 MHz by an electromagnetic coil.
a. The maximum restoring force that can be applied to the disk without breaking it is 40,000 N. What is the maximum oscillation amplitude that won't rupture the disk?
b. What is the disk's maximum speed at this amplitude?

63. || A compact car has a mass of 1200 kg. When empty, the car bounces up and down on its springs 2.0 times per second. What is the car's oscillation frequency when it is carrying four 70 kg passengers?

64. ||| A car with a total mass of 1400 kg (including passengers) is driving down a washboard road with bumps spaced 5.0 m apart. The ride is roughest—that is, the car bounces up and down with the maximum amplitude—when the car is traveling at 6.0 m/s. What is the spring constant of the car's springs?

65. || A 500 g air-track glider attached to a spring with spring con- INT stant 10 N/m is sitting at rest on a frictionless air track. A 250 g glider is pushed toward it from the far end of the track at a speed of 120 cm/s. It collides with and sticks to the 500 g glider. What are the amplitude and period of the subsequent oscillations?

66. |||| Tarzan, who has a mass of 80 kg, holds onto the end of a vine INT that is at a 12° angle from the vertical. He steps off his branch and, just at the bottom of his swing, he grabs onto his chimp friend Cheetah, whose mass is 40 kg. What is the maximum angle the rope reaches as Tarzan swings to the other side?

67. ||| In the Pirate Boat ride at the INT amusement park, riders swing back and forth in a pendulum-like "boat." The distance from the boat to the pivot point is 13 m, and the maximum angle the boat reaches is 40°.

a. What is the maximum speed the boat attains?
b. What is the apparent weight of a 55 kg rider at the bottom of the arc?

68. || The typical American man has a leg length of 0.85 m and BIO walks at a speed of 1.4 m/s. A giraffe's legs are 1.8 m long. At what speed do you expect a giraffe to walk? **Hint:** An animal's speed is proportional to the length of its legs times the frequency of its strides.

69. |||| Suppose that during each step, the leg of the student in BIO Example 14.11 swings through a total distance of 2.0 m. At the end of the swing, this foot rests on the ground for 0.2 s before the other leg begins its swing.
a. At what speed does this student walk? (Think carefully about how far forward the student moves at each step.)
b. The swinging leg reaches its maximum speed at the bottom of its arc. How many times faster is this maximum leg speed (measured with respect to the ground) than the average walking speed?

70. |||| A physical pendulum consists of a uniform rod that can swing INT freely from one end, with a small, heavy bob attached to the other end. If the length of the rod is 1.35 m, and the mass of the bob and the rod are both 1.0 kg, what is the period of this pendulum?

71. ||| The pendulum on a grandfather clock has a period of 2.00 s. If the clock is not wound, the pendulum's amplitude begins to decay at a rate of 0.53% each pendulum period.
a. What is the time constant of this pendulum?
b. What percentage of the pendulum's energy is lost each period?

72. |||| Orangutans can move by brachiation, swinging like a pen- BIO INT dulum beneath successive handholds. If an orangutan has arms that are 0.90 m long and repeatedly swings to a 20° angle, taking one swing immediately after another, estimate how fast it is moving in m/s.

73. |||| An infant's toy has a 120 g wooden animal hanging from INT a spring. If pulled down gently, the animal oscillates up and down with a period of 0.50 s. His older sister pulls the spring a bit more than intended. She pulls the animal 30 cm below its equilibrium position, then lets go. The animal flies upward and detaches from the spring right at the animal's equilibrium position. If the animal does not hit anything on the way up, how far above its equilibrium position will it go?

74. |||| A jellyfish can propel BIO itself with jets of water pushed out of its bell, a flexible structure on top of its body. The elastic bell and the water it contains function as a mass-spring system, greatly increasing efficiency. Normally, the jellyfish emits one jet right after the other, but we can get some insight into the jet system by looking at a single jet thrust. Figure P14.74 shows a graph of the motion of one point in the wall of the bell for such a single jet; this is the pattern of a damped oscillation. The spring constant for the bell can be estimated to be 1.2 N/m.

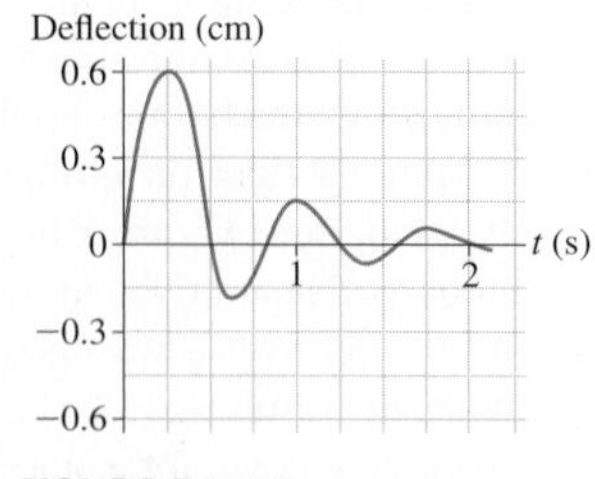

FIGURE P14.74

a. What is the period for the oscillation?
b. Estimate the effective mass participating in the oscillation. This is the mass of the bell itself plus the mass of the water.
c. Consider the peaks of positive displacement in the graph. By what factor does the amplitude decrease over one period? Given this, what is the time constant for the damping?

75. |||| A 200 g oscillator in a vacuum chamber has a frequency of 2.0 Hz. When air is admitted, the oscillation decreases to 60% of its initial amplitude in 50 s. How many oscillations will have been completed when the amplitude is 30% of its initial value?

76. ||| While seated on a tall bench, extend your lower leg a small BIO amount and then let it swing freely about your knee joint, with no muscular engagement. It will oscillate as a damped pendulum. Figure P14.76 is a graph of the lower leg angle versus time in such an experiment. Estimate (a) the period and (b) the time constant for this oscillation.

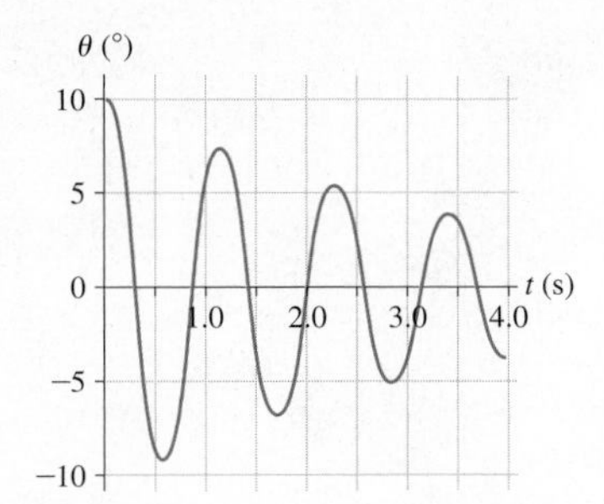

FIGURE P14.76

MCAT-Style Passage Problems

The Spring in Your Step BIO

In Chapter 10, we saw that a runner's Achilles tendon will stretch like a spring and then rebound, storing and returning energy during a step. We can model this as the simple harmonic motion of a mass-spring system. When the foot rolls forward, the tendon spring begins to stretch as the weight moves to the ball of the foot, transforming kinetic energy into elastic potential energy. This is the first phase of an oscillation. The spring then rebounds, converting potential energy to kinetic energy as the foot lifts off the ground. The oscillation is fast: Sprinters running a short race keep each foot in contact with the ground for about 0.10 second, and some of that time corresponds to the heel strike and subsequent rolling forward of the foot.

77. | We can make a static measurement to deduce the spring constant to use in the model. If a 61 kg woman stands on a low wall with her full weight on the ball of one foot and the heel free to move, the stretch of the Achilles tendon will cause her center of gravity to lower by about 2.5 mm. What is the spring constant?
A. 1.2×10^4 N/m B. 2.4×10^4 N/m
C. 1.2×10^5 N/m D. 2.4×10^5 N/m

78. | If, during a stride, the stretch causes her center of mass to lower by 10 mm, what is the stored energy?
A. 3.0 J B. 6.0 J
C. 9.0 J D. 12 J

79. | If we imagine a full cycle of the oscillation, with the woman bouncing up and down and the tendon providing the restoring force, what will her oscillation period be?
A. 0.10 s B. 0.15 s
C. 0.20 s D. 0.25 s

80. || Given what you have calculated for the period of the full oscillation in this model, what is the landing-to-liftoff time for the stretch and rebound of the sprinter's foot?
A. 0.050 s B. 0.10 s
C. 0.15 s D. 0.20 s

Web Spiders and Oscillations BIO

All spiders have special organs that make them exquisitely sensitive to vibrations. Web spiders detect vibrations of their web to determine what has landed in their web, and where.

In fact, spiders carefully adjust the tension of strands to "tune" their web. Suppose an insect lands and is trapped in a web. The silk of the web serves as the spring in a spring-mass system while the body of the insect is the mass. The frequency of oscillation depends on the restoring force of the web and the mass of the insect. Spiders respond more quickly to larger—and therefore more valuable—prey, which they can distinguish by the web's oscillation frequency.

Suppose a 12 mg fly lands in the center of a horizontal spider's web, causing the web to sag by 3.0 mm.

81. | Assuming that the web acts like a spring, what is the spring constant of the web?
A. 0.039 N/m B. 0.39 N/m
C. 3.9 N/m D. 39 N/m

82. | Modeling the motion of the fly on the web as a mass on a spring, at what frequency will the web vibrate when the fly hits it?
A. 0.91 Hz B. 2.9 Hz
C. 9.1 Hz D. 29 Hz

83. | If the web were vertical rather than horizontal, how would the frequency of oscillation be affected?
A. The frequency would be higher.
B. The frequency would be lower.
C. The frequency would be the same.

84. | Spiders are more sensitive to oscillations at higher frequencies. For example, a low-frequency oscillation at 1 Hz can be detected for amplitudes down to 0.1 mm, but a high-frequency oscillation at 1 kHz can be detected for amplitudes as small as 0.1 μm. For these low- and high-frequency oscillations, we can say that
A. The maximum acceleration of the low-frequency oscillation is greater.
B. The maximum acceleration of the high-frequency oscillation is greater.
C. The maximum accelerations of the two oscillations are approximately equal.

15 Traveling Waves and Sound

This bat's ears are much more prominent than its eyes. It appears that hearing is a much more important sense than sight for bats. How does a bat use sound waves to locate prey?

LOOKING AHEAD ▸▸

Traveling Waves

Shaking one end of the spring up and down causes a disturbance—a **wave**—to travel along the spring.

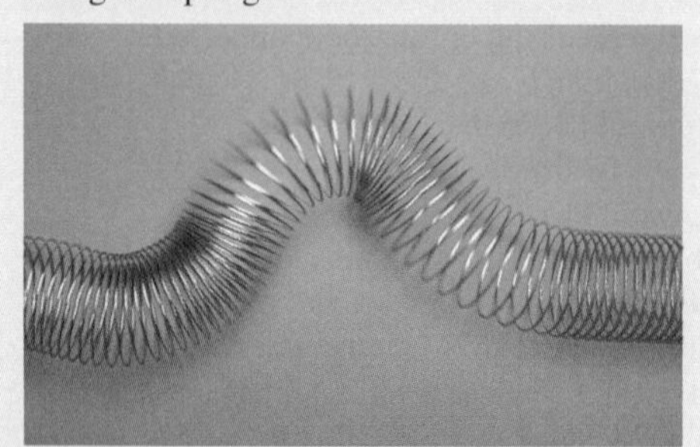

You'll learn the **wave model** that describes phenomena ranging from light waves to earthquake waves.

Describing Waves

A wave moves out from the vibrating rod. The shape of the wave is *sinusoidal,* a form we've seen for oscillations.

The terms and equations used to describe waves are closely related to those for oscillations, as you'll see.

Energy and Intensity

All waves carry energy. The energy of a laser beam can be used to burn away unwanted parts of the cornea in eye surgery.

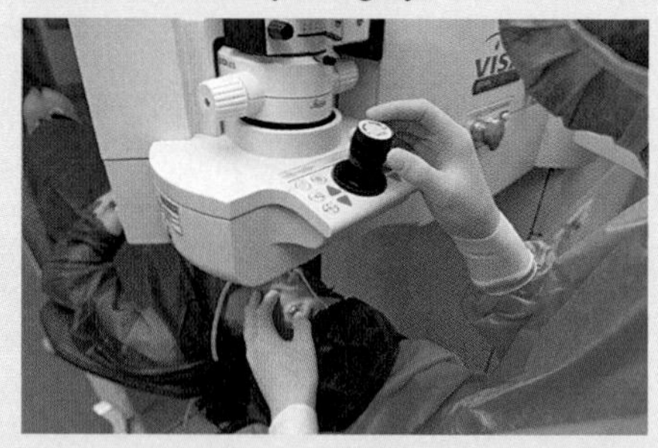

You'll learn to calculate **intensity,** a measure of how spread out or concentrated a wave's energy is.

GOAL **To learn the basic properties of traveling waves.**

LOOKING BACK ◂◂

Simple Harmonic Motion

In Chapter 14, you learned to use the terminology of simple harmonic motion to describe oscillations.

The terms you learned—such as period, frequency, and amplitude—apply to descriptions of wave motion as well.

STOP TO THINK

A wooden toy hangs from a spring. When you pull it down and release it, it reaches the highest point of its motion after 1.0 s. What is the frequency of the oscillation?

A. 2.0 Hz
B. 1.5 Hz
C. 1.0 Hz
D. 0.5 Hz

15.1 The Wave Model

The *particle model* that we have been using since Chapter 1 allowed us to simplify the treatment of motion of complex objects by considering them to be particles. Balls, cars, and rockets obviously differ from one another, but the general features of their motions are well described by treating them as particles. As we saw in Chapter 3, a ball or a rock or a car flying through the air will undergo the same motion. The particle model helps us understand this underlying simplicity.

In this chapter we will introduce the basic properties of waves with a **wave model** that emphasizes those aspects of wave behavior common to all waves. Although sound waves, water waves, and radio waves are clearly different, the wave model will allow us to understand many of the important features they share.

The wave model is built around the idea of a **traveling wave,** which is an organized disturbance that travels with a well-defined wave speed. This definition seems straightforward, but we must understand several new terms to gain a complete understanding of the concept of a traveling wave.

Mechanical Waves

Mechanical waves are waves that involve the motion of a substance through which they move, the **medium.** For example, the medium of a water wave is the water, the medium of a sound wave is the air, and the medium of a wave on a stretched string is the string.

As a wave passes through a medium, the atoms that make up the medium are displaced from equilibrium, much like pulling a spring away from its equilibrium position. This is a **disturbance** of the medium. The water ripples of FIGURE 15.1 are a disturbance of the water's surface.

FIGURE 15.1 Ripples on a pond are a traveling wave.

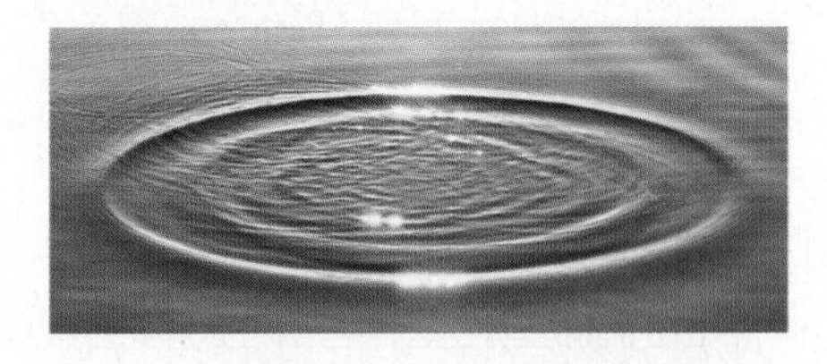

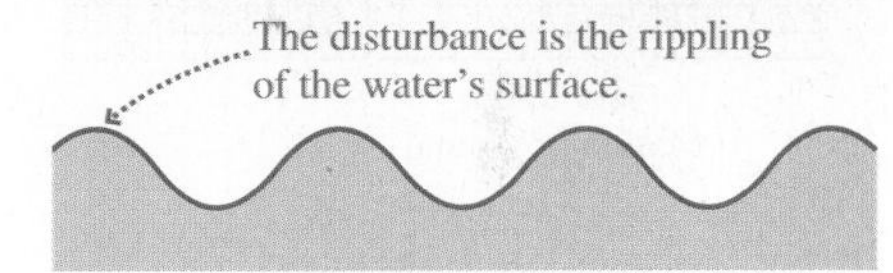
The disturbance is the rippling of the water's surface.
The water is the medium.

A wave disturbance is created by a *source.* The source of a wave might be a rock thrown into water, your hand plucking a stretched string, or an oscillating loudspeaker cone pushing on the air. Once created, the disturbance travels outward through the medium at the **wave speed** v. This is the speed with which a ripple moves across the water or a pulse travels down a string.

The disturbance propagates through the medium, and a wave does transfer *energy,* but **the medium as a whole does not travel!** The ripples on the pond (the disturbance) move outward from the splash of the rock, but there is no outward flow of water. Likewise, the particles of a string oscillate up and down but do not move in the direction of a pulse traveling along the string. **A wave transfers energy, but it does not transfer any material or substance outward from the source.**

Electromagnetic and Matter Waves

Mechanical waves require a medium, but there are waves that do not. Two important types of such waves are electromagnetic waves and matter waves.

Electromagnetic waves are waves of an *electromagnetic field.* Electromagnetic waves are very diverse, including visible light, radio waves, microwaves, and x rays. Electromagnetic waves require no material medium and can travel through a vacuum; light can travel through empty space, though sound cannot. At this point, we have not defined what an "electromagnetic field" is, so we won't worry about the precise nature of what is "waving" in electromagnetic waves. The wave model can describe many of the important aspects of these waves without a detailed description of their exact nature. We'll look more closely at electromagnetic waves in Chapter 25, once we have a full understanding of electric and magnetic fields.

One of the most significant discoveries of the 20th century was that material particles, such as electrons and atoms, have wave-like characteristics. We will learn how a full description of matter at an atomic scale requires an understanding of such **matter waves** in Chapter 28.

You may have been at a sporting event in which spectators do "The Wave." The wave moves around the stadium, but the spectators (the medium, in this case) stay right where they are. This is a clear example of the principle that a wave does not transfer any material.

Transverse and Longitudinal Waves

Most waves fall into two general classes: *transverse* and *longitudinal.* For mechanical waves, these terms describe the relationship between the motion of the particles that carry the wave and the motion of the wave itself, as the following table shows.

Two types of wave motion

A transverse wave

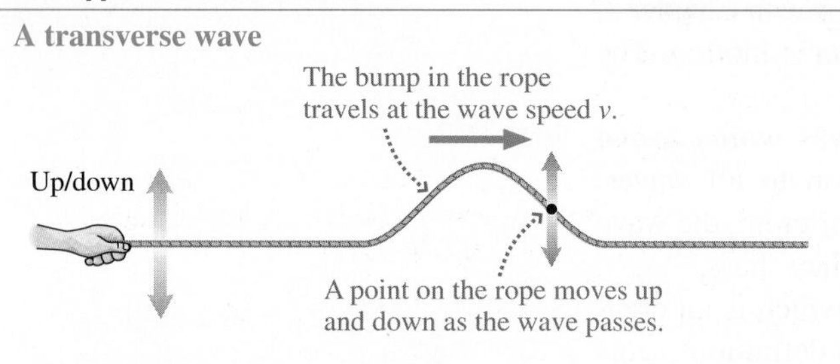

For mechanical waves, a **transverse wave** is a wave in which the particles in the medium move *perpendicular* to the direction in which the wave travels. Shaking the end of a stretched string up and down creates a wave that travels along the string in a horizontal direction while the particles that make up the string oscillate vertically.

A longitudinal wave

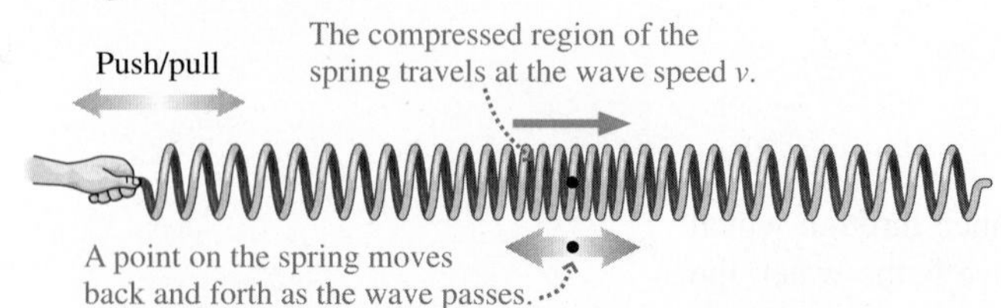

In a **longitudinal wave,** the particles in the medium move *parallel* to the direction in which the wave travels. Quickly moving the end of a spring back and forth sends a wave—in the form of a compressed region—down the spring. The particles that make up the spring oscillate horizontally as the wave passes.

FIGURE 15.2 Different types of earthquake waves.

The passage of a longitudinal P wave expands and compresses the ground. The motion is parallel to the direction of travel of the wave.

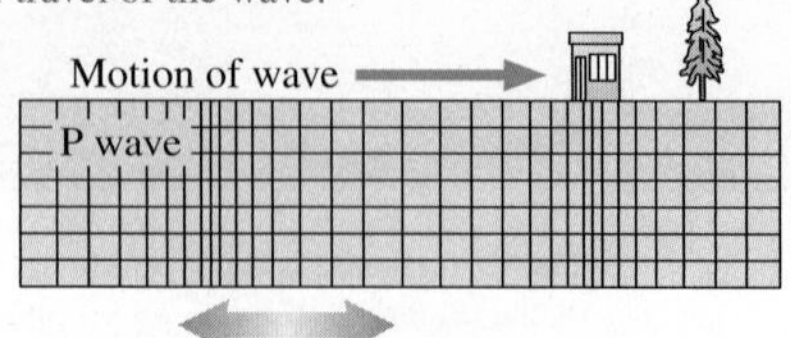

The passage of a transverse S wave moves the ground up and down. The motion is perpendicular to the direction of travel of the wave.

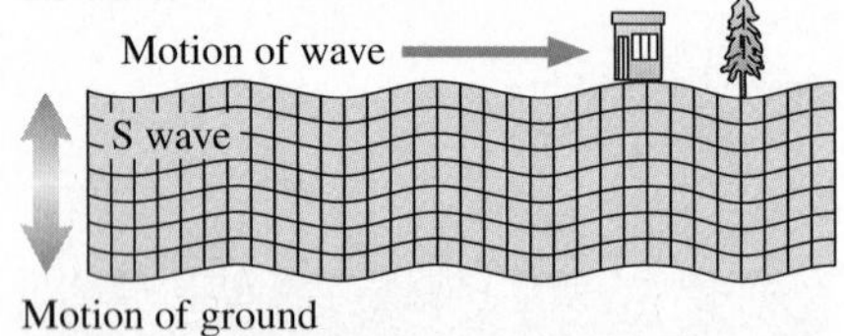

A video to support a section's topic is embedded in the eText.

Video Traveling Waves

The rapid motion of the earth's crust during an earthquake can produce a disturbance that travels through the earth. The two most important types of earthquake waves are S waves (which are transverse) and P waves (which are longitudinal), as shown in FIGURE 15.2. The longitudinal P waves are faster, but the transverse S waves are more destructive. Residents of a city a few hundred kilometers from an earthquake will feel the resulting P waves as much as a minute before the S waves, giving them a crucial early warning.

STOP TO THINK 15.1 Spectators at a sporting event do "The Wave," as shown in the photo on the preceding page. Is this a transverse or longitudinal wave?

15.2 Traveling Waves

When you drop a pebble in a pond, waves travel outward. But how does this happen? How does a mechanical wave travel through a medium? In answering this question, we must be careful to distinguish the motion of the wave from the motion of the particles that make up the medium. The wave itself is not a particle, so we cannot apply Newton's laws to the wave. However, we can use Newton's laws to examine how the medium responds to a disturbance.

Waves on a String

FIGURE 15.3 shows a transverse *wave pulse* traveling to the right along a stretched string. Imagine watching a little dot on the string as a wave pulse passes by. As the pulse approaches from the left, the string near the dot begins to curve. Once the string

FIGURE 15.3 The motion of a string as a wave passes.

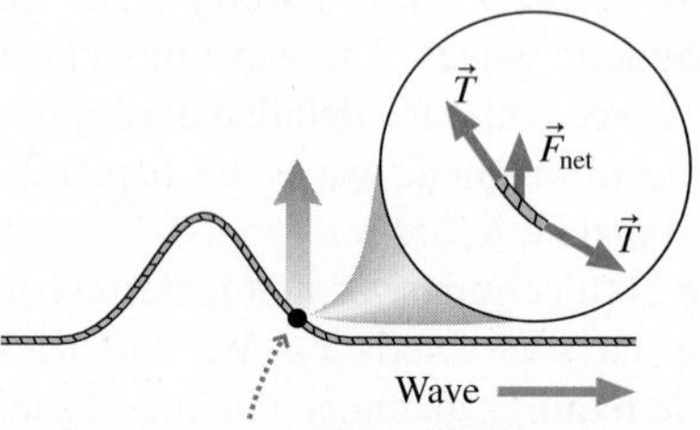

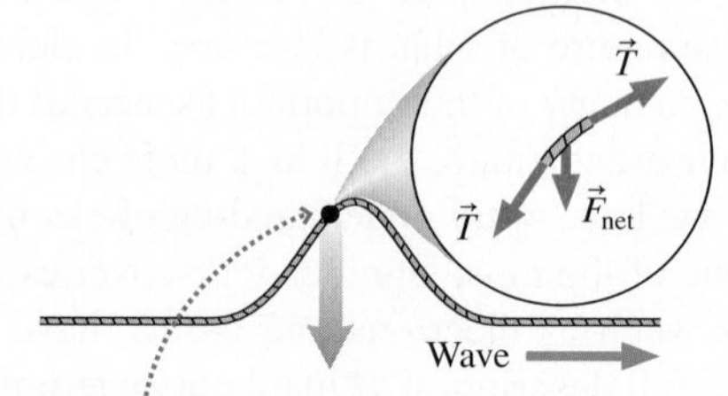

curves, the tension forces pulling on a small segment of string no longer cancel each other. As the wave passes, the curvature of the string leads to a net force that first pulls each little piece of the string up and then, after the pulse passes, back down. Each point on the string moves perpendicular to the motion of the wave, so **a wave on a string is a transverse wave.**

No new physical principles are required to understand how this wave moves. The motion of a pulse along a string is a direct consequence of the tension acting on the segments of the string. An external force created the pulse, but **once started, the pulse continues to move because of the internal dynamics of the medium.**

Sound Waves

Next, let's see how a sound wave in air is created using a loudspeaker. When the loudspeaker cone in FIGURE 15.4a moves forward, it compresses the air in front of it, as shown in FIGURE 15.4b. The *compression* is the disturbance that travels forward through the air. This is much like the sharp push on the end of the spring on the preceding page, so **a sound wave is a longitudinal wave.** We usually think of sound waves as traveling in air, but sound can travel through any gas, through liquids, and even through solids.

Just as the motion of a wave on a string is determined by the physics of a string under tension, the motion of a sound wave in air is determined by the physics of gases that we explored in Chapter 12. Once created, the wave in Figure 15.4b will propagate forward; its motion is entirely determined by the properties of the air.

FIGURE 15.4 A sound wave produced by a loudspeaker.

(a) The loudspeaker cone moves in and out in response to electrical signals.

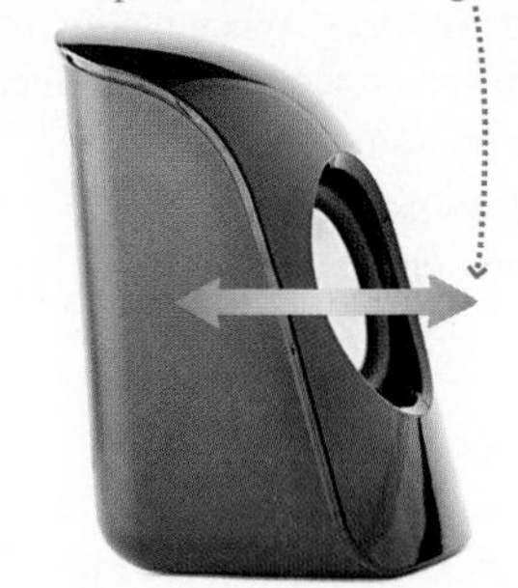

(b) The loudspeaker receives an electrical pulse and pushes out sharply. This creates a compression of the air.

Compression

v_{sound}

Speaker

Molecules

This compression is the disturbance. It travels at the wave speed v_{sound}.

Video Figure 15.4

Wave Speed Is a Property of the Medium

The above discussions of waves on a string and sound waves in air show that the motion of these waves depends on the properties of the medium. In fact, we'll see that **the wave speed does not depend on the shape or size of the pulse, how the pulse was generated, or how far it has traveled**—only the medium that carries the wave. Let's examine this idea in more detail for both waves on a string and sound waves.

What properties of a string determine the speed of waves traveling along the string? The only likely candidates are the string's mass, length, and tension. Because a pulse doesn't travel faster on a longer and thus more massive string, neither the total mass m nor the total length L is important. Instead, the speed depends on the mass-to-length *ratio,* which is called the **linear density** μ of the string:

$$\mu = \frac{m}{L} \tag{15.1}$$

Linear density characterizes the *type* of string we are using. A fat string has a larger value of μ than a skinny string made of the same material. Similarly, a steel wire has a larger value of μ than a plastic string of the same diameter.

How does the speed of a wave on a string vary with the tension and the linear density? Using what we know about forces and motion, we can make some predictions:

- A string with a greater tension responds more rapidly, so the wave will move at a higher speed. **Wave speed increases with increasing tension.**
- A string with a greater linear density has more inertia. It will respond less rapidly, so the wave will move at a lower speed. **Wave speed decreases with increasing linear density.**

A full analysis of the motion of the string leads to an expression for the speed of a wave that shows both of these trends:

$$v_{string} = \sqrt{\frac{T_s}{\mu}} \tag{15.2}$$

Wave speed on a stretched string with tension T_s and linear density μ

The subscript s on the symbol T_s for the string's tension will distinguish it from the symbol T for the *period* of oscillation.

Sensing water waves BIO The African clawed frog has a simple hunting strategy: It sits and "listens" for prey animals. It detects water waves, not sound waves, using an array of sensors called the *lateral line organ* on each side of its body. An incoming wave reaches different parts of the organ at different times, and this allows the frog to determine where the waves come from.

Every point on a wave pulse travels with the speed given by Equation 15.2. We can increase the wave speed either by *increasing* the string's tension (making it tighter) or by *decreasing* the string's linear density (making it skinnier). We'll examine the implications for stringed musical instruments in Chapter 16.

EXAMPLE 15.1 When does the spider sense its lunch? BIO

All spiders are very sensitive to vibrations. An orb spider will sit at the center of its large, circular web and monitor radial threads for vibrations created when an insect lands. Assume that these threads are made of silk with a linear density of 1.0×10^{-5} kg/m under a tension of 0.40 N, both typical numbers. If an insect lands in the web 30 cm from the spider, how long will it take for the spider to find out?

STRATEGIZE When the insect hits the web, a wave pulse will be transmitted along the silk fibers. The speed of the wave depends on the properties of the silk.

PREPARE The speed of a wave on a string is given by Equation 15.2.

SOLVE First, we determine the speed of the wave:

$$v = \sqrt{\frac{T_s}{\mu}} = \sqrt{\frac{0.40\ \text{N}}{1.0 \times 10^{-5}\ \text{kg/m}}} = 200\ \text{m/s}$$

The time for the wave to travel a distance $d = 30$ cm to reach the spider is

$$\Delta t = \frac{d}{v} = \frac{0.30\ \text{m}}{200\ \text{m/s}} = 1.5\ \text{ms}$$

ASSESS Spider webs are made of very light strings under significant tension, so the wave speed is quite high and we expect a short travel time—important for the spider to quickly respond to prey caught in the web. Our answer makes sense.

What properties of a gas determine the speed of a sound wave traveling through the gas? It seems plausible that the speed of a sound pulse is related to the speed with which the molecules of the gas move—faster molecules should mean a faster sound wave. In ◀◀ SECTION 12.2, we found that the typical speed of an atom of mass *m*, the root-mean-square speed, is

$$v_{\text{rms}} = \sqrt{\frac{3k_B T}{m}}$$

where k_B is Boltzmann's constant and T the absolute temperature in kelvin. A thorough analysis finds that the sound speed is slightly less than this rms speed, but has the same dependence on the temperature and the molecular mass:

$$v_{\text{sound}} = \sqrt{\frac{\gamma k_B T}{m}} = \sqrt{\frac{\gamma R T}{M}} \qquad (15.3)$$

Sound speed in a gas at temperature T

v_{sound} / T SQUARE ROOT p. 401

In Equation 15.3, R is the gas constant, M is the molar mass (kg per mol), and γ is a constant that depends on the gas: $\gamma = 1.67$ for monatomic gases such as helium, $\gamma = 1.40$ for diatomic gases such as nitrogen and oxygen, and $\gamma \approx 1.3$ for a triatomic gas such as carbon dioxide or water vapor.

Certain trends in Equation 15.3 are worth mentioning:

- The speed of sound in air (and other gases) increases with temperature. **For calculations in this chapter, you can use the speed of sound in air at 20°C, 343 m/s, unless otherwise specified.**
- At a given temperature, the speed of sound increases as the molecular mass of the gas decreases. Thus the speed of sound in room-temperature helium is faster than that in room-temperature air.
- The speed of sound doesn't depend on the pressure or the density of the gas.

TABLE 15.1 lists the speeds of sound in various materials. The bonds between atoms in liquids and solids result in higher sound speeds in these phases of matter. Generally, sound waves travel faster in liquids than in gases, and faster in solids than in liquids.

TABLE 15.1 The speed of sound

Medium	Speed (m/s)
Air (0°C)	331
Air (20°C)	343
Helium (0°C)	970
Ethyl alcohol	1170
Water	1480
Human tissue (ultrasound)	1540
Lead	1200
Aluminum	5100
Granite	6000
Diamond	12,000

The speed of sound in a solid depends on its density and stiffness. Light, stiff solids (such as diamond) transmit sound at very high speeds.

EXAMPLE 15.2 **The speed of sound on Mars**

On a typical Martian morning, the very thin atmosphere (which is almost entirely carbon dioxide) is a frosty $-100°C$. What is the speed of sound on such a morning? At what approximate temperature would the speed be double this value?

STRATEGIZE Equation 15.3 gives the speed of sound in terms of the molar mass and the absolute temperature. We will assume that the atmosphere is composed of pure CO_2.

PREPARE The molar mass of CO_2 is the sum of the molar masses of the constituents: 12 g/mol for C and 16 g/mol for each O, giving $M = 12 + 16 + 16 = 44\text{ g/mol} = 0.044\text{ kg/mol}$. The absolute temperature is $T = 173\text{ K}$. Carbon dioxide is a triatomic gas, so $\gamma = 1.3$.

SOLVE The speed of sound with the noted conditions is

$$v_{\text{sound}} = \sqrt{\frac{\gamma RT}{M}} = \sqrt{\frac{(1.3)(8.31\text{ J/mol}\cdot\text{K})(173\text{ K})}{0.044\text{ kg/mol}}} = 210\text{ m/s}$$

Rather than do a separate calculation to determine the temperature required for the higher speed, we can make an argument using proportionality. The speed is proportional to the square root of the temperature, so doubling the speed requires an increase in the temperature by a factor of 4 to about 690 K, or about 420°C.

ASSESS The speed of sound doesn't depend on pressure, so even though the atmosphere is "thin" we needn't adjust our calculation. The speed of sound is lower for heavier molecules and colder temperatures, so we expect that the speed of sound on Mars, with its cold, carbon-dioxide atmosphere, will be lower than that on earth, just as we found.

Electromagnetic waves, such as light, travel at a much higher speed than do mechanical waves. As we'll discuss in Chapter 25, all electromagnetic waves travel at the same speed in a vacuum. We call this speed the **speed of light,** which we represent with the symbol c. The value of the speed of light in a vacuum is

$$v_{\text{light}} = c = 3.00 \times 10^8\text{ m/s} \tag{15.4}$$

This is almost one million times the speed of sound in air! At this speed, light could circle the earth 7.5 times in one second.

NOTE ▶ The speed of electromagnetic waves is lower when they travel through a material, but this value for the speed of light in a vacuum is also good for electromagnetic waves traveling through air. ◀

EXAMPLE 15.3 **How far away was the lightning?**

During a thunderstorm, you see a flash from a lightning strike. 8.0 seconds later, you hear the crack of the thunder. How far away did the lightning strike?

STRATEGIZE Two different kinds of waves are involved, with very different wave speeds. The flash of the lightning generates light waves; these waves travel from the point of the strike to your position almost instantaneously. The strike also generates the sound waves that you hear as thunder; these waves travel much more slowly.

PREPARE The time for light to travel 1 mile (1610 m) is

$$\Delta t = \frac{d}{v} = \frac{1610\text{ m}}{3.00 \times 10^8\text{ s}} = 5.37 \times 10^{-6}\text{ s} \approx 5\ \mu\text{s}$$

We are given the time to an accuracy of only 0.1 s, so it's clear that we can ignore the travel time for the light flash! The delay between the flash and the thunder is simply the time it takes for the sound wave to travel.

SOLVE We assume that the speed of sound has its room-temperature (20°C) value of 343 m/s. During the time between seeing the flash and hearing the thunder, the sound travels a distance

$$d = v\Delta t = (343\text{ m/s})(8.0\text{ s}) = 2.7 \times 10^3\text{ m} = 2.7\text{ km}$$

ASSESS This seems reasonable. As you know from casual observations of lightning storms, an 8-second delay between the flash of the lightning and the crack of the thunder means a strike that is close but not too close. A few km seems reasonable.

Video What the Physics? Speed of Sound

Distance to a lightning strike Sound travels approximately 1 km in 3 s, or 1 mi in 5 s. When you see a lightning flash, start counting seconds. When you hear the thunder, stop counting. Divide the result by 3, and you will have the approximate distance to the lightning strike in kilometers; divide by 5 and you will have the approximate distance in miles.

STOP TO THINK 15.2 Suppose you shake the end of a stretched string to produce a wave. Which of the following actions will increase the speed of the wave down the string? There may be more than one correct answer; if so, give all that are correct.

A. Move your hand up and down more quickly as you generate the wave.
B. Move your hand up and down a greater distance as you generate the wave.
C. Use a heavier string of the same length, under the same tension.
D. Use a lighter string of the same length, under the same tension.
E. Stretch the string tighter to increase the tension.
F. Loosen the string to decrease the tension.

15.3 Graphical and Mathematical Descriptions of Waves

Describing waves and their motion takes a bit more thought than describing the motion of particles. An object described by the particle model, such as a car or a runner, is at a particular position x at a single instant of time t. This means we can describe a particle using a function of time, such as $x(t)$ or $v(t)$. Unlike a particle, a wave is not localized at a single point; it exists—and is changing with time—throughout the entire medium through which it travels. Thus to describe waves mathematically, we will need to use functions of both time *and* space, such as $y(x, t)$.

Before starting in on a mathematical description of waves, however, we want to get a good *visual* understanding of wave motion. To do so, we'll study two kinds of wave graphs: *snapshot graphs,* which depict the shape of the wave at one instant of time, and *history graphs,* which show the motion over time of a single point of the wave.

NOTE ► The analysis that follows is for a wave on a string, which is easy to visualize, but the results apply to any traveling wave. ◄

FIGURE 15.5 A snapshot graph of a wave pulse on a string.

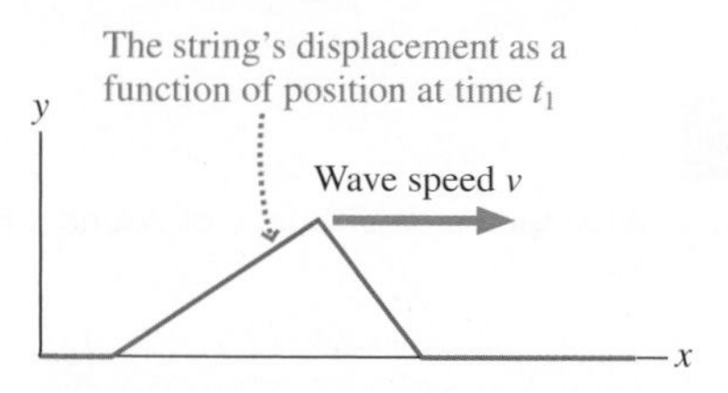
The string's displacement as a function of position at time t_1
Wave speed v
y
x

FIGURE 15.6 A sequence of snapshot graphs shows the wave in motion.

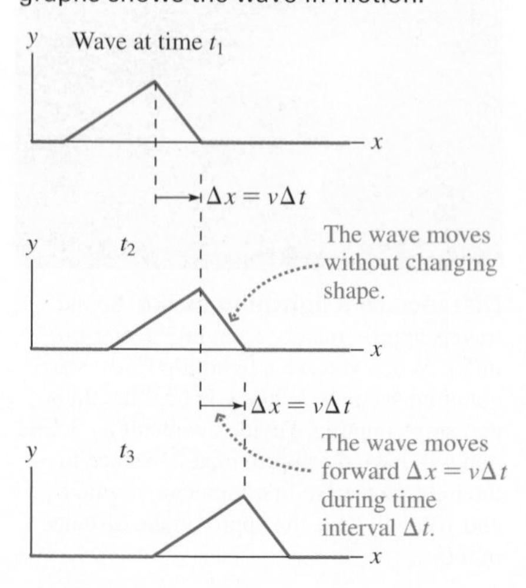
Wave at time t_1
$\Delta x = v\Delta t$
t_2
The wave moves without changing shape.
$\Delta x = v\Delta t$
t_3
The wave moves forward $\Delta x = v\Delta t$ during time interval Δt.
y
x

Snapshot and History Graphs

Consider the wave pulse shown moving along a stretched string in FIGURE 15.5. (We will consider somewhat artificial triangular and square wave pulses in this section to clearly show the edges of the pulse.) The graph shows the string's displacement y at a particular instant of time t_1 as a function of position x along the string. This is a "snapshot" of the wave, much like what you might make by using a camera to take a picture of the wave at the time t_1. A graph that shows the wave's displacement as a function of position at a single instant of time is called a **snapshot graph.**

As the wave moves, we can take more snapshots. FIGURE 15.6 shows a sequence of snapshot graphs as the wave of Figure 15.5 continues to move. These are like successive frames from a video, reminiscent of the sequences of pictures we saw in Chapter 1. The wave pulse moves forward a distance $\Delta x = v\,\Delta t$ during each time interval Δt; that is, the wave moves at a constant speed.

A snapshot graph shows the motion of the *wave*, but that's only half the story. Now we want to consider the motion of the *medium*. FIGURE 15.7a shows four snapshot graphs of a wave as it travels on a string. In each of the graphs we've placed a dot, located at position x_1, at one point on the string. As the wave travels horizontally, the dot moves vertically because a wave on a string is a transverse wave. We use the vertical positions of the dots from the snapshot graphs to construct the graph in FIGURE 15.7b, which shows the motion of this one point on the string. We call this a **history graph** because it shows the history—the time evolution—of a particular point in the medium. The snapshot graphs are pictures of the waves at particular instants in time, but the history graph isn't: It's a record of the motion of *one point* in the medium over time.

Notice that the snapshot graphs show the steeper edge of the wave on the *right* but the history graph has the steeper edge on the *left*. As the wave moves toward the dot, the steep leading edge of the wave causes the dot to rise quickly. Later, the shallower trailing edge causes the dot to descend more slowly. On the history graph, as on any displacement-versus-time graph, earlier times are to the left and later times are to the right. The rapid rise when the wave hits the dot is at an early time and so appears on the left side of the Figure 15.7b history graph, whereas the slow descent of the dot occurs later and so appears to the right side of the graph. This reversal will not be present in all cases, though; you'll need to consider each situation individually.

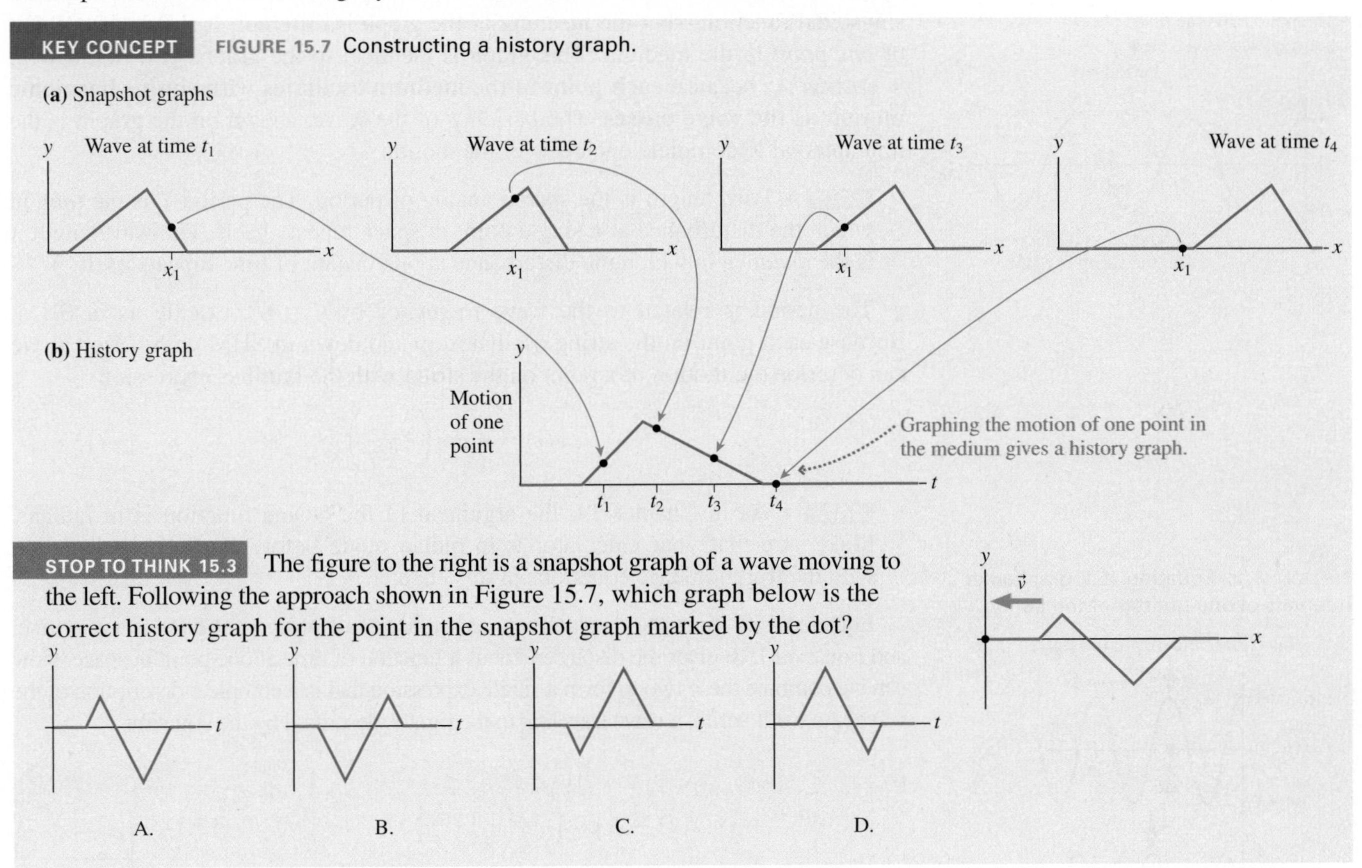

FIGURE 15.7 Constructing a history graph.

STOP TO THINK 15.3 The figure to the right is a snapshot graph of a wave moving to the left. Following the approach shown in Figure 15.7, which graph below is the correct history graph for the point in the snapshot graph marked by the dot?

A. B. C. D.

The Mathematical Description of Sinusoidal Waves

Waves can come in many different shapes, but for the mathematical description of wave motion we will focus on a particular shape, the **sinusoidal wave.** This is the type of wave produced by a source that oscillates with simple harmonic motion. A loudspeaker cone that oscillates in SHM radiates a sinusoidal sound wave.

The pair of snapshot graphs in FIGURE 15.8 shows two successive views of a string carrying a sinusoidal wave, revealing the motion of the wave as it moves to the right. We define the **amplitude** A of the wave to be the maximum value of the displacement. The crests of the wave—the high points—have displacement $y_{crest} = A$, and the troughs—the low points—have displacement $y_{trough} = -A$. Because the wave is produced by a source undergoing SHM, which is periodic, the wave is periodic as well. As you move from left to right along the wave frozen in time in the top snapshot graph of Figure 15.8, the disturbance repeats itself over and over. The distance spanned by one cycle of the motion is called the **wavelength** of the wave. Wavelength is symbolized by λ (lowercase Greek lambda) and, because it is a length, it is measured in units of meters. The wavelength is shown in Figure 15.8 as the distance between two crests, but it could equally well be the distance between two troughs. As time passes, the wave moves to the right; comparing the two snapshot graphs in Figure 15.8 makes this motion apparent.

FIGURE 15.8 Snapshot graphs show the motion of a sinusoidal wave.

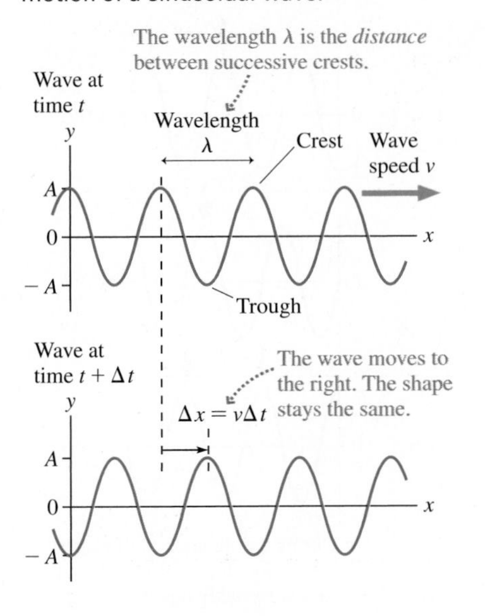

The snapshot graphs of Figure 15.8 show that the wave, at one instant in time, is a sinusoidal function of the distance x along the wave, with wavelength λ. At the time represented by the top graph of Figure 15.8, the displacement is given by

$$y(x) = A\cos\left(2\pi\frac{x}{\lambda}\right) \tag{15.5}$$

FIGURE 15.9 A history graph shows the motion of a point on a string carrying a sinusoidal wave.

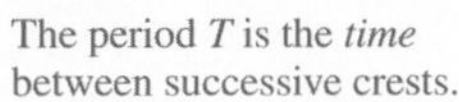
The period T is the time between successive crests.

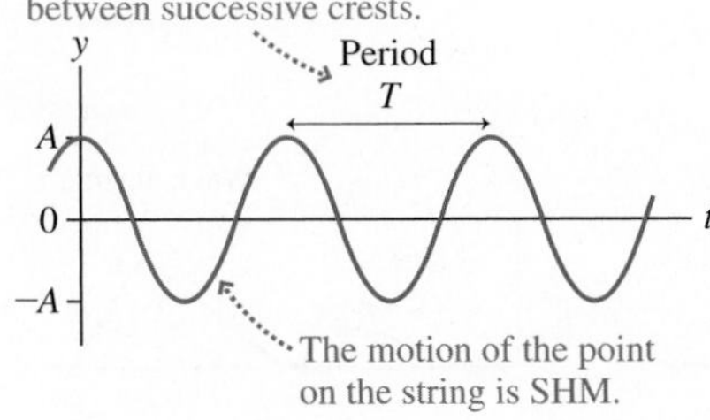
y
Period
T
A
0
−A
t
The motion of the point on the string is SHM.

Next, let's look at the motion of a point in the medium as this wave passes. FIGURE 15.9 shows a history graph for a point on a string as the sinusoidal wave of Figure 15.8 passes by. This graph has exactly the same shape as the snapshot graphs of Figure 15.8—it's a sinusoidal function—but the meaning of the graph is different: It shows the motion of *one point* in the medium. This graph is identical to the graphs you first saw in ◀◀ SECTION 14.3 because **each point in the medium oscillates with simple harmonic motion as the wave passes.** The *period* T of the wave, shown on the graph, is the time interval to complete one cycle of the motion.

NOTE ▸ Wavelength is the spatial analog of period. The period T is the *time* in which the disturbance at a single point in space repeats itself. The wavelength λ is the *distance* in which the disturbance at one instant of time repeats itself. ◂

The period is related to the wave *frequency* by $T = 1/f$, exactly as in SHM. Because each point on the string oscillates up and down in SHM with period T, we can describe the motion of a point on the string with the familiar expression

$$y(t) = A\cos\left(2\pi\frac{t}{T}\right) \tag{15.6}$$

NOTE ▸ As in Chapter 14, the argument of the cosine function is in radians. Make sure that your calculator is in radian mode before starting a calculation with the trigonometric equations in this chapter. ◂

FIGURE 15.10 Equation 15.7 graphed at intervals of one-quarter of the period.

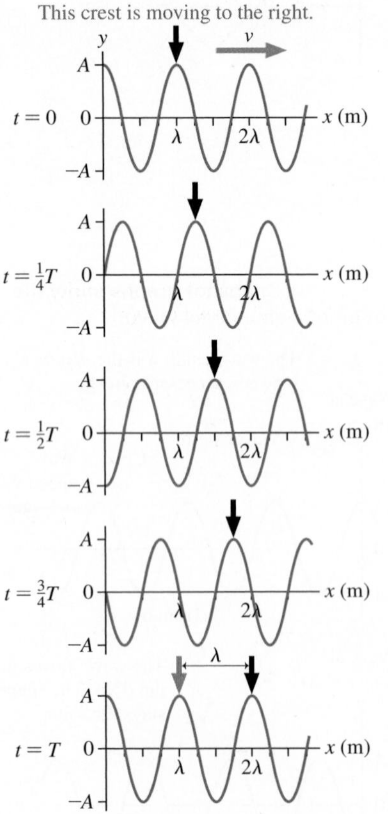
This crest is moving to the right.
y
v
A
0
−A
x (m)
λ
2λ
t = 0
t = ¼T
t = ½T
t = ¾T
t = T
During a time interval of exactly one period, the crest has moved forward exactly one wavelength.

Equation 15.5 gives the displacement as a function of position at one instant in time, and Equation 15.6 gives the displacement as a function of time at one point in space. How can we combine these two to form a single expression that is a complete description of the wave? As we'll verify, a wave traveling to the right is described by the equation

$$y(x, t) = A\cos\left(2\pi\left(\frac{x}{\lambda} - \frac{t}{T}\right)\right) \tag{15.7}$$

Displacement of a traveling wave moving to the right with amplitude A, wavelength λ, and period T

SINUSOIDAL p. 491

The notation $y(x, t)$ indicates that the displacement y is a function of the *two* variables x and t. We must specify both where (x) and when (t) before we can calculate the displacement of the wave.

We can understand why this expression describes a wave traveling to the right by looking at FIGURE 15.10. In the figure, we have graphed Equation 15.7 at five instants of time, each separated by one-quarter of the period T, to make five snapshot graphs. The crest marked with the arrow represents one point on the wave. As the time t increases, so does the position x of this point—the wave moves to the right. One full period has elapsed between the first graph and the last. During this time, the wave has moved by one wavelength. And, as the wave has passed, each point in the medium has undergone one complete oscillation.

For a wave traveling to the left, we have a slightly different form:

$$y(x, t) = A\cos\left(2\pi\left(\frac{x}{\lambda} + \frac{t}{T}\right)\right) \tag{15.8}$$

Displacement of a traveling wave moving to the left

NOTE ▸ A wave moving to the right (the $+x$-direction) has a $-$ in the expression, while a wave moving to the left (the $-x$-direction) has a $+$. Remember that the sign is *opposite* the direction of travel. ◂

EXAMPLE 15.4 Determining the rise and fall of a moving boat

A boat is moving to the right at 5.0 m/s with respect to the water. An ocean wave is moving to the left, opposite to the motion of the boat. The waves have 2.0 m between the top of the crests and the bottom of the troughs. The period of the waves is 8.3 s, and their wavelength is 110 m. At one instant, the boat sits on a crest of the wave. 20 s later, what is the vertical displacement of the boat?

STRATEGIZE Because the wave is moving to the left, the wave height as a function of position and time is given by Equation 15.8. From the boat's speed, we will find its position x at time $t = 20$ s, and then use Equation 15.8 to find its height.

PREPARE We begin with a visual overview, as in FIGURE 15.11. Let $t = 0$ be the instant the boat is on the crest, and draw a snapshot graph of the traveling wave at that time. The boat begins at a crest of the wave, so we see that the boat can start the problem at $x = 0$.

FIGURE 15.11 Visual overview for the boat.

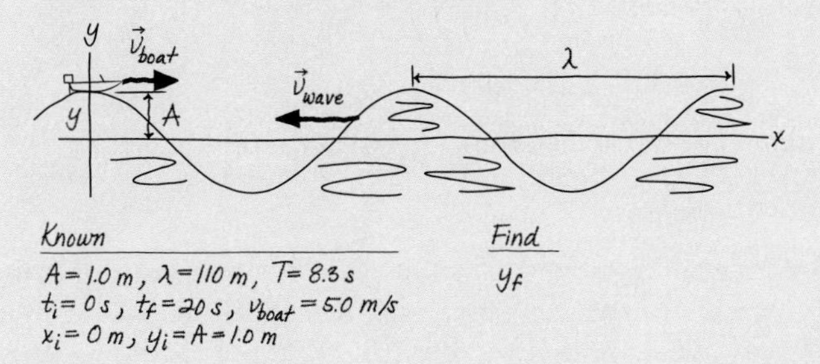

The distance between the high and low points of the wave is 2.0 m; the amplitude is half this, so $A = 1.0$ m. The wavelength and period are given in the problem.

SOLVE The boat is moving to the right at 5.0 m/s. At $t_f = 20$ s the boat is at position

$$x_f = (5.0\text{ m/s})(20\text{ s}) = 100\text{ m}$$

We need to find the wave's displacement at this position and time. Substituting known values for amplitude, wavelength, and period into Equation 15.8, for a wave traveling to the left, we obtain the following equation for the wave:

$$y(x, t) = (1.0\text{ m})\cos\left(2\pi\left(\frac{x}{110\text{ m}} + \frac{t}{8.3\text{ s}}\right)\right)$$

At $t_f = 20$ s and $x_f = 100$ m, the boat's displacement on the wave is

$$y_f = y(\text{at } 100\text{ m}, 20\text{ s}) = (1.0\text{ m})\cos\left(2\pi\left(\frac{100\text{ m}}{110\text{ m}} + \frac{20\text{ s}}{8.3\text{ s}}\right)\right)$$
$$= -0.42\text{ m}$$

Don't forget that your calculator must be in radian mode when you make your final computation!

ASSESS The final displacement is negative—which means that the boat is in a trough of a wave, not underwater.

The Fundamental Relationship for Sinusoidal Waves

In Figure 15.10, one critical observation is that the wave crest marked by the arrow has moved one full wavelength between the first graph and the last. That is, **during a time interval of exactly one period T, each crest of a sinusoidal wave travels forward a distance of exactly one wavelength λ.** Because speed is distance divided by time, the wave speed must be

Video Traveling Waves Problems

$$v = \frac{\text{distance}}{\text{time}} = \frac{\lambda}{T} \tag{15.9}$$

Because $f = 1/T$, it is customary to write Equation 15.9 in the form

$$v = \lambda f \tag{15.10}$$

Relationship between velocity, wavelength, and frequency for sinusoidal waves

Although Equation 15.10 has no special name, it is *the* fundamental relationship for sinusoidal waves. When using it, keep in mind the *physical* meaning that a wave moves forward a distance of one wavelength during a time interval of one period.

At this point, it's worthwhile to bring together the details about wave motion we've learned in order to review and to make connections.

SYNTHESIS 15.1 Wave motion

Wave speed is determined by the medium.

For a wave on a stretched string:

$$v_{string} = \sqrt{\frac{T_s}{\mu}}$$

String tension (N)

String mass divided by length (kg/m)

For a sound wave in a gas:

$$v_{sound} = \sqrt{\frac{\gamma k_B T}{m}}$$

Temperature (K)

Molecular mass (kg)

For example, at room temperature in air:

$$v_{sound}(20\ °C) = 343\ m/s$$

In a vacuum or air, the speed of light is

$$v_{light} = 3.00 \times 10^8\ m/s$$

A sinusoidal wave moves one wavelength in one period, giving the fundamental relationship

$$v = \lambda f = \frac{\lambda}{T}$$

Speed of the wave (m/s)

Wavelength (m)

Frequency of the wave (Hz)

Period of the wave (s)

As a sinusoidal wave travels, each point in the medium moves with simple harmonic motion.

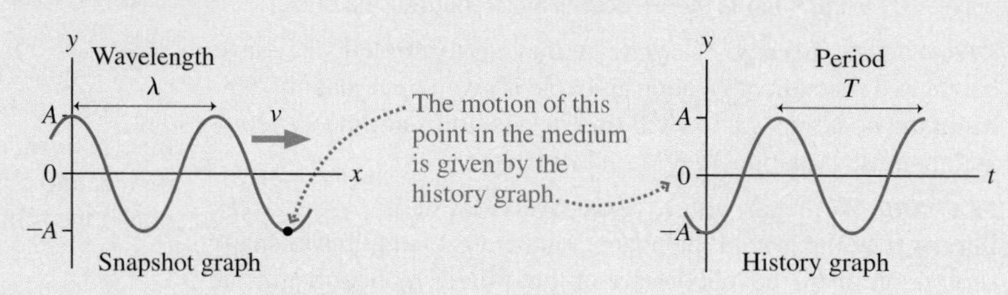

The displacement of a traveling sinusoidal wave is given by the equations

$$y(x,t) = A\cos\left(2\pi\left(\frac{x}{\lambda} - \frac{t}{T}\right)\right)$$

Wave moving to the right

$$y(x,t) = A\cos\left(2\pi\left(\frac{x}{\lambda} + \frac{t}{T}\right)\right)$$

Wave moving to the left

These expressions are appropriate for a wave that is at its maximum ($y = A$) at the origin ($x = 0$) at $t = 0$.

EXAMPLE 15.5 Writing the equation for a wave

A sinusoidal wave with an amplitude of 1.5 cm and a frequency of 100 Hz travels at 200 m/s in the positive x-direction. Write the equation for the wave's displacement as it travels.

STRATEGIZE The wave is moving to the right, so its equation will have the form of Equation 15.7. We will need the amplitude A, the wavelength λ, and the period T.

PREPARE The problem statement gives the characteristics of the wave: $A = 1.5\ cm = 0.015\ m$, $v = 200\ m/s$, and $f = 100\ Hz$.

SOLVE The amplitude is given in the problem statement. To find the wavelength, we can use the fundamental relationship for sinusoidal waves:

$$\lambda = \frac{v}{f} = \frac{200\ m/s}{100\ Hz} = 2.0\ m$$

The period can be calculated from the frequency:

$$T = \frac{1}{f} = \frac{1}{100\ Hz} = 0.010\ s$$

With these values in hand, we can write the equation for the wave's displacement:

$$y(x,t) = (0.015\ m)\cos\left(2\pi\left(\frac{x}{2.0\ m} - \frac{t}{0.010\ s}\right)\right)$$

ASSESS If the speed of a wave is known, you can use the fundamental relationship for sinusoidal waves to find the wavelength if you are given the frequency, or the frequency if you are given the wavelength. You'll often need to do this in the early stages of problems that you solve.

STOP TO THINK 15.4 Three waves travel to the right with the same speed. Which wave has the highest frequency? All three graphs have the same horizontal scale.

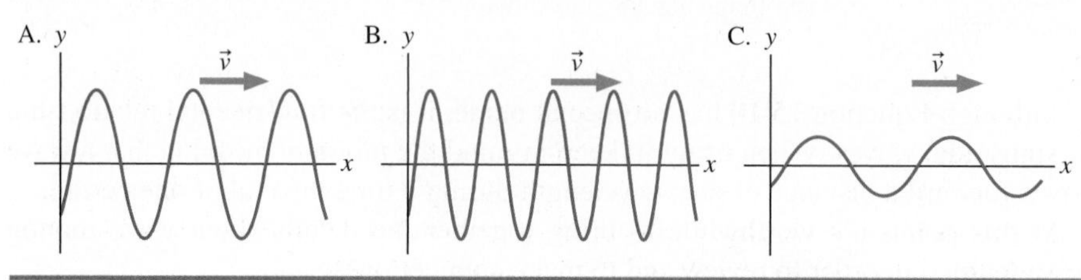

15.4 Sound and Light Waves

Think about how you are experiencing the world right now. Chances are, your senses of sight and sound are hard at work, detecting and interpreting light and sound waves from the world around you.

Sound Waves

We saw in Figure 15.4 how a loudspeaker creates a sound wave. If the loudspeaker cone moves with simple harmonic motion, it will create a sinusoidal sound wave, as illustrated in FIGURE 15.12. Each time the cone moves forward, it pushes the air molecules closer together, creating a region of higher pressure. A half cycle later, as the cone moves backward, the air has room to expand and the pressure decreases. These regions of higher and lower pressure are called **compressions** and **rarefactions,** respectively.

As Figure 15.12 suggests, it is often most convenient and informative to think of a sound wave as a pressure wave. As the graph of the pressure shows, the pressure oscillates sinusoidally around the atmospheric pressure p_{atmos}. This is a snapshot graph of the wave at one instant of time, and the distance between two adjacent crests (two points of maximum compression) is the wavelength λ.

When the wave reaches your ear, the oscillating pressure causes your eardrums to vibrate. This vibration is transferred through your inner ear to the cochlea, where it is sensed, as we learned in Chapter 14. Humans with normal hearing are able to detect sinusoidal sound waves with frequencies between about 20 Hz and 20,000 Hz, or 20 kHz. Low frequencies are perceived as a "low pitch" bass note, while high frequencies are heard as a "high pitch" treble note.

FIGURE 15.12 A sound wave is a pressure wave.

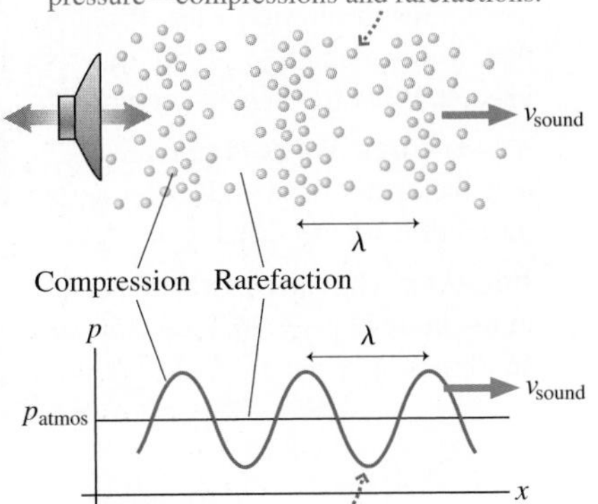
The loudspeaker cone moves back and forth, creating regions of higher and lower pressure—compressions and rarefactions.
v_{sound}
λ
Compression
Rarefaction
p
λ
v_{sound}
p_{atmos}
x
The sound wave is a pressure wave. Compressions are crests; rarefactions are troughs.

EXAMPLE 15.6 Range of wavelengths of sound

What are the wavelengths of sound waves at the limits of human hearing and at the midrange frequency of 500 Hz? Notes sung by human voices are near 500 Hz, as are notes played by striking keys near the center of a piano keyboard.

STRATEGIZE We will solve for the wavelengths using the fundamental relationship $v = \lambda f$.

PREPARE We will do our calculation at room temperature, 20°C, so we will use $v = 343$ m/s for the speed of sound.

SOLVE For the three wavelengths, the fundamental relationship gives

$$f = 20\text{ Hz}: \quad \lambda = \frac{v}{f} = \frac{343\text{ m/s}}{20\text{ Hz}} = 17\text{ m}$$

$$f = 500\text{ Hz}: \quad \lambda = \frac{v}{f} = \frac{343\text{ m/s}}{500\text{ Hz}} = 0.69\text{ m}$$

$$f = 20\text{ kHz}: \quad \lambda = \frac{v}{f} = \frac{343\text{ m/s}}{20 \times 10^3\text{ Hz}} = 0.017\text{ m} = 1.7\text{ cm}$$

ASSESS The wavelength of a 20 kHz note is only 1.7 cm. At the other extreme, a 20 Hz note has a huge wavelength of 17 m! A wave moves forward one wavelength during a time interval of one period, and a wave traveling at 343 m/s can move 17 m during the $\frac{1}{20}$ s period of a 20 Hz note.

It is well known that dogs are sensitive to high frequencies that humans cannot hear. Other animals also have quite different ranges of hearing than humans; some examples are listed in TABLE 15.2.

Elephants communicate over great distances with vocalizations at frequencies far too low for us to hear. There are animals that use frequencies well above the range of our hearing as well. High-frequency sounds are useful for *echolocation*—emitting a pulse of sound and listening for its reflection. Bats, which generally feed at night, rely much more on their hearing than their sight. They find and catch insects by echolocation, emitting loud chirps whose reflections are detected by their large, sensitive ears. The frequencies that they use are well above the range of our hearing; we call such sound **ultrasound.** Why do bats and other animals use high frequencies for this purpose?

In Chapter 19, we will look at the *resolution* of optical instruments. The finest detail that your eye—or any optical instrument—can detect is limited by the wavelength of light. Shorter wavelengths allow for the imaging of smaller details.

The same limitations apply to the acoustic image of the world made by bats. In order to sense fine details of their surroundings, bats must use sound of very short wavelength (and thus high frequency). Other animals that use echolocation, such as porpoises, also produce and sense high-frequency sounds.

TABLE 15.2 Range of hearing for selected animals

Animal	Range of hearing (Hz)
Elephant	<5–12,000
Owl	200–12,000
Human	20–20,000
Dog	30–45,000
Mouse	1000–90,000
Bat	2000–100,000
Porpoise	75–150,000

Sound travels very well through tissues in the body, and the reflections of sound waves from different tissues can be used to create an image of the body's interior. The fine details necessary for a clinical diagnosis require the short wavelengths of ultrasound. You have certainly seen ultrasound images taken during pregnancy, where the use of x rays is clearly undesirable.

EXAMPLE 15.7 Ultrasonic frequencies in medicine BIO

To make a sufficiently detailed ultrasound image of a fetus in its mother's uterus, a physician has decided that a wavelength of 0.50 mm is needed. What frequency is required?

STRATEGIZE We need to remember that the speed of sound in tissue is different from the speed in air.

PREPARE The speed of ultrasound in the body is given in Table 15.1 as 1540 m/s.

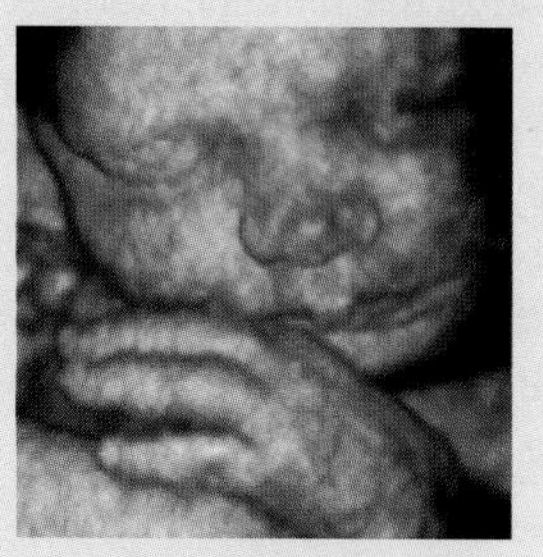

Computer processing of an ultrasound image shows fine detail.

SOLVE We can use the fundamental relationship for sinusoidal waves to calculate

$$f = \frac{v}{\lambda} = \frac{1540 \text{ m/s}}{0.50 \times 10^{-3} \text{ m}} = 3.1 \times 10^6 \text{ Hz} = 3.1 \text{ MHz}$$

ASSESS This is a reasonable result. Clinical ultrasound uses frequencies in the range of 1–20 MHz. Lower frequencies have greater penetration; higher frequencies (and thus shorter wavelengths) show finer detail.

Light and Other Electromagnetic Waves

A light wave is an electromagnetic wave, an oscillation of the electromagnetic field. Other electromagnetic waves, such as radio waves, microwaves, and ultraviolet light, have the same physical characteristics as light waves even though we cannot sense them with our eyes. As we saw earlier, all electromagnetic waves travel through a vacuum (or air) with the same speed: $c = 3.00 \times 10^8$ m/s.

The wavelengths of light are extremely short. Visible light is an electromagnetic wave with a wavelength (in air) of roughly 400 nm ($= 400 \times 10^{-9}$ m) to 700 nm ($=700 \times 10^{-9}$ m). Each wavelength is perceived as a different color. Longer wavelengths, in the 600–700 nm range, are seen as orange or red light; shorter wavelengths, in the 400–500 nm range, are seen as blue or violet light.

FIGURE 15.13 shows that the visible spectrum is a small slice out of the much broader **electromagnetic spectrum.** We will have much more to say about light and the rest of the electromagnetic spectrum in future chapters.

FIGURE 15.13 The electromagnetic spectrum from 10^6 Hz to 10^{18} Hz.

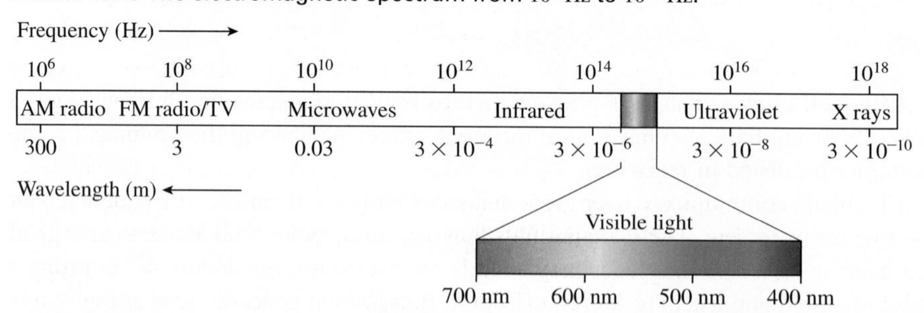

EXAMPLE 15.8 Finding the frequency of microwaves

The wavelength of microwaves in a microwave oven is 12 cm. What is the frequency of the waves?

STRATEGIZE The frequency is related to the wavelength by the fundamental relationship $f = v/\lambda$.

PREPARE Microwaves are electromagnetic waves, so their speed is the speed of light, $c = 3.00 \times 10^8$ m/s.

SOLVE Using the fundamental relationship, we find

$$f = \frac{v}{\lambda} = \frac{3.00 \times 10^8 \text{ m/s}}{0.12 \text{ m}} = 2.5 \times 10^9 \text{ Hz} = 2.5 \text{ GHz}$$

ASSESS This is a high frequency, but the speed of the waves is also very high, so our result seems reasonable.

We've just used the same equation to describe ultrasound and microwaves, which brings up a remarkable point: The wave model we've been developing applies equally well to all types of waves. Features such as wavelength, frequency, and speed are characteristics of waves in general. At this point we don't yet know any details about electromagnetic fields, yet we can use the wave model to say some significant things about the properties of electromagnetic waves.

The 2.5 GHz frequency of microwaves in an oven that we found in Example 15.8 is similar to the frequencies of other devices. Many cordless phones work at a frequency of 2.4 GHz, and cell phones at just under 2.0 GHz. Although these frequencies are close to those of the waves that heat your food in a microwave oven, the *intensity* is much less—which brings us to the topic of the next section.

STOP TO THINK 15.5 Comparing two different types of electromagnetic waves, infrared and ultraviolet, we can say that

A. Infrared has a longer wavelength and higher frequency than ultraviolet.
B. Infrared has a shorter wavelength and higher frequency than ultraviolet.
C. Infrared has a longer wavelength and lower frequency than ultraviolet.
D. Infrared has a shorter wavelength and lower frequency than ultraviolet.

15.5 Energy and Intensity

A traveling wave transfers energy from one point to another. The sound wave from a loudspeaker sets your eardrum into motion. Light waves from the sun warm the earth and, if focused with a lens, can start a fire. The *power* of a wave is the rate, in joules per second, at which the wave transfers energy. As you learned in ◀◀ SECTION 10.10, power is measured in watts. A person singing or shouting as loud as possible is emitting energy in the form of sound waves at a rate of about 1 W, or 1 J/s. In this section, we will learn how to characterize the power of waves. A first step in doing so is to understand how waves change as they spread out.

Circular, Spherical, and Plane Waves

If you take a photograph of ripples spreading on a pond and mark the location of the *crests* on the photo, your picture looks like FIGURE 15.14a. The lines that locate the crests are **wave fronts,** and they are spaced precisely one wavelength apart. A wave like this is called a **circular wave,** a two-dimensional wave that spreads across a surface.

Although the wave fronts are circles, you would hardly notice the curvature if you observed a small section of the wave front very far away from the source. The wave fronts would appear to be parallel lines, still spaced one wavelength apart and traveling at speed v as in FIGURE 15.14b.

Many waves, such as sound waves or light waves, move in three dimensions. Loudspeakers and lightbulbs emit **spherical waves.** The crests of the wave form a series of spherical shells separated by the wavelength λ. In essence, the waves are three-dimensional ripples. It is useful to draw wave-front diagrams such as Figure 15.14a, but now the circles are slices through the spherical shells locating the wave crests.

If you observe a spherical wave far from its source, the small piece of the wave front that you can see is a little patch on the surface of a very large sphere. If the radius of the sphere is large, you will not notice the curvature and this patch of the wave front appears to be a plane. FIGURE 15.15 illustrates the idea of a **plane wave.**

FIGURE 15.14 The wave fronts of a circular or spherical wave.

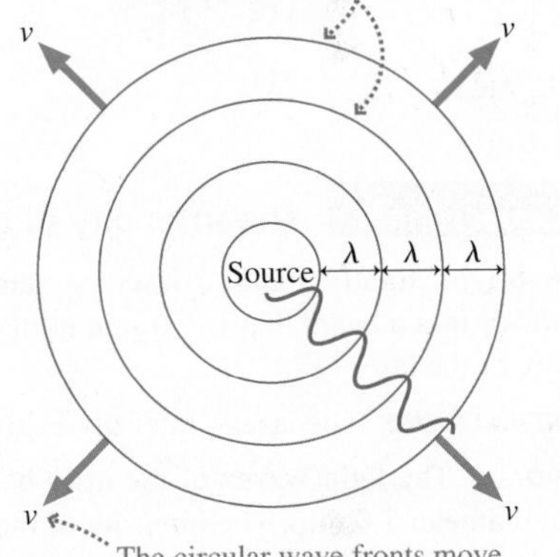

FIGURE 15.15 A plane wave.

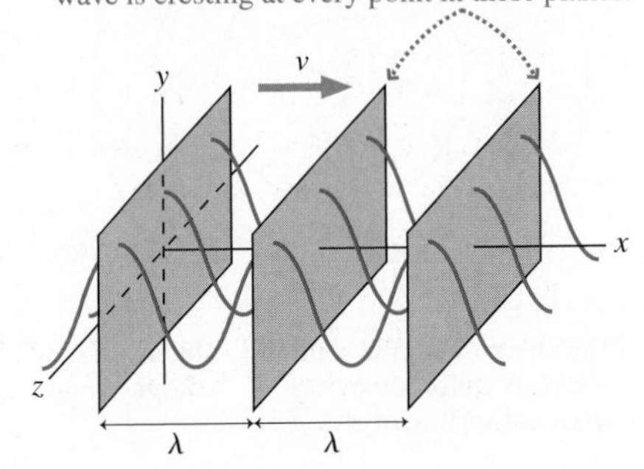

Very far from the source, small segments of spherical wave fronts appear to be planes. The wave is cresting at every point in these planes.
y
v
x
z
λ
λ

Power, Energy, and Intensity

Imagine doing two experiments with a lightbulb that emits 2 W of visible light. In the first, you use this single lightbulb to illuminate a large lecture hall. In the second experiment, you use the same lightbulb to illuminate a small closet. The power emitted by the bulb is the same in both cases, but, as you know, the walls of the closet appear much brighter than those of the lecture hall because the same power is concentrated in a much smaller area. We say that the light striking the closet walls is more *intense* than the light striking the walls of the lecture hall. Similarly, a loudspeaker that beams its sound forward into a small area produces a louder sound in that area than a speaker of equal power that radiates the sound in all directions. Quantities such as brightness and loudness depend not only on the rate of energy transfer, or power, but also on the *area* that receives that power.

FIGURE 15.16 Plane waves of power P impinge on area a.

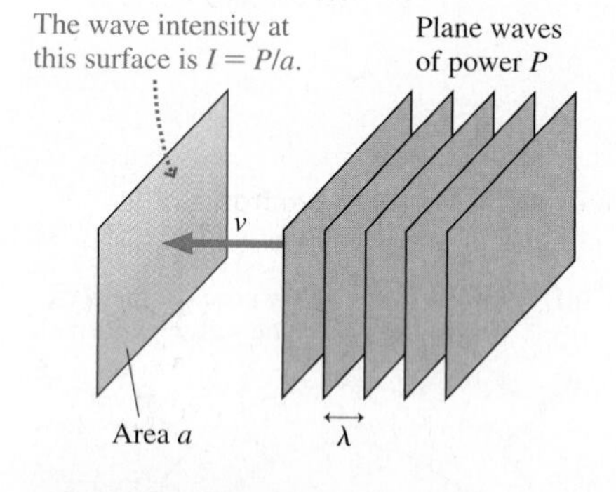

FIGURE 15.16 shows a wave impinging on a surface of area a. The surface is perpendicular to the direction in which the wave is traveling. This might be a real physical surface, such as your eardrum or a solar cell, but it could equally well be a mathematical surface in space that the wave passes right through. If the wave has power P, we define the **intensity** I of the wave as

$$I = \frac{P}{a} \tag{15.11}$$

The SI units of intensity are W/m^2. Because intensity is a power-to-area ratio, a wave concentrated onto a small area has a higher intensity than a wave of equal power that is spread out over a large area.

NOTE ▸ In this chapter we will use a for area to avoid confusion with amplitude, for which we use the symbol A. ◂

EXAMPLE 15.9 The intensity of a laser beam

A bright, tightly focused laser pointer emits 1.0 mW of light power into a beam that is 1.0 mm in diameter. What is the intensity of the laser beam?

STRATEGIZE The laser's intensity is given by Equation 15.11.

SOLVE The light waves of the laser beam pass through a circle of diameter 1.0 mm. The intensity of the laser beam is

$$I = \frac{P}{a} = \frac{P}{\pi r^2} = \frac{0.0010\ \text{W}}{\pi(0.00050\ \text{m})^2} = 1300\ \text{W/m}^2$$

ASSESS This intensity is roughly equal to the intensity of sunlight at noon on a summer day. Such a high intensity for a low-power source may seem surprising, but the area is very small, so the energy is packed into a tiny spot. You know that the light from a laser pointer won't burn you but you don't want the beam to shine into your eye, so an intensity similar to that of sunlight seems reasonable.

FIGURE 15.17 A source emitting uniform spherical waves.

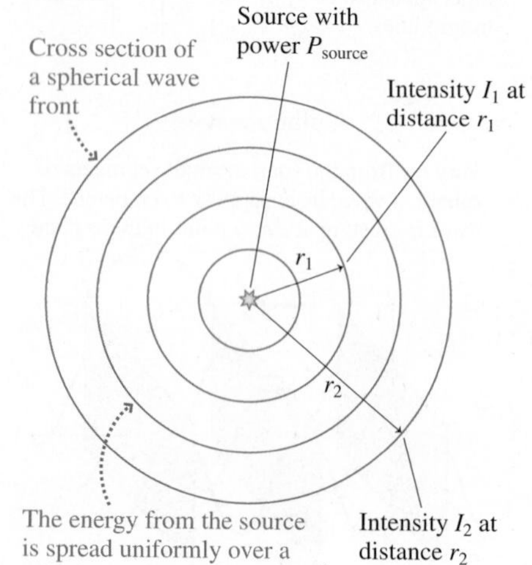

Sound from a loudspeaker and light from a lightbulb become less intense as you get farther from the source. This is because spherical waves spread out to fill larger and larger volumes of space. If a source of spherical waves radiates uniformly in all directions, then, as FIGURE 15.17 shows, the power at distance r is spread uniformly over the surface of a sphere of radius r. The surface area of a sphere is $a = 4\pi r^2$, so the intensity of a uniform spherical wave is

$$I = \frac{P_{source}}{4\pi r^2} \tag{15.12}$$

Intensity at distance r of a spherical wave from a source of power P_{source}

INVERSE-SQUARE p. 191

The inverse-square dependence of r is really just a statement of energy conservation. The source emits energy at the rate of P joules per second. The energy is spread over a larger and larger area as the wave moves outward. Consequently, the *energy per area* must decrease in proportion to the surface area of a sphere.

If the intensity at distance r_1 is $I_1 = P_{source}/4\pi r_1^{\,2}$ and the intensity at r_2 is $I_2 = P_{source}/4\pi r_2^2$, then you can see that the intensity *ratio* is

$$\frac{I_1}{I_2} = \frac{r_2^2}{r_1^2} \tag{15.13}$$

You can use Equation 15.13 to compare the intensities at two distances from a source without needing to know the power of the source.

NOTE ▶ Wave intensities are strongly affected by reflections and absorption. Equations 15.12 and 15.13 apply to situations such as the light from a star and the sound from a firework exploding high in the air. Indoor sound does *not* obey a simple inverse-square law because of the many reflecting surfaces. ◀

The better to hear you with BIO The great grey owl has its ears on the front of its face, hidden behind its facial feathers. Its round face works like a radar dish, collecting the energy of sound waves over a large area and "funneling" it into the ears. This allows owls to sense very quiet sounds.

EXAMPLE 15.10 Intensity of sunlight on Mars

The intensity of sunlight on the earth's surface is approximately 1000 W/m^2 at noon on a summer day. Mars orbits at a distance from the sun approximately 1.5 times that of earth.

a. Assuming similar absorption of energy by the Martian atmosphere, what would you predict for the intensity of sunlight at noon during the Martian summer?
b. The Sojourner rover, an early Mars rover, had a rectangular array of solar cells approximately 0.60 m long and 0.37 m wide. What is the maximum solar energy this array could capture?
c. If we assume a solar-to-electric conversion efficiency of 18%, typical of high-quality solar cells, what is the maximum useful power the solar cells could produce?

STRATEGIZE We aren't given the power emitted by the sun or the distances from the sun to earth and the sun to Mars, but we can find the intensity at the surface of Mars by using ratios of distances and intensities. Once we know the intensity, we will use the area of the solar array to determine the energy that can be captured.

PREPARE The ratio $(r_{earth}/r_{Mars})^2$ is equal to $(1/1.5)^2$.

SOLVE a. According to Equation 15.13,

$$I_{Mars} = I_{earth}\frac{r_{earth}^2}{r_{Mars}^2} = (1000\ \text{W/m}^2)\left(\frac{1}{1.5}\right)^2 = 440\ \text{W/m}^2$$

b. The area of the solar array is

$$a = (0.60\ \text{m})(0.37\ \text{m}) = 0.22\ \text{m}^2$$

If the solar cells are turned to face the sun, the full area can capture energy. We can rewrite Equation 15.11 to give the power capture by the solar array in terms of area and intensity:

$$P = Ia = (440\ \text{W/m}^2)(0.22\ \text{m}^2) = 100\ \text{W}$$

That is, the cells capture solar energy at the rate of 100 joules per second.

c. Not all of this energy is converted to electricity. Multiplying by the 18% efficiency, we get

$$P = (100\ \text{W})(0.18) = 18\ \text{W}$$

ASSESS The intensity of sunlight is quite a bit less on Mars than on earth, as we would expect given the greater distance of Mars from the sun. And the efficiency of solar cells is modest, so the final result—though small—seems reasonable. Indeed, the peak power from the rover's solar array was just less than this.

STOP TO THINK 15.6 A plane wave, a circular wave, and a spherical wave all have the same intensity. Each of the waves travels the same distance. Afterward, which wave has the highest intensity?

A. The plane wave B. The circular wave C. The spherical wave

15.6 Loudness of Sound

Ten guitars playing in unison sound only about twice as loud as one guitar. Generally, **increasing the sound intensity by a factor of 10 results in an increase in perceived loudness by a factor of approximately 2.** The difference in intensity between the quietest sound you can detect and the loudest you can safely hear is a factor of 1,000,000,000,000! A normal conversation has 10,000 times the sound intensity of a whisper, but it sounds only about 16 times as loud.

The loudness of sound is measured by a quantity called the **sound intensity level.** Because of the wide range of intensities we can hear, and the fact that the

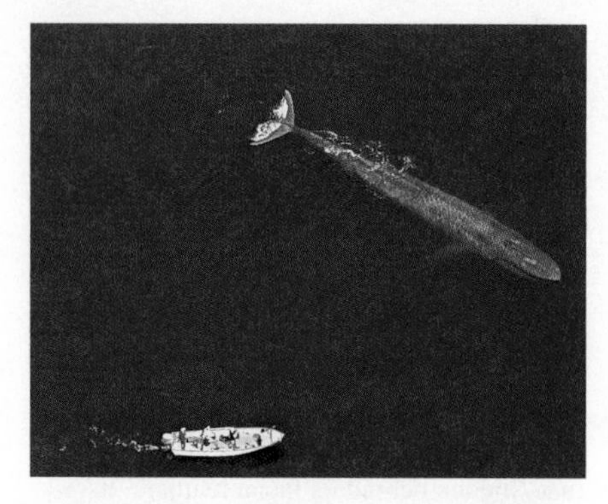

The loudest animal in the world BIO The blue whale is the largest animal in the world, up to 30 m (about 100 ft) long, weighing 150,000 kg or more. It is also the loudest. At close range in the water, the 10–30 second calls of the blue whale would be intense enough to damage tissues in your body. Their loud, low-frequency calls can be heard by other whales hundreds of miles away.

difference in perceived loudness is much less than the actual difference in intensity, the sound intensity level is measured on a *logarithmic scale.* In this section we will explain what this means. The units of sound intensity level (i.e., of loudness) are *decibels,* a word you have likely heard.

The Decibel Scale

There is a lower limit to the intensity of sound that a human can hear. The exact value varies among individuals and with frequency, but an average value for the lowest-intensity sound that can be heard in an extremely quiet room is

$$I_0 = 1.0 \times 10^{-12}\ \mathrm{W/m^2}$$

This intensity is called the *threshold of hearing.*

It's logical to place the zero of our loudness scale at the threshold of hearing. All other sounds can then be referenced to this intensity. To create a loudness scale, we define the *sound intensity level,* expressed in **decibels** (dB), as

$$\beta = (10\ \mathrm{dB})\log_{10}\left(\frac{I}{I_0}\right) \tag{15.14}$$

Sound intensity level in decibels for a sound of intensity I

β is the lowercase Greek letter beta. The decibel is named after Alexander Graham Bell, inventor of the telephone. Sound intensity level is dimensionless, since it's formed from the ratio of two intensities, so decibels are actually just a *name* to remind us that we're dealing with an intensity *level* rather than a true intensity.

Equation 15.14 takes the base-10 logarithm of the intensity ratio I/I_0. As a reminder, logarithms work like this:

If you express a number as a power of 10 . . . . . . the logarithm is the exponent.

$$\log_{10}(1000) = \log_{10}(10^3) = 3$$

Right at the threshold of hearing, where $I = I_0$, the sound intensity level is

$$\beta = (10\ \mathrm{dB})\log_{10}\left(\frac{I_0}{I_0}\right) = (10\ \mathrm{dB})\log_{10}(1) = (10\ \mathrm{dB})\log_{10}(10^0) = 0\ \mathrm{dB}$$

The threshold of hearing corresponds to 0 dB, as we wanted.

We can find the intensity from the sound intensity level by taking the inverse of the $\log_{10}$ function. Recall, from the definition of the base-10 logarithm, that $10^{\log(x)} = x$. Applying this to Equation 15.14, we find

$$I = (I_0)10^{(\beta/10\ \mathrm{dB})} \tag{15.15}$$

TABLE 15.3 lists the sound intensity levels and intensities for a number of typical sounds. Notice that the sound intensity level increases by 10 dB each time the actual intensity increases by a factor of 10. For example, the sound intensity level increases from 70 dB to 80 dB when the sound intensity increases from $10^{-5}\ \mathrm{W/m^2}$ to $10^{-4}\ \mathrm{W/m^2}$. A 20 dB increase in the sound intensity level means a factor of 100 increase in intensity; 30 dB a factor of 1000. We found earlier that sound is perceived as "twice as loud" when the intensity increases by a factor of 10. In terms of decibels, we can say that the apparent loudness of a sound doubles with each 10 dB increase in the sound intensity level.

The range of sounds in Table 15.3 is very wide; the top of the scale, 130 dB, represents 10 trillion times the intensity of the quietest sound you can hear. Vibrations of this intensity will injure the delicate sensory apparatus of the ear and cause pain. Exposure to less intense sounds also is not without risk. A fairly short exposure to 120 dB can cause damage to the hair cells in the ear, but lengthy exposure to sound intensity levels of over 85 dB can produce damage as well.

TABLE 15.3 Sound intensity levels and intensities of common environmental sounds

Sound	β (dB)	I (W/m^2)
Threshold of hearing	0	1.0×10^{-12}
Person breathing, at 3 m	10	1.0×10^{-11}
A whisper, at 1 m	20	1.0×10^{-10}
Classroom during test, no talking	30	1.0×10^{-9}
Residential street, no traffic	40	1.0×10^{-8}
Quiet restaurant	50	1.0×10^{-7}
Normal conversation, at 1 m	60	1.0×10^{-6}
Busy traffic	70	1.0×10^{-5}
Vacuum cleaner, for user	80	1.0×10^{-4}
Niagara Falls, at viewpoint	90	1.0×10^{-3}
Pneumatic hammer, at 2 m	100	0.010
Home stereo at max volume	110	0.10
Rock concert	120	1.0
Threshold of pain	130	10

Before we begin to look at examples that use sound intensity level, let's summarize what we've learned about the energy carried by waves.

SYNTHESIS 15.2 Wave power and intensity

We have different measures of the energy carried by waves.

The **intensity** of a wave is the power per square meter of area:

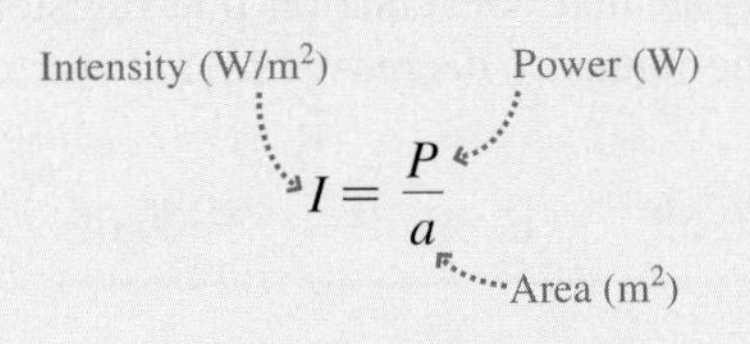

When a wave hits a surface, the power absorbed in area a is

$$P = Ia$$

Spherical waves spread over the surface of a sphere of increasing size.

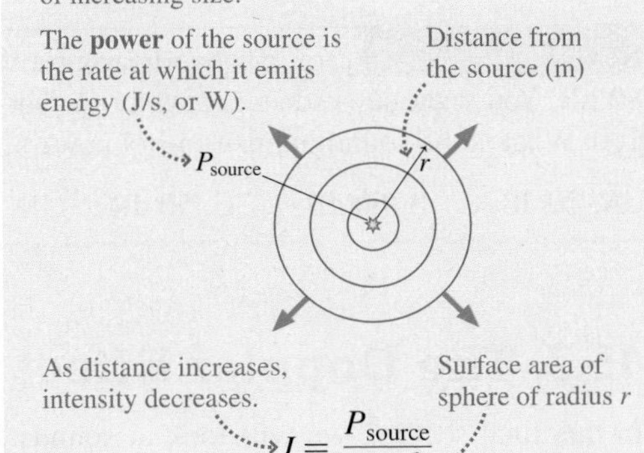

As distance increases, intensity decreases.

Surface area of sphere of radius r

$$I = \frac{P_{source}}{4\pi r^2}$$

We compute **sound intensity level** from relative intensity:

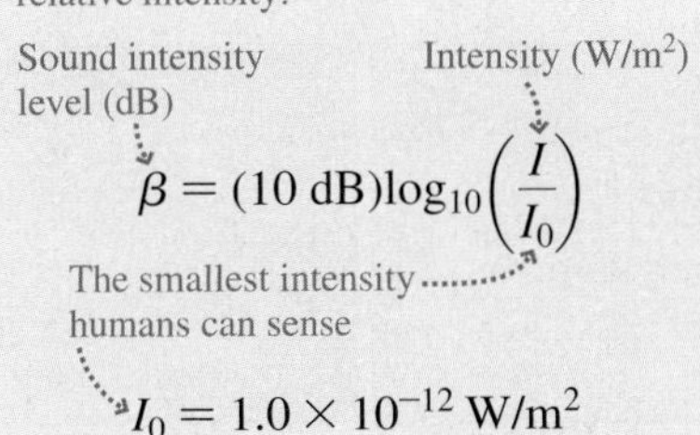

Sound intensity level (dB)
Intensity (W/m²)
$\beta = (10\text{ dB})\log_{10}\left(\dfrac{I}{I_0}\right)$
The smallest intensity humans can sense
$I_0 = 1.0 \times 10^{-12}\text{ W/m}^2$

We can also compute intensity from sound intensity level:

$$I = (I_0)10^{(\beta/10\text{ dB})}$$

EXAMPLE 15.11 Finding the loudness of a shout

A person shouting at the top of his lungs emits about 1.0 W of energy as sound waves. What is the sound intensity level 1.0 m from such a person?

STRATEGIZE We assume that the shouting person emits a spherical sound wave. Synthesis 15.2 gives the details for the decrease in intensity with distance and the calculation of the resulting sound intensity level.

SOLVE At a distance of 1.0 m, the sound intensity is

$$I = \frac{P}{4\pi r^2} = \frac{1.0\text{ W}}{4\pi(1.0\text{ m})^2} = 0.080\text{ W/m}^2$$

Thus the sound intensity level is

$$\beta = (10\text{ dB})\log_{10}\left(\frac{0.080\text{ W/m}^2}{1.0 \times 10^{-12}\text{ W/m}^2}\right) = 110\text{ dB}$$

ASSESS This is quite loud (compare with values in Table 15.3), as you might expect.

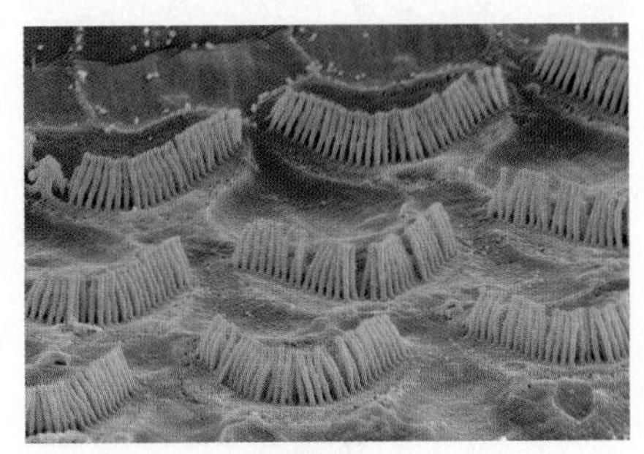

Hearing hairs BIO This electron microscope picture shows the hair cells in the cochlea of the ear that are responsible for sensing sound. Even very tiny vibrations transmitted into the fluid of the cochlea deflect the hairs, triggering a response in the cells. Motion of the basilar membrane by as little as 0.5 nm, about 5 atomic diameters, can produce an electrical response in the hair cells. With this remarkable sensitivity, it is no wonder that loud sounds can damage these structures.

NOTE ► In calculating the sound intensity level, be sure to use the $\log_{10}$ button on your calculator, not the natural logarithm button. On many calculators, $\log_{10}$ is labeled LOG and the natural logarithm is labeled LN. ◄

EXAMPLE 15.12 How far away can you hear a conversation?

The sound intensity level 1.0 m from a person talking in a normal conversational voice is 60 dB. Suppose you are outside, 100 m from the person speaking. If it is a very quiet day with minimal background noise, are you able to hear him or her?

STRATEGIZE We know how sound intensity changes with distance, but not sound intensity level, so we need to break this problem into three steps. First, we will convert the sound intensity level at 1.0 m into intensity, using Equation 15.15. Second, we will compute the intensity at a distance of 100 m, using Equation 15.13. Finally, we will convert this result back to a sound intensity level, using Equation 15.14, so that we can judge the loudness.

PREPARE We denote the intensity at 1 m by $I(1\text{ m})$ and that at 100 m by $I(100\text{ m})$.

SOLVE The intensity is

$$I(1\text{ m}) = (1.0 \times 10^{-12}\text{ W/m}^2)10^{(60\text{ dB}/10\text{ dB})} = 1.0 \times 10^{-6}\text{ W/m}^2$$

Continued

The intensity at 100 m can be found using Equation 15.13:

$$\frac{I(100\text{ m})}{I(1\text{ m})} = \frac{(1\text{ m})^2}{(100\text{ m})^2}$$

$$I(100\text{ m}) = I(1\text{ m})(1.0\times10^{-4}) = 1.0\times10^{-10}\text{ W/m}^2$$

This intensity corresponds to a sound intensity level

$$\beta = (10\text{ dB})\log_{10}\left(\frac{1.0\times10^{-10}\text{ W/m}^2}{1.0\times10^{-12}\text{ W/m}^2}\right) = 20\text{ dB}$$

This is above the threshold for hearing—about the level of a whisper—so it can be heard on a very quiet day.

ASSESS The sound is well within what the ear can detect. However, normal background noise is rarely less than 40 dB, which would make the conversation difficult to decipher. Our result thus seems reasonable based on experience. The sound is at a level that can theoretically be detected, but it is much quieter than the ambient level of sound, so it's not likely to be noticed.

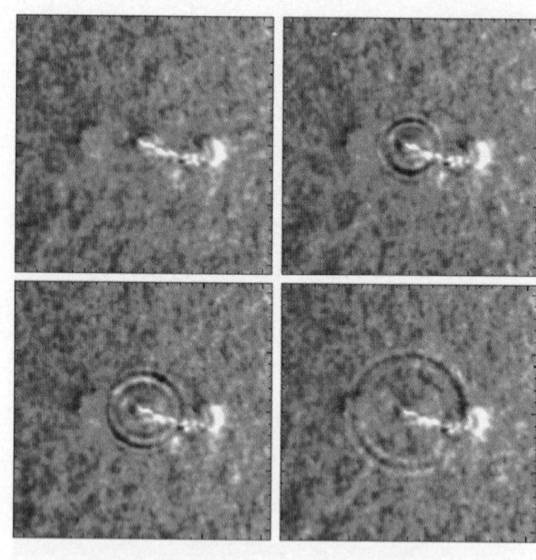

Catching a wave on the sun The Doppler effect shifts the frequency of light emitted from the sun's surface. If the surface is rising, the light is shifted to higher frequencies; these positions are shown darker. If the surface is falling, the light is shifted to lower frequencies and is shown lighter. This series of images of the sun's surface shows a wave produced by the disruption of a solar flare.

STOP TO THINK 15.7 You are overhearing a very heated conversation that registers 80 dB. You walk some distance away so that the intensity decreases by a factor of 100. What is the sound intensity level now?

A. 70 dB B. 60 dB C. 50 dB D. 40 dB E. 30 dB F. 20 dB

15.7 The Doppler Effect and Shock Waves

In this final section, we will look at sounds from moving objects and for moving observers. You've likely noticed that the pitch of an ambulance's siren gets higher as it approaches you and drops as it goes past you. A higher pitch suddenly becomes a lower pitch. This change in frequency, which is due to the motion of the ambulance, is called the *Doppler effect.* A more dramatic effect happens when an object moves faster than the speed of sound. The crack of a whip is a *shock wave* produced when the tip moves at a *supersonic* speed. These examples—and much of this section—concern sound waves, but the phenomena we will explore apply generally to all waves.

Sound Waves from a Moving Source

FIGURE 15.18a shows a source of sound waves moving away from Pablo and toward Nancy at a steady speed v_s. The subscript s indicates that this is the speed of the source, not the speed of the waves. The source is emitting sound waves of frequency f_s as it travels. Part a of the figure shows the positions of the source at times $t = 0$, T, $2T$, and $3T$, where $T = 1/f_s$ is the period of the waves.

FIGURE 15.18 A motion diagram showing the wave fronts emitted by a source as it moves to the right at speed v_s.

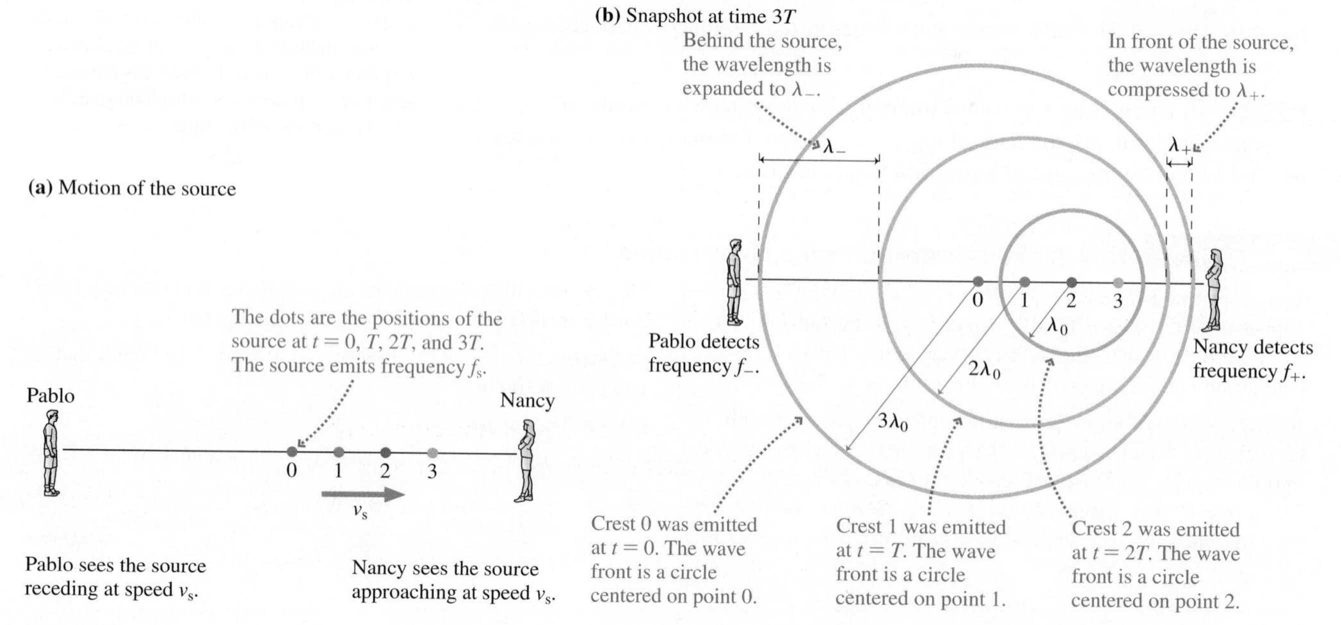
(a) Motion of the source
The dots are the positions of the source at $t = 0, T, 2T$, and $3T$. The source emits frequency f_s.
Pablo
Nancy
0 1 2 3
v_s
Pablo sees the source receding at speed v_s.
Nancy sees the source approaching at speed v_s.
(b) Snapshot at time $3T$
Behind the source, the wavelength is expanded to λ_-.
In front of the source, the wavelength is compressed to λ_+.
λ_-
λ_+
Pablo detects frequency f_-.
Nancy detects frequency f_+.
λ_0
$2\lambda_0$
$3\lambda_0$
Crest 0 was emitted at $t = 0$. The wave front is a circle centered on point 0.
Crest 1 was emitted at $t = T$. The wave front is a circle centered on point 1.
Crest 2 was emitted at $t = 2T$. The wave front is a circle centered on point 2.

After a wave crest leaves the source, its motion is governed by the properties of the medium. The motion of the source cannot affect a wave that has already been emitted. Thus each circular wave front in FIGURE 15.18b is centered on the point from which it was emitted. You can see that the wave crests are bunched up in the direction in which the source is moving and are stretched out behind it. The distance between one crest and the next is one wavelength, so the wavelength λ_+ that Nancy measures is *less* than the wavelength $\lambda_s = v/f_s$ that would be emitted if the source were at rest. Similarly, λ_- behind the source is larger than λ_s.

These crests move through the medium at the wave speed v. Consequently, the frequency $f_+ = v/\lambda_+$ detected by the observer whom the source is approaching is *higher* than the frequency f_s emitted by the source. Similarly, $f_- = v/\lambda_-$ detected behind the source is *lower* than frequency f_s. This change of frequency when a source moves relative to an observer is called the **Doppler effect.** The frequency heard by a stationary observer depends on whether the observer sees the source approaching or receding:

Approaching source

$$f_+ = \frac{f_s}{1 - v_s/v}$$

Frequency of the source (Hz): f_s. Speed of the *waves* (m/s): v. Speed of the *source* of the waves (m/s): v_s.

The observed frequency is *increased*.

Receding source

$$f_- = \frac{f_s}{1 + v_s/v} \qquad (15.16)$$

The observed frequency is *decreased*.

As expected, $f_+ > f_s$ (the frequency is higher) for an approaching source because the denominator is less than 1, and $f_- < f_s$ (the frequency is lower) for a receding source.

EXAMPLE 15.13 How fast are the police driving?

A police siren has a frequency of 550 Hz as the police car approaches you and 450 Hz after it has passed you and is moving away. How fast are the police traveling?

STRATEGIZE The siren's frequency is altered by the Doppler effect. The frequency is f_+ as the car approaches and f_- as it moves away. We will write two equations for these frequencies and solve for the speed of the police car, v_s.

SOLVE Because our goal is to find v_s, we rewrite Equations 15.16 as

$$f_s = \left(1 + \frac{v_s}{v}\right)f_- \quad \text{and} \quad f_s = \left(1 - \frac{v_s}{v}\right)f_+$$

Subtracting the second equation from the first, we get

$$0 = f_- - f_+ + \frac{v_s}{v}(f_- + f_+)$$

Now we can solve for the speed v_s:

$$v_s = \frac{f_+ - f_-}{f_+ + f_-}v = \frac{100 \text{ Hz}}{1000 \text{ Hz}} 343 \text{ m/s} = 34 \text{ m/s}$$

ASSESS This is pretty fast (about 75 mph) but reasonable for a police car speeding with the siren on.

A Stationary Source and a Moving Observer

Suppose the police car in Example 15.13 is at rest while you drive toward it at 34 m/s. You might think that this is equivalent to having the police car move toward you at 34 m/s, but there is an important difference. Mechanical waves move through a medium, and the Doppler effect depends not just on how the source and the observer move with respect to each other but also on how they move with respect to the medium. The frequency heard by a moving observer depends on whether the observer is approaching or receding from the source:

Approaching observer

$$f_+ = \left(1 + \frac{v_o}{v}\right)f_s$$

Speed of the *observer* (m/s): v_o. Frequency of the source (Hz): f_s. Speed of the *waves* (m/s): v.

The observed frequency is *increased*.

Receding observer

$$f_- = \left(1 - \frac{v_o}{v}\right)f_s \qquad (15.17)$$

The observed frequency is *decreased*.

The greater the distance to a galaxy, the faster it moves away from us. This photo shows galaxies at distances of up to 12 billion light years. The great distances imply large red shifts, which make the light from the most distant galaxies appear distinctly red.

The Doppler Effect for Light Waves

If a source of light waves is receding from you, the wavelength λ_- that you detect is longer than the wavelength λ_s emitted by the source. Because the wavelength is shifted toward the red end of the visible spectrum, the longer wavelengths of light, this effect is called the **red shift.** Similarly, the light you detect from a source moving toward you is **blue shifted** to shorter wavelengths. For objects moving at normal speeds, this is a small effect; you won't see a Doppler shift of the flashing light of the police car in the previous example!

Distant galaxies *all* have a distinct red shift—all distant galaxies are moving away from us. How can we make sense of this observation? The most straightforward explanation is that the galaxies of the universe are *all* moving apart from each other. Extrapolating backward in time brings us to a point when all the matter of the universe—and even space itself, according to the theory of relativity—began rushing out of a primordial fireball. Many observations and measurements since have given support to the idea that the universe began in a *Big Bang* about 14 billion years ago.

Frequency Shift on Reflection from a Moving Object

A wave striking a barrier or obstacle can *reflect* and travel back toward the source of the wave, a process we will examine more closely in the next chapter. For sound, the reflected wave is called an *echo.* A bat uses echolocation to determine the distance to a flying insect by measuring the time between the emission of an ultrasonic chirp and the detection of the echo from the insect. But if the bat is to catch the insect, it's just as important to know where the insect is going—its velocity. The bat can figure this out by noting the *frequency shift* in the reflected wave, another application of the Doppler effect.

Suppose a sound wave of frequency f_s travels toward a moving object. The object would see the wave's frequency Doppler shifted to a higher frequency f_+, as given by Equation 15.17. The wave reflected back toward the sound source by the moving object is also Doppler shifted to a higher frequency because, to the source, the reflected wave is coming from a moving object. Thus the echo from a moving object is "double Doppler shifted." If the object's speed v_o is much lower than the wave speed v $(v_o \ll v)$, the frequency shift of waves reflected from a moving object is

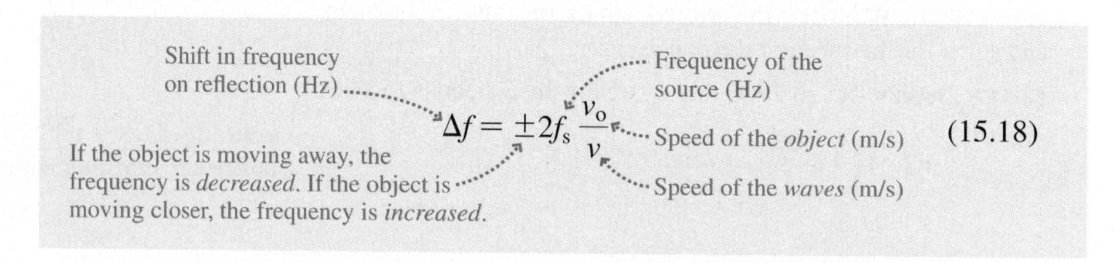

$$\Delta f = \pm 2 f_s \frac{v_o}{v} \quad (15.18)$$

Notice that there is no shift (i.e., the reflected wave has the same frequency as the emitted wave) when the reflecting object is at rest.

Frequency shift on reflection is observed for all types of waves. Radar units emit pulses of radio waves and observe the reflected waves. The time between the emission of a pulse and its return gives an object's position. The change in frequency of the returned pulse gives the object's speed. This is the principle behind the radar guns used by traffic police, as well as the Doppler radar images you have seen in weather reports.

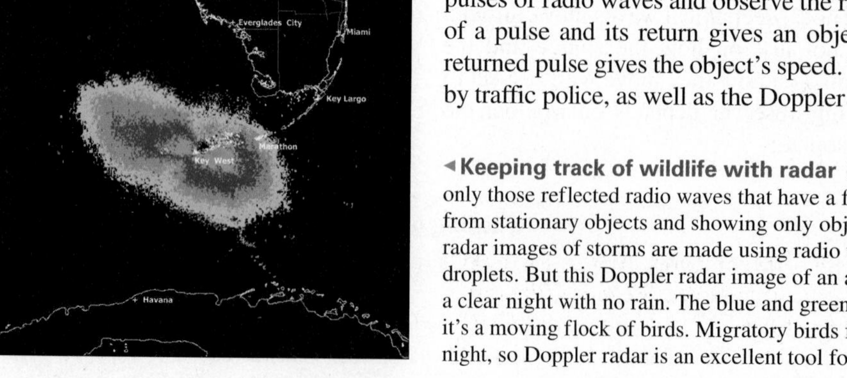

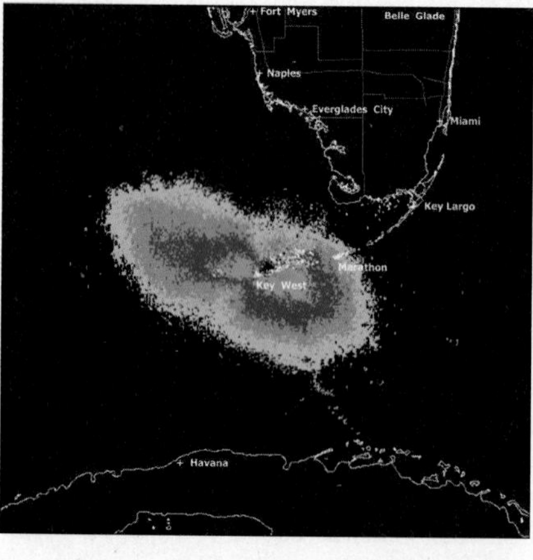

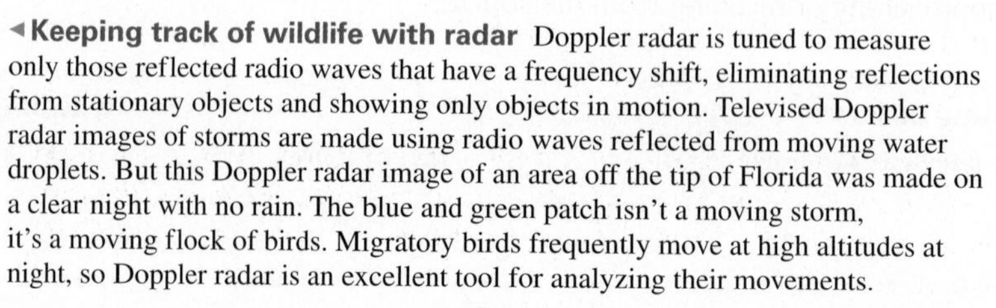

◀ **Keeping track of wildlife with radar** Doppler radar is tuned to measure only those reflected radio waves that have a frequency shift, eliminating reflections from stationary objects and showing only objects in motion. Televised Doppler radar images of storms are made using radio waves reflected from moving water droplets. But this Doppler radar image of an area off the tip of Florida was made on a clear night with no rain. The blue and green patch isn't a moving storm, it's a moving flock of birds. Migratory birds frequently move at high altitudes at night, so Doppler radar is an excellent tool for analyzing their movements.

EXAMPLE 15.14 The Doppler blood flowmeter BIO

If an ultrasound source is pressed against the skin, the sound waves reflect off tissues in the body. Some of the sound waves reflect from blood cells moving through arteries toward or away from the source, producing a frequency shift in the reflected wave. A biomedical engineer is designing a *Doppler blood flowmeter* to measure blood flow in an artery where a typical flow speed is known to be 0.60 m/s. What ultrasound frequency should she use to produce a frequency shift of 1500 Hz when this flow is detected?

STRATEGIZE The ultrasound is reflecting from a moving object—the blood—so Equation 15.18 will apply.

PREPARE In Equation 15.18, v_0 is the speed of the blood, or 0.60 m/s, and v is the speed of sound in human tissue, or 1540 m/s.

SOLVE We can rewrite Equation 15.18 to calculate the frequency of the emitter:

$$f_s = \left(\frac{\Delta f}{2}\right)\left(\frac{v}{v_{blood}}\right)$$

The values on the right side are all known. Thus the required ultrasound frequency is

$$f_s = \left(\frac{1500\ \text{Hz}}{2}\right)\left(\frac{1540\ \text{m/s}}{0.60\ \text{m/s}}\right) = 1.9\ \text{MHz}$$

ASSESS Doppler units to measure blood flow in deep tissues actually work at about 2.0 MHz, so our answer is reasonable.

If we add a frequency-shift measurement to an ultrasound imaging unit, like the one in Example 15.7, we have a device—called *Doppler ultrasound*—that can show not only structure but also motion. *Doppler ultrasound* is a very valuable tool in cardiology because an image can reveal the motion of the heart muscle and the blood in addition to the structure of the heart itself.

Shock Waves

When sound waves are emitted by a moving source, the frequency is shifted, as we have seen. Now, let's look at what happens when the source speed v_s exceeds the wave speed v and the source "outruns" the waves it produces.

Earlier in this section, in Figure 15.18, we looked at a motion diagram of waves emitted by a moving source. FIGURE 15.19 is the same diagram, but in this case the source is moving faster than the waves. This motion causes the waves to overlap. (Compare Figure 15.19 to Figure 15.18.) The amplitudes of the overlapping waves add up to produce a very large amplitude wave—a **shock wave.**

Anything that moves faster than the speed of sound in air will create a shock wave. FIGURE 15.20a is a specialized photograph of a **supersonic** (faster than the speed of sound) jet airplane. The photograph is enhanced to show the shock waves that emanate from the bow, wings, and tail of the aircraft. This shock wave travels along with the jet. If the jet flew over you, the passing of the shock wave would produce a **sonic boom,** a loud noise that sounds like a clap of thunder. The crack of a whip is a sonic boom as well, though on a much smaller scale.

FIGURE 15.19 Waves emitted by a source traveling faster than the speed of the waves in a medium produce a shock wave.

The source of waves is moving to the right at v_s. The positions at times $t = 0, t = T, t = 2T, \ldots$ are marked.

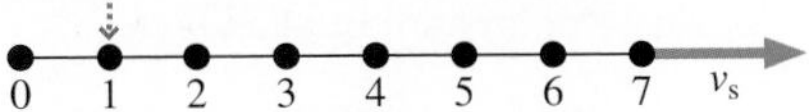

At each point, the source emits a wave that spreads out. A snapshot is taken at $t = 7T$.

Crest 3 was emitted at $t = 3T$, $4T$ before the snapshot. The wave front is a circle of radius 4λ centered on point 3.

4λ

0 1 2 3 4 5 6 7 v_s

The source speed is greater than the wave speed, so during the time $t = 4T$ since crest 3 was emitted, the source has moved farther than crest 3.

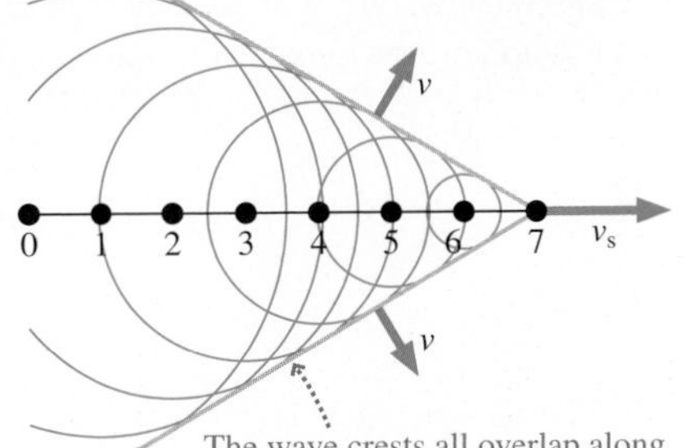

The wave crests all overlap along these lines. The overlapping crests make a V-shaped shock wave that travels at the wave speed.

FIGURE 15.20 Extreme and everyday examples of shock waves.

Shock waves are not necessarily an extreme phenomenon. A boat (or even a duck!) can easily travel faster than the speed of the water waves it creates; the resulting wake, with its characteristic "V" shape as shown in FIGURE 15.20b, is really a shock wave.

STOP TO THINK 15.8 Amy and Zack are both listening to a source of sound waves that is moving to the right. Compare the frequencies each hears.

A. $f_{Amy} > f_{Zack}$
B. $f_{Amy} = f_{Zack}$
C. $f_{Amy} < f_{Zack}$

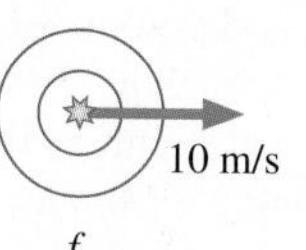

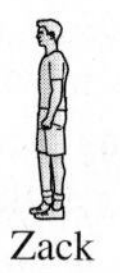

INTEGRATED EXAMPLE 15.15 Shaking the ground

Earthquakes are dramatic slips of the earth's crust. You may feel the waves generated by an earthquake even if you're some distance from the *epicenter*, the point where the slippage occurs. Earthquake waves are usually complicated, but some of the long-period waves shake the ground with motion that is approximately simple harmonic motion. FIGURE 15.21 shows the vertical positions of the ground recorded at a distant monitoring station following an earthquake that hit Japan in 2003. This particular wave traveled with a speed of 3500 m/s.

a. Was the wave transverse or longitudinal?
b. What were the wave's frequency and wavelength?
c. What were the maximum speed and the maximum acceleration of the ground during this earthquake wave?
d. Intense earthquake waves produce accelerations greater than the free-fall acceleration of gravity. How does this wave compare?

FIGURE 15.21 The vertical motion of the ground during the passage of an earthquake wave.

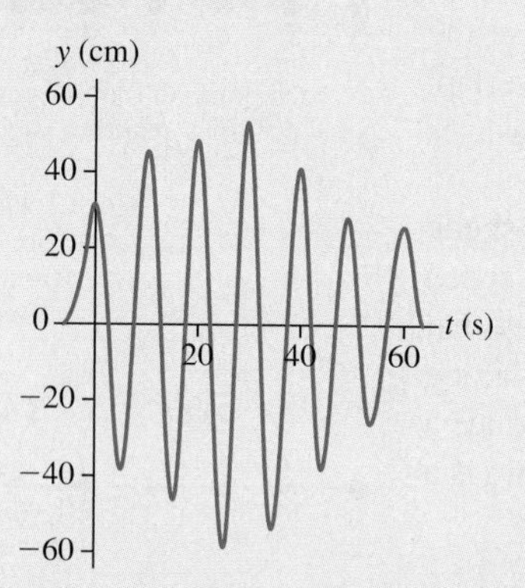

STRATEGIZE Figure 15.21 is a history graph showing the motion in the y-direction (i.e., the *vertical* motion) of one point of the medium—the ground. The graph is approximately sinusoidal, so we will model the motion of the ground as simple harmonic motion and the traveling wave as a sinusoidal wave. Consequently, we will use the fundamental relationships for sinusoidal waves to relate the wavelength, frequency, and speed.

PREPARE If we look at the middle portion of the wave shown in Figure 15.21, we can estimate the amplitude to be 0.50 m (it varies, but this is a reasonable average over a few cycles). We can also see from the graph that 6 cycles of the oscillation took 60 s, so the period was $T = 10$ s.

SOLVE

a. As the graph shows, the motion of the ground was vertical, perpendicular to the horizontal motion of the wave traveling along the ground, so this was a transverse wave.
b. The period was 10 s, so the frequency was $f = 1/T = 1/10\text{ s} = 0.10$ Hz. The speed of the wave was 3500 m/s, giving a wavelength of

$$\lambda = \frac{v}{f} = \frac{3500\text{ m/s}}{0.10\text{ Hz}} = 35{,}000\text{ m} = 35\text{ km}$$

c. The motion of the ground was simple harmonic motion with frequency 0.10 Hz and amplitude 0.50 m. We can compute the maximum speed and acceleration using relationships from Chapter 14:

$$v_{max} = 2\pi fA = (2\pi)(0.10\text{ Hz})(0.50\text{ m}) = 0.31\text{ m/s}$$
$$a_{max} = (2\pi f)^2 A = [2\pi(0.10\text{ Hz})]^2(0.50\text{ m}) = 0.20\text{ m/s}^2$$

d. This was a reasonably gentle earthquake wave, with an acceleration of the ground much less than the free-fall acceleration of gravity:

$$a_{max}\text{ (in units of } g) = \frac{0.20\text{ m/s}^2}{9.8\text{ m/s}^2} = 0.020g$$

ASSESS The wavelength is quite long, as we might expect for such a fast wave with a long period. Given the relatively small amplitude and long period, it's no surprise that the maximum speed and acceleration are relatively modest. You'd certainly feel the passage of this wave, but it wouldn't knock buildings down. This value is less than the acceleration for the sway of the top of the building in Example 14.3 in Chapter 14!

SUMMARY

GOAL To learn the basic properties of traveling waves.

GENERAL PRINCIPLES

The Wave Model

This model is based on the idea of a **traveling wave,** which is an organized disturbance traveling at a well-defined **wave speed** v.

- In **transverse waves** the particles of the medium move *perpendicular* to the direction in which the wave travels.

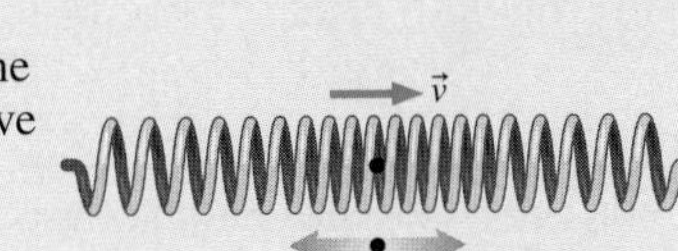

- In **longitudinal waves** the particles of the medium move *parallel* to the direction in which the wave travels.

A wave transfers energy, but there is no material or substance transferred.

Mechanical waves require a material **medium**. The speed of the wave is a property of the medium, not the wave. The speed does not depend on the size or shape of the wave.

- For a **wave on a string,** the string is the medium.

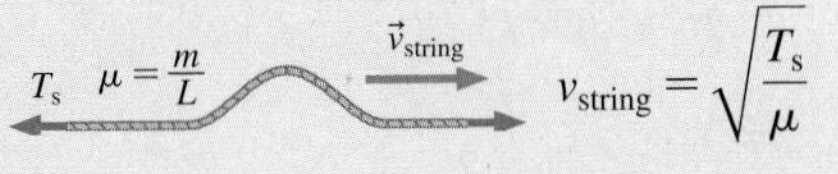

$$v_{string} = \sqrt{\frac{T_s}{\mu}}$$

- A **sound wave** is a wave of compressions and rarefactions of a medium such as air.

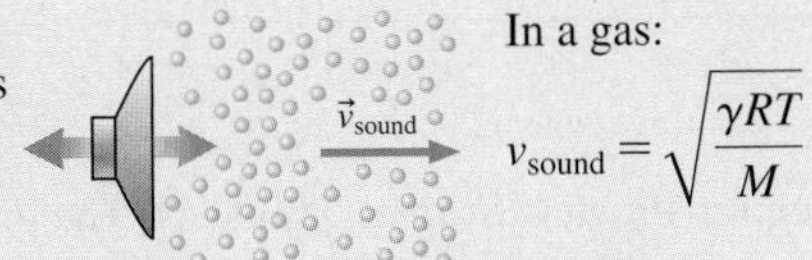

In a gas:

$$v_{sound} = \sqrt{\frac{\gamma RT}{M}}$$

Electromagnetic waves are waves of the electromagnetic field. They do not require a medium. All electromagnetic waves travel at the same speed in a vacuum, $c = 3.00 \times 10^8$ m/s.

IMPORTANT CONCEPTS

Graphical representation of waves

A **snapshot graph** is a picture of a wave at one instant in time. For a periodic wave, the **wavelength** λ is the distance between crests.

Fixed t:

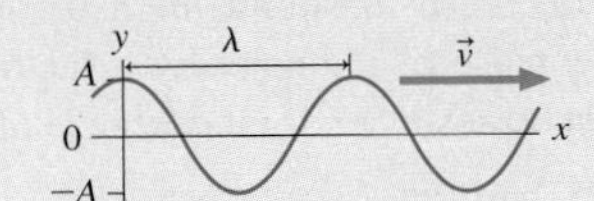

A **history graph** is a graph of the displacement of one point in a medium versus time. For a periodic wave, the **period** T is the time between crests.

Fixed x:

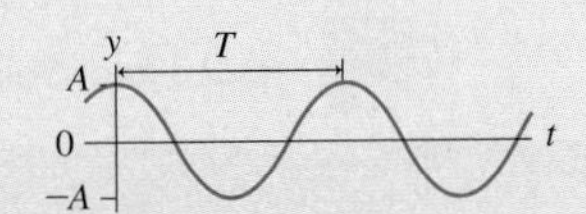

Mathematical representation of waves

Sinusoidal waves are produced by a source moving with simple harmonic motion. The equation for a sinusoidal wave is a function of position and time:

$$y(x, t) = A\cos\left(2\pi\left(\frac{x}{\lambda} \pm \frac{t}{T}\right)\right)$$

+: wave travels to left

−: wave travels to right

For sinusoidal waves:

$$T = \frac{1}{f} \qquad v = \lambda f$$

The **intensity** of a wave is the ratio of the power to the area:

$$I = \frac{P}{a}$$

For a **spherical wave** the power decreases with the surface area of the spherical **wave fronts:**

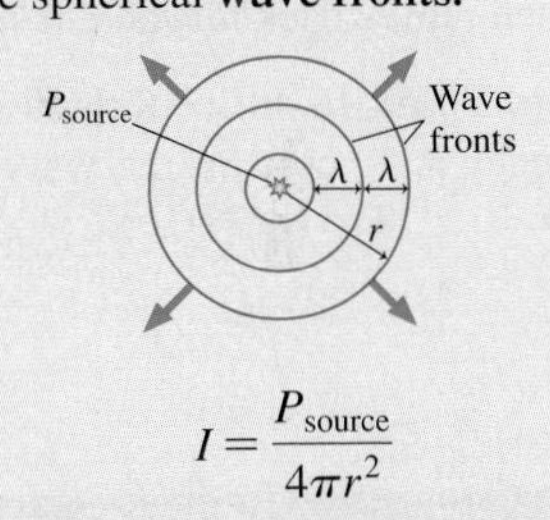

$$I = \frac{P_{source}}{4\pi r^2}$$

APPLICATIONS

The loudness of a sound is given by the **sound intensity level.** This is a logarithmic function of intensity and is in units of **decibels** (dB).

The sound intensity level, in dB, is given by

$$\beta = (10 \text{ dB})\log_{10}\left(\frac{I}{I_0}\right)$$

where $I_0 = 1.0 \times 10^{-12}$ W/m² is the *threshold of hearing,* the quietest sound that we can hear. A sound at the threshold of hearing corresponds to 0 dB.

The **Doppler effect** is a shift in frequency when there is relative motion of a wave source (frequency f_s, wave speed v) and an observer.

Moving source, stationary observer:

Receding source:

$$f_- = \frac{f_s}{1 + v_s/v}$$

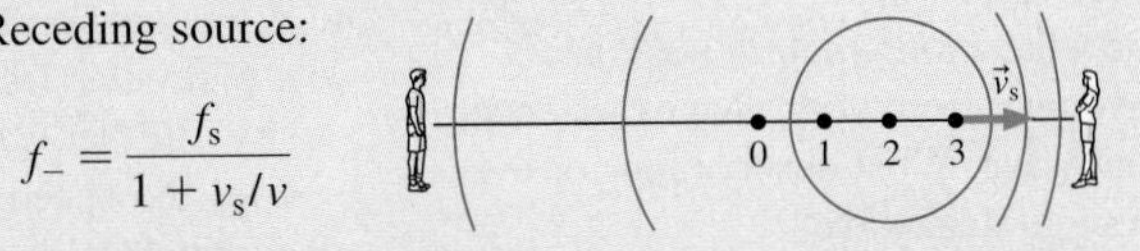

Approaching source:

$$f_+ = \frac{f_s}{1 - v_s/v}$$

Moving observer, stationary source:

Approaching the source: $f_+ = \left(1 + \frac{v_o}{v}\right)f_s$

Moving away from the source: $f_- = \left(1 - \frac{v_o}{v}\right)f_s$

Reflection from a moving object:

For $v_o \ll v$, $\Delta f = \pm 2f_s\frac{v_o}{v}$

When an object moves faster than the wave speed in a medium, a **shock wave** is formed.

Learning Objectives After studying this chapter, you should be able to:

- Use the wave model to understand the basic properties of waves. *Conceptual Questions 15.1, 15.2; Problem 15.8*
- Calculate the speeds of waves on strings and the speeds of sound waves. *Conceptual Questions 15.4, 15.5; Problems 15.1, 15.2, 15.3, 15.5, 15.7*
- Use the graphical and mathematical representations of waves. *Conceptual Questions 15.10, 15.11; Problems 15.9, 15.14, 15.16, 15.17, 15.18*
- Solve problems about sound and light waves. *Conceptual Questions 15.12, 15.13; Problems 15.24, 15.26, 15.27, 15.28, 15.30*
- Apply the ideas of energy and power to waves. *Conceptual Question 15.14; Problems 15.31, 15.32, 15.35, 15.36, 15.37*
- Calculate the loudness of sound. *Conceptual Question 15.16; Problems 15.40, 15.41, 15.42, 15.43, 15.44*
- Understand and use the Doppler effect. *Conceptual Questions 15.17, 15.18; Problems 15.51, 15.52, 15.53, 15.54, 15.55*

STOP TO THINK ANSWERS

Chapter Preview Stop to Think: D. The toy goes from the lowest point to its highest point in 1.0 s. This is one-half of a full oscillation, so the period is 2.0 s, giving a frequency of 0.5 Hz.

Stop to Think 15.1: Transverse. The wave moves horizontally through the crowd, but individual spectators move up and down, transverse to the motion of the wave.

Stop to Think 15.2: D, E. Shaking your hand faster or farther will change the shape of the wave, but this will not change the wave speed; the speed is a property of the medium. Changing the linear density of the string or its tension will change the wave speed. To increase the speed, you must decrease the linear density or increase the tension.

Stop to Think 15.3: A. As the wave moves toward the origin, this point moves up at a constant speed for a short distance, reverses and moves down for a greater distance until the displacement is negative, and then moves back to where it started. Graph A shows this motion.

Stop to Think 15.4: B. All three waves have the same speed, so the frequency is highest for the wave that has the shortest wavelength. (Imagine the three waves moving to the right. The one with the crests closest together has the crests passing by most rapidly.)

Stop to Think 15.5: C. Figure 15.13 shows that infrared is to the left of the visible light region of the spectrum and ultraviolet is to the right. The upper scale on the diagram shows that infrared thus has a lower frequency than ultraviolet. All electromagnetic waves travel at the same speed, so infrared must have a longer wavelength.

Stop to Think 15.6: A. The plane wave does not spread out, so its intensity will be constant. The other two waves spread out, so their intensity will decrease.

Stop to Think 15.7: B. If the intensity decreases by a factor of 10, this corresponds to a decrease in sound intensity level of 10 dB. $100 = 10 \times 10$, so a decrease of intensity by a factor of 100 corresponds to two 10 dB decreases, for a total decrease of 20 dB.

Stop to Think 15.8: C. The source is moving toward Zack, so he observes a higher frequency. The source is moving away from Amy, so she observes a lower frequency.

 Video Tutor Solution Chapter 15

QUESTIONS

Conceptual Questions

1. a. In your own words, define what a *transverse wave* is.
   b. Give an example of a wave that, from your own experience, you know is a transverse wave. What observations or evidence tells you this is a transverse wave?
2. a. In your own words, define what a *longitudinal wave* is.
   b. Give an example of a wave that, from your own experience, you know is a longitudinal wave. What observations or evidence tells you this is a longitudinal wave?
3. The wave pulses shown in Figure Q15.3 travel along the same string. Rank in order, from largest to smallest, their wave speeds v_1, v_2, and v_3. Explain.

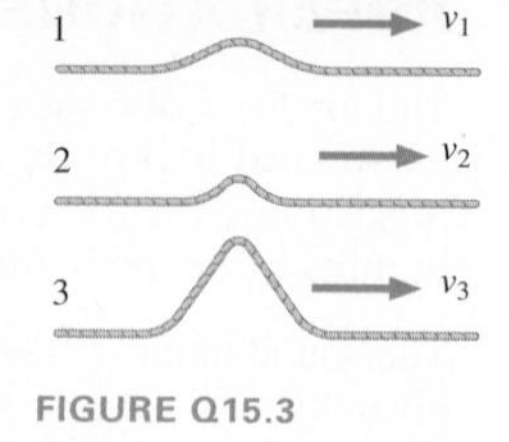

FIGURE Q15.3

4. A loudspeaker emits a sound wave at a particular frequency. If the intensity of the wave is doubled, how does the speed of the wave change? What if the frequency is doubled?

Problem difficulty is labeled as | (straightforward) to ||||| (challenging). INT Problems labeled integrate significant material from earlier chapters; BIO are of biological or medical interest.

The eText icon indicates when there is a video tutor solution available for the chapter or for a specific problem. To launch these videos, log into your eText through Mastering™ Physics or log into the Study Area.

5. A wave pulse travels along a string at a speed of 200 cm/s. What will be the speed if:
   a. The string's tension is doubled?
   b. The string's mass is quadrupled (but its length is unchanged)?
   c. The string's length is quadrupled (but its mass is unchanged)?
   d. The string's mass and length are both quadrupled?
   Note that parts a–d are independent and refer to changes made to the original string.
6. BIO Harbor seals, like many animals, determine the direction from which a sound is coming by sensing the difference in arrival times at their two ears. A small difference in arrival times means that the object is in front of the seal; a larger difference means it is to the left or right. There is a minimum time difference that a seal can sense, and this leads to a limitation on a seal's direction sense. Seals can distinguish between two sounds that come from directions 3° apart in air, but this increases to 9° in water. Explain why you would expect a seal's directional discrimination to be worse in water than in air.
7. Home professionals like realtors use ultrasonic measuring tools to quickly measure the size of rooms. These instruments work by sending out a pulse of ultrasonic sound, then measuring the time it takes for this pulse to reflect off a wall and return to the device. If you used such a device on a particularly hot day, would you measure the length of a given room to be longer or shorter than on a cool day?
8. When in air, a waterproof speaker emits sound waves that have a frequency of 1000 Hz. When the speaker is lowered into water, does the frequency of the sound increase, decrease, or remain the same? Does the wavelength of the sound increase, decrease, or remain the same?
9. Figure Q15.9 shows a history graph of the motion of one point on a string as a wave traveling to the left passes by. Sketch a snapshot graph for this wave.

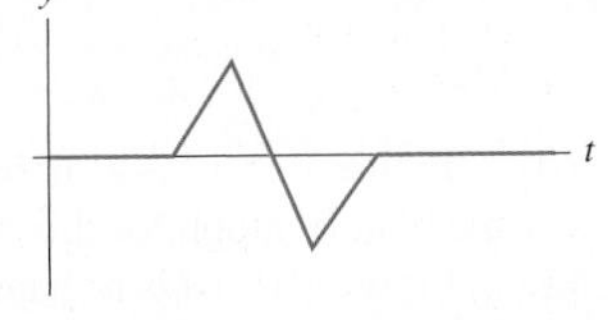

FIGURE Q15.9

10. Figure Q15.10 shows a history graph *and* a snapshot graph for a wave pulse on a string. They describe the same wave from two perspectives.
    a. In which direction is the wave traveling? Explain.
    b. What is the speed of this wave?

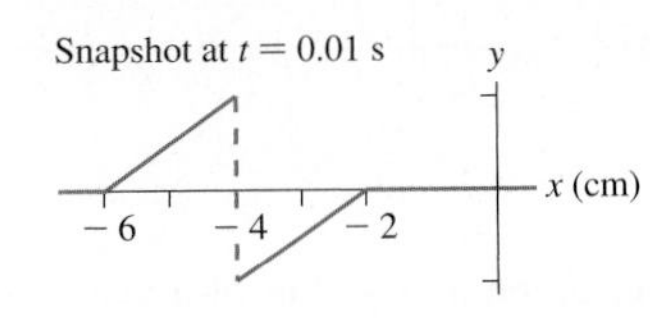

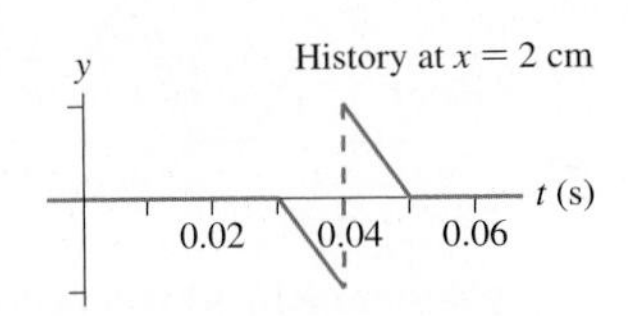

FIGURE Q15.10

11. Rank in order, from largest to smallest, the wavelengths λ_1 to λ_3 for sound waves having frequencies $f_1 = 100$ Hz, $f_2 = 1000$ Hz, and $f_3 = 10{,}000$ Hz. Explain.
12. BIO Bottlenose dolphins use echolocation pulses with a frequency of about 100 kHz, higher than the frequencies used by most bats. Why might you expect these water-dwelling creatures to use higher echolocation frequencies than bats?

13. BIO Some bat species have auditory systems that work best over a narrow range of frequencies. To account for this, the bats adjust the sound frequencies they emit so that the returning, Doppler-shifted sound pulse is in the correct frequency range. As a bat increases its forward speed, should it increase or decrease the frequency of the emitted pulses to compensate?
14. A laser beam has intensity I_0.
    a. What is the intensity, in terms of I_0, if a lens focuses the laser beam to 1/10 its initial diameter?
    b. What is the intensity, in terms of I_0, if a lens defocuses the laser beam to 10 times its initial diameter?
15. A wave pulse travels along a horizontal string. As the pulse passes a point on the string, the point moves vertically up and then back down again. How does the vertical speed of the point compare to the speed of the wave?
16. The volume control on a stereo is designed so that three clicks of the dial increase the output by 10 dB. How many clicks are required to increase the power output of the loudspeakers by a factor of 100?
17. A bullet can travel at a speed of over 1000 m/s. When a bullet is fired from a rifle, the actual firing makes a distinctive sound, but people at a distance may hear a second, different sound that is even louder. Explain the source of this sound.
18. You are standing at $x = 0$ m, listening to seven identical sound sources described by Figure Q15.18. At $t = 0$ s, all seven are at $x = 343$ m and moving as shown below. The sound from all seven will reach your ear at $t = 1$ s. Rank in order, from highest to lowest, the seven frequencies f_1 to f_7 that you hear at $t = 1$ s. Explain.

1 → 50 m/s, speeding up
2 → 50 m/s, steady speed
3 → 50 m/s, slowing down
4 At rest
50 m/s, speeding up ← 5
50 m/s, steady speed ← 6
50 m/s, slowing down ← 7

FIGURE Q15.18

19. Police radar guns work by measuring the change in frequency of a reflected pulse of electromagnetic waves. A gun reads correctly only if a car is moving directly toward or away from the person making the measurement. Explain why this is so.

Multiple-Choice Questions

20. || An AM radio listener is located 5.0 km from the radio station. If he is listening to the radio at the frequency of 660 kHz, how many wavelengths fit in the distance from the station to his house?
    A. 11 B. 22 C. 33 D. 132
21. BIO || The probe used in a medical ultrasound examination emits sound waves in air that have a wavelength of 0.12 mm. What is the wavelength of the sound waves in the patient?
    A. 0.027 mm B. 0.12 mm C. 0.26 mm D. 0.54 mm
22. BIO | Ultrasound can be used to deliver energy to tissues for therapy. It can penetrate tissue to a depth approximately 200 times its wavelength. What is the approximate depth of penetration of ultrasound at a frequency of 5.0 MHz?
    A. 0.29 mm B. 1.4 cm C. 6.2 cm D. 17 cm

23. | A sinusoidal wave traveling on a string has a period of 0.20 s, a wavelength of 32 cm, and an amplitude of 3 cm. The speed of this wave is
A. 0.60 cm/s. B. 6.4 cm/s. C. 15 cm/s. D. 160 cm/s.

24. || Two strings of different linear density are joined together and pulled taut. A sinusoidal wave on these strings is traveling to the right, as shown in Figure Q15.24. When the wave goes across the boundary from string 1 to string 2, the frequency is unchanged. What happens to the velocity?
A. The velocity increases.
B. The velocity stays the same.
C. The velocity decreases.

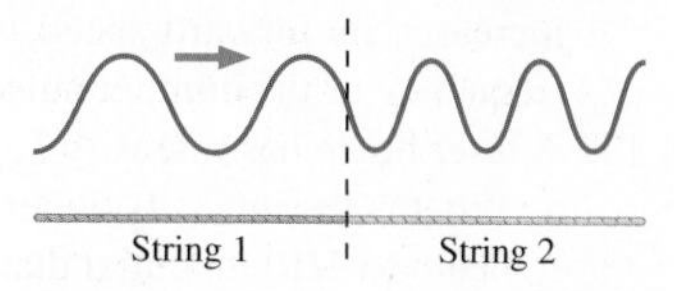

FIGURE Q15.24

25. || You stand at $x = 0$ m, listening to a sound that is emitted at frequency f_s. Figure Q15.25 shows the frequency you hear during a four-second interval. Which of the following describes the motion of the sound source?
A. It moves from left to right and passes you at $t = 2$ s.
B. It moves from right to left and passes you at $t = 2$ s.
C. It moves toward you but doesn't reach you. It then reverses direction at $t = 2$ s.
D. It moves away from you until $t = 2$ s. It then reverses direction and moves toward you but doesn't reach you.

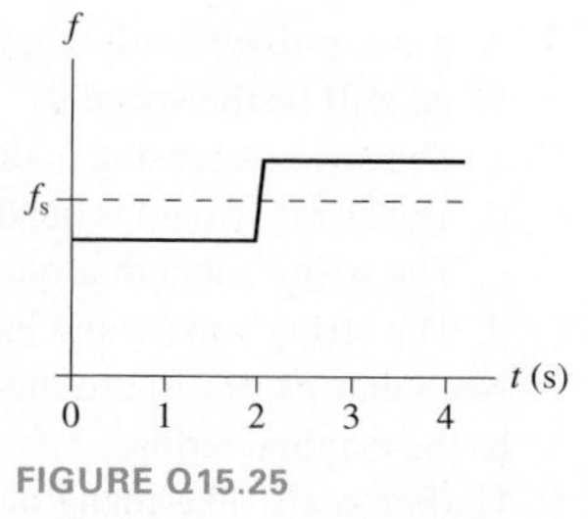

FIGURE Q15.25

PROBLEMS

Section 15.1 The Wave Model

Section 15.2 Traveling Waves

1. || The wave speed on a string under tension is 200 m/s. What is the speed if the tension is doubled?
2. || The wave speed on a string is 150 m/s when the tension is 75.0 N. What tension will give a speed of 180 m/s?
3. || The back wall of an auditorium is 26.0 m from the stage. If you are seated in the middle row, how much time elapses between a sound from the stage reaching your ear directly and the same sound reaching your ear after reflecting from the back wall?
4. ||| Bats sense objects in the dark by echolocation, in which they emit very short pulses of sound and then listen for their echoes off the objects. A bat is flying directly toward a wall 50 m away when it emits a pulse. 0.28 s later it receives the pulse. What is the bat's speed? BIO
5. ||| A scientist measures the speed of sound in a monatomic gas to be 449 m/s at 20°C. What element does this gas consist of?
6. | A medical ultrasound imaging system sends out a steady stream of very short pulses. To simplify analysis, the reflection of one pulse should be received before the next is transmitted. If the system is being used to create an image of tissue 12 cm below the skin, what is the minimum time between pulses? How many pulses per second does this correspond to? BIO
7. || An earthquake 45 km from a city produces P and S waves that travel outward at 5000 and 3000 m/s, respectively. Once city residents feel the shaking of the P wave, how much time do they have before the S wave arrives?

Section 15.3 Graphical and Mathematical Descriptions of Waves

8. || A stationary boat in the ocean is experiencing waves from a storm. The waves move at 56 km/h and have a wavelength of 160 m, both typical values. The boat is at the crest of a wave. How much time elapses until the boat is first at the trough of a wave?
9. || Figure P15.9 is a snapshot graph of a wave at $t = 0$ s. Draw the history graph for this wave at $x = 6$ m, for $t = 0$ s to 6 s.

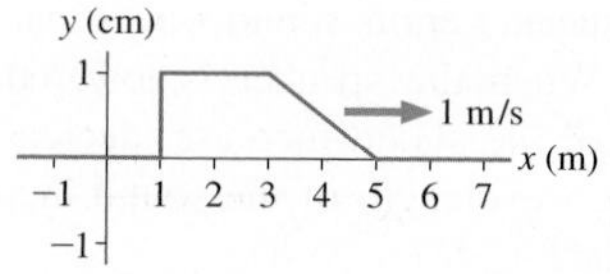

FIGURE P15.9

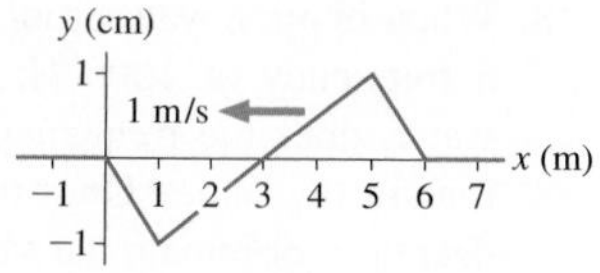

FIGURE P15.10

10. || Figure P15.10 is a snapshot graph of a wave at $t = 2$ s. Draw the history graph for this wave at $x = 0$ m, for $t = 0$ s to 8 s.
11. | Figure P15.10 is a snapshot graph of a wave at $t = 2$ s. What is the maximum speed reached by a particle on the string?
12. ||| Figure P15.12 is a history graph at $x = 0$ m of a wave moving to the right at 1 m/s. Draw a snapshot graph of this wave at $t = 1$ s.

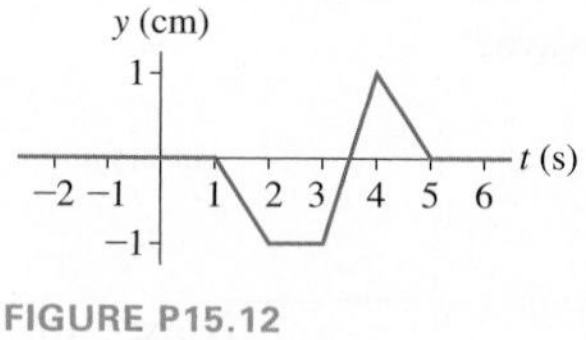

FIGURE P15.12

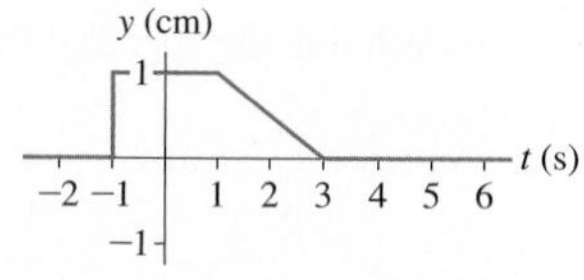

FIGURE P15.13

13. ||| Figure P15.13 is a history graph at $x = 2$ m of a wave moving to the left at 1 m/s. Draw the snapshot graph of this wave at $t = 0$ s.
14. ||| Figure P15.14 shows history graphs of two different points on a string as a wave pulse moves along the string. The blue curve is the history graph for the point at $x = 1.0$ cm, and the green curve is for the point at $x = -3.0$ cm. What is the velocity (including the correct sign for its direction) of this wave?

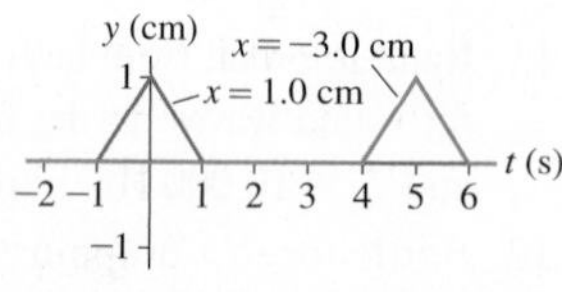

FIGURE P15.14

15. | A sinusoidal wave has period 0.20 s and wavelength 2.0 m. What is the wave speed?
16. | A sinusoidal wave travels with speed 200 m/s. Its wavelength is 4.0 m. What is its frequency?
17. || The motion detector used in a physics lab sends and receives 40 kHz ultrasonic pulses. A pulse goes out, reflects off the object being measured, and returns to the detector. The lab temperature is 20°C.
    a. What is the wavelength of the waves emitted by the motion detector?
    b. How long does it take for a pulse that reflects off an object 2.5 m away to make a round trip?
18. | The displacement of a wave traveling in the positive x-direction is $y(x, t) = (3.5 \text{ cm}) \times \cos(2.7x - 92t)$, where x is in m and t is in s. What are the (a) frequency, (b) wavelength, and (c) speed of this wave?
19. ||| A traveling wave has displacement given by $y(x, t) = (2.0 \text{cm}) \times \cos(2\pi x - 4\pi t)$, where x is measured in cm and t in s.
    a. Draw a snapshot graph of this wave at $t = 0$ s.
    b. On the same set of axes, use a dotted line to show the snapshot graph of the wave at $t = 1/8$ s.
    c. What is the speed of the wave?
20. | Figure P15.20 is a snapshot graph of a wave at $t = 0$ s. What are the amplitude, wavelength, and frequency of this wave?

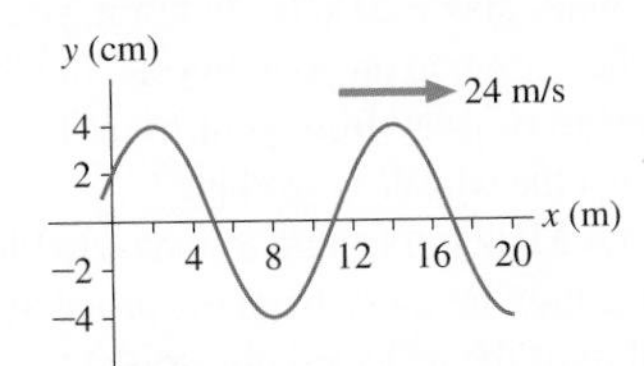

FIGURE P15.20

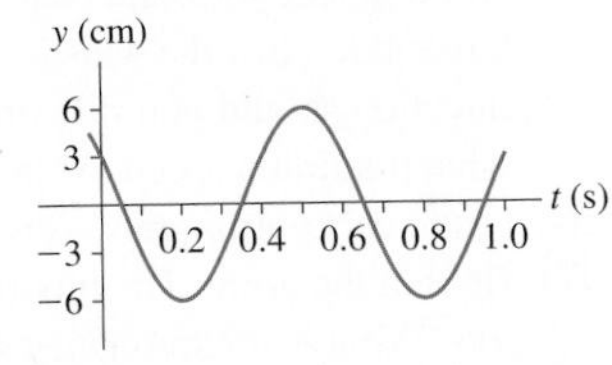

FIGURE P15.21

21. | Figure P15.21 is a history graph at $x = 0$ m of a wave moving to the right at 2 m/s. What are the amplitude, frequency, and wavelength of this wave?
22. || Figure P15.22 shows snapshot (left) and history (right) graphs for a wave traveling on a string. The snapshot graph shows the wave at $t = 0$ s; the history graph shows the displacement of the point on the string at $x = 1.5$ cm, indicated by the dot in the snapshot graph. What is the speed of the wave, and in which direction is it traveling?

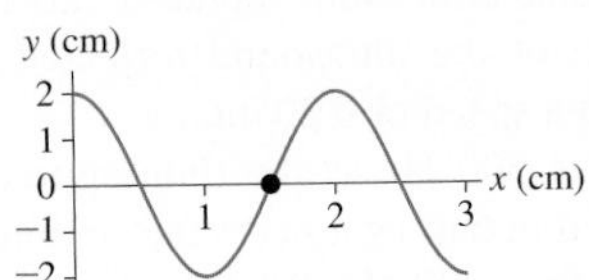

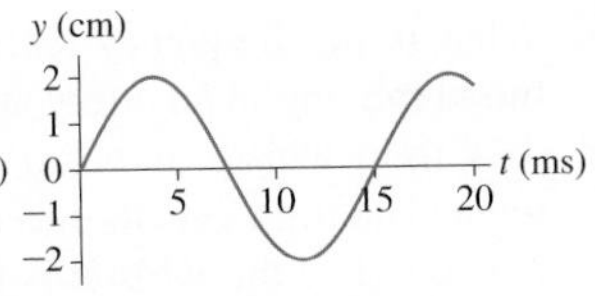

FIGURE P15.22

23. |||| A sinusoidal wave moving to the left has a wavelength of 5.0 cm and a frequency of 50 Hz. At $t = 0$ s, the wave has a crest at $x = 0$ cm. What is the earliest time after $t = 0$ s at which there is a crest at the position $x = 3.0$ cm?

Section 15.4 Sound and Light Waves

24. | People with very good pitch discrimination can very quickly BIO determine what note they are listening to. The note on the musical scale called C_6 (two octaves above middle C) has a frequency of 1050 Hz. Some trained musicians can identify this note after hearing only 12 cycles of the wave. How much time does this correspond to?
25. | A dolphin emits ultrasound at 100 kHz and uses the timing BIO of reflections to determine the position of objects in the water. What is the wavelength of this ultrasound?
26. | Elephants can communicate over distances as far as 6 km by BIO using very low-frequency sound waves. What is the wavelength of a 10 Hz sound wave emitted by an elephant?
27. || a. What is the frequency of blue light that has a wavelength of 450 nm?
    b. What is the frequency of red light that has a wavelength of 650 nm?
28. | Research vessels at sea can create images of their surroundings by sending out sound waves and measuring the time until they detect echoes. This image of a shipwreck on the ocean bottom was made from the surface with 600 kHz ultrasound.

    a. What was the wavelength?
    b. How deep is the shipwreck if echoes were detected 0.42 s after the sound waves were emitted?
29. | A bullet shot from a rifle travels at 1000 m/s. What is the elapsed time between when the bullet strikes a target 500 m away, and when the sound of the gunshot reaches the target?
30. || a. An FM radio station broadcasts at a frequency of 101.3 MHz. What is the wavelength?
    b. What is the frequency of a sound source that produces the same wavelength in 20°C air?

Section 15.5 Energy and Intensity

31. || Sound is detected when a sound wave causes the eardrum BIO INT to vibrate (see Figure 14.26). Typically, the diameter of the eardrum is about 8.4 mm in humans. When someone speaks to you in a normal tone of voice, the sound intensity at your ear is approximately 1.0×10^{-6} W/m^2. How much energy is delivered to your eardrum each second?
32. ||| At a rock concert, the sound intensity 1.0 m in front of the BIO INT bank of loudspeakers is 0.10 W/m^2. A fan is 30 m from the loudspeakers. Her eardrums have a diameter of 8.4 mm. How much sound energy is transferred to each eardrum in 1.0 second?
33. ||| The intensity of electromagnetic waves from the sun is 1.4 kW/m^2 just above the earth's atmosphere. Eighty percent of this reaches the surface at noon on a clear summer day. Suppose you model your back as a 30 cm $\times$ 50 cm rectangle. How many joules of solar energy fall on your back as you work on your tan for 1.0 h?
34. |||| A sun-like star is barely visible to naked-eye observers on earth when it is a distance of 7.0 light years, or 6.6×10^{16} m, away. The sun emits a power of 3.8×10^{26} W. Using this information, at what distance would a candle that emits a power of 0.20 W just be visible?
35. ||| A large solar panel on a spacecraft in earth orbit produces 1.0 kW of power when the panel is turned toward the sun. What power would the solar cell produce if the spacecraft were in orbit around Saturn, 9.5 times as far from the sun?
36. ||| Solar cells convert the energy of incoming light to electric energy; a good quality cell operates at an efficiency of 15%. Each person in the United States uses energy (for lighting, heating, transportation, etc.) at an average rate of 11 kW. Although sunlight varies with season and time of day, solar energy falls on the United States at an average intensity of 200 W/m^2. Assuming you live in an average location, what total solar-cell area would you need to provide all of your energy needs with energy from the sun?

37. || LASIK eye surgery uses pulses of laser light to shave off tissue from the cornea, reshaping it. A typical LASIK laser emits a 1.0-mm-diameter laser beam with a wavelength of 193 nm. Each laser pulse lasts 15 ns and contains 1.0 mJ of light energy. BIO
    a. What is the power of one laser pulse?
    b. During the very brief time of the pulse, what is the intensity of the light wave?
38. || Using a dish-shaped mirror, a *solar cooker* concentrates the sun's energy onto a pot for cooking. A cooker with a 1.5-m-diameter dish focuses the sun's energy onto a pot with a diameter of 25 cm. Given that the intensity of sunlight is about 1000 W/m^2,

    a. How much solar power does the dish capture?
    b. What is the intensity at the base of the pot?
39. | The world's most powerful laser is the LFEX laser in Japan. It can produce a 2 petawatt (2×10^{15} W) laser pulse that last for 1 ps. The laser is focused onto a small spot that is 30 μm in diameter. What is the light intensity within this spot?

Section 15.6 Loudness of Sound

40. | What is the sound intensity level of a sound with an intensity of 3.0×10^{-6} W/m^2?
41. ||| What is the sound intensity of a whisper at a distance of 2.0 m, in W/m^2? What is the corresponding sound intensity level in dB?
42. ||| The record for the world's loudest burp is 109.9 dB, measured at a distance of 2.5 m from the burper. Assuming that this sound was emitted as a spherical wave, what was the power emitted by the burper during his record burp? BIO
43. || The sound intensity from a jack hammer breaking concrete is 2.0 W/m^2 at a distance of 2.0 m from the point of impact. This is sufficiently loud to cause permanent hearing damage if the operator doesn't wear ear protection. What are (a) the sound intensity and (b) the sound intensity level for a person watching from 50 m away?
44. ||| A concert loudspeaker suspended high off the ground emits 35 W of sound power. A small microphone with a 1.0 cm^2 area is 50 m from the speaker. What are (a) the sound intensity and (b) the sound intensity level at the position of the microphone?
45. ||| The African cicada is the world's loudest insect, producing a sound intensity level of 107 dB at a distance of 0.50 m. What is the intensity of its sound (in W/m^2) as heard by someone standing 3.0 m away? BIO
46. ||| From a distance of 4.0 m, a bystander listens to a jackhammer breaking concrete. How far would he need to move from the jackhammer so that its perceived loudness decreases by a factor of 8?
47. || A rock band playing an outdoor concert produces sound at 120 dB 5.0 m away from their single working loudspeaker. What is the sound intensity level 35 m from the speaker?
48. ||| Your ears are sensitive to differences in pitch, but they are not very sensitive to differences in intensity. You are not capable of detecting a difference in sound intensity level of less than 1 dB. By what factor does the sound intensity increase if the sound intensity level increases from 60 dB to 61 dB? BIO
49. || 30 seconds of exposure to 115 dB sound can damage your hearing, but a much quieter 94 dB may begin to cause damage after 1 hour of continuous exposure. You are going to an outdoor concert, and you'll be standing near a speaker that emits 50 W of acoustic power as a spherical wave. What minimum distance should you be from the speaker to keep the sound intensity level below 94 dB? BIO
50. ||| When you speak, your voice sounds 10 dB louder to someone standing directly in front of you than to someone at the same distance but directly behind you. What is the ratio of the intensity of your voice for someone in front of you to the intensity for someone behind you? BIO

Section 15.7 The Doppler Effect and Shock Waves

51. | An opera singer in a convertible sings a note at 600 Hz while cruising down the highway at 90 km/h. What is the frequency heard by
    a. A person standing beside the road in front of the car?
    b. A person standing beside the road behind the car?
52. || An osprey's call is a distinct whistle at 2200 Hz. An osprey calls while diving at you, to drive you away from her nest. You hear the call at 2300 Hz. How fast is the osprey approaching? BIO
53. || A whistle you use to call your hunting dog has a frequency of 21 kHz, but your dog is ignoring it. You suspect the whistle may not be working, but you can't hear sounds above 20 kHz. To test it, you ask a friend to blow the whistle, then you hop on your bicycle. In which direction should you ride (toward or away from your friend) and at what minimum speed to know if the whistle is working?
54. | An echocardiogram uses 4.4 MHz ultrasound to measure blood flow in the aorta. The blood is moving away from the probe at 1.4 m/s. What is the frequency shift of the reflected ultrasound? BIO
55. | A friend of yours is loudly singing a single note at 400 Hz while driving toward you at 25.0 m/s on a day when the speed of sound is 340 m/s.
    a. What frequency do you hear?
    b. What frequency does your friend hear if you suddenly start singing at 400 Hz?
56. ||| The frequency of light emitted from hydrogen present in the Andromeda galaxy has been found to be 0.10% higher than that from hydrogen measured on earth. Is this galaxy approaching or receding from the earth, and at what speed?
57. ||| A Doppler blood flow unit emits ultrasound at 5.0 MHz. What is the frequency shift of the ultrasound reflected from blood moving in an artery at a speed of 0.20 m/s? BIO
58. ||| A train whistle is heard at 300 Hz as the train approaches town. The train cuts its speed in half as it nears the station, and the sound of the whistle is then 290 Hz. What is the speed of the train before and after slowing down?

General Problems

59. ||| At the 18 km cruising altitude of Concorde, a passenger aircraft that flew at twice the speed of sound, the temperature was −57°C. What was the Concorde's cruising speed?
60. ||| A 2.0-m-long string is under 20 N of tension. A pulse travels the length of the string in 50 ms. What is the mass of the string?
61. ||| A female orb spider has a mass of 0.50 g. She is suspended from a tree branch by a 1.1 m length of 0.0020-mm-diameter silk. Spider silk has a density of 1300 kg/m^3. If you tap the branch and send a vibration down the thread, how long does it take to reach the spider? BIO INT

62. || A spider spins a web with silk threads of density 1300 kg/m^3 BIO and diameter 3.0 μm. A typical tension in the radial threads of such a web is 7.0 mN. Suppose a fly hits this web. Which will reach the spider first: the very slight sound of the impact or the disturbance traveling along the radial thread of the web?

63. ||| In 2003, an earthquake in Japan generated 1.1 Hz waves INT that traveled outward at 7.0 km/s. 200 km to the west, seismic instruments recorded a maximum acceleration of $0.25g$ along the east-west axis.
   a. How much time elapsed between the earthquake and the first detection of the waves?
   b. Was this a transverse or a longitudinal wave?
   c. What was the wavelength?
   d. What was the maximum horizontal displacement of the ground as the wave passed?

64. ||| A coyote can locate a sound source with good accuracy by BIO comparing the arrival times of a sound wave at its two ears. Suppose a coyote is listening to a bird whistling at 1000 Hz. The bird is 3.0 m away, directly in front of the coyote's right ear. The coyote's ears are 15 cm apart.
   a. What is the difference in the arrival times of the sound at the left ear and the right ear?
   b. What is the ratio of this times difference to the period of the sound wave?

   Hint: You are looking for the difference between two numbers that are nearly the same. What does this near equality imply about the necessary precision during intermediate stages of the calculation?

65. ||| A wave travels along a steel string with a speed of 22 m/s. At some point along the string, its diameter doubles. What is the speed of the wave in this thicker part?

66. || The string in Figure INT P15.66 has a linear density of 7.2×10^{-5} kg/m. What is the speed of a wave on this string?

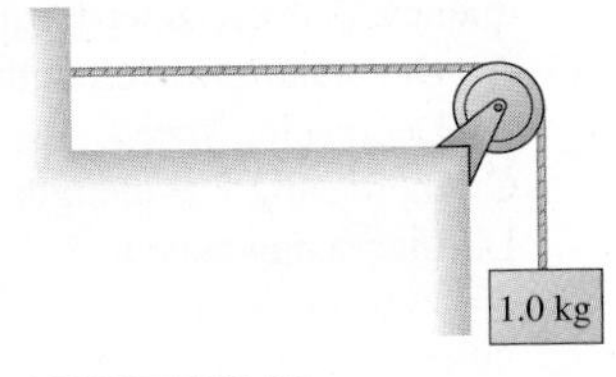

FIGURE P15.66

67. || Low-frequency vertical BIO oscillations are one possible cause of motion sickness, with 0.30 Hz having the strongest effect. Your boat is bobbing in place at just the right frequency to cause you the maximum discomfort. The water wave that is bobbing the boat has crests that are 30 m apart.
   a. What is the speed of the waves?
   b. What will be the boat's vertical oscillation frequency if you drive the boat at 5.0 m/s in the direction of the oncoming waves?

68. ||| Figure P15.68 shows two snapshot graphs taken 10 ms apart, INT with the blue curve being the first snapshot. What are the (a) wavelength, (b) speed, (c) frequency, and (d) amplitude of this wave?

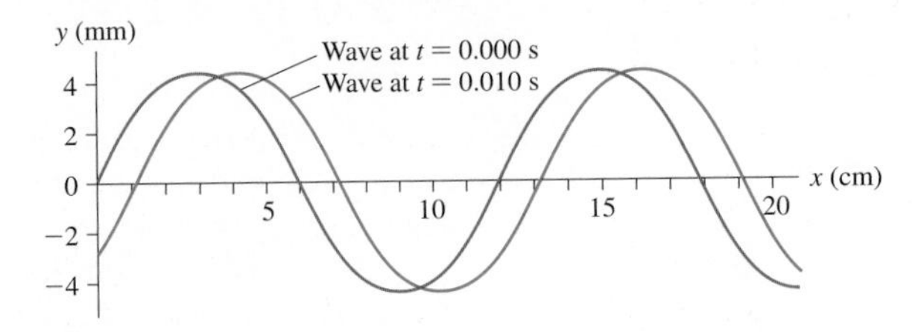

FIGURE P15.68

69. ||| The pressure in a sound wave in steel is given by $p(x, t) = p_{atm} + p_0 \cos(2.4x - (1.4 \times 10^4)t)$, where p_{atm} is atmospheric pressure, p_0 is the amplitude of the wave, x is in m, and t in s. What are the speed and frequency of this wave?

70. ||| A wave on a string is described by $y(x, t) = (3.0\text{ cm}) \times \cos[2\pi(x/(2.4\text{ m}) + t/(0.20\text{ s}))]$, where x is in m and t in s.
   a. In what direction is this wave traveling?
   b. What are the wave speed, frequency, and wavelength?
   c. At $t = 0.50$ s, what is the displacement of the string at $x = 0.20$ m?

71. | Write the y-equation for a wave traveling in the negative x-direction with wavelength 50 cm, speed 4.0 m/s, and amplitude 5.0 cm.

72. ||| A point on a string undergoes simple harmonic motion as INT a sinusoidal wave passes. When a sinusoidal wave with speed 24 m/s, wavelength 30 cm, and amplitude of 1.0 cm passes, what is the maximum speed of a point on the string?

73. ||| The threshold of hearing—the lowest-intensity sound a person BIO can hear—depends on, among many factors, the frequency of the sound. During a hearing exam, the hearing specialist creates an *audiogram*, a graph of the patient's hearing threshold (in dB) versus sound frequency; thresholds above 20 dB indicate hearing loss. For the audiogram shown in Figure P15.73, what is the ratio of the sound intensity (in W/m^2) of the faintest sound the patient can hear at 1 kHz to that of the faintest sound she can hear at 6 kHz?

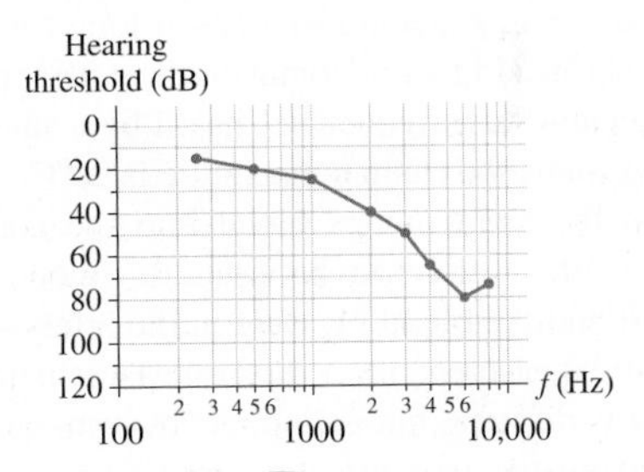

FIGURE P15.73

74. | The total power consumption by all humans on earth is approximately 10^{13} W. Let's compare this to the power of incoming solar radiation. The intensity of radiation from the sun at the top of the atmosphere is 1380 W/m^2. The earth's radius is 6.37×10^6 m.
   a. What is the total solar power received by the earth?
   b. By what factor does this exceed the total human power consumption?

75. | A dark blue cylindrical bottle is 22 cm high and has a INT diameter of 7.0 cm. It is filled with water. The bottle absorbs 60% of the light that shines on it as it lies on its side in the noonday sun, with intensity 1000 W/m^2. By how much will the temperature of the water increase in 5.0 min if there's negligible heat loss to the surrounding air?

76. || Assume that the opening of the ear canal has a diameter BIO of 7.0 mm. For this problem, you can ignore any focusing of energy into the opening by the pinna, the external folds of the ear.
   a. How much sound power is "captured" by one ear at 0 dB, the threshold of hearing?
   b. How much energy does one ear "capture" during 1 hour of listening to a lecture delivered at a conversational volume of 60 dB?

77. | The sound intensity 50 m from a wailing tornado siren is 0.10 W/m^2. What is the sound intensity level 300 m from the siren?

Watch Video Solution Problems 15.79 and 15.81

78. | One of the loudest sound generators ever created is the INT Danley Sound Labs Matterhorn. When run at full power, the device uses an input power of 40,000 W. The output power registers 94 dB at 250 m. What is the efficiency of this device—that is, the ratio of output power to input power?

79. ||| A harvest mouse can detect sounds below the threshold of BIO human hearing, as quiet as -10 dB. Suppose you are sitting in a field on a very quiet day while a harvest mouse sits nearby. A very gentle breeze causes a leaf 1.5 m from your head to rustle, generating a faint sound right at the limit of your ability to hear it. The sound of the rustling leaf is also right at the threshold of hearing of the harvest mouse. How far is the harvest mouse from the leaf?

80. |||| A speaker at an open-air concert emits 600 W of sound power, radiated equally in all directions.
   a. What is the intensity of the sound 5.0 m from the speaker?
   b. What sound intensity level would you experience there if you did not have any protection for your ears?
   c. Earplugs you can buy in the drugstore have a noise reduction rating of 23 decibels. If you are wearing those earplugs but your friend Phil is not, how far from the speaker should Phil stand to experience the same loudness as you?

81. | A physics professor demonstrates the Doppler effect by tying INT a 600 Hz sound generator to a 1.0-m-long rope and whirling it around her head in a horizontal circle at 100 rpm. What are the highest and lowest frequencies heard by a student in the classroom? Assume the room temperature is 20°C.

82. ||||| When the heart pumps blood into the aorta, the *pressure* BIO INT *gradient*—the difference between the blood pressure inside the heart and the blood pressure in the artery—is an important diagnostic measurement. A direct measurement of the pressure gradient is difficult, but an indirect determination can be made by inferring the pressure difference from a measurement of velocity. Blood is essentially at rest in the heart; when it leaves and enters the aorta, it speeds up significantly and—according to Bernoulli's equation—the pressure must decrease. A doctor using 2.5 MHz ultrasound measures a 6000 Hz frequency shift as the ultrasound reflects from blood ejected from the heart.
   a. What is the speed of the blood in the aorta?
   b. What is the difference in blood pressure between the inside of the heart and the aorta? Assume that the patient is lying down and that there is no difference in height as the blood moves from the heart into the aorta.

MCAT-Style Passage Problems

Echolocation BIO

As discussed in the chapter, many species of bats find flying insects by emitting pulses of ultrasound and listening for the reflections. This technique is called *echolocation*. Bats possess several adaptations that allow them to echolocate very effectively.

83. | Although we can't hear them, the ultrasonic pulses are very loud. In order not to be deafened by the sound they emit, bats can temporarily turn off their hearing. Muscles in the ear cause the bones in their middle ear to separate slightly, so that they don't transmit vibrations to the inner ear. After an ultrasound pulse ends, a bat can hear an echo from an object a minimum of 1 m away. Approximately how much time after a pulse is emitted is the bat ready to hear its echo?
   A. 0.5 ms B. 1 ms C. 3 ms D. 6 ms

84. | Bats are sensitive to very small changes in frequency of the reflected waves. What information does this allow them to determine about their prey?
   A. Size B. Speed C. Distance D. Species

85. | Some bats have specially shaped noses that focus ultrasound echolocation pulses in the forward direction. Why is this useful?
   A. Increasing intensity reduces the time delay for a reflected pulse.
   B. The energy of the pulse is concentrated in a smaller area, so the intensity is larger; reflected pulses will have a larger intensity as well.
   C. Increasing intensity allows the bat to use a lower frequency and still have the same spatial resolution.

86. | Some bats utilize a sound pulse with a rapidly decreasing frequency. A decreasing-frequency pulse has
   A. Decreasing wavelength.
   B. Decreasing speed.
   C. Increasing wavelength.
   D. Increasing speed.

16 Superposition and Standing Waves

The didgeridoo is a musical instrument played by aboriginal tribes in Australia. It consists of a hollow tube that is open at both ends. How can the player get a wide range of notes out of such a simple device?

LOOKING AHEAD ➤

Superposition

Where the two water waves meet, the motion of the water is the sum, or **superposition,** of the waves.

You'll learn how this **interference** can be constructive or destructive, leading to larger or smaller amplitudes.

Standing Waves

The superposition of waves on a string can lead to a wave that oscillates in place—a **standing wave.**

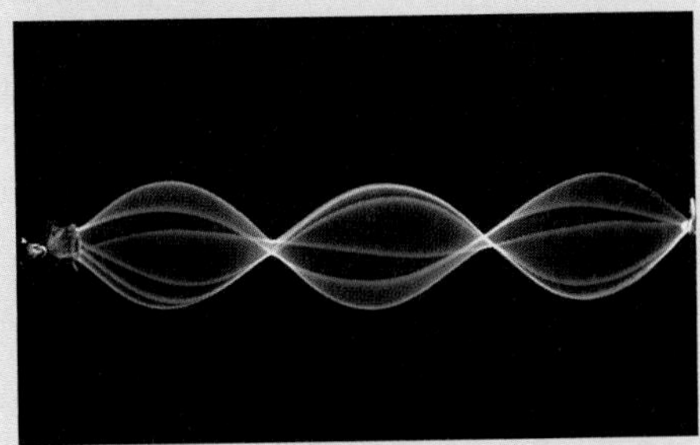

You'll learn the patterns of standing waves on strings and standing sound waves in tubes.

Speech and Hearing

Changing the shape of your mouth alters the pattern of standing sound waves in your vocal tract.

You'll learn how your vocal tract produces, and your ear interprets, different mixes of waves.

GOAL To use the idea of superposition to understand the phenomena of interference and standing waves.

LOOKING BACK ◀

Traveling Waves

In Chapter 15 you learned the properties of traveling waves and relationships among the variables that describe them.

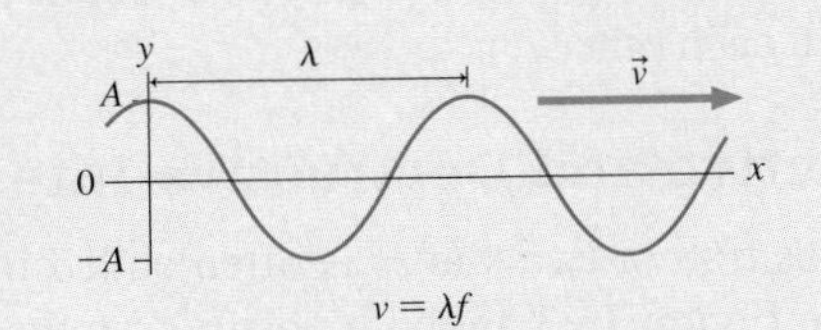

$v = \lambda f$

In this chapter, you'll extend the analysis to understand the interference of waves and the properties of standing waves.

STOP TO THINK

A 170 Hz sound wave in air has a wavelength of 2.0 m. The frequency is now doubled to 340 Hz. What is the new wavelength?

A. 4.0 m
B. 3.0 m
C. 2.0 m
D. 1.0 m

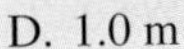

16.1 The Principle of Superposition

FIGURE 16.1a shows two baseball players, Alan and Bill, at batting practice. Unfortunately, someone has turned the pitching machines so that pitching machine A throws baseballs toward Bill while machine B throws toward Alan. If two baseballs are launched at the same time and with the same speed, they collide at the crossing point and bounce away. Two baseballs cannot occupy the same point of space at the same time.

FIGURE 16.1 Two baseballs cannot pass through each other. Two waves can.

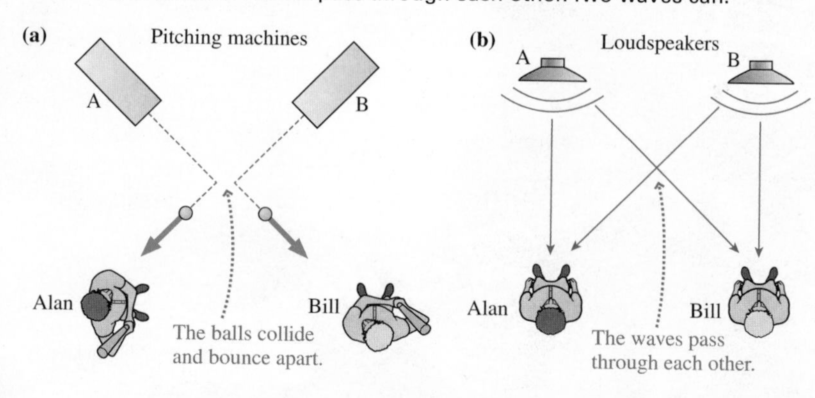

FIGURE 16.2 Two waves on a stretched string pass through each other.

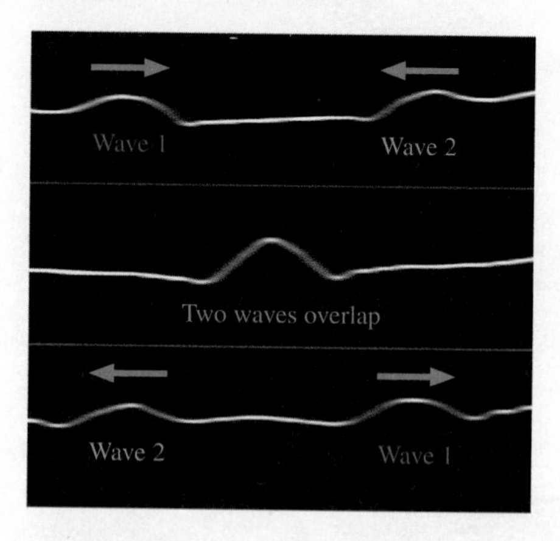

But unlike baseballs, sound waves *can* pass directly through each other. In FIGURE 16.1b, Alan and Bill are listening to the stereo system in the locker room after practice. Both hear the sound from each speaker perfectly, without any distortion or missing sound. Evidently the sound wave that travels from speaker A toward Bill passes right through the wave that travels from speaker B toward Alan, with no effect on either wave. This is a basic property of waves.

What happens to the medium at a point where two waves are present simultaneously? What is the displacement of the medium at this point? FIGURE 16.2 shows a sequence of photos of two waves traveling along a stretched string. In the first photo, the waves are approaching each other. In the second, the waves overlap; you can see that the displacement of the string is larger than it was for either of the individual waves. A careful measurement reveals that the displacement is the sum of the displacements of the two individual waves. In the third frame, the waves have passed through each other and continue on as if nothing had happened.

This result is not limited to stretched strings; the outcome is the same whenever two waves of any type pass through each other. This is known as the *principle of superposition:*

FIGURE 16.3 The superposition of two waves on a string as they pass through each other.

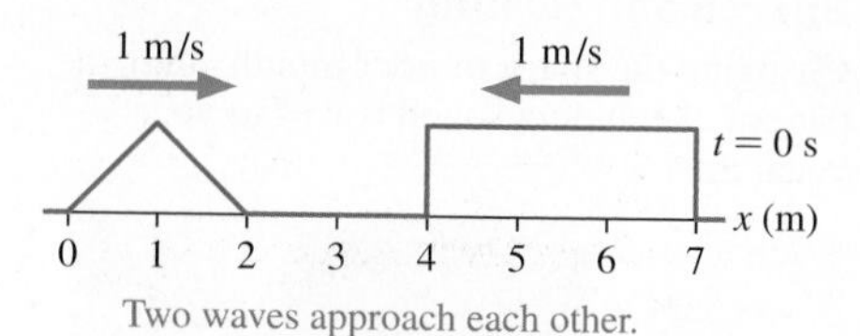

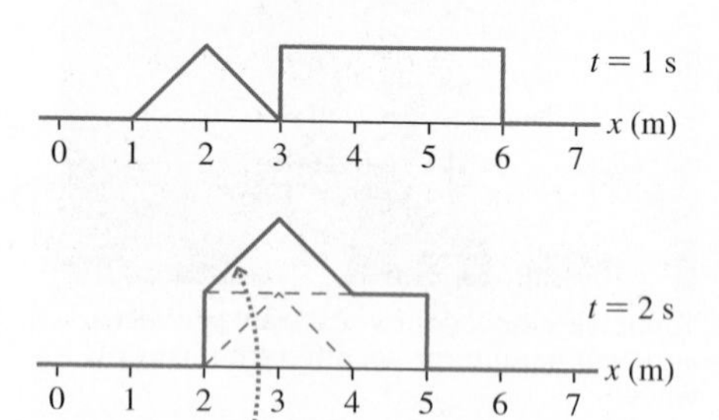

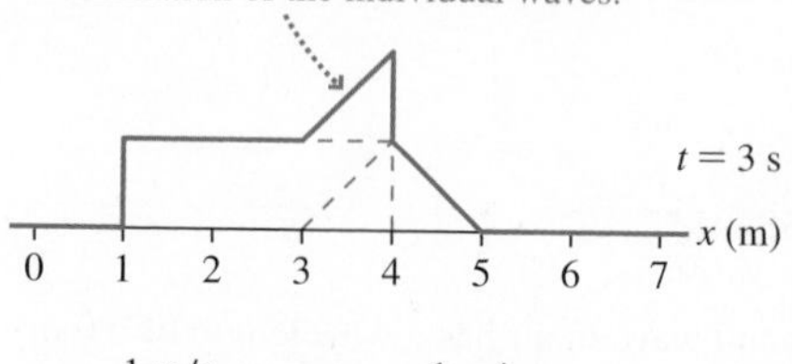

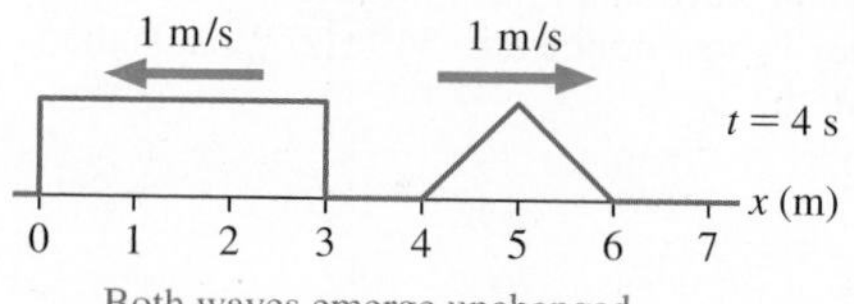

> **Principle of superposition** When two or more waves are *simultaneously* present at a single point in space, the displacement of the medium at that point is the sum of the displacements due to each individual wave.

Let's illustrate this principle with an idealized example. FIGURE 16.3 shows five snapshot graphs taken 1 s apart of two waves traveling at the same speed (1 m/s) in opposite directions along a string. The displacement of each wave is shown as a dashed line. The solid line is the sum *at each point* of the two displacements at that point. This is the displacement that you would actually observe as the two waves pass through each other.

Constructive and Destructive Interference

The superposition of two waves is often called **interference.** The displacements of the waves in Figure 16.3 are both positive, so the total displacement of the medium where they overlap is larger than it would be due to either of the waves separately. We call this **constructive interference.**

FIGURE 16.4 shows another series of snapshot graphs of two counterpropagating waves—waves traveling in opposite directions—but this time one has a negative displacement. The principle of superposition still applies, but now the displacements are opposite to each other. The displacement of the medium where the waves overlap is *less* than it would be due to either of the waves separately. We call this **destructive interference.**

In the series of graphs in Figure 16.4 the displacement of the medium at $x = 3.5$ m is always zero. The positive displacement of the wave traveling to the right and the negative displacement of the wave traveling to the left always exactly cancel at this spot. The complete cancellation at one point of two waves traveling in opposite directions is something we will see again. When the displacements of the waves cancel, where does the energy of the wave go? We know that the waves continue on unchanged after their interaction, so no energy is dissipated. Consider the graph at $t = 1.5$ s. There is no net displacement at any point of the medium at this instant, but the *string is rapidly moving vertically.* The energy of the waves hasn't vanished—it is in the form of the kinetic energy of the medium.

FIGURE 16.4 Two waves with opposite displacements produce destructive interference.

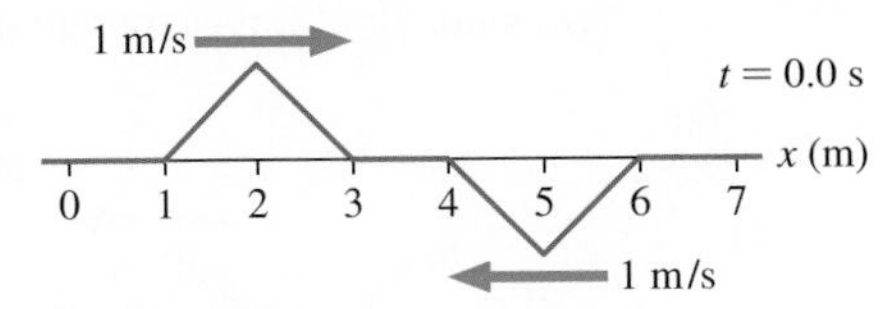

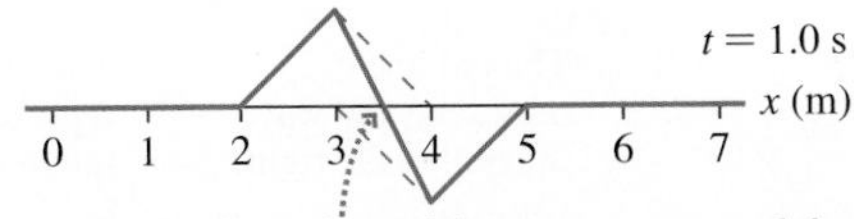

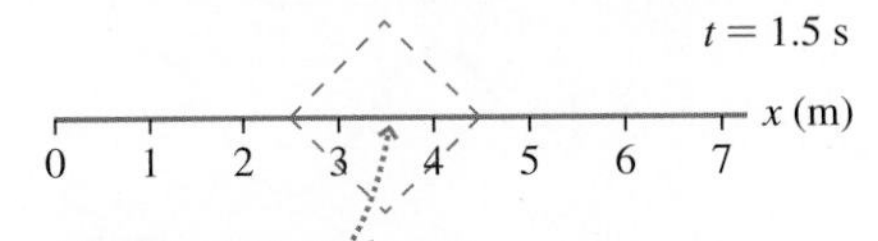

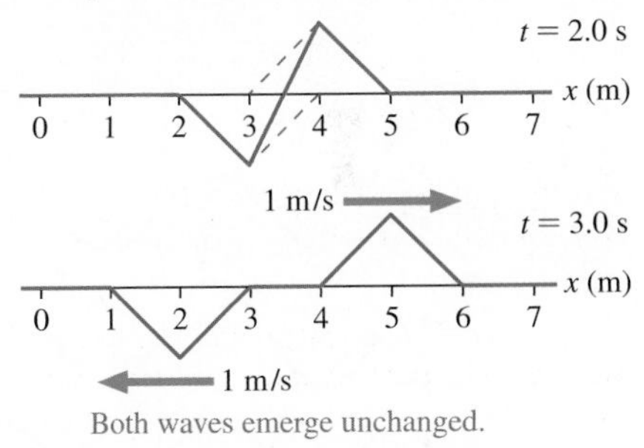

STOP TO THINK 16.1 Two pulses on a string approach each other at speeds of 1 m/s. What is the shape of the string at $t = 6$ s?

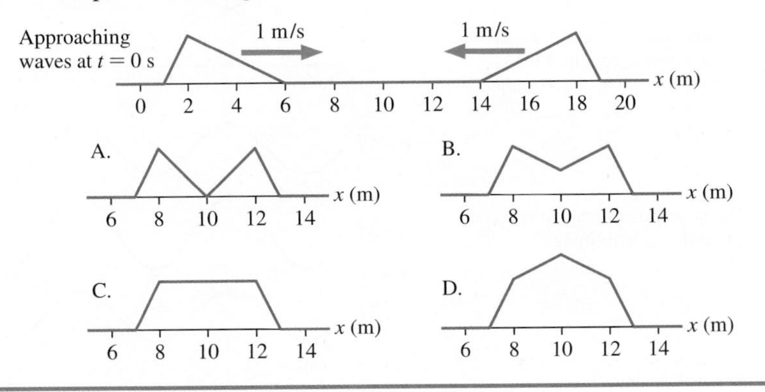

A video to support a section's topic is embedded in the eText.

Video What the Physics? Destructive Interference

16.2 Standing Waves

When you pluck a guitar string or a rubber band stretched between your fingers, you create waves. But how is this possible? There isn't really anywhere for the waves to go, because the string or the rubber band is held between two fixed ends. FIGURE 16.5 shows a strobe photograph of waves on a stretched elastic cord. This is a wave, though it may not look like one because it doesn't "travel" either right or left. Waves that are "trapped" between two boundaries, like those in the photo or on a guitar string, are what we call *standing waves.* **Individual points on the string oscillate up and down, but the wave itself does not travel.** It is called a **standing wave** because the crests and troughs "stand in place" as it oscillates. As we'll see, a standing wave isn't a totally new kind of wave; it is simply the superposition of two traveling waves moving in opposite directions.

FIGURE 16.5 The motion of a standing wave on a string.

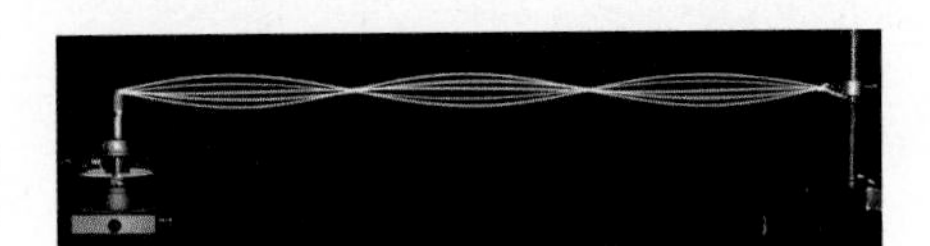

Superposition Creates a Standing Wave

Suppose we have a string on which two sinusoidal waves of equal wavelength and amplitude travel in opposite directions, as in FIGURE 16.6a (next page). When the waves meet, the displacement of the string will be a superposition of these two waves. FIGURE 16.6b shows nine snapshot graphs, at intervals of $\frac{1}{8}T$, of the two waves as they move through each other. The red and orange dots identify particular crests of each of the waves to help you see that the red wave is traveling to the right and the orange wave to the left. At *each point,* the net displacement of the medium is found by adding the red displacement and the orange displacement. The resulting blue wave is the superposition of the two traveling waves.

FIGURE 16.6 Two sinusoidal waves traveling in opposite directions.

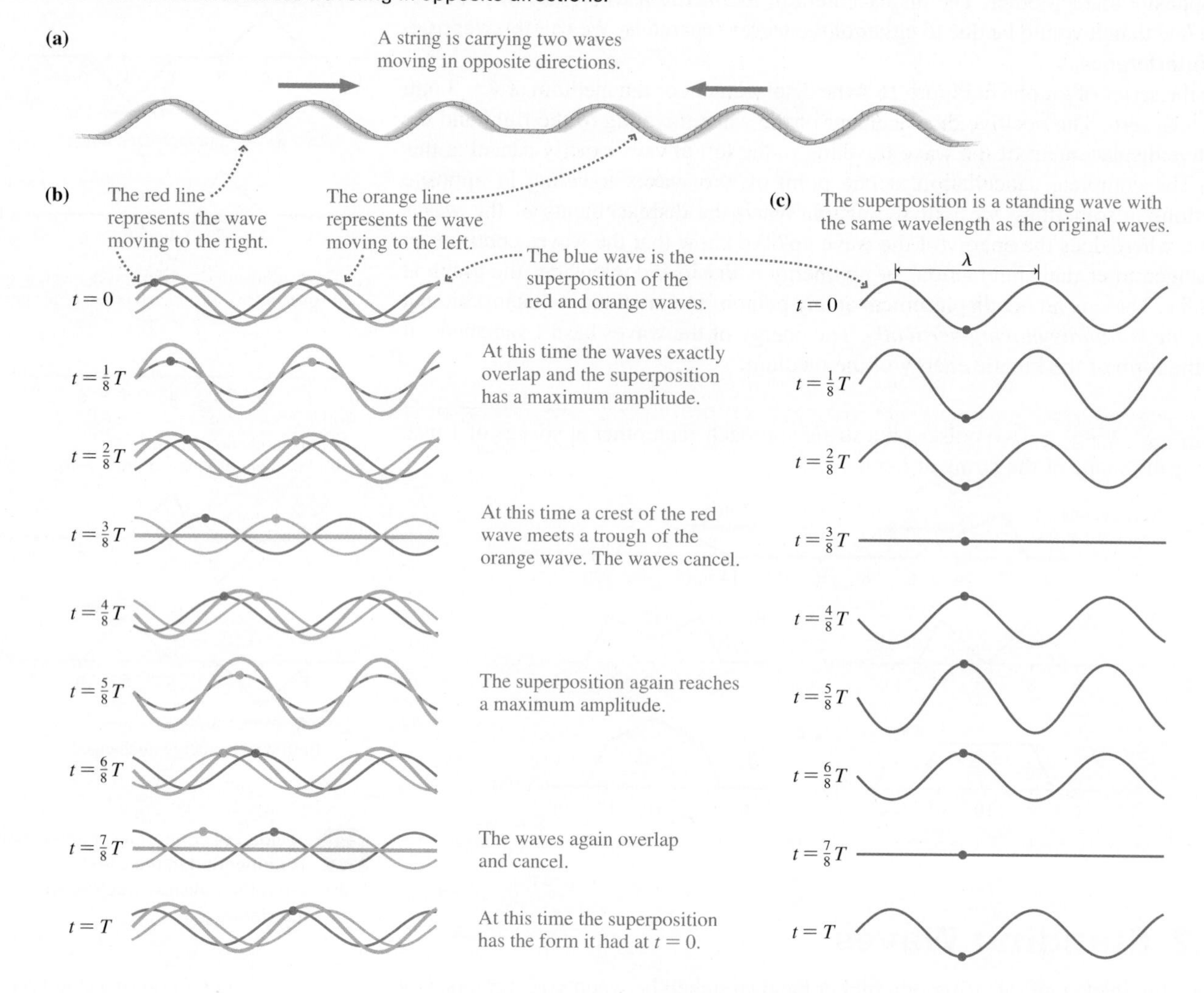
(a)
A string is carrying two waves moving in opposite directions.
(b)
The red line represents the wave moving to the right.
The orange line represents the wave moving to the left.
The blue wave is the superposition of the red and orange waves.
(c)
The superposition is a standing wave with the same wavelength as the original waves.
λ
t = 0
t = 1/8 T
At this time the waves exactly overlap and the superposition has a maximum amplitude.
t = 2/8 T
t = 3/8 T
At this time a crest of the red wave meets a trough of the orange wave. The waves cancel.
t = 4/8 T
t = 5/8 T
The superposition again reaches a maximum amplitude.
t = 6/8 T
t = 7/8 T
The waves again overlap and cancel.
t = T
At this time the superposition has the form it had at t = 0.

FIGURE 16.6c shows just the superposition of the two waves. This is the wave that you would actually observe in the medium. The blue dot shows that the wave in Figure 16.6c is moving neither right nor left. The superposition of the two counter-propagating traveling waves is a standing wave. Notice that the wavelength of the standing wave, the distance between two crests or two troughs, is the same as the wavelengths of the two traveling waves that combine to produce it.

Nodes and Antinodes

In FIGURE 16.7 we have superimposed the nine snapshot graphs of Figure 16.6c into a single graphical representation of this standing wave. The graphs at different times overlap, much like the photos of the string at different times in the strobe photograph of Figure 16.5. The motion of individual points on the standing wave is now clearly seen. A striking feature of a standing-wave pattern is points that *never move!* These points, which are spaced $\lambda/2$ apart, are called **nodes.** Halfway between the nodes are points where the particles in the medium oscillate with maximum displacement. These points of maximum amplitude are called **antinodes,** and you can see that they are also spaced $\lambda/2$ apart. This means that **the wavelength of a standing wave is twice the distance between successive nodes or successive antinodes.**

FIGURE 16.7 Superimposing multiple snapshot graphs of a standing wave clearly shows the nodes and antinodes.

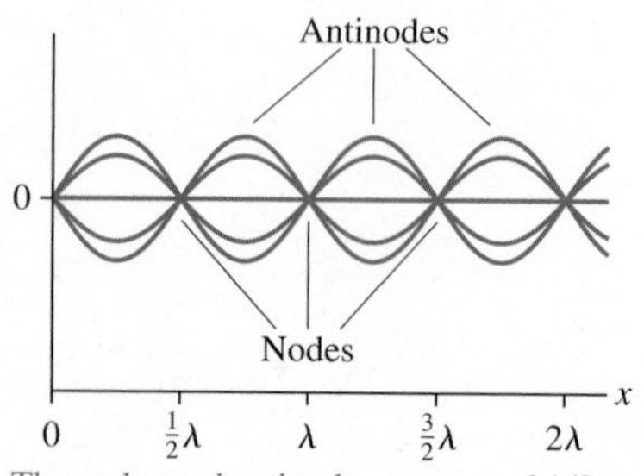
Antinodes
0
Nodes
x
0
½λ
λ
3/2λ
2λ
The nodes and antinodes are spaced $\lambda/2$ apart.

It seems surprising and counterintuitive that some particles in the medium have no motion at all. This happens for the same reason we saw in Figure 16.4: The two waves exactly offset each other at that point. Look carefully at the two traveling waves in Figure 16.6b. You will see that the nodes occur at points where at *every instant* of time the displacements of the two traveling waves have equal magnitudes but *opposite signs.* Thus the superposition of the displacements at these points is always zero—they are points of destructive interference. The antinodes have large displacements. They correspond to points where the two displacements have equal magnitudes and the *same sign* at all times. Constructive interference at these points gives a displacement twice that of each individual wave.

The intensity of a wave is largest at points where it oscillates with maximum amplitude. FIGURE 16.8 shows that the points of maximum intensity along the standing wave occur at the antinodes; the intensity is zero at the nodes. For a standing sound wave, the loudness varies from zero (no sound) at the nodes to a maximum at the antinodes and then back to zero. The key idea is that **the intensity is maximum at points of constructive interference and zero at points of destructive interference.**

FIGURE 16.8 Intensity of a standing wave.

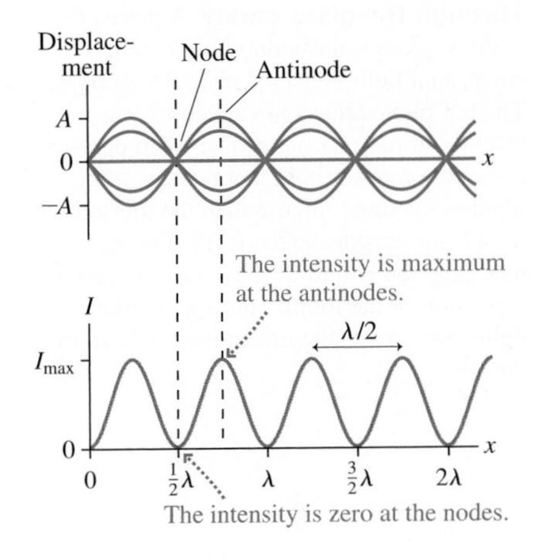

EXAMPLE 16.1 Setting up a standing wave

Two children hold an elastic cord at each end. Each child shakes her end of the cord 2.0 times per second, sending waves at 3.0 m/s toward the middle, where the two waves combine to create a standing wave. What is the distance between adjacent nodes?

SOLVE The distance between adjacent nodes is $\lambda/2$. The wavelength, frequency, and speed are related as $v = \lambda f$, as we saw in Chapter 15, so the wavelength is

$$\lambda = \frac{v}{f} = \frac{3.0\ \text{m/s}}{2.0\ \text{Hz}} = 1.5\ \text{m}$$

The distance between adjacent nodes is $\lambda/2$ and thus is 0.75 m.

STOP TO THINK 16.2 A standing wave is set up on a string. A series of snapshots of the wave are superimposed to produce the diagram at right. What is the wavelength?

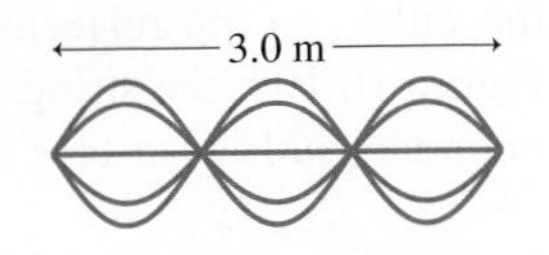

A. 6.0 m B. 4.0 m C. 3.0 m D. 2.0 m E. 1.0 m

16.3 Standing Waves on a String

The oscillation of a guitar string is a standing wave. A standing wave is naturally produced on a string when both ends are fixed (i.e., tied down), as in the case of a guitar string or the string in the photo of Figure 16.5. We also know that a standing wave is produced when there are two counterpropagating traveling waves. But you don't shake both ends of a guitar string to produce the standing wave! How do we actually get two traveling waves on a string with both ends fixed? Before we can answer this question, we need a brief explanation of what happens when a traveling wave encounters a boundary or a discontinuity.

Reflections

We know that light reflects from mirrors; it can also reflect from the surface of a pond or from a pane of glass. As we saw in Chapter 15, sound waves reflect as well; that's how an echo is produced. To understand reflections, we'll look at waves on a string, but the results can be applied to other waves as well.

Suppose we have a string that is attached to a wall or other fixed support, as in FIGURE 16.9. The wall is what we will call a *boundary*—it's the end of the medium. When the pulse reaches this boundary, it reflects, moving away from the wall. *All* the wave's energy is reflected; hence **the amplitude of a wave reflected from a boundary is unchanged.** Figure 16.9 shows that the amplitude doesn't change when the pulse reflects, but the pulse is inverted.

FIGURE 16.9 A wave reflects when it encounters a boundary.

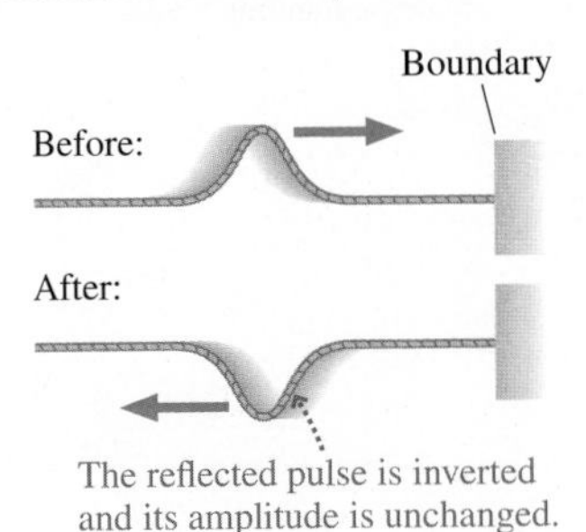

Through the glass darkly A piece of window glass is a discontinuity to a light wave, so it both transmits and reflects light. The left photo shows the windows in a brightly lit room at night. The small percentage of the interior light that reflects from windows is more intense than the transmitted light from outside, so the reflection dominates and the windows show a mirror-like reflection of the room. Turning the room lights out reveals the transmitted light from outside.

Waves also reflect from what we will call a *discontinuity,* a point where there is a change in the properties of the medium. FIGURE 16.10a shows a discontinuity where a string with a larger linear density on the left connects with a string with a smaller linear density on the right. The tension is the same in both strings, so the wave speed is slower on the left, faster on the right. Whenever a wave encounters a discontinuity, some of the wave's energy is *transmitted* forward and some is *reflected* backward. Because energy must be conserved, both the transmitted and the reflected pulses have a smaller amplitude than the initial pulse in this case.

FIGURE 16.10 The reflection of a wave at a discontinuity.

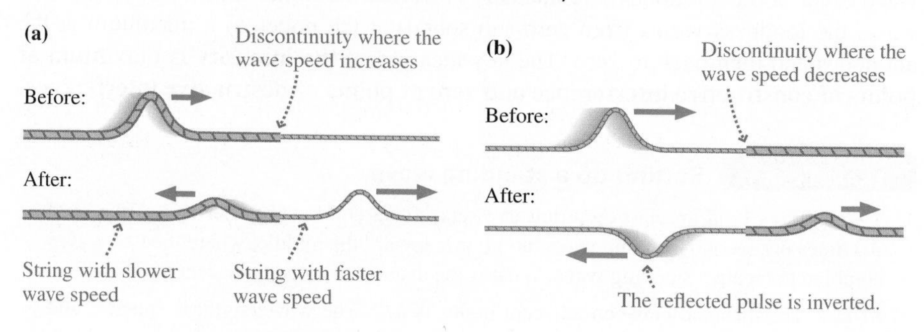

In FIGURE 16.10b, an incident wave encounters a discontinuity at which the wave speed decreases. Once again, some of the wave's energy is transmitted and some is reflected.

In Figure 16.10a, the reflected pulse is right-side up. The string on the right is light and provides little resistance, so the junction moves up and down as the pulse passes. This motion of the string is like the original "snap" of the string that started the pulse, so the reflected pulse has the same orientation as the original pulse. In Figure 16.10b, the string on the right is more massive, so it looks more like the fixed boundary in Figure 16.9 and, as in that situation, the reflected pulse is inverted.

Creating a Standing Wave

FIGURE 16.11 Reflections at the two boundaries cause a standing wave on the string.

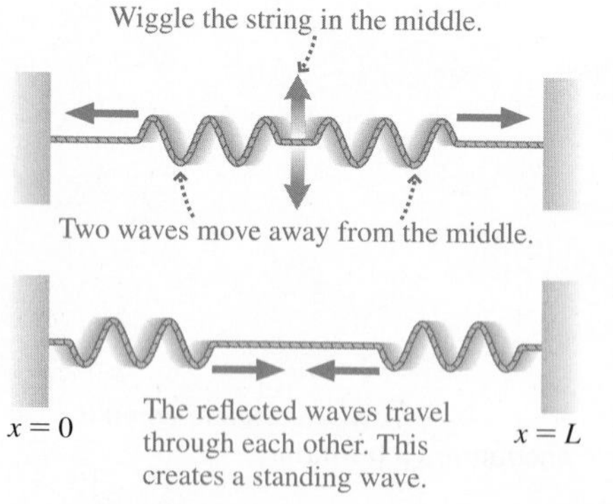

Now that we understand reflections, let's look at how to create a standing wave. FIGURE 16.11 shows a string of length L that is tied at $x = 0$ and $x = L$. This string has *two* boundaries where reflections can occur. If you wiggle the string in the middle, sinusoidal waves travel outward in both directions and soon reach the boundaries, where they reflect. The reflections at the ends of the string cause two waves of *equal amplitude and wavelength* to travel in opposite directions along the string. As we've just seen, these are the conditions that cause a standing wave!

What kind of standing waves might develop on the string? There are two conditions that must be met:

- Because the string is fixed at the ends, the displacements at $x = 0$ and $x = L$ must be zero at all times. Stated another way, we require nodes at both ends of the string.
- We know that standing waves have a spacing of $\lambda/2$ between nodes. This means that the nodes must be equally spaced.

FIGURE 16.12 shows the first three possible waves that meet these conditions. These are called the standing-wave **modes** of the string. To help quantify the possible waves, we can assign a **mode number** m to each. The first wave in Figure 16.12, with a node at each end, has mode number $m = 1$. The next wave is $m = 2$, and so on.

NOTE ▸ Figure 16.12 shows only the first three modes, for $m = 1$, $m = 2$, and $m = 3$. But there are many more modes, for all possible values of m. ◂

The distance between adjacent nodes is $\lambda/2$, so the different modes have different wavelengths. For the first mode in Figure 16.12, the distance between nodes is the length of the string, so we can write

$$\lambda_1 = 2L$$

The subscript identifies the mode number; in this case $m = 1$. For $m = 2$, the distance between nodes is $L/2$; this means that $\lambda_2 = L$. Generally, for any mode m the wavelength is given by the equation

$$\lambda_m = \frac{2L}{m} \qquad m = 1, 2, 3, 4, \cdots \tag{16.1}$$

Wavelengths of standing-wave modes of a string of length L

FIGURE 16.12 The first three possible standing waves on a string of length L.

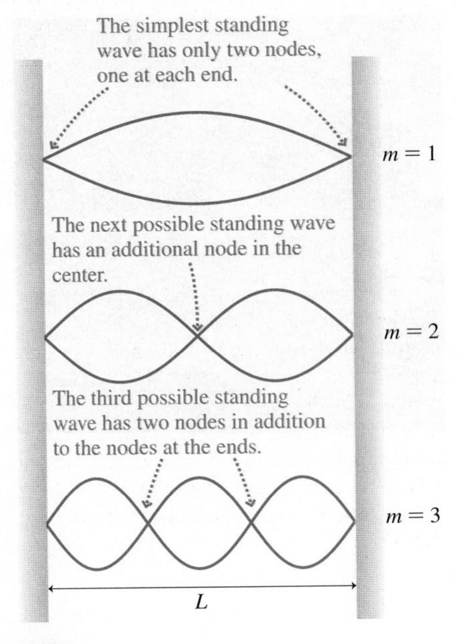
The simplest standing wave has only two nodes, one at each end.
m = 1
The next possible standing wave has an additional node in the center.
m = 2
The third possible standing wave has two nodes in addition to the nodes at the ends.
m = 3
L

Video Figure 16.12

These are the only possible wavelengths for standing waves on the string. **A standing wave can exist on the string *only* if its wavelength is one of the values given by Equation 16.1.**

NOTE ▶ Other wavelengths, which would be perfectly acceptable wavelengths for a traveling wave, cannot exist as a *standing* wave of length L because they do not meet the constraint of having a node at each end of the string. ◀

If standing waves are possible for only certain wavelengths, then only specific oscillation frequencies are allowed. Because $\lambda f = v$ for a sinusoidal wave, the oscillation frequency corresponding to wavelength λ_m is

$$f_m = \frac{v}{\lambda_m} = \frac{v}{2L/m} = m\left(\frac{v}{2L}\right) \qquad m = 1, 2, 3, 4, \ldots \tag{16.2}$$

Frequencies of standing-wave modes of a string of length L

You can tell by looking at a particular standing wave what mode number it corresponds to. FIGURE 16.13 shows the first three standing-wave modes for a wave on a string fixed at both ends, with mode numbers, wavelengths, and frequencies labeled. **The mode number m is equal to the number of antinodes of the standing wave,** so you can determine the mode by counting the number of antinodes (*not* the number of nodes). Once you know the mode number, you can use the formulas for wavelength and frequency to determine these values.

KEY CONCEPT FIGURE 16.13 Possible standing-wave modes of a string fixed at both ends.

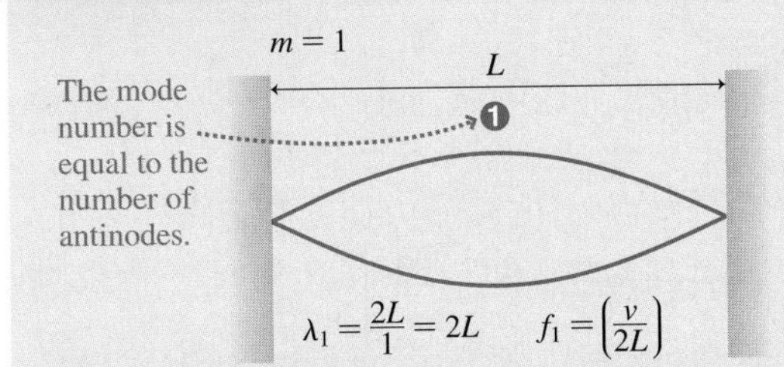
m = 1
L
The mode number is equal to the number of antinodes.
1
$\lambda_1 = \frac{2L}{1} = 2L$
$f_1 = \left(\frac{v}{2L}\right)$

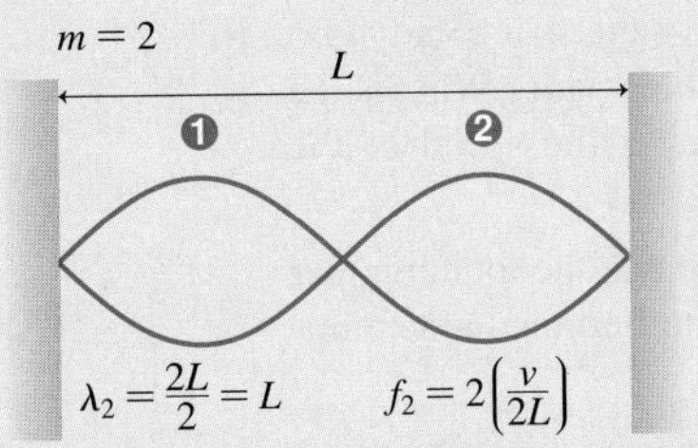
m = 2
L
1
2
$\lambda_2 = \frac{2L}{2} = L$
$f_2 = 2\left(\frac{v}{2L}\right)$

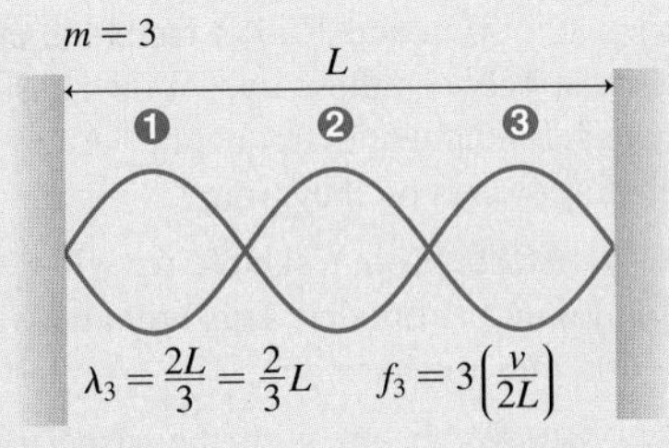
m = 3
L
1
2
3
$\lambda_3 = \frac{2L}{3} = \frac{2}{3}L$
$f_3 = 3\left(\frac{v}{2L}\right)$

STOP TO THINK 16.3 A 2.0-m-long string carries a standing wave as in the figure at right. Extend the pattern and the formulas shown in Figure 16.13 to determine the mode number and the wavelength of this particular standing-wave mode.

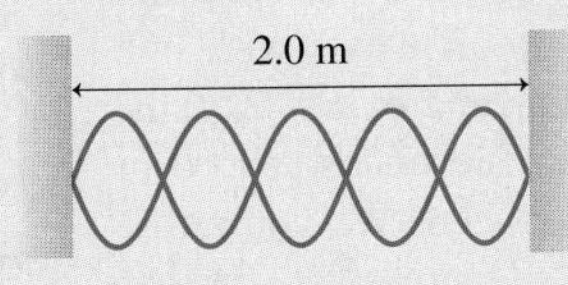
2.0 m

A. $m = 6, \lambda = 0.67$ m B. $m = 6, \lambda = 0.80$ m C. $m = 5, \lambda = 0.80$ m
D. $m = 5, \lambda = 1.0$ m E. $m = 4, \lambda = 0.80$ m F. $m = 4, \lambda = 1.0$ m

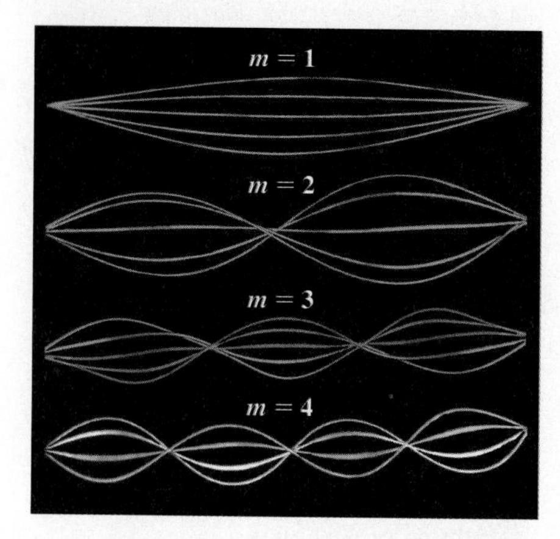

FIGURE 16.14 Resonant modes of a stretched string.

In ◀◀ SECTION 14.7, we looked at the concept of *resonance*. A mass on a spring has a certain frequency at which it "wants" to oscillate. If the system is driven at its resonance frequency, it will develop a large amplitude of oscillation. A stretched string will support standing waves, meaning it has a series of frequencies at which it "wants" to oscillate: the frequencies of the different standing-wave modes. We can call these **resonant modes,** or more simply, **resonances.** A small oscillation of a stretched string at a frequency near one of its resonant modes will cause it to develop a standing wave with a large amplitude. FIGURE 16.14 shows photographs of the first four standing-wave modes on a string, corresponding to four different driving frequencies.

NOTE ▶ When we draw standing-wave modes, as in Figure 16.13, we usually show only the *envelope* of the wave, the greatest extent of the motion of the string. The string's motion is actually continuous and goes through all intermediate positions as well, as we see from the time-exposure photographs of standing waves in Figure 16.14. ◀

The Fundamental Frequency and Higher Harmonics

An inspection of the frequencies of the first three standing-wave modes of a stretched string in Figure 16.13 reveals an interesting pattern. The first mode has frequency

$$f_1 = \frac{v}{2L} \qquad (16.3)$$

f_1 vs. v: PROPORTIONAL p. 39; f_1 vs. L: INVERSE p. 116

We call this the **fundamental frequency** of the string. All of the other modes have frequencies that are multiples of this fundamental frequency. We can rewrite Equation 16.2 in terms of the fundamental frequency as

$$f_m = mf_1 \qquad m = 1, 2, 3, 4, \ldots \qquad (16.4)$$

The allowed standing-wave frequencies are all integer multiples of the fundamental frequency. This sequence of possible frequencies is called a set of **harmonics.** The fundamental frequency f_1 is also known as the *first harmonic,* the $m = 2$ wave at frequency f_2 is called the *second harmonic,* the $m = 3$ wave is called the *third harmonic,* and so on. The frequencies above the fundamental frequency, the harmonics with $m = 2, 3, 4, \ldots$, are referred to as the **higher harmonics.**

EXAMPLE 16.2 Identifying harmonics on a string

A 2.50-m-long string vibrates as a 100 Hz standing wave with nodes at 1.00 m and 1.50 m from one end of the string and at no points in between these two. Which harmonic is this? What is the string's fundamental frequency? And what is the speed of the traveling waves on the string?

STRATEGIZE From a sketch, we will determine the total number of nodes and antinodes. This will allow us to find the mode number and thus the harmonic.

PREPARE We begin with the visual overview in FIGURE 16.15, in which we sketch this particular standing wave and note the known and unknown quantities. We set up an x-axis with one end of the string at $x = 0$ m and the other end at $x = 2.50$ m. The ends of the string are nodes, and there are nodes at 1.00 m and 1.50 m as well, with no nodes in between. We know that standing-wave nodes are equally spaced, so there must be other nodes on the string, as shown in Figure 16.15a. Figure 16.15b is a sketch of the standing-wave mode with this node structure.

FIGURE 16.15 A visual overview of the string.

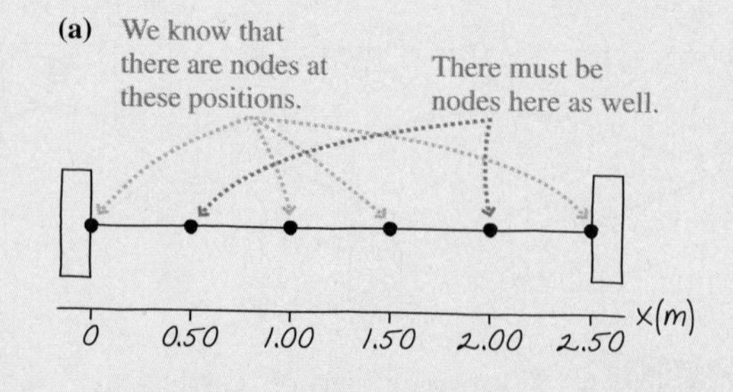

(b) The mode with nodes at these positions looks like this.

Known
$L = 2.50$ m
$f_m = 100$ Hz
Find
m, f_1, v

SOLVE We count the number of antinodes of the standing wave to deduce the mode number; this is mode $m = 5$. This is the fifth harmonic. The frequencies of the harmonics are given by $f_m = mf_1$, so the fundamental frequency is

$$f_1 = \frac{f_5}{5} = \frac{100\text{ Hz}}{5} = 20\text{ Hz}$$

The wavelength of the fundamental mode is $\lambda_1 = 2L = 2(2.50\text{ m}) = 5.00\text{ m}$, so we can find the wave speed using the fundamental relationship for sinusoidal waves:

$$v = \lambda_1 f_1 = (5.00\text{ m})(20\text{ Hz}) = 100\text{ m/s}$$

ASSESS We can calculate the speed of the wave using any possible mode, which gives us a way to check our work. The distance between successive nodes is $\lambda/2$. Figure 16.15 shows that the nodes are spaced by 0.50 m, so the wavelength of the $m = 5$ mode is 1.00 m. The frequency of this mode is 100 Hz, so we calculate

$$v = \lambda_5 f_5 = (1.00\text{ m})(100\text{ Hz}) = 100\text{ m/s}$$

This is the same speed that we calculated earlier, which gives us confidence in our results.

Stringed Musical Instruments

Stringed musical instruments, such as the guitar, the piano, and the violin, all have strings that are fixed at both ends and tightened to create tension. A disturbance is generated on the string by plucking, striking, or bowing. The disturbance creates standing waves on the string.

In Chapter 15, we saw that the speed of a wave on a stretched string depends on T_s, the tension in the string, and μ, the linear density, as $v = \sqrt{T_s/\mu}$. Combining this with Equation 16.3, we find that the fundamental frequency is

$$f_1 = \frac{v}{2L} = \frac{1}{2L}\sqrt{\frac{T_s}{\mu}} \qquad (16.5)$$

Standing waves on a bridge This photo shows the Tacoma Narrows suspension bridge on the day in 1940 when it experienced a catastrophic oscillation that led to its collapse. Aerodynamic forces caused the amplitude of a particular resonant mode of the bridge to increase dramatically until the bridge failed. In this photo, the red line shows the original line of the deck of the bridge. You can clearly see the large amplitude of the oscillation and the node at the center of the span.

When you pluck or bow a string, you initially excite a wide range of frequencies. However, resonance sees to it that the only frequencies that persist are those of the possible standing waves. The string will support a wave of the fundamental frequency f_1 plus waves of all the higher harmonics f_2, f_3, f_4, and so on. Your brain interprets the sound as a musical note of frequency f_1; the higher harmonics determine the *tone quality,* a concept we will explore later in the chapter.

For instruments like the guitar or the violin, the strings are all the same length and under approximately the same tension. The strings have different frequencies because they differ in linear density. The lower-pitched strings are "fat" while the higher-pitched strings are "skinny." This difference changes the frequency by changing the wave speed.

CONCEPTUAL EXAMPLE 16.3 **Tuning and playing a guitar**

Although a guitar's strings are all the same length, a player can produce many different notes by pressing the string against frets, metal bars on the neck of the guitar, as shown in FIGURE 16.16. The fret becomes the new end of the string, making the effective length shorter.

FIGURE 16.16 Guitar frets.

a. A guitar player plucks a string to play a note. He then presses down on a fret to make the string shorter. Does the new note have a higher or lower frequency?

b. The frequency of one string is too low. (Musically, we say the note is "flat.") How must the tension be adjusted to bring the string to the right frequency?

REASON

a. The fundamental frequency, the note we hear, is $f_1 = (1/2L)\sqrt{T_s/\mu}$. Because f_1 is inversely proportional to L, decreasing the string length increases the frequency.

b. Because f is proportional to the square root of T_s, the player must increase the tension to increase the fundamental frequency.

ASSESS If you watch a guitarist, you can see that he plays higher notes by moving his fingers to shorten the strings.

EXAMPLE 16.4 **Setting the tension in a guitar string**

The fifth string on a guitar plays the musical note A, at a frequency of 110 Hz. On a typical guitar, this string is stretched between two fixed points 0.640 m apart, and this length of string has a mass of 2.86 g. What is the tension in the string?

STRATEGIZE Equation 16.5 relates the frequency to the string tension.

PREPARE Strings sound at their fundamental frequency, so 110 Hz is f_1.

Continued

SOLVE The linear density of the string is

$$\mu = \frac{m}{L} = \frac{2.86 \times 10^{-3}\ \text{kg}}{0.640\ \text{m}} = 4.47 \times 10^{-3}\ \text{kg/m}$$

We can rearrange Equation 16.5 for the fundamental frequency to solve for the tension in terms of the other variables:

$$T_s = (2Lf_1)^2\mu = [2(0.640\ \text{m})(110\ \text{Hz})]^2(4.47 \times 10^{-3}\ \text{kg/m}) = 88.6\ \text{N}$$

ASSESS If you have ever strummed a guitar, you know that the tension is quite large, so this result seems reasonable. If each of the guitar's six strings has approximately the same tension, the total force on the neck of the guitar is a bit more than 500 N.

Standing Electromagnetic Waves

FIGURE 16.17 A laser contains a standing light wave between two parallel mirrors.

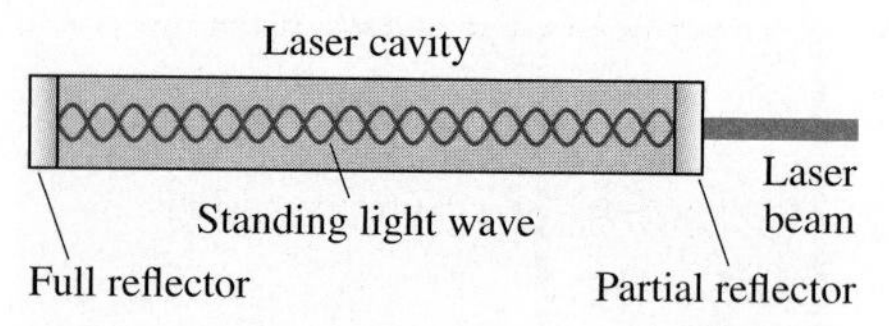

The standing-wave descriptions we've found for a vibrating string are valid for any transverse wave, including an electromagnetic wave. Standing light waves can be established between two parallel mirrors that reflect the light back and forth. The mirrors are boundaries, analogous to the boundaries at the ends of a string. This is how a laser works. The two facing mirrors in FIGURE 16.17 form a *laser cavity*.

Because the mirrors act exactly like the points to which a string is tied, the light wave must have a node at the surface of each mirror. (To allow some of the light to escape the laser cavity and form the *laser beam,* one of the mirrors lets some of the light through. This doesn't affect the node.)

Microwave modes A microwave oven uses microwaves with a wavelength of about 12 cm to heat food. The inside walls of a microwave oven are reflective to microwaves, so we have the correct conditions to set up a standing wave. A standing wave has high intensity at the antinodes and low intensity at the nodes, so your oven has hot spots and cold spots, as we see in this thermal image showing the interior of an oven. A turntable in a microwave oven keeps the food moving so that no part of your dinner remains at a node or an antinode.

EXAMPLE 16.5 Finding the mode number for a laser

A helium-neon laser emits light of wavelength $\lambda = 633$ nm. A typical cavity for such a laser is 15.0 cm long. What is the mode number of the standing wave in this cavity?

STRATEGIZE Just like a standing wave on a string, the standing light wave has nodes at each end, so Equation 16.1 applies.

SOLVE The standing light wave in a laser cavity has a mode number m that is roughly

$$m = \frac{2L}{\lambda} = \frac{2 \times 0.150\ \text{m}}{633 \times 10^{-9}\ \text{m}} = 474{,}000$$

ASSESS The wavelength of light is very short, so we'd expect there to be many nodes within the length of the cavity. A high mode number seems reasonable.

STOP TO THINK 16.4 The illustration to the right shows a standing wave on a string. Which of the modes shown below (on the same string) has twice the frequency of the original wave?

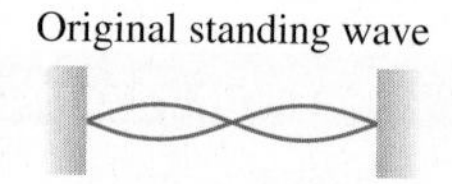

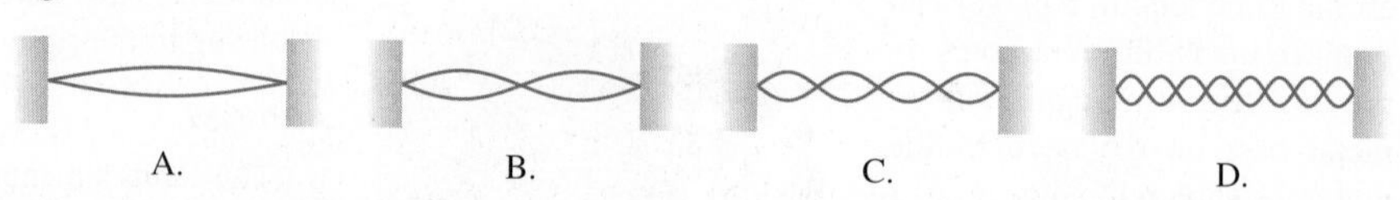

16.4 Standing Sound Waves

Video Standing Sound Waves

Wind instruments like flutes, trumpets, and didgeridoos work very differently from stringed instruments. The player blows air into one end of a tube, producing standing waves of *sound* that make the notes we hear.

You saw in ◀◀ SECTION 15.4 that a sound wave is a longitudinal pressure wave. The air molecules oscillate back and forth parallel to the direction in which the wave is traveling, creating *compressions* (regions of higher pressure) and *rarefactions* (regions of lower pressure). Consider a sound wave confined to a long, narrow column of air, such as the air in a tube. A wave traveling down the tube eventually reaches the end, where it encounters the atmospheric pressure of the surrounding environment. This is a discontinuity, much like the small rope meeting the big rope in Figure 16.10b. Part of the wave's energy is transmitted out into the environment, allowing you to hear the sound, and part is reflected back into the tube. Reflections at both ends of the tube create waves traveling both directions inside the tube, and their superposition, like that of the reflecting waves on a string, is a standing wave.

We start by looking at a sound wave in a tube open at both ends. What kind of standing waves can exist in such a tube? Because the ends of the tube are open to the atmosphere, the pressure at the ends is fixed at atmospheric pressure and cannot vary. This is analogous to a stretched string that is fixed at the end. As a result, **the open end of a column of air must be a node of the pressure wave.**

FIGURE 16.18a shows a column of air open at both ends. We call this an *open-open tube.* Just as the antinodes of a standing wave on a string are points where the string oscillates with maximum displacement, the antinodes of a standing sound wave are where the pressure has the largest variation, creating, alternately, the maximum compression and the maximum rarefaction. In Figure 16.18a, the air molecules squeeze together on the left side of the tube. Then, in FIGURE 16.18b, half a cycle later, the air molecules squeeze together on the right side. The varying density creates a variation in pressure across the tube, as the graphs show. FIGURE 16.18c combines the information of Figures 16.18a and 16.18b into a graph of the pressure of the standing sound wave in the tube.

As the standing wave oscillates, the air molecules "slosh" back and forth along the tube with the wave frequency, alternately pushing together (maximum pressure at the antinode) and pulling apart (minimum pressure at the antinode). This makes sense, because sound is a longitudinal wave in which the air molecules oscillate parallel to the tube.

FIGURE 16.18 The $m = 2$ standing sound wave inside an open-open column of air.

(a) At one instant

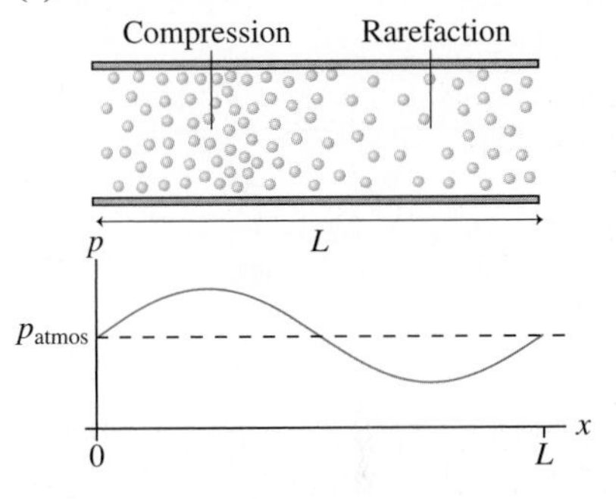

(b) Half a cycle later

The shift between compression and rarefaction means a motion of molecules along the tube.

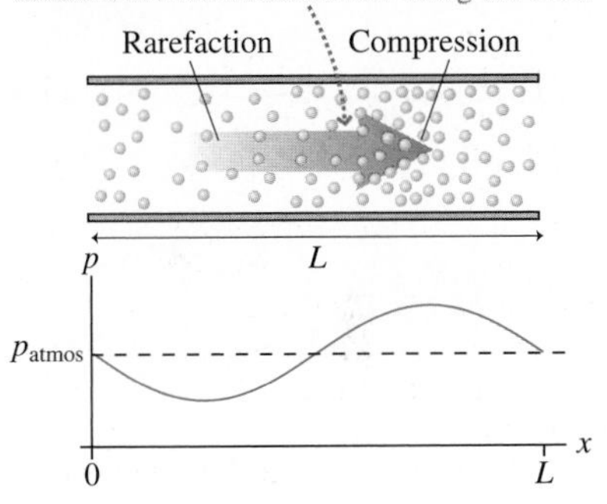

(c) At the ends of the tube, the pressure is equal to atmospheric pressure. These are nodes.

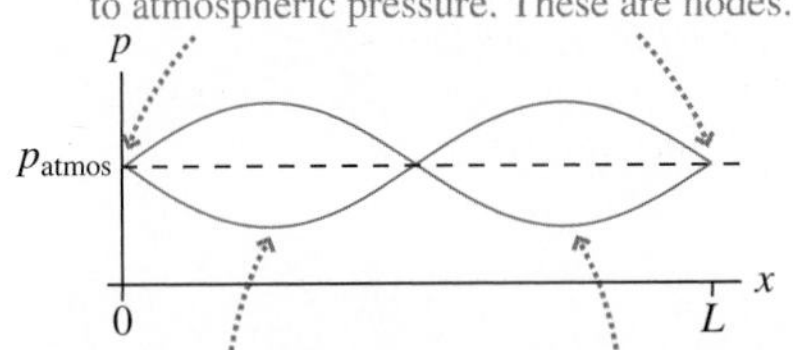

At the antinodes, each cycle sees a change from compression to rarefaction and back to compression.

NOTE ▶ The variation in pressure from atmospheric pressure in a real standing sound wave is much smaller than Figure 16.18 implies. When we display graphs of the pressure in sound waves, we won't generally graph the pressure p. Instead, we will graph Δp, the variation from atmospheric pressure. ◀

Many musical instruments, such as a flute, can be modeled as open-open tubes. The flutist blows across one end to create a standing wave inside the tube, and a note of this frequency is emitted from both ends of the flute. The possible standing waves in tubes, like standing waves on strings, are resonances of the system. A gentle puff of air across the mouthpiece of a flute can cause large standing waves at these resonant frequencies.

Other instruments work differently from the flute. A trumpet or a clarinet is a column of air open at the bell end but *closed* by the player's lips at the mouthpiece. To be complete in our treatment of sound waves in tubes, we need to consider tubes that are closed at one or both ends. At a closed end, the air molecules can alternately rush toward the wall, creating a compression, and then rush away from the wall, leaving a rarefaction. Thus **a closed end of an air column is an antinode of pressure.**

FIGURE 16.19 on the next page shows graphs of the first three standing-wave modes of a tube open at both ends (an open-open tube), a tube closed at both ends (a *closed-closed tube*), and a tube open at one end but closed at the other (an *open-closed tube*), all with the same length L. These are graphs of the pressure wave, with a node at open ends and an antinode at closed ends. The standing wave in the closed-closed tube looks like the wave in the open-open tube except that the positions of the nodes and antinodes are interchanged. In both cases there are m half-wavelength segments between the ends; thus the wavelengths and frequencies of an open-open tube and a closed-closed tube are the same as those of a string tied at both ends:

$$\lambda_m = \frac{2L}{m} \qquad f_m = m\left(\frac{v}{2L}\right) = mf_1 \qquad m = 1, 2, 3, 4, \ldots \tag{16.6}$$

Wavelengths and frequencies of standing sound wave modes in an open-open or closed-closed tube

FIGURE 16.19 The first three standing sound wave modes in columns of air with different ends. These graphs show the pressure variation in the tube.

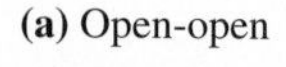
(a) Open-open
The ends of the open-open tube are nodes, so possible modes have a node at each end.
L

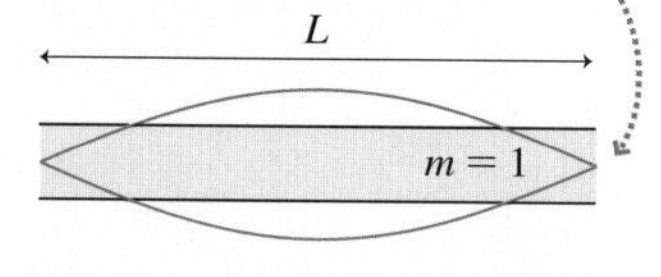
m = 1

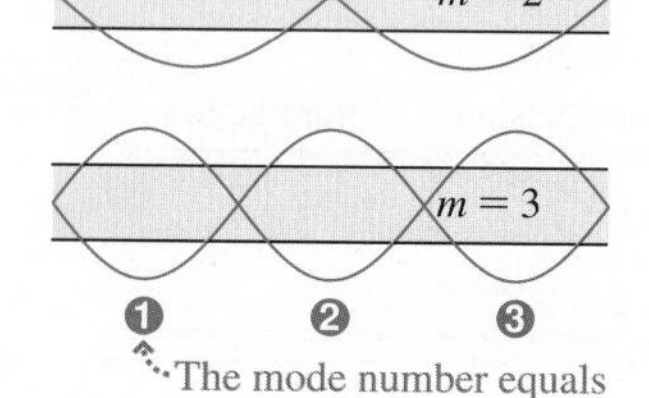
m = 2
m = 3
1
2
3
The mode number equals the number of antinodes.

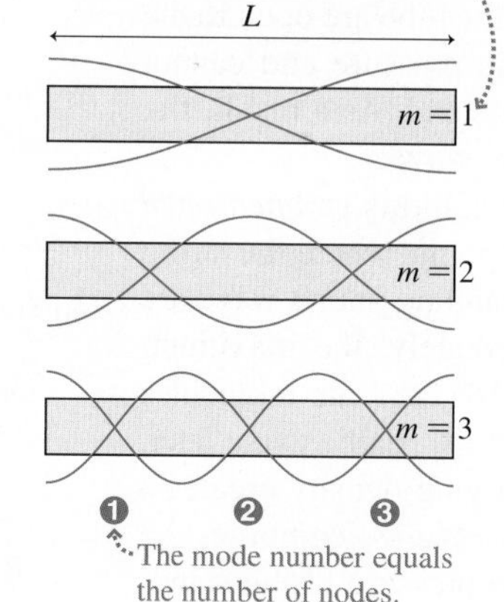
(b) Closed-closed
The ends of the closed-closed tube are antinodes, so possible modes have an antinode at each end.
L
m = 1
m = 2
m = 3
1
2
3
The mode number equals the number of nodes.

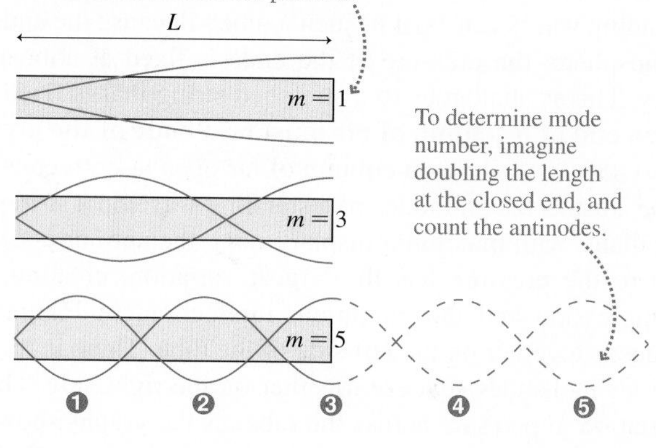
(c) Open-closed
The open-closed tube has a node at one end and an antinode at the other. Only odd-numbered modes are possible.
L
m = 1
m = 3
m = 5
To determine mode number, imagine doubling the length at the closed end, and count the antinodes.
1
2
3
4
5

The open-closed tube is different, as we can see from Figure 16.19c. The $m = 1$ mode has a node at one end and an antinode at the other, and so has only one-quarter of a wavelength in a tube of length L. The $m = 1$ wavelength is $\lambda_1 = 4L$, twice the $m = 1$ wavelength of an open-open or a closed-closed tube. Consequently, **the fundamental frequency of an open-closed tube is half that of an open-open or a closed-closed tube of the same length.**

The wavelength of the next mode of the open-closed tube is $4L/3$. Because this is $1/3$ of λ_1, we assign $m = 3$ to this mode. The wavelength of the subsequent mode is $4L/5$, so this is $m = 5$. In other words, an open-closed tube allows only odd-numbered modes. Consequently, the possible wavelengths and frequencies are

$$\lambda_m = \frac{4L}{m} \qquad f_m = m\left(\frac{v}{4L}\right) = mf_1 \qquad m = 1, 3, 5, 7, \ldots \tag{16.7}$$

Wavelengths and frequencies of standing sound wave modes in an open-closed tube

NOTE ▶ Because sound is a pressure wave, the graphs of Figure 16.19 are *not* "pictures" of the wave as they are for a string wave. The graphs show the pressure variation versus position x. The tube itself is shown merely to indicate the location of the open and closed ends, but the diameter of the tube is *not* related to the amplitude of the wave. ◀

At this point, it's worthwhile to compare and contrast the details we've seen for different types of standing waves. We can also formulate a general strategy for solving standing-wave problems.

◀ **Fiery interference** In this apparatus, a speaker at one end of the metal tube emits a sinusoidal wave. The wave reflects from the other end, which is closed, to make a counter-propagating wave and set up a standing sound wave in the tube. The tube is filled with propane gas that exits through small holes on top. The burning propane allows us to easily discern the nodes and the antinodes of the standing sound wave.

SYNTHESIS 16.1 Standing-wave modes

There are only two sets of frequency/wavelength relationships for standing-wave modes.

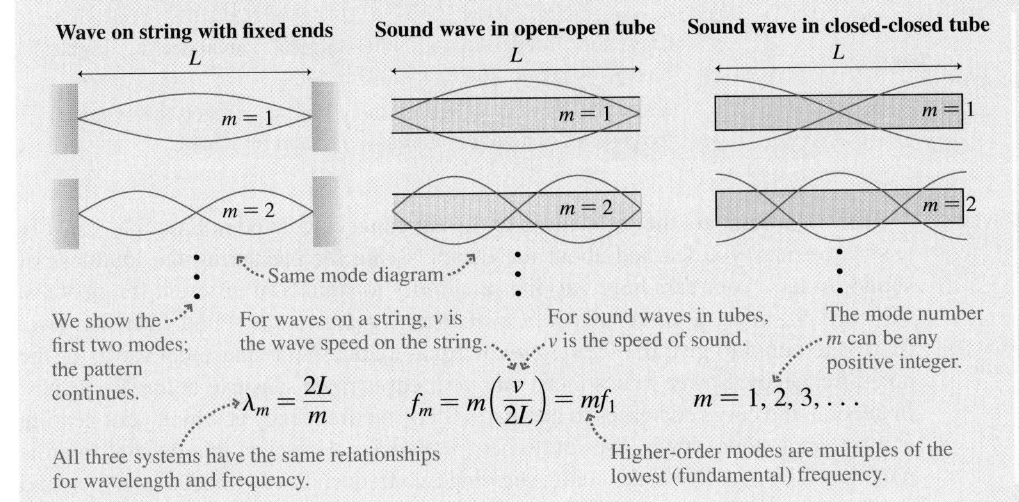

Wave on string with fixed ends
Sound wave in open-open tube
Sound wave in closed-closed tube
L
m = 1
m = 2
Same mode diagram
We show the first two modes; the pattern continues.
For waves on a string, v is the wave speed on the string.
For sound waves in tubes, v is the speed of sound.
The mode number m can be any positive integer.
$\lambda_m = \frac{2L}{m}$
$f_m = m\left(\frac{v}{2L}\right) = mf_1$
$m = 1, 2, 3, \ldots$
All three systems have the same relationships for wavelength and frequency.
Higher-order modes are multiples of the lowest (fundamental) frequency.

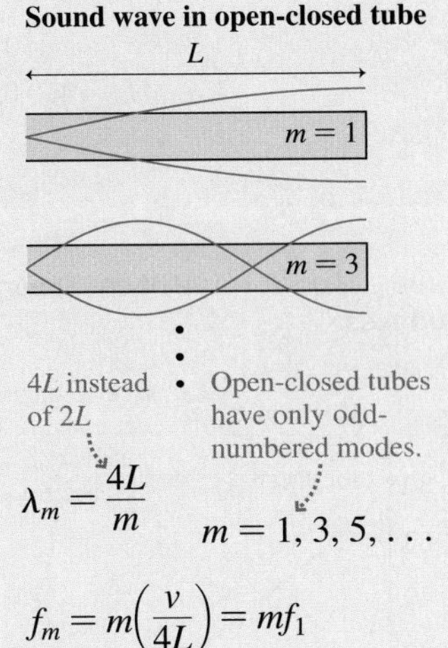

Sound wave in open-closed tube
L
m = 1
m = 3
4L instead of 2L
Open-closed tubes have only odd-numbered modes.
$\lambda_m = \frac{4L}{m}$
$m = 1, 3, 5, \ldots$
$f_m = m\left(\frac{v}{4L}\right) = mf_1$

PROBLEM-SOLVING APPROACH 16.1 Standing waves

We can use the same general approach for any standing wave.

STRATEGIZE Standing waves occur whenever a wave is confined between two boundaries. In this case, only certain modes, with particular wavelengths and frequencies, can exist.

PREPARE

- For sound waves, determine what sort of pipe or tube you have: open-open, closed-closed, or open-closed.
- For string or light waves, the ends will be fixed points.
- For other types of standing waves, such as electromagnetic or water waves, the mode diagram will be similar to one of the cases outlined in Synthesis 16.1. You can then work by analogy with waves on strings or sound waves in tubes.
- Determine known values: length of the tube or string, frequency, wavelength, positions of nodes or antinodes.
- It may be useful to sketch a visual overview, including a picture of the relevant mode.

SOLVE Once you have determined the mode diagram, you can use the appropriate set of equations in Synthesis 16.1 to find the wavelength and frequency. These equations work for any type of wave.

ASSESS Does your final answer seem reasonable? Is there another way to check on your results? For example, the frequency times the wavelength for any mode should equal the wave speed—you can check to see that it does.

EXAMPLE 16.6 Resonances of the ear canal BIO

The eardrum, which transmits vibrations to the sensory organs of your ear, lies at the end of the ear canal. As FIGURE 16.20 shows, the ear canal in adults is about 2.5 cm in length. What frequency standing waves can occur within the ear canal that are within the range of human hearing? The speed of sound in the warm air of the ear canal is 350 m/s.

STRATEGIZE The ear canal acts as a tube that supports standing sound waves.

PREPARE We proceed according to the steps in Problem-Solving Approach 16.1. We can treat the ear canal as an open-closed tube: open to the atmosphere at the external end, closed by the eardrum at the other end. The possible standing-wave modes appear as in Figure 16.19c. The length of the tube is 2.5 cm.

FIGURE 16.20 The anatomy of the ear.

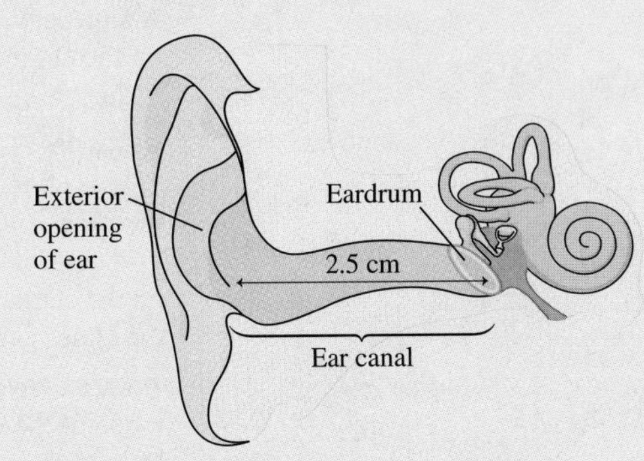

Exterior opening of ear
Eardrum
2.5 cm
Ear canal

Continued

SOLVE Equation 16.7 gives the allowed frequencies in an open-closed tube. We are looking for the frequencies in the range 20 Hz–20,000 Hz. The fundamental frequency is

$$f_1 = \frac{v}{4L} = \frac{350\text{ m/s}}{4(0.025\text{ m})} = 3500\text{ Hz}$$

The higher harmonics are odd multiples of this frequency:

$$f_3 = 3(3500\text{ Hz}) = 10{,}500\text{ Hz}$$
$$f_5 = 5(3500\text{ Hz}) = 17{,}500\text{ Hz}$$

These three modes lie within the range of human hearing; higher modes are greater than 20,000 Hz.

ASSESS The ear canal is short, so we expect the resonant frequencies to be high; our answers seem reasonable.

FIGURE 16.21 A curve of equal perceived loudness.

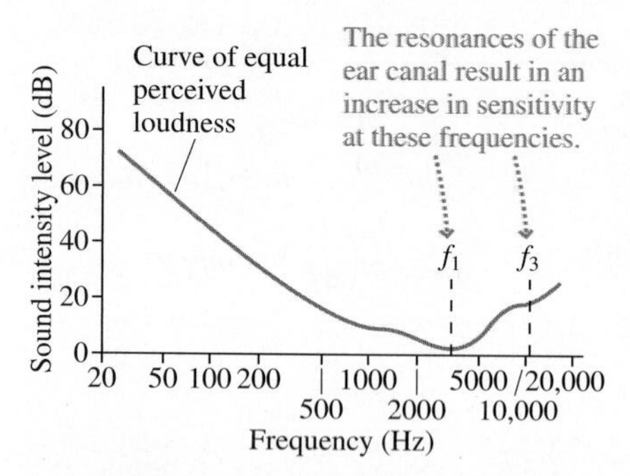

How important are the resonances of the ear canal calculated in Example 16.6? In ◀◀ SECTION 15.6, you learned about the decibel scale for measuring the loudness of sound. In fact, your ears have varying sensitivity to sounds of different frequencies. FIGURE 16.21 shows a curve of *equal perceived loudness,* the sound intensity level (in dB) required to give the *impression* of equal loudness for sinusoidal waves of the noted frequency. Lower values mean that your ear is more sensitive at that frequency. In general, the curve decreases to about 1000 Hz, the frequency at which your hearing is most acute, then slowly rises at higher frequencies. However, this general trend is punctuated by two dips in the curve, showing two frequencies at which a quieter sound produces the same perceived loudness. As you can see, these two dips correspond to the resonances f_1 and f_3 of the ear canal. Incoming sounds at these frequencies produce a larger oscillation, resulting in an increased sensitivity to these frequencies.

Wind Instruments

With a wind instrument, blowing into the mouthpiece creates a standing sound wave inside a tube of air. The player changes the notes by using her fingers to cover holes or open valves, changing the effective length of the tube. The first open hole becomes a node because the tube is open to the atmosphere at that point. The fact that the holes are on the side, rather than literally at the end, makes very little difference. The length of the tube determines the standing-wave resonances, and thus the musical note that the instrument produces.

Many wind instruments have a "buzzer" at one end of the tube, such as a vibrating reed on a saxophone or clarinet, or the musician's vibrating lips on a trumpet or trombone. Buzzers like these generate a continuous range of frequencies rather than single notes, which is why they sound like a "squawk" if you play on just the mouthpiece without the rest of the instrument. When the buzzer is connected to the body of the instrument, most of those frequencies cause little response. But the frequencies from the buzzer that match the resonant frequencies of the instrument cause the buildup of large amplitudes at these frequencies—standing-wave resonances. The combination of these frequencies makes the musical note that we hear.

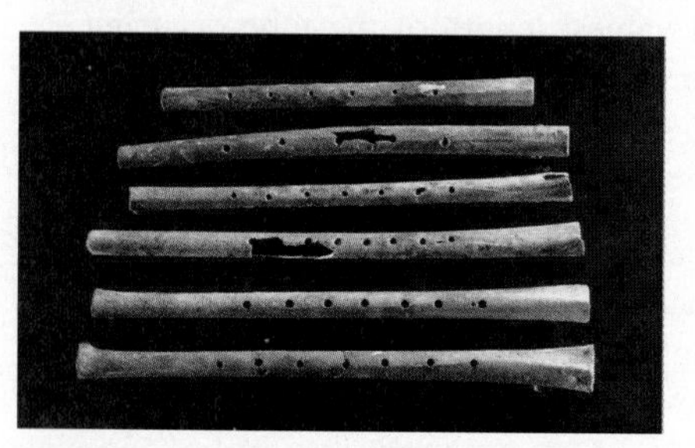

Truly classical music The oldest known musical instruments are bone flutes from burial sites in central China. The flutes in the photo are up to 9000 years old and are made from naturally hollow bones from crowned cranes. The positions of the holes determine the frequencies that the flutes can produce. Soon after the first flutes were created, the design was standardized so that different flutes would play the same notes—including the notes in the modern Chinese musical scale.

CONCEPTUAL EXAMPLE 16.7 **Comparing the flute and the clarinet**

A flute and a clarinet have about the same length, but the lowest note that can be played on the clarinet is much lower than the lowest note that can be played on the flute. Why is this?

REASON A flute is an open-open tube; the frequency of the fundamental mode is $f_1 = v/2L$. A clarinet is open at one end, but the player's lips and the reed close it at the other end. The clarinet is thus an open-closed tube with a fundamental frequency $f_1 = v/4L$. This is about half the fundamental frequency of the flute, so the lowest note on the clarinet has a much lower pitch. In musical terms, it's about an octave lower than the flute.

ASSESS A quick glance at Synthesis 16.1 shows that the wavelength of the lowest mode of the open-closed tube is longer than that of the open-open tube, so we expect a lower frequency for the clarinet.

A clarinet has a lower pitch than a flute because its lowest mode has a longer wavelength and thus a lower frequency. But the higher harmonics are different as well. An open-open tube like a flute has all of the harmonics; an open-closed tube like a clarinet has only those with odd mode numbers. This gives the two instruments a very different tone quality—it's easy to distinguish the sound of a flute from that of a clarinet. We'll explore this connection between the harmonics an instrument produces and its tone quality in the next section.

EXAMPLE 16.8 The importance of warming up

Wind instruments have an adjustable joint to change the tube length. Players know that they may need to adjust this joint to stay in tune—that is, to stay at the correct frequency. To see why, suppose a "cold" flute plays the note A at 440 Hz when the air temperature is 20°C.

a. How long is the tube? At 20°C, the speed of sound in air is 343 m/s.
b. As the player blows air through the flute, the air inside the instrument warms up. Once the air temperature inside the flute has risen to 32°C, increasing the speed of sound to 350 m/s, what is the frequency?
c. At the higher temperature, how must the length of the tube be changed to bring the frequency back to 440 Hz?

STRATEGIZE A flute is an open-open tube with fundamental frequency $f_1 = v/2L$. We know the speed of sound v and the frequency f_1, so we can find the tube length L. A higher speed of sound will then lead to a higher frequency. Finally, using the higher speed of sound, we can find the new length of the tube that will give the required 440 Hz frequency.

SOLVE

a. At 20°C, the length corresponding to 440 Hz is

$$L = \frac{v}{2f_1} = \frac{343 \text{ m/s}}{2(440 \text{ Hz})} = 0.390 \text{ m}$$

b. As the speed of sound increases, the frequency changes to

$$f_1(\text{at } 32°\text{C}) = \frac{350 \text{ m/s}}{2(0.390 \text{ m})} = 449 \text{ Hz}$$

c. To bring the flute back into tune, the length must be increased to give a frequency of 440 Hz with a speed of 350 m/s. The new length is

$$L = \frac{v}{2f_1} = \frac{350 \text{ m/s}}{2(440 \text{ Hz})} = 0.398 \text{ m}$$

Thus the flute must be increased in length by 8 mm.

ASSESS A small change in the absolute temperature produces a correspondingly small change in the speed of sound. We expect that this will require a small change in length, so our answer makes sense.

STOP TO THINK 16.5 A tube that is open at both ends supports a standing wave with harmonics at 300 Hz and 400 Hz, with no harmonics between. What is the fundamental frequency of this tube?

A. 50 Hz B. 100 Hz C. 150 Hz D. 200 Hz E. 300 Hz

16.5 Speech and Hearing BIO

When you hear a note played on a guitar, it sounds very different from the same note played on a trumpet. And you have perhaps been to a lecture in which the speaker talked at essentially the same pitch the entire time—but you could still understand what was being said. Clearly, there is more to your brain's perception of sound than pitch alone. How do you tell the difference between a guitar and a trumpet? How do you distinguish between an "oo" vowel sound and an "ee" vowel sound at the same pitch?

FIGURE 16.22 The frequency spectrum and a graph of the sound wave of a guitar playing a note with fundamental frequency 262 Hz.

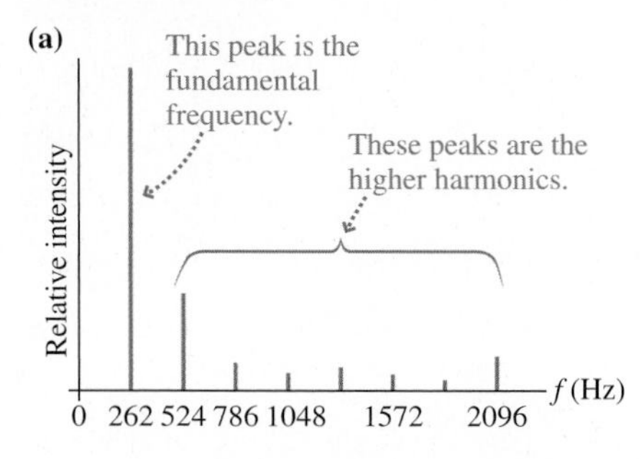
(a)
This peak is the fundamental frequency.
These peaks are the higher harmonics.
Relative intensity
f (Hz)
0 262 524 786 1048 1572 2096

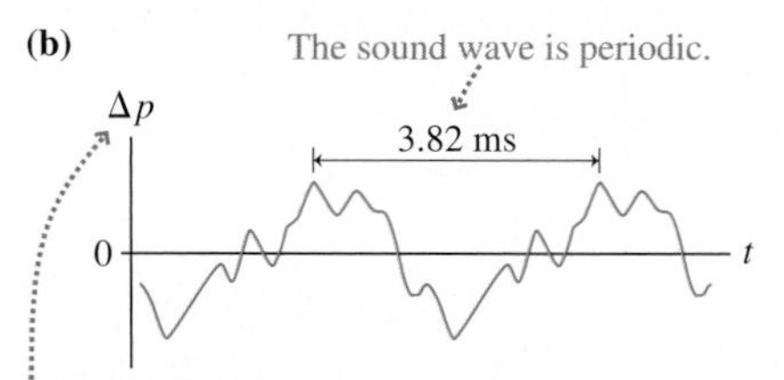
(b)
The sound wave is periodic.
Δp
3.82 ms
0
t
The vertical axis is the change in pressure from atmospheric pressure due to the sound wave.

The Frequency Spectrum

To this point, we have pictured sound waves as sinusoidal waves, with a well-defined frequency. In fact, most of the sounds that you hear are not pure sinusoidal waves. Most sounds are a mix, or superposition, of different frequencies. For example, we have seen how certain standing-wave modes are possible on a stretched string. When you pluck a string on a guitar, you generally don't excite just one standing-wave mode—you simultaneously excite many different modes.

If you play the note "middle C" on a guitar, the fundamental frequency is 262 Hz. There will be a standing wave at this frequency, but there will also be standing waves at the frequencies 524 Hz, 786 Hz, 1048 Hz, . . . , all the higher harmonics predicted by Equation 16.4.

FIGURE 16.22a is a bar chart showing all the frequencies present in the sound of the vibrating guitar string. The height of each bar shows the relative intensity of that harmonic. A bar chart showing the relative intensities of the different frequencies is called the **frequency spectrum** of the sound.

When your brain interprets the mix of frequencies from the guitar in Figure 16.22a, it identifies the fundamental frequency as the *pitch.* 262 Hz corresponds to middle C, so you will identify the pitch as middle C, even though the sound consists of many different frequencies. Your brain uses the higher harmonics to determine the **tone quality,** which is also called the *timbre.* The tone quality—and therefore the higher harmonics—is what makes a middle C played on a guitar sound quite different from a middle C played on a trumpet. The frequency spectrum of a trumpet would show a very different pattern of the relative intensities of the higher harmonics.

The sound wave produced by a guitar playing middle C is shown in FIGURE 16.22b. The sound wave is periodic, with a period of 3.82 ms that corresponds to the 262 Hz fundamental frequency. **The higher harmonics don't change the period of the sound wave; they change only its shape.** The sound wave of a trumpet playing middle C would also have a 3.82 ms period, but its shape would be entirely different.

The didgeridoo, the musical instrument shown at the start of the chapter, consists of a hollow stem or branch. It is an open-closed tube like a clarinet or trumpet. The player blows air through his lips pressed at the end, but he also adds sounds from his vocal cords. This changes the mix of standing-wave modes that are produced, permitting a wide range of sounds.

FIGURE 16.23 The frequency spectrum from the vocal cords, and after passing through the vocal tract.

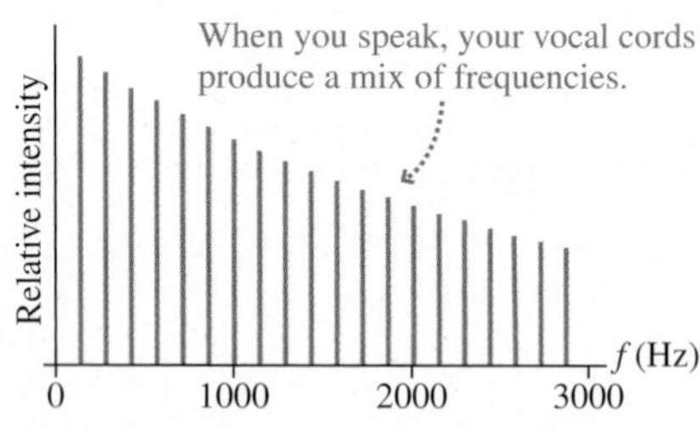
(a) Frequencies from the vocal cords
When you speak, your vocal cords produce a mix of frequencies.
Relative intensity
f (Hz)
0 1000 2000 3000

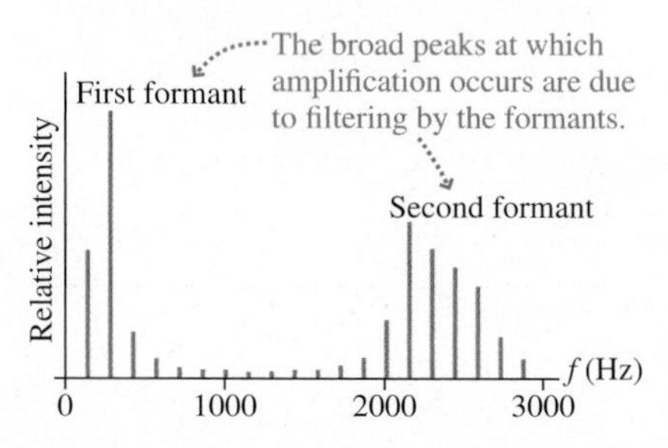
(b) Actual spoken frequencies (vowel sound "ee")
When you form your vocal tract to make a certain vowel sound, it increases the amplitudes of certain frequencies and suppresses others.
The broad peaks at which amplification occurs are due to filtering by the formants.
First formant
Second formant
Relative intensity
f (Hz)
0 1000 2000 3000

Vowels and Formants

Try this: Keep your voice at the same pitch and say the "ee" sound, as in "beet," then the "oo" sound, as in "boot." Pay attention to how you reshape your mouth as you move back and forth between the two sounds. The two vowel sounds are at the same pitch, but they sound quite different. The difference in sound arises from the difference in the higher harmonics, just as for musical instruments. As you speak, you adjust the properties of your vocal tract to produce different mixes of harmonics that make the "ee," "oo," "ah," and other vowel sounds.

Speech begins with the vibration of your vocal cords, stretched bands of tissue in your throat. The vibration is similar to that of a wave on a stretched string. In ordinary speech, the average fundamental frequency for adult males and females is about 150 Hz and 250 Hz, respectively, but you can change the vibration frequency by changing the tension of your vocal cords. That's how you make your voice higher or lower as you sing.

Your vocal cords produce a mix of different frequencies as they vibrate—the fundamental frequency and a rich mixture of higher harmonics. If you put a microphone in your throat and measured the sound waves right at your vocal cords, the frequency spectrum would appear as in FIGURE 16.23a.

There is more to the story though. Before reaching the opening of your mouth, sound from your vocal cords must pass through your vocal tract—a series of hollow cavities including your throat, mouth, and nose. The vocal tract acts like a series of tubes, and, as in any tube, certain frequencies will set up standing-wave resonances.

The rather broad standing-wave resonances of the vocal tract are called **formants.** Harmonics of your vocal cords at or near the formant frequencies will be amplified; harmonics far from a formant will be suppressed. FIGURE 16.23b shows the formants of an adult male making an "ee" sound. The filtering of the vocal cord harmonics by the formants is clear.

You can change the shape and frequencies of the formants, and thus the sound you make, by changing the shape and length of your vocal tract. You do this by changing your mouth opening and the shape and position of your tongue. The first two formants for an "ee" sound are at roughly 270 Hz and 2300 Hz, but for an "oo" sound they are 300 Hz and 870 Hz. The much lower second formant of the "oo" emphasizes midrange frequencies, making a "calming" sound, while the more strident sound of "ee" comes from enhancing the higher frequencies.

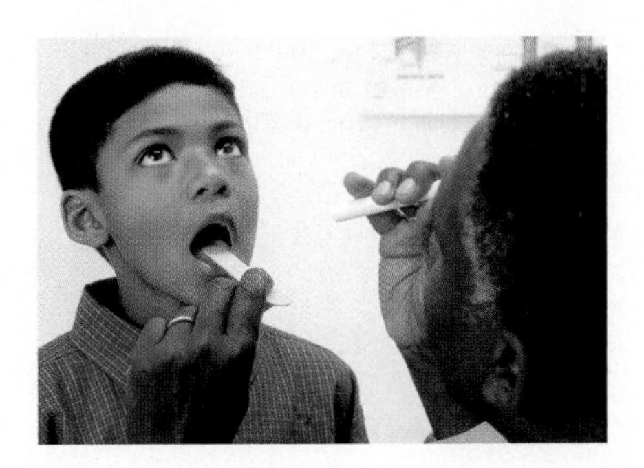

Saying "ah" BIO Why, during a throat exam, does a doctor ask you to say "ah"? This particular vowel sound is formed by opening the mouth and the back of the throat wide—giving a clear view of the tissues of the throat.

CONCEPTUAL EXAMPLE 16.9 High-frequency hearing loss BIO

As you age, your hearing sensitivity will decrease. For most people, the loss of sensitivity is greater for higher frequencies. The loss of sensitivity at high frequencies may make it difficult to understand what others are saying. Why is this?

REASON It is the high-frequency components of speech that allow us to distinguish different vowel sounds. A decrease in sensitivity to these higher frequencies makes it more difficult to make such distinctions.

ASSESS This result makes sense. In Figure 16.23b, the lowest frequency is less than 200 Hz, but the second formant is over 2000 Hz. There's a big difference between what we hear as the pitch of someone's voice and the frequencies we use to interpret speech.

STOP TO THINK 16.6 If you speak at a certain pitch, then hold your nose and continue speaking at the same pitch, your voice sounds very different. This is because

A. The fundamental frequency of your vocal cords has changed.
B. The frequencies of the harmonics of your vocal cords have changed.
C. The pattern of resonance frequencies of your vocal tract has changed, thus changing the formants.

FIGURE 16.24 Interference of waves from two sources.

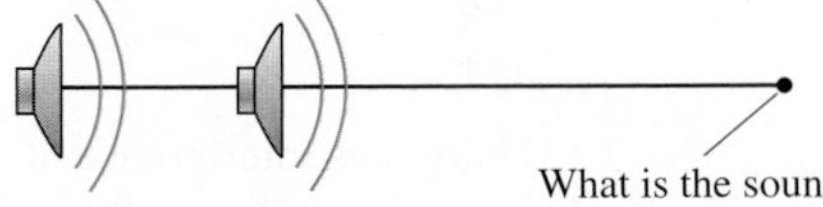
(a) Two sound waves overlapping along a line
What is the sound at this point?
Speaker 2
Speaker 1

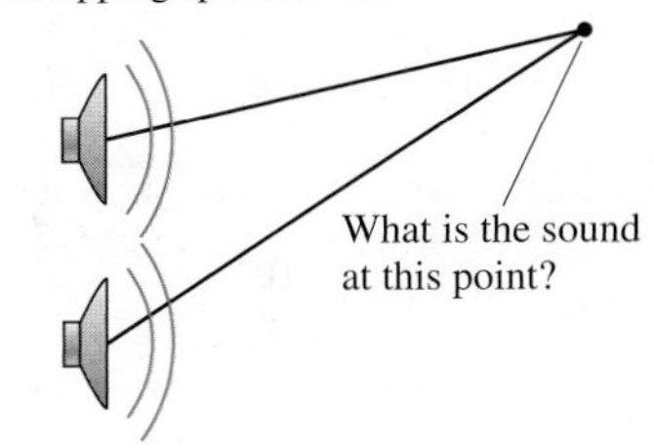
(b) Two overlapping spherical sound waves
What is the sound at this point?

16.6 The Interference of Waves from Two Sources

Perhaps you have seen headphones that offer "active noise reduction." When you turn on the headphones, they produce sound that somehow *cancels* noise from the external environment. How does adding sound to a system make it quieter?

We began the chapter by noting that waves, unlike particles, can pass through each other. Where they do, the principle of superposition tells us that the displacement of the medium is the sum of the displacements due to each wave acting alone. Consider the two loudspeakers in FIGURE 16.24, both emitting sound waves with the same frequency. In Figure 16.24a, sound from loudspeaker 2 passes loudspeaker 1, then two overlapped sound waves travel to the right along the x-axis. What sound is heard at the point indicated with the dot? And what about at the dot in Figure 16.24b, where the speakers are side by side? These are two cases we will consider in this section. Although we'll use sound waves for our discussion, the results apply to all waves.

Interference Along a Line

FIGURE 16.25 shows traveling waves from two loudspeakers spaced exactly one wavelength apart. The graphs are slightly displaced vertically from each other so that you can see what each wave is doing, but the *physical situation* is one in which the waves

FIGURE 16.25 Constructive interference of two waves traveling along the x-axis.

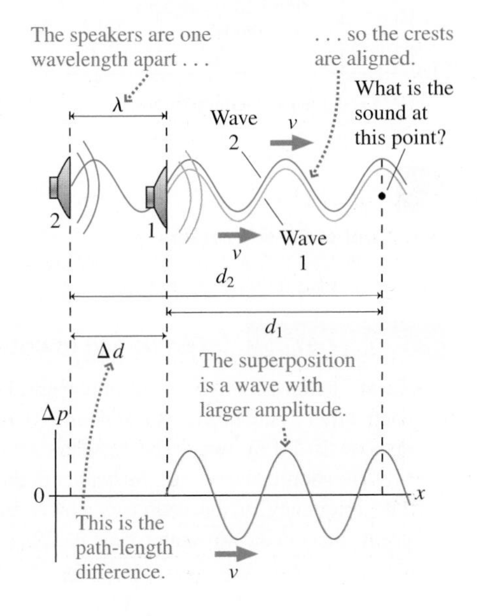
The speakers are one wavelength apart . . .
. . . so the crests are aligned.
What is the sound at this point?
λ
Wave 2
v
2
1
Wave 1
v
d2
d1
Δd
The superposition is a wave with larger amplitude.
Δp
0
x
This is the path-length difference.
v

are traveling *on top of* each other. We assume that the two speakers emit sound waves of identical frequency f, wavelength λ, and amplitude A.

According to the principle of superposition, at every point along the line the net sound pressure wave will be the sum of the pressures from the individual waves. Because the two speakers are separated by one wavelength, the two waves are aligned crest-to-crest and trough-to-trough. Waves aligned this way are said to be **in phase;** waves that are in phase march along "in step" with each other. The superposition is a traveling wave with wavelength λ and twice the amplitude of the individual waves. This is constructive interference.

NOTE ▸ Textbook illustrations like Figure 16.25 can be misleading because they're frozen in time. The net sound wave in Figure 16.25 is a *traveling* wave, moving to the right with the speed of sound v. It differs from the two individual waves only by having twice the amplitude. This is not a standing wave with nodes and antinodes that remain in one place. ◂

If d_1 and d_2 are the distances from loudspeakers 1 and 2 to a point at which we want to know the combined sound wave, their difference $\Delta d = d_2 - d_1$ is called the **path-length difference.** It is the *extra* distance traveled by wave 2 on the way to the point where the two waves are combined. In Figure 16.25, we see that constructive interference results from a path-length difference $\Delta d = \lambda$. But increasing Δd by an additional λ would produce exactly the same result, so we will also have constructive interference for $\Delta d = 2\lambda$, $\Delta d = 3\lambda$, and so on. In other words, **two waves will be in phase and will produce constructive interference any time their path-length difference is a whole number of wavelengths.**

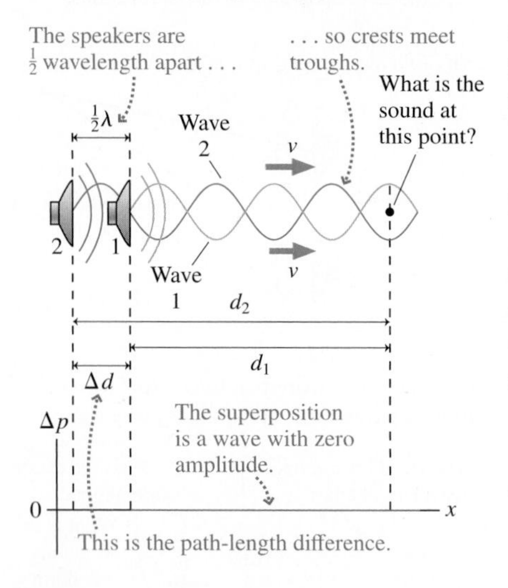

FIGURE 16.26 Destructive interference of two waves traveling along the x-axis.

In FIGURE 16.26, the two speakers are separated by half a wavelength. Now the crests of one wave align with the troughs of the other, and the waves march along "out of step" with each other. We say that the two waves are **out of phase.** When two waves are out of phase, they are equal and opposite at every point. Consequently, the sum of the two waves is zero *at every point.* This is destructive interference.

The destructive interference of Figure 16.26 results from a path-length difference $\Delta d = \frac{1}{2}\lambda$. Again, increasing Δd by an additional λ would produce a picture that looks exactly the same, so we will also have destructive interference for $\Delta d = 1\frac{1}{2}\lambda$, $\Delta d = 2\frac{1}{2}\lambda$, and so on. That is, **two waves will be out of phase and will produce destructive interference any time their path-length difference is a whole number of wavelengths plus half a wavelength.**

In summation, for two identical sources of waves, constructive interference occurs when the path-length difference is

$$\Delta d = m\lambda \qquad m = 0, 1, 2, 3, \ldots \tag{16.8}$$

and destructive interference occurs when the path-length difference is

$$\Delta d = \left(m + \frac{1}{2}\right)\lambda \qquad m = 0, 1, 2, 3, \ldots \tag{16.9}$$

Video Out-of-Phase Speakers

NOTE ▸ The path-length difference needed for constructive or destructive interference depends on the wavelength and hence the frequency. If one particular frequency interferes destructively, another may not. ◂

EXAMPLE 16.10 Interference of sound from two speakers

Susan stands directly in front of two speakers that are in line with each other. The farther speaker is 6.0 m from her; the closer speaker is 5.0 m away. The speakers are connected to the same 680 Hz sound source, and Susan hears the sound loud and clear. The frequency of the source is slowly increased until, at some point, Susan can no longer hear it. What is the frequency when this cancellation occurs? Assume that the speed of sound in air is 340 m/s.

STRATEGIZE Susan hears the sound "loud and clear" at the initial frequency of 680 Hz, which indicates constructive interference with $\Delta d = m\lambda_{\text{initial}}$. From the information given, we can find Δd and λ_{initial} and thus m. Destructive interference will occur when the frequency is increased (and the wavelength decreased) until $\Delta d = (m + \frac{1}{2})\lambda_{final}$; this will determine the final wavelength λ_{final} and hence the final frequency.

PREPARE We start with a visual overview of the situation, as shown in FIGURE 16.27. (The speakers are actually in line with each other, but we've displaced them vertically for clarity.) The sound waves from the two speakers overlap at Susan's position. The path-length difference—the extra distance traveled by the wave from speaker 1—is just the difference in the distances from the speakers to Susan's position. In this case,

$$\Delta d = d_2 - d_1 = 6.0 \text{ m} - 5.0 \text{ m} = 1.0 \text{ m}$$

Initially, with a 680 Hz tone and a 340 m/s sound speed, the wavelength is

$$\lambda_{initial} = \frac{v}{f} = \frac{340 \text{ m/s}}{680 \text{ Hz}} = 0.50 \text{ m}$$

FIGURE 16.27 Visual overview of the loudspeakers.

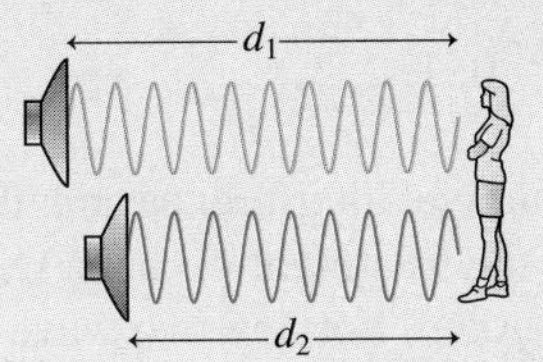
d_1
d_2

Known
$d_1 = 6.0$ m
$d_2 = 5.0$ m
$v = 340$ m/s
Find
f

SOLVE We can find the value of m from

$$m = \frac{\Delta d}{\lambda_{initial}} = \frac{1.0 \text{ m}}{0.50 \text{ m}} = 2$$

We see that m is a whole number, as expected for constructive interference. As the frequency increases and the wavelength decreases, at some point the criterion for destructive interference, $\Delta d = (m + \frac{1}{2})\lambda_{final} = 2.5\lambda_{final}$, will be met. We thus have

$$\lambda_{final} = \frac{\Delta d}{2.5} = \frac{1.0 \text{ m}}{2.5} = 0.40 \text{ m}$$

This corresponds to a frequency of

$$f = \frac{340 \text{ m/s}}{0.40 \text{ m}} = 850 \text{ Hz}$$

ASSESS 850 Hz is an increase of 170 Hz from the original 680 Hz, an increase of one-fourth of the original frequency. This makes sense: Originally, 2 cycles of the wave "fit" in the 1.0 m path-length difference; now, 2.5 cycles "fit," an increase of one-fourth of the original.

Another interesting and important case of interference, illustrated in FIGURE 16.28, occurs when one loudspeaker emits a sound wave that is *the exact inverse* of the wave from the other speaker. If the speakers are side by side, so that $\Delta d = 0$, the superposition of these two waves will result in destructive interference; they will completely cancel. This destructive interference does not require the waves to have any particular frequency or any particular shape.

Headphones with *active noise reduction* use this technique. A microphone on the outside of the headphones measures ambient sound. A circuit inside the headphones produces an inverted version of the microphone signal and sends it to the headphone speakers. The ambient sound and the inverted version of the sound from the speakers arrive at the ears together and interfere destructively, reducing the sound intensity. In this case, *adding* sound results in a *lower* overall intensity inside the headphones!

FIGURE 16.28 Opposite waves cancel.

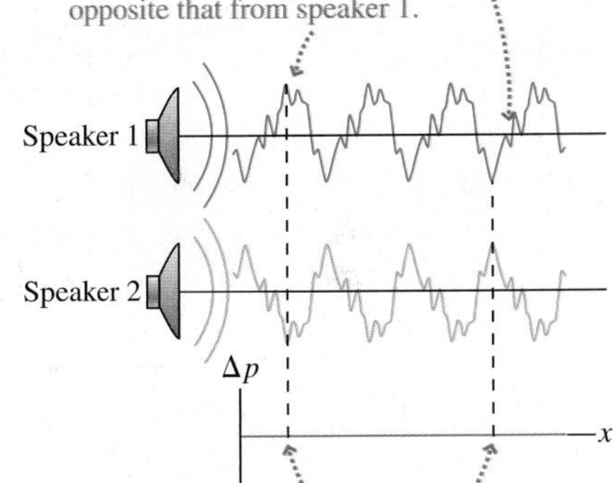
At each position, the wave from speaker 2 is opposite that from speaker 1.
Speaker 1
Speaker 2
Δp
x
The superposition is thus a wave with zero amplitude at all points.

Interference of Spherical Waves

Interference along a line illustrates the idea of interference, but it's not very realistic. In practice, sound waves from a loudspeaker or light waves from a lightbulb spread out as spherical waves. FIGURE 16.29 shows a wave-front diagram for a spherical wave. Recall that the wave fronts represent the *crests* of the wave and are spaced by the wavelength λ. Halfway between two wave fronts is a trough of the wave. What happens when two spherical waves overlap? For example, imagine two loudspeakers emitting identical sound waves radiating in all directions. FIGURE 16.30 on the next page shows the wave fronts of the two waves. This is a static picture, of course, so you have to imagine the wave fronts spreading out as new circular rings are born at the speakers. The waves overlap as they travel, and, as was the case in one dimension, this causes interference.

Consider a particular point like that marked by the red dot in Figure 16.30. The two waves each have a crest at this point, so there is constructive interference here. But at other points, such as that marked by the black dot, a crest overlaps a trough, so this is a point of destructive interference.

FIGURE 16.29 A spherical wave.

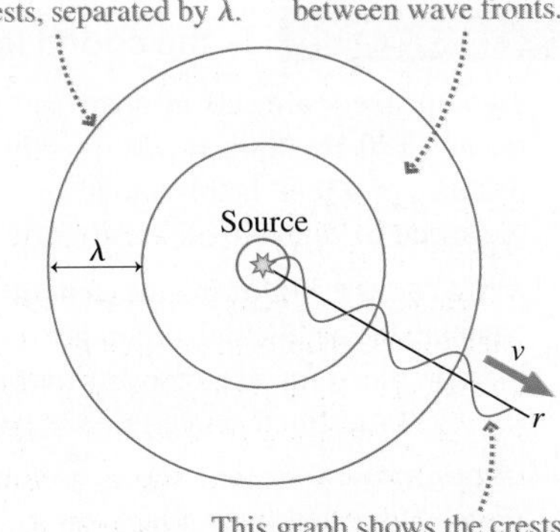
The wave fronts are crests, separated by λ.
Troughs are halfway between wave fronts.
Source
λ
v
r
This graph shows the crests and troughs of the wave.

FIGURE 16.30 The overlapping wave patterns of two sources.

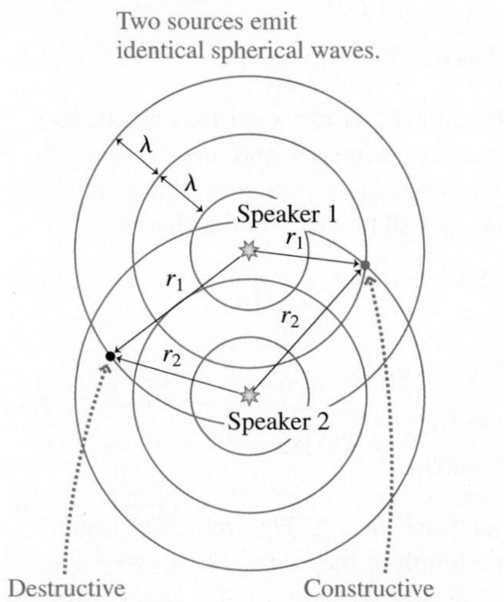

Two sources emit identical spherical waves.

Destructive interference occurs where a crest overlaps a trough.

Constructive interference occurs where two crests overlap.

Notice—simply by counting the wave fronts—that the red dot is three wavelengths from speaker 2 ($r_2 = 3\lambda$) but only two wavelengths from speaker 1 ($r_1 = 2\lambda$). The path-length difference of the two waves arriving at the red dot is $\Delta r = r_2 - r_1 = \lambda$. That is, the wave from speaker 2 has to travel one full wavelength more than the wave from speaker 1, so the waves are in phase (crest aligned with crest) and interfere constructively. You should convince yourself that Δr is a *whole number of wavelengths* at every point where two wave fronts intersect.

Similarly, the path-length difference at the black dot, where the interference is destructive, is $\Delta r = \frac{1}{2}\lambda$. As with interference along a line, destructive interference results when the path-length difference is a whole number of wavelengths plus half a wavelength.

Thus the general rule for determining whether there is constructive or destructive interference at any point is the same for spherical waves as for waves traveling along a line. For identical sources, constructive interference occurs when the path-length difference is

$$\Delta r = m\lambda \qquad m = 0, 1, 2, 3, \ldots \tag{16.10}$$

Destructive interference occurs when the path-length difference is

$$\Delta r = \left(m + \frac{1}{2}\right)\lambda \qquad m = 0, 1, 2, 3, \ldots \tag{16.11}$$

The conditions for constructive and destructive interference are the same for spherical waves as for waves along a line. And the treatment we have seen for sound waves can be applied to any wave, as we have noted. For any two wave sources, the following Tactics Box sums up how to determine whether the interference at a point is constructive or destructive.

TACTICS BOX 16.1 Identifying constructive and destructive interference

1. Identify the path length from each source to the point of interest. Compute the path-length difference $\Delta r = |r_2 - r_1|$.
2. Find the wavelength, if it is not specified.
3. If the path-length difference is a whole number of wavelengths ($\lambda, 2\lambda, 3\lambda, \ldots$), crests are aligned with crests and there is constructive interference.
4. If the path-length difference is a whole number of wavelengths plus a half wavelength ($1\frac{1}{2}\lambda, 2\frac{1}{2}\lambda, 3\frac{1}{2}\lambda, \ldots$), crests are aligned with troughs and there is destructive interference.

Exercises 9, 10 

NOTE ▸ Keep in mind that interference is determined by Δr, the path-length *difference*, not by r_1 or r_2. ◂

EXAMPLE 16.11 Is the sound loud or quiet?

Two speakers are 3.0 m apart and play identical tones of frequency 170 Hz. Sam stands directly in front of one speaker at a distance of 4.0 m. Is this a loud spot or a quiet spot? Assume that the speed of sound in air is 340 m/s.

STRATEGIZE Following the steps of Tactics Box 16.1, we need to compute the path-length difference, $r_2 - r_1$, and determine whether it is of the form $m\lambda$ (constructive interference, loud sound) or $(m + \frac{1}{2})\lambda$ (destructive interference, quiet sound).

PREPARE FIGURE 16.31 shows a visual overview of the situation, showing the positions of and path lengths from each speaker.

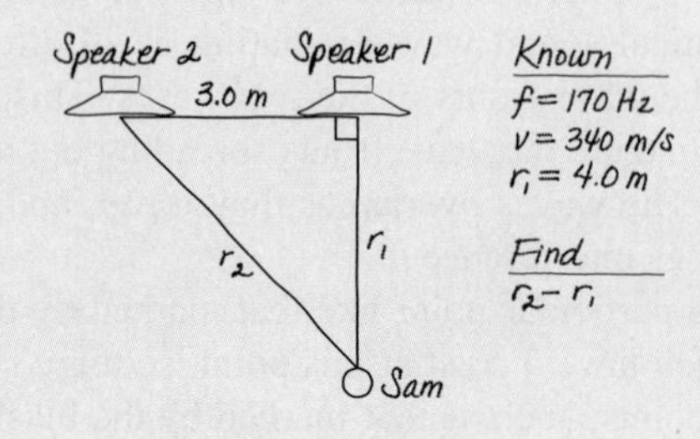

FIGURE 16.31 Visual overview of two speakers.

SOLVE We start by computing the path-length difference. r_1, r_2, and the distance between the speakers form a right triangle, so we can use the Pythagorean theorem to find

$$r_2 = \sqrt{(4.0\text{ m})^2 + (3.0\text{ m})^2} = 5.0\text{ m}$$

Thus the path-length difference is

$$\Delta r = r_2 - r_1 = 1.0\text{ m}$$

Next, we compute the wavelength:

$$\lambda = \frac{v}{f} = \frac{340\text{ m/s}}{170\text{ Hz}} = 2.0\text{ m}$$

The path-length difference is $\frac{1}{2}\lambda$, so this is a point of destructive interference. Sam is at a quiet spot.

You are regularly exposed to sound from two separated sources: stereo speakers. When you walk across a room in which a stereo is playing, why don't you hear a pattern of loud and soft sounds? First, we don't listen to single frequencies. Music is a complex sound wave with many frequencies, but only one frequency at a time satisfies the condition for constructive or destructive interference. Most of the sound frequencies are not affected. Second, reflections of sound waves from walls and furniture make the situation much more complex than the idealized two-source picture in Figure 16.30. Sound wave interference can be heard, but it takes careful selection of a pure tone and a room with no hard, reflecting surfaces. Interference that's rather tricky to demonstrate with sound waves is easy to produce with light waves, as we'll see in the next chapter.

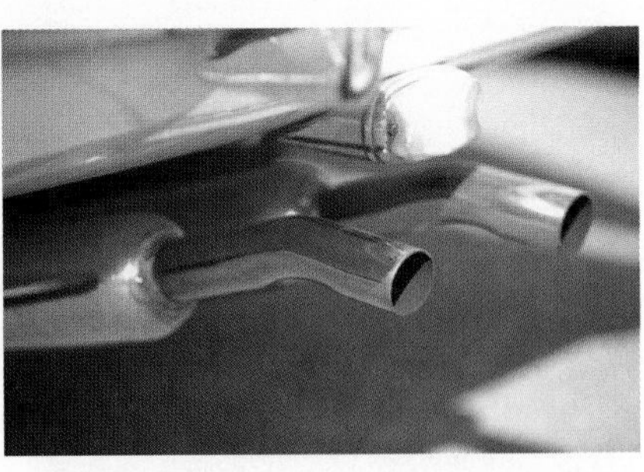

Taming and tuning exhaust noise
The wide section of pipe at the end of this exhaust system is a *resonator*. It's a tube that surrounds the exhaust pipe. One end is open to the exhaust system and the other end (the end seen here) is closed. Sound waves in the exhaust enter, reflect from the closed end, and reenter the exhaust pipe. If the length of the tube is just right, the reflected wave is out of phase with the sound wave in the pipe, producing destructive interference. The resonator is tuned to eliminate the loudest frequencies from the engine, but other frequencies will produce constructive interference, enhancing them and giving the exhaust a certain "note."

STOP TO THINK 16.7 These speakers emit identical sound waves with a wavelength of 1.0 m. At the point indicated, is the interference constructive, destructive, or something in between?

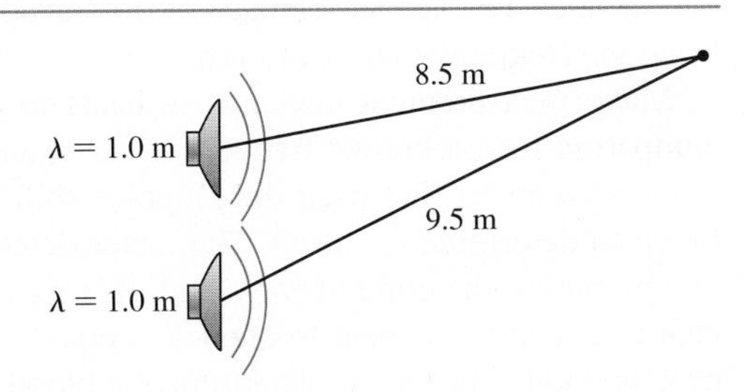
8.5 m
$\lambda = 1.0$ m
9.5 m
$\lambda = 1.0$ m

16.7 Beats

Suppose two sinusoidal waves are traveling toward your ear, as shown in FIGURE 16.32. The two waves have the same amplitude but slightly different frequencies: The orange wave has a slightly higher frequency (and thus a slightly shorter wavelength) than the red wave. This slight difference causes the waves to combine in a manner that alternates between constructive and destructive interference. Their superposition, drawn in blue below the two waves, is a wave whose amplitude shows a

FIGURE 16.32 Creating beats by the superposition of two sound waves with slightly different frequencies.

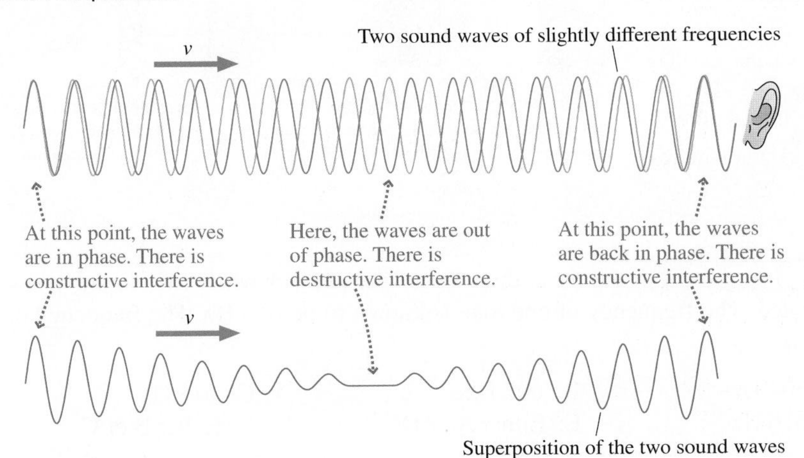
Two sound waves of slightly different frequencies
v
At this point, the waves are in phase. There is constructive interference.
Here, the waves are out of phase. There is destructive interference.
At this point, the waves are back in phase. There is constructive interference.
v
Superposition of the two sound waves

periodic variation. As the waves reach your ear, you will hear a single tone whose intensity is *modulated.* That is, the sound goes up and down in volume, loud, soft, loud, soft, . . . , making a distinctive sound pattern called **beats.**

Suppose the two waves have frequencies f_1 and f_2 that differ only slightly, so that $f_1 \approx f_2$. A complete mathematical analysis would show that the air oscillates against your eardrum at frequency

$$f_{osc} = \frac{1}{2}(f_1 + f_2)$$

This is the *average* of f_1 and f_2, and it differs little from either since the two frequencies are nearly equal. Further, the intensity of the sound is modulated at a frequency called the *beat frequency:*

$$f_{beat} = |f_1 - f_2| \tag{16.12}$$

The beat frequency is the *difference* between the two individual frequencies.

FIGURE 16.33 is a history graph of the wave at the position of your ear. You can see both the sound wave oscillation at frequency f_{osc} and the much slower intensity oscillation at frequency f_{beat}. Frequency f_{osc} determines the pitch you hear, while f_{beat} determines the frequency of the loud-soft-loud modulations of the sound intensity.

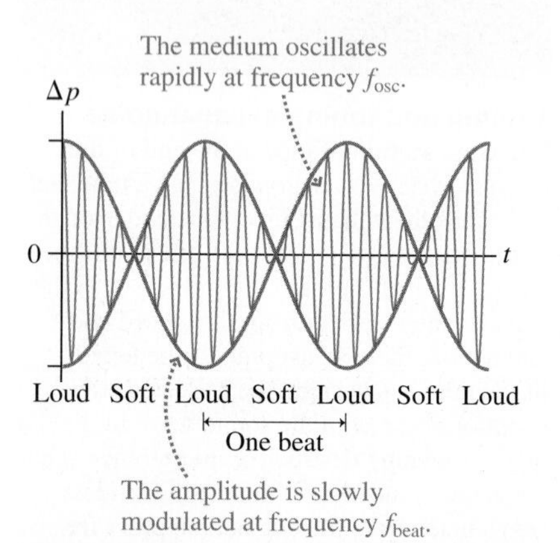
The medium oscillates rapidly at frequency f_{osc}.
Δp
0
t
Loud Soft Loud Soft Loud Soft Loud
One beat
The amplitude is slowly modulated at frequency f_{beat}.

FIGURE 16.33 The modulated sound of beats.

Musicians can use beats to tune their instruments. If one flute is properly tuned at 440 Hz but another plays at 438 Hz, the flutists will hear two loud-soft-loud beats per second. The second flutist is "flat" and needs to shorten her flute slightly to bring the frequency up to 440 Hz.

Many measurement devices use beats to determine an unknown frequency by comparing it to a known frequency. For example, Chapter 15 described a Doppler blood flowmeter that used the Doppler shift of ultrasound reflected from moving blood to determine its speed. The meter determines this very small frequency shift by combining the emitted wave and the reflected wave and measuring the resulting beat frequency. The beat frequency is equal to the shift in frequency on reflection, exactly what is needed to determine the blood speed.

EXAMPLE 16.12 Detecting bats using beats BIO

The little brown bat is a common bat species in North America. It emits echolocation pulses at a frequency of 40 kHz, well above the range of human hearing. To allow observers to "hear" these bats, the bat detector shown in FIGURE 16.34 combines the bat's sound wave at frequency f_1 with a wave of frequency f_2 from a tunable oscillator. The resulting beat frequency is isolated with a filter, then amplified and sent to a loudspeaker. To what frequency should the tunable oscillator be set to produce an audible beat frequency of 3 kHz?

STRATEGIZE The oscillator's frequency needs to be close to that of the bat's pulses, so that the resulting beat frequency is low enough for humans to hear.

SOLVE The beat frequency is $f_{beat} = |f_1 - f_2|$, so the oscillator frequency and the bat frequency need to *differ* by 3 kHz. An oscillator frequency of either 37 kHz or 43 kHz will work nicely.

ASSESS These two frequencies are close to 40 kHz, a requirement to have a clear beat frequency.

FIGURE 16.34 The operation of a bat detector.

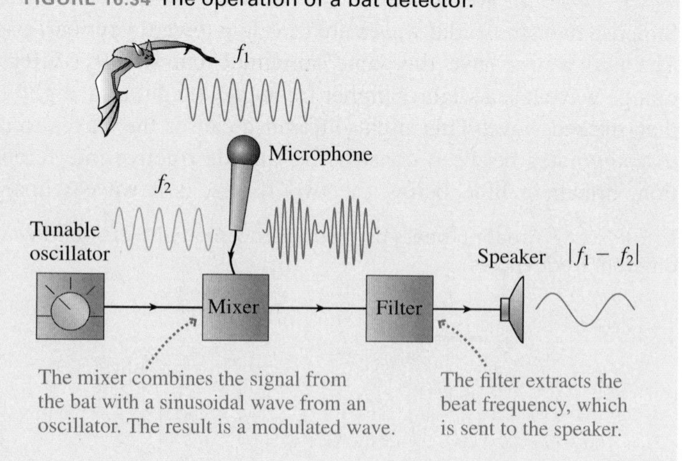
f_1
Microphone
f_2
Tunable oscillator
Speaker $|f_1 - f_2|$
Mixer
Filter
The mixer combines the signal from the bat with a sinusoidal wave from an oscillator. The result is a modulated wave.
The filter extracts the beat frequency, which is sent to the speaker.

STOP TO THINK 16.8 You hear three beats per second when two sound tones are generated. The frequency of one tone is known to be 610 Hz. The frequency of the other is

A. 604 Hz
B. 607 Hz
C. 613 Hz
D. 616 Hz
E. Either A or D
F. Either B or C

INTEGRATED EXAMPLE 16.13 The size of a dog determines the sound of its growl BIO

The sounds of the human vocal system result from the interplay of two different oscillations: the oscillation of the vocal cords and the standing-wave resonances of the vocal tract. A dog's vocalizations are based on similar principles, but the canine vocal tract is simpler than that of a human. When a dog growls or howls, the shape of the vocal tract is essentially a tube closed at the larynx and open at the lips.

All dogs growl at a low pitch because the fundamental frequency of the vocal cords is quite low. But the growls of small dogs and big dogs differ because they have very different formants. The frequency of the formants is determined by the length of the vocal tract, and the vocal-tract length is a pretty good measure of the size of a dog. A larger dog has a longer vocal tract and a correspondingly lower-frequency formant, which sends an important auditory message to other dogs.

The masses of two different dogs and the frequencies of their formants are given in TABLE 16.1.

a. What are the approximate vocal-tract lengths of these two dogs? Assume that the speed of sound is 350 m/s at a dog's body temperature.
b. Growls aren't especially loud; at a distance of 1.0 m, a dog's growl is about 60 dB—the same as normal conversation. What is the acoustic power emitted by a 60 dB growl?
c. The lower-pitched growl of the Doberman will certainly sound more menacing, but which dog's 60 dB growl sounds louder to a human? **Hint:** Assume that most of the acoustic energy is at frequencies near the first formant; then look at the curve of equal perceived loudness in Figure 16.21.

TABLE 16.1 Mass and acoustic data for two growling dogs

Breed	Mass (kg)	First formant (Hz)	Second formant (Hz)
West Highland Terrier (Westie)	8.0	650	1950
Doberman	38	350	1050

STRATEGIZE We will model the dog's vocal tract as an open-closed tube.

PREPARE FIGURE 16.35 shows an idealized model of the vocal tract. The formants correspond to the standing-wave resonances of this system. The first two modes are shown in the figure. The first formant corresponds to $m = 1$; the second formant corresponds to $m = 3$ because an open-closed tube has only odd harmonics. Knowing the frequencies of the formants allows us to determine the length of the tube that produces them.

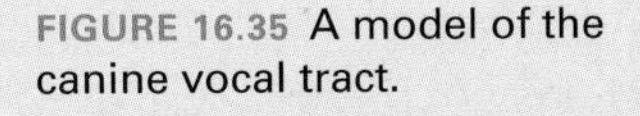

FIGURE 16.35 A model of the canine vocal tract.

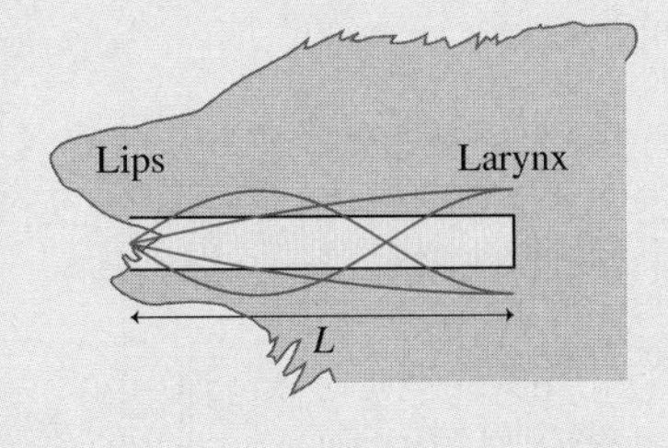

For the question about loudness, we can determine the sound intensity from the sound intensity level. Knowing the distance, we can use this value to determine the emitted acoustic power.

SOLVE

a. We can use the frequency of the first formant to find the length of the vocal tract. The standing-wave frequencies of an open-closed tube are given by

$$f_m = m\frac{v}{4L} \qquad m = 1, 3, 5, \ldots$$

Rearranging this to solve for L, with $m = 1$, we find

$$L(\text{Westie}) = m\frac{v}{4f_m} = (1)\frac{350\text{ m/s}}{4(650\text{ Hz})} = 0.13\text{ m}$$

$$L(\text{Doberman}) = m\frac{v}{4f_m} = (1)\frac{350\text{ m/s}}{4(350\text{ Hz})} = 0.25\text{ m}$$

The length is greater for the Doberman, as we would predict, given the relative masses of the dogs.

b. Equation 15.15 lets us compute the sound intensity for a given sound intensity level:

$$I = (1.0\times 10^{-12}\text{ W/m}^2)10^{(\beta/10\text{ dB})}$$

$$= (1.0\times 10^{-12}\text{ W/m}^2)10^{(60\text{ dB}/10\text{ dB})} = 1.0\times 10^{-6}\text{ W/m}^2$$

This is the intensity at a distance of 1.0 m. The sound spreads out in all directions, so we can use Equation 15.12, $I = P_{\text{source}}/4\pi r^2$, to compute the power of the source:

$$P_{\text{source}} = I\cdot 4\pi r^2 = (1.0\times 10^{-6}\text{ W/m}^2)\cdot 4\pi(1.0\text{ m})^2 = 13\ \mu\text{W}$$

A growl may sound menacing, but that's not because of its power!

c. Figure 16.21 is a curve of equal perceived loudness. The curve steadily decreases until reaching $\approx$3500 Hz, so the acoustic power needed to produce the same sensation of loudness decreases with frequency up to this point. That is, below 3500 Hz your ear is more sensitive to higher frequencies than to lower frequencies. Much of the acoustic power of a dog growl is concentrated at the frequencies of the first two formants. The Westie's growl has its energy at higher frequencies than that of the Doberman, so the Westie's growl will sound louder—though the lower formants of the Doberman's growl will make it sound more formidable.

ASSESS The lengths of the vocal tract that we calculated—13 cm (5 in) for a small terrier and 25 cm (10 in) for a Doberman—seem reasonable given the size of the dogs. We can also check our work by looking at the second formant, corresponding to $m = 3$; using this harmonic in the second expression in Equation 16.7, we find

$$L(\text{Westie}) = m\frac{v}{4f_m} = (3)\frac{350\text{ m/s}}{4(1950\text{ Hz})} = 0.13\text{ m}$$

$$L(\text{Doberman}) = m\frac{v}{4f_m} = (3)\frac{350\text{ m/s}}{4(1050\text{ Hz})} = 0.25\text{ m}$$

This exact match to our earlier calculations gives us confidence in our model and in our results.

Video Harmonics and Voices

SUMMARY

GOAL To use the idea of superposition to understand the phenomena of interference and standing waves.

GENERAL PRINCIPLES

Principle of Superposition

The displacement of a medium when more than one wave is present is the sum of the displacements due to each individual wave.

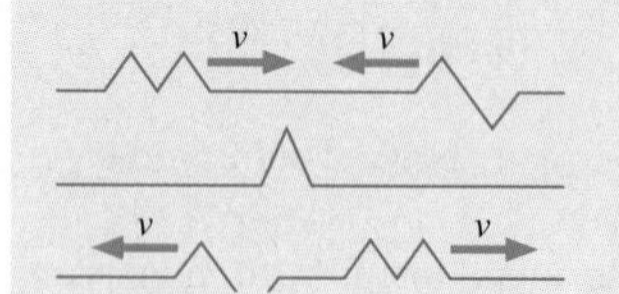

Interference

The superposition of two or more waves into a single wave is called interference.

Constructive interference occurs when crests are aligned with crests and troughs with troughs; the waves are *in phase*. It occurs when the path-length difference Δd is a whole number of wavelengths.

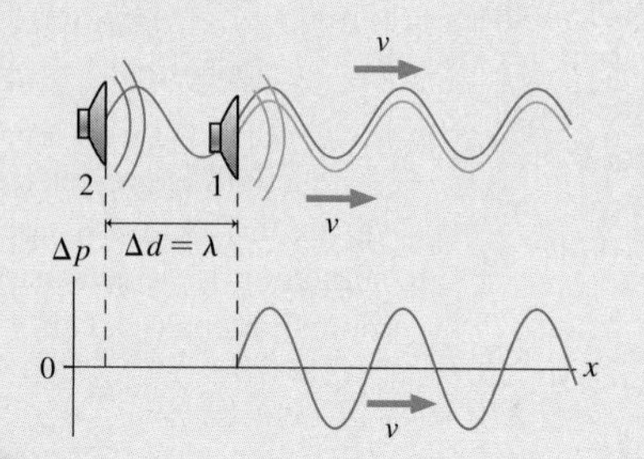

Destructive interference occurs when crests are aligned with troughs; the waves are *out of phase*. It occurs when the path-length difference Δd is a whole number of wavelengths plus half a wavelength.

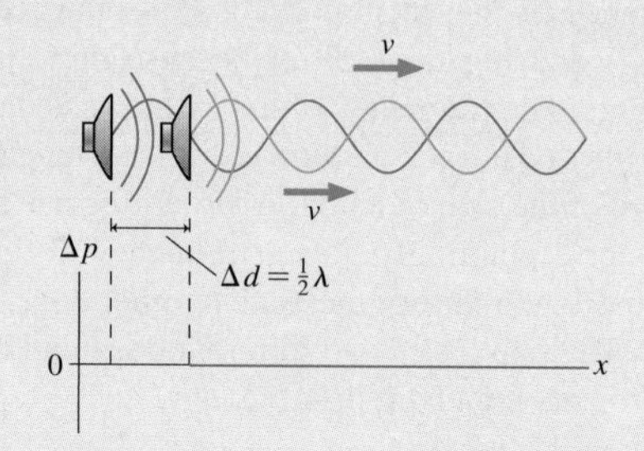

IMPORTANT CONCEPTS

Standing Waves

Two identical traveling waves moving in opposite directions create a standing wave.

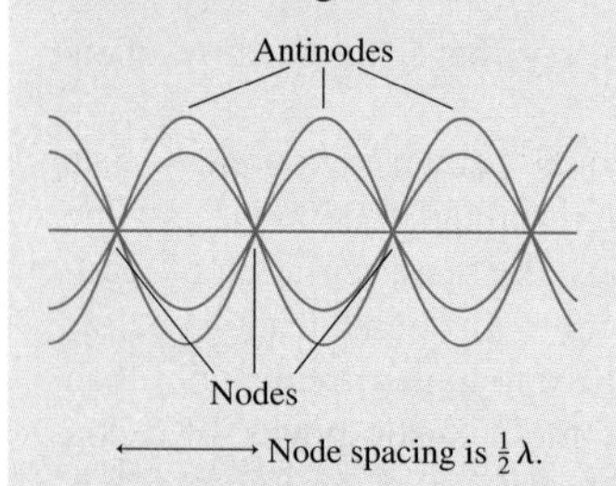

The boundary conditions determine which standing-wave frequencies and wavelengths are allowed. The allowed standing waves are **modes** of the system.

A **standing wave on a string** has a node at each end. Possible modes:

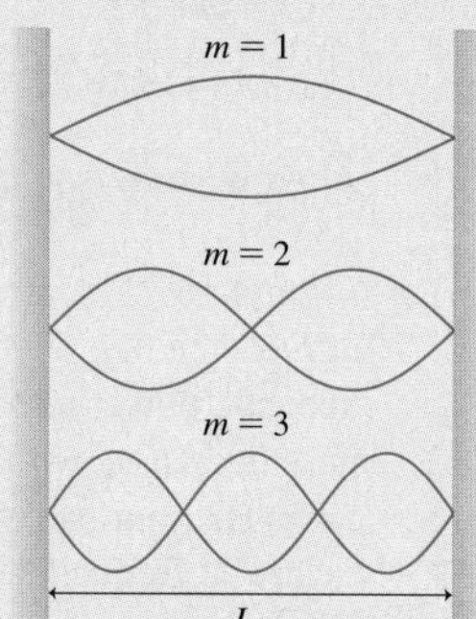

$$\lambda_m = \frac{2L}{m} \qquad f_m = m\left(\frac{v}{2L}\right) = mf_1$$

$m = 1, 2, 3, \ldots$

A **standing sound wave in a tube** can have different boundary conditions: open-open, closed-closed, or open-closed.

Open-open

$$f_m = m\left(\frac{v}{2L}\right)$$

$m = 1, 2, 3, \ldots$

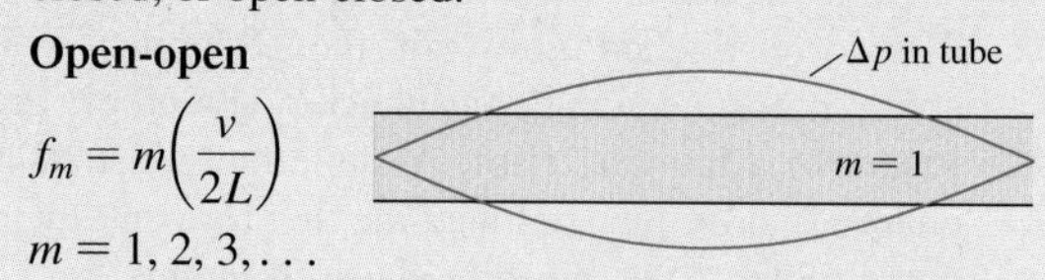

Closed-closed

$$f_m = m\left(\frac{v}{2L}\right)$$

$m = 1, 2, 3, \ldots$

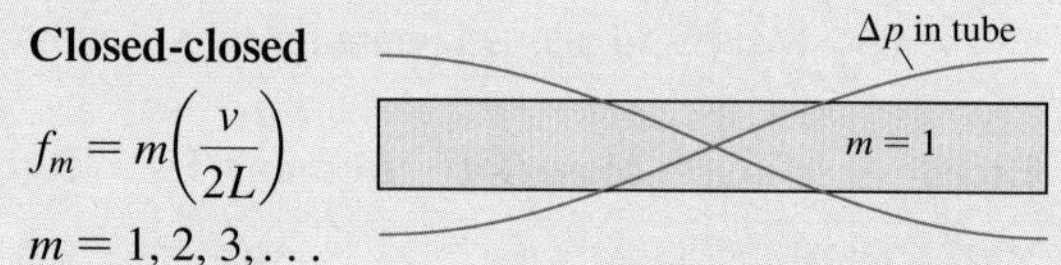

Open-closed

$$f_m = m\left(\frac{v}{4L}\right)$$

$m = 1, 3, 5, \ldots$

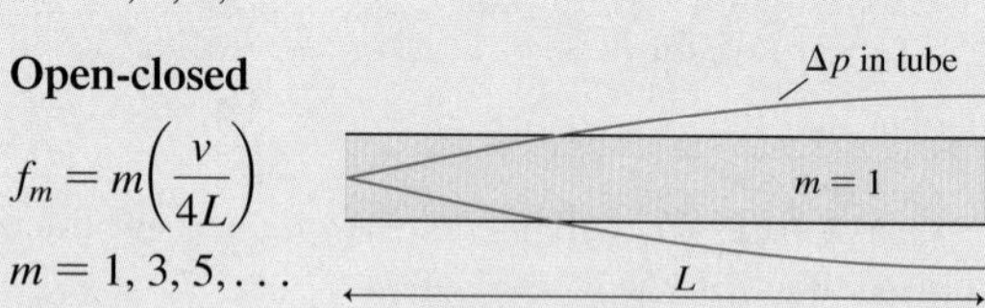

APPLICATIONS

Beats (loud-soft-loud-soft modulations of intensity) are produced when two waves of slightly different frequencies are superimposed.

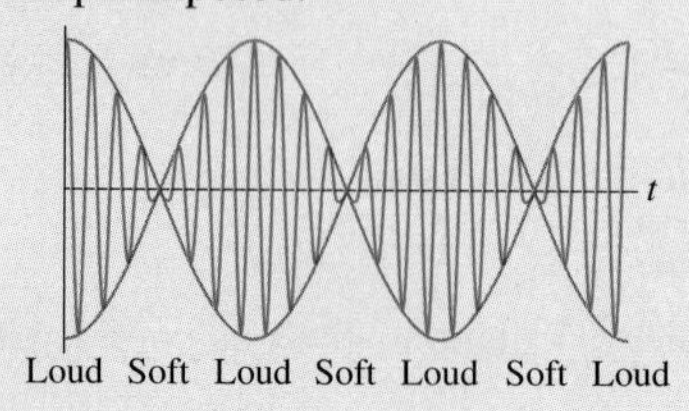

$$f_{\text{beat}} = |f_1 - f_2|$$

Standing waves are multiples of a **fundamental frequency,** the frequency of the lowest mode. The higher modes are the higher **harmonics.**

For sound, the fundamental frequency determines the perceived **pitch;** the higher harmonics determine the **tone quality.**

Our vocal cords create a range of harmonics. The mix of higher harmonics is changed by our vocal tract to create different vowel sounds.

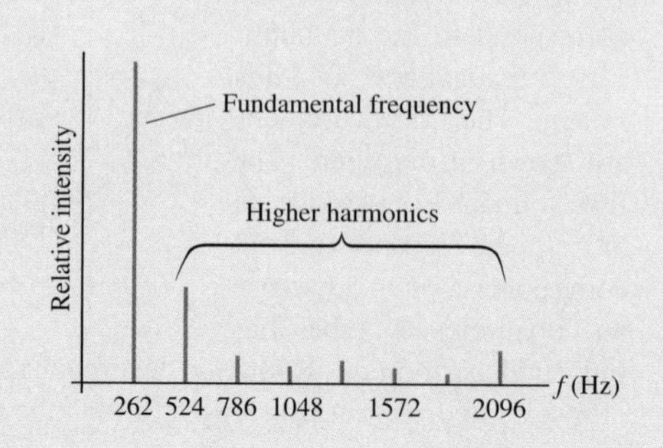

Learning Objectives After studying this chapter, you should be able to:

- Understand and apply the principle of superposition. *Conceptual Question 16.2; Problems 16.1, 16.2, 16.3, 16.4, 16.5*
- Solve problems about standing waves on a string. *Conceptual Questions 16.10, 16.16; Problems 16.8, 16.9, 16.12, 16.15, 16.16*
- Solve problems about standing sound waves. *Conceptual Questions 16.9, 16.13; Problems 16.20, 16.22, 16.23, 16.26, 16.32*
- Apply wave concepts to speech and hearing. *Conceptual Questions 16.14, 16.17; Problems 16.33, 16.34, 16.35, 16.36*
- Calculate the constructive and destructive interference of waves from two sources. *Conceptual Questions 16.3, 16.11; Problems 16.37, 16.40, 16.42, 16.43, 16.44*
- Calculate the beat frequency of two simultaneous waves. *Problems 16.45, 16.46, 16.47, 16.48, 16.49*

STOP TO THINK ANSWERS

Chapter Preview Stop to Think: D. The frequency, wavelength, and speed are related by $v = \lambda f$. The wave speed is the speed of sound, which stays the same, so doubling the frequency means halving the wavelength to 1.0 m.

Stop to Think 16.1: C. The figure shows the two waves at $t = 6$ s and their superposition. The superposition is the *point-by-point* addition of the displacements of the two individual waves.

Stop to Think 16.2: D. There is a node at each end of the string and two nodes in the middle, so the distance between nodes is 1.0 m. The distance between successive nodes is $\lambda/2$. so the wavelength is 2.0 m.

Stop to Think 16.3: C. There are five antinodes, so the mode number is $m = 5$. The equation for the wavelength then gives $\lambda = 2L/m = 2(2.0 \text{ m})/5 = 0.80$ m.

Stop to Think 16.4: C. Standing-wave frequencies are $f_m = mf_1$. The original wave has frequency $f_2 = 2f_1$ because it has two antinodes. The wave with frequency $2f_2 = 4f_1$ is the $m = 4$ mode with four antinodes.

Stop to Think 16.5: B. 300 Hz and 400 Hz are not f_1 and f_2 because 400 Hz $\neq 2 \times 300$ Hz. Instead, both are multiples of the fundamental frequency. Because the difference between them is 100 Hz, we see that $f_3 = 3 \times 100$ Hz and $f_4 = 4 \times 100$ Hz. Thus $f_1 = 100$ Hz.

Stop to Think 16.6: C. Holding your nose does not affect the way your vocal cords vibrate, so it won't change the fundamental frequency or the harmonics of the vocal-cord vibrations. But it will change the frequencies of the formants of the vocal system: The nasal cavities will be open-closed tubes instead of open-open tubes. The frequencies that are amplified are altered, and your voice sounds quite different.

Stop to Think 16.7: Constructive interference. The path-length difference is $\Delta r = 1.0 \text{ m} = \lambda$. Interference is constructive when the path-length difference is a whole number of wavelengths.

Stop to Think 16.8: F. The beat frequency is the difference between the two frequencies.

Video Tutor Solution Chapter 16

QUESTIONS

Conceptual Questions

1. Light can pass easily through water and through air, but light will reflect from the surface of a lake. What does this tell you about the speed of light in air and in water?

2. A small wave pulse and a large wave pulse approach each other on a string; the large pulse is moving to the right. Some time after the pulses have met and passed each other, which of the following statements is correct? (More than one answer may be correct.) Explain.
   A. The large pulse continues unchanged, moving to the right.
   B. The large pulse continues moving to the right but is smaller in amplitude.
   C. The small pulse is reflected and moves off to the right with its original amplitude.
   D. The small pulse is reflected and moves off to the right with a smaller amplitude.
   E. The two pulses combine into a single pulse moving to the right.

Problem difficulty is labeled as | (straightforward) to ||||| (challenging). INT Problems labeled integrate significant material from earlier chapters; BIO are of biological or medical interest.

The eText icon indicates when there is a video tutor solution available for the chapter or for a specific problem. To launch these videos, log into your eText through Mastering™ Physics or log into the Study Area.

3. You are listening to music from a loudspeaker. Then a second speaker is turned on. Is it possible that the music you now hear is quieter than it was with only the first speaker playing? Explain.
4. A guitarist finds that the pitch of one of her strings is slightly flat—the frequency is a bit too low. Should she increase or decrease the tension of the string? Explain.
5. BIO Certain illnesses inflame your vocal cords, causing them to swell. How does this affect the pitch of your voice? Explain.
6. If you're swimming underwater and knock two rocks together, you will hear a very loud noise. But if your friend above the water knocks two rocks together, you'll barely hear the sound. Explain.
7. Figure Q16.7 shows a standing sound wave in a tube of air that is open at both ends.
   a. Which mode (value of m) standing wave is this?
   b. Is the air vibrating horizontally or vertically?

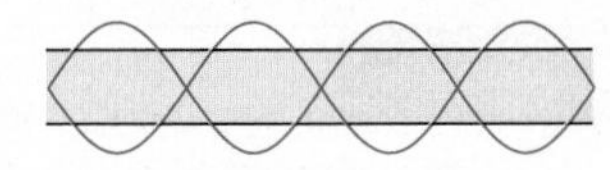

FIGURE Q16.7

8. A typical flute is about 66 cm long. A piccolo is a very similar instrument, though it is smaller, with a length of about 32 cm. How does the pitch of a piccolo compare to that of a flute?
9. Some pipes on a pipe organ are open at both ends, others are closed at one end. For pipes that play low-frequency notes, there is an advantage to using pipes that are closed at one end. What is the advantage?
10. A washtub bass is a simple instrument consisting of a string stretched between a pole and a metal washtub. The musician can play different notes by pulling back more or less on the pole. Explain how this tuning method works.
11. You are standing directly between two loudspeakers playing identical 350 Hz tones. Describe what you will hear if you start walking along a line perpendicular to the line between the speakers.
12. If you pour liquid in a tall, narrow glass, you may hear sound with a steadily rising pitch. What is the source of the sound, and why does the pitch rise as the glass fills?
13. BIO When you speak after breathing helium, in which the speed of sound is much greater than in air, your voice sounds quite different. The frequencies emitted by your vocal cords do not change since they are determined by the mass and tension of your vocal cords. So what *does* change when your vocal tract is filled with helium rather than air?
14. Figure Q16.14 shows frequency spectra of the same note played on a flute (modeled as an open-open tube) and on a clarinet (a closed-open tube). Which figure corresponds to the flute, and which to the clarinet? Explain.

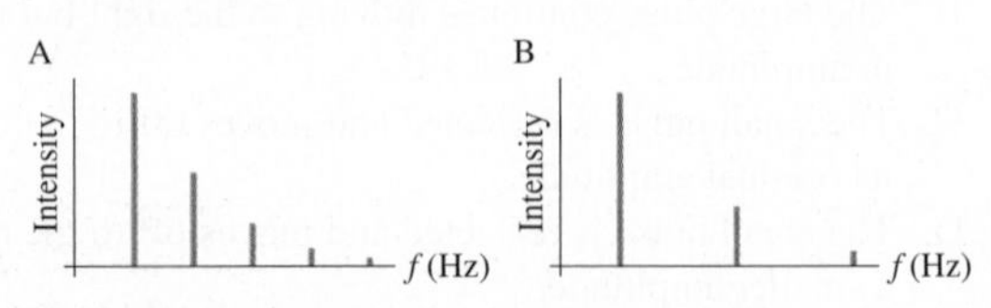

FIGURE Q16.14

15. A synthesizer is a keyboard instrument that can be made to sound like a flute, a trumpet, a piano—or any other musical instrument. Pressing a key sets the fundamental frequency of the note that is produced. Other settings that the player can adjust change the sound quality of the synthesizer without altering the fundamental frequency. How is this change accomplished? What is being adjusted?
16. Figure Q16.16 shows a standing wave on a string. Point A on the string is shown at time $t = 0$. In terms of the wave's period T, what are the earliest times after $t = 0$ that points B, C, and D reach the positions shown?

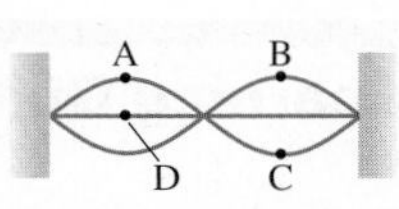

FIGURE Q16.16

17. BIO A small boy and a grown woman both speak at approximately the same pitch. Nonetheless, it's easy to tell which is which from listening to the sounds of their voices. How are you able to make this determination?

Multiple-Choice Questions

Questions 18 through 20 refer to the snapshot graph of Figure Q16.18.

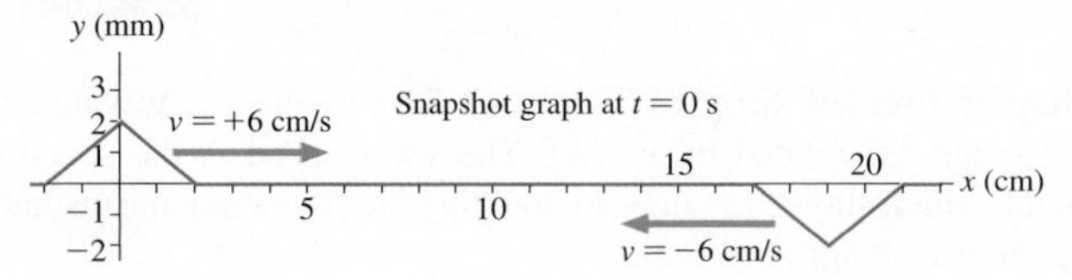

FIGURE Q16.18

18. | At $t = 1$ s, what is the displacement y of the string at $x = 7$ cm?
   A. −1.0 mm B. 0 mm C. 0.5 mm
   D. 1.0 mm E. 2.0 mm
19. | At $x = 3$ cm, what is the earliest time that y will equal 2 mm?
   A. 0.5 s B. 0.7 s C. 1.0 s
   D. 1.5 s E. 2.5 s
20. | At $t = 1.5$ s, what is the value of y at $x = 10$ cm?
   A. −2.0 mm B. −1.0 mm C. −0.5 mm
   D. 0 mm E. 1.0 mm
21. | In a tube, standing-wave modes are found at 200 Hz and 400 Hz. The tube could *not* be
   A. Open-closed. B. Open-open.
   C. Closed-closed. D. It could be any of these.
22. | A student in her physics lab measures the standing-wave modes of a tube. The lowest frequency that makes a resonance is 20 Hz. As the frequency is increased, the next resonance is at 60 Hz. What will be the next resonance after this?
   A. 80 Hz B. 100 Hz C. 120 Hz D. 180 Hz
23. | An organ pipe is tuned to exactly 384 Hz when the temperature in the room is 20°C. Later, when the air has warmed up to 25°C, the frequency is
   A. Greater than 384 Hz. B. 384 Hz.
   C. Less than 384 Hz.
24. | BIO Resonances of the ear canal lead to increased sensitivity of hearing, as we've seen. Dogs have a much longer ear canal—5.2 cm—than humans. What are the two lowest frequencies at which dogs have an increase in sensitivity? The speed of sound in the warm air of the ear is 350 m/s.
   A. 1700 Hz, 3400 Hz B. 1700 Hz, 5100 Hz
   C. 3400 Hz, 6800 Hz D. 3400 Hz, 10,200 Hz
25. | The frequency of the lowest standing-wave mode on a 1.0-m-long string is 20 Hz. What is the wave speed on the string?
   A. 10 m/s B. 20 m/s C. 30 m/s D. 40 m/s
26. | Suppose you pluck a string on a guitar and it produces the note A at a frequency of 440 Hz. Now you press your finger down on the string against one of the frets, making this point the new end of the string. The newly shortened string has 4/5 the length of the full string. When you pluck the string, its frequency will be
   A. 350 Hz B. 440 Hz C. 490 Hz D. 550 Hz

PROBLEMS

Section 16.1 The Principle of Superposition

1. | Figure P16.1 is a snapshot graph at $t = 0$ s of two waves on a taut string approaching each other at 1 m/s. Draw six snapshot graphs, stacked vertically, showing the string at 1 s intervals from $t = 1$ s to $t = 6$ s.

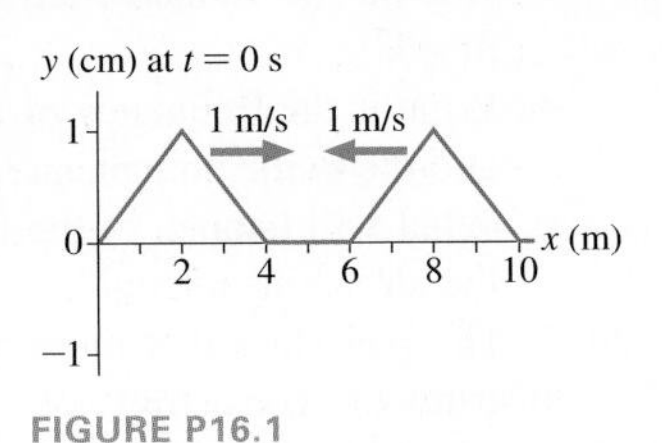

FIGURE P16.1

2. ||| Figure P16.2 is a snapshot graph at $t = 0$ s of two waves approaching each other at 1 m/s. Draw two snapshot graphs, stacked vertically, showing the string at $t = 2$ s and 3 s.

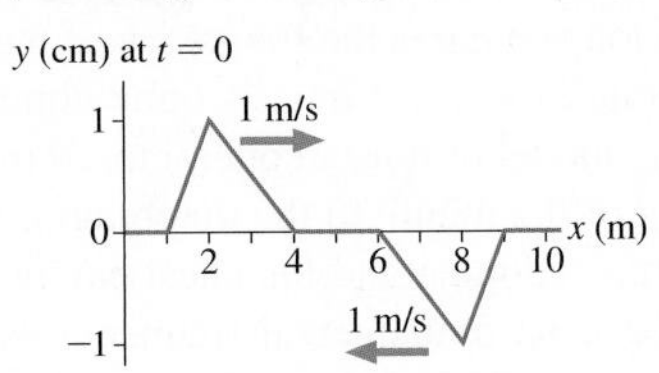

FIGURE P16.2

3. ||| Figure P16.2 is a snapshot graph at $t = 0$ s of two waves approaching each other at 1 m/s. Draw a history graph of the point of the string at $x = 4$ cm.

4. || Figure P16.4a is a snapshot graph at $t = 0$ s of two waves on a string approaching each other at 1 m/s. At what time was the snapshot graph in Figure P16.4b taken?

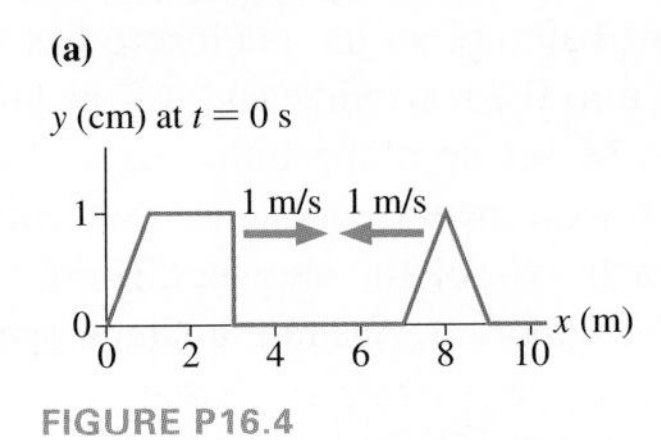

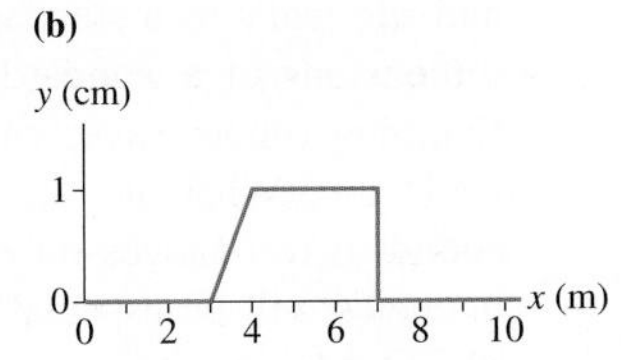

FIGURE P16.4

5. | Figure P16.5 is a snapshot graph at $t = 0$ s of two waves on a string approaching each other at 1 m/s. List the values of the displacement of the string at $x = 5.0$ cm at 1 s intervals from $t = 0$ s to $t = 6$ s.

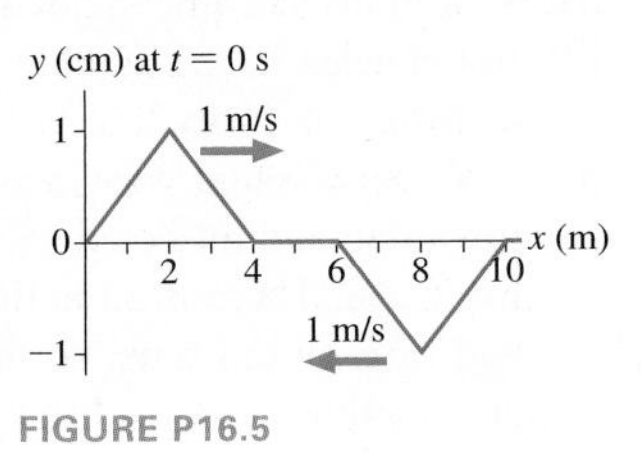

FIGURE P16.5

Section 16.2 Standing Waves

Section 16.3 Standing Waves on a String

6. |||| Figure P16.6 is a snapshot graph at $t = 0$ s of a pulse on a string moving to the right at 1 m/s. The string is fixed at $x = 3$ m. Draw a snapshot graph of the string at time $t = 1.25$ s.

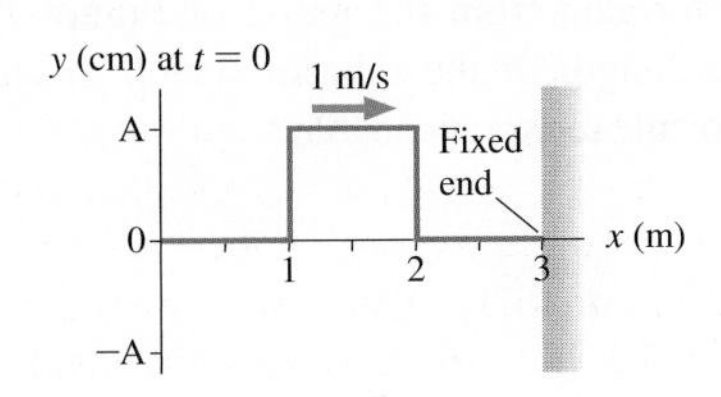

FIGURE P16.6

7. ||| At $t = 0$ s, a small "upward" (positive y) pulse centered at $x = 6.0$ m is moving to the right on a string with fixed ends at $x = 0.0$ m and $x = 10.0$ m. The wave speed on the string is 4.0 m/s. At what time will the string next have the same appearance that it did at $t = 0$ s?

8. ||| You are holding one end of an elastic cord that is fastened to a wall 3.0 m away. You begin shaking the end of the cord at 3.5 Hz, creating a continuous sinusoidal wave of wavelength 1.0 m. How much time will pass until a standing wave fills the entire length of the string?

9. ||| A 2.0-m-long string is fixed at both ends and tightened until the wave speed is 40 m/s. What is the frequency of the standing wave shown in Figure P16.9?

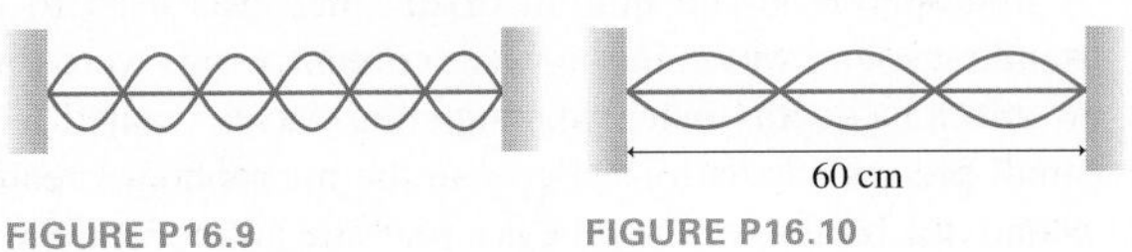

FIGURE P16.9 FIGURE P16.10

10. || Figure P16.10 shows a standing wave oscillating at 100 Hz on a string. What is the wave speed?

11. || A bass guitar string is 89 cm long with a fundamental frequency of 30 Hz. What is the wave speed on this string?

12. || The four strings of a bass guitar are 0.865 m long and are tuned to the notes G (98 Hz), D (73.4 Hz), A (55 Hz), and E (41.2 Hz). In one bass guitar, the G and D strings have a linear mass density of 5.8 g/m, and the A and E strings have a linear mass density of 26.8 g/m. What is the total force exerted by the strings on the neck?

13. || a. What are the three longest wavelengths for standing waves on a 240-cm-long string that is fixed at both ends?
    b. If the frequency of the second-longest wavelength is 50.0 Hz, what is the frequency of the third-longest wavelength?

14. ||| The G string on a guitar is 59 cm long and has a fundamental frequency of 196 Hz. A guitarist can play different notes by pushing the string against various frets, which changes the string's length. The first fret from the neck gives A♭ (207.65 Hz); the second fret gives A (220 Hz). How far apart are the first and second frets?

15. || A guitar string with a linear density of 2.0 g/m is stretched between supports that are 60 cm apart. The string is observed to form a standing wave with three antinodes when driven at a frequency of 420 Hz. What are (a) the frequency of the fifth harmonic of this string and (b) the tension in the string?

16. || Some guitarists like the feel of a set of strings that all have the same tension. For such a guitar, the G string (196 Hz) has a mass density of 0.31 g/m. What is the mass density of the A string (110 Hz)?

17. || The lowest note on a grand piano has a frequency of 27.5 Hz. The entire string is 2.00 m long and has a mass of 400 g. The vibrating section of the string is 1.90 m long. What tension is needed to tune this string properly?

18. | An experimenter finds that standing waves on a string fixed at both ends occur at 24 Hz and 32 Hz, but at no frequencies in between.
    a. What is the fundamental frequency?
    b. Draw the standing-wave pattern for the string at 32 Hz.

Watch Video Solution Problem 16.27

19. || Ocean waves of wavelength 26 m are moving directly toward a concrete barrier wall at 4.4 m/s. The waves reflect from the wall, and the incoming and reflected waves overlap to make a lovely standing wave with an antinode at the wall. (Such waves are a common occurrence in certain places.) A kayaker is bobbing up and down with the water at the first antinode out from the wall. How far from the wall is she? What is the period of her up-and-down motion?

Section 16.4 Standing Sound Waves

20. | The lowest frequency in the audible range is 20 Hz. What are the lengths of (a) the shortest open-open tube and (b) the shortest open-closed tube needed to produce this frequency?
21. |||| To study the physical basis of underwater hearing in frogs, scientists used a vertical tube filled with water to a depth of 1.4 m. A microphone at the bottom of the tube was used to create standing sound waves in the water column. Frogs were lowered to different depths where the standing waves created large or small pressure variations. Because the microphone creates the sound, the bottom of the tube is a pressure antinode; the water's surface, fixed at atmospheric pressure, is a node. BIO
    a. What is the fundamental frequency of this water-filled tube?
    b. A frog sits on a platform located 0.28 m from the bottom. What is the lowest frequency that would result in a sound node at this point?
22. | The vuvuzela is a simple horn, typically 0.65 m long, that fans use to make noise at sporting events. What is the frequency of the fundamental note produced by a vuvuzela?

23. | The world's longest organ pipe, in the Boardwalk Hall Auditorium in Atlantic City, is 64 feet long. What is the fundamental frequency of this open-open pipe?
24. | An organ pipe is made to play a low note at 27.5 Hz, the same as the lowest note on a piano. Assuming a sound speed of 343 m/s, what length open-open pipe is needed? What length open-closed pipe would suffice?
25. || The speed of sound in room temperature (20°C) air is 343 m/s; in room temperature helium, it is 1010 m/s. The fundamental frequency of an open-closed tube is 315 Hz when the tube is filled with air. What is the fundamental frequency if the air is replaced with helium? INT
26. | *Parasaurolophus* was a dinosaur whose distinguishing feature was a hollow crest on the head. The 1.5-m-long hollow tube in the crest had connections to the nose and throat, leading some investigators to hypothesize that the tube was a resonant chamber for vocalization. If you model the tube as an open-closed system, what are the first three resonant frequencies? Assume a speed of sound of 350 m/s. BIO

27. || A drainage pipe running under a freeway is 30.0 m long. Both ends of the pipe are open, and wind blowing across one end causes the air inside to vibrate.
    a. If the speed of sound on a particular day is 340 m/s, what will be the fundamental frequency of air vibration in this pipe?
    b. What is the frequency of the lowest harmonic that would be audible to the human ear?
    c. What will happen to the frequency in the later afternoon as the air begins to cool?
28. | The pan flute is a musical instrument consisting of a number of closed-end tubes of different lengths. When the musician blows over the open ends, each tube plays a different note. The longest pipe is 0.33 m long. What is the frequency of the note it plays?

29. | Although the vocal tract is quite complicated, we can make a simple model of it as an open-closed tube extending from the opening of the mouth to the diaphragm, the large muscle separating the abdomen and the chest cavity. What is the length of this tube if its fundamental frequency equals a typical speech frequency of 200 Hz? Assume a sound speed of 350 m/s. Does this result for the tube length seem reasonable, based on observations on your own body? BIO
30. | You know that you sound better when you sing in the shower. This has to do with the amplification of frequencies that correspond to the standing-wave resonances of the shower enclosure. A shower enclosure is created by adding glass doors and tile walls to a standard bathtub, so the enclosure has the dimensions of a standard tub, 0.75 m wide and 1.5 m long. Standing sound waves can be set up along either axis of the enclosure. What are the lowest two frequencies that correspond to resonances on each axis of the shower? These frequencies will be especially amplified. Assume a sound speed of 343 m/s.
31. |||| A child has an ear canal that is 1.3 cm long. At what sound frequencies in the audible range will the child have increased hearing sensitivity? BIO
32. || When a sound wave travels directly toward a hard wall, the incoming and reflected waves can combine to produce a standing wave. There is an antinode right at the wall, just as at the end of a closed tube, so the sound near the wall is loud. You are standing beside a brick wall listening to a 50 Hz tone from a distant loudspeaker. How far from the wall must you move to find the first quiet spot? Assume a sound speed of 340 m/s.

Section 16.5 Speech and Hearing

33. || The first formant of your vocal system can be modeled as the resonance of an open-closed tube, the closed end being your vocal cords and the open end your lips. Estimate the frequency of the first formant from the graph of Figure 16.23, and then estimate the length of the tube of which this is a resonance. Does your result seem reasonable? BIO

34. || When you voice the vowel sound in "hat," you narrow the BIO opening where your throat opens into the cavity of your mouth so that your vocal tract appears as two connected tubes. The first is in your throat, closed at the vocal cords and open at the back of the mouth. The second is the mouth itself, open at the lips and closed at the back of the mouth—a different condition than for the throat because of the relatively larger size of the cavity. The corresponding formant frequencies are 800 Hz (for the throat) and 1500 Hz (for the mouth). What are the lengths of these two cavities? Assume a sound speed of 350 m/s.
35. ||| The first and second formants when you make an "ee" vowel BIO INT sound are approximately 270 Hz and 2300 Hz. The speed of sound in your vocal tract is approximately 350 m/s. If you breathe a mix of oxygen and helium (as deep-sea divers often do), the speed increases to 750 m/s. With this mix of gases, what are the frequencies of the two formants?
36. ||| Figure P16.36 shows BIO the two lowest resonances recorded in the vocal tract of the eastern towhee, a small songbird.
    a. Is this bird's vocal tract better modeled as an open-open tube or an open-closed tube?
    b. Estimate the length of the towhee's vocal tract.

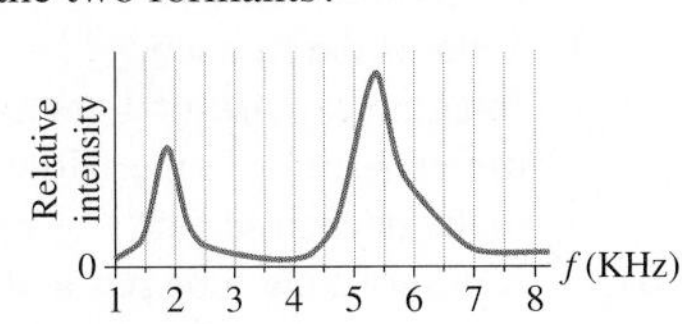

FIGURE P16.36

Section 16.6 The Interference of Waves from Two Sources

37. ||| Two loudspeakers emit identical sound waves along the x-axis. The sound at a point on the axis has maximum intensity when the speakers are 20 cm apart. The sound intensity decreases as the distance between the speakers is increased, reaching zero at a separation of 30 cm.
    a. What is the wavelength of the sound?
    b. If the distance between the speakers continues to increase, at what separation will the sound intensity again be a maximum?
38. ||| Two loudspeakers in a 20°C room emit 686 Hz sound waves along the x-axis. What is the smallest distance between the speakers for which the interference of the sound waves for a point on the axis is destructive?
39. |||| Two loudspeakers, 1.0 m apart, emit sound waves with the same frequency along the positive x-axis. Victor, standing on the axis to the right of the speakers, hears no sound. As the frequency is slowly tripled, Victor hears the sound go through the sequence loud-soft-loud-soft-loud before becoming quiet again. What was the original sound frequency?
40. || In noisy factory environments, it's possible to use a loudspeaker to cancel persistent low-frequency machine noise at the position of one worker. The details of practical systems are complex, but we can present a simple example that gives you the idea. Suppose a machine 5.0 m away from a worker emits a persistent 80 Hz hum. To cancel the sound at the worker's location with a speaker that exactly duplicates the machine's hum, how far from the worker should the speaker be placed? Assume a sound speed of 340 m/s.
41. |||| Two identical loudspeakers separated by distance d emit 170 Hz sound waves along the x-axis. As you walk along the axis, away from the speakers, you don't hear anything even though both speakers are on. What are three possible values for d? Assume a sound speed of 340 m/s.
42. | Figure P16.42 shows the circular wave fronts emitted by two sources. Make a table with rows labeled P, Q, and R and columns labeled r_1, r_2, Δr, and C/D. Fill in the table for points P, Q, and R, giving the distances as multiples of λ and indicating, with a C or a D, whether the interference at that point is constructive or destructive.

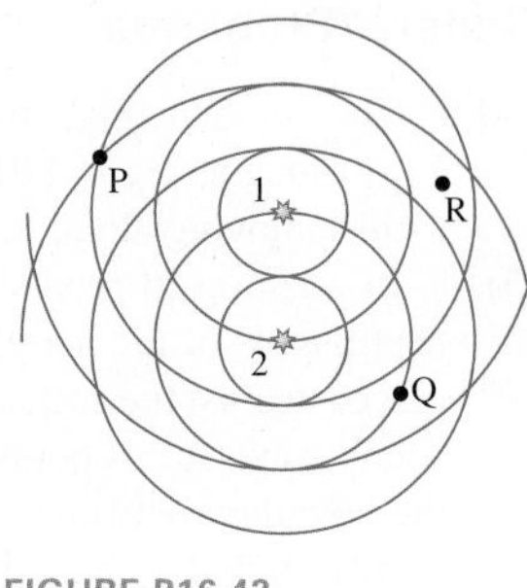

FIGURE P16.42

43. ||| Two identical loudspeakers 2.0 m apart are emitting 1800 Hz sound waves into a room where the speed of sound is 340 m/s. Is the point 4.0 m directly in front of one of the speakers, perpendicular to the line joining the speakers, a point of maximum constructive interference, perfect destructive interference, or something in between?
44. || Two identical loudspeakers 2.0 m apart are emitting sound waves into a room where the speed of sound is 340 m/s. Abby is standing 5.0 m in front of one of the speakers, perpendicular to the line joining the speakers, and hears a maximum in the intensity of the sound. What is the lowest possible frequency of sound for which this is possible?

Section 16.7 Beats

45. ||| Musicians can use beats to tune their instruments. One flute is properly tuned and plays the musical note A at exactly 440 Hz. A second player sounds the same note and hears that her instrument is slightly "flat" (that is, at too low a frequency). Playing at the same time as the first flute, she hears two loud-soft-loud beats per second. What is the frequency of her instrument?
46. ||| A student waiting at a stoplight notices that her turn signal, which has a period of 0.85 s, makes one blink exactly in sync with the turn signal of the car in front of her. The blinker of the car ahead then starts to get ahead, but 17 s later the two are exactly in sync again. What is the period of the blinker of the other car?
47. | Two strings are adjusted to vibrate at exactly 200 Hz. Then the tension in one string is increased slightly. Afterward, three beats per second are heard when the strings vibrate at the same time. What is the new frequency of the string that was tightened?
48. ||| Figure P16.48 shows the superposition of two sound waves. What are their frequencies?

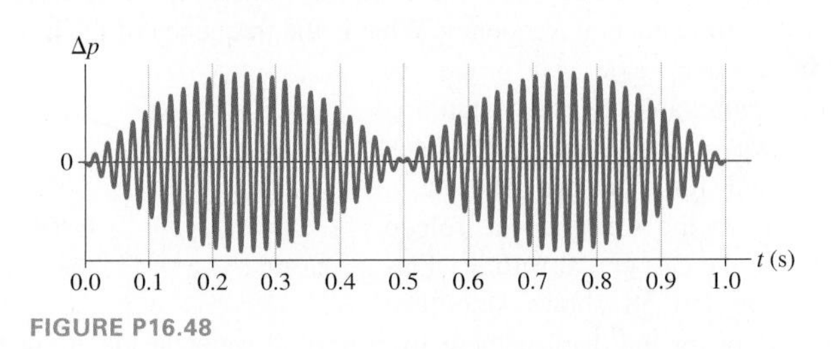

FIGURE P16.48

49. || A flute player hears four beats per second when she compares her note to a 523 Hz tuning fork (the note C). She can match the frequency of the tuning fork by pulling out the "tuning joint" to lengthen her flute slightly. What was her initial frequency?

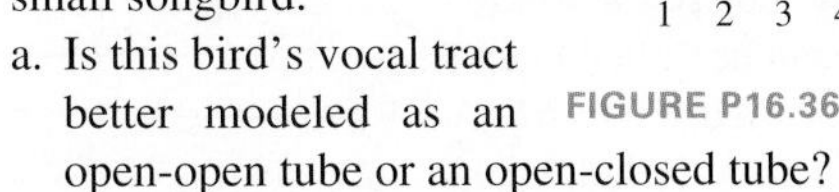

Watch Video Solution Problems 16.57 and 16.65

General Problems

50. | The fundamental frequency of a standing wave on a 1.0-m-long string is 440 Hz. What would be the wave speed of a pulse moving along this string?

51. ||| In addition to producing images, ultrasound can be used to heat tissues of the body for therapeutic purposes. An emitter is placed against the surface of the skin; the amplitude of the ultrasound wave at this point is quite large. When a sound wave hits the boundary between soft tissue and bone, most of the energy is reflected. The boundary acts like the closed end of a tube, which can lead to standing waves. Suppose 0.70 MHz ultrasound is directed through a layer of tissue with a bone 0.55 cm below the surface. Will standing waves be created? Explain.
BIO INT

52. ||| An 80-cm-long steel string with a linear density of 1.0 g/m is under 200 N tension. It is plucked and vibrates at its fundamental frequency. What is the wavelength of the sound wave that reaches your ear in a 20°C room?
INT

53. |||| Tendons are, essentially, elastic cords stretched between two fixed ends; as such, they can support standing waves. These resonances can be undesirable. The Achilles tendon connects the heel with a muscle in the calf. A woman has a 20-cm-long tendon with a cross-section area of 110 mm^2. The density of tendon tissue is 1100 kg/m^3. For a reasonable tension of 500 N, what will be the resonant frequencies of her Achilles tendon?
BIO INT

54. |||| Two loudspeakers directly face each other 30 m apart, with the left speaker positioned at $x = 0$ m. The pressure of the sound wave emitted by the left speaker is described by the equation $\Delta p_L = p_0 \cos(1.90x - 630t)$, while that from the right speaker is given by $\Delta p_R = p_0 \cos(1.90x + 630t)$, where x is measured in m and t in s. What is the point nearest to the left speaker at which there is a node in the sound wave?
INT

55. |||| Spiders may "tune" strands of their webs to give enhanced response at frequencies corresponding to the frequencies at which desirable prey might struggle. Orb web silk has a typical diameter of 0.0020 mm, and spider silk has a density of 1300 kg/m^3. To give a resonance at 100 Hz, to what tension must a spider adjust a 12-cm-long strand of silk?
BIO INT

56. || A particularly beautiful note reaching your ear from a rare Stradivarius violin has a wavelength of 39.1 cm. The room is slightly warm, so the speed of sound is 344 m/s. If the string's linear density is 0.600 g/m and the tension is 150 N, how long is the vibrating section of the violin string?
INT

57. || A 12 kg hanging sculpture is suspended by a 90-cm-long, 5.0 g steel wire. When the wind blows hard, the wire hums at its fundamental frequency. What is the frequency of the hum?
INT

58. || Lake Erie is prone to remarkable *seiches*—standing waves that slosh water back and forth in the lake basin from the west end at Toledo to the east end at Buffalo. Figure P16.58 shows smoothed data for the displacement from normal water levels along the lake at the high point of one particular seiche. 3 hours later the water was at normal levels throughout the basin; 6 hours later the water was high in Toledo and low in Buffalo.

FIGURE P16.58

a. What is the wavelength of this standing wave?
b. What is the frequency?
c. What is the wave speed?

59. ||||| A guitar player can change the frequency of a string by "bending" it—pushing it along a fret that is perpendicular to its length. This stretches the string, increasing its tension and its frequency. The G string on a guitar is 64 cm long and has a tension of 74 N. The guitarist pushes this string down against a fret located at the center of the string, which gives it a frequency of 392 Hz. He then bends the string, pushing with a force of 4.0 N so that it moves 8.0 mm along the fret. What is the new frequency?
INT

60. ||| A carbon-dioxide laser emits infrared light with a wavelength of 10.6 μm.
INT
a. What is the length of a tube that will oscillate in the $m = 100{,}000$ mode?
b. What is the frequency?
c. Imagine a pulse of light bouncing back and forth between the ends of the tube. How many round trips will the pulse make in each second?

61. || A 40-cm-long tube has a 40-cm-long insert that can be pulled in and out, as shown in Figure P16.61. A vibrating tuning fork is held next to the tube. As the insert is slowly pulled out, the sound from the tuning fork creates standing waves in the tube when the total length L is 42.5 cm, 56.7 cm, and 70.9 cm. What is the frequency of the tuning fork? The air temperature is 20°C.

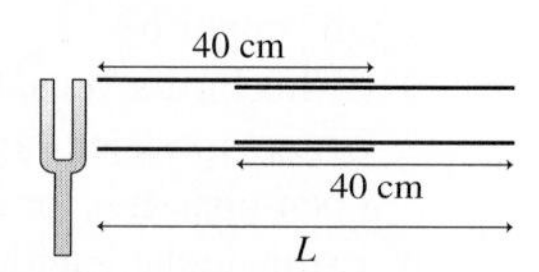

FIGURE P16.61

62. || The width of a particular microwave oven is exactly right to support a standing-wave mode. Measurements of the temperature across the oven show that there are cold spots at each edge of the oven and at three spots in between. The wavelength of the microwaves is 12 cm. How wide is the oven?

63. ||| Two loudspeakers located along the x-axis as shown in Figure P16.63 produce sounds of equal frequency. Speaker 1 is at the origin, while the location of speaker 2 can be varied by a remote control wielded by the listener. He notices maxima in the sound intensity when speaker 2 is located at $x = 0.75$ m and 1.00 m, but at no points in between. What is the frequency of the sound? Assume the speed of sound is 340 m/s.

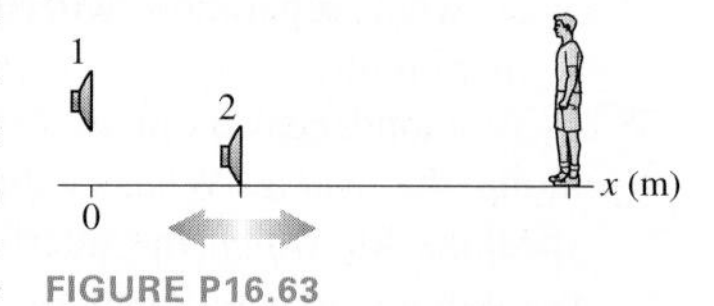

FIGURE P16.63

64. | Two loudspeakers 42.0 m apart and facing each other emit identical 115 Hz sinusoidal sound waves in a room where the sound speed is 345 m/s. Susan is walking along a line between the speakers. As she walks, she finds herself moving through loud and quiet spots. If Susan stands 19.5 m from one speaker, is she standing at a quiet spot or a loud spot?

65. ||| You are standing 2.50 m directly in front of one of the two loudspeakers shown in Figure P16.65. They are 3.00 m apart and both are playing a 686 Hz tone in phase. As you begin to walk directly away from the speaker, at what distances from the speaker do you hear a *minimum* sound intensity? The room temperature is 20°C.

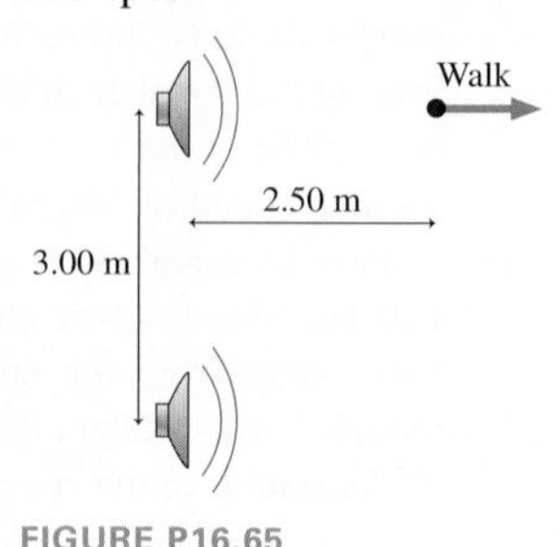

FIGURE P16.65

66. ||| Two loudspeakers, 4.0 m apart and facing each other, play identical sounds of the same frequency. You stand halfway between them, where there is a maximum of sound intensity. Moving from this point toward one of the speakers, you encounter a minimum of sound intensity when you have moved 0.25 m.
    a. What is the frequency of the sound?
    b. If the frequency is then increased while you remain 0.25 m from the center, what is the first frequency for which that location will be a maximum of sound intensity?
67. || Piano tuners tune pianos by listening to the beats between the harmonics of two different strings. When properly tuned, the note A should have the frequency 440 Hz and the note E should be at 659 Hz. The tuner can determine this by listening to the beats between the third harmonic of the A and the second harmonic of the E. A tuner first tunes the A string very precisely by matching it to a 440 Hz tuning fork. She then strikes the A and E strings simultaneously and listens for beats between the harmonics. What beat frequency indicates that the E string is properly tuned?
68. ||| A flutist assembles her flute in a room where the speed of sound is 342 m/s. When she plays the note A, it is in perfect tune with a 440 Hz tuning fork. After a few minutes, the air inside her flute has warmed to where the speed of sound is 346 m/s.
    a. How many beats per second will she hear if she now plays the note A as the tuning fork is sounded?
    b. How far does she need to extend the "tuning joint" of her flute to be in tune with the tuning fork?
69. || Police radars determine speed by measuring the shift of radio waves reflected by a moving vehicle. They do so by determining the beat frequency between the reflected wave and the 10.5 GHz emitted wave. Some units can be calibrated by using a tuning fork; holding a vibrating fork in front of the unit causes the display to register a speed corresponding to the vibration frequency. A tuning fork is labeled "55 mph." What is the frequency of the tuning fork? INT
70. || A Doppler blood flowmeter emits ultrasound at a frequency of 5.0 MHz. What is the beat frequency between the emitted waves and the waves reflected from blood cells moving away from the emitter at 0.15 m/s? BIO INT
71. | An ultrasound unit is being used to measure a patient's heartbeat by combining the emitted 2.0 MHz signal with the sound waves reflected from the moving tissue of one point on the heart. The beat frequency between the two signals has a maximum value of 520 Hz. What is the maximum speed of the heart tissue? BIO INT

MCAT-Style Passage Problems

Harmonics and Harmony

You know that certain musical notes sound good together—harmonious—whereas others do not. This harmony is related to the various harmonics of the notes.

The musical notes C (262 Hz) and G (392 Hz) make a pleasant sound when played together; we call this consonance. As Figure P16.72 shows, the harmonics of the two notes are either far from each other or very close to each other (within a few Hz). This is the key to consonance: harmonics that are spaced either far apart or very close. The close harmonics have a beat frequency of a few Hz that is perceived as pleasant. If the harmonics of two notes are close but not too close, the rather high beat frequency between the two is quite unpleasant. This is what we hear as dissonance. Exactly how much a difference is maximally dissonant is a matter of opinion, but harmonic separations of 30 or 40 Hz seem to be quite unpleasant for most people.

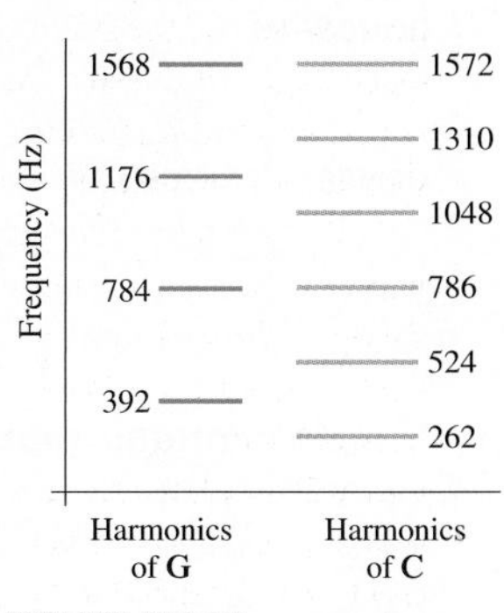

FIGURE P16.72

72. | What is the beat frequency between the second harmonic of G and the third harmonic of C?
    A. 1 Hz B. 2 Hz C. 4 Hz D. 6 Hz
73. | Would a G-flat (frequency 370 Hz) and a C played together be consonant or dissonant?
    A. Consonant B. Dissonant
74. | An organ pipe open at both ends is tuned so that its fundamental frequency is a G. How long is the pipe?
    A. 43 cm B. 87 cm C. 130 cm D. 173 cm
75. | If the C were played on an organ pipe that was open at one end and closed at the other, which of the harmonic frequencies in Figure P16.72 would be present?
    A. All of the harmonics in the figure would be present.
    B. 262, 786, and 1310 Hz
    C. 524, 1048, and 1572 Hz
    D. 262, 524, and 1048 Hz

PART IV SUMMARY Oscillations and Waves

KNOWLEDGE STRUCTURE IV

BASIC GOALS

How can we describe oscillatory motion? What are the distinguishing features of waves? How does a wave travel through a medium? What happens when two waves meet?

CROSS-CUTTING CONCEPTS

The mathematics describing oscillations are *universal,* describing a mass on a spring, a gibbon swinging from a branch, and the speed of a walking person. Similarly, the same theory of waves works for stringed musical instruments, ground motion during an earthquake, and the warming energy of the sun.

GENERAL PRINCIPLES

Simple harmonic motion occurs when a linear restoring force acts to return a system to equilibrium. The period of simple harmonic motion depends on physical properties of the system but not the amplitude. All sinusoidal waves, whether water waves, sound waves, or light waves, have the same functional form. When two waves meet, they pass through each other, combining where they overlap by superposition.

Simple harmonic motion

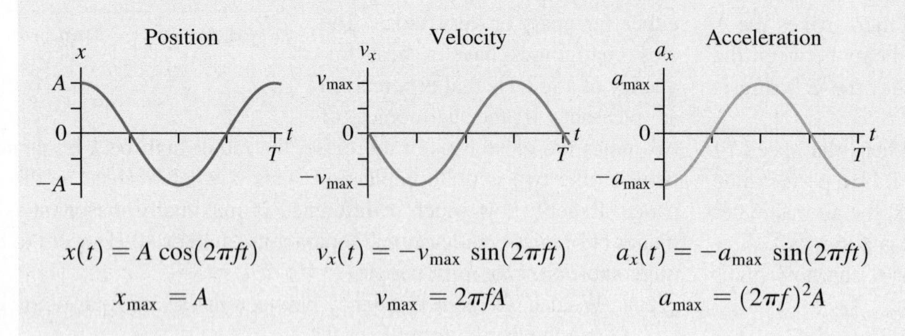

$$x(t) = A\cos(2\pi f t) \qquad v_x(t) = -v_{max}\sin(2\pi f t) \qquad a_x(t) = -a_{max}\sin(2\pi f t)$$

$$x_{max} = A \qquad v_{max} = 2\pi f A \qquad a_{max} = (2\pi f)^2 A$$

The frequency of a mass on a spring depends on the spring constant and the mass:

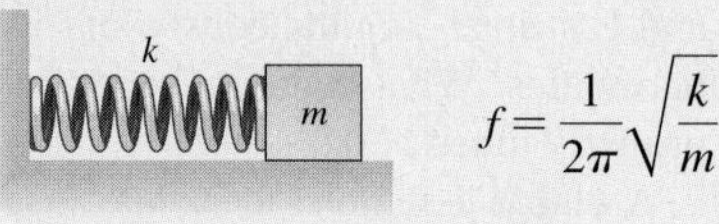

$$f = \frac{1}{2\pi}\sqrt{\frac{k}{m}}$$

The frequency of a pendulum depends on its length and the free-fall acceleration:

$$f = \frac{1}{2\pi}\sqrt{\frac{g}{L}}$$

Traveling waves

A **mechanical wave** travels through a medium at a well-defined wave speed v.

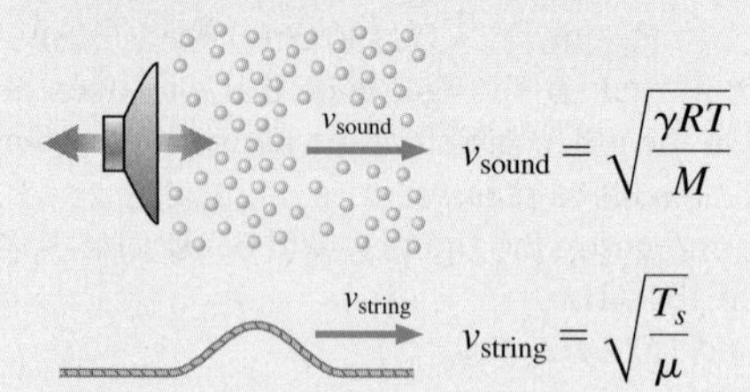

$$v_{sound} = \sqrt{\frac{\gamma RT}{M}} \qquad v_{string} = \sqrt{\frac{T_s}{\mu}}$$

Light waves travel through a vacuum at speed $c = 3.00 \times 10^8$ m/s.

A **snapshot graph** shows the wave at a particular instant in time:

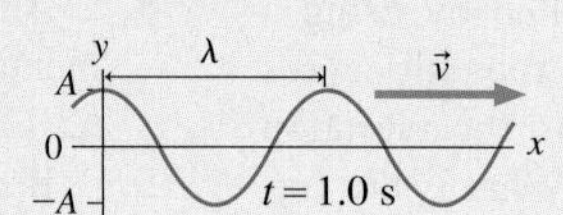

A **history graph** shows the displacement of the wave versus time at one point in the medium:

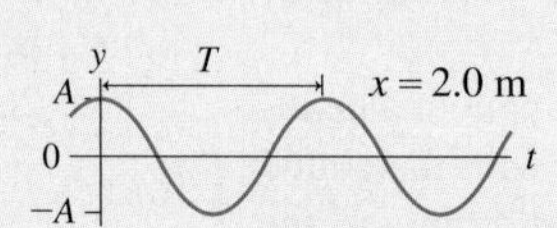

The speed v, wavelength λ, period T, and frequency f of a wave are related:

$$T = \frac{1}{f} \qquad v = \lambda f$$

A sinusoidal traveling wave is described by the equation

$$y = A\cos\left(2\pi\left(\frac{x}{\lambda} \pm \frac{t}{T}\right)\right)$$

+: wave travels to the left

−: wave travels to the right

Superposition and interference

The superposition of two waves is the sum of the displacements of the two individual waves.

Constructive interference occurs when crests are aligned with crests and troughs with troughs. Constructive interference occurs when the path-length difference is a whole number of wavelengths.

Destructive interference occurs when crests are aligned with troughs. Destructive interference occurs when the path-length difference is a whole number of wavelengths plus half a wavelength.

Standing waves

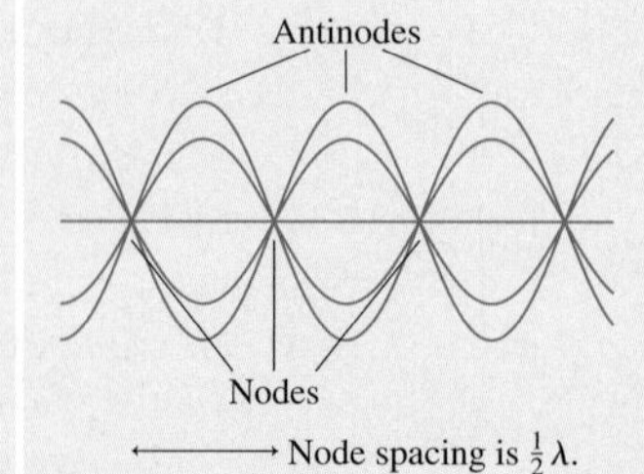

Standing waves are due to the superposition of two traveling waves moving in opposite directions.

The boundary conditions determine which standing-wave frequencies and wavelengths are allowed. The allowed standing waves are modes of the system.

For a string of length L, the modes have wavelength and frequency

$$\lambda_m = \frac{2L}{m} \qquad f_m = m\left(\frac{v_{string}}{2L}\right) = m f_1 \qquad m = 1, 2, 3, 4, \ldots$$

Waves in the Earth and the Ocean

In December 2004, a large earthquake off the coast of Indonesia produced a devastating water wave, called a *tsunami,* that caused tremendous destruction thousands of miles away from the earthquake's epicenter. The tsunami was a dramatic illustration of the energy carried by waves.

It was also a call to action. Many of the communities hardest hit by the tsunami were struck hours after the waves were generated, long after seismic waves from the earthquake that passed through the earth had been detected at distant recording stations, long after the possibility of a tsunami was first discussed. With better detection and more accurate models of how a tsunami is formed and how a tsunami propagates, the affected communities could have received advance warning. The study of physics may seem an abstract undertaking with few practical applications, but on this day a better scientific understanding of these waves could have averted tragedy.

Let's use our knowledge of waves to explore the properties of a tsunami. In Chapter 15, we saw that a vigorous shake of one end of a rope causes a pulse to travel along it, carrying energy as it goes. The earthquake that produced the Indian Ocean tsunami of 2004 caused a sudden upward displacement of the seafloor that produced a corresponding rise in the surface of the ocean. This was the *disturbance* that produced the tsunami, very much like a quick shake on the end of a rope. The resulting wave propagated through the ocean, as we see in the figure.

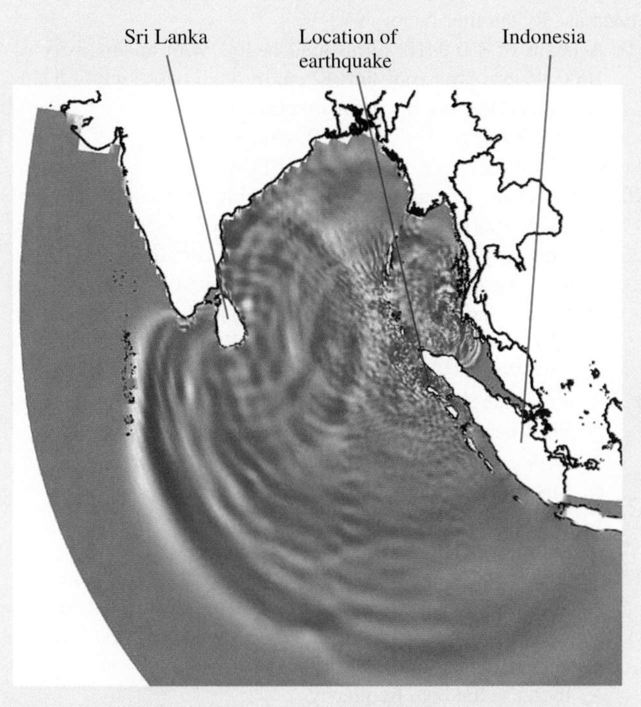

One frame from a computer simulation of the Indian Ocean tsunami three hours after the earthquake that produced it. The disturbance propagating outward from the earthquake is clearly seen, as are wave reflections from the island of Sri Lanka.

This simulation of the tsunami looks much like the ripples that spread when you drop a pebble into a pond. But there is a big difference—the scale. The fact that you can see the individual waves on this diagram that spans 5000 km is quite revealing. To show up so clearly, the individual wave pulses must be very wide—up to hundreds of kilometers from front to back.

A tsunami is actually a "shallow water wave," even in the deep ocean, because the depth of the ocean is much less than the width of the wave. Consequently, a tsunami travels differently than normal ocean waves. In Chapter 15 we learned that wave speeds are fixed by the properties of the medium. That is true for normal ocean waves, but the great width of the wave causes a tsunami to "feel the bottom." Its wave speed is determined by the depth of the ocean: The greater the depth, the greater the speed. In the deep ocean, a tsunami travels at hundreds of kilometers per hour, much faster than a typical ocean wave. Near shore, as the ocean depth decreases, so does the speed of the wave.

The height of the tsunami in the open ocean was about half a meter. Why should such a small wave—one that ships didn't even notice as it passed—be so fearsome? Again, it's the *width* of the wave that matters. Because a tsunami is the wave motion of a considerable mass of water, great energy is involved. As the front of a tsunami wave nears shore, its speed decreases, and the back of the wave moves faster than the front. Consequently, the width decreases. The water begins to pile up, and the wave dramatically increases in height.

The Indian Ocean tsunami had a height of up to 15 m when it reached shore, with a width of up to several kilometers. This tremendous mass of water was still moving at high speed, giving it a great deal of energy. A tsunami reaching the shore isn't like a typical wave that breaks and crashes. It is a kilometers-wide wall of water that moves onto the shore and just keeps on coming. In many places, the water reached 2 km inland.

The impact of the Indian Ocean tsunami was devastating, but it was the first tsunami for which scientists were able to use satellites and ocean sensors to make planet-wide measurements. An analysis of the data has helped us better understand the physics of these ocean waves. We won't be able to stop future tsunamis, but with a better knowledge of how they are formed and how they travel, we will be better able to warn people to get out of their way.

PART IV PROBLEMS

The following questions are related to the passage "Waves in the Earth and the Ocean" on the previous page.

1. Rank from fastest to slowest the following waves according to their speed of propagation:
   A. An earthquake wave B. A tsunami
   C. A sound wave in air D. A light wave
2. The increase in height as a tsunami approaches shore is due to
   A. The increase in frequency as the wave approaches shore.
   B. The increase in speed as the wave approaches shore.
   C. The decrease in speed as the wave approaches shore.
   D. The constructive interference with the wave reflected from shore.
3. In the middle of the Indian Ocean, the tsunami referred to in the passage was a train of pulses approximating a sinusoidal wave with speed 200 m/s and wavelength 150 km. What was the approximate period of these pulses?
   A. 1 min B. 3 min
   C. 5 min D. 15 min
4. If a train of pulses moves into shallower water as it approaches a shore,
   A. The wavelength increases.
   B. The wavelength stays the same.
   C. The wavelength decreases.
5. The tsunami described in the passage produced a very erratic pattern of damage, with some areas seeing very large waves and nearby areas seeing only small waves. Which of the following is a possible explanation?
   A. Certain areas saw the wave from the primary source, others only the reflected waves.
   B. The superposition of waves from the primary source and reflected waves produced regions of constructive and destructive interference.
   C. A tsunami is a standing wave, and certain locations were at nodal positions, others at antinodal positions.

The following passages and associated questions are based on the material of Part IV.

Deep-Water Waves

Water waves are called *deep-water waves* when the depth of the water is much greater than the wavelength of the wave. The speed of deep-water waves depends on the wavelength as follows:

$$v = \sqrt{\frac{g\lambda}{2\pi}}$$

Suppose you are on a ship at rest in the ocean, observing the crests of a passing sinusoidal wave. You estimate that the crests are 75 m apart.

6. Approximately how much time elapses between one crest reaching your ship and the next?
   A. 3 s
   B. 5 s
   C. 7 s
   D. 12 s
7. The captain starts the engines and sails directly opposite the motion of the waves at 4.5 m/s. Now how much time elapses between one crest reaching your ship and the next?
   A. 3 s
   B. 5 s
   C. 7 s
   D. 12 s
8. In the deep ocean, a longer-wavelength wave travels faster than a shorter-wavelength wave. Thus, a higher-frequency wave travels ______ a lower-frequency wave.
   A. Faster than
   B. At the same speed as
   C. Slower than

Attenuation of Ultrasound BIO

Ultrasound is absorbed in the body; this complicates the use of ultrasound to image tissues. The intensity of a beam of ultrasound decreases by a factor of 2 after traveling a distance of 40 wavelengths. Each additional travel of 40 wavelengths results in a decrease by another factor of 2.

9. A beam of 1.0 MHz ultrasound begins with an intensity of 1000 W/m^2. After traveling 12 cm through tissue with no significant reflection, the intensity is about
   A. 750 W/m^2 B. 500 W/m^2
   C. 250 W/m^2 D. 125 W/m^2
10. A physician is making an image with ultrasound of initial intensity 1000 W/m^2. When the frequency is set to 1.0 MHz, the intensity drops to 500 W/m^2 at a certain depth in the patient's body. What will be the intensity at this depth if the physician changes the frequency to 2.0 MHz?
    A. 750 W/m^2 B. 500 W/m^2
    C. 250 W/m^2 D. 125 W/m^2
11. A physician is using ultrasound to make an image of a patient's heart. Increasing the frequency will provide
    A. Better penetration and better resolution.
    B. Less penetration but better resolution.
    C. More penetration but worse resolution.
    D. Less penetration and worse resolution.
12. A physician is using Doppler ultrasound to measure the motion of a patient's heart. The device measures the beat frequency between the emitted and the reflected waves. Increasing the frequency of the ultrasound will
    A. Increase the beat frequency.
    B. Not affect the beat frequency.
    C. Decrease the beat frequency.

Measuring the Speed of Sound

A student investigator is measuring the speed of sound by looking at the time for a brief, sinusoidal pulse from a loudspeaker to travel down a tube, reflect from the closed end, and reach a microphone. The apparatus is shown in Figure IV.1a; typical data recorded by the microphone are graphed in Figure IV.1b. The first pulse is the sound directly from the loudspeaker; the second pulse is the reflection from the closed end. A portion of the returning wave reflects from the open end of the tube and makes another round trip before being detected by the microphone; this is the third pulse seen in the data.

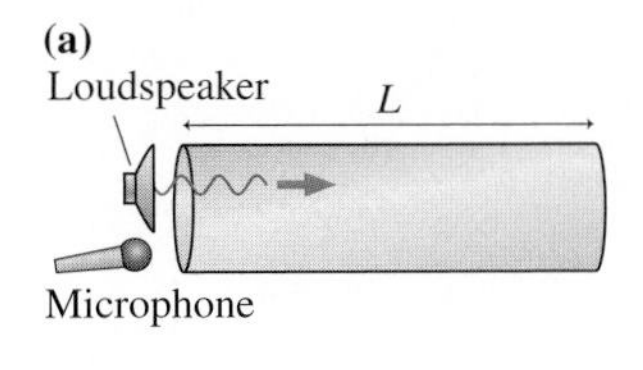

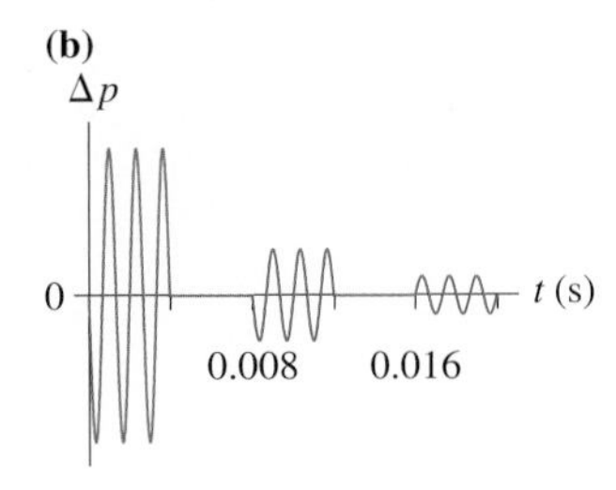

FIGURE IV.1

13. What was the approximate frequency of the sound wave used in this experiment?
    A. 250 Hz
    B. 500 Hz
    C. 750 Hz
    D. 1000 Hz
14. What can you say about the reflection of sound waves at the ends of a tube?
    A. Sound waves are inverted when reflected from both open and closed tube ends.
    B. Sound waves are inverted when reflected from a closed end, not inverted when reflected from an open end.
    C. Sound waves are inverted when reflected from an open end, not inverted when reflected from a closed end.
    D. Sound waves are not inverted when reflected from either open or closed tube ends.
15. What was the approximate length of the tube?
    A. 0.35 m
    B. 0.70 m
    C. 1.4 m
    D. 2.8 m
16. An alternative technique to determine sound speed is to measure the frequency of a standing wave in the tube. What is the wavelength of the lowest resonance of this tube?
    A. $L/2$
    B. L
    C. $2L$
    D. $4L$

In the Swing

A rope swing is hung from a tree right at the edge of a small creek. The rope is 5.0 m long; the creek is 3.0 m wide.

17. You sit on the swing, and your friend gives you a gentle push so that you swing out over the creek. How long will it be until you swing back to where you started?
    A. 4.5 s   B. 3.4 s
    C. 2.2 s   D. 1.1 s
18. Now you switch places with your friend, who has twice your mass. You give your friend a gentle push so that he swings out over the creek. How long will it be until he swings back to where he started?
    A. 4.5 s   B. 3.4 s
    C. 2.2 s   D. 1.1 s
19. Your friend now pushes you over and over, so that you swing higher and higher. At some point you are swinging all the way across the creek—at the top point of your arc you are right above the opposite side. How fast are you moving when you get back to the lowest point of your arc?
    A. 6.3 m/s   B. 5.4 m/s
    C. 4.4 m/s   D. 3.1 m/s

Additional Integrated Problems

20. BIO The jumping gait of the kangaroo is efficient because energy is stored in the stretch of stout tendons in the legs; the kangaroo literally bounces with each stride. We can model the bouncing of a kangaroo as the bouncing of a mass on a spring. A 70 kg kangaroo hits the ground, the tendons stretch to a maximum length, and the rebound causes the kangaroo to leave the ground approximately 0.10 s after its feet first touch.
    a. Modeling this as the motion of a mass on a spring, what is the period of the motion?
    b. Given the kangaroo mass and the period you've calculated, what is the spring constant?
    c. If the kangaroo speeds up, it must bounce higher and farther with each stride, and so must store more energy in each bounce. How does this affect the time and the amplitude of each bounce?
21. BIO A brand of earplugs reduces the sound intensity level by 27 dB. By what factor do these earplugs reduce the acoustic intensity?
22. BIO Sperm whales, just like bats, use echolocation to find prey. A sperm whale's vocal system creates a single sharp click, but the emitted sound consists of several equally spaced clicks of decreasing intensity. Researchers use the time interval between the clicks to estimate the size of the whale that created them. Explain how this might be done.

    Hint: The head of a sperm whale is complex, with air pockets at either end.

Mathematics Review

Algebra

Using exponents: $a^{-x} = \dfrac{1}{a^x}$ $\quad a^x a^y = a^{(x+y)}$ $\quad \dfrac{a^x}{a^y} = a^{(x-y)}$ $\quad (a^x)^y = a^{xy}$

$a^0 = 1$ $\quad a^1 = a$ $\quad a^{1/n} = \sqrt[n]{a}$

Fractions: $\left(\dfrac{a}{b}\right)\left(\dfrac{c}{d}\right) = \dfrac{ac}{bd}$ $\quad \dfrac{a/b}{c/d} = \dfrac{ad}{bc}$ $\quad \dfrac{1}{1/a} = a$

Logarithms: Natural (base e) logarithms: If $a = e^x$, then $\ln(a) = x$ $\quad \ln(e^x) = x$ $\quad e^{\ln(x)} = x$

Base-10 logarithms: If $a = 10^x$, then $\log_{10}(a) = x$ $\quad \log_{10}(10^x) = x$ $\quad 10^{\log_{10}(x)} = x$

The following rules hold for both natural and base-10 algorithms:

$\ln(ab) = \ln(a) + \ln(b)$ $\quad \ln\left(\dfrac{a}{b}\right) = \ln(a) - \ln(b)$ $\quad \ln(a^n) = n\ln(a)$

The expression $\ln(a + b)$ cannot be simplified.

Linear equations: The graph of the equation $y = ax + b$ is a straight line. a is the slope of the graph. b is the y-intercept.

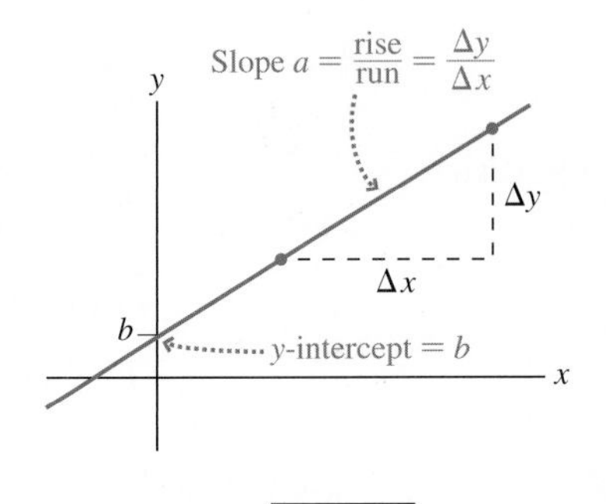

Proportionality: To say that y is proportional to x, written $y \propto x$, means that $y = ax$, where a is a constant. Proportionality is a special case of linearity. A graph of a proportional relationship is a straight line that passes through the origin. If $y \propto x$, then

$$\frac{y_1}{y_2} = \frac{x_1}{x_2}$$

Quadratic equation: The quadratic equation $ax^2 + bx + c = 0$ has the two solutions $x = \dfrac{-b \pm \sqrt{b^2 - 4ac}}{2a}$.

Geometry and Trigonometry

Area and volume:

Rectangle
$A = ab$

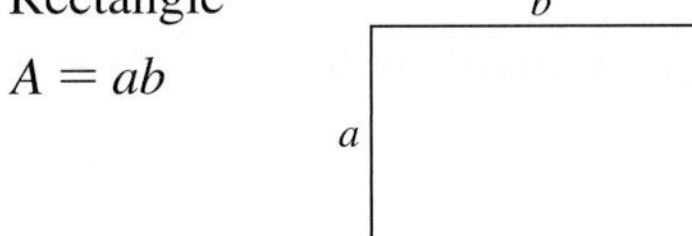

Rectangular box
$V = abc$

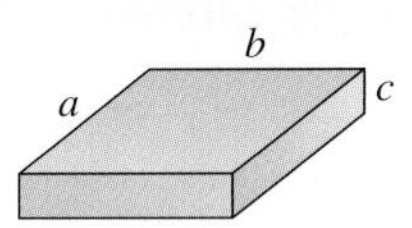

Triangle
$A = \frac{1}{2}ab$

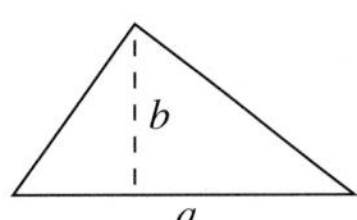

Right circular cylinder
$V = \pi r^2 l$

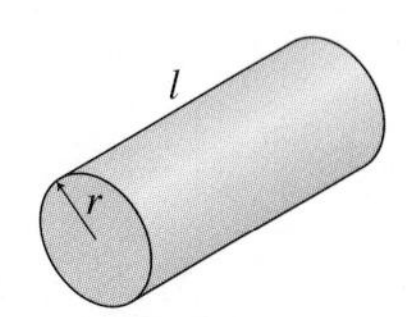

Circle
$C = 2\pi r$
$A = \pi r^2$

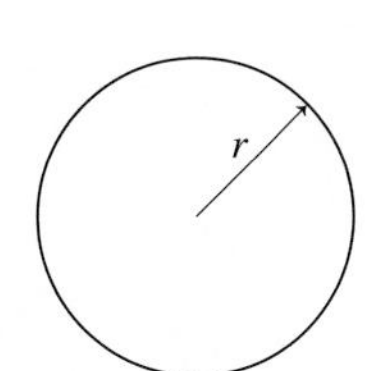

Sphere
$A = 4\pi r^2$
$V = \frac{4}{3}\pi r^3$

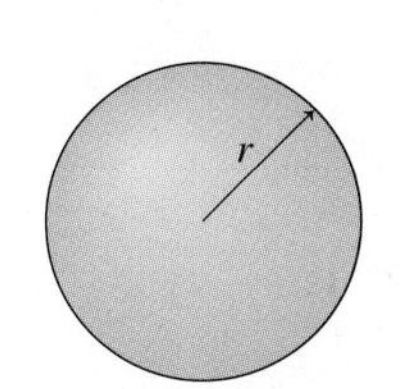

Arc length and angle: The angle θ in radians is defined as $\theta = s/r$.
The arc length that spans angle θ is $s = r\theta$.
$2\pi \text{ rad} = 360°$

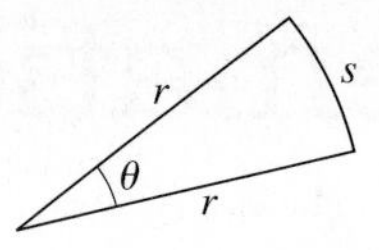

Right triangle: Pythagorean theorem $c = \sqrt{a^2 + b^2}$ or $a^2 + b^2 = c^2$

$$\sin\theta = \frac{b}{c} = \frac{\text{far side}}{\text{hypotenuse}} \qquad \theta = \sin^{-1}\left(\frac{b}{c}\right)$$

$$\cos\theta = \frac{a}{c} = \frac{\text{adjacent side}}{\text{hypotenuse}} \qquad \theta = \cos^{-1}\left(\frac{a}{c}\right)$$

$$\tan\theta = \frac{b}{a} = \frac{\text{far side}}{\text{adjacent side}} \qquad \theta = \tan^{-1}\left(\frac{b}{a}\right)$$

In general, if it is known that the sine of an angle θ is x, so $x = \sin\theta$, then we can find θ by taking the *inverse sine* of x, denoted $\sin^{-1}x$. Thus $\theta = \sin^{-1}x$. Similar relationships apply for cosines and tangents.

General triangle: $\alpha + \beta + \gamma = 180° = \pi \text{ rad}$

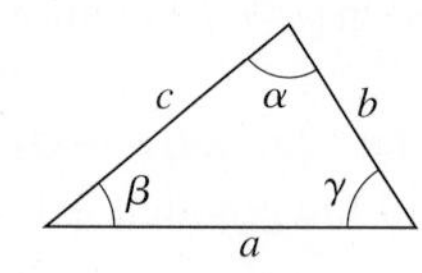

Identities:

$$\tan\alpha = \frac{\sin\alpha}{\cos\alpha} \qquad \sin^2\alpha + \cos^2\alpha = 1$$

$$\sin(-\alpha) = -\sin\alpha \qquad \cos(-\alpha) = \cos\alpha$$

$$\sin(2\alpha) = 2\sin\alpha\cos\alpha \qquad \cos(2\alpha) = \cos^2\alpha - \sin^2\alpha$$

Expansions and Approximations

Binomial approximation: $(1+x)^n \approx 1 + nx \quad \text{if} \quad x \ll 1$

Small-angle approximation: If $\alpha \ll 1$ rad, then $\sin\alpha \approx \tan\alpha \approx \alpha$ and $\cos\alpha \approx 1$.

The small-angle approximation is excellent for $\alpha < 5°$ (≈ 0.1 rad) and generally acceptable up to $\alpha \approx 10°$.

Periodic Table of Elements

Key: Atomic number — 27; Symbol — Co; Atomic mass — 58.9

Period																		
1	1 H 1.0																	2 He 4.0
2	3 Li 6.9	4 Be 9.0											5 B 10.8	6 C 12.0	7 N 14.0	8 O 16.0	9 F 19.0	10 Ne 20.2
3	11 Na 23.0	12 Mg 24.3											13 Al 27.0	14 Si 28.1	15 P 31.0	16 S 32.1	17 Cl 35.5	18 Ar 39.9
4	19 K 39.1	20 Ca 40.1	21 Sc 45.0	22 Ti 47.9	23 V 50.9	24 Cr 52.0	25 Mn 54.9	26 Fe 55.8	27 Co 58.9	28 Ni 58.7	29 Cu 63.5	30 Zn 65.4	31 Ga 69.7	32 Ge 72.6	33 As 74.9	34 Se 79.0	35 Br 79.9	36 Kr 83.8
5	37 Rb 85.5	38 Sr 87.6	39 Y 88.9	40 Zr 91.2	41 Nb 92.9	42 Mo 95.9	43 Tc [98]	44 Ru 101.1	45 Rh 102.9	46 Pd 106.4	47 Ag 107.9	48 Cd 112.4	49 In 114.8	50 Sn 118.7	51 Sb 121.8	52 Te 127.6	53 I 126.9	54 Xe 131.3
6	55 Cs 132.9	56 Ba 137.3	71 Lu 175.0	72 Hf 178.5	73 Ta 180.9	74 W 183.9	75 Re 186.2	76 Os 190.2	77 Ir 192.2	78 Pt 195.1	79 Au 197.0	80 Hg 200.6	81 Tl 204.4	82 Pb 207.2	83 Bi 209.0	84 Po [209]	85 At [210]	86 Rn [222]
7	87 Fr [223]	88 Ra [226]	103 Lr [262]	104 Rf [267]	105 Db [268]	106 Sg [271]	107 Bh [274]	108 Hs [277]	109 Mt [278]	110 Ds [281]	111 Rg [281]	112 Cn [285]	113 Nh [286]	114 Fl [289]	115 Mc [289]	116 Lv [293]	117 Ts [294]	118 Og [294]

Transition elements: columns 21 Sc through 30 Zn (groups 3–12).

Inner transition elements

Lanthanides 6	57 La 138.9	58 Ce 140.1	59 Pr 140.9	60 Nd 144.2	61 Pm 144.9	62 Sm 150.4	63 Eu 152.0	64 Gd 157.3	65 Tb 158.9	66 Dy 162.5	67 Ho 164.9	68 Er 167.3	69 Tm 168.9	70 Yb 173.0
Actinides 7	89 Ac [227]	90 Th 232.0	91 Pa 231.0	92 U 238.0	93 Np [237]	94 Pu [244]	95 Am [243]	96 Cm [247]	97 Bk [247]	98 Cf [251]	99 Es [252]	100 Fm [257]	101 Md [258]	102 No [259]

An atomic mass in brackets is that of the longest-lived isotope of an element with no stable isotopes.

Answers

Chapter 1

Answers to odd-numbered multiple-choice questions

19. A
21. C
23. B
25. C
27. B
29. A

Answers to odd-numbered problems

1.

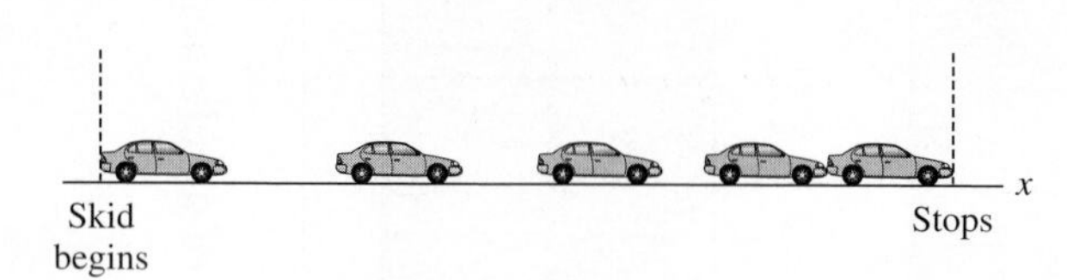

3.

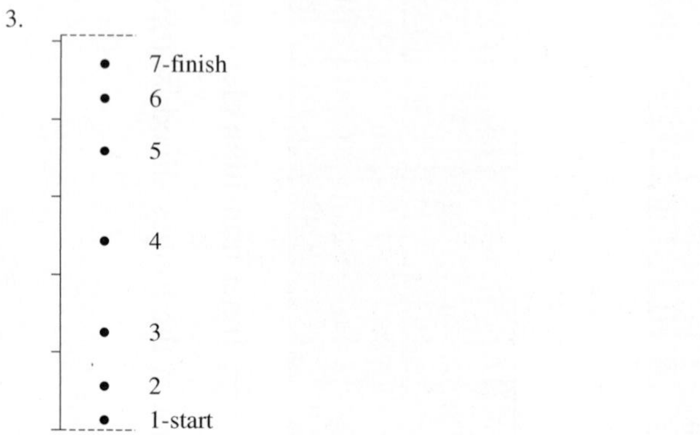

5. a. 2 mi b. 2 mi
7. −22 m
9. 800 m
11. Bike, boy, cat, toy car
13. −1.0 m/s
15. a. Second 100 m b. 9.88 m/s
17. a. 0.20 m b. 20 m/s c. 27 m/s
19. a. 3 b. 3 c. 3 d. 2
21. a. 846 b. 7.9 c. 5.77 d. 13.1
23. 8.8480×10^3 m
25. We get 6×10^{-9} m/s; your answer should be close.
27. 487 m
29. 1.16×10^3 ft
31. 71 m, 45° south of west
33. 38 km
35. 6.1 m
37. 177 m
39. 150 cm
41. a. 4.6 m b. 6.1 m
43.

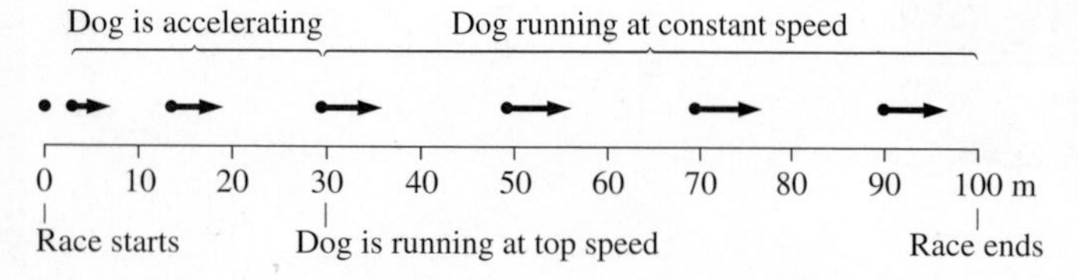

45.

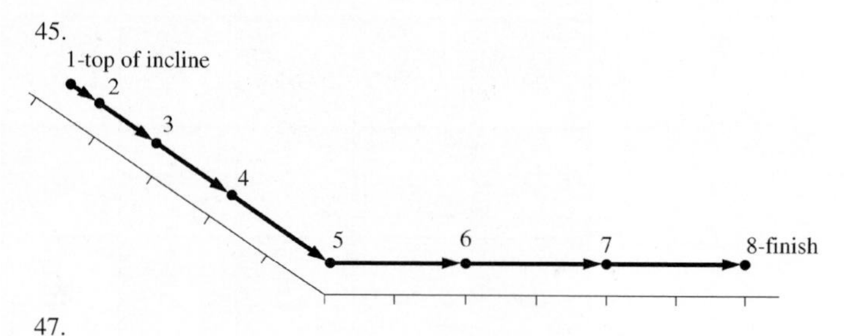

47.

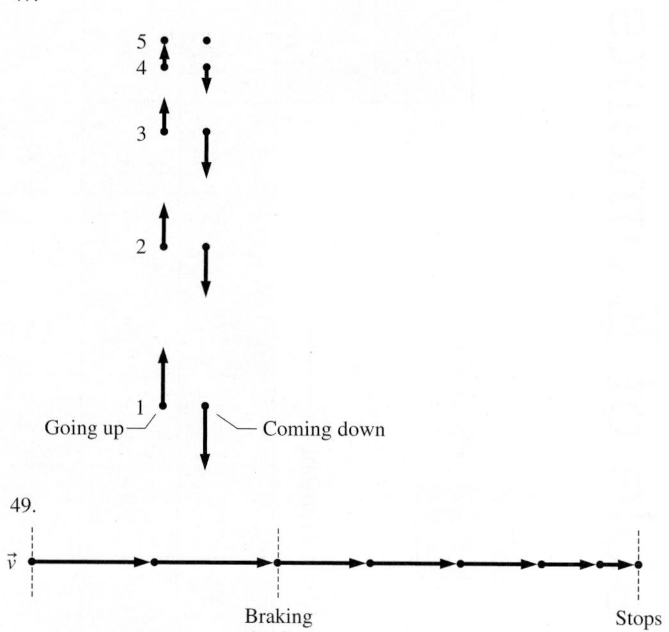

49.

55. 2.5×10^9 s
57. 69 mph
59. 9.3% longer
61. 3.0×10^4 m/s
63. 26 mph
65. a. 25 km b. 0 km c. 28 m/s
67. a. 22 mph
69. a. AB and CD b. All segments c. AB and CD
71. a. 21° b. 990m c. 4.1 m/s
73. a. 220 m b. 220 m c. 260 s
75. a. 1.41 km b. 1.06 m/s
77. A

Chapter 2

Answers to odd-numbered multiple-choice questions

15. C
17. C
19. D
21. B
23. A
25. C
27. B

Answers to odd-numbered problems

1. b.

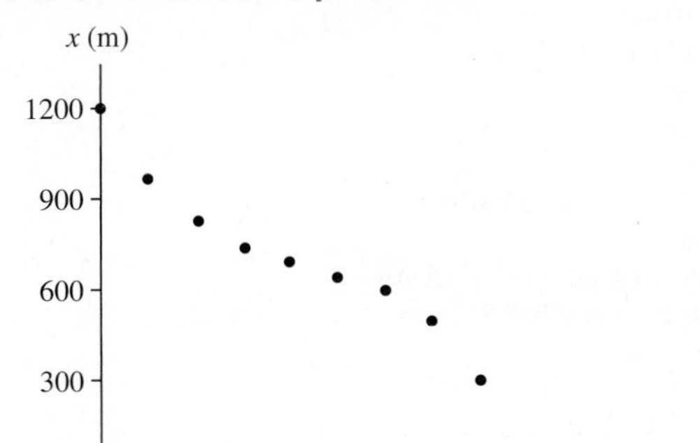

3. a. 8 s b.

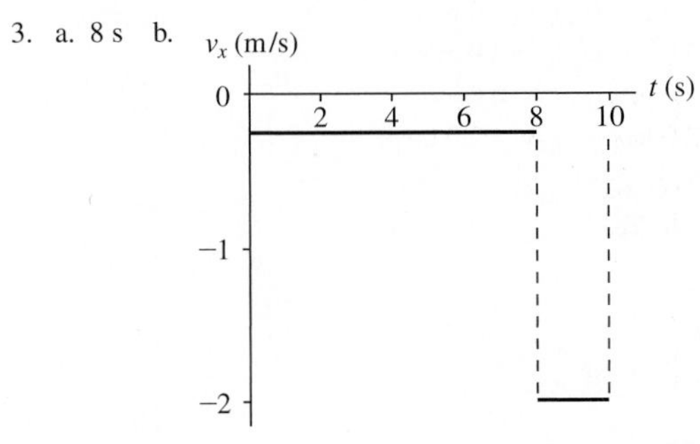

5. a.

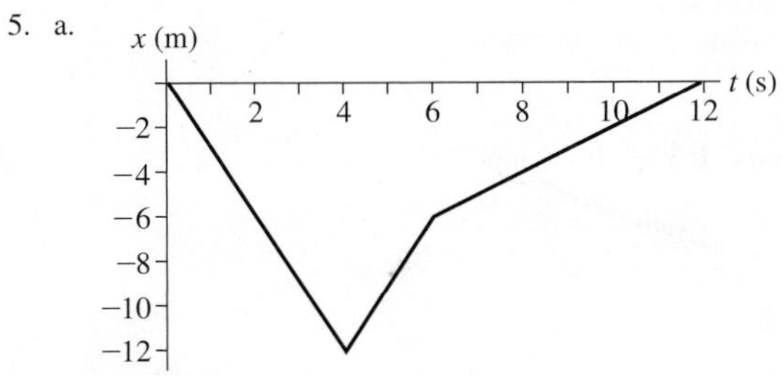

b. 0 m

7.

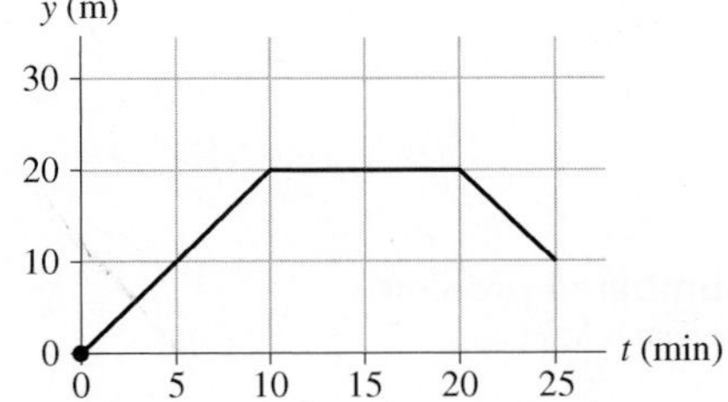

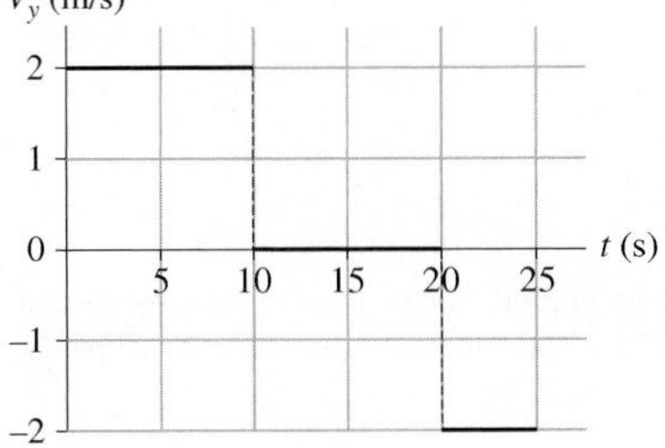

9. 0.43 s
11. a. Beth b. 20 min
13. 4.3 min
15. 12 m/s
17. a.

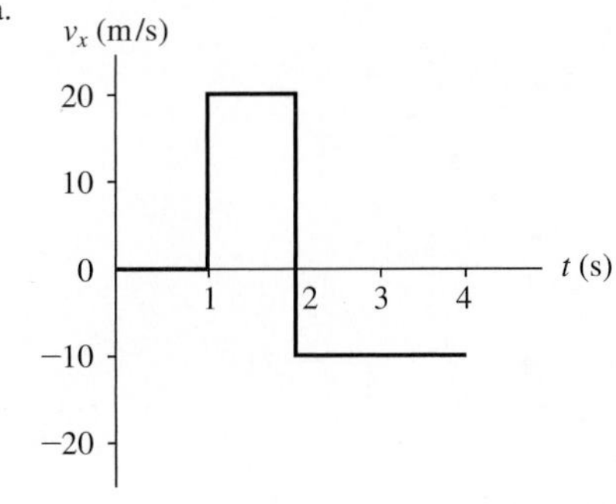

b. Yes, at $t = 2$ s

19. a. 26 m, 28 m, 26 m b. Yes, at $t = 3$ s
21.

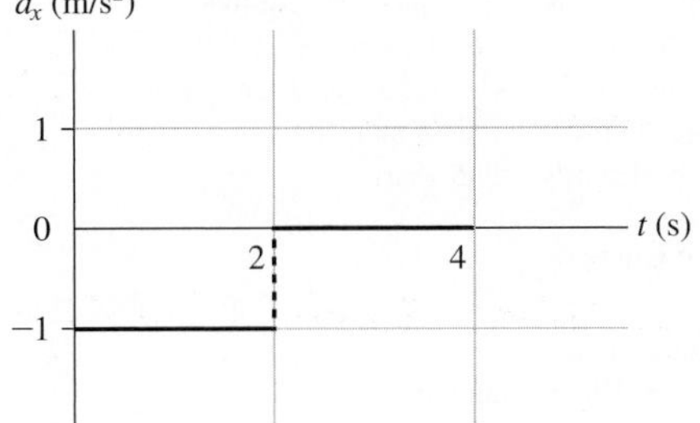

23.

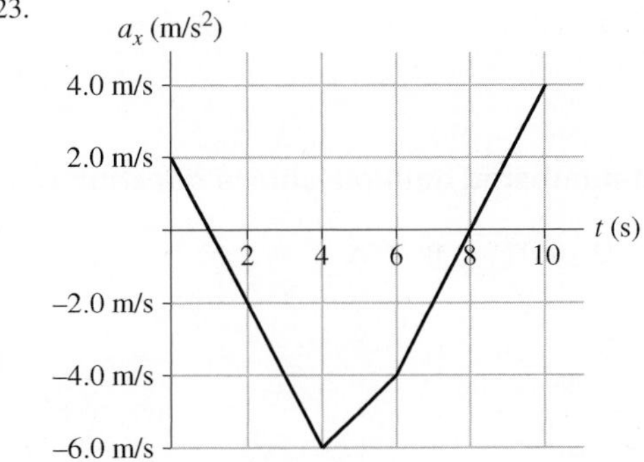

25. 6.1 m/s^2, 2.5 m/s^2, 1.5 m/s^2
27. Trout
29. a. 2.7 m/s^2 b. $0.27g$ c. 134 m, 440 ft
31. a. 15 m/s b. 0.22 m
33. 2.8 m/s^2
35. 37 m/s^2
37. Yes
39. a. 5 m b. 22 m/s
41. a. 2.8 s b. 31 m
43. 10.0 s
45. 0.18 s
47. 52 m
49. 1.1 s
51. 4.6 m/s
53. 7.7 m/s
55. a. 3.0 s b. 15 m/s c. −31 m/s, −35 m/s
57. 57 mph

59. b.

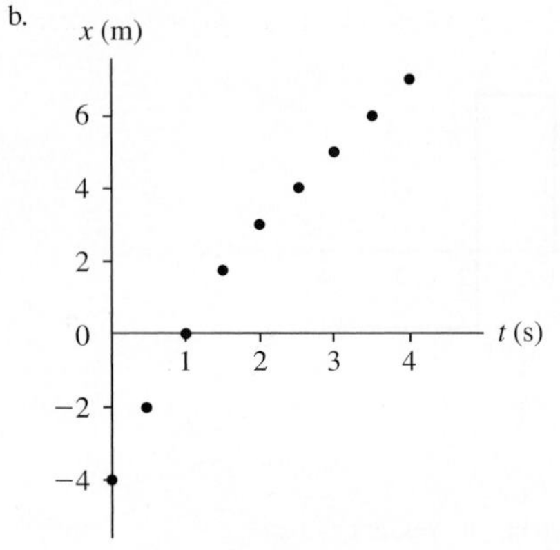

c. 4 m d. 4 m e. 4 m/s f. 2 m/s g. -2 m/s^2

61. a. 2.3 m/s^2, $0.23g$ b. 35 s c. 4.2 km
63. a. 16 m/s$^2 = 1.6g$ b. 5 m/s$^2 = 0.5g$ c. 2 cm
65. a. $\approx$1.0 cm b. 35 m/s^2 c. 0.84 m/s
67. a. 1000 m/s^2 b. 1.0 m/s c. 5.1 cm
69. 3.2 s
71. a. 1.7 s b. -9.9 m/s
73. 110 m
75. a. 4.1 s b. Equal speeds
77. Man: 10.2 s; horse: 10.7 s; man wins
79. Honda wins by 1.0 s
81. 5.5 m/s^2
83. B

Chapter 3

Answers to odd-numbered multiple-choice questions

21. C
23. a. B b. E c. C d. D e. B f. A
25. A
27. C
29. C
31. E

Answers to odd-numbered problems

5. 8.0 m
7. 87 m/s
9. a. $d_x = 71$ m, $d_y = -71$ m b. $v_x = 280$ m/s, $v_y = 100$ m/s
   c. $a_x = 0.0$ m/s^2, $a_y = -5.0$ m/s^2
11. a. 45 m/s, 63° b. 6,3 m/s^2, $-72°$
13. 7.1 km, 7.5 km
15. 25 m
17. 530 m
19. 1.3 s
25. Vector C, to the right
27. a. d. 10 m

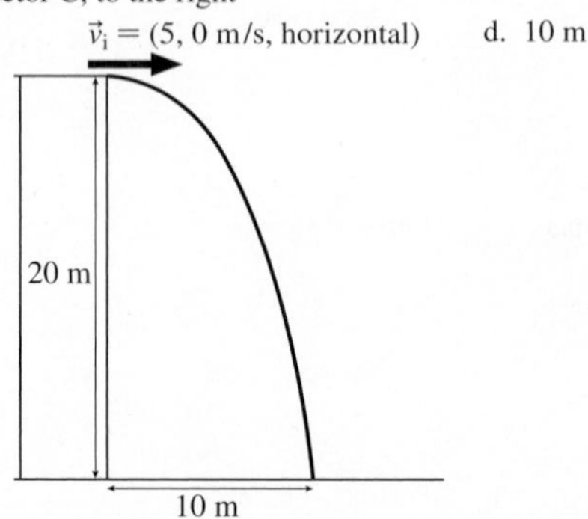

29. 1.1 m
31. a. 10 m/s b. 50°
33. a. 12 m/s b. 12 m/s
35. 6.8 m
37. 10 m/s^2, $1.0g$
39. 2.7×10^{-3} m/s^2
41. a. 4.0 m/s^2 b. 32 m/s^2
43. 27 m
45. Ball 1: 1.5 m/s; ball 2: 15 m/s
47. Ball 1: 15 m/s; ball 2: 5 m/s
49. 2.0 km/h
51. a. (8, 7) c. $\sqrt{113}$, 41°
53. a. $(-2, 2)$ b. 2.8, 135°
55. 2.4×10^3 ft
57. a. 26 m/s b. 70°
59. a. 63 m b. 7.1 s
61. a. $(v_x)_0 = 2.0$ m/s $(v_y)_0 = 4.0$ m/s
   $(v_x)_1 = 2.0$ m/s $(v_y)_0 = 2.0$ m/s
   $(v_x)_2 = 2.0$ m/s $(v_y)_0 = 0.0$ m/s
   $(v_x)_3 = 2.0$ m/s $(v_y)_0 = -2.0$ m/s
   b. 2.0 m/s^2 c. 63°
63. a. 6.0 m/s b. Yes
65. 6.0 m
67. 0.82 m
69. Yes, by 1.0 m
71. 16° from the vertical
73. 6.1 m
75. 41° south of west
77. a. 7.2° south of east b. 2.5 h
79. 30 s
81. 5.0 m/s
83. a. 2.9 m/s^2, $0.30g$ b. 57 mph
85. B
87. C

Chapter 4

Answers to odd-numbered multiple-choice questions

21. C
23. D
25. D
27. C
29. C
31. B

Answers to odd-numbered problems

1. First is rear-end; second is head-on
5.

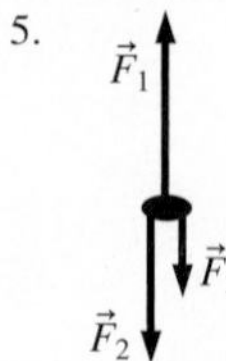

7. Gravity by ground
9. Weight, normal force by ground, kinetic friction force by ground
11. Weight, normal force by slope, kinetic friction force by slope
13. $m_1 = 0.080$ kg and $m_3 = 0.50$ kg
15. 3.0 m/s^2
17. a. 16 m/s^2 b. 4.0 m/s^2 c. 8.0 m/s^2 d. 32 m/s^2
19. $0.71g$
21. 0.25 kg
23. a. No b. 10 m/s^2 to the left
25. 6.6 N
27. 1.1×10^3 N

33.

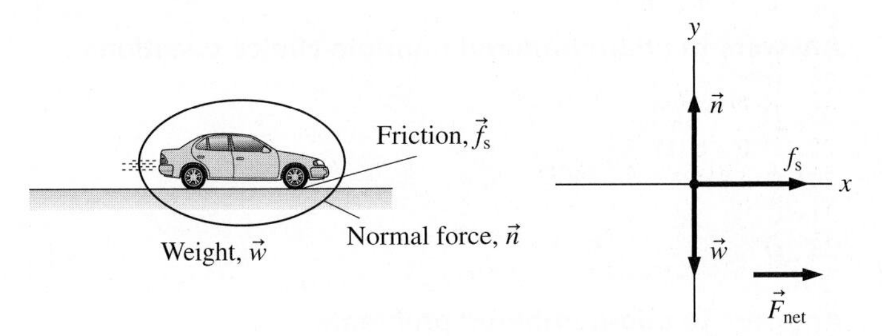

35.

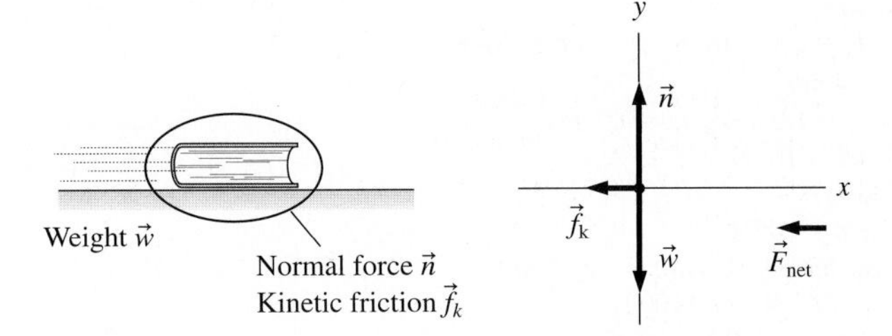

37.

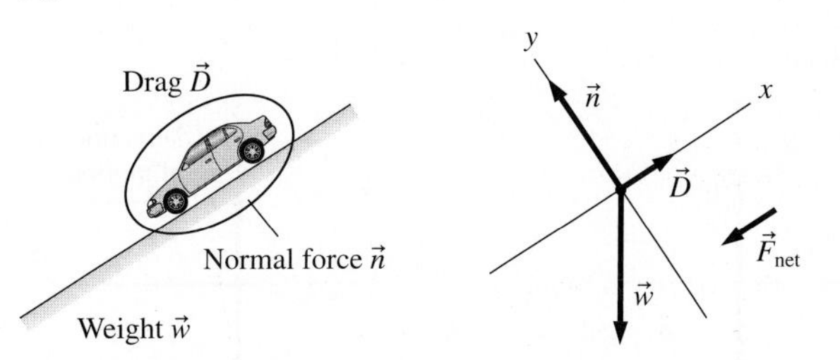

39.

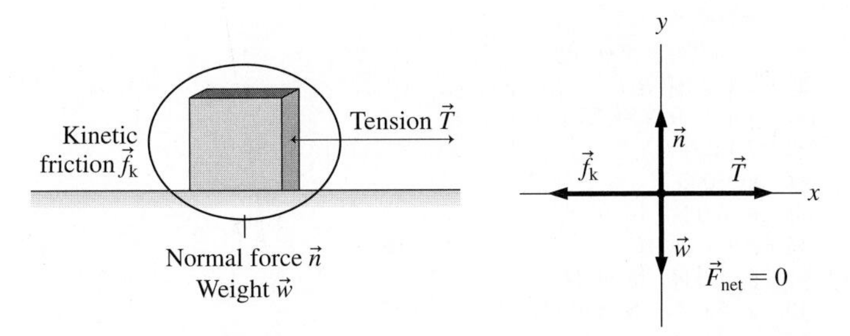

41.

43.

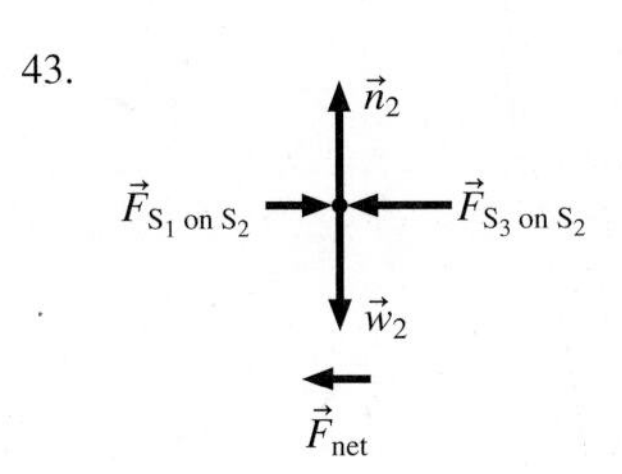

45. Normal force of road on car and of car on road; friction force of road on car and of car on road

47.

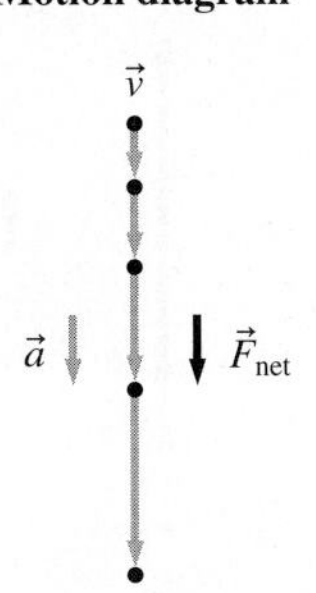

49.

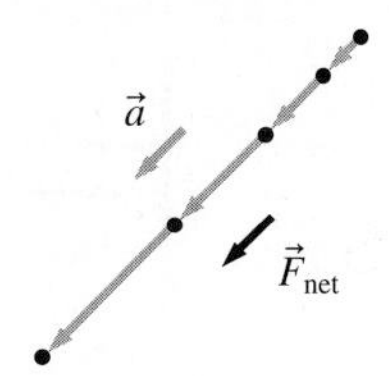

51.

$\vec{F}_{net} = \vec{0}$ N

53.

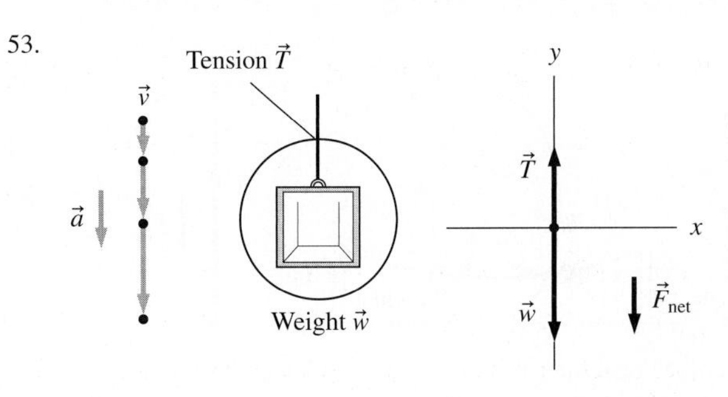

55.

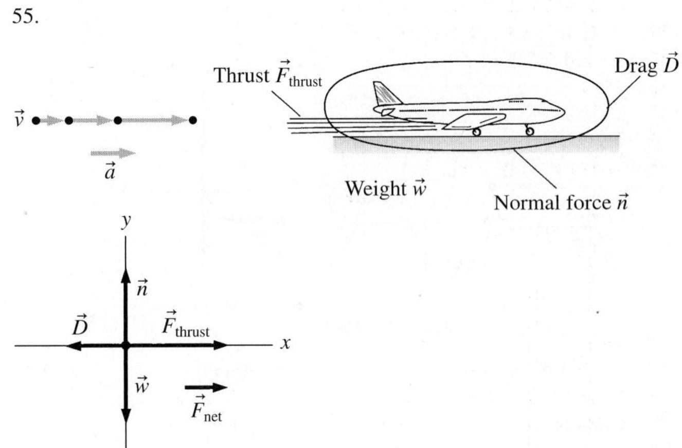

57.

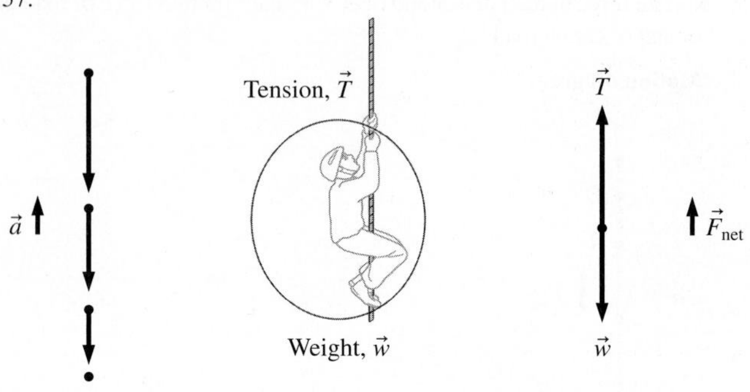

59.

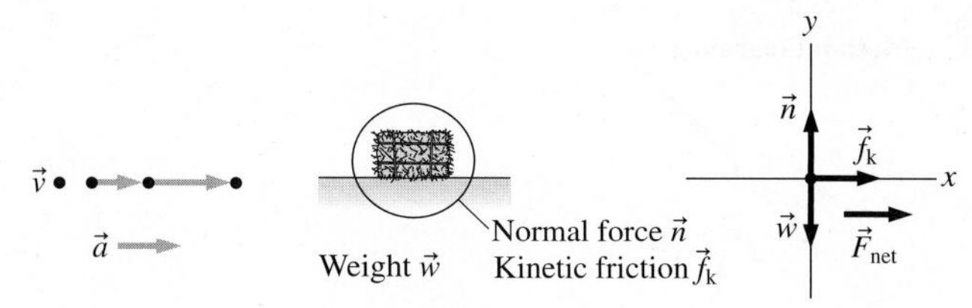

61.

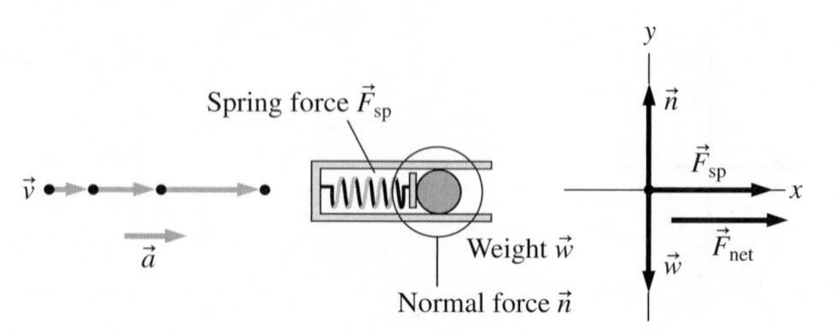

63.

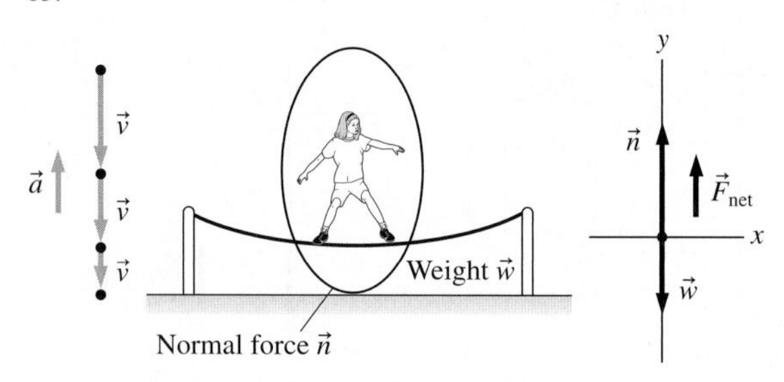

65.

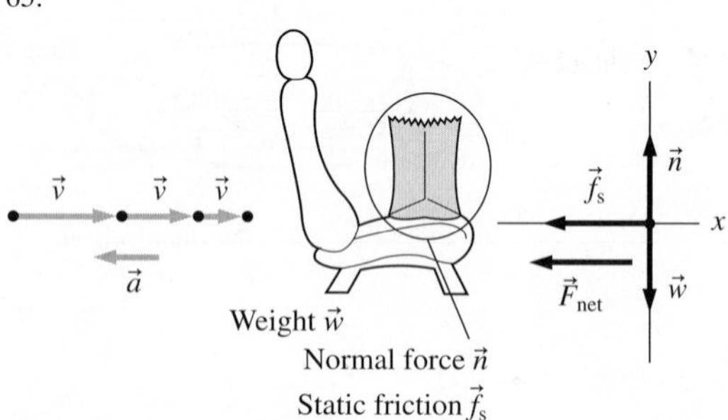

67. a. Ground on greyhound b. 320 N c. 80 m
69. b. 160 N c. 1/5
71. 0.084 N
73. Greater on the way up
75. C
77. C

Chapter 5

Answers to odd-numbered multiple-choice questions

23. B
25. D
27. a. B b. D
29. C
31. C
33. D

Answers to odd-numbered problems

1. $T_1 = 87$ N, $T_2 = 50$ N
3. $T_l = 3.6 \times 10^3$ N, $T_r = 3.9 \times 10^3$ N
5. 49 N
7. $T_l = 4.1 \times 10^2$ N, $T_r = 5.0 \times 10^2$ N
9. 2.4 m/s
11. $a_x = 1.0$ m/s^2, $a_y = 0.0$ m/s^2
13. 1.2×10^2 N
15. 310 N
17. 3.8×10^2 N
19. a. 590 N b. 740 N c. 590 N
21. a. 780 N b. 1100 N
23. 1000 N, 740 N, 590 N
25. b. 180 N
27. 0.25
29. 140 m
31. a.

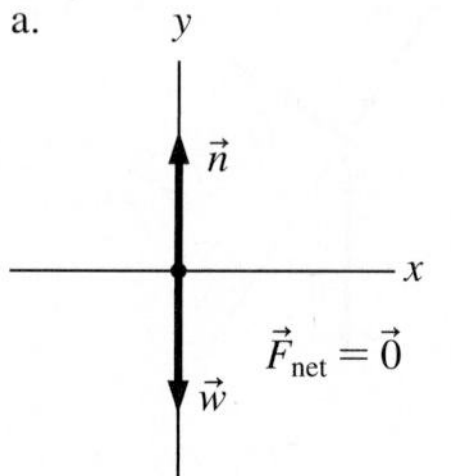

b.

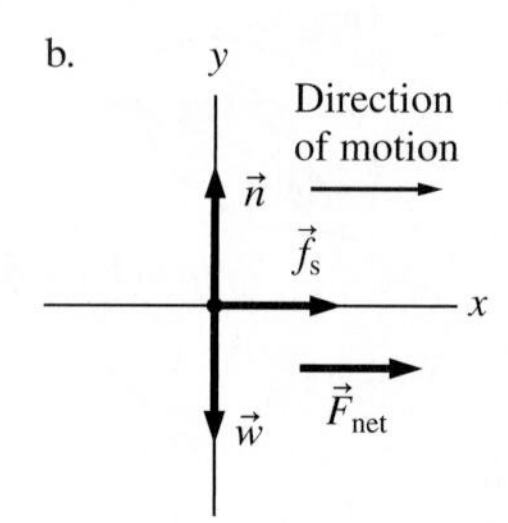

33. ≈ 500 s, ≈ 3000 m
35. 2.4×10^4 N
37. 4.6×10^2 s ≈ 7.7 min
39. 7.4 m/s^2
41. a. 3000 N b. 3000 N
43. a. 6.0 N b. 10 N
45. 3.9×10^2 N
47. a. 1.0 N b. 50 N
49. a. 530 N b. 5300 N
51. $F_{net}(1\text{ s}) = 8$ N, $F_{net}(4\text{ s}) = 0$ N, $F_{net}(7\text{ s}) = -12$ N
53. a. 490 N b. 740 N
55. a. 490 N b. 240 N
57. 3200 N, 5 times actual weight
59. a. 6800 N b. 1.4×10^6 N c. 11 times, 2300 times
61. a. 5.2 m/s^2 b. 1000 kg
63. 160 N
65. 60 m
67. ≈ 36 mph
69. a. -10.2 m/s^2 b. 35 m/s
71. a. 11 kg b. 57 kg
73. 0.98 kg
75. Stay at rest
77. $T_1 = 17$ N, $T_2 = 27$ N
79. 14 N
81. a. 220 N b. 440 N
83. Down at 0.93 m/s^2
85. C
87. B

Chapter 6

Answers to odd-numbered multiple-choice questions

21. D
23. E
25. B
27. A
29. A
31. D

Answers to odd-numbered problems

1. 3.9 m/s
3. a. 0.56 rev/s b. 1.8 s
5. a. 2.3 Hz b. 31 m/s c. 4.5×10^2 m/s^2
7. a. 3.0×10^4 m/s b. 6.0×10^{-3} m/s^2
9. a. 5.1g b. 57%
11. 0.56 Hz
13. $T_3 > T_1 = T_4 > T_2$
15. 1.8
17. 9400 N, toward center, static friction
19. a. 1700 m/s^2 b. 240 N
21. 13 m/s
23. 270 N
25. 4.7×10^2 N
27. 20 m/s
29. 78%, or just over 3/4
31. a. 1.8×10^4 m/s^2 b. 4.4×10^3 m/s^2
33. 99 min
35. 1/2
37. 6.0×10^{-4}
39. 3.9 m/s^2
41. a. 3.53×10^{22} N b. 1.99×10^{29} N c. 0.564%
43. a. 3.77 m/s^2 b. 25.9 m/s^2
45. 4.0
47. 280 y
49. 92 min, 7700 m/s
51. 0.039 au
53. 11 h
57. 9.5×10^3 N
59. 12%
61. 1.1
63. 7.4 m/s
65. No
67. 5.4 m/s
69. 22 m/s
71. 2400 m
73. a. 3.0×10^{24} kg b. 0.89 m/s^2
75. (12 cm, 0 cm)
77. 6.5×10^{23} kg
79. 0.48 m/s
81. C
83. A

Chapter 7

Answers to odd-numbered multiple-choice questions

19. C
21. D
23. A
25. A
27. D
29. D

Answers to odd-numbered problems

1. a. $\theta = \pi/2$ b. $\theta = 0$ c. $\theta = 4\pi/3$
3. 1.7×10^{-3} rad/s
5. a. 1.3 rad, 72° b. 3.9×10^{-5} rad/s
7. 3.0 rad
9. a. 12.5 rad b. −2.5 rad/s
11. a. 0.105 rad/s b. 0.00105 m/s
13. 11 m/s
15. 960 rev
17. a. 160 rad/s^2 b. 50 rev
19. $\tau_1 < \tau_2 = \tau_3 < \tau_4$
21. −0.20 N · m
23. 28.3 N
25. 7.5 N · m
27. 5.5 N · m
29. 2.5 N · m
31. −4.9 N · m
33. a. 34 N · m b. 24 N · m
35. $x = 0.0333$ m, $y = 0.0333$ m
37. 2.93 mm
39. 7.2×10^{-7} kg · m^2
41. 4.5 cm
43. 1.8 kg
45. 6.0×10^{-3} N · m
47. 8.0 N · m
49. 0.047 N · m
51. 17 rad/s^2
53. 0.11 N · m
55. 0.50 s
57. a. 14 rad/s b. 11 m/s c. 7.9 m/s
59. 2.0 m
61. b.

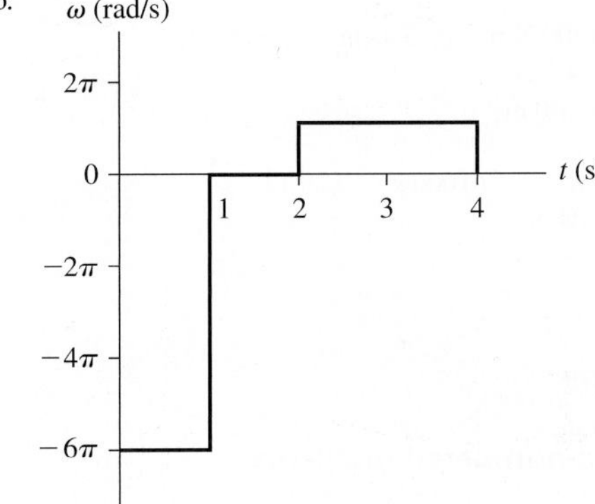

63. 55 rotations
65. −0.94 N · m
67. 7.5 cm
69. a. 1.4 kg · m^2 b. Increase
71. 3.5 rad/s^2
73. 180 s
75. −0.28 N · m
77. 2.2 N
79. B
81. B
83. B

Chapter 8

Answers to odd-numbered multiple-choice questions

19. A
21. B
23. D
25. B
27. B

Answers to odd-numbered problems

1. Right 470 N; left 160 N
3. 15 cm
5. 140 N
7. 160 N
9. 1.4 m
11. 590 N
13. 98 N
15. 16°
17. 50°
19. 0.93 m
21. 0.96 N
23. 0.30 N/m
25. a. 390 N/m b. 18 cm
27. a. 0.13 m b. 0.08 m
29. 98 N/m
31. a. 2 mm b. 0.25 mm
33. 2.0 kg
35. 16 cm
37. 0.0078%
39. 1.0 mm
41. a. 3.8 mN b. 9 mm
43. 3.3%
45. 78 N
47. 1100 N
49. a. 340 N b. 8.6
51. $F_1 = 750$ N, $F_2 = 1000$ N
53. 350 N
55. 1.0 m
57. a. 49 N b. 1500 N/m c. 3.4 cm
59. a. 1.0 cm b. 4.5 N
61. a. 2300 N b. 1.0 m
63. 0.25 = 25%
65. a. 4.8×10^{-2} m^2 b. 70,000 kg c. 0.14
67. a. 206 N b. 84 N
69. B
71. C

Part I Problems

Answers to odd-numbered problems

1. A
3. A
5. C
7. B
9. D
11. A
13. B
15. D
17. B
19. A
21. b. 2.8 N

Chapter 9

Answers to odd-numbered multiple-choice questions

23. C
25. A
27. A
29. C

Answers to odd-numbered problems

1. 15 m/s
3. 2.6 kg · m/s
5. 60 N
7. 1500 N
9. 0.50 s
11. 960 N
13. −110 N
15. a. −19 kN b. −280 kN
17. a. 5.08 kg · m/s b. 3400 N
19. 0.205 m/s
21. 0.31 m/s
23. 1.4 m/s
25. 0.47 m/s
27. 4.8 m/s
29. 5.5 cm/s
31. 2.3 m/s
33. 1.0 kg
35. 1.7 m/s at 45° north of east
37. (−2 kg · m/s, 4 kg · m/s)
39. 14 m/s at 45° north of east
41. 0.025 kg · m^2/s, into the page
43. 1.3 rev/s
45. 2.8×10^{34} kg · m^2/s
47. 0.20 s
49. 930 N
51. 3.8 m/s
53. 12 kg · m/s, left; 15° above the horizontal
55. a. 0.59 m/s d. Yes
57. 1.1 m/s
59. The quarterback
61. 6.3 m
63. 1.7 m/s, 55° below the x-axis
65. a. 8.0 m/s b. 100g
67. 20 m/s downward
69. 2.0 m/s
71. 1.5×10^7 m/s, forward
73. 0.85 m/s, 72° below the x-axis
75. 400 rpm
77. 0
79. 18 rev/s, clockwise
81. B
83. C

Chapter 10

Answers to odd-numbered multiple-choice questions

25. C
27. C
29. C
31. C

Answers to odd-numbered problems

1. a. −30 J b. 30 J
3. Rope 1: 0.919 kJ; rope 2: 0.579 kJ
5. a. 0 J b. −43 J c. 43 J
7. −220 J
9. 110 km/h
11. a. 14 m/s b. Factor of 4
13. 4.0 m/s
15. 0.86 kJ
17. 0.049 J
19. 1.9 kg · m^2
21. 6.2 m
23. 14 kJ
25. 10 J
27. 9.7 J
29. 16.5 kJ
31. 6.2 kJ
33. 110 N

35. a. 5.1 m b. 10 m/s c. 22 m/s
37. 4.5 m/s
39. a. Yes b. 14 m/s
41. 3.1 kJ
43. 26 m
45. 17 kJ
47. 24 m/s
49. a. Kinetic energy to gravitational potential energy b. 1.0 m c. 4.4 m/s
51. a. Right b. 17.3 m/s c. $x = 1.0$ m and 6.0 m
53. 0.08 nm
55. 0.86 m/s and 2.9 m/s
57. 1/2
59. a. 1.8×10^2 J b. 59 W
61. a. 9.5×10^5 J b. 2.0×10^4 W
63. 45 kW
65. 34 min
67. a. 500 N b. 1.7×10^4 W
69. 2.0×10^4 W
71. 91 kJ
73. a. 20 J b. 0.74 J c. 13 J d. 6.3 J
75. a. 2.0 mJ b. 1/2
77. 4.2 m/s
79. 5.8 m
81. a. 0.34 m
83. a. $\sqrt{\frac{(m + M)\,kd^2}{m^2}}$ b. 2.0×10^2 m/s c. 99.8%
85. a. 1.7 m/s b. 2/3
87. a. 2.6 m/s b. 33 cm
89. 1.3 m/s
91. D
93. B
95. B
97. B
99. B

Chapter 11

Answers to odd-numbered multiple-choice questions

33. D
35. C
37. D

Answers to odd-numbered problems

1. 1.7 MJ
3. 3.3%
5. 8 LEDs, 8.0 W
7. 2060 Cal
9. 370,000 J = 89,000 cal = 89 Cal
11. 490 Cal
13. ≈ 1.4 km
15. 2.0 h
17. 710 m
19. 1800 reps
21. a. 196 J b. 15,680 J c. 0.0094 donuts
23. 1200 km
25. 2.9×10^{-2} g
27. −290°C, −452°F
29. 1.3
31. 700 J from the system
33. 200 J
35. 10 J, increases
37. 0.40
39. 5000 J
41. 67%
43. 47%, 35%
45. 1.5
47. a. 200 J b. 250 J
49. 7.3
51. 6.3 kW
53. a. Refrigerator c b. Refrigerator b
55. 1 slice
57. a. 670 Cal b. 1200 W
59. 830 W
61. a. 80 W b. 5 people
63. a. 950 Cal b. 46 W, 54 W less
65. 230 K
67. a. 110 m b. 32%
69. 92
71. a. 86 MJ b. 2.8 mi^2
73. A
75. A
77. D
79. A

Part II Problems

Answers to odd-numbered problems

1. A
3. B
5. D
7. C
9. A
11. C
13. D
15. B
17. C
19. B
21. D
23. 0.22 m/s to the west
25. a. 3.0 m/s b. 120 N c. 60 J

Chapter 12

Answers to odd-numbered multiple-choice questions

31. D
33. B
35. A
37. B

Answers to odd-numbered problems

1. Carbon
3. 3.5×10^{24}
5. 0.024 m^3
7. 313°C
9. −9.3°C
11. 11°C
13. 1.1 N
15. 47.9 psi
17. 350 m/s
19. 8.9×10^8 molecules
21. 700 L
23. 6.0°C
25. 5.5×10^3 m^3
27. 1.6 atm
29. a. 9500 kPa
31. a. 98 cm^3
33. a. Isochoric b. 910 K, 300 K
35. a. Isobaric b. 120°C c. 0.0094 mol
37. a. 2.5 cm^3 b. Water to the bubble
39. 35 psi

41. 32 psi
43. 4.7 m
45. 0.5 cm
47. 0.83%
49. 13 kJ
51. a. 36°C b. 3000 J
53. 2300 W
55. 71 min
57. 12 W, 250 Cal/d
59. 14°F
61. 76°C
63. 28°C
65. 0.218 kg
67. 610 g
69. a. 91 J b. 140°C
71. 9.0°C
73. 830 W
75. 6.0 W
77. 4.7×10^{-4} m^2
79. 300 W
81. 8 ms
83. No
85. 60 J
87. a. 80 J b. 1.3 W
89. 0.12°C
91. 2.4×10^6 kg/min
93. 0.54 kg/h
95. 3.8 kg/h
97. a. 83 J/(kg · K) b. 2.0×10^5 J/kg
99. 60 J
101. a. 810 J b. −490 J c. 0 J
103. a. 320 J b. 0.13
105. a. 84 W b. 180 W c. Radiation d. Chilly
107. A
109. B
111. B

Chapter 13

Answers to odd-numbered multiple-choice questions

33. A
35. D
37. C
39. A
41. B
43. D

Answers to odd-numbered problems

1. 1200 kg/m^3
3. 14.1 kg
5. 11 kg/m^3
7. 1100 atm
9. 3.8 kPa
11. 0.10 atm
13. 3.2 km
15. 88,000 Pa
17. a. 1.1×10^5 Pa b. 4400 Pa for both A–B and A–C
19. 3.68 mm
21. 0.82 m
23. a. 3.0 cm b. 3.2 cm
25. 1.9 N
27. 1065 kg/m^3
29. 63 kg
31. 0.96 N
33. 3.2 m/s
35. a. 1.0 m/s, 16 m/s b. 3.1×10^{-4} m^3/s
37. 110 kPa
39. 3.0 m/s
41. 1.8×10^5 Pa
43. 260 Pa
45. 3200 Pa
47. 3.1 kPa
49. a. 2.8×10^{-4} m/s b. 3.5×10^2 Pa
51. 17 Pa
53. 4.8×10^{23} atoms
55. 27 cm
57. 750 kg/m^3
59. 44 N
61. a. 0.38 N b. 20 m/s
63. 2×10^{-3} m/s
65. a. $p_2 = p_{atmos}$ b. 4.6 m
67. 4.1 cm
69. 8.2×10^5 Pa
71. 8.8×10^{-5} m^3/s
73. B
75. C

Part III Problems

Answers to odd-numbered problems

1. A
3. B
5. D
7. B
9. C
11. A
13. C
15. D
17. A
19. 300 N
21. a. 0.071 mol b. 6.7×10^{-4} m^3 c. 0.019 mol

Chapter 14

Answers to odd-numbered multiple-choice questions

23. a. B b. C c. C
25. a. B b. B c. B d. E e. E
27. B
29. B

Answers to odd-numbered problems

1. 2.3 ms
3. 0.80 s, 1.3 Hz
5. 40 N
7. a. 13 cm b. 9.0 cm
9. a. 20 cm b. 0.25 Hz
11. 2.0×10^{-2} m
13. a. 0.2 m/s^2 b. 1/50
15. a. 0.67 m/s b. 5.2 m/s^2
17. 1.6 mm
19. a. $U = 1/4E$ $K = 3/4E$ b. $A/\sqrt{2}$
21. a. 2.83 s b. 1.41 s c. 2.00 s d. 1.41 s
23. a. 2.0 cm b. 0.63 s c. 5.0 N/m d. 20 cm/s e. 1.0×10^{-3} J f. 15 cm/s
25. a. 0.50 s b. 5.5 cm c. 70 cm/s d. 0.049 J
27. a. 4.00 s b. 5.66 s c. 2.83 s d. 4.00 s
29. 35.7 cm
31. 3.67 m/s^2
33. 8.7×10^{-2} kg · m^2
35. 1.9 s
37. a. 1.24 s b. 97 steps/min

39. 10.0 s
41.

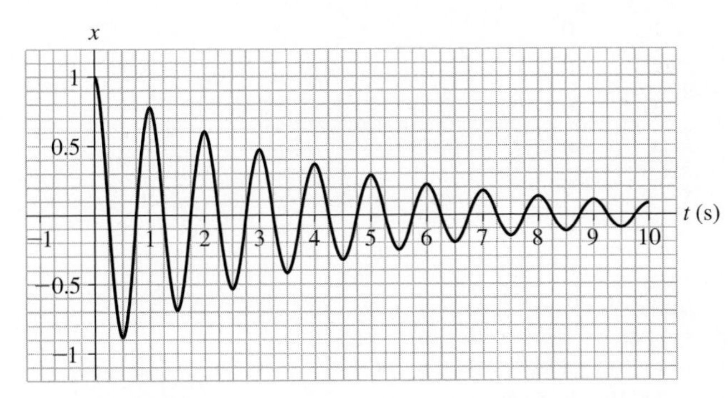

43. a. 5.0 kHz b. 4.8×10^{-3} s
45. 2 Hz, 0.25 s
47. 2.8 s
49. 250 N/m
51. 1.1 m/s
53. a. 0.169 kg b. 0.565 m/s
55. 2.8 Hz
57. 5.86 m/s^2
59. a. 0.25 Hz, 3.0 s b. 6.0s, 1.5 s c. 9/4
61. a. 6.40 cm b. 28.3 cm/s
63. 1.8 Hz
65. 11 cm. 1.7 s
67. a. 7.9 m/s b. 8.0×10^2 N
69. a. 1.02 m/s b. 3.2 times
71. a. 376 s b. 1.06%
73. 73 cm
75. 240 oscillations
77. D
79. A
81. A
83. C

Chapter 15

Answers to odd-numbered multiple-choice questions

21. D
23. D
25. D

Answers to odd-numbered problems

1. 280 m/s
3. 0.076 s
5. Neon
7. 6.0 s
9.

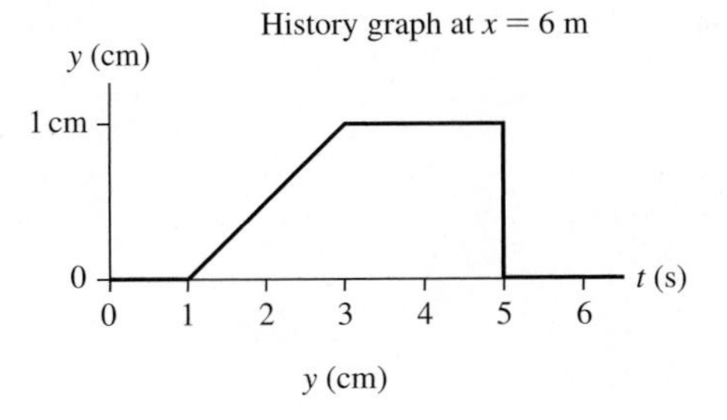

11. 1.0 cm/s
13.

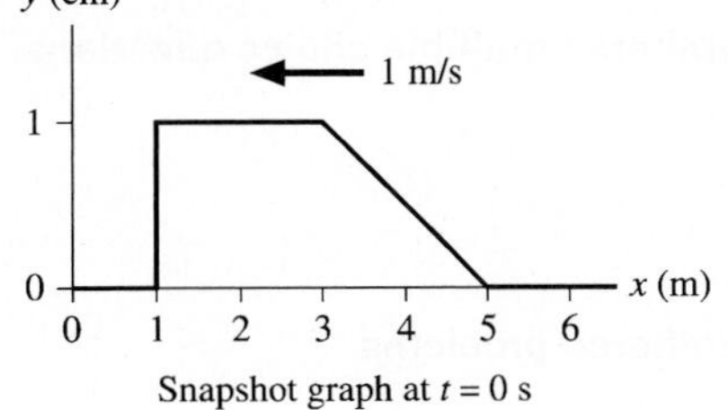

15. 10 m/s
17. a. 8.6 mm b. 15 ms
19. a., b.

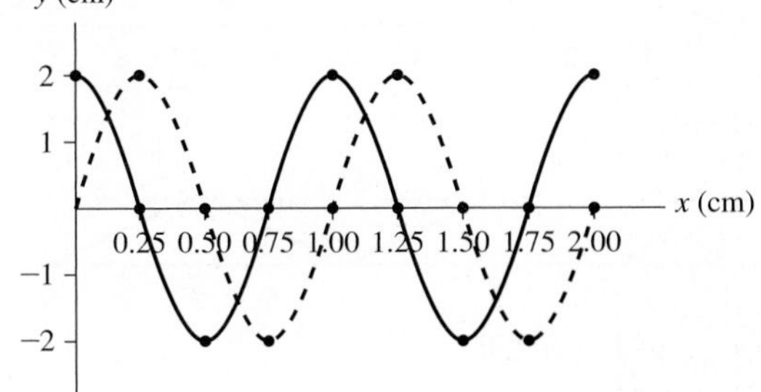

c. 2.0 m/s

21. 6.0 cm, 1.7 Hz, 1.2 m
23. 8.0 ms
25. 15 cm
27. a. 6.7×10^{14} Hz b. 4.6×10^{14} Hz
29. 0.96 s
31. 55 pJ
33. 600 kJ
35. 11 W
37. a. 6.7×10^1 W b. 8.5×10^{10} W/m^2
39. 2.8×10^{24} W/m^2
41. 25 pW/m^2, 14 dB
43. a. 3.2 mW/m^2 b. 95 dB
45. 1.4 mW/m^2
47. 103 dB
49. 49 m
51. a. 650 Hz b. 560 Hz
53. 16 m/s away
55. a. 431 Hz b. 429 Hz
57. 1.3 kHz
59. 589 m/s
61. 1.0 m/s
63. a. 29 s b. Longitudinal c. 6.4 km d. 5.1 cm
65. 11 m/s
67. a. 9 ms b. 0.47 Hz
69. 5.8×10^3 m/s, 2230 Hz
71. $y(x, t) = (5.0\text{ cm})\cos\left[2\pi\left(\frac{x}{50\text{ cm}} + \frac{t}{0.125\text{ s}}\right)\right]$
73. 3.2×10^{-6}
75. 0.78°C
77. 94 dB
79. 4.7 m
81. 620 Hz, 580 Hz
83. D
85. B

Chapter 16

Answers to odd-numbered multiple-choice questions

19. A
21. A
23. A
25. D

Answers to odd-numbered problems

1.

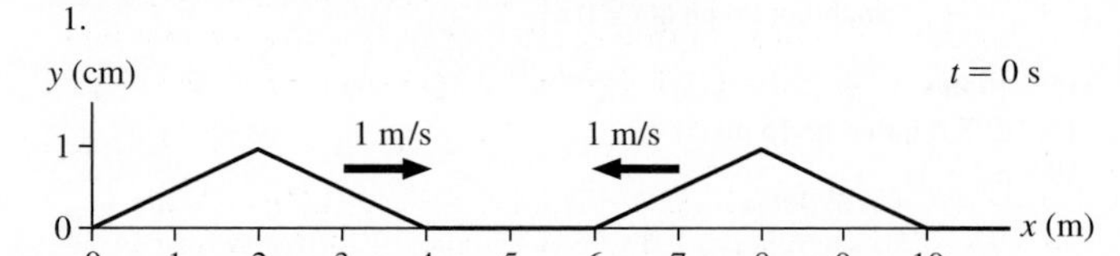

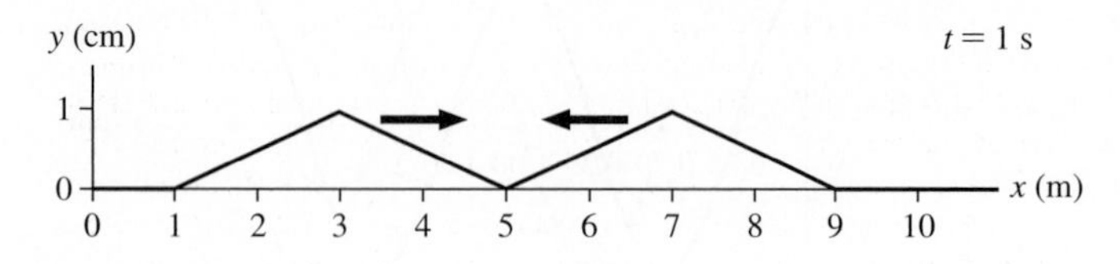

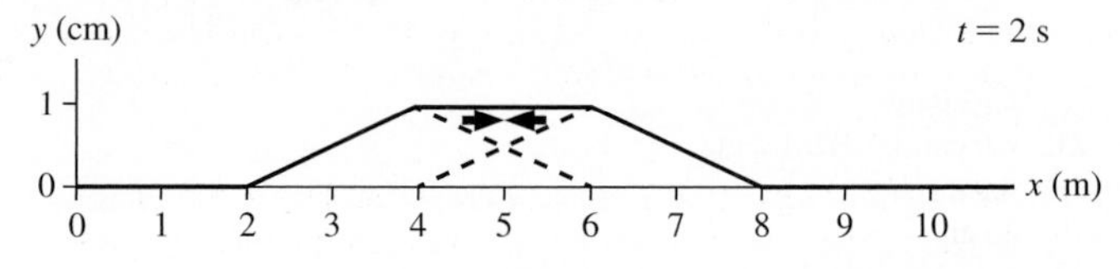

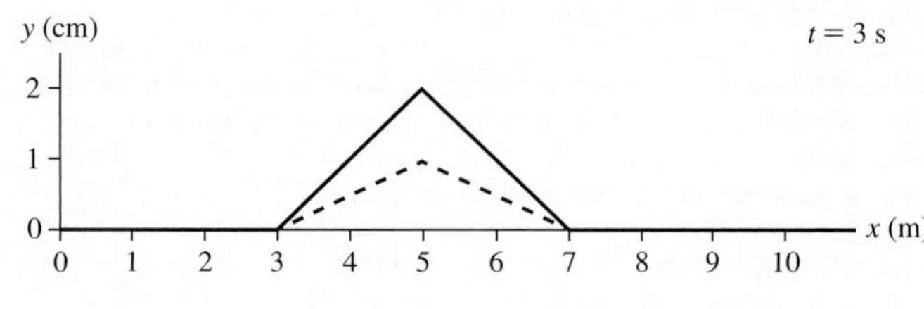

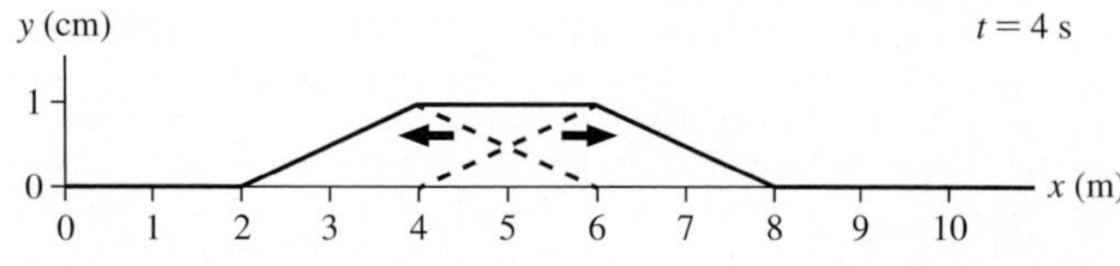

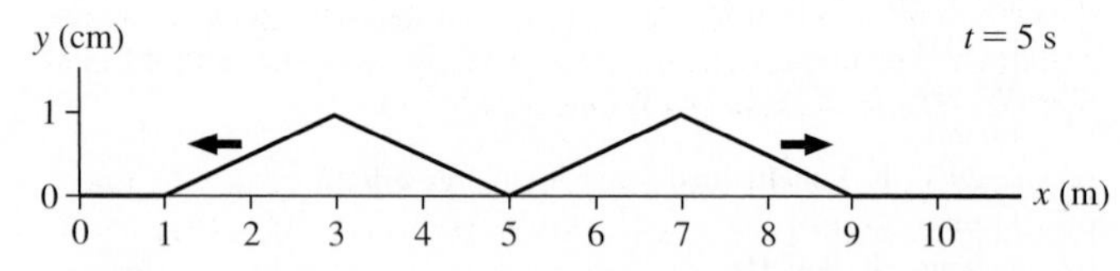

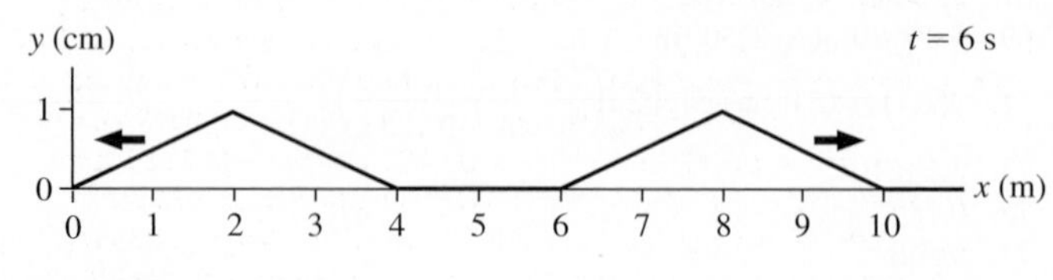

3.

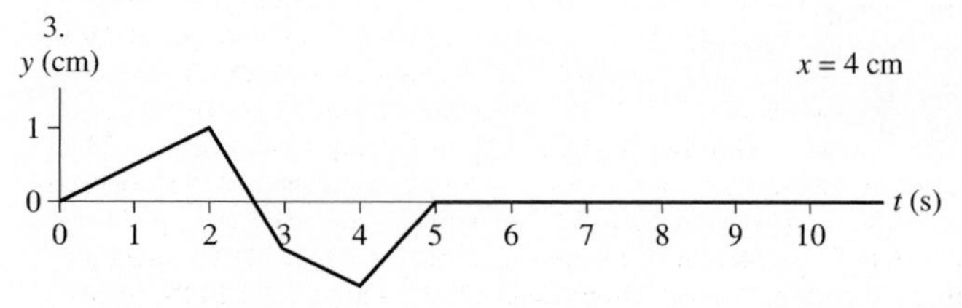

5.

t(s)	y(cm)
0	0
1	0
2	0
3	0
4	0
5	0
6	0

7. 5.0 s
9. 60 Hz
11. 53 m/s
13. a. 4.80 m, 2.40m, 1.60 m b. 75.0 Hz
15. a. 700 Ha b. 56 N
17. 2180 N
19. 13 m, 5.9 s
21. a. 2.6×10^2 Hz b. 1.3 kHz
23. 8.8 Hz
25. 928 Hz
27. a. 5.67 Hz b. 22.7 Hz c. The frequency will decrease.
29. 44 cm
31. 6700 Hz
33. 290 Hz, 0.30 m
35. 580 Hz, 4900 Hz
37. a. 20 cm b. 40 cm
39. 515 Hz
41. 1.0 m, 3.0 m, 5.0 m
43. Perfectly destructive
45. 438 Hz
47. 203 Hz
49. 527 Hz
51. Yes
53. 160 Hz, 320 Hz, 480 Hz, . . .
55. 2.4×10^{-6} N
57. 81 Hz
59. 408 Hz
61. 1210 Hz
63. 1400 Hz
65. 2.98 m, 5.62 m, 17.88 m
67. 2 Hz
69. 1750 Hz
71. 0.20 m/s
73. B
75. B

Part IV Problems

Answers to odd-numbered problems

1. $v_{light} > v_{earthquake} > v_{sound} > v_{tsunami}$
3. D
5. B
7. B
9. C
11. B
13. C
15. C
17. C
19. C
21. 500

Credits

FM P. iii top Sally Knight/Pearson Education; **p. iii middle** John Eisele/ Colorado State University; **bottom** Stuart Field; **p. xii** Ethan Daniels/ Shutterstock; **p. xx** Jim Zipp/Science Source; **p. xxi left** Gabriel12/ Shutterstock; **right** Kate Grishakova/Shutterstock; **p. xxii left** Arno Vlooswijk/TService/Science Source; **right** Andy Harmer/Alamy Stock Photo; **p.xxiii** Juniors Bildarchiv GmbH/Alamy Stock Photo; **p. xxiv** CBCK Christine/Alamy Stock Photo; **p. xxv** Lisa Werner/Alamy Stock Photo.

PART I OPENER P. 2 Interfoto/Zoology/Alamy Stock Photo.

CHAPTER 1 P. 4 top Jim Zipp/Science Source; **bottom left** Faltermaier/ Vario images, Inh. Susanne Baumgarten e.K./Alamy Stock Photo; **bottom right** Johnny Greig/Science Source; **p. 5 top left** Theo Allofs/Corbis/Getty Imagers; **top right** Thomas Peter/Reuters; **bottom left** Richard Megna/ Fundamental Photographs, NYC.; **bottom right** Steve Allen/Getty Images; **p. 8 bottom left** Chris Carroll/Corbis/Getty Images; **p. 14 top** Stuart Field; **center** Stuart Field; **bottom** Clive Rose/Getty Images; **p. 16** NASA Earth Observatory; **p. 18 top** Niels Van Kampenhout/Alamy Stock Photo; **center** Vikram Hoshing/Ephotocorp/Alamy Stock Photo; **p. 19 top left** Darren Baker/Corbis/Getty Images; **p. 19 top center** Michael Burrell/Alamy Stock Photo; **top right** C Squared Studios/Photodisc/Getty images; **p. 19 middle** Khaled Desouki/AFP/Getty Images; **bottom** Balfour Studios/Alamy Stock Photo; **p. 20** Alan Schein Photography/Corbis/Getty Images; **p. 23** Mehau Kulyk/Science Source; **p. 28 Problem 22** Jupiterimages/Stockbyte/Getty Images; **p. 29 Problems 41–42** Philip Dalton/Alamy Stock Photo; **p. 30 Problem 68** Sebastian Kaulitzki/Alamy Stock Photo.

CHAPTER 2 P. 32 top Eduard KyslynskyyShutterstock; **p. 32 middle, left to right** Jhorrocks/E+/Getty Images; Steve Bloom Images/Alamy Stock Photo; EpicStockMedia/Shutterstock; **p. 41** Milos Muller/Shutterstock; **p. 44** Texas Transportation Institute; **p. 47** NASA; **p. 50** Vincent Munier/ Nature Picture Library; **p. 52** Kuritafsheen/RooM the Agency/Alamy Stock Photo; **p. 55** NASA; **p. 56** Mike Brinson/The Image Bank/Getty Images; **p. 58** Steve Bloom Images/Alamy Stock Photo; **p. 62 Question 3** Jim Barber/Shutterstock; **p. 66 Problem 28** Krzysztof Odziomek/Shutterstock; **p. 67 Problem 40** Courtesy Hövding; **p. 68 Problem 45** Richard Megna/ Fundamental Photographs, NYC; **p. 69 Problem 66** EcoPrint/Shutterstock; **Problem 68** Mirko Graul/Shutterstock.

CHAPTER 3 P. 71 top Martin Harvey/Alamy Stock Photo; **p. 71 middle, left to right** Erik Isakson/Tetra Images/Alamy Stock Photo; Kyle Krause/ Photolibrary/Getty Images; Kristy-Anne Glubish/Design Pics Inc./Alamy Stock Photo; **p. 71 bottom right** PCN Chrome/PCN Photography/Alamy Stock Photo; **p. 74** Tom Uhlman/Alamy Stock Photo; **p. 80** Scott Markewitz/ The Image Bank/Getty Images; **p. 81** ML Harris/The Image Bank/ Getty Images; **p. 83** Ivan Chudakov/Shutterstock; **p. 85** Richard Megna/ Fundamental Photographs, NYC.; **p. 89** Gustoimages/Science Source; **p. 90** Don Henry/Alamy Stock Photo; **p. 92** Phil Noble/Reuters; **p. 93** David Madison/Allsport Concepts/Getty Images; **p. 95** Jerry Young/Dorling Kindersley, Ltd.; **p. 98 Question 7** Andy Lyons/Getty Images; **Question 14** Jupiterimages/Stockbyte/Getty Images; **p. 100 Problem 15** Don Mason/ Corbis/Getty Images; **Problem 18** Mike Dobel/Alamy Stock Photo; **p. 101 Problem 42** frogmo9/Fotolia; **p. 103 Problem 70** Kai Pfaffenbach/Reuters.

CHAPTER 4 P. 105 top MarclSchauer/Shutterstock; **p. 105 left to right** Wittek, R./Arco Images GmbH/Alamy Stock Photo; Bernd Friedel/123RF; PCN Photography/Alamy Stock Photo; **bottom** Dlewis33/iStock/Getty Images; **p. 106 left** NASA; **p. 107 top, left to right** National Highway Traffic Safety Administration; **p. 107 left, top to bottom** Tony Freeman/ PhotoEdit Inc.; Dmitry Kalinovsky/Shutterstock; Comstock/Stockbyte/ Getty Images; Dorling Kindersley, Ltd.; StudioOneNine/Shutterstock; Creative Digital Vision/Pearson Education; **p. 110** Sportpoint/Shutter- stock; **p. 113** D.P. Wilson/FLPA/Science Source; **p. 117** Richard Megna/ Fundamental Photographs, NYC.; **p. 118** Stan Honda/AFP/Getty Images; **p. 121** Roger Bacon/REUTERS/Alamy Stock Photo; **p. 123** Joppo/Shut- terstock; **p. 125** Robert salthouse/Alamy Stock Photo; **p. 128 Question 7** Richard Megna/Fundamental Photographs, NYC.; **Question 19** Rein Taf- enau/Alamy Stock Photo; **p. 129 Problem 3** Longview/The Image Bank/ Getty Images.

CHAPTER 5 P. 134 top Nick Garbutt/Nature Picture Library/Alamy Stock Photo; **p. 134 middle, left to right** Digital Vision/Getty Images; Eric Schrader/Pearson Education, Inc.; Radius Images/Alamy Stock Photo; **p. 135** Cesar Rangel/AFP/Getty Images; **p. 141** NASA; **p. 142** Erin Patrice O'Brien/The Image Bank/Taxi/Getty Images; **p. 144 top** NASA; **bottom** NASA; **p. 148 top** Nulinukas/Shutterstock; **bottom** Roman Krochuk/Shut- terstock; **p. 151 top** Eric Schrader/Pearson Education, Inc.; **bottom** Marc Xavier/Fotolia; **p. 168 Problem 2** Dumrath/F1online digitale Bildagentur GmbH/Alamy Stock Photo; **p. 169 Problem 5** Martin M Rotker/Science Source/Getty Images; **Problem 9** Sport In Pictures/Alamy Stock Photo; **p. 170 Problem 20** J. D. Pooley/AP Images; **p. 173 Problem 73** Richard Megna/Fundamental Photographs, NYC.; **Problem 74** Martyn Hayhow/ AFP/Getty Images.

CHAPTER 6 P. 175 top Gabriel12/Shutterstock; **middle, left to right** Blue Images/Corbis/Getty Images; Matthew Jacques/Shutterstock; NASA Earth Observing System; **bottom right** Tom Carter/Alamy Stock Photo; **p. 176** Kristy-Anne Glubish/Design Pics Inc./Alamy Stock Photo; **p. 177** Val lawless/Shutterstock; **p. 178** Mark A Leman/The Image Bank/Getty Images; **p. 182 top** Mark Scheuern/Alamy Stock Photo; **bottom** Robert Laberge/Getty Images; **p. 183 top** Matt Perrin Sport/Alamy Stock Photo; **bottom** Richard Megna/Fundamental Photographs, NYC.; **p. 184** FPG/ Hulton Archive/Getty Images; **p. 185** Susumu Nishinaga/Science Source; **p. 186 top** Hubble Heritage Team/NASA; **middle** Wacky Racing/Pictures- byrob/Alamy Stock Photo; **p. 187** Richard Megna/Fundamental Photo- graphs, NYC.; **p. 189 top** NASA; **bottom** M.G.M /Album/M.G.M/Album/ Newscom; **p. 192** Ralph White/NASA; **p. 193** NASA; **p. 195** Jet Propulsion Laboratory/NASA; **p. 198 Question 3** Mike Fuchslocher / Alamy Stock Photo; **p. 199 Question 15** Jupiterimages/Stockbyte/Getty Images; **p. 200 Problem 8** Rafa Irusta/Shutterstock; **Problem 10** Images-USA/Alamy Stock Photo; **p. 201 Problem 14** Todd Korol/Reuters; **Problem 18** Exra Shaw/Getty Images; **Problem 25** Barry Lewis/Alamy Stock Photo; **p. 203 Problem 63** Frilet Patrick/Hemis.fr/Hemis/Alamy Stock Photo; **Problem 64** Dwphotos/Shutterstock; **p. 204 Problem 79** NASA.

CHAPTER 7 P. 205 top Anthony Collins / Alamy Stock Photo; **p. 205 middle, left to right** Hadrian/Shutterstock; Dmbaker/123RF; Oleg Mikhaylov/Shutterstock; **bottom right** Graficart.net/Alamy Stock Photo; **p. 206** BMJ/Shutterstock; **p. 211** Dennis D. Potokar/Science Source; **p. 216 left top** PhotoAbility/Alamy Stock Photo; **left bottom** Chris Mole/Shut- terstock; **p. 216** Stock Image/Pixland/Alamy Stock Photo; **p. 218** Sputnik/ Alamy Stock Photo; **p. 219** Image 100/Alamy Stock Photo; **p. 222** Franck Fife/Afp Photo/Newscom; **p. 224** Stuart Field; **p. 226** Tiffany Brown/AP Images; **p. 228** Cordier Sylvain/Hemis.fr/Alamy Stock Photo; **p. 231 top** Richard Megna/Fundamental Photographs, NYC.; **bottom** John Mead/ Science Source; **p. 232** Danilo Calilung/Alamy Stock Photo; **p. 236 Ques- tion 4** Kari Marttila/Alamy Stock Photo; **Question 12** Richard Megna/ Fundamental Photographs, NYC.; **Question 14** Alex Milan Tracy/Sipa USA/Newscom; **Question 15** Dennis Frates/Alamy Stock Photo; **p. 237 Questions 27–29** Atsushi Tomura/Aflo Sport/Newscom; **p. 238 Problem 14** SCDNR's Southeastern Regional Taxonomic Center (SERTC).; **p. 240 Problem 52** ZUMA Press, Inc./Alamy Stock Photo; **p. 241 Problem 60** Harry Todd/Fox Photos/Hulton Archive/Getty Images; **p. 242 Problem 72** Lucian Coman/Shutterstock; **Problem 78** Images by J.Edwards/D. Whitaker/M.Laskwaski/A.Acosta; **p. 243 Problem 81** John Terence Turner/ Alamy Stock Photo.

CHAPTER 8 **P. 244 top** Eugene Mynzul/Shutterstock; **p. 244 middle, left to right** Larin Andrey Vladimirovich/Shutterstock; Christopher Cassidy/Alamy Stock Photo; Helen Birkin/Shutterstock; **bottom right** Mr Doomits/Shutterstock; **p. 247** James Kay/SCPhotos/Alamy Stock Photo; **p. 251 top** Stuart Field; **bottom** Richard Megna/Fundamental Photographs, NYC.; **p. 252** Ted Kinsman/Science Source; **p. 256** Javier Larrea/Age fotostock/Getty images; **p. 257** Harry Taylor/Dorling Kindersley, Ltd.; **p. 258 Figure 8.25 left** Arco/G. Lacz/Arco Images GmbH/Alamy Stock Photo; **right** Courtesy of Ellen Meiselman/Flickr; **Figure 8.26 (a)** John Eisele/Colorado State University; **p. 261** Adam Pearlstein/Pearson Education, Inc.; **p. 264 Question 4** Uz Foto /Shutterstock; **p. 265 Question 5, left** Fizkes/Shutterstock; **Question 5, right** Fuse/Corbis/Getty Images; **Question 13** Mint Images/REX/Shutterstock; **Question 14, left and right** John Eisele/Colorado State University/Colorado State University Photography; **p. 267 Problem 16** Richie Chan/Shutterstock; **p. 269 Problem 45** Ratmaner/Shutterstock; **Problem 46** Ivan Smuk/123RF; **Problem 50** Chris Curtis/Shutterstock; **p. 270 Problem 62** Kai Pfaffenbach/Reuters;

PART I SUMMARY **P. 273** Bill Schoening, Vanessa Harvey, REU Program/NOA/AURA/NSF; **p. 274 left** Owen Humphreys/AP Images; **right** SPL/Science Source.

PART II OPENER **P. 276** Jim Zipp/Science Source.

CHAPTER 9 **P. 278 top** James Hager/Robertharding/Alamy Stock Photo; **middle, left to right** Pete Fontaine/WireImage/Getty Images; INTERFOTO/Alamy Stock Photo; Dave Fleetham/Design Pics/Perspectives/Getty Images; **bottom** iStock/Getty Images; **p. 279** Brian Jones; **p. 281** Freebilly/Shuterstock; **p. 284** Petr Zurek/Alamy Stock Photo; **p. 288** Richard Megna/Fundamental Photographs, NYC; **p. 292** Georgette Douwma/Nature Picture Library; **p. 295** Charlie Edwards/Photodisc/Getty images; **p. 297** Emery Wells/Image Bank Film Signature/Getty Images; **p. 298 top** NASA; **bottom** Ivan Kuzmin/Alamy Stock Photo; **p. 299 left** Stockfolio/Alamy Stock Photo; **right** Kostas Tsironis/AP Images; **p. 302 Question 12** NASA; **Question 18** Vespasian/Alamy Stock Photo; **p. 305 Problem 44** Macor/123RF; **p. 306 Problem 52** Anthony Pierce/Alamy Stock Photo.

CHAPTER 10 **P. 309 top** Kate Grishakova/Shutterstock; **middle, left to right** Muratart/Shutterstock; Chris Rout/Alamy Stock Photo; Kali9/iStock/Getty Images; **p. 310 left to right** Doug Martin/Science Source; David Kleyn/Alamy Stock Photo; Erik Isakson/Tetra Images/Alamy Stock Photo; **p. 311 top, left to right** Danita Delimont/Gallo Images/Getty Images; Nicholas Toh/Shutterstock; Photo Researchers, Inc/Alamy Stock Photo; **p. 311 left, top to bottom** Stuart Franklin/Getty Images; Spencer Grant/AGE Fotostock; Scott Sroka/National Geographic/Getty Images; David Gray/Reuters; **p. 312 left to right** Kimimasa Mayama/REUTERS/Alamy Stock Photo; Gary Morrison/The Image Bank/Getty Images; Gart Buss/The Image Bank/Getty Images; **p. 315 top left** Al Grillo/AP Images; **bottom right** Danny Beier/Panther Media/AGE Fotostock; **p. 317** InsectWorld/Shutterstock; **p. 320** NASA; **p. 321** lassedesignen/Shutterstock; **p. 322** Meredith Davenport/REUTERS/Alamy Stock Photo; **p. 325 top right** Brian Jones; **middle left** Chris Salvo/The Image Bank/Getty Images; **middle right** Chip Forelli/The Image Bank/Getty Images; **p. 328** High Speed Nature/Alamy Stock Photo; **p. 340** Ted Kinsman/Science Source; **p. 341 top** Neil Setchfield/Planet Images/Getty Images; **bottom** David Madison/The Image Bank/Getty Images; **p. 344** Charles D. Winters/Science Source; **p. 348 Problem 18** Tom Heaton/REX/Shutterstock; **p. 349 Problem 40** Paul Miguel/Alamy Stock Photo; **p. 351 Problem 70** Kristel Segeren/Shutterstock; **p. 353 Problem 88** Paulo Oliveira/Alamy Stock Photo.

CHAPTER 11 **P. 354 top** Mitchell Funk/The Image Bank/Getty Images; **middle, left to right** Dirima/Shutterstock; Tony McConnell/Science Source; Robas/Vetta/Getty Images; **p. 355 top, left to right** Brian Jones; Willard Clay/The Image Bank/Getty Images; David Woodfall/The Image Bank/Getty Images; **bottom, left to right** Brian Jones; Steve Gorton/Dorling Kindersley, Ltd.; Steve Allen/Getty Images; **p. 356** Eye35.pix/Alamy Stock Photo; **p. 357** Christina Richards/Shutterstock; **p. 359** Pearson Education, Inc.; **p. 360** Science Photo Library/Alamy Stock Photo; **p. 361 top** Asgeir Helgestad/Nature Picture Library; **bottom** Jurgen Freund/Nature Picture Library; **p. 363** MaximImages archive/Alamy Stock Photo; **p. 364** CJM Photography/Alamy Stock Photo; **p. 365 top** Richard Megna/Fundamental Photographs, NYC. ; **bottom** W. Phillips, NIST; **p. 368** Richard Megna/Fundamental Photographs, NYC.; **p. 370 left to right** Chris Ladd/The Image Bank/Getty Images; Tony McConnell/Science Source; Brian Jones; **p. 373** Wagan Corporation; **p. 378** Basement Stock/Alamy Stock Photo; **p. 380** Andy Sacks/The Image Bank/Getty Images; **p. 382** Eye35.pix/Alamy Stock Photo; **p. 383 Question 5** Tom Reichner/Shutterstock; **p. 384 Question 20** Richard Megna/Fundamental Photographs, NYC.; **p. 386 Problems 24–26** Dennis Jacobsen/Shutterstock; **Problem 41** David Nunuk/All Canada Photos/Alamy Stock Photo; **p. 387 Problem 54** Brian Jones; **p. 388 Problem 63** Anthony Kaminju/Reuters; **p. 389 Problem 72** Martin Harvey/Science Source;

PART II SUMMARY **P. 391 left** Scott Camazine/Science Source; **right** National Optical Astronomy Observatories/Science Source; **p. 392** Rex/Shutterstock; **p. 393** NASA Earth Observing System.

PART III OPENER **P. 394–395** Vladsilver/Shutterstock.

CHAPTER 12 **P. 396 top** Arno Vlooswijk/TService/Science Source; **middle, left to right** Nui7711/Fotolia; Steve Hamblin/Alamy Stock Photo; NHPA/PhotoShot/Superstock; **bottom** B.A.E. Inc./Alamy Stock Photo; **p. 397 top** Wolfelarry/Fotolia; **middle** Coprid/Fotolia; **bottom** Image Source/IS020/Alamy Stock Photo; **p. 398** Richard Megna/Fundamental Photographs, NYC.; **p. 400** STS-129 Crew, Expedition 21 Crew/NASA; **p. 401** JPL/NASA; **p. 403** Tom McHugh/Science Source; **p. 404** Adam36/Fotolia; **p. 411** U of Illinois At Urbana-Champaign, Dept. of Atmospheric Sciences; **p. 412** Richard Choy/Photolibrary/Getty Images; **p. 414** TongRo Images Inc/Getty Images; **p. 416 top** Kenneth B. Storey/Carleton University; **bottom** Pete Orelup/Flickr RF/Getty Images; **p. 418 top, left to right** Alessandro Trovati/AP Images; Stephen Oliver/Alamy Stock Photo; Tom Brakefield/Photodisc/Getty Images; **middle** Pinkomelet/Shutterstock; **p. 423 left to right** Cn Boon/Alamy Stock Photo; Gary S. Settles & Jason Listak/Science Source; Pascal Goetgheluck/Science Source; Brian McEntire/Shutterstock; **p. 424 top** Gillian Merritt/Alamy Stock Photo; **bottom** Kwest/Shutterstock; **p. 425 top** Tony McConnell/Science Source; **bottom** 2000 John Hopkins University/Applied Physics Laboratory LLC; **p. 426** Ted Kinsman/Science Source; **p. 429 top** Robert Hamilton/Alamy Stock Photo; **middle** Power and Syred/Science Source; **p. 430** Brian Jones; **p. 433 Question 12** Eric Schrader/Pearson Education, Inc.; **p. 434 Question 27** Matthew Johnston/Alamy Stock Photo; **p. 437 Question 55** Brian Jones; **p. 438 Problem 75** B. Mauck et al., "Thermal windows on the trunk of hauled-out seals: hot spots for thermoregulatory evaporation?" J Exp Biol 206, 1727-1738, doi: 10.1242/?jeb.00348; **Problem 81** Eye of Science/Science Source; **p. 439 Problem 95** Four Oaks/Shutterstock.

CHAPTER 13 **P. 441 top** David Tipling Photo Library/Alamy Stock Photo; **middle, left to right** Hank Shiffman/Shutterstock; Attila Volgyi/Xinhua/Alamy Stock Photo; Andrey Popov/Shutterstock; **bottom** Christopher Gardiner/Shutterstock; **p. 445** Woods Hole Oceanographic Institution/Oceanus Magazine; **p. 447** NOAA; **p. 451** Harald Sund/The Image Bank/Getty Images; **p. 453 top** Altrendo Images/Getty Images; **bottom** Vladimir Wrangel/Shutterstock; **p. 455** Alix/Science Source; **p. 456** Don Farrall/Photodisc/Getty Images; **p. 457** Culture-images GmbH/Alamy; **p. 459** Ed Endicott / Alamy Stock Photo; **p. 460** Zephyr/Science Source; **p. 463** Monkey Business Images/Shutterstock; **p. 465 top** Martin Harvey/Alamy Stock Photo; **bottom** DuCane Medical Imaging Ltd./Science Source; **p. 466** David M. Phillips/Science Source; **p. 468 top** Rusla Ruseyn/Shutterstock; **middle** Vincent Hazat/PhotoAlto/Alamy Stock Photo; **p. 472 Question 21** Vicspacewalker/Shutterstock; **Question 24** Suzanne Oliver; **p. 474 Problem 12** Reinhard Dirscherl/Alamy Stock Photo; **p. 475 Problem 22** Denis-Huot/Nature Picture Library/Alamy Stock Photo;

PART III SUMMARY **P. 480** Egill Bjarnason/Alamy Stock Photo.

PART IV OPENER **P. 482–483** D-Robert Franz/Image State/Alamy Stock Photo.

CHAPTER 14 **P. 484 top** Tom McHugh/Science Source; **middle, left to right** With Love Photographs/Flickr RF/Getty Images; Richard Megna/Fundamental Photographs, NYC.; Wan Xiang/Newscom; **p. 486 top** Oscar Burriel/Science Source; **bottom** CORBIS RF/AGE Fotostock; **p. 490** Randall Fung/Corbis/Getty Images; **p. 493** Serge Kozak/AGE Fotostock; **p. 494** Barry Bland/Alamy Stock Photo; **p. 496** NASA; **p. 497** H. Craighead/Cornell Univ.; **p. 498** Aris Messinis/AFP/Getty Images; **p. 499** Tim Wright/Corbis Documentary/Getty Images; **p. 502** Drubig-photo/Fotolia; **p. 503** Steve P./Alamy Stock Photo; **p. 504** Laszlo Podor/Alamy Stock Photo; **p. 505** Tom Williams/Roll Call Photos/Newscom; **p. 507** Sirtravelalot/Shutterstock; **p. 511 Question 16** Image Werks Co.,Ltd./Alamy Stock Photo; **Question 19** Leo Mason/Popperfoto/Getty Images; **p. 514 Problem 37** Josep Pena/Fotolia; **p. 515 Problem 46** John Henshall/Alamy Stock Photo; **p. 516 Problem 67** MJTH/Shutterstock; **p. 517 Problem 81** Ted Horowitz/Corbis/Getty Images.

CHAPTER 15 **P. 518 top** Andy Harmer/Alamy Stock Photo; **middle, left to right** Stuart Field; Richard Megna/Fundamental Photographs, NYC; Aldo Liverani/Andia/Alamy Stock Photo; **bottom** Brian Jones; **p. 519 top** Olga Selyutina/Shutterstock; **bottom** Eddy Lemaistre/Corbis Sport/Getty Images; **p. 521 top** Aleksandr Belugin/Alamy Stock Photo; **bottom** James Gerhold/Photolibrary/Getty Images; **p. 523** Yevgeniy11/Shutterstock; **p. 530** B. Benoit/Science Source; **p. 533** BMJ/Shutterstock; **p. 534** Mark Carwardine/Photolibrary/Getty Images; **p. 535** Steve Gschmeissner/Science Source; **p. 536** NASA; **p. 538 top** NASA; **bottom** NOAA/Science Source; **p. 539 left** NASA; **right** Robert Harding/Robert Harding World Imagery/Alamy Stock Photo; **p. 543 Question 12** David M Schrader/Fotolia; **Question 19** John Roman/Shutterstock; **p. 545 Problem 28** Monitor National Marine Sanctuary/NOAA; **p. 546 Problem 38** Sihasakprachum/Shutterstock.

CHAPTER 16 **P. 549 top** Martin Philbey/Redferns/Getty Images; **middle, left to right** Amana images inc./Alamy Stock Photo; Richard Megna/Fundamental Photographs, NYC.; Wayhome studio/Shutterstock; **p. 550** Brian Jones; **p. 551** Pearson Education, Inc.; **p. 554** Brian Jones; **p. 556** Brian Jones; **p. 557 top** Stuart Field; **bottom** Photodisc/Getty Images; **p. 558** Field, Stuart; **p. 560** Dave Maiullo as seen in "THAT PHYSICS SHOW" Off-Broadway www.thatphysicsshow.com; **p. 562** Courtesy Brookhaven National Laboratory; **p. 565** ImageState/Alamy Stock Photo; **p. 569** Davidf/iStock/Getty Images; **p. 571** Utekhina Anna/Shutterstock; **p. 573 Question 1** Dale W. Brown/Shutterstock; **p. 574 Question 10** Jim West/Alamy Stock Photo; **p. 576 Problem 22** Allstar Picture Library/Alamy Stock Photo; **Problem 26** Joe Tucciarone/Science Source; **Problem 28** PCJones/Alamy Stock Photo; **p. 578 Problem 59** ldutko/Shutterstock;

PART IV SUMMARY **P. 581** Courtesy D.Vatvani, Deltares, the Netherlands.; **p. 583** Zoonar GmbH/Alamy Stock Photo.

Index

For users of the two-volume edition, pages 1–583 are in Volume 1 and pages 584–1093 are in Volume 2.

A

B

C

D

F

G

H

I

N

Q

R

T

U

V

W

X

Y

Z